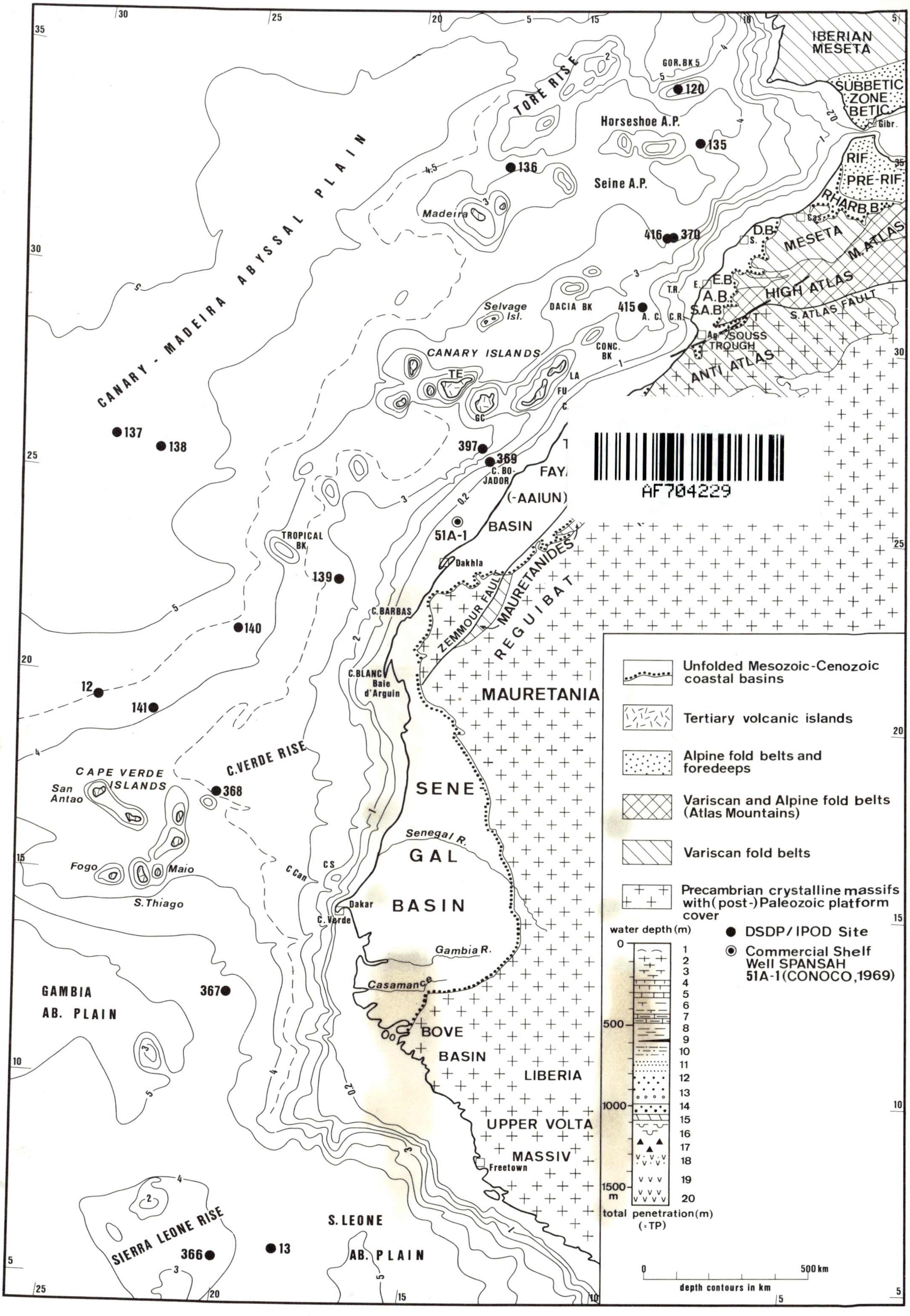

IBERIAN MESETA
SUBBETIC ZONE BETIC
Gibr.
RIF
PRE-RIF
RHARB B.
MESETA
M. ATLAS
HIGH ATLAS
S. ATLAS FAULT
SOUSS TROUGH
ANTI ATLAS
GOR. BK
120
TORE RISE
Horseshoe A.P.
135
136
Seine A.P.
Madeira
416
370
CANARY - MADEIRA ABYSSAL PLAIN
Selvage Isl.
DACIA BK
415
CONC. BK
CANARY ISLANDS
TE
GC
LA
FU
137
138
397
369
C. BOJADOR
(-AAIUN)
BASIN
51A-1
TROPICAL BK
Dakhla
139
C. BARBAS
140
ZEMMOUR FAULT
MAURETANIDES
REGUIBAT
C. BLANC
Baie d'Arguin
12
141
MAURETANIA
C. VERDE RISE
CAPE VERDE ISLANDS
San Antao
368
SENE-
Senegal R.
GAL
Fogo
Maio
S. Thiago
C Can
CS
Dakar
C. Verde
BASIN
Gambia R.
Casamance
367
GAMBIA AB. PLAIN
BOVE BASIN
LIBERIA
UPPER VOLTA
MASSIV
Freetown
SIERRA LEONE RISE
366
13
S. LEONE AB. PLAIN
Unfolded Mesozoic-Cenozoic coastal basins
Tertiary volcanic islands
Alpine fold belts and foredeeps
Variscan and Alpine fold belts (Atlas Mountains)
Variscan fold belts
Precambrian crystalline massifs with (post-) Paleozoic platform cover
DSDP/IPOD Site
Commercial Shelf Well SPANSAH 51A-1 (CONOCO, 1969)
water depth (m)
total penetration (m) (=TP)
500 km
depth contours in km

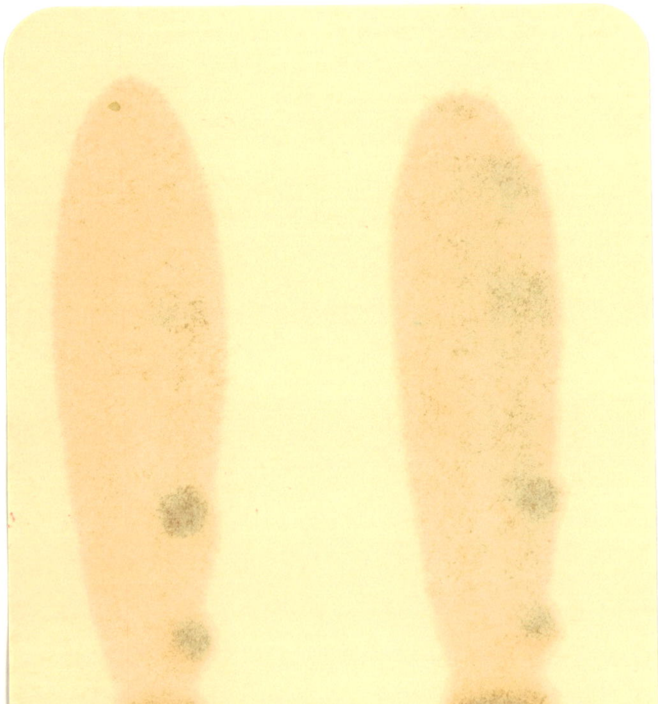

Geology of the Northwest African Continental Margin

Edited by
U. von Rad K. Hinz M. Sarnthein E. Seibold

With 325 Figures

Springer-Verlag
Berlin Heidelberg New York 1982

Dr. Ulrich von Rad
Prof. Dr. Karl Hinz
Bundesanstalt für Geowissenschaften und Rohstoffe
Stilleweg 2, 3000 Hannover 51, FRG

Prof. Dr. Michael Sarnthein
Prof. Dr. Eugen Seibold
Geologisch-Paläontologisches Institut der Universität
Olshausenstraße 40/60, 2300 Kiel, FRG

Front cover motif: Prof. P. Wurster, Bonn

ISBN 3-540-11257-X Springer-Verlag Berlin Heidelberg New York
ISBN 0-387-11257-X Springer-Verlag New York Heidelberg Berlin

Library of Congress Cataloging in Publication Data. Main entry under title: Geology of the northwest African continental margin. Bibliography: p. Includes index. 1. Geology – Africa, Northwest. 2. Continental margins – Africa, Northwest. I. Rad, Ulrich von. QE339.N67G46
556.4 81-18550 AACR2.

Offsetprinting: Beltz Offsetdruck, Hemsbach/Bergstr.
Bookbinding: Konrad Triltsch, Graphischer Betrieb, Würzburg.
2132/3130-543210

Preface

The Northwest African and the conjugate Northeast American continental margins provide excellent examples of mature passive continental margins which record in their thick sediment cover the birth and evolution of the Atlantic Ocean Basins during the past 200 million years. Due to a dense net of single- and multichannel seismic reflection lines, numerous Deep Sea Drilling Project (DSDP) sites and petroleum exploration wells, and the well-known onshore geology, the Northwest African margin is one of the best-documented margins of the world. A particular advantage of this margin is that its coastal basins are well exposed and accessible for detailed stratigraphic studies and onshore-offshore correlations, whereas the Mesozoic-Cenozoic marginal basins off eastern North America underlie only the present continental shelf and slope.

During the past decade, the Northwest African margin was studied in great detail, particularly by the Woods Hole Oceanographic Institution, Lamont-Doherty Geological Observatory, Imperial College London, Shell Internationale Petroleum Maatschappij (Den Haag), Kiel University, and the Bundesanstalt für Geowissenschaften und Rohstoffe (BGR) at Hannover. A number of important synthesis papers summarized these results; e.g., Summerhayes et al., 1971; Dillon and Sougy, 1974; Seibold and Hinz, 1974; Uchupi et al., 1976; Lehner and de Ruiter, 1977 (all quoted in the references of Seibold, this Vol.). During the late 1970's, the International Phase of Ocean Drilling (IPOD), including its pre- and post-site surveys, achieved a major breakthrough for our understanding of the evolution of the Northwest African margin. GLOMAR CHALLENGER visited this margin six times; in 1968/69 (Leg 3), in 1970 (Legs 13/14 with Sites 120, 135-141), in 1975 (Leg 41, Sites 366-370), in 1976 (Leg 47A with Site 397 and Leg 50 with Sites 415/416), and, most recently, in 1981 (Leg 79). We include in this volume a short report from the shipboard party with the latest results of the Mazagan Slope Sites 544-547, data which could no longer be incorporated into the other papers of our volume, or in the front cover figure. Continued and increasing interest in the offshore petroleum potential of the North Atlantic continental margins helped also to stimulate new scientific exploration programs for the 1980's, such as the Ocean Margin Drilling Project with GLOMAR EXPLORER.

The present volume was conceived by the German IPOD group at Kiel in May 1979. It was agreed that the timing was appropriate for a comprehensive synthesis volume summarizing the major new results on this margin in view of the pertinent literature of the past decade. Most papers were open for discussion during a special small symposium at Bonn-Bad Godesberg in October 1980.

The synthesis volume concentrates on geophysical, geological, paleontological, and geochemical studies obtained during the preparation and evaluation of DSDP Legs 14, 41, 47A, 50 and 79, as well as on the structure and Mesozoic-Cenozoic evolution of the onshore coastal basins and the offshore continental margin. It includes also deep crustal geophysical surveys, geological-volcanological studies of the Canary and Cape Verde Islands, paleoenvironmental research based on Neogene to Quaternary cores, and a comparison of the Northwest African margin with its counterpart off eastern North America.

During the past 12 years, the Deutsche Forschungsgemeinschaft (DFG) in Bonn-Bad Godesberg has funded all METEOR cruises and most of the German geoscientific research projects which concentrated on Northwest Africa and its margin. We are especially grateful to Dr. F. Goerlich (Secretary of the Earth Sciences Division of the DFG) and to Dr. H. Beiersdorf (coordinator of the German IPOD contribution at BGR) for their continuous advice and encouragement. The Bundesminister für Forschung und Technologie (BMFT; Federal Minister of Research and Technology, Bonn) supported our continental margin projects with R/V VALDIVIA and BGR's continental margin study group at Hannover, the reports of which formed the nucleus of this synthesis volume (cf. papers no.3, 4, 6, 7, 9). DFG and BMFT together financed the Initial Contribution for the German participation in IPOD.

The JOIDES Planning Committee and the JOIDES Advisory Panel on Passive Margins also supported our idea to compile a regional synthesis volume summarizing all DSDP data from the Northwest African margin.

Dr. Hohm, Mrs. Grössl, Mrs. Schüll, and Mrs. von dem Bussche from the Springer Verlag (Heidelberg) were very helpful and cooperative. The artwork on the front cover of the book is from Professor P. Wurster (Bonn).

Each paper was reviewed by at least one of the editors and one or several peers. On behalf of all authors we should like to thank all colleagues who contributed to the quality of the papers by careful reviews. Last but not least, we are most grateful to all those who helped to make this volume possible, from the captains and crews of our research vessels to the marine and laboratory technicians, translators, and secretaries who went to great trouble to produce camera-ready manuscripts. In order to speed up publication and to keep the price of the book as low as possible, we decided to use the technique of setting the manuscripts in typewriter composition by the authors. Hence, we ask the reader to excuse any printing errors, inconsistencies, and the less-than-perfect appearance of some of our manuscripts.

Hannover and Kiel, August 1981

Ulrich von Rad
Karl Hinz
Michael Sarnthein
Eugen Seibold

Contents

Comparison of the Northwest African and Northeast American Margins

Evolution of Volcanism

Cretaceous Stratigraphy, Sedimentation and Paleoenvironment

Cenozoic Stratigraphy, Sedimentation and Paleoenvironment

Inorganic/Organic Geochemistry

Contributors

You will find the addresses at the beginning of the respective contributions

Introduction

1 The Northwest African Continental Margin - An Introduction

Eugen Seibold[1]

ABSTRACT

The Northwest African continental margin has many features which parallel those of the Northeast American continental margin, both in structure and evolution. However, up to the present time, many aspects of orogenic and volcanic activity and of climatic and oceanographic evolution have differed.

1. Introduction

The Northwest African continental margin, and the contingent Northeast American continental margin are of special scientific interest, as they belong to the oldest passive margins of the modern oceans. Land geology, geophysical data and deep sea drilling results indicate that rifting and sea floor spreading began in late Triassic to mid-Jurassic times. As a result, thick sediment accumulations have developed on most parts of these mature passive continental margins. As a negative consequence, their early evolution is documented by deeply buried rocks and hence hard to evaluate. However, as a positive consequence, one can retain the hope that commercial hydrocarbon reserves can still be discovered in the thick sediment prisms of the continental slope and rise.

Despite their common origin, both margins developed fundamental differences during their history, such as:

- the structure of the continental margins, including orogenic and volcanic activities;
- the paleoclimate of the margins, as well as of the continental hinterland;
- the paleo-oceanography of the adjacent regions of the North Atlantic Ocean (for a detailed comparison of the NW African with the NE American margin see Jansa and Wiedmann, this Vol.).

This volume deals with the Northwest African continental margin between the Azores-Gibraltar Fracture Zone (35°N) and Vema Fracture Zone (10°N; see Figs.1,4). It covers the geology of the onshore coastal basins, the structure and origin of the continental margin, as well as the development of volcanism and the paleoenvironment evolution during Mesozoic and Cenozoic times.

2. History of Investigation

Portuguese sailors were the first to explore the area and this was to mark the beginning of a new epoch. Not only was there new regional exploration, but also experience and direct observations replaced and corrected medieval ideas and speculations, which originated from ancient literature, such as Ptolemeus' geography.

A major historical event was Gil Eannes' discovery in 1434 that Cape Bojador could be passed. Prior to this time, it was common belief

1 Geologisch-Paläontologisches Institut der Universität Kiel
Olshausenstraße 40-60
2300 Kiel, FRG

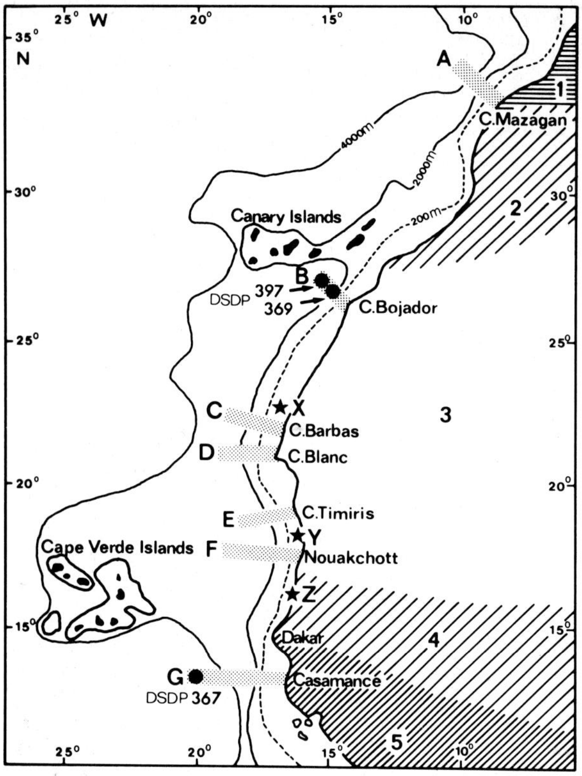

Fig.1
General overview of Northwest Africa and its continental margin with location of profiles A-G (see Fig.2) and present position of climatic zones. 1 Mediterranean scrub, warm-temperature, winter rains; 2 steppe, hot, summer dry; 3 desert, hot, dry; 4 steppe, hot, winter dry; 5 savannah, tropical, winter dry

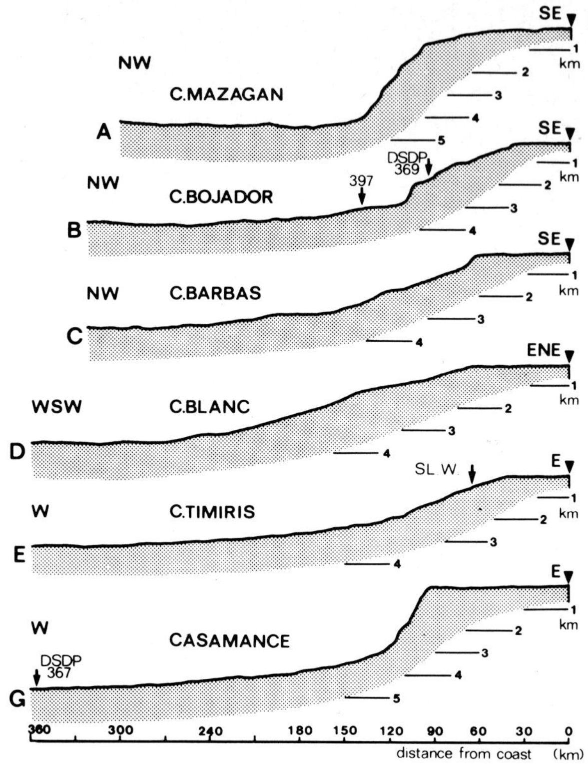

Fig.2
Selected morphological cross-sections of the Northwest African continental margin (for location see Fig.1)

that south of this cape sailing would be extremely dangerous, if not impossible, because of shallow, muddy, even viscid offshore waters and dramatic and incalculable currents. It was predicted that, more to the south, there would be more desert-like, burned and uninhabitable land, without rivers and vegetation. Therefore, the discovery of the Senegal River mouth and of green coastal regions around Cape Verde (sic) in 1444 was sensational. Sierra Leone was discovered just before Prince Henry the Navigator died in 1460. He was the extraordinary personality stimulating these expeditions and advocating the idea to open the seaway to India. It took the Portuguese, with their excellent navigation and mapping, approximately 30 years to outline a realistic coastline from Morocco to Cape Palmas. The contours of a continent began to come into focus as emerging from an inpenetrable fog.

Since the 1950s, again after roughly 30 years, geology and geophysics have acquired a rough idea of offshore structures and sediments. Existing isolated data were presented and discussed during a Symposium Geology of the East Atlantic Continental Margin, held in Cambridge (UK) in March 1970. It was sponsored by SCOR (Scientific Committee on Oceanic Research), CMG (Commission for Marine Geology of the International Union of Geological Sciences) and UNESCO. The results were published by F.M. Delaney (1971). It was prepared by SCOR Working Group 31 (K.O. Emery - Woods Hole, A.P. Lisitsin-Moscow, A.S. Laughton-Wormley, G. Boillot-

Rennes, H. Holtedahl-Bergen E.S.W. Simpson-Cape Town, E. Seibold-Kiel, Chairman). After the Symposium, the group made the following comments to the sponsoring organizations and to the active workers in the field:

"1) We recognise that investigations already undertaken on a national and international basis, as shown by the Symposium, have mainly been concentrated on the continental shelf, especially off Europe, and only patchy work has been done on the slope and rise, and that knowledge of the deeper structure of the margin is limited. We therefore recommend that national efforts be aimed at the major geographical gaps in knowledge, extended to the slope and rise, and be intensified, whenever possible. Such investigations in many cases critically depend upon international cooperation.

2) The major scientific and economic problems, such as the genesis of continental margins and the sedimentary processes on them, especially require an international interdisciplinary approach.

The reconstruction of the North Atlantic, prior to continental separation, places the poorly-known structure of the Northwest African continental margin against the well-studied margin of eastern North America...

Similarly, sedimentary processes on the margins, which have operated since the separation of the continents, have been studied only in very limited areas. This is an important problem, too, because of the latitude range, the varying oceanographic conditions, environments of deposition and the economic potential. Therefore, we recommend, as a first step, that a systematic reconnaissance study of the East Atlantic Continental Margin of the African continent be made...

3) We strongly support a concept of regional working meetings to consolidate the knowledge acquired in key areas...

4) We recommend that SCOR and IUGS encourage the establishment of national committees with countries which do not have them at present, to establish and maintain contact with international programmes.

5) We recognise the significant contributions already made by commercial companies and governmental agencies in furthering the knowledge of the Atlantic margin of Africa, through the partial release of their geological or geophysical data. In the interest of science, we urge those organisations which have further relevant information to make this available as far as possible for publication.

6) We note with interest the geotraverse across the North Atlantic from Cape Hatteras to Northwest Africa planned by ESSA" (Seibold and Emery in Delany, 1971).

These recommendations resulted in numerous international activities. In geophysics, the most important was the 1971-74 Geophysical-Geological Study of the Eastern Continental Margin, a project of the International Decade of Ocean Exploration (IDOE) which covered the margin between Cape Town and Portugal. Furthermore, it was agreed during the Symposium that German efforts should be concentrated between Morocco and Sierra Leone. Thus the investigation of the structure, stratigraphy and paleoenvironment of the Northwest African continental margin became one of the main targets of German oceanographic and marine geoscientific research during the past decade. Reviews of these activities and of the cruises of R.V. Meteor 8/9 - 1967, 25-1971, 30-1973, 39-1975, 46-1977 and R.V. Valdivia (10-1975) were published by Seibold and Hinz (1974, 1976).

The most systematic survey was realized by R.V. Atlantis II. The results and a very detailed bibliography were given by Uchupi et al. (1976). Luyendyk and Bunce (1973) published geophysical results off Morocco. Members of the Lamont-

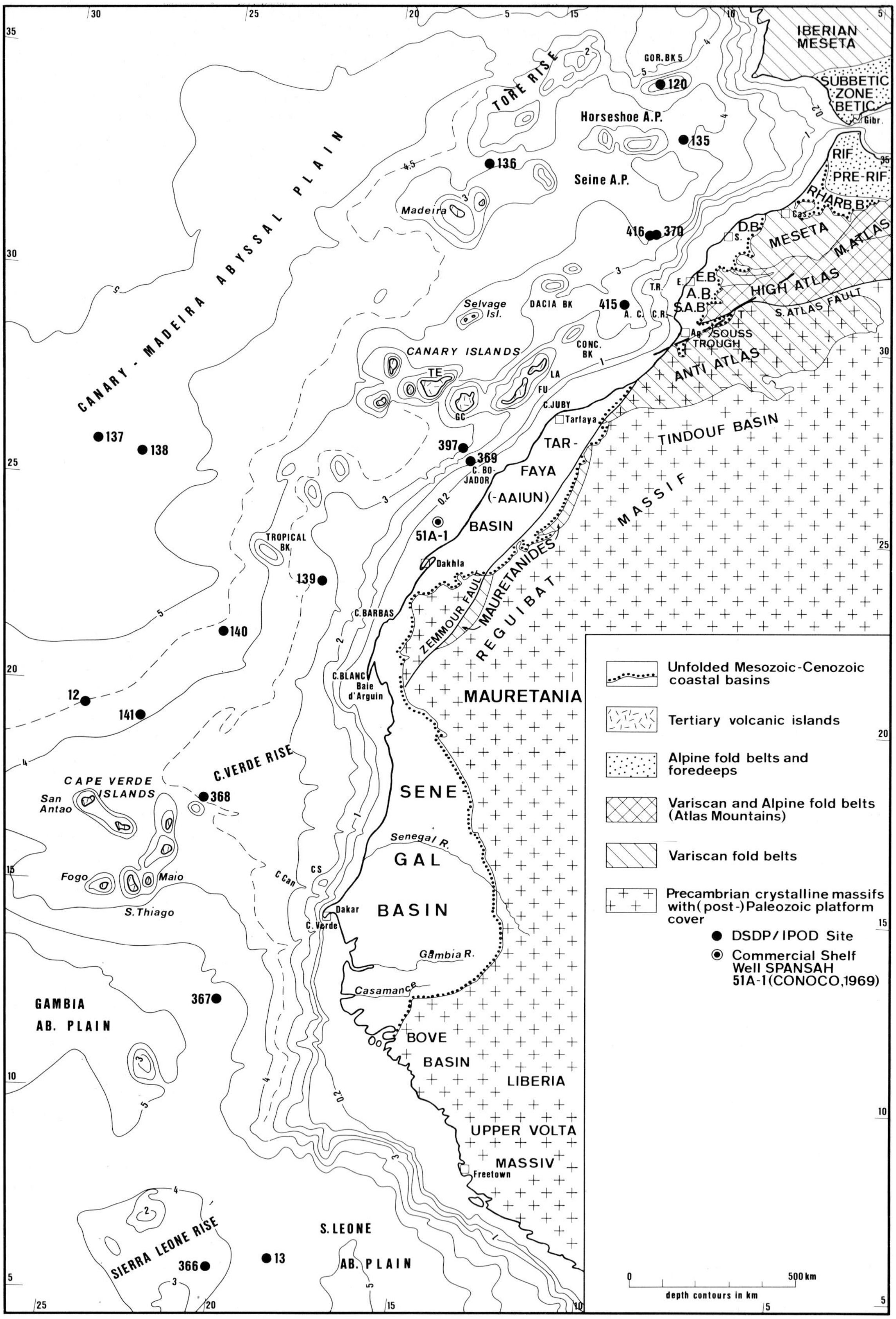
IBERIAN MESETA
SUBBETIC ZONE BETIC
Gibr.
RIF
PRE-RIF
RHARB B.
MESETA
M. ATLAS
HIGH ATLAS
S. ATLAS FAULT
SOUSS TROUGH
ANTI ATLAS
TINDOUF BASIN
MASSIF
REGUIBAT
MAURETANIDES
ZEMMOUR FAULT
MAURETANIA
TAR-FAYA (-AAIUN) BASIN
SENE-GAL BASIN
Senegal R.
Gambia R.
Casamance
BOVE BASIN
LIBERIA
UPPER VOLTA MASSIV
Freetown
Dakar
C. Verde
C. BLANC
Baie d'Arguin
C. BARBAS
Dakhla
C. BOJADOR
C. JUBY
Tarfaya
51A-1
GOR. BK
Horseshoe A.P.
Seine A.P.
TORE RISE
Madeira
CANARY - MADEIRA ABYSSAL PLAIN
Selvage Isl.
DACIA BK
CONC. BK
CANARY ISLANDS
TE
GC
LA
FU
TROPICAL BK
C. VERDE RISE
CAPE VERDE ISLANDS
San Antao
Fogo
Maio
S. Thiago
C Can
CS
GAMBIA AB. PLAIN
SIERRA LEONE RISE
S. LEONE AB. PLAIN
120
135
136
416
370
415
397
369
137
138
139
140
12
141
368
367
366
13
Unfolded Mesozoic-Cenozoic coastal basins
Tertiary volcanic islands
Alpine fold belts and foredeeps
Variscan and Alpine fold belts (Atlas Mountains)
Variscan fold belts
Precambrian crystalline massifs with (post-)Paleozoic platform cover
DSDP/IPOD Site
Commercial Shelf Well SPANSAH 51A-1 (CONOCO, 1969)
0
500 km
depth contours in km

Doherty Geological Observatory evaluated numerous cruises off Northwest Africa (Pitman and Talwani, 1972; Rabinowitz, 1974; Jacobi et al., 1975; Hayes and Rabinowitz, 1975; Jacobi, 1976; Embley and Jacobi, 1977; Jacobi and Hayes, this Vol.).

Several French cruises concentrated on sediments and their organic content, as ORGON III, 1976. Summerhayes and his coworkers reported on British activities in 1971 and 1976. Many of the geophysical results from these cruises, together with some commercial seismic profiles and results (e.g., Beck and Lehner, 1974; Lehner and de Ruiter, 1977), were used to prepare drill sites with the "Glomar Challenger" of the Deep Sea Drilling Project (DSDP), and the International Phase of Ocean Drilling (IPOD). Certainly, this marks a real breakthrough to a better knowledge of the evolution of the Northwest African continental margin (Leg 2: Peterson, Edgar et al., 1970; Leg 3: Maxwell, Von Herzen et al., 1970; Leg 14: Hayes, Pimm et al., 1972; Leg 41: Lancelot, Seibold et al., 1978; Leg 47A: von Rad, Ryan et al., 1979; Leg 50: Lancelot, Winterer et al., 1980).

From 1978 to 1980, the Bundesanstalt für Geowissenschaften und Rohstoffe (BGR, Hannover) evaluated and compiled a large amount of published and unpublished geophysical and geological data on the structure, geological development and hydrocarbon potential of the Northwest African continental margin. Some of the results of this BGR continental margin study group are now published in this volume by Hinz et al., Ranke et al., and Wissmann.

←

Fig.3
Geological sketch map of North West Africa and bathymetry of its continental margin (modified and updated from various sources by U. von Rad). For sources and abbreviations see Front Cover Figure

3. General Morphology

Shelf widths off Northwest Africa exceed 120 km only off wide coastal basins bordered by stable massifs as off the Reguibat High north and south of Cape Blanc, and the Guinea Arch/Bove Basin/Upper Volta Massif in the south. Minimum shelf widths ($<$ 20 km) occur off Cape Rhir and off Cape Verde (Fig.3). Generally, the shelfs are only 40-60 km wide. Shelf breaks occur around 110 - 150 m waterdepth.

The continental slope mostly passes gradually to the rise in water depths ranging from 1500 m to 4000 m. Generally, slope angles vary between 1° and 3°, but off Morocco or off Casamance they may reach 5°-6°. Due to young tectonism or erosional sculpturing of buried Mesozoic carbonate platforms (e.g. Mazagan Escarpment), the slopes may locally reach inclinations up to 40° (Fig.2).

Between the Straits of Gibraltar and the marginal Tafelny Plateau numerous, mainly NW directed small submarine canyons dissect the continental slope. For a length of at least 200 km, the complex, huge Agadir Canyon system extends from the shelf break to the upper rise to a water depth of at least 3200 m. It separates the Tafelny Plateau in the north from the North Tarfaya margin to the south which is also sculptured by many sub-recent small submarine canyons cutting into the outer shelf and upper slope.

South of Cape Bojador to about 25°N, the continental slope is dissected by many canyons, slope valleys and gullies, indicating erosional destruction over wide areas (Seibold and Hinz, 1974; von Rad and Wissmann, this Vol.). South of this area, no real canyons were found up to Cape Blanc. From there to southern Senegal, canyons and other slope incisions are again widespread. These striking differences seem to be the result of different former sediment supply from land.

An unexplained morphological feature is the 70 m high slope wall in a water depth of 500 m between $19^{o}40$'N and Cape Verde.

Echosounder and seismic profiles, together with sediment cores, indicate that the area is widely influenced by mass movements off the Sahara. They normally occur in water depths below 2000 m and occasionally the slides, slumps and debris flows may extend for several hundreds of kilometers onto the continental rise. The last mass movements occurred about 10,000 years ago, when the sea level rise was very fast (Pflaumann et al., this Vol.). The morphology of the continental slope and rise is also influenced by diapirism. Many examples are found off Morocco and to a minor degree off Senegal and Mauretania (Hinz et al., Wissmann, Jacobi and Hayes, all this Vol.).

The continental rise is less than 100 to 1500 km wide, disturbed by the above-mentioned mass movements, by deep-water currents and the continuation of large canyons crossing the rise as channels, sometimes accompanied by levees (Embley, 1976; Embley and Jacobi, 1977; Jacobi, 1976; Seibold and Hinz, 1974; Sarnthein et al., this Vol.).

A general bathymetric map of our area was published by Uchupi (1977), and sectors by Egloff (1972), Marshall et al. (1977), Laughton et al. (1975), Hinz et al. Ranke et al., Jacobi and Hayes, all this Vol.; see also Fig.3).

4. Surface Sediments

Presently, currents and surface waves intensively affect the shelf sediments by erosion, reworking or retarding deposition. Therefore, they greatly vary from coarse sands to muds and are generally coarser near the shelf edge. Carbonate contents vary, too. Carbonates are essentially of biogenic origin, both recent or as relict material, and dominate off the Sahara. South of Cape Blanc terrigenous eolian and fluviatile supply dilutes the carbonates. Off Morocco terrigenous sediments are also widespread, because of higher rainfall and other factors (Summerhayes et al., 1976). Near Cape Blanc biogenic shelf carbonates indicate a boundary between temperate and warm water communities (Lees and Buller, 1972; Lees, 1975; Koopmann et al., 1979).

Grain sizes decrease with water depth on the continental slope. A northward directed undercurrent off Senegal-Mauretania causes some anomalies due to winnowing out fines in several hundred meters water depths and concentrating them in adjacent deeper waters. Stirring particles up by benthic organisms certainly assists in transporting them. Continental slope carbonates are characterized by planktonic foraminifera in the sand fraction and coccoliths in the fine silt to clay fraction. Carbonate contents decrease from about 50% - 70% in the north to less than 30% in the south, caused by the same factors as on the shelf.

Climatic zonation and sorting effects cause a decrease of illite from north to south and an increase of montmorollinite in the same direction and from shallow to deep water (Lange, 1975).

The continental slope and rise off Northwest Africa received less than 100 mm of sediment in 1000 years. During glacial periods, and particularly during the early warm stages, sedimentation rates increased by a factor of 2. For general information on surface sediments see Lisitzin, 1969; Seibold and Fütterer, 1980; and Sarnthein et al., this Vol.

During arid periods with lowered sea level, eolian-sand turbidites were formed (Sarnthein and Diester-Haass, 1977). Indicators for arid versus humid climatic conditions in the African hinterland are, partly controversially, discussed in Lutze et al. (1979), Diester-Haass (1979), Sarnthein (1979), and in Sarnthein et al., 1980 (see also Sarnthein et al., this Vol.; Diester Haass and Chamley, this Vol.).

As a special feature off Northwest Africa, coastal upwelling should be mentioned. It occurs throughout the year between 20°N and 25°N. During summer, up to Portugal, and during winter, down to Sierra Leone. However, in contrast to Southwest

Africa, only a small fraction of the high organic input from surface waters reaches the sediment. Slope and rise sediments contain 0.3% - 4% organic carbon (Seibold, 1980; Diester-Haass and Müller, 1979).

5. General Structure

Northwest Africa shows three elongated onshore coastal basins: (1) the Atlas (Essaouira-Agadir) Basin (="Atlas Gulf" of Wurster and Stets, this Vol.) with the Moroccan Meseta and High Atlas as hinterland, (2) the Aaiun-Tarfaya Basin as a successor of the Paleozoic Tindouf Basin, and (3) the Senegal-Mauretania Basin preceded by the Paleozoic Taudeni Basin.

These onshore basins are separated by Precambrian Highs, the Antiatlas High in the north, together with the Agadir or Tizi n'Test fault trend, which has been active since the Paleozoic, the Reguibat High in the center and the Guinea Arch in the south (Reyre, 1966; Choubert, 1968; Dillon and Sougy, 1974; Caratini and Coumes, 1979). On the continental margin, these basins merge.

The general structure, therefore, is similar to the Northeast American margin, where basins are partly separated by platforms as seaward protrusions of the pre-Jurassic continental crust (Folger et al., 1979).

Many details of the correlation between the basins on either side of the Atlantic and between the American and African basins are unknown. The same is true with the internal basin patterns. This is partly due to differing opinions regarding the original situation before the Atlantic break-up, as to the relations between initial transform faults and basin boundaries, and simply because of lack of data. For pre-drift reconstructions see Bullard et al., 1965; Drake et al., 1968; Le Pichon et al., 1971, 1977; Rabinowitz, 1974; Dewey et al., 1973; Van Houten, 1977; Klitgord and Schouten, 1980.

The oceanic basement of the Northwest African margin is situated in the magnetic quiet zone. Its boundary lies about 500 km offshore, i.e. less than off North America. Seaward, magnetic anomalies 22 to 25 of the M-Sequence border this zone, indicating an approximate age of 150 - 155 m.y. of this oldest basement (Fig.4; see Hinz et al., this Vol.; Roeser, this Vol.).

At least a dozen transform faults running approximately west-east with an approximate average spacing of 200 km were observed. Offsets are relatively small, less than 100 km, even across the Canary Lineament (Hayes and Rabinowitz, 1975; Jacobi and Hayes, this Vol.).

The character of the crust underneath the magnetic quiet zone and the boundaries between oceanic and continental crust are still a matter of discussion. Off Morocco, refraction seismic studies and gravity modeling (Weigel et al., this Vol.; Hinz et al., this Vol.) show that the continental crust thins by about 20 km in the transition from the coastline to the deep sea. The largest thickness reduction takes place between the outer shelf and upper slope (Fig.5). The observed anomalous crust below the Moroccan continental margin is interpreted by the above-mentioned authors to be attenuated continental or "rift-stage" crust (see also the magnetic slope anomalies of Roeser, this Vol.).

Despite much progress in the field of thermal models of continental margin development, the driving forces of subsidence are still poorly understood. Subsidence began locally during the rift stage, i.e., as early as Triassic; regionally, later with the onset of drifting, during Early to Middle Jurassic times. Off Northwest Africa, Jurassic subsidence rates were determined as 80 - 120 m/m.y., about the same or even up to 140 m/m.y. in the Early Cretaceous, and declining to about 15 m/m.y. or even less during Cenozoic (Rona, 1970; von Rad and Einsele, 1980; Ranke et al, this Vol.). Under normal conditions, all these figures are smaller than shelf sedimentation rates, but continental

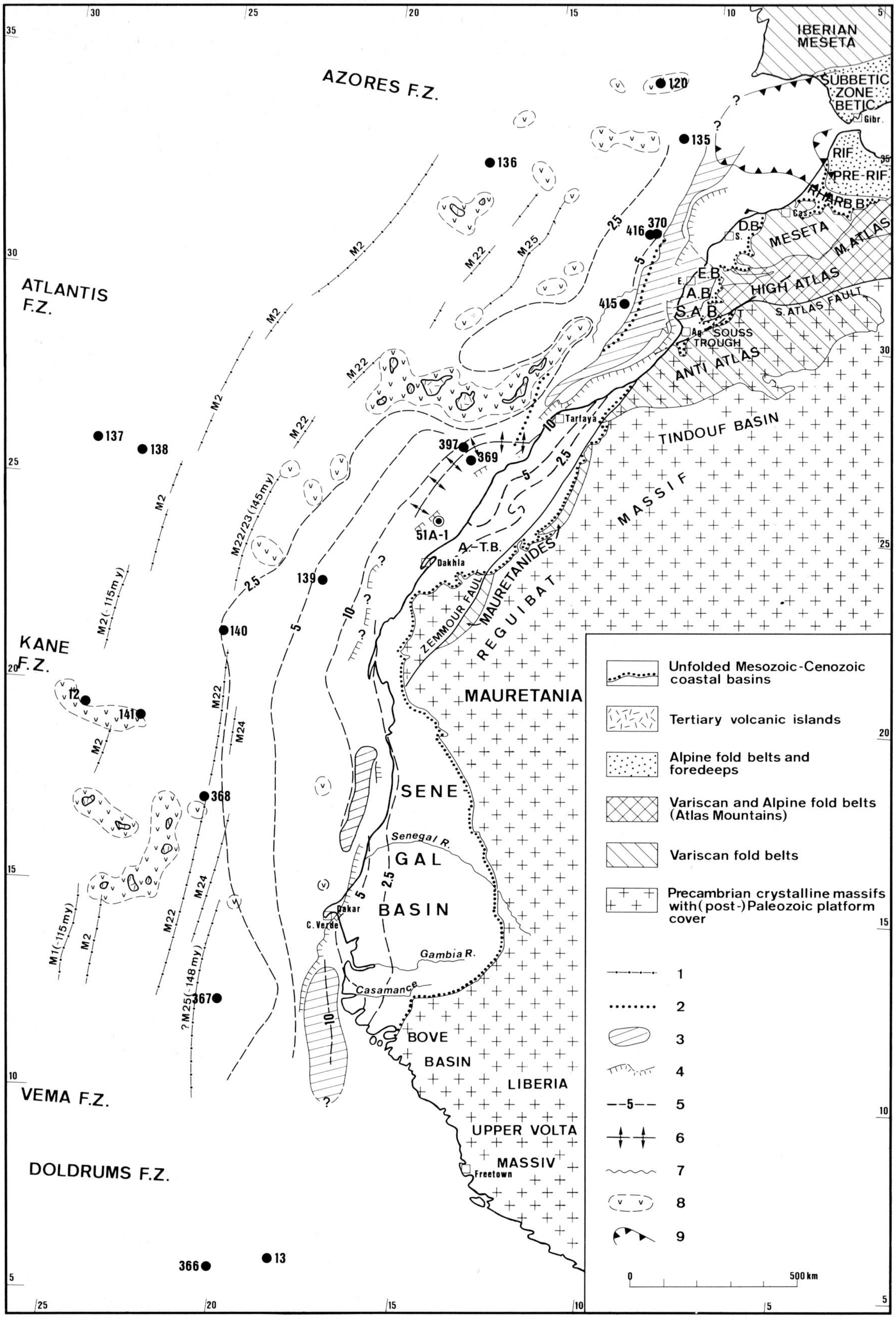
IBERIAN MESETA
SUBBETIC ZONE
BETIC
Gibr.
RIF
PRE-RIF
RHARB B.
MESETA
M. ATLAS
HIGH ATLAS
S. ATLAS FAULT
SOUSS TROUGH
ANTI ATLAS
TINDOUF BASIN
MASSIF
REGUIBAT
MAURETANIDES
ZEMMOUR FAULT
MAURETANIA
SENE GAL BASIN
Senegal R.
Gambia R.
Casamance
Dakar
C. Verde
BOVE BASIN
LIBERIA
UPPER VOLTA
MASSIV
Freetown
Tarfaya
Dakhla
A.-T.B.
AZORES F.Z.
ATLANTIS F.Z.
KANE F.Z.
VEMA F.Z.
DOLDRUMS F.Z.
M2(~115my)
M1(~115my)
M22/23(145my)
?M25(~148my)
Unfolded Mesozoic-Cenozoic coastal basins
Tertiary volcanic islands
Alpine fold belts and foredeeps
Variscan and Alpine fold belts (Atlas Mountains)
Variscan fold belts
Precambrian crystalline massifs with (post-)Paleozoic platform cover
1
2
3
4
5
6
7
8
9
0
500 km

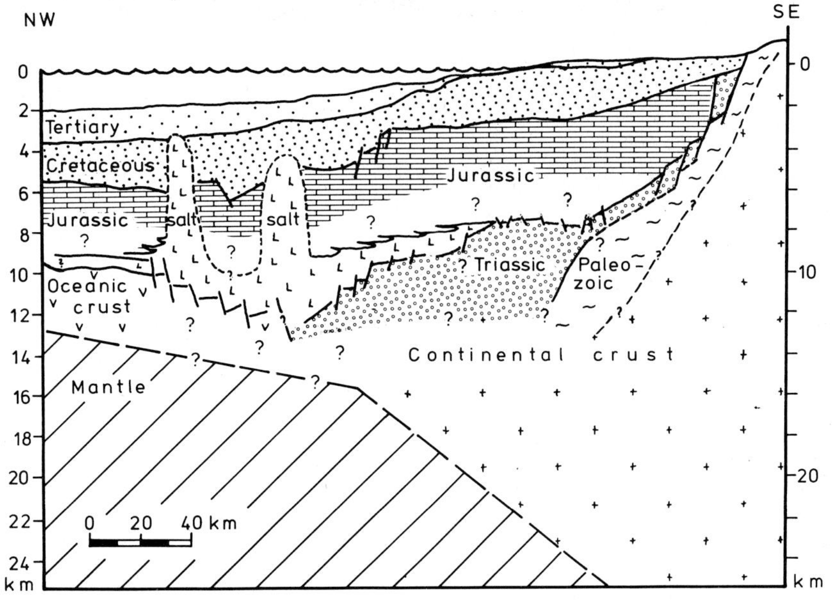

Fig.5
Schematic cross-section of the Southwest Moroccan continental margin (northern Tarfaya Basin). Highly generalized (especially landward of coast and below Jurassic) from reflection/refraction seismic and gravimetric data, well information and published sources by Hinz, Dostmann and Fritsch (this Vol.)

Fig.4
Structural sketch map of North West Africa and its continental margin. Sources for onshore geology see Front Cover Figure. Sources for offshore structure: 2-7 BGR continental margin study (Hinz et al., unpublished reports 1979/80); 1 magnetic anomalies according to Perry et al., 1980 (Naval Research Lab. Acoustics Division Map); M25 off Morocco according to Schouten and Klitgord (in Hinz et al., this Vol.); 2 magnetic slope anomaly S1 (after Roeser, this Vol.); 3 approximate limit of diapiric structures of probable evaporite origin (Late Triassic/Early Jurassic salt); 4 outer edge of Late Jurassic/Early Cretaceous carbonate platform; 5 approximate 2.5 km (=2 s), 5 km, 10 km sediment isopachs; 6 continental slope anticline off Western Sahara (roughly coinciding with depocenter of marginal basin with 10 to >14 km thick sediment fill); 7 outer limit of upper Cretaceous gravity slides; 8 volcanic masses of volcanic islands and selected seamounts; 9 outer limit of Miocene nappes and olistostromes of the Rif and Betic Mountains

slope sedimentation rates rarely exceed the figures given for the Jurassic.

After Lehner and de Ruiter (1977), oceanic basement of Jurassic age is at a depth of 5 - 6 km below sea level in regions of thin overburden in the eastern Atlantic, but at 12 - 15 km depth in marginal basins (see Fig.5). This is also the estimated maximum sediment thickness off Cape Bojador (von Rad and Arthur, 1979; Ranke et al., this Vol.). Off Morocco up to more than 5 km are reported, off Senegal up to more than 12 km (Fig.4). The same order of magnitude holds true for the Northeast American margin (Folger et al., 1979).

6. Genesis of the Northwest African Continental Margin

6.1 General Remarks

The evolution of passive continental margins depends on many factors. The relationship of time (age), sea level fluctuations, rates of subsidence, and of sediment supply - both from the continents and the oceans - are the more important factors. Sediment contribution is influenced by paleocirculation patterns (e.g., continental upwelling) and by oceanic and continental climates, as demonstrated by the upbuilding of carbonate reefs, the outbuilding of deltas, and the diminished sedimentation rates off arid hinterlands.

Margin destruction is caused mainly by turbidity currents, other gravitational mass movements, and deep geostrophic contour currents. These are normally the result of sea level changes or changes of the oceanic circulation. They were due to the opening or closure of new deep-water pathways to the adjacent basins, or to global climatic changes. This holds true for all passive margins. Additionally, off Northwest Africa, Tertiary volcanism has also to be considered. For the following, we can only give some highlights of the combination of these factors off Northwest Africa since the opening of the central North Atlantic.

6.2 Triassic to Liassic Early Rift History

The earliest stages of the Northwest African continental margin evolution are deeply buried. Direct observations on land and geophysical results, however, allow a rough sketch of a possible scenario for the Triassic (Lancelot and Winterer, 1980; Todd and Mitchum, 1977; Van Houten, 1977). Pre-drift reconstructions of the oldest central North Atlantic agree in a juxtaposition of Morocco and Nova Scotia, with parallel Hercynian structures on both sides (Le Pichon et al., 1977; Klitgord and Schouten, 1977; Jansa and Wiedmann, this Vol.). Triassic grabens on the Moroccan mainland were filled with red detrital sediments and lagoonal clays. Possibly, first marine evaporites were formed in some special basins at the end of Triassic. During this time, - basalt flows and doleritic sills were formed.

The evolution from rift phases to drift apparently occurred during the late Liassic (about 170 m.y. ago). A marine transgression, possibly entering from the north, and not from the Tethys, marks the beginning of the North Atlantic Ocean (Bhat et al., 1975; Lancelot and Winterer, 1980). Reviews of the paleobathymetry of the central Atlantic are given by Sclater et al. (1977) and of the paleoenvironment by Berggren and Hollister (1974) and Thiede (1979). During the Early Jurassic, thick evaporites were deposited (Fig. 5; Hinz et al., this Vol.). Off Morocco, possibly a regression during Early Dogger supported their formation. Off Senegal and Mauritania similar evaporites occur (Fig.4; Burke, 1975).

6.3 Mid- to Late Jurassic Transgression and Carbonate Buildup

After the initial rifting, the main event was a circum-Atlantic transgression coming from the Tethys during Callovian/Oxfordian times. Direct indications from rocks in the Atlantic are available only from now on, as dredged shallow-water carbonates from the Mazagan Escarpment (Renz et al., 1975; Wissmann and von Rad, 1979) or Oxfordian deep water carbonates from DSDP Sites 367 off West Africa and Sites 99, 100, 105 and Callovian sediments from Site 533 off Northeast America. During this period in the southern part of the central North Atlantic, oceanic crust was formed, i.e., shortly after the drift had begun off Morocco. Comparisons with the Northeast American Margin are mainly based on Rona (1970) Jansa et al. (1975,1979) Schlee et al. (1976) Sheridan (1976) and Folger et al. (1979).

On both sides of the central North Atlantic, carbonate platforms, together with algal and coral reefs, were formed all along the margins (Fig.4). It is surprising that these buildups partly influenced the morphology up to recent times, as demonstrated by steep slopes off Mazagan or Casamance (Figs.2,3). Initial cooling of the lithosphere caused a continued rapid subsidence. High deposition rates, however, kept pace, resulting in more than 5 km thick, shallow-water carbonate sequences.

As a consequence, the platforms, together with the advancing transgression, blocked terrestrial input to the young basins on both sides of the newly formed Mid-Atlantic Ridge (for subsidence and evolution of water depths, see Berger and von Rad 1972; Sclater et al., 1977). Therefore, deep-water argilaceous, but pelagic, reddish limestones with aptychi and ammonites were deposited above the Calcite Compensation Depth (CCD), comparable with the ammonitico rosso of the Mediterranean Alpine fold belt (Bernoulli and Jenkyns, 1974). Off Morocco, distal turbidites of Tithonian to Hauterivian age occur (Lancelot and Winterer, 1980). Off Senegal, the ideal symmetric evolution in the western and eastern Atlantic basins continued through the Neocomian (Lancelot and Seibold, 1978).

6.4 Early Cretaceous Regression and Delta Formation

During the Early Cretaceous the High Atlas began to rise and coastal basins continued to subside, but the carbonate platform buildup continued, too. Finally, these platforms were drowned (Schlager, 1980) and overwhelmed by fluvial deltaic and shelf sediments, similar to off North America. Generally, this process seems to be time-transgressive toward the south on both contingent margins (termination of carbonate buildup at Georges Bank and off Cape Hatteras during mid- Cretaceous times; at Blake Plateau in Barremian times, in the Bahamas carbonate buildup up to Recent). Off Morocco, however, carbonate buildup terminated in the Late Cretaceous to early Paleogene at the Mazagan Plateau, in Early Cretaceous off the Essaouira Basin, and in earliest Cretaceous time off the Tarfaya Basin (Hinz et al., this Vol.). Off Mauretania carbonate platforms were formed until Aptian, off Northern Senegal until Albian, and off Southern Senegal until Tertiary times (Lehner and de Ruiter, 1977). Lower Cretaceous Wealden-type deltas were widespread around the Northern and Central Atlantic and were especially well studied within our area at Cape Bojador (Einsele and von Rad, 1979; von Rad and Arthur 1979; Wiedmann et al., this Vol.; Ranke et al., this Vol.). Seaward, these sequences grade into laminated prodelta mudstones.

Off Morocco, deep-water marlstones and claystones contain redeposited shallow-water material from the drowned carbonate platform and from the Atlas and Meseta hinterlands as distal turbidites (Schlager, 1980; Robertson and Bernoulli this Vol.).

Off Senegal, Barremian variegated claystones occur, indicating an abrupt shallowing of the CCD after the deposition of carbonates since the Oxfordian (Lancelot and Seibold, 1978; Thierstein, 1979).

Off North America, material prograding seaward over the carbonate shelf edges resulted in extensive sedimentation on the continental slope and rise.

6.5 Mid-Cretaceous Anoxic Events

Similar relations between the eastern and western Atlantic basins exist in mid-Cretaceous (Barremian to Cenomanian) times . Carbonaceous "black shales" were recovered in both basins, again masked by terrigenous input and partly influenced by gravitational sliding and deep water currents off Morocco.

The paleoenvironment favorable to the preservation of these huge amounts of organic matter and the sources of it are an interesting problem (see for example, Schlanger and Jenkyns, 1976; Lancelot and Seibold, 1978). The CCD was close to the basin bottoms. We have to assume partly restricted circulation with reducing conditions in the water column, at least above sediment layers with high organic contents originating from marine organisms (Tissot et al., 1979).

The landward extent of these sediments is extremely important, because they represent excellent hydrocarbon source rocks. However, up to now, very little is known, except for exploration wells in the Senegal basin underneath the shelf southeast of Cape Bojador and some indications in the slope Site 369. In outcrops of the Aaiun-Tarfaya Basin, laminated bituminous marls were observed even in the late Cenomanian to Turonian and attributed to regional coastal upwelling and/or an expanded oxygen-minimum zone (Einsele and Wiedmann, 1975; Wiedmann et al., this Vol.). Even less known is the possible contact of these "black" shales and marls with the former carbonate reef slopes.

6.6 Late Cretaceous Sedimentation

Several causes seem to be responsible for the surprisingly thin upper Cretaceous in the North American and African Basins:

- Cenomanian and other transgressions, partly invading the Sahara reduced the terrigenous sediment supply to the Atlantic.

- Shallowing of the CCD, certainly related to the transgression, resulted in a dissolution of organic carbonates. Commonly green terrigenous clays and silty clays, with almost no carbonates, were deposited. Distal turbidites and other processes caused distinct cycles therein. Off Morocco, gravitational sliding was observed and associated with the Late Cretaceous emergence of the Western High Atlas mountains (Lancelot and Winterer, 1980; Price, 1980).

- Additionally to the reduced sedimentation, bottom currents removed this material within the basins and especially along the continental contours. The correlation of these negative events is an open problem, because hiatuses between Cenomanian and Paleogene are widespread in the American basin and off Cape Bojador. At Site 397, the sedimentary record even from late Hauterivian through Oligocene is missing (Arthur et al., 1979). 1-2 km of lower slope and upper rise sediments were removed there and caused the 100 m.y. hiatus. Major erosional events possibly occurred during Late Cretaceous and Early Tertiary, and most dramatically during the Oligocene (von Rad and Wissmann, this Vol.).

Deep-water pathways opened from the northernmost Atlantic by the separation of Greenland and Spitsbergen (for 25 m.y.) and caused strong western boundary currents. Since the Early Tertiary, the Gulf Stream reduced direct sediment input to the deep sea.

Possibly, the opening of South Atlantic deep water pathways and climatic changes around Antartica were responsible for the deep-sea erosion on the African slope and rise. Much more work, however, has to be done to decipher the importance and origin of hiatuses, so effectively brought into discussion by Moore and Heath (1977) and Vail et al. (1977).

6.7 Paleogene Sedimentation and Erosion

Paleogene sections off North Africa consist of silty clays and claystones together with turbidites off Morocco. As off North America, Eocene cherts and porcellanites indicate a major change in the silica budget, possibly related to drastically changed paleocirculation patterns and to a land climate favoring silica input and triggering siliceous bioproduction in oceanic surface waters. Off Northwest Africa, high contents of palygorskite in Late Cretaceous and Paleogene sediments might indicate recycling of material, deposited under subtropical brackish conditions by wind or river transport into the sea or precipitation during diagenesis from silica, Mg- and Al-rich alkaline solutions in the deep sea environment (Riech and von Rad, 1979).

There are indications that formation of submarine canyons began at least since the Paleocene. However, Oligocene erosion was especially important. A global sea level fall of several hundred meters (Vail et al., 1977) may have caused slumping, turbidity currents, and canyon cutting on the continental slopes and increased deep water circulation (see also Diester-Haass and Chamley, this Vol., for the Oligocene climatic and eustatic history). Additionally, pre-Antarctic bottom water circulation was intensified (Kennett, 1977).

Slopes, oversteepened by these processes as off Cape Bojador, are an exciting example for causing rapid sedimentation of flysch-type sequences in a "passive" margin setting (Arthur and von Rad, 1979).

6.8 Neogene History of Sedimentation and Volcanism

Generally, the Northwest African shelves and slopes were covered only by a thin veneer of Neogene terrigenous sediments, both because of more arid climates and of bypassing of material to the continental rise. More humid conditions in northeast America favored direct terrigenous input to the rise, accentuated by supply from the North caused by western boundary currents (Sarnthein et al., this Vol.). Climate and currents were not the only assymetric features of the eastern and western basins:

(1) Since Early Tertiary in Morocco north of the South Atlas Fault, folding and faulting occurred when the European and African plates finally collided. Price (1980) postulated an influence of the Late Cretaceous to Tertiary Atlas orogenetic events for the margin sediments off central Morocco. Hinz et al. (this Vol.), however, suggest that these sediments were only influenced by gravity tectonics due to salt diapirism. During Mid to Late Tertiary, the Rif and High Atlas Mountains were uplifted. Therefore, Tertiary turbidites are widespread off Morocco (especially in Site 370/ 416)

(2) Off Northeast America, volcanic activity, as on Bermuda or the New England seamounts together with the regional uplift of the Bermuda Rise, started prior to the Early Cenozoic.

Off Northwest Africa, precursors of volcanism might have been active since mid-Cretaceous times (Robertson and Bernoulli, this Vol.). Growth and uplift of offshore features, however, did not begin before the late Paleogene and culminated during the Miocene both in the Cape Verde Islands (Grunau et al., 1975; Lancelot and Seibold, 1978) and in the Canary Islands archipelago (Schmincke, 1976, and this Vol.; McDougall and Schmincke, 1977; Schmincke and von Rad, 1979).

These events added volcanoclastics to the marine sediments. However, the uplift of rises and island chains had even more important consequences for sedimentation. As an example, turbidity currents originating off northern Senegal were diverted to the south by the Cape Verde Rise rising during the Miocene. Regionally, the evolution of the Canary and Cape Verde Island areas certainly influenced strength and directions especially of deep water currents. The most dramatic erosional events apparently occurred before the Neogene. Nevertheless, minor hiatuses of a few m.y. were observed, particularly in the carefully studied Neogene section of DSDP Site 397 off Cape Bojador (von Rad, Ryan et al., 1979) and Site 366 (Sarnthein et al., this Vol.).

As off North America, Late Cretaceous through Paleogene erosion and nondeposition changed to rapid Miocene deposition in the deep basins off Northwest Africa. From middle Miocene until Recent times, nannofossil marls and oozes prevail because of a major lowering of the CCD, a possible higher fertility of surface water masses and a reduced terrigenous supply by gradually more accentuated arid climatic phases in the Sahara (Sarnthein et al., this Vol.).

Acknowledgements. I should like to thank U. von Rad (Hannover) for supplementing and improving a first draft of this manuscript, and K. Hinz (Hannover) and U. von Rad for providing Figs. 3-5. K. Hinz, U. von Rad and M. Sarnthein (Kiel) kindly reviewed the manuscript.

REFERENCES

Arthur MA, von Rad U, Cornford C, McCoy FW, Sarnthein M (1979) Evolution and sedimentary history of the Cape Bojador continental margin, northwestern Africa. In: von Rad U, Ryan WBF, et al., Init Repts Deep Sea Drilling Project 47 (1) Washington (U.S. Gov Printing Office): 773-816

Beck RH, Lehner P (1974) Oceans, new frontier in exploration. Amer Assoc Petroleum Geologists Bull 58: 376-395

Berger WH, von Rad U (1972) Cretaceous and Cenozoic sediments from the Atlantic Ocean. In: Hayes DE, Pimm AC et al Init Repts of the Deep Sea Drilling Project 14, Washington (U.S. Government Printing Office): 787-954

Berggren WA, Hollister CD (1974) Paleogeography, paleobiogeography and the history of circulation in the Atlantic Ocean. Soc Econ Paleont Mineral, Spec Publ 20: 126-186

Bernoulli D, Jenkyns HC (1974) Alpine, Mediterranean, and Central Atlantic Mesozoic facies in relation to the early evolution of the Tethys. In: Dott RH, Shaver RH (eds), Modern and Ancient Geosynclinal Sedimentation: Soc Econ Paleont Mineral, Spec Publ 19: 129-160

Bhat H, McMillan NJ, Aubert J, Porthault B, Surin M (1975) North American and African drift - the record in Mesozoic coastal plain rocks, Nova Scotia and Morocco. In: Yorath CJ, et al (eds), Canadian Continental Margins and offshore Petroleum Potential, Mem Can Soc Petrol Geol 4: 375-389

Burke K (1975) Atlantic evaporites formed by evaporation of water spilled from Pacific, Tethyan and Southern oceans. Geology 3: 613-616

Bullard EC, Everett JE, Gilbert Smith A (1965) The fit of the continents around the Atlantic. In: A Symposium on Continental Drift, Phil Trans Roy Soc London 258A (no 1088): 41-51

Caratini C, Coumes F (1979), Le cadre géologique, structural et sedimentaire. Rev Inst Francais du Pétrole 34 (6): 858-872

Choubert G (1968) International tectonic map of Africa, 1: 5,000,000. Assoc African Geol Surveys and UNESCO, Paris

Delany FM (ed) (1971) ICSU-SCOR Working Party 31, Symposium Cambridge. The Geology of the East Atlantic Continental Margin, 1, general and economic papers Rep No 70/13, Inst Geol Sci, London 176 (1971), Africa; Rep No 70/16 209 pp (1970)

Dewey IF, Pitman CC, Ryan WBF, Bonnin J (1975) Plate tectonics and the evolution of the Alpine system. Geol Soc Amer Bull 84: 3137-3180

Diester-Haass L (1979) Indicators of continental climates in marine sediments. A reply. "Meteor"-Forsch Ergebn C 31: 55-58

Diester-Haass L, Müller PJ (1979) Processes influencing sand fraction composition and organic matter content in surface sediments off West Africa (12°-19^oN), Meteor Forsch-Ergebn Reihe C 31: 21-47

Dillon WP, Sougy JMA (1974) Geology of West Africa and Canary and Cape Verde Islands. In: Nairn AEM, Stehli FG (eds), 2, New York (Plenum Press): 315-390

Drake CL, Ewing M, Sutton GH (1959) Continental margins and geosynclines: the east coast of North America, north of Cape Hatteras, In: Physics and Geochemistry of the Earth, 3, New York (Pergamon Press): 110-198

Egloff J (1972) Morphology of the ocean basins seaward of northwest Africa: The Canary Islands to Monrovia, Liberia. Amer Ass Petroleum Geol Bull 56: 694-706

Einsele G, von Rad U (1979) Facies and paleoenvironment of Lower Cretaceous sediments at DSDP Site 397 and in the Aaiun Basin (Northwest Africa). In: von Rad U, Ryan WBF et al Init Repts Deep Sea Drilling Project 47 (1) Washington (U.S. Gov Print Office): 559-578

Einsele G, Wiedmann J (1975), Faunal and sedimentological evidence for upwelling in the Upper Cretaceous coastal basin of Tarfaya, Morocco. Ninth Internat Congress of Sedimentology Nice, Thème 1: 67-72

Embley RW (1976) New evidence for occurrence of debris flow deposits in the deep sea. Geology 4: 371-374

Embley RW, Jacobi R (1977) Distribution and morphology of large submarine sediment slides and slumps on Atlantic continental margins. In: Richards F (ed), Marine Slope Stability, Mar Geotechn, 2: 205-228

Folger DW, Dillon WP, Grow JA, Klitgord KD, Schlee JS (1979) Evolution of the Atlantic continental margin of the United States. In: Talwani M, Hay WW, Ryan WBF (eds), Deep Drilling Results in the Atlantic Ocean: Continental Margins and Paleoenvironment, Washington (Am Geophys Union): 87-108

Grunau HR, Lehner P, Cleintaur MR, Allenbach P, Bakker G (1975) New radiometric ages and seismic data from Fuerteventura (Canary Islands), Maio (Cape Verde Islands) and Sao Tomé (Gulf of Guinea). In: Progress in Geodynamics, Amsterdam (Roy Neth Acad Arts Sci): 90-118

Hayes DE, Pimm AC et al. (1972) Initial Reports Deep Sea Drilling Project Washington DC, 14: 975 pp

Hayes DE, Rabinowitz PD (1975) Mesozoic magnetic lineations and the magnetic quiet zone off northwest Africa. Earth Planet Sci Lett 28: 105-115

Hinz K, Seibold E (1980) Passive Continental Margins: NW Africa, North Atlantic, NW Australia. In: Closs H, et al (eds), Mobile Earth, International Geodynamics Project, Final Report, F.R.G.: 127-138

Jacobi RD (1976) Sediment slides on the northwestern continental margin of Africa. Marine Geol 22: 157-173

Jacobi RD, Rabinowitz PD, Embley RW (1975) Sediment waves on the Moroccan continental rise. Marine Geol 19: 61-68

Jansa LF, Enos P, Tucholke BE, Gradstein FM, Sheridan RE (1979) Mesozoic-Cenozoic sedimentary formations of the North American Basin, Western North Atlantic. In: Talwani M, Hay WW, Ryan WBF (eds), Deep Drilling Results in the Atlantic Ocean: Continental Margins and Paleoenvironment, Washington (Am Geophys Union): 1-57

Jansa LF, Wade JA (1975) Geology of the continental margin off Nova Scotia and Newfoundland. Offshore Geology of Eastern Canada, Geol Survey of Canada, Paper 74-30 (2): 258 pp

Kennett JP (1977) Cenozoic evolution of Antarctic glaciation, the Circum-Antarctic Ocean, and their impact on global paleoceanography. J Geophys Res 82: 3843-3860

Klitgord KD, Schouten H (1977) The Onset of Sea Floor Spreading from Magnetic Anomalies. Symposium of the geological development of New York Bight Proceedings, Palisades NY (Lamont-Doherty Geological Observatory)

Koopmann B, Lees A, Piessens P, Sarnthein M (1979) Skeletal carbonate sands and wind-derived silty marls off the Sahara coast: Baie du Levrier, Arguin Platform, Mauretania. "Meteor" Forsch-Ergebn C 30: 15-57

Lancelot Y, Seibold E et al (1978) Initial Reports of the Deep Sea Drilling Project Washington DC 41: 1259 pp

Lancelot Y, Winterer EL et al (1980) Initial Reports of the Deep Sea Drilling Project Washington, DC 50: 868 pp

Lange H (1975) Herkunft und Verteilung von Oberflächensedimenten des westafrikanischen Schelfs und Kontinentalhanges. "Meteor" Forsch Ergebn C 22: 61-84

Laughton AS, Roberts DG, Graves P (1975) Bathymetry of the northeast Atlantic: Mid-Atlantic Ridge to southwest Europa. Deep-Sea Research 22: 791-810

Lees A (1975) Possible influence of salinity and temperature on modern shelf carbonate sedimentation. Marine Geol 19: 159-198

Lees A, Buller AT (1972) Modern temperate-water and warm- water shelf carbonate sediments contrasted. Marine Geol 13: 67-73

Lehner P, de Ruiter PAD (1977) Structural history of the Atlantic margin of Africa. Amer Assoc Petroleum Geol Bull 61: 961-981

Le Pichon X, Fox PJ (1971) Marginal offsets, fracture zones, and the early opening of the North Atlantic. J Geophys Res 76 (no 26): 6294-6308

Le Pichon X, Sibuet JC and Francheteau C (1977) The fit of the continents around the North Atlantic Ocean. Tectonophysics 38: 169-209

Lisitzin AP (ed) (1969) Atlas of Atlantic Ocean sediments. Akad Nauk SSSR (russ.), Moscow

Lutze GF, Sarnthein M, Koopmann B, Pflaumann U, Erlenkeuser H, Thiede J (1979) Meteor Cores 12309: Late Pleistocene reference section for interpretation of the Neogene of Site 397. In: von Rad U, Ryan WBF et al, Init Repts Deep Sea Drilling Project Washington (US Gov Printing Office) 47: 727-740

Maxwell AE, von Herzen RP, Andrews JE, Boyce RE, Milow ED, Hsü KJ, Percival SF, Saito T (1970) Summary and conclusions. In: Maxwell AE et al, Init Repts Deep Sea Drilling Project Washington (US Gov Printing Office) 5: 441-474

McDougall I, Schmincke HU (1977) Geochronology of Gran Canaria, Canary Islands: age of shield building and other magnetic phases. Bull Volcanology 40: 1-21

Moore TC and Heath GR (1977) Survival of deep-sea sedimentary sections. Earth Planet Sci Lett, 37: 71-80

ORGON (1979) La géochimie organique des sédiments marins profonds. Mission ORGON 3, Arnould M, Pelet R (eds) Edit Centre Nat Rech Scient, Paris

Peterson MNA, Edgar NT, von der Borch CC, Rex RW (1970) Cruise Leg summary and discussion. In: Peterson MNA et al Init Rept Deep Sea Drilling Project Washington DC (U.S. Gov Print Office) 2:413

Pitman WC III, Talwani M (1972) Sea-floor spreading in the North Atlantic. Geol Soc Am Bull 83: 619-646

Price J (1980) Gravity tectonics on a passive margin: Deep Sea Drilling Project Site 415 in relation to regional seismic data. In: Lancelot Y, Winterer EL et al, Init Reps Deep Sea Drilling Project, Washington DC (US Gov Printing Office, 50: 759-772

Rabinowitz PD (1974) The boundary between oceanic and continental crust in the western North Atlantic. In: Burk CA, Drake CL (eds) The Geology of Continental Margins New York, Heidelberg, Berlin (Springer): 391-408

Renz O, Imlay R, Lancelot Y, Ryan WBF (1975) Ammonite-rich Oxfordian limestones from the base of the continental slope off Northwest Africa. Eclogae geol Helv 68: 431-448

Reyre D (1964) Bassins sedimentaires du littoral africain, Symp Ass Serv Géol Afr, New Delhi

Riech V, von Rad U (1979) Silica diagenesis in the Atlantic Ocean: diagenetic potential and transformations. In: Talwani M, Hay WW, Ryan WBF (eds), Deep Drilling Results in the Atlantic Ocean: Continental Margins and Paleoenvironments, Washington (Am Geophys Union): 315-340

Rona PA (1970) Comparison of continental margins of eastern North America at Cape Hatteras and northwestern Africa at Cape Blanc. Amer Assoc Petrol Geol Bull 54(1): 129-157

Sarnthein M (1979) Indicators of continental climates in marine sediments, a discussion. "Meteor" Forsch-Ergeb C 31: 49-51

Sarnthein M, Diester-Haass (1977) Eolian-sand turbidites. J Sed Petr 47: 868-890

Sarnthein M, Seibold E, Rognon P (eds) (1980), Sahara and surrounding seas - sediments and climatic changes. Palaeoecology of Africa and the Surrounding Islands, Symposium Mainz, Rotterdam (Balkema), 408 pp

Schlager W (1980) Mesozoic calciturbidites in Deep Sea Drilling Project hole 416A- Recognition of a drowned carbonate platform. In: Lancelot Y, Winterer EL et al, Init Repts Deep Sea Drilling Project 50, Washington (US Gov Printing Office) 50: 733-750

Schlanger SO, Jenkyns HC (1976) Cretaceous oceanic anoxic events, causes and consequences. Geol Mijnbow 55: 179-184

Schlee J, Behrendt JC, Robb JM, Mattick RE, Taylor PT, Lawson BJ (1976) Regional geologic framework off northeastern United States. Amer Assoc Petroleum Geol Bull 60: 926-951

Schmincke HU (1976) The geology of the Canary Islands. In: Kunkel G (ed) Biogeography and ecology in the Canary Islands, Den Haag (Junk): 77-184

Schmincke HU, von Rad U (1979) Neogene evolution of Canary Islands volcanism inferred from ash layers and volcanoclastic sandstones of DSDP Site 397 (Leg 47A). In: von Rad U, Ryan WBF et al, Init Repts Deep Sea Drilling Project, Washington (US Gov Printing Office) 47: 703-726

Sclater JG, Hellinger S, Tappscott C (1977) The paleobathymetry of the Atlantic Ocean from the Jurassic to Present.J Geol 85: 509-552

Seibold E (in press) Sediments in upwelling areas, particularly off Northwest Africa. Rapports et Procès Verbaux Series ICES, Kopenhagen

Seibold E, Fütterer D (in press) Sediment dynamics on the NW African continental margin. In: Heezen Memorial Volume, Wiley and Chichester

Seibold E, Hinz K (1974) Continental slope construction and destruction, West Africa. In: Burk CA, Drake CHL (eds), The Geology of Continental Margins, New York, Heidelberg, Berlin (Springer): 179-196

Seibold E, Hinz K (1976) German cruises to the continental margin of North West Africa in 1975: general reports and preliminary results from "Valdivia" 10 and "Meteor" 39. "Meteor" Forsch-Ergebn C 35: 47-80

Sheridan RE (1976) Sedimentary basins of the Atlantic margin of North America. Tectonophysics 36: 113-132

Summerhayes CP, Milliman JD, Briggs SR, Bee AG, Hogan C q (1976) Northwest African shelf sediments: influence of climate and sedimentary processes. J Geol 84: 277-300

Summerhayes CP, Nutter AH, Tooms JS (1971) Geological structure and development of the continental margin of Northwest Africa. Marine Geol 11: 1-27

Thiede J (1979) History of the North Atlantic Ocean: evolution of an asymmetric zonal paleoenvironment in a latitudinal ocean basin. In: Talwani M, Hay WW, Ryan WBF (eds), Deep Drilling Results in the Atlantic Ocean: Continental Margins and Paleoenvironment, Washington (Am Geophys Union): 275-298

Thierstein HR (1979) Paleoceanographic implications of organic carbon and carbonate distribution in Mesozoic deep-sea sediments. In: Talwani M, Hay WW, Ryan WBF (eds), Deep Drilling Results in the Atlantic Ocean: Continental Margins and Paleoenvironment, Washington (Am Geophys Union): 249-274

Tissot B, Deroo G, Herbin JP (1979) Organic matter in Cretaceous sediments of the North Atlantic: Contribution to sedimentology and paleogeography, In: Talwani M, Hay WW, Ryan WBF (eds), Deep drilling results in the Atlantic Ocean: Continental margins and paleoenvironment. Washington (Am Geophys Union): 312-374

Todd RG, Mitchum Jr RM (1977) Identification of Upper Triassic, Jurassic, and Lower Cretaceous seismic sequences in the Gulf of Mexico and offshore West Africa. In: Vail P R et al, Seismic stratigraphy and global changes of sea level, Amer Assoc Petroleum Geol Memoir 26: 145-163

Uchupi E (1971) Bathymetric Atlas of the Atlantic, Caribbean and Gulf of Mexico, Woods Hole Ocean Inst Techn Rep 71-72

Uchupi E, Emery KO, Bowin CO, Phillips JD (1976) Continental margin of Western Africa: Senegal to Portugal. Amer Assoc Petroleum Geologists Bull 60: 809-878

Vail PR, Mitchum, Jr. RM, Thompson III S (1977) Global cycles of relative changes of sea level. In: Vail PR et al Seismic stratigraphy and global changes of sea level, Amer Assoc Petroleum Geol Memoir 26: 83-98

van Houten FB (1977) Triassic-Liassic deposits of Morocco and eastern North America: a comparison. Amer Ass Petr Geol Bull 61: 79-99

von Rad U, Arthur MA (1979) Geodynamic, sedimentary and volcanic evolution of the Cape Bojador continental margin (NW Africa). In: Talwani M, Hay WW, Ryan WBF (eds), Deep Drilling Results in the Atlantic Ocean: Continental Margins and Paleoenvironment: Washington (Am Geophys Union):-187-203

von Rad U, Einsele G (1980) Mesozoic-Cainozoic subsidence history and palaeobathymetry of the northwest African continental margin (Aaiun Basin to DSDP Site 397). In: Kent P, Laughton AS, Roberts DG, Jones EWJ (eds) The Evolution of Passive Continental Margins in the Light of Recent Deep Drilling Results Phil Trans Roy Soc London A, 294: 37-50

von Rad U, Ryan WBF, et al. (1979) Initial Reports of the Deep Sea Drilling Project, 47, part 1, Washington DC: 835 pp

Wissmann G, von Rad U (1979) Seismic structure, continental basement, and Mesozoic sediments from the Mazagan Plateau off Morocco. "Meteor" Forsch-Ergebn C 31: 1-20

Structure and Geodynamic Evolution of the Continental Margin

2 Preliminary Results from DSDP Leg 79 Seaward of the Mazagan Plateau off Central Morocco

Karl Hinz[1], Edward L. Winterer[2], Peter O. Baumgartner[2], Martin J. Bradshaw[3], James E. T. Channel[4], Michel Jaffrezo[5], Lubomir F. Jansa[6], Robert Mark Leckie[7], Johnnie N. Moore[8], Jürgen Rullkötter[9], Carl Schaftenaar[10], Torsten H. Steiger[11], Vassil Vuchev[12], and George E. Wiegand[13]

1. Introduction

The region of the Mazagan Plateau (Figure 1) offers unique access for Glomar Challenger to document the Jurassic environment of rifting and the early subsidence history of a segment of passive margin bordering the proto-Atlantic. The edge of the Upper Jurassic carbonate platform bulges seaward here from its normal position beneath the present Moroccan continental shelf, and crops out along a spectacular escarpment that falls about 1 km to a broad slope leading to the abyssal sea floor (Figures 2 and 7). The escarpment reveals a section of Mesozoic strata, near the base of which Oxfordian algal limestone has been dredged. With a series of holes drilled near the escarpment, we planned to piece together a composite section that would record the sequence of depositional environments and subsidence history of this thick carbonate platform, including the timing of platform drowing and the installation of pelagic conditions. The results of drilling here bear strongly on the early evolution of the deeply buried conjugate margin on the North American side.

Seaward of the Mazagan slope is a band about 50 km wide where diapirs punch upward into a pile of sediments about 4 km thick (Figure 1). Dredges and cores on the slopes of one of these structural highs only about 15 km from the foot of the escarpment (Figure 2), recovered fragments of sheared granite. We believed this structure might thus bring close to the sea floor very old sediments, dating to the earliest stage of Atlantic evolution, in the early Jurassic or late Triassic. The dredged fragments of granite suggest that this particular structure might not be a salt diapir but a Hercynian basement high, overlain by "syn-rift" to early "post-rift" clastic and restricted (evaporitic or carbonaceous) sediments. The nature of the evaporites that appear in the

1 Bundesanstalt für Geowissenschaften und Rohstoffe, Stilleweg 2, 3000 Hannover 51, FRG

2 Scripps Institution of Oceanography, La Jolla, CA, USA

3 University of Aston, Birmingham, Great Britain

4 Lamont-Doherty Geological Observatory, Palisades, USA

5 Université Pierre et Marie Curie, Paris, France

6 Geological Survey of Canada, Dartmouth, Nova Scotia, Canada

7 University of Colorado, Boulder, CO, USA

8 University of Montana, Missoula, MT, USA

9 Kernforschungsanlage Jülich, 5170 Jülich, FRG

10 Texas A and M University College Station, TX, USA

11 Institut für Paläontologie und Historische Geologie, 8000 München, FRG

12 Kansas Geological Survey, Lawrence, KS, USA

13 Florida State University Tallahassee, FL, USA

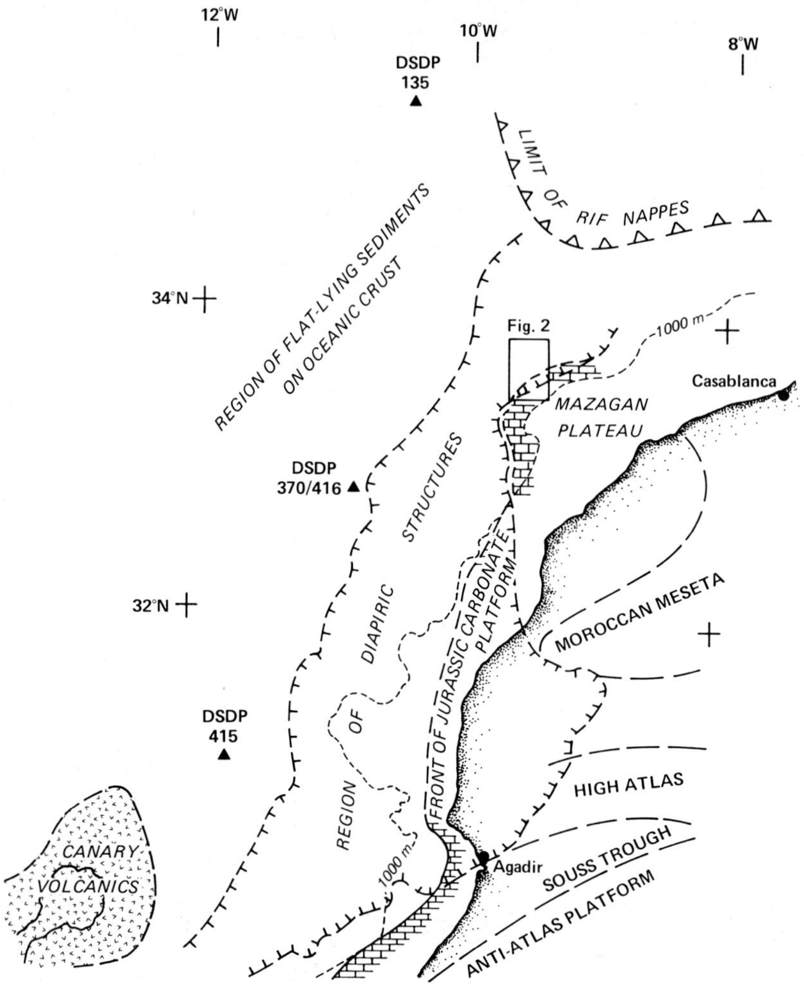

Fig. 1. Map showing the regional setting of the Mazagan Plateau (see also back cover figure)

diapir province is an important unsolved problem, and clues as to their age and as to whether they formed in a deep or shallow basin are badly needed.

Leg 79 began in Las Palmas, in the Canary Islands, on 15 April 1981 and ended in Brest on 30 May 1981, 45 days later. Thirty four days were spent at 4 drill sites, 2 days in drydock in Lisbon, and 9 days in transit. We drilled at 4 sites (Figure 2):

Site 544, located on the seaward (NE) margin of the structural high from which granitic fragments were dredged, at a place where acoustic basement is covered by only about 100 meters of pelagic sediments.

Site 545, located near the foot of the steep part of the Mazagan Escarpment at a place where older Mesozoic strata are blanketed by only a thin wedge of Cenozoic slope sediments, and where seismic data suggested that the drill might reach basement at perhaps 1000 meters depth.

Site 546, located on the next structural high seaward of Site 544. This site was chosen after drilling at Site 544 discovered gneissic basement, in order to test a structure that appeared from seismic data to be more certainly diapiric. Permission to drill was given by DSDP on the grounds that acoustic basement at Site 546 (as at Site 544) is above the level of the surrounding deep sea floor, and is covered by only a thin mantle of pelagic sediments.

Site 547, located down the plunge of the same structure drilled at Site 544, at a place where the

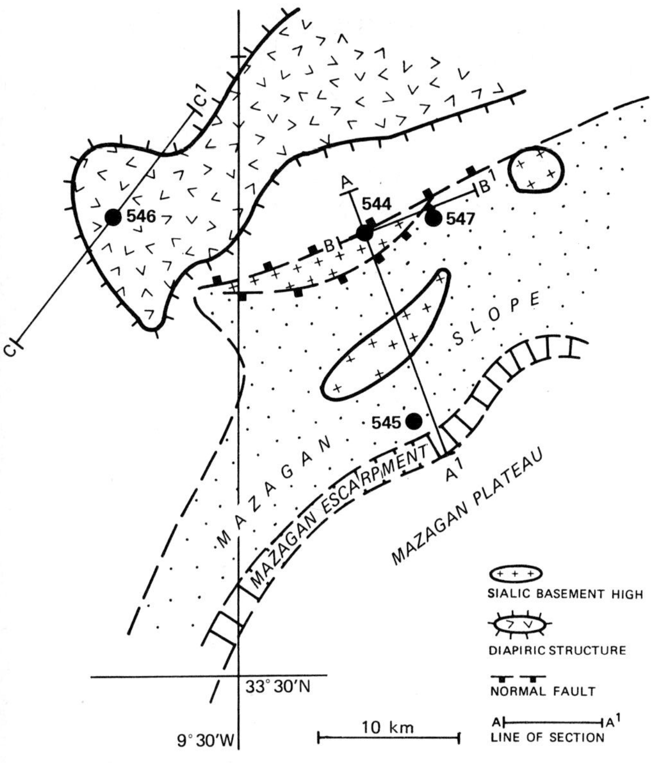

Fig. 2. Mazagan Escarpment area, showing sites drilled during Leg 79, location of lines of cross sections, and major geologic features (cf. also Hinz, Dostmann and Fritsch, this Vol.)

sedimentary section overlying acoustic basement is much thicker than at Site 544.

Results from each of the sites arc described in the following section, and graphic logs are given in Figures 3,4,5 and 6. Geologic cross-sections through the drill sites are shown as Figures 7,8 and 9.

Because of the preliminary nature of this report which was assembled only two months after the end of the cruise from the Leg 79 summary, we have not quoted any literature. The interested reader is referred to the referencens quoted in Hinz, Dostmann and Fritsch; Jansa and Wiedmann; and Seibold (all, this Vol.).

2. Stratigraphic results of Sites 544-547

2.1 Site 544 (latitude: 33°46.0'N; longitude 9°24.3'W; water depth: 3607 meters; sub-bottom depth: 235.0 meters; recovered section: 83.1 meters).

Three holes were drilled at Site 544; at the first hole (544) we took only a single core, and we

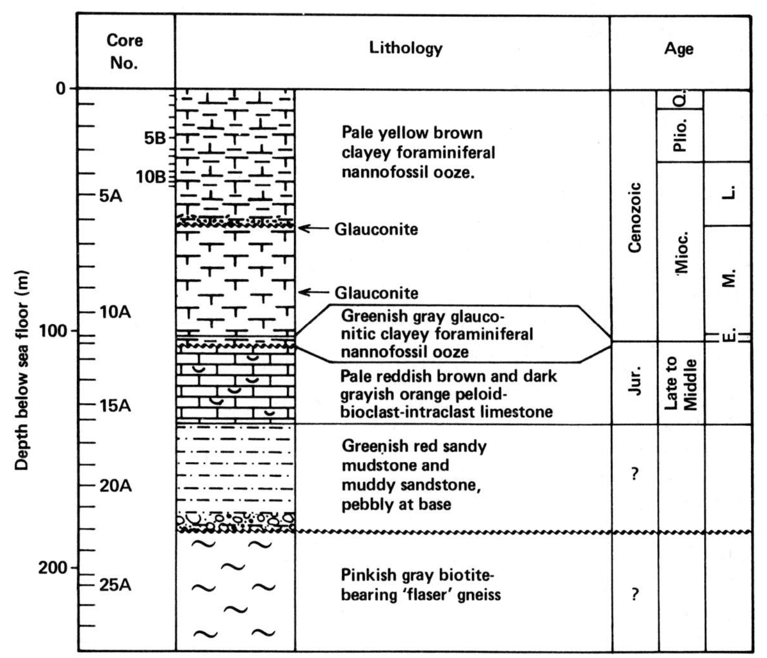

Fig. 3. Lithologic column drilled at Site 544

judged the thickness of sediment overlying basement was too little to allow burial of the bottomhole assembly. We moved the ship about 250 meters uphill to the southeast and drilled Hole 544A to basement. Hole 544B obtained 39.3 meters of hydraulic piston cores from the Neogene at the same location as Hole 544A.

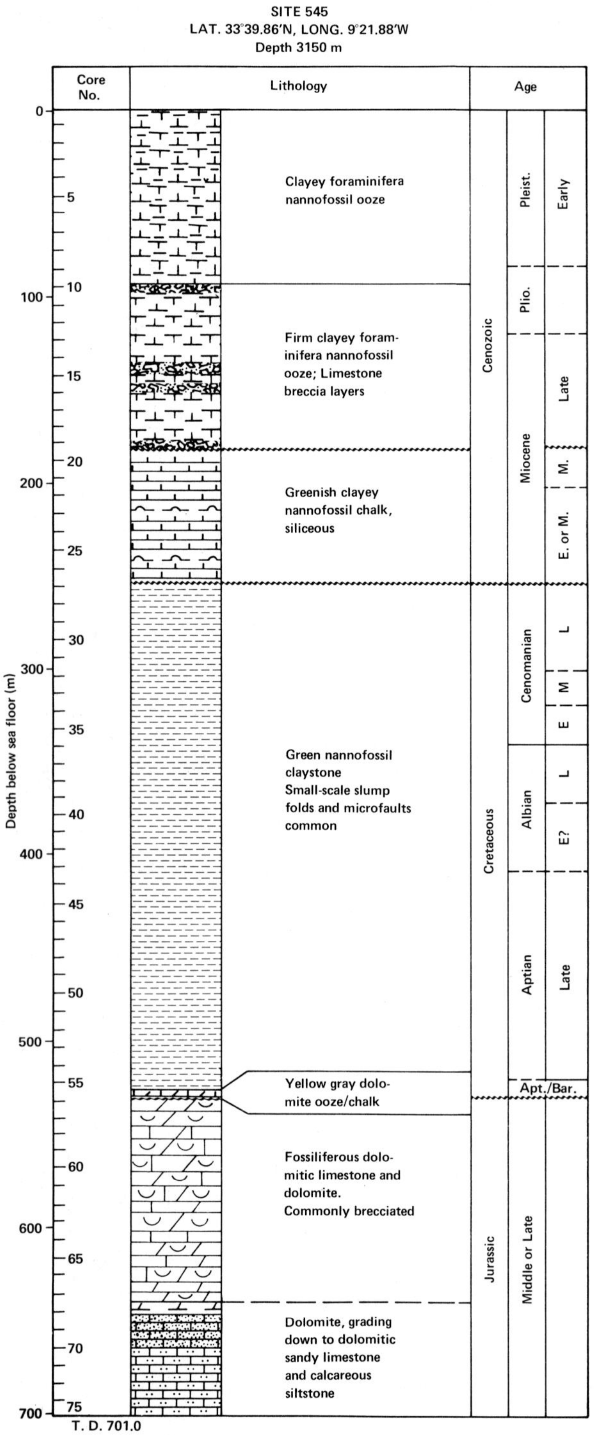

Fig. 4. Lithologic column drilled at Site 545

The following sequence was drilled at Hole 544A (Figure 3).

(1) Neogene sediments (0-105.8 meters)

0-56 meters: Pale yellowish brown clayey foraminiferal nannofossil ooze, early Pleistocene to late Miocene (Zone N17). A hiatus separates these beds from the next unit below.

56-102.6 meters: Greenish clayey foraminiferal nannofossil ooze and chalk. Middle Miocene (N14) to late Early Miocene (N9).

102.6-103.8 meters: Greenish glauconitic nannofossil chalk and nannofossil-bearing claystone. Early Miocene (N7). A major unconformity separates the Neogene from the Jurassic.

(2) Mesozoic rocks (103.8-184.2 meters)

103.8-139.2 meters: Limestone: pale reddish-brown and orange skeletal intraclast peloid packstone with patches of grainstone. "Posidonia", Protoglobigerina, ammonites, echinoids, crinoids, ostracods, vagile foraminifers, terebratulids, bivalves, sponge fragments, serpulids, bryozoans, Fe/Mn encrusted hardgrounds. Oxfordian (?).

139.2-184.2 meters: Grayish-red sandy mudstone and muddy sandstone. Pebbly at base. Unknown age.

184.2-235.0 meters: Pinkish-gray biotite-bearing granitic "flaser" gneiss. Unknown age.

2.2 Site 545 (latitude: 33°39.9'N, longitude 9°21.9'W; water depth: 3150 meters; sub-bottom depth: 701 meters; recovered section: 354 meters). The following sequence was drilled (Figure 4):

0-86 meters: Yellowish-brown clayey foraminiferal nannofossil ooze. Early Pleistocene.

86-180 meters: Yellowish-brown firm clayey nannofossil ooze with a few layers of breccia, with clasts of carbonate platform rocks of Jurassic age. One clast of Senonian pelagic limestone encloses peloids of calpionellid (Berriasian or Tithonian) limestone. Pliocene to late Miocene (N16/17). Rests unconformably on:

180-253 meters: Greenish clayey radiolarian-bearing nannofossil chalk. Middle Miocene (N15) to middle (NN5 or 6) or early Miocene (N5/6). Rests unconformably on:

253-531 meters: Green nannofossil claystone and claystone with abundant slump folds, microfaults and cut by low-angle slide surfaces that result in both repetition and omission of stratigraphic section. The basal few meters is dolomitized nannofossil chalk. Late Cenomanian through early Aptian/late Barremian. Rests unconformably on:

531-635 meters: Dolomitized limestone with shallow-water, high-energy characteristics, and containing coral debris, algal fragments, miliolids, molluscs, echinoderms and oolites. Locally brecciated, with iron-stained fissure fillings. Middle or Late Jurassic, but age could not be more precisely determined aboard ship. Grades downward to:

635-701 meters: Dolomitized sandy limestone with ammonites. The sandy component, which includes angular quartz and fresh feldspar grains up to granule size, increases to 90% in the deepest core. An ammonite from a core at 638 meters (identified by O. Renz shortly after the termination of the leg as Sowerbiceras sp.) indicates an Oxfordian age.

2.3 Site 546 (latitude: 33°46.7'N, longitude: 9°33.9'W; water depth: 3992 meters; sub-bottom depth: 192 meters; recovered section: 118.5 meters).

The following sequence was drilled (Figure 5):

0-134 meters: Yellowish-brown slightly clayey foraminiferal nannofossil ooze. Late Pleistocene to late Miocene (N17). Rests unconformably on:

134-147 meters: Greenish clayey stiff nannofossil ooze and reddish brown clay. Glauconite is common. Middle Miocene (N10-13).

147-149 meters: Greenish and grayish-red sandy clay with layers rich in gypsum crystals. The red layers are barren, but the greenish layers contain Early Cretaceous and Late Jurassic radiolarians and Early Cretaceous nannofossils, mixed with middle Miocene radiolarians, foraminifers, and nannofossils.

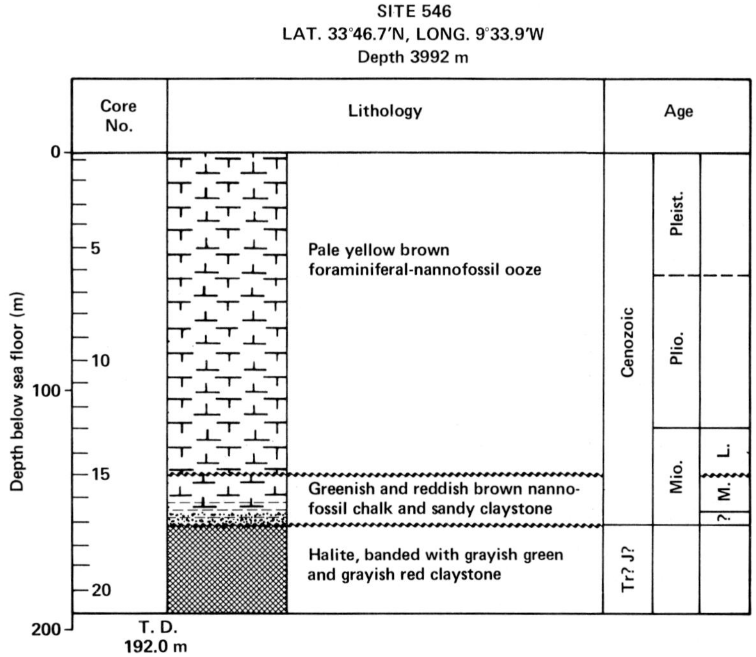

Fig. 5. Lithologic column drilled at Site 546

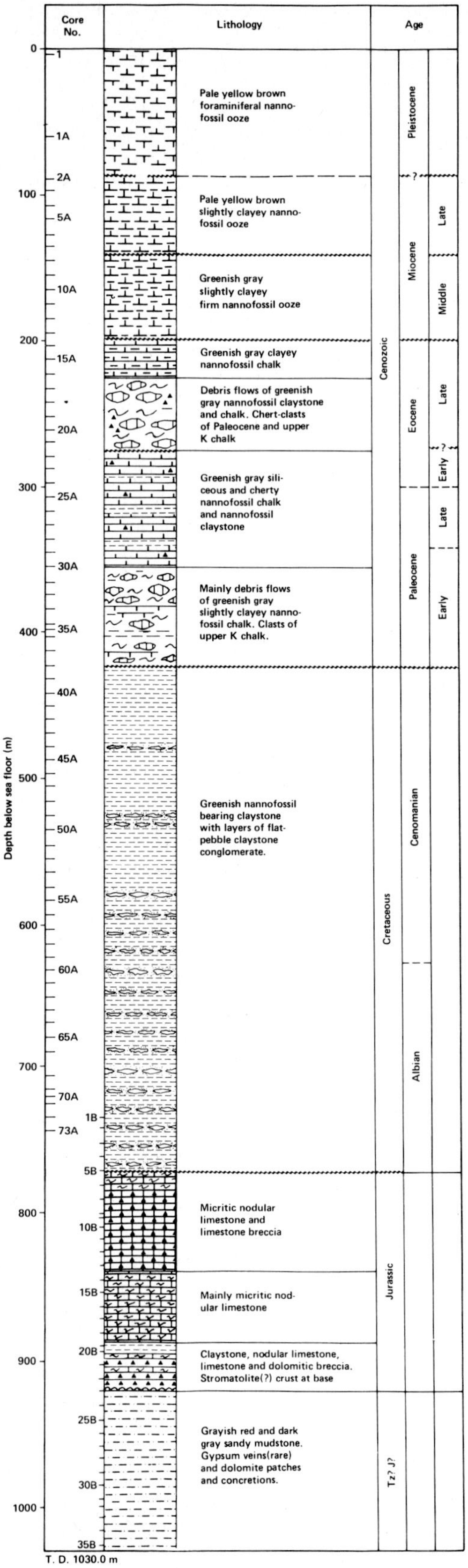

149-156.5 meters: Barren grayish-red gypsiferous sandy clay.

156.5-192 meters: Halite, banded with grayish-red, grayish-green, and gray clayey streaks and cut by rare veins of anhydrite. Most banding dips gently, but single unbroken core lengths show considerable variation in dips, suggesting folding. Some limpid halite crystals are as much as 1 cm in diameters.

The salinity of interstitial waters increased steadily from 34 to 217 per mil in the Neogene sediments above the halite, but no dissolved hydrocarbons could be detected.

A geologic cross-section through the site (Figure 9) shows the diapiric structure and documents an early salt-pillow stage and several episodes of diapiric rise. It also shows a possible solution basin over the top of the structure in which the sediments with mixed Cretaceous and Miocene fossils were deposited.

2.4 Site 547 (latitude: 33°46.8'N, longitude: 9°21.0'W; water depth: 3940.5 meters; sub-bottom depth: 1030 meters; recovered section: 649 meters). This was a re-entry site; at Hole 547, the bit was washed through the upper 32.5 meters of section to test conditions for washing in the re-entry cone and 16-inch surface casing. Hole 547A was spot-cored to 89 meters and continuously cores to 744.5 meters. Hole 547B, the re-entry hole, was continuously cored from 724.5 to 1030 meters, with 3 re-entries.

The following sequence was drilled (Figure 6):

0-141 meters: Pale yellowish-brown slightly clayey foraminiferal nannofossil ooze and nannofossil ooze. Quaternary (N23) to late Miocene (N17). Rests unconformably on:

141-205 meters: Greenish-gray nannofossil ooze. Middle Miocene (N15) to early Miocene (N4).

Fig. 6. Lithologic column drilled at Site 547

205-225 meters: Greenish-gray slightly clayey nannofossil ooze and chalk. Late Eocene. Rests unconformably on:

225-269 meters: Conglomerate beds (interpreted as debris flows), made of clasts (up to 4.2 meters, but typically 5 cm in diameter), largely of claystone, nannofossil claystone, clayey chalk, nannofossil chalk and porcelanite. The clasts float in a nannofossil chalk matrix. Contains resedimented Eocene, Paleocene and Late Cretaceous fossils. The youngest fossils are of late Eocene age.

269-355 meters: Greenish-gray clayey nannofossil chalk, with common radiolarians and porcelanite nodules. Early Eocene to early Paleocene (P1, NP5).

355-422 meters: Mainly debris flows of greenish-gray nannofossil chalk. Probably early Paleocene (P1), but reworked Maestrichtian chalk is the most common material. Rests unconformably (at the regional seismic "red" reflector) on:

422-773 meters: Grayish-green nannofossil-bearing claystone and mudstone with common intraformational flat-pebble mudstone conglomerate layers, slump structures, and repeated sections suggesting slide sheets. Middle Cenomanian (*Rotalipora cushmani* Zone) to late Albian (*Ticinella breggiensis* Zone).

773-775 meters: Greenish-gray claystone, pale greenish white micritic limestone and graded pebbly limestone beds. Foraminifers in the upper claystone beds are Early Cretaceous (Valanginian-Aptian), and calpionellids in the limestone pebble beds are late Tithonian.

775-791 meters: Red and green nodular limestone and limestone pebble and breccia beds.

The nodular limestone resembles Upper Jurassic nodular limestone units known in outcrop sections in Tethys. Ossicles of the pelagic crinoid *Saccocoma* occur in these beds along with radiolarians, aptychi, and occasional phragomocones of ammonites. An aptychus identified by O. Renz as *Lamellaptychus sparcilamellosus* (Kimmeridgian) occurs at 784.2 meters. *Saccocoma* occur in the upper part of the unit.

The breccia and pebbly beds, which are both mud- and grain-supported, are typically 10-30 cm thick, and the clasts, up to 15 cm in size, are reddish and greenish biomicrite and peloidal grainstone, some with coated grains, oolites and skeletal debris.

791-795 meters: Beneath a 7-cm thick layer of dark reddish-brown claystone are red and greenish nodular limestones as above, but with abundant shells of the pelagic lamellibranch *"Posidonia" (Bositra)*, some of which are current sorted.

795-846 meters: Limestone breccia, as above, with minor reddish and greenish nodular limestone interbeds. The breccia beds, 20-40 cm thick, are clast-supported and the clasts, up to 18 cm in diameter, are light gray limestone containing oncolites, ooids, coated grains, *Bositra*, sponge spicules, fragments of thick-shelled molluscs, and ammonites. The beds are separated by *in situ* crusts. Detrital quartz appears at 805 meters and increases in abundance downward, accompanied by clasts of feldspathic sandstone. An ammonite at 821.5 meters was identified by O. Renz as *Lytoceras* aff. *fimbriatum* (Middle Jurassic-Oxfordian).

846-891 meters: Mainly yellowish-brown nodular limestone, with micritic limestone nodules floating in a darker brown clay matrix. A 25-cm layer of burrowed black shale, containing nearly 7% organic carbon occurs at 847.1 meters. No age-diagnostic fossils could be identified aboard ship.

891-924 meters: Olive-black and olive-gray mudstone, dolomitic limestone-mudstone breccia, calcareous wackestone breccia. Near the base is a layer of laminated micrite with fenestral fabric, resembling stromatolites, and at the very base a bed of dolomitized limestone breccia. The claystone at 891 meters contains a very diversified and well-preserved benthic foraminiferal fauna, which, for

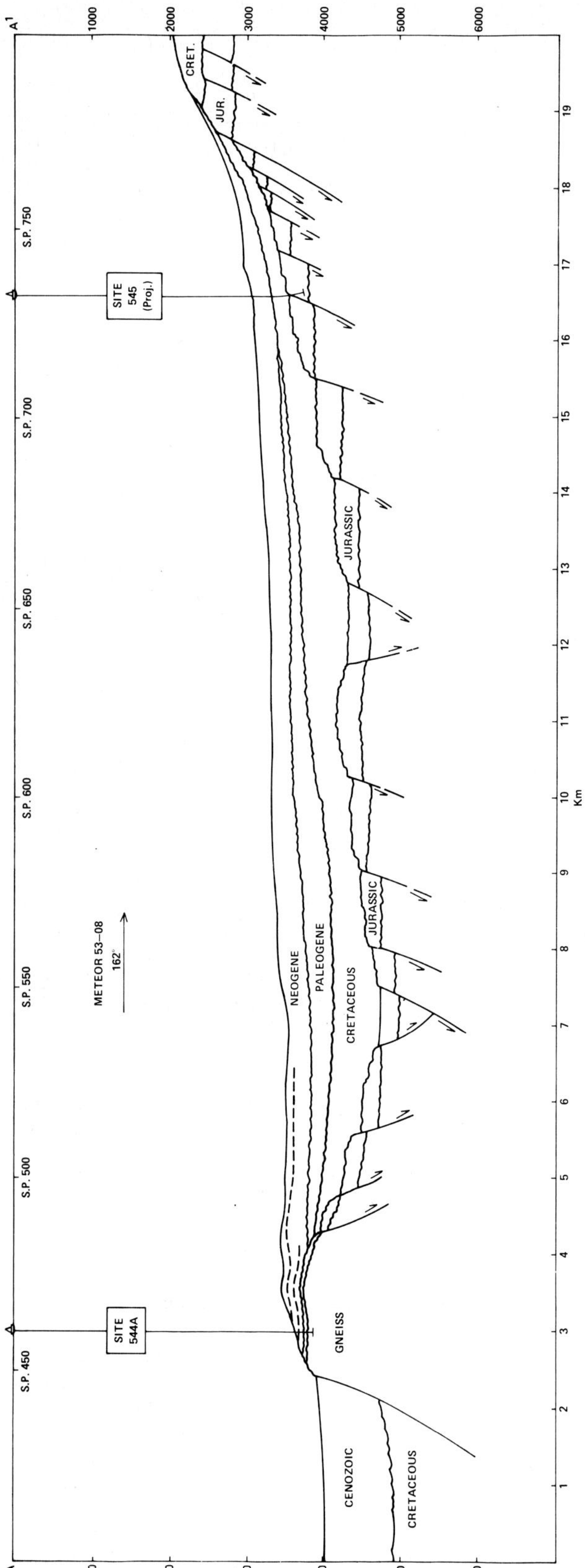

want of adequate references, could not be dated aboard ship. Nannofossils very tentatively dated as Pliensbachian occur in this unit.

924-1030 meters: Grayish-red sandy mudstone, interbedded in the upper part with dark gray sandy mudstone. Sand grains and granules of quartz and feldspar occur as floating grains and pods or ill-defined laminae, constituting 5-20% of the rock. A few veins of gypsum were cored and dolomitic concretions and patches occur. At the very base of the cored section, from 1027.5-1030 meters are abundant poorly-rounded pebbles of dolomite. No fossils, other than a few plant fibers, were found in this unit aboard ship. Seismic data suggest the drill was very close to basement when we were forced to cease drilling because the bit would not turn when weight was applied.

3. Tentative interpretations

Until we have better paleontological control in the Jurassic from shore-based studies, interpretations must be tentative; nonetheless, a few remarks on how the data from the individual sites fit together are warranted.

(1) Neogene sedimentation was mainly hemipelagic, and was above the CCD at all sites. A regional hiatus between Messinian and Serravalian beds occurs at all sites, and corresponds to the hiatus at this level at other sites on the continental margin of the eastern Atlantic.

(2) Paleogene sedimentation on the Mazagan slope was punctuated by episodes of debris flows that carried blocks of older Paleogene and Upper Cretaceous hemipelagic sediments. These

Fig. 7. Geologic cross-section along Meteor seismic profile 53-08, through Sites 544 and 547 (projected into line) (cf. also Hinz, Dostmann and Fritsch, this Vol.)

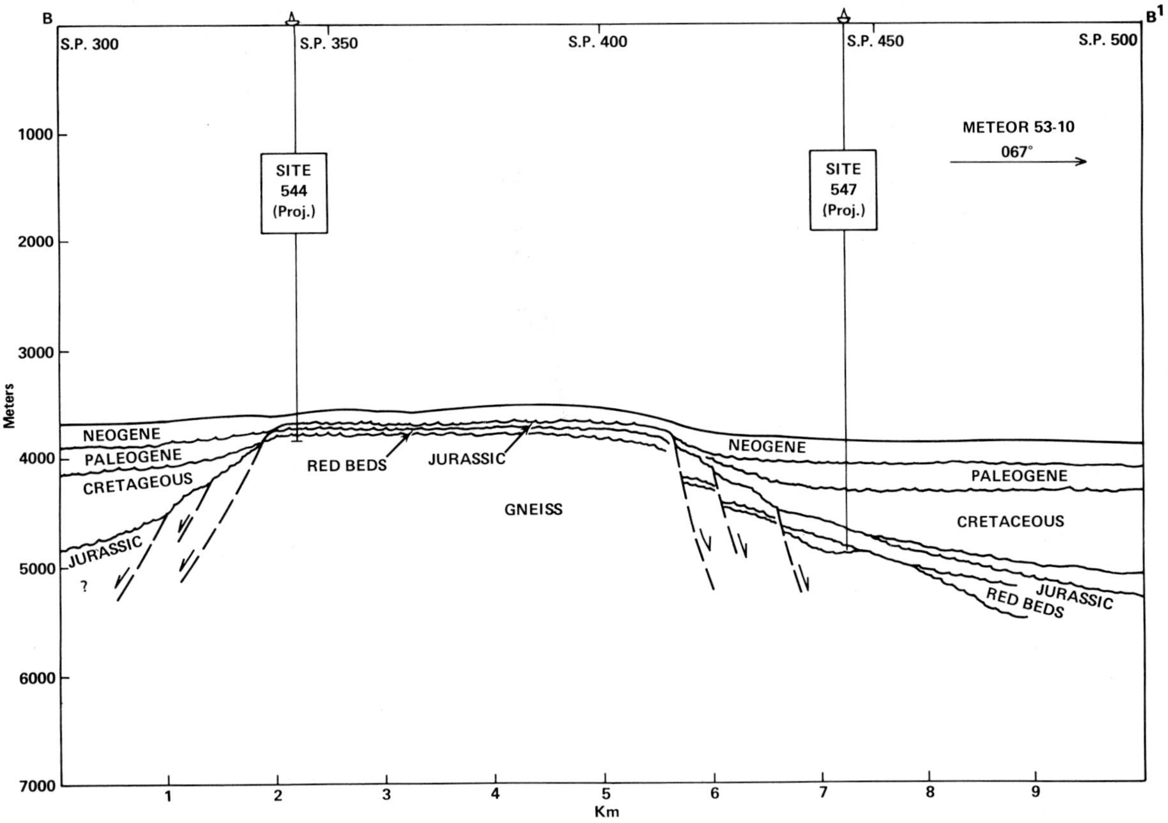

Fig. 8. Geologic cross section along Meteor seismic profiles 53-10, through Sites 544 and 547 (projected into line)

flows probably originated high on the Mazagan Escarpment of even on the outer parts of the plateau, and carried coarse debris at least 15-20 km seaward. Paleogene sediments are moderately siliceous, containing both radiolarians and porcelanite nodules, and thus contrast with the Neogene, which contains virtually no biogenous silica.

(3) Cretaceous sedimentation is remarkably like that at an analogously located site (415) drilled on Leg 50 low on the continental slope, about 200 km SW of the Mazagan area. As at most sites in the eastern Atlantic, the post-Cenomanian Upper Cretaceous is entirely absent (except as clasts in Cenozoic debris flows) and the Middle Cretaceous is represented by claystone deposits replete with small- and large-scale slump structures, microfaults and intraformational conglomerate beds, and is riven by low - angle faults - probably the bases of gravity-driven slide sheets - that repeat section.

Taken together, the Cretaceous and Cenozoic strata record episodic defacing of the Mazagan slope, with successively older rocks being laid bare: the Middle Cretaceous conglomerates are intraformational; the Paleogene debris flows carry older Cenozoic and Upper Cretaceous clasts, and the Neogene breccias contain Jurassic clasts.

(4) Jurassic sedimentation shows great diversity of facies. Tithonian and Kimmeridgian rocks are of pelagic facies (calpionellid limestone and nodular limestone) but include limestone breccia with clasts of probable Oxfordian and Dogger age. Oxfordian rocks include *in situ* shallow-water and perhaps moderate-depth limestone at

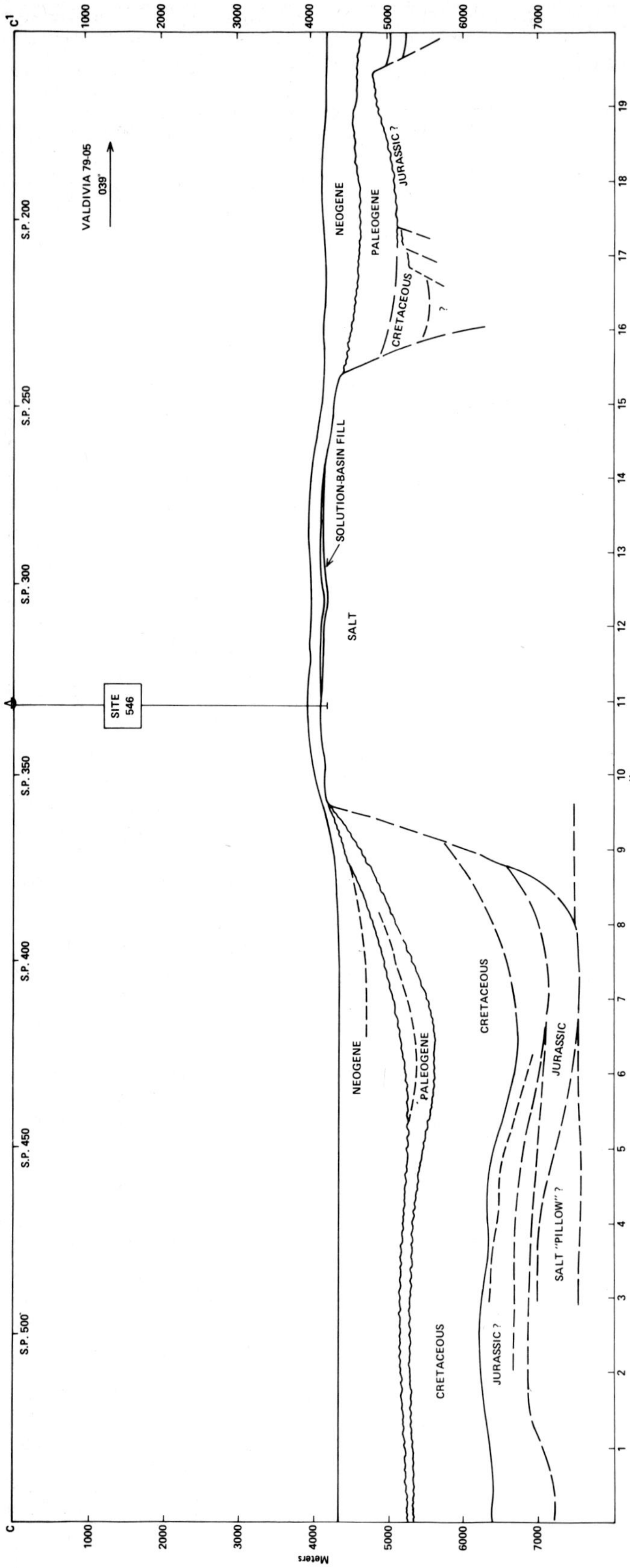

Fig. 9. Geologic cross-section along Valdivia seismic profile 79-05, through Site 546

Sites 544 and 545, and resedimented (deeper water?) breccias at Site 547. Somewhat sandy limestone of Oxfordian age occurs at Site 545. The lower part of the carbonate section at Site 547 shows very shallow-water features, but is dated only very tentatively as Liassic.

The diversity of lithofacies suggests a diversity of environments, and the contrasts in the Jurassic rocks between Sites 544 and nearby Site 547 suggest that there was considerable relief during Oxfordian times. The Jurassic succession at Sites 544, 545 and 547 all show facies reflecting decreasing terrigenous detrital influence and increasing water depth upward from the base.

The unconformity separating Tithonian/Berriasian limestone from Aptian claystone contrasts strongly with the situation at Site 416, about 80 km to the SW in the Morocco Basin, where the Lower Cretaceous consists of about 700 meters of flysch.

(5) Cross-section A-A' (Figure 7) shows many normal faults, but this is but one of many possible interpretations of the structural style. We show the faults as cutting the Jurassic but not the Cretaceous; it is possible that there are fewer faults, and that they predate much of the Jurassic carbonate accumulation.

(6) Continental red-bed sediments were drilled below Jurassic limestone at Sites 544 and 547, and they may be present at Site 545 below the level cored; the deepest rocks sampled there are calcareous sandstone. The red beds are lithologically similar to Triassic (?) red beds onshore in Morocco. The Mazagan red beds include slumps, breccias and mud flow deposits, suggesting nearby relief.

(7) Seismic data show diapiric structures extending seaward some 50 km from Site 546, but mixed with other structural highs not clearly diapiric. It is possible that sialic horsts and salt diapirs are interspersed in this province; indeed, the diapirism may have been abetted by the faulting that accompanied attenuation of the seaward edge of the continental crust. Only drilling of some of the candidate sialic blocks will establish for certain whether the salt was depoted on oceanic, or continental crust, or on some mixture of these two crustal types.

We suspect an age correspondence between the salt at Site 546 and the red beds at 544 and 547. The gypsum at 547 suggests aridity, and the map relations suggest the red beds may be a marginal facies bordering the salt. The implications of this possibility are profound, since the structural relief between the original salt layer next to the diapir at Site 546, and the red beds at Site 544 is about 4 km.

(8) The gneissic basement rocks at Site 544 are overlain by a very thin stratigraphic section, with an unconformity separating Oxfordian (?) from Miocene beds. The relations are similar to those on "outer highs", so common on other passive margins at the boundary between oceanic and continental crust.

3 The Continental Margin of Morocco: Seismic Sequences, Structural Elements and Geological Development

Karl Hinz, Hans Dostmann, and Jürgen Fritsch[1]

ABSTRACT

This paper discusses seismic sequences, structural elements and the geological development of the Moroccan continental margin from the North Tarfaya segment in the south, northward to the Tafelney Plateau, the Essaouira segment, the Mazagan Plateau, and the Pre-Rif segment.

The dominating structural unit of the Moroccan continental margin is a NNE-trending zone of salt diapirs. This zone starts southeast of Fuerteventura and continues to the north beneath the Pre-Rif nappes. The age of the salt is probably Late Triassic/earliest Jurassic. In the offshore areas, salt movement started during the Early Cretaceous. The western boundary of the diapiric province is sharp and locally fault-controlled. In the south, within the North Tarfaya segment, the diapiric zone is 10-45 km seaward of the Upper Jurassic carbonate platform edge, an additional prominent structural unit of the continental margin of Morocco. The diapiric zone broadens within the Tafelney Plateau and Essaouira segment and continues into the Southwestern Moroccan Coastal Basin.

The western boundary of the Moroccan salt diapiric province matches well with the eastern boundary of the diapiric zone of Nova Scotia (Canada). This means that before separation of the African and North American plates, one uniform evaporite basin existed. This basin broadened during the early development of the margin when the crust was uplifted, eroded, thinned and stretched by unknown crustal and subcrustal processes. The observed anomalous crust beneath the Moroccan continental slope and deep sea is interpreted to be attenuated continental crust which is contaminated by basaltic intrusions. In our opinion, the distinct western boundary of the Moroccan salt diapiric province marks the site where actual separation of Morocco and Nova Scotia began.

Following the separation (about Sinemurian), carbonates were deposited in the uniformly subsiding basin. About 45 million years after deposition of red beds and occurrence of basaltic volcanism in the Southwestern Moroccan Coastal Basin, more rapid subsidence occurred in the Middle Jurassic of those parts of the basin underlain by a contaminated thinned continental crust. This led to the formation of a continental slope, carbonate banks and carbonate platforms. Carbonate platform formation terminated in the earliest Cretaceous in the North Tarfaya segment, in the Early Cretaceous in the Essaouira segment and in the Late Cretaceous/?Early Tertiary on the Mazagan Plateau.

Subsidence continued until the Middle Cretaceous in the Moroccan continental margin segments with exception of the Mazagan Plateau. The Tafelney Plateau was formed during the Late Cretaceous as a result of the epeirogenetic uplift

1 Bundesanstalt für Geowissenschaften und Rohstoffe, Stilleweg 2
3000 Hannover 51, FRG

of the Western High Atlas and intensive salt diapirism. This stage of diapirism, initiated by the epeirogenetic uplift and associated with large-scale gravity glidings, caused the development of complex seismic structures in the outer parts of the Tafelney Plateau and Essaouira segment.

Major erosional events occurred between the Cenomanian and the Eocene as well as during the Oligocene and Miocene.

1. Introduction

The continental margin of Northwest Africa is one of the main working areas of German marine sciences. A review of the German activities and results off Northwest Africa up to 1975 was given by Seibold and Hinz (1974 and 1976).

Since 1977 three additional cruises have been carried out (METEOR cruises no. 46/1977 and no. 53/1980; VALDIVIA-Westafrica cruise 1979) with the objectives to fill gaps from the previous cruises and to contribute to postsite and presite surveys for GLOMAR CHALLENGER-IPOD Drilling, Leg 41, 47, 50 and 79.

The geophysical data from these cruises (Fritsch et al., 1978; Wissmann and von Rad, 1979; Hinz and Seibold, 1980) and of previous reconnaissance surveys by N.V. LADY GLORITTA (Beck and Lehner, 1974; Lehner and Ruiter, 1977) and R.V. ATLANTIS II (Uchupi and Emery, 1974; Uchupi et al., 1976), plus the results of IPOD deep sea drilling (Lancelot and Seibold, 1977; Lancelot and Winterer, 1980) indicate that the old Moroccan continental margin offers the only opportunity to study the Early and Middle Jurassic paleoenvironment, the early subsidence history of the proto-Atlantic, and the old features presently believed to be characteristic of a rifted margin (FUSOD Report, 1977) by GLOMAR EXPLORER drilling.

The paper discusses the seismic sequences, structural elements, and the physiography of the Moroccan continental margin between latitudes 28° and 35°N. We shall rely on our geophysical data plus integration of these data with geological information from the western Moroccan sedimentary basins.

2. Physiography and Geological Setting

The Moroccan margin between 28°N and 35°N has a varied physiography (Fig. 1).
The shelf width varies between 13 km at C. Ghir to about 80 km south of latitude 29°N; on average the width is 40 km.

In general, the shelf break occurs at the 150 m depth contour. Disregarding small areas with local highs of 10 m to 30 m (see also McMaster and La Chance, 1968; Summerhayes et al., 1971; Robb, 1971), the sea floor in the shelf region has a flat relief with a seaward slope of 0°1' to 0°3' (Tooms et al., 1971; McMaster and La Chance 1968).

The continental slope and rise extend from the shelf break to 1500 m depth in the south, east of the island of Fuerteventura, and to about 4000 m in the north, north of C. Cantin.

Physiographically the continental margin off Morocco can be subdivided into five segments from the south to the north: In the south, the North Tarfaya segment (segment 1) extends from latitude 28°N to C. Ghir. The seafloor slopes smoothly seaward at an angle of 1.3° to 1.9°. Recent canyons occur in the region of the outer shelf and the upper continental slope. The complex Agadir Canyon system, which extends into the deep sea, separates the North Tarfaya segment from the Tafelney Plateau/Terrace (segment 2), a large marginal plateau of about 3500 km^2 (as outlined by the 1000 m water depth contour). The Tafelney Plateau is situated along the seaward extension of the Western High Atlas.

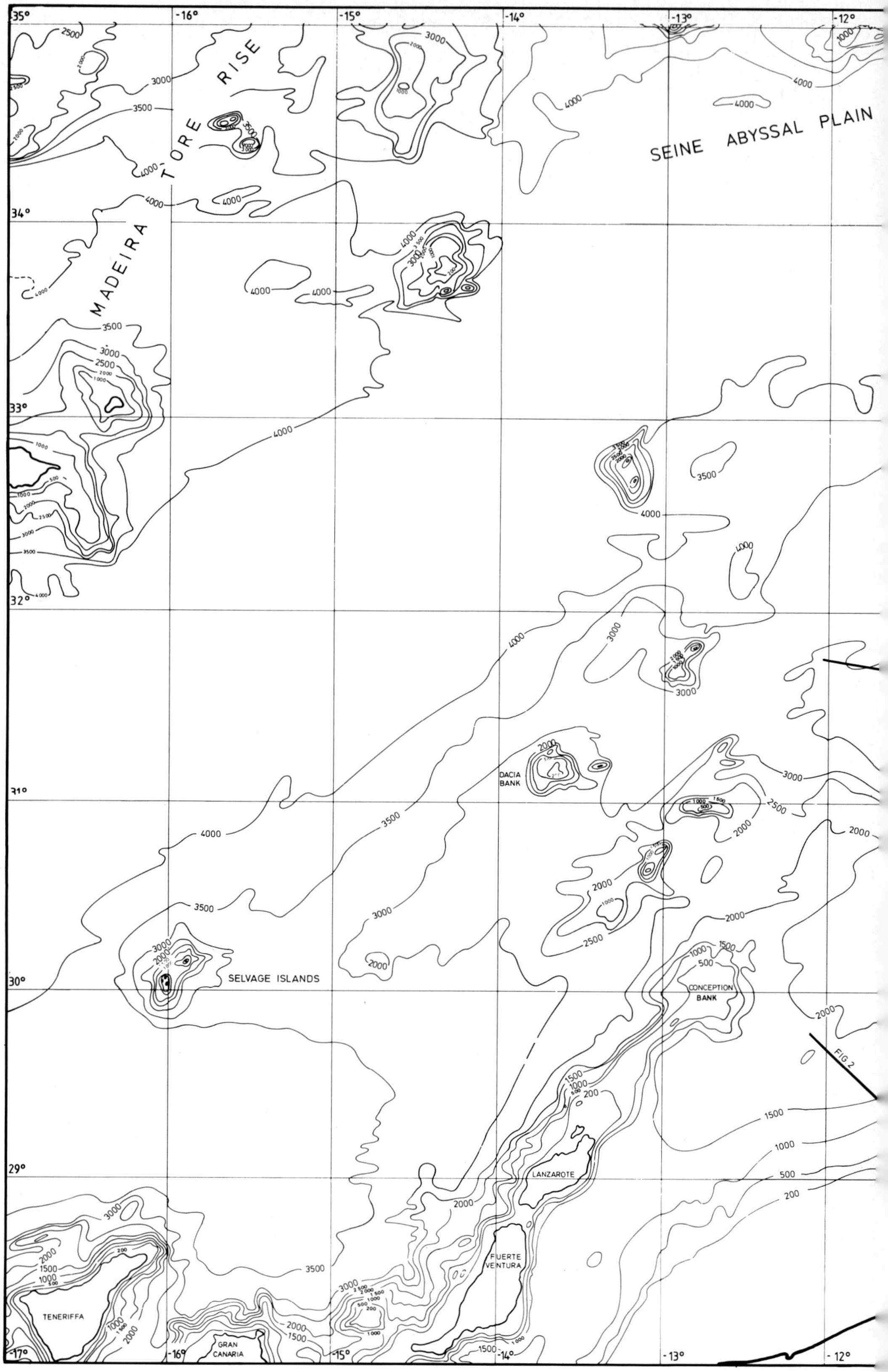

Fig. 1
Bathymetric map of Morocco Sources: Echosoundings of the METEOR cruises no. 9/1967, 39/1975, 46/1977, 53/1980, the VALDIVIA West Africa cruise 1979;

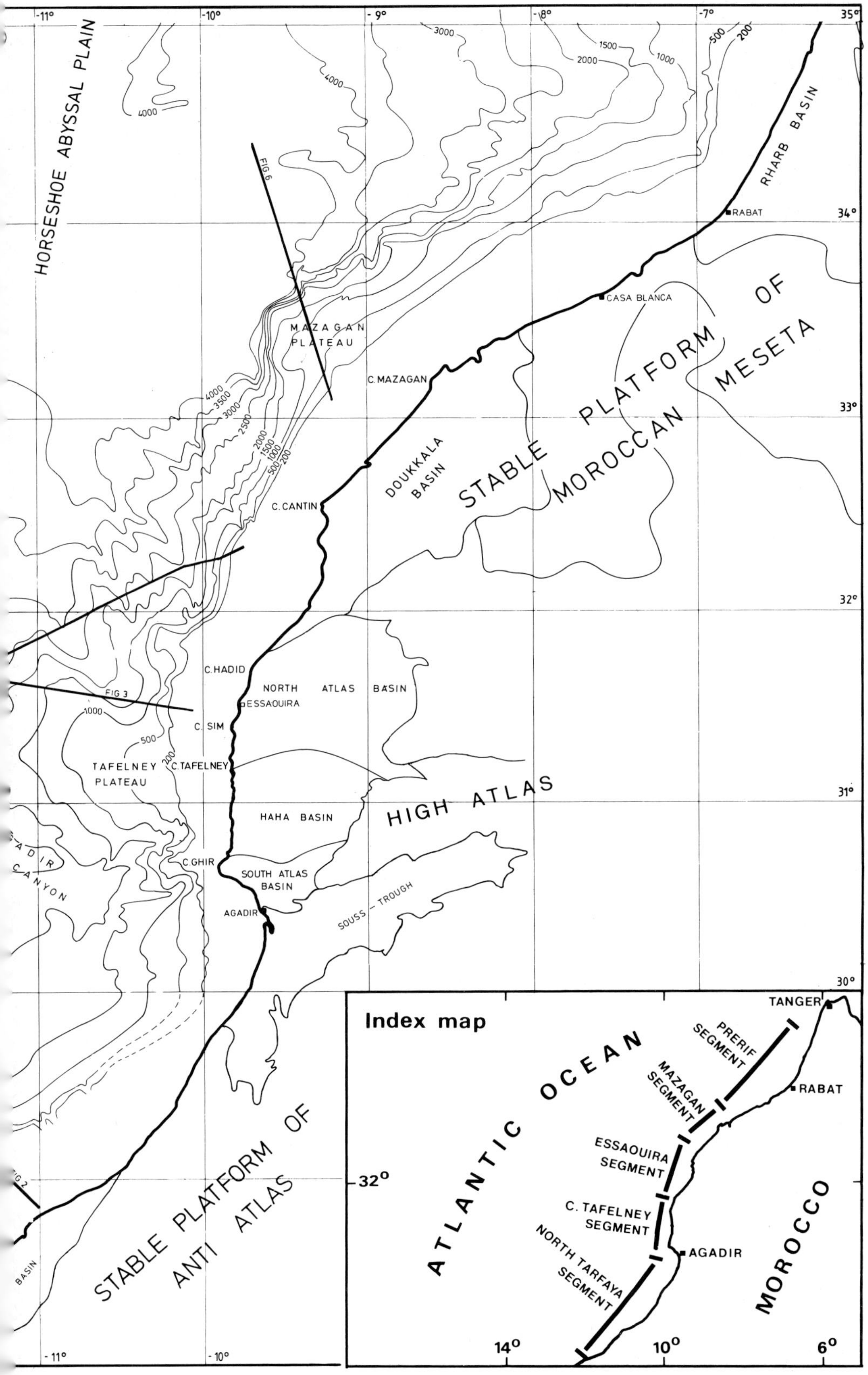

Bathymetric map Ocean Atlantique, sheets IGL 8080 + 8081, 1:1,000,000, 1975; Carte Générale Bathymétrique des Océans - GEBCO 104, 1:1,000,000, 1972; Uchupi et al., 1976; Jacobi et al., 1975

The Essaouira segment (segment 3), located between the Tafelney Plateau/Terrace and the Mazagan Plateau, is distinguished by an irregular sea floor relief with slopes from 1.4° to 3°. Numerous canyons within this segment strike northwest, across the continental slope.

The Mazagan Plateau (segment 4) lies off the Moroccan Meseta between 33°N and 34°N. The plateau, which covers an area of about 1600 km^2 at depths ranging from 500 m to 1800 m, drops steeply into the deep sea to the west and southwest at slopes up to 38°.

North of the Mazagan Plateau, in the seaward extension of the Rharb Basin, lies the Pre-Rif segment (segment 5); here the continental slope broadens to about 250 km. North of latitude 35°N the average slope is 0.85°.

Conspicuous physiographic provinces seaward of the continental slope are:

- the northeast trending submarine East Canary Ridge, which runs from the eastern Canary Islands Fuerteventura and Lanzarote to the Conception Bank and ends in a chain of seamounts trending north-south;
- a submarine ridge running ENE from the Selvagen Islands over the Dacia Bank;
- the Madeira-Tore Rise and to the east of it, an east-west trending ridge containing the Ampere Bank.

The Atlantic coastal region of Morocco (Ager, 1974; Ambroggi, 1963; Choubert and Marcais, 1952; Faure-Muret and Choubert, 1971; Robb, 1971) is subdivided from north to south into the following geological units:

- the Rif, an Alpine-type orogene distinguished by numerous overthrusts;
- the Rharb Basin, a Molasse-like trough between the Rif and the Moroccan Meseta;
- the Doukkala Basin; a Mesozoic-Tertiary series, up to 1000 m thick, overlies the Moroccan Coastal Meseta Platform;
- the Southwestern Moroccan Basin, bounded in the north by the Moroccan Coastal Meseta, in the east by the Variscan-Paleozoic complex of the High Atlas and in the south by the Anti Atlas. Three basins are grouped under the name of the Southwestern Moroccan Basin: The North Atlas Basin, also called the Essaouira Basin or Mogador Syncline, which is distinguished by salt diapirs; the Haha Basin, also known as the Western High Atlas or the High Central Zone, which is an uplifted block with outcropping Early Cretaceous to Jurassic sediments; the South Atlas Basin, also called Agadir Basin, which together with the Tertiary Souss Trough adjacent to the south, forms a sub-basin of the North Atlas Basin;
- the stable Anti Atlas Platform, a Precambrian high with interposed Variscan metamorphic rocks belonging to the Westafrican craton.

Between the volcanic East Canary Ridge in the southwest and the Moroccan coastal region a broad marginal sedimentary basin, characterized by diapiric structures, exists (Beck and Lehner, 1974; Uchupi and Emery, 1974; Uchupi et al., 1976; Hinz and Seibold, 1980; Lancelot and Winterer, 1980); it will be discussed next.

3. Seismic Stratigraphy and Geological Structure of the Moroccan Continental Margin Segments

The seismic sequences recognized in the different physiographical segments of the Moroccan continental margin are best described by means of typical seismic transects. The location of the transects is indicated in Fig. 1.

3.1 North Tarfaya Segment

A typical profile across the North Tarfaya segment shows a seaward thickening sedimentary wedge intruded by salt beneath the continental slope Fig. 2).

The oldest recognizable seismic reflection layer is the sequence MO (Morocco)-5, the top of which, in

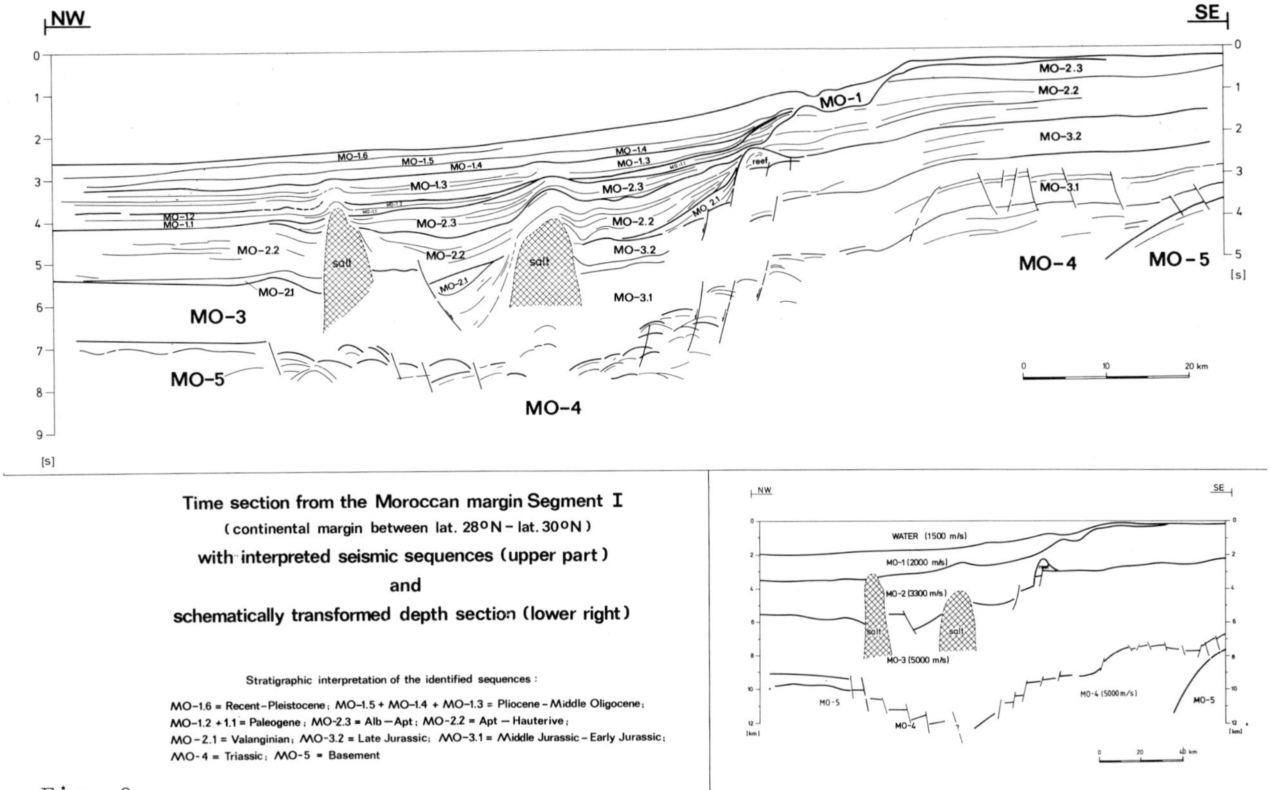

Fig. 2
Time section from the Northern Tarfaya segment with interpreted seismostratigraphic sequences and schematically transformed depth section. Location of the profile is indicated in Fig.1

the present shelf area, is represented by a strong reflector that dips seaward at about 3°. According to extrapolated borehole information, the upper boundary of MO-5 corresponds to Paleozoic metamorphic basement; it plunges to a depth of more than 10 km beneath the continental slope. The top of the Paleozoic sequence MO-5 was not observable under the present continental slope even at recording times of 8 s (TWT). The seismics cannot resolve whether the acoustic basement observed at the northwest end of the line is the same top of Paleozoic rocks inferred for the shelf region or whether it is oceanic basalt.

Beneath the shelf and upper continental slope, the MO-4 sequence is characterized by a sub-parallel reflection pattern as it progressively onlaps the top of the MO-5. Under the present lower continental slope the top of the MO-4 is distinguished by a prominent diffraction pattern, which indicates intensive faulting. In the extreme northwest part of the line, sequence MO-4 appears to terminate at faults against an indicated basement complex. Considering extrapolated onshore data (Choubert and Marcais, 1952; van Houten, 1977) the sedimentary sequence MO-4 is interpreted as Triassic. From the seismic data it is obvious that beneath the lower continental slope region, the MO-4 sequence contains sufficient evaporites to form diapirs locally.

Like the underlying Triassic sequence, sequence MO-3.1 and MO-3.2 possess a sub-parallel reflection pattern in the present shelf area; this indicates uniform sedimentation on a uniformly subsiding shelf. According to offshore borehole information (Vail et al., 1977, part 2 and part 8), the MO-3.2 sequence in the shelf region consists mainly of Late Jurassic shallow water sediments (Puerto Cansado Formation). The MO-3.1 sequence is interpreted as Early and Middle Jurassic.

Under the present upper continental slope, the top of the Late Jurassic MO-3.2 sequence slopes seaward at 10° to 20°. According to Vail et al. (1977) this slope marks the edge of a Late Jurassic carbonate platform and represents the continental slope at the turn of the Jurassic/Cretaceous (Tithonian/-Berriasian). At the edge of the carbonate platform, biolithic structures up to 5 km wide can be recognized by a discontinuous reflection pattern and diffractions on some sections. These structures are interpreted as reefs or bioherms; boreholes on the Moroccan outer shelf (Vail et al., 1977, page 160) have shown them to be composed of reef-building corals, oolites, and calcareous algae of Late Jurassic (Tithonian/Berriasian) age.

In contrast, the Late Jurassic carbonates, drilled to the west of these structures contain deep-water planktonic organisms. From the difference in relief between the Late Jurassic sediments containing shallow water organisms and their contemporaneous deep sea equivalents (Vail et al., 1977; Bhat et al., 1975), water depth seaward of the former shelf edge was 2500 m in Tithonian/Berriasian times. Estimates from the seismic data gave us paleowater depths for the Late Jurassic in the North Tarfaya segment of between 2300 m and 2700 m. The Late Jurassic continental shelf margin edge, which today lies at a depth of between 1700 m and 2800 m, was situated in Tithonian/Berriasian times about 10 km to 20 km seaward of the present shelf edge.

From the reflection seismic data it can be deduced that the sequences MO-4 (interpreted as Triassic) and MO-3.1 (interpreted as Early and Middle Jurassic) were deposited in a fairly uniform subsiding basin, which earlier had been subdivided by pre-Triassic troughs and swells. At about the end of the deposition of sequence MO-3.1, possibly at the beginning of the Dogger, a more rapid and fault-associated subsidence of the seaward part of the basin started, which led to the formation of the continental slope. After these movements, the North Tarfaya segment was not affected by any major faulting, apart from that connected with salt tectonics.

The sedimentation on the Late Jurassic-earliest Cretaceous carbonates began with the sequence MO-2.1, which is distinguished by landward onlap and seaward downlap. The sequence MO-2.1, which Vail et al., (1977) consider to be formed of deep-water clastic sediments (Valanginian) could only be detected south of latitude 29°N in the reflection seismic data within the North Tarfaya segment.

The overlying sequences MO-2.2 and MO-2.3 show a parallel to subparallel reflection pattern in the shelf region. In contrast, beneath the upper continental slope, a seaward dipping, and mainly landward onlapping reflection pattern is exhibited, which indicates deposition in deep water on the former continental slope. Sequences MO-2.2 and MO-2.3 are separated by an erosional unconformity. Drilling results indicate that sequences MO-2.2 (interpreted as HauterivianAptian) and MO-2.3 (interpreted as Aptian-Albian) consist of deltaic sandstones, clay and claystone of the Tan Tan Formation. The erosional unconformity between sequences MO-2.2 and MO-2.3, which is usually clearly observable in the North Tarfaya segment, could be related to a global lowstand of the sea level in the Middle Aptian, postulated by Vail et al. (1977).
The thin Middle Cretaceous Aguidir Formation (50 m to 360 m thick), which has been penetrated by drillholes on the shelf (Faure-Muret and Choubert, 1971) could not be recognized in the reflection seismic data beyond the shelf.

After a phase of intensive erosion (probably during the Early Eocene) during which most of the Late Cretaceous sediments were eroded in the North Tarfaya segment, there followed deposition of the Tertiary sequence MO-1.
The Tertiary sediments are divided by unconformities into a number of

subsequences. In the North Tarfaya segment at least three erosional unconformities are observable:

- an Early Eocene unconformity, which forms the boundary between the Cretaceous and Tertiary;
- an Oligocene unconformity, which forms the base of subsequence MO-1.3;
- a Miocene unconformity, which marks the base of sub- sequence MO-1.4.

3.2 Tafelney Plateau Segment

In contrast to the North Tarfaya segment, all the seismic sections from the Tafelney Plateau segment, situated between latitudes 30°45'N and 31°30'N, show a complex structural style, which can be divided from west to east into (Fig. 3, for location see Fig. 1):

- a zone with flat-lying Cenozoic to Mesozoic sediments;
- a north-south trending zone with complex seismic structures which apparently encloses the Tafelney Plateau;
- the Tafelney Plateau area, with anticlinal structures.

In the western part of the area the following designation of seismic markers has been made by the shipboard party of IPOD-Leg 50 (Lancelot et al., 1977):

In DSDP site 415 the brown horizon represents the top of a series of cemented sand-marl layers of the lower Miocene. When compared with the North Tarfaya segment the brown horizon would be correlated with a Miocene unconformity on the shelf (Robb, 1971) and the base of sequence MO-1.4. Although not sampled, the sequence MO-1.3, which can be distinguished at DSDP site 415, is interpreted as lower Miocene to Oligocene. The Neogene sediments are composed of marls, chalk and oozes (Lancelot et al., 1977). The thin Paleogene sediments which underlie sequence MO-1.3 are divided by a further unconformity into the sequences MO-1.1 and MO-1.2. In DSDP site 415 the Paleogene sediments consist mainly of mudstone with some interbedded chert and marlstone with limestone.

The red horizon has been defined according to the stratigraphical classification of the shipboard party of Leg 50 (Lancelot et al., 1977) as the boundary between early Paleocene pelagic clays and late Paleocene turbidites. Because of its uncertain stratigraphic classification, the red horizon could be correlated either with the Early Eocene erosional unconformity observed in the North Tarfaya segment or with a Cretaceous (97 m.y.) unconformity caused by a global sea-level fall postulated by Vail et al. (1977). We shall treat it in the following as the Tertiary/Cretaceous boundary.

Beneath the red horizon is a seaward thinning sequence (sequence MO-2.4, Fig.3, western part) up to 0.8 s (reflection time) thick, and distinguished by a discontinuous to chaotic reflection pattern. Sequence MO-2.4 represents an allochthonous layer which in its upper part is weakly folded. The sequence is composed of marls with clays and limestones of Cenomanian to Albian age (Site Reports, leg 50; Lancelot et al., 1976).

The orange horizon represents the base of the upper Cretaceous allochthonous sequence MO-2.4.

Based on the results of DSDP site 416 (Lancelot et al., 1976), which is already situated in the Essaouria segment, the green horizon was interpreted as the top of a flysch sequence, composed mainly of turbidites of Tithonian to Aptian age.

The blue horizon is distinguished by several consecutive marked reflection maxima and minima. It very probably represents the top of the Late Jurassic deep-water carbonates drilled in the North Tarfaya segment (Vail et al., 1977).

The yellow horizon approximates the top of the sedimentary sequence MO-4, interpreted as Triassic, which apparently terminates in the west against a basement complex (sequence MO-5), possibly composed of oceanic basalts. The reflection horizons and sequences which are clearly correlatable in the outer, seaward part of the Tafelney Pla-

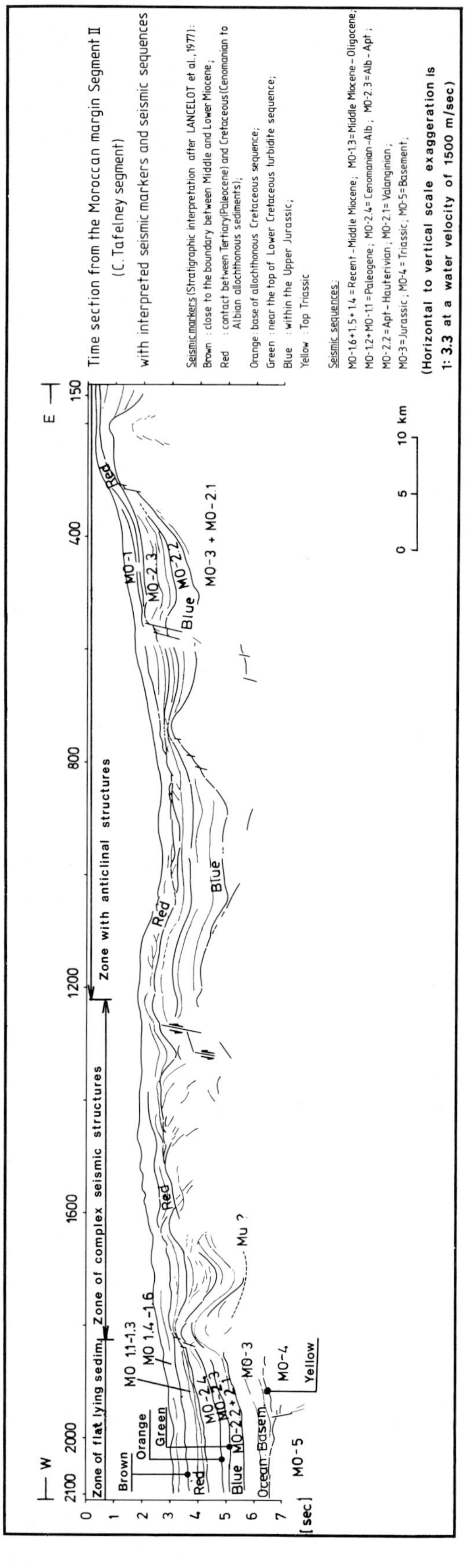
Time section from the Moroccan margin Segment II
(C. Tafelney segment)
with interpreted seismic markers and seismic sequences
Seismic markers (Stratigraphic interpretation after LANCELOT et al., 1977):
Brown : close to the boundary between Middle and Lower Miocene;
Red : contact between Tertiary (Paleocene) and Cretaceous (Cenomanian to Albian allochthonous sediments);
Orange: base of allochthonous Cretaceous sequence;
Green : near the top of Lower Cretaceous turbidite sequence;
Blue : within the Upper Jurassic;
Yellow : Top Triassic
Seismic sequences:
MO-1.6+1.5+1.4 = Recent - Middle Miocene; MO-1.3 = Middle Miocene - Oligocene; MO-1.2+MO-1.1 = Paleogene; MO-2.4 = Cenomanian-Alb; MO-2.3 = Alb - Apt; MO-2.2 = Apt - Hauterivian; MO-2.1 = Valanginian; MO-3 = Jurassic; MO-4 = Triassic; MO-5 = Basement;
(Horizontal to vertical scale exaggeration is 1: 3.3 at a water velocity of 1500 m/sec)
Zone of flat lying sedim.
Zone of complex seismic structures
Zone with anticlinal structures
W
E
[sec]
Ocean Basem.
0 5 10 km

Fig. 3

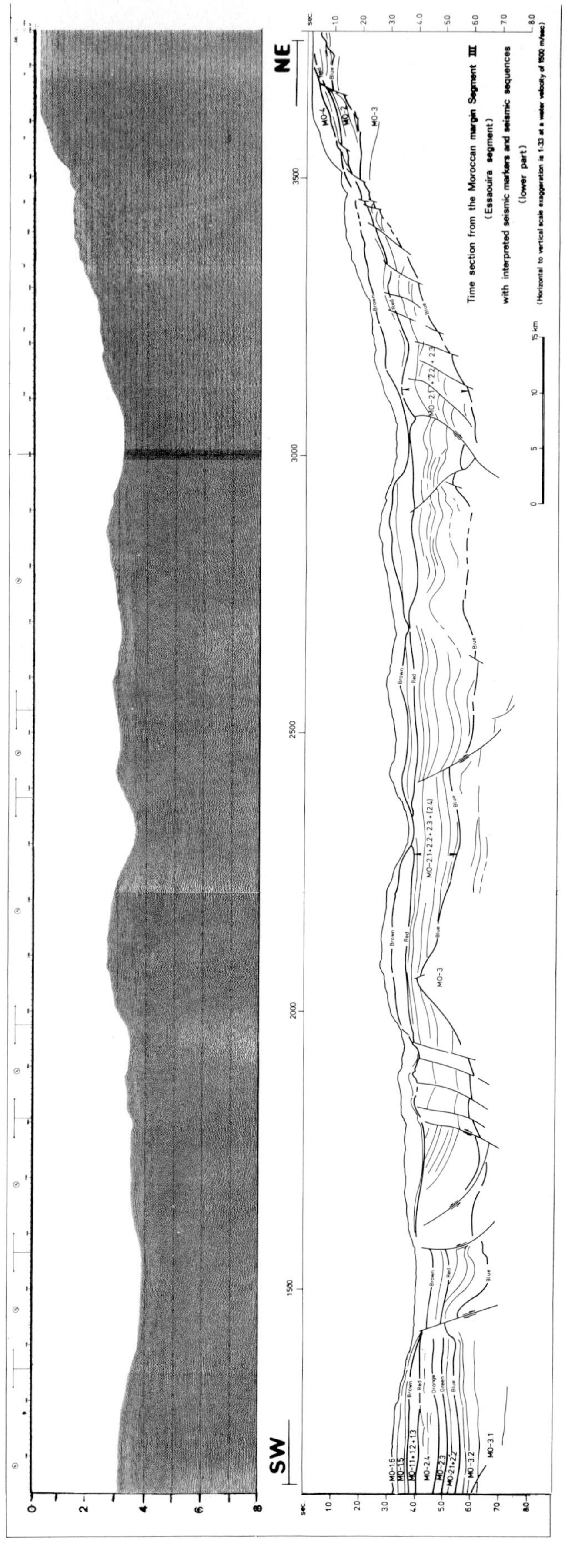
SW
NE
Time section from the Moroccan margin Segment III
(Essaouira segment)
with interpreted seismic markers and seismic sequences
(lower part)
(Horizontal to vertical scale exaggeration is 1:3.3 at a water velocity of 1500 m/sec)
0 5 10 15 km

Fig. 4

teau segment are, apart from the red horizon, not recognizable in an approximately 30-km-wide strip defined as the "zone of complex seismic structures" (Fig.3). In the reflection seismic records of some lines across this zone, a diffraction pattern with landward curving hyperbolas is visible beneath the red horizon. Such a reflection pattern could point to a melange formed of gravitational overthrusts, the distal part of which was drilled in site 415.

It is striking that in contrast to the brown reflection horizon, the red horizon shows little relief. This clearly indicates that the red reflection horizon has only been weakly deformed since its deposition. In other seismic lines from the "zone of complex seismic structures", diapiric structures rise high enough to deform the sea floor.

The area bordering the "zone of complex seismic structures" to the east is defined by numerous anticlinal structures, some of which appear to have been originated in the Early Cretaceous. The sequence MO-2.2, interpreted as Hauterivian-Aptian, thins in some areas (Fig. 3, SP 400-900) against the flanks of the anticlines by onlap onto the reflector blue, thereby indicating a continuing doming in the Early Cretaceous. In other areas (e.g., Fig. 3, SP 900-1200), the sequence MO-2.3 (inferred as Aptian-Albian) has been deformed, thus indicating movement during the Late Cretaceous. The anticlines observed beneath the present shelf lie in the seaward continuation of the Oufta anticline and the Djebel Amsittène and further onshore east-west trending synclinal and anticlinal structures (Robb, 1971, see also Fig. 9).

It can be assumed that the nearshore structures in the Tafelney Plateau area are the direct continuation of the above-mentioned onshore structures, that were formed as the result of the alpine-type deformation of the Western High Atlas in the Tertiary. Deep salt domes beneath some anticlines observed in the shelf region are suggested by diffraction pattern at the core of the folds.
Salt diapirs within the Tafelney Plateau segment are 5-km to 10-km-wide structures that can be observed on the VALDIVIA-profiles in the zone lying directly to the east of the "zone of complex seismic structures". These structures penetrate the Upper Jurassic horizon blue, doming the red horizon and occasionally bow up the seafloor.

The epeirogenic uplift of the Western High Atlas during Late Cretaceous times (Faure-Muret and Choubert, 1971; Robb, 1971; Bhat et al., 1975) led to the formation of the present-day Tafelney Plateau, as indicated by the elevated level of the time contours of the Upper Jurassic blue horizon (Fig.10). The emergence of the former continental margin appears to have been associated with block faulting during Late Cretaceous, the formation of the Tafelney Plateau, and the development of salt diapirs. In the border areas of the present-day Tafelney Plateau the diapiric movements appear to have been especially intensive in the Late Cretaceous. Our interpretation is that the "zone of complex seismic structures" formed as a part of large-scale gravity sliding and intensive salt movement triggered by uplift of the Tafelney Plateau segment in the Late Cretaceous.

After several periods of intensive erosion in the post-Cenomanian-Eocene period (horizon red), there followed the deposition of Cenozoic sediments, which are also cut by several unconformities. In the Tafelney Plateau segment the brown horizon/unconformity is widely present. According to the available data, the trend of the Agadir

◄◄Fig. 3
Time section from the northern margin of the Tafelney Plateau segment with interpreted seismostratigraphic sequences. Location of the profile is indicated in Fig. 1

◄Fig. 4
Reflection seismic record of METEOR-line M 46-09 crossing the Essaouira segment with interpreted seismostratigraphic sequences. Location of the profile is indicated in Fig. 1

Canyon on the southern edge of the Tafelney Plateau was tectonically influenced by the formation of the Tafelney Plateau.

3.3 Essaouira Segment

The Essaouira continental margin segment lies in the seaward continuation of the Essaouira Basin between latitudes 31°30'N and 33°N. METEOR-line M 46-09 (Fig. 4, for location see Fig. 1 and 9), which crosses the Essaouira segment in a NE-SW direction has been chosen as a typical transect for discussion of the geological structures of this continental margin segment.

As was the Tafelney segment, the Essaouira segment is distinguished by a north-south trending structural boundary (SP 1500, Fig. 4), lying between longitudes 10°40'W and 11°10'W, which is associated with a positive magnetic anomaly (Roeser, this vol.). In the Essaouira segment this structural boundary also coincides with a positive free air gravity anomaly.

West of this boundary are more than 3 s (reflection time) thick, flatlying sediments that have been stratigraphically subdivided by extrapolation of the results of DSDP sites 415, 416, 370 (Lancelot et al., 1977). The stratigraphic classification of the seismic marker horizons has been given in the previous section.

Immediately east of the structural boundary lies a narrow zone (SP 1500-2000, Fig. 4), where intensely deformed sediments overlie the gently folded and faulted blue horizon (interpreted as top of Upper Jurassic deep-water carbonates). The intensive deformation of the Cretaceous sediments is explained by gravitational processes, which were probably triggered by salt movement during Late Cretaceous times.

In the landward adjoining zone (SP 2000-3200, Fig.4) the deepest recognizable reflection horizon blue is folded and faulted. The overlying Cretaceous sediments have been locally (e.g., SP 2500-3000) more intensely deformed than the blue horizon, again an indication of gravitational tectonics and/or mud diapirism in the Cretaceous. With the available reflection seismic data one cannot always clearly decide whether the anticlinal structures have been formed exclusively by halokinesis, or whether tectonic doming/ folding played the dominant role. It is clear that the anticlinal structures are post-Late Jurassic, because the blue horizon is always either domed or locally pierced. These facts, together with the observation of flat-lying reflection horizons, underneath the blue horizon (see SP 2200-2500, Fig. 4), allow one to draw the conclusion that evaporites must have been very important in the formation of anticlinal structures in the Essaouira segment.

In addition to the main NE-SW-trending anticlinal structures, most probably formed by salt movement, many smallerscale deformations are visible in the seismic records. They are characterized by diffraction patterns, reflection-free areas and by east or west dipping sequences of reflectors. They possibly formed by gravitational processes and/or mud diapirism. Further, they have only been observed in the deep water area between the present 1000 m isobath and the marked structural boundary to the west. In contrast, the large anticlinal structures interpreted by us as being halokinetic in origin extend clearly to the shelf area.

In Fig. 4 (SP 3200-3700) the blue horizon shows a relatively steep seaward dip. Beneath the present outer shelf it is flat and displaced by faults. The relatively steep slope of the blue horizon is interpreted to be an ancient paleoslope of Late Jurassic to earliest Cretaceous age. From the sections available, the relatively steep slope of horizon blue is only recognized in the northern part of the Essaouira segment. From this it can be concluded that the edge of the Jurassic carbonate platform lies mainly beneath the present shelf.

In the Essaouira segment the thickness of the Cretaceous sequences is

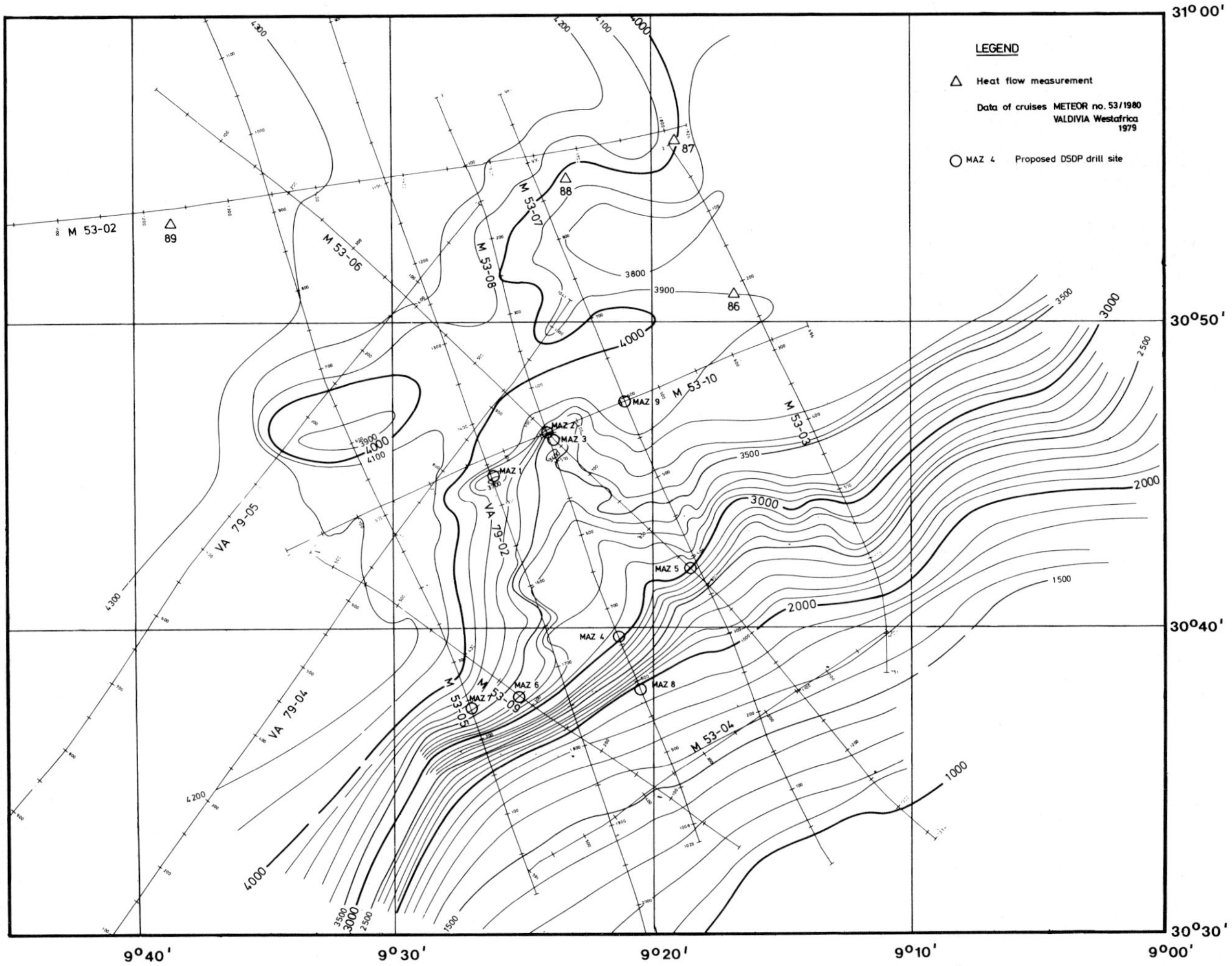

Fig. 5
Bathymetric map of the Mazagan Plateau segment with location of geophysical lines, heat flow stations and proposed IPOD drilling sites for Leg 79. Contour interval 100 m

strongly reduced by erosion beneath the present slope and outer shelf.

3.4 Mazagan Plateau Segment

The Mazagan Plateau, about 1600 km^2 at 500 m to 1500 m water depth (Fig.1 and 5), lying seaward from the Moroccan Meseta, is the dominant feature of the Mazagan Plateau segment, which adjoins the Essaouira segment to the north.

Line VA-7902 has been chosen as a typical section from the Mazagan Plateau segment. The migrated seismic section and the seismostratigraphic interpretation of this line is shown in Fig.6. The results of the DSDP sites 135, 370 and 416 have been extrapolated along with the results of seafloor sampling (Renz et al., 1975; Bee, 1973; Wissmann and von Rad, 1979) for the seismostratigraphic interpretation.

Sequence Ma-3 is the oldest recognizable complex on the outer Mazagan Plateau. Since algal and foraminiferal shallow-water limestones of Oxfordian to Kimmeridgian age have been dredged from the steep slope at depths of 2600 m to 3300 m, sequence Ma-3 in the Mazagan Plateau area very probably

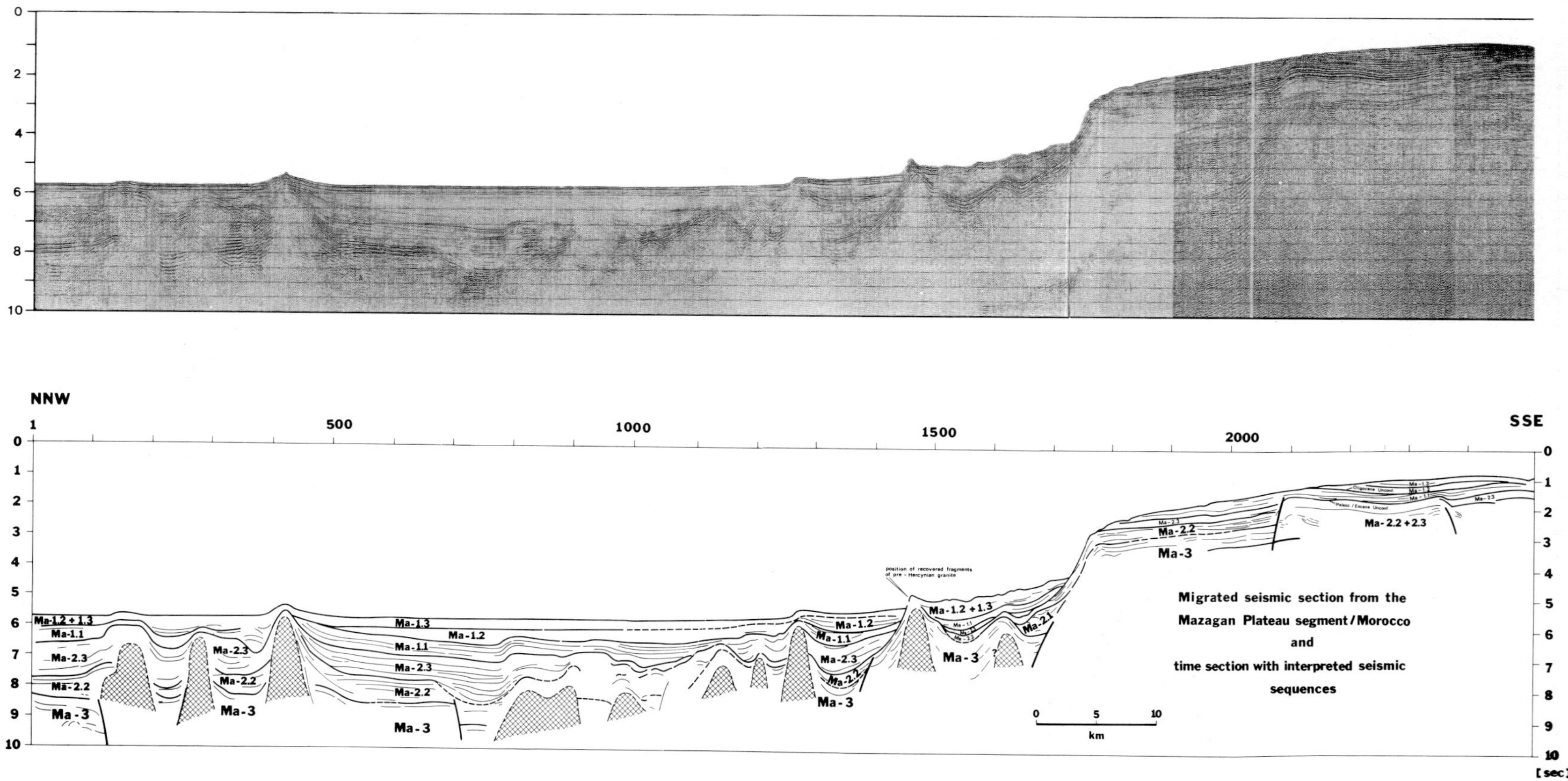

Fig. 6
Migrated seismic section from the Mazagan Plateau segment and time section with interpreted seismostratigraphic sequences. Location of the profile is indicated in Fig. 1. Stratigraphic interpretation of identified sequences: Ma-1.3 = Lower Pliocene-Recent, Ma-1.2 = Miocene, Ma-1.1 Eocene-Oligocene, Ma-2.3=Upper Aptian/Albian - Lower Eocene, Ma-2.2 = Hauterivian - Lower Aptian, Ma-2.1 = Valanginian, Ma-3 = Upper Jurassic

represents Late Jurassic carbonate platform sediments.

The overlying sequence Ma-2.2 is approximately 0.5 s (reflection time) thick, and is distinguished by a sub- parallel reflection pattern. The top of this sequence is marked in the outer plateau area by an unconformity; extrapolating the results of Bee (1973), the unconformity is Cenomanian in age.

In the outcrop region of sequences Ma-2.2 and Ma-2.3 on the steep slope middle Aptian quartz-bearing siliceous organism-rich nanno-marls (Wissmann and von Rad, 1979), as well as glauconitic marls and clastic sediments of the Eocene (Renz et al., 1975), have been dredged or sampled by piston coring. According to Wissmann and von Rad (1979) the middle Aptian sediments were deposited in an outer shelf environment. From these findings, sequence Ma-2.2 has been interpreted as Lower Cretaceous and sequence Ma-2.3 as Upper Cretaceous. The top of sequence Ma-2.3 is also defined by a clear unconformity, probably formed during a global regression in the Paleocene and/or Eocene.

The seismic records from the Mazagan Plateau show structures exhibiting either a discontinuous reflection pattern or mound-like structures (e.g., SP 2080-2370, Fig.6), outlined by strong top-reflectors. Both types of structure have been interpreted as bioherms/-reefs. In the outer plateau area, adjacent to the steep slope, the top of these structures are at a depth between 2300 m and 2700 m in the uppermost part of the Upper Jurassic sequence Ma-3. In contrast, the top of the observed mound structure further landward lies directly under the Paleocene/-Eocene unconformity at a depth of 1200 m to 1400 m.
Assuming an age of 130 x 10^6 years for the top of sequence Ma-3 and an age of 60 x 10^6 or 53 x 10^6 years respectively for the Paleocene/-Eocene unconformity, it follows that the subsidence of the Mazagan Plateau was not uniform with respect to either time or space. The parts of the plateau lying directly next to the steep slope, subsided about 1100 m to 1300 m in the Cretaceous time at a rate of 14.2 to 18.5 m/m.y. whereas the more landward situated parts remained at a shallow water depth until the early Paleocene. From Jurassic to Late Cretaceous times, the Mazagan Plateau was a stable swell that gradually began to subside in the Late Cretaceous. The shelf edge has migrated since this time (i.e., in the last 100 m.y.) about 60 km landwards.

The transition from the Mazagan Plateau to the deep sea is sharp and probably associated with faults. The adjacent deep sea basin is 75 km to 100 km wide and contains diapirs. Within this basin the following seismostratigraphic units were observed:
Sequence Ma-3 is the oldest unit identified; the upper part of it is interpreted to be a Late Jurassic deep water series, comparable with the Cat Gap formation of the North American Atlantic Basin (Jansa et al., 1979). A sedimentary wedge (sequence Ma-2.1), observed only on line VA-7902 (SP 1640-1730, Fig.6), is characterized by landward onlap and seaward downlap and is thought to represent part of a Valanginian deep water sequence of clastic sediments.

Sequence Ma-2.2, which overlies the Jurassic deep water carbonates, is distinguished by a low-frequency subparallel, strongly discontinuous pattern of reflections. Sequence Ma-2.2 is interpreted to consist of pelagic Valanginian/Hauterivian to Aptian marls and limestones (white chalky limestones unit of Jansa and Macqueen, 1978) that grades landward to a larger proportion of turbidites. The sequence Ma-2.2 thins toward the steep slope of the Mazagan Plateau. Apparently sediments of the Lower Cretaceous were intensively eroded during the Aptian either by contour currents, or as the result of global sea-level changes.

During mid-Cretaceous times halokinetic movements started, especially along the landward-side of the deep sea part of the Mazagan Plateau segment (SP 1000-1650, Fig.6). These movements led to a division of the deep sea basin into small troughs and swells. It was in this warped area that sequence Ma-2.3 was deposited. From the results of DSDP sites 370 and 416 sequence Ma-2.3 is interpreted as a terrigenous turbidite equivalent of the Cretaceous black shales.

After an erosional interval sequence Ma-1.1 was deposited. As a result of DSDP sites 370 and 416, we surmise that sequence Ma-1.1 could be composed of Eocene to Oligocene terrigenous marls. In the Oligocene intensive erosion occurred in both the plateau and deep sea areas, followed by the deposition of the seismically transparent sequences Ma-1.2 and Ma-1.3.

The nature of the structure seen in Fig. 6 between SP 1400 and 1500 has been subject of much discussion because granitic rock fragments were found in piston corer samples of VEMA cruise VA-30-13 and in dredge samples of the VALDIVIA-Westafrica cruise in 1979. The structure has been interpreted as slivers of foundered continental basement, pushed upward along landward dipping thrust faults and later as an autochthonous basement ridge (Wissmann and von Rad, 1979). The time migrated seismic sections of METEOR cruise no. 53, 1980, and of the VALDIVIA-Westafrica cruise, 1979, imply that very probably this structure is a salt dome similar to the adjacent diapiric structures. The granitic fragments found are interpreted as being debris of the continental crystalline basement of the Mazagan Plateau transported by piercing salt.[1]

3.5 Pre-Rif Segment

In the seaward continuation of the Rharb Basin and the Rif Province, a layered sequence of Plio-Pleistocene age lies beneath the sea floor (Watkins and Hope, 1979). This sequence which is more than 3.5 s (reflection time) thick beneath the present shelf, overlies a seismically chaotic sequence interpreted to be the seaward continuation of the Pre-Rif nappes. On its southern boundary the Pre-Rif nappes overlie a low-frequency, disturbed reflection marker, which is locally domed, indicating deep-seated Mesozoic salt structures. Figure 7 shows a time contour map to the top of the Pre-Rif nappes.

4. Deep Crustal Structure of the Moroccan Continental Margin

The continental margin of Morocco is thought to have been formed during the separation from Nova Scotia/North America about 200 m.y. ago (van Houten, 1977). Although the Africa-North America plate motions have been intensively studied (Le Pichon et al., 1977), there is as yet no agreement on the position of the ocean-continent boundary and on the nature of the lower crust underneath the present slope and rise.

In addition to sono-buoy refraction seismic measurements (Hoskins et al., 1974), which due to inadequate penetration do not contribute to the solution of the problem, refraction seismic studies (Fig.8) have been carried out on three lines off Morocco (Weigel et al., 1978; Goldflam et al., 1980; Hinz, 1977; Thiessen, 1979; Weigel et al., this vol.).

On line M 39-I, lying in the North Tarfaya segment and extending from the Conception Bank eastward to the outer shelf near Ifni, a 5 km thick layer with a velocity of 5.8 km/s underlies 6-7 km thick sedimentary rocks with velocities of 2.2 km/s, 3 km/s and 4.9 km/s. This layer is in turn underlain by a complex with a velocity of 7.1 km/s.

On the lines M 46-101 and M 46-102, lying in the Essaouira segment, a layer with vp 5.8 km/s underlying a 5-6 km sedimentary sequence (velocities of 1.7 km/s to 2 km/s and 3.7 km/s to 4 km/s) has been observed outside the diapiric province, i.e., directly west of longitude 11°W. Beneath this layer also exists a deeper layer with a velocity of 7.3 km/s.

1 Recent drilling (HINZ K., WINTERER E.L. et al., this Vol.) encountered granitic gneiss. The structure probably is a rotational sialic block moved along listric faults

Fig. 7
Time contour map of the top of the Pre-Rif nappes in ms (reflection time)

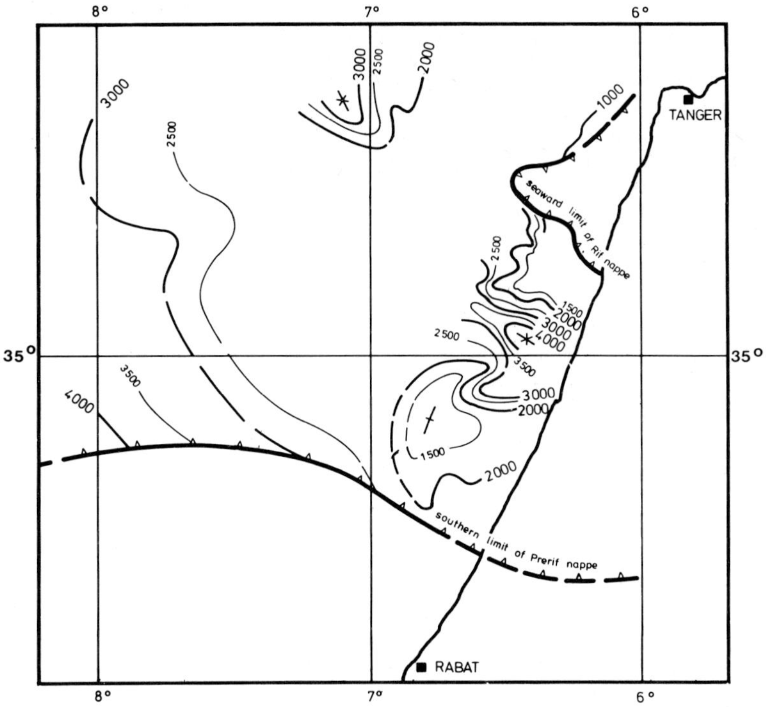

The 5.8 km/s-layer observed in both the North Tarfaya and the Essaouira segments most probably represents the top of the crystalline basement. It is highly probable that this layer represents continental crust in the shelf and continental slope areas, because Paleozoic metamorphics have been penetrated in the borehole IFNI 1. In the deep water area this layer could also be old oceanic crust, i.e., oceanic layer 2 c (Houtz and Ewing, 1976) according to the observed velocity.

The nature and composition of the unit with vp = 7.1 km/s to 7.3 km/s observed in both the Moroccan continental margin segments is unknown. For the area of the continental slope/ outer shelf, this unit could represent continental crust contaminated by basaltic intrusions. During the latest Triassic and earliest Liassic, basalt flows and dolerite sills were emplaced in the Southwestern Moroccan Basin (van Houten, 1977; Wurster, 1980; Brown, 1980). In the deep water areas, seaward of the diapiric province, this clearly defined deep crustal layer with velocities of 7.3 km/s (the western part of lines M 46-101 and M 46-102) could be explained by intrusion of lighter mantle derivates (Drake and Nafe, 1968) into old oceanic crust.

In the Moroccan coastal region, a crustal thickness of 30 km was determined on a line running from Agadir to Oued Sem (Makris and Demnati, 1979). Considering the results of the refraction lines previously discussed, the crust thins by about 20 km in the transition from the coast to the deep sea.

This observation is in principle also confirmed by gravity modeling, which was carried out on three selected transects across the North Tarfaya segment, the Essaouira segment and the Mazagan Plateau segment. The location of these transects is shown in Fig.8. The results of the gravity modeling are presented in Fig.8.

Both the results of the refraction seismic surveys and of the gravity modeling show that the continental crust from the coast to the deep sea thins constantly. The largest reduction in thickness takes place beneath the outer shelf and slope. The processes responsible for the observed crustal thinning are unknown. Sub-aerial erosion during

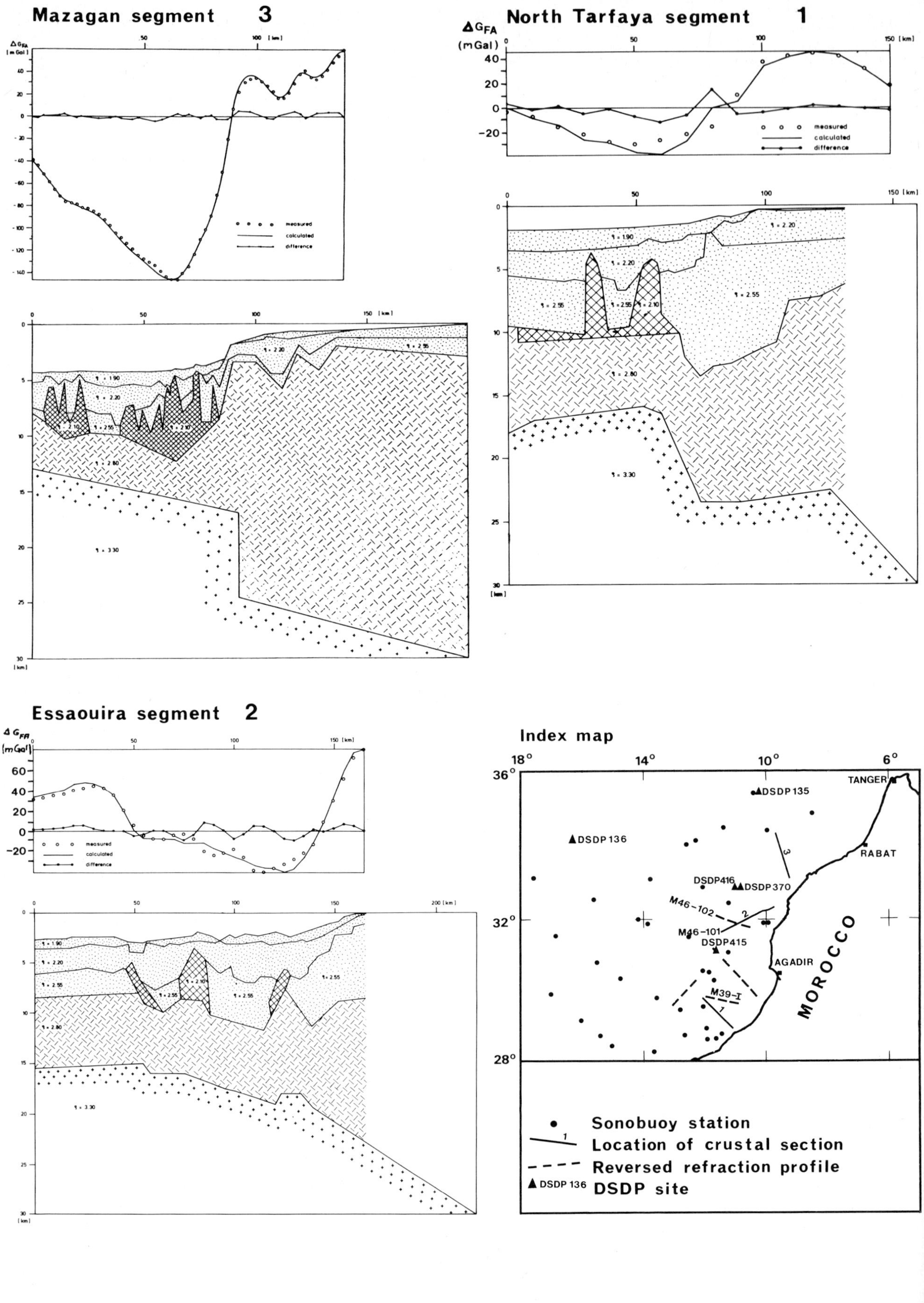
Mazagan segment 3
North Tarfaya segment 1
Essaouira segment 2
ΔG_FA (mGal)
measured
calculated
difference
[km]
ρ = 1.90
ρ = 2.20
ρ = 2.55
ρ = 2.10
ρ = 2.80
ρ = 3.30
Index map
18°
14°
10°
6°
36°
32°
28°
TANGER
RABAT
AGADIR
MOROCCO
DSDP 135
DSDP 136
DSDP416
DSDP370
DSDP415
M46-102
M46-101
M39-I
Sonobuoy station
Location of crustal section
Reversed refraction profile
DSDP site

initial uplift and crustal stretching are suggested to occur (Artemjev and Artyushkov, 1971; Sleep, 1971; De Chapral et al., 1978).

5. Geological and Tectonic Development of the Moroccan Continental Margin

As determined geophysically, structural elements of the Moroccan continental margin are shown in Fig.9.

The dominant structural unit is a NNE-SSW trending zone of salt diapirs, which runs northward from east of the island of Fuerteventura (Canary Islands) and can be traced to the southern margin of the Pre-Rif nappes. The age of the salt is probably Late Triassic to Liassic and started to migrate in the Early Cretaceous, continuing up to the present.

The western boundary of the salt diapir zone, which can be traced for about 800 km, is sharply defined and locally associated with faults. It is parallel to the shelf edge at a distance of 100 km to 150 km from the present coast and is locally displaced by 10 km to 20 km in an east-west direction; the displacement may indicate the existence of fracture zones.

In the southern area of the North Tarfaya segment, the diapiric zone is 35 km to 90 km wide and lies about 10 km to 45 km seaward of the Upper Jurassic carbonate platform. In the area of the Tafelney and Essaouira segments the zone becomes wider and extends into the Southwestern Moroccan Coastal Basin.

Fig. 8
Schematic crustal sections across the continental margin of the North Tarfaya segment (1), the Essaouira segment (2), and the Mazagan segment (3). Location of the crustal sections derived from gravity modeling, and the location of sonobuoy refraction seismic stations and reversed refraction seismic lines are indicated in the Index map and Fig. 9

In the area of the Mazagan Plateau segment, clearly identifiable diapirs are directly next to the edge of the Late Jurassic carbonate platform. At present it cannot be discounted that the cores of some of the biohermic structures found in the Mesozoic platform sediments of the Mazagan Plateau may have formed over salt pillows.

In the Southwestern Moroccan Coastal Basin the salt diapir zone overlies Paleozoic basement, in contrast to the deep offshore area, where it overlies a thinned crust.

The eastern boundary of the salt diapir province off Nova Scotia/-Canada mapped from BGR seismic data correlates convincingly well with the western boundary of the Moroccan diapir province (see insert on Fig.9). From this we believe that before separation of the Africa-North America plates a single evaporite basin existed, which is preserved in the Southwestern Moroccan Basin and beneath the Nova Scotian margin. This basin has been enlarged to its present size (see insert on Fig.9) by crustal and sub-crustal processes (e.g., upwelling of asthenosphere; uplift, stretching and thinning of the crust, volcanism; later subsidence as the result of lithospheric cooling) that initiated continental separation leading to the drift stage.

We interpret the thin crust observed beneath the diapir zone in offshore Morocco as an attenuated continental crust which has been contaminated by basaltic intrusions. It is analogous to the Scotian shelf where Paleozoic metamorphic basement flanks a seaward zone of thinned continental (?) crust. In our opinion, the distinct western boundary of the Moroccan diapir province marks the zone where the actual separation of Morocco and Nova Scotia occurred. Linear magnetic anomalies, similar to those off Nova Scotia, have been observed within the offshore Moroccan diapir province (Fig.9 and Roeser, this vol.).

After actual separation in the Sinemurian, presumably carbonates were deposited in the uniformly sub-

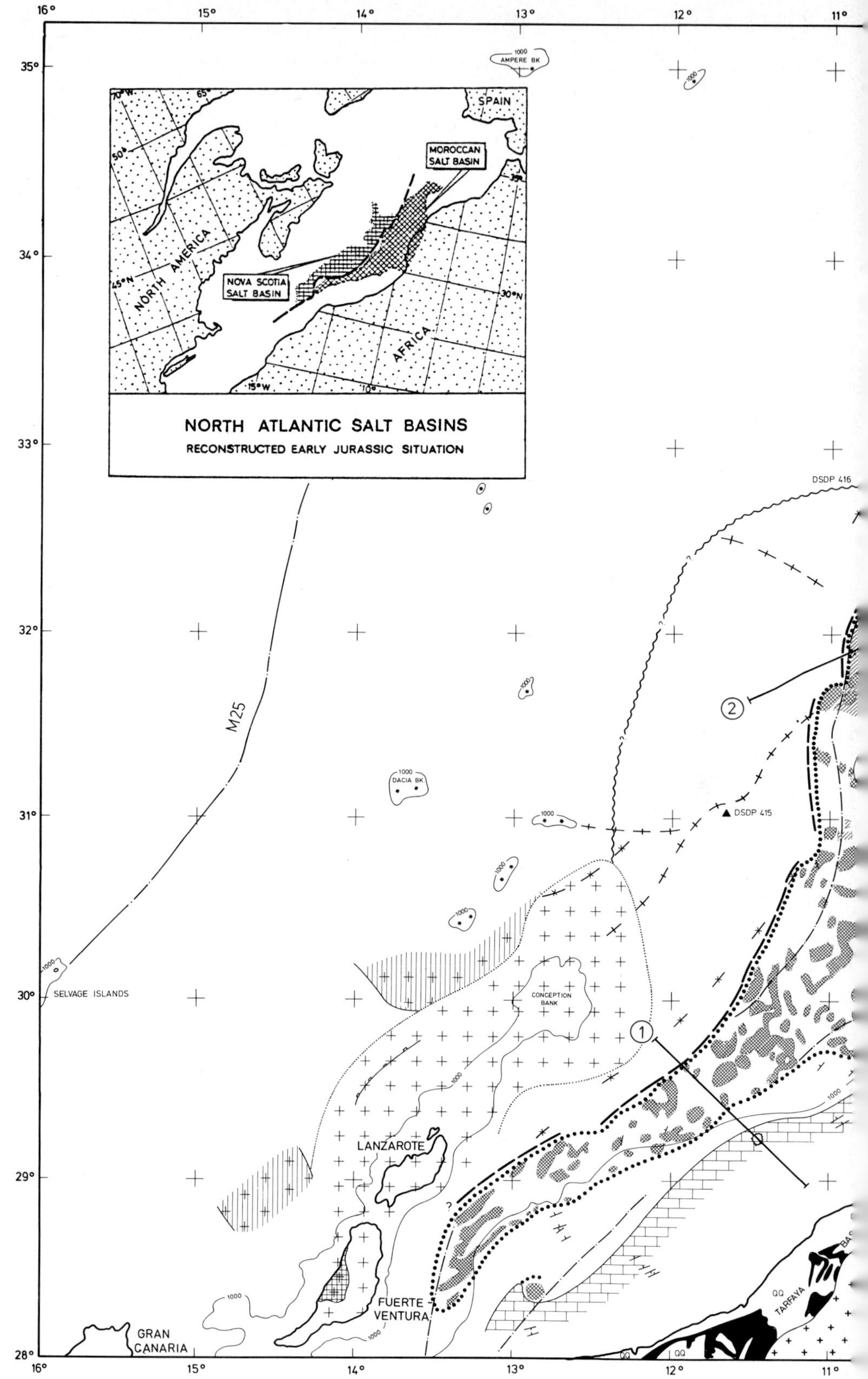

Fig. 9
Structural map of the Moroccan continental margin and adjacent Northwest Africa

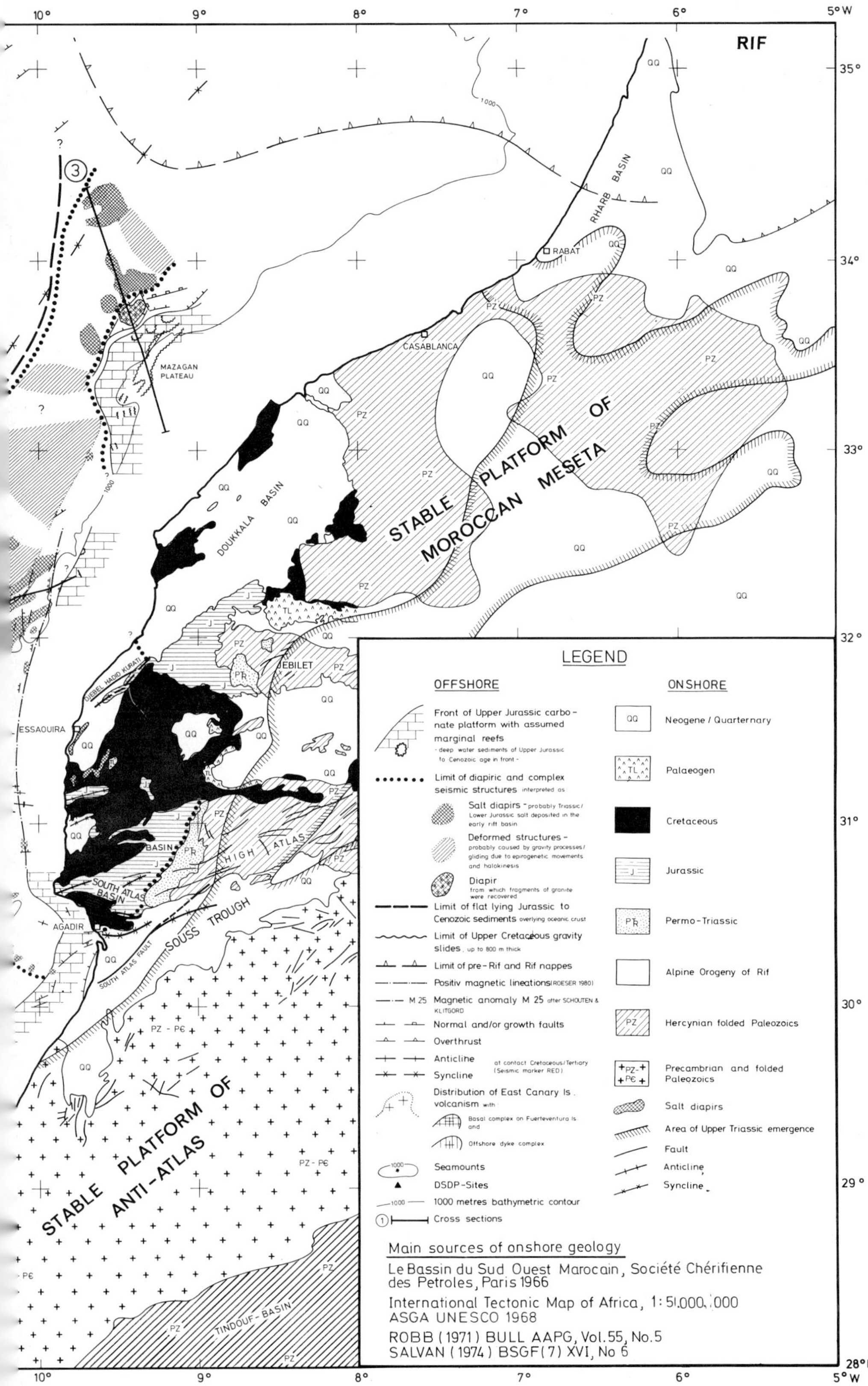

10°
9°
8°
7°
6°
5°W
35°
34°
33°
32°
31°
30°
29°
28°N
RIF
RHARB BASIN
RABAT
CASABLANCA
MAZAGAN PLATEAU
DOUKKALA BASIN
STABLE PLATFORM OF MOROCCAN MESETA
JEBILET
ESSAOUIRA
DJEBEL HADID KURATI
BASIN
HIGH ATLAS
SOUTH ATLAS BASIN
AGADIR
SOUSS TROUGH
SOUTH ATLAS FAULT
STABLE PLATFORM OF ANTI-ATLAS
TINDOUF-BASIN
QQ
PZ
TL
J
PR
PZ - P€
1000
3
LEGEND
OFFSHORE
Front of Upper Jurassic carbonate platform with assumed marginal reefs
- deep water sediments of Upper Jurassic to Cenozoic age in front -
Limit of diapiric and complex seismic structures interpreted as
Salt diapirs - probably Triassic/Lower Jurassic salt deposited in the early rift basin
Deformed structures - probably caused by gravity processes/gliding due to epirogenetic movements and halokinesis
Diapir from which fragments of granite were recovered
Limit of flat lying Jurassic to Cenozoic sediments overlying oceanic crust
Limit of Upper Cretaceous gravity slides, up to 800 m thick
Limit of pre-Rif and Rif nappes
Positiv magnetic lineations (ROESER 1980)
M 25 Magnetic anomaly M 25 after SCHOUTEN & KLITGORD
Normal and/or growth faults
Overthrust
Anticline
Syncline
at contact Cretaceous/Tertiary (Seismic marker RED)
Distribution of East Canary Is. volcanism with
Basal complex on Fuerteventura Is. and
Offshore dyke complex
Seamounts
DSDP-Sites
1000 metres bathymetric contour
Cross sections
ONSHORE
Neogene / Quaternary
Palaeogen
Cretaceous
Jurassic
Permo-Triassic
Alpine Orogeny of Rif
Hercynian folded Paleozoics
Precambrian and folded Paleozoics
Salt diapirs
Area of Upper Triassic emergence
Fault
Anticline
Syncline
Main sources of onshore geology
Le Bassin du Sud Ouest Marocain, Société Chérifienne des Petroles, Paris 1966
International Tectonic Map of Africa, 1:5.000.000 ASGA UNESCO 1968
ROBB (1971) BULL AAPG, Vol.55, No.5
SALVAN (1974) BSGF (7) XVI, No 6

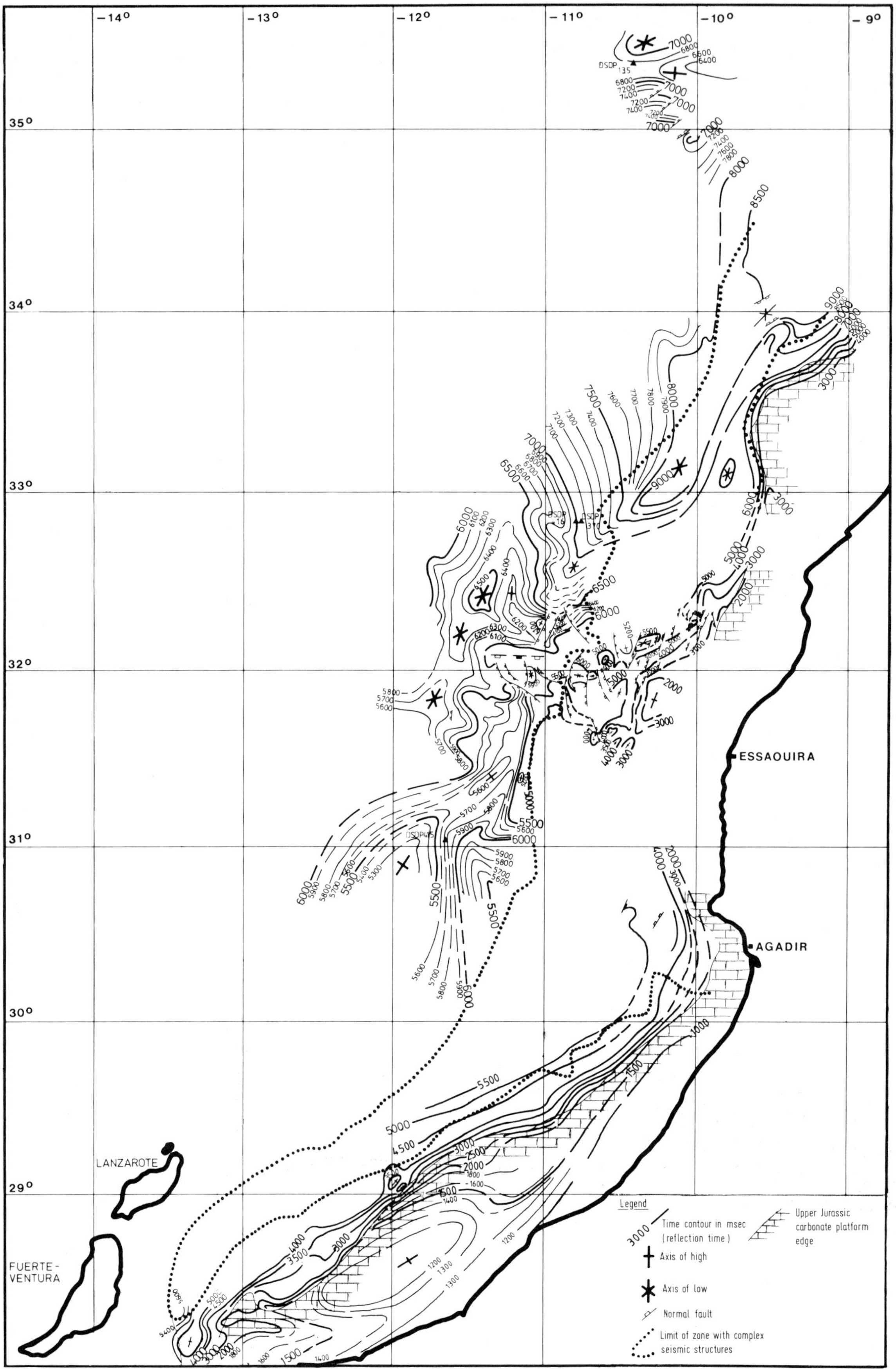

-14°
-13°
-12°
-11°
-10°
-9°
35°
34°
33°
32°
31°
30°
29°
DSDP 135
DSDP 416
DSDP 370
DSDP 415
ESSAOUIRA
AGADIR
LANZAROTE
FUERTE-VENTURA
Legend
3000 Time contour in msec (reflection time)
Axis of high
Axis of low
Normal fault
Limit of zone with complex seismic structures
Upper Jurassic carbonate platform edge

siding basin. About 45 m.y. years after deposition of Triassic red beds over a Hercynian unconformity plus basaltic volcanism in the Southwestern Moroccan Basin, more rapid subsidence took place during Middle Jurassic in the seaward part of the basin, as the thinned and intensively contaminated lithosphere cooled and faulted. This tectonic event, associated with a marine regression in the Bajocian (Jansa et al., 1979), led to the formation of a paleoslope in the Jurassic. For the Tithonian/Berriasian paleowater depths between 2300 m and 2700 m have been estimated from seismic data in the North Tarfaya segment as described previously.

After the formation of a continental slope mainly carbonates built platforms in the various segments of the Moroccan continental margin. Carbonate platform buildups terminated between Agadir and Tarfaya at the beginning of the Early Cretaceous by deposition of the Tan Tan delta. In the Essaouira segment carbonate platform buildup continued into the Early Cretaceous; on the Mazagan Plateau, it continued into the Late Cretaceous; or possibly even Early Tertiary.

Subsidence continued until the Middle Cretaceous in the Moroccan continental margin segments with the exception of the Mazagan Plateau. The Mazagan Plateau was a stable block in a shallow-water environment until the Late Cretaceous, which gradually subsided since that time.

Individual segments developed in a similar manner, although differences occurred in sediment supply, subsidence, and intensity of erosion. The epeirogenic uplift of the Western High Atlas in the Late Cretaceous brought an end of major subsidence of the Moroccan margin. As a result of the uplift of the Western High Atlas, the Tafelney Plateau was formed. The formation of the Tafelney Plateau associated with faulting initiated intensive salt diapirism. This salt diapirism led to large-scale gravity glidings thereby causing development of the complex seismic structures in the outer parts of the Tafelney Plateau and Essaouira segments.

Major erosional events occurred between the Cenomanian and Eocene, during the Oligocene and Miocene in all segments of the Moroccan continental margin.

As a result of the Rif-orogenesis, nappes and olisthostromes slid downslope in the Miocene to mask pre-Paleogene structures of the Pre-Rif segment.

Acknowledgements. The authors are greatly indebted to Prof. Dr. F. Bender, President of BGR, Prof. Dr. Dürbaum, Dr. Stober, Dr. Wilckens (BMFT) and Dr. Goerlich (DFG) for promoting and supporting these investigations.

The assistance of Dr. P. Kehrer and Dr. E.H.K. Kempter in the compilation and analysis of onshore and well data, by Dr. H. Meyer in the seismic processing and by the marine geophysical staff of BGR in collecting geophysical data is gratefully acknowledged.

We are grateful to the Direction de la Géologie, Ministère de l'Energie et des Mines, Morocco, for the permission to carry out the research.

We thank Dr. J.S. Schlee, Dr. D.E. Hayes and Dr. Ph. D. Rabinowitz for critically reviewing this manuscript and offering constructive comments.

The research was financed by the Deutsche Forschungsgemeinschaft (DFG) and the Bundesministerium für Forschung und Technologie (BMFT).

Fig.10
Time contour map of horizon "BLUE" (= top of Late Jurassic carbonates) in ms (reflection time)

References

Addel-Monem A, Watkins ND, Gast PW (1971) Potassium-argon ages, volcanic stratigraphy, and geomagnetic polarity history of the Canary Islands: Lanzarote, Fuerteventura, Gran Canaria, and La Gomera. Americ J Sci 271: 490-521

Ager DV (1974) The western High Atlas of Morocco and their significance in the history of the North Atlantic. Geol Association (London) Proc Vol 85 No 1: 23-41

Ambroggi R (1963) Etude géologique du versant meridional du Haut Atlas occidentale et de la plaine du Souss. Notes Mem Serv Geol Moroc No 157: 321 pp 181 figs

Artemjev M and Artyuskov EV (1971) Structure and isostasy of the Baikal Rift and the mechanism of rifting. Geophysical Res 76: 1197-1211

Beck RH (1972) The oceans, the new frontier in Exploration. Austral Petrol Expl Ass J Vol 12 Part 2: 1-21

Beck RH Lehner P (1974) Oceans, new frontier in Exploration. Bull Am Assoc Petrol Geol 58: 376-395

Beck RH, Lehner P, Diebold P, Bakker G, Doust H (1975) New geophysical data on key problems of global tectonics. 9th World Petroleum Congr (Tokyo) PD 1(1)

Bee AG (1973) The marine geochemistry and geology of the Atlantic continental shelf of central Morocco. Ph thesis Roy School of Mines Imperial College (London): 268 pp

Behrens M, Krumsiek K, Meyer DE., Schäfer A, Siehl A, Stets J, Thein J, Wurster P (1978) Sedimentationsabläufe im Atlas Golf (Kreide-Küstenbecken Morokko). Geol Rundschau 67 2: 424-453

Beurlen K (1974): Die geologische Entwicklung des Atlantischen Ozeans. Geotekt Forsch 46: 1-69

Bhat H, McMillan NJ, Aubert J, Porthault B, Surin M (1975) North American and African drift - the record in Mesozoic coastal plain rock, Nova Scotia and Morocco. in: Yorath CJ et al. (eds) Canadian continental margins and offshore petroleum exploration. Mem Can Soc Petrol Geol 4: 375-389

Brown RH (1980) Triassic rocks of Argana Valley, Southern Morocco, and their regional structural implications. AAPG Bull Vol 64 No 7: 988-1003

Buttkus B, Harjes HP, Hinz E, Hubral P (1977) Auswertung reflexionsseismischer Messungen der METEOR-Fahrt Nr. 39. Unveröff Bericht der BGR 10. Jan. 1977 Hannover

Choubert G, Marcais J. (1952) Geologie du Maroc: apercu structural. 19th Internat Geol Congr, (Alger) Mon Rég Ser 3 Vol 6: 9-73

Choubert G, Faure-Muret A, Hottinger L (1966) Apercu géologique du bassin côtier de Tarfaya. Notes et mémoires du service géologique No 175 Tome I, stratigraphie: 7-219 (Rabat, Maroc)

Cornford C, Rullkötter J, Welte D (1978) Akkumulation organischen Materials aus dem Kontinentalabhang und -fuß des östlichen Nordatlantik. Ergebnisse von DSDP Sites 397, 398, 415 und 416. Abstract of paper presented at Kolloquium des "DFG- Schwerpunktprogramms Deep Sea Drilling Projekt" (Hannover, 4-6 April, 1978)

Cornford C, Rullkötter J, Welte D (1978) Organic geochemistry of DSDP Leg 47 A, site 397 eastern North Atlantic:organic petrography and extractable hydrocarbons. Initial Reports of the Deep Sea Drilling Project, Vol 47 (1):511-522

Dash BP, Bosshard E (1969) Seismic and gravity investigations around the western Canary Islands. Earth and Planetary Science Letters Vol 7 No. 2: 169-177

De Chapral, Guenoch O, Montadert L, Roberts DG (1978) Rifting, crustal attenuation and subsidence in the Bay of Biscay. Nature 275: 706-711

Dillon WP (1974): Structure and development of the southern Morocco continental shelf. Marine Geology 16: 121-143

Dillon WP, Sougy JMA (1974) Geology of West Africa and Canary and Cape Verde Islands. in: Nairn, AEM and Stehli, FG: The ocean basins and margins. Vol 2, The North Atlantic (New York, Plenum Press): 315-390

Dow WD (1977) Petroleum source beds on continental slopes and rises. AAPG Continuing Education Course; Short Course "Geology of continental margins" (Washington)

Drake CL, Nafe JE (1968) The transition from ocean to continent from seismic refraction data. Geophysical Monograph No 12 Amer Geophys Union: 174-186

Embley RW (1976) New evidence for occurrence of debris flow deposits in the deep sea. Geology Vol 4: 371-374

Evans I et al. (1974): Jurassic sedimentation in the High Atlas mountains of Morocco during early rifting of Africa and North America. Geology Vol 2 No. 6: 295-296

Faure-Muret A, Choubert G (1971) Le Maroc. Domaine rifaine et atlasique. In Tectonique de l'Afrique U.N.E.S.C.O.: 17-46 (Paris)

Fritsch J (1977): Bericht über den Fahrtabschnitt M 46/2 Casablanca - Las Palmas der METEOR-Westafrika-Fahrt 1977 (Reise Nr. 46). Unpubl BGR report Archiv Nr. 79019 Dezember 1977 (Hannover)

Fritsch J, Hinz K, von Rad U, Roeser H, Weigel W, Wissmann G (1978) Ergebnisse geowissenschaftlicher Untersuchungen der VALDIVIA-Westafrika-Fahrt VA-10/1975. Forschungsbericht M78-03 des Bundesministerium für Forschung und Technologie BMFT-FB M78-03: 37 pp 22 Fig.

Fusod (1977) The Future of Scientific Ocean Drilling. A report by an ad-hoc subcommittee of the JOIDES Executive Committee. Appendix 4: 32-49 (Univers. Washington, Seattle, Washington 98195)

Fuster JM, Cendrero A, Gastesi P, Ibarrola E, Ruiz JL (1968 a) Geology and volcanology of the Canary Islands, Fuerteventura. Inst. "Lucas Mallada" Madrid, Intern Symp Volcanology (Tenerife Sept. 1968) Spec Rub: 239 pp

Fuster JM, Santin SF, Sagredo A (1968 b): Geology and volcanology of the Canary Islands, Lanzarote. Inst. "Lucas Mallada" Madrid, Intern Sym Volcanology (Tenerife, Sep. 1968) Spec Pub: 117 pp

Gastesi A (1973) Is the Betancuria Massif, Fuerteventura, Canary Islands, an uplifted piece of oceanic crust? Nature Phys Sciences Vol. 26 Dec 17/24: 102-104

Giesel W, Seibold E (1968) Sedimentechogramme vom iberomarokkanischen Kontinentalrand. "Meteor"-Forschungsergebnisse C No 1: 53-75

Goldflam P, Hinz K, Weigel W, Wissmann G (1980) Some feature of the northwest African margin and magnetic quiet zone. Phil Trans R Soc (London) A 294: 87-96

Grunau HR, Lehner P, Cleintuar MR, Allenbach P, Bakker G. (1975) New radiometric ages and seismic data from Fuerteventura (Canary Islands), Maio (Cape Verde Islands) and Sao Tome (Gulf of Guinea. In Progress in Geodynamics, Roy Soc Neth Acad Arts and Sci: 89-118

Hansen W (1979) Eine Untersuchung der Amplitudenverteilung der seismischen Wellen bei den Messungen des F.S. "VALDIVIA" 1975 vor der Küste Mauritaniens, Northwest-Afrika. Hamburger Geophysikalische Einzelschriften Reihe A Heft 39 (Hamburg) G.M.L. Wittenborn Söhne, 2 Hamburg 13

Hayes DE and Rabinowitz PD (1975) Mesozoic magnetic lineations and the magnetic quiet zone off Northwest Africa. Earth and Planetary Science Letters 28: 105-115

Hinz K (1970) Neuere geophysikalische Ergebnisse aus dem Bereich des äußeren Schelfs und Kontinentalabfalls. Erdöl und Kohle Erdgas Petrochemie 23 Jahrg: 70-73

Hinz K (1977) Bericht über den Fahrtabschnitt M 46/1 Hamburg - Casablanca 8.10.77 - 5.11.77 der "METEOR"-Westafrika-Fahrt 1977 (Reise Nr. 46). Unpubl BGR report Archiv Nr. 78906 Nov. 1977 (Hannover)

Hinz K, Amann K and Schendl G (1975) Erdölexploration im tiefen Wasser. Erdöl und Kohle Erdgas Petrochemie Bd 28 H 4: 175-182

Hinz K and Seibold E (1980) Passive Continental Margins: NW-Africa,

North Atlantic, NW-Australia. In: Mobile Earth Internat Geodynamics Project Final Report of the Fed Rep Germany: 127-138 (Harald Boldt Verlag, Boppard)

Hoskins H, Rogers CU, Wod AO (1974): Data report of oblique reflection-refraction radio-sonobuoy profiles on the African Atlantic continental margin (R/V Atlantis II cruises 67 and 75). Distributed as WHOI Reference No 74-34 (Woods Hole Oceanographic Institution)

Houtz E, Ewing J (1976) Upper crustal structure as a function of plate age. Journal of Geophysical Research Vol 81 No 14: 2490-2496

Jansa LF, Gardner JV, Dean WS (1977) Mesozoic sequences of central North Atlantic. Initial Reports of the Deep Sea Drilling Project Vol XLI: 991-1031 (Washington U.S. Government Printing Office)

Jansa LF, Macqueen RW (1978) Stratigraphy and Hydrocarbon Potential of the Central North Atlantic Basin. Geoscience Canada Vol 5 No. 4: 176-183

Jansa LF, Enos P, Tucholke BE, Gradstein FM, Sheridan RE (1979) Mesozoic-Cenozoic sedimentary formations of the North American Basin; Western North Atlantic. in Maurice Ewing Series 3: Deep Drilling Results in the Atlantic Ocean continental margins and Paleoenvironment. Americ Geophys Union (Washington D.C.)

Jacobi RD, Rabinowitz PD, Embley RW (1975) Sediment waves on the Moroccan continental rise. Marine Geology 19: 61-67

Keen CE, Hyndman RD (1979) Geophysical Review of the continental margins of eastern and western Canada. Canadian Journal of Earth Sciences Vol 16 No. 3 Part 2: 712-747

Kendrick JW, Hood A, Castano J (1977) Petroleum generating potential of sediments from Leg 41 DSDP. in: Lancelot Y and Seibold E (ed.): Initial Reports of the Deep Sea Drilling Project Vol 41: 817-819

Lancelot Y, Winterer EL, Bosellini A, Boutefeu AG, Boyce RE, Cepek P, Fritz D, Galimov EM, Taguchig K, Vincent E, Westberg J (1976) Initial core descriptions, Deep Sea Drilling Project, Leg 50. Prepared for the National Science Foundation under contract C-482 by the University of California

Lancelot Y, Seibold E, Gardner JV (1977) Geophysical observations collected underway on GLOMAR CHALLENGER-Leg 41. Initial Reports of the Deep Sea Drilling Project Vol. XLI: 1135-1179 (Washington U.S. Government Printing Office)

Lancelot Y, Seibold E (1977) The Evolution of the Central Northeastern Atlantic - Summary of Results of Leg 41. Initial Reports of the Deep Sea Drilling Project Vol XLI: 1215-1245 (Washington U.S. Government Printing Office)

Lancelot Y, Winterer EL and the shipboard party of Leg 50 (1977) In the Atlantic: documenting early rifting. Geotimes 22: 24-27

Lancelot Y and Winterer EL (1980) Evolution of the Moroccan oceanic basin and adjacent continental margin - A synthesis. Initial Reports of the Deep Sea Drilling Project: 801-821 (Washington U.S. Government Printing Office)

Lehner P, de Ruiter PAD (1976) Africa's Atlantic margin typified by string of basins. The Oil and Gas Journal Nov 8: 252-266

Lehner P, de Ruiter PAC (1977) Structural history of Atlantic margin of Africa. Am Assoc Petrol Geol Bull Vol 61 No 7: 961-981

Le Pichon X, Sibuet J-C, Francheteau J (1977) The fit of the continents around the North Atlantic Ocean. Tectonophysics 38: 169-209

Luyendyk BP, Bunce ET (1973) Geophysical Study of the Northwest African margin of Morocco. Deep-Sea Res Vol 20: 537-549

Markis J, Demnati A (1979) Deep seismic soundings in Morocco and a crust and upper mantle model deduced from seismic and gravity data. in press

Martinis B, Visintin V (1966) Données géologiques sur le basin sédimentaire côtier de Tarfaya (Maroc méridional). In: Reyre D. (ed.) 1966: Sedimentary basins of the African coasts (symposium): 13-26 Assoc of African Geological Surveys, (Paris)

McMaster RL, La Chance TP (1968) Seismic reflection studies on northwestern African continental shelf, Strait of Gibraltar to Mauritania. Bull Am Assoc Petrol Geol Vol 52 No 12: 2387-2396

Michard A (1976) Elements de Geologie marocaine. Notes et Mem Ser Geol du Maroc No 252: 408 pp

Pratsch J-C (1974): Westafrika, Atlantik, Kontinentaldrift. Compendium 74/75 Ergänzungsband der Zeitschrift Erdöl und Kohle; Erdgas; Petrochemie: 114-122

Price I (1979) Gravity tectonics on a passive margin: Deep Sea Drilling Project Site 415 in relation to regional seismic data. Initial Reports of the Deep Sea Drilling Project 50: 759-771 (Washington U.S. Gov Print Office)

Renz O, Imlay R, Lancelot Y, Ryan WBF (1975) Ammoniterich Oxfordian limestones from the base of the continental slope off Northwest Africa. Eclogae geol Helv Vol 68/2: 431-448

Robb JM (1971) Structure of the continental margin between Cape Rhir and Cape Sim, Morocco, Northwest Africa. Bull Amer Assoc Petrol Geol Vol 55, No 5: 643-650

Robertson AHF and Stillman CJ (1979) Submarine volcanic and associated sedimentary rocks of the Fuerteventura Basal Complex, Canary Islands. Geol Mag 116, 3: 203-214

Roeser HA, Hinz K, Plaumann S (1971) Continental margin structure in the Canaries. Inst Geol Sci Report 70/16: 27-36

Roeser HA (1972) Geophysikalische Untersuchungen im östlichen Kanarenraum. Dissertation Georg August Universität (Göttingen)

Ronov AB (1958) Organic carbon in sedimentary rocks. Geochemistry No 5: 510-536

Rothe P (1964) Fossile Straußeneier auf Lanzarotte. Natur und Museum 94: 175-178

Rothe P (1968) Mesozoische Flysch-Ablagerungen auf der Kanaren-insel Fuerteventura. Geol Rundschau 58: 314-322

Salvan HM (1974) Les series saliféres du Trias marocain; caractères géneraux et possibilites d'interpretation: Bull Soc Geol France 7. ser, 16: 724-731

Schmincke H-U (1976) The geology of the Canary Islands. in: Kunkel G. (ed): Biogeography and Ecology in the Canary Islands. Junk (The Hague): 67-184

Scholle P (1977): Deep-water carbonate sedimentology, diagenesis and petroleum potential. Am Assoc Petrol Geologists, Program and Abstracts: 50-51

Seibold E and Hinz K (1974) Continental Slope Construction and Destruction, West-Africa. - In: The Geology of Continental Margins, (ed) Burk CA and Drake CL: 179-196 (Springer Verlag New York Heidelberg Berlin)

Seibold E, Hinz K (1976) German cruises to the continental margin of Northwest Africa in 1975: General reports and preliminary results from "VALDIVIA" 10 and "METEOR" 39. "Meteor" Forschungsergebnisse C No 25: 47-80.

Sleep NH (1971) Thermal effects of the formation of Atlantic continental margins by continental break up. Geophys Journ of the Royal Astronom Society 24: 325-350

Stillman CJ, Robertson AHF (1977) The dyke swarm of the Fuerteventura Basal Complex, Canary Islands. Abstr Geol Soc Lond Newsletter 6, 8

Summerhayes CP, Nutter AH Tooms JS (1971) Geological structure and development of the continental margin of Northwest Africa. Marine Geology Vol 11 No 1: 1-25

Summerhayes CP, Milliman JD, Briggs SR, Bee AG, Hogan C (1976): Northwest African shelf sediments: Influence of climate and sedimentary processes. Journ of Geology Vol. 84: 277-300

Thiessen J (1979) Seismische und gravimetrische Untersuchungen der Krustenstruktur am marokkanischen Kontinentalrand. Diplomarbeit, Institut für Geophysik der Universität Hamburg.

Tooms JS, Summerhayes CP, McMaster RL (1971) Marine geological studies on the northwest African margin: Rabat - Dakar. Report No 70/16 Institute of Geological Sciences. ICSU/ SCOR Working Party 31 Symposium 4 Africa: 13-25 (Cambridge 1970)

Uchupi E, Emery KO (1974) Seismic reflection, magnetic, and gravity profiles of the Eastern Atlantic continental margin and adjacent deep sea floor: II Congo Canyon (Rep. of Zaire) to Lisbon (Portugal). Woods Hole Oceanogr Inst Ref No. 74-19 14 sheets

Uchupi E, Emery KO, Bowin CO, Phillip JD (1976): The continental margin off Western Africa: Senegal to Portugal. Bull Am Assoc Petrol Geol 60 No. 5: 809-878.

Vail PR, Mitchum RM, Todd RG, Widmier JM, Thompson S, Sangree JB, Bubb JN, Hatlelid WG (1977) Seismic stratigraphy and global changes of Sea Level. Seismic Stratigraphy - application to hydrocarbon exploration Memoir 26 Am Ass Petrol Geol: 49-212 (Tulsa, Oklahoma, USA)

Vail PR, Mitchum RM, Shipley TH, Buffler RT (1977) Unconformities of the North Atlantic. Preprint for publication in Philosophical Transaction of the Royal Society "The Evolution of Passive Continental Margins in the Light of Recent Drilling Results" October 19-20, 1977 Symposium, London

van Houten FB (1977) Triassic-Liassic deposits of Morocco and Eastern North America: Comparison. AAPG Bull Vol 61 No 1: 79-99

von Rad U, Arthur MA (1979) Geodynamic, sedimentary, and volcanic evolution of the Cape Bojador continental margin (NW Africa). In: Talwani et al. (eds): Deep drilling results in the Atlantic Ocean: Continental margins and paleoenvironment, M Ewing Series 3, Amer Geophys Union: 187-203

Warme JE et al. (1974): Structural framework and depositional history of the High Atlas Mountains, Morocco. AAPG/SEPM Ann Mtg Abs Vol 1: 95 pp

Watkins JS, Hoppe KW (1979) Seismic reflection reconnaissance of the Atlantic margin of Morocco. In: Maurice Ewing Series 3; Deep Sea Drilling Results in the Atlantic Ocean continental margins and Paleoenrivonment. Americ. Geophys. Union, Washington D.C. 2: 205-217

Weigel W, Goldflam P, Hinz K (1978) The crustal structure of the Conception Bank. Marine Geophys Researches 3: 381-392

Welte DH (1979) Hydrocarbon source rocks in deep sea sediments. Paper OTC 3408 presented at the 11th Annual Offshore Technology Conference in Houston April 30 - May 3 1979: 457-461

Wiedmann J, Butt A, Einsele G (1978) Vergleich von marokkanischen Kreide-Küstenaufschlüssen und Tiefseebohrungen (DSDP): Stratigraphie, Paläoenvironment und Subsidenz an einem passiven Kontinentalrand. Geol Rundschau 67 2: 454-508

Wissmann G, von Rad U (1979): Seismic structure, continental basement, and Mesozoic sediments from the Mazagan Plateau off Morocco. - "Meteor" Forsch. Erg. Reihe C No 31: 1-20

Wurster P (1980) Comments on "Rifting and its geological environment" by Ryan WB, Fox PJ Phil Trans Royal Soc London A 294: 121-122

4 Magnetic Anomalies in the Magnetic Quiet Zone off Morocco

Hans A. Roeser[1]

ABSTRACT

Magnetic data are presented from seven geophysical surveys which the Bundesanstalt für Geowissenschaften und Rohstoffe (BGR) has carried out off the Atlantic coast of Morocco since 1967. The observed elongated anomalies in the magnetic quiet zone parallel in part the Mesozoic M-anomalies which are observed further offshore, and in part the present shelf edge. The most westward anomaly is assumed to mark the initial rift zone, whereas the other anomalies lie on crust of continental origin. They may reveal a pattern of crustal weakness in this area.

1. Introduction

Since 1967, the Bundesanstalt für Geowissenschaften und Rohstoffe (BGR) has carried out geophysical surveys off the Atlantic coast of Northwest Africa. Geomagnetic measurements were made on the following cruises:
METEOR 8 (1967), METEOR 9 (1967), VALDIVIA 10 (1975), METEOR 39 (1975), METEOR 46 (1977), VALDIVIA 79 (1979), METEOR 53 (1980)

One of the main objectives of these integrated geophysical investigations was to obtain information on the early rift stage of development of the Atlantic Ocean. This paper presents the magnetic data of BGR surveys which lie off Morocco between 28°and 34° N (see also Fig. 2 on back cover).

As the main emphasis of the surveys was placed on seismic investigations, the profile lines were rarely located so as to satisfy the requirements of gravimetry and magnetics. In some places the lines are closely spaced and may even cross each other at small angles, whereas in other places there are large gaps. Generally, the lines are shorter than is desirable for the magnetic interpretation.

The BGR surveys lie in the "magnetic quiet zone" off West Africa (Heirtzler and Hayes, 1967) where magnetic anomalies appeared to be very small and irregular or even absent. However, to a certain degree, the term "magnetic quiet zone" is a question of scale. Plotting the anomalies at a larger than normal vertical scale considerably reduces the width of the magnetic quiet zone off West Africa. The following discussion mainly concerns the zone between this reduced quiet zone and the Moroccan coast.

2. Processing of the Magnetic Data

As the surveys span more than one decade there are considerable differences in the data acquisition and processing techniques that have been used.

In 1967 celestial navigation, supported by radar fixes in the neighborhood of the Canary Islands, was used. Adjustment of the positions and deadreckoning was manually achieved.

In the later cruises integrated navigation systems were employed with satellite positioning and Doppler sonar. In METEOR cruise no. 39, deadreckoning errors were corrected by a smoothing process. For the other cruises the satellite positions were assumed to be exact and the Doppler sonar deadreckoning was corrected by the addition of a linear function of time which increased between consecutive satellites from 0 to the observed satellite update.

1 Bundesanstalt für Geowissenschaften und Rohstoffe, Stilleweg 2, 3000 Hannover 51, FRG

Errors of up to 5 nm may occur in the 1967 surveys and of up to 1 nm in the later surveys.

The total intensity of the magnetic field was measured in 1967 with a Varian V-4931DR proton magnetometer and on all the other surveys with a Geometrics G801/3 proton magnetometer. The magnetic field of the ship was removed with the help of the Bullard-Mason formula (Bullard and Mason, 1961) in its first approximation

$$F_{Head} = -a * \cos(\alpha)$$

where α is the course of the ship relative to magnetic north. The coefficient a was determined by the apparent changes of the magnetic field during sharp changes of course and was found to vary between 5 and 20 nT depending on the ship and the length of the towing cable.

The removal of the regional trend is based on the international geomagnetic reference fields, namely IGRF 1965 (IAGA, 1969) for the 1967 cruises and IGRF 1975 (IAGA, 1975) for the other cruises. As usual, the removal of the pure IGRF left considerable mean deviations from the zero level. These were reduced by the additional subtraction of 100 nT for METEOR cruise no. 9 and 920 nT - 30 nT * latitude (latitude in degrees) for the other cruises. The profiles of METEOR cruise no. 9 lie between 27°and 29° N and in this area the two functions do not differ by more than 50 nT. This is considered to be insignificant for the interpretation presented here.

The diurnals were only removed for METEOR cruise no. 9 with data from the geomagnetic observatory of Tenerife.

3. Discussion

3.1 Previous Work

The knowledge of the magnetic field off West Africa is poor in comparison to that

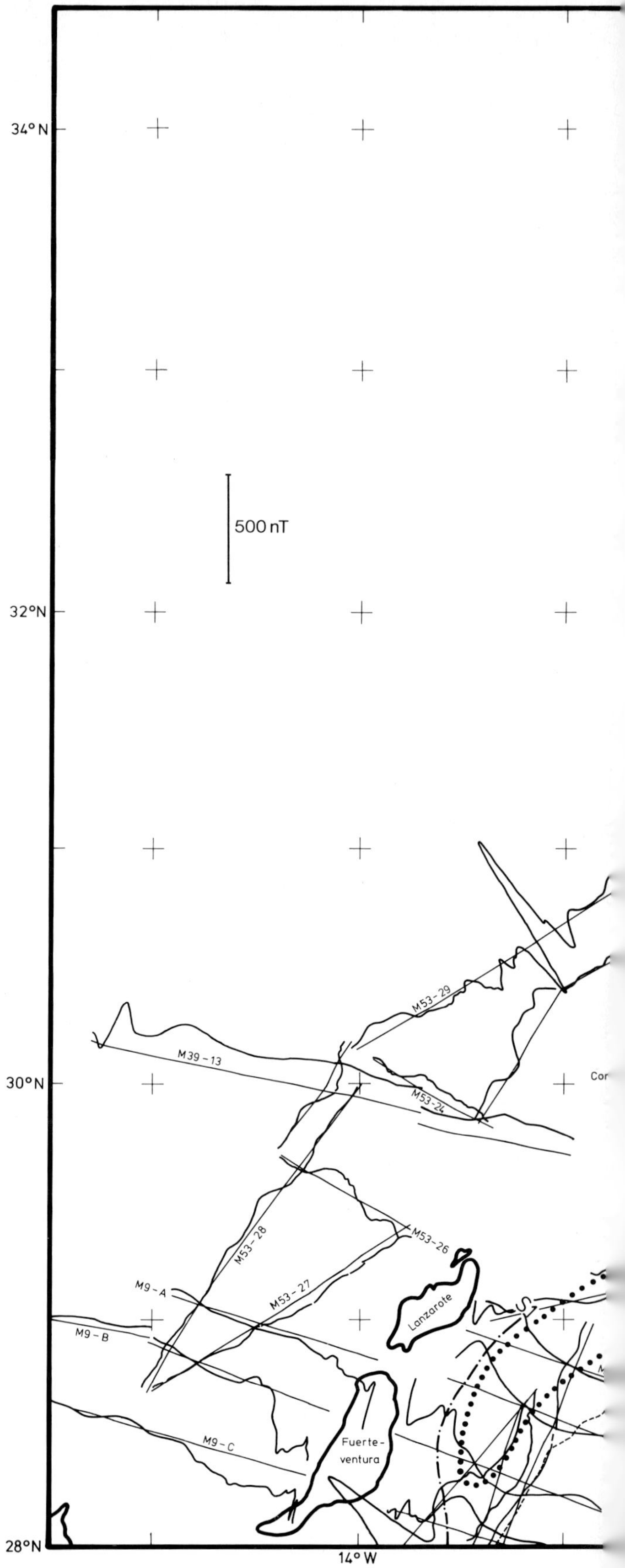

Fig. 1
Magnetic anomalies plotted perpendicular to the tracks
Dotted line limit of diapiric and complex seismic structures after Hinz et al., this Vol.;
Thin lines profiles with profile numbers;
Heavier lines magnetic anomaly curves;
Dashed-dotted lines magnetic anomalies S1 to S4

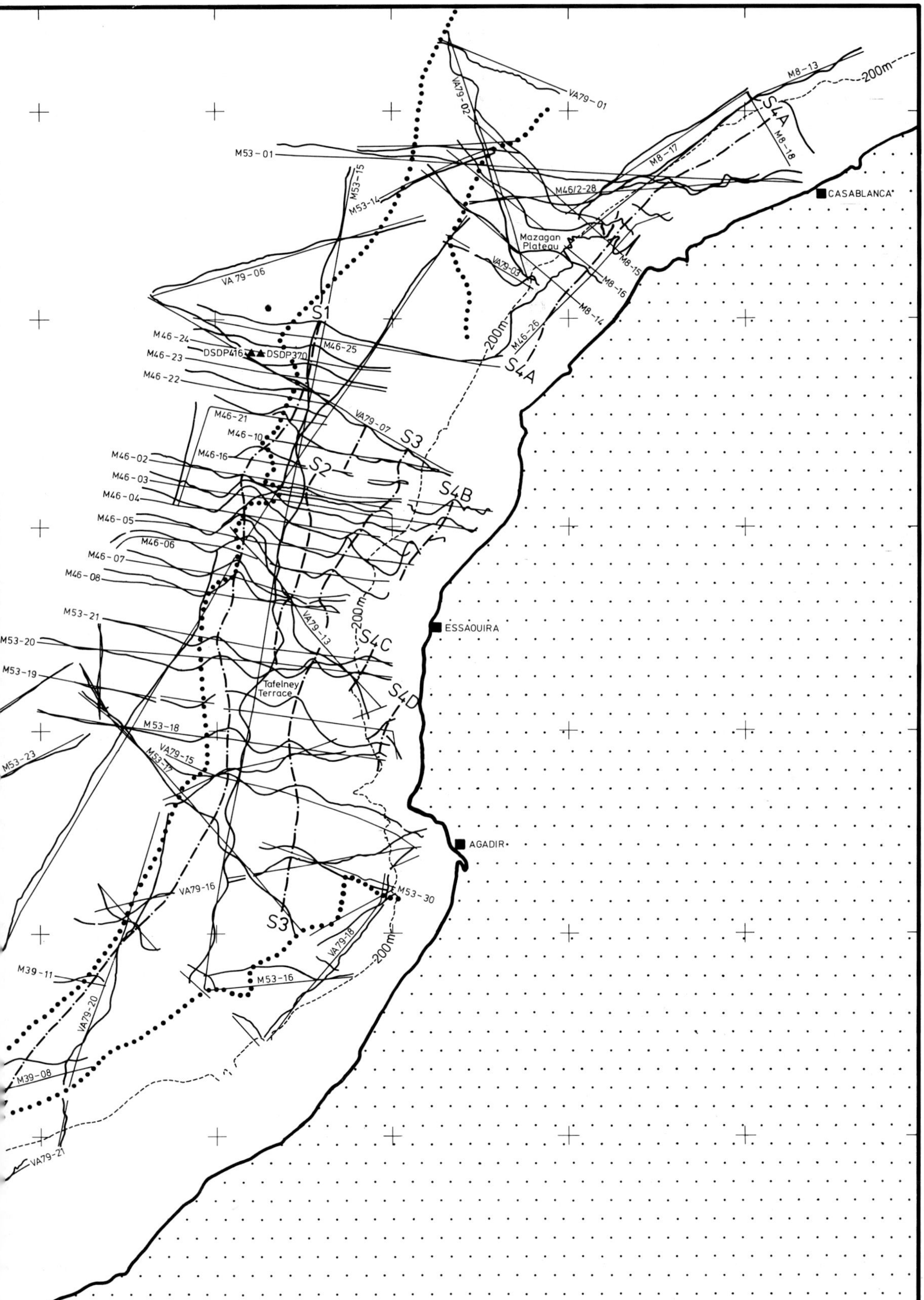

12°W
10°W
8°W
CASABLANCA
ESSAOUIRA
AGADIR
Mazagan Plateau
Tafelney Terrace
DSDP416
DSDP370
200m
S1
S2
S3
S4A
S4B
S4C
S4D
M53-01
M53-15
M53-14
VA79-01
VA79-02
VA79-03
VA 79-06
VA79-07
VA79-13
VA79-15
VA79-16
VA79-18
VA79-20
VA79-21
M8-13
M8-14
M8-15
M8-16
M8-17
M8-18
M46/2-28
M46-26
M46-24
M46-25
M46-23
M46-22
M46-21
M46-10
M46-16
M46-02
M46-03
M46-04
M46-05
M46-06
M46-07
M46-08
M53-21
M53-20
M53-19
M53-18
M53-17
M53-23
M53-30
M53-16
M39-11
M39-08

of the Atlantic coast of North America (Klitgord and Behrendt, 1977, 1979; Behrendt and Klitgord, 1980; Schouten and Klitgord, 1977). In a comprehensive study, Hayes and Rabinowitz (1975) mapped the Mesozoic M-anomalies off West Africa between 10°and 35°N. East of the M-sequence in few places near the coast they recognized a discontinuous anomaly which they suggested might be the analog to the East Coast Magnetic Anomaly (ECMA) at the North American Atlantic coast (Taylor et al., 1968). The magnetic field east of the M-anomalies and offshore from Morocco has been investigated by Dillon (1974), Roeser et al. (1971) and Roeser (1972). Roeser et al. (1971) defined an elongated anomaly between the Canary Islands and Africa that is parallel to the North American East Coast Anomaly. Between Tarfaya and Agadir, offshore from the Aaiun Basin, the Anti-Atlas mountains and the Souss Trough, Dillon (1974) observed some weak and irregular anomalies of unknown origin. Luyendyk and Bunce (1972) presented four lines in the area under discussion, however, their anomaly scale was too small and they saw only a "near-zero amplitude characteristic".

3.2 General Description of the Magnetic Anomalies

Figure 1 shows the magnetic anomalies plotted along track. All of the magnetic data is not included in areas of dense coverage and where it is strongly disturbed by temporal variations of the Earth´s magnetic field.

For ease of reference the anomalies are numbered. The most prominent anomaly is S1 (S for "slope") which lies at the western boundary of a region of diapiric and complex seismic structures (Hinz et al., this Vol.). West of S1 only small anomalies are observed except for those anomalies that are related to the volcanic apron of the Canary Islands and the seamounts in the northeastern prolongation of Fuerteventura and Lanzarote. Some additional elongated anomalies east of S1 are designated as S2, S3 and S4. These anomalies are only roughly parallel to each other and it is not always clear whether they are continuous across the gaps between the surveyed lines. Some anomalies with different trends also occur. They are not shown in Fig. 1 because the necessary profiles for reasons of clarity had to be omitted from the presentation.

The character of the anomalies varies along strike. This can be best seen on lines M46-08 and M46-02 for anomaly S1 and on line M46-05 for anomaly S3. The peak of S3 on line M46-05 shows that here the source of the anomaly must lie much shallower than on the other lines. Further north, S3 decreases and cannot be traced across the gap between lines VA79-07 and M46-25.

The degree of smoothness of the anomalies exhibits a noteworthy variation. On the Tafelney Terrace the anomalies S1 and S3 are smoother than to the north of it on the lines M46-10 to M46-08. Further to the north, on lines M46-25 to M46-22, S1 is smoother than elsewhere in the survey area.

Between lines M53-20 and M46-16 anomaly S2 exactly parallels anomaly S1. South of this region S2 and S3 are replaced by only one anomaly which is tentatively assumed to be the prolongation of S3.

The four anomalies S4A to S4D lie on the shelf just east of the shelf break. Anomalies S4A, S4B and S4C are presumably continuous across the area over a distance of 350 km. S4D is slightly offset relative to S4C.

3.3 Comparison of S1 and the East Coast Magnetic Anomaly

Figures 2 and 3 show some essential magnetic features of the east coast of North America, viz. the East Coast Magnetic Anomaly and the Blake Spur Anomaly together with the Mesozoic anomalies M2 and M22, and of the West African coast, viz. the anomaly S1 and also the anomalies M2 and M22. These figures demonstrate convincingly that anomaly S1 is the analog of the East Coast Magnetic Anomaly which parallels the Atlantic coast of the U.S.A. and Canada near the continental slope (therefore the alternate term "slope anomaly"). If S1 and the East Coast Anomaly mark corresponding events of the sea-floor spreading history of the North Atlantic, then sea-floor spreading was symmetrical to within a few percent up until magnetic anomaly M22. In addition to their positions, the shapes of the anomalies are in agreement when one accounts for the differences in their strikes and the inclination of the magnetic field of the earth at their locations. This has been shown for one line by Roeser (1972). It has not been possible to decide whether these slope anomalies lie on oceanic or on continental crust (on either the American or the African margins of the North Atlantic).

Until this study, it was not clear whether the slope anomaly is as common on the African as on the American margin because it had been observed in only a few

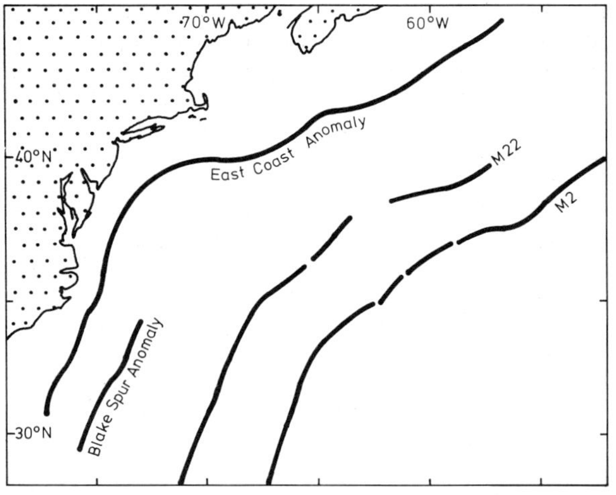

Fig. 2. East Coast Magnetic Anomaly and the Mesozoic anomalies M2 and M22 off the Atlantic coast of North America between 28° and 45°N. (After Emery et al., 1970 and Vogt and Einwich, 1979)

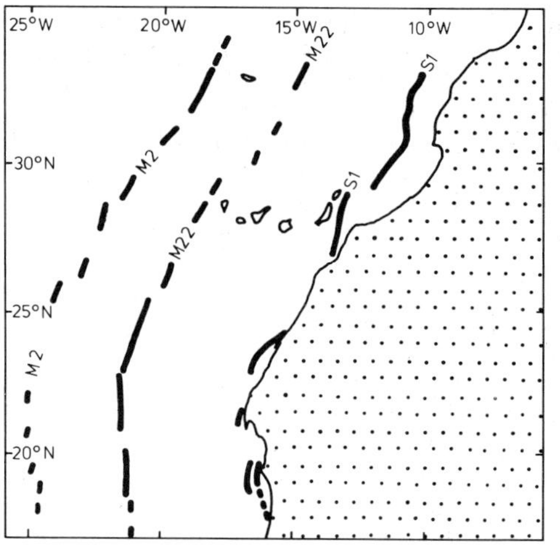

Fig. 3. Anomaly S1 and the Mesozoic anomalies M2 and M22 offshore Northwest Africa (M-anomalies after Hayes and Rabinowitz, 1975)

places and seemed to be absent in others. The BGR surveys show the existence of this anomaly over a distance of 450 km at a place where it had been assumed to be absent. This indicates that the slope anomaly is as common and continuous at the West African coast as it is at the North American Atlantic coast. However, on the West African side it may lie partly on the shelf or even in the coastal regions.

3.4 Seismic Statements and Magnetic Alternatives

The seismic data from the numerous BGR surveys provide a very detailed picture of the crustal structure in the area between anomaly S1 and the Moroccan coast (Hinz et al., this Vol.). These authors conclude that the crust in this region is of continental origin with strong contamination by basaltic intrusions. Typically, the basement lies at a depth of about 10 km or more. The layering of the sediments is severely disturbed by diapiric structures which are interpreted to be salt domes. The salt is probably of Upper Triassic/lowermost Jurassic age, and started to move in the Early Cretaceous.

Do the magnetic data support this view? The configuration of the anomalies appears consistent with a sea-floor spreading origin, and this would conflict with the continental origin of the crust as interpreted from the seismic data. However, there is strong evidence against a sea-floor spreading origin for the anomalies, and this will be discussed in the following sections.

3.5 A Basic Consideration

Sea-floor spreading anomalies are caused by a magnetic layer with stripes of normal and inverse magnetization. Although the widths of the individual stripes can vary over a wide range, the mean width of the stripes with normal magnetization should be the same as that for the stripes with inverse magnetization (excluding some extremely long intervals of constant polarity). The anomaly picture which is produced by a layer with these stripes should therefore statistically not change when we invert the sign of the anomalies. Indeed this is the case when we investigate sufficiently long sequences of sea-floor spreading anomalies. However, it is not true for anomalies S1 to S4. This can be seen by simply turning Fig. 1 upside down. Only in a very few places are negative anomalies more pronounced than their positive counterparts.

Although this is not proof against a sea-floor spreading origin for the anomalies, this heuristic indication should not be overlooked.

3.6 Coincidences with Diapiric Structures

At the present stage of knowledge, the diapiric structures mainly show an irregular pattern (Hinz et al., Fig. 9). This is also the case in areas with dense seismic coverage, so it is not the result of insufficient coverage. The most prominent structure extends 80 km in a NNE direction from 31.5° N, 10.5° W. Between lines M46-08 and M46-02 anomaly S3 coincides exactly with this structure. This indicates a relationship between the diapiric structure and the magnetic body which produces anomaly S3.

Model calculations reveal a depth of 2 km below sealevel for the uppermost parts of the magnetic body on line M46-05; that is in the uppermost part of the diapir and high above the acoustic basement in the vicinity. As the reflection seismic data show that the diapir has deformed Tertiary layers, it is assumed that at least partly the magnetic bodies were emplaced in Tertiary time. They may have triggered the rise of the salt at this line. The nature of the magnetic rocks is not known. Brown (1980) reports that coastal salt anticlines expose Upper Triassic salt, mudstone, and basalt in their cores. Although he does not give more details, the context indicates a Late Triassic age for these basalts. This age seems unlikely in the case of the S3 body, for, if indeed the body was emplaced at the beginning of sea-floor spreading or earlier, it would have been high above sea level in Jurassic and Early Cretaceous time. Presumably it would not have survived subaerial erosion over such a long period. It is possible, however, that anomaly S3 delineates an old zone of weakness which was associated with magmatic activity over a long time during the evolution of the continental margin.

Independent of the open questions it is certain that anomaly S3 is not a sea-floor spreading anomaly although the parallelism to S1 suggests that.

3.7 Anomalies S2 and S4

Anomaly S4 lies on the shelf; the distance from the shelf break varies between 0 and 10 km. On some lines the uppermost parts of the magnetic source rocks of S4 cannot lie deeper than 3 km below sea level; southeast of the Mazagan Plateau they even reach the sea-bed. These anomalies demonstrate that here the shelf break is not only a topographic feature but that it has also deeper structures associated with it. This is not the case for most of the shelf breaks of the North Atlantic. The reason for the unusual behavior between 30°and 34 N may be that sedimentation could not keep pace with the subsidence of the margin.

S2 is weaker and smoother than the other anomalies and has no relationship with topographic or structural features.

3.8 Anomaly S1

There is no doubt that anomaly S1 can be followed from 33° to 31° N, and the few existing lines indicate its continuity as far as 29°N where a westward offset of 80 km seems to occur. It then continues south to 27°N as a very prominent anomaly between the Canary Islands and the African coast (Roeser et al., 1971).

Although in general S1 parallels the shelf edge there are remarkable changes in the distance from the anomaly center to the shelf edge. At 33°N the distance is 100 km, but it decreases to 80 km between 32° and 30.8° N. After an increase to 110 km it again decreases to only 30 km at the 29° N offset. Immediately south of this offset the distance is 60 km, and at 27.5° N S1 crosses the shelf edge.

The depth to the uppermost parts of the anomaly sources varies over a wide range, and this is considered to be an important clue to the answer to the question of whether the anomaly is of sea-floor spreading origin or not. Seismic and gravimetric data indicate that in the vicinity of anomaly S1 off Morocco the basement lies at a depth of 7 - 9 km. The uppermost parts of the block that causes anomaly S1 lie at 5 to 10 km; 7 km seems to be a typical value. Obviously in the reflection seismic records we see directly the bodies which cause anomaly S1. S1 probably marks the line of the initial breakup. Modeling of S1 with the observed bodies proved difficult. That is in part due to the quite complex structures and possibly also to inhomogeneities of the magnetization of the rocks.

S1 cannot be only the edge effect from a plate which has homogeneous magnetization and underlies the quiet zone. As the magnetization in the quiet zone is assumed to be normal (Hayes and Rabinowitz, 1975), the edge anomaly should have the opposite sign of the observed anomaly. Furthermore, the shape of the anomaly, especially the minimum along its western flank, indicates that the amount of magnetic material immediately west of S1 is much smaller than below S1 itself. An explanation for this may be that in the first stage of sea-floor spreading more magnetic material is produced or that it rises to a higher level than later on. Further detailed investigations may reveal most important information on the beginning of the rift stage.

3.9 Faults and Anomaly Offsets

The apparent offset of anomaly S1 east of Lanzarote corresponds to a similar offset of anomaly M22 at 30°N (Fig. 2). North of this offset S1 nearly parallels the African shelf break, whereas M22 and M2 are much more linear. Obviously the spreading pattern became progressively more simple during the early opening of the Atlantic. As well as the 60 km offset of anomaly S1 east of Lanzarote, there may also be an offset of about 10 km at line M46-10. Anomaly S2 is also offset at this location.

Another 10 km offset may exist between lines M46-08 and M53-21 at the northern edge of the Tafelney Terrace which, as indicated previously, also separates anomalies with different degrees of smoothness. It is not clear why the anomalies are smoother over the Tafelney Terrace than over the southern part of the Essaouira Segment.

Because of the lack of information on the shelf region at present it is impossible to relate these offsets to faults at land (e.g., Brown, 1980). Likewise the correlation with offsets in the M-anomalies mapped by Hayes and Rabinowitz (1975) would require investigations in the magnetic quiet zone west of anomaly S1.

4. Conclusions

Hinz et al. (1981) concluded that S1 marks the initial rifting zone. This is confirmed by the magnetic interpretation. S1 is the analog to the East Coast Magnetic Anomaly off the U.S. coast. Because of the lower sedimentation rate off Morocco the magnetic source rocks lie much shallower here, and it is possible to see them in the reflection seismic records.

The elongated magnetic anomalies east of S1 lie on initially continental crust. They may be due to basic intrusions into the thinned and faulted crust. The only available age information is that the salt diapir connected with S2 deforms Tertiary layers at a place where the magnetic rocks are in the uppermost part of the structure.

The magnetic anomalies may reflect an instability pattern in the crust and thus help to reveal one of the possible responses of continental crust to the forces it experiences during the development of a continental margin (Bott 1979).

Vogt and Einwich (1979) propose a spreading rate of 1.22 cm/y for the magnetic quiet zone. Then sea-floor spreading would have started about 180 m.y. ago. It is not clear whether the Jurassic constant polarity interval can be extended that far back. On the other hand, spreading rates may be as large as 2.38 cm/y, a rate which was derived by the same authors for the interval between anomalies 33 and M0. Then the beginning of sea-floor spreading could be dated to 160 m.y. before present which according to Larson and Hilde (1975) would fall into the normal polarity interval. A possible edge effect from a plate with normal polarity underlying the magnetic quiet zone cannot be observed.

Even though anomalies S2 to S4 are not sea-floor spreading anomalies, it cannot be excluded by magnetics alone that the crust in this whole area was initially of oceanic character. However, the joint geological and geophysical information excludes a sea-floor spreading origin of the crust east of S1.

Acknowledgements. The BGR surveys were supported mainly by the Deutsche Forschungsgemeinschaft and the Bundesministerium für Forschung und Technologie.

I thank the many colleagues who were involved in these investigations, especially K. Hinz, the leader of the BGR marine geophysics group, and P. Kewitsch, the technical engineer for the BGR gravity/magnetics equipment since 1970. Further I wish to thank K. Hinz and P. Symonds (BMR/Canberra) who reviewed the paper thoroughly. Discussions with J. Behrendt, H. Dostmann, J. Fritsch, J. Hanisch, K. Hinz and U. von Rad are gratefully acknowledged.

REFERENCES

Behrendt JC, Klitgord KD (1980) High sensitivity aeromagnetic survey of the U.S. Atlantic continental margin. Geophysics 45: 1813-1846

Bott MHP (1979) Subsidence mechanism at passive continental margins. In: Watkins JS et al (eds) Geological and geophysical investigations of continental margins. AAPG Mem 29: 3-9

Brown RH (1980) Triassic rocks of Argana Valley, southern Morocco, and their regional structural implications. AAPG Bull 64: 988-1003

Bullard EC, Mason RG (1961) The magnetic field astern of a ship. Deep-sea Res 8: 20-27

Dillon WP (1974) Structure and development of the western Moroccan continental shelf. Mar Geol 16: 121-143

Emery KO, Uchupi E, Phillips JD, Bowin CO, Bunce ET, Knott ST (1970) Continental rise off eastern North America. Am Assoc Petrol Geol Bull 54: 44-108

Hayes DE, Rabinowitz PD (1975) Mesozoic magnetic lineations and the magnetic quiet zone off northwest Africa. Earth Planet Sci Lett 28: 105-115

Heirtzler JR, Hayes DE (1967) Magnetic boundaries in the North Atlantic Ocean. Science 157: 185-187

Hinz K, Dostmann H, Fritsch J (1981) The continental margin off Morocco - Seismic sequences, structural elements and the geologic development. This volume

IAGA Commission 2 Working Group 4 (1969) The International Geomagnetic Reference Field 1965.0. J Geomagn Geoelectr 21:569-571

IAGA Division I Study Group, Geomagnetic Reference Fields (1975) International Geomagnetic Reference Field 1975. J Geomagn Geoelectr 27: 437-439

Klitgord KD, Behrendt JC (1977) Aeromagnetic anomaly map of the U.S. Atlantic continental margin. US Geol Surv Misc Field Stud MF-913, 2 sheets, scale 1:1 000 000

Klitgord KD, Behrendt JC (1979) Basin structure of the U.S. Atlantic margin. In: Watkins JS et al (eds) Geological and geophysical investigations of continental margins. AAPG Mem 29: 85-112

Larson RL, Hilde TWC (1975) A revised time scale of magnetic reversals for the Early Cretaceous and Late Jurassic. J Geophys Res 80: 2586-2594

Luyendyk BP, Bunce ET (1972) Geophysical study of the nothwest African margin off Morocco. Deep-sea Res 20: 537-549

Roeser HA (1972) Geophysikalische Untersuchungen im östlichen Kanarenraum. PhD Thesis, Univ Goettingen

Roeser HA, Hinz K, Plaumann S (1971) Continental margin structure in the Canaries. In: Delany FM (ed) The geology of the east Atlantic continental margin, 2. Africa. Rep no 70/16 Inst Geol Sci, London; pp 27-36

Schouten H, Klitgord KD (1977) Map showing Mesozoic magnetic anomalies; western North Atlantic. US Geol Surv Misc Field Stud Map MF-915, scale 1:2 000 000

Taylor PT, Zietz I, Dennis LS (1968) Geologic implications of aeromagnetic data for the eastern continental margin of the United States. Geophysics 33: 755-780

Vogt PR, Einwich AM (1979) Magnetic anomalies and sea-floor spreading in the western North Atlantic, and a revised calibration of the Keathley (M) geomagnetic reversal chronology. In: Init Rep Deep Sea Drilling Project 43: 857-876

5 Atlas and Atlantic - Structural Relations

Johannes Stets and Paul Wurster[1]

ABSTRACT

– The High and Middle Atlas mountains originated from continental rift structures, the history of which lasted from Triassic until Turonian times.
– The formation of the Atlas Mountains finally was caused by inversion of the earlier rift structure. In the western High Atlas inversion began during the Late Cretaceous and lasted until Tertiary and Quaternary times. Uplift in the central parts of the rising mountain chains was accompanied by downwarping of subsequent marginal basins along the southern and northern flanks of the High Atlas.
– Vertical movements, starting with subsidence and followed by uplift, dominated the structural evolution of the Atlas mobile belts, which belong entirely to the African plate.
– The Atlas Rift at first formed an integral part of the Triassic North Atlantic rift system, along which Pangea began to break up. This means that during the initial stages the Atlantic furrow and the Atlas Rift were equal structures. Later on the main Atlantic rift developed to an oceanic structure, whereas the Atlas Rift failed to do so.

Intracratonic mountain building processes and ocean genesis may occasionally have the same origin. Our results on the evolution of the High Atlas imply that vertical movements should not be neglected in future discussions on global tectonics.

1. Introduction

The High Atlas and the Middle Atlas Mountains of Morocco are parts of the complicated Northwest African Atlas system, comprising also the Rif, Tell, and Sahara Atlas. By the Betic and the Calabrian arch, connections exist to the Alpine chains of western Europe. Without any doubt, Northwest Africa participated in strike-slip movements, in drifting and collision processes of the western Mediterranean. If we accept the term "alpine" for all of them, the High Atlas was also affected by alpine geodynamics. On the other hand, origin and development of the High Atlas seem rather different from the genesis of the Alps.

Another approach is possible from the west. Here the High Atlas mountain ranges obviously end at the passive margin of the Atlantic Ocean. A continuation toward the nearby Canary Islands is not evident. If we rearrange the contingent margins of the central Atlantic, we find no equivalents of the Atlas system in Northeast America, although relations existed during the Paleozoic with respect to the tectonic history of the Appalachians and the West African Variscan belts. We will not discuss here the problems of a Proto-Atlantic, although in the western High Atlas structural levels of the Precambrian and Variscan orogenies are also well exposed. Our discussion will be restricted to the relations between the Atlas mountains and the Atlantic Ocean during the Mesozoic and the Cainozoic.

Theories of "modern global tectonics" postulate mainly lateral plate movements to be responsible for the formation of ocean basins and orogenetic belts. The generation of new ocean floor at spreading axes induces the drifting of continents, after crustal separation is achieved following periods of rifting. On the other hand, orogenetic belts are often associated with the subduction of old oceanic crust at active continental margins. According to this model, ocean basins and orogenetic belts are quite different structures. They were formed by antagonistically working processes: fracturing and faulting caused by extension on one hand, folding and thrusting caused by crustal shortening on the other. Last but not least, the resulting physiographic features are absolutely contrasting.

Classical geological hypotheses explain continental uplift, synclines, aulacogens and graben structures to be caused by slow epirogenetic or taphrogenetic movements. Geosynclines and mountain chains of alpine type on the other side are distinct stages in the course of orogenetic evolution. According to these aspects continental or intracratonic graben structures are contrasting features with respect to mobile orogenetic belts and high mountain chains.

According to former and more recent theories the two adjacent global structures dealt with in this paper, i.e., the High Atlas and the Atlantic Ocean, developed in quite different ways:

1 Geologisches Institut der Universität Bonn, Nussallee 8, D–5300 Bonn 1, FRG

- Initial rifting, crustal separation, formation of oceanic crust at a mid-oceanic spreading axis, sea-floor spreading, and subsidence of the continental margins are the predominant factors which control the development of the Atlantic Ocean during the Mesozoic and the Cainozoic.
- Mainly vertical movements in an intraplate continental position are responsible for the origin and the evolution of the High and Middle Atlas Mountains at the same time. Hence, the Atlas Mountains were not formed at the convergent boundary between two plates.

In spite of all these differences, concerning feature, structural evolution, and geological interpretation, there must be some relationship between the High Atlas and the Atlantic, besides neighborhood, or etymological and mythical analogies. In order to find these out, it will be useful to summarize the geological history of the High Atlas, as well as of the Atlantic Ocean from the beginning. Block diagrams, drawn for this purpose, may illustrate the different stages at comparable times.

The ideas and concepts presented in this paper are the results of geological investigations in the western High Atlas of Morocco. As the authors were involved in a research program on Cretaceous sedimentology and paleogeography of the Atlantic coastal basins of Morocco, they had to formulate a structural model of basin development of this region. This model is valid for the Cretaceous and corresponds with the paleogeographic and tectonic evolution of the Atlantic Ocean, as well as of the Atlas system during Mesozoic and Cainozoic times, because significant changes in structural evolution and associated sedimentation in this area took place in post-Variscan time.

Studies of the literature and of geological maps – especially the compendium on the elements of Moroccan geology by Michard (1976) and the excellent geological maps of Morocco at a scale of 1 : 500 000 – provided additional knowledge about the pre- and post-Cretaceous formations and the crustal deformation of Morocco.

2. Structural Setting of Morocco

Concerning the problems dealt with here, a short review on the geological and structural setting of Morocco seems to be useful. Morocco and the adjacent Atlantic province are characterized by the following structural units (Fig. 1; see also back cover figure):

- African Platform,
- High and Middle Atlas mountain belts,
- Moroccan Meseta and High Plateaus,
- Salt diapir province off the Moroccan west coast (including parts of the onshore coastal basin),
- Rif Atlas fold belt.

With respect to the problems discussed in this paper, the Rif fold belt is not included, as no direct contact exists with the High and Middle Atlas.

2.1 African Platform

The northern part of the African Platform, the Anti-Atlas, borders the High Atlas at its southern margin. It contains rock series of the Precambrian, which are exposed in domal uplifts and are elsewhere covered by domed and slightly folded Lower Cambrian series. Toward the south and the east an increasingly thick sedimentary cover forms folded transition zones, and further south occupies the Paleozoic Tindouf and Reggane Basins, as well as the Sahara Basin farther to the east. Thus the Anti-Atlas forms a structural high at the northern border of the African Platform (Fig. 2).

Mesozoic beds are not preserved throughout the Anti-Atlas except in its eastern part. There, a thin blanket of Upper Cretaceous to Quaternary sediments covers the Hamada-du-Guir. The sedimentary record of the Mesozoic in the High Atlas region indicates that the Anti-Atlas remained a structural high throughout that time. It was a source area for detrital supply into the adjacent troughs. Very slow epirogenetic movements took place only in this region. It has remained stable since the Paleozoic.

2.2 High Atlas and Middle Atlas

The western and central High Atlas border the African Platform directly (Fig. 2) and attain spectacular heights of more than 4000 m (Jbel Toubkal 4165 m, Jbel Mgoun 4071 m). Their width measures at least 50 km southeast of Marrakech, and up to 100 km in the central part between Midelt and Ksar-es-Souk. This narrow mountain range trends eastnortheast-westsouthwest to east-west, that is parallel to the Anti-Atlas structural high. Near the small town of Beni-Mellal the narrower Middle Atlas branches off in a more northeasterly direction.

In the western High Atlas, south of Marrakech ("massif ancien": Michard, 1976), the Precambrian socle, like that of the Anti-Atlas, is exposed in the Jbel Toubkal region and in its neighborhood. Toward the west and northwest thick Cambro-Ordovician and Carboniferous series are exposed that were intruded by Variscan granites (Schaer, 1964; H. and G. Termier, 1971). East and southeast of the Jbel Toubkal area thinner Cambro-Ordovician to Carboniferous series with a different facies are exposed. They are similar to those of the eastern Anti-Atlas between Ouarzazate and Ksar-es-Souk.

A bundle of northeast-southwest trending faults crosses the High Atlas at the "Tizi-n'Test" ("accident du Nfis", "accident du Tizi-n'Test": Ambroggi and Neltner, 1952; Laville, 1975 in Michard, 1976). This fault system is suggested to be a late-Variscan strike-slip fault of enormous extent and of a complex history (Proust, Petit and Tapponier, 1977). Strike-slip movements of greater extent, especially in the Mesozoic cover, are not yet evident. Indeed, this structural element or geosuture is of great importance with respect to pre-Triassic geology of Morocco, as it separates the more rigid older Anti-Atlas from the mobile Variscan belt. This is obvious in the Jbel Toubkal region: In general, the whole Paleozoic sequence underwent higher deformation during the Variscan orogeny to the northwest than to the southeast of this fault system.

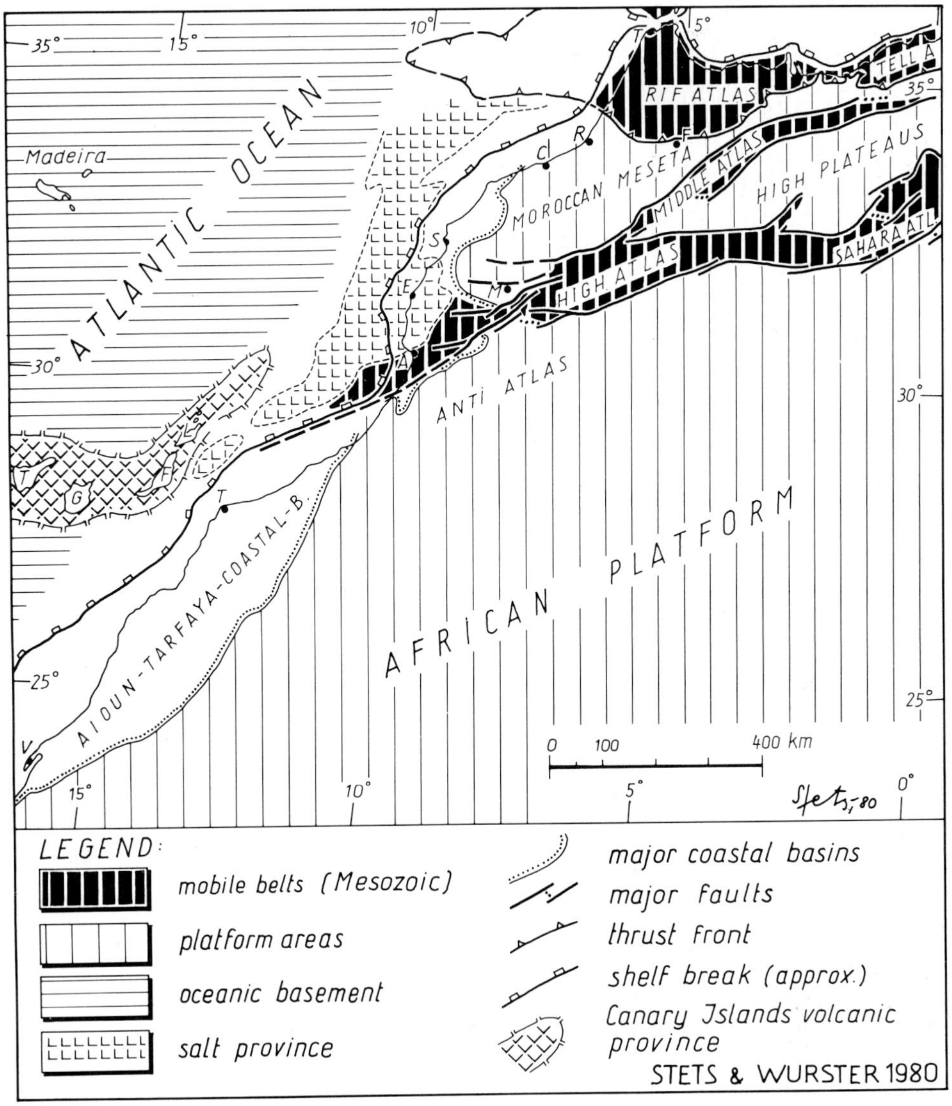

Fig. 1. Regional relationship of Mesozoic mobile belts, platform areas, coastal basins, and continental margin in Northwest Africa. Morocco: **A** Agadir; **C** Casablanca; **E** Essaouira; **F** Fes; **M** Marrakech; **R** Rabat; **S** Safi; **T** Tarfaya; **V** Villa Cisneros; Canary Islands: **F** Fuerteventura; **G** Gran Canaria; **L** Lanzarote; **T** Tenerife

Significant changes occurred in post-Variscan time, when the Paleozoic sequence was covered by Permian to Upper Triassic sediments. The total thickness of the Mesozoic sequence varies from less than 1000 m to more than 3000 m. A maximum is recorded from the Atlas Basin (Agadir-Essaouira- or Haha Basin: Duffaud et al., 1966) in the west, and from the central High Atlas in the east (du Dresnay, 1971, 1972).

The structural deformation pattern of the central High Atlas and of the Middle Atlas, where the Mesozoic cover is still intact, is accentuated by eastnortheast-westsouthwest, east-west, and northeast-southwest running, partly intersecting crests of narrow anticlinal structures (de Sitter, 1952). A similar oriented fault pattern is represented in the central part of the western High Atlas ("massif ancien"), where the Variscan basement is exposed. Tectonic interactions of the Variscan socle and of its younger sedimentary cover are very impressive (Wegmann, 1961; Schaer, 1967; Michard, 1976; Laville et al., 1977). According to this setting we assume a rhomboidal block pattern of the socle, caused by more or less steeply dipping faults, also in regions that now are covered by younger sediments.

The northern and southern flanks of the High Atlas are accompanied by narrow marginal basins, filled with Upper Cretaceous and younger sediments: for example the Haouz Basin in the north and the "sillon préafricain" in the south. The most important of these basins from the structural point of view is the Souss Basin (Ambroggi, 1963), which opens toward the Atlantic

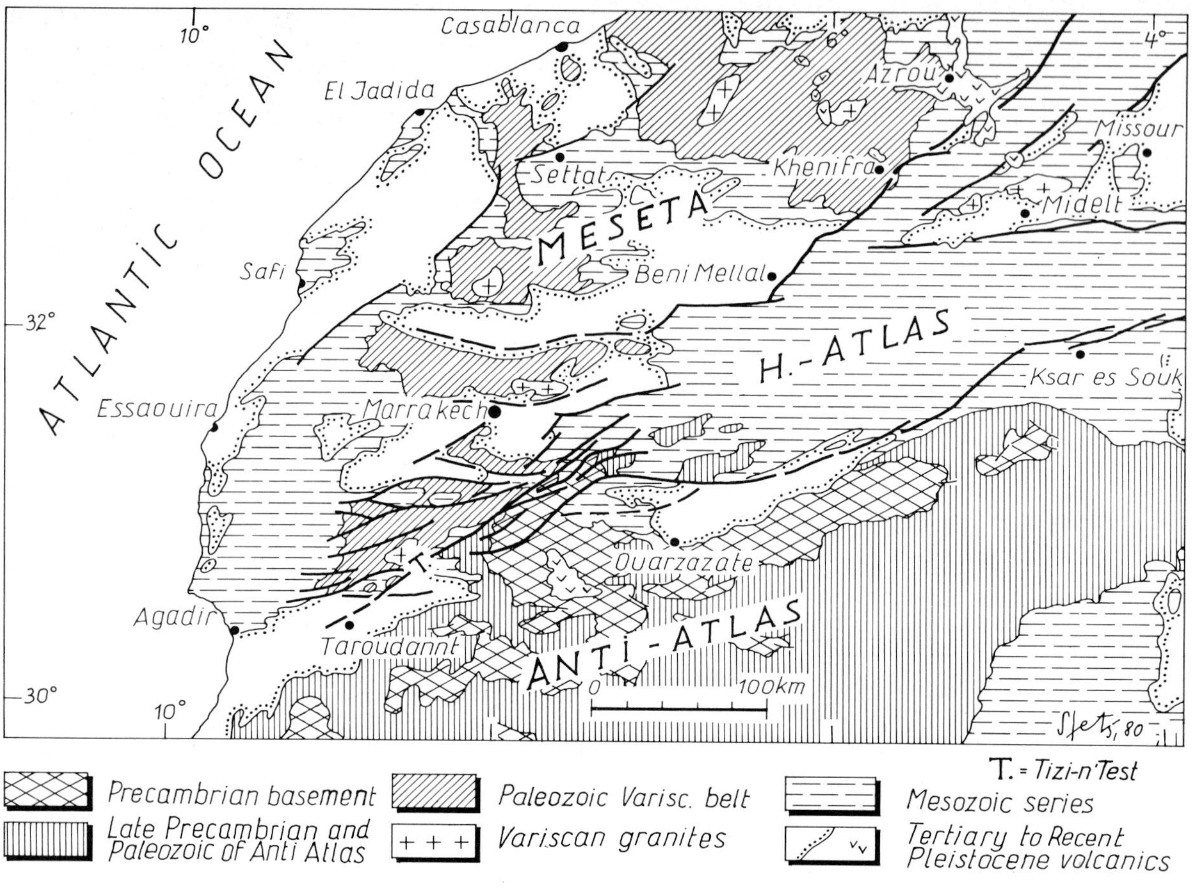

Fig. 2. Geological setting of the investigated area in Morocco; basic map for Figs. 5 and 6

Ocean south of Agadir (Fig. 2). Thrust faults generally separate the High Atlas from these marginal basins and from the stable platforms in the north and in the south ("accident nord- et sudatlasique": P. and L. Russo, 1934; Ambroggi and Neltner, 1952; Choubert and Marçais, 1952; Choubert and Faure-Muret, 1962; Michard, 1976). These thrust faults cannot be traced for longer distances (Proust et al., 1977), but are set off en echelon, thus forming the flanks of the High Atlas and the Middle Atlas as well. Especially north of Agadir no "South Atlas Fault" or "Agadir Fault" (Rod, 1962), as cited in many recent papers, can be traced either from geological or by geophysical evidence (Dahmani, Demnati, Makris pers. comm.). The main bordering fault, separating the High Atlas from the Anti-Atlas in the west, is the "Tizi-n'Test Fault", which crosses the Souss Basin toward the southwest.

2.3 Moroccan Meseta and High Plateaus

North and Northwest of the High Atlas and of the Middle Atlas the Paleozoic socle of the Moroccan Meseta is exposed in many places: Jbilet, Rehamna, northern Meseta. Very thick Paleozoic series of Early Cambrian to Carboniferous age were studied by numerous authors (Michard, 1976). Northeast-southwest trending faults and fold structures of varying intensity characterize the tectonic style of this area, which was mainly affected by the Variscan orogeny. A number of granites pierced the Paleozoic sedimentary series. In the subsurface a Precambrian basement is assumed, perhaps similar to that of the African Platform.

A thin, locally lacking Mesozoic blanket covers the Paleozoic socle. This sedimentary cover was not highly affected by fault tectonics or by folding. The block of the Jbilet Mountains was only raised as a horst structure in continuation of the High Atlas east-west trending fault systems. East of Safi and Casablanca northeast-southwest running faults downfault blocks and basins toward the Atlantic continental margin: Doukkala, Chaouia (Destombes and Jeanettes, 1966).

The Morocco-Oranian Meseta ("Meseta marocooranaise") north of the High Atlas and southeast of the Middle Atlas is a part of the High Plateaus ("Hauts Plateaux"), stretching farther east toward Algeria. In its westernmost part ("Haute-Moulouya") granitic and highly metamorphic rocks are exposed in domal uplifts (Emberger, 1965), forming a stable platform. The rocks within this area may also have been deformed during the Variscan orogeny. The crystalline basement is here covered by a thin blanket of Mesozoic sediments, which also varies in thickness. This cover was not affected by folding or faulting (Michard, 1976). With respect to the post-Variscan sedimentary history and to basin analysis, the Mesetas reacted as stable platform areas like the African Platform throughout Mesozoic and Cainozoic times.

2.4 Salt Diapir Province

Offshore surveys confirmed the existence of a highly complicated salt diapir province off the Moroccan west coast (Hinz et al., this Vol.; Beck and Lehner, 1974; Uchupi et al., 1976; Le Pichon et al., 1977; Lehner and de Ruiter, 1977). Between Agadir and Essaouira, where the High Atlas joins the Atlantic Ocean, this salt diapir province extends further landward, thus indicating an indirect relationship between Atlas and Atlantic (Fig. 1). At its southern boundary this salt diapir province is bordered by a northeast-southwest trending fault line. It can be traced through the Souss Basin toward the above mentioned Tizi-n'Test Fault, thus separating the stable African Platform in the southeast from the adjacent mobile belt in the northwest. The seaward boundary of this salt diapir province passes east of the Canary Islands Fuerteventura and Lanzarote and ends near latitude 28° N. Oceanic crust lies west of these two islands, as Drake and Ewing (1968) have postulated. The salt diapir province hides the transition zone between the Atlantic Ocean on one side and the westernmost High Atlas and the Moroccan Meseta on the other. Because of this structural setting of the deeper subsurface the continental margin is here not easily revealed in detail even by seismic research (Hinz et al., this Vol.; Weigel et al., this Vol.).

2.5 Conclusions

The structural pattern of the different tectonic units of Morocco (Figs. 1, 2) can be defined as follows: During Mesozoic and Cainozoic time the Anti-Atlas, the Moroccan and the Morocco-Oranian Mesetas remained as isolated stable blocks in a constant, rather high position. High Atlas and Middle Atlas reacted as narrow mobile belts. Since the Triassic they intersect the stable continental platform which was created and consolidated during the Variscan orogeny. Block movements and deformation processes within the mobile belts went on rather slowly but continuously, as will be shown below. The structural units (stable platforms and mobile belts) end abruptly at the Atlantic continental margin, due to passive margin formation. Now the geological evolution of the High Atlas can be revealed in context with the Atlantic and northwest African framework.

3. Atlas Rift

Thickness and facies of the Lower to Mid-Cretaceous in the High Atlas belt (Behrens et al., 1978) show two important facts:

- The thickness of Cretaceous sediments changes abruptly near to border faults, if the mountain range is crossed from north to south (Wurster and Stets, this Vol., Fig. 5),
- lithology and facies indicate that during Cretaceous times continental, lagoonal and shallow-marine environments prevail over the whole area. No deep-sea facies existed throughout that time.

This means that accumulation and subsidence continued to remain in equilibrium. Thickness of the sedimentary basin fill therefore may indicate the approximate amount of subsidence, and can be used for structural basin analysis.

Two diagrammatic sections across the High Atlas (Fig. 3) show the structural development of the basin perpendicular to its axis. The uniform Turonian carbonate series (Mid-Cretaceous) represents the best marker horizon in the field. It can also be taken as a reference for reconstruction. In addition, Jurassic and Triassic series, which overlie the Paleozoic basement unconformably, are indicated in the sections.

As a result, the High Atlas mobile belt is revealed as an area of significant subsidence in the center, flanked on either side by the structural highs of the Moroccan Meseta in the north and the Anti-Atlas in the south. Both are characterized by thin or lacking sedimentary cover, thus emphasizing the platform character. The thickness of the Cretaceous sediments increases abruptly from the margins toward the interior of the basin. These changes indicate the syngenetic character of the normal faults at the margins. The pre-Cretaceous series show the same trend. Within the trough, the different stages of the Cretaceous show also similarly conspicuous variations in thickness. We interpret this phenomenon by differentiated movements of rhomboidal-shaped tilting blocks, downfaulted during sedimentation. We suggest that a system of horst-and-graben structures, consisting of such blocks, split the socle within the Atlas belt. These results led us to the idea that during Early to Mid-Cretaceous times this narrow High Atlas basin was a complicated graben structure or even a rift.

In the westernmost High Atlas, where the Mesozoic cover is still more complete and better known by well data, cross-sections from the Triassic up to the Tertiary can be drawn. The maps and data given by Duffaud et al. (1966) offer a most reliable basis for a very similar reconstruction as given for the Cretaceous. The results are shown in Fig. 4:

– *Triassic*: The lowermost diagram shows striking thickness variations for the Triassic, ranging from zero in the Anti-Atlas to more than 6000 m in the Argana basin ("couloir d'Argana"; Ambroggi, 1963; Demaison, 1965; Duffaud et al., 1966; Salvan, 1968; Tixeront, 1973; Van Houten, 1976). In agreement with the French and American authors, the best interpretation of these data is a Triassic rift system in the High Atlas region (Van Houten, 1977; Van Houten and Brown, 1977; Mattauer et al., 1977; Brown, 1980). Differentiated subsidence of individual blocks in the interior of the basin caused thickness changes at short distances.

– *Jurassic*: During the Jurassic variations in thickness are not so evident here as in the Triassic, but the same tendency is clearly visible.

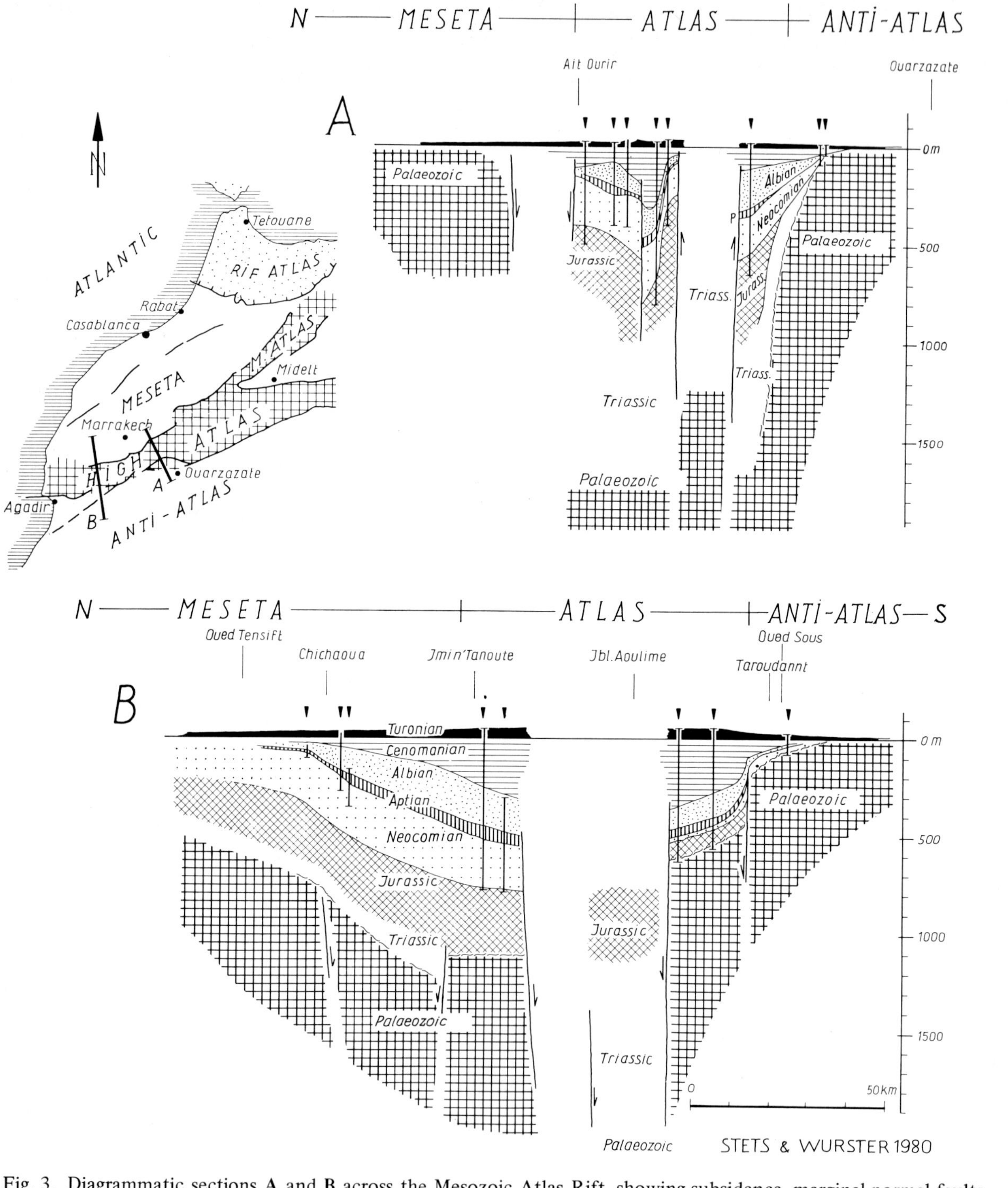

Fig. 3. Diagrammatic sections **A** and **B** across the Mesozoic Atlas Rift, showing subsidence, marginal normal faults, and role of syngenetic fault pattern in the interior of the rift

– *Cretaceous*: In Early to Mid-Cretaceous times subsidence continued and sedimentation went on in the interior of the rift. Our interpretation has been drawn by thick dashed lines in Fig. 4. It does not agree with the opinion of the french authors (Duffaud et al., 1966; Choubert and Faure-Muret, 1962), who assume a central paleogeographic high in the interior of the High Atlas basin. However, correlation of lithology and facies in many profiles from the northern and southern border of the High Atlas shows clearly that up to the Mid-

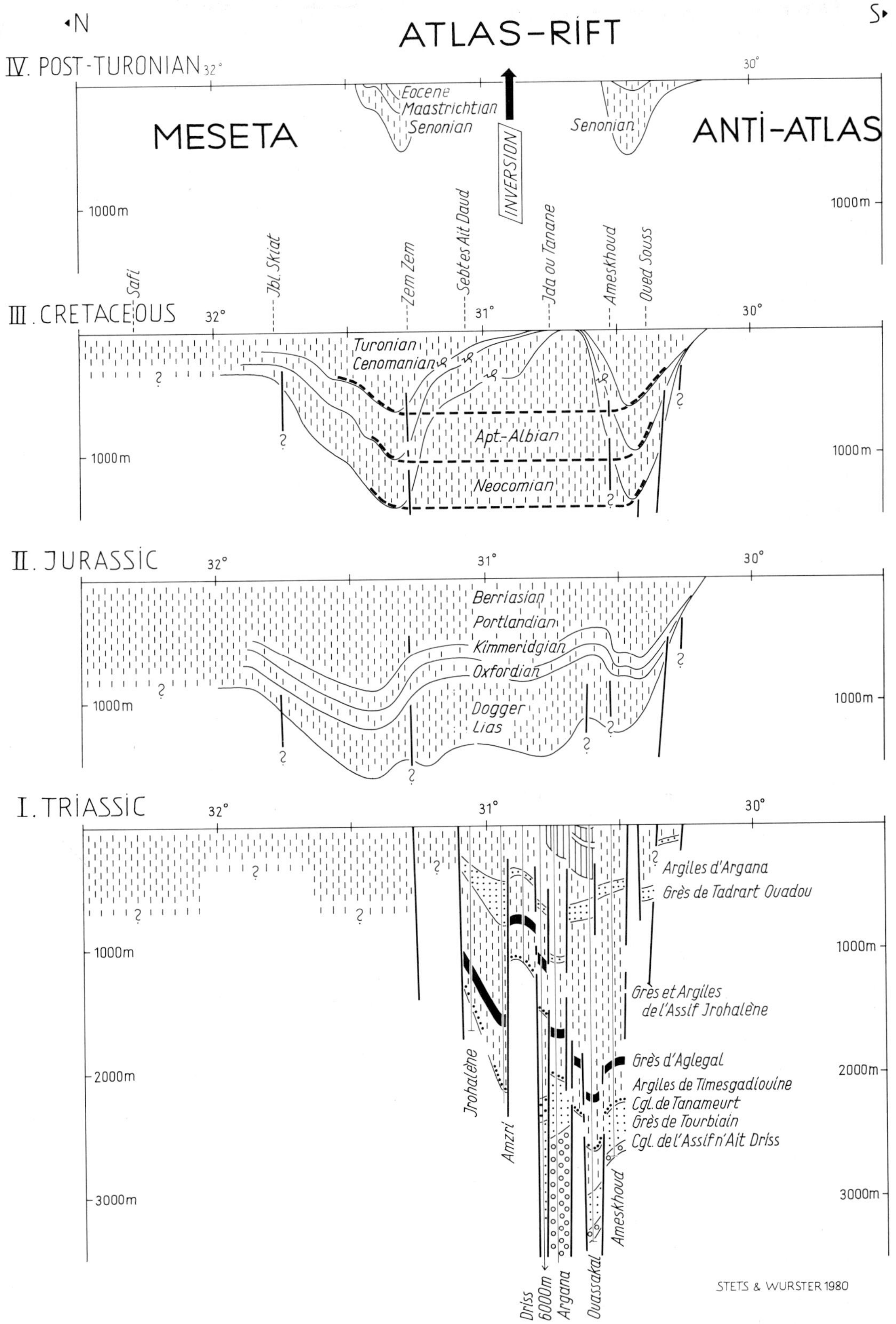

Fig. 4. Diagrammatic sections across the Mesozoic Atlas Rift between meridians 9° and 10° 30'; compiled from maps and data of Duffaud et al. (1966); for contradictional interpretation of Cretaceous section see text

Cretaceous a continuously sinking basin must have existed: Sediment thickness increases toward the center of the basin, and ammonites reveal that during Aptian to Albian times the sea entered the rift far to the east. Therefore the funnel-shaped Atlas Gulf must have been characterized by a marine environment of a greater extent than before (Behrens et al., 1978; Wurster and Stets, this Vol.). During the Turonian, subsidence came to an end and the rifting process died out. In Late Cretaceous and Tertiary times, an inversion took place, and the Atlas mobile belt began to rise.

The structural situation cited above can be represented in a generalized map of that Mesozoic graben or rift stage. This map contains the approximate amounts of subsidence, summarized from the Triassic up to the Mid-Cretaceous (Fig. 5). Data for the Cretaceous are taken from our own field work, combined with those given by Choubert and Faure-Muret (1962). Excellent data of du Dresnay (1971, 1972) and Duffaud et al. (1966) elucidated the Jurassic paleogeography. Data for the Triassic paleogeography are derived from Tixeront (1973), Van Houten (1977), and Van Houten and Brown (1977).

Stable platform conditions in the Anti-Atlas and on the Mesetas are indicated by thin or lacking sedimentary cover. In between the mobile Atlas belts are situated. The interior of the Atlas Rift is characterized by the rhomboidal syngenetic block pattern, simplified in Fig. 5. Only the most active faults are shown. At the margins the syngenetic normal faults, bordering the Atlas Rift and the platform areas, are clearly traceable. They are not obvious in the field, but can be well recognized in the sedimentary record. Proust (1962) already mentioned the action of syngenetic normal faults at the northern and southern flanks of the High Atlas in the Triassic and in the Jurassic. We are now able to prove them also in the Cretaceous.

In the region of Agadir and Essaouira the Atlas Basin is characterized by a maximum thickness of sediments of more than 3000 m. It developed since the Triassic, containing thick deposits of red clastics and of evaporites (Tixeront, 1973; Van Houten, 1977; Brown, 1980). In the Jurassic and Early to Mid-Cretaceous marine, lagoonal, and fluvial sediments were deposited. The thickness of the whole Mesozoic series decreases toward the east (central part of the western High Atlas, "massif ancien"). Remnants of the Triassic cover near the Tizi-n'Test and north of Taroudannt, as well as the Cretaceous sedimentary record, show that in this transition zone subsidence may have reached more than 1000 m.

In the Jbel Toubkal region south and southeast of Marrakech the reconstruction is quite complicated and seems quite doubtful, as now the Mesozoic cover is found only in relics. Most of it was removed by erosio-

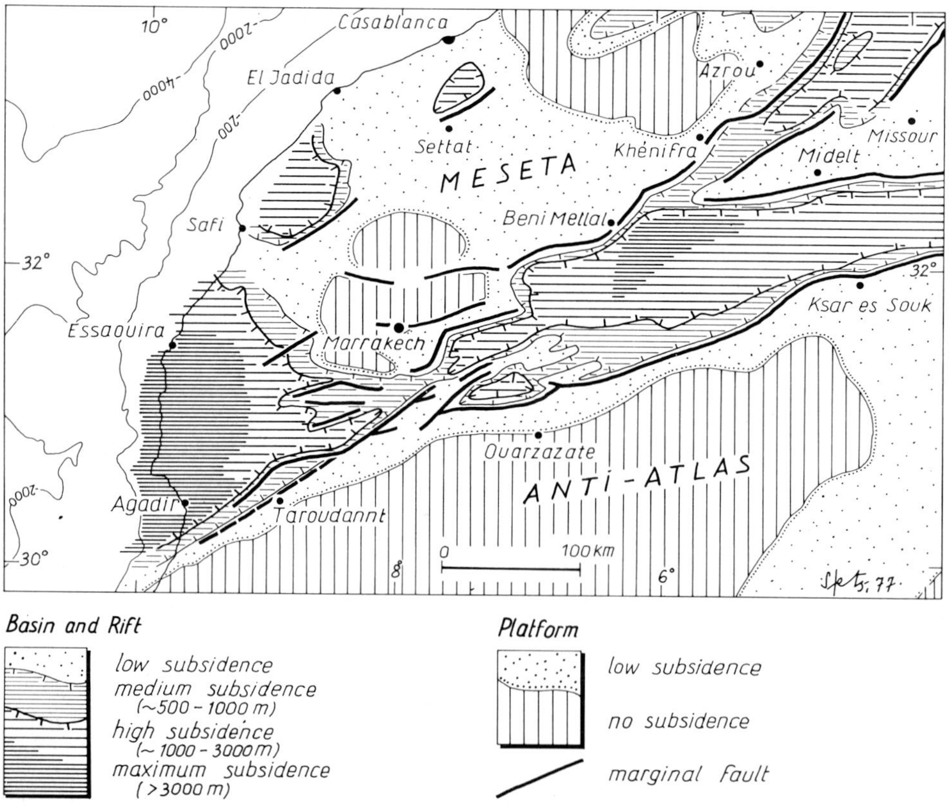

Fig. 5. Map of the Mesozoic Atlas Rift (Triassic-Turonian); the mobile belt is bordered by syngenetic normal faults; for geological framework see Fig. 2

nal processes. Local highs and basins can be traced, depending on the syngenetic fault block pattern. Here the total thickness may not have reached 1000 m. A narrow horst existed, trending northeast-southwest and plunging toward the northeast. The bundle of faults crossing the High Atlas at the Tizi-n'Test may have caused this structural high. In the Early Cretaceous (Apto-Albian) a facies change can be observed along this structure, but the sedimentary history records no trace of severe strike-slip movements along this fault at that time. Nevertheless, this ancient geosuture prevents the development of a uniform rift system throughout the High Atlas from the Atlantic in the west to the Sahara Atlas in the east.

Another area of high subsidence during the Triassic and the Jurassic lies farther to the east in the central High Atlas (Van Houten, 1977; du Dresnay, 1971, 1975; Choubert and Faure-Muret, 1962). Concerning this area, du Dresnay has formulated for the first time the idea of a Jurassic rift structure. Recent research work in the central High Atlas (Schaer and Persoz, 1976), and close to its northern and southern border (Laville et al., 1977; Laville, 1978) indicated that rifting processes may have died out here in the Late Jurassic, and that deformation already began in Late Jurassic to Early Cretaceous times. Contemporaneous intrusion of magmas and effusion of lavas is also known from this region at the same time. Equivalents are not recorded from the western High Atlas, nor did our investigations give any evidence in that region. Accordingly, subsidence of the central part of the Atlas Rift came to an end earlier than in the western High Atlas. In the Cretaceous higher amounts of subsidence are evident only in the area south and southeast of Beni-Mellal, where the Middle Atlas branches off. There they are restricted to synclincal basins, showing major changes in thickness and lithofacies.

4. Atlas Mountain Range

The maps of Duffaud et al. (1966), the data of Ambroggi (1963) and Michard (1976), as well as the results of our own field work at the northern and southern flanks of the western High Atlas (Below, 1976; Brix, 1976; Schmitz, 1977; Wellpott, 1977; Behrens et al., 1978) imply that in the Turonian basin subsidence failed in the Atlas Rift. The thickness of the up to 80-m-thick carbonate rock series of this stage near the Atlantic decreases only slowly toward the east and toward the margins. These rocks cover the mobile belt, as well as parts of the stable platform areas, i.e., the Mesetas and the northern margin of the Anti-Atlas. No abrupt thickness changes of these sediments could be detected.

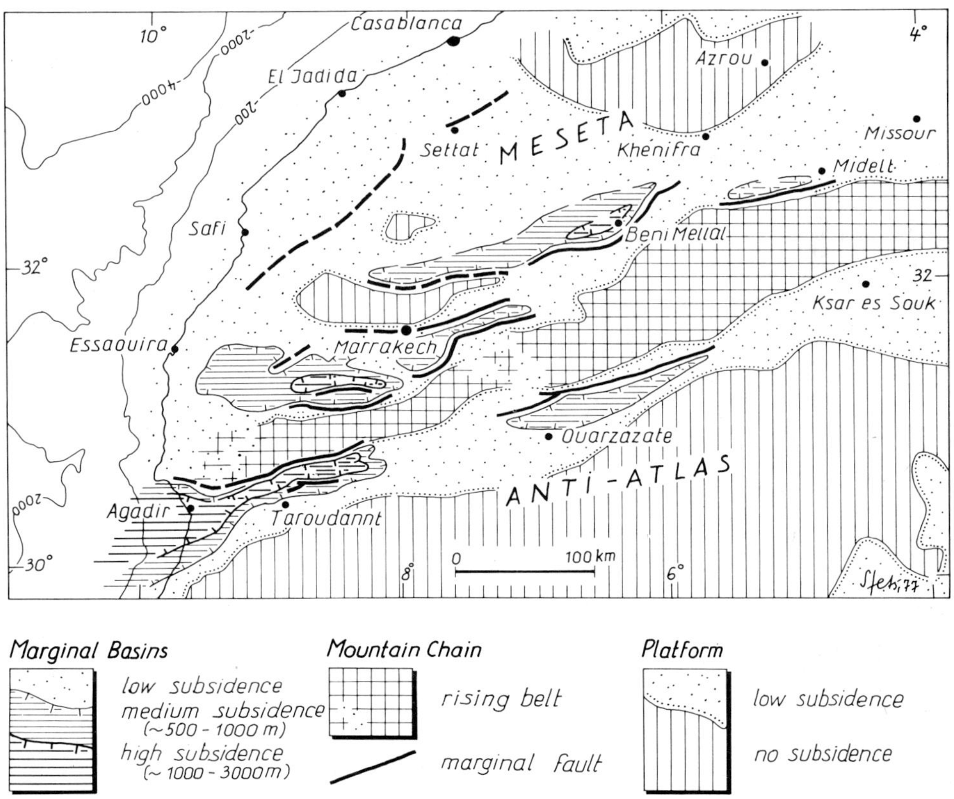

Fig. 6. Post-Turonian inversion of the former Atlas Rift; subsidence of subsequent marginal basins along the southern and northern flanks of the High Atlas is indicated; for geological framework see Fig. 2

During Late Cretaceous to Early Tertiary times the mobile western High Atlas belt began to rise. Structural inversion took place. It is already indicated in the uppermost diagrammatic section of Fig. 4. The thickness of different Upper Cretaceous rock series decreases toward the interior of the belt. Pebbles within some of the series are derived from the rising mountain chain. Changes in content and composition of heavy minerals (Brix, 1976, 1981) also indicate an important change of the structural evolution.

All data now available allow us to construct another map, showing the average amount of subsidence at that time (Fig. 6). While the stable blocks of the African Platform and of the Mesetas remained in their rather high position, the western High Atlas belt began to rise. The uplift of the central High Atlas, which partly began in Late Jurassic time, continued. This means that the whole mountain belt was then involved in a structural inversion process.

The uplift of the mobile belts was accompanied by the formation of subsequent marginal troughs (Voigt, 1963) along the northern and southern flanks of the High Atlas. Beginning in the Late Cretaceous, they were at first filled with alternating shallow-marine and lagoonal sediments in the west, passing into continental series farther to the east. In the Neogene, thick continental clastics of molasse type accumulated.

The subsequent marginal troughs partly communicate with each other. Especially in the north they are set off en echelon. The thickness of the sediments shows that basin subsidence generally did not reach the same high values in the marginal troughs as in the pre-Turonian rift. They may not exceed about 1000 m, except in the western part of the "sillon préafricain", that is in the Souss Basin.

The marginal faults, which favored the rifting process in Triassic to Mid-Cretaceous times, were transformed to steeply dipping thrust faults. Later they tilted onto the foreland and turned into complicated overthrusts, favoring the formation of nappes at the "accidents nord-et sudatlasiques".

In the beginning, the transition from subsidence to uplift was inconspicuous. As a result, the former Atlas Rift rose slowly but continuously. Later the mountain building process became more accentuated, and a spectacular mountain chain formed. There are no indications for a strong orogenetic deformation. Only inversion took place, and the original pattern of the rhomboidal blocks remained more or less undisturbed. Only tendency and direction of movements changed.

Further field work and structural investigations are necessary to reconstruct the mountain building processes in more detail. At present, we are convinced that the High Atlas is no orogenetic belt of alpine or appalachian type, piled up mostly by horizontal stress, and perhaps caused by some sort of subduction. We assume a special intracratonic or intracontinental mountain-building process without vigorous stress deformation. The pre-existing rhomboidal fault block pattern, initiating and favoring the rifting process, now controlled the uplift of individual blocks, as well as of the whole High Atlas mobile belt. Folds and fractures of the Mesozoic cover within the mountain range are significantly characterized by this pattern. Folding and fracturing during the mountain-building process did not exceed that of the ancient Variscan orogeny. High deformation is confined only to narrow, elongated zones, accompanying the faults, which separate the internal blocks. Obviously, the High Atlas Mesozoic sequence is nearly undeformed, and no traces of stress are represented. It is indeed a surprising fact that this orogenetic belt rose from a rift only by uplift.

5. Evolution of the High Atlas

Summarizing our considerations and results, the evolution of the High Atlas can be illustrated by a series of block diagrams (Fig. 7). The diagrams had to be generalized and exaggerated, as necessary, when models and theories are evolved.

I. *Triassic*: The lowermost block shows an initial stage of continental rifting during the Triassic. Normal faults and rather steep escarpments bordered the original graben structure of the Atlas Rift, symbolizing syngenetic movements. Detritus, nourishing broad alluvial and fluvial fans, is derived from the adjacent high platforms, i.e., the Moroccan Meseta and the Anti-Atlas. In deep depressions, mud plains, sabkhas and salt lakes originated. Outpouring lava flows finally covered the whole bottom of the continuously subsiding rift.

II. *Jurassic*: During the Jurassic, the supply of continental detritus continued. According to a rising sea level, the open sea entered into the rift depression. The Atlantic invaded from the west, the Mediterranean from the east. Lowlands of the Moroccan Meseta were inundated as well. At the flanks of the rift patch reefs built upward. Over the steep escarpments turbidity currents cascaded down to the bottom of the trough, as subsidence was locally not compensated by sedimentation at that time. This model is valid especially for the central Atlas Rift.

III. *Cretaceous*: During Early to Mid-Cretaceous times subsidence of the Atlas Rift continued. The sea inundated the adjacent platforms when it reached its highest level (Wurster and Stets, this Vol.). Littoral belts mark the previous marginal escarpments. During regressions, fluvial and deltaic fans prograded into the gulf from the east and from the south. Continental deposits and shallow-marine sediments interfinger in the west. After Turonian times subsidence of the rift area declined and finally came to an end.

IV. *Tertiary*: Starting in Late Cretaceous times the buried Atlas Rift began to rise. The marginal normal faults were transformed into thrust faults. After reaching the surface, the filling of the Mesozoic trough was gradually eroded. On each flank of the rising mountain chain subsequent marginal troughs originated. Erosion products that came down from the rising belt nourished alluvial and fluvial fans, filled the marginal basins, and covered also the neighboring platform areas. At the margins of the rising belt, slices of the Mesozoic sedimentary pile were cut off and thrust onto the foreland. Partly

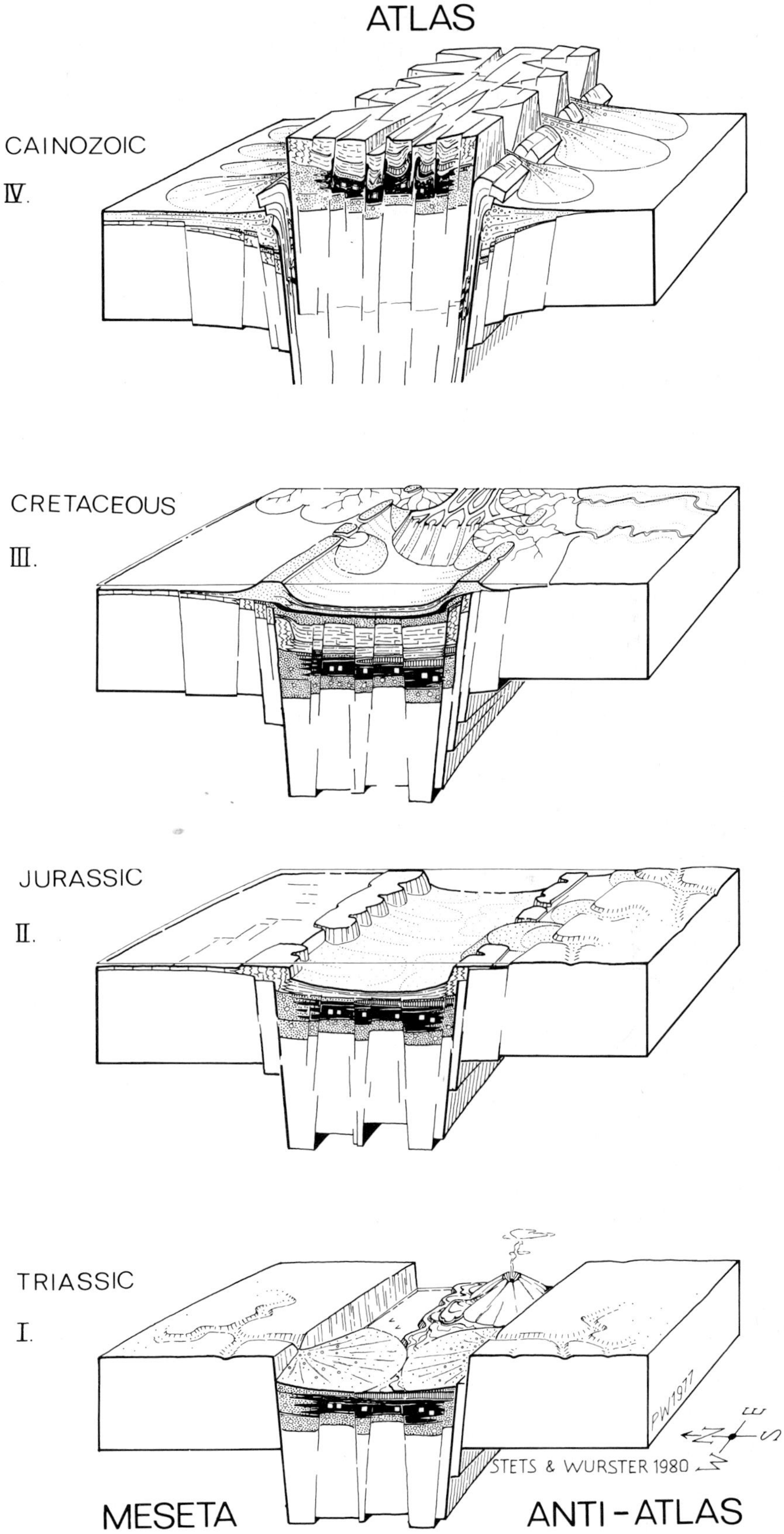

Fig. 7. Block diagrams illustrating the geological evolution of the High Atlas

they were overturned and glided as gravity nappes onto the younger series. The tectonic pattern of the interior of the mountain belt, consisting of the rhomboidal blocks, remained rather undeformed. The main impression, gained in the field and by the studies of maps and sections, is that the transformation of the Atlas Rift toward a mountain chain happened only by inversion processes. Lateral movements, implying strong crustal shortening, cannot be observed.

6. Evolution of the Atlantic Ocean

The basic steps in the evolution of the Atlantic Ocean are also traced in a series of block diagrams (Fig. 8). They are based on the well established knowledge about origin and formation of the Atlantic, gained by fitting reconstructions, by geophysical data, by deep-sea drilling and by geochronological data, published in numerous papers.

I. *Triassic*: During the Triassic a major rift system began to transsect Pangea. A broad bundle of rift zones, containing rhomboidal fault block patterns, tilted blocks, and horst-and-graben structures, originated. The Atlantic rift zone is part of this system. The subsiding troughs were filled mainly with continental detrital sediments and evaporites, which later on were covered by widespread lava flows.

II. *Jurassic*: By Mid-Jurassic time crustal separation was already achieved in the central Atlantic. A narrow proto-oceanic basin opened, as sea-floor spreading started after a Mid-Atlantic spreading ridge had been formed. In the course of time, the basin grew wider and deeper by drifting and subsidence. Volcanic activity was only confined to the open ocean. At and upon the highly faulted continental margins salinar troughs, carbonate platforms, and widespread littoral zones were situated. In front of the margins, deep-sea sediments accumulated.

III. *Cretaceous*: In Cretaceous times, the process of drifting and subsidence continued. The bottom of the widening ocean basin and the shelf areas were already completely separated environments at that time.

Caused by this evolution Triassic continental troughs, Late Triassic to Early Jurassic evaporite basins, Upper Jurassic carbonate platforms, and Cretaceous sedimentary wedges can be found at both contingent margins of the Atlantic Ocean in North America and Northwest Africa. To-day this ocean measures about 6000 km in width between Cape Hatteras and Cap Blanc. For the past 150 million years the average half-spreading rate amounts to 2 cm per year. The result of this continuous evolution during the Mesozoic and further in the Cainozoic is a large ocean, mainly controlled by horizontal movements. In contrast, the neighboring Atlas mountain range seems to be only a midget.

7. Structural Relations

To find out the relations between Atlas and Atlantic, we must go back to the initial stages. Numerous data have been assembled on the break-up of Pangea and on the opening of the Atlantic Ocean, especially on the best fit of the continents around the North Atlantic (Le Pichon et al., 1977). In the region under discussion, it seems difficult to solve this problem, as the continental margin consists of systems of step faults heading against the dip and of horst-and-graben structures (Dillon, 1974). They are mostly covered by sedimentary wedges, prograding toward the Atlantic Ocean on both sides or by carbonate platforms (Sheridan, 1974; Beck and Lehner, 1974; Jansa and Wade, 1975; Schlee et al., 1976; Behrens et al., 1978; Wiedmann et al., 1978; von Rad and Einsele, 1980; Hinz et al., this Vol.). Nature and origin of the margin therefore have to be revealed indirectly as the relations between Atlas and Atlantic.

The type of the northwest African continental margin was classified as a "divergent margin" by Thompson (1976), characterized by passive subsidence without earthquakes or considerable volcanism, except in the rift stage, and during early sea-floor spreading. Basalts, indicating a rift stage, are only found in the Late Triassic to Early Jurassic north of Agadir (Michard, 1976). Olivine and quartz-tholeiitic basalts are recorded from the High Atlas mobile belt (Manspeizer et al., 1978) at the same time emphasizing that rifting also took place there.

Following the classification of "extended ocean margins" of Thompson (1976), it can be stated that the continental margin off West Morocco is "evaporite controlled" in places. The thick evaporites of the salt diapir province (Fig. 1), stretching farther landward into the Atlas Rift, are assumed to be of Late Triassic to Early Jurassic age (Beck and Lehner, 1974; Uchupi et al., 1976; Le Pichon et al., 1977; Hinz et al., this Vol.). They filled the early Atlantic rift just where both rift structures, Atlas and Atlantic, met. This type of margin is restricted to the offshore area west of the High Atlas and of the Moroccan Meseta. Toward the southeast, it is cut off by the geosuture, which separates the Anti-Atlas from the High Atlas mobile belt, and which extends farther northeast to the Tizi-n'Test Fault.

South of that geosuture, Triassic evaporites, to a minor extent, are only recorded from the Aaiun (Tarfaya) Basin (El-Aioun Basin: Querol, 1966; Van Houten, 1977), but no salt diapirism has been observed off the West Sahara south of 29° N. There, the continental margin is built up of a wedge-shaped clastic cover, which is the prevailing structure of the deeper subsurface (Wiedmann et al., 1978; von Rad et al., 1979; von Rad and Einsele, 1980; Ranke et al., this Vol.). Therefore it seems obvious that the Atlas Rift branched off the Atlantic rift near the Canary Islands. Farther to the north, the salt diapir province extends along the Moroccan west coast, trending parallel to the margin that is parallel to the eastern border of the ancient rift (Fig. 1, 9).

Considering the pre-drift position of Africa, North America and Iberia, we are able to outline the original relations between Atlas and Atlantic (Fig. 9). As Burke

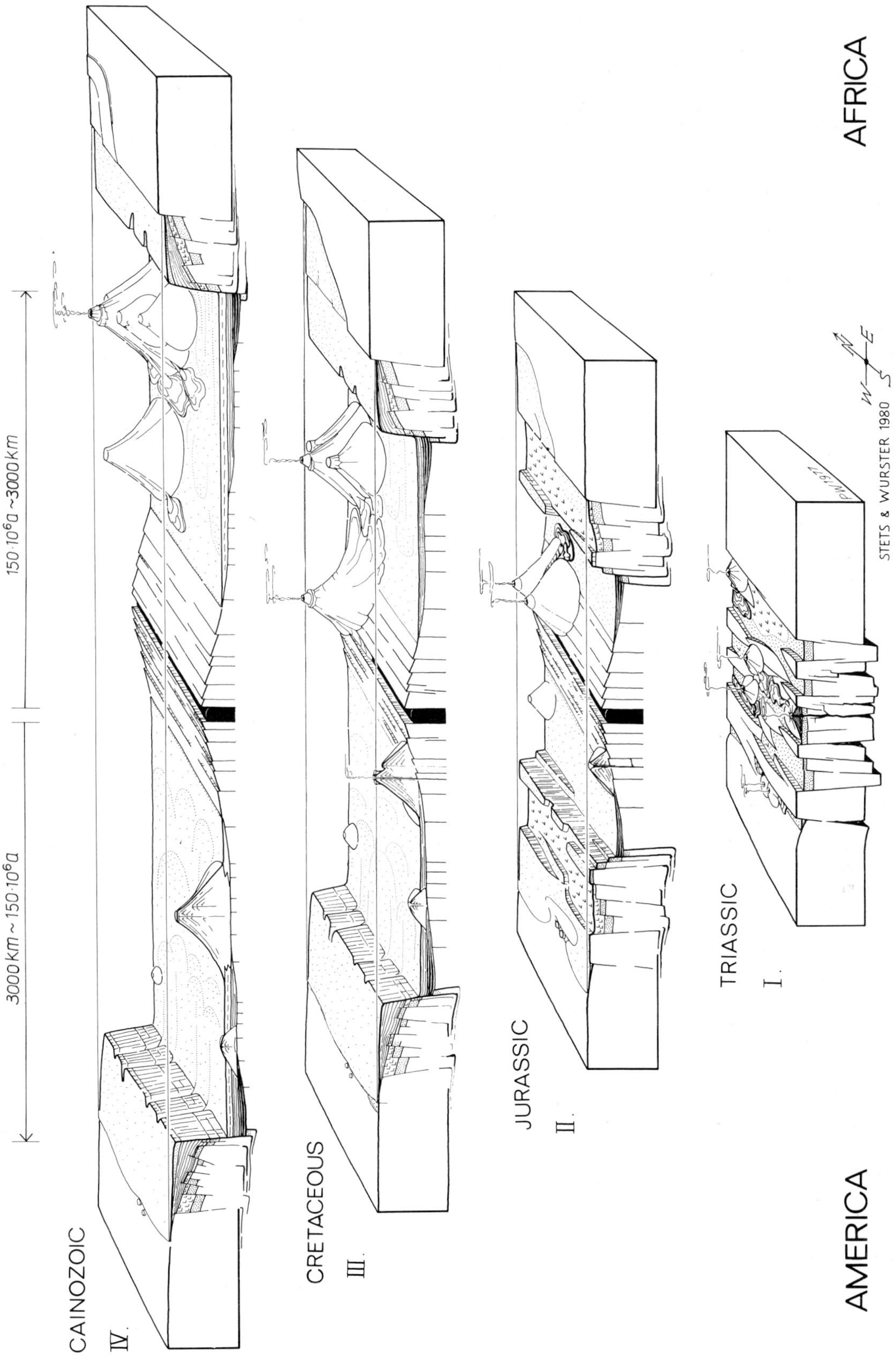

Fig. 8. Block diagrams illustrating the geological evolution of the Atlantic Ocean

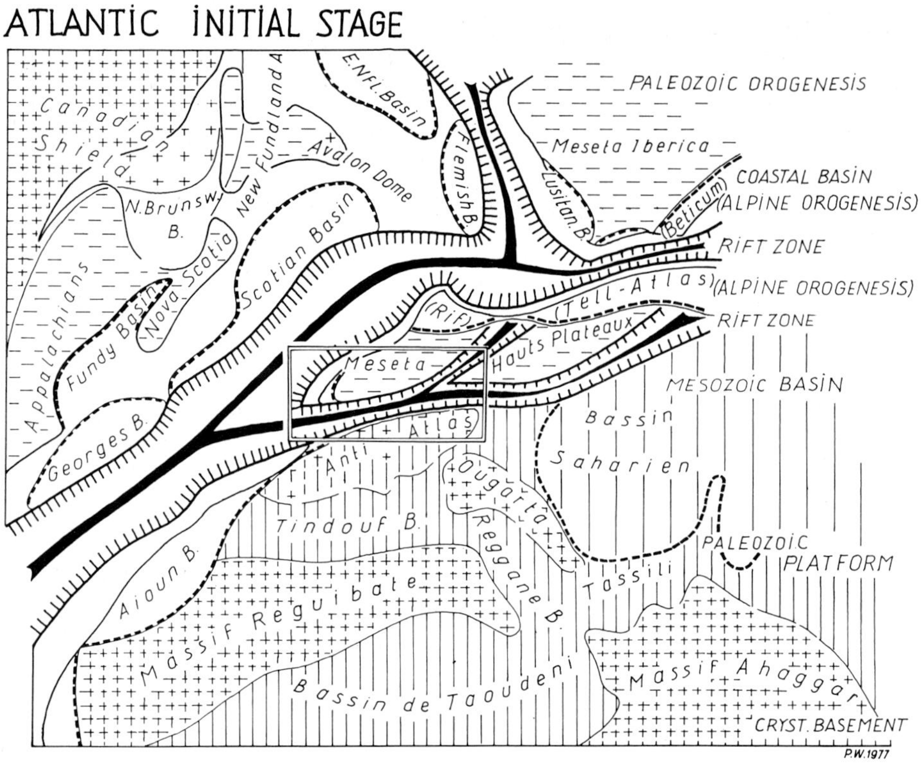

Fig. 9. Atlantic and Atlas system; initial stage of the main rift structures, Mesozoic basins and cratons; for investigated area see Figs. 2, 5 and 6

(1976) showed, the main Atlantic furrow was composed of many different graben structures, rifts, and troughs, which opened successively and initially operated in a different manner, at different places, and at different times. According to him, rifting in the region under discussion started in the Triassic to Early Jurassic. Cousminer and Manspeizer (1977) argue that the major phase of rifting began in the Carnian stage at almost the same time, when the Atlas Rift originated. Most of these dispersed rift and fracture zones along the Atlantic furrow failed, and gave rise only to Mesozoic troughs, basins, or sedimentary wedges. The final dividing line between the future two continents evidently was closer to Africa than to America. As the North Atlantic complex rift zone originated successively from south to north, the adventive branches may be considered as potential pathways for continental break-up, later separation, and drifting. From our point of view, an important bifurcation occurred at the latitude of the Canary Islands, where the Atlas Rift as a spectacular branch runs via the present Agadir toward the east, just between the Anti-Atlas and the Moroccan Meseta.

Another bifurcation occurred near Beni-Mellal, one branch taking its way along the central High Atlas and the Sahara Atlas, the other along the Middle Atlas (Fig. 9). Similar bifurcations or triple junction positions can be assumed at the latitude of the Azores Islands between the Rif-Grand Banks area and the Iberian plate.

We postulate a joint and synchronous Triassic origin of both rift systems, Atlas and Atlantic. The impulses of crustal movements finally were the same: break-up of Pangea was caused by lateral extension. Pre-existing geosutures, as for example the Appalachian front and the Tizi-n'Test fault line, determined the trend of separation. Finally, the main Atlantic rift developed to an oceanic rift and succeeded in drifting, whereas the Atlas Rift failed. Otherwise Morocco and Algeria would be Canadian provinces to-day!

Striking differences between the Atlas and the Atlantic became evident during the Jurassic and the Cretaceous. Sea-floor spreading and subsidence of the continental margins began to control the structural evolution of the oceanic Atlantic rift (Thompson, 1976; Sclater et al., 1977; Le Pichon et al., 1977). During the Cretaceous, continuous sedimentation took place along the border of the African Platform in the Aaiun coastal basin (Wiedmann et al., 1978; von Rad and Einsele, 1980; Ranke et al., this Vol.), and along the Moroccan west coast. In the intracontinental Atlas Rift, subsidence continued during the Cretaceous without significant horizontal extension.

Thus, contrasting the Atlantic evolution, the Atlas Rift shows no perceptible indication for major horizontal drifting or strike-slip movements. Cumulative subsidence in the Atlas Rift as well as at the passive East Atlantic margin resulted in a deep re-entrant Atlas Gulf in Early Cretaceous time. Oscillation of the sea level and differentiated tilting of blocks within the rift zone controlled facies distribution and environmental patterns as well (Wurster and Stets, this Vol.). The rift remained in its previous state until the Turonian, by which time it ceased to subside.

Further striking divergencies in the evolution of the Atlas and Atlantic are increasingly evident by the post-Turonian inversion of the Atlas Rift, when a genuine mountain chain originated. This "orogenetic" evolution definitely contrasts with the continuous spreading of the Atlantic ocean basin. This horizontal extension was accompanied by subsidence at the passive continental margins. The vertical uplift of the High and Middle Atlas mountain chains on the other hand demonstrates rather antagonistic movements. Therefore we suggest that vertical movements should be more considered than it has been done up to now.

In the Haha coastal area, between Agadir and Essaouira, the Northwest African passive continental margin incorporates the westernmost parts of the former Atlas Rift and later mountain range. Within this region, uplift of the Atlas mobile belt and downwarping of the continental margin overlap. Therefore subsidence and uplift compensate each other, and the Atlas coastal basin behaves like a huge tilting block, which conceals the western extension of the Atlas system toward the Atlantic.

Acknowledgments. We are greatly indebted to Professor Dr. Hans Closs, Hannover, who stimulated our project as an onshore contribution to IPOD/DSDP sites off the Northwest African continental margin. Field work, done in 1975–1978, was largely supported by the German Science Foundation (DFG, grants No. Wu 32/15–19). Further thanks are due to Drs. Hilali, Boudda, Bensaid, and Dahmani (Ministère de l'Energie et des Mines, Division de la Géologie, Service de la Carte Géologique), Rabat, who facilitated the field work in the High Atlas in every respect. For critical reading of the manuscript and for discussions we thank Dr. Peter Ziegler, The Hague. Other reviews made by Drs. U. von Rad and K. Hinz, Hannover, and Dr. A. Schäfer, Bonn, provided helpful comments and improved the English style. Recently Dr. R. du Dresnay, Rabat, contributed effective advice concerning the Mesozoic sequence.

REFERENCES

Ambroggi R (1963) Etude géologique du versant méridional du Haut Atlas occidental et de la plaine du Souss. Notes Mém Serv Géol Maroc 157: 321 pp

Ambroggi R, Neltner L (1952) Haut Atlas et Anti-Atlas. Généralités. 19 Congr Géol Int Alger 1952, Livre-Guide Maroc 12: 65 pp

Arthaud F, Matte Ph (1975) Le décrochement tardihercynien du sudouest de l'Europe. Géométrie et essai de reconstitution des conditions de la déformation. Tectonophysics 25: 139–171

Beck RH, Lehner P (1974) Oceans, new frontier in exploration. AAPG Bull 58: 376–395

Behrens M, Krumsiek K, Meyer DE, Schäfer A, Siehl A, Stets J, Wurster P (1978) Sedimentationsabläufe im Atlas-Golf (Kreide Küstenbecken Marokko). Geol Rundsch 67: 424–453

Below R (1976) Geologische Aufnahmen am Südrand des Hohen Atlas bei Taroudannt etc. Diplomarb, Bonn, 89 pp (unpubl)

Boigk H (1968) Gedanken zur Entwicklung des Niedersächsischen Tektogens. Geol Jahrb 85: 861–900

Brix M (1976) Geologische Aufnahmen am Südrand des Hohen Atlas bei Taroudannt (Marokko) etc. Diplomarb, Bonn, 78 pp (unpubl)

Brix M (1981) Schwermineralanalyse und andere sedimentologische Untersuchungen als Beitrag zur Rekonstruktion der strukturellen Entwicklung des westlichen Hohen Atlas/Marokko. Dissertation, Bonn, 247 pp

Brown RH (1980) Triassic Rocks of Argana Valley, Southern Morocco, and their Regional Structural Implications. AAPG Bull 64: 988–1003

Burke K (1976) Development of graben associated with the initial ruptures of the Atlantic Ocean. Tectonophysics 36: 93–112

Burke K (1978) Evolution of continental rift systems in the light of plate tectonics. In: Ramberg JB, Neumann E-R (eds) Tectonics and Geophysics of Continental Rifts. Reidel, Dordrecht/Holland, pp 1–9

Choubert G, Faure-Muret A (1962) Evolution du domaine Atlasique Marocain dépuis les temps Paléozoiques. In: Durand-Delga M (ed) Livre à la mémoire P Fallot, Mém H-S Soc Géol Fr 1: 447–527

Choubert G, Marçais J (1952) Aperçu structural. In: Géologie du Maroc. 19 Congr Géol Int Alger 1952, Monogr Rég, 3 Sér Maroc 6: 9–73

Cousminer HL, Manspeizer W (1977) Autunian and Carnian palynoflorules. Contribution to the chronology and tectonic history of the Moroccan pre-Atlantic borderland. In: Swain FM (ed) Stratigraphic micropaleontology of Atlantic basins and borderlands. Elsevier, Amsterdam, pp 185–204

Demaison GJ (1965) The Triassic salt in the Algerian Sahara. In: Salt basins around Africa. Inst Petrol, London, pp 91–100

Destombes J, Jeanette A (1966) Mémoire explicatif de la Carte géotechnique de la Méséta cotière a l'Est de Casablanca etc. Notes Mém Serv Géol Maroc 180: 104 pp

Dillon WP (1974) Structure and development of the southern Moroccan continental shelf. Mar Geol 16: 121–143

Drake CL, Ewing FJ (1968) The continental margin of the eastern United States. Can J Earth Sci 5: 993–1010

Dresnay R du (1964) Les discontinuités de sédimentation pendant le Jurassique dans la partie orientale du domaine atlasique Marocaine etc. C R Mém 1 Coll Jurassique Luxembourg 1962: 899–912

Dresnay R du (1971) Extension et développement des phénomènes récifaux jurassiques dans le domaine atlasique marocain, particulièrement au Lias moyen. Bull Soc Géol Fr 13 (7): 46–56

Dresnay R du (1972) Relations paléogéographiques entre sillons atlasiques (Haut et Moyen Atlas et plateforme des Hauts Plateaux (Maroc oriental) pendant l'Aalénien et le Bajocien. Mém B R G M Fr 75: 147–161

Dresnay R du (1975) Influence of late Hercynian structural heritage and syntectonic events on Jurassic sedimentation in the marine trough of central and eastern High Atlas of Morocco. 9 Int Congr Sedimentol Nice 1975 Thème 4,1: 103–108

Duffaud MF (red) et al (1966) Le basin du Sud-ouest Marocain. In: Reyre D (ed) Bassins sédimentaires du littoral Africain. Ass Serv Géol Africain, Paris, p 5–12

Emberger A (1965) Introduction a l'étude des minéralisations plombifères de la Haute-Moulouya. Notes Mém Serv Géol Maroc 181: 167–174

Houten van FB (1976) Late Variscan nonmarine deposits, northwestern Africa: implications for pre-drift North Atlantic reconstruction. Am J Sci 276/6: 671–693

Houten van FB (1977) Triassic-Liassic deposits of Morocco and Eastern North America. AAPG Bull 61: 79–99

Houten van FB, Brown RH (1977) Latest Paleozoic – Early Mesozoic paleogeography, northwestern Africa. J Geol 85: 143–156

Jansa LF, Wade JA (1975) Geology of the continental margin of Nova Scotia and Newfoundland. Geol Surv Can Pap 74–30,2: 51–105

Lajat D, Biju-Duval B, Connard R, Letouzey J, Winnock E (1975) Prolongement dans l'Atlantique de la partie externe de l'Arc bético-rifain. Bull Soc Géol Fr 17 (7): 481–485

Lancelot Y, Winterer EL (1980) Evolution of the Moroccan Oceanic Basin and adjacent continental margin – a synthesis. Init Rep Deep Sea Drilling Project 50: 801–821

Laville E (1977) L'évolution sédimentaire et tectonique de la couverture jurassique de la région d'Immouzzer de Marmoucha (Moyen Atlas – Maroc). Bull Soc Géol Fr 19 (7): 1151–1158

Laville E (1978) Incidence des jeux successifs d'un âccident synsédimentaire sur les structures plicatives du versant nord du Haut Atlas central (Maroc). Bull Soc Géol Fr 20 (7): 329–337

Laville E, Lesage JL, Seguret M (1977) Géométrie, cinématique (dynamique) de la tectonique atlasique sur le versant sud du Haut-Atlas marocain. Bull Soc Géol Fr 19 (7): 527–539

Lehner P, de Ruiter PAD (1977) Structural history of the Atlantic margin of Africa. AAPG Bull 61: 961–981

Le Pichon X, Sibuet JS, Franchetau J (1977) The fit of the continents around the North Atlantic Ocean. Tectonophysics 38: 169–209

Linden van der WJM (1978) Passive continental margins and intracratonic rifts, a comparison. In: Ramberg JB, Neumann E-R (eds) Tectonics and Geophysics of Continental Rifts. Reidel, Dordrecht/Holland, p 221–227

Manspeizer W, Puffer JH, Cousminer HL (1978) Separation of Morocco and eastern North America: A Triassic-Liassic stratigraphic record. Geol Soc Am Bull 89: 901–920

Mattauer M, Tapponier P, Proust F (1977) Sur les mécanismes de déformation des chaînes atlasiques du Maroc. Bull Soc Géol Fr 19 (7): 521–526

Michard A (1976) Eléments de géologie marocaine. Notes Mém Serv Géol Maroc 252: 408pp

Neumann E-R, Ramberg JB (1978) Paleorifts – concluding remarks. In: Ramberg JB, Neumann E-R (eds) Tectonics and Geophysics of Continental Rifts. Reidel, Dordrecht/Holland, pp 409–424

Proust F (1962) Tectonique de socle par failles inverses, en liaison avec d'anciennes failles normales dans le Haut Atlas (Maroc). C R Somm Soc Géol Fr 1962: 9–10

Proust F, Petit J-P, Taponnier P (1977) L'accident du Tizi-n'Test et le rôle des décrochements dans la tectonique du Haut Atlas occidental (Maroc). Bull Soc Géol Fr 19 (7): 541–551

Querol R (1966) Regional geology of the Spanish Sahara. In: Reyre D (ed) Bassins sédimentaires du littoral Africain. Assoc Serv Géol Afr, Paris, pp 27–39

Rad von U, Einsele G (1980) Mesozoic-Cainozoic subsidence history and palaeobathymetry of the northwest African continental margin (Aaiun Basin to DSDP site 397). Philos Trans R Soc London Ser A 294: 37–50

Rad von U, Cepek P, von Stackelberg U, Wissmann G, Zobel B (1979) Cretaceous and Tertiary sediments from the northwest African slope (Dredges and cores supplementing DSDP results). Mar Geol 29: 273–312

Rod E (1962) Fault pattern, northwest corner of Sahara shield. AAPG Bull 46: 529–552

Russo P, Russo L (1934) Le grand accident sud-atlasien. Bull Soc Géol Fr 4 (5): 375–384

Salvan HM (1968) L'évolution du problème des évaporites et ses conséquences sur l'interprétation des gisements marocains. Mines Géol Rabat 27: 5–30

Schaer J-P (1964) Aspects de la tectonique dans le bloc occidental du massif ancien du Haut Atlas. C R Acad Sci Paris 258: 2353–2356

Schaer J-P (1967) Interférences entre les structures du socle et celles de la couverture dans le Haut-Atlas marocain. In: Etages tectoniques, Colloq Neuchâtel 1966. La Bacconière, Neuchâtel, pp 297–306

Schaer J-P, Persoz F (1976) Aspects structuraux et pétrographiques du Haut-Atlas calcaire de Midelt (Maroc). Bull Soc Géol Fr 18 (7): 1239–1250

Schlee J, Behrendt JC, Grow JA, Robb JM, Mattick RE, Taylor PT, Lawson BJ (1976) Regional geologic framework off northeastern United States. AAPG Bull 60: 926–951

Schmitz J (1977) Geologische Aufnahmen am Südrand des Hohen Atlas bei Taroudannt etc. Diplomarb, Bonn, 119 pp (unpubl)

Sclater JC, Hellinger St, Tapscott Ch (1977) The paleobathymetry of the Atlantic Ocean from the Jurassique to the present. J Geol 85: 519–552

Sheridan RE (1974) Atlantic continental margin of North America. In: Burk CA, Drake CL (eds) The geology of the continental margins. Springer, Berlin Heidelberg New York, pp 391–407

Sitter LU de (1952) Plissement croisé dans le Haut-Atlas. Geol Mijnbouw 14: 277–282

Stets J, Wurster P (1977) Atlas and Atlantic – structural relations, EOS 58: 908 (Abstr only)

Stets J, Wurster P (1981) Zur Strukturgeschichte des Hohen Atlas. Geol Rundsch (in press)

Termier H, Termier G (1971) Le massif granito-dioritique du Tichka (Haut-Atlas occidental Maroc). Les régions et les terrains. Notes Mém Serv Géol Maroc 216: 240 pp

Thompson ThL (1976) Plate tectonics in oil and gas exploration of continental margins. AAPG Bull 60: 1463–1501

Tixeront M (1973) Lithostratigraphie et minéralisations cuprifères et uranifères stratiformes, syngénétiques et familières des formations détritiques permo-triasiques du Couloir d'Argana, Haut Atlas occidental (Maroc). Notes Mém Serv Géol Maroc 249: 147–177

Uchupi E, Emery KO, Bowin CO, Phillips JO (1976) Continental margin off western Africa: Senegal to Portugal. AAPG Bull 60: 809–878

Voigt E (1963) Über Randtröge vor Schollenrändern und ihre Bedeutung im Gebiet der Mitteleuropäischen Senke und angrenzender Gebiete. Z Dtsch Geol Ges 114 (1962): 378–418

Wegmann CE (1961) Le Jura plissé dans la perspective des études sur la comportement des socles. In: Durand-Delga M (ed) Livre à la mémoire P Fallot. Mém H-S Soc Géol Fr 2: 99–104

Wellpott A (1977) Geologische Aufnahmen am Südrand des Hohen Atlas bei Taroudannt (Marokko) etc. Diplomarb, Bonn: 85 pp (unpubl)

Wiedmann J, Butt A, Einsele G (1978) Vergleich von marokkanischen Kreide-Küstenaufschlüssen und Tiefseebohrungen (DSDP): Stratigraphie, Paläoenvironment und Subsidenz an einem passiven Kontinentalrand. Geol Rundsch 67: 454–508

Ziegler PA (1978a) Northwestern Europe: Tectonics and basin development. Geol Mijnbouw 57: 589–626

Ziegler PA (1978b) North Sea Rift and Basin Development. In: Ramberg JB, Neumann E-R (eds) Tectonics and Geophysics of Continental Rifts 2. Reidel, Dordrecht/Holland, pp 249–277

6 Stratigraphy, Facies and Tectonic Development of the On- and Offshore Aaiun-Tarfaya Basin – A Review

Ulrich Ranke, Ulrich von Rad, and Gerd Wissmann[1]

ABSTRACT

Based mainly on published data, we attempt a synthesis of the stratigraphy, facies and tectonic evolution of the onshore Aaiun-Tarfaya Basin and its offshore extension, the West Saharan Marginal Basin. Basement rocks are Precambrian, and folded Paleozoic sediments (Mauritanides belt): they dip gently westward and are overlain by a seaward thickening wedge of Mesozoic to Cenozoic continental to shallow-marine sediments. Jurassic to Cretaceous sediments extend from the onshore basin to the present shelf and upper slope, where they are more than 12 km thick.

In the onshore basin Triassic clastic rocks, evaporites, and basalt sills and lower to middle Jurassic evaporites and carbonates are overlain by a 1-2 km thick sequence of upper Jurassic neritic carbonates (Puerto Cansado Formation). These formations document high subsidence rates (80-100 m/m.y.) which were compensated by carbonate buildup after a major transgression, that coincided with the Liassic to Oxfordian breakup of Pangaea.

The subsidence rates increased slightly (110-140 m/m.y.) during the Early Cretaceous, when 1-4 km of regressive continental to marine-deltaic, Wealden-type sediments were deposited as part of two major delta systems (Tan Tan/Jreibichat Formation).

Restricted mid-Cretaceous sediments (Calcaires d' Aguidir) are overlain by up to 1.0 km of transgressive Late Cretaceous shallow-marine carbonates (Lebtaina Formation) and similar Paleogene sediments (Samlat Formation). The subsidence and accumulation rates decreased considerably during this time. Upper Cretaceous to Paleogene sediments were truncated over a wide area during several erosional cycles.

Because of uplift and erosion, the Neogene is thin or missing in the onshore coastal basin; an exception is the 1000 m thick sequence of ?deltaic to estuarine claystones in the northwestern Aaiun Basin. Sediment bypassed the present shelf and slope during this time, to accumulate in a new depocenter as gravitative and hemipelagic deposits along the uppermost continental rise.

1. Introduction

After the Senegal Basin, the Aaiun-Tarfaya Basin plus its offshore extension, the West Saharan Marginal Basin, is the largest marginal basin off Northwest Africa (see front cover figure). It is filled by a thick sequence of Mesozoic, terrigenous to shallow-marine sediments, and a thin veneer of Tertiary shallow-marine deposits (Figs.1,2). Together they are more than 14 km thick (see Chap. 2.21). According to Klitgord and Behrendt's (1979) reconstruction, the matching counterpart basin

1 Bundesanstalt für Geowissenschaften und Rohstoffe, Stilleweg 2, 3000 Hannover 51, FRG

along the North American margin during pre-drift time (175 m.y. BP) is the Georges Bank Basin (cf. also Jansa and Wiedmann, this Vol.). However, certain tectonic features and the basin geometry of the "Aaiun-West Saharan Marginal Basin" are more reminiscent of the Baltimore Canyon Trough off New Jersey (Van Houten, 1977; Schlee and Fritsch, in press).

Both conjugate areas are old, mature passive margins which contain in their thick sediment sequence the record of the rift stage and the post-breakup evolution of the Atlantic Ocean, including the subsidence history of the margin during the past 200 m.y. The Mesozoic evolution of the "Aaiun-West Saharan Marginal Basin" can be well compared with that of the contingent Northeast American margin (Arthur et al., 1979; Jansa and Wiedmann, this Vol.); however, the Tertiary history shows marked differences because of a different tectonic and paleooceanographic setting.

The main objective of this paper, which is partly based on an internal BGR report (Ranke, 1979), is to attempt a brief synthesis of the pertinent published information on the onshore Aaiun-Tarfaya Basin between 22° and 29°N. Our summary is mainly based on the stratigraphic data from 48 oil exploration wells by the "Departamento de Investigaciones Petroliferas de AUXINI (quoted as AUXINI, 1969; see Figs. 1,5) and on the stratigraphical and tectonic compilations of Querol (1966), Choubert et al. (1966), Ratschiller (1969), Viotti (1966), Martinis and Visintin (1966), and Wiedmann et al. (1978).

In addition, we shall discuss some aspects of the morphology, stratigraphy and structure of the offshore extension of the Aaiun-Tarfaya Basin under the present shelf and slope, a subject covered in more detail in von Rad and Wissmann (this Vol.). For the offshore area we use the results of DSDP Sites 137-139 (Berger and von Rad, 1972), 369 (Lancelot, Seibold et al., 1977), 397 (von Rad, Ryan et al., 1979), and of the shelf well Spansah 51A-1 (CONOCO, 1969; von Rad and Einsele, 1980), in addition to published and unpublished seismic data from BGR cruises and other sources.

Finally, we shall comment briefly on the tectonic evolution and subsidence history of the Cape Bojador continental margin during the past 200 million years (von Rad and Einsele, 1980; von Rad and Arthur, 1979).

2. Geology of the Aaiun-Tarfaya Basin

2.1 General remarks, geological setting, morphology

The Aaiun-Tarfaya Basin has been named Tarfaya Basin in its northern part, and (El) Aaiun (or Rio del Oro) Basin in the area of the former Spanish Sahara. It belongs to a string of elongate coastal or marginal basins which formed during Mesozoic and Cenozoic seaward of the stable West African craton (Dillon and Sougy, 1974; Lehner and de Ruiter, 1976).

The onshore part of the coastal basin has a length of about 700 km (N-S) and a width of 100 to 250 km in the east-west direction. It covers an area of about 125,000 km^2 and contains a sediment volume of about 625,000 km^3. The on- and offshore Aaiun-Tarfaya Basin covers an area of 460,000 km^2 and contains a sediment volume of about 3.7 million km^3. Hence, more than 80% of the total basin fill is contained under the present continental shelf, slope, and uppermost rise (Figs. 1,4,9A) at the broad "West Saharan Marginal Basin".

The landward part of the Aaiun-Tarfaya-Basin is bounded by the Anti Atlas Mountains to the northeast and by the Precambrian Reguibat Massif and the Paleozoic fold belt of the Mauritanides to the southeast and south. A swell in the northern extension of this fold belt separates the Aaiun-Tarfaya Basin from its precursor, the Tindouf Basin, which is filled by undeformed Paleozoic sediments (see back cover figure). Towards the west, the coastal basin is open to the sea and its oceanward boundary

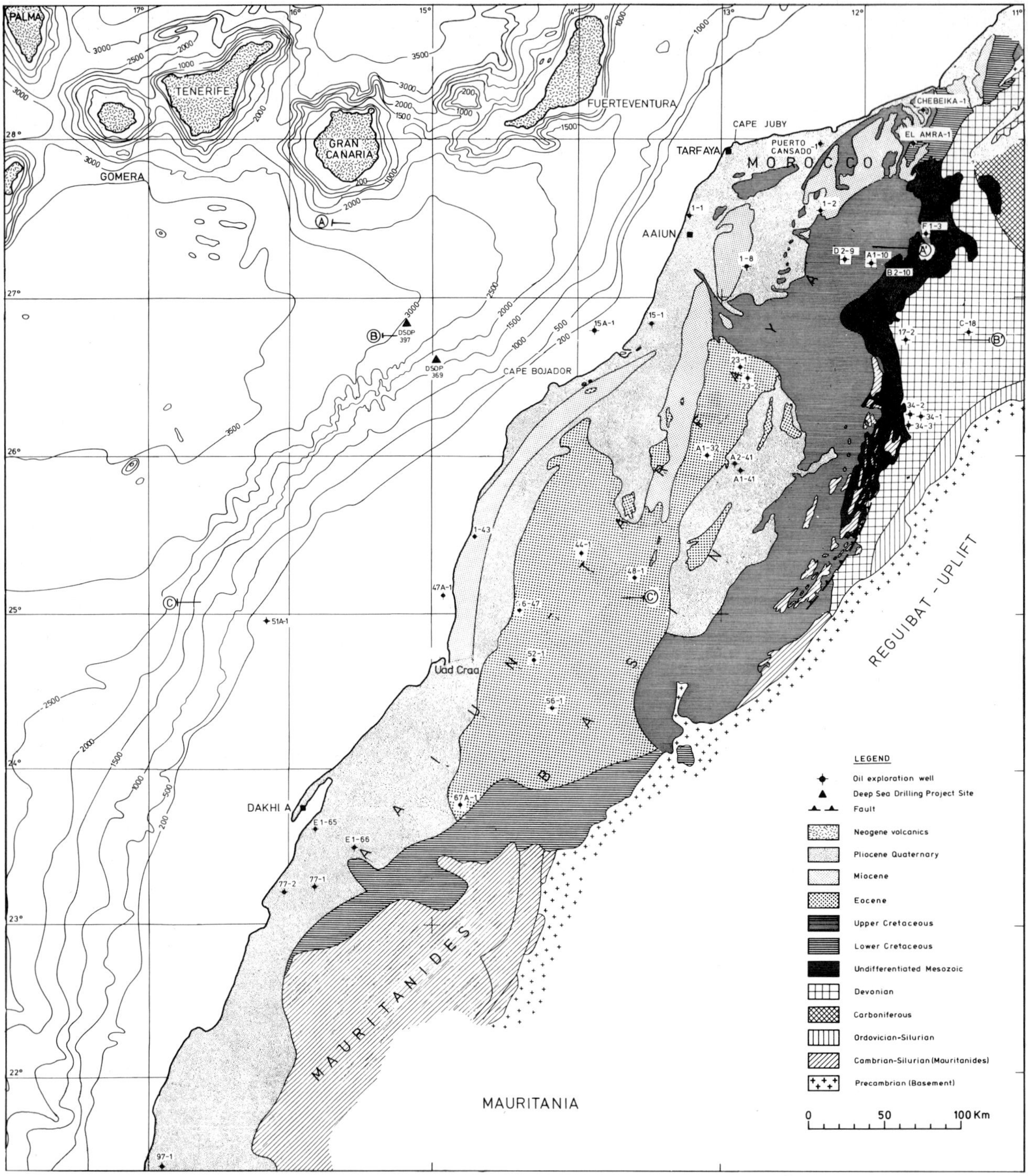

Fig. 1. Geology of the Aaiun-Tarfaya Basin and bathymetry of the adjacent West Saharan continental margin after International Tectonic Map of Africa 1:5,000 000 (Choubert et al., 1968), Querol (1966), AUXINI (1969), and Ratschiller (1970). Bathymetry after various sources (GEBCO sheets, Uchupi et al., 1976; unpublished BGR-surveys); for a more detailed bathymetric map of the Cape Bojador margin see von Rad and Wissmann (this Vol)

Fig.2. Generalized lithostratigraphy of the onshore Aaiun-Tarfaya Basin (sources as in Fig.1)

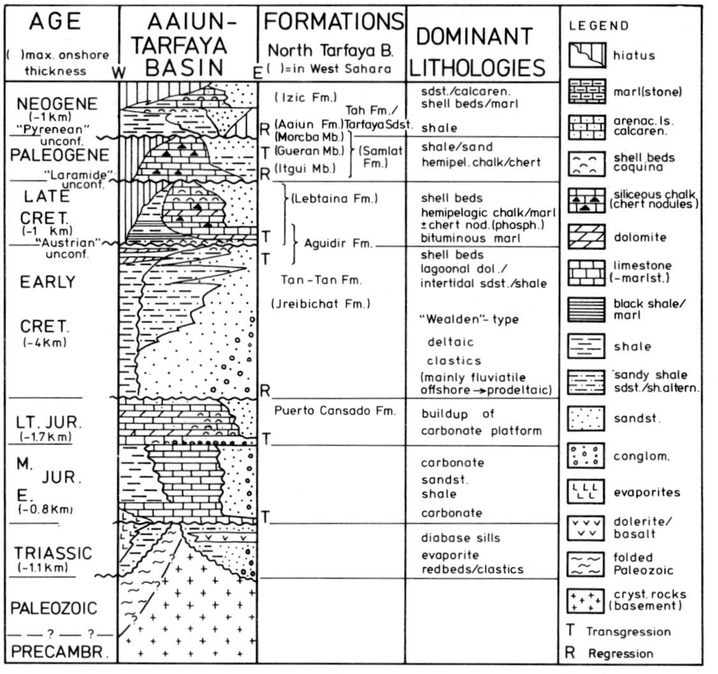

is arbitrarily defined by the 2.5 km (= 2s) sediment isopach (Fig.4). The offshore marginal basin is bounded to the north by the Canary Archipelago (Schmincke, this Vol.), whereas the southern boundary to the Senegal Basin is at about 22°N latitude, where the Reguibat Massif extends almost to the coastline (see front cover figure).

The morphology of the West Saharan continental shelf, slope and upper rise (Fig.1) has been described and interpreted by Seibold and Hinz (1974), von Rad et al. (1979), and von Rad and Wissmann (this Vol.). The main features are a southward widening shelf (25-60 km), a shelf-break at about 110 m, a 60-100 km wide, comparatively steep slope indented by numerous lower slope canyons and merging into a broad continental rise.

2.2 Stratigraphy of the onshore Aaiun-Tarfaya Basin

2.21 Basement definition

The Aaiun-Tarfaya Basin is filled by Mesozoic and Cenozoic continental to shallow-marine sediments overlying basement of Precambrian and/or Paleozoic age (Figs.2,3).

The basement definition differs from Hinz (1970), who correlated a 6 km/s layer at 9 km depth under the shelf (station METEOR 9-39) and at 9.2 to 9.8 km depth under the continental rise (M 9-38 and M 9-41) with acid crystalline rocks underlying Upper Paleozoic sediments that extend seaward from the continent into the deep sea.

Hinz (1970) interpreted the 6 km/s layer with continental lithofacies, as this velocity was considered too large for oceanic layer 2 (4.6-5.4 km/s) rocks. However Grunau et al. (1975) do show a velocity of 6 km/s for oceanic basement on a multi-

Fig.3. East-west continental margin sections across the Tarfaya-Aaiun Basin and the West Saharan continental margin (for location see Fig.1). Data sources for onshore geology see Fig.1. Offshore: basement depth after Roeser et al. (1971) and unpublished BGR refraction/gravity data; stratigraphy and structure after DSDP Sites 369 and 397, Spansah 51A-1 (CONOCO, 1969), and various published and unpublished seismic reflection data

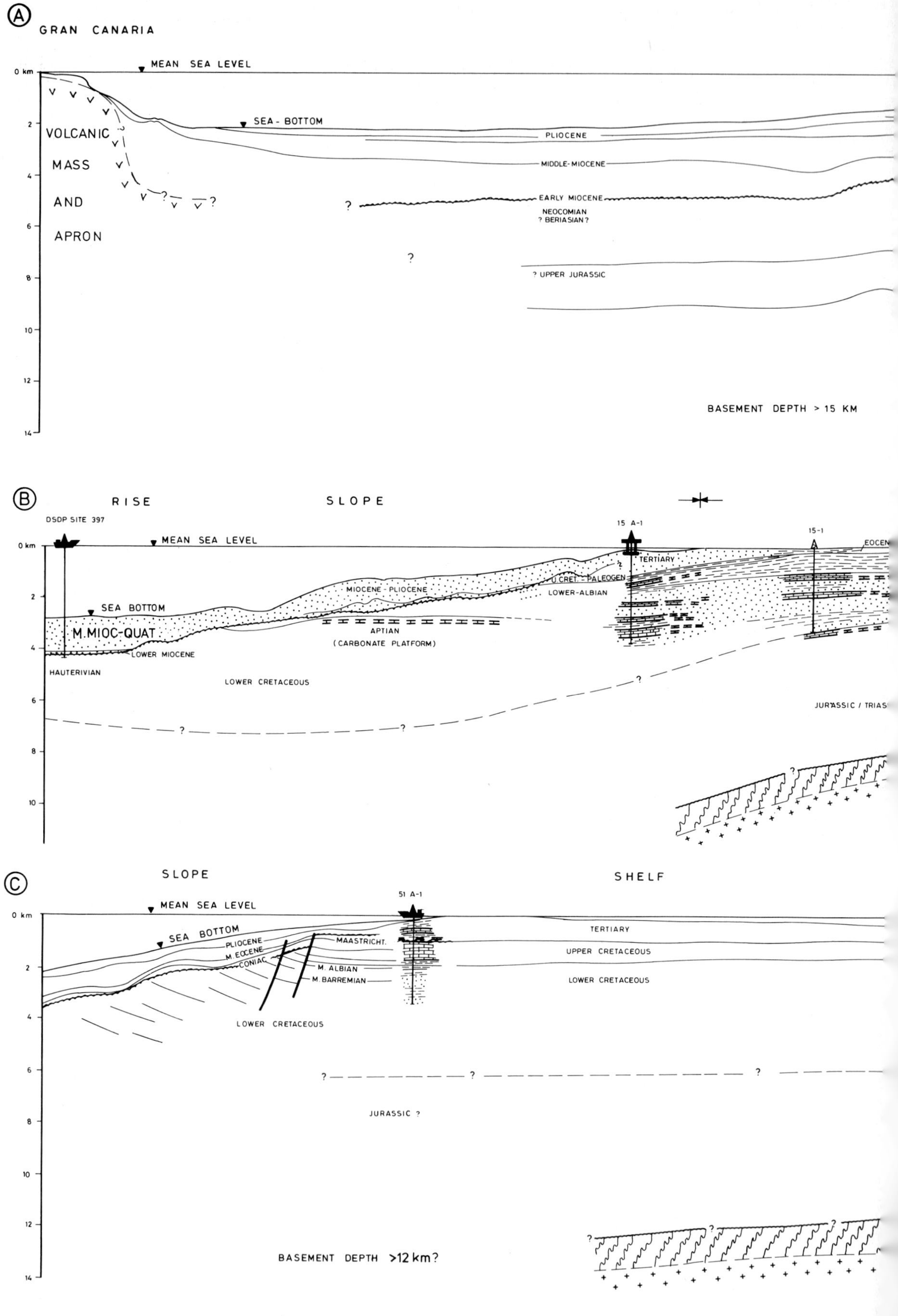

Fig. 3 (legend see p. 89)

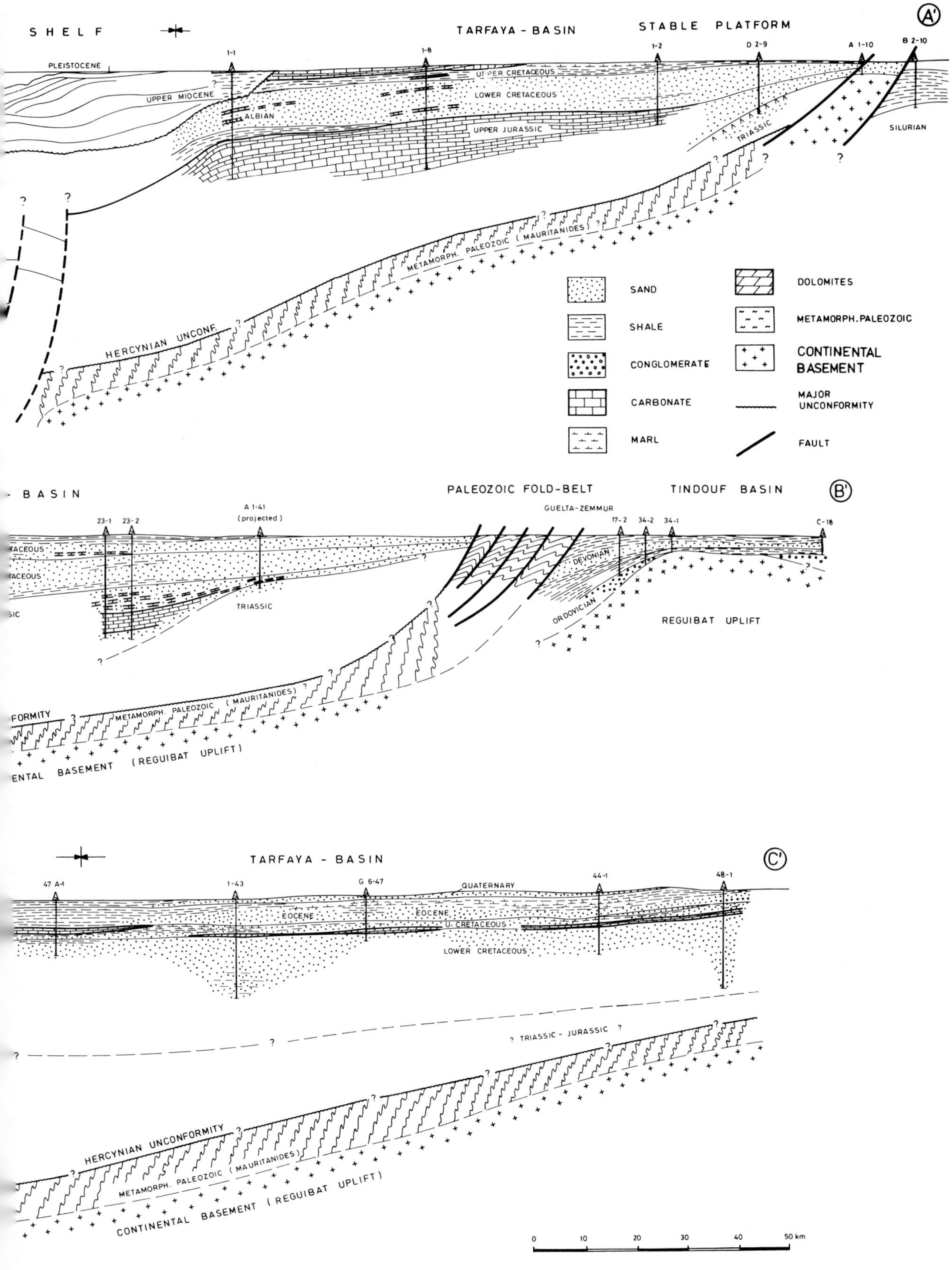
SHELF
TARFAYA - BASIN
STABLE PLATFORM
A'
PLEISTOCENE
UPPER MIOCENE
ALBIAN
UPPER CRETACEOUS
LOWER CRETACEOUS
UPPER JURASSIC
TRIASSIC
SILURIAN
METAMORPH. PALEOZOIC (MAURITANIDES)
HERCYNIAN UNCONF.
SAND
SHALE
CONGLOMERATE
CARBONATE
MARL
DOLOMITES
METAMORPH. PALEOZOIC
CONTINENTAL BASEMENT
MAJOR UNCONFORMITY
FAULT
BASIN
PALEOZOIC FOLD-BELT
TINDOUF BASIN
B'
A 1-41 (projected)
GUELTA-ZEMMUR
DEVONIAN
TRIASSIC
ORDOVICIAN
REGUIBAT UPLIFT
METAMORPH. PALEOZOIC (MAURITANIDES)
ENTAL BASEMENT (REGUIBAT UPLIFT)
TARFAYA - BASIN
C'
QUATERNARY
EOCENE
U. CRETACEOUS
LOWER CRETACEOUS
? TRIASSIC - JURASSIC ?
HERCYNIAN UNCONFORMITY
METAMORPH. PALEOZOIC (MAURITANIDES)
CONTINENTAL BASEMENT (REGUIBAT UPLIFT)
0 10 20 30 40 50 km

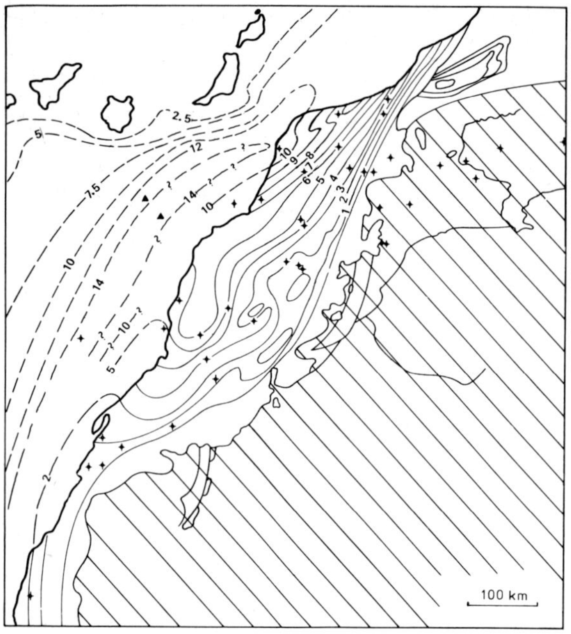

Fig.4. Basement depth of Aaiun-Tarfaya Basin (in km) according to aeromagnetic survey conducted 1959 (Choubert et al., 1968). Offshore marginal basin: depth contours = very approximate sediment isopachs in km above continental basement, contaminated and/or oceanic crust after unpublished BGR refraction seismic/-gravity data. Hatched area: Precambrian Reguibat Massif and Paleozoic Mauretanides. For petroleum exploration wells and DSDP sites see Fig.1

channel seismic profile west of Lanzarote, in an area where crust created by seafloor-spreading is generally believed to exist (Goldflam et al., 1980).

Later recalculations of the METEOR refraction seismic data by Richter (unpublished BGR data, 1975) changed the basement depths slightly to 7.1 km under the shelf and 10.5 km or 11.5 km under the rise. Multichannel reflection seismic records obtained in the area of the METEOR refraction survey show at 12 km depth under the shelf (Fig. 3A) basement above which high-amplitude onlapping reflectors dip westward to depths of more than 14 km (calculated from interval velocities). Even though basement lies here beneath the 8s-sweep of the seismic record, the overlying smoothly prograding sediments testify to the absence of a large-scale topography of the basement layer, which had already been deduced by Hinz (1970) from lines of equal velocity connecting the METEOR refraction stations. The apparent inconsistency between refraction and reflection depths cannot be reconciled with the available data. However since other multichannel lines crossing the West Saharan continental margin further south also reach 15 km depths to basement, these more modern data are favoured over the 1967 METEOR refraction data.

Because of a linear magnetic anomaly, Roeser et al. (1971) implied that the crust between Gran Canaria and the African coast consisted of basic rocks with oceanic character. They estimated the magnetic layer to lie beneath the observed 6 km/s layer at about 10-15 km depth. An extrapolation of the measured geothermal gradient of 3.5°C/100 m of DSDP Site 397 (Ryan, in von Rad, Ryan et al., 1979, p. 72) to the Curie temperature range of oceanic basalts (550°-600°C) shows that this temperature range may indeed only be reached between 16 and 17 km. Thus the basement depths from magnetic modeling may also support the reflection seismic depths of "more than 14 km".

2.22 Triassic

Several exploration wells in the central Aaiun-Tarfaya Basin (e.g. A1-41, A2-41, D2-9, B2-10, 56-1, 23-2) penetrated continental rock sequences. Evaporites, redbeds, reddish sandstones, conglomerates and volcanic rocks of Triassic age overlie either metamorphic and folded rock sequences of the Mauritanides on a Hercynian unconformity, or they overlie crystalline rocks of the Precambrian Reguibat Massif (Choubert et al., 1966; AUXINI, 1969; Dillon and Sougy, 1974). In the Chebeika-1 well a 1 km thick siltstone/sandstone/claystone sequence of

?Triassic age is interrupted by a 265 m thick diabase sill complex (Fig.5; Martinis and Visintin, 1966). These doleritic flood basalts resemble typical early-rift basalts, erupted during the early tensional stages of graben formation (cf. the coeval doleritic basalts in the High Atlas area and those of the Newark Series in New Jersey). These grabens were rapidly filled by coarse terrigenous debris during the initial rift stage of the early Atlantic (Lehner and de Ruiter, 1977; Van Houten, 1977). The marked NNE-trending zone of upper Triassic to lower Liassic salt diapirs which can be traced along most of the Moroccan margin (Hinz et al., this Vol.; see back cover figure) is not developed in the Aaiun-West Saharan Marginal Basin south of the Canary Islands. However, most drillholes did not penetrate below a seaward thickening wedge of Jurassic to Cretaceous clastic and carbonate rocks overlain by a Cenozoic sequence; the units can be followed seaward to the present continental slope (Fig.3,5,9A).

2.23 Jurassic

The oldest sediments, which can be safely dated, are carbonate rocks that indicate an early Jurassic marine transgression onto the Triassic rift sediments or evaporites. Lower and middle Jurassic sequences are present only in the northern part of the coastal basin. The Puerto Cansado-1 well penetrated 640 m of lower to middle Jurassic silty sandstones, limestones, dolomitic limestones, and dolomites. These rocks are overlain by almost 1500 m of upper Jurassic neritic marly limestones and calcarenites, intercalcated with marl, shale and sandstone ("Puerto Cansado Formation"). The presence of a conglomerate at the base of the upper Jurassic indicates a major Callovian to Oxfordian transgression which was followed by major carbonate buildup, especially during the Oxfordian, Kimmeridgian, and Tithonian.

In the central and western coastal basin, the upper Jurassic Puerto Cansado Formation consists of alternating carbonates, marls, shales, and evaporites. Near the eastern margin of the basin the marine carbonates grade into a nearshore-littoral and continental facies, in which the Triassic cannot be differentiated from the Jurassic (Fig.6).

The Jurassic carbonate platform which underlies large parts of the Northwest African continental outer shelf and upper slope, especially off Morocco and Senegal (see back cover figure), is less well developed in the central Aaiun-Tarfaya Basin. North off the northern Tarfaya Basin (north of Cape Juby), however, it can be observed in seismic records as a steep (10-20°) "carbonate front" (Mitchum and Vail, 1977; Hinz et al., this Vol.). This paleo-shelf edge lies about 10-20 km seaward of the present shelf edge. Beyond the downfaulted edge of this platform, deep-water carbonates accumulated. Near the top of the upper Jurassic there is an increasing number of sandy layers, which must be derived from a (?) rising hinterland in the south and east. These clastics indicate a renewed regressional cycle near the Jurassic/Cretaceous boundary. Off the northern Tarfaya Basin carbonate buildup ended in the earliest Cretaceous, earlier than in the Senegal Basin and much earlier than at the Mazagan Plateau.

The maximum thickness of Triassic plus Jurassic sediments in the onshore Aaiun-Tarfaya Basin is more than 2 km (Querol, 1966), and under the present shelf and slope more than 3 km - possibly ranging up to 8 km (Wissmann in von Rad et al., 1979; see Fig.9). However, these estimates are based only on limited seismic refraction information and hence are speculative. It is possible that folded upper Paleozoic strata of the Mauritanides are included in the section between the poorly defined "basement" and the base of the Cretaceous; their exclusion would decrease the estimated "Jurassic" thickness values considerably.

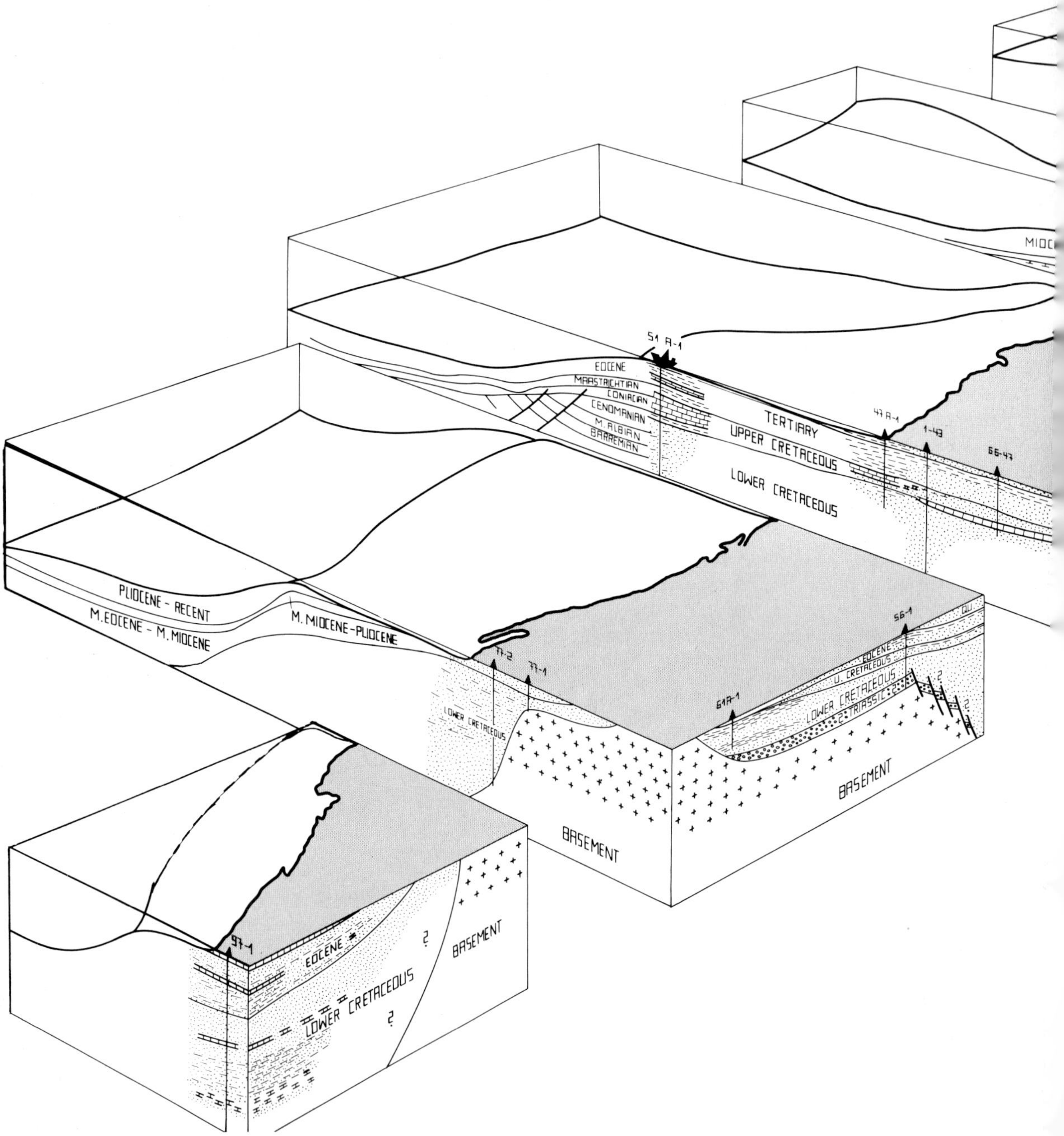

Fig.5. Block diagrams with schematic E-W and N-S cross-sections of the Aaiun-Tarfaya Basin and its adjacent continental margin. Data source and location of oil exploration wells see Fig.1

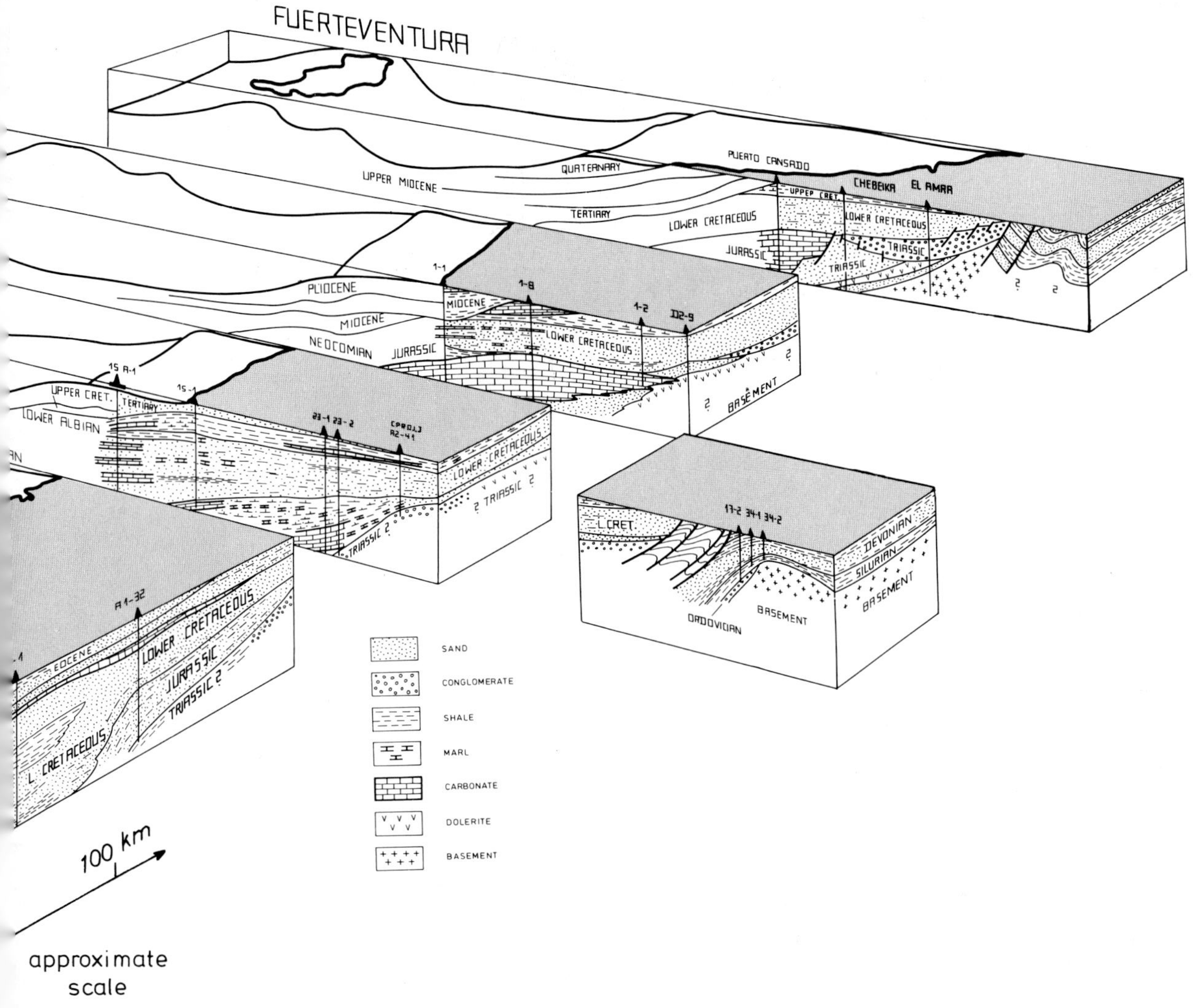

2.24 Early Cretaceous

During Early Cretaceous, especially Neocomian-Barremian times, a thick sequence of continental to marine-deltaic clastic sediments (Sables de Tan Tan or Jreibichat Formation) was deposited in the Aaiun-Tarfaya Basin. It terminated the buildup of the carbonate platform (Martinis and Visintin, 1966; Ratschiller, 1970; see Fig.2). When compared to the Jurassic, the facies boundaries shifted westwards (Fig.7). In the central basin only barren clastic sediments were deposited in an alluvial plain environment. Paralic (lagoonal/intertidal) to littoral and shallow- marine deltaic ("delta topsets") sediments are observed in the area between the present shoreline and shelfbreak, whereas the present slope and upper rise is underlain by rapidly deposited (?) deltaic foreset and laminated prodelta sediments (for a three-dimensional facies model see Einsele and von Rad, 1979). The lower Cretaceous deltaic sequence (1-4 km thick) was accumulated during and after a major global Valanginian regression about 132

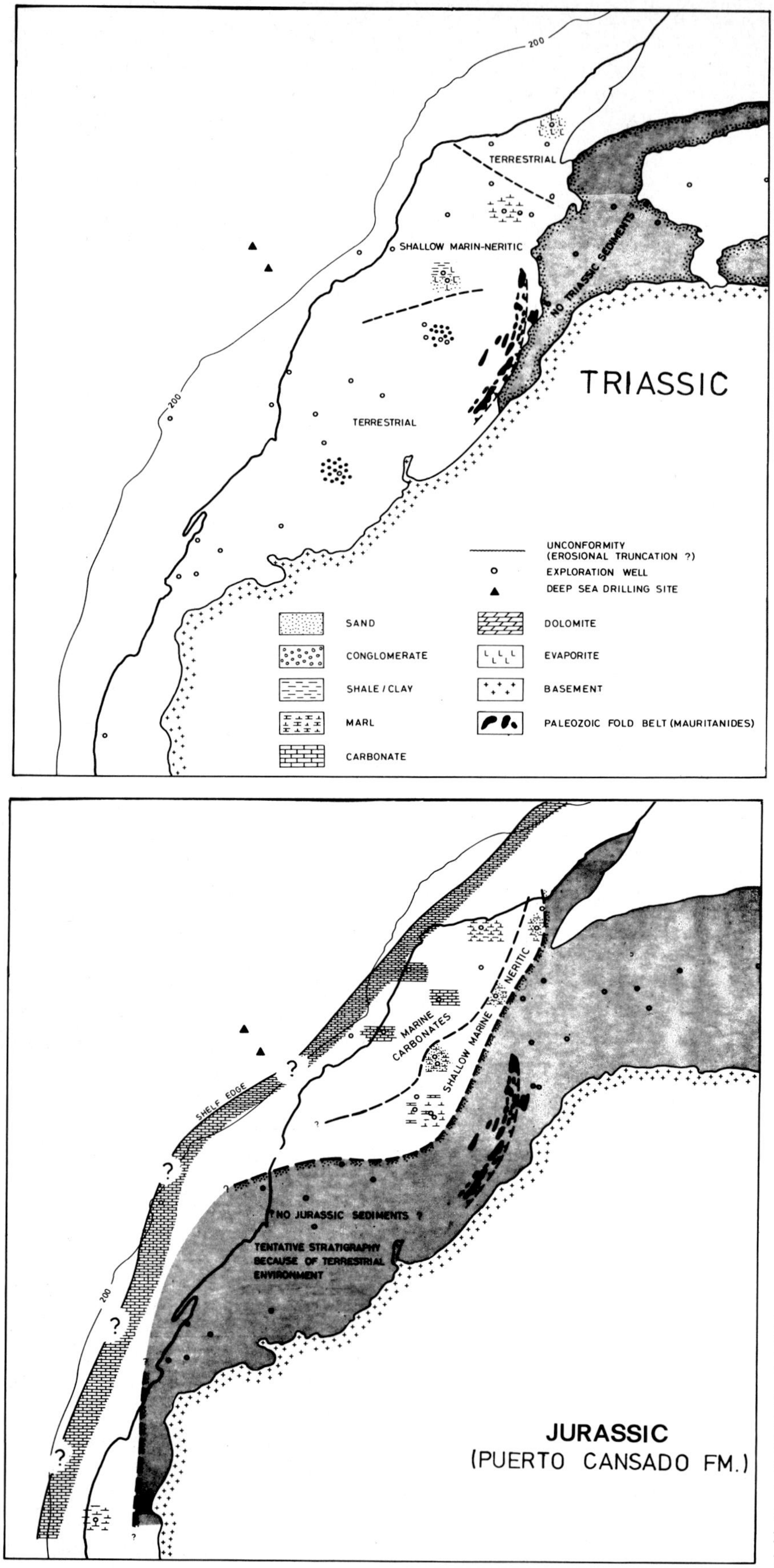

Fig.6
Facies distribution and paleogeography of the Aaiun- Tarfaya Basin during Triassic and Jurassic times. Present coastline and -200 m contour for comparison

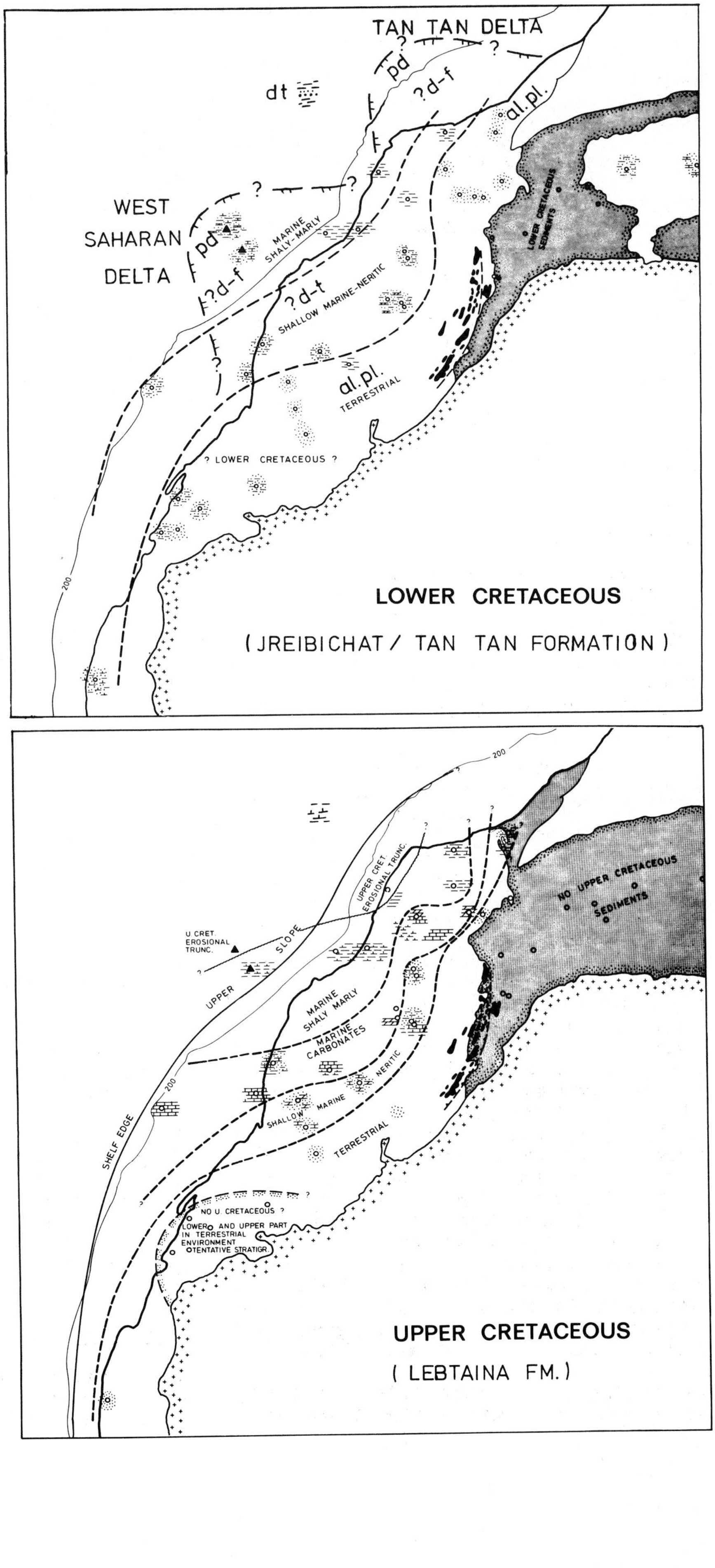

Fig.7
Facies distribution and paleogeography of the Aaiun-Tarfaya Basin during Early and Late Cretaceous times.- Abbreviations: al.pl. = abyssal plain, d-t = delta topsets, d-f = delta foresets, pd = prodelta, dt = distal turbidites (Fuerteventura: Robertson and Bernoulli, this Vol.)

m.y. ago (Vail et al., 1977). Two major progradational lower Cretaceous "Wealden"-type delta systems can be distinguished in the area of the Aaiun-Tarfaya Basin: the Tan-Tan delta system (Cape Juby-Fuerteventura) to the north (Martinis and Visintin, 1966; Price 1980; Robertson and Bernoulli; this Vol.), and the Cape Bojador delta system (central Aaiun Basin - Cape Bojador - Site 397) to the south (Einsele and von Rad, 1979). Both deltas may have been part of one huge coalescing Tan Tan - Cape Bojador delta system. For a discussion of the occurrence of more or less coeval "Wealden" type sequences around the North Atlantic see von Rad and Arthur (1979).

2.25 Late Cretaceous

In the Aaiun-Tarfaya Basin the regressive fluvial delta plain facies is overlain by transgressive fine-grained lagoonal, tidal, and shallow-marine muds, sands and coquinas (Aptian-Albian Calcaires d' Aguidir Formation, Martinis and Visintin, 1966; Einsele and von Rad, 1979). During the late Albian to early Cenomanian transgressions, water depth increased continuously and the facies zones migrated landward (Wiedmann et al, 1978; Einsele and Wiedmann, this Vol.). Inner to outer shelf depths were reached during Cenomanian to Turonian times (Fig.7). An expanded oxygen minimum zone and/or regional upwelling generated laminated organicrich marls, equivalents of the "mid-Cretaceous" black shales, in the late Cenomanian and Turonian (Wiedmann et al., 1978; Einsele and Wiedmann, this Vol.). The upper Cretaceous sediments are comparatively thin, hemipelagic marls and carbonates, characterized by waning terrigenous influx and an increased plankton input. During the Santonian and Campanian, Ostrea-type coquinas and silty marl-and claystones reappeared in the northern Tarfaya Basin; they indicate shallower (inner shelf to beach) paleodepths leading in the latest Cretaceous to emergence of the area (Wiedmann et al, 1978).

According to Ratschiller (1970), the shallow-marine upper Cretaceous (to Eocene) Lebtaina Formation (300-600 m) unconformably overlies the continental lower Cretaceous Jreibichat Formation in the northern Aaiun Basin. The Lebtaina Formation consists of marlstones interbedded with siliceous and/or phosphatic chalks and with several horizons of chert concretions and phosphorite. Eastward, this formation grades into the continental Tebaila and Briber facies, also of Late Cretaceous age (Fig.2).

The maximum thickness of the upper Cretaceous-marine rocks, in the Aaiun-Tarfaya Basin, is ca. 1000 m (898 m in Spansah 51A-1). In general, one can see a land- and seaward decrease of thickness from a maximum under the present shelf. A marked erosional truncation can be observed in the northwestern offshore Aaiun-Tarfaya Basin between Cape Juby and Cape Bojador (Daora 1-1 to Site 397; Figs. 3,5,7). Here not only upper Cretaceous, but also the overlying Paleogene and part of the underlying lower Cretaceous strata were eroded (von Rad and Wissmann, this Vol.). According to seismic information (Hinz, 1979; Wissmann, 1979) and the results of DSDP Site 397 (Arthur et al., 1979), the erosion was especially extensive in the area of the present lower continental slope and upper rise off Cape Bojador (Figs. 7,8,9A). The reason for the erosion of large parts of the upper Cretaceous strata is still enigmatic. Arthur et al (1979) and von Rad and Arthur (1979) invoke a conspicuous erosional event during late Oligocene to earliest Miocene which was caused by the coincidence of a major regression with intensified slumping, canyon incision, and bottom water circulation (? pre-Antarctic Bottom Water). A more complex, multi-phase mid-Cretaceous to Cenozoic history of slope rejuvenation and destruction is inferred by von Rad and Wissmann (this Vol.) from geophysical and geological evidence.

2.26 Paleogene

The Paleogene Samlat Formation unconformably overlies upper Cretaceous strata (Fig.2). The Paleocene

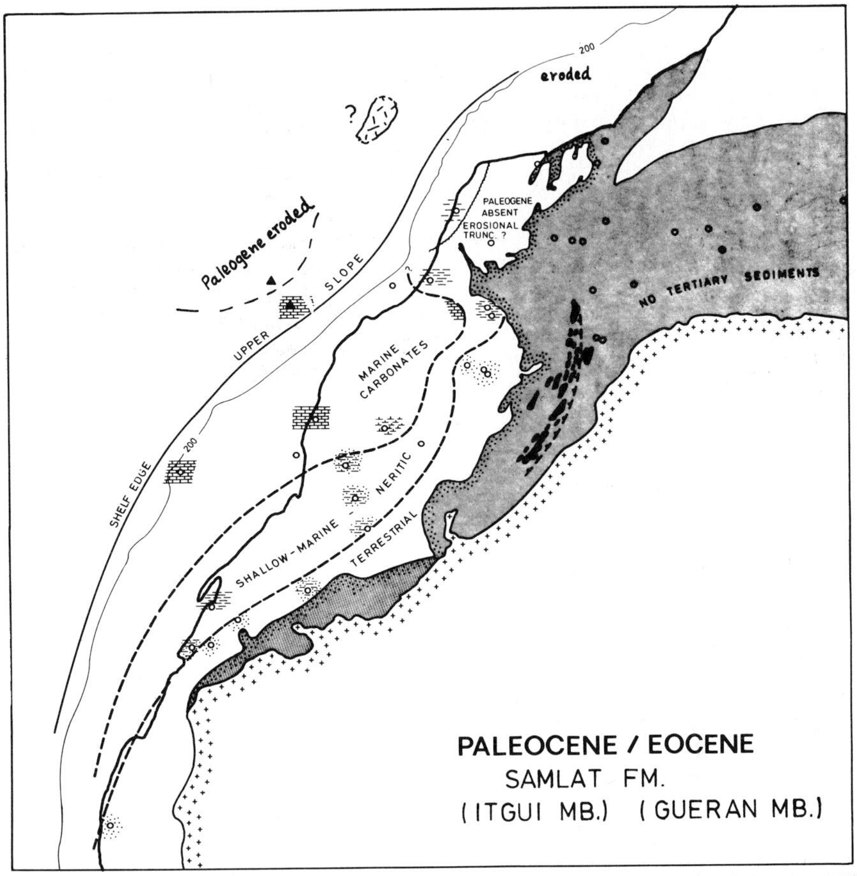

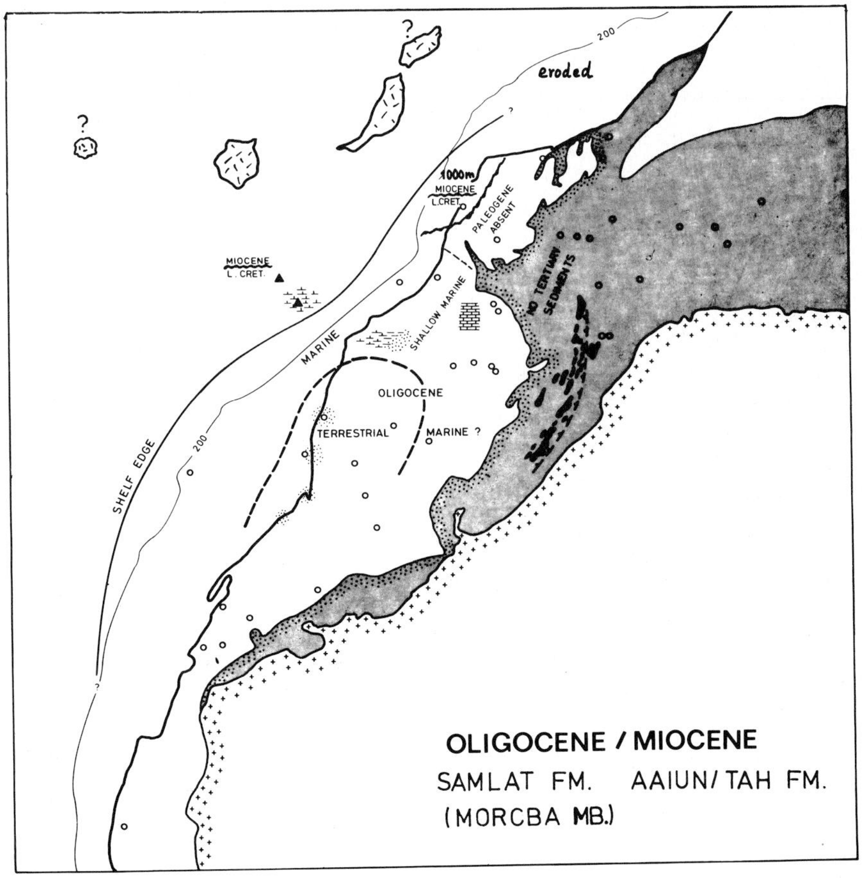

Fig.8
Facies distribution and paleogeography of the Aaiun- Tarfaya Basin during Paleocene/Eocene and Oligocene/Miocene times

Itgui Member of the Samlat Formation (Ratschiller, 1970) is 60-300 m thick and consists of predominantly marine sediments, including siliceous chalks with chert concretions. The Eocene Gueran Member is lithologically similar and reaches thicknesses of up to 800 m. A major input of clastic sediments during the Eocene indicates a regression which may coincide with a global sea level lowstand during mid-Eocene time (Querol, 1966; Vail et al., 1977). Eocene strata are exposed over large areas in the central and southern Aaiun Basin (Fig.1).

The Oligocene (to Miocene) Morcba Member of the Samlat Formation is as much as 300 m thick, characterized by continental sandstones and conglomerates, and exposed only in the southern Aaiun Basin (Figs.1,2,-8; Ratschiller, 1970). Elsewhere (especially in the northern coastal basin) the Oligocene is missing because of widespread erosion during this regressive period. Major Oligocene hiatuses (including parts of the early Miocene) can be observed in wells 1-1, 15-1, 1-32, 43-1, 56-1 and Puerto Cansado-1.

In the central Aaiun-Tarfaya Basin, the thickness of Paleogene sediments exceeds 1000 m. Towards the shelf and slope, this thickness decreases to 640 m (Spansah 51A-1) and about 200 m (Site 369; with hiatuses).

2.27 Neogene

The Neogene rocks are thin (< 100 m) and only exposed in the western Aaiun-Tarfaya Basin (Figs.2,8). In general, the transgressive upper Miocene Aaiun Formation is less than 50 m thick and consists of sandy limestones, calcarenites, sandstones and oyster beds. In the northern Tarfaya Basin the Miocene Tah Formation (Argiles de Tah) consists of 35 m of shallow-marine sandy claystones which thicken considerably seawards and southwards.

In the vicinity of the Daora 1-1 well, a 1000 m thick Mio-Pliocene sequence of ?deltaic/estuarine claystones with sandy intercalations overlies unconformably the lower Cretaceous Tan Tan formation (AUXINI, 1969), signifying a 80-100 m.y. hiatus, very similar to that at the upper rise Site 397 (Figs.3, 5).

The Izic Formation of late Miocene to Pliocene age consists mainly of arenaceous limestones and calcarenites, as much as 80 m in thickness (Ratschiller, 1970). Landwards, it grades into a continental facies (the Erbeig Facies) and northward into the Pliocene to Quaternary Tarfaya Sandstone (grès de Tarfaya; Martinis and Visintin, 1966).

2.3 Stratigraphy of the offshore Aaiun-West Saharan Basin

Since the stratigraphy of the Cape Bojador or West Saharan Marginal Basin has been outlined by Lancelot and Seibold (1978), von Rad and Einsele (1980), and Arthur et al. (1979), we will summarize here only the most important facts.

The ocean-continent boundary is obscure under the West Saharan margin, because it is overlain by up to 8 km of pre-Cretaceous and up to >6 km of Cretaceous and Tertiary strata. Distinct or modified oceanic crust with velocities of 5.9-6.7 km/s overlying a shallow (11-15 km) Moho can be observed under the Canary Islands Tenerife, Gran Canaria, Fuerteventura and Lanzarote (Banda et al, in press). Therefore the ocean-continent boundary must lie somewhere between Lanzarote-Fuerteventura and the continent.

The Early Cretaceous is represented by a very thick regressive Wealden-type deltaic facies ranging from alluvial plain to estuarine/intertidal sediments under the present shelf (Spansah 51A-1), to distal prodelta muds (DSDP Site 397; Einsele and von Rad, 1979).

Upper Cretaceous transgressive marine sediments are much thinner and only preserved landward of the intermediate slope (Figs.3,7,9A). Their thickness decreases from the shelf (898 m in Spansah 51A-1) to the slope (60 m in Site 369, with hiatuses) indicating a seaward thinning and pinch-out of individual strata because of reduced terrigenous supply (Hinz, 1979; von Rad and Einsele, 1980).

Only a few hundred meters of Paleogene sediments were deposited without any progradation above a widespread hiatus at the Cretaceous-Tertiary boundary. They are also truncated seaward of the intermediate slope (Figs.3,9A).

In general, the Neogene is extremely thin under the present shelf and slope, because most sediment bypassed this area on its way to the new depocenter on the uppermost rise. Beneath the upper rise, an oversteepened late Oligocene-early Miocene erosional surface is covered by 600 m of rapidly deposited lower Miocene mass flow units which are overlain in turn by a 700 m thick drape of middle Miocene to Quaternary hemipelagic nanno chalks and marls (von Rad and Einsele, 1980).

3. Structure

Two major tectonic directions control the Aaiun-Tarfaya Basin and its hinterland: the WSW-ENE "Atlas direction" of the bordering Precambrian to Paleozoic Reguibat and Anti Atlas Highs, and the NNE-SSW direction of the Mesozoic-Cenozoic "NW Africa Geosyncline" (Querol, 1966). The early Mesozoic basin fill overlies folded Cambrian to Carboniferous strata of the Mauritanides belt or Precambrian crystalline rocks on a Hercynian unconformity.

In general, the sedimentary strata in the eastern Aaiun- Tarfaya Basin are horizontal. Seawards they are progressively more inclined to the west without forming a marked hingeline (Fig.3). The northern Tarfaya Basin has a monoclinal structure that trends NW and WNW; dipping strata are interrupted by a gentle updoming near Puerto Cansado-1. Geophysical data suggest deeply buried normal faulting, lowering the basin floor westward (Martinis and Visintin, 1966).

Tectonic uplift of the Reguibat and Anti Atlas Highs is evident in Early Cretaceous times, when major deltaic systems pushed the receding coastline further to the west.

During the Eocene/Oligocene, when the African and Eurasian plates collided and alpine-type movements folded the Atlas Mountains, the Reguibat and Anti Atlas Highs experienced only uplift and erosion, as documented by slight unconformities and by coarse terrigenous input into the adjacent Aaiun-Tarfaya Basin.

Under the upper continental slope the horizontal layering of the Mesozoic rocks changes into a more blockfaulted appearance as blocks with landward dipping strata are downfaulted along growth faults, especially apparent in rocks of the lower Cretaceous deltaic sequence.

Along the lower continental slope, a "slope anticline" has been described by Hinz et al. (1974) which is not due to alpine-type compressional tectonics (Uchupi et al., 1976), but is probably due to differential loading of the landward thickening Cretaceous sediment wedge and to roll-over structures between antithetic growth faults (Hinz, 1979; Wissmann, 1979; Arthur et al., 1979; von Rad and Wissmann, this Vol.).

4. Tectonic evolution and Subsidence History

The Aaiun-West Saharan Basin developed from a rift to a marginal basin which started to subside after late Paleozoic orogenic movements and a post-Hercynian (Permian-Triassic) erosional phase of uplift. It formed at the same time as most of the basins of the NNE-SSW trending "Northwest African Geosyncline" (Querol, 1966). Rifting started in the Late Triassic and continued until Liassic times. Breakup of Pangaea and separation of the North American and African plates started probably in late Triassic to Liassic times.

The early to middle Jurassic subsidence rates were small (< 20 m/m.y. at Puerto Cansado-1), and the basin

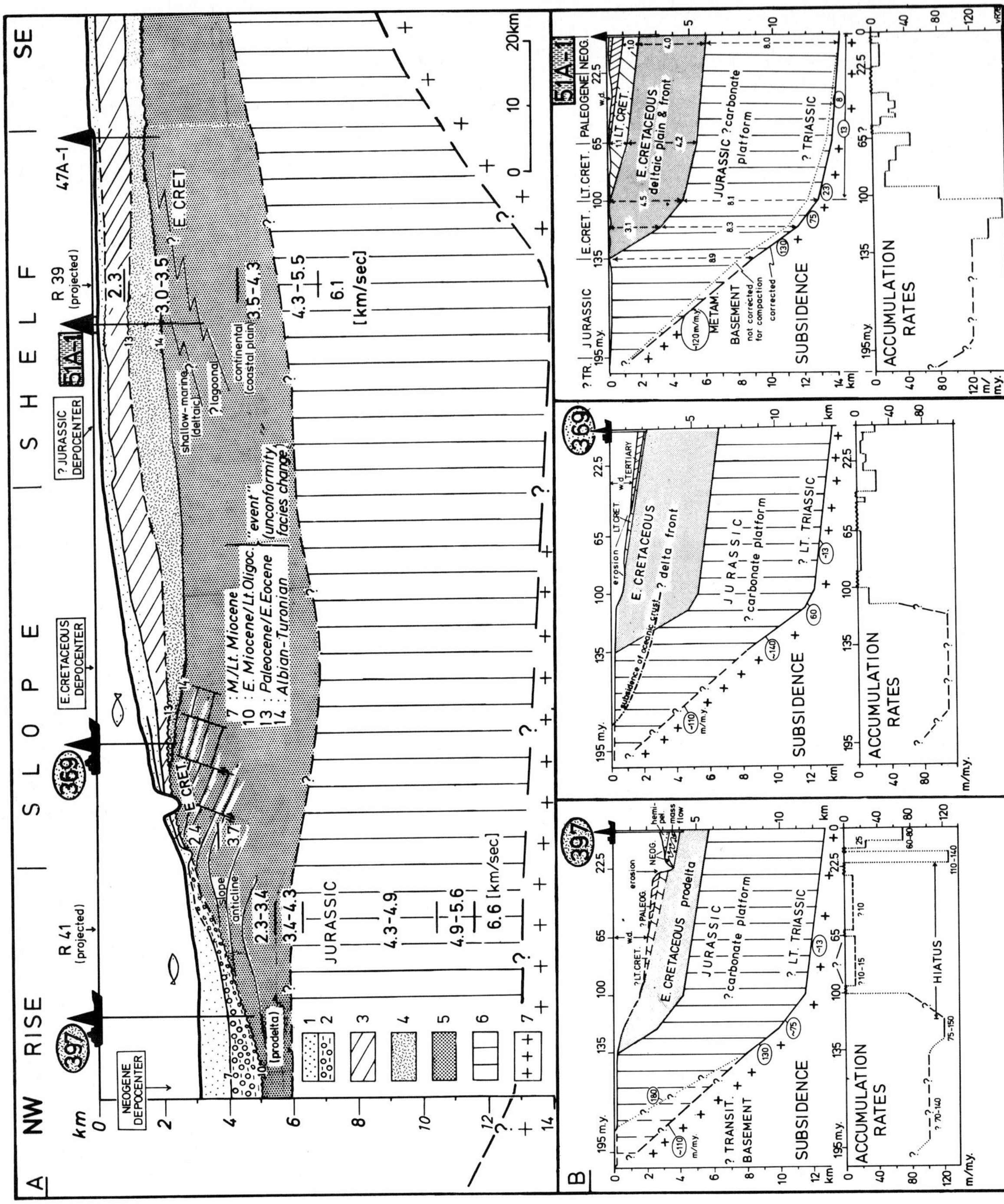

Fig.9. A: Highly generalized SE-NW cross-section from the outer Aaiun Basin to the upper rise off Cape Bojador (drilling sites indicated in Fig. 1). Basement depth and Jurassic/Cretaceous boundary is very speculative and interpreted from refraction and multichannel seismic profiles. 1 Neogene (mainly hemipelagic), 2 early Neogene on slope and rise (mainly allochthonous), 3 Paleogene, 4 Late Cretaceous, 5 Early Cretaceous, 6 Jurassic, 7 "Basement". B: Mesozoic-Cenozoic subsidence rates (m/m.y.), changes in water depth and accumulation rates at three drilling sites from the shelf (51A-1), slope (369), and upper rise (397). A and B after von Rad and Einsele (1980) and von Rad and Arthur (1979)

was filled by redbeds, evaporites, flood basalts (Chebeika-1), and silty-sandy claystones in a continental to paralic environment.

During Late Jurassic, especially Oxfordian to Kimmeridgian times, the subsidence rates increased rapidly (Puerto Cansado-1:100 m/m.y.; 47A-1, 43-1:80 m/m.y.). We infer similar uniform subsidence rates for the offshore area (Spansah 51A-1, Fig.9B; von Rad and Einsele, 1980). This strong subsidence was compensated by carbonate buildup, so the water depth landward of the marked platform edge remained very shallow (e.g. oolites in Puerto Cansado-1!). The seaward part of the basin subsided more rapidly along normal faults and led to a conspicuous paleoslope and formation of a deep-water carbonate basin (Hinz et al., this Vol.).

The great thickness of rapidly accumulated Wealden-type deltaic clastics, deposited during Neocomian to Barremian times off Cape Bojador as delta plain (51A-1), delta foreset (369) or prodelta deposits (397) induced an increase of subsidence to 110-140 m/m.y. (Fig.9), whereas the Tan-Tan Formation deposited in the onshore northern Tarfaya Basin (Puerto Cansado-1) subsided only by about 40 m/m.y., similar to the overlying Aguidir Formation of Aptian-Albian age.

During the Late Cretaceous and Cenozoic, the subsidence off Cape Bojador slowed down to intermediate (23 m/m.y.: Late Cretaceous at 51A-1) and finally very small values (8 m/m.y.: Tertiary at 51A-1). The accumulation rates lagged behind the subsidence rates to result in a gradual deepening of the outer continental margin to its present depth (Fig.9).

During most of the Neogene, the onshore Aaiun-Tarfaya Basin experienced uplift and erosion (von Rad and Einsele, 1980). At the upper continental rise off Cape Bojador (Site 397), however, a 1-2 km deep erosional scarp was formed, and the lower slope receded due to several Cretaceous to Oligocene erosional events (von Rad and Wissmann, this Vol.). In the early Miocene, the oversteepened lower slope was rapidly covered by mass flows, until equilibrium conditions were gradually established (Arthur and von Rad, 1979).

Acknowledgements

This study was supported by the Bundesministerium für Forschung und Technologie (BMFT) and the Deutsche Forschungsgemeinschaft.
We are very grateful for critical comments by Drs. G.Einsele, K.Hinz, and J.Schlee who kindly reviewed our paper.

REFERENCES

Arthur MA, von Rad U, Cornford C, McCoy FC, Sarnthein M (1979) Evolution and sedimentary history of the Cape Bojador continental margin, Northwestern Africa. In: Ryan WBF, von Rad U, et al. Init Reps Deep Sea Drilling Project 47 (1): 773-816

Arthur MA, von Rad U (1979) Early Neogene base-of-slope sedimentation at Site 397: sequential evolution of gravitative mass transport processes and redeposition along the Northwest African passive margin. In: Ryan WBF, von Rad U et al. Init Reps Deep Sea Drilling Project 47 (1): 603-639

AUXINI (Departamento de Investigaciones Petroliferas de AUXINI, 1969) Correlacion estratigraphica de los sondeos perforados en el Sahara espanol. Bol Geol Minero, Madrid 83: 235-251

Banda E, Danobeitia JJ, Surinach E, Ansorge J (in press) Features of crustal structure under the Canary Islands. Earth Plan Sci Letters

Berger WH, von Rad U (1972) Cretaceous and Cenozoic sediments from the Atlantic Ocean. In: Hayes DE, Pimm AC et al. Init Reps Deep Sea Drilling Project 14: 787-954

Choubert G, FaureMuret A, Hottinger L (1966) Apercu géologique du bassin cotier de Tarfaya. Notes et Mém Serv Géol Maroc (I): 7-106

Choubert G, Faure-Muret A, Sougy J (1968) Carte tectonique internationale de l'Afrique 1:5,000 000, Sheet 1. Assoc Afr Geol Surv, UNESCO, Paris

CONOCO (1969) Spansah 51A-1, offshore Spanish Sahara. Final Report by Continental Oil Company of Spain, Spanish Gulf Oil Company, Compania Espanol de Petroleas S.A. Exploration Department, unpublished report and logs

Dillon RS, Sougy JHA (1974) Geology of West Africa and Canary and Cape Verde Islands. In: Nairn AEM, Stehli FG (eds) The ocean basins and margins, 2. The North Atlantic: New York (Plenum Press): 315-390

Einsele G, von Rad U (1979) Facies and paleoenvironment of Early Cretaceous sediments in DSDP Site 397 and in the Aaiun Basin (NW Africa). In: von Rad U, Ryan WBF et al., Init Reps Deep Sea Drilling Project 47(1): 559-577

Goldflam P, Hinz K, Weigel W, Wissmann G (1980) Some features of the northwest African margin and magnetic quiet zone. Phil Trans R Soc Lond A 294: 87-96

Grunau HR, Lehner P, Cleintuar MR, Allenbach P, Bakker G (1975) New radiometric ages and seismic data from Fuerteventura (Canary Islands), Maio (Cape Verde Islands) and Sao Tomé (Gulf of Guinea). In: Borradaille JI et al (eds) Progress in geodynamics (Amsterdam, 3-4 April 1975) Koninkl Nederlandse Akad Wetensch: 90-118

Hinz K, (1970) Neuere geophysikalische Ergebnisse aus dem Bereich des äusseren Schelfs und Kontinentalabfalls. Erdöl und Kohle, Erdgas, Petrochemie 23 (2): 70-73

Hinz K (1979) Seismic sequences of Cape Bojador, Northwest Africa. In: von Rad U, Ryan WBF et al., Init Reps Deep Sea Drilling Project 47(1): 485-489

Hinz K, Seibold E, Wissmann G (1974) Continental slope anticline and unconformities off West Africa. "Meteor" Forsch Ergebn C, no 17: 67-73

Klitgord KD, Behrendt JC (1979) Basin Structure of the U.S. Atlantic margin. In: Watkins JS et al (eds), Geological and geophysical investigations of continental margins. Amer Assoc Petrol Geol Mem 29: 85-112.

Lancelot Y, Seibold E et al (1977), Initial Reports of the Deep Sea Drilling Project, v 41, Washington (US Gov Print Off), 1259 pp.

Lancelot Y, Winterer EL (1980) Evolution of the Moroccan oceanic basin and adjacent continental margin - a synthesis. In: Lancelot Y, Winterer EL et al Init Reps Deep Sea Drilling Project 50: 801-833

Lehner P, de Ruiter PAD (1977) Structural history of the Atlantic margin of Africa. Amer Assoc Petrol Geol Bull 61: 961-981

Martinis B, Visintin V (1966) Données géologiques sur le bassin sédimentaire cotier de Tarfaya (Moroc Méridional). In: Reyre D (ed) Sedimentary Basins of the African Coast, part 1 (Atlantic coast), Ass Afr Geol Surv, UNESCO: 13-26

Mitchum Jr RM, Vail PR (1977) Seismic stratigraphic interpretation procedure. In: Vail PR et al (eds), Seismic stratigraphy and global changes of sea level, Amer Assoc Petrol Geol Memoir 26: 135-144

Querol R (1966) Regional geology of the Spanish Sahara. In: Reyre D (ed) Sedimentary basins of the African coast, part 1 (Atlantic coast), Assoc Afr Geol Surv, UNESCO: 27-39

Price I (1980) Provenance of the Jurassic-Cretaceous flysch,Deep Sea Drilling Project Sites 370 and 416. In: Lancelot Y, Winterer EL et al, Init Reps Deep Sea Drilling Project 50: 751-757

Ranke U (1979) Bau, geologische Entwicklung und Kohlenwasserstoffhöffigkeit des Kontinentalrandes vor Nordwestafrika. Tarfaya-Becken, Teil 1, Festland und Schelf. Unpubl Rep Bundesanst Geowiss Rohst Hannover, 52 pp, 9 Figs, 15 Encl

Ratschiller LK (1970) Lithostratigraphy of the northern Spanish Sahara. Memorie Museo Tridentino Sci Trento 18 (1): 1-80

Roeser HA, Hinz K, Plaumann S (1971) continental margin structure in the Canaries. In: Delany FM (ed). The Geology of the East Atlantic continental Margin, Inst of Geological Sciences Rep 70 (16): 27-36

Schlee JS, Fritsch J (in press) Seismic stratigraphy of the Georges Bank Basin complex, offshore New England. In: J S Watkins et al (eds), Hedberg Research Conference, Amer Assoc Petrol Geol Memoir

Seibold E, Hinz K (1974) Continental slope construction and destruction, West Africa. In: Burk CA, Drake CL (eds), The geology of continental margins. New York (Springer): 179-196

Uchupi E, Emery KO, Bowin CO, Phillips JO (1976) Continental margin off Western Africa. Amer Assoc Petrol Geol Bull 60: 809-878

Vail PR, Mitchum Jr RM, Thompson III S (1977) Global cycles of relative changes of sea level. In: Vail PR et al (eds) Seismic stratigraphy and global changes of sea level, Amer Assoc Petrol Geol Memoir 26: 83-98

Van Houten FB (1977) Triassic-Liassic deposits of Morocco and eastern North America: Comparison. Amer Assoc Petroleum Geol Bull 61 (1): 79-99

Viotti C (1966) Resultats stratigraphiques du sondage Puerto Cansado 1 du bassin cotier de Tarfaya. Notes et Mem Serv Géol Maroc 175: 225-249

von Rad U, Arthur MA (1979) Geodynamic, sedimentary, and volcanic evolution of the Cape Bojador continental margin (NW Africa). In: Talwani M et al (eds), Deep drilling results in the Atlantic Ocean: Continental margins and paleoenvironment, M Ewing Series 3, Amer Geophys Union: 187-203

von Rad U, Cepek P, von Stackelberg U, Wissmann G, Zobel B (1979) Cretaceous and Tertiary sediments from the Northwest African slope (dredges and cores supplementing DSDP results). Mar Geol 29: 273-312

von Rad U, Einsele G (1980) Mesozoic-Cainozoic subsidence history and paleobathymetry of the northwest African continental margin (Aaiun Basin to DSDP Site 397). In: Kent P et al. (eds). The evolution of passive continental margins in the light of recent deep drilling results, Phil Trans Royal Soc (London) A 294: 37-50

von Rad U, Ryan WBF et al (1979) Initial Reports of the Deep Sea Drilling Project, v. 47(1), 835 pp

Wiedmann J, Butt A, Einsele G (1978) Vergleich von marokkanischen Kreide-Küstenaufschlüssen und Tiefseebohrungen (DSDP): Stratigraphie, Paläoenvironment und Subsidenz an einem passiven Kontinentalrand. Geol Rundschau 67 (2): 454-508

Wissmann G (1979) Cape Bojador slope, an example for potential pitfalls in seismic interpretation without the information of outer margin drilling. In: von Rad U, Ryan WBF et al Init Reps Deep Sea Drilling Project 47(1): 491-499

7 Cretaceous-Cenozoic History of the West Saharan Continental Margin (NW Africa): Development, Destruction and Gravitational Sedimentation

Ulrich von Rad and Gerd Wissmann[1]

ABSTRACT

The most important constructional phases of the mature passive West Saharan continental margin involved the buildup of a thick Jurassic carbonate platform, overwhelmed by up to 4 km of Early Cretaceous Wealden-type deltaic sediments. Roll-over structures between antithetic growth faults produced the "lower slope anticline" and a convex Cretaceous paleoslope. We recognized no salt diapirs, but local piercement structures probably caused by mobilized Early Cretaceous prodeltaic shale. During late Cretaceous to Tertiary times the continental slope was hardly prograded. During Tertiary (especially Neogene) times the depocenter shifted to the upper rise.

Evidence from seismic profiles correlated with DSDP, commercial well, dredge, and land geological data allow the differentiation of the complex, multi-phase mid-Cretaceous to Cenozoic history of slope destruction. At the following times slope canyons were excavated and regional unconformities formed: (1) mid- Cretaceous (upper slope; precursor valleys); (2) early Senonian; (3) Paleocene (pronounced unconformity); (4) late Eocene (local); (5) mid-to late Oligocene (major lower slope erosion; up to 100 m.y. hiatus; major canyon cutting phase);(6) mid- to late Miocene (local); (7) Pliocene (shelf and upper slope; deeply incised Uad Craa Channel); (8) late Pleistocene (canyon reactivation, slope gullies, and incisions).

Some of these accentuated erosional phases (especially 5 and 8) can be correlated with intensified Atlantic bottom water circulation from climatic cooling during periods of exceptionally low sea level causing hiatuses on the shelf and increased mass wasting at the slope (Fig.10). A very large sediment volume (7,500 -15,000 km^3 off Cape Bojador) was removed during the mid-Tertiary sculpturing of the lower slope, probably by enhanced contour currents (pre-AABW?) and associated gravitative mass transport. In total 900 km^3 of sediments were excavated by slope canyons, most of which do not extend to the shelfbreak. Both processes caused a gradual retreat of the base-of-slope line and accentuated the convex shape of the slope in this area.

During early to mid-Miocene times, massive slope failure led to the gravitative deposition of olistostromes, debris flows, turbidites and slumps on the oversteepened lowermost paleoslope, until an equilibrium gradient was achieved (Figs. 11,12). During the late Pleistocene the same processes caused giant translational slides with up to >100 m high slide scars (Figs. 1,8). The triggering of these massive slides at this very time and setting (passive margin, low gradient and sedimentation rates) is still enigmatic.

---

1 Bundesanstalt für Geowissenschaften und Rohstoffe, Stilleweg 2, 3000 Hannover 51, FRG

1. Introduction

The West Saharan continental slope and rise is the offshore continuation of the Aaiun-Tarfaya coastal basin (Fig.1) and is part of the mature Atlantic-type continental margin off NW Africa. This margin is built up by >10 km thick Mesozoic and Cenozoic sediments, deposited after the early Jurassic continental breakup between the African and North American plates. By now this area has become one of the best documented passive margins: this is due to a dense net of reflection and refraction seismic profiles from a number of BGR expeditions with R/V "Meteor" and "Valdivia" (Hinz et al., 1974; Seibold and Hinz, 1974, 1976). These data are supplemented by geophysical data from Woods Hole Oceanographic Institution (Uchupi et al., 1976), Lamont Doherty Geological Observatory, and commercial surveys by the petroleum industry (e.g. Lehner and de Ruiter, 1977; Mitchum and Vail, 1977). Two DSDP sites, one at the intermediate slope (Site 369; Lancelot, Seibold et al., 1977), and one at the uppermost rise off Cape Bojador (Site 397; von Rad, Ryan et al., 1979) brought a major breakthrough for our understanding of the geological evolution of this continental margin. Several presite surveys helped to prepare these drilling sites by correlating seismic sequences from coastal basins via the shelf to the slope and rise (Hinz, 1979; Wissmann, 1979). Cretaceous and Tertiary sediments dredged and cored in erosional "windows" cut by deeply incised slope canyons (von

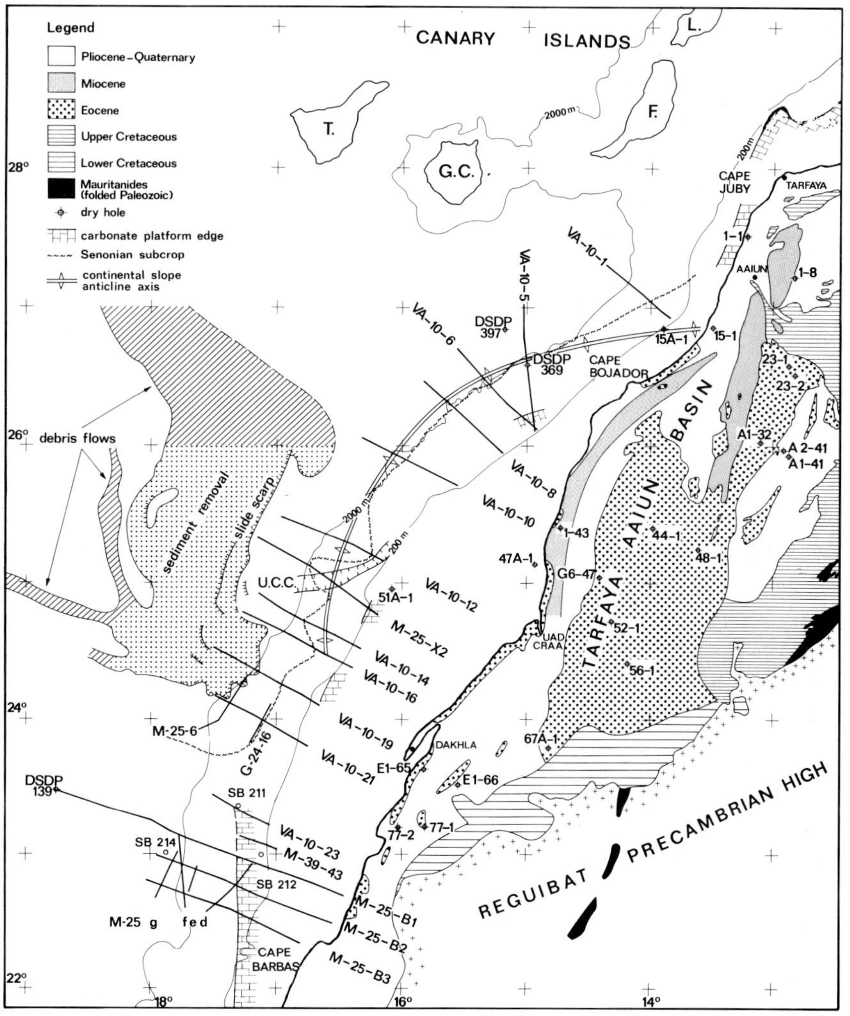

Fig.1
Sketch map of the West Saharan continental margin with surface geology of Tarfaya-Aaiun Basin, location of oil exploration wells (AUXINI, 1969), DSDP sites, and BGR seismic profiles from "Meteor" (M25, M39) and "Valdivia" (VA-10) cruises (see also Ranke et al., this Vol.). Carbonate platform edge north of 28°N after Hinz et al. (this Vol.); areal extent of huge Quaternary slides and debris flows after Embley (1976) with slide scarps modified by BGR data.
SB "Atlantis II" sonobuoys (Hoskins et al., 1974) UCC = Uad Craa Channel (up to 1000 m of Pliocene erosion); T Tenerife, GC Gran Canaria; F Fuerteventura; L Lanzerote

Rad et al., 1979) supplemented the stratigraphic and paleoenvironmental information of the two DSDP Sites and of several commercial on- and offshore wells from the adjacent Aaiun- Tarfaya Basin (Fig.3; von Rad and Einsele, 1980).

The general geodynamic, sedimentary and volcanic evolution of the West Saharan margin between Cape Bojador and Uad Craa has been summarized by von Rad and Arthur (1979), von Rad and Einsele (1980) and Ranke et al. (this Vol.). Since the West Saharan continental slope and upper rise

Fig.3
Generalized lithostratigraphy of DSDP Sites 139, 370/416, 397 and 369; and commercial shelf and Aaiun (-Tarfaya) Basin wells Spansah 51A-1, Daora 1-1, Tidewater 56-1 (after AUXINI, 1969), Alisio 15A-1 (after Connan, 1974; Querol, 1970). For location see Fig.1. Encircled numbers: sedimentation rates (m/m.y.). Note different vertical scales for DSDP and commercial wells

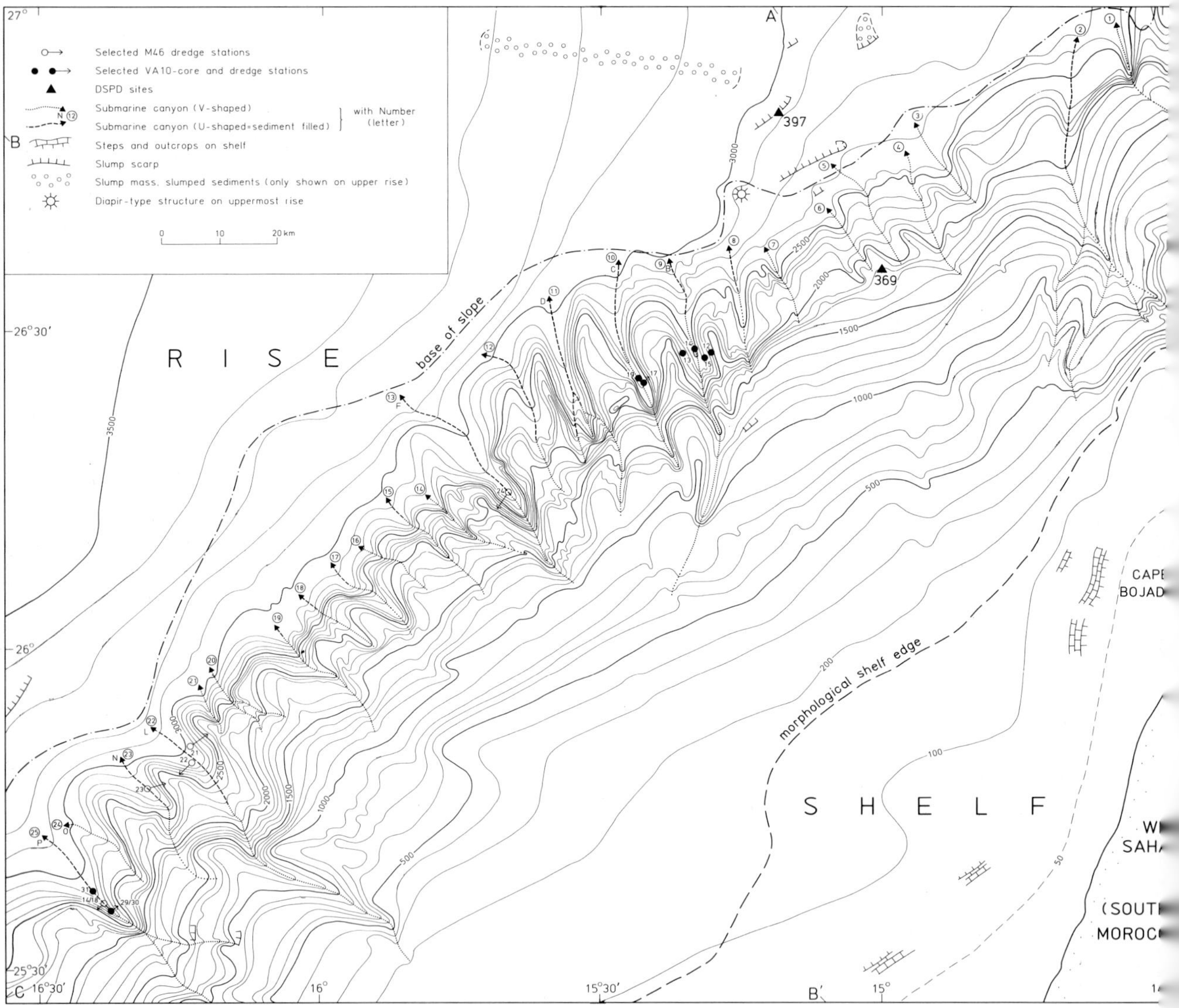

Fig.2
Bathymetry and morphology of the northern West Saharan continental margin (100 m contours). Compiled by the authors from all bathymetric and seismic profiles available at the BGR

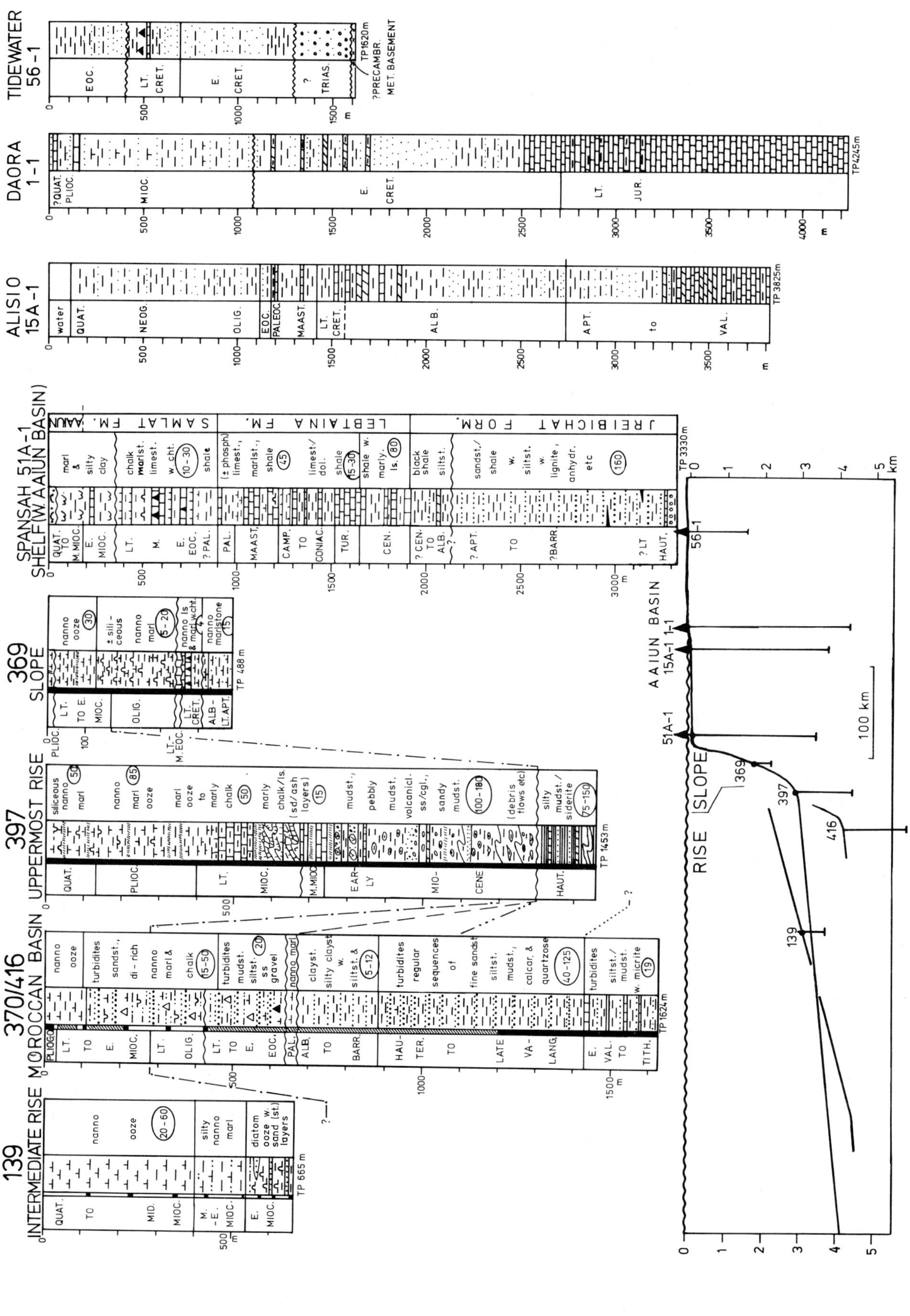
TIDEWATER 56-1
DAORA 1-1
ALISIO 15A-1
SPANSAH 51A-1 SHELF (W. AAIUN BASIN)
369 SLOPE
397 UPPERMOST RISE
370/416 MOROCCAN BASIN
139 INTERMEDIATE RISE
SAMLAT FM.
LEBTAINA FM.
JREIBICHAT FORM.
AAIUN BASIN
SLOPE
RISE
100 km
TP 1620 m
TP 4245 m
TP 3825 m
TP 3330 m
TP 488 m
TP 1453 m
TP 1624 m
TP 665 m
?PRECAMBR. MET. BASEMENT

were shaped during the past 135 m.y. from previous paleoslopes by a number of discrete constructional and destructional phases (cf. Seibold and Hinz, 1974; Schlee et al., 1979), the main objective for this paper is to decipher the temporal and spatial development of these erosional and depositional events from the morphology, seismic stratigraphy, and the geological record of the DSDP Sites, deep-sea dredges and commercial shelf wells.

2. Morphology

To find dredge locations in the numerous canyons, the bathymetry of the continental slope off the northern West Sahara has been mapped in more detail (Fig.2). The margin off Cape Bojador is characterized by a comparatively narrow (25 km) shelf which widens southward to 55 km. Beyond the shelf-break at 110 m (90 - 150 m) a comparatively gentle continental slope extends down to a water depth of 2700 - 3300 m, where it grades into the near-horizontal uppermost rise (see also Ranke et al., this volume). In Fig.2 the rise-slope boundary is defined by the proximal onlap of nearsurface rise sediments onto the continental slope. The uppermost continental rise is characterized by small (5 - 20 m), marked slump scars and sediment-filled deep-sea channels extending from some submarine canyons, but lacks any deep-sea fans.

Off Cape Bojador, the continental slope is 60 km (north) to 100 km (south) wide and shows a convex profile with a moderate inclination (2°) in its upper part and an over-steepened (4°) lower slope underlain by a "slope anticline" (Fig.7). This is remarkable, since most continental slopes of the world (including our southern profiles; see Figs. 4,6,12) have a concave profile, with inclinations decreasing downslope (Shepard, 1973). The lower slope is conspicuously dissected by numerous, deeply incised "slope canyons". The intercanyon areas are characterized by a hummocky relief produced by numerous small-scale slumps and gullies. These small-scale features which according to seismic evidence comprise only the uppermost tens of meters of Plio-Pleistocene sediments, are likewise concentrated in the oversteepened part below a water depth of 1000 m.

About 30% of the 250 km long front of the lower continental slope off Cape Bojador, shown in Fig.2, is eroded by 25 submarine canyons. The marked dissection especially of the lower slope is a very conspicuous, unique feature of this part of the West Saharan continental margin. The major canyons are 20 - 60 km long and have incision depths of 200 - 900 m; according to seismic profiles and dredge results, they are genuine submarine canyons with V-profiles and steep rocky walls (Shepard and Dill, 1966) Beyond a water depth of 2000 - 2900 m along the axis, these canyons are frequently U-shaped with a fill of several hundred meters of younger sediments. The smaller (100-200 m) incisions and gullies can be classified as submarine valleys. These incisions have been briefly described by Rust and Wienecke (1973) and Seibold and Hinz (1974), and will be covered in more detail by Wissmann and von Rad (in press). Most submarine canyons originate at the lower and intermediate slope. Only two major canyons (Canyons 9 and 13) have their heads at about 1000 m water depth and three (Canyons 2,22-25) at about 300 m water depth. Only Canyon 2 reaches the shelf edge and might have received shelf and nearshore sediment by longshore drift during lowered sea level stages, as usual for most normal submarine canyons around the world. None of these submarine canyons can be traced to adjacent land canyons or river mouths, and most canyons did not receive shelf-derived sediment material.

3. Continental Margin Structure

The following notes document the late Mesozoic and Cenozoic structure of the offshore Aaiun-Tarfaya Basin, as revealed by BGR single-trace seismic profiles from cruises METEOR-25 and 39, and VALDIVIA-10 (Fig. 1; Seibold, 1972; Seibold and Hinz, 1976) and they supplement data presented in von Rad et al.

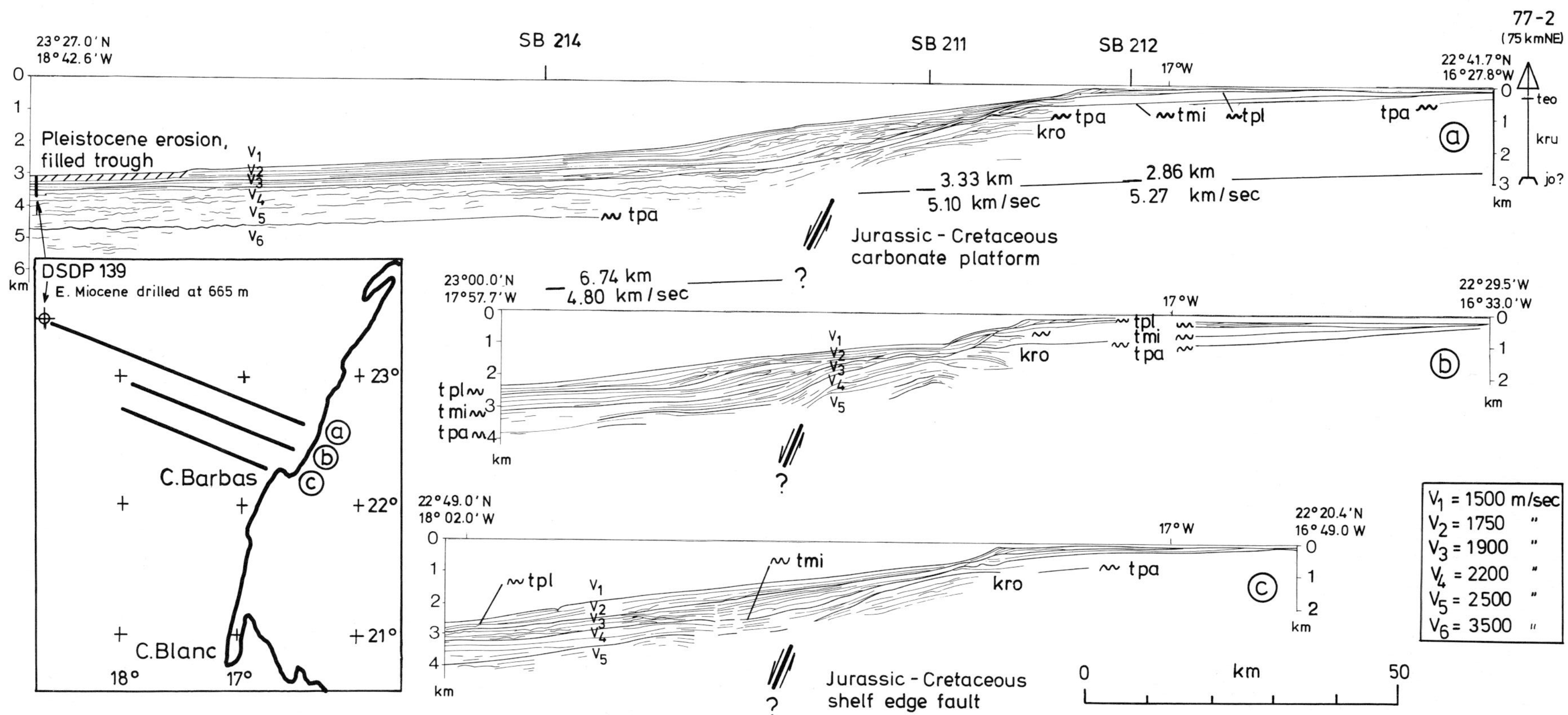

Fig.4

Depth-converted line drawings of seismic profiles across the continental margin near Cape Barbas, aligned along 17°W. Upper profile is correlated with DSDP Site 139 and projected onshore well 77-2; depths to high velocity refractor in sonobuoys SB 211, 212, 214 from Hoskins et al. (1974). V_1 to V_6 interval velocities for seismic sequences; kro upper Cretaceous rocks; tpa, tmi, tpl Paleocene, Miocene, Pliocene erosion surfaces

(1979), Wissmann (1979), Hinz (1979), and Wissmann and von Rad (in press), which cover mainly the northern part of the West Saharan continental slope.

We interpret the seismic profiles with the aid of DSDP and commercial well data (AUXINI, 1969; Connan, 1974; AAPG Bull., annual world oil developments; Ranke et al., this Vol.), the geological information compiled in Fig. 1 (Choubert et al., 1968; Querol, 1966; Ratschiller, 1970), and rock samples (von Rad et al., 1979). We will now discuss the regional structure of the margin from south to north.

3.1 Southern West-Saharan Margin (off Reguibat High)

Most profiles do not cross the shelf, where multiple reflections severely reduce the seismic penetration. Only three profiles north of Cape Barbas run closely inshore, because they were intended to find a possible offshore continuation of the Reguibat Precambrian High (Fig.1).

The oldest formations exposed at the shoreline in the southern Aaiun Basin are Eocene regressive, coarse sandstones which truncate upper Cretaceous rocks to rest directly on the lower Cretaceous (Querol, 1966; Fig.1). These sandstones grade laterally to the northwest into siliceous and calcareous shales, as encountered in the Spansah 51a-1 offshore well (Fig.3). Thus the geological evidence points to an extended phase of erosion from Paleocene to middle Eocene times. Since only thin remnants of Paleocene sediments have been found in the Spansah well (CONOCO, 1969) and in the north. part of the Aaiun Basin (Querol, 1966), whereas Eocene rocks have been frequently recovered by drilling (Fig.3) and dredging (von Rad et al., 1979), the climax of erosion seems to have been in the Paleocene.

In METEOR profiles M-25-B1, B2 and B3 (Fig.1) several strong westward-dipping reflectors can be differentiated from the horizontal shelf multiples. Figure 4 shows tracings of these profiles, which were converted to depths with assumed interval velocities equal or slightly higher than those computed from normal move-out in industrial profiles of the area. The deepest reflector under the shelf (tpa) seems to have been formed during the Paleogene erosion phase, because its eastward continuation projects to Eocene shore formations which were drilled in well 77-2 (AUXINI, 1969). We interpret the westward dipping reflectors overlying the Paleogene unconformity as Miocene (tmi) and Pliocene (tpl) unconformities and trace them to their outcrops on the shelf, where the truncated sequences have been sampled (Summerhayes et al., 1971, 1976).

The Paleogene unconformity generally marks the top of an upper Cretaceous paleo-slope westward of the present shelf break. No rocks older than Cretaceous could be seen on the seismic records. However, a correlation of late Jurassic continental rocks at 2703 m in well 77-2 (AUXINI, 1969) with the depth of sonobuoy high-velocity refractors (> 5 km/s) suggests the existence of a Jurassic-Early Cretaceous carbonate platform under the shelf (Fig.4). Our interpretation agrees with Jurassic carbonate facies velocities from seismic data off Morocco (K.Hinz, personal communication), but not with Hoskins et al. (1974) and Uchupi et al. (1976, p. 825, line 122), who associated these refractions with basement.

We correlate the western boundary of the carbonate platform with the increased gradient or paleo-slope of the upper Cretaceous sediments below the Paleogene unconformity, under the assumption that the rocks with a velocity of 4.8 km/s under sonobuoy station SB 214 represent Jurassic deep-water limestones. The depth difference between the platform and deep-water carbonates is partly attributed to a shelf edge fault, which was active in Jurassic to Cretaceous times, similar to that along the Moroccan continental margin (Hinz et al., this Vol.).

A comparison of the thicknesses of the Tertiary sequences under the shelf and slope (Fig.4) shows that

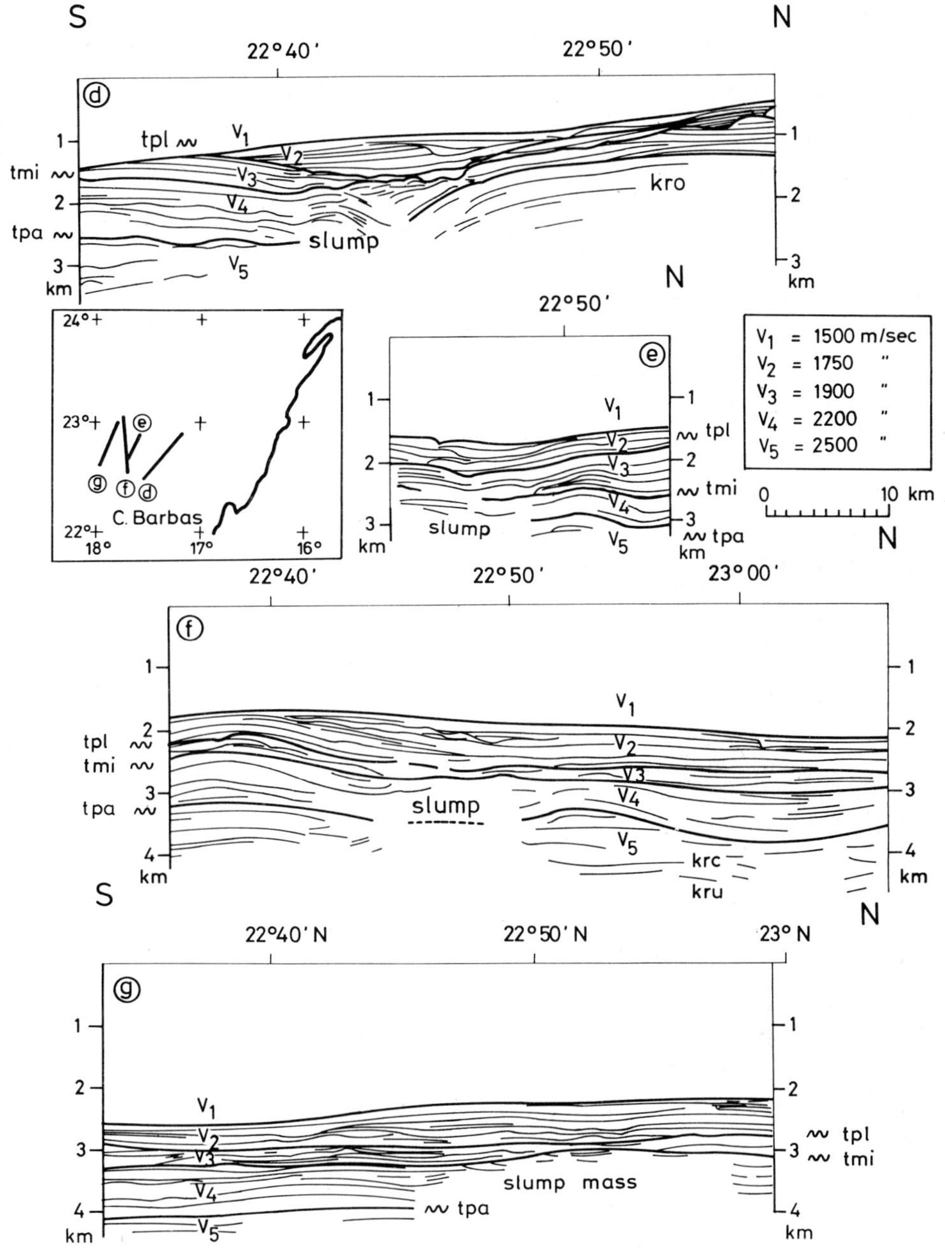

Fig.5

Depth-converted line drawings of seismic slope-parallel (N - S) profiles off Cape Barbas. For abbreviations see Fig. 4; krc Cenomanian; kru Lower Cretaceous rocks. Note Pliocene channel in the shallowest profile. Slump marks a transparent zone with some diffractions, where well-bedded pre-middle Miocene reflectors are discontinued

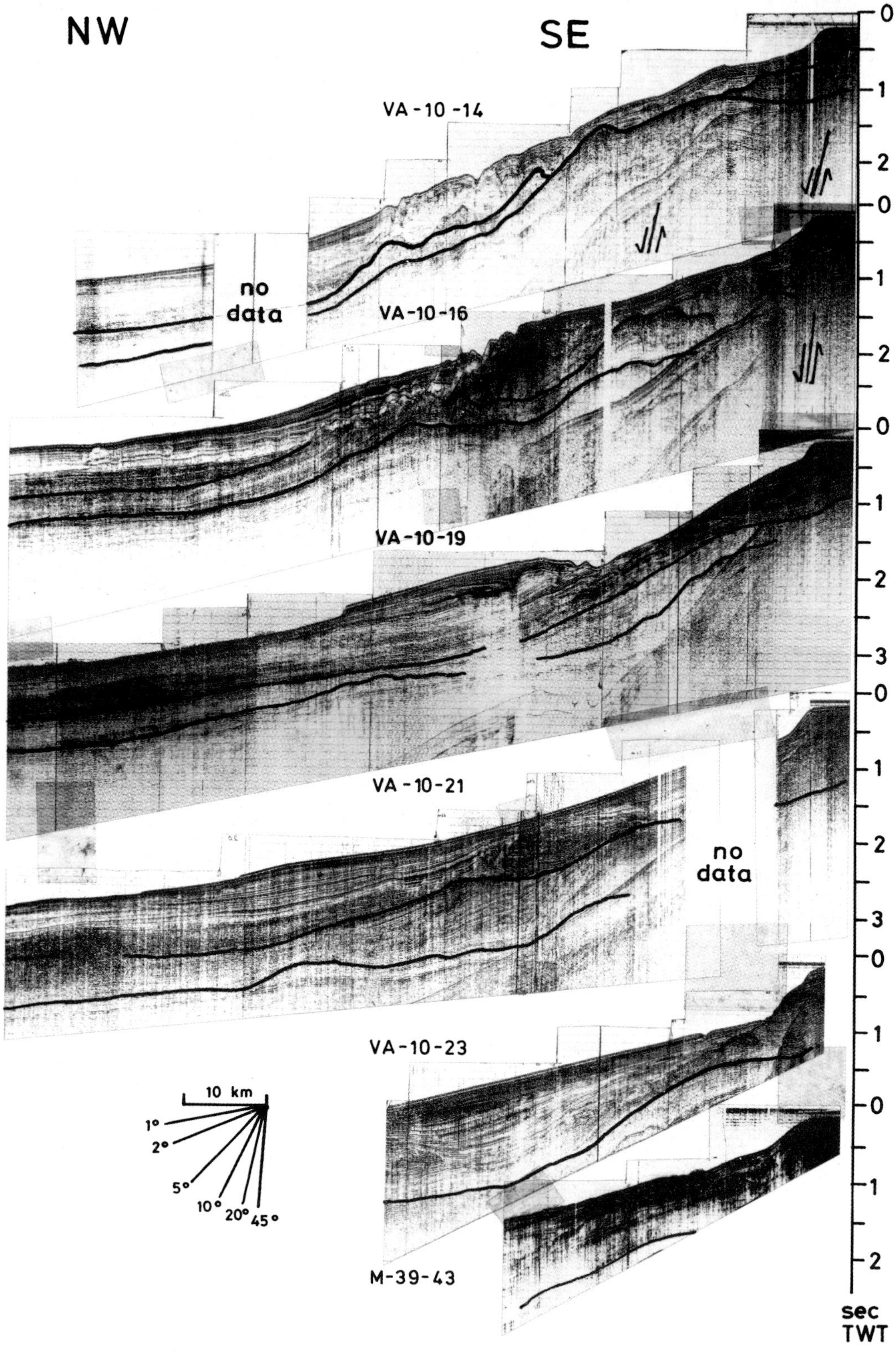

Fig. 6 (legend see p. 116)

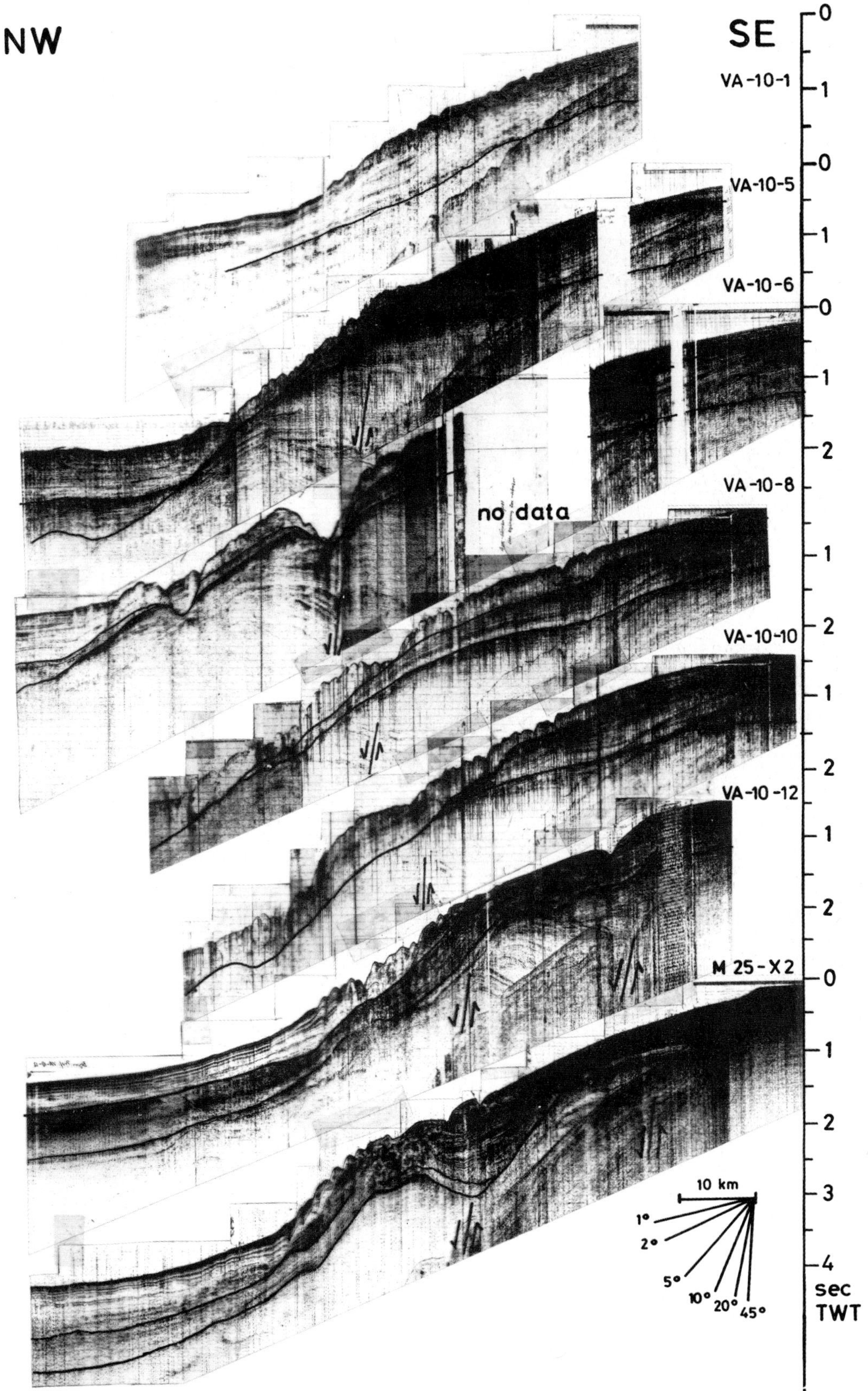

Fig. 7 (legend see p. 116)

the Tertiary depocenter was seaward of the Cretaceous slope. There the younger sequences lap with a general westward progradation onto the unconformities bounding the older formations. The Miocene and Pliocene unconformities erosionally truncate older rocks, locally cutting down to the Paleocene unconformity and forming channels under the upper slope. The turbidity currents depositing the early Miocene coarse sands of Site 139 (Sarnthein and Diester-Haass, 1977) might have been funnelled through these channels. Several Pleistocene unconformities shown in the high-resolution single-channel profiles were not specifically marked, with the exception of a broad Pleistocene erosional channel around Site 139, which was not recognized in the spot-cored drilling record.

An alignment of coast-parallel profiles along $22^{\circ}40$'N (Fig.5) clearly shows Pleistocene erosion surfaces and slide scarps. The easternmost profile (Fig.5d) demonstrates the rapid thickness variation of the Tertiary units and pronounced channel formation during Miocene and Pliocene times. We ascribe a SE - NW trending zone of chaotic sediment layering to Paleogene slump masses, but a fault or fracture zone (Rona, 1970) origin could only be excluded with deeper-penetration seismic records.

3.2 Central West-Saharan Margin (23°N - 25°N)

The alignment of slope-normal profiles in Figs. 6 and 7 allows a synoptic overview of the Tertiary evolution of slope sedimentation and erosion.

Two conspicuous unconformities (D1 and D2) show up in single-trace seismic profiles of the West Saharan margin (Hinz et al., 1974; Uchupi et al., 1976). The deeper one (D1=tpa in Figs.4,5) marks the base of the Paleogene section. In many places on the slope the Paleogene section, including this unconformity, has been eroded by younger erosional events of regional importance.

Hinz et al. (1974) interpreted the younger discontinuity D2 as an Oligocene to early Miocene hiatus. In more recent interpretations D2 only forms a minor sub-unconformity within the larger Quaternary to middle Eocene upper slope/shelf sequence CB 1a of Hinz (1979). Wissmann (1979) points out that the "slope-D2" is not merely a lithological change within the early Miocene from nannofossil to more siliceous marls (Lancelot, Seibold et al., 1978), but also a surface of erosional truncation, at least locally under the continental

Fig. 6 (see p. 114)
Original SE - NW seismic profiles across the West Saharan margin (southern part) aligned along shelfbreak. Depending on seismic penetration either one (D2 Middle Miocene) or two (an upper D2, and a lower D1 Paleocene/Eocene) unconformities are outlined. Horizontal scale is approximate, as ship speed varied slightly. On profile VA-10-19 (near 1.8 s) note contour current channeling and possible bottom simulating clathrate horizon above west-dipping Neogene horizon horizons, but lack of faulting as

Fig. 7 (see p. 115)
Original SE - NW seismic profiles across the West Saharan margin (northern part). Merging of unconformities under the slope and uppermost rise north of VA-10-16 (Fig. 6) and a major late Oligocene erosional event drilled in Site 397 presents clear age assignments for upper rise - lower slope reflectors. Note interpreted growth faults with rollover structures, probably fault controlled Uad Craa Channel (M25-X2) filled with >1000 m of post-Miocene sediments, hummocky slump surface prevalent on lower continental slope, and northward flattening of shelf break

slope. Below the lower slope and upper rise off Cape Bojador a major mid-Tertiary (?Oligocene) unconformity (see Fig.11A) underlies the Neogene sequence CB 1b of Hinz (1979). In general, the unconformities D1 and D2 subdivide the continental margin sediments into a Paleogene (D1 to D2) and Neogene (above D2) section.

The southern profiles of Fig. 6 still show a linear to concave seafloor topography, which is smooth and undisturbed over long distances. The shelf edge is clearly defined by an upper slope escarpment. The alternation of constructional and destructional processes along the continental slope results in an extremely variable physiography; In profile VA-10-23 we observe this escarpment truncating horizontal Miocene strata (von Rad et al., 1979) and a broad trough filled with landward onlapping and seaward downlapping late Miocene sediments; only 16 km further south (in profile M 39-43) the escarpment is buried under prograded sediments.

Watkins and Hoppe (1979) interpreted the broad positive ocean floor feature in the middle of profile VA-10-19 as a prominent slump. Older sediments are indeed laid bare upslope of this feature and middle Miocene diatom mud was recovered in a piston core from 965 m water depth (von Rad et al., 1979). However, we do not believe that this sedimentary unit has slumped to its present position, as we neither see the tensional faults at the head of the postulated slump (Fig.4 of Watkins and Hoppe, 1979), nor any compressional elevations at its toe. We see, however, post-middle-Miocene to Recent channeling (possibly by contour currents) along the eastern side of the sediment block which does not extend to the neighboring profiles. On the other hand, we agree with these authors that mid-Tertiary olistostromes do exist along the West-Saharan margin. The juxtaposition of horizontally layered strata with westward offlapping younger sediments at 2 s reflection time on profiles VA-10-21 and 16 proves that such mass sediment movements took place in Miocene time. These interpreted olistostromes are represented by unlayered, seismically transparent zones under the upper rise, which are up to 300 m thick and possess an undulating surface (VA-10-21 : western termination of D2).

Todays downslope motion of Neogene sediments created various positive and negative mass wasting structures: slumps and mass flows, slide scars, and channels.

We used the 100 m high slide scar on the seaward end of the sediment block of profile VA-10-19 and three slide scars visible in the lower portion of profile VA-10-21 to modify the area of sediment removal (Fig.1), outlined by Embley (1976). The concentration of Pleistocene slide scars between profiles VA-10-21 and VA-10-16 (Fig.6) must be due to more local factors, such as regional tectonics and increased sediment input.

Hummocky sea bottom topography and seismically transparent heaps on the sea-floor on the western end of profile VA-10-19 indicate the area of deposition of slumps or debris flows that originated at such slide scars.

Figure 8 shows a seismic section along the strike of the slope toward DSDP Site 139 with numerous less pronounced unconformities, that are not outlined in the profiles of Fig.6 and 7. Early Miocene slumps, slide scars, and channels are clearly visible. The slide scars of today's ocean bottom (see also Embley and Jacobi, 1977) are more than 20° steep and have been sampled by piston cores, which dated the unconformable glide plane as "mid-Pleistocene" (see 4.7).

Figure 9 presents a multichannel line which runs upslope and parallel to the profile of Fig. 8 and shows Early Cretaceous to Neogene unconformities. Broad and rather shallow precursor valleys had already been excavated in mid-Cretaceous times at those locations, where deep canyons were incised during the Paleogene.

The formation of large channels (VA-10-23, Fig.6; and the Uad Craa Channel in M-25-X2, Fig.7) traver-

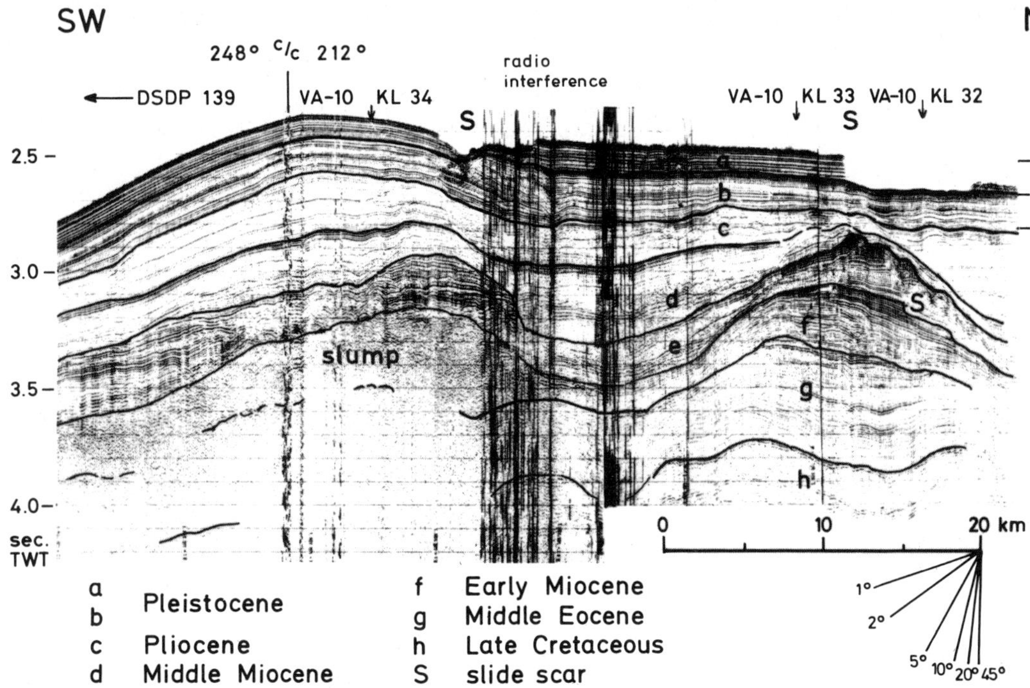

Fig.8
High resolution slope-parallel air gun profile M-25-6 with location of piston cores KL 32 - 34, Pleistocene and Miocene slide scars (S), and Miocene and Late Cretaceous channeling on lower slope. The early Miocene slump shown has a seismic structure comparable to the slump on Fig. 5

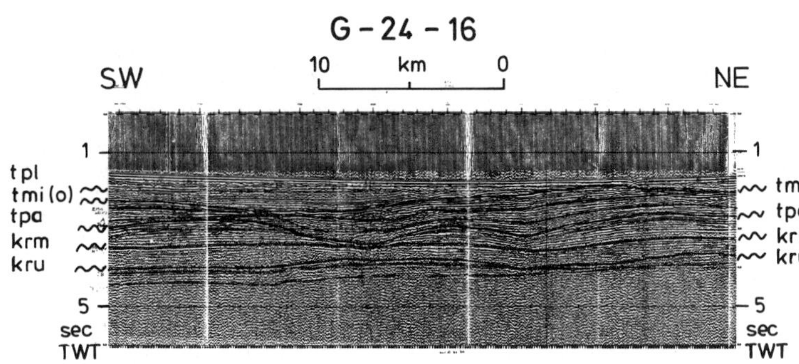

Fig.9
24-channel Gulf profile 24-16 (kindly provided by J.S. Watkins) located east of profile M-25-6. The profile shows separate Cretaceous and Tertiary erosional surfaces which merge in other locations. Deep and broad channels on the upper slope were first formed in Paleocene. Abbreviations as in Figs. 4 and 5, krm mid-Cretaceous, tmi(o) upper Miocene

sing the slope obliquely in Miocene times still warrants explanation. Fault control of the Cretaceous substratum, gigantic slumps, erosion by tidal currents, or, most likely, offshore extension of an onshore paleo-drainage system (Uad Craa) via canyons are proposed as possible causes.

3.3 Northern West-Saharan Margin (25° - 28°N)

As pointed out by Hinz et al. (1974) and Uchupi et al. (1976), the unique feature of the West Saharan continental slope are the "folded" strata under the Paleogene unconformity D1. The axis of the anticlinal ridge has been mapped in Fig.1: its northern termination and eastward deflection has been demonstrated in METEOR profile M 39-14 by Wissmann (1979), and its southern end is here interpreted to be at 24°N, 100 km further north than in Uchupi et al. (1976). GSI-Line A1L (von Rad, Ryan et al., 1979, p.76) proved that there exists no basement ridge under the sediment anticline, as postulated by Lancelot and Seibold (1978, p.346). Uchupi et al. (1976, p. 846) believed that the slope anticline was formed by "Pyreneic (late Eocene) epeirogenesis as a result of long instability along the boundary between the stable African shield and the Atlantic Ocean". However, the limit of Alpine orogenic activity seems to lie further north at the South Atlas Flexure (Nairn et al., 1980; Querol, 1966, his Fig.5). Furthermore our data and Line A1L point to systems of Cretaceous growth faults with tremendous rollover structures rather than Eocene folding and generally support the differential subsidence model of Lancelot and Seibold (1978). We observe a seaward migrating series of growth faults in the lower to middle Cretaceous sediments of the West Saharan slope. This phenomenom is related to spillover of deltaic sediments which accumulate at higher rates than can be accomodated by the subsiding block seaward of the growth fault (Bruce, 1973).

It is not clear whether salt pillows are involved in the deformation of the Cretaceous sediments. Although Triassic/Jurassic evaporites are shown offshore between Dakhla and Cape Bojador by Uchupi et al. (1976, Fig. 35) and Guieu (1976) and local subcircular uplifts are mentioned by King in Querol (1966), no salt diapirs were observed in any of the profiles available to us. Locally, diapiric upwarps are present under the Paleogene unconformity D1 at the foot of the continental slope as in profiles M 25-X2, VA-10-12 (Fig.7) and in KANE-9 profile (20.8.1968, 24°°; Lowrie and Escowitz, 1969). Tentatively, we interpret these upwarps as plugs of an Early Cretaceous prodeltaic shale unit (Einsele and von Rad, 1979), mobilized by the motion of the growth faults in the overlying Cretaceous marls (Site 369 and 51a-1), similar to the Akata Formation of the Niger Delta (Egbogah and Lambert-Aikhionbare, 1980). The asymmetrical piercement structure at 1 s two-way travel time on profile VA-10-16 (Figs.6, 12) might be an example of ?Miocene plastic deformation of shale by differential loading (Coleman and Garrison, 1977).

Uchupi et al. (1976, their Fig. 7) interpret surface undulations on the lower slope (e.g. M 25-X2) as slumps and think that the underlying downward- and seaward-curved interfaces between groups of subbottom reflectors represent the soles of these slumps. Since these slump planes do not extend below the Miocene unconformity D2, the stepped sea-floor morphology could have been due to rotational peripheral slumps, similar to those described in a smaller scale from Mississippi mouth bar slopes (Coleman and Garrison, 1977), although the West Saharan Pleistocene rates of carbonate-dominated hemipelagic sedimentation are orders of magnitude smaller (Seibold and Hinz, 1974). Profiles VA-10-16 and 12 give evidence that the slumping leads to a general thinning of the sedimentary column (Almagor and Wiseman, 1977).

In Fig.7 prograding Cenozoic sediments fill channels (e.g., VA-10-6) and cover Cretaceous fault blocks (VA-10-12, M 25-X2) to form a convex sea-floor, the relief of which is currently being reduced by the rotational slumps described. Since the post-early Miocene slope sediments are thinner in the northern profiles, the convex shape of the ocean bottom seems to be mainly due to the paleo-topography formed by the tilted Cretaceous fault blocks, and not due to increased progradation by Tertiary terrigenous sediments.

4. Development of the West Saharan Margin

Figure 10 shows the Mesozoic-Cenozoic evolution of the West Saharan continental margin in the context of the history of global sea level changes, Atlantic paleooceanography, paleoclimate, and Canary

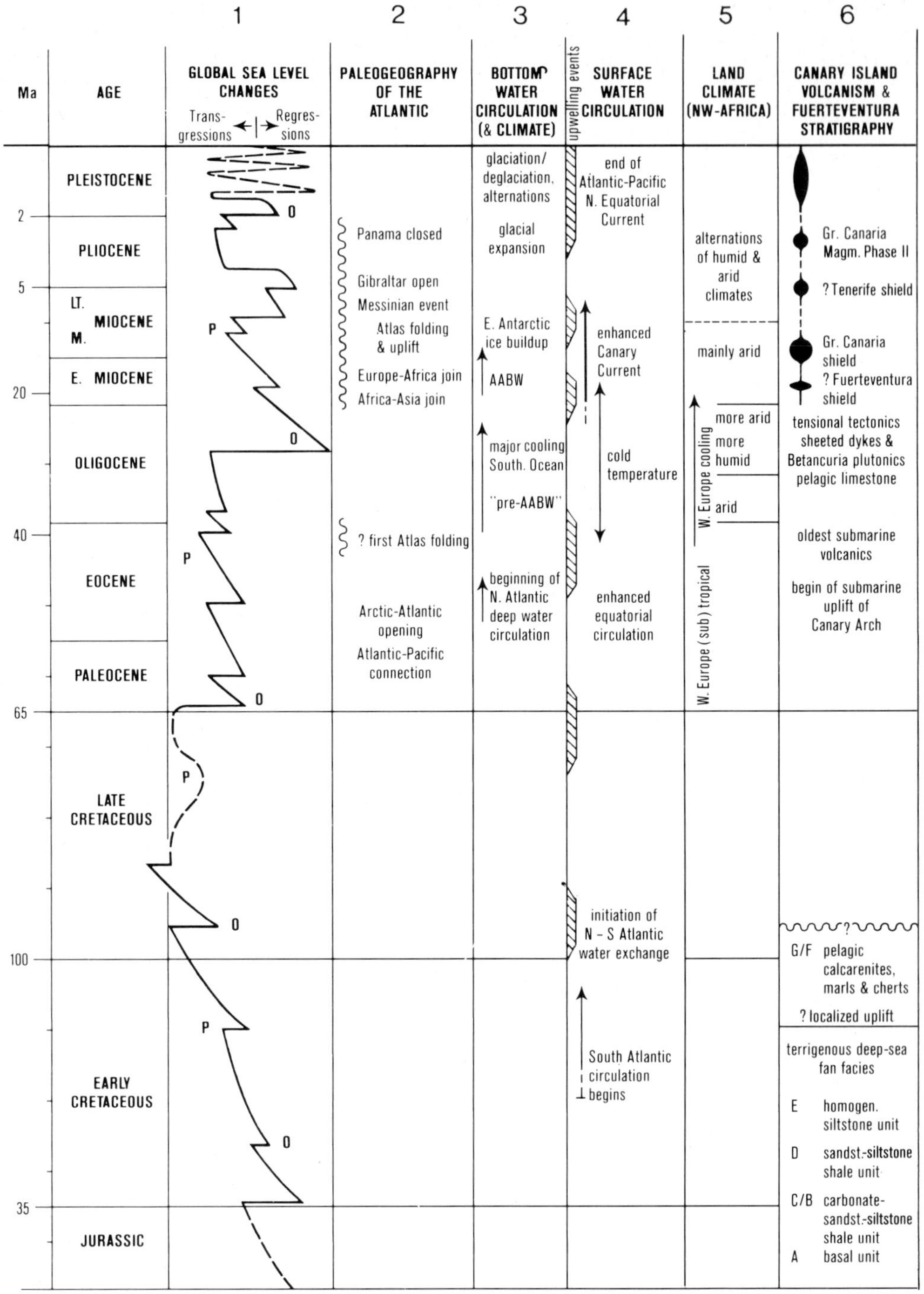

Fig.10
Schematic overview of the development of the Cape Bojador continental margin (rise, slope, shelf) during the past 150 m.y. Global sea level fluctuations, paleogeography and paleooceanography of the Atlantic, land climate and history of Canary Island volcanism are shown for comparison. Subsidence and accumulation rates in m/m.y. after von Rad and Einsele (1980). P "polytaxic" (= thalassocratic); O "oligotaxic" (= epeirocratic) times after Fischer and

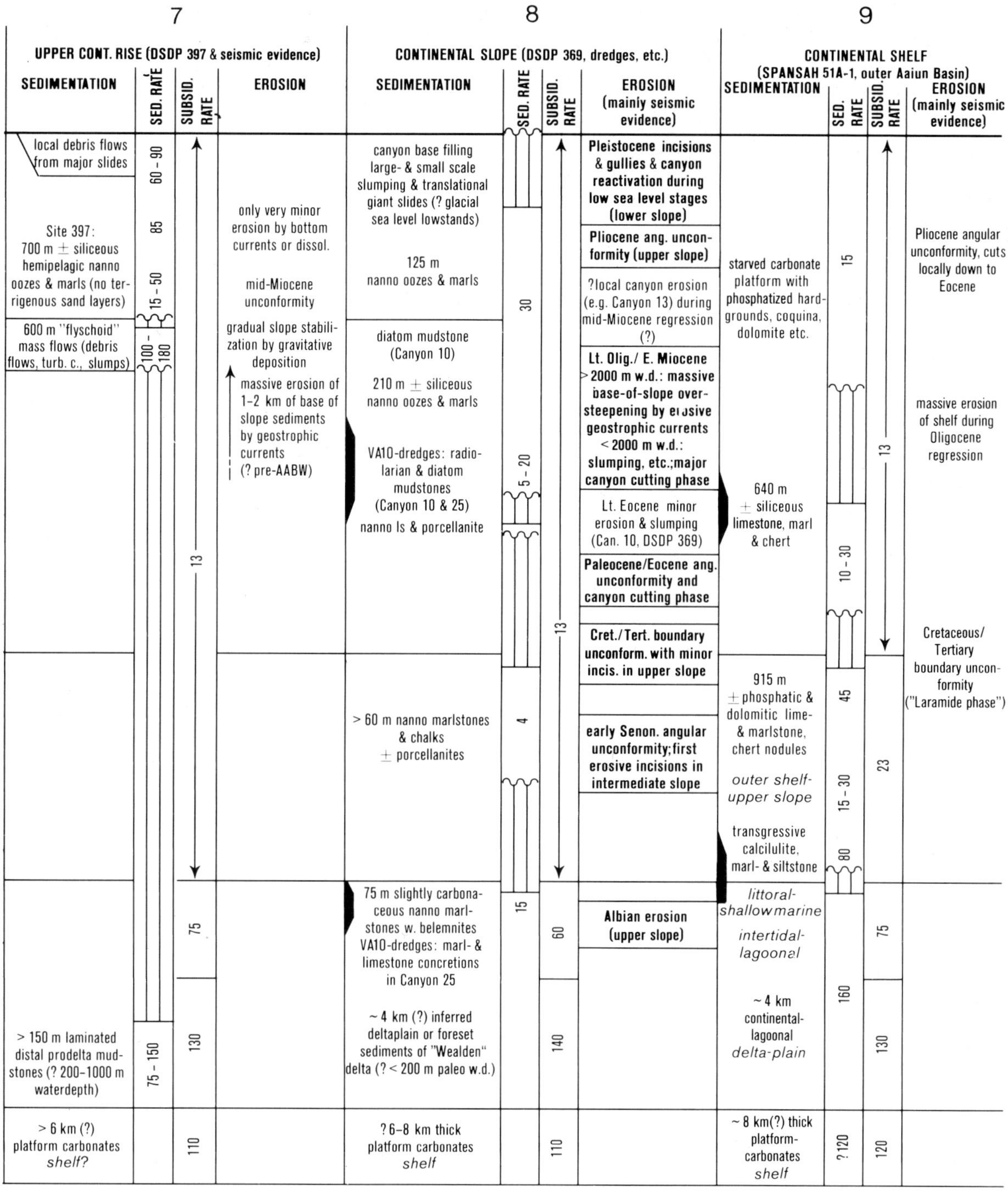

Arthur (1977). Sources for 1 Vail et al., 1977; Pitman, 1978; 2-4 Bergren and Hollister, 1974; Lancelot and Seibold, 1978; Arthur et al., 1979. 5 Sarnthein in Arthur et al., 1979; Diester-Haass, 1979. 6 Schmincke, 1976; McDougall and Schmincke, 1977; Schmincke and von Rad, 1979; Robertson and Stillman, 1979; Watkins and Hoppe, 1979. 7 von Rad, Ryan et al., 1979. 8 Lancelot, Seibold et al., 1978; von Rad et al., 1979. 9 CONOCO, 1969; von Rad and Einsele, 1980

Island volcanism (see also von Rad and Arthur, 1979; Arthur et al., 1979). The main topic of this figure, however, is a schematic overview of the temporal and spatial development of the depositional and erosional episodes at the West Saharan continental rise, slope, and shelf (see also Figs. 11 and 12). A strong correlation exists between sea-level stands, area of shelf seas, bottom water circulation, and global erosional cycles: during thalassocratic or -"polytaxic" times (e.g., 15,45,80 and 110 m.y. ago) sea level was high (transgression), the area of shelf seas large, the climate warm and equable, and the bottom water circulation sluggish with continuous pelagic sedimentation (Fischer and Arthur, 1977). During epeirocratic or "oligotaxic" times, especially in the glacial Pleistocene late Oligocene (30 m.y. ago) and Paleocene (60 m.y. ago), regressions caused hiatuses in previous shelf seas; at the same times the climate was colder with sharp latitudinal and vertical temperature gradients which led to an intensified bottom water circulation, periodic backcutting of the continental slope, as well as increased excavation of submarine canyons and mass gravity transport to the continental rise.

4.1 Jurassic to Early Cretaceous Construction of the Continental Margin

In comparison with the continental margins off Senegal, Mauritania (Wissmann, this Vol.) and off central Morocco (Hinz et al., this Vol.), we have very little information on the Jurassic history of the 180 m.y. old West Saharan margin. According to seismic evidence, we assume a thickness of about 4 - 5 km of Jurassic (to Early Cretaceous) carbonates (v=4 - 5 km/s) underlying a thick Early Cretaceous deltaic section under the present shelf (Fig. 10). However, the quality of the BGR seismic records in this area precludes any further resolution of the deeply buried Jurassic sequence or of distinct marginal reefs at the seaward platform edge, as observed further north. Following the worldwide Oxfordian "Atlantic" transgression (Lancelot and Winterer, 1980) carbonates were built up on a tectonically stable, but rapidly subsiding (80 - 120m/m.y.; von Rad and Einsele, 1980) shelf. The thick sequence of Late Jurassic shallow-water limestones and marls (Puerto Cansado Formation), penetrated in the western Aaiun Basin (e.g. Daora 1-1; Fig.3) grades eastward into a much thinner, more continental facies. Thus the shoreline of the Late Jurassic Atlantic lay about 100 - 200 km landward of Cape Bojador, whereas the outer edge of the carbonate platform extended a few tens of km seaward of the present shelf edge (Fig.1). The Oxfordian continental margin off central and southwestern Morocco had already a relief of 2-3 km between the top of the carbonate platform and the adjacent abyssal plains (Lancelot and Winterer, 1980; Hinz et al., this Vol.).

During a short-lived Valanginian transgression, the carbonate platforms off central Morocco were drowned (Schlager, 1980). Due to the following Hauterivian regression the carbonate supply to the deep-sea fans was gradually replaced by rapidly deposited, up to 4-km-thick "Wealden-type" deltaic sediments representing the Tan-Tan Formation in the Tarfaya Basin (Figs.3,10,11; cf. von Rad and Arthur, 1979). This huge delta system which moved the coastline about 100 km seaward comprised a complete sequence of continental alluvial plain deposits (Aaiun Basin), paralic to shallow-marine sediments (present coast to Spansah 51A-1), delta foreset (?Site 369) and prodelta sediments (Site 397; Einsele and von Rad, 1979), to distal turbidites on outer submarine fans and adjacent abyssal plains (Hauterivian flysch on Fuerteventura: Robertson and Bernoulli, this Vol.). The postulated diapiric piercement structures below the present lower slope (Fig. 12) are probably later upwarps of these prodeltaic sediments. The increased rates of clastic sedimentation which are observed all around the North Atlantic Basin (von Rad and Arthur, 1979, their Fig. 4) were probably due to increased uplift and denudation of the hinterland (here the Antiatlas

Mountains and Reguibat Massive) and a change to a more humid, temperate, moderately warm climate during a global regression (Chamley, 1979).

4.2 Mid-Cretaceous to Paleogene History of Margin Construction and Erosion

During the Aptian/Albian and Cenomanian eustatic transgressions the coastline migrated landward and the terrigenous clastic sediments were replaced by marine claystones (Aguidir Formation) grading seaward into pelagic carbonate oozes. Multichannel seismic records, correlated with the shelf well Spansah 51A-1 (Fig.3; CONOCO, 1969; von Rad and Einsele, 1980), indicate that already during Albian times parts of the upper continental slope were eroded (Fig.12). At the same time the Fuerteventura flysch underwent localized uplift, deep-marine erosion and redeposition (Robertson and Stillman, 1979). An expanded oxygen minimum layer and/or regional upwelling conditions produced "mid-Cretaceous" laminated, organic-rich sediments. In deeply incised slope canyons we dredged Aptian to late Albian, finely laminated, pelagic foraminiferal nanno marlstones with up to 2.5% organic carbon, and concretionary layers and lenses of fibrous calcite, fish debris, phosphorite, pyrite concentrations, pointing to an oxygen-depleted environment along the upper slope or outer shelf (Wissmann and von Rad, in press). Laminated, olive black nanno marlstones of late Aptian age were also described from the slope Site 369 (Lancelot, Seibold, et al., 1978).

After a major Turonian transgression the western Aaiun Basin had subsided to water depths of several hundred meters. At that time a continental slope was shaped, the location of which has not changed very much for the past 90 m.y. (see also Fig. 4; Wiedmann et al., 1978).

In the area of the upper and intermediate slope (near Canyon 25; Fig.2) we recognize a marked angular unconformity of early Senonian age which locally cuts into lower Cretaceous beds and might have been produced by the first meridional contour-following bottom currents (Figs. 10 and 12). During this time the first canyon-type incisions with thalweg depths of about 50 m and valley widths of 200 - 400 m formed along the intermediate slope.

The major Late Cretaceous transgressions of West Africa resulted in diminished terrigenous supply and the continental slope was hardly prograded during Late Cretaceous to Eocene times. Hemipelagic chalks and marl was deposited at low subsidence and sedimentation rates (Site 369, 51A-1; see Fig.10). The high contents of siliceous and calcareous plankton and phosphatic concretions suggest high plankton fertility. Upwelling and oxygen-minimum conditions occurred during late Paleogene times along the upper continental slope (von Rad et al., 1979).

A further angular unconformity (tpa in Figs. 4,5,9), which might correspond to the Laramide phase of the Atlas orogeny, separates the upper Cretaceous from the overlying Paleocene section. Also at this time (? late Paleocene) the Agadir Canyon (Lancelot and Winterer, 1980) and several West Saharan submarine canyons were eroded for the first time.

An important Eocene angular unconformity (D1) is present below the West-Saharan continental slope (Figs. 6,7, 10,12; Hinz et al., 1974; Wissmann, 1979). At the same time, when the Atlas Mountains experienced major folding and uplift (post-Lutetian or Pyrenean phase), the onshore Aaiun-Tarfaya Basin was also exposed to subaerial erosion (Querol, 1966). However, we do not see any evidence of alpine-type deformation in the seismic records of the West Saharan margin.

During late Eocene times Paleogene and Cretaceous strata were eroded and many slope canyons initiated, especially in the area between Canyons 22 and 25 (Figs. 2,7; von Rad et al., 1979, their Fig. 4A).

Thus a very large part of the Late Cretaceous to Paleogene record was removed by different erosional cycles, especially at the outer continental margin.

4.3 Mid-Tertiary Erosion

The rapid late Eocene-early Oligocene global cooling led to widespread Antarctic glaciation and the production of cold deep Antarctic Bottom Water (AABW). During late Oligocene times (22-25 m.y.) the Drake Passage opened between Antarctica and South America-Africa resulting in an unrestricted circum-Antarctic Current and a pre-AABW-current in the Southwest Atlantic (Kennett, 1977; Van Andel et al., 1978; McCoy and Zimmerman, 1977). Possibly a branch of this bottom current reached the NE Atlantic through gaps in the Mid-Atlantic Ridge (von Stackelberg et al., 1979) and touched the lower continental slope off NW Africa (von Rad and Arthur, 1979). Apparently this northward-flowing eastern geostrophic contour-following current was active for long time intervals and scoured the lowermost continental slope and uppermost rise. Since Site 369 has a full Oligocene record, this contour current did not affect the intermediate and upper slope above 2000 m water depth (Fig.11). Locally, the erosion reached down below the Paleogene and lower Senonian unconformities which crop out below the Neogene sediments along this escarpment (Figs.6,7,12) and produced a 100 m.y. hiatus at Site 397.

This conspicuous erosional event along the lower margin coincides with a major global late Oligocene regression which resulted in a sea-level drop beyond the present shelf edge and a widespread Oligocene hiatus under the continental shelf (well 51A-1; Figs.3,10). The over-steepening of the Oligocene lower continental slope by the scour of enhanced geostrophic currents and the exceptionally low sea level stand triggered periodic rotational slumps of metastable or critically overloaded slope sediments (Arthur and von Rad, 1979). Volcano-tectonic seismic shocks from the early Canary volcanism might have triggered the massive slope failure. Major slumps and slides transformed into huge debris flows and high-density turbidity currents which deepened already existing slope canyons or eroded new ones. In many cases these canyons cut deeply beyond the previous Eocene, Paleocene, or Senonian unconformities (see von Rad et al.,1979, their Fig. 4A).

4.4 Early Neogene Gravitational Sedimentation at the Uppermost Rise

Direct witnesses of the gravitative sedimentation are the Oligocene slumps in DSDP Site 369 (Lancelot, Seibold, et al., 1975) and late Eocene to Oligocene slump breccias derived from steep walls of submarine canyons (VA10-19KL; von Rad et al., 1979). An especially thick flyschoid sequence of early Miocene debris flows, turbidites and slumps cascaded down from the shelf and slope off Cape Bojador and accumulated on the erosional surface at the uppermost rise (Fig.11). Subaqueous mass flow deposits gradually decreased of over-steepened (8°) Oligocene gradient of the lowermost continental slope, until equilibrium conditions were achieved (Fig.11; Arthur and von Rad, 1979). The local presence of pebbles of Aptian to Paleogene age in the early Miocene debris flows of DSDP Site 397 proves that these strata were exposed along the over-steepened continental slope at the end of the Oligocene erosional period, although mainly early Miocene sediments were reworked by these gravitative processes.

A local, thin, middle Miocene debris flow, which, however, contained a very heterogenous mixture of Albian to Miocene pebbles directly overlying late Aptian nanno marls, was discovered in a lower slope core (VA 10-15 KL; von Rad et al., 1979). A large-scale early Miocene olistostrome of Miocene age can be seismically detected on the lower continental slope of Uad Craa (Figs.6,12). At the intermediate rise (Site 139) thick proximal "eolian sand turbidites" were deposited during an early Miocene (23 - 20 m.y.) low sea-level stage, when vigorous offland trade winds moved the Saharan sand dunes directly to the

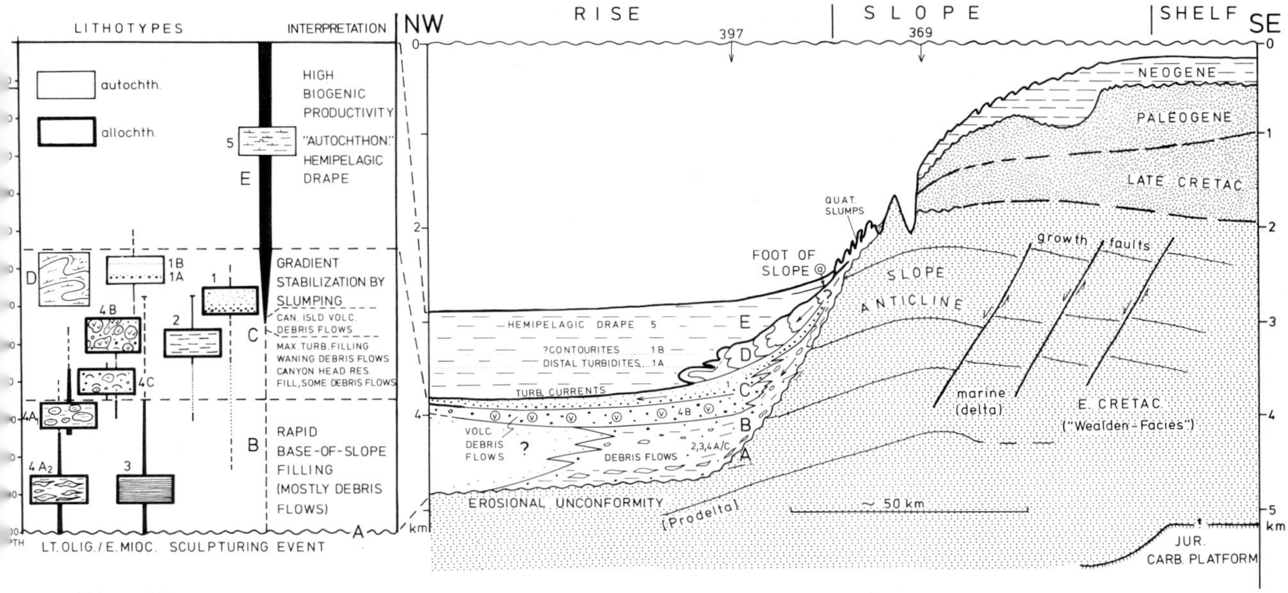

Fig.11
Schematic cross-section of continental margin off Cape Bojador. Left diagram shows spatial and temporal evolution of slope destruction and gravitative sedimentation on upper rise (Site 397, von Rad and Arthur, 1979). A,B,C,D,E designate major erosional and depositional cycles in both graphs, numbers (1A,1B,2,3,4A,4B,4C) indicate allochthonous lithofacies types (Arthur and von Rad, 1979); a proximal onlap

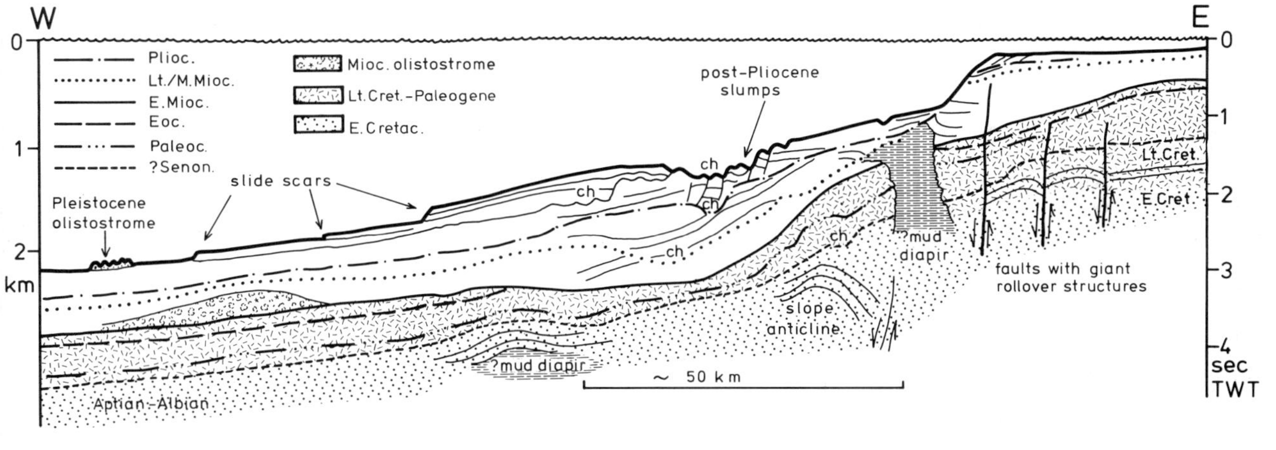

Fig.12
Schematic profile of the continental margin off Uad Craa (25°N). Note inferred mud diapirs in lower Cretaceous sediments; Miocene and Quaternary olistostromes; and channels (ch) eroding Recent, early Quaternary, Pliocene, Miocene, and Late Cretaceous strata (? contour-following currents). For more detailed seismic information see Figs.6 and 8

shelfbreak (Sarnthein and Diester-Haass, 1977), and turbidity currents funnelled the sediments through channels across the slope (see Chap. 3.1).

The accentuated vertical, lateral, and headward erosion of submarine canyons, especially along the lower continental slope, resulted in a distinct retreat of the base of the slope line (Seibold and Hinz, 1974). Roughly, the sediment volume removed by all submarine canyons (Fig.2) from the continental slope is about 900 km^3. This corresponds only to 10% of the total volume of sediment stripped from the base of the continental slope during the mid-Tertiary (and earlier) erosional events (7,500 - 15,000 km^3, von Rad and Arthur, 1979). Since the heads of most canyons are located on the intermediate slope below 1000 m (Fig.2), we invoke deep marine (and not shelf-related) erosional processes over lengthy time intervals for the formation of these canyons.

During the Neogene most terrigenous sediments bypassed the Mesozoic depocenter under the continental slope and accumulated at the uppermost rise. Therefore the low sedimentation rates on the slope lagged behind the already low subsidence rates (Fig.10), causing a gradual deepening of the outer margin.

4.5 Late Neogene Hemipelagic Sedimentation

The sculpturing of the lower continental slope by enhanced eastern contour currents ceased abruptly about 20 m.y. ago. It is possible that the margin-sweeping contour currents were blocked or diverted by the gradual uplift of the Cape Verde Rise (Lancelot and Seibold, 1978) and the submarine precursor of the Canary Arch (Schmincke and von Rad, 1979) during early to middle Miocene times. By early late Miocene all redeposition had ceased and a more or less continuous hemipelagic sedimentation followed: high plankton fertility under upwelling conditions and an abundant supply of fine terrigenous material resulted in high accumulation rates (Fig.-10,11).

Smaller canyons were still excavated during late Neogene times (Fig.10). An example is Canyon 13 on the intermediate slope. This erosional event was perhaps influenced by a mid-Miocene sea level lowstand around 11 m.y.B.P..The intercanyon areas, however, were hardly eroded during that time, and the submarine canyons along the lower slope were gradually filled.

A marked Pliocene angular unconformity can be recognized under the shelf and upper slope (Figs.4,5,8, 9,12,). Locally, this unconformity cuts down into Eocene strata. Off Uad Craa a huge channel (up to 1 km deep and up to 14 km wide), running obliquely across the upper continental slope (Fig.1) was excavated during Pliocene times and consequently completely filled by Plio-Pleistocene sediments (Figs.5,7). It is possible that this canyon system grades into a thin, morphologically evident suprafan on the upper rise.

4.6 Pleistocene Canyon Cutting and Giant Slides

Whereas most submarine canyons around the world developed during Pleistocene low sea-level stages (Shepard and Dill, 1966), the first excavation and main erosion of the West Saharan canyons date back to Paleogene or even Late Cretaceous times. However, we assume that the U-shaped canyons which during the late Neogene were partly buried by soft or semiconsolidated hemipelagic sediments, were re-excavated during glacial low sea level stages (Fig.10; Wissmann and von Rad, in press). Apparently the more pluvial climate which prevailed in the Western Sahara during and shortly after these glacial periods and during the beginning Holocene transgressions (13 000-9 000 y. ago; M. Sarnthein, personal communication) caused increased terrigenous input, and enhanced canyon erosion by turbidity currents and slides. Numerous small incisions, gullies, and broad valleys along the lower continental slope are all products of Pleistocene erosion. In the intercanyon areas small-scale shallow slumps were caused by the instability of locally oversteepened slopes.

The largest surficial translational slides so far discovered on the Atlantic sea floor were described by Embley and Jacobi (1977) from the lower West Saharan slope between 26 and 24°N (Fig.1). The large-scale mass gravity transport involved more than 600 km^3 of sediments which avalanched downslope in a few brief slump intervals. Downslope of these slide scars we find 10-20 m thick debris flows (olistostromes) which spill out for several hundreds of km onto the rise apron over an area of 18,000 km^2 (Embley, 1976).

The sheer magnitude of these mass wasting processes is conspicuous, especially if we realize that these slope failures were triggered on a tectonically quiet "passive" margin with slope gradients of only 1.5° which was not influenced by major earthquakes (no recorded epicenters between 1600 and 1978), salt diapirism, orogenetic movements, or high terrigenous sedimentation rates and ensuing porewater overpressure (as off the Mississippi delta). Our VALDIVIA profiles show several NNE - SSW trending slide scars between 1700 and 1970 m water depth with up to 20° steep steps of 100 - 150 m which we sampled by piston cores (Figs.8,12). The core at the foot of the lowermost slide scar (VA 10-32 KL) penetrated 232 cm of soft late Quaternary ooze, unconformably underlain by semiconsolidated, partly brecciated nanno marl ooze of "mid-Pleistocene" age (upper part of zone NN19, ca. 0.5 - 1.2 m.y. ago, P. Čepek, personal communication). This unconformity represents the glide plane of the 100 m thick slide mass which had moved downslope later than 500 000 years ago, most probably during one of the last Quaternary glacial low sea level stages. The slumps off Senegal-Mauritania were triggered during the late Glacial rapid sea-level rise between 12,000 and 10,000 years B.P. (Pflaumann et al., this Vol.).

The cause of this large-scale gravity sliding at this location is still enigmatic. Figures 6 and 12 show slope-parallel channels cut during Quaternary times into the continental slope southeast of the slide scars. Therefore we suggest that local undercutting of the slope and support removal by enhanced contour-following bottom currents and possibly methane-induced high porewater pressure (high biogenic methane content: ?clathrates in Site 397, 200 - 550 m subbottom) during late Quaternary glacial stages (Summerhayes et al., 1979; Dingle, 1980) might have contributed to the massive slope failure which triggered progressive deep-water slides and debris flows.

Finally, the Holocene transgression ended all gravitative and canyon-cutting processes. Today, the strong south-southwestward directed Canary Current and the aridity of the hinterland prevents any significant sedimentation on the West Saharan shelf, whereas the velocities of the semi-permanent north-directed counter-current along the upper continental slope are hardly sufficient for the erosion of sediments.

5. Comparison with Destructional Phases at the Northwest Atlantic Continental Slope

In general, the continental margin off NE America experienced a similar multi-phase history of slope construction and destruction as its NW African counterpart. The oldest erosional unconformity at the lower slope off the United States is also of mid-Cretaceous age (Schlee et al., 1979). In the area of the New England submarine canyons, precursor valleys had already been excavated from the Cretaceous escarpment prior to Campanian/Maastrichtian time, and later deepened in pre-mid-Eocene and Pleistocene times (Ryan et al., 1978). The present slope off the eastern United States is an erosional surface, cut into Paleogene and Late Cretaceous strata during various Cenozoic regressions, which caused a retreat of the shelfbreak of 5 - 30 km (10 - 300 km) as compared with the Late Cretaceous (late Jurassic/Early Cretaceous) shelfbreaks (Schlee et al., 1979). A marked diachronous angular unconformity of late Eocene to early Miocene age ("Horizon A^u") was formed by a precursor of the Western Boundary Undercurrent along the foot of the NE American slope; this catastrophic erosion removed about $2x10^5 km^3$ of sediments, especially off Cape Hatteras (Tucholke and Mountain, 1979). The interaction of the Gulf Stream and the

Western Boundary Undercurrent, which originated suddenly after the mid-Tertiary subsidence of the Greenland-Iceland-Faroer Ridge, has eroded Neogene sediments or inhibited sedimentation on Blake Plateau (McGregor, 1978). Whereas the southerly flowing, erosive contour currents of Arctic bottom water along the American margin are well known, the northward flowing contour currents along the East Atlantic margin have a different source (Antarctic Bottom Water?) and timing (since mid-Oligocene?), and sculptured the margin only locally (especially off the northern West Sahara and central Morocco). However, we must admit that these boundary currents which today are extremely weak are still poorly defined and speculative (cf. Lancelot and Winterer, 1980).

Similar Plio-Pleistocene mass wasting phenomena, as off the West Sahara, were observed off New England by Mac Ilvaine and Ross (1979): they detected slump scars that vary in length from tens of meters to several km and in relief from 1 m to hundreds of m. During glacial periods of rapid deposition these gravitational processes were the dominant erosional agents on the continental slope. During interglacial stages, however, biological and hydrodynamic processes became the dominant sedimentary processes. According to McGregor (1978), a Pliocene unconformity, similar to that observed in our profiles, became the slip surface for a large submarine slide (68 km^2, 11 km^3) off the Delaware coast.

Acknowledgements. This research was supported by the Deutsche Forschungsgemeinschaft (DFG) and the Bundesministerium für Forschung und Technologie (BMFT). We are very grateful to Dr. J. Watkins (Pittsburgh) for his permission to publish a GULF multichannel profile and to Drs. K. Hinz (Hannover), C.P. Summerhayes (Houston) and E.Uchupi (Woods Hole) for reviewing our paper. Professor Dr. E. Seibold (Kiel) and Dr. L. Diester-Haass (Würzburg) reviewed a previous draft of this paper.

REFERENCES

Almagor G, Wiseman G (1977) Analysis of submarine slumping in the continental slope off the southern coast of Israel. Mar Geotechn 2: 349-388

Arthur MA, von Rad U (1979) Early Neogene base-of-slope sedimentation at Site 397: sequential evolution of gravitative mass transport processes and redeposition along the Northwest African passive margin. In: Ryan WBF, von Rad U et al. Init Repts Deep Sea Drilling Project 47(1): 603-639

Arthur MA, von Rad U, Cornford C, McCoy FC, Sarnthein, M (1979) Evolution and sedimentary history of the Cape Bojador continental margin, Northwestern Africa. In: Ryan WBF, von Rad U, et al. Init Repts Deep Sea Drilling Project 47(1): 773-816

AUXINI (Departamento de Investigaciones Petroliferas de AUXINI, 1969) Correlacion estratigraphica de los sondeos perforados en el Sahara espanol. Bol Geol Minero, Madrid 83: 235-251

Bergren WA, Hollister CD (1974) Paleogeography and the history of circulation in the Atlantic region. In: Hay WW (ed) Studies in Paleooceanography. Soc Econ Petrol Min Spec Publ 20: 126-186

Bruce CH (1973) Pressured shale and related sediment deformation: Mechanism for development of regional contemporaneous faults. Am Assoc Petrol Geol Bull 57 (5): 878-886

Chamley H (1979) North Atlantic clay sedimentation and paleoenvironment since late Jurassic. In: Talwani M et al (eds) Deep Drilling results in the Atlantic Ocean: continental margins and paleoenvironment. M. Ewing Ser 3, Am Geophys Union, Washington DC; pp 342-361

Choubert G, Faure-Muret A, Sougy J (1968) Carte tectonique internationale de l'Afrique 1: 5,000 000, Sheet 1. Assoc Afr Geol Surv, UNESCO, Paris

Coleman JM, Garrison LE (1977) Geological aspects of marine slope stability, northwestern Gulf of Mexico. Mar Geotechn 2: 9-44

Connan J (1974) Time-temperature relation in oil genesis. Am Assoc Petrol Geol Bull 58: 2516-2521

CONOCO (1969) Spansah 51A-1, offshore Spanish Sahara. Final Report by Continental Oil Company of Spain, Spanish Gulf Oil Company, Compania Espanol de Petroleas S.A. Exploration Department, unpublished report and logs

Diester-Haass L (1979) DSDP Site 397: climatological, sedimentological, and oceanographic changes in the Neogene autochthonous sequence. In: von Rad U, Ryan WBF et al Init Repts Deep Sea Drilling Project 47(1): 647-670

Dingle RV (1980) Large allochthonous sediment masses and their role in the construction of the continental slope and rise off southwestern Africa. Mar Geol 37: 333-354

Einsele G, von Rad U (1979) Facies and paleoenvironment of Early Cretaceous sediments in DSDP Site 397 and in the Aaiun Basin (NW Africa). In: von Rad U, Ryan WBF et al., Init Repts Deep Sea Drilling Project 47(1): 559-577

Embley RW (1976) New evidence for occurrence of debris flow deposits in the deep sea. Geology 4: 371-374

Embley RW, Jacobi RD (1977) Distribution and morphology of large submarine sediment slides and slumps on Atlantic continental margins. In: Richards AF (ed) Marine slope stability. Mar Geotechn 2:205-228

Egbogah EO, Lambert-Aikhionbare DO (1980) Possible new oil potentials of the Niger delta. Oil Gas J 78 (n 15): 176-184

Fischer AG, Arthur MA (1977) Secular variations in the pelagic realm. In: Cook HE, Enos P (eds) Deep water carbonate environments, Soc Econ Paleont Min Spec Publ 25: 19-50

Guieu G (1976) Les bassins à evaporites de la marge ouest- africaine du Maroc a l'Angola. Dep Géol Univ Dakar Rep no 32

Hinz K (1979) Seismic sequences of Cape Bojador, Northwest Africa. In: von Rad U, Ryan WBF et al., Init Repts Deep Sea Drilling Project 47(1): 485-489

Hinz K, Seibold E, Wissmann G (1974) Continental slope anticline and unconformities off West Africa. "Meteor" Forsch Ergebn C 17: 67-73

Hoskins H, Rogers CU, Woo AO (1974) Data report of oblique reflection-refraction radio-sonobuoy profiles on the African Atlantic continental margin (R.V. ATLANTIS II cruise 67 and 75). Woods Hole Ocean Inst Rep 74-34: 40 pp

Kennett JP (1977) Cenozoic evolution of Antarctic glaciation, the circum-Antarctic Ocean and their impact on global paleooceanography. J Geophys Res 82: 3843-3860

Lancelot Y, Seibold E (1978) The evolution of the central northeastern Atlantic-Summary of results of DSDP Leg 41. In: Lancelot Y, Seibold E et al., Init Repts Deep Sea Drilling Project 41: 1215-1245

Lancelot Y, Seibold E et al (1978), Initial Reports of the Deep Sea Drilling Project, vol 41. U.S. Government Printing Off Washington, 1259 pp

Lancelot Y, Winterer EL (1980) Evolution of the Moroccan oceanic basin and adjacent continental margin - a synthesis. In: Lancelot Y, Winterer EL et al Init Repts Deep Sea Drilling Project 50: 801-833

Lehner P, de Ruiter PAD (1977) Structural history of the Atlantic margin of Africa, Am Assoc Petrol Geol Bull 61: 961-981

Lowrie A, Escowitz E (1969) Kane 9: global ocean floor and research data series. US Government Printing Office Washington (DC) 1: 976 pp

Mac Ilvaine JC, Ross DA (1979) Sedimentary processes on the continental slope of New England. J Sedim. Petrol 49 (2): 563-574

McCoy FW, Zimmerman HB (1977) A history of sediment lithofacies in the South Atlantic Ocean. In: Supko PR, PerchNielsen K et al. Init Repts Deep Sea Drilling Project 39: 1047-1080

McDougall I, Schmincke HU (1977) Geochronology of Gran Canaria, Canary Islands: age of shield building volcanism and other magmatic phases, Bull Volcanol 40: 1-21

McGregor B (1978) Seismic reflection profiles of the United-States east coast continental

margin, Natl Ocean At Agency Tech Rep ERL 398-AOML 28: 17 pp
Mitchum Jr RM, Vail PR (1977) Seismic stratigraphic interpretation procedure. In: Vail PR et al (eds), Seismic stratigraphy and global changes of sea level, Am Assoc Petrol Geol Mem 26: 135-144
Nairn EM, Noltimier HC, Nairn B (1980) Surface magnetic survey of the Souss Basin, southwestern Marocco: evaluation of the tectonic role postulated for the Agadir and Tarfaya Fault Zones and the South Atlas Flexure. Tectonophysics 64: 235-248
Querol R (1966) Regional geology of the Spanish Sahara. In: Reyre D (ed) Sedimentary basins of the African coast; Symposium, part 1, Atlantic coast. Assoc Afr Geol Surv UNESCO: 27-39
Pitman III WC (1978) Relationship between eustacy and stratigraphic sequences of passive margins. Geol Soc Am Bull 89: 1389-1403
Ratschiller LK (1970) Lithostratigraphy of the northern Spanish Saharan. Mem Mus Tridentino Sci Trento 18 (1): 1-80
Robertson AHF, Stillman CJ (1979) Late Mesozoic sedimentary rocks of Fuerteventura, Canary Islands: implications for West African continental margin evolution. J Geol Soc London 136: 47-60
Rona PA (1970) Comparison of continental margins of eastern North America at Cape Hatteras and northwestern Africa at Cap Blanc. Am Assoc Petrol Geol Bull 54: 129-157
Rust U, Wienecke F (1973) Bathymetrische und geomorphologische Bearbeitung von submarinen "Einschnitten" im Seegebiet vor Westafrika. Ein methodischer Versuch. Münchner Geogr Abh Geogr Inst Univ München 9:53-67
Ryan WBF, Cita MB, Miller EL, Hanselmann D, Nesteroff WD, Hecker B, Nibbelink M (1978) Bedrock geology in New England submarine canyons. Oceanol Acta 1 (2): 233-254
Sarnthein M, Diester-Haass L (1977) Eolian sand turbidites off Northwest-Africa. J Sediment Petr 47: 868-890
Schlager W (1980) Mesozoic calciturbidites in Deep Sea Drilling Project Hole 416A: recognition of a drownwed carbonate platform. In: Lancelot Y, Winterer EL et al Init Repts Deep Sea Drilling Project 50: 733-749
Schlee JS, Dillon WP, Grow JA (1979) Structure of the continental slope off the eastern United States. In: Doyle LJ, Pilkey OH (eds) Geology of continental slopes, Soc Econ Paleont Min Spec Publ 27: 95-117
Schmincke HU (1976) The geology of the Canary Islands. In: Kunkel G (ed) Biogeography and ecology in the Canary Islands, W.Junk, The Hague pp 67-184
Schmincke HU, von Rad U (1979) Neogene evolution of Canary Island volcanism inferred from ash layers and volcaniclastic sandstones of DSDP Site 397 (Leg 47A). In: von Rad U, Ryan WBF et al., Init Repts Deep Sea Drilling Project 47(1): 703-725
Seibold E (1972) Cruise 25/1971 of R.V. "Meteor": continental margin of West Africa General report and preliminary results. "Meteor" Forsch Ergebn C 10: 17-38
Seibold E, Hinz K (1974) Continental slope construction and destruction, West Africa. In: Burk CA, Drake CL (eds) The geology of continental margins. Springer New York Heidelberg Berlin, pp 179-196
Seibold E, Hinz K (1976) German cruises to the continental margin of North West Africa in 1975: General reports and preliminary results from "Valdivia" 10 and "Meteor" 39. "Meteor" Forsch Ergebn C 25: 47-80
Shepard FP (1973) Submarine geology, 3rd edn Harper & Row New York, 517 pp
Shepard FP, Dill RF (1966) Submarine canyons and other sea valleys. Rand McNalley & Co Chicago, 381 pp
Summerhayes CP, Bornhold BD, Embley RW (1979) Surficial slides and slumps on the continental slope and rise of South West Africa: a reconnaissance study. Mar Geol 31: 265-277
Summmerhayes CP, Milliman JD, Briggs SR, Bee AG, Hogan C (1976) Northwest African shelf sediments: influence of climate and sedimentary processes. J Geol 84: 277-300

Summerhayes CP, Nutter AH, Tooms JS (1971) Geological structure and development of the continental margin of Northwest Africa. Mar Geol 11: 1-25

Tucholke BE, Mountain GS (1979) Seismic stratigraphy, lithostratigraphy and paleosedimentation patterns in the North American Basin. In: Talwani M et al (eds) Deep Drilling results in the Atlantic Ocean: continental margins and paleoenvironment, M Ewing Ser 3. Am Geophys Union: 59-86

Uchupi E, Emery KO, Bowin CO, Phillips JO (1976) Continental margin off Western Africa. Am Assoc Petrol Geol Bull 60: 809-878

Vail PR, Mitchum Jr RM, Thompson III S (1977) Global cycles of relative changes of sea level. In: Vail PR et al (eds) Seismic stratigraphy and global changes of sea level Am Assoc Petrol Geol Mem 26: 83-98

Van Andel T, Thiede J, Sclater JG, Hay WW (1978) Depositional history of the South Atlantic Ocean during the last 125 m.y. J Geol 85: 651-698

von Rad U, Arthur MA (1979) Geodynamic, sedimentary, and volcanic evolution of the Cape Bojador continental margin (NW Africa). In: Talwani M et al (eds) Deep drilling results in the Atlantic Ocean: Continental margins and paleoenvironment, M Ewing Ser 3, Am Geophys Union: 187-203

von Rad U, Cepek P, von Stackelberg U, Wissmann G, Zobel B (1979) Cretaceous and Tertiary sediments from the Northwest African slope (dredges and cores supplementing DSDP results). Mar Geol 29: 273-312

von Rad U, Einsele G (1980) Mesozoic-Cainozoic subsidence history and paleobathymetry of the northwest African continental margin (Aaiun Basin to DSDP Site 397). Phil Trans Royal Soc London Ser A 294: 37-50

von Rad U, Ryan WBF et al (1979) Initial Reports of the Deep Sea Drilling Project, 47(1): 835 pp

von Stackelberg U, von Rad U, Zobel B (1979) Asymmetric sedimentation around Great Meteor Seamount (North Atlantic). Mar Geol 33: 117-132

Watkins JS, Hoppe KW (1979) Seismic reflection reconnaissance of the Atlantic margin of Morocco. In: Talwani M et al (eds) Deep drilling results in the Atlantic Ocean: continental margins and paleoenvironment, M Ewing Ser 3; Am Geophys Union: 205-217

Wiedmann J, Butt A, Einsele G (1978) Vergleich von marokkanischen Kreide-Küstenaufschlüssen und Tiefseebohrungen (DSDP): Stratigraphie, Paläoenvironment und Subsidenz an einem passiven Kontinentalrand. Geol Rundschau 67 (2): 454-508

Wissmann G (1979) Cape Bojador slope, an example for potential pitfalls in seismic interpretation without the information of outer margin drilling. In: von Rad U, Ryan WBF et al Init Repts Deep Sea Drilling Project 47(1): 491-499

Wissmann G, von Rad U (in press) Morphologie und Struktur der submarinen Canyons am Kontinentalhang vor der West- Sahara (NW Afrika). "Meteor" Forsch Ergebn Reihe C

8 Deep Seismic Structure (Mauritania and Central Morocco)

W. Weigel[1], Gerd Wissmann[2], and P. Goldflam[1]

ABSTRACT

This paper presents two crustal sections through the Moroccan and Mauritanian continental margin in the vicinity of the Canary and Cape Verde Islands. The sections are based mainly on joint refraction surveys of the Institut für Geophysik (Hamburg) and the Federal Institute for Geosciences and Natural Resources (BGR, Hannover) and additional information from published refraction work in the survey areas.

On both sections rift basins were crossed, in which evaporites had formed diapirs. A 7.1 - 7.3 km/s layer could be followed from the rise to the shelf in both areas, while Moho arrivals could only be observed over the continental rise of Mauritania. Here the typical oceanic layer 2 seems to be thin or missing in a more than 100 km wide zone of transitional crust, in which the 7.1 km/s layer probably forms the top of the crystalline basement. From this structural relationship a model is postulated, which interprets that prominent 7.1 km/s refraction horizon as the cooled and solidified top of the original elongated mantle plume that lead to the splitting up and contributed to the drifting of the continents. A thick layer with a velocity of 5.8 km/s overlying the postulated plume surface off Morocco could then either constitute > 100 km of thinned continental crust or massive carbonate sediments. From the refraction results clear indications for an oceanic crust exist only more than 100 km west of the present continental slope.

The lithological interpretation of the seismic velocities is complicated by the presence of high velocity (vp > 5 km/s) carbonate platform rocks under the continental slope and of volcanic rocks (intrusions on the shelf, sea-mounts and islands on the rise and slope).

1. Introduction

The evolution of passive continental margins with its relation to sea-floor spreading and plate tectonics has been studied geologically and geophysically for more than ten years. Especially the continental margin of North West Africa was the object of several expeditions, some of which partly served as DSDP/IPOD pre-site surveys, of the German vessels "METEOR" and "VALDIVIA" in the period 1967-1980 (Closs et al., 1969; Seibold and Hinz, 1976). On four cruises (Fig. 1) the deep crustal structure off Mauritania ("VALDIVIA"-cruise 10) and off Morocco ("METEOR"-cruises 39, 46 and 53) has been studied by explosion refraction seismics (Weigel et al., 1977; Weigel et al., 1978; Thiessen, 1979; Hansen, 1979; Goldflam et al., 1980; Gebhardt and Weigel, 1981; Wissmann, 1979; Hinz et al., this Vol.), partly combined with refraction seismic investigations in Morocco (Makris, 1981) and on Lanzarote (Pavia, 1977).

The history of the continental margin surely can be linked to the geological events on the bordering continental regions. From the results presented in this paper it is obvious, that in the important transition zone between ocean and continent, informations from the deeper crust are sparse. Nevertheless, an evolutionary model has been derived.

The main geological structures of Northwest Africa are

1. offshore volcanic features, such as the Canary Islands (Bosshard and Macfarlane,

---

1 Institut für Geophysik, Universität Hamburg, 2000 Hamburg, FRG

2 Bundesanstalt für Geowissenschaften und Rohstoffe, Stilleweg 2, 3000 Hannover 51, FRG

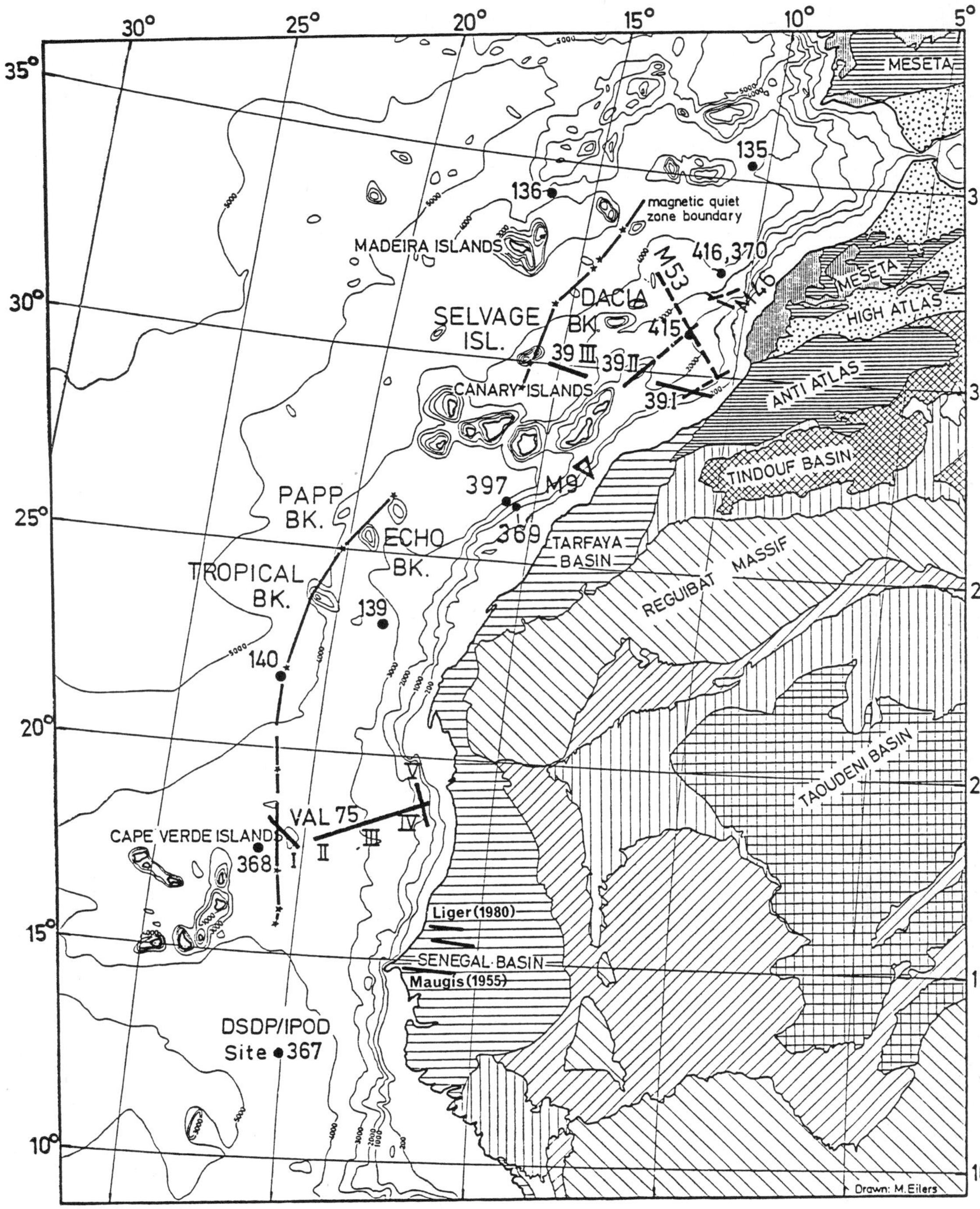

Fig. 1. Deep seismic sounding on the continental margin of NW-Africa by IfG Hamburg and BGR Hannover.
Solid lines: refraction seismic profiles (I, II...).
VAL 75 : VALDIVIA cruise 10/1975, I : Cape Verde Rise
II-V: Continental margin off Mauritania
M 39 : METEOR cruise 39/1975, 39 I : Continental margin off Morocco
39 II : Conception Bank (in strike profile)
39 III: Conception Bank - Selvagem Grande
M 9 : METEOR cruise 9/1967
dashed lines: M 46/1977, M 53/1980 (Thiessen,1979; Gebhardt and Weigel, 1981)

1970; Schmincke,1979; Schmincke, this Vol.) and East Canary Ridge (Weigel et al., 1978; Watkins and Hoppe, 1979), the Selvage Islands, the Dacia, Papp, Echo and Tropical Bank and the Cape Verde Islands (Dash et al., 1976). Ages reported for some of these structures range from Cretaceous to Miocene (Robertson and Bernoulli, this Vol.; Schmincke, this Vol.) and are therefore younger than the crust surrounding the volcanic intrusions;
2. the marginal basins such as the Senegal-Mauritania, Tarfaya-Aaiun and South Atlas basins, characterized by diapiric or wall like salt structures with sediments of several thousand meters thickness (Luyendik and Bunce, 1973; Uchupi et al., 1976; Kent et al., 1980; Hinz et al., this Vol; Ranke et al., this Vol.);
3. the Precambrian Reguibat Massif, the Anti Atlas of Paleozoic age and the High Atlas belonging to the Variscan and Alpine fold belts (Stets and Wurster, this Vol.).

(Compare also front and back cover figures.)

The crustal structure of the continental margin has been studied on refraction profiles which all are situated within the magnetic quiet zone (Fig. 1).

All refraction seismograms (demonstrated in Figs. 3A-5B, 7-14A) were produced by detonating explosives of 25 - 200 kg charges, which were dropped from the research vessels in distances of about one nautical mile. The seismic signals were received by bottom hydrophones (water depth less than 1 000 m) and suspended hydrophones (water depth more than 1 000 m) which were cable connected to moored buoys. After preamplification and passing a low-frequency (9.5 cps) modulator, the signals were transmitted by high frequency transmission to the research vessel, where they were recorded on paper - for controlling the seismic efficiency - and on magnetic tape. The mean distances between the buoys on one profile are about 60 km. Positioning relied on satellite navigation and the direct wave of water sound.

Each seismogram has been filtered by a band pass filter between 2 and 20 Hz and the amplitude was adjusted to keep an optimal signal to noise ratio. All seismograms of one profile section are placed together in seismogram sections (Figs. 3-5, 7-14) reduced by the compression velocity (Vp) of 6.0 km/s. Reversed seismogram sections (Figs. 10 A,B) sometimes show differences in positions for the same shot, since the shot positions do not always coincide with the line connecting two receiver buoys. The correlation lines of vp are mainly straight lines, which point to a layered structure and to the absence of strong velocity gradients. The presented crustal sections (Figs. 6,16) have been computed by a ray tracing method (Fig. 15).

2. Results and interpretation

2.1. Morocco - Selvagem Grande

The refraction results of this region are summarized in Table 1 and the calculated crustal structure models in Fig. 6. The models show, that the different geological provinces, the marginal basin off Morocco (M 39/I), the Conception Bank (M 39/II), and the deep sea region west of the East Canary Ridge (M 39/III), are also expressed by the deep crustal structure. From the seismogram sections (Figs.3A-5B), three refractors within the crust are obvious. The first arrivals on section 3B are supported by multiple refracted events. The low signal to noise ratio in Fig. 4 seems to be caused by energy dissipation from continental slope irregularities in the sediments.

Under the continental margin (M 39/I) the existence of more than 7 km of sediments could be demonstrated. The 4.9 km/s layer is interpreted by Hinz et al. (this Vol.) as carbonates and evaporites that built up in a subsiding rift basin.

The 5.8 km/s layer rises to about 5000 m under the shelf and may constitute the Paleozoic basement, which crops out onshore. The interpretation of this layer in Table 1 was left open, since a sedimentary or basaltic composition could not be excluded. The refractor comes to lie at a depth of 8.5 km, whereas the crystalline basement sequence MO-5 of Hinz et al. (this Vol.) plunges to a depth of more than 10 km beneath the continental slope. Velocities for the metamorphic Paleozoic or the Precambrian bottom hole formation encountered in the Ifni Marine 1 exploratory well at 2001 meter depth (AAPG World Oil Developments, 1976) are not known.

On the other hand the 5.6 km/s layer of the METEOR 9 stations south of the Canary Islands (Fig. 2A) most probably represents sediments (Roeser et al., 1971; Ranke et al., this Vol.)

Fig. 2
A. Compressional-velocity distribution in the continental margin region off Morocco and the Canary Islands (Interpretation compare Table 1)
B. Compressional-velocity distribution in the continental margin region off Mauritania (Interpretation compare Table 2)

--- Magnetic Quiet Zone Boundary

sediments

carbonates

A

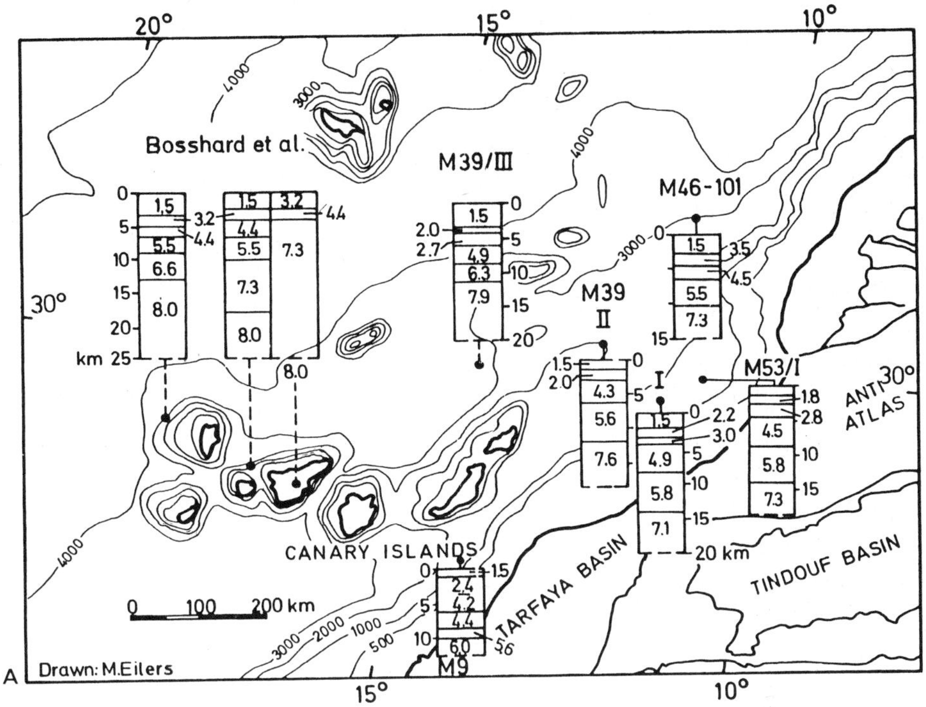

B

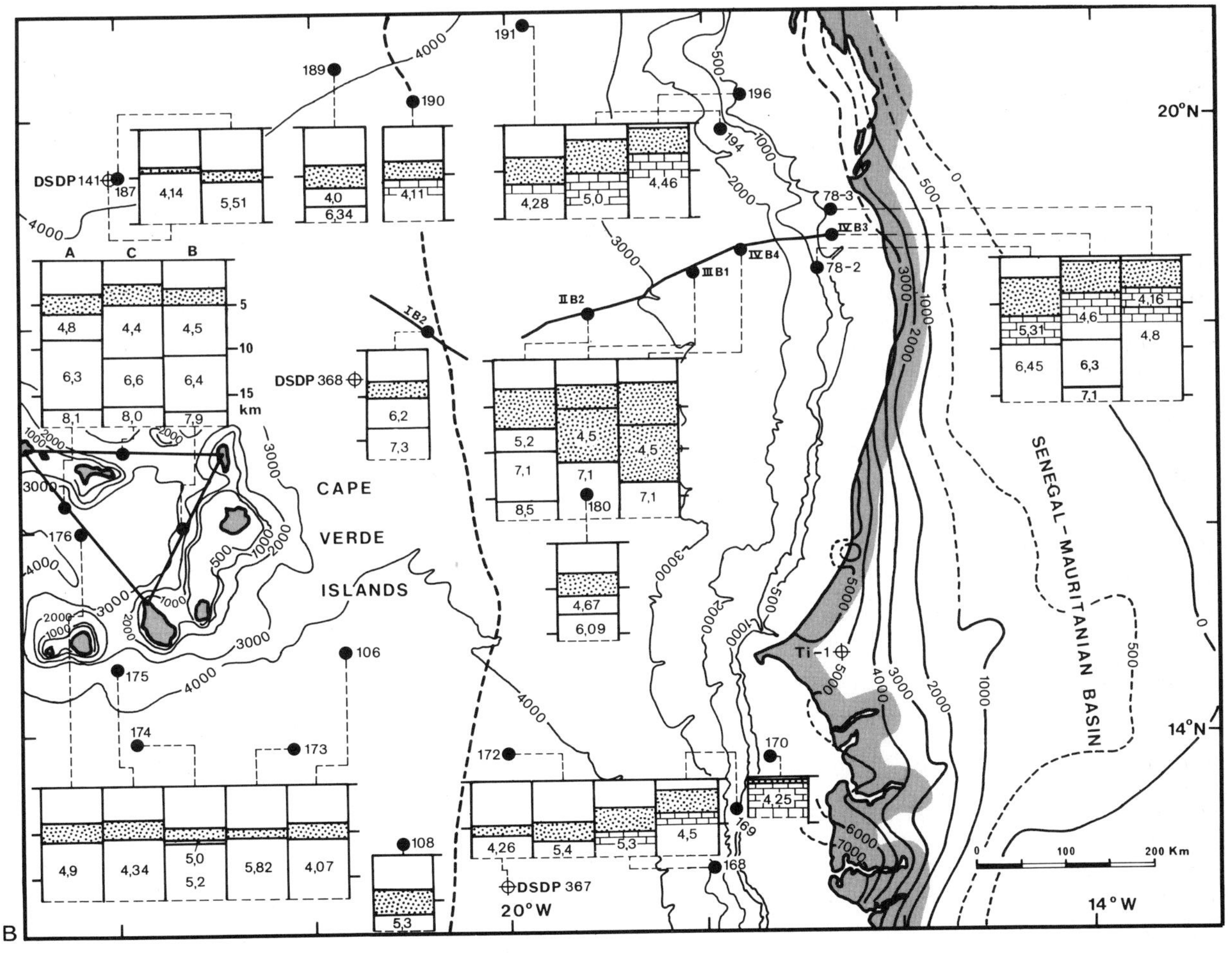

We interpret the 7.1 - 7.3 km/s refractor (Figs. 6, 2A) as generated from mantle material which cooled and contracted and led to the formation of the continental margin sediment prism. No normal mantle velocity near 8.0 km/s has been observed. Because of this we assume that such a velocity will be reached in much greater depth, where it could not represent a first order discontinuity, but at most a zone of a small velocity gradient. From these results it is obvious that in the more than 200 km wide region of the sedimentary basin between the East Canary Ridge and the Morocco shelf no typical oceanic crust is present, even if the 5.8 km/s layer should constitute pillow lavas.

The very clear seismogram sections of the Conception Bank region (39/II) are presented in Weigel et al. (1978). They show a high signal to noise ratio which decreases as the sediment cover increases on the flanks of the Bank. The p-velocities observed are mainly associated with volcanic material which relates the Conception Bank to the East Canary Ridge (Banda et al., in press). The Conception Bank formed as an isolated unit in or on the edge of older crust, which had been produced by sea-floor spreading. A Pliocene/Miocene origin of the Conception Bank is indicated by its thrusting through Miocene, Pliocene and post-Pliocene sedimentary deposits, which could be stratigraphically identified by correlation to DSDP Sites 415 and 416 (Hinz et al., this Vol.).

In Figures 5A and 5B (M 39/III) two seismogram sections from the deep sea region west of the Conception Bank are presented. They show three apparent crustal velocities which have been identified as oceanic layers 2, 3 and the crust/mantle boundary.

Our data support the existence of a typical oceanic crust with a clear "Moho" west of the East Canary Ridge, where it had also been found by other authors (Bosshard and Macfarlane, 1970; Grunau et al., 1975; Uchupi et al., 1976).

2.2. Mauritania - Cape Verde Rise

Some results of the VALDIVIA-10/1975 refraction seismic profiles off Mauritania have already been reported (Weigel and Wissmann, 1977; Goldflam et al., 1980). New data such as the two ship (NOAA DISCOVERER, ISELIN) refraction measurements taken in 1972 between the Cape Verde Islands and Mauritania by Dash and Rona were not available for comparison with our work.

The main purpose of this paragraph is to present new information about the opposite ends of the VALDIVIA profile over the Cape Verde Rise and the shelf of Mauritania. A review of the interpretation over the central part of the profile is given with a documentation of seismograms for the complete crustal section presented. The discussion of the VALDIVIA-10/1975 profile will lead from the deep sea crustal structure of the Cape Verde Rise to the continental shelf of Mauritania and the lithological interpretation of the refraction velocities is compared to the results of sonobuoys along three east-west traverses offshore the Senegal-Mauritania Basin (Fig. 2B):
The refraction results of the VALDIVIA-10/1975 profiles are summarized in Table 2 and the calculated structural models in Fig. 16.

2.2.1. Cape Verde Archipelago and Rise (Profile I)

The seismogram section (Fig. 7) shows very clear refraction arrivals with high signal/noise ratios and small damping rates, which point to hard rock discontinuities and a thin sediment cover. The true velocities observed in two directions are 6.2 and 7.3 km/s. By reason of the great water depth sedimentary velocities could not be observed, but are known from reflection profiles (Table 2).

The results of profile I over the Cape Verde Rise (Fig. 7) will be compared with the records obtained by Dash et al. (1976) from the recording stations on the islands of Sao Vicente, Santiago and Sal in the Cape Verde archipelago. What is the nature of the 6.2 km/s layer at 5 km depth under profile I-buoy B2 and does it represent the same material as the 6.3 to 6.6 km/s layer at ca. 10 km depth under the Cape Verde archipelago? Velocities of 6.0 to 6.6 km/s have also been encountered under many islands of the volcanic Leeward West Indies (Lesser Antilles Island Arc) and there have been partly interpreted as pieces of Atlantic oceanic crust (Eschenbrenner et al., 1980). While such elevated velocities seem to be commonly associated with volcanic provinces, obviously the geographical position and the geological environment must be considered in an explanation, as similar velocities of 6.3 to 6.45 km/s have been found in the more continental setting of the Mauritania slope parallel to the 1000 m isobath at about 10 km depth (Weigel and Wissmann, 1977; EXPLORA-1978 sonobuoy 2 in Fig. 2B).

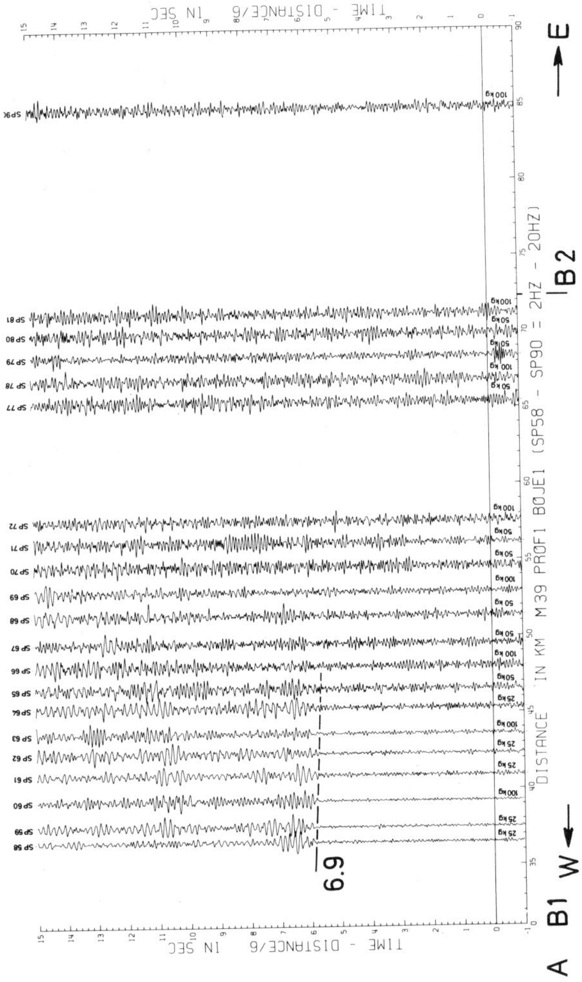

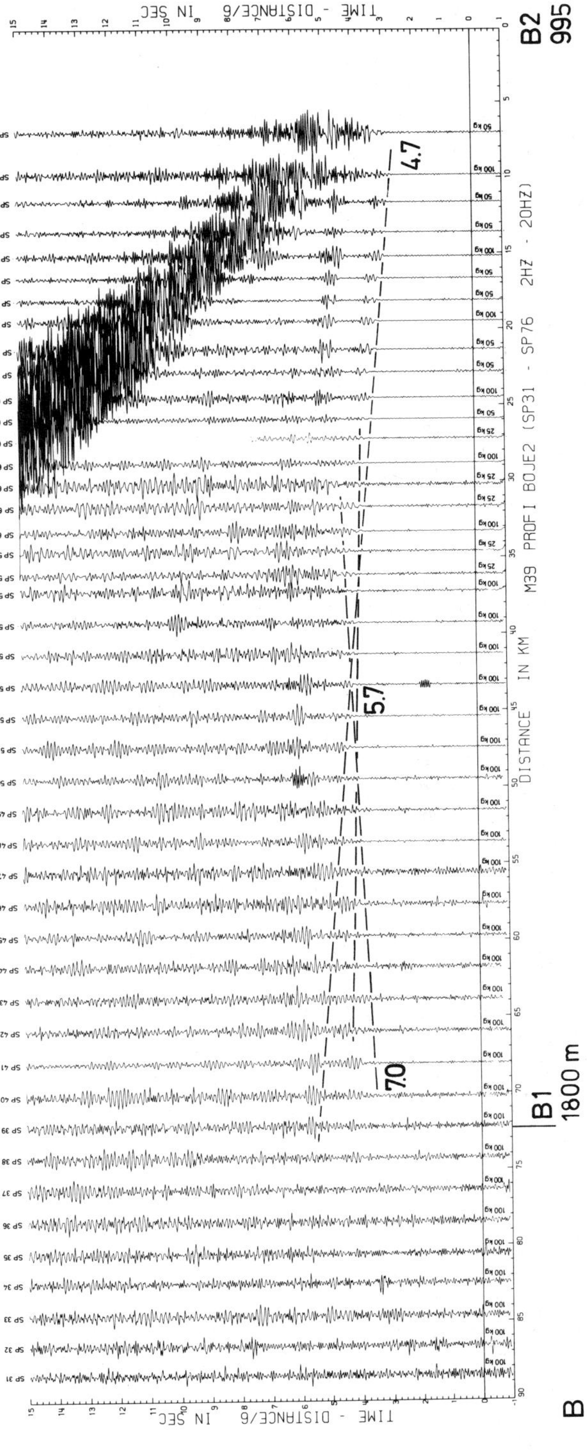

Fig.3A. Seismogram section M 39/I, observed from E to W relative to B1. Clear apparent velocity of 6.9 km/s
B. Seismogram section M 39/I, observed from E to W relative to B2 (5.7 km/s also identified by second arrivals)

Dash et al. (1976) infer 3-6 km basaltic pillow lava (4.4 - 4.8 km/s) overlying 6-8 km of "high density plutonic (typically gabbroic)" rocks (6.3 - 6.6 km/s). As a velocity of 6.3 km/s is typical for oceanic layer 3 and as the magnetic quiet zone (MQZ) boundary occurs landward of the islands (Fig. 1), the original composition of the crust underlying the Cape Verde archipelago was probably oceanic.

This argument should also be valid for the western part of VALDIVIA-10 profile I, which crosses the MQZ boundary over the Cape Verde Rise (Fig. 1). However the model calculated from the arrivals recorded by the BGR anchored buoys B1 to B3 shows no obvious crustal change across the MQZ: a 6.2 km/s layer dips from 5 km in the northwest to 6 km depth in the southeast (Fig. 16).

Applying the velocity interpretation of Dash et al. (1976) "plutonic (typically gabbroic) oceanic layer 3 rocks" would therefore exist 5 km shallower under the Cape Verde Rise than under the Cape Verde archipelago (Fig. 2B). In the following we shall discuss this discrepancy.

Analyses of rock samples show both structures to be genetically related (Grunau et al., 1975; Natland, 1977). So far no conclusive evidence has been presented that "the building of the Cape Verde archipelago began at an early stage in the Mesozoic" (Dash et al., 1976). On Maio a submarine volcanic pile may have produced Late Cretaceous tuffs (Robertson and Bernoulli, this Vol.) and reached sea level during the Miocene.

A 19 m.y. old, 12.5 m thick olivine-gabbro sill (Duncan and Jackson, 1977) was drilled in DSDP site 368. It intrudes into Albian to Turonian black shales and together with many more sills seen in the reflection seismic profiles (e.g. METEOR-25 on p. 235 of Lancelot, Seibold et al., 1977), it testifies to the volcano-tectonic origin of the Cape Verde Rise. These Early Miocene intrusive events somewhat predated the Middle Miocene uplift of the rise of more than 1000 metres, which was deduced from changes of the CCD in the sedimentological record (Lancelot, Seibold et al., 1977, p. 235). This uplift is seen as a

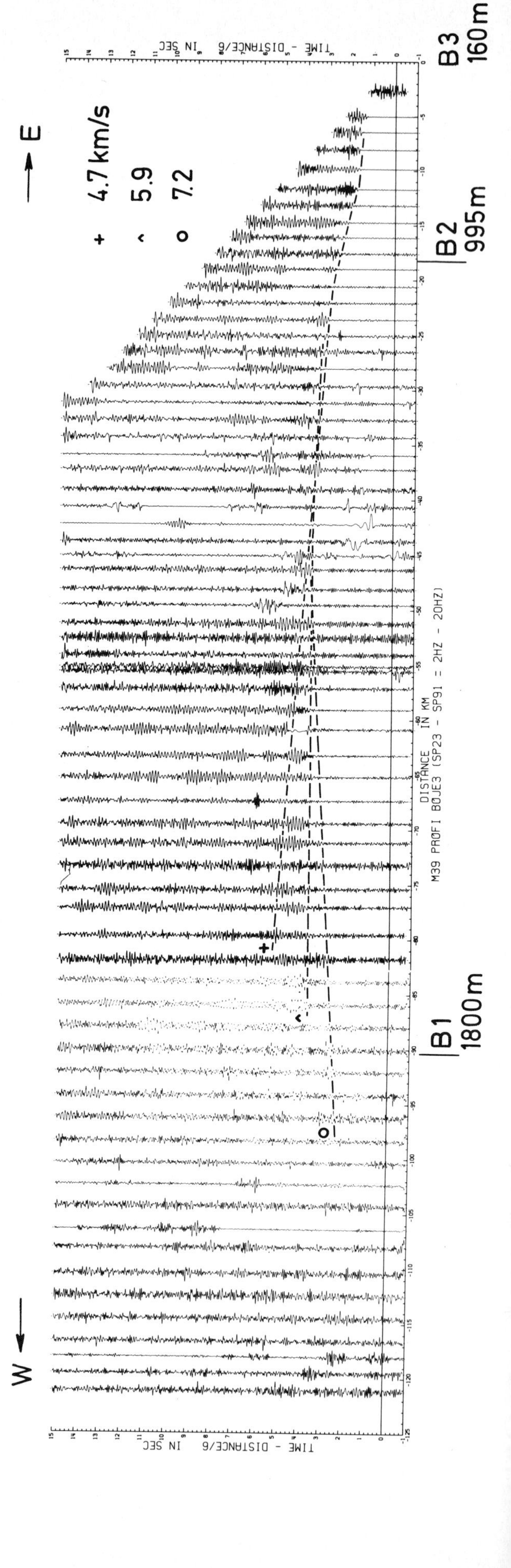

Fig. 4. Seismogram section M 39/I, observed from E to W relative to B3. Seismic energy weak: influenced by 2-8 km thick sediment cover and salt diapirs (Hinz, this Vol.)

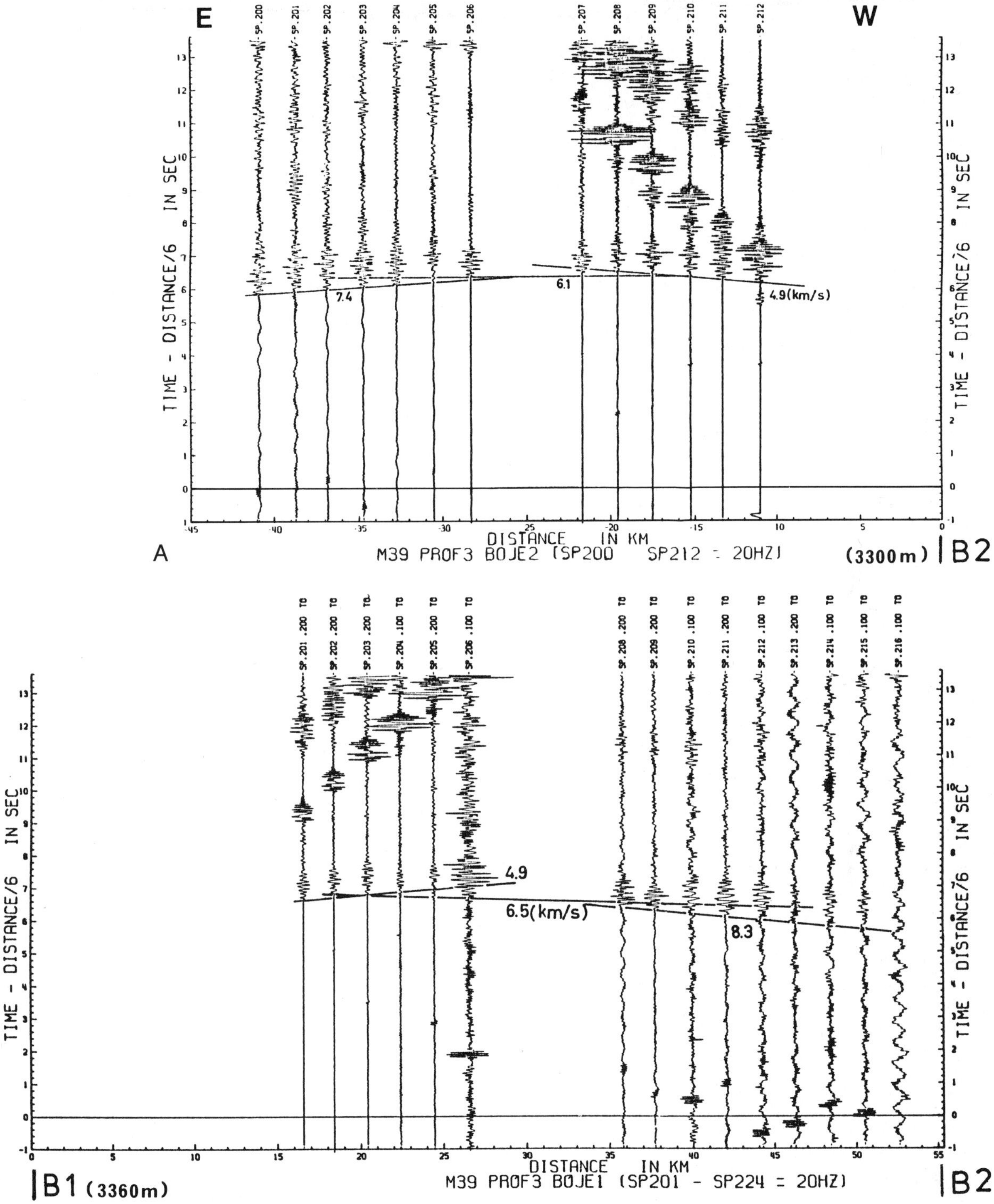

Fig. 5A. Seismogram section M 39/III, observed from W to E relative to B2
B. Seismogram section M 39/III, deep sea region, observed from E to W relative to B1. 8.3 km/s: apparent velocity corresponding to 7.4 km/s in Fig. 5A

local basement high underneath the Cape Verde Rise on a SHELL multichannel seismic profile (Lancelot, Seibold et al., 1977, p. 234) and it took place contemporaneous with the early Neogene volcanic activity that built the Cape Verde Islands.

The velocities measured in the sills of Site 368 (3.6 and 4.6 km/s) are lower than the 6.2 km/s of the VALDIVIA refraction profile I. But the 5 to 6 km depth calculated for this refraction layer coincides with the depth to basement on the SHELL profile. Therefore we interpret the 6.2 km/s layer to represent original oceanic layer 2 basalt, which was altered by Neogene intrusions of "plutonic, typically gabbroic" rocks of higher velocity.

These rocks under the Cape Verde Rise are not equivalent to the 6.3 to 6.6 km/s layer under the Cape Verde Islands and we think there they were not "depressed (from 5 to about 10 km depth) as a result of the loading of accumulated materials of the islands (Dash et al., 1976)".

The latter argument was to explain the "depression of the Moho from oceanic to transitional (16-17 km) depth", but our VALDIVIA-10 profile II to the east of the Cape Verde Rise shows that the Moho (8.5 km/s) has a probably regional depth of 16 km even in an area without volcanic structures.

Transitional Moho depths of 13-22 km were also calculated from gravimetric studies on the Cape Verde Islands (Mendes-Victor, 1970). The values of 17 km for Sao Nicolau and 18 km for Boa Vista in particular support the results of 16.2 and 16.7 km for the refraction profiles C and B of Dash et al. (1976).

Our data on the Cape Verde Rise do not reveal the depth to the Moho, as the BGR buoys did not record clear arrivals beyond 60 km, while in the time-distance plots of Dash et al. (1976) Moho first arrivals start to appear only in a distance of 60 km from the receivers. But if we assume a regional Moho depth of 16 km, the higher velocities of our Cape Verde Rise model (Fig. 16) at comparatively shallower depth than under the Cape Verde islands, call for earlier Moho arrivals, which we cannot see in our data. From this we conclude that here the crust/mantle boundary does not exist as a first order discontinuity.

We interpret the 7.3 km/s layer of profile I as the intrusion metamorphized equivalent of the 6.3 - 6.6 km/s oceanic layer 3 under the islands. This interpretation of the refraction seismic data agrees with the reflection seismic interpretation of the basalts drilled in DSDP Site 367 and correlated as a reflector to Site 368, as flat oceanic basement basalts (Goldflam et al., 1980).

This flat basement horizon was correlated to 28 sonobuoy stations offshore the Senegal Basin (Hoskins et al., 1974) and a mean velocity for oceanic layer 2A was calculated: $v_p = 4.95 \pm 1.16$ km/s. Within the more restricted province of the Cape Verde Islands volcanism the basement velocity and the variation is lower, namely

$$v_p = 4.59 \pm 0.25 \text{ km/s}$$

This statistic based on reflection seismic profiles reconfirms our interpretation of the 6.2 km/s layer under the Cape Verde Rise as altered Middle Jurassic (150 m.y.) oceanic basement, which had cooled and normally subsided to a depth of 6 km, before it was uplifted by 1 km in the Miocene.

2.2.2. Mauritania continental slope (profiles II, III, IV)

The seismogram sections of the deep sea region of the continental rise (Fig. 8 A,B) are characterized by first arrivals from the upper crust (5.0 - 5.2 km/s), the lower crust (7.1 - 7.2 km/s) and the upper mantle (8.3 - 8.8 km/s), which is also indicated by PmP reflections near B 1 in Fig. 8 A. Further east towards the slope off Mauritania the p-velocities in the upper crust change to lower values and a more differentiated structure.

The most prominent horizon is that with the apparent velocities of 6.9 - 7.2 km/s, which is observed with very clear phase onsets from the western part of the rise to the continental shelf over a distance of more than 300 km (Figs. 8 A - 14). The high signal to noise ratio and the correlation by straight lines characterizes this horizon as a first order discontinuity with a smooth morphology. Mantle first arrivals

Fig. 6. Crustal structure from raytracing models: Continental margin off Morocco - Conception Bank - Selvagem Grande. Travel-times reduced by 6 km/s. Dashed line in M 39/I from reflection results

- Table 1 -

Velocity Interpretation (Fig. 6)

Profiles METEOR 39/I-III: Selvagem Grande - Conception Bank - Continental Margin

39III: Conception Bank - Selvagem Grande: oceanic crust

	Vp (km/s)	Geological interpretation	
water:	1.51		
sediments:	2.0	unconsolidated	layer 1
	2.7	consolidated sediments	
rocks:	4.9	pillow basalt + massive basalt	layer 2
	6.3	gabbro, metagabbro, metabasalt	layer 3
	7.9	mantle	

39 II: Conception Bank: Tertiary volcanism

	Vp (km/s)	Geological interpretation
water:	1.51	
sediments:	2.5	mainly volcanic material from reflection results
rocks:	4.3	pillow basalt
	5.6	massive basalt (intruded)
	7.6	mantle material (reduced velocity from high temperatures)

39 I: Continental Margin: continental and transitional crust

	Vp (km/s)	Geological interpretation
water:	1.51	
sediments:	2.2	Tertiary - Quaternary
	3.0	Cretaceous
	4.9	Lower Jurassic - Cretaceous *) (including rift sediments and evaporites)
rocks:	5.8	Paleozoic-Precambrian basement: (thinned and contaminated continental crust *) (sediments or basalts also possible, see Fig. 2A:M9)
	7.1	plume surface or "transitional" rocks (Figs. 17, 18A,B)

*) HINZ et al., this vol.

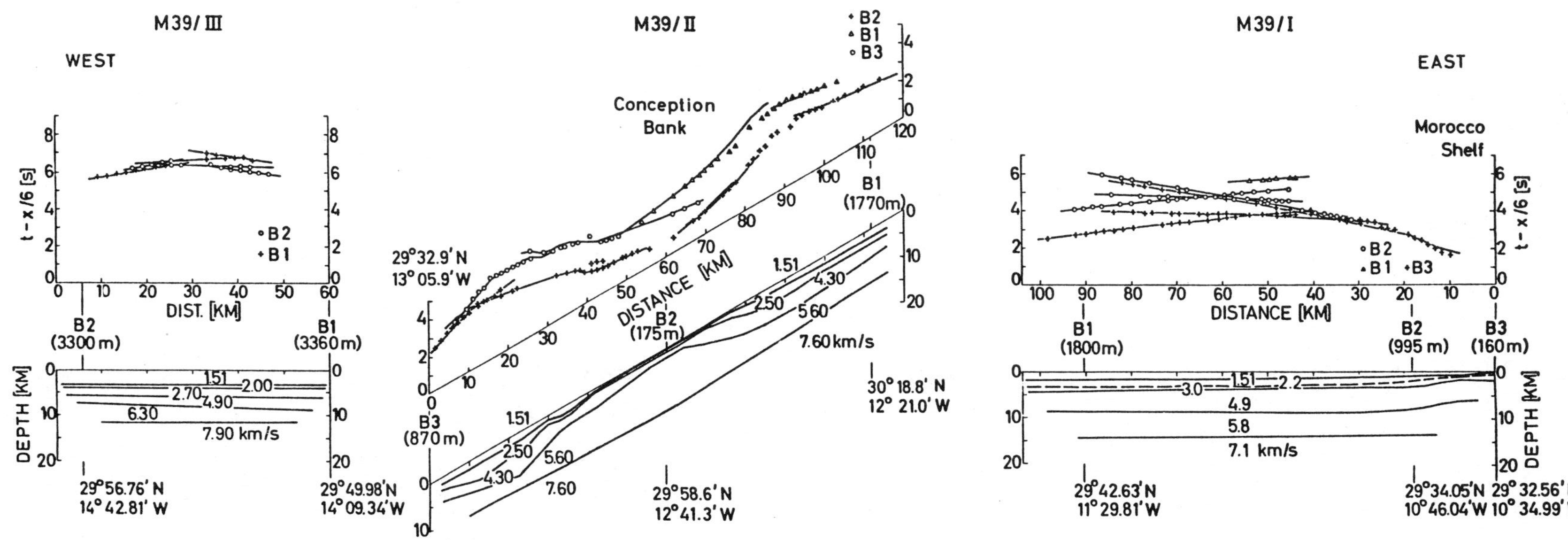

are only clearly identified on sections 8 A and 8 B, where the crossover distance (7.1/8.8 and 7.1/8.3) is about 42 km, less than the length of the profile sections 9 A, B; 10A, B; 12 A, B; 13 A, B; 14 A, B. Therefore we state that the transition to the mantle is neither represented by a discontinuity nor a zone of a high velocity gradient but rather by a zone of a low velocity gradient. The velocity of 8.5 km/s (Fig.12B) from second arrivals cannot unambiguously be related to the "Moho".

The outstanding feature of the VALDIVIA-10/1975 refraction profile is the well-documented existence of a 7.1 km/s layer dipping continuously from 9 km under the continental rise to 14 km under the shelf of Mauritania.

In a preliminary report (Weigel and Wissmann, 1977) it was stated, that "even 400 km west of the continental slope no typical oceanic crust was found". This statement referred to a "higher than normal" oceanic layer 3 velocity of 7.1 km/s overlying a mantle layer with 8.5 km/s.

The mantle could here only be observed over the western part of profile II up to 300 km away from the Mauritanian coastline, where it dips east from 14 to 16 km depth (Figs. 8 A and B, 16).

Examination of the literature since 1973 indicates that velocities of 7.1 km/s in oceanic crust are not so atypical after all, probably a consequence of the use of improved equipment (Houtz, 1977). First a basal layer with velocities of 7.2 - 7.6 km/s between layer 3 and mantle (Moores and Jackson, 1974; Fox et al., 1973) was recognized, which in later statistical studies about the structure and constitution of oceanic crust became an integral part of layer 3 during a process of redefinition and adaption of older terms to a steadily widening data base (Christensen and Salisbury, 1975; Trehu et al., 1976).

Even though in a review of ten years of marine geology and geophysics from 1967-1977 Le Pichon (1977) still maintained that there was no generally accepted standard structure for layer 3, it now seems ob-

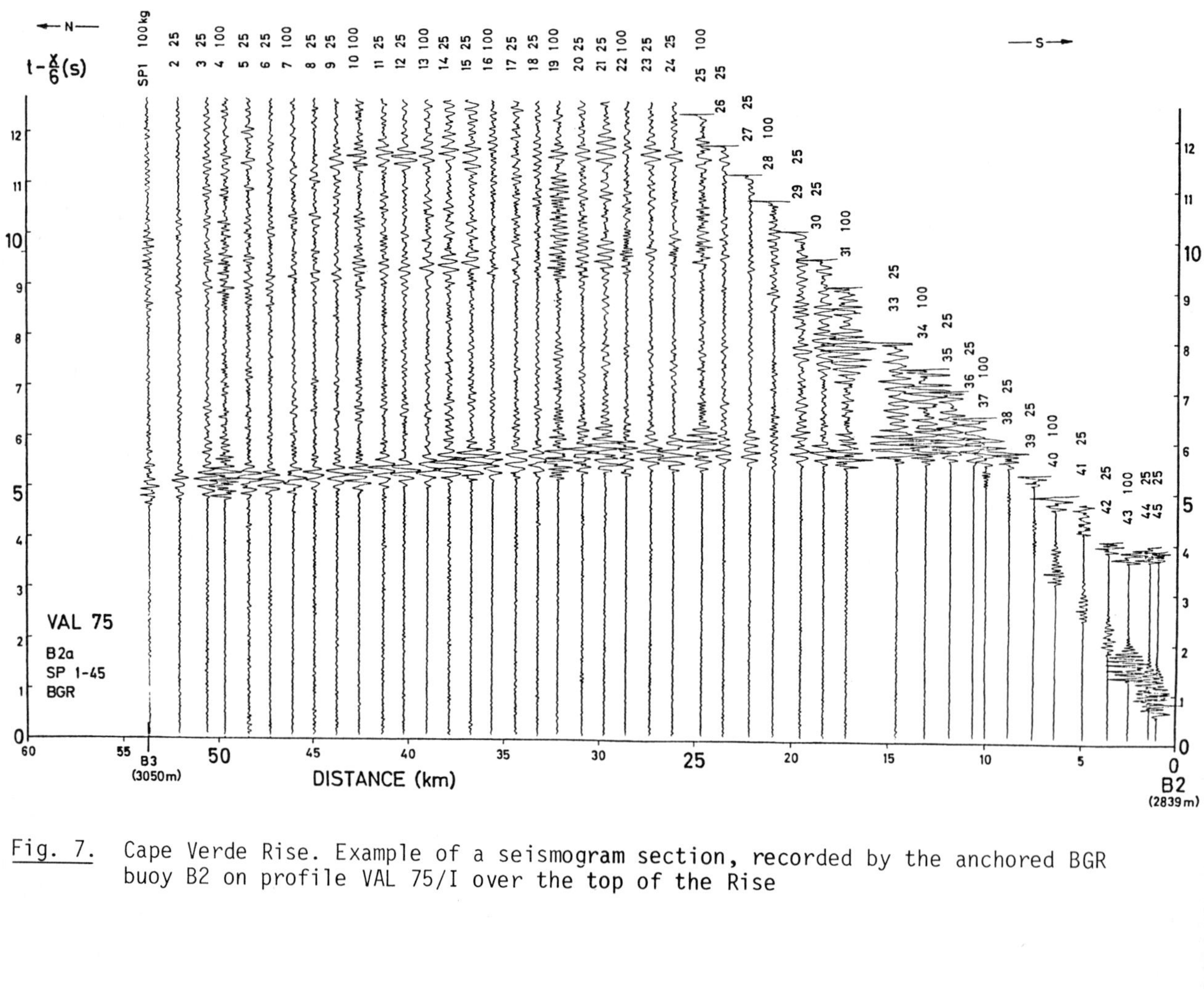

Fig. 7. Cape Verde Rise. Example of a seismogram section, recorded by the anchored BGR buoy B2 on profile VAL 75/I over the top of the Rise

Fig. 8A. Seismogram section VAL 75/II, observed from E to W relative to B2. 8.8. km/s: apparent mantle p-velocity

Fig. 8B. Seismogram section VAL 75/II, observed from W to E relative to B1. 8.3 km/s: apparent mantle p-velocity

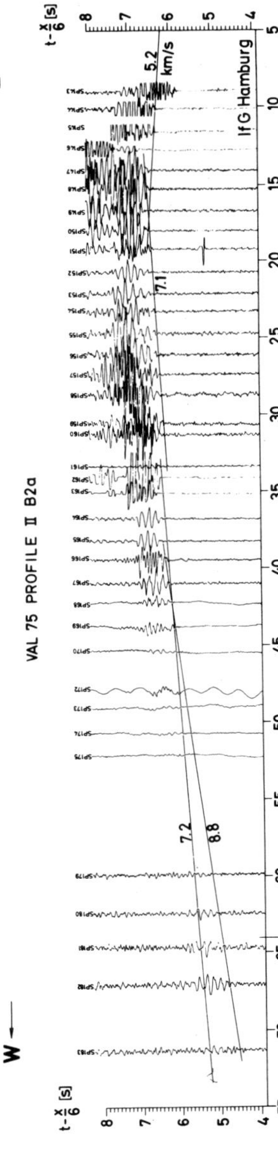

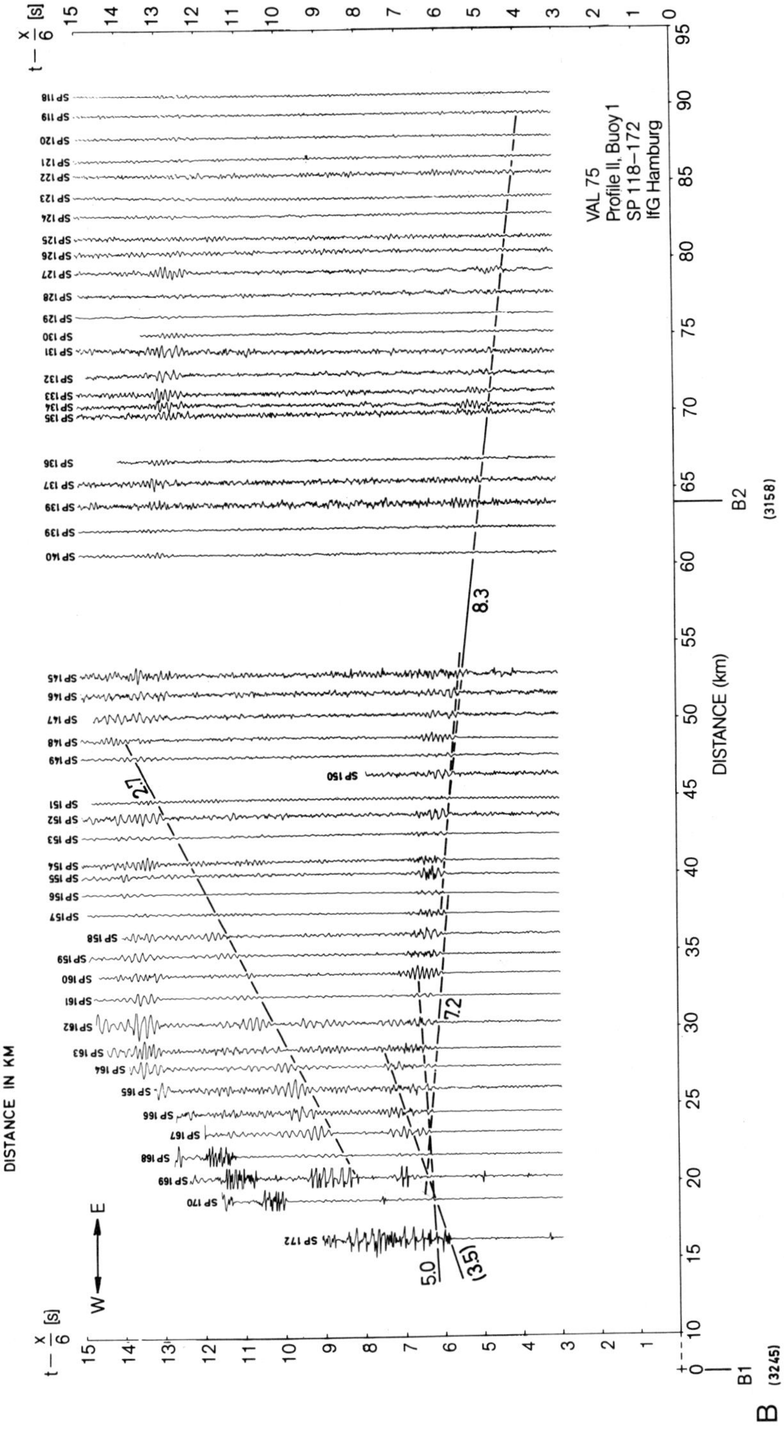

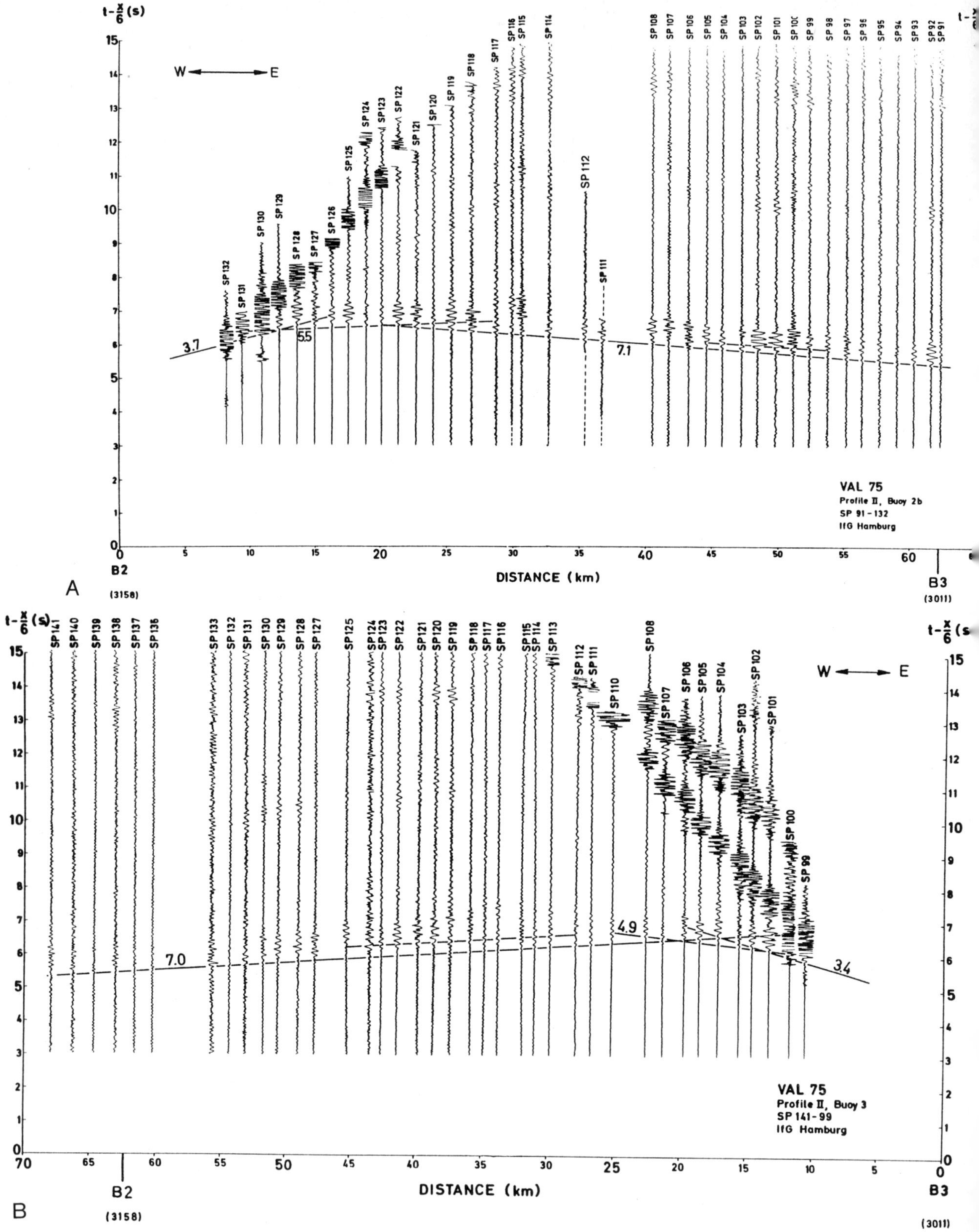

Fig. 9A. Seismogram section VAL 75/II, observed from W to E relative to B2
B. Seismogram section VAL 75/II, observed from E to W relative to B3

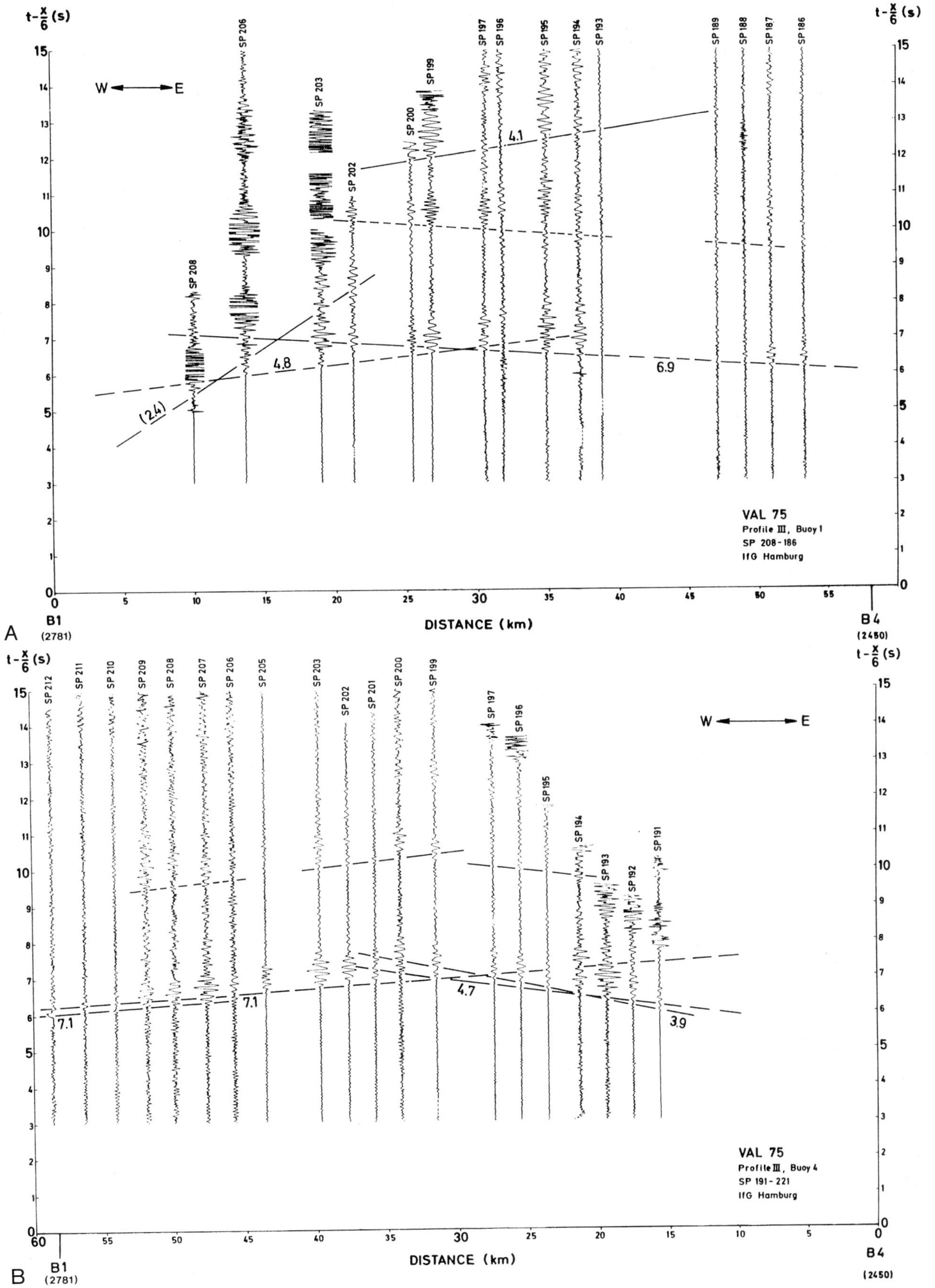

Fig. 10A. Seismogram section VAL 75/III, observed from W to E relative to B1
B. Seismogram section VAL 75/III, observed from E to W relative to B4

vious that the "often cited two-layer model of 5 ± 0.5 km/s basement (Layer 2) on a 6.75 ± 0.25 km/s layer 3 is an over-simplification" (Houtz and Ewing, 1976): Drilling in the North Atlantic of the uppermost half kilometre (Hall and Robinson, 1979) and in situ sampling of the plutonic foundation of oceanic crust in the Carribean Cayman Trough spreading centre (Caytrough, 1979) have, for example, predicted a significant lateral heterogeneity of the crustal material.

The improved understanding of and the increased range of layer 3 velocities plus new data led us to modify our original concept of the crustal structure offshore Mauritania:

A multichannel profile (EXPLORA 78-48) obtained by the BGR in 1978 along the VALDIVIA-10 refraction profile of 1975 clearly shows oceanic basement (vp = 5.2 km/s). It extends up to 100 km further east from the point where oceanic crust is last defined by the presented refraction results, before it is obscured by the first sea floor multiple reflection. There the basement reaches the depth (9-10 km) of the east dipping refraction layer of 7.1 km/s. Therefore oceanic crust with a decreasing layer 2 thickness and a "higher than normal layer 3 velocity" exists at least 200 km from the Mauritanian coast. The boundary between oceanic crust generated by sea floor spreading and originally continental crust,

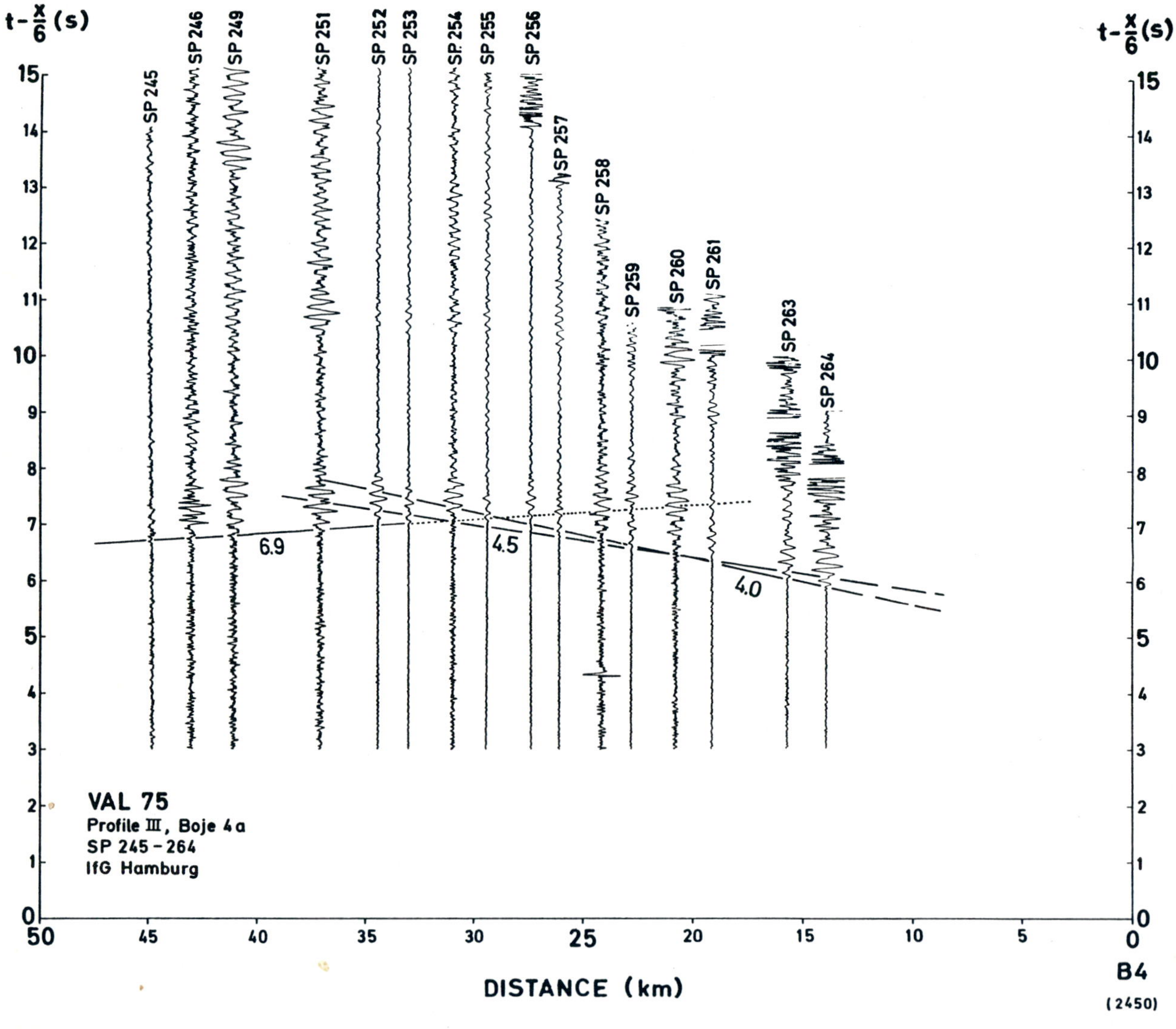

Fig. 11. Seismogram section VAL 75/III, observed from E to W relative to B4

Fig. 12

A. Seismogram section VAL 75/IV, observed from W to E relative to B4. High velocity 7.7 km/s caused mainly by updipping seafloor

B. Seismogram section VAL 75/IV, observed from E to W relative to B2 on the continental slope: max. seafloor inclination in the interval 0-30 km. Weak indication for mantle arrivals (8.5 km/s)

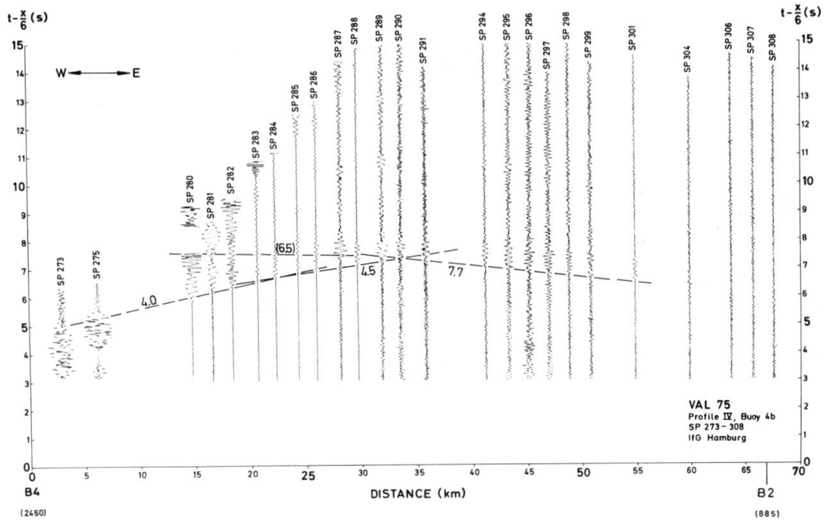

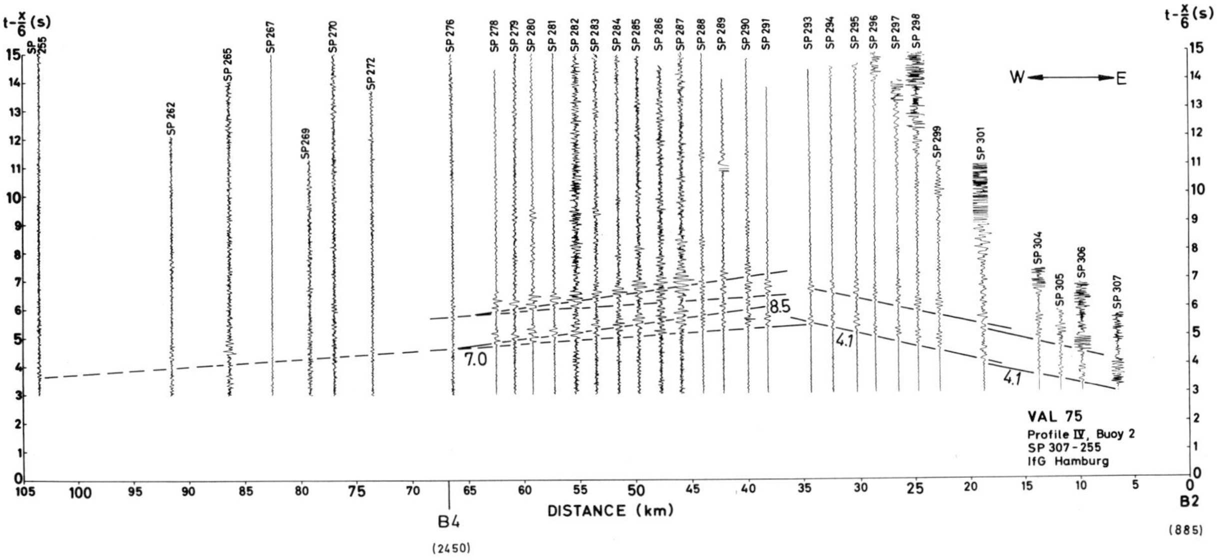

perhaps thinned and contaminated, must lie within this interval.

2.2.3. Mauritania upper slope and continental shelf (profiles IV, V)

The seismogram sections of Fig. 12 A - 14 A present informations about the continental slope and shelf region. The apparent velocities (Figs. 12 A, B; 13 A, B) are highly influenced by the dips of the sea bottom and deeper layers (compare the ray path model Fig. 15). On Fig. 14 A a p-velocity of 6.3 km/s occurs which is supported by strong second refraction arrivals originated by multiple reflected waves between sea surface and sea bottom.

The 7.1 km/s layer is also well documented under the upper slope and shelf (Figs. 14 A and B, 13 A): the interpretation there depends on the significance given to the 6.3 km/s layer above it. The latter layer was first only recognized over a 10 km short interval near the critical distance of the 7.1 km/s layer (Fig. 14 A) in the northern part of profile V. It was then tentatively associated with Paleozoic/Precambrian basement (Weigel and Wissmann, 1977) at 11 km depth.

In 1978 the sonobuoy 78-2 also recorded at 9.4 km depth an updip velocity of 6.45 km/s to the southeast of profile V,

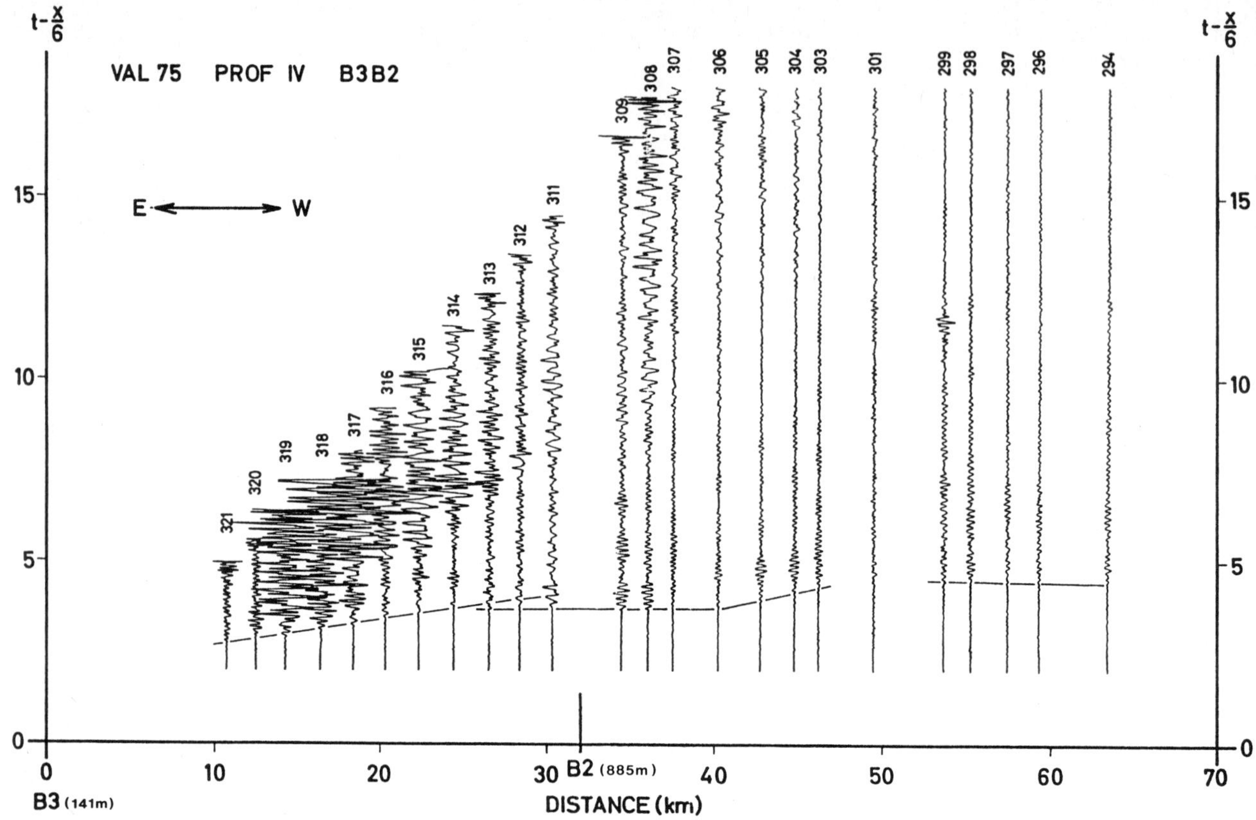

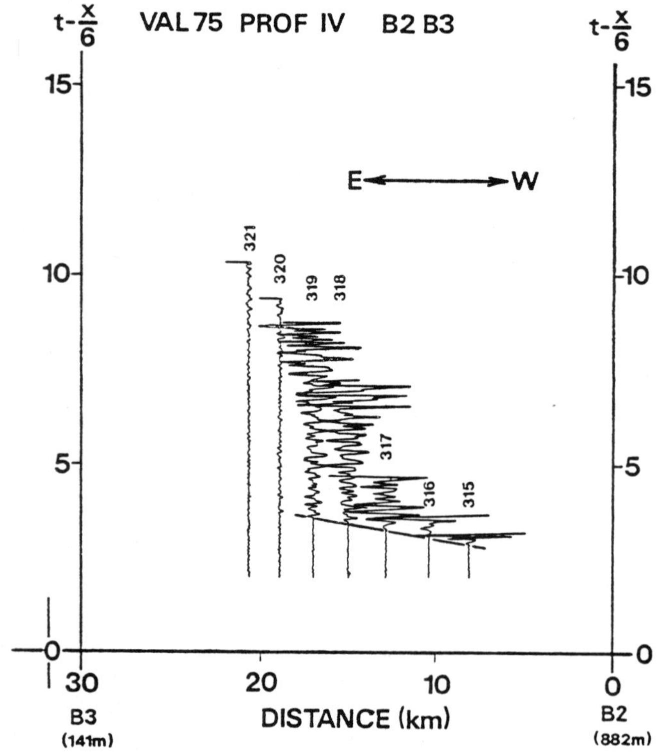

Fig. 13A. Seismogram section VAL 75/IV, observed from E to W down the slope relative to B3

B. Seismogram section VAL 75/IV, observed from W to E up the slope relative to B2. Note sudden loss of seismic energy (faults?, thick sediments) in shot records 320 and 321

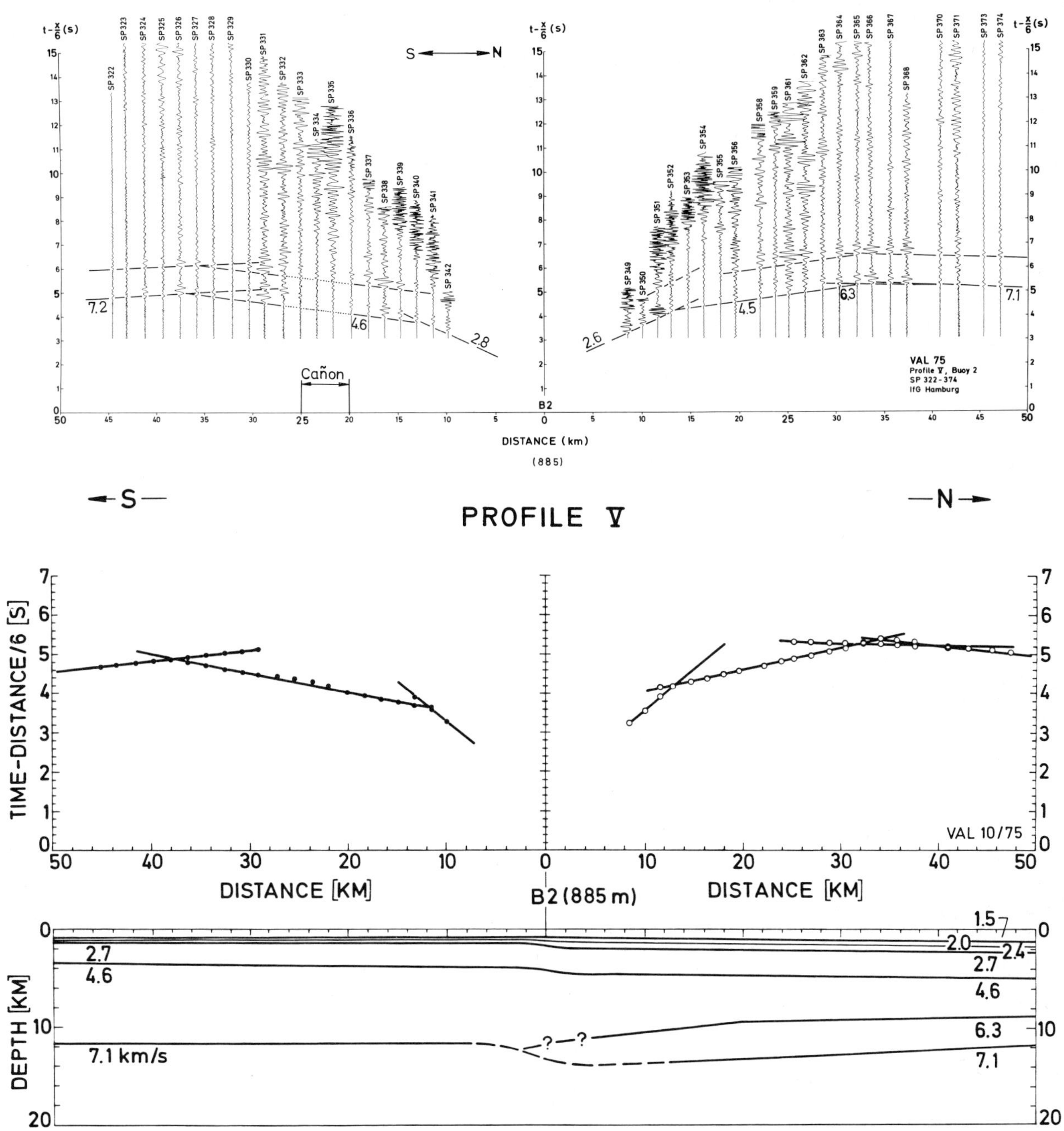

Fig. 14A. Seismogram section VAL 75/V, observed by B2 to N and S, in strike direction of the continental slope along 1000 m isobath

B. Crustal structure in strike direction of the continental slope

while sonobuoy 78-3 to the northeast found a downdip velocity of 4.8 km/s at 5.2 km depth (Fig. 2 B). Because of the large difference in depth these two layers are probably not related. On the EXPLORA 48-channel profiles 78-48 and 50 the deepest correlatable reflectors at about 5 km depth dip seaward at 15° - 20 ° thus cautioning against an uncritical acceptance of the sonobuoy apparent velocities. Overlapping refraction data give another clue about the existence of a 6.3 km/s layer. Buoy 3 (Fig. 13 A,B) recorded such a layer that dips from 7.5 to 11 km depth (Fig. 16).

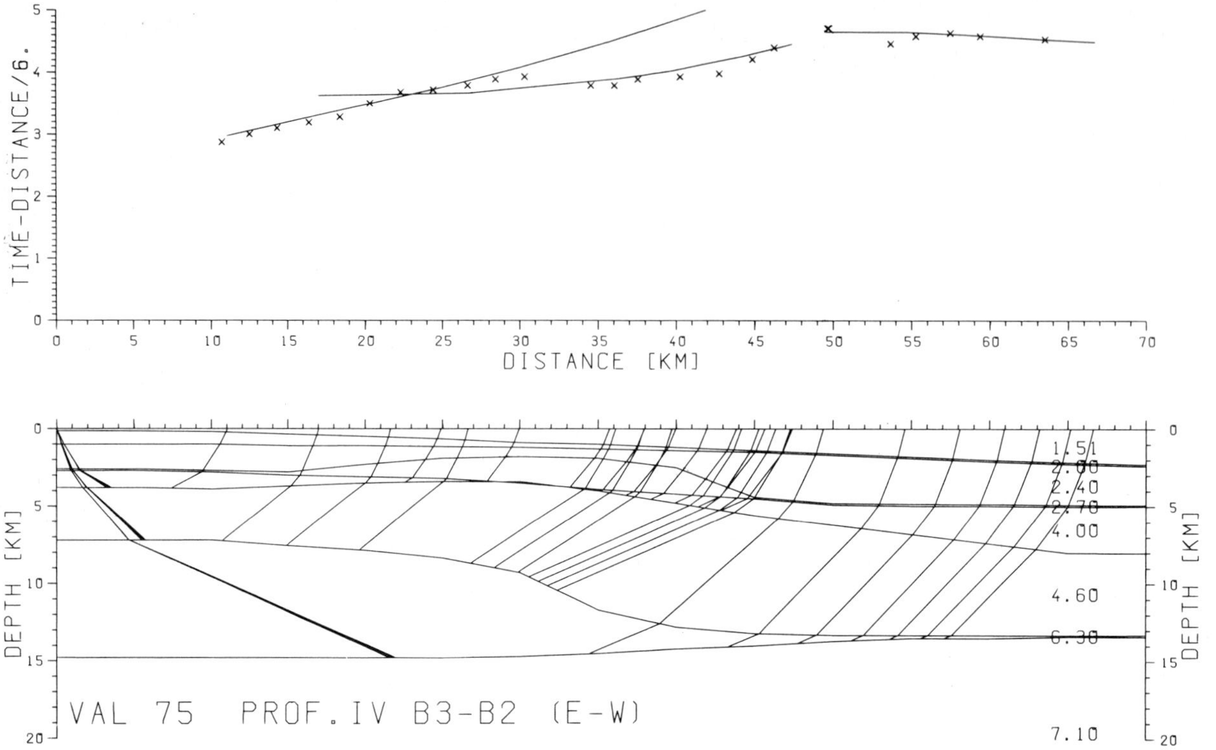

Fig. 15. Ray path model for shelf buoy B3. Continuation of 7.1-refractor under the upper slope

Under the Cape Verde Rise we associated velocities of 6.0 - 6.3 km/s with volcanic rocks (see 2.2.1.) and it has been suggested (Lehner, pers. communication) that some of the diapiric bodies off Mauritania (including the feature interpreted as Cretaceous carbonate front in Fig. 7 of Goldflam et al., 1980) are in fact igneous intrusions.

We will now discuss the interpretation of the 6.3 km/s layer as either basement or igneous intrusion with additional data. We will also consider some observations of the basement depth under the onshore basin for comparison with the 6.3 km/s refractor offshore (Fig. 16).

Magnetic data give some support to the interpretation as intrusions. Separated by 30 km, two magnetic lineations (Rona et al., 1970), convex to the west, curve around Cape Mirik (19°23'N). The southern termination of the western magnetic anomaly seems to coincide with a north-south striking, voluminous intrusion, which was crossed by profile VA-10-66 (Goldflam et al., 1980) and possibly 78-48 (Wissmann, this Vol.).

The three separate observations of a 6.3 - 6.45 km/s layer around 60-80 km from the Mauritanian coast all lie between these two magnetic anomalies. The landward anomaly near 17°30'N continues onshore (Liger, 1980) as the "positive West-African coast anomaly" approximately along the 5000 m basement depth contour (A.S.G.A., 1968; Fig. 2B) to M'bour south of Dakar. Even onshore it seems to be related to intrusive bodies along a zone of crustal weakness, which marks the eastern boundary of intermediate type crust, produced during the rifting phase of the Atlantic (Liger, 1980). This interpretation of the magnetic anomalies is supported by their position along the eastern border of the coastal gravity gradient (Crenn and Rechenmann, 1965) and by onshore refraction data in northern Senegal:

a) Spectral analysis of the magnetic and gravity anomalies located the sources in around 7 km depth, which is about 2 km more than shown on the Tectonic Map of Africa (Fig. 2B).
b) Seven kilometers depth to basement have also been measured on two refraction seismic profiles (Fig. 1), striking E-W to the north and south of the Leona dome (15°45'N). They show faulted basement with velocities between 5750

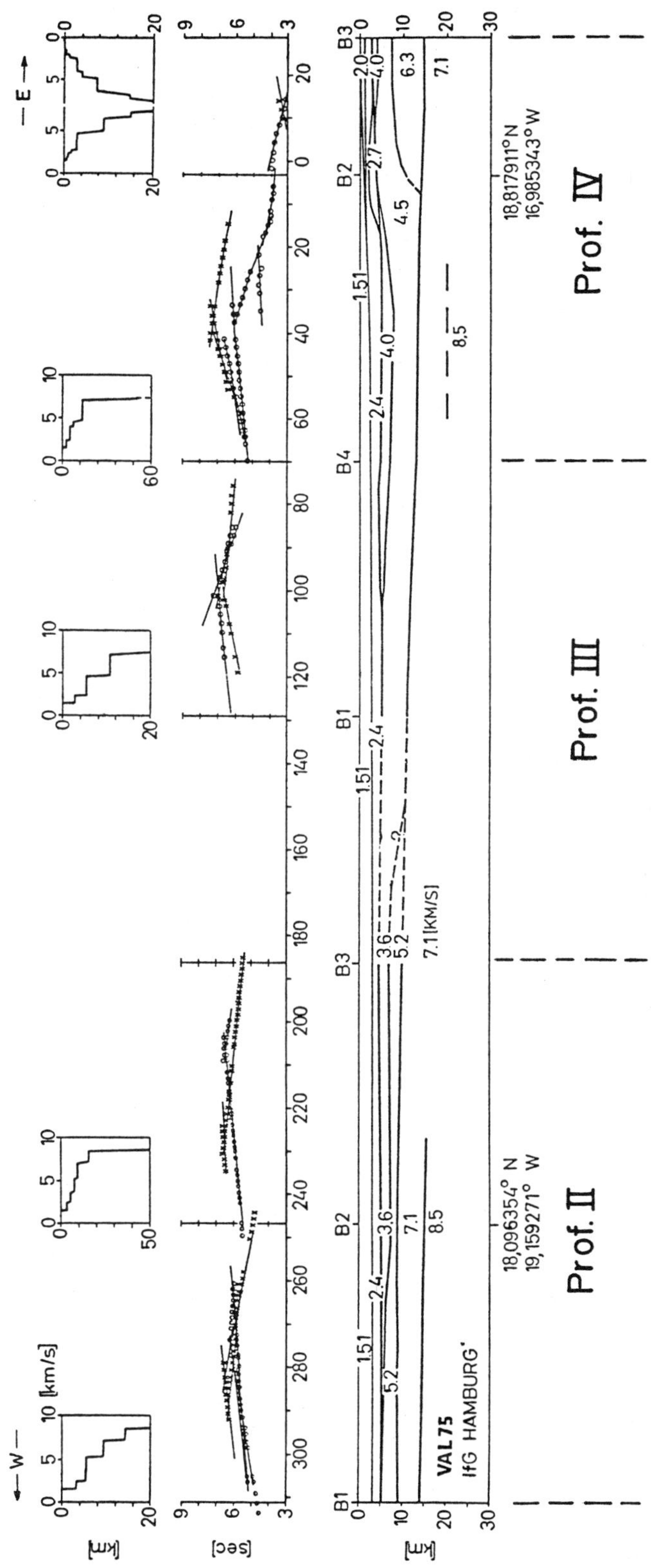

Table VAL 75

Prof	Buoy	Longitude	Latitude	Water depth (m)
I	B1	20,399239°W	17,638208°N	3214
	B2	20,816036°W	17,963584°N	2839
	B3	21,228993°W	18,210838°N	3050
II	B1	19,738574°W	17,931505°N	3245
	B2	19,159271°W	18,096354°N	3158
	B3	18,590685°W	18,251121°N	3011
III	B1	18,103973°W	18,486428°N	2781
	B4	17,615083°W	18,730068°N	2450
IV	B2	16,985343°W	18,817511°N	885 (b)
	B3	16,685542°W	18,863162°N	141 (b)

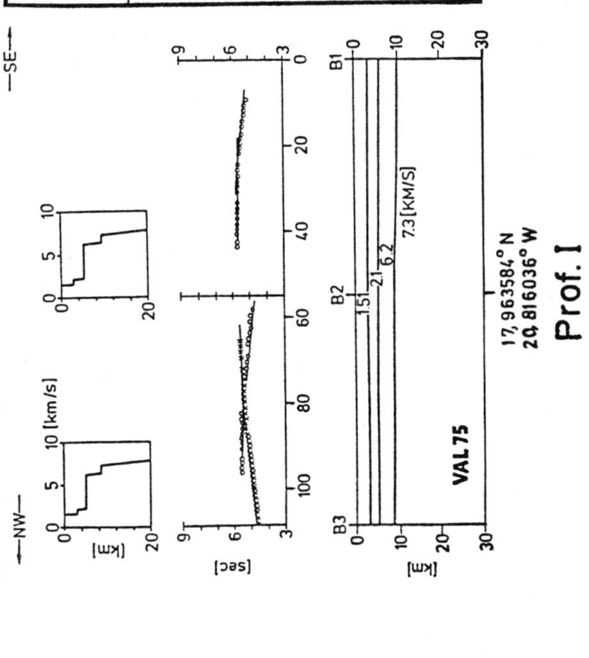

Fig. 16. Crustal structure from ray tracing models: Cape Verde Rise and the continental margin area off Mauritania.
Travel-times reduced by 6 km/s; all distances in km

and 6300 m/s plunging with increasing dip to depths of 7000 m near the coast (Liger, 1980). Westward dipping sediment layers with velocities between 4850 and 5600 m/s lap onto the basement at about 4 km depth. Similarly in a 1953 refraction survey of northern Senegal, a fast marker horizon of 5.9 km/s taken to be basement by Maugis (1955) was drilled as ? Mid-Cretaceous carbonate in 3 km depth by the stratigraphic deep hole Tienaba-1 (Hedberg, 1956) 70 km east of Dakar (Fig. 2).

The compilation of these data serves to illustrate that a petrological interpretation of seismic velocities especially in areas of complicated structure and facies changes (continental margins) can only be achieved by a combination of different geophysical methods. For example, high refraction velocities along three traverses offshore the Senegal Basin (Fig. 2B) have been compared and correlated with the aid of reflection seismic records to either consolidated sediments, basement or carbonates. From the available data therefore it is not definite, whether the 6.2 km/s layer under the shelf represents basement or igneous intrusions.

The restricted occurrence of this layer east of the 1000 m isobath, the approximate coincidence with the basement depths of the Tectonic Map of Africa (after they are increased by 2000 m as proven in Senegal) and the existence of the same basement velocities under the onshore basin favour the 6.2 km/s layer as continental (? Paleozoic/Precambrian) granodioritic (Smithson, 1978) basement.

Accepting this interpretation, we now consider the 7.1 km/s layer, which passes continuously from oceanic under continental crust (Fig. 16). The interpretation as gabbroic oceanic layer 3 may be upheld only to the line, where oceanic layer 2 can be identified on reflection seismic records.

However, velocities near 7 km/s are also found in continental crust and they are typical for gabbroic rocks to (> 7 km/s) pyroxene granulites (Fountain et al.,1975; Smithson and Brown, 1977).

Considering the lateral and vertical heterogeneity of the deep continental crust revealed by reflection seismic studies (COCORP; Smithson, 1978), a continuous 7.1 km/s layer may mean a first order discontinuity in the composition of oceanic and continental crust.

The discussed interpretation of the VALDIVIA-10/1975 refraction profile forms the basis of a gravitational model for the Mauritanian continental margin (Fig. 17), an improved version of an earlier model (in Fritsch et al., 1978). A deeply subsided basin filled with rift sediments (evaporites, carbonates) exists under the lower continental slope and is well expressed by a gravity anomaly.

3. Conclusions

The evolution of the continental margin off Morocco and Mauritania will be explained from the presented seismic crustal

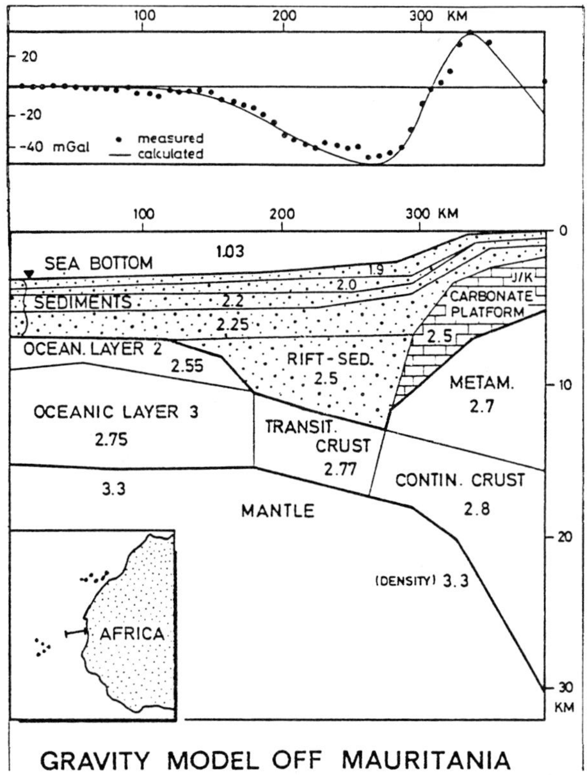

Fig. 17. Schematic crustal section across the continental margin of Mauritania. Density polygons derived from the seismic model Fig. 16, densities adjusted to fit gravity data (after Fritsch et al., 1978). Surface of elongated mantle plume corresponds to surface of transitional crust and continental crust. Top of mantle under shelf and slope only chosen to fit onshore gravity anomalies

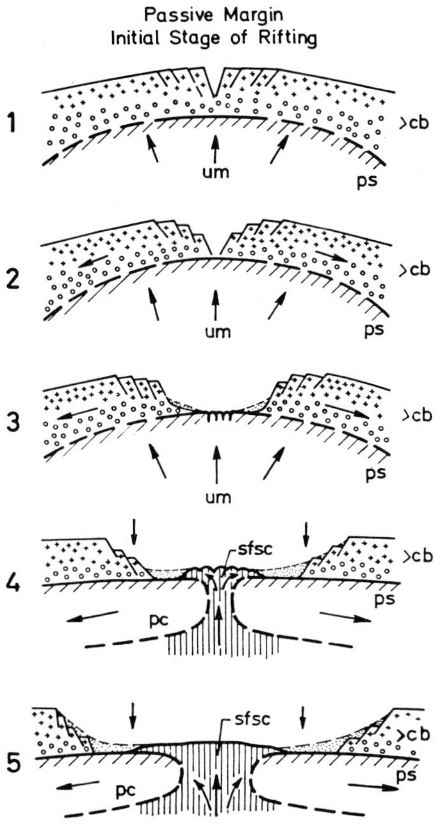

Fig. 18A. Schematic model for the evolution of the passive continental margin off NW Africa based on the presented refraction seismic models (our emphasis on phases 3,4).
cb : continental block
um : updoming mantle (elongated plume along the rift zone)
ps : plume surface
pc : plume crust
sfsc: sea-floor spreading generated crust

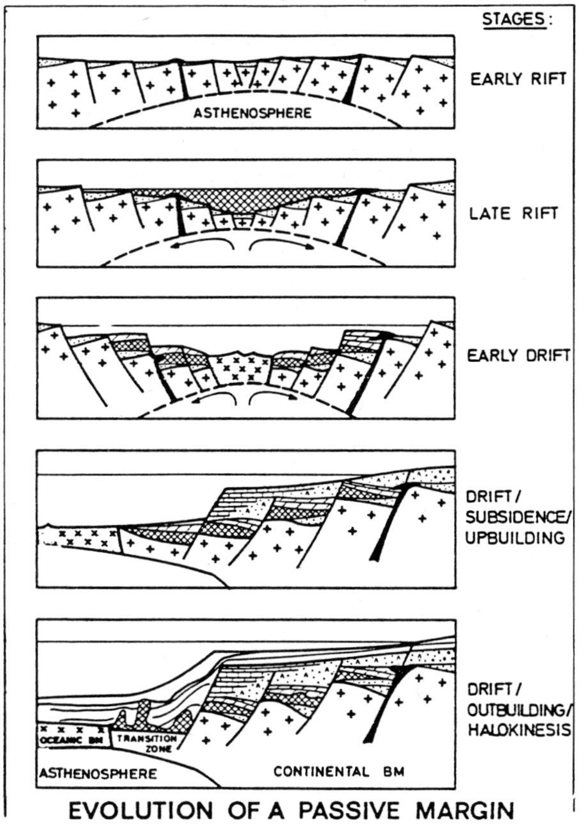

Fig. 18B. Evolution model for the passive margin off Morocco. Adapted after Hinz et al. (this Vol.) from DSDP Leg 50 Initial Reports by introducing a crustal transition zone at the foot of the continental sialic blocks

sections (Fig. 6, 16) by rifting of the African and American continental blocks, which started about 200 million years ago.

It is here assumed, that prior to brittle failure of the upper crust (Montadert et al., 1979) an upwelling zone of mantle material extended several thousands of kilometers along the direction of the later mid oceanic rift zone. As a precursor to the spreading process this zone was not restricted to only a narrow region in a transverse direction, but stretched far into the present ocean and land areas.

This assumption is supported by the presented crustal sections, all of which show compressional seismic wave velocities of 7.1 - 7.3 km/s in a depth of about 12 km under the continental rise dipping down to about 15 km under the continental slope off Morocco as well as under the slope off Mauritania. This prominent seismic horizon has been observed in the marginal basin of Morocco over more than 200 km between the outer shelf and the East Canary Ridge. Off Mauritania this horizon stretches without interruption from the shelf to the Cape Verde Rise over a distance of more than 500 km. Therefore this marked horizon plays an important role for the evolution of the NW-African passive continental margin.

From the presented crustal sections a model for the evolution of the continental margins off Mauritania and Morocco has been derived (Fig. 18A). By this conception the marked horizon with the p-velocity of 7.1. - 7.3 km/s, which continues under the shelf (Fig. 15, 16) and then perhaps to 25 km and more under the continent (Makris, 1981), is identified as a solidified former mantle plume surface. The presented schematic model (Fig. 18 A) demonstrates five phases of two (Cochran, 1981) successive stages.

A: Rifting Stage:

1. Updoming of mantle material (um) along a linearly far extended upwelling zone (rift-plume), with uplifting being effective hundreds of kilometers perpendicular to this zone. Breaking of the continental lithosphere block (cb) along this zone, origin of faults. ps: plume surface.
2. Separation of the continental blocks - by gravitational sliding on the hot and viscous plume surface, perhaps driven by convection under the continental blocks.
3. Further separation of the blocks by sliding. Cooling of the upper plume by crustal water convection (Lister, 1980) and development of a brittle zone. Begin of subsidence and breaking of the solidified plume surface (ps). Sedimentation down the continental flanks onto the new basement ps.

B: Spreading Stage:

4. Further separation and subsidence, origin of the marginal basin and the continental slope which can extend landward (compare Figs. 6 and 16). Thinning of the upper crust by stretching and subcrustal melting associated with contamination may occur. - Beginning of sea-floor spreading in a smaller zone of uprising magma. Flood basalts cover ps. - sfsc: sea-floor spreading generated oceanic crust.
5. Continental block (cb) and plume crust (pc) are coupled into one unit, which further separates from the opposite continental block by sea-floor spreading within a small zone (initiating fast cooling compared with former slow cooling of the plume). ps builds the oceanic basement in the rift basin region and dips down to the continent as early flank of the upwelling zone.

The early stage of rifting is characterized by an updoming of mantle material and a decrease of density and p-velocity during the updoming process. The material seems to be partially molten and of high viscosity and temperature. In regions of active rifting, compressional velocities around 7.0 - 7.6 km/s are often to be found: for example in the Red Sea, in the Afar rift (Berckhemer et al., 1975), under the Mid Atlantic Ridge (Forsyth, 1977), at the "hot spot" Iceland and frequently in active volcanic areas like Tenerife (Bosshard and Macfarlane,1970). These zones are of elevated temperatures compared to passive continental margins and contain low pressure zones where the material often rises to the surface.

The origin of passive continental margins by rifting processes has recently been reviewed (Meissner, 1980; Bott, 1980). The continents are breaking generally in the upper reaches of the upwelling zone and slide apart, more or less driven by the forces of gravity and additionally by convective forces in the asthenosphere under the continental block.

In our interpretation it is assumed that the surface of the upwelling zone (p_s in Fig. 18 A) builds the young basement of the juvenile ocean. Then the cooling phase begins, accompanied by contraction and an increase of density and p-velocity. In some areas, especially under epicontinental seas, pore water accelerates the cooling process (Lister, 1980) and the subsidence. The plume surface forms a discontinuity or a zone of a high velocity gradient. Velocity depth functions controlled by synthetic seismogram analyses (Hansen, 1979) prove the existence of such a discontinuity off Mauritania.

Near the top of the plume the compressional velocity should rise from a value less than 7.0 km/s in the heated, low pressure phase to more than 7.0 km/s after cooling and subsidence. Lower velocities which may lead to shadow zones in the observed refraction travel time curves, may still exist at greater depth: perhaps an indication for the phenomenon, that in the transitional "upwelling zone" no deeper refractor than that with the velocity 7.1 km/s could be observed by first arrivals.

A rising elongated plume and the sea-floor spreading mechanism are related processes. In both cases material rises from the mantle into a higher level and expands by heating and reduced pressure; but the plume extends over a large region and builds an approximately continuous surface. This could explain the magnetic quiet zone (Cochran, 1981) compared with the magnetized zone of a "pure sea-floor spreading generated crust" following the "plume phase".

If we date the magnetic quiet zone boundary between the Canary and Cape Verde Islands as about 155 m.y. (Pitman and Talwani, 1972) and the onset of separation of the continents as about 180-200 m.y., then the "plume phase" lasted for about 25-45 m.y. before the margin consolidated (100 m.y. in Kanes et al., 1973). This time in the Jurassic to Early Cretaceous coincides with the phase of high sinking rates of the continental margin (Hinz et al., this Vol.) perhaps influenced by sea water cooling of the crust. That the crust inside the magnetic quiet zone need not be sea-floor spreading generated agrees with Roeser's (this Vol.) interpretation of magnetic continental slope anomalies offshore Morocco as associated with magmatic injections into an old zone of weakness.

Likewise in the Mauritanian margin region far west of the coast no separation into the typical oceanic layer components 2 A, 2 B, 3 A, 3 B can be proven by the refraction results. Mainly layer 2 B and layer 3 B material seems to cause refractions. This is interpreted as a special magma differentiation occurring during the transition time from rifting (or from the "plume phase") to the "spreading phase". The eastward thinning of the 5.2-layer (Fig. 16) may indicate a lavaflow from a first spreading center located perhaps between B 1 and B 2. A continuation of the 5.2-layer (layer 2) to the continental slope is not clear. If layer 2 is mainly responsible for the magnetic lineations, their absence in the rift basin may also indicate a poorly developed basaltic layer 2. However even in that region, where layer 2 exists, no clear magnetic anomalies are evident off Mauritania. A thinning or absence of oceanic layer 2 near the continental margin has also been observed in other parts of the world ocean:

In the Gulf of Mexico a refraction layer of 7.0 - 7.2 km/s at 12 km depth constitutes the oceanic basement under evaporites and lithified carbonates with 5.0 km/s (Ladd et al., 1976). Orlenok (1976) interpreted Atlantic basement velocities ranging from 4.0 to 7.2 km/s to indicate a substantial heterogeneity in the composition of basement rocks. Keen and Hyndman (1979) report a 10 km wide zone of transitional basement between oceanic and continental crust offshore Nova Scotia in which a 7.4 km/s layer lies beneath the sediments (3.6 to 4.4 km/s).

However seaward of the U.S. East Coast Magnetic Anomaly in the area of the Baltimore Canyon Trough and in approximate juxtaposition to Mauritania a 7.1 - 7.2 km/s refraction layer occurs underneath a normal thickness (1-2 km) of oceanic layer 2 basement, identified in multichannel reflection seismic profiles. The 7.1 km/s layer is identified as oceanic layer 3, which here dips to 15 km depth under the shelf (Grow, 1980). A basement transition zone has not been interpreted there, but based on the reflection records seems not to be wider than 10 km.

The general evolution of a passive continental margin as demonstrated in Fig. 18 A need not progress uniformely in different geological provinces: Sedimentation onto the cooling plume surface (stage 3) begins with the deposition of evaporites and continental clastics (Fig. 16) off Mauritania. Off Morocco (Fig. 18 B) a thinned and contaminated continental crust (v_p = 5.8 km/s, Hinz et al., this Vol.) formed the subsiding base onto which evaporites were deposited (Lancelot and Winterer, 1980).

If this interpretation is correct, then rifting off Morocco began near the eastern border of the East Canary Ridge which early in the continental margin evolution must have been a zone of crustal weakness to be later intruded by Tertiary volcanic masses. West of this zone oceanic crust was generated.

If however the 5.8 km/s-velocity is interpreted as carbonate sediments, then the evolution occurred in the same way as off Mauritania. If on the other hand the layer is interpreted as basalts over the whole region of observation (Fig. 6) then the rift plume must have split up and sea floor spreading began already under the continental slope. But for this latter interpretation we have no strong indications.

An abrupt juxtaposition of oceanic sea floor spreading generated crust and continental crust as postulated off Sierra Leone (Sheridan et al., 1969) and off New Jersey, US coast (Sheridan et al., 1979) cannot be firmly stated from our refraction data.

Within a 200 km wide zone of the marginal basins offshore Morocco and Mauritania no typical "Moho" with compressional velocity first arrivals around 8 km/s was observed, neither under the Conception Bank (East Canary Ridge) nor under the Cape Verde Rise. The reason may be a vertical mixing and equalization process in the uprising mantle material, which leads to a low velocity gradient and produces a high gradient or discontinuity (7.1 - 7.3 km/s) only near the surface. So we find a clear Moho discontinuity only in "undisturbed" regions of "pure" continental or sea floor spreading generated crust.

Most of the published continental margin crustal sections lack information about the deep structure just under the slope, shelf and coastal basin, the transition zone from the present ocean areas to the continental block. In all cases it seems not to be clear, whether the 7.1 - 7.3 km/s marker-horizon continues under the continental block or whether it ends under the continental slope.

To answer this question combined on- and offshore refraction and deep-penetration, expanded-spread reflection seismic studies can shed more light on the crustal and upper mantle structure in the transition region from ocean to continent.

Acknowledgements

This study was financially supported by the Bundesministerium für Forschung und Technologie (BMFT) under grant MF 0196 and by the Deutsche Forschungsgemeinschaft.

The authors wish to thank all persons that participated in the cruises and contributed to the collection of data at sea, especially Mr. Herber and Mr. Richter, who serviced the IfG-telemetry-buoys and the shooting crew of the BGR led by Dr. Geyh. Thanks are also due to the officers and crew of the research vessels VALDIVIA and METEOR and to the Moroccan and Mauritanian authorities, who supported the German marine research in the economic zones of their countries' continental margin areas.

We are very grateful for critical comments by Drs. von Rad and Hinz and for a discussion with Prof. Closs, who drew our attention to specialized literature about petrology and geochemistry of continental rifts, which we had not considered in our more structurally oriented review of seismic data.

REFERENCES

Association of African Geological Societies and UNESCO (1968) International Tectonic map of Africa, scale 1 : 5,000,000. Paris, sheet 1

Banda E, Danobeitia JJ, Surinach E, Ansorge J (1981) Features of Crustal Structure under the Canary Islands (in press in Earth and Planetary Science Letters)

Berckhemer H, Baier B, Bartelsen H, Behle A, Burckhard H, Gebrande H, Makris J, Menzel H, Miller H, Vees R (1975) Deep seismic soundings in the Afar region and on the Highland of Ethiopia. Inter Union Commission on Geodynamics, Scientific report No. 14

Bosshard E, Macfarlane DJ (1970) Crustal structure of the Western Canary Islands from seismic refraction and gravity data. J Geophys Res 75: 4901-4918

Bott MHP (1980) Problems of passive margins from the viewpoint of the geodynamics project: a review. Phil Trans R Soc Lond A 294: 5-16

+Caytrough (1979) Geological and Geophysical Investigation of the Midcayman Rise Spreading Center: Initial Results and Observations. In: Talwani M et al. (eds) Deep Drilling Results in the Atlantic Ocean: Ocean Crust. M Ewing Series 2: 66-93

Christensen NI, Salisbury MH (1975) Structure and constitution of the lower oceanic crust. Rev Geophys Space Phys 13:57-86

Closs H, Dietrich G, Hempel G, Schott W, Seibold E (1969) Atlantische Kuppenfahrten 1967 mit dem Forschungsschiff Meteor. "Meteor" Forsch-Ergebn A, 5: 1-77

Cochran JR (1981) The Gulf of Aden: Structure and Evolution of a young ocean basin and continental margin. J Geophys Res 86 B 1, n 1 : 263-287

Crenn Y and Rechenmann J (1965) Mesures gravimétriques et magnétiques au Sénégal et en Mauritanie occidentale. Cah ORSTOM, sér Géophys, n 6: 59 p

Dash BP, Ball MM, King GA, Butler LW, Rona PA (1976) Geophysical investigation of the Cape Verde Archipelago. J Geophys Res 81, n 29: 5249-5259

Duncan RA, Jackson ED (1977) Geochronology of basaltic rocks recovered by DSDP Leg 41, Eastern Atlantic Ocean. In: Lancelot Y, Seibold E et al. Init Reps Deep Sea Drilling Project 41: 1113-1118

Eschenbrenner S, Dorel J, Viode J-P (1980) Coupes Sismiques des Structures Superficielles dans les Petites Antilles - II: Martinique. Pageoph 118: 807-821

Forsyth DW (1977) The evolution of the upper mantle beneath mid-ocean ridges. Tectonophysics 38: 89-118

Fountain DM, Carlson RL, Salisbury MH, Christensen NI (1975) Possible lower crustal rocks recovered on Leg 31 by Deep Sea Drilling in the Phillipine Sea. Mar Geol 19: M 75-M 80

Fox PJ, Schreiber E, Peterson JJ (1973) The Geology of the Oceanic Crust: Compressional Wave Velocities of Oceanic Rocks. J Geophys Res 78, n 23: 5155-5172

Fritsch J, Hinz K, von Rad U, Roeser H, Weigel W, Wissmann G (1978) Ergebnisse geowissenschaftlicher Untersuchungen der VALDIVIA-Westafrika Fahrt VA-10/ 1975. Forschungsbericht M 78-03 des Bundesministeriums für Forschung und Technologie, BMFT-FB M 78-03, 37 p

Gebhardt V, Weigel W (1981) METEOR-Expedition 53, 1980: Erste refraktionsseismische Ergebnisse vom Kontinentalrand vor Marokko. Paper presented on the annual meeting of the DGG, Heidelberg, 30.3.-3.4.1981

Goldflam P, Hinz K, Weigel W, Wissmann G (1980) Some features of the Northwest African margin and magnetic quiet zone. Phil Trans R Soc Lond A 294: 87-96

Grow JA (1980) Deep structure and evolution of the Baltimore Canyon Trough in the vicinity of the COST No. B-3 well. In Scholle PA (ed) Geological studies of the COST No. B-3 well, United States Mid-Atlantic Continental Slope Area. Geol Survey Circ 833: 117-132

Grunau HR, Lehner P, Cleintuar MR, Allenbach P, Bakker G (1975) New radiometric ages and seismic data from Fuerteventura (Canary Islands) and Sao Tomé (Gulf of Guinea). In: Progress in Geodynamics Symposium, April 3-4. Koninkligke Nederlandse Akademie van Wetenschappen: 90-118

Hall JM, Robinson PT (1979) Deep Crustal Drilling in the North Atlantic Ocean. Science 204, n 4393: 573-586

Hansen W (1979) Eine Untersuchung der Amplitudenverteilung der seismischen Wellen bei den Messungen des F.S. VALDIVIA 1975 vor der Küste Mauretaniens, Nordwest-Afrika. Hamburger Geophysikalische Einzelschriften, Reihe A, Heft 39, Hamburg, GML Wittenborn Söhne, 2 Hamburg 13

Hedberg HD (1955) Developments in Foreign Petroleum Fields - Africa. AAPG Bull 40: 1607

Hoskins H, Rogers CU, Woo AO (1974) Data report of oblique reflection - refraction radio-sonobuoy profiles on the African Atlantic continental margin (R/V ATLANTIS II cruises 67 and 75). Woods Hole Oceanogr Inst Ref No 74-34: 40 p

Houtz RE (1977) Interval Velocities in Crustal Layers From Sonobuoy Records of Mantle Reflections. J Geophys Res 82, n 33: 5395-5400

Houtz RE, Ewing J (1976) Upper Crustal Structure as a Function of Plate Age. J Geophys Res 81, n 14: 2490-2498

Kanes WH, Saadi M, Ehrlich E, Alem A (1973) Moroccan Crustal Response to Continental Drift. Science 180: 950-952

Keen CE, Hyndman RD (1979) Geophysical review of the continental margins of eastern and western Canada. Can J Earth Sci 16, n 3 (part 2): 712-747

Kent P, Laughton AS, Roberts DG, Jones EWJ (1980) The evolution of passive continental margins in the light of recent deep drilling. Royal Society discussion meeting held on 19 and 20 October 1977. Phil trans Roy Soc London, ser A, 294, n 1409: 1-208

Ladd JW, Buffler RT, Watkins JS, Worzel JL (1976) Deep seismic reflection results from the Gulf of Mexico. Geology 4, n 7: 365-368

Lancelot Y, Winterer EL (1980) Evolution of the Moroccan Oceanic basin and adjacent continental margin - A synthesis. In Lancelot Y, Winterer EL Init Reps of DSDP 50: 801-821

Lancelot Y, Seibold E et al (1977) Initial Reports of the Deep Sea Drilling Project, 41, Washington (U.S. Gov Print Off): 1259 pp

Le Pichon X (1977) 1967-77: Ten years of marine geology and geophysics. Nature 267: 765-769

Liger J-L (1980) Structure profonde du bassin côtier sénégalo-mauritanien, interprétation de données gravimétriques et magnétiques. Travaux des Lab. des Sciences de la Terre, ser B, 16: 158 pp

Lister CRB (1980) Heat flow and hydrothermal circulation. Ann Rev Earth Planet Sci, 8: 95-117

Luyendik BP, Bunce ET (1973) Geophysical study of the Northwest African margin off Morocco. Deep-Sea Res Vol 20: 537-549

Makris J (1981) Seismic studies in Morocco (abstr). 11th Colloquium of African Geology, The Open University, Milton Keynes, April 13-15, 1981

Maugis P (1955) Etudes de préreconnaissance pétrolière dans le bassin sédimentaire du Sénégal. Extrait de Rev de l'Inst Français du Pétrole, 10, n 5: 99-128

Meissner R (1980) The development of continental margins - a consequence of plumes intruding a variable viscosity tectonosphere (in press)

Mendes-Victor LA (1970) L'interprétation des mesures gravimétriques et magnétiques aux îles du Cap Vert et la théorie de l'expansion des fonds océaniques. Thèse à l'Université de Strasbourg: 198 pp

Montadert L, de Charpel O, Roberts D, Guennoc P, Sibuet J (1979) Northeast Atlantic Passive Margins: Rifting and Subsidence Processes. In Talwani M et al (eds) Deep Drilling Results in the Atlantic Ocean: Continental Margins and Paleoenvironment. M Ewing Series 3: 154-186

Moores EM, Jackson ED (1974) Ophiolites and oceanic crust. Nature 250, 5462: 136-138

Natland J (1977) Composition of basaltic rocks recovered at Sites 367 and 368, Deep Sea Drilling Project, near the Cape Verde Islands. In: Lancelot Y, Seibold E et al Initial Reps of DSDP, 41: 1107-1112

Pavia J (1977) Comparacion de la estructura de la corteza en Canarias y Baleares por refraccion sismica. Tesis de Licenciatura, University of Barcelona: 88 p

Orlenok VV (1976) Neomobilism in the light of data on the structure of Atlantic bottom sediments. Geotectonics 9, n 6:397-402

Pitman III W, Talwani M (1972) Sea-Floor spreading in the North Atlantic. Geol Soc of Am Bull 83: 619-646

Roeser HA, Hinz K, Plaumann S (1971) Continental margin structure in the Canaries. In Delany FM (ed) The geology of the East Atlantic continental margin. Great Britain Inst Geol Sci Rep 70/16: 27-36

Rona PA, Brakl J, and Heirtzler JR (1970) Magnetic Anomalies in the northeast Atlantic between the Canary and Cape Verde Islands. J Geophys Res 75: 7412-7420

Schmincke HU (1979) Age and crustal structure of the Canary Islands. J Geophys 46: 217-224

Seibold E, Hinz K (1976) German cruises to the continental margin of North West Africa in 1975: General reports and preliminary results from "VALDIVIA" 10 and "METEOR" 39. "Meteor" Forsch-Ergebn C, No 25: 47-80

Sheridan RE, Houtz RE, Drake CL, Ewing M (1969) Structure of Continental Margin off Sierra Leone, West Africa. J Geophys Res 74, n 10: 2512-2530

Sheridan RE, Grow JA, Behrendt JC, Bayer KC (1979) Seismic refraction study of the continental edge off the eastern United States. Tectonophysics, 59: 1-26

Smithson SB (1978) Modeling continental crust: structural and chemical constrainst. Geophys Res Letters 5, n 9: 749-752

Smithson SB, Brown SK (1977) A model for lower continental crust. Earth Planet Sci Letters: 134-144

Thiessen J (1979) Seismische und gravimetrische Untersuchungen der Krustenstruktur am marokkanischen Kontinentalrand. Diplomarbeit, Institut Geophys Universität Hamburg

Uchupi E, Emery KO, Bowin CO, and Phillips JD (1976) Continental margin off Western Africa: Senegal to Portugal. AAPG Bull 60, n 5: 809-878

Watkins JS, Hoppe KW (1979) Seismic reflection reconnaissance of the Atlantic margin of Morocco. In Talwani M et al (eds) Deep drilling results in the Atlantic Ocean: continental margins and paleoenvironment, M Ewing Ser 3; Am Geophys Union: 205-217

Weigel W, Wissmann G (1977) A first crustal section from seismic observations west of Mauritania. In Publ Inst Geophys Pol Acad Sci A4, 115: 369-380

Weigel W, Goldflam P, Hinz K (1978) The crustal structure of Conception Bank. Mar Geophys Res 3: 381-392

Wissmann G (1979) Cape Bojador slope, an example for potential pitfalls in seismic interpretation without the information of outer margin drilling. In: von Rad U, Ryan WBF et al Init Reps Deep Sea Drilling Project 47 (1): 491-499

9 Stratigraphy and Structural Features of the Continental Margin Basin of Senegal and Mauritania

Gerd Wissmann[1]

ABSTRACT

Published geological and geophysical information was compiled and considered with some gravimetric and seismic data of the BGR to obtain a synthesis of the stratigraphy and structure of the Senegal-Mauritania Basin:

Refraction and gravimetric data show the deepest part (ca. 14 km) of the basin to lie under the lower continental slope. It is flanked in the west by an oceanic hinge line in Jurassic basalts with hundred-meter fault throws. In the east faults of a continental hinge line offset continental blocks up to a kilometer. This structural boundary of the continent had been active until the Tertiary and formed listric growth faults in the Cretaceous section north of Dakar.

Doleritic rocks occurring in Senegal are not associated with rift tectonics of the Atlantic ocean formation. Triassic rift sediments have not been found onshore south of the Precambrian Reguibat High, however during this time and until the Middle Jurassic a peneplanation surface was eroded in eastern Senegal (? rift onset unconformity).

The base of the Jurassic section and the time of the first marine transgression are not known, but regional subsidence of the continental margin is assumed to have started during the Liassic. An almost continuous succession of Jurassic to Cenomanian platform carbonates forms a regressive Mesozoic shelf edge with a 4 km high escarpment under the southern continental slope. This most prominent structural feature deepens to the north and is less pronounced off Mauritania. Paleozoic? to Triassic diapirs, originating from an evaporite layer underneath the carbonates at 8 to 10 km depth off Casamance, there occur landward of the Mesozoic shelf edge, while off Mauritania ?Liassic diapirs grew from > 6 km depth on the seaward side.

A thick Cretaceous-Tertiary basin offshore Mauritania, with a regional northward deepening of sediment layers, thins in a southern and western direction. Its eastern border is drawn by a group of N-S striking faults along the 200 m-isobath.

Extensive deltaic wedges developed during the Cretaceous off Mauritania and possibly drowned the Mesozoic carbonate platform. Another Cretaceous to Paleogene carbonate platform exists further north off Cape Blanc in the form of a capital letter L. Its southern limb possibly marks the eastern extension of the Kane Fracture Zone, which is tentatively interpreted to extend into the onshore region of the Arguin Platform.

After the Early Jurassic extensional tectonic phase a second major tectonic phase during the Late Cretaceous established fault controlled anticlines and volu-

1 Bundesanstalt für Geowissenschaften und Rohstoffe, Stilleweg 2, 3000 Hannover 51, FRG

minous igneous intrusions, documented by semi-circular gravity anomalies, in the coastal region of the basin.

During the early Tertiary carbonate deposition reappeared on the present shelf, while the coast line regressed, as an epicontinental sea flooded almost the whole basin.

In the Neogene a prograding sedimentation pattern dominated again. Renewed activity on deep coastal faults especially in the Miocene and Early Quaternary uplifted deep water sediments and culminated in an intensive alcaline volcanism on the Cape Vert peninsula, while sea mounts were constructed on the continental slope. Early Miocene slumps and turbidite deposits under the lower continental slope point to the activity of mass wasting processes. The Middle Miocene shelf edge lies under the upper continental slope. During the Pliocene/-Pleistocene numerous canyons on the continental margin, which coalesce into two major channel systems on the continental rise, cut into the sediments that had outlasted several paleogene and neogene erosional cycles.

1. Introduction

The Senegal-Mauritania Basin is the largest marginal basin off Northwest Africa. Its onshore part alone extends over an area of about 340 000 km^2 in Mauritania, Senegal, Gambia and Guinea-Bissao (see front cover figure). It is filled by a >7000 m thick sequence of Mesozoic, terrigeneous to shallow marine sediments (Castelain, 1965). In the reported maximum onshore thickness of the Cenozoic section (ca. 1500 m: Castelain, 1965) Paleogene rocks dominate with more than 1200 m of sediments. Together an accumulation of more than 8000 m (Sociéte Africaine des Pétroles (1961): schematic cross-section of the Senegal basin near Dakar in Castelain, 1965) of almost uninterrupted Upper Jurassic to Quaternary sediments are known.

Compared to the Aaiun-Tarfaya Basin north of 21°N latitude (cf. Ranke et al., this Vol.) the Senegal-Mauritania Basin has probably undergone the same Mesozoic evolution, but their histories during the Tertiary differ. While at least in the southern coastal part of the Aaiun-Tarfaya Basin the Paleogene sediments still reach the same thicknesses as in Senegal, the prominent Neogene erosional unconformity documented off the West-Sahara (von Rad and Wissmann, this Vol.) is lacking in the Senegal-Mauritania Basin.

According to Klitgord and Behrendt's (1979) reconstruction there is not one, but several matching counterpart basins along the North American margin during pre-drift time (175 m.y. BP): the Carolina Trough (Dillon and Paull, 1978) the Southeast Georgia Embayment (Dillon et al., 1979) and the Blake Plateau (Shipley et al., 1978; Paull and Dillon, 1980).

On the other side of the Atlantic few structural features (most of them based on gravity anomalies: Liger and Roussel, 1979) are known at least in the onshore Senegal-Mauritania Basin, which would justify a differentation into several subbasins. Off Mauritania the offshore part of this large West African basin is covered with more than 10 km of sediments (Weigel et al., this Vol.) and very little information is available about the basement configuration under the continental slope. Offshore Gambia and Casamance in southern Senegal a thick sequence of Mesozoic shelf carbonates overlying a ?Liassic evaporite layer at 8 to 10 km depth masks seismic reflections from the underlying crust and possible Mesozoic rift grabens even on commercial oil company profiles (Lehner and de Ruiter , 1977).

The main objective of this paper is therefore to present new information about the Mesozoic stratigraphy and structure of the deep offshore Senegal-Mauritania marginal basin under the continental slope as interpreted on BGR multichannel seismic profiles obtained off Mauritania. Previously published single channel profiles from this area (Mc Master and Lachance, 1968; Lowrie and Escowitz, 1969; Collette et al., 1969; Seibold and Hinz, 1974; Uchupi et al., 1976) rarely penetrate beneath the Cenozoic sediments. They are considered in a brief synthesis of all pertinent available data on the Senegal-Mauritania Basin, but the seismic profile spacing in the continental margin area is still so large and the seismic penetration so variable that it seemed unfeasible to prepare structure-contour and isopach maps of seismic sequences. This was attempted by Wissmann over the continental rise with reflection time maps for basement, the top of mid-Cretaceous black shales and the Eocene marker horizon A in Goldflam et al. (1980). Because of a lack of data only typical structural features of the Senegal-Mauritania basin will be exemplified in the following.

2. Geology of the Senegal-Mauritania Basin

2.1 Geological setting

The northern boundary of the Senegal-Mauritania Basin is formed by a piece of the Precambrian African craton, the Reguibat High. Along the eastern and southern boundary a Paleozoic foldbelt stretches more than 1800 km from Southern Morocco to Sierra Leone: the Mauritanides (Sougy, 1962; Machens, 1966; Serie d'Akjout of 350-270 m.y., Choubert and Faure-Muret, 1971). In latest- to post-Devonian time (Lecorche and Sougy, 1978) the Mauritanides were thrust onto the cratonic crust or onto Early Paleozoic sediments of intracratonic basins. The Hercynian fold-belt is nowhere wider than 200 km and it shows a westward increasing degree of metamorphism. South of 13°N the Mauritanides split into a NS- and a SW-striking mountain-chain: the Rokelides (Dia et al., in press).

The tectonic framework of the Mauritanides changes from overturned east-facing folds in the south to several generations of "thinskin" allochthonous nappes with large transport distances in the north. To the west the Paleozoic basement dips under the Mesozoic-Tertiary Senegal-Mauritania Basin. In eastern Senegal this basement has been peneplained by the ? Triassic niveau d' aplanissement 1 (Michel, 1973; p. 322) possibly a "rift onset unconformity" formed prior to the formation of the Atlantic.

Onset of rifting in other parts of Northwest Africa was commonly associated with Triassic to Liassic igneous activity (Bertrand and Prioton, 1975). Thick (200 m) doleritic sills and dykes also occur in Senegal (Michel, 1973, p. 75-78; Maugis, 1955, p. 123; Hebrard, 1978, p. 92). One hundred kilometer from the coast the oil well D 1 drilled diorites in 4103 m depth. However in Senegal these igneous rocks have been related to either the Hercynian orogeny, as they intrude Paleozoic sediments as young as Upper Carboniferous and metamorphic rocks of eastern and southern Senegal (Michel, 1973) or to the Neogene volcanism of the Cape Vert peninsula (Maugis, 1955, p. 125).

2.2 Basement

The western extension of the folded and partly metamorphic Paleozoic basement under the onshore sediment basin is only known indirectly and within a certain depth range, because of limitations in the geophysical methods employed (see also Weigel et al., this Vol.). For instance high velocity (5.9 km/s) seismic refractors, first charted as basement in ca. 3 km depth, were later drilled as Lower Cretaceous dolomites at about 3000 m in the first stratigraphic well in Senegal, Tienaba 1 (Maugis, 1955). In a zone between 15° and 16° W, the gradient of the by 3° west-dipping basement increases (Elouard, 1975), and basement is no longer reached by geoelectric measurements (Mathiez and Huot, 1966) west of this apparent basement hingeline.

A basement high (dôme de Guier) resulting from vertical crustal movements generally uplifting the continent in the latest Cretaceous was drilled between lakes Guier and Rkiz in Senegal (Trenous and Michel, 1971). From oil exploration investigations along the coastline the depth to basement is quoted as about 3000-5000 m in Mauritania (Crenn and Rechenmann 1965, p. 45 and 47) and 6000-8000 m in Senegal (Elouard, 1975).

Basement depth contours under the onshore northern Mauritanian Basin between 18°N and 21°N in Hebrard (1978) call for a sediment thickness between 2000 and 3000 m along the coast line.

2.3 Mesozoic-Cenozoic basin fill

2.3.1 General remarks

The surface geology does not allow many deductions about the Mesozoic-Tertiary basin, as this is almost completely covered by a Quaternary basin with marine sediments in Mauritania. Michel (1973) presents two-cross-sections for Senegal derived from surface geology. The oldest sediments onlapping the Mauritanides are Eocene marly chalks in the upper Senegal valley and clayey sandstones (Michel and Pascal, 1967). Some Maastrichtian sandy limestones also crop out over the Thies anticline, which was uplifted by the volcanism on the Cape Vert penisula (Fig. 1). Stratigraphic data from deep wells are also very sparse:

The Tienaba-1 well (Maugis, 1955), reached Aptian-Albian lagoonal limestones; later wells tested an Upper Jurassic carbonate platform (Michel, 1973, p. 79) uplifted by Oligocene (Crevola, 1978) to Recent volcanism east of Dakar (cross-section by Ayme, 1965). Results of deep stratigraphic wells are not known for onshore Mauritania, where only some shallow water wells reaching Maastrichtian sands above granitic basements could be correlated in a N-S profile (Fig. 2; Ortlieb, 1975; Hebrard, 1978).

Since the Jurassic marine sediments of the widening Atlantic Ocean transgressed onto a smoothly subsiding Paleozoic basement. They attained their maximum eastward extent during Late Cretaceous and Paleogene

Fig 1 Geology onshore and bathymetry offshore the Senegal-Mauritania Basin (compiled by Ranke, BGR). West of the Senegal River the stratigraphically complex, originally marine Mio-/Pliocene series of the Continental terminal (Tessier et al., 1975) have been uncovered to show the large extent of the Eocene epicontinental sea. ►

seamount mentioned in chapter 3.1

Ds-1, D-1, Ti-1 = onshore wells

M = Motounsi; BL = Bou Lanouar; I = Idini; L = Legat

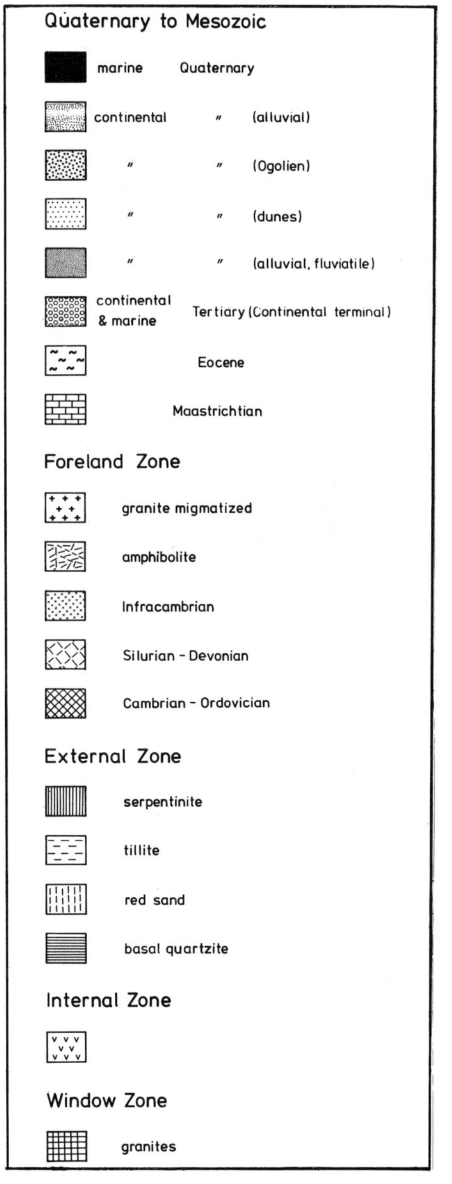

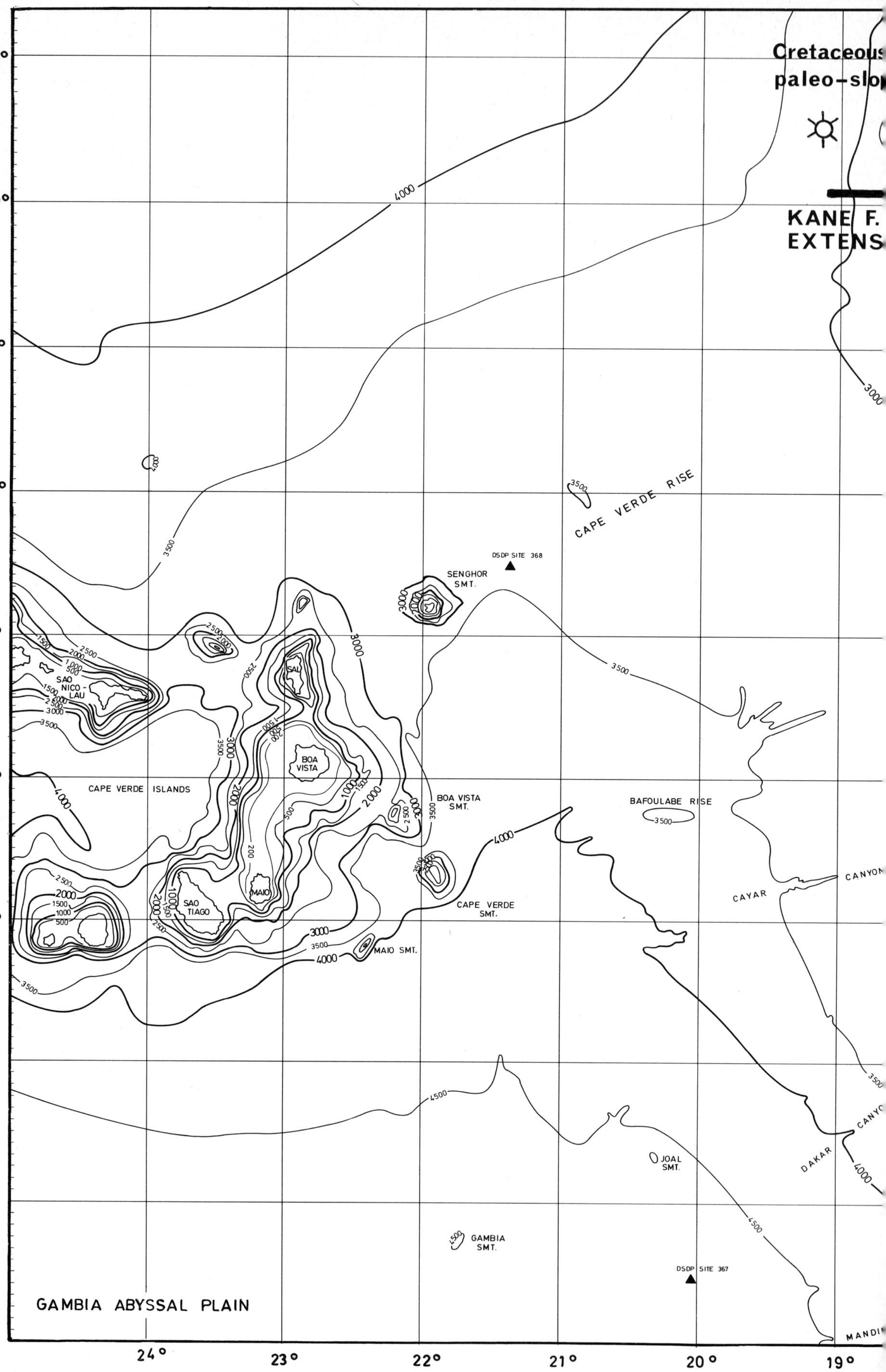

Fig. 1 (legend see p. 163)

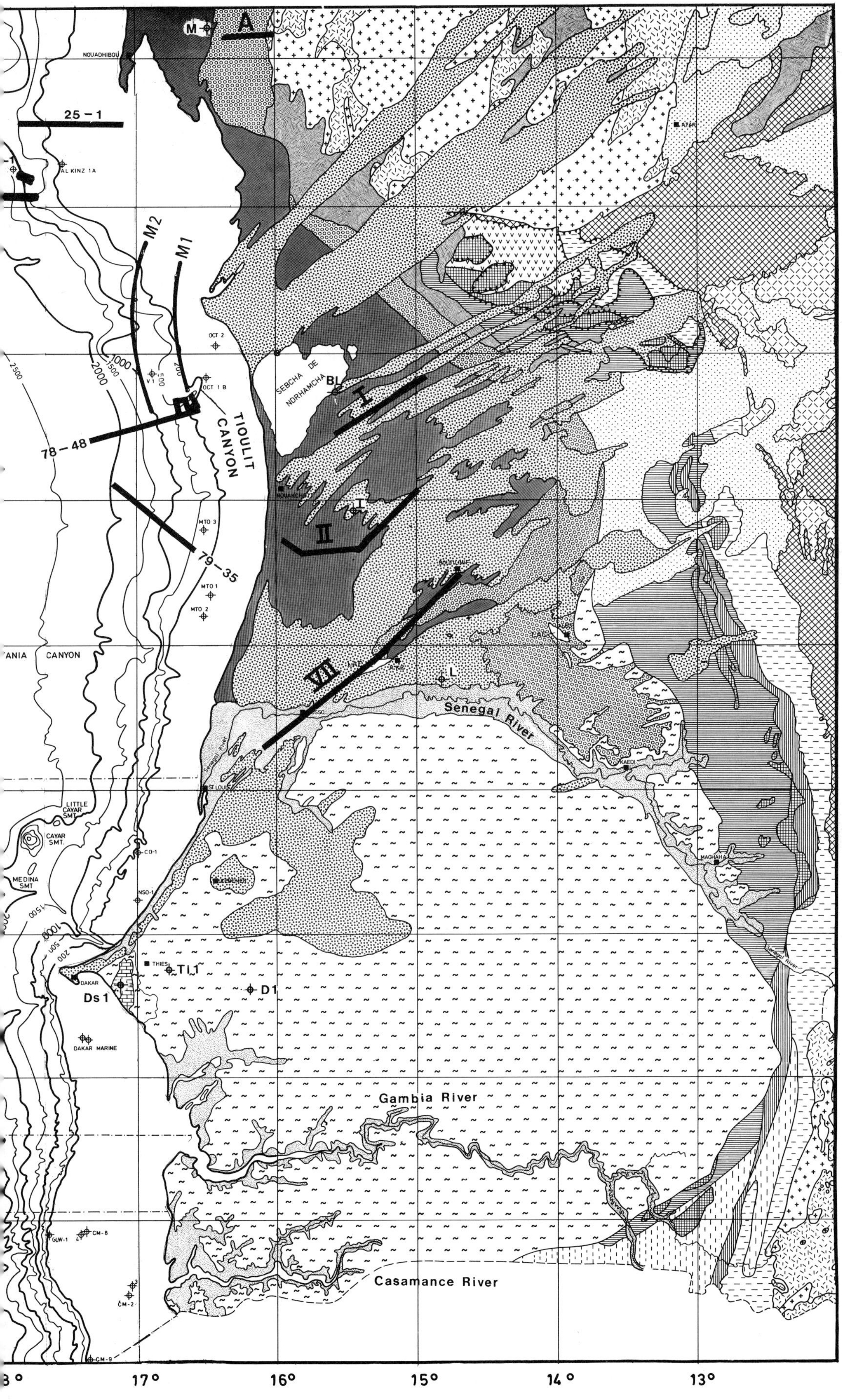

A
NOUADHIBOU
25 – 1
AL KINZ 1A
M2
M1
ATAR
OCT 2
SEBCHA DE NDRHAMCHA
BL
I
OCT 1 B
TIOULIT CANYON
78 – 48
II
MTO 3
79 – 35
MTO 1
MTO 2
VII
L
Senegal River
TANIA CANYON
LITTLE CAYAR SMT.
CAYAR SMT.
MEDINA SMT
ST LOUIS
KAEDI
MAGHAMA
THIES
Ti 1
DAKAR
Ds 1
D1
DAKAR MARINE
Gambia River
Casamance River
CM-8
CM-2
CM-9
17°
16°
15°
14°
13°

times in an embayment of Senegal outlined by Flicoteaux et al. (1975), who also present an isopach map down to the Upper Cretaceous.

2.3.2 Stratigraphy

The following evolution of the onshore sediment basin is based mainly on Elouard (1975) and Michel (1973):

Triassic rift sediments as found in the Newark basin of the USA and interpreted under the Southeast Georgia Embayment (Dillon et al., 1979) have only been drilled in the onshore Aaiun-Tarfaya Basin (Ranke et al., this Vol.), but they have not been reported in the Senegal-Mauritania Basin.

Upper Jurassic shallow water oolithic carbonate platform sediments are the oldest Mesozoic sediments known. They were drilled between 3250 and 4010 m depth in the N'Diass 1 (Ds 1) well east of Dakar (Castelain, 1965). Because they were uplifted by Tertiary volcanism and tectonics they lie about 2 km higher than sediments of the same age under the conjugate American continental margin (shallowest level under the Blake Plateau: 5.5 km). Besides, this Jurassic shelf edge lies on the American side of the Atlantic under 1 to 2 km of sea water, whereas on the African side the oldest sedimentological record of the Atlantic opening could be drilled onshore.

Early Cretaceous rocks (775 m in D 1, 1319 m in Ds 1) present a sandy facies in the east, with interbedded carbonates near the coast.

In the Middle Cretaceous the sediments become finer grained: Albian to Lower Cenomanian (600-1150 m) gray shales and claystones change with siltstones containing gypsum and lignite. The epicontinental sea advances further to the east. Transgressive Upper Cenomanian sands and gravels, frequently showing crossbedding structures, diminish eastwards in thickness from 600 to 150 m (Michel, 1973; cross-sections p. 62).

In the Turonian the sediments (< 100 m) pass from sands again into gray shales. During the Senonian sands, gravels and conglomerates (60-300 m) are deposited in the eastern basin, while clays and carbonates (600 m) accumulate in the present coastal region.

Maastrichtian sediments were recognized under the whole coastal basin. Their thickness is reported as 1800 m in Senegal (Carte geol. de Senegal, 1967) and around 400 m in Mauritania, however in contrast to Senegal their stratigraphic classification was not quite certain in Mauritania.

Paleocene limestones between 40 and 120 m thick lie discordantly on the Upper Cretaceous. Their facies shows an increasing sand content toward the basin margin in the east.

Early Eocene (Ypresian) sandy to chalky sediments with some glauconite and phosphate horizons are cropping out in large parts of Senegal, but not in Mauritania. Their facies changes from fully marine-neritic at the coast over littoral to continental from west to east. Their thickness decreases in the same direction from up to 400 m near the coast NE of Rufisque to their complete absence along the eastern basin margin. Because of a general uplift of the continent the Lower Eocene transgression did not reach as much inland as during the Paleocene.

Middle Eocene (Lutetian) crops out in many locations on river banks in the Senegal-Mauritanian basin. Continuing basement subsidence resulted in a wide transgression, which reached the eastern basin margin and eroded the Paleocene sediments there (Elouard, 1975). In profile this transgression is documented by a series of sandy sediments near the base grading into chalky dolomitic layers near the top of the middle Eocene; from fully marine carbonates with glauconite and phosphate-rich sediments in the western and central onshore basin the facies changes to detritic-continental input at the basin margin. According to palynomorphs (Dieng, 1965) a humide

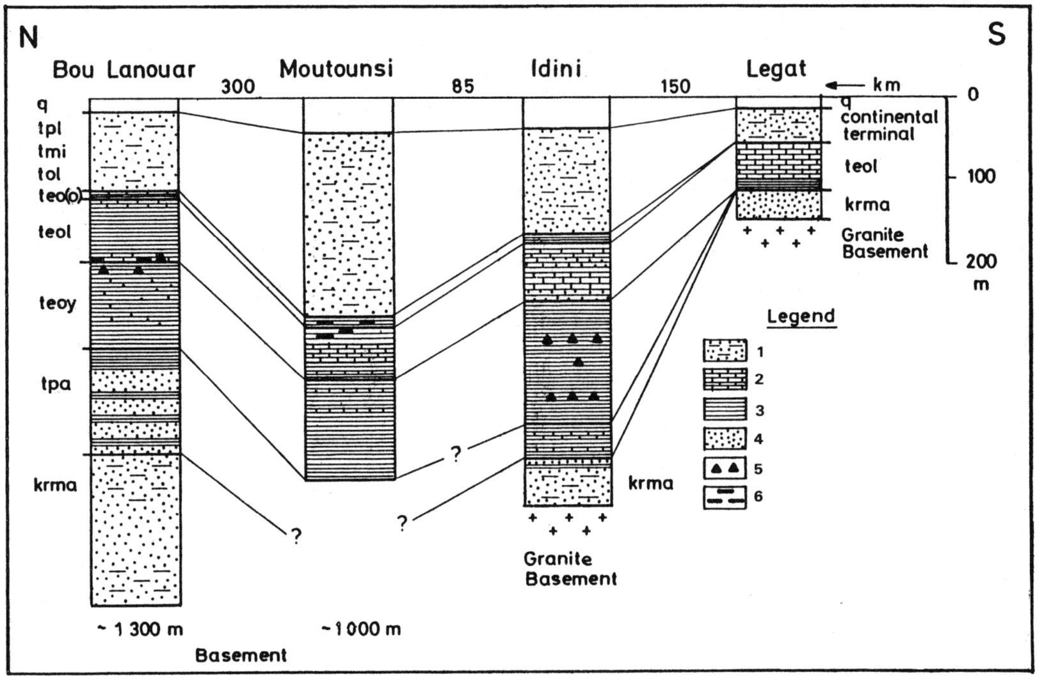

Fig. 2 North-south correlation of deep water wells in Mauritania after Elouard (1975). Another well (21°55'N, 17°03'W) near Nouadhibou bottoming at 418 m depth in Lutetian carbonates and shales is mentioned by Hebrard (1978). Locations shown in Fig.1, distance between wells in km indicated on top.
q = Quaternary, tpl = Pliocene, tmi = Miocene, tol = Oligocene, teo(o) = upper Eocene, teol = Lutetian, teoy = Ypresian, tpa = Paleocene, krma = Maastrichtian. 1 = clayey sandstone, 2 = carbonate, 3 = shale, 4 = sand, 5 = chert layers, 6 = lignite layers

tropical climate existed in the Senegal-Mauritania Basin during the middle Eocene, which became more arid towards the end of the Eocene and within the Oligocene.

Late Eocene lagoonal dark sands with lignite and pyrite layers and clay exist in ca. 10-175 m thick beds only in three embayments on the coast (Elouard, 1975).

During the Oligocene most of the basin was uplifted and several small local sedimentation centers were established that existed into the Miocene. Near Nouadhibou about 300 m of marls and chalks overlying chalky sands were found, which had been deposited in very shallow water. In a paleogeographical reconstruction of coast-lines since the Paleocene Flicoteaux et al. (1975) show how the different marine transgressions in the middle Eocene, the early Oligocene and the earliest and late Miocene form steadily receding gulfs in the Casamance and Gambia river regions.

Pliocene marine sediments were not recognized in Senegal and only in a small zone around Cape Timrik (19°20'N) in Mauritania (Michel, 1973, p. 80). They may be incorporated in a series of variegated clayey and sandy, strongly ferruginous layers, which represent the end of the Tertiary humide climate and the beginning of Quaternary arid sedimentation: "the Continental terminal" (Tessier et al., 1975). With thicknesses between 20 and 300 m the "Continental terminal" levels the paleorelief and covers almost the whole basin in Senegal. In northern Mauritania the structural relationships of the originally marine ? Mio-/Pliocene sediments constituting the

"Continental terminal" with the overlying Quaternary deposits are complicated (Ortlieb, 1975).

Early Quaternary crossbedded littoral dune deposits form the seafloor on the 5 to 15 m shallow Arguin Bank (Ortlieb, 1975; p. 50). In the Quaternary four transgressions (sandy-chalky sediments) and regressional phases (lacustrine sediments, sebkahs and dunes) can be distinguished mainly in the Mauritanian basin (Hebrard, 1978).

The following notes are intented to present a review of the offshore parts of the Senegal-Mauritania continental margin basin, updated by results of BGR marine geophysical surveys (Seibold and Hinz, 1976; Fritsch et al. 1978; Goldflam et al. 1980).

3. Offshore Senegal-Mauritania Basin

3.1 Bathymetry

The bathymetry of the Northwest African continental margin (Fig. 1) is compiled mainly from GEBCO-charts, supplemented in special areas by detailed surveys of the BGR (North Senegal-Cayar Seamount, Tioulit Canyon at 19°N off Mauritania) and other institutions (CSS Baffin survey off Sengal; Bay d' Arguin, University of Louvain). Because of a high profile density, the Canadian Senegal Offshore Survey (Ruffman et al., 1977) is the centerpiece of the presented bathymetric map, whereas the BGR data were used mainly off Senegal (Schlüter in Seibold and Hinz, 1974; Edelmann, 1979) and off Mauritania (Fig. 3). An area, where information is sparse lies on the continental margin off the Bay d' Arguin (northern Mauritania), which probably is dissected by numerous deep submarine canyons. Two channels of 15 m depth enclosing the Arguin Bank in the northwest and southeast (Hebrard, 1978; p. 42) probably lead to the head of such postulated canyons, which have also been indicated by indentations of the 500 m-isobath in Rona (1971). A more detailed bathymetric and microphysiographic map of the southern Mauritanian and Senegal margin is given by Jacobi and Hayes (this Vol.).

The 4000 m isobath outlines the Gambia abyssal plain open to the SW and the Cape Verde-Madeira abyssal plain, which opens to the NW. Both are separated by the volcanic Cape Verde Islands Archipelago and the Cape Verde Rise adjoining them in the NE:

The western and central islands of the Cape Verde Group rise 5-7 km above the surrounding ocean floor, one third of their elevation lying above sea-level. The older eastern islands (cf. Robertson and Bernoulli, this Vol.) which are denudated by erosion, sit on a broad, probably volcanic submarine platform and do not reach more than 400 m above sea-level (Klug, 1973). The central eastern island, Boa Vista, possesses an eastern moat filled with ca. 400 m of post-Middle Miocene sediments, which lap onto the Miocene volcanic apron on the slope. The erosional unconformity forming the base of the moat may indicate that strong erosive pre-Middle Miocene bottom currents (Jacobi and Hayes, this Vol.) were not completely deflected westwards around the Cape Verde Islands. Another example for the volcanic element structuring the bathymetry is a Miocene intrusion under the Bafoulabe Rise (Ruffman et al., 1977), which therefore was not built by accretionary sedimentary processes.

Hebrard (1978) drew attention to an isolated ? Tertiary seamount near 20°31'N and 19°08'W. Though this seamount is said to rise from 3297 m to a depth of 16 m on the International Hydrographic Office Map A 1 (4. ed., 1958), to 18 m on the London Hydrographic Office Map 3848 (1972) and to 20 m on the Russian bathymetry maps of the Atlantic Ocean (1 : 10 x 10^6, 1971; 1 : 20 x 10^6, 1974), its presence is not documented on the 1981 GEBCO plotting sheet n. 129 of this area (pers. comm. Becker, German Hydrographic Institute). Two seismic profiles in about 16 km distance give no definite clues about the seamounts presence: the eastern upslope profile shows no trace of a volcanic body, while

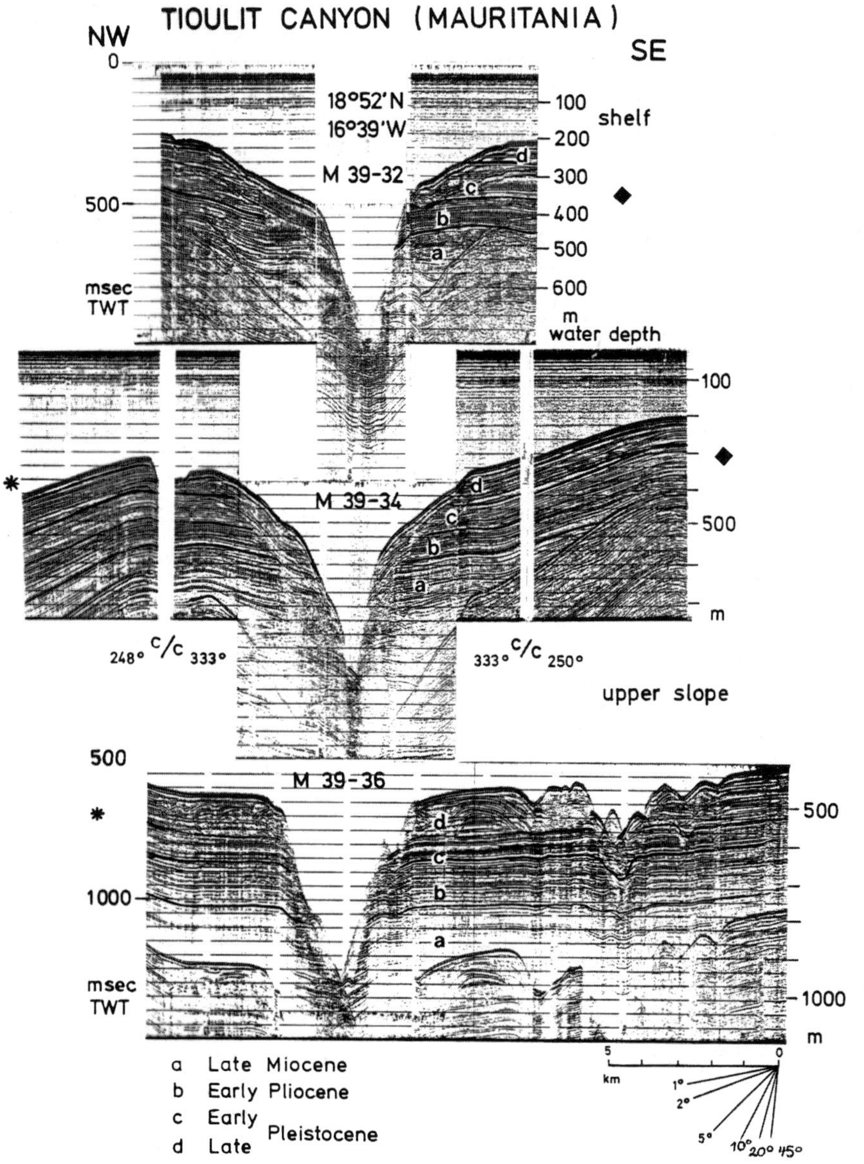

Fig. 3 Seismic profiles shot with a small air-gun at high repetition rates across the Tioulit canyon. Locations of profiles indicated in Fig. 1. Rhomboids and stars show, where the profile continues: the changing course headings are given in degrees. Age of sequences represented by letters a to d. The name of the canyon was chosen (Seibold and Hinz, 1976) from a small settlement on the coast near the canyon head. The same canyon is named Tikatane canyon in Hebrard (1978)

the western downslope profile exhibits chaotic bedding or slumps in Pliocene sediments, which could have been caused by a Miocene/Pliocene intrusion located to the east of the profile.

The continental slope shows a variable width: off southern Senegal it is relatively small and was shaped by the growth of a Mesozoic carbonate platform now lying under the shelf edge, rather than by progradation of sediment carried by the onshore rivers, as suggested by Ruffman et al. (1977). In the north the continental slope broadens to 2 or 3 times of the width in the south because of the presence of the Cayar seamount. The steeper continental slope off southern Senegal is associated with a broader-than-normal shelf region (80 km off Casamance): the shelf attains similar widths only off the Arguin Platform in northern Mauritania (Piessens and Chabot, 1977). The southern broad shelf is covered by active deltas and drowned valleys with shelf bypass-channels of the Casamance river (Ruffman et al., 1977).

The smaller sediment contribution of the Senegal River is carried by a nearshore current to the head of the Cayar Canyon the head of which is only 1 km away from the coast (Dietz et al., 1968; Seibold and Hinz, 1974). The concentration of canyons along the Senegal continental slope is rather constant (10 canyons/100 km, Ruffman et al., 1977); probably, it extends north to Mauritania, although the canyons are unsufficiently documented there.

The smaller canyons cannot be distinguished beyond the foot of the continental slope, the larger ones empty into a system of two deep sea-channels, recognizable by indentations of the 4500 m isobath. These channels funnel and transport suspended sediments onto the abyssal plain (ponded turbidites near DSDP Site 367 in Lancelot, Seibold et al., 1977; Goldflam et al., 1980). As these canyons do not cut deeper than into Pliocene sediments in their distal parts, they seem to be of Pleistocene, or at most Pliocene age.

3.2 Structure of the continental margin basin

Onshore gravity measurements indicate an increasing sediment thickness between 14°30'N and 17°30'N along a 400 km N-S trending band of negative anomalies (Fig. 4) from 15°30'W towards the coast together with a corresponding modification of the crust from 16°W westwards (Crenn and Rechenmann, 1965, p. 46). This interpretation would mean that in a distance of 50 km east of the coastline the thickness of the continental crust is already reduced compared to "normal" continental crust, which extends to 30 km depth (Liger and Roussel, 1979 a).

In the coastal zone of Senegal a positive Bouguer gravity anomaly ($>$ 40 mgal) is explained by thinned continental crust, while oceanic crust is interpreted along the meridian of Dakar (17°30'W) (Liger and Roussel, 1979 b). The same interpretation could apply to the positive gravity anomaly along the Mauritanian coast (Crenn and Rechenmann, 1965; Blot et al., 1962).
It is very likely that the local ellipsoidal positive anomalies of 60-80 mgal in Mauritania are caused by volcanic features like the Senonian intrusive batholith of Leona (Pascal and Michel, 1967) and the Oligocene (Crevola, 1978) to Pleistocene (Cantagrel et al., 1976, Crevola, 1980) volcanics of the Cape Vert peninsula.

If isostasy is assumed along the continental margin, the area integrals over the seaward negative and landward positive gravity anomaly constituting the "margin effect" must be equal. Therefore the position of the zero-milligal free-air anomaly contour delineates approximately the continent/ocean boundary (Fritsch et al., 1978). The continent/ocean boundary thus defined runs under the continental slope of Senegal and Mauritania (Fig. 4; Fritsch et al., 1978). Because of a strong gravity gradient the continental block has a better defined edge off southern Senegal than further north. There dense volcanic bodies of the Cape Vert peninsula and the Cayar seamount pull the zero-contour to the west of the probable continent/ocean boundary.

Off the Arguin Platform the zero-milligal contour (as well as the reflection time contours to basement are left-laterally offset 100 km to the west, possibly outlining a fault along a paleo-Kane-Fracture Zone (Twigt, 1980; p. 104). This interpretation assumes continental crust which juts out far to the west of Nouadhibou approximately along the 2000 m

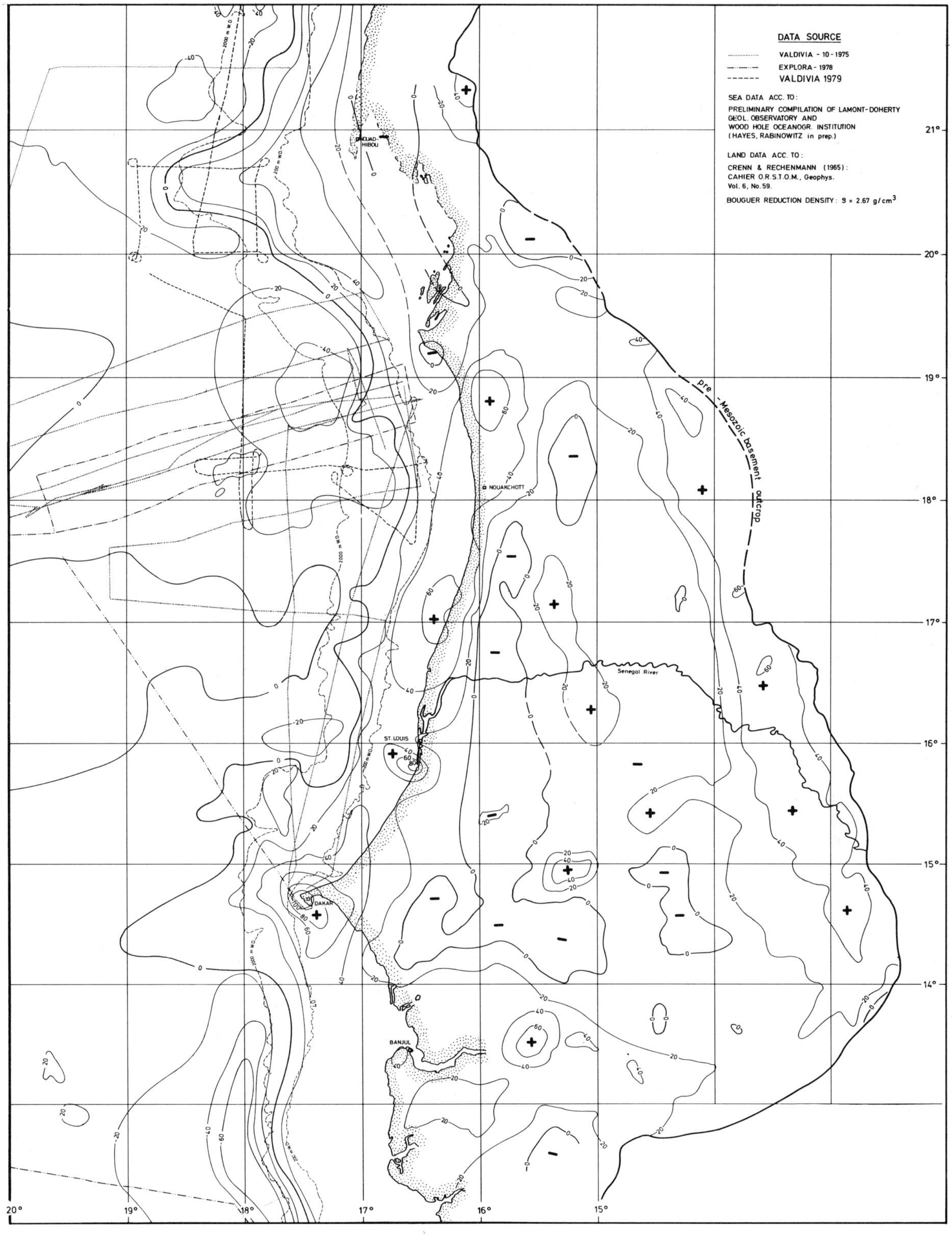

Fig. 4 Gravity anomalies over the Senegal-Mauritania Basin. Onshore Bouguer anomalies and offshore Free-Air anomalies are depicted, to reach a common reference datum at sea level (Fritsch et al., 1978)

isobath. The postulated ? Tertiary seamount at 20°31'N, 19°08'W may testify continuing activity along this fault.

Where this paleo-Kane-Fracture Zone extends onshore, it corresponds to a belt of E-W trending gravity lows of -10 to - 20 mgals (axis Chami-Atar in Hebrard, 1978, p. 24), which cuts the N-S striking gravity high of the Mauritanides at 20°30'N.
This E-W gravity minimum is interpreted as a series of granitic intrusions (= 2.62 g/cm^3) of lighter density than the Paleozoic Mauritanides (= 2.85 g/cm^3) (Crenn and Rechenmann, 1965, p. 31).

The north-south trending BGR-profile VALDIVIA 79-26 crosses the Kane-Fracture Zone Extension somewhere along the 18°W meridian (Fig. 5). Basement was only tentatively identified on this profile, but there seems to exist no obvious basement topography, like ridges or valleys, which are normally associated with fracture zones in younger oceanic crust. The only sign of the postulated Kane-Fracture Zone is a zone of fault-disturbed sediments showing little offset, which may indicate lateral movement. Since the southern part of VALDIVIA 79-26 profile lies within the Mesozoic Magnetic Quiet Zone (MQZ), it is however curious that a fracture zone should disturb Paleogene to Neogene sediments. As a Tertiary transform motion along a fracture zone in Jurassic crust is impossible, the reflection seismic observation could perhaps be explained by a left lateral strike slip to left oblique normal fault. The postulated extension of the fracture zone indicated by some bathymetric, gravimetric and reflection seismic features is by no means sufficiently documented. The presented interpretation should stimulate the discussion of the observed data and serve to foster research about the eastern continuation (Twigt, 1980) of Atlantic fracture zones into oldest oceanic crust, now covered by a thick pile of Mesozoic and Cenozoic sediments.

The continent/ocean boundary defined by the zero-milligal free-air anomaly lies between the 500 m and 2000 m isobaths along the Senegal-Mauritanian continental slope. This basic framework of the continental margin basin is also supported by seismic measurements. As the deep crustal structure based on refraction seismic is discussed by Weigel et al. (this Vol.), the following notes shall concentrate on reflection seismic evidence. The penetration of reflection seismic data is obscured by intrusive sills and strongly reflecting volcanoclastic sediments on the Senegal continental slope north of Dakar and by strong multiple reflections off Mauritania, but the following observations could be made:

A Triassic (Templeton, 1971) to Early Jurassic rift basin is located between an oceanic (Uchupi and Austin, 1979) and a continental hingeline:

a) Almost flat-lying oceanic-layer-2 basalts, correlated to DSDP Site 367 (Goldflam et al., 1980), can be observed dipping landward to about 8 sec TWT (about 10 km) off Mauritania and 7 sec (= 8 km; Uchupi et al., 1976) off Southern Senegal, where they are down-faulted to the east along small normal faults of an "oceanic basement hinge line".

b) A low frequency-two-phase reflector, which represents the deepest unconformity off Senegal, is stepped down to the West along Cretaceous growth faults of the coast-parallel Cayar Fault Zone. A series of larger faults along the Mauritanian shelf edge delineates the rim of the buried carbonate platform. Both faults systems probably describe the trace of a "continental basement hinge line" even though the basement topography cannot be clearly seen on the seismic records (Fig. 6,7,8).

The difference lies in the amount of throw: while continental blocks have been down-faulted by several hundreds of meters, the oceanic basalt is rather more flexured supplemented by minor faults with throws in the order of 100 meters.

Between them lies either a part, or in the extreme case the total fill of the rift graben which originated

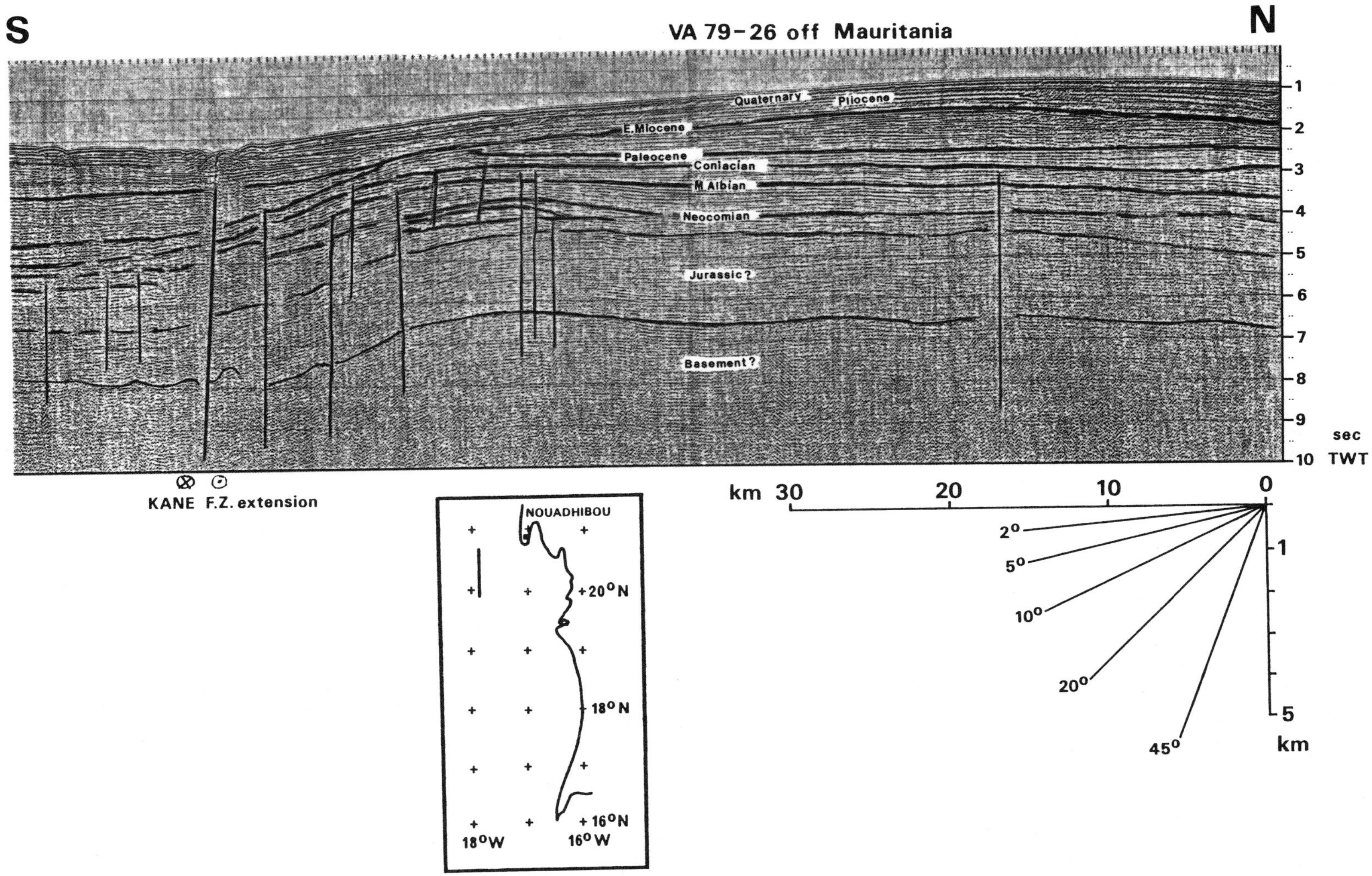

Fig. 5 Seismic reflection profile VALDIVIA 79-26 crossing the postulated paleo-KANE-Fracture-Zone. Basement is only tentatively identified as the deepest unconformity hidden among the many multiple reflections. Stratigraphic correlations from DSDP 368, 367 and Al Kinz 1-A (Rona, 1973)

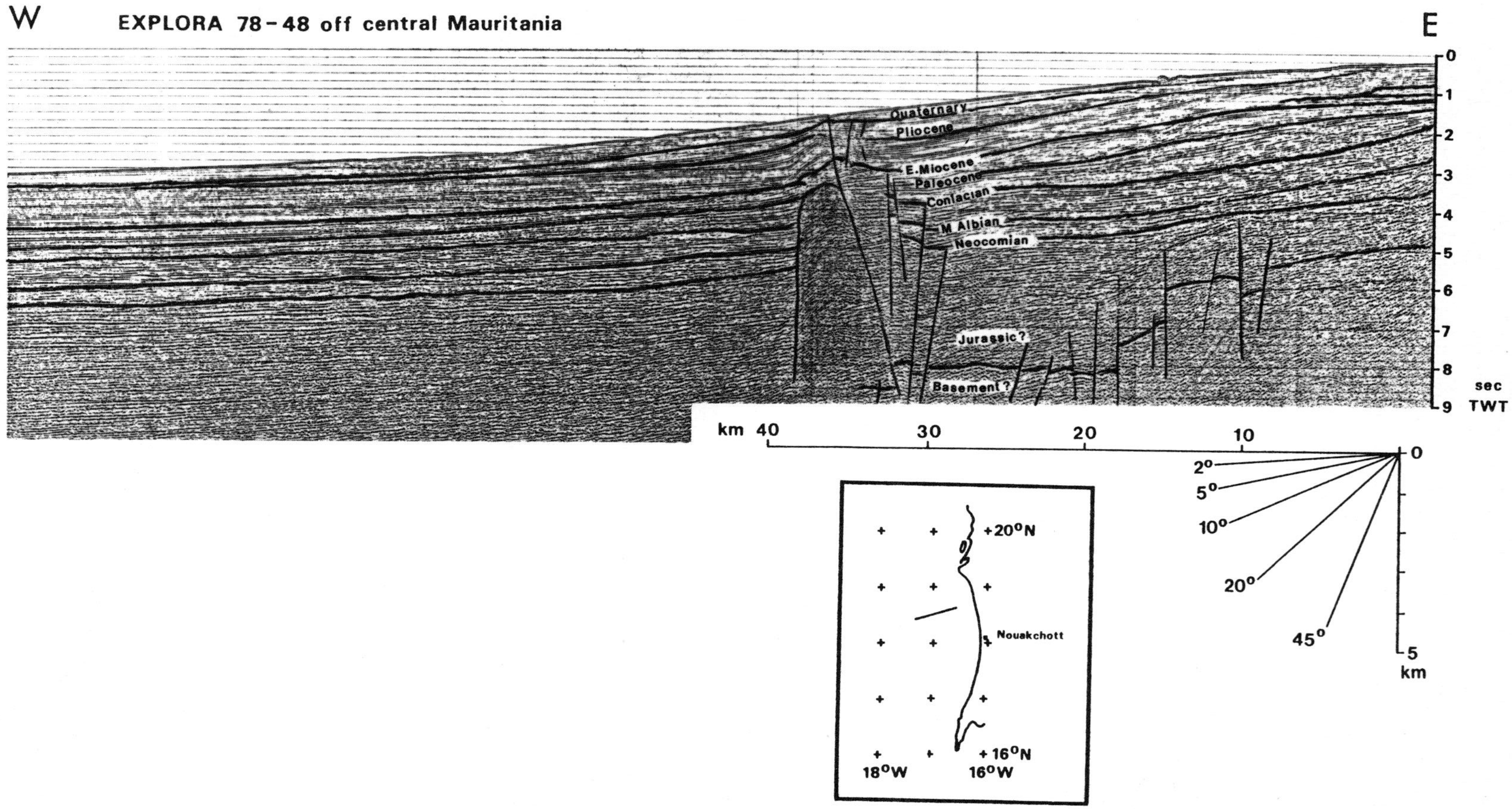

Fig. 6 Seismic reflection profile EXPLORA 78-48 off central Mauritania. Basement ist only tentatively identified as a deep, faulted, low frequency, two phase reflector. Stratigraphy correlated from DSDP 368, 367 becomes uncertain to the east of the salt overhang along a fault or perhaps volcanic plug (Lehner, pers. comm.) at km 35

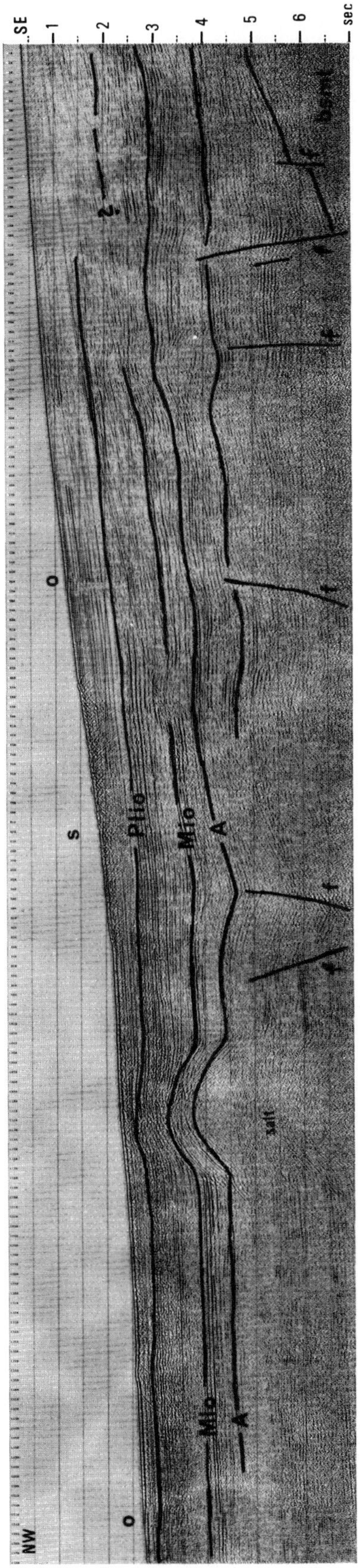

Fig. 7 Seismic reflection profile VALDIVIA 79-35 crossing the zone of diapirs (see Fig. 9). Strong amplitude, low frequency reflectors (? carbonates) lapping onto an inclined surface at 5-6 sec under the upper slope define the basement. Undulations of deeper reflectors probably caused by normal faults. Note sediment slides and olistostromes at seafloor

before the onset of Atlantic ocean floor spreading (Weigel et al., this Vol.).

The floor of the marginal rift graben thus defined between an "oceanic and continental hinge line" lies between 12 to 14 km depth under the upper continental slope and cannot be seen on the available reflection seismic records. However the lack of layer-2 basalt velocities in the refraction seismic profiles over this zone indicates that rift sediments may have been directly deposited on lower crustal rocks with a layer 3 velocity (v_p = 7.1 km/sec; Weigel et al., this Vol.).

Diapiric structures (Fig. 9) indicate that evaporites were deposited on the subsiding continental basement blocks in a narrow elongate zone between 16°N and 19°N. This salt is probably of Early Jurassic age (Lehner and de Ruiter, 1977). The area of salt diapirs lies opposite the paleo-estuary of the Senegal River which ran northward to the Sebkha Ndramcha (18°30'N) during Mio-/Pliocene (Michel, 1973, p. 80) and may have provided the sediment overburden for halokinesis.

After the main post-rift subsidence phase during the Jurassic to Early Cretaceous a second major phase of vertical tectonics during the Late Cretaceous (Senonian) continued the disintergration of the continental basement blocks. Intrusive activity (Leona Dome batholith south of St. Louis; Michel and Pascal, 1967) accompanied the faulting of the continental margin, where Upper Cretaceous beds climb 3-4 km in zones of synsedimentary faults from the shelf to the coast. If compared to the Senegalese margin, the Mauritanian margin seems to be less tectonically structured.

North of the suggested Kane Fracture Zone Extension, an anticline of Lower Cretaceous to Paleocene sediments marks the position of the Cretaceous continental slope in the form of a westward convex arc which on the south adjoins the E-W Fracture-Zone Extension. This anticline (Fig. 5) was drilled by the A-1 well (Rachwal and Deste-

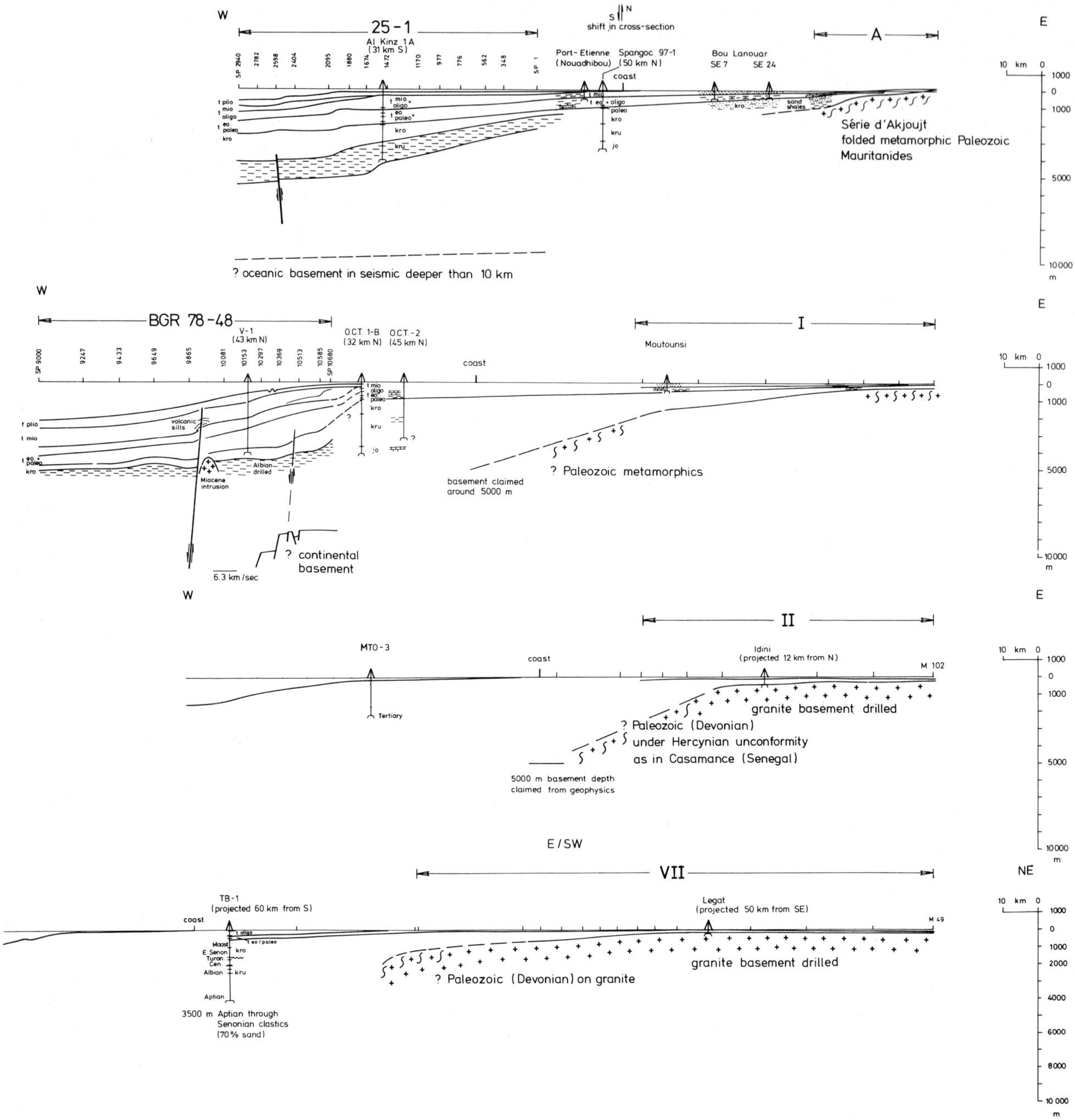

Fig. 8 Geological cross-sections through the Mauritanian continental margin constructed from seismic profiles offshore, oil wells (Rona, 1973; AAPG World Oil Developments) and geoelectric soundings (Mathiez and Huot, 1966) onshore

fano, 1980). The Cape Blanc anticline does not consist of Cretaceous growth fault roll-over structures as the prominent slope anticline off the West-Sahara (von Rad and Wissmann, this Vol.). Since the West Saharan slope anticline also terminates near 24°30'N (von Rad and Wissmann, this Vol.), it is probably a structure independant of the Cape Blanc anticline.

The Cretaceous paleoslope has also been recognized off Cape Blanc

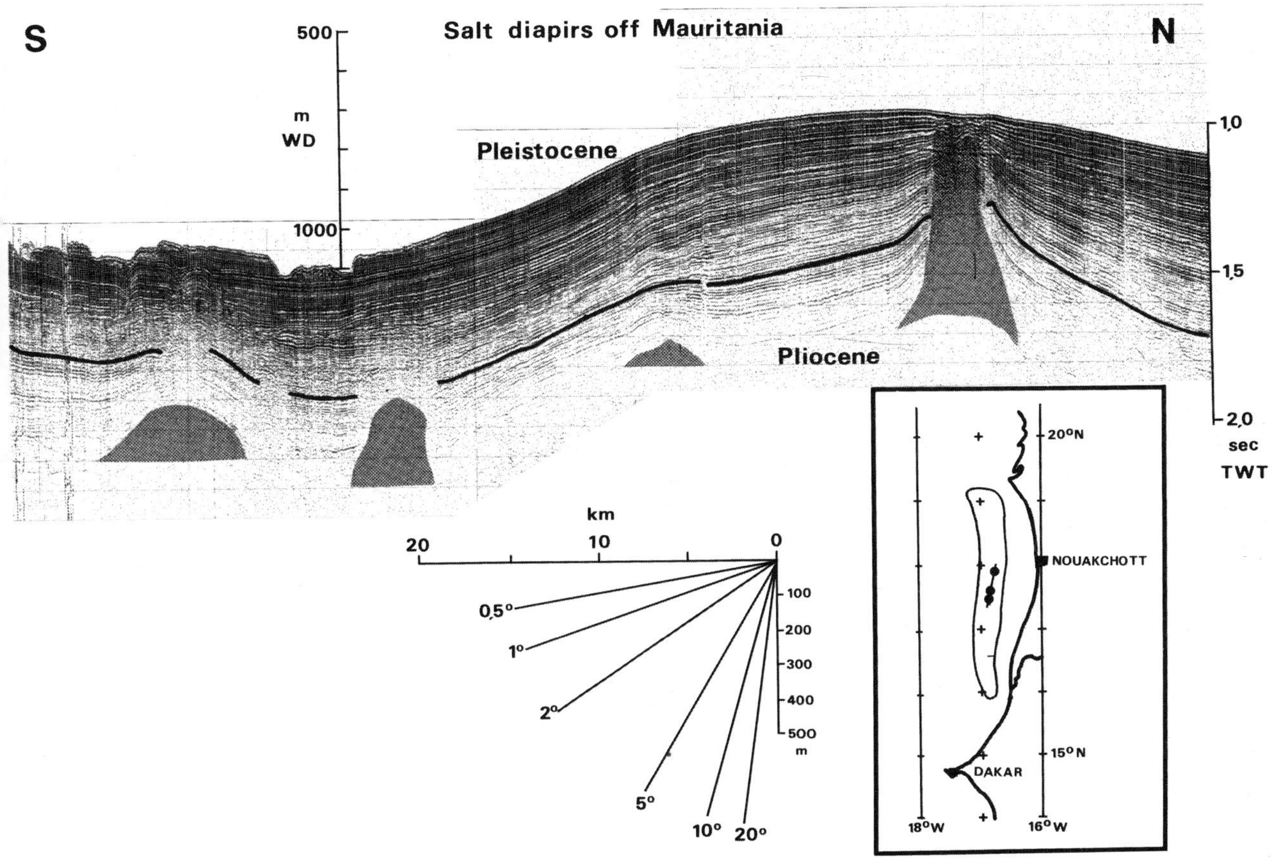

Fig. 9 Seismic reflection profile METEOR 25-E 2 in the elongate zone of Early Jurassic salt diapirs off Mauritania. Thickened reflector is interpreted to separate Pliocene from Pleistocene sediments. Note caved-in cap rock on the northern diapir

(Fig. 10, Seibold and Hinz, 1976). Low frequency, high amplitude Cretaceous reflectors terminating against the crest of the anticline (Fig. 10c) probably represent layers of a carbonate bank complex. A middle Miocene shelf edge (Fig.10 a, c) had receded landward 70 km relative to the Late Cretaceous/Paleocene shelf edge.

The stratigraphy of the seismic sequences was tied to the deep basin DSDP Sites 368 and 139 and to the Al-Kinz 1-A and O.C.T. 1-B wells on the Mauritanian shelf reported by Rona (1973). The single-trace profiles a and b of Fig. 10 lack seismic penetration beneath the Paleocene unconformity, but show high amplitude, continuous and well-laminated seismic sequences which are interpreted as Miocene and Pliocene turbidites. Intercalated between these turbidites there are seismically diffuse zones under the lower continental slope. They probably constitute Miocene slump deposits, which slid down from the upper slope and left steep slide scars behind (cf. von Rad and Wissmann, this Vol.).

Gravitational redeposition of large sediment masses as olistostromes and slides (Fig. 7) also takes place in Pleistocene to Recent sediments. During the last 20 000 years the terrigeneous material in the deep sea sediments was mainly supplied by the offshore blowing Harmattan-winds: in a 1000 km broad zone adjoining the coast an average amount of nearly 50 x 10^6 tons of eolian sediments per year (Koopmann, 1979) were deposited.

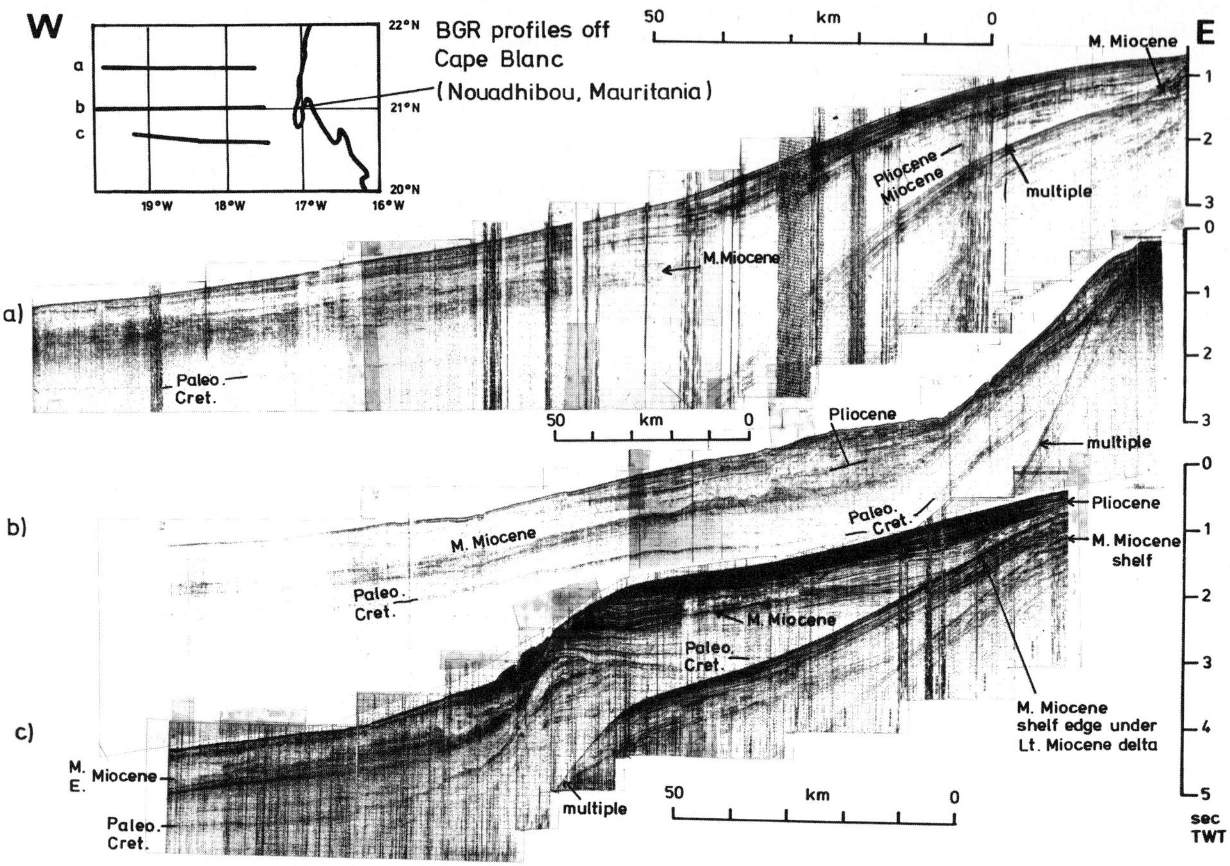

Fig. 10 Seismic reflection profiles off Cape Blanc

Acknowledgements

The author wants to thank Dr. J.S. Watkins for providing the Gulf oil company seismic profile 25-1. The profile is shown with the author's interpretation as a line drawing in Fig. 8 of this paper.

Drs. von Rad and Hinz critically read and improved drafts of this paper.

REFERENCES

Aymé JM (1965) The Senegal salt basin. In: Salt basins around Africa. London Inst Petroleum: 83-90

Bertrand H, Prioton JM (1975) Les dolerites marocaines et l'ouverture de l'Atlantique, Etude pétrologique et geochimique. Dissertation Université Claude Bernard-Lyon

Blot C, Crenn Y, Rechenmann J (1962) Eléments apportées par la gravimétrie à la connaissance de la tectonique profonde du Sénégal. Comptes rendus 254: 1131-1133

Cantagrel J-M, Lappartient J-R, Tessier F (1976) Nouvelles données géochronologique sur le volcanisme ouest-africain. 4ème Réun ann Sci Terre, Paris, 13-16 avril 1976: p 93

Castelain J (1965) Apercu stratigraphique et micropaléontologique du bassin du Sénégal, historique de la découverte paléontologique. BRGM Mémoire n 32, Colloque International de Micropaléontologie (Dakar, 6-11 mai 1963): 135-161

Choubert G, Faure-Muret A (1971) Tectonique de l'Afrique. Paris: 372-378

Collette BJ, Ewing JI, Lagaay RA, Truchan M (1969) Sediment distribution in the oceans: The Atlantic between 10° and 19°N. Mar Geol 7: 279-345

Crenn Y, Rechenmann J (1965) Mesures gravimétriques et magnétiques au Sénégal et au Mauritanie Occidentale. Cahiers ORSTOM géophysique, Paris, n 6

Crevola G (1978) Volcanisme de la presqu'ile du Cap-Vert. In Trav Lab Sci Terre, Rapport bisanuelle d'activité 1976-1978, Saint Jerome Marseille: 32

Crevola G (1980) Données nouvelles sur le volcanisme quaternaire dans la tête de la presqu'île du Cap Vert, Sénégal. CR Acad Sc Paris ser D 291, n 7:617-620

Dia O, Lecorche J-P, Le Page A, Les relations entre les differents troncons des Mauritanides de la Mauritanie nordoccidentale à la frontière de Guinée. Rev Géograph phys et de Géol dynam: (in press)

Dia O, Lecorche, J-P, Le Page (1979) The principal relationship between northern, central and southern regions of the main Mauritanides belt. In: Résumés du 10e Colloque de Géologie africaine, Montpellier, 25-27 avril 1979: 113-114

Dieng M (1965) Contribution à l'étude géologique du Continental terminal du Sénégal. Rap Bur Rech Géol Minières Dakar DAK 65-A-27: 176 p

Dietz RS, Knebel JJ, Somers LH (1968) Cayar submarine canyon. Geol Soc America Bull 79: 1821-1828

Dillon WP, Paull CK (1978) Interpretation of multichannel seismic-reflection profiles of the Atlantic continental margin off the coasts of South Carolina and Georgia. US Geol Survey, Miscell Field Studies Map MF-936

Dillon WP, Paull CK, Buffler RT, Fail JP (1979) Structure and development of the Southeast Georgia Embayment and Northern Blake Plateau: Preliminary Analysis. In Watkins et al. (eds), Geological and geophysical Investigations of Continental Margins. Amer Assoc Petrol Geol Mem 29: 27-41

Edelmann P (1979) Tiefenkarte westafrikanischer Kontinentalabhang 15°-17° N, 16°-18°40'W, 1 : 500 000. Diplomarbeit Geograph Inst Uni Kiel

Elouard P (1975) Formations sédimentaires de Mauritanie atlantique. In: Notice explicative de la carte géologique à 1 : 10^6 de la Mauritanie. Editions BRGM, Paris: 171-233

Flicoteaux R, Lappartient J-R. Trenous J-Y (1975) Contribution à l'étude du Tertiaire du bassin sénégalo-mauritanien (au sud de la vallée du Sénégal). Evolution paléogéographique et structurale. 3e Réunion and Sci Terre, Montpellier, 23-25 avril, p 150

Fritsch J, Hinz K, von Rad U, Roeser H, Weigel W, Wissmann G (1978) Ergebnisse geowissenschaftlicher Untersuchungen der VALDIVIA Westafrika Fahrt VA 10/1975. BMFT-Forschungsbericht M 78-03: 60 pp

Goldflam P, Hinz K, Weigel W, Wissmann G (1980) Some features of the northwest African margin and magnetic quiet zone. Phil Trans R Soc Lond A, 294: 87-96

Hebrard L (1978) Contribution a l'étude géologique du quaternaire du littoral mauritanien entre Nouakchott et Nouadhibou 18°-21° latitude Nord, participation à l'étude des désertifications du Sahara. Docum Lab Géol Fac Sci Lyon n 71: 210 pp

Hedberg HD (1955) Developments in Foreign Petroleum Fields - Africa. AAPG Bull 40: 1607

Klitgord KD, Behrendt JC (1979) Basin structure of the U.S. Atlantic margin. In: Watkins JS et al. (eds), Geological and geophysical investigations of continental margins. Amer Assoc Petrol Geol Mem 29: 85-112

Klug H (1973) Beiträge zur Geographie der mittelatlantischen Inseln. Geograph Inst Uni Kiel, 39: 169-205

Koopmann B (1979) Saharastaub in den Sedimenten des subtropisch-tropischen Nordatlantik während der letzten 20 000 Jahre. Uni Kiel Diss: 107 pp

Lancelot Y, Seibold E et al. (1977) Initial Rep of DSDP 41, 1259 p. Washington DC US Government Printing Office

Lecorche J-P, Sougy J (1978) Les Mauritanides, Afrique orientale - Essai de Synthese. In: Tozer ET and Schenk PE (eds) Caledonian-Appalachian orogen of the North Atlantic region. Geol Survey Canada Paper 78-13: 231-239

Lehner P, de Ruiter PAC (1977) Structural history of Atlantic Margin of Africa. Amer Assoc Petrol Geol Bull 61, n 7: 961-981

Le Page A (1979) Les principales unités structurales des Mauritanides dans la région de Bakel (Sénégal oriental) et leur signification. 10éme Colloq Intern Géol Afric, Montpellier 25-27 avril 1979: 117-118

Liger JL, Roussel J (1979 a) Etude gravimétrique du bassin côtier, profond sénégalais. Résumés du 10e Colloque de Géologie Africaine, Montpellier, 25-27 avril, 1979: 130-131

Liger JL, Roussel J (1979 b) Etude gravimétrique du bassin côtier profond du Sénégal. Rev Géol dyn et Géograph physique 21. n 5: 419-427

Lowrie A, Escowitz E (1969) Kane 9, Global ocean floor analysis and research data series 1, Washington DC, US Govt Printing Office: 976 p

Machens E (1966) Zur geotektonischen Entwicklung von Westafrika. Z deut geol Ges 116: 589-597

Mathiez JP, Huot G (1966) Prospection géophysique et recherches d'eaux souterraines. Exemples d'application en Afrique Occidentale - Comité Inter-Africain d'Etudes Hydrauliques, Paris, 154 p

Maugis P (1955) Etudes de préreconnaissance pétroliére dans le bassin sédimentaire du Sénégal. Bull de la Dir Fed Mines et Géol, Dakar, 19: 99-128

Michel P, Pascal M (1967) Carte géologique des formations sédimentaires de la vallée du Sénégal. In: Notice explicative de la carte géologique au 1/200 000 "Saint-Louis", Bureau de Recherches Géol et Min, Dakar: 8-9

Michel P (1973) Les bassins des fleuves Sénégal et Gambie. Etude géomorphologique. Tome 1-3. - Mémoires ORSTOM n 63, Paris

Nahon D (1976) Cuirasses ferrugineuses et encroutements calcaires au Sènègal occidental et en Mauritanie, systemes évolutifs: géochimie, structures, relais et coexistence. Sciences géologique Inst Géol Université Strasbourg, Mémoire n 44: 232 pp

Ortlieb L (1975) Recherches sur les formations plio-quaternaires du littoral ouest-saharien (28°30'-20°40' lat. N). Trav et documents de l'ORSTOM n 48: 267 pp

Pascal M, Michel P (1967) Etude géologique: Tectonique. In Notice explicative de la carte géologique au 1/200 000 "Louga" Bureau de Recherches Géol et Min, Dakar: 5-14

Paull CK, Dillon WP (1980) Structure, stratigraphy and geologic history of Florida-Hatteras shelf and inner Blake Plateau. Amer Assoc Petrol Geol Bull 64, n 3: 339-358

Piessens P, Chabot AG (1977) Bathymetry and sediments of the Arguin Platform, Mauritania, West Africa. Mém Inst Géol Uni Louvain 29: 369-379

Rachwal CA, Destefano ER (1980) Petroleum Developments in Central and Southern Africa in 1979, AAPG Bull 64: p 1789

Rona PA (1971) Bathymetry off central northwest Africa. Deep Sea Res 18: 321-327

Rona PA (1973) Relations between Rates of Sediment Accumulation on Continental Shelves, Sea-Floor Spreading, and Eustacy Inferred from the Central North Atlantic. Geol Soc America Bull 84: 2851-2872

Ruffman A, Meagher LJ, Stewart McG (1977) Bathymetry of the continental shelf and margin of Senegal and the Gambia, v 1. Dep Fisheries and Environment, Canada: 21-98

Seibold E, Hinz K (1976) German cruises to the continental margin of North West Africa in 1975: General reports and preliminary results from VALDIVIA 10 and METEOR 39. Meteor Forsch Ergebn C 25: 47-80

Seibold E, Hinz K (1974) Continental slope construction and destruction, West Africa. In: Burk CA and Drake CL (eds) The geology of continental margins. Springer, New York: 179-196

Shipley TH, Buffler RT, Watkins JS (1978) Seismic stratigraphy and geologic history of Blake Plateau and adjacent Western Atlantic Continental Margin. Amer Assoc Petrol Geol Bull 62, n 5: 792-812

Sougy J (1962) West African Fold Belt. Geol Soc America Bull 73: 871-876

Templeton RSM (1971) The geology of the continental margin between Dakar and Cape Palmas. In: Delany FM (ed) The geology of the East Atlantic continental margin: Great Britain Inst Geol Sci Rept 70/16: 47-60

Tessier F, Flicoteaux R, Lappartient J-R, Nahon D, Triat J-M (1975) Reform of the concept of "Continental terminal" in the coastal sedimentary basins of West Africa. Trav Lab Sci Terre, Ser A, n 8: 6 p

Trenous J-Y, Michel P (1971) Etude de la structure du dôme de Guier (Sénégal nord-occidental) Bull Soc géol France 13, n 1-2: 133-139

Twigt W (1980) Magnetic anomalies over fracture zones in the central North Atlantic Ocean. Geologica Ultraiectina, Mededelingen Inst Aardwetenschappen Rijksuniversität Utrecht 26: 125 p

Uchupi E, Austin JA (1979) The stratigraphy and structure of the Laurentian cone region. Can J Earth Sci 16: 1726-1752

10 Bathymetry, Microphysiography and Reflectivity Characteristics of the West African Margin Between Sierra Leone and Mauritania

Robert D. Jacobi[1] and Dennis E. Hayes[2]

Abstract

This report maps and analyzes the bathymetry and seafloor reflectivity characteristics along the northwest African margin between 3° and 23°N. The bathymetric trends strongly suggest that the morphology associated with the mid-ocean ridge (MOR) spreading system strikes north throughout the area mapped (north of ∿10°N). The system of equatorial fracture zones defines a dominant E-W morphologic fabric in the southwestern portion of the map area, and appears to extend through the northeast-trending Sierra Leone Rise in the southeastern part of the area. The Sierra Leone Rise is an aseismic system of plateaus probably formed at the MOR crest during a phase of excess basalt generation.

The reflectivity characteristics and microphysiography show that both downslope processes (turbidity flows and sediment slides) and along-slope processes (geostrophic bottom currents) have been important in shaping the present morphology of the continental rise. Turbidity flows and sediment slides are sparse south of 10°N; north of 10°N numerous turbidity flow pathways occur on the continental slope and rise. The Cayar Canyon System defines the largest pathway which extends from Senegal to the Gambia Abyssal Plain (900 km). Sediment slides are extremely common and occur on the continental slope and rise and at the base of isolated seamounts, islands, and ridges. The longest and largest sediment slide extends from the shelf break near Senegal to southeast of the Cape Verde Islands (570+ km). The largest slide complex is adjacent to the largest turbidity flow system. We believe rapid deposition by turbidity flows may provide an important mechanism for generating unstable slope and rise deposits that later fail as sediment slides.

Strong geostrophic bottom currents are not presently active in much of the map area. However, in the recent past, such currents probably flowed in the basins along the continental margin. In the basins north of the Kane Gap, a north-flowing bottom current operated that had two likely sources: Antarctic bottom water (AABW) flowing eastward through fracture zones, and bottom water flowing north through the Kane Gap. In the Sierra Leone Basin, a counter-clockwise gyre of bottom water was probably sustained by eastward-flowing AABW in the south.

1 Department of Geological Sciences, State Univ. New York at Buffalo, 4240 Ridge Lea Road, Amherst, N.Y. 14226, USA

2 Lamont-Doherty Geological Observatory, Palisades, NY 10964, USA

1. Introduction

The present bathymetry of the Northwest African continental margin results from a complex interaction of many processes. The igneous fabric generated by the seafloor spreading process provides the basic control on the morphology; this is subsequently modified by sedimentation and erosion processes. Seafloor spreading and related processes formed abyssal hills, fracture zone ridges and troughs, aseismic ridges, off-ridge volcanic islands and seamounts. Near-bottom sedimentation processes result in a multitude of physiographic features in the sedimentary cover, including (1) sediment slide complexes; (2) canyon/channel/distributary systems caused by downslope movement of currents and sediments; and (3) sediment waves and furrows caused by along-slope movement of currents.

Studies of 3.5 and 12 kHz echograms have shown increasingly that certain ocean-bottom echoes can be correlated with depositional/erosional processes acting at the seafloor. Thus, in order to study both the basement features and the sedimentary features on the northwest African margin, we utilized both airgun, seismic reflection profiles (∿100Hz), and echograms (3.5 kHz and 12 kHz), and secondarily, cores, bottom photographs, and near-bottom temperature data.

Two important products of these studies are maps of the bathymetry (Fig. 1), and reflectivity characteristics and microphysiography (Fig. 2). Figs. 1, 2 have been generalized from our more detailed ones (Jacobi and Hayes, in prep.) because oversized figures could not be included in this volume. The bathymetric contours are given in units of nominal fathoms because much of the data base could only be obtained in these units. Nominal fathoms are reflection time units based upon an assumed value for the velocity of sound in water. In our case, one nominal fathom assumes a sound velocity of 800 fm/s and is therefore exactly equivalent to 0.0025 s of two-way reflection time (1 fm = 1.828 m).

Selected 100 fm contours are shown in Fig. 1 to try to illustrate the principal morphologic fabric and gradients but to avoid unnecessary detail. Because all contours are not shown and some are shown as discontinuous ones, "nonconsecutive contours" are occasionally present in Fig. 1.

Generalized maps of the region have been published previously by Egloff (1972), Emery et al. (1975) and Uchupi et al. (1976). Bathymetric maps of smaller areas were contoured by Krause (1974), Dietz et al. (1968), Rona (1971), Fox (1972), Schlee et al. (1974) and Hobart et al. (1975). The detailed bathymetric map of Ruffman et al. (1977) drew heavily from preliminary versions of our maps of this area. Jacobi (1976) and Embley and Jacobi (1977a,b) studied in detail the microphysiography of a small part of the continental margin north of Senegal. Additional structural and sedimentological studies have been carried out by Collette et al.(1969), Rona (1970), Schneider and Johnson (1970), Behrendt et al. (1974), Schlee et al. (1974), Seibold and Hinz (1974), McMaster et al. (1975), and Wissmann (this Vol.).

2. Bathymetry

2.1 Fracture Zone and Spreading Ridge Morphology

The morphology associated with spreading mid-ocean ridges (MOR) dominates the northwestern map area, while the nearly orthogonal morphologic fabric associated with fracture zones dominates the southwestern map area. In the southwest, the major fracture zones trend approximately E-W and are separated by distances of approximately 100 km. Two very large fracture zones (Doldrums F.Z. and Four North F.Z.) were recognized and discussed by Heezen and Tharp (1965), Heezen et al. (1969), Fox (1972) and Cochran (1973). Several smaller fracture zones commonly are present within each degree of latitude. The trends of these minor fractures are not well known.

In the southwestern map area ship tracks that have an E-W orientation show a "second-order" fabric not readily apparent on N-S tracks. These relatively short-wavelength features very probably are related to the topographic fabric of the spreading ridge. The contours shown in Fig. 1 only outline the major E-W trending basement blocks bounded by fracture zones and do not define the smaller scale morphology that parallels the ridge crest.

The overall fracture zone morphology becomes less dominant north of approximately 12°N; however, some fracture zones still dramatically affect basement level there (e.g., 16°N, 29°W). Except in a few areas (e.g., 17°N, 29°W; 19°N, 27°W) the trend of these fracture zones is difficult to determine.

Similarly, the trend of the spreading-ridge morphology is not well defined. Uchupi (1971), Egloff (1972), and Uchupi et al. (1976) show abyssal hills (spreading-ridge morphology) with a northeast trend.

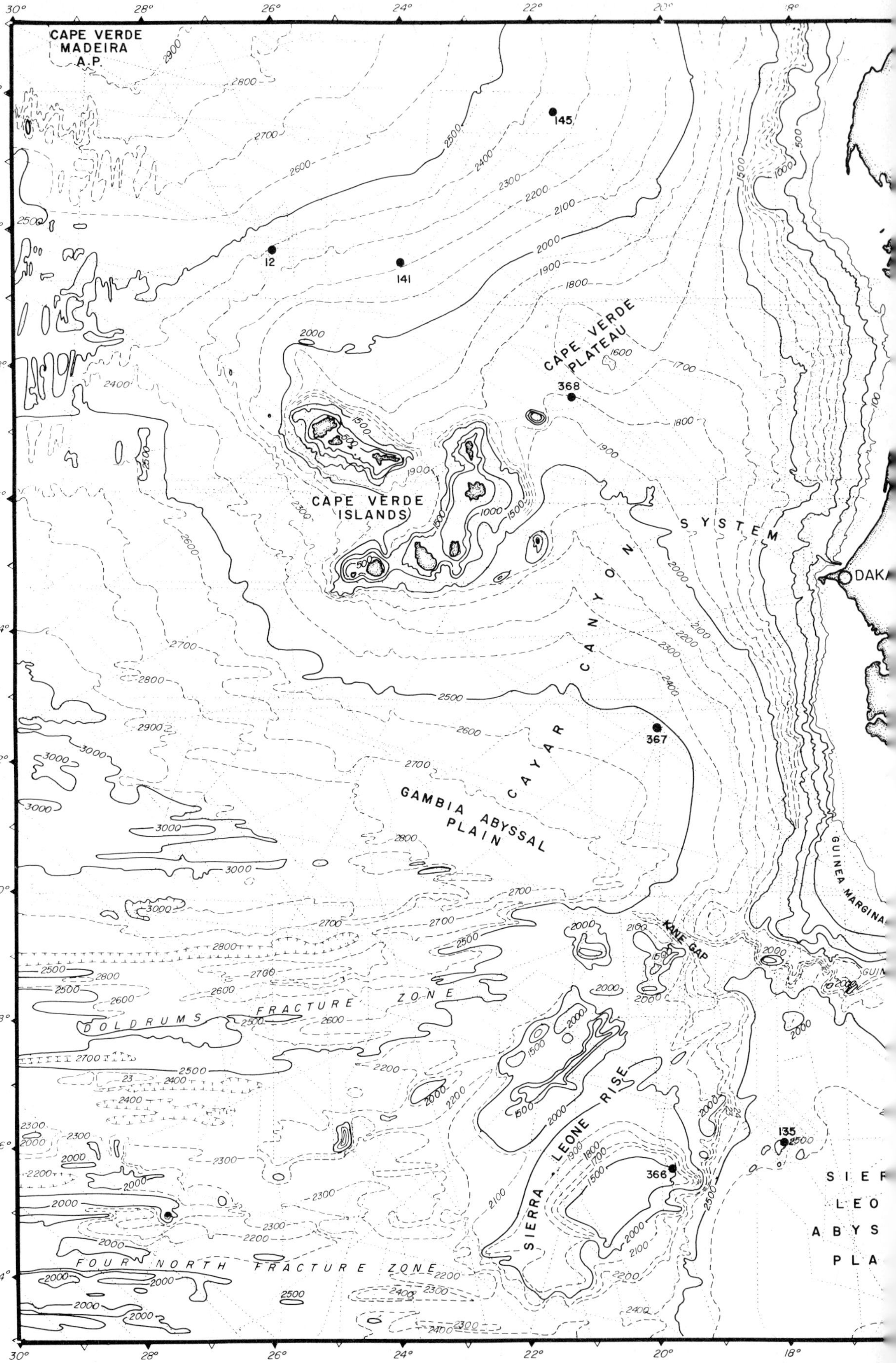
CAPE VERDE MADEIRA A.P.
CAPE VERDE PLATEAU
CAPE VERDE ISLANDS
CAYAR CANYON SYSTEM
GAMBIA ABYSSAL PLAIN
KANE GAP
GUINEA MARGINAL
DOLDRUMS FRACTURE ZONE
SIERRA LEONE RISE
SIERRA LEONE ABYSSAL PLAIN
FOUR NORTH FRACTURE ZONE
DAKA
145
12
141
368
367
366
135

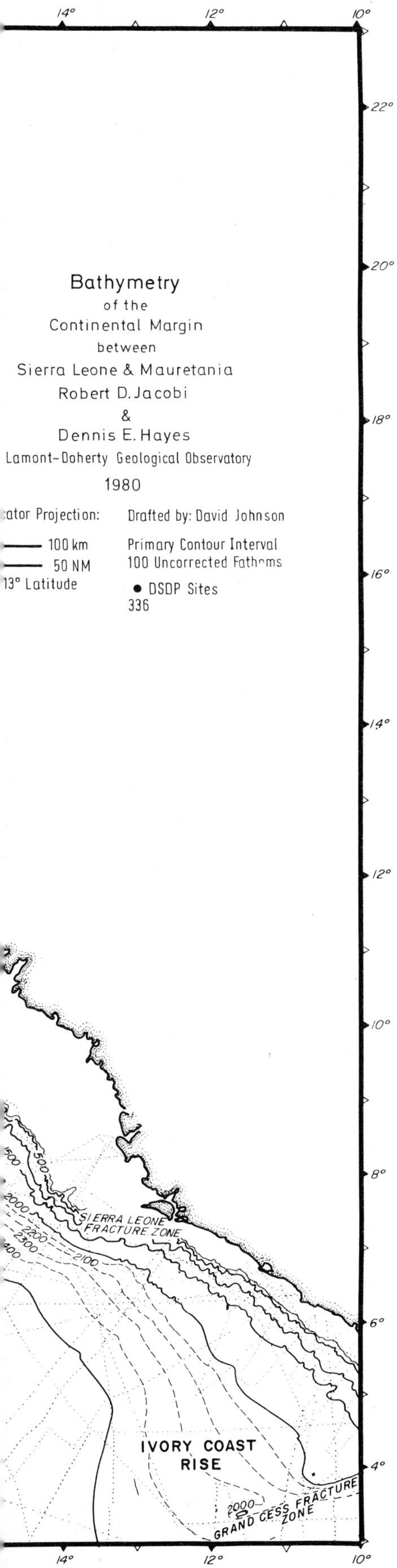

However, we believe the true trend of the spreading-ridge morphology is just slightly east of the north based on the trend of a few well-mapped abyssal hills (at 17°-18°N, 29°W; 22.5°N, 29.2°W; 22°N, 29°W; 20°N, 26°W) and fracture zones (at 17°N, 20°N and 22°N).

In the eastern map area, the location of fracture zones is difficult to determine because of the thick overlying prism of sediment on the continental rise. However, large offsets in the continental shelf break provide indirect evidence for the location and likely trend of ancient major fracture zones. The most dramatic example of such a "break" is at the intersection of the Guinea Fracture Zone with the continental margin at 9°N. The actual trend of the Guinea Fracture Zone has been a matter of contention. Krause (1964), Le Pichon and Fox (1971), and Le Pichon et al. (1977) suggested that the Guinea Fracture Zone extends west-northwest or northwest from the offset of the shelf break, whereas Heezen and Tharp (1965), Templeton (1971), and Emery et al. (1975) suggested that the fracture zone extends west from the same offset in the shelf break. The magnetics data of McMaster et al. (1970) support the interpretation of an E-W trend. The bathymetric trends mapped in Fig. 1 suggest the presence of a nearly continuous east-west fracture zone extending from the Guinea Marginal Plateau through the Sierra Leone Rise to the Doldrums Fracture Zone in the west.

Fig. 1
Bathymetry of the North Atlantic off northwest Africa between Sierra Leone and Mauritania. Primary contour interval is 500 nominal fms. Selected 100 fm also shown. Note that here one nominal fathom equals 1/400 s of two-way reflection time; one fm equals 1.8288 m. Primary data sources: soundings collected by the Lamont-Doherty Geological Observatory ships (R/V VEMA and R/V ROBERT D. CONRAD) and the U.S. Naval Oceanographic Office. In addition, the coastal charts published by the Defense Mapping Agency/Topographic Center were also utilized.
Please note: for DSDP Site "145" read 140, for Site "135" read 13

30°
28°
26°
24°
22°
20°
18°
CAPE VERDE
MADEIRA
A.P.
22°
20°
18°
16°
14°
12°
10°
8°
6°
4°
140
12
141
CAPE VERDE
PLATEAU
368
CAPE VERDE
ISLANDS
SYSTEM
CANYON
CAYAR
DAKA
367
GAMBIA ABYSSAL
PLAIN
GUINEA MARGINAL
DOLDRUMS FRACTURE ZONE
SIERRA LEONE RISE
366
SIER
LEO
ABYS
PLAI
FOUR NORTH FRACTURE ZONE
SL

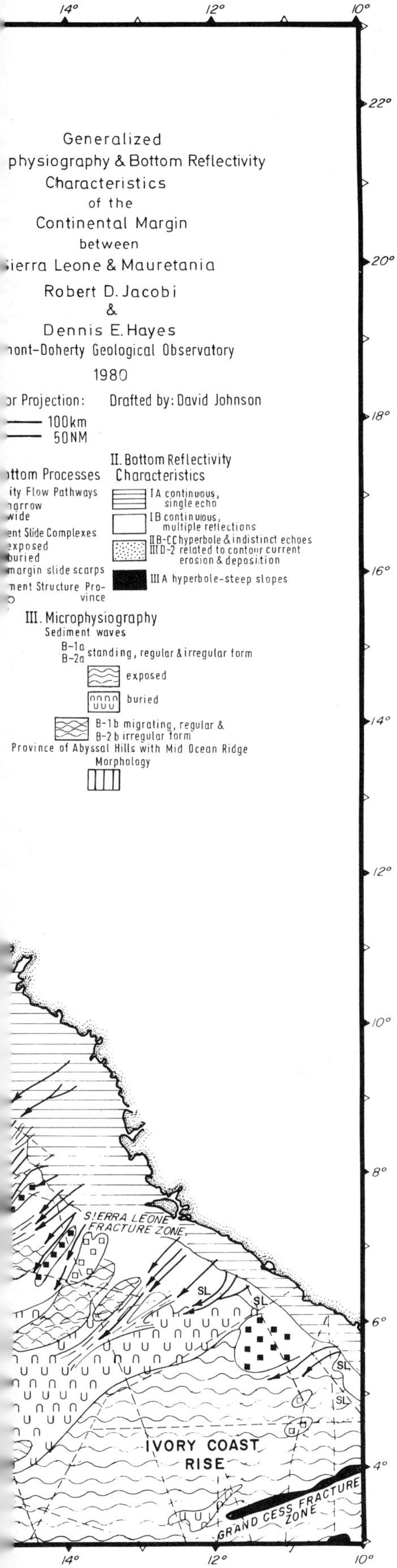

Another prominent offset of the shelf break occurs at 7°N where magnetic anomalies along the continental margin suggest that an east-northeast trending fracture zone intersects the margin (McMaster et al., 1973; McMaster et al., 1975). Emery et al. (1975) extended this fracture zone through the Sierra Leone Rise, although very little evidence exists concerning its continuity in the Sierra Leone Basin. The presence of a basement ridge at 5.3°N, 18.5°W, and a basement level change with associated seamounts, are consistent with the hypothesized west-trending fracture zones in the western Sierra Leone Basin.

A prominent offset of the continental margin near 4°N defines the northeast-trending Grand Cess Fracture Zone (Behrendt et al., 1974; Mascle and Sibuet, 1974). Less prominent offsets of the shelf break occur at 11°N, 12.5°N, 16°N, and 20°N, but their relationship to fracture zones is not yet known.

2.2 North-South Atlantic Fossil Plate Boundary

It is difficult to locate precisely the fossil plate boundary between the eastern North Atlantic and eastern South Atlantic. Magnetic anomalies from seafloor spreading are not easily identified because the amplitudes of these N-S trending anomalies are very low near the equator. In addition, the closely spaced equatorial fracture zones themselves have relatively high-amplitude magnetic anomalies. The difference in trends between fracture zones in the North Atlantic and those in the South Atlantic can provide

Fig. 2
Microphysiography and reflectivity characteristic map of the North Atlantic off northwest Africa between Sierra Leone and Mauritania. Lamont-Doherty Geological Observatory 3.5 kHz echograms and seismic reflection profiles, 12 kHz echograms also were utilized where 3.5 kHz echograms were unavailable or of poor quality. Additional data is from the U.S. Naval Oceanographic Office (unpubl. data); Lowrie and Escowitz (1969) and Hayes and Pimm et al. (1972)

some important constraints on the problem. The observed difference in trend between the Doldrums Fracture Zone at 8°N (with an east-northeast trend) and a fracture zone at 11.7°N, 25°W (with an E-SE trend) suggests to us that the fossil plate boundary between the North and South Atlantic probably strikes west to W-NW through a point near 10°N, 22°W. This hypothesis is also based upon the observation that deeper seafloor characterizes the area immediately north of ∿10°N.

2.3 Sierra Leone Rise

The Sierra Leone Rise is composed of a series of plateaus with superimposed seamounts. The trend of the eastern and western boundaries of the plateaus, individual seamounts, and seamount chains generally is northeast (Fig. 1). We are suggesting that the southern boundary of the Sierra Leone Rise coincides with the 4°N Fracture Zone (see Fig. 1). The northern boundary of the Sierra Leone Rise lies near the intersection of the Guinea Fracture Zone (Egloff, 1972) at the continental shelf. Offsets of the western boundary of the Sierra Leone Rise at 6°, 7°, and 9°N, and perhaps at 8°N coincide with adjacent fracture zones. East-west transverse ridges also may occur on the Sierra Leone Rise at 4°N, 7°N, 8.5°N, and 9°N. These offsets and transverse ridges suggest that the fracture zones extend eastward through the Sierra Leone Rise and control the north and south boundaries of the various basement blocks within the Sierra Leone Rise. Consideration of the bathymetry data and the synthetic fracture zones generated from Rabinowitz and LaBrecque's (1979) early pole of rotation strongly suggest that the Sierra Leone Rise is not comprised of a series of South Atlantic fracture zones.

The origin of the Sierra Leone Rise remains speculative. However, since no tectonism or volcanism occurred at DSDP site 366 (5.6°N, 19.8°W) in Cenozoic times (Lancelot et al., 1975), we hypothesize that the bedrock of the plateaus was generated at the mid-oceanic ridge. Sibuet and Mascle (1977) speculated that the Sierra Leone Rise marks the fossil plate boundary between the "main" South Atlantic and a mini-plate in the Sierra Leone Basin. Their mini-plate hypothesis cannot be tested with our present data because of the lack of clearly defined magnetic lineations, fracture zones and MOR morphology in the Sierra Leone Basin.

2.4 Cape Verde Islands

Ballard and Hemler (1967) suggested that the Cape Verde Islands were fault-bounded on the south and east. The northern and southern boundaries of both the eastern and western groups of the Cape Verde Islands all lie along possible eastward extensions of major fracture zones. Therefore, the zones of weakness defined by the fracture zones probably partially controlled the Miocene and younger uplift and volcanism of the region.

The age of volcanism (generally post-dating the generation of the underlying ocean floor), the nature of the volcanism (e.g., Dillon and Sougy, 1974) and the tectonic trends (especially the northeastern trend of the Cape Verde Plateau) suggest that the islands record an off-axis "hot spot" trace.

3. Microphysiography and Reflectivity Characteristics

We divide echoes returned from the seafloor into two major types: (1) those that define microphysiographic features, i.e., small features on the seafloor that are relatively accurately reproduced on standard 3.5 kHz and 12 kHz echograms (e.g., long wavelength sediment waves and large canyons); and (2) those echoes from other smaller features on the seafloor that are not portrayed accurately but which exhibit unique characteristics, and as such serve to define reflectivity characteristics provinces. Reflection hyperbolae, for example, are one group of these echo types that we will refer to as "reflectivity characteristics." This dual classification obviously gives rise to some overlap; for example, Damuth (1975) includes sediment waves as a subset of "hyper-

Table 1. Sea-Floor Reflectivity Characteristics

Type			Sketch of Echo Type	Example of Echo Type in Fig	Description of Echo Type	Physiographic Provinces in Which Echo Type is Common	Possible Origin And/Or Associated Deposit	Selected References
This Paper	Damuth and Hayes (1977)	Symbol on Fig. 2						
IA	IA			3	Continuous, strong, sharp echo with no sub-bottom reflectors	Continental shelves	Thin deposits of sand and gravel over consolidated or semi-consolidated sediments	Damuth (1975)
IB	IB			5	Continuous, strong, sharp echo with continuous sub-bottom reflectors	Continental slope, rise and abyssal plain, absent in areas of proximal turbidites	In regions of distal turbidites, reflectors are commonly turbidites	Ryan & Heezen (1965), Kermabon et al. (1969), Akal (1972), Embley (1975)
							In pelagic environments reflectors from variations in density and/or calcium carbonate content	Embley (1975)
							commonly relatively few sand/silt beds are associated with echo	Damuth (1975)
IIA	II,B			4, 7	Continuous, strong, prolonged echo; commonly no sub-bottom reflectors, except near boundaries of echo type, where discontinuous sub-bottoms occur	Turbidity-flow pathways (eg., channels) on continental slope and rise	Echo is actually very small hyperbolae probably the result of small, somewhat regular features caused by turbidity flows (e.g. erosional furrows or or flutes)	Ewing et al. (1973), Embley (1975)
							Relatively high % of sand/silt beds associated with echo	Damuth (1975)

Table 1. Sea-Floor Reflectivity Characteristics (cont'd)

Type			Sketch of Echo Type	Example of Echo Type in Fig	Description of Echo Type	Physiographic Provinces in Which Echo Type is Common	Possible Origin And/Or Associated Deposit	Selected References
This Paper	Damuth and Hayes (1977)	Symbol on Fig. 2						
IIB			II B 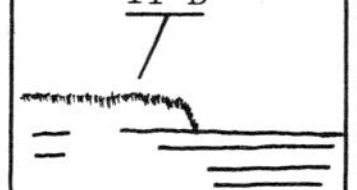	5, 7, 8	Weak, slightly prolonged echo commonly forming a transparent wedge that is elevated w.r.t. "normal" sea floor	Occur downslope from slide scar on continental slope and rises, at base of seamounts. Occur both within and without canyons and channels.	Generally related to debris flow material, esp. if slide scar or toe of material is observable.	Embley (1975, 1976), Embley and Jacobi (1977b), Jacobi (1976)
		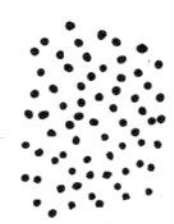	II B - cc 	11	Slightly prolonged, variable strength echo forming transparent wedges with varying thicknesses, sub-bottoms common.	Occur in areas with no observable slide scar, esp. fracture zones.	In some areas (e.g., fracture zones) where scars and toes do not not occur, thick turbidites or contour current deposits may cause echo, although a debris flow is also possible (see also echo type III D-3)	
IIIA	IIIA			9, 11	Large hyperbolae with varying vertex elevations (∿10-100m) + λ	Continental slope, canyon walls, abyssal hills, seamounts and fracture-zone topography	Steep, rugged slopes	Damuth (1975)
IIIC	IIIC	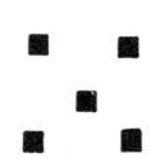	III C - 1 	7, 8	Large hyperbolae with varying vertex elevation (>40m), some hyperbolae enclose zones of multiple sub-bottom reflectors	Continental slope and rise, slope and base of seamounts, fracture zones, abyssal hills	"Hummocky material" - slide material near slide scar. Hummocks with internal reflectors are probable slide blocks (olistoliths), or, in some areas, portions of section that did not slide	Jacobi (1976) Flood et al. (1979)

Table 1. Sea-Floor Reflectivity Characteristics (cont'd)

Type: This Paper	Type: Damuth and Hayes (1977)	Symbol on Fig. 2	Sketch of Echo Type	Example of Echo Type in Fig	Description of Echo Type	Physiographic Provinces in Which Echo Type is Common	Possible Origin And/Or Associated Deposit	Selected References
III C (cont'd)	III C (cont'd)	■ ■ ■ ■ ■	III C - 2 	6, 7, 8	Large hyperbolae with varying vertex elevation (<40m), no sub-bottom reflectors	Same as above	"Blocky material" - slide material downslope from III C-1 and upslope from II B. Apparently material was less rigid than III C-1 and more rigid than II B	Embley (1975), Jacobi (1976), Flood et al. (1979)
		■ ■ ■ ■ ■	III C - 3 	6	Large hyperbolae, slightly varying vertex elevations, regular-irregular λ. Commonly associated with II A along channels. Forms areas of raised topography.	Along channels on continental rise.	Source unclear: either a result of bed forms along levees (small sediment waves) or slide material that moved down a channel and is presently being reworked.	
		○ ○ ○ ○ ○	III C - 4 		Large hyperbolae, slightly varying vertex elevations, irregular λ. Associated with both I B and II A echo types. Forms areas of raised relief. Generally associated with piercement structures observed on seismic reflection profiles.	Generally on continental rise.	Source is probably disturbed (deformed and uplifted) sediments resulting from the rising piercement structures, although a source similar to that for III C-2 and III C-3 is possible in many cases.	Embley and Jacobi (1977a)

Table 1. Sea-Floor Reflectivity Characteristics (cont'd)

Type: This Paper	Type: Damuth and Hayes (1977)	Type: Symbol on Fig. 2	Sketch of Echo Type	Example of Echo Type in Fig	Description of Echo Type	Physiographic Provinces in Which Echo Type is Common	Possible Origin And/Or Associated Deposit	Selected References
III D		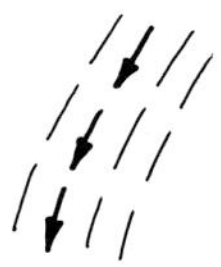	III D - 1 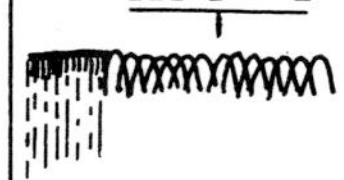	4	Very regular, discrete hyperbolae with vertices tangent to seafloor or a sub-bottom reflector, commonly associated with zones of IIA.	Zones of III D-1 occur primarily on continental rise approximately parallel and adjacent to (or within) channel systems.	Source is probable bed forms associated with turbidity flows, especially flutes.	
		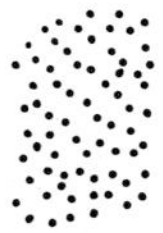	III D - 2 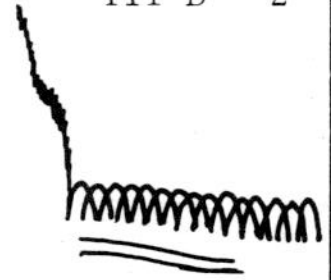	9, 10	Same as above except associated with IIIA of seamount or continental slope or IB/IIA of continental rise. Commonly associated with condensed or missing sedimentary section.	Zones of III D-2 are parallel to bathymetric contours on continental rise or occur around seamounts.	Source is generally furrows caused by narrow zones of erosion or non-deposition of contour currents. Furrows are parallel to current flow.	Laughton (1962) Heezen et al. (1966) Hollister and Heezen (1972), Hollister et al. (1974) Flood and Hollister (1974), Lonsdale and Spiess (1977)
		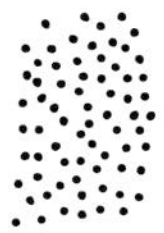	III D - 3 	11	Same as above, except associated with II B - cc.	Same as II B - cc fracture zones, some isolated seamounts.	If IIB - cc is caused by contour current activity, III D - 3 is probably related to ripples or furrows caused by the contour current.	
		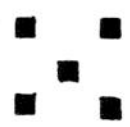	III D - 4 	5	Same as above, except associated with II B.	Same as II B - continental slope and rise and at base of seamounts.	Sources probably small bumps on top surface of debris flow deposit.	

Table 2. Microphysiographic features

Type: This Paper	Type: Damuth and Hayes (1977)	Type: Symbol on Fig. 2	Sketch of Echo Type	Example of Echo Type in Fig	Description of Echo Type	Physiographic Provinces in Which Echo Type is Common	Possible Origin And/Or Associated Deposit	Selected References
A				3	Locally planar sea-floor	Continental shelf, slope, rise, and abyssal plain		Egloff (1972) Damuth & Hayes (1977), Embley (1975)
B		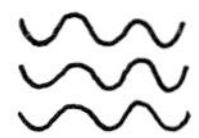	B - 1a 	12	Standing (climbing) sediment waves with regular form (λ and amplitude).	Echoes form zones approximately parallel to bathymetric contours on continental rise and less commonly to channel systems on continental rise; form zones around seamounts	Contour currents, and less commonly turbidity flows	Bryan & Markl (1966), Heezen et al. (1966) Schneider et al. (1967), Fox et al. (1968), Heezen & Johnson (1969), Markl et al. (1970), Ewing et al. (1971), Egloff (1972), Kenyon & Belderson (1973), Hollister et al. (1974), Jacobi et al. (1975), Damuth & Hayes (1977), Embley & Langseth(1977) Lonsdale & Spiess (1977), Damuth (1979), Flood (1978, 1980)
		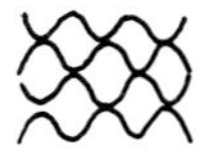	B - 1b 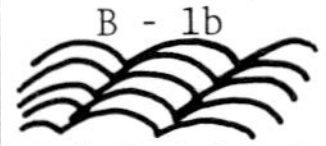	12	Migrating sediment waves with regular form (λ)			
		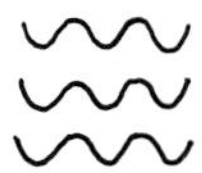	B - 2a 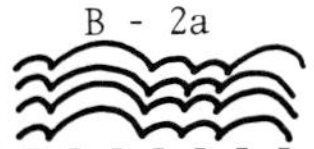	13	Standing (climbing) sediment waves with irregular form	Same as above, except zones more common along channel systems than parallel to continental rise bathymetric contours	Contour currents and/or turbidity flows	
		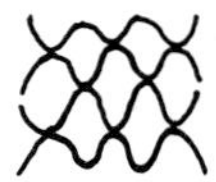	B - 2b 	13	Migrating sediment waves with irregular form			

Table 2. Microphysiographic features

Type: This Paper	Type: Damuth and Hayes (1977)	Symbol on Fig. 2	Sketch of Echo Type	Example of Echo Type in Fig	Description of Echo Type	Physiographic Provinces in Which Echo Type is Common	Possible Origin And/Or Associated Deposit	Selected References
C			C - 1 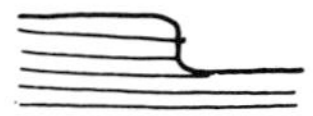	5, 7	Erosional scarp (with no displacement of underlying strata)	Continental slope and rise, seamounts abyssal hills and feature zones	(1) Exposed portion of a slide scar (especially if cross-section is open-ended or box-shaped), (2) steep wall of deep-sea channel.	Embley (1975, 1976), Jacobi (1976), Embley and Jacobi (1977b), Flood et al. (1979)
			C - 2 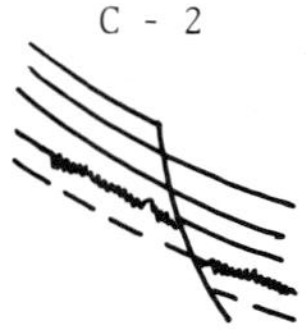		Fault scarp (displaced sub-bottom reflectors occur beneath surface expression of fault)	Same as above	Faulting caused by instability of sedimentary sequence because of (1) oversteepened slope caused by (a) contour current erosion downslope (b) sediment sliding downslope (c) high sediment input; (2) tectonic activity in basement	Jacobi (1976), Embley and Jacobi (1977b)
		SL	C - 3a 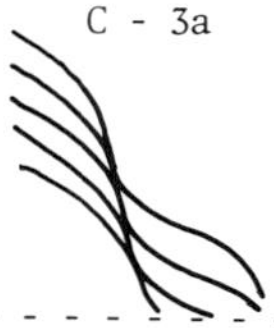	8	Sharp monoclinal flexure. Note difference in form WRT B-1b and B-2b. C-3a generally has a long gentle slope and a short steep slope	Same as above	Ductile faulting resulting from causes similar to C-2	
		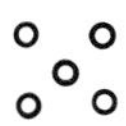	C - 3b 		Sharp monoclinal flexure (located beneath flexure is is piercement structure observed on seismic reflection profiles)	Generally on continental rise	Piercement structure may act as a tectonic dam and/or may actively deform the sediment sequence	

bolae" and hence a reflectivity characteristic, whereas most sediment waves in this paper are classified as microphysiography.

The origin and areal distribution of many of these echo types have been studied extensively in other areas (e.g., Flood and Hollister, 1974; Embley, 1975; Jacobi, 1976; Damuth and Hayes, 1977). Tables 1,2 summarize our compilation of the relevant information concerning the echo classification used, including: (1) examples, (2) map distribution, (3) probable origin, and (4) selected references. Careful inspection of these tables reveals that the origin of certain echo types is unequivocal (e.g., strong, prolonged single echoes, type IIA); in contrast, the origin of some other echo types is unclear (e.g., prolonged echo type IIB-cc) and must therefore be regarded as speculative.

Several echo types may occur in combination because they are associated with a common near-bottom process, such as turbidity flows or contour currents. Figure 2 illustrates the areal distribution of echo types that define: (1) three near-bottom processes (turbidity, flows, sediment slides and piercement processes); (2) selected reflectivity characteristics provinces (single echo type IIID-2 combined with weak echo type IIB-cc); and (3) microphysiographic features (migrating and standing sediment waves type B; sediment slide scarps type C-1; and abyssal hills). In the following section we briefly review the recognition, origin, and distribution of the various echo types associated with each of the near-bottom processes. We also discuss the areal distribution and significance of each of the near-bottom processes that we believe has been important in the map area.

4. Near-Bottom Processes

4.1 Turbidity Flow Pathways

The echo types common to turbidity flow pathways are summarized in Table 3. The most characteristic echo types associated with the pathways are erosional scarps (C-1; microphysiographic

Table 3

Bottom Process	Location	Echo Types	Figure	Comments
Turbidity Flows	Continental slope	III A	4	Most common because of steep-walled canyon
		B & C	4	A few echoes observed
	Continental rise a) Channel	II A, III D-1; C-1	4	As downslope gradient decreases, morphology becomes subdued. Wall of channel can be C-1 or a gentle slope with no scarp Fig. 4).
	b) Levee and areas away from channel	II A	4	From probable spill-over and sheet flow of turbidity flows
		B	13	B-2b more common than other types of waves.[a] If both B-1 and B-2 occur on same levee, B-1 occur generally closer to the channel. B echoes occur only rarely on the lower parts of deep sea fans.
	Areas with turbidity flow deposits return echoes observed on seismic reflection profiles that are very strong, slightly prolonged, planar reflections.			

[a]The abundance and location of these irregular waves may be a function of relatively sporadic and irregular flow of turbidity as Kenyon and Belderson (1973) suggested for similar waveforms influenced by contour current deposits

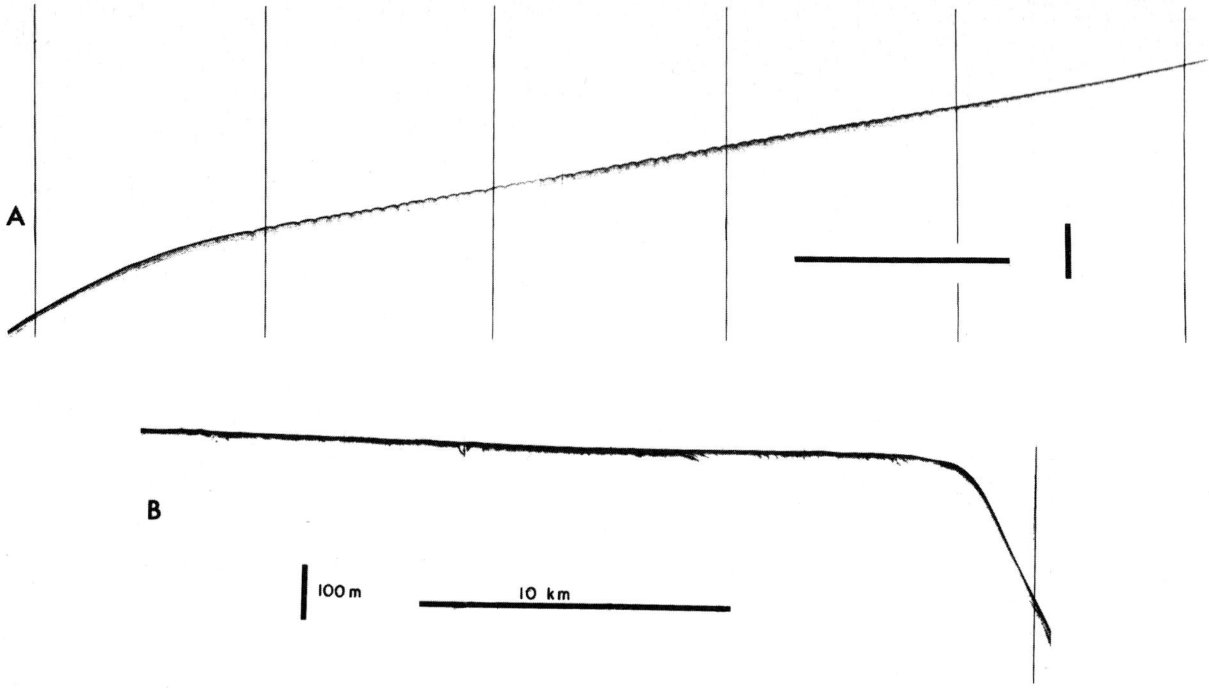

Fig. 3. Type IA distinct, continuous echoes with very few or no sub-bottoms that are located on the continental shelf

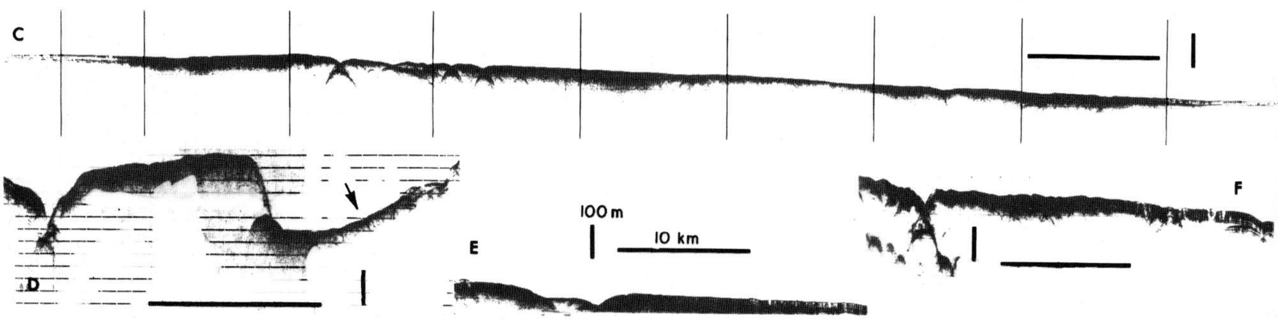

Fig. 4. Type IIA continuous, strong, prolonged echo, commonly without sub-bottoms. These echoes imply turbidity flow pathways. Note the variable sub-bottom extent of Type IIA echoes in echograms C, E, F. Echogram D has Type IIID-1 echoes (marked by arrow) in a possible slide scal now utilized by turbidity flows

features) and strong, prolonged echoes (type IIA; reflectivity characteristic). For examples see profiles in Fig. 4. Weak echoes (type IIB), large, irregularly spaced hyperbolae (type IIIC-2) and some small, regularly spaced hyperbolae (type IIID-1) also occur along many of the turbidity flow pathways (e.g., Profile D, Fig. 4 and Profile Q, Fig. 6). Such echoes occur north of the Cayar Canyon System and also near the Cape Verde Islands. These echoes have an uncertain origin, but are the result of either sediment slide deposits or levees. South of the Cape Verde Islands a definite slide scar, adjacent to the channel, occurs upslope from the deposit which gives rise to the type IIB and III echoes (Fig. 2). In general, we believe most of the type IIB and irregular hyperbolae result from sediment slides following approximately the same path as the turbidity flows. It is difficult even on 3.5 kHz echograms to distinguish between the subtypes of IIID echoes. Therefore, some turbidity flow pathways mapped in the Sierra Leone Rise area may be misidentified.

Sediment waves (echo type B) caused by contour currents commonly occur in the same general area as sediment waves produced by turbidity flows. It is usually difficult to discern which near-bottom process caused many of the waves. In this study area, sediment waves that are observed on levees are assumed to be part of the turbidity flow regime. The observed reversal in the migration direction of the sediment waves on opposing sides of the levee at 8°N, 15.5°W (Profile UU, Fig. 13) provides additional evidence to substantiate this assumption. Sediment waves adjacent to prominent levees that have a form and orientation similar to those on the crest of the levee also are thought to be caused by turbidity flow. However, two criteria suggest that the major sediment wave provinces mapped in this area are caused directly by contour currents. Firstly, ponded turbidites (with planar-bedding) are presently burying the sediment wave province in the southeastern map area (∿15°W, 5°N; Profile RR, Fig. 13). This strongly suggests that at least recent turbidity flows do not create or maintain most of the sediment waves in that province.

Secondly, within the two major sediment wave provinces south of the Cape Verde Islands and west of Dakar, sediment waves generally trend perpendicular to the turbidity flow pathways. The sediment waves in these provinces probably were not generated by turbidity flow, as the observed trend of sediment waves along levees appears to be generally parallel or slightly oblique to the turbidity flow pathways.

The distribution of the major turbidity flow pathways is shown in Fig. 2; such pathways occur along the entire continental slope south of about 20°N and north of the Guinea Marginal Plateau (10°N). South of the Guinea Marginal Plateau relatively few turbidity flow pathways appear to be active at present, and they are the shortest major pathways in the study area, probably because of the narrow continental rise there.

Major turbidity flow pathways also extend from the Cape Verde Islands northward to the Cape Verde-Madeira Abyssal Plain and southwestward to the nearby fracture zone valleys. A few channels also extend away from isolated seamounts and ridges. These turbidity flows may be generated from sediment slides.

The most notable canyon system in the area is the Cayar Canyon System (Dietz et al., 1968; Egloff, 1972). This system extends about 900 km from the continental slope near 15°N to a small deep-sea fan adjacent to the Gambia Abyssal Plain. This abyssal plain is very narrow and much of the turbidity flow continues westward past the abyssal plain and infills various fracture zone troughs (Fig. 2). Dietz et al. (1968) suggested that southerly-directed longshore currents provide the source materials of the turbidity flows in the Cayar Canyon.

4.2 Sediment Slide

Slide scars and redeposited slide material are observed on both 3.5 kHz records and low-frequency seismic reflection records. The echo types commonly associated with sediment slides are summarized in Table 4. "Erosional scarps" (echo type C-1) formed by the upturned edges of slide scars bound the zones of removal and are commonly 50 m high, but range up to 250 m high (Figs. 5, 7). Except for the erosional scarps, the slide-scars generally parallel prominent sub-bottom reflectors, even over successive crests and troughs of sediment waves (Profile G, Fig. 5). In these areas the sliding sediment appears to transfer very little energy across the slide scar to the autochthonous sediment because the sliding sediment did not erode down through the prominent reflector over the crests of the sediment waves. Fault scarps from rotating slump blocks (echo type C-2) occur only rarely on the northwest African margin. In many of the slide zones multiple slide scars and the complex depositional morphology indicate multiple slide movements (Jacobi, 1976).

The deposited slide material includes "hummocky" slide material (echo type IIIC-1, Fig. 7); "blocky" slide material (echo type IIIC-2, Figs. 5, 6, 7); and debris flow material (echo types IIB and IIID-4, Figs. 5, 7). These echo types illustrate an in-

Table 4

Bottom Processes	Location (progressively in a downslope sequence)	Echo Types	Figure	Comments
Sediment slides	Farthest upslope echo of sediment slide complex; on continental slopes, rises, seamounts and fracture zone walls	C-1	5	Except for boundaries of the slide scar where it cuts upsection, the slide scar follows strong reflectors, even where the reflector has considerable local relief due to bottom features such as sediment waves.
		C-2		The zone of removal is rarely bounded by C-2 rather than C-1.
	Generally near C-1	III C-I	7,8	Olistoliths do not appear to have travelled long distances on this continental margin.
	Near C-1 if no III C-1, otherwise downslope from III C-1	III C-2	6,7,8	Cores in areas returning III C-2 and II B echoes usually have highly disturbed sediment (Jacobi, 1976).
	Near C-1 if no III C-1 and 2; otherwise downslope from III C-2	II B and III D-4	5,7,8	
	Downslope of III C-2 and II B	II A	4,7	Volume considerations suggest a lack of extensive turbidity flows generated by debris flows off the Northwest African margin (Jacobi, 1976). However, younger turbidity flows commonly utilize the morphologic depression caused by the slide scar.

creasing liquefaction downslope (hummocky→blocky→debris flow). Highly deformed sediments are observed in cores raised from areas of blocky material and debris flow material (Embley, 1976; Jacobi, 1976; Embley and Jacobi, 1977b; Jacobi and Mrozowski, 1979). Middleton and Hampton (1973) suggest that turbidity flows should be generated at the leading edge of debris flows. Such turbidity flows are extremely difficult to differentiate from more recent ones that merely utilize an existing sediment slide depression. On the northwest African margin mass balance considerations of the volume of material removed vs. the volume of slide material deposited suggest that turbidity flows generated from debris flows are of minor volumetric importance (Jacobi, 1976).

Deposited sediment slide material that subsequently has been eroded partially or buried is generally difficult to recognize. Eroded blocky slide material commonly returns sporadic type IIIC hyperbolic echoes interspersed with echoes returned from bottom-current pathways (types IIA and IIID). Irregular sediment waves (B-2a) observed downslope from a type C-1 or C-2 scarp may result from the hemipelagic sediment which mantles hummocky and blocky slide material (i.e., drape folds; 21.5°N, 18.5°W, Fig. 2; Profiles S-V, Fig. 8). Alternatively, these irregular waves may be "slump folds" caused by downslope creep of the sediments (Embley and Jacobi, 1977b).

Type IIIC hyperbolae which appear at depth beneath the waves (Profiles U, V; Fig. 8) may be the result of buried slide material or a rough decollement surface.

Major zones of sediment slides occur on the continental margin north of Dakar, Senegal, between 15°N and 18°N (Fig. 7; Seibold and Hinz, 1974; Jacobi, 1976; Embley and Jacobi, 1977b). These slides extend westward from near the shelf break to at least

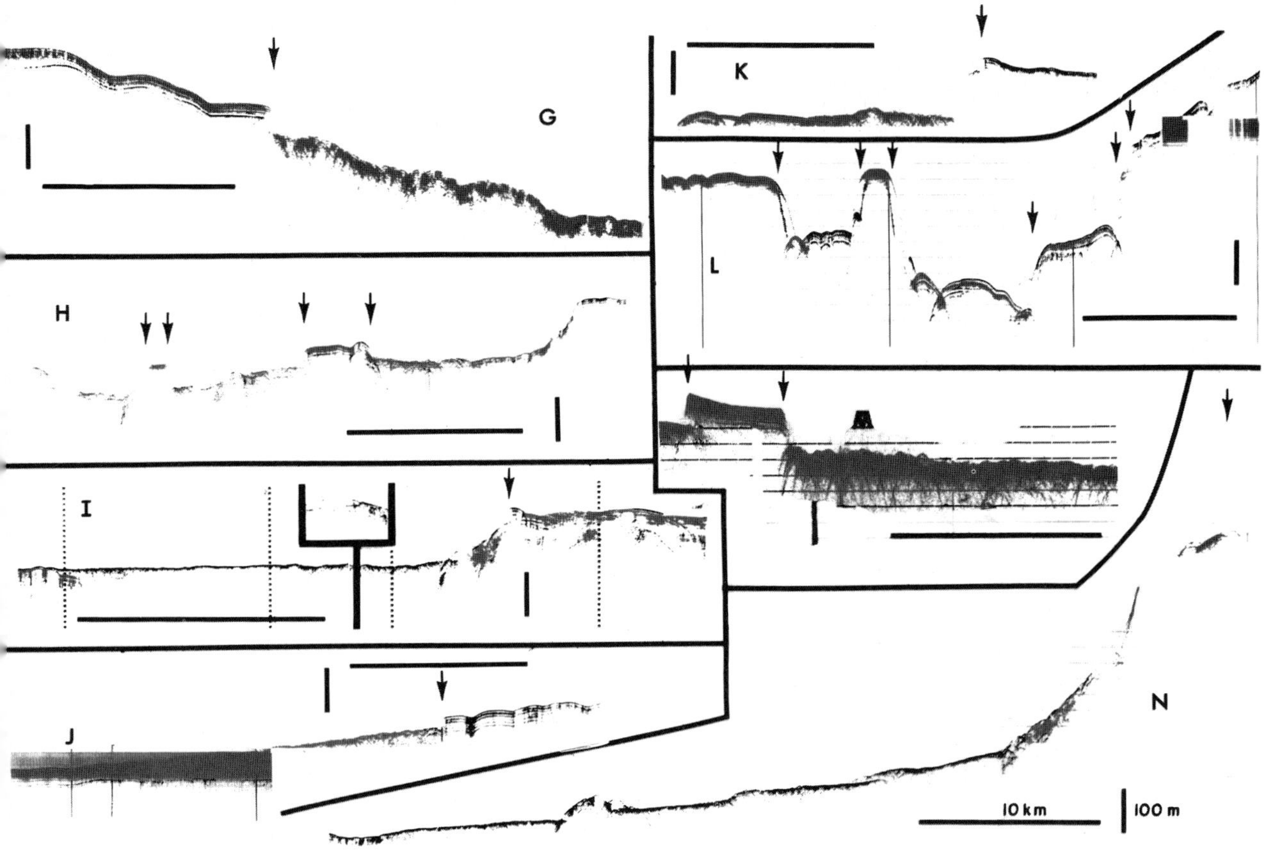

Fig. 5. Type IB, IIB, IIIC-2, C-1 echoes - slide scars and slide material. Type IB continuous, strong echo with multiple reflectors in nondisturbed parts (e.g., left of echogram G). Type IIB indistinct weak echo with few hyperbolae (probable debris flow deposits) seen at left of center on echograms H, I, J, and N. Insert in echogram I (located in Doldrums Fracture Zone) shows limited extent of debris flow deposit when ship turned at 90° to remainder of profile. Type IIIC-2 hyperbolae (blocky slide material) are seen in echograms G, M and right of center in profiles J, K, M, and N. Type C-1 scarps (slide scars) observed on most echograms (marked by arrows); note box-shaped cross-section of slide scar in echogram L and note parallelism of slide scar to layering in sediment waves in echogram G

21°W, and may contribute material along the Cayar Canyon as far south as 14°N (Fig. 2 and Profile Q, Fig. 6). Between 15.5°N and 14°N, much of the previously deposited slide material is partially reworked by younger turbidity flows in the Cayar Canyon System.

Several large slide zones that occur well north of Dakar are now either buried or manifested primarily as slump folds (e.g., echo type B-2a at 22.4°N, 18.3°W; see discussion above). Relatively few slides or turbidity flow pathways occur on the continental margin south of the Guinea Fracture Zone. Slide material is also evident near the base of seamounts (e.g., 17.5°N, 26.5°W; 9°N, 19.5°W; Cape Verde Islands; Sierra Leone Rise; Profile I, Fig. 5).

Although some of the sedimentary deposits in fracture zones that return type IIID-3 echoes (hyperbolae) and IIB-cc echoes may have slides as their ultimate source (Profile I, Fig. 5), others may be the result of contour current deposition (see discussion below).

The observation that abundant sediment slides occur in areas of abundant

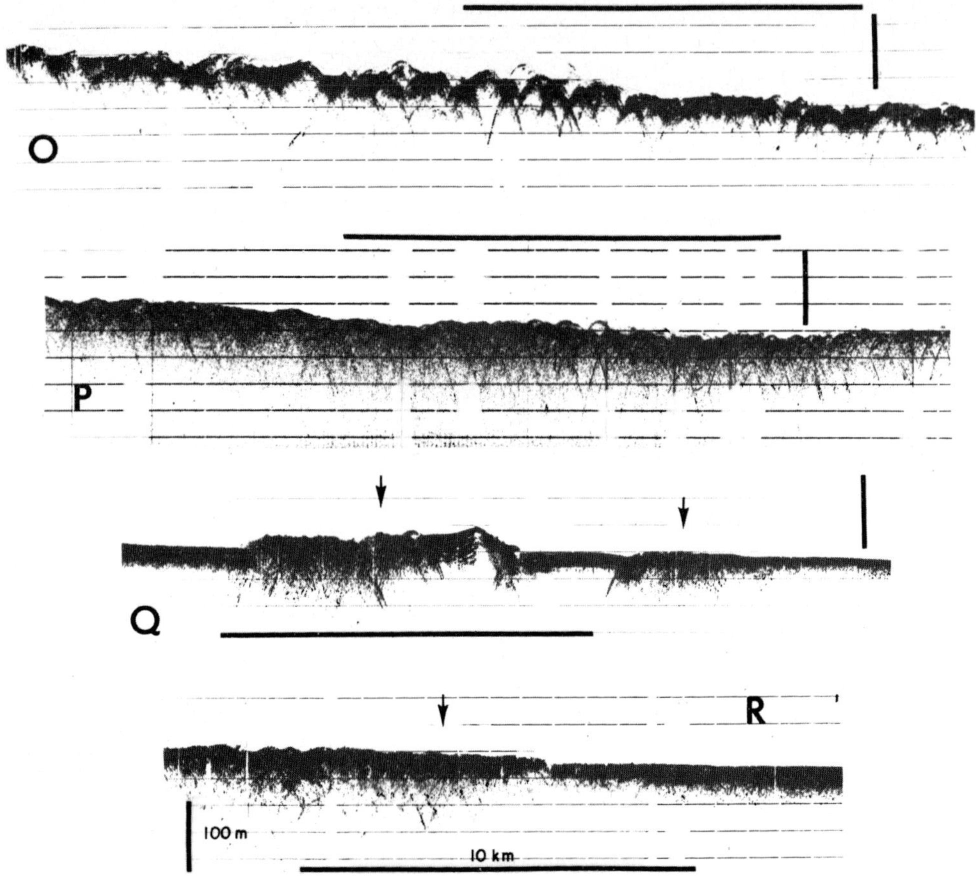

Fig. 6. Type IIIC-2 hyberbolae - blocky slide material. Note elevation of blocky material (arrow) with respect to undisturbed sediment (at right) in echogram R. IIIC-3 hyperbolae on echogram Q (elevated areas marked by arrows) are probably slide material, but could be part of a levee related to the turbidity flow pathways shown at left and center

turbidity flows and that few sediment slides occur in areas of sparse turbidity flows suggests a causal interrelationship. We suggest that the rapid deposition resulting from turbidity flows may give rise to unstable deposits due to associated excess pore pressures. Such deposits would then be especially susceptible to failure when subjected to additional loading by normal sedimentation processes, eustatic sea level changes, or severe storms. Although earthquakes could also trigger movement of such unstable deposits, this particular region is presently aseismic and probably has been for some time. Abbott et al. (1981) and Embley and Jacobi (1977b) have proposed similar mechanisms for other slide areas.

4.3 Contour Currents

Areas of contour current activity are represented on 3.5 kHz echograms as moats (Profile GG, Fig. 10); erosion surfaces that crosscut underlying reflectors (Fig. 10, all profiles); sediment waves (echo type B, Figs. 12, 13); hyperbolae (echo types IIID-2; Profiles Z-DD, Fig. 9); IIID-3 (Profile EE, Fig. 10); and IIB-cc echoes (all profiles, Fig. 11). Moats and erosional furrows result from fast bottom water flow (∿15 cm/s) whereas sediment waves and drift deposits (type IIB-cc) are generated by slower moving bottom water (∿10 cm/s). Analysis of 100 Hz seismic profiler records reveals that areas affected by contour currents commonly have rela-

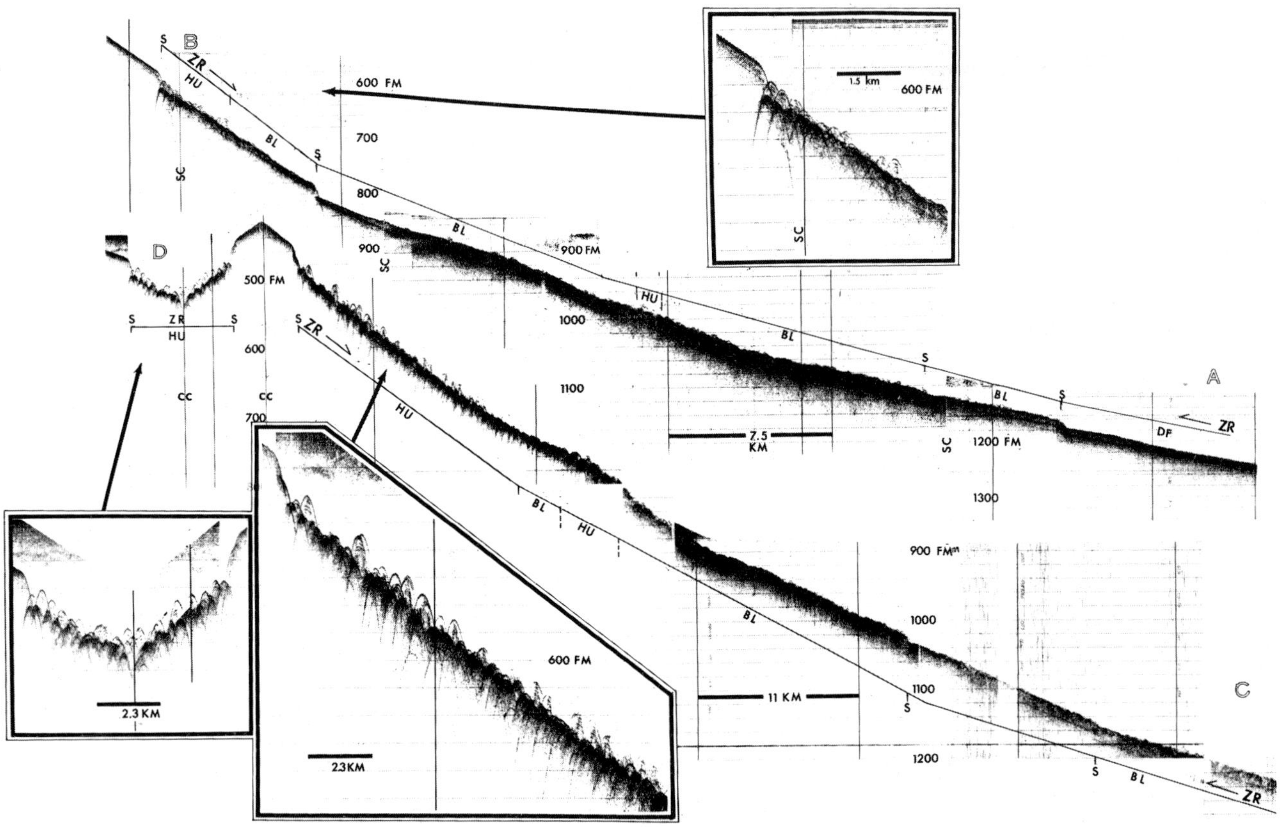

Fig. 7. Type IIB, IIIC-1, IIIC-2, hyperbolae and indistinct, weak echoes. Profiles of two sediment slides north of Dakar, from Jacobi (1976). BL blocky slide material (type IIIC-2); HU hummocky slide material (type IIIC-1); S slide scar (type C-1); ZR zone of removal of sediment during sliding. Note internal reflectors in the large hummocks

tively thick and unconformable transparent sedimentary sections.

North of the Kane Gap, sediment waves and erosional furrows are the primary features caused by contour currents. Sediment waves occur locally around seamounts, on the Cape Verde Plateau, and in the northwestern map area (e.g., 19.5°N, 29°W; Fig. 2). Two large sediment wave provinces occur on the continental rise south of the Cape Verde Islands and west of Senegal at 13.5°W, 19.5°W. The depth of the upper reaches of both sediment wave provinces is somewhat variable but it occurs approximately at the slope/rise boundary ($\sim$1700 fm$\approx$3100 m). However, the lower boundary of both provinces lies consistently near the 2500-2550 fm (4570-4650 m) isobaths where a marked decrease in the gradient of the continental rise occurs (Fig. 1). The provinces consist of both standing and migrating sediment waves (Fig. 12) whose orientations may be either oblique or parallel to the contours. Uchupi et al. (1976) suggested that all the sediment waves found west of Dakar, Senegal have a slump origin. Because the sediment waves found nearest to Dakar (Fig. 2) trend parallel to the regional contours, are irregular and have a type C3a morphology (Profile W, Fig. 8), they may be slump folds. However, we believe most of the sediment waves are not slump folds because the sediment waves trend obliquely to the regional bathymetric contours and are associated with a moat around a basement peak (e.g., at 13.5°N, 20°W; Fig. 2, and Profile OO, Fig. 12). Furthermore, in some of the troughs between migrating sediment waves, subbottom reflectors onlap discordantly the adjacent upslope sediment waves. The observed geometry of reflectors

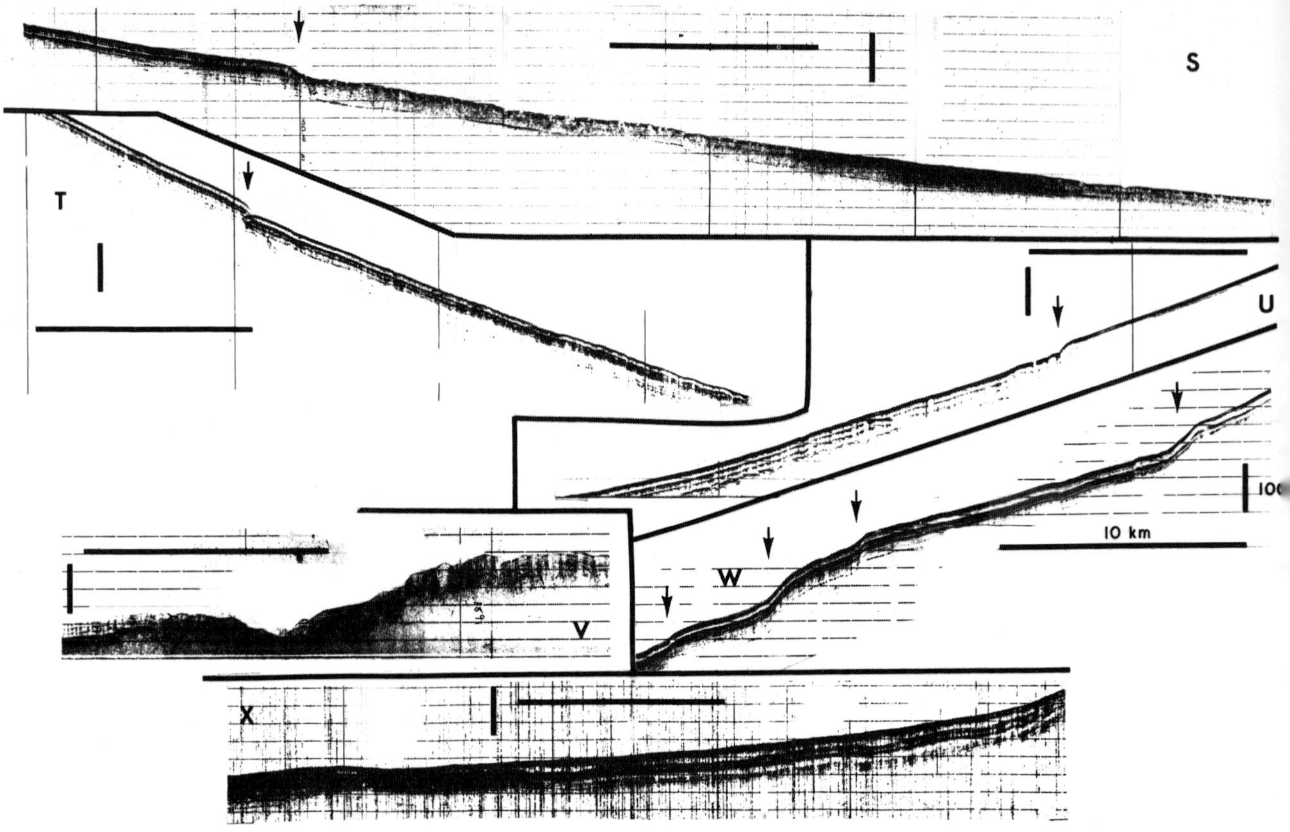

Fig. 8. Type IIB, IIIC-1, IIIC-2 echoes buried by type IB echoes; type C-3 scarps. Echogram S shows buried type IIB and IIIC-2 echoes (debris flow and blocky material) downslope from a buried C-1 scarp (marked by arrow). These buried echoes result in "sediment waves" in the overlying material. Echograms T and U show possible buried IIIC-2 echoes downslope from possible burried C-1 scarps (marked by arrows). See text for alternate hypothesis concerning origin of hyperbolae in profiles T and U. Echogram V shows possible buried IIIC-1 echoes (hummocky material). Echogram W shows C-3 scarps caused by slumping along ductile faults which are marked with arrows. Echogram X shows buried type IIIB echoes (debris flow material) disposed in several layers (e.g., 10 m below surface in center of profile). Note that in echograms S, T, and V the transparent layer beneath the hyperbolae is either nonexistent or much thinner upslope from the buried scarps

within individual sediment waves indicates clearly that reflectors were not originally continuous and then later faulted (Profile PP, Fig. 12).

The oblique strike of the sediment waves with respect to the bathymetric contours suggests a northerly flow of bottom water west of Senegal, and a westerly flow south of the Cape Verde Islands. Our inspection of bottom photographs and studies of potential bottom temperatures indicate the presence of relatively sluggish bottom water in this area at present. Therefore, many of the observed sediment waves were probably generated by paleocontour currents.

The sediment waves west of Senegal overlie a strong sub-bottom reflector observed in seismic reflection profiles at 0.5 s (2-way travel time) below the seafloor. This reflector represents a Middle-Late Miocene unconformity (Lancelot et al., 1975). Sheridan et al. (1974) and Tucholke and Mountain (1979) suggested that on the Blake-Bahama Rise and in the Blake-Bahama Basin (western North Atlantic) strong, erosional con-

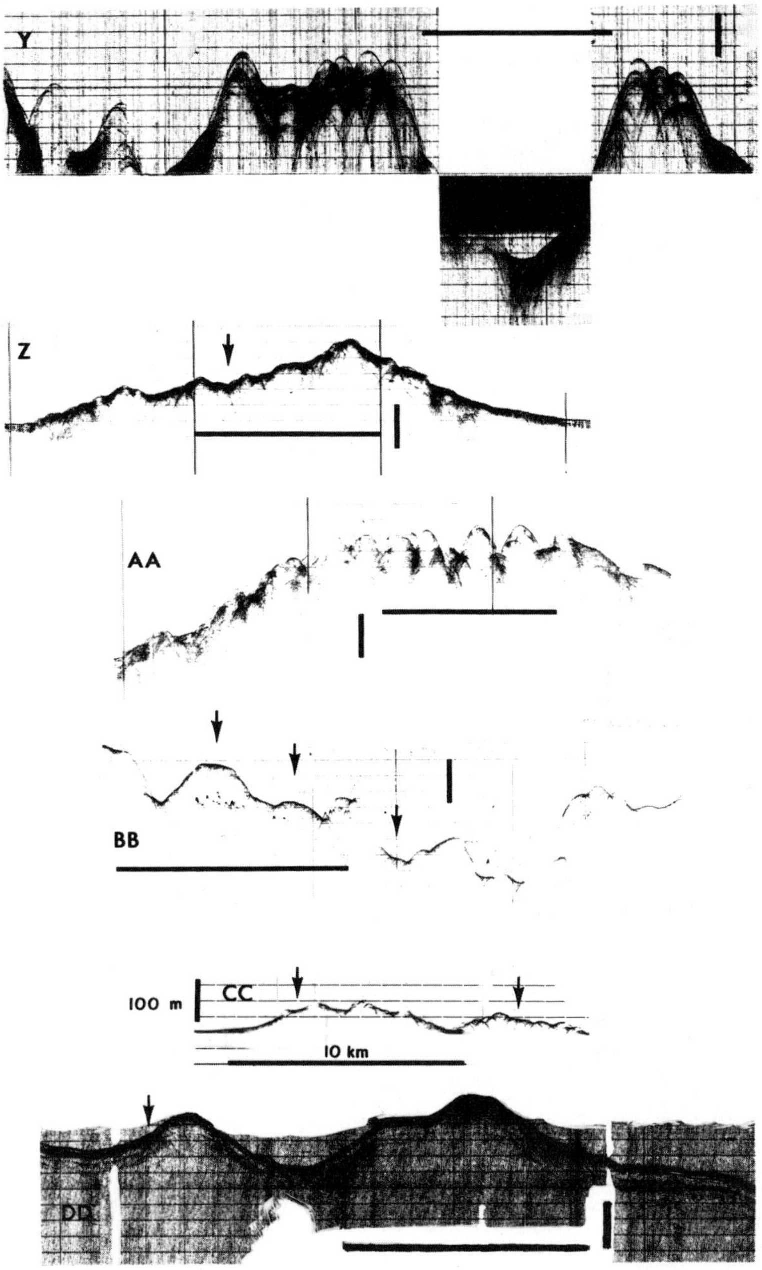

Fig. 9
Type IIID-2 hyperbolae and IIA prolonged echoes. Echograms Y and AA show good IIA echoes resulting from basement-controlled steep slopes (basement covered with pelagic sediments). The high reflectivity (IIA and some IIID-2) on the upper three profiles is suggestive of bottom water flow around the peaks, especially evident where IIA echoes grade into IIID echoes (arrow in profile Z). Type IIID-2 hyperbolae are displayed in lower three profiles (representative examples marked by arrows). Note possible slide complex at right of AA

tour currents diminished in their intensity in the Miocene, thus resulting in deposition and the development of sediment waves. If a similar situation had occurred in the Cape Verde area, the gradual weakening of bottom water flow could account for the observation that migrating sediment waves change upsection to standing waves.

Erosional furrows produced by contour currents (echo type IIID-2) occur around and on seamounts in the northwestern map area (Profiles BB, CC; Fig. 9). These features do not occur generally in water depths greater than 2500 fm (4570 m), although a few seamounts (e.g., 21°N, 28°W; 19.5°N, 26.4°W) have erosional moats at greater depths. A clockwise flow of bottom water around the Cape Verde Islands is suggested by the orientation of sediment waves and drift deposits, moats, and by the presence of erosional furrows. Because the erosional furrows return echo hyperbolae that are tangent to the surface of the seafloor, we be-

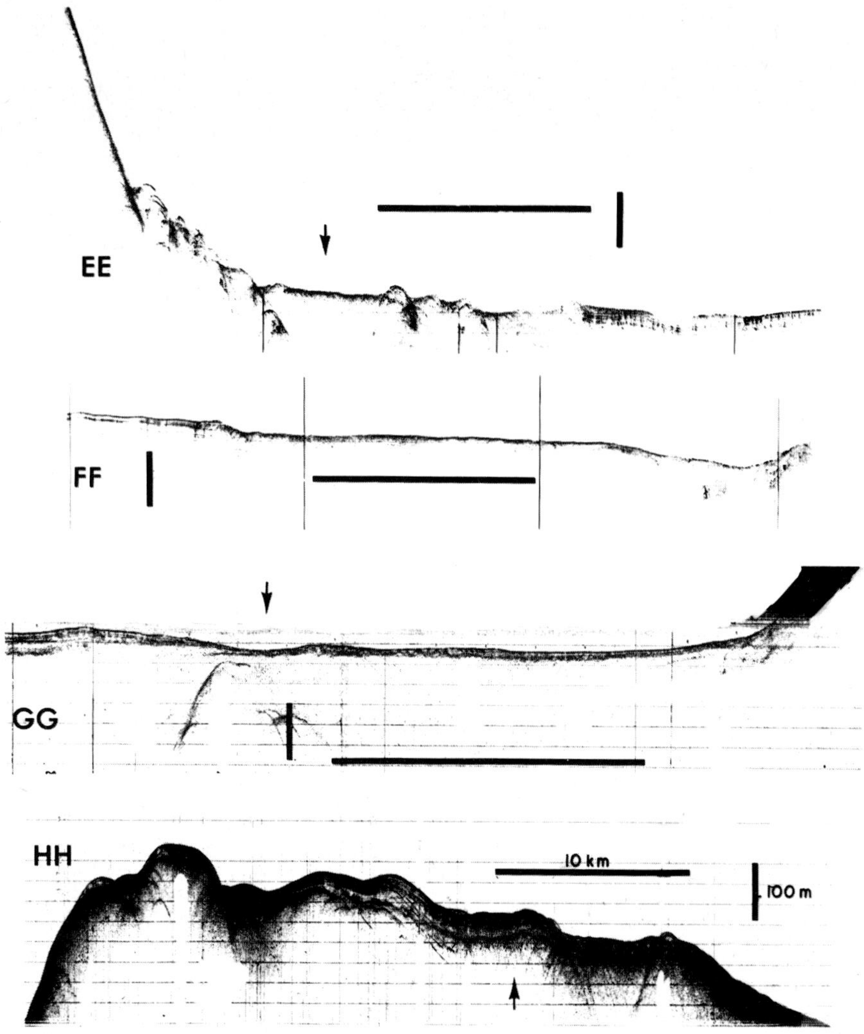

Fig. 10
Moats with Type IIID-2 hyperbolae and IIA prolonged echoes. Type IIID-2 hyperbolae (marked by arrows in profiles EE and HH) result from bottom current activity. In all cases the sedimentary section thins toward the seamount because of either thinner beds or actual disappearance of reflector units. In profile GG note thinning of sedimentary section (at arrow) near a seamount located off the ship's track and recorded on the echogram as a side echo

lieve that bottom water either is presently active in this area or was active in the recent past.

As most of the current features suggest an overall northward flow of bottom water, the source of the bottom water is probably Antarctic Bottom Water (AABW) flowing from both the Sierra Leone Basin and the Guiana Basin (western Atlantic). It is significant that the greatest depth (2500-2550 fm≈ 4570-4650 m) of most current related features in this area is essentially the sill depth of the primary passages by which the AABW enters the Cape Verde Basin from the Sierra Leone Basin and the Guiana Basin.

Hyperbolae (type IIID-2) and indistinct echoes (type IIA) occur in a zone at Kane Gap (9.5°N, 19.8°W; Fig. 2). A moat is located immediately south of Kane Gap and is slightly deeper than the sill depth of the gap; another moat occurs just east of the Kane Gap at 9.2°N, 18.8°W. The presence of these moats and the disposition of sediment

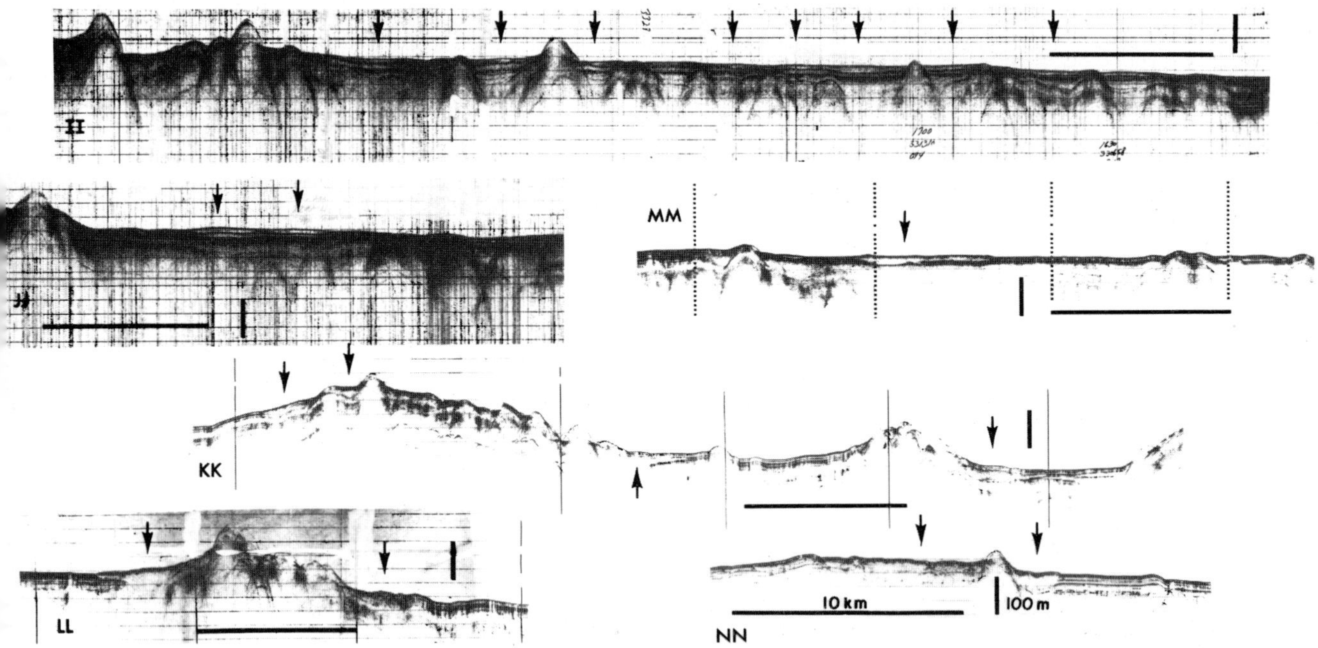

Fig. 11. Type IIID-2 hyperbolae, and IIB-cc indistinct echoes (arrows mark general location of best examples). The deposits are either debris flow deposits reworked by contour currents or are contour-current drift deposits. Type IIIA hyperbolae are basement features. On echogram KK note greater sediment fill on south side of ridge (to left) and greater erosion (type IIA and IIID-2 echoes amongst the IIIA hyperbolae) or nondeposition on north side. On echogram KK type IIB-cc echoes located at center and right are buried. Note gradation from type IIB-cc echo to type IIID-3 hyperbolae in apparently the same deposit (left two arrows in profile KK, left arrow in profiles NN and LL)

waves and hyperbolic echoes suggest that the Kane Gap still serves as a conduit for AABW flow. This conclusion is strongly supported by bottom photographs and potential temperature profiles that indicate a layer of bottom water 320 m thick in flowing through the Kane Gap (Hobart et al., 1975).

In the southwestern map area sediment waves, type IIID-3 hyperbolae and type IIB indistinct echoes commonly occur in the floors of fracture zones and along the lower slopes of their adjacent walls (Profile I, Fig. 5; Fig. 11). Type IIID-3 hyperbolae also occur locally along the walls of the fracture zones, especially the south walls (Profiles Z, AA, DD, Fig. 9; Profile HH, Fig. 10). Echo type IIID-2 is notably lacking from the floors of fracture zone troughs (Profiles Z, DD; Fig. 9). The source of the IIID-3 hyperbolae and the IIB indistinct echoes is not clear. The presence of nearby slide scars (e.g., Profile I, Fig. 5) suggests that some of the type IIB echoes are the result of debris flow deposits. However, we believe that contour-following currents created many of the deposits returning these echo types, especially in the Doldrums Fracture Zone and along the south side of the 4°N Fracture Zone, for the following reasons: (1) These deposits are associated with erosional features (e.g., moats and bevelled upper surfaces of irregular bedding) and sediment waves. (2) Some deposits are not associated with any obvious bathymetric high that could provide a debris flow source. (3) Commonly, several deposits occur superimposed and form slight bathymetric highs (Profiles II, JJ; Fig. 11), whereas debris flows seek local bathymetric lows. (4) The "stacked" deposits commonly have a lensing nature which is not common in many debris flow deposits. (5) Echo-

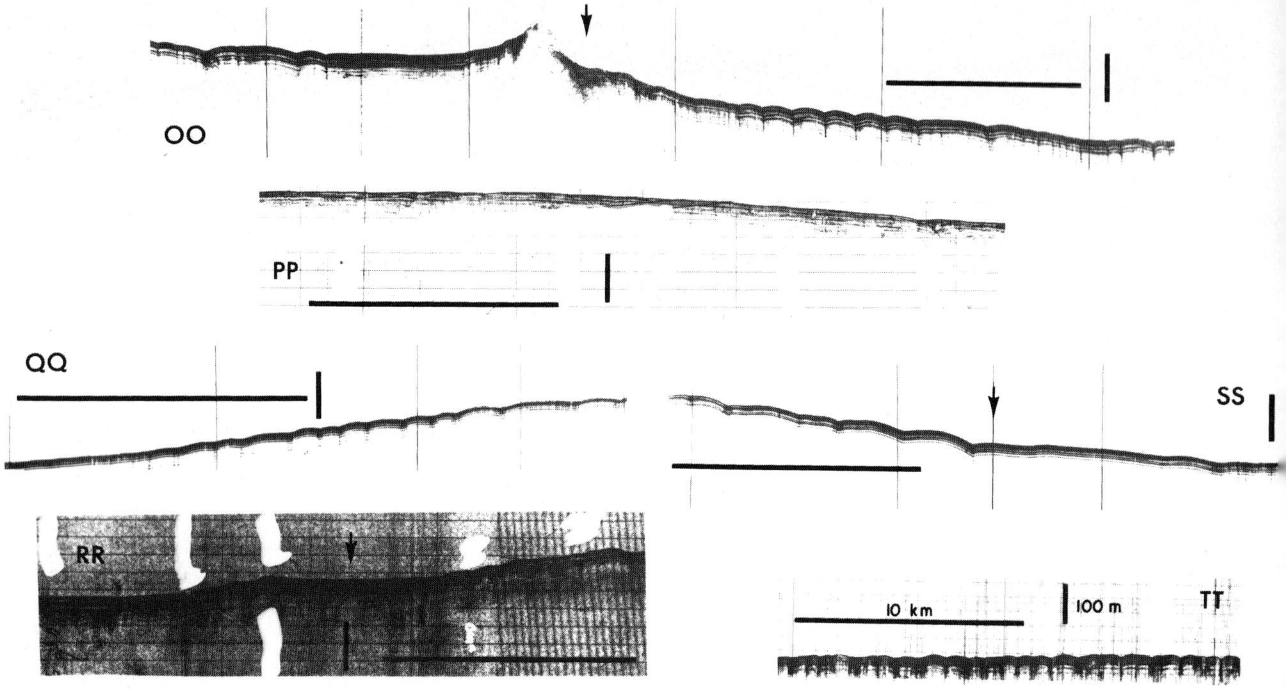

Fig. 12. Type B-1a and B-1b echoes - standing and migrating sediment waves with regular form. Echograms OO and PP show migrating waves (B-1b) and the other echograms show standing waves (B-1a). In echogram OO note erosion features (IIA and IIID-2, marked by arrow) on downslope side of the seamount and absence of sediment waves on upslope (lee) side of seamount. On echogram PP note reversal of apparent migration direction of sediment waves as apparent slope of the ocean floor changes. Echogram SS shows the apparent wavelength change as a result of ship's course change (marked by arrow). Echogram RR shows ponded sediments over apparently relict sediment waves (arrow marks area of ponded sediment

grams taken across fracture zones show that these deposits commonly are absent in depressions below the general level of the fracture zone sill depth, unlike deposits of debris flows and turbidity flows that typically form in the bathymetric lows.

Overall, these relationships incicate that these deposits were probably created by contour currents. They may represent the planar bed form common to the lower part of the upper flow regime. The orientation of a few sediment waves and the stronger erosion on the south walls of fracture zones, suggests that the bottom water flows (or recently flowed) eastward. The eastward flow of AABW is consistent with bottom potential temperature studies of Wüst (1955).

Several provinces of sediment waves occur on the Sierra Leone Rise. The largest province is located on a plateau at ∿20.5°W, 5°N (Figs. 1 and 2). Sediment waves in this province are distinctive because they are recorded on echograms as double-crested waves (Profile XX, Fig. 13), suggesting they may have crescent-shaped crest lines; their origin is unclear.

Three major sediment wave provinces are mapped in the Sierra Leone Basin (Fig. 2). Sediment waves in the eastern province are associated both with levees of turbidity flow pathways (Profile UU, Fig. 13) and with contour currents (see earlier turbidity flow discussion). The sediment waves formed by contour currents are partially buried by more recent planar-bedded turbidites (Fig. 2; Profile RR, Fig. 12). The southwestern boundary of the buried sediment waves is defined by the 2600 fm isobath (4755 m); the northeastern boundary has a variable depth. The oblique strike of individual sediment

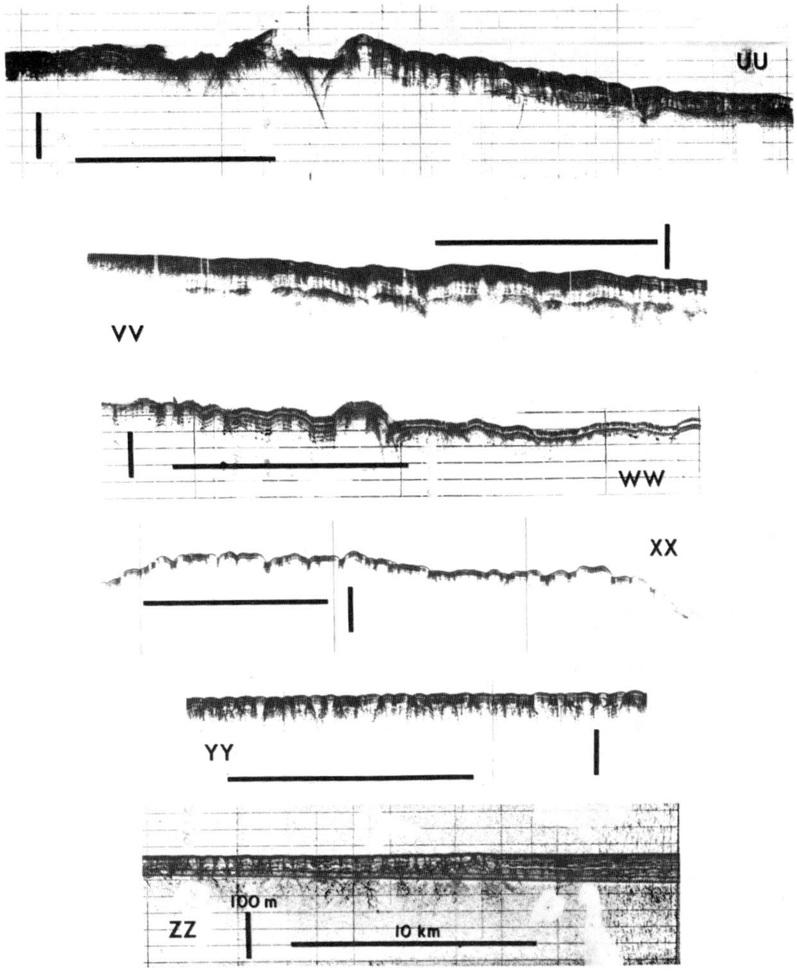

Fig. 13
Type B-2a and B-2b echoes - standing and migrating sediment waves with irregular form. Echograms UU and VV show migrating sediment waves (B-2b) on levees. The other echograms have standing sediment waves with irregular form. Note IIa echoes (prolonged) in channels on echograms UU and VV. Echogram XX shows "double" wave (typical of Sierra Leone Rise) that may suggest slightly crescent-shaped dunes. Echogram ZZ shows "sediment waves" that may be related to upward-migration of fluids in the sediments

waves with respect to the regional bathymetric contours suggests a possible northward flow of contour currents in the eastern province.

In the area of the Grand Cess Fracture Zone, short wavelength, regular sediment waves (type B-1; Profiles YY, ZZ, Fig. 13) are superimposed on long wavelength sediment waves with irregular form (type B-2). On seismic reflection profiles, piercement structures appear to cause the longer wavelength sediment waves, whereas some of the short wavelength sediment waves may be related to upward migration of fluids in the sediments (Profile ZZ, Fig. 13).

Another province of sediment waves is located along the west margin of the Sierra Leone Basin (5°N, 19.5°W; Fig. 2); the lower boundary of this province is also defined by the 2600 fm (4755 m) isobath. Erosional features (echo type IIID-2 hyperbolae and IIA indistinct echoes) occur along the western boundary of this province and are probably primarily caused by contour currents. Seamounts east of this province (e.g., 6°N, 18°W) have associated sediment waves to a depth of ∿2600 fm (4755 m); possible "drift" deposits occur on the south side of some of the seamounts (e.g., 5.3°N, 18.3°W). Most of these seamounts have type IIID-2 hyperbolae, especially on their north sides, and seismic reflection profiles reveal much thinner sedimentary sections on their north sides. These observations of erosion on the north sides of seamounts and deposition on the south sides, suggest southward flow of bottom water along the western margin. The bottom current pattern for the Sierra Leone Basin appears to be a counter clockwise gyre.

4.4 Piercement Structures

Piercement structures commonly cause irregular sediment waves (type B-2;

e.g., 19.5°N, 22.5°W), scarps (type C-3b) and doming of the sedimentary section. Rona 1969, 1970) believed most piercement structures in the map area are the result of salt diapirism, whereas Pautot et al. (1970) suggested that some of the structures are basement peaks penetrating the sedimentary cover. A DSDP hole drilled on a piercement structure at 19.4°N, 24°W (site 141) encountered basalt overlain by unfossiliferous clay at an anomalously shallow sub-bottom depth (Hayes and Pimm et al., 1972). Lancelot and Embley (1977) suggested that the piercement structures can form on basement highs by pelagic mud squeezing up along lines of least pressure within the sediments. In the Cape Verde Basin, numerous piercement structures are located on basement highs and some have associated C-3b scarps. Several of these piercement structures are oriented in north-south bands (e.g., 19°N, 23°W) and form effective tectonic dams to downslope movement of sediments.

Piercement structures are observed commonly beneath the deposits found in the northern part of the map area that return IIIC-4 echoes or, less commonly 11B echoes. These deposits are surrounded by younger, bathymetrically lower turbidites and hemipelagic sediments. A piston core from one such elevated deposit at 22.5°N, 25°W contains a chaotic pebbly mud with sediment of Middle Miocene age. Embley and Jacobi (1977a) suggested that diapirism vertically displaced these relatively old sediments and gave rise to the pebbly mud.

In scattered areas (where the hemipelagic and turbidite section overlying the pelagic mud is thin; e.g., 13.2°N, 24.8°W) some of the previously identified "piercement structures" are simply pelagic mud disposed in drape folds over basement peaks.

5. SUMMARY

From this study of the bathymetry and microphysiography one can begin to appreciate the complex interaction of the multiple processes that have shaped the northwest African continental margin. The basement morphology remains evident in the western portion of the area. Seafloor spreading resulted in the formation of N-S "abyssal hills" which are preserved in the northwest. The Cape Verde Islands resulted from Miocene uplift and Miocene-Recent volcanism, and the islands appear to have trends both parallel and oblique to the older fracture zones. The northeast trends of the Cape Verde Islands and Plateau may be related to "off-axis volcanism."

Near-bottom processes are responsible for many of the morphologic features observed in the eastern map area. In the past, the most dominant "force" shaping the northwest African continental rise was contour currents with sediment input from turbidity flow and sediment slides. At present, however, the strength of the contour currents here has diminished and turbidity flows and slides are now the dominant near-bottom processes.

Turbidity flows carved many canyons and channels on the continental slope and rise north of the Guinea Marginal Plateau but relatively few south of the Guinea Marginal Plateau. The Cape Verde Islands, isolated seamounts, and fracture zone ridges are also sediment sources for turbidity flows. Turbidity flows are gradually extending the bounddaries of the abyssal plains and the lower continental rise by supplying detritus to submarine fans, abyssal plains, and fracture zone valleys west of the present abyssal plains. The relatively small influence of turbidity flows south of the Guinea Marginal Plateau suggests a relatively low sediment supply. This low sediment supply is also suggested by the relatively small buildup of the continental shelf (e.g., the continental shelf is narrower and the continental slope is steeper and narrower in the south).

Sediment slides also are presently modifying the shape of the continental margin by removing large amounts of material from the continental slope and upper rise and redepositing the material lower on the rise. The longest slide complex extends 570+ km from the continental shelf break near 16°N, 17°W, to the lower continental rise near 13.5°N, 21.5°W. Sediment slides also occur on the flanks of isolated seamounts and the Cape Verde Islands, and along fracture zones. The largest slides and greatest concentration of slides occur on the continental slope and rise northward from Dakar, Senegal. We believe it is most significant that the longest slide is adjacent to the largest turbidity flow system (Cayar Canyon System). Furthermore, sediment slides are in general more abundant in areas that have abundant turbidity flows. Relatively few sediment slides are present south of the Guinea Marginal Plateau where turbidity flows are also much less prevalent. We conclude that many sediment slides in this area may have resulted from the creation of unstable deposits by the rapid sedimentation associated with turbidity flows.

In pre-Middle Miocene time, strong contour-following bottom currents eroded many areas on the continental rise and resulted in a prominent Middle Miocene unconformity. These currents began to diminish in intensity in Middle Miocene time, with the result that sediment waves were then deposited along much of this continental rise.

The source of the near-bottom currents was very likely AABW flowing eastward from the western Atlantic through the troughs of the major fracture zones. Fracture zone troughs south of the Sierra Leone Rise guided the AABW directly into the Sierra Leone Basin. Bottom water in fracture zone troughs north of 10°N entered the Cape Verde Basin directly, but AABW also entered the Cape Verde Basin through the Kane Gap from the Sierra Leone Basin. The geostrophic currents generally flowed north along the continental margin north of the Sierra Leone Rise and were deflected westward around the Cape Verde Islands. The sill depths of major bathymetric gaps apparently serve to control the maximum depth to which bottom current processes have been effective on this margin.

ACKNOWLEDGEMENTS. We would like to thank the Captain, Officers and Crew, and Chief Scientists of the R.V. VEMA and ROBERT D. CONRAD for their assistance in gathering the data. We thank John Damuth and Robert Embley for useful discussions, and David Johnson for drafting Figs. 1 and 2. The manuscript was critically reviewed by John Damuth and Roger Flood. This study was supported by National Science Foundation Grant OCE-76-21786 and Office of Naval Research contract N00014-75-C-0210.

REFERENCES

Abbott D, Menke W, Hobart M, Anderson R (1981) Evidence for excess pore pressures in southwest Indian Ocean sediments. J Geophys Res 86: 1813-1827

Akal T (1972) The relationship between the physical properties of underwater sediments that affect bottom reflection. Mar Geol 13: 251-266

Ballard JA, Hemler LG (1967) Structure of the Cape Verde Rise (abstr). Am Geophys Union Trans 50: 210

Behrendt JC, Schlee J, Robb JM, Silverstein MK (1974) Structure of the continental margin of Liberia, West Africa. Geol Soc Am Bull 85: 1143-1158

Belousov IM, Valvashko GM, Gorodnitsky AM, Shreider AA, Sedov AP (1975) Studies of the structure of low-anomalous magnetic fields in the Atlantic Ocean. OKEAHOMORNA 15: 95-101

Bryan GM, Markl RG (1966) Microtopography of the Blake-Bahama region. Tech Rep 8 (CU-8-66-Nobsr 85077) Lamont Geol Obs

Cochran JR (1973) Gravity and magnetic investigations in the Guinea Basin, Western Equatorial Atlantic. Geol Soc Am Bull 84: 3249-3268

Collette BJ, Ewing JI, Lagaay RA, Truchan M (1969) Sediment distribution in the oceans: The Atlantic between 10° and 19°N. Mar Geol 7: 279-345

Damuth JE (1975) Echo character of the western equatorial Atlantic floor and its relationship to the dispersal and distribution of terrigenous sediments. Mar Geol 18: 17-45

Damuth JE (1979) Migrating sediment waves created by turbidity currents in the northern South China Basin. Geology 7: 520-523

Damuth JE, Hayes DE (1977) Echo character of the East Brazilian continental margin and its relationship to sedimentary processes. Mar Geol 24: 73-95

Dietz RS, Knebel HJ, Somers LH (1968) Cayar submarine canyon. Geol Soc Am Bull 79: 1821-1828

Dillon WP, Sougy JMA (1974) Geology of West Africa and Canary and Cape Verde Islands. In: Nairn A, Stehli F (eds) The Ocean Basins and Margins, vol 2, The North Atlantic. Plenum Press New York, pp 315-390

Egloff J (1972) Morphology of ocean basin seaward of northwest Africa:

Canary Islands to Monrovia, Liberia. Am Assoc Petrol Geol 56(4): 694-706
Embley RW (1975) Studies of deep sea sedimentation processes using high frequency seismic data. PhD Thesis, Columbia Univ (unpublished)
Embley RW (1976) New evidence for debris flow deposits in the deep sea. Geology 4: 371-374
Embley RW, Jacobi RD (1977a) The occurrence of Middle Miocene sediment in a piston core on the Cape Verde Rise and its relationship to piercement structures. Am Assoc Petrol Geol 61: 2004-2009
Embley RW, Jacobi RD (1977b) Distribution and morphology of large submarine slides and slumps on Atlantic continental margins. Marc Geotech 2: 205-228
Embley RW, Langseth MG (1977) Sedimentation processes on the continental rise of northeastern South America. Mar Geol 25: 279-297
Emery KO, Uchupi E, Phillips J, Bowin C, Mascle J (1975) Continental margin off western Africa: Angola to Sierra Leone. Am Assoc Petrol Geol 59: 2209-2265
Ewing M, Eittreim S, Ewing J, Le Pichon X (1971) Sediment transport and distribution in the Argentine Basin, 3, Nepheloid layer and processes of sedimentation. In: Ahrens LH, Press F, Runcorn SK, Urey HC (eds) Phys Chem Earth 8: 49-78
Ewing M, Embley RW, Shipley TH (1973) Observations of shallow layering utilizing the pingerprobe echo-sounding system. Mar Geol 14: M55-M63
Flood, RD (1978) Studies of deep-sea sedimentary microtopography in the North Atlantic Ocean. Unpubl PhD Thesis, Woods Hole Oceanographic Institute/Massachusetts Institute of Technology, 395 pp
Flood RD (1980) Deep-sea sedimentary morphology: Modelling and interpretation of echo-sounding profiles. Mar Geol 38: 77-92
Flood RD, Hollister CD (1974) Current-controlled topography on the continental margin off the eastern United States. In: Burke CA, Drake CL (eds) The geology of continental margins. Springer Berlin Heidelberg New York, pp 197-206
Flood RD, Hollister CD, Lonsdale P (1979) Disruption of the Feni sediment drift by debris flow from Rockall Bank. Mar Geol 32: 311-334
Fox PJ (1972) The geology of some Atlantic fracture zones, Caribbean escarpments and the nature of the oceanic basement and crust. PhD Thesis, Columbia Univ, 357 p
Fox PJ, Heezen BC, Harian AM (1968) Abyssal antidunes. Nature (London) 220(5166): 470-472
Hayes DE, Rabinowitz PD (1975) Mesozoic magnetic lineations and the magnetic quiet zone off northwest Africa. Earth Planet Sci Lett 28: 105-115
Hayes DE, Pimm AC, Beckmann JP, Benson WE, Berger WH, Roth PE, Supko PR, von Rad U (1972) Init Rep Deep Sea Drilling Project XIV: 970 pp
Heezen BC, Johnson GL (1969) Mediterranean undercurrent and microphysiography west of Gibraltar. Bull Inst Oceanog 67: 95
Heezen BC, Tharp M (1965) Tectonic fabric of the Atlantic and Indian oceans and continental drift. Philos Trans R Soc London 258: 90-106
Heezen BC, Gerard RD, Tharp M.(1964) The Vema Fracture Zone in the equatorial Atlantic. Bull Geol Soc Am 69: 733-739
Heezen BC, Hollister CD, Ruddiman WF (1966) Shaping of the continental rise by deep geostrophic contour currents. Science 152(3721): 502-508
Heezen BC, Fox PJ, Johnson GL, Tharp M, Ballard A (1969) A fracture zone at 8°N. Abstr Trans Geophys Union 50: 211
Hobart MA, Bunce ET, Sclater JG (1975) Bottom water flow through the Kane Gap, Sierra Leone Rise, Atlantic Ocean. J Geophys Res 80: 5083-5088
Hollister CD, Heezen BC (1972) Geologic effects of ocean bottom currents; western North Atlantic. In: Gordon AL (ed) Studies in physical oceanography. Gordon and Breach, London, pp 37-66
Hollister CD, Flood RD, Johnson DA, Lonsdale P, Southard JB (1974) Abyssal furrows and hyperbolic echo tracks on the Bahama Outer Ridge. Geology 2: 395-400
Jacobi RD (1976) Sediment slides on the northwestern continental margin of Africa. Mar Geol 22: 157-173
Jacobi RD, Mrozowski CL (1979) Sediment studies and sediment waves in the Bonin Trough, western Pacific. Mar Geol 29: M1-M9

Jacobi RD, Rabinowitz PD, Embley RW (1975) Sediment waves on the Moroccan continental rise. Mar Geol 19: M61-M67

Kenyon NH, Belderson RH (1973) Bedforms of the Mediterranean undercurrent observed with side-scan sonar. Sediment Geol 9: 77-99

Kermabon A, Gehin C, Blavier P, Tonarelli B (1964) Acoustic and other physical properties of deep sea sediments in the Tyrrhenian abyssal plain. Mar Geol 7: 129-145

Krause DC (1964) Guinea fracture zone in the Equatorial Atlantic. Science 146: 57-59

Lancelot Y, Embley RW (1977) Evidence for sediment diapirism in the deep sea. Am Assoc Petrol Geol Bull 61: 1991-2000

Lancelot Y, Seibold E, Cepek P, Dean WE, Evemeer V, Gardner JV, Jansa LF, Johnson D, Krasheninnikov V, Pflaumann U, Rankin JG, Trabant P (1975) The eastern North Atlantic (Leg 41). Geotimes 20: 18-21

Laughton AS (1962) Discrete hyperbolic echoes from an otherwise smooth deep-sea floor. Deep Sea Res 9: 218-219

Le Pichon X, Fox PJ (1971) Marginal offsets, fracture zones and the early opening of the North Atlantic. J Geophys Res 76: 6294-6308

Le Pichon X, Sibuet, JC, Francheteau J (1977) The fit of the continents around the North Atlantic Ocean. Tectonophysics 38: 169-209

Lonsdale P, Speiss FN (1977) Abyssal bedforms explored with a deeply towed instrument package. Mar Geol 23: 57-75

Lowrie A, Escowitz E (1969) Kane 9. US Naval Oceanogr Office, 971p

Markl RG, Bryan GM, Ewing JI (1970) Structure of the Blake-Bahama Outer Ridge. J Geophys Res 75(24): 4539-4555

Mascle J, Sibuet JC (1974) New pole for early opening of South Atlantic. Nature (London) 252: 464-465

McMaster RL, de Boer J, Ashraf A (1970) Magnetic and seismic reflection studies on continental shelf off Portuguese Guinea, Guinea and Sierra Leone, West Africa. Am Assoc Petrol Geol 54: 158-167

McMaster RL, Ashraf A, de Boer J (1973) Transverse continental margin fracture zone off Sierra Leone. Nature (London) Phys Sci 244: 93-94

McMaster RL, Christoffersen E, Ashraf A (1975) Structural framework of continental shelf and slope off southwestern Sierra Leone, West Africa. Am Assoc Petrol Geol 59: 2161-2171

Middleton GV, Hampton MA (1973) Sediment gravity flows: Mechanics of flow and deposition. In: Middleton G, Bouma A (eds) Turbidites and deep water sedimentation. SEPM short course, pp 1-38

Morgenstern N (1967) Submarine slumping and the initiation of turbidity currents. In: Richards A (ed) Marine geotechnique. U Illinois Press, Urbana ILL, pp 189-220

Pautot G, Auzende J, Le Pichon X (1970) Continuous deep-sea salt layer along North Atlantic margins related to early phase of rifting. Nature (London) 227: 351-354

Rabinowitz PD, LaBrecque J (1979) The Mesozoic South Atlantic Ocean and Evolution of its continental margins. J Geophys Res 84: 5973-6002

Rona PA (1969) Possible salt domes in the deep Atlantic off North-west Africa. Nature (London) 224: 141-143

Rona PA (1970) Comparison of continental margins of eastern North America at Cape Hatteras and northwestern Africa at Cap Blanc. Am Assoc Petrol Geol 54: 129-157

Rona PA (1971) Bathymetry off central northwest Africa. Deep Sea Res 18: 321-327

Ruffman A, Neagher LJ, Stewart JMcq (1977) Bathymetry of the continental shelf and margin of Senegal and the Gambia, West Africa. In: CSS BAFFIN offshore survey Senegal and the Gambia, 1, Dep Fish Environ, Ottawa

Ryan WBF, Heezen BC (1965) Ionian Sea submarine canyons and the 1908 Messina Turbidity Current. Bull Geol Soc Am 76: 915-932

Schlee J, Behrendt JC, Robb JM (1974) Shallow structure and stratigraphy of Liberian continental margin. Am Assoc Petrol Geol 58: 708-728

Schneider ED, Johnson GL (1970) Deep-Ocean diapir occurrences. Am Assoc Petrol Geol 54: 2151-2169

Schneider ED, Fox PJ, Hollister CD, Needham HD, Heezen BC (1967) Further evidence of contour currents in the western North Atlantic. Earth Planet Sci Lett 2: 351-359

Seibold E, Hinz K (1974) Continental slope construction and destruction, West Africa. In: Burke CA, Drake CL (eds) The geology of continental margins. Springer, Berlin Heidelberg New York, pp 179-196

Sheridan RE, Golovchenko X, Ewing JI (1974) Late Miocene turbidite horizon in Blake-Bahama Basin. Am Assoc Petrol Geol 58: 1797-1805

Sibuet JC, Mascle J (1977) Origin and evolution of Ceara and Sierra Leone Rises (Equatorial Atlantic). Abstr Trans Am Geol Union 58: 504

Templeton RSM (1971) The geology of the continental margin between Dakar and Cape Palmas. In: Delany FM (ed) The geology of the East Atlantic continental margin. G B Inst Geol Sci Rep 70/16: 47-60

Tucholke BE, Mountain GS (1979) Seismic stratigraphy, lithostratigraphy and paleosedimentation patterns in the North American Basin. In: Talwani M, Hay W, Ryan W (eds) Deep drilling results in the Atlantic Ocean: Continental margins and paleoenvironment. Maurice Ewing Ser 3. Am Geophys Union, Washington DC, pp 58-86

Uchupi E (1971) Bathymetric Atlas of the Atlantic, Caribbean, and Gulf of Mexico. Ref No 71-72 Woods Hole Oceanogr Inst, Mass

Uchupi E, Emery KO, Bowin CO, Phillips JD (1976) Continental margin off Western Africa: Senegal to Portugal. Am Assoc Petrol Geol 60: 809-878

Wüst G (1955) Stromgeschwindigkeit en in tiefen und Bodenwasser des Atlantischen Ozeans auf grand dynamishen Berechnung der Deutschen Atlantischen Expedition 1925-1927. Pap Mar Biol Oceanogr Suppl Deep Sea Res 3: 373-397

Comparison of the Northwest African and Northeast American Margins

11 Mesozoic-Cenozoic Development of the Eastern North American and Northwest African Continental Margins: A Comparison

Lubomir F. Jansa[1] and Jost Wiedmann[2]

Abstract

During the last decade much information has been accumulated on the evolution of the Mesozoic-Cenozoic basins bordering the North Atlantic. This paper ties together the available seismic and stratigraphic data from the conjugate margins of eastern North America and Northwest Africa and contributes to a better understanding of the history of the North Atlantic Ocean with its passive continental margins. Due to the differing actual bathymetric position of the basins at the western and eastern oceanic margins, the available stratigraphic data from these margins are not entirely concordant. These data are mainly based on drilling activity in the west, and both drilling and onshore-based studies in the east.

Continental fit and oceanic fracture zones enable us to correlate from north to south the following basin pairs: the Scotian Basin on the Canadian shelf in the west, with the Agadir-Essaouira (or Atlas) Basin of Central Morocco ("northern transect") in the east; the Georges Bank Basin on the northeastern U.S. shelf with the Tarfaya-Aaiun Basin of Southern Morocco ("central transect") and, finally, the Blake Plateau Basin of the southeastern U.S. shelf with the Mauritania-Senegal Basin on the African margin ("southern transect"). The Baltimore Canyon Trough on the central U.S. shelf has, however, no exact corresponding African basin; it is placed against the southernmost portion of the Tarfaya-Aaiun Basin and the Reguibat High and is discussed here within the scope of the "central transect".

With regard to the initiation of the oceanic evolution it is important, that the positioning of the later North Atlantic is by no means a fortuitous occurrence. Most of the basins considered are bordered towards the continents by Paleozoic foldbelts having been consolidated at differing ages (Acadian, Hercynian, Alleghenian), but by similar tectonic stress. Most of the Paleozoic structures parallel the later break-up of the Afro-American plate or the actual continental margins. The corresponding Paleozoic troughs are to be considered as a first attempt of building a proto-Atlantic.

After erosion took place during the late Paleozoic, the history of the North Atlantic begins in the Triassic or even in the late Permian with the first phase of continental graben evolution and graben filling. Information about this graben event is restricted to the northern and central transect. The pre-Jurassic history of the southern transect is not yet detected.

The next evolutionary step is that of initial rifting which can be likewise dated in the northern and central transects as being of late Triassic to early Jurassic age. It is associated at both margins with the deposition of evaporites and with the appearance of diabasic,

1 Geological Survey of Canada, Bedford Institute of Oceanography, Dartmouth, N.S., Canada

2 Institut für Geologie und Paläontologie, Universität Tübingen, Sigwartstraße 10, 7400 Tübingen, FRG

basaltic and doleritic volcanics. Regional disconformities in the Scotian and Essaouira-Agadir Basins are related with Early Cimmerian tectonics.

The next depositional break can be dated as being late Liassic, when open marine conditions were established. Also the first ocean spreading can be related to a late Lower Jurassic age.

The Middle Jurassic can be regarded as time of installation of carbonate platforms and carbonate banks, mainly continuing into the Upper Jurassic, but everywhere competing with the landward construction of deltaic complexes. In the basins of the northern transect the installation of platform carbonates is associated with a first phase of reef development. Throughout nearly all basins, we have a second phase of reef evolution at the top of the Jurassic. It is locally linked with the formation of hardgrounds, short-term emersion or disconformities due to widespread Cimmerian tectonics. Occasionally, however, reef evolution may have persisted through most of Jurassic time, probably on rotated fault-blocks, and thus forming the "slope anticlines" or the Essaouira-Agadir and Tarfaya-Aaiun Basins. Therefore, this time is intimately related to the formation of the future shelf and continental slope.

At Callovian time, the first global sea level rise can be traced through all the basins.

The Late Cimmerian tectonics can be considered as the source of the next environmental change. Due to continental uplift, the detritic input rapidly increases and governs the early Cretaceous silicoclastic regime through the basins. The carbonate platforms were mainly covered by large delta plains. Coeval steepening of the relief leads to the deposition of turbidites at the Moroccan slope and rapidly increasing subsidence of the oceanic basin to autochthonous pelagic sedimentation below CCD. Locally, i.e. on the outer shelfs of the Blake Plateau Basin and the southern Senegal Basin the deposition of platform carbonates continues through the Lower Cretaceous.

With the mid-Cretaceous, correlations across the North Atlantic become difficult due to the installation of differing sedimentary and oceanographic regimes at both margins. In the Moroccan coastal basins the Middle and Upper Cretaceous evolution is characterized by mainly inner shelf conditions. The accumulation of thick shell beds is related to continuous subsidence and low terrigenous input. Three transgressive peaks are the expression of global sea level rises at the late Aptian, Turonian and Campanian. The Turonian transgressive peak is associated with the deposition of cherts and black shales, especially in the Tarfaya-Aaiun Basin. These deposits can be explained by first coastal upwelling. This idea is supported by the Upper Cretaceous hiatus detected in the DSDP sites drilled in the Moroccan Basin and related to contour currents and the installation of a new oceanographic regime at the northwestern African margin. Moreover, black shales are unknown from the Senegal Basin and from the conjugate American basins.

In the silty background sedimentation of the Upper Cretaceous of the Senegal coastal basin, the Turonian and Campanian transgressive pulses can be recognized, as also in the shaly background sedimentation of the Scotian Basin. They are, however, less distinct or unknown from the southernmost American shelf basins, where a progressive deepening towards the Cretaceous-Tertiary boundary is to be recognized.

With this boundary the evolution at both North Atlantic margins becomes completely divergent. The uplift of the northwestern African coastal basins is initiated by first Subhercynian tectonics in the Tarfaya Basin. It continues with the Laramid uplift of the Moroccan Meseta, the Tarfaya-Aaiun and the Senegal Basin, leading to the deposition of phosphatic nearshore sediments. Towards the outer shelf and oceanic basin these sediments grade into chert-bearing clays and can, thus, be regarded as another expression of coastal upwelling at the northwestern African margin during Paleocene and Eocene.

At this time, subsidence of the Agadir-Essaouira Basin still continues, but is transformed into the uplift of the western High Atlas

Ranges by the Pyrennean orogeny. The final uplift of the Northwest African coastal basins can be related to Savic tectonics between Paleo- and Neogene. Most of the tectonic events at the African margin can be correlated with salt motion in the subsurface.
Neogene marine incursions into the Northwest African coastal basin are only of random importance.

In the Northeast American shelf basins, however, subsidence continues into the Paleogene and even into the Neogene. In the Scotian Basin considerable amounts of glacigene clastics accumulated during the Neogene. The Cretaceous-Tertiary boundary is camouflaged in nearly all the basins by a wide-spread Paleogene hiatus, which can be related to the installation of contour currents during the early Tertiary, leading to erosion at the continental slope and shelf.

Thus, the evolution of the contiguous basins across the North Atlantic is found to be intimately linked during their Triassic, Jurassic and early Cretaceous history, but increasingly separated during late Cretaceous and Tertiary time. During the early history a series of coeval events can be recognized and possibly related to discontinuous stages of rifting and spreading. Later on the evolution of the marginal basins is found to be a complicated interplay of subsidence rates, tectonic activity, terrigenous influx, sea level alterations and local oceanographic conditions.

1. Introduction

The North American and the contingent Northwest African margins provide excellent examples of mature passive continental margins which evolved during Mesozoic to Cenozoic times. The objective of this paper is to compare the geologic history of both conjugate margins, using: (1) all available offshore oil exploration plus deep sea drilling data and reflection seismic data on the eastern North American margin; (2) onshore and offshore geology, as well as seismic data from the Northwest African margin to see the similarity or dissimilarity in the margin development, and (3) arrive at a better understanding of the plate tectonic processes and basin development in the North Atlantic (cf. also Seibold, this Vol.).

The problem of determining the past geologic history of the two contingent margins is closely determined by our ability to reconstruct the paleoposition of the continental plates. Earlier continental fits (Choubert, 1935; Bullard et al., 1965; Le Pichon and Fox, 1971; Choubert et al., 1972; Talwani and Eldholm, 1972; Le Pichon et al., 1977) have been refined by matching the fracture zones across the North Atlantic. The latitudinal positions of the conjugate margin have thus been determined with a precision of about 100 km (Schouten and Klitgord, 1980). The directions of the fracture zones define the directions of the opening ("flow lines"), but they do not indicate, how far the continental crust extends or what is the best limit for the continental fit.

The validity of the geophysical reconstructions can be tested by geologic methods. Several such comparisons between Northwest Africa and North America have been attempted (see, for example, Rona, 1970; Bhat et al., 1975; Jansa and Wade, 1975; Lancelot and Winterer, 1980; Schlee, 1980).

The geologic data on the eastern North American margin have been provided by more than 78 oil exploratory wells located on the Scotian Shelf and by 5 deep stratigraphic boreholes located on the Atlantic shelf off the United States. The drilling confirmed that the geologic development of the margin is more complex than was indicated by the earlier geophysical studies. However, despite the intensive exploration of the eastern North American margin during the last ten years (Schlee et al., 1976; Folger et al., 1979; Jansa et al., 1979; Sheridan and Enos, 1979; Tucholke and Mountain, 1979), the continental rise and lower slope remain virtually unexplored, and thus many of the outstanding geologic problems, such as the character of the early rift deposits, the origin of salt, the nature and location of the oceanic-continental crust boundary, and the timing of the North Atlantic opening remain unresolved.

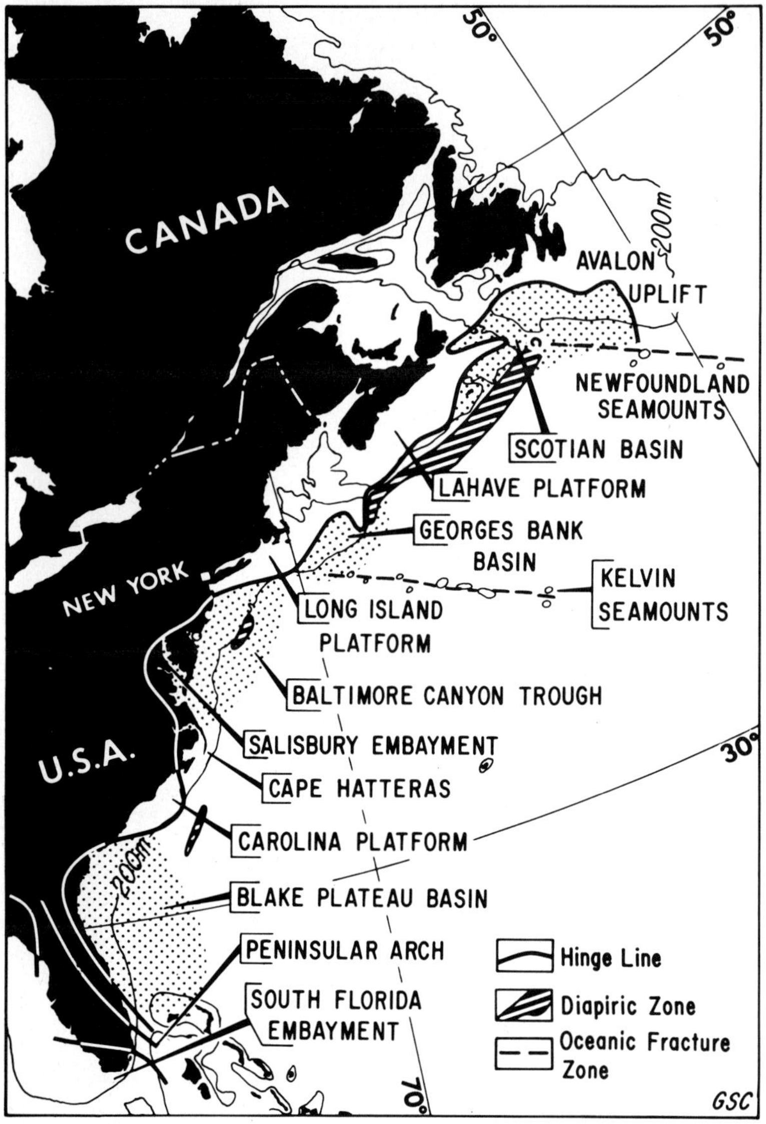

Fig. 1. Location map with major structural elements of the Northeast American margin. Location of the major sedimentary basins is shown by dotted pattern

Less drilling activity has been done on the Northwest African shelf, from where almost no data were published. The only exception is the commercial well Spansah 51A-1, drilled in the Tarfaya shelf by CONOCO in 1969 (cf. von Rad and Einsele, 1980; Arthur et al., 1979). However, significant scientific data were generated on the Northwest African slope and rise by the Deep Sea Drilling Project (Hayes, Pimm et al., 1972; Lancelot, Seibold et al., 1978; von Rad, Ryan et al., 1979; Lancelot, Winterer et al., 1980; Hinz, Winterer et al., this Vol.). This unique information about the development of the slope and rise is not available on the North American margin. Another unique feature of the NW African sedimentary basins is the fact that parts of these basins are exposed and thus accessible for detailed onshore stratigraphic and tectonic studies. First attempts of detailed

Fig. 2. Location map of sedimentary basins, DSDP sites and interpreted seismic profiles along the Northwest African margin. 1 Mazagan Plateau; 2 Essaouira Basin; 3 Northern Tarfaya Basin; 4 Central Senegal Basin; 5 Southern Senegal Basin

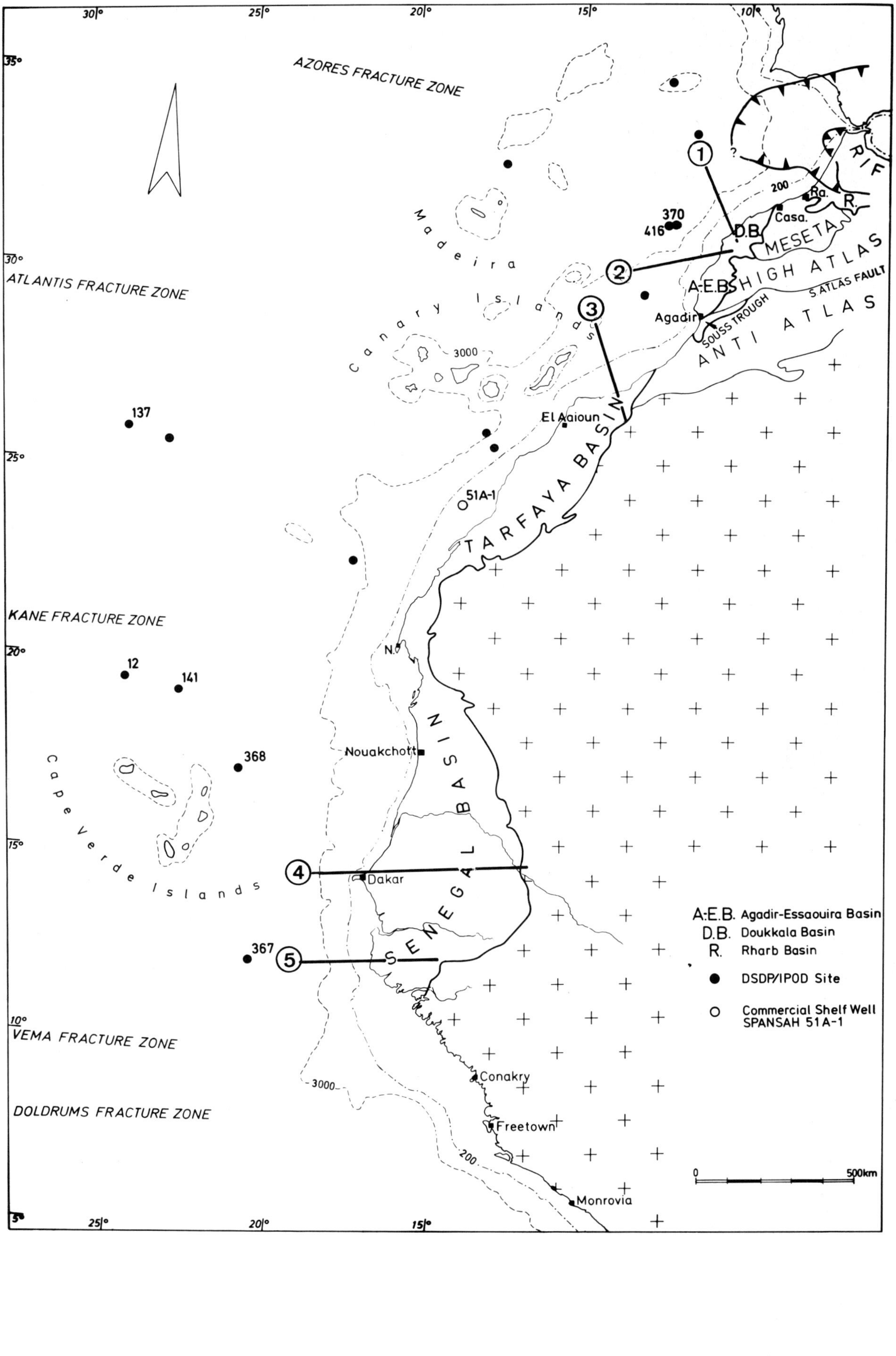

AZORES FRACTURE ZONE
ATLANTIS FRACTURE ZONE
KANE FRACTURE ZONE
VEMA FRACTURE ZONE
DOLDRUMS FRACTURE ZONE
Madeira
Canary Islands
Cape Verde Islands
370
416
137
12
141
368
367
51A-1
3000
200
RIF
Ra.
R.
Casa.
D.B
MESETA
A-E.B
HIGH ATLAS
S.ATLAS FAULT
SOUSS TROUGH
ANTI ATLAS
Agadir
El Aaioun
TARFAYA BASIN
N.
Nouakchott
SENEGAL BASIN
Dakar
Conakry
Freetown
Monrovia
A-E.B. Agadir-Essaouira Basin
D.B. Doukkala Basin
R. Rharb Basin
DSDP/IPOD Site
Commercial Shelf Well SPANSAH 51A-1
0
500km

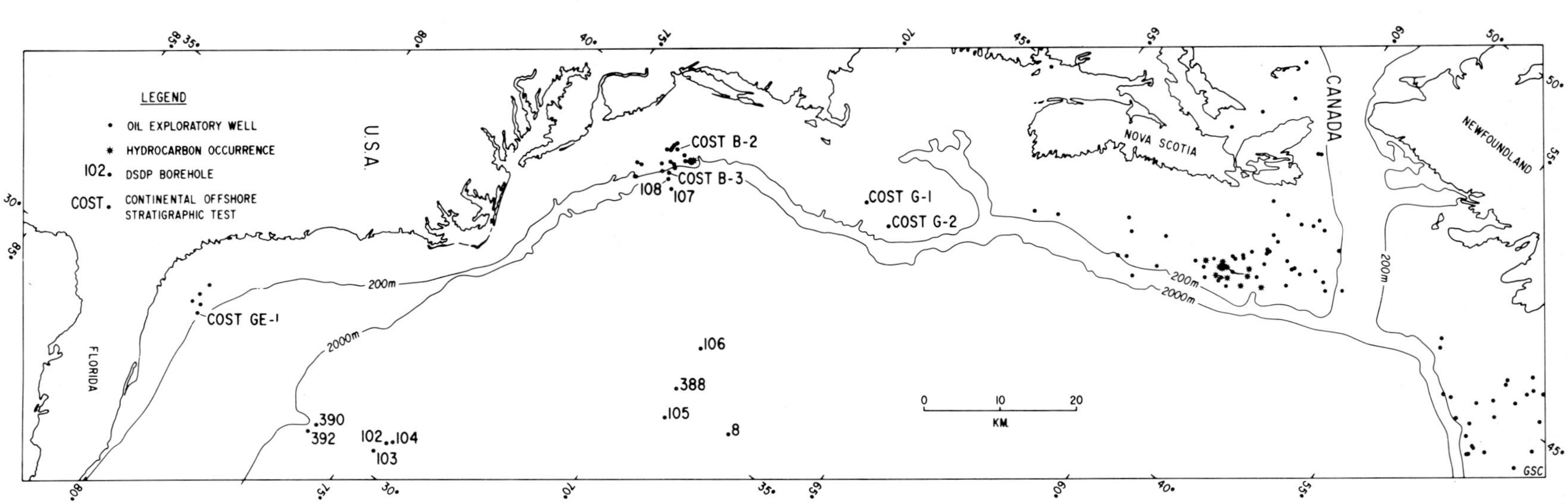

Fig. 3. General location map showing the coastline and edge of the Northeast American shelf (200 m isobath), location of offshore oil exploratory wells, continental offshore stratigraphic test wells (COST) and DSDP boreholes

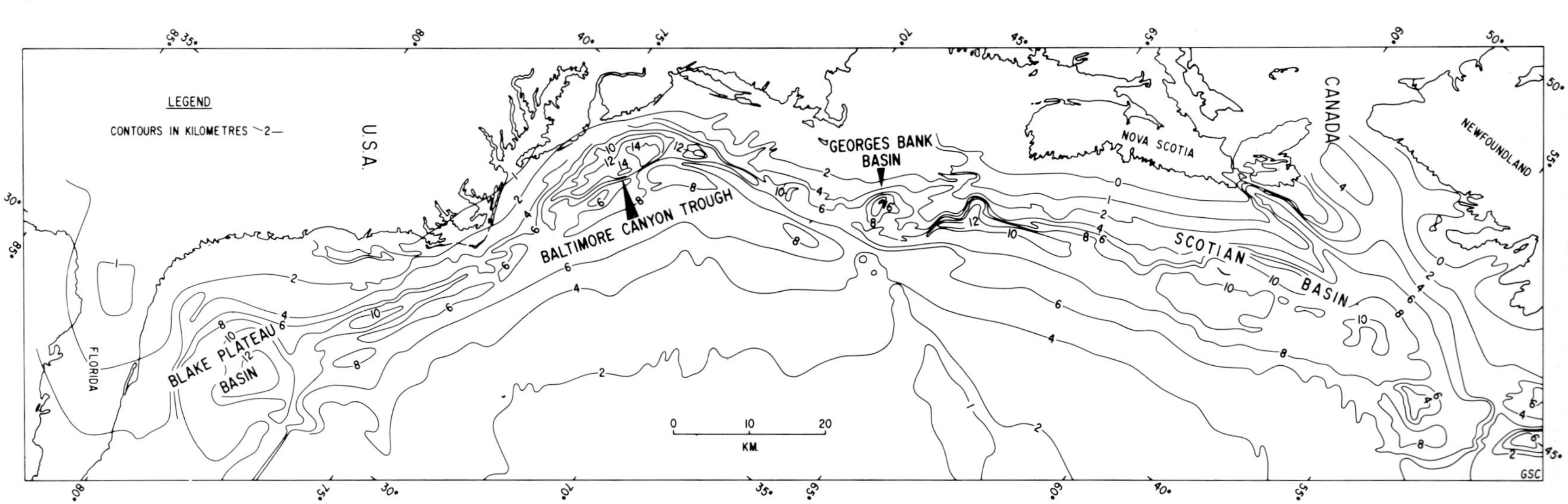

Fig. 4. Isopach map of the Mesozoic-Cenozoic sediments of the eastern North American continental margin (compiled from Jansa and Wade, 1975; Klitgord and Behrendt, 1979; Dillon et al., 1979; Tucholke et al., in press; Schlee and Jansa, in press)

onshore-offshore correlation have been made by Wiedmann et al. (1978a), Einsele and von Rad (1979), and Arthur et al. (1979).

2. Tectonic Framework

2.1. Northeastern American Margin

The eastern North American margin has been built over the fractured eastern edge of the North American continent during the opening of the North Atlantic Ocean. Infilling of the basins (Fig. 1) and over the basement highs has formed a sedimentary terrace which abuts against the continent. The top of this terrace forms the present continental shelf, which is from 2 to 540 km wide. The shelf has its greatest width off the southeastern Grand Banks and is narrowest adjacent to southern Florida (Fig. 3). Several Mesozoic-Cenozoic sedimentary basins underlie the shelf (Fig. 4) and are separated from one another by structural highs (basement highs, ridges, platforms). From the north to the south the basins are: the East Newfoundland Basin, Scotian Basin, Georges Bank Basin, Baltimore Canyon Trough, and Blake Plateau Basin (Figs. 1, 4).

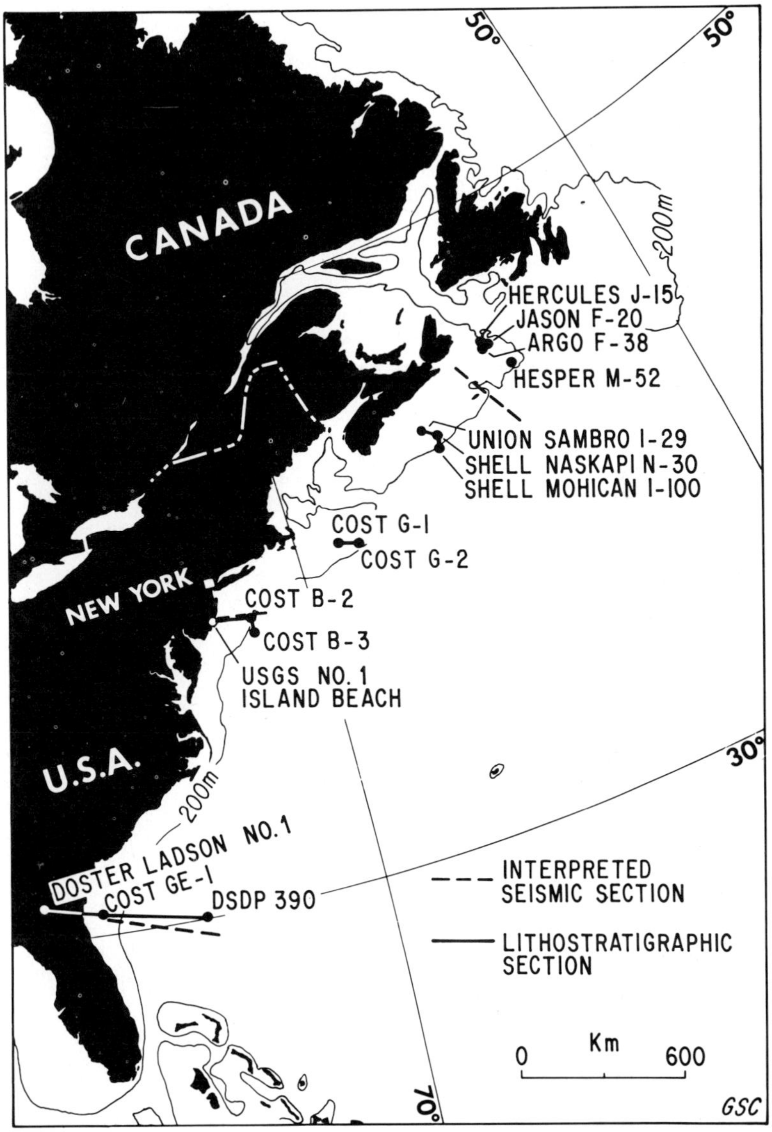

Fig. 5. Location map of stratigraphic sections, interpreted seismic profiles and wells at the eastern North American margin, discussed in this paper

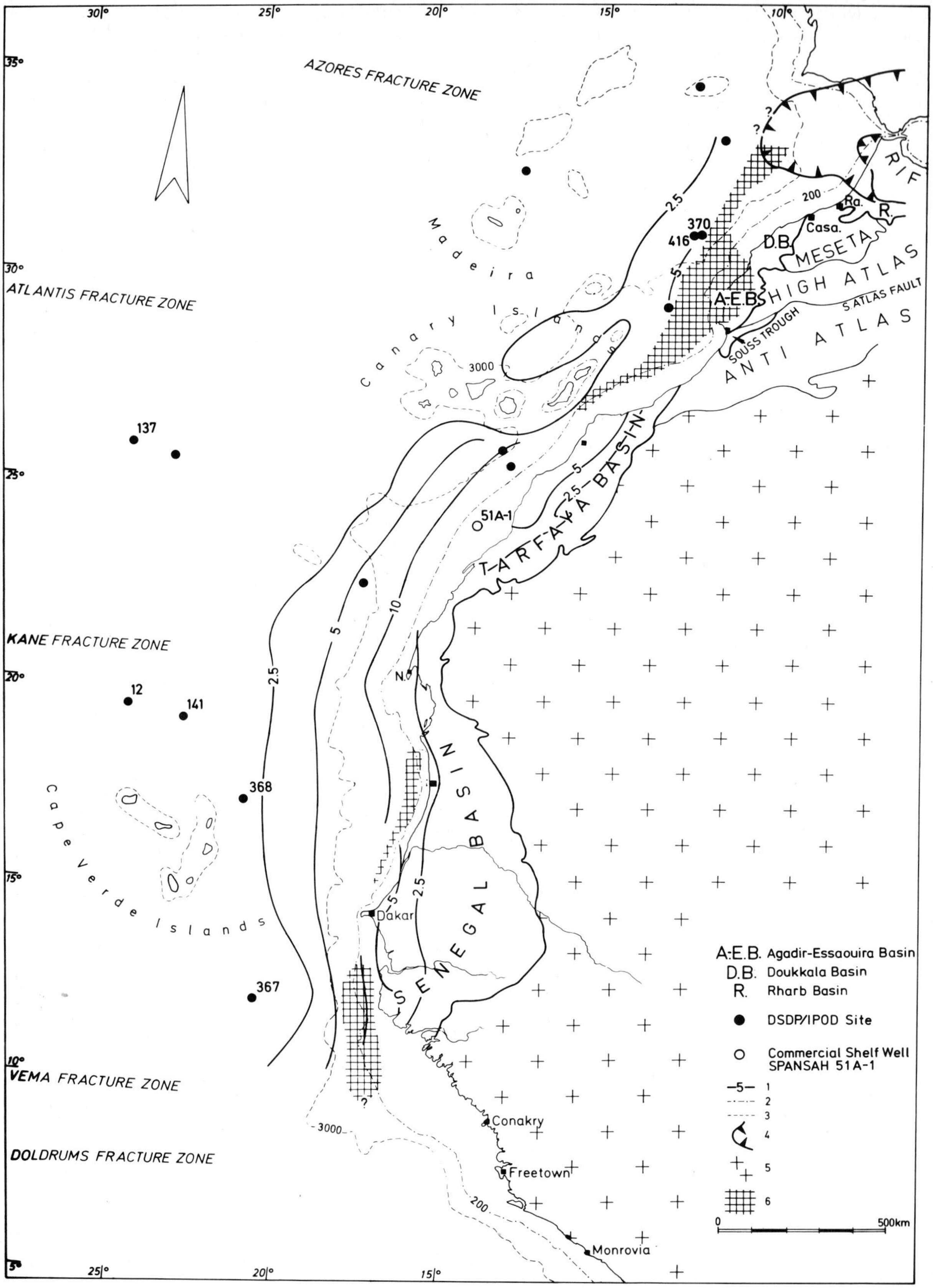

AZORES FRACTURE ZONE
ATLANTIS FRACTURE ZONE
KANE FRACTURE ZONE
VEMA FRACTURE ZONE
DOLDRUMS FRACTURE ZONE
Madeira
Canary Islands
Cape Verde Islands
RIF
MESETA
HIGH ATLAS
ANTI ATLAS
S. ATLAS FAULT
SOUSS TROUGH
TARFAYA BASIN
SENEGAL BASIN
Casa.
Ra.
D.B.
A-E.B.
R.
Dakar
Conakry
Freetown
Monrovia
N.
51A-1
137
12
141
368
367
370
416
3000
200
A-E.B. Agadir-Essaouira Basin
D.B. Doukkala Basin
R. Rharb Basin
DSDP/IPOD Site
Commercial Shelf Well SPANSAH 51A-1
0 500km

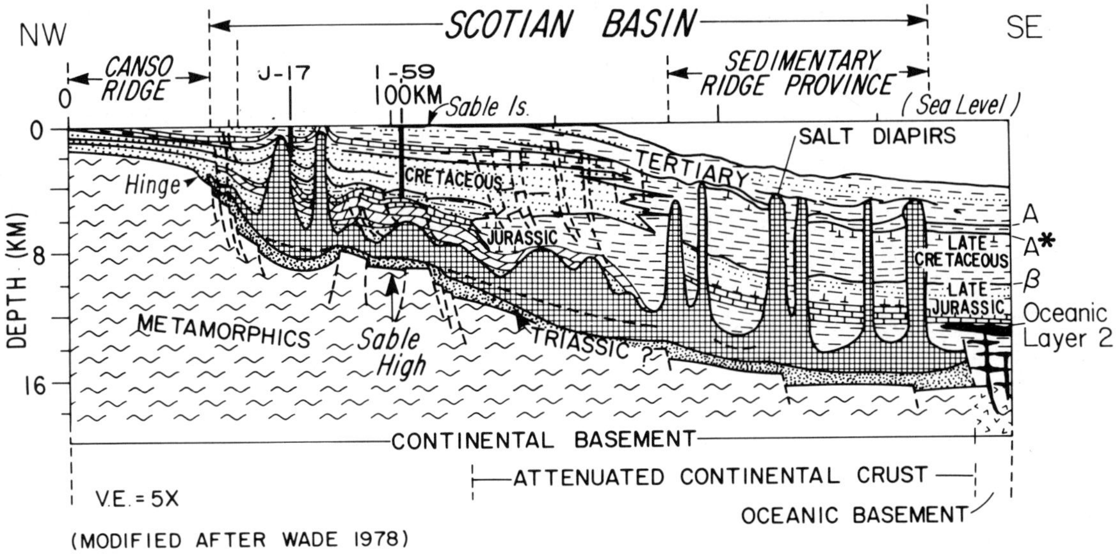

Fig. 7. Geologic interpretation of a multichannel reflection seismic section across the continental margin off Nova Scotia (modified after Wade, 1978; for location of section see Fig. 5)

The location of the offshore sedimentary basins and basement highs reflects the Paleozoic structural trend of the Appalachian fold belt (H. Williams, 1978), with the basins located seaward of the structural embayments and the platforms being seaward of the promontories. Thus, the Paleozoic structural trends in part determined the location of the Mesozoic-Cenozoic marginal basins (Schlee et al., 1976).

The shape and geometry of the sedimentary basins and the interpreted reflection seismic sections (Figs. 7, 10, 12) show that the basins are actually structural monoclines (mio- and eugeosynclines of King, 1975), which are bordered on the landward side by the continent, but are open towards the ocean. Because of this structural characteristic it would be more correct to classify them as half-basins.

The Scotian Basin (Fig. 7), is built over a broad structural embayment located over both the divergent margin seaward of Nova Scotia and a transform margin south of Newfoundland (Sherwin, 1973; Jansa and Wade, 1975; Parsons, 1975).

The basin is about 800 km wide and is bordered on its northeast side by the Avalon Uplift and on the southern side by the LaHave Platform (Jansa and Wade, 1975). Sediments, Triassic to Recent in age, attain thicknesses in excess of 11 km beneath the outer shelf and upper continental slope (Fig. 7). Off Nova Scotia the area of the maximum thickness coincides in part with the 'Sedimentary Ridge Province' which is formed by elongated diapiric ridges of salt or shale in a zone 64 km wide and 640 km long (Figs. 1, 7). This zone is located offshore of the carbonate banks front.

The sediments in the basin thin off to the southwest over the LaHave Platform, which separates the Scotian Basin from the much smaller (200 km) Georges Banks Basin. An isopach map reveals an irregular pattern of the basin morphology.

Fig. 6. Location and isopach map of Mesozoic-Cenozoic sedimentary basins and diapiric zones along the Northwest African margin (from back cover figure). 1 Isopachs of Mesozoic-Cenozoic sediments (in km); 2 200 m isobath, largely coinciding with the shelf edge; 3 3000 m isobath; 4 Nappes of Rif-Betic Systems; 5 Continental basement, Hercynican platforms and fold belts; 6 Diapiric zones

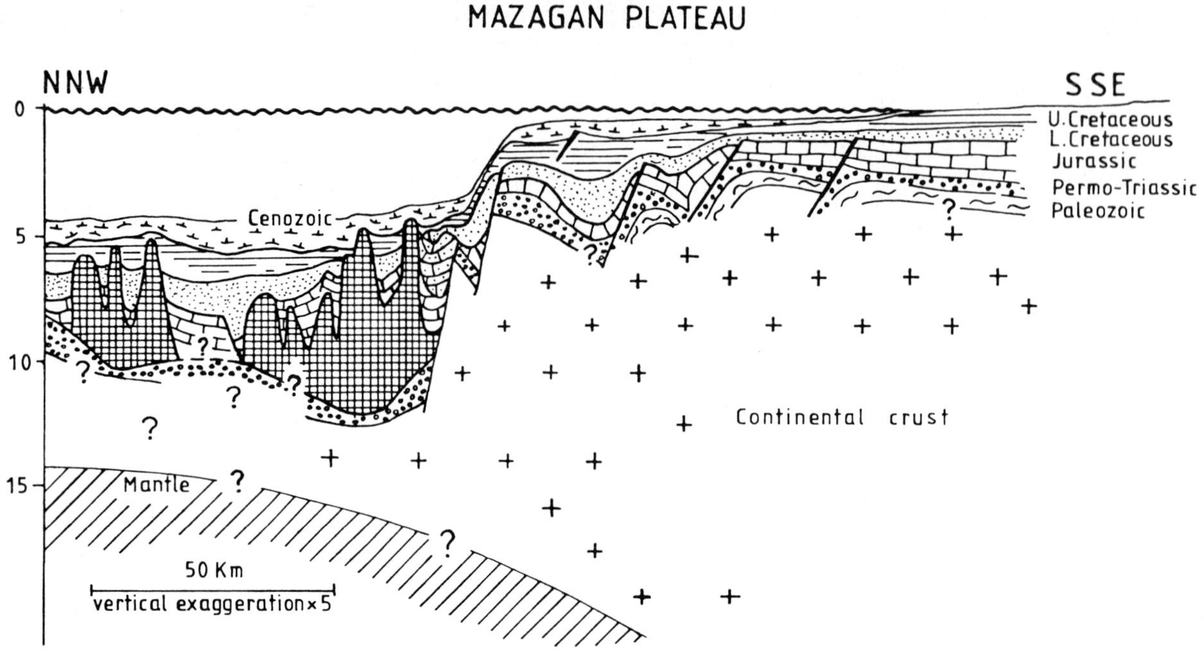

Fig. 8. Geologic interpretation of a multichannel reflection seismic section across the Mazagan Platform (seismic data from Hinz et al., this Vol.) and new drilling data from DSDP Leg 79. (For location see Fig. 2, 1)

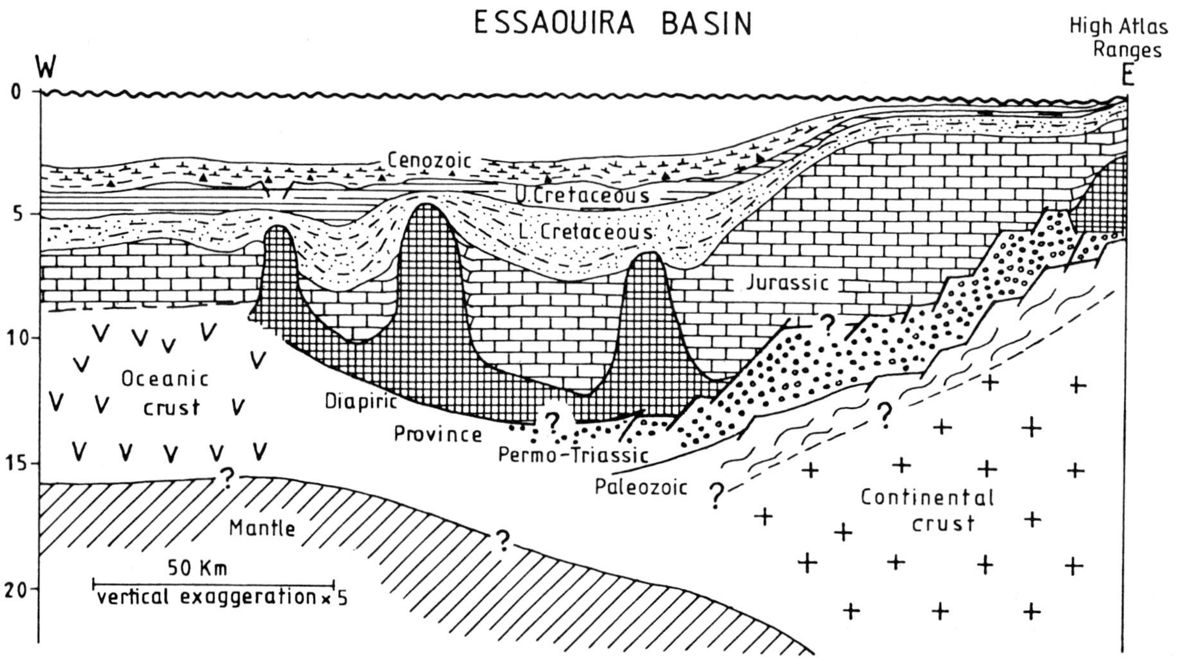

Fig. 9. Geologic interpretation of a multichannel reflection seismic section across the continental margin of the Essaouira Basin (seismic data from Hinz et al.,this Vol.). For location see Fig. 2, 2

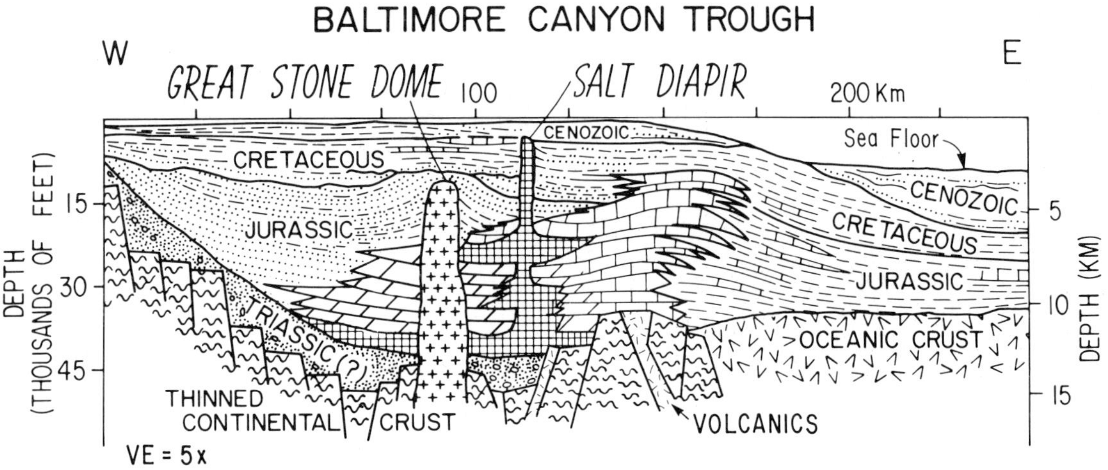

Fig. 10. Schematic section through the Baltimore Canyon Though, interpreted from several multichannel seismic reflection profiles (modified after Schlee and Jansa, in press). The Great Dome is probably not a volcanic plug (as indicated in this figure), but rather an area of frequent volcanic sills and dykes intruding the sedimentary sequence in the general region outlined as the "Great Stone Dome". The oceanic-continental crust boundary is highly speculative and its exact position is unknown. For location see Fig. 5

The basement blocks are much larger and less deeply buried than in the Scotian Basin, and, therefore, isolate the oldest sedimentary infilling as sub-basins (Schlee and Jansa, in press). A broad wedge of sediments as much as 12 km thick blankets these older deposits and smoothes out the basement morphology.

The Baltimore Canyon Trough (Fig. 10), which is the next basin to the south, is separated from the Georges Bank Basin at the north by the Long Island Platform and from the Blake Plateau Basin on the south by the Carolina Platform. The Salisbury Embayment is the onshore expression of the Baltimore Canyon Trough. The width of this basin is about 400 km. Similar to the Scotian Basin, this basin has also the shape of a wide embayment and may have been built over the transcurrent margin. The deepest part of the basin is an elongated east-west oriented trough which parallels the Long Island Platform. The main part of the Baltimore Canyon Trough is outlined by a zone of rapid thickening (6 to 10 km isopach) beneath the continental shelf. The deepest part of the basin contains over 14 km of Triassic to Recent sedimentary rocks (Schlee, 1981).

The Baltimore Canyon Trough terminates near Cape Hatteras. To the south it is separated from the Blake Plateau Basin by the Carolina Platform (Klitgord and Behrendt, 1979), which is covered by 2 to 4 km of sediments (Fig. 12). The Carolina Platform also separates two different sedimentary provinces of the margin - the mainly clastic depositional province to the north from the mainly carbonate province to the south. That part of the Blake Plateau Basin which underlies the present shelf has been referred to as the Southeast Georgia Embayment by Dillon et al. (1979).To the south the Blake Plateau Basin is bordered by the Florida Platform (Eardley, 1951). Seaward of the Blake Plateau, south of latitude 30° N, the ocean floor drops abruptly to form the Blake Escarpment, whereas at the southern part of the plateau the transition to the abyssal plain is less abrupt. The vertical elevation of the Escarpment above the abyssal plain is about 3200 m. The Blake Plateau Basin is similar to the other marginal basins and has an embayment

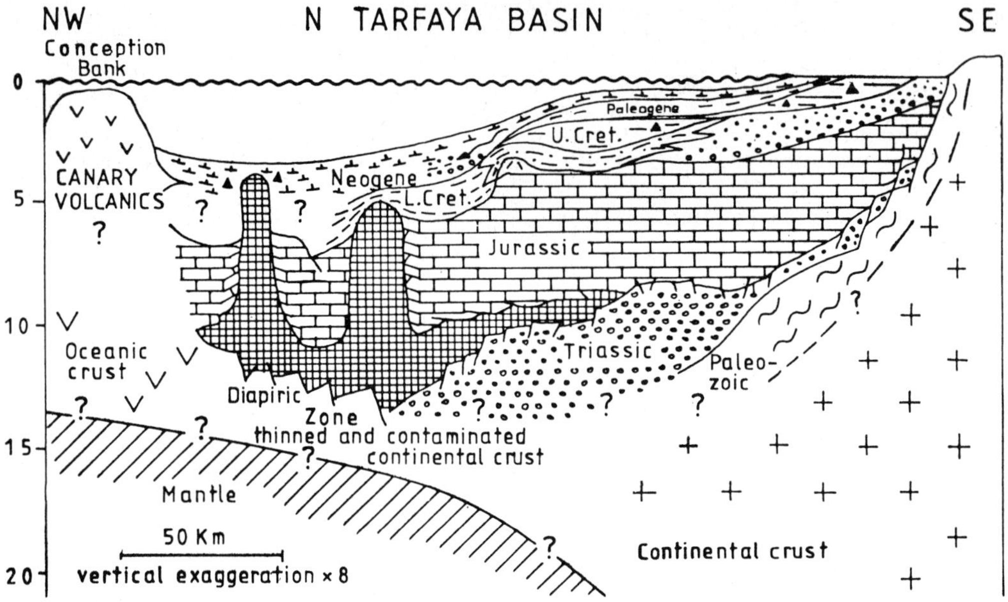

Fig. 11. Geologic interpretation of a multichannel reflection seismic section across the continental margin of the northern Tarfaya Basin (seismic data from Hinz et al., this Vol.). For location see Fig. 2, 3

shape. The basin is filled by more than 12 km of thick Mesozoic-Cenozoic deposits (Klitgord and Behrendt, 1979). The geologic history of the Blake Plateau Basin is more complex than that of the other basins, as will be discussed later in detail. Most of the basin is underlain by predominantly shallow-water platform carbonates, that grade into pelagic sediments when the margin subsided during Aptian-Albian (Benson, Sheridan et al., 1978). The near-shore part of the basin experienced much lower subsidence and was a site of mixed clastic-carbonate deposition during Early Cretaceous and Tertiary (Amato and Bebout, 1978).

The morphological difference between the present shelf (200 m) and the Blake Plateau, which is covered by 600 to 100 m of water, results from Tertiary changes in the paleocirculation pattern, with the submarine currents inhibiting deposition of sediments over the plateau, despite its continued subsidence.

2.2. Northwest African Margin

The northwestern continental margin of Africa (Figs. 2, 6) shows striking similarities to the matching Northeast American margin, but also important differences. Similar to Northeast America is the presence of a comparable number of sedimentary coastal basins alternating with structural highs. From north to south these basins are (1) the Rif-Betic Basin located on both sides of the Strait of Gibraltar and its southernmost outlyer, the Rharb Basin; (2) the Essaouira-Agadir (or Atlas) Basin; (3) the Tarfaya-Aaiun Basin and finally (4) the Mauritania-Senegal Basin. The interpretation of the northernmost Rif-Betic Basin on or near the boundary of the Eurasian and African plates is too problematic to be included in the present context. How the interfingering of the eastern North Atlantic with the western termination of the Mediterranean Tethys, west of the Alboran Massive, was established, is still a matter of discussion.

The separation of the Mesozoic to Cenozoic coastal basins by either Precambrian basement, Hercynian platforms or fold belts, forming the eastern borderland, at the same time, is directly comparable with Northeastern America. The Northwest African basins are also following Hercynian structures. The Hercynian

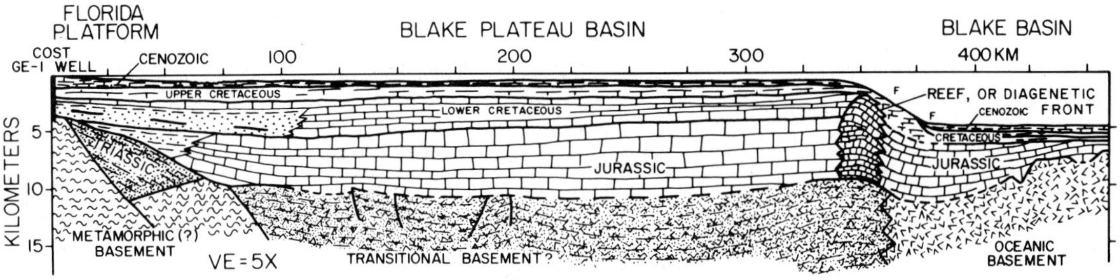

Fig. 12. Interpreted multichannel seismic section across the Blake Plateau Basin indicating the thickness, age and rock types (modified after Schlee et al., 1979). For location of the section see Fig. 5

fold belts of the Moroccan Meseta, High and Anti Atlas and Mauritanides can be considered as a first attempt to establish a proto-Atlantic rift during late Paleozoic time.

The separation of the mobile and mainly unfolded Mesozoic-Cenozoic coastal basins by stable onshore highs is documented also by a considerable offshore extension of the submerged Jurassic carbonate platform. Only the Essaouira-Agadir Basin continues towards the east into the mobile belt of the High Atlas with Hercynian and Alpidic history during which the coastal trough possibly communicated at times with the extended continental Sahara Basin (cf. Stets and Wurster, Wurster and Stets, this Vol.).

But nevertheless the Northwest African basins are comparable to those of the opposite margin and are also structural monoclines ("miogeosynclines") or half-basins rather than true basins. They are open towards the ocean. They differ, however, considerably in their actual position. Their sedimentary fills comprise the actual shelf, but they extend also onto the continent. The onshore width of the basins ranges from 100 km in the Essaouira-Agadir Basin to 500 km in the Senegal Basin.

The continental shelf (with less than 200 m water depth) ranges from a minimum width of 10 km near Dakar to a maximum of 210 km off Guinea, the average value of 40 km being considerably less than off northeast America. The Mesozoic-Cenozoic sediments of the Northwest African basins are mainly unfolded, but there are diapiric structures in nearly all basins being related to the occurrence of Triassic-Early Jurassic evaporites. These diapiric provinces parallel the actual shelf (see backcover figure and Fig. 6) and correspond (Fig. 1) with the Northeast American basins (Van Houten, 1977).

The Essaouira-Agadir or Atlas Basin (Figs. 6, 8, 9) is oriented northeast-southwest (cf. "Atlas Gulf" of Stets and Wurster, this Vol.). Towards the north it is bordered by the Mazagan Plateau (Fig. 8) which forms the offshore continuation of the Moroccan Meseta. Towards the south it is separated from the adjacent Tarfaya-Aaiun Basin by the Agadir Canyon system which is to some extent the offshore extension of the Cenozoic Souss Trough, and the Anti Atlas High. The sediments, ranging from Permo-Triassic to Recent, attain thicknesses of 8 km and have their maximum near the present-day coastline. A distinctive Jurassic platform can be recognized offshore and its edge largely coincides with the actual shelf edge in this area. In oceanward direction they are paralleled by the diapiric zone measuring 900 km in length and 200 km in maximum width (Hinz et al., this Vol.). In the centre of the basin the diapiric zone grades into the shelf and even into the onshore part of the basin.

The Tarfaya-Aaiun Basin (Figs. 6, 11) is bordered towards the north by the stable Anti Atlas High, towards the east and south by the Precambrian Reguibat High with its Paleozoic cover, i.e. the fold belt of the Hercynian Mauritanides. In total the basin measures about 1000 km in length and 500 km in maximum width. Across the Reguibat Arch, between Cape Barbas and Cape Blanc it continuously grades into the Senegal Basin.

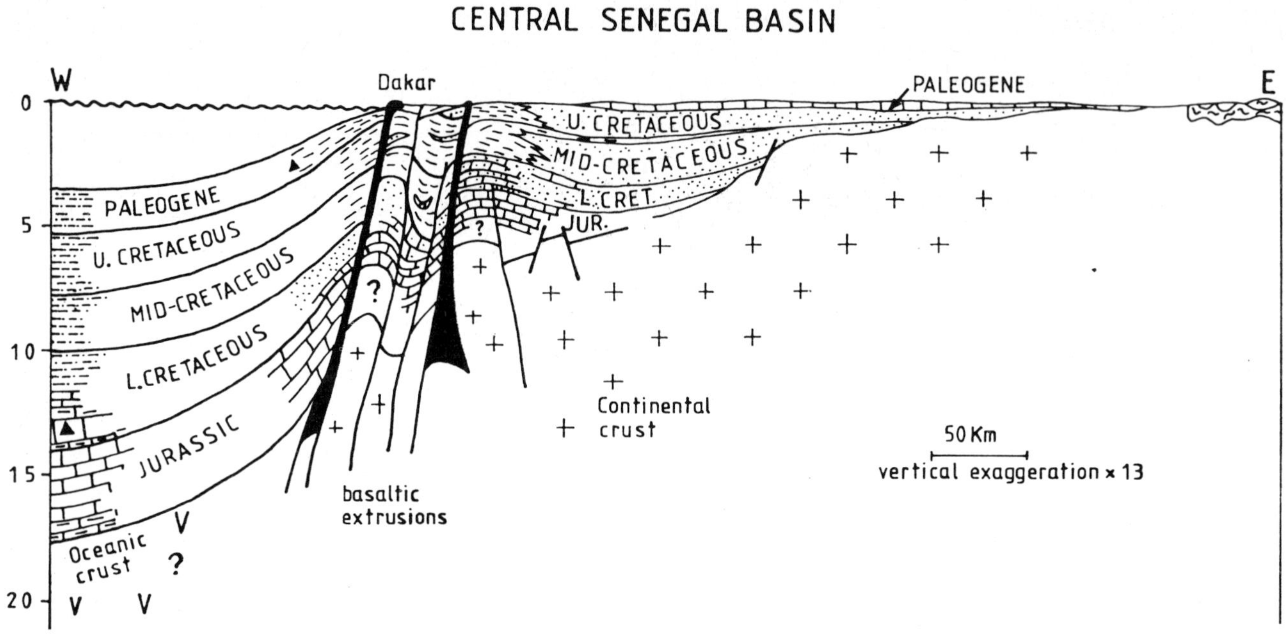

Fig. 13. Geological section through the central Senegal Basin (after Spengler et al., 1966). For location see Fig. 2, 4

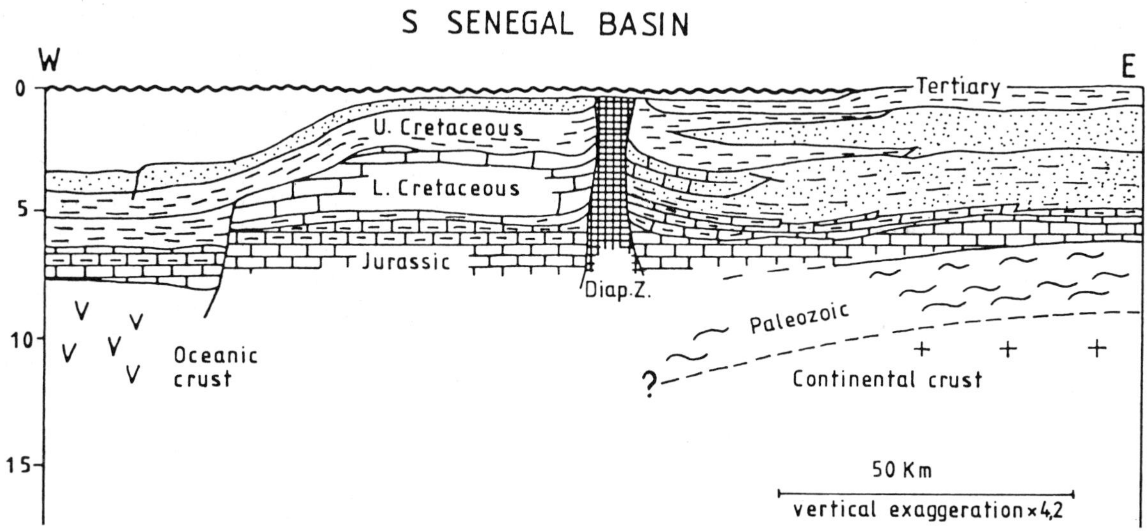

Fig. 14. Geologic interpretation of a multichannel reflection seismic profile across the southern Senegal Basin (after Schlee, 1980). For location see Fig. 2, 5

Sediments ranging from Triassic (?) to Recent attain thicknesses of more than 15 km, which can be recognized on the inner shelf and towards present-day's coast line in the north, whereas towards the south they appear below the outer shelf and slope. The edge of the Jurassic carbonate platform mainly follows the shelf edge (see backcover figure; Hinz et al., Ranke et al., this Vol.). The Jurassic platform is widest at the western extension of the Anti Atlas in the north and west of Cape Barbas in the south. The diapiric zone of the Morocco Sea continues only into the northern part of the Tarfaya Basin and has not yet been proved south of Tarfaya (Fig.6).

The southernmost Senegal Basin (Figs. 6, 13, 14) measures 1400 km in length and has a maximum width of 1000 km between the Precambrian basement in the east and the interfingering and overlying Cape Verde volcanics in the west. The southern boundary can be defined by the Vema Fracture Zone and the southernmost occurrence of ?Triassic-?Jurassic evaporites. The Recent shelf edge again largely follows the edge of the Jurassic-Lower Cretaceous platform both realizing their maximum width at the southern margin of the Senegal Basin at Gambia and the Casamance River embayments. Sediments of more than 10 km thickness can be dated from Late Jurassic to Recent (see Wissmann, this Vol.). In the northern Senegal Basin the continental slope is accompanied by a narrow seam of diapirs - about 400 km long and 30 km wide, mainly coinciding with the centre of deposition (see backcover figure). In the southern basin, however, deposition centre and the southern diapiric zone - of about 400 to 100 km - are integrated into the prograding shelf.

3. Continent-Ocean Crust Boundary

The character and location of the boundary between oceanic and continental crust along the eastern North American margin remains unknown, and none of the data presently available solve the problem. Drilling on the shelf has established that Paleozoic slightly metamorphosed, mainly Cambro-Ordovician, and less older sedimentary rocks, minor volcanics, and Devonian granites underlie the Mesozoic sedimentary deposits of the shelf. Such rocks were penetrated in the Scotian Basin (Jansa and Wade, 1975; McIver, 1972) and Georges Banks (COST G-1 well; Amato and Bebout, 1978).

Seismic profiles running westward from the quiet magnetic zone boundary, where the crust is oceanic, as documented by deep sea drilling (Hollister, Ewing et al., 1972; Lancelot, Seibold et al., 1978), do not show any change in the character of the acoustic basement, all the way to the area of the lower continental rise and up to the diapiric Sedimentary Ridge Province of Nova Scotia, and to a similar zone of diapirs off the Blake Plateau (Grow et al., 1979). Since it has not been established which characteristic on the reflection seismic profiles should be equivalent to the transition between the oceanic and continental crust along the passive continental margins, there is then the possibility that this boundary remains undetected on this profile. However, if we assume that the unchanged characteristic of the ocean basement reflector indicates the extension of oceanic crust westward, then the western limit of the oceanic crust can be placed somewhere between the Blake Spur and East Coast Magnetic Anomalies (Schouten and Klitgord, 1977; Shipley et al., 1978). Using the sedimentation rates as established for the lower part of Site 391 (Benson, Sheridan et al., 1978) and the depth to basement derived from the multichannel seismic profile published by Shipley et al. (1978), the age of the Blake Spur magnetic anomaly is known to be approximately 165 m.y.(Sheridan, Gradstein et al., in press). Thus, the age of the formation of oldest oceanic crust and the initiation of the continental spreading process in the southern part of the central North Atlantic can be estimated approximately at 170 to 175 m.y. This conclusion is consistent with estimates of Jansa and Wade (1975) and Van Houten (1977), who placed the initial North Atlantic opening into late Lower Jurassic, based on geological evidence.

If the above conclusion is correct, the developing North Atlantic Basin with its margins should behave as a structural unity until late Liassic times and the marginal basins should have similar geologic histories.

From the Northwest African margin no better data are available to locate and date the boundary between oceanic and continental crust. A crustal thickness of about 30 km has been calculated by Makris and Demnati (in press) at the coastal area of Morocco between Agadir and Oued Zem. A rapid crustal thinning to 20 km thickness underneath the actual coastline can be assumed from refraction seismic lines described by Hinz et al., and Weigel et al. (this Vol.) and Fig. 9. The rapid reduction of continental crust continues towards the shelf and slope, where the top of the crystalline basement is related to the seismic 5.8 km/s-layer (Hinz et al., this Vol.), if correlated with Paleozoic metamorphics drilled by the continental well IFNI 1. In the deep ocean, the oceanic layer 2c (Houtz and Ewing, 1976) shows similar velocities, here interpreted as old oceanic crust. Problems arise, however, in the interpretation of a deeper layer of vp = 7.1-7.3 km/s, located inside and outside the diapiric zone of the Morocco Basin, which could be either contaminated continental or oceanic crust as well (for a more detailed discussion see Weigel et al., Wissmann; both this Vol.).

The age of initial drift and thus the formation of the oldest oceanic crust cannot be determined with comparable precision as at the Northeast American margin. The oldest magnetic anomaly off Morocco (M25), running from Selvage Island to Seine Abyssal Plain, with an absolute age of 150 m.y., is still 20-25 m.y. too young for the late Lower Jurassic age of early spreading (cf. also Roeser, this Vol.).

4. Stratigraphy and Sedimentary History

The most extensively drilled sedimentary basin on the eastern North American margin is the Scotian Basin, with about 71 drill holes completed to the present day. Only 5 deep stratigraphic wells were drilled on the U.S. coast (Fig. 3). Thus, the stratigraphy of the margin as discussed here is mainly based on the drilling and reflection seismic data from the Scotian Basin to which the stratigraphic development of the U.S. marginal basins is compared.

The stratigraphy of the Mesozoic-Cenozoic sequences of the Scotian Basin was the subject of several papers (McIver, 1972; Jansa and Wade, 1975; Sherwin, 1973; Ascoli, 1976; Given, 1977; Eliuk, 1978; Barss et al., 1979). Therefore, only a very brief general account of the data is given here. Results of the Continental Offshore Stratigraphic Test (COST) wells off the U.S. coast are published in a series of open file reports by the U.S. Geological Survey. These reports contain geologic and operational summaries of the wells, including detailed biostratigraphy and rock description. Data from these reports were used extensively here to establish the general characteristic of the stratigraphic framework for the U.S. east coast marginal basins.

Considering the northwest African offshore basins, as mentioned above, the data were available from only one commercial shelf well (SPANSAH 51A-1; von Rad and Arthur, 1979). Therefore , we mainly refer to the DSDP drilling sites reproduced in Fig. 2. Onshore data are mainly taken from Ambroggi, 1963; Duffaud et al., 1966; Martinis and Visintin, 1966; Spengler et al., 1966; Choubert et al., 1967; Viotti,1967; AUXINI, 1969; Ratschiller, 1971; Dillon and Sougy, 1974; Wiedmann et al., 1978a, 1978b, this Vol. Moreover, a considerable amount of seismic information is available (McMaster and Lachance, 1968; Summerhayes et al., 1971; Uchupi et al., 1976; Lehner and Ruiter, 1977; Hinz, 1979; Watkins and Hoppe, 1979, Hinz et al., this Vol.).

For better comparison, the stratigraphical description of the Essaouira-Agadir Basin immediately follows that of Scotian Basin, description of Tarfaya-Aaiun Basin that of Georges Bank Basin and Baltimore Canyon Trough and, finally, the consideration of Senegal Basin that of Blake Plateau Basin.

For a better understanding of the stratigraphic discussion, representative stratigraphic columns of the Scotian (Fig. 15) and Atlas Basin (Fig. 16) are reproduced.

4.1. Northern Transect

4.1.1. Scotian Basin (Figs. 15, 17)

A relatively complete succession of upper Triassic and Jurassic sediments overlies the Paleozoic basement comprised of slightly metamorphosed Cambro-Ordovician sedimentary

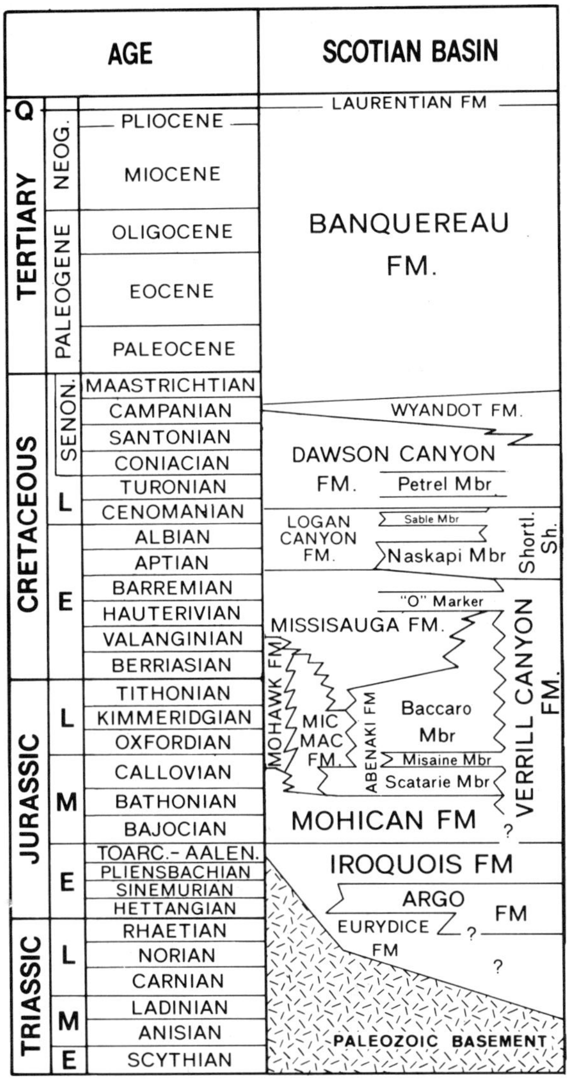

Fig. 15. Type stratigraphic column of the Scotian Basin and formation names used

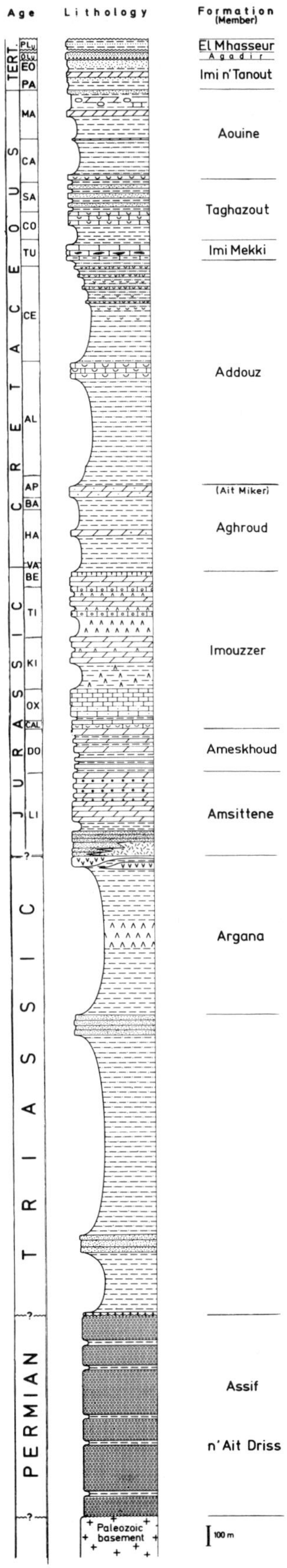

Fig. 16. Type stratigraphic column of ► the Atlas Basin and formation names used

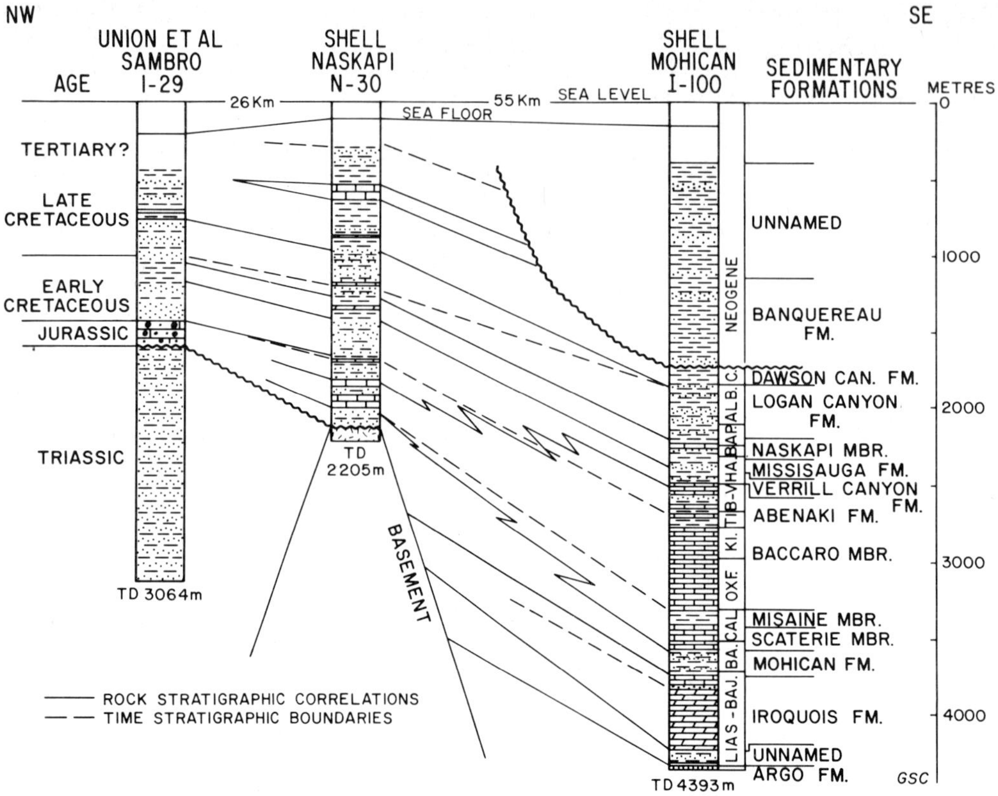

Fig. 17. Lithologic cross-section across the southern part of the Scotian Basin (for location see Fig. 5). The section shows the predominance of clastics near to the updip margin of the basin, whereas carbonates occur in the offshore part of the margin. Note the unconformity in the ?early Lower Jurassic and the deep ?Oligocene-Miocene erosion near to the recent shelf edge

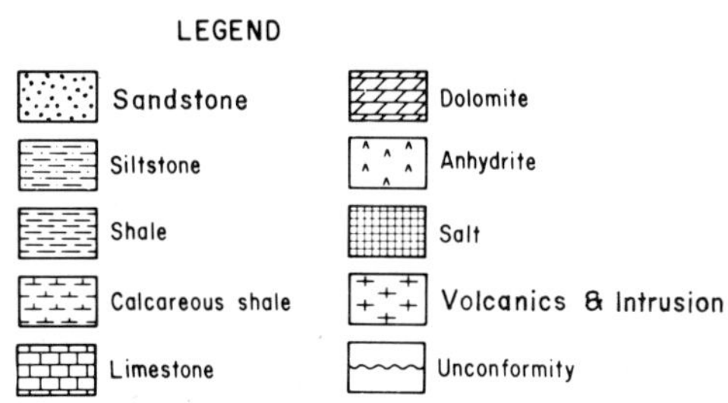

rocks, Devonian granites and unmetamorphosed Carboniferous sediments (Jansa and Wade, 1975; McIver, 1972). The oldest post-Paleozoic rocks which are represented by reddish silty shales, sandstones, and minor conglomerates have not been dated, but it is assumed that they are Triassic. In the Bay of Fundy, these redbeds are more than 3400 m thick (Jansa et al., 1980). In the Scotian Basin the evaporites were penetrated only on the upper landward flank of the basin (Argo Formation, McIver, 1972).

Evaporites of the Argo Formation are 780 m thick and were dated by palynology as Rhaetian to Hettangian-Sinemurian (Barss et al., 1979). The redbeds intercalated with the salt are thought to have been deposited in a very shallow, semi-restricted basin. The Sedimentary Ridge Province, which is located beneath the lower continental slope and offshore of the edge of the carbonate platform (Fig. 7) was interpreted by Jansa and Wade (1975) to be formed by either diapiric salt or shale. The salt in these diapirs could be older than the salt drilled at the upper flank of the Scotian Basin.

Towards the shallower part of the basin, seismic profiles indicate that the redbed-evaporite sequence is separated from the overlying strata by a conspicuous unconformity, which Jansa and Wade (1975) correlated with the Early Cimmerian tectonic event of Stille (1924). The tectonic significance of this unconformity is conjectural. The unconformity is not recognizable in the deeper part of the basin, where deposition may have continued without interruption, but a similar unconformity has been recently recognized in the Baltimore Canyon Trough (Schlee and Jansa, in press).

The redbeds and evaporites are overlain with a sharp contact by a ?Sinemurian to Bajocian carbonate unit, 600 m thick, named the Iroquois Formation by McIver (1972) (Fig. 17). The carbonates are fine to micro-crystalline dolomites with anhydrite and stromatolitic beds in the lower part of the sequence, which grade upward into secondary dolomitized oolitic and skeletal limestones. Changes in the carbonate composition indicate deepening of the depositional environment from semi-restricted tidal flat to an open marine subtidal environment during late Early Jurassic. The lower Jurassic carbonate sequence locally extends beneath the present continental slope, where it is assumed to be composed by argillaceous carbonates. Jansa (in press) suggested that the lower Jurassic carbonates formed a ramp similar to the present Persian Gulf (Purser, 1973).

A clastic sequence up to 264 m thick, redefined as Mohican Formation by Given (1977), overlies the lower Jurassic sequence of the Iroquois Formation (Figs. 15, 17). The occurrence of quartz grains forming nuclei in the oolites of the uppermost part of the Iroquois carbonates indicates initiation of a slow regression as early as in late Liassic times. The clastics of the Mohican Formation are alternating fine to coarse-grained sandstones and variegated shales. Occasional thin algal limestones are intercalated between the clastics in wells located in a distal part of the shelf. The clastics were deposited under continental and marginal marine depositional conditions, during approximately late Bajocian-Bathonian time (Barss et al., 1979). The deposition of this clastic wedge has been correlated with the rejuvenation of the source area by Jansa and Wade (1975) as related to the initiation of the drift of continents in the North Atlantic.

Carbonate sedimentation in the Scotian Basin has been resumed during late Bathonian when a transgressive blanket of high energy carbonates, oolitic and sandy oncolitic grainstones and packestones of the Scatarie Member of the Abenaki Formation (McIver, 1972) were deposited along the outer part of the margin.

A brief ingression during the Callovian deposited a more than 100 m thick shaly unit (the Misaine Shale Member), thus temporarily disrupting carbonate deposition in the basin. The Callovian shale has a sharp contact with the overlying Oxfordian-Berriasian carbonate sequence (the Baccaro Member of the Abenaki Formation). Carbonate deposition became more and more regionally restricted to the outer shelf during late Jurassic, with offshore banks developing near the paleoshelf edge at the end of the Jurassic.

The composition of the upper Jurassic carbonate sequences differs according to their position on the carbonate platform. The high-energy carbonates which occur near the paleo-shelf edge during Oxfordian-Early Kimmeridgian give away to biomicrites, argillaceous limestones, and calcareous shales deposited in an outer shelf environment during the Kimmeridgian. This lithologic change reflects a minor late Jurassic transgression. During the late Kimmeridgian, a new regressive pulse is marked by the decrease in shaliness, and the strata mainly comprise biomicrite and skeletal wackestone containing pelagic biota such as calpionnellids, ammonites, and pelagic foraminifera (Jansa et al., 1980). At this time isolated coral-stromatoporoid bioherms have also developed near the shelf-edge (Eliuk, 1978). The intensity of the regression increased during the late Tithonian and carbonate deposition in the basin terminated between the Tithonian-Berriasian-Valanginian, when the carbonate banks were covered by prograding clastics (Figs. 7, 17). A sharp boundary between the top of the limestones and the overlying clastics, the reddish colouration associated with the hematization of the glauconite at the boundary, and the occurrence of diplaced Tithonian limestone blocks into Valanginian deposits (Jansa et al., 1980) indicate at least minor tectonic disturbance of the Late Cimmerian phase near the shelf edge area.

Deposits coeval to the late Jurassic carbonates near the basin margin were continental clastics (sands, variegated shales and minor coal beds) of the Mohawk Formation (Given, 1977) and the marginal marine clastics of the MicMac Formation (Fig. 15). The MicMac Formation is composed of shale, frequently reddish coloured, intercalated with well sorted sandstones and thin limestone beds. This formation is more than 1200 m thick near the Sable Island area and it is Callovian to Tithonian in age (Barss et al., 1979). The third late Jurassic lithofacies (Verril Canyon Formation) are medium grey calcareous shales deposited in the front of the carbonate banks. Only the uppermost part of this formation was penetrated by the wells, and the detailed characteristics of this unit remain unknown.

The Early Cretaceous tectonic episode resulted in a regional change of deposition from predominantly carbonate shelf into silicoclastic, paralic deposition, represented by the Missisauga and Logan Canyon Formations. Both of these formations comprise poorly to moderately sorted, fine to conglomeratic sandstones intercalated with minor siltstone, shale, and coal beds. A brief marine ingression during the Hauterivian caused development of a few oolitic limestone beds intercalated with sandstones (Fig. 17). A more widespread transgression is noticeable in the Aptian, when a shale unit of about 300 m thickness, named the Naskapi shales, had been deposited in the basin. This marine shale separates the Missisauga and Logan Canyon clastics, both of which are about 1200 m thick, depending on the position in the basin (Fig. 15). The clastic wedges are Berriasian-Cenomanian in age and thin in an offshore direction (Fig. 17). During the late Cenomanian the next transgression pulse flooded the coastal plain and the sand accumulation ceased, except in the most marginal areas of the basin. In the basin, the glauconitic siltstones and shales were deposited in a middle to outer shelf environment during the Late Cretaceous. The shale deposition was interrupted twice by a period of a limestone and/or chalk sedimentation. The first limestone deposition occurred during Turonian and the second one during late Santonian-Maastrichtian. During the latter the Wyandot chalk has been deposited (McIver, 1972). Both these periods are explained to indicate periods of rising sea level. The Wyandot chalk has been deposited in an environment corresponding to the outer shelf-upper continental slope. Chalk deposition ceased during early Maastrichtian, but upper slope depositional conditions persisted up to the Eocene. During this time mainly mudstones were deposited as a prograding wedge which filled the area and built the present shelf configuration (Figs. 15, 17). A renewed period of coarse clastic deposition during the late Miocene to Pleistocene is related to glaciation and a lowered sea level, when most of the shelf was subaerially, exposed.

4.1.2. Essaouira-Agadir or Atlas Basin (Figs. 16, 18)

In the last decades much information has been accumulated on the post-Hercynian evolution of the onshore part of this basin, mainly by Choubert et al, 1962; Ambroggi, 1963; Duffaud et al., 1966; Ager, 1974; Van Houten, 1977; Behrens et al., 1978; Wiedmann et al., 1978a; Stets and Wurster, Wurster and Stets, Stamm and Thein, Behrens and Siehl, all this Vol.

In the western part of the High Atlas Ranges the Paleozoic basement is disconformably covered by almost 4000 m thick red conglomerates, sandstones and clays of presumably Permian age (Assif n'Ait Driss Formation) (Fig. 16). Sediments of this early stage of graben evolution

Fig. 18. Cretaceous onshore-offshore correlation between the Essaouira and Agadir subbasins and the DSDP Sites 370/416 and 137 (after Wiedmann et al., 1978a)

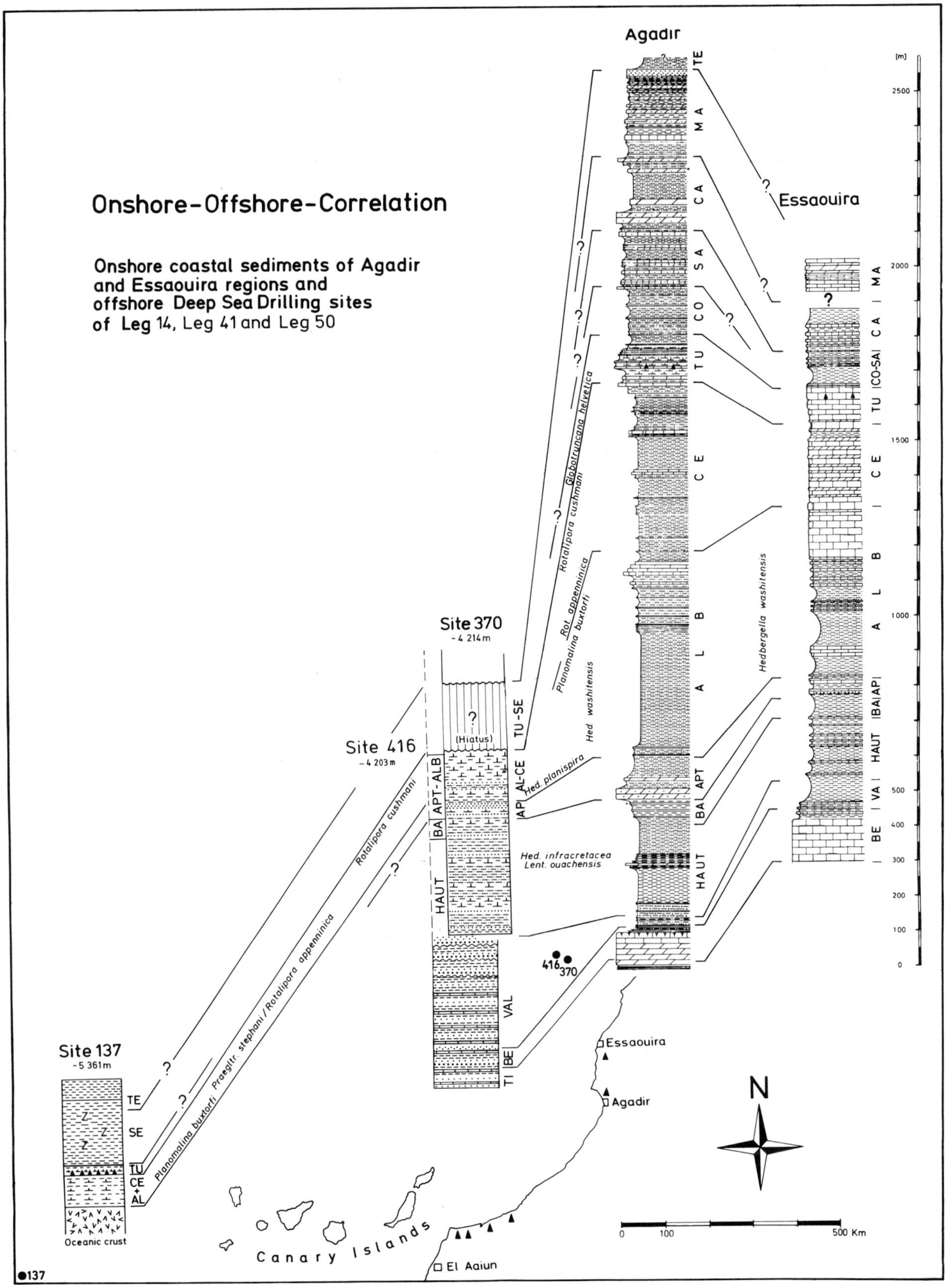
Onshore-Offshore-Correlation
Onshore coastal sediments of Agadir and Essaouira regions and offshore Deep Sea Drilling sites of Leg 14, Leg 41 and Leg 50
Agadir
Essaouira
Site 370
-4 214m
Site 416
-4 203m
Site 137
-5 361m
(Hiatus)
Oceanic crust
Globotruncana helvetica
Rotalipora cushmani
Rot. appenninica
Planomalina buxtorfi
Hed. washitensis
Hed. planispira
Hed. infracretacea
Lent. ouachensis
Hedbergella washitensis
Praegltr. stephani / Rotalipora appenninica
Planomalina buxtorfi
Canary Islands
Essaouira
Agadir
El Aaiun
N
500 Km
[m]
●137

are accumulated in the central part of the western High Atlas Ranges near Argana (Duffaud et al., 1966).

The Assif n'Ait Driss valley exhibits maximum thicknesses of these basal red beds and may serve as type-locality of the equivalent formation. This formation is again disconformably overlain by red clays, sandstones, marls and sands, passing into red clays and evaporites towards the top, having thicknesses between 1000 and 3500 m and including doleritic volcanics in their upper portion. This red sequence is named Argana Formation and is of Triassic to early Liassic age. The Argana evaporites continue into the actual shelf and slope and could be continuous with the mainly offshore diapiric province (Fig. 6). According to Hinz et al. (this Vol.) their western boundary matches the eastern boundary of the coeval diapiric zone of the Scotian Basin (Hinz, in prep.) (Fig. 1). The subsequent reef limestones of Djebel Amsittene (Duffaud et al., 1966) were deposited in the marginal part of the slowly subsiding basin. They interfinger with and are followed again by a series of red sandstones and clays, continuing into evaporites, especially dolomites. This transgressive sequence may be named Amsittene Formation and is of middle to late Liassic age (Fig. 16). The basal limestone member can be attributed to an Upper Sinemurian-Pliensbachian age. It is thus comparable with the Iroquois Formation of Scotian Basin and includes all sediments of the initial rifting stage. The environment changes progressively from sublittoral via tidal to open marine conditions. In thickness the Amsittene Formation exceeds 400 m. It is covered by about 220 m of conglomerates, red sandstones and clays and again by dolomites, attributed to early to middle Dogger and regarded as Ameskhoud Formation. It resembles the Mohikan Formation of Scotian Basin showing also distinctive regressive characters.

During late Middle Jurassic time a stable carbonate platform (Fig.9) became established and continued through the complete Upper Jurassic. The carbonate platform has been drilled seaward of the Mazagan Plateau during DSDP Leg 79 (see Hinz, Winterer et al., this Vol.). Within this period of continuous subsidence 700-800 m of well-bedded limestones, marls, calcarenites and dolomites were deposited and are treated here as Imouzzer Formation. Correlation with the coeval Abenaki Formation of the Scotian Basin is evident. In the central portion of the Agadir-Essaouira Basin, northeast of Agadir, three regressive phases can be recognized during the deposition of the Imouzzer Formation, i.e. the Lower Oxfordian calcarenite, and the Lower Kimmeridgian and Tithonian red clays and evaporites. A fourth regression is documented by a final hardground of Berriasian age, closing up the platform sedimentation in the Agadir part of the basin (Figs. 9, 16, 18). The transgressive Callovian portion of the Imouzzer Formation is characterized by high-energy biosparitic and oolitic limestones, while the higher limestone levels show a reef-like facies with nerineid gastropods, corals and brachiopods. The alternating redbeds and evaporites thin towards the present coastline and grade into reef limestones, e.g. at Cape Rhir.

The paleoshelf edge is buried by the continental slope deposits. The drastic change from platform sedimentation to detrital influx at the base of the Cretaceous can be related to tectonic activity at the Jurassic-Cretaceous boundary (late Cimmerian movements), and in consequence to uplifting source areas in the eastern High Atlas and Anti Atlas (Schlager, 1980). The top of the platform limestones exhibits besides the mentioned hardgrounds also local erosional surfaces and is finally buried by about 400 m of clays, marls and clastics of the Upper Berriasian to Lower Aptian Agroud Formation. The same change in depositional environment from carbonate shelf to silicoclastic paralic deposition is known from the Scotian Basin. Increasing subsidence from Berriasian to late Barremian times is documented by progressive pelagic influences and interpreted as to indicate inner to midshelf environmental conditions (Wiedmann et al., this Vol.). This continuous trend is only interrupted by a short episode of redbed sedimentation during the mid-Hauterivian. The upper portion of the Agroud Formation is also equivalent to the next regressive phase during which 40 m of dolomites, sandstones with slumping and redbeds containing charophytes, were deposited. They were extensively studied by Einsele (this Vol.). They are treated here as Ait Mikr Member and are of Lower Aptian age (Figs. 16, 18).

The subsequent transgression is first of all combined with stratigraphic condensation (late Aptian) and leads finally to the deposition of silty marls and thick oyster shell beds during Albian and Cenomanian time. The 1100 m thick sequence is named here Addouz Formation. The inner to mid-shelf deposits are followed by 130 m of well bedded micritic limestones and chert layers alternating with black shales, representing the peak of the mid-Cretaceous transgression. This Turonian Imi Mekki Formation can be related to an outer shelf to slope environment (Wiedmann et al., this Vol.). Despite the fact that these sediments are of random importance for the Agadir-Essaouira Mesozoic, both black shales and chert layers are of widespread distribution and were, if not eroded, recognized in several DSDP sites of the eastern North Atlantic (i.e. Sites 137, ?135, 13). They even permit correlation of this event into the Rif-Betic Basin.

During the following Coniacian and Santonian time the environment drastically changed again to nearshore conditions. The Almodraga Formation (Figs. 16, 18) is represented by a sequence of 300 m of Astarte shell beds, marls, deltaic sands and sandstones. It is followed by the final Cretaceous transgression being Campanian to Maastrichtian in age and documented by about 470 m of marls, dolomites, chalk and concretionary limestones of the Aouine Formation. Their fossil content point again to an inner to outer shelf environment. Phosphatic sands close the Cretaceous System and may be regarded as the proof of the worldwide regression at the Cretaceous-Tertiary boundary. The mainly marine conditions of the Essaouira-Agadir Coastal Basin drastically change towards the east into continental red sedimentation (Behrens et al., 1978; Stets and Wurster, this Vol.). Only the late Aptian and the Turonian strata preserve their marine characteristics and the latter may even connect with the epicontinental Trans-Saharan Sea (Stamm and Thein, this Vol.).

Due to increasing separation during the Cretaceous only the main trends of sedimentation are comparable with the Cretaceous of the Scotian Basin: the predominance of clastic sedimentation during the Early Cretaceous, the Aptian transgression as well as the Upper Cretaceous transgressive pulses leading to the deposition of Turonian limestones and the Wyandot Chalk of late Santonian to Maastrichtian age.

Paleocene and Eocene are well represented in the zone of Imi n' Tanout. This Imi n'Tanout Formation (Fig. 16) shows reduced thicknesses of 150 m. At first, it is transgressive, starting with Paleocene pelagic marls and phosphatic sands, passing into dolomites, sandstones and red marls of Eocene age. The Upper Oligocene Agadir Formation overlies the Eocene disconformably with only a few meters of conglomerates and sandstones. This is the expression of first tectonic activity (Pyrennean Orogeny) in the Essaouira-Agadir Basin, followed by the main orogeny after the Oligocene (Savic Orogeny). The subsequent marine near-shore marls and sandstones of the El Mhasseur Formation are attributed to the early Pliocene which is followed by a final unconformity by late Pliocene marine to continental sandstones of varying thicknesses.

Offshore information is mainly based on DSDP Sites 370/416 in the deep Moroccan Basin and 415 at the continental rise near the termination of the Agadir Canyon (Lancelot, Winterer et al., 1980). Important features of the sedimentary evolution at these sites are (1) the presence of Late Jurassic (Tithonian) to Early Cretaceous (Hauterivian) turbidites as a proof of a pre-existing relief; (2) deposition of Cretaceous turbidites and hemipelagic claystones (Barremian to Albian) below CCD as a proof of early subsidence; (3) an Upper Cretaceous hiatus as a first proof of the erosion by bottom currents; (4) deposition of a reduced Tertiary section (500-600 m) above CCD, starting with Paleocene nanno marls, passing into Eocene mudstone with chert (Site 415) suggesting an increase of productivity of surface

Table 1. Evolutionary events in Mesozoic-Cenozoic basin development at the

	SCOTIAN BASIN	GEORGES BANK BASIN	BALTIMORE CANYON TROUGH
QUATERNARY NEOGENE PALEOGENE	 *subsidence* *hiatus*	 *reduced subsidence* *+ sedimentation* *hiatus* *?Laramid tectonics*	 clays *transgressive pulse* *regressive pulse* clays *transgressive pulse* *hiatus* *Laramid tectonics*
UPPER CRETACEOUS	*shelf construction* chalk, clays *transgressive pulse*	ooze *transgressive pulse*	calcareous shales *transgressive pulse*
MID-CRETACEOUS	limestones, silicoclastics *transgressive pulse* *deltaic fans*	shales *transgressive pulse*	*regressive pulse* clays *transgressive pulse*
LOWER CRETACEOUS	*Austrian tectonics* silicoclastics *deltaic fans* *platform drowning* *Late Cimmerian tectonics*	silicoclastics *deltaic fans* *? Late Cimmerian tectonics*	silicoclastics *deltaic fans - non-marine* *reef phase*
UPPER JURASSIC	*Second reef phase* *carbonate platform* *'Atlantic' transgression* *First reef phase*	*carbonate platform and banks with clastic tongues* *'Atlantic' transgression*	*carbonate platform with clastic tongues*
MIDDLE JURASSIC	*early drift* clastics *regressive pulse*		
LOWER JURASSIC	*mid-Cimmerian tectonics* evaporites, dolomites *late rifting*	evaporites, dolomites clastics *late rifting*	
TRIASSIC PERMIAN	*Early Cimmerian tectonics* alcaline volcanics redbeds *graben filling* *early rifting*		*Early Cimmerian tectonics*

Northeast American and Northwest African continental margins

AKE PLATEAU BASIN	SENEGAL BASIN	TARFAYA-AAIUN BASIN	ESSAOUIRA-AGADIR BASIN
litic limestones alk environment	*volcanic activity* *uplift* inner shelf limestones chert, phosphates *?upwelling* *Laramid tectonics*	*volcanic activity* *uplift* *random marine incursion* *Pyrennean tectonics* phosphates *?upwelling* *Laramid tectonics*	*final uplift* *random marine incursion* *main uplift* *Savic tectonics* *continental phase* *Pyrennean tectonics* pelagic sediments *transgressive pulse*
ogressive deepening	pelagic sediments *transgressive pulse*	shell beds	shell beds *regressive pulse*
dstone-clay environment *ré-Gosavian tectonics*	*regressive pulse* clays *transgressive pulse*	black shales, cherts *main transgression* *+ upwelling* shell bed accumulation	black shales, cherts *main transgression* *+ upwelling* shell bed accumulation
litic limestones rine clastics *ansgressive pulse* astics hydrite *n-marine phase*	calcarenites *deltaic phase*	*Austrian tectonics* clastics *deltaic phase* *Late Cimmerian tectonics*	*condensation event* *initial ingression* clastics *deltaic phase* *initial diapirism* *platform drowning* *Late Cimmerian tectonics*
rbonate platform	*carbonate platform* *'Atlantic' transgression*	*reef phase* *carbonate platform*	*Second reef phase* *carbonate platform* *'Atlantic' transgression*
		evaporites diabasic volcanics *late rifting*	*early drift* redbeds *deltaic phase* *First reef phase*
		redbeds *graben filling* *early rifting*	evaporites alcaline volcanism *late rifting*
			Early Cimmerian tectonics evaporites alcaline volcanics redbeds *early rifting* *Palatinate tectonics* redbeds *graben filling* *early rifting*

waters and possibly upwelling; (5) another large hiatus at Site 415, ranging from Late Eocene to Early Miocene and possibly related to the installation of a modern system of deep oceanic circulation; at the same time continuous turbidite sedimentation at Site 370/416, where the Eocene is disconformably resting on Paleocene; and, (6) finally, congruent sedimentation of nanno marls at both sites during Late Miocene and Pliocene.

If we compare the evolution of the Essaouira-Agadir Basin with the Scotian Basin (Table 1), it becomes obvious that both regions were intimately related during the early stage of rifting (Triassic), the evaporite event (Triassic-Early Liassic) and largely also during the buildup of the carbonate which is mainly restricted to the late Jurassic.

With regard to time variation of the carbonate buildup at the Moroccan margin we refer to Seibold, this Vol. Moreover, new results of DSDP Leg 79 from the Mazagan Plateau provided us with exciting new information on the Jurassic-Early Cretaceous early rift, early drift and subsidence history of this continental margin (see Hinz, Winterer et al., this Vol.).

Also the drastic change from the drowning platform to the sedimentation of silicoclastic fan deposits masking these platforms and the slope during the Early Cretaceous coincides to a large extent. During mid-Cretaceous time, however, differing regimes were installed on both sides of the Atlantic. Sedimentation of Turonian black shales and cherts, related to increasing plankton productivity and a first coastal upwelling, is restricted to the Moroccan margin. Only the two global late Cretaceous transgressive peaks (Turonian, Campanian) are recognizable by similar trends to chalk deposition in both regions. During the Tertiary, however, the separation is complete and no common depositional trend can be recognized.

4.2. Central Transect

4.2.1. Georges Bank Basin (Fig. 19)

The stratigraphic information for the Georges Bank Basin is provided by two COST wells drilled in this basin (Amato and Simonis, 1980; Amato and Bebout, 1980). The COST G-2 well was drilled near the present shelf edge and the COST G-1 well westward from the COST G-2 well, closer to the basin margin (Fig. 3). The COST G-1 well bottomed in Paleozoic dolomites underlain by upper Cambrian graphitic phyllites radiometrically dated at 450 to 550 m.y. Of all the COST wells drilled on the eastern North American margin only the COST G-2 well penetrated rocks old enough to core a thick section of early Jurassic evaporites and carbonates. Unfortunately, these strata are unfossiliferous and thus lithological correlation with the Scotian Basin is used to indicate the probable age of the rocks.

The oldest sediment encountered in the COST G-2 well is a salt (Fig. 19) which can be traced seismically to the diapiric province of the Scotian Basin (Wade, 1978) and thus can be tentatively correlated with the Argo Formation of the Scotian Basin. The salt is overlain by a 2502 m thick sequence of fine to micro-crystalline dolomites, irregularly alternating with beds of anhydrite, dark grey shale, and light to dark brown-grey micritic limestones. Oolitic and pelleted structures are poorly recognizable in some of the dolomites. Anhydrite beds come up to 3 m thickness and are more common in the upper half than in the lower half of this sequence. This minor lithological change occurs near the base of the Middle Jurassic boundary. The top of the unit is paleontologically dated as pre-Callovian (Bebout, 1980) and it is assumed here that it is Bathonian in age. This dolomite anhydrite-limestone sequence is in general correlative with the Iroquois Formation of the Scotian Shelf, with the upper part of the unit in the COST G-2 well being the facies equivalent to the lower part of the Mohican Formation (Figs. 19, 22).

The COST G-1 well which is located updip of the COST G-2 well contains considerably more lower Jurassic clastics, but the upper part of the Iroquois Formation has a similar development to that in the COST G-2 well.

The depositional conditions in the Georges Bank Basin were more vari-

Fig. 19. Lithologic cross-section of the Georges Bank Basin (for location see Fig. 5). The column next to the ages in COST G-2 well shows correlation with the lithostratigraphic units of the Scotian Basin. Note the seaward lithofacies change to carbonates

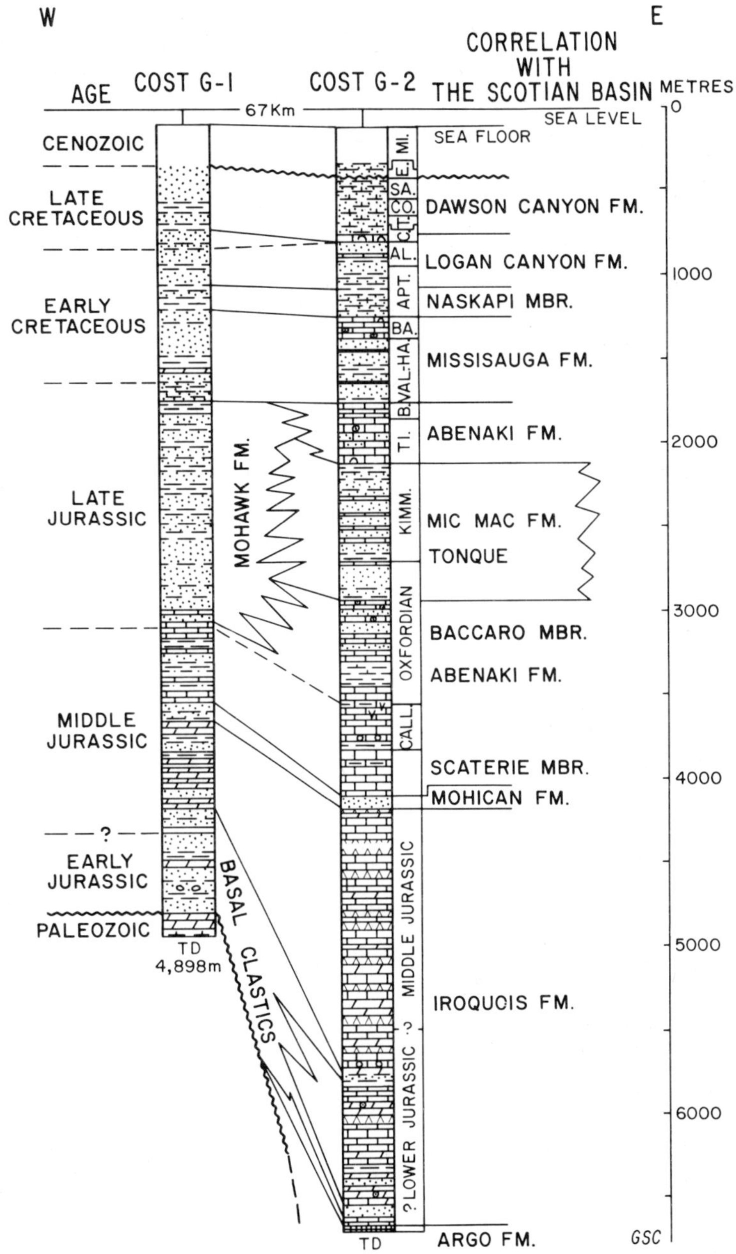

able during Early Jurassic than in The Scotian Basin and fluctuated from semi-restricted, hypersaline lagoonal, to a normal open marine, shallow-water environment. The late Middle Jurassic was a period of clastics deposition in the basin similar to that in the Scotian Basin. The clastics in the distal part of the basin (COST G-2 well) are very thin and are overlain by shallow-water carbonates (oolitic, pelleted, and skeletal). Limestone deposition in the outer shelf area was interrupted by a prograding wedge of clastics during Kimmeridgian-Tithonian, but the deposition of the limestones has been resumed again in the late Tithonian and continued into Berriasian. In the updip part of the basin (COST G-1 well) the deposition of terrestrial and nearshore clastics predominated for most of the late Jurassic (Fig. 19).

The Cretaceous strata in both COST wells have a similar development to synchronous deposits of the Scotian Basin. However, the Upper Cretaceous

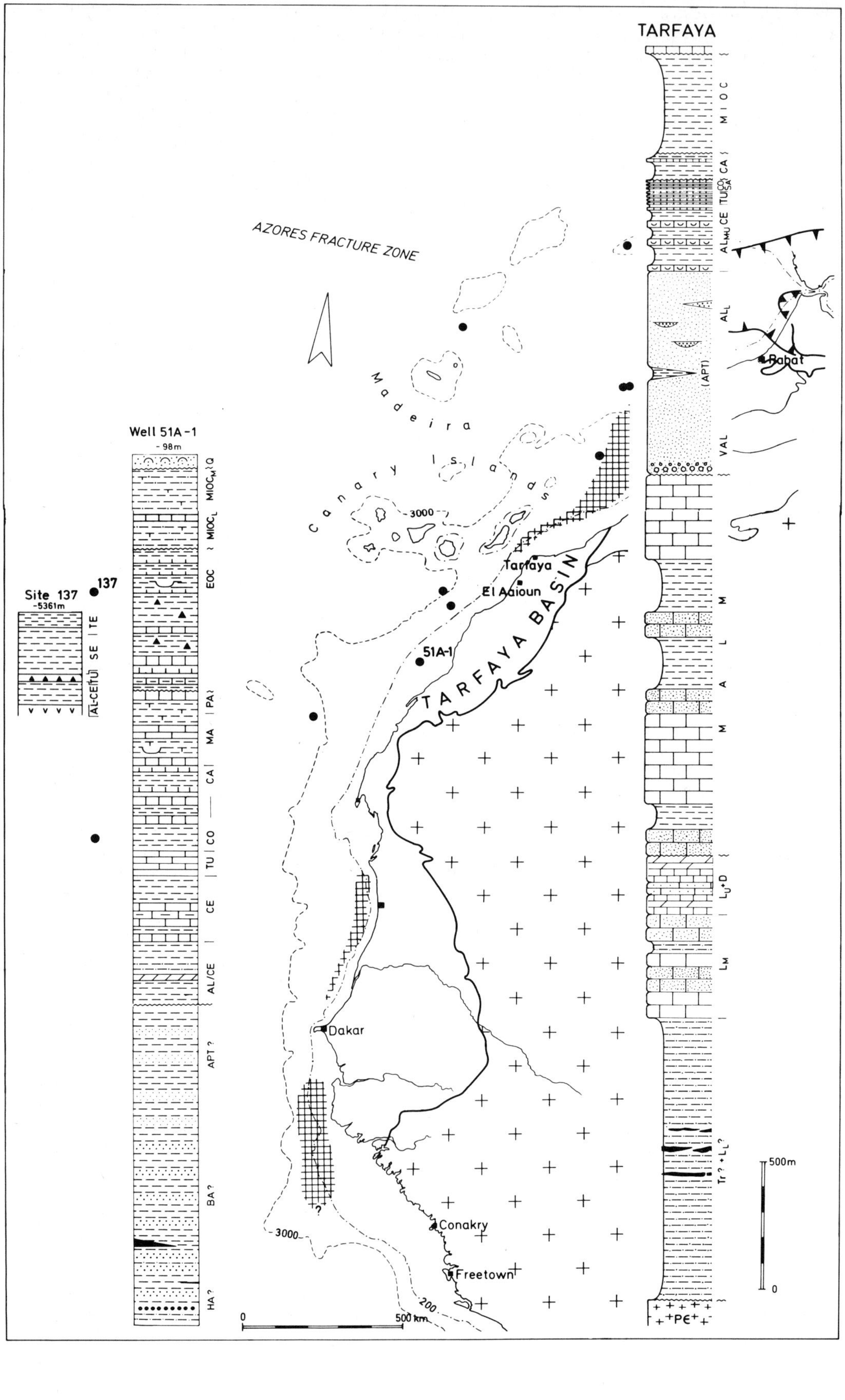
TARFAYA
AZORES FRACTURE ZONE
Madeira
Canary Islands
3000
Tarfaya
El Aaioun
TARFAYA BASIN
51A-1
Rabat
Dakar
Conakry
Freetown
3000
200
0
500 km
Well 51A-1
-98m
Site 137
-5361m
137
500m
0
MIOC
CA
TU CO SA
CE
AL
VAL
(APT)
M A L M
L U + D
L M
Tr ? + L L ?
PЄ
Q
EOC
PA
MA
AL/CE
APT ?
BA ?
HA ?
SE
TE

sediments are much thinner in the Georges Bank Basin than in the Scotian Basin (see Figs. 19, 22). The Logan Canyon equivalent in the COST G-2 well is developed in a more shaly facies and contains intercalated skeletal limestone beds, which reflect not only the position of the well in the distal part of the basin but also quite less availability of coarse-grained clastics in this basin during Middle Cretaceous. Except for 10 m of Paleocene marine shale, no Tertiary rocks were recovered by the wells. The Tertiary-Quaternary is less than 500 m thick in the COST G-2 well (Amato and Simonis, 1980) in comparison to more than 2 km of similar deposits at the outer shelf in the Scotian Basin, indicating slower subsidence of the Georges Bank Basin during late Cretaceous-Tertiary.

4.2.2. Baltimore Canyon Trough (Fig. 21)

The stratigraphic development of the Mesozoic-Cenozoic strata in this basin was tested by two COST wells: the COST B-2 well (Smith, Amato et al., 1976; Scholle, 1977) and COST B-3 well (Amato and Simonis, 1979). The COST B-2 well is located updip from the COST B-3 well, with the latter well being located on the present continental slope, in water depths of 818 m (Fig. 3).

Both COST wells bottomed in a Kimmeridgian clastic sequence (Fig. 21), comprised of interbedded sandstones, siltstones, shale, and coal. The shale in the updip part of the basin is variegated grey, brown and red, whereas in the more distal part of the basin it is grey. The other difference is that in the distal part of the basin thin beds of oolitic limestones are intercalated with the clastics. According to the lithologic composition the updip sequence is developed in a similar lithofacies, and it is synchronous to the Mohawk Formation as well, and the downdip section is correlatable with the MicMac Formation of the Scotian Basin. The facies development suggests that during the late Jurassic the alluvial plain in the Baltimore Canyon Trough was more than 120 km wide and extended almost up to the present shelf edge.

A 500 m thick zone of Tithonian shallow-water carbonates is intercalated with clastics in the MicMac Formation equivalent of the COST B-3 well. This carbonate wedge suggests that this area was located landward from a major offshore carbonate bank (Fig. 10) which is identifiable on an USGS seismic profile, line 2 (Grow et al., 1979). The seismic section indicates that the development of the carbonate bank in this region continued into the Early Cretaceous, with a 'reef' having developed during the terminal stage of the carbonate build up growth. The carbonate bank is equivalent to the Abenaki Formation in the Scotian Basin (Fig. 21). The lower Cretaceous of both Baltimore Canyon Trough COST wells is characterized by deposition of nonmarine and deltaic clastics. As might be predicted, the clastics in the distal part of the basin (COST B-3 well) contain thicker shaly intervals and more numerous limestone beds than occur in the updip well (Fig. 21).

A sharp lithologic boundary and an important hiatus separates mid-Cretaceous clastics from mainly clayey late Cenomanian to Tertiary strata. The latest Cretaceous and Tertiary of the Baltimore Canyon Trough shows less similarity with the coeval sedimentary sequences of the Scotian Basin than do the underlying Jurassic and lower Cretaceous sediments. In the Baltimore Canyon Trough a thick sandy unit occurring in the Coniacian-Santonian suggests a brief regression pulse in this area. Also, the upper Cretaceous chalk is not developed in this basin, but instead calcareous shale was deposited. An unconformity separates the Maastrichtian beds from the Eocene, comprised of intercalated calcareous clay, limestone and chalk, all deposited in a slope environment (Scholle, 1979). Evidence for a late Oligocene regression is present in both wells based on foraminiferal biota

Fig. 20. Onshore-offshore correlation of Tarfaya-Aaiun Basin with SPANSAH shelf well 51A-1 and DSDP Site 137

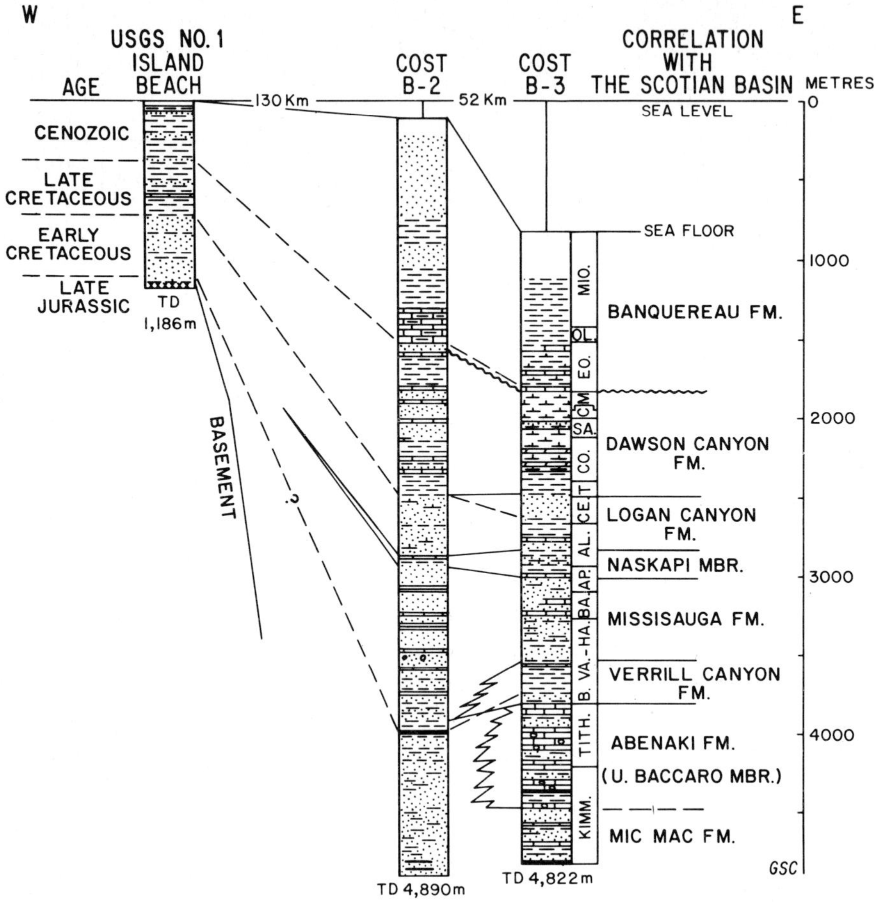

Fig. 21. Lithologic cross-section in the Baltimore Canyon Trough (for location see Fig. 5). Note the correlation of major lithologic units with the Scotian Basin

(Adinolfi and Jacobson, 1979), but it is not evident lithologically. This brief regression was followed by a rise in sea level and deposition of prograding wedges of clay on the shelf during Miocene time, and by alternating shale-sand sequences in the more marginal parts of the basin.

4.2.3. Tarfaya-Aaiun Basin (Fig. 20)

Stratigraphic information about this basin is mainly based on Martinis and Visintin, 1966; Querol, 1966; Choubert et al., 1967; Viotti, 1967; Ratschiller, 1971; Van Houten, 1977; Wiedmann et al., 1978a, 1978b. It is, however, supplemented by a certain number of onshore wells and the commercial shelf well SPANSAH 51A-1 (see also Ranke et al., this Vol.).

The Chebeika-1 well in the northern coastal part of the onshore basin penetrated 4200 m of Mesozoic sediments to bottom finally into the Precambrian basement. The lower part of the Mesozoic was represented by 1100 m of red silts, sandstones and clays, including a diabasic layer of considerable thickness in their upper portion (Fig. 20). This sequence may be treated as Chebeika Formation and can easily be compared with the above mentioned Triassic to early Liassic Argana Formation of the Atlas Basin; but a Permian age of their lower beds or of the basal sediments in the central basins cannot be denied. No evaporites have been recognized in the onshore well; but from the available seismic data (Hinz et al., this

Vol.) it is obvious that the central Moroccan diapiric province continues into the northern Tarfaya Basin and is here well established beneath the lower continental slope (Figs. 6, 20).

Another problem of the Chebeika-1 well is the separation of these Triassic-Liassic clastics from the disconformably following sandstones, sands and red clays of the Tantan Formation. This formation is generally considered to represent the Lower Cretaceous (?Valanginian to Lower Albian), but may as well include the continental development of the Jurassic (Fig. 20). In a distance of less than 100 km,the Puerto Cansado-1 well drilled more than 2000 m of Jurassic sediments without arriving at the base.

Further subdivision of Puerto Cansado Formation is difficult. The lower 400 m consist of sandstones and subordinated limestones of late Lower Jurassic and may be equivalent to the Amsittene Formation further north. The following 230 m of limestone, dolomites and subordinated sandstones may correspond to the middle Jurassic (and the Ameskhoud Formation? of the Essaouira-Agadir Basin). The Upper Jurassic (including the Berriasian) is represented by 1500 m of algal limestones, marly or detritic limestones, marls and sandstones, which are easy to compare with the northern Imouzzer Formation. An unconformity has been detected at the base of this sequence.

The Upper Jurassic part of the Puerto Cansado Formation formed the shelf platform. The paleoshelf edge is burried underneath the actual slope off Central Morocco. It is marked by a distinctive edge of a carbonate platform which seems to represent a reef or bioherm body of Tithonian-Berriasian age (Hinz et al., this Vol.).

Comparison with the Georges Bank is interesting. Here the oldest Mesozoic has not been penetrated. In the COST G-2 well the sequence starts with a considerable thickness of the early to middle Jurassic Iroquois Formation, represented by dolomites, limestones, and anhydrite. This sequence resembles the middle Jurassic carbonate member of the Puerto Cansado Formation. The equivalents of the late Jurassic Abenaki Formation include an important sequence of clastics of about Kimmeridgian age between shallow-water carbonates at the base and the top. Almost the same sequence is found in the upper section of the Puerto Cansado Formation.

The change from platform carbonate to terrigenous sediments is even more pronounced as it is in the northern basins. Even in the Tarfaya Basin this change is possibly related to Late Cimmerian tectonics, documented by unconformities at the base of the Tantan Formation (Martinis and Visintin, 1963). The latter is of variable lithology and thickness: in the El Amra-1 well 2300 m of conglomerates and sandstones of alluvial plains were drilled, resting unconformably on Precambrian basement; the Puerto Cansado-1 well penentrated nearly 1400 m of sands, sandstones and clays, here unconformably on Jurassic; this means that the nearly 3000 m thick sequence of the Tantan Formation drilled at Oued Chebeika (Martinis and Visintin, 1963; Fig. 20) in part may belong to the Jurassic. The Tantan Formation crops out in the northern part of the basin. Their age is determined by the preceding and following sediments dated to range from Valanginian to lower Albian. The late Aptian transgression of the northern basins is also to be recognized in the northernmost outlyers of the basin, where the Wealden-type Tantan deltaic sedimentation is interfingering with marine late Aptian carbonates (Choubert et al., 1967; von Rad and Arthur, 1979). The marine equivalents of the Tantan Formation were found below the present-day shelf (SPANSAH 51A-1 well) and at the uppermost continental rise at DSDP Site 397 (Einsele and von Rad, 1979; Wiedmann, 1979) and are at the same time the oldest sediments recovered by offshore drilling. The thick continental series of the onshore alluvial plains grade - on the present shelf - into a likewise thick series of Wealden-like deltaic sandstones and siltstones and, finally, on the continental rise into prodelta muds with sideritic layers. In the abyssal area of DSDP Sites 137 and 138, however, Albian-Cenomanian sequences were found to cover immediately the oceanic basement (Fig. 20).

Comparable widespread deltaic systems prograding over Jurassic

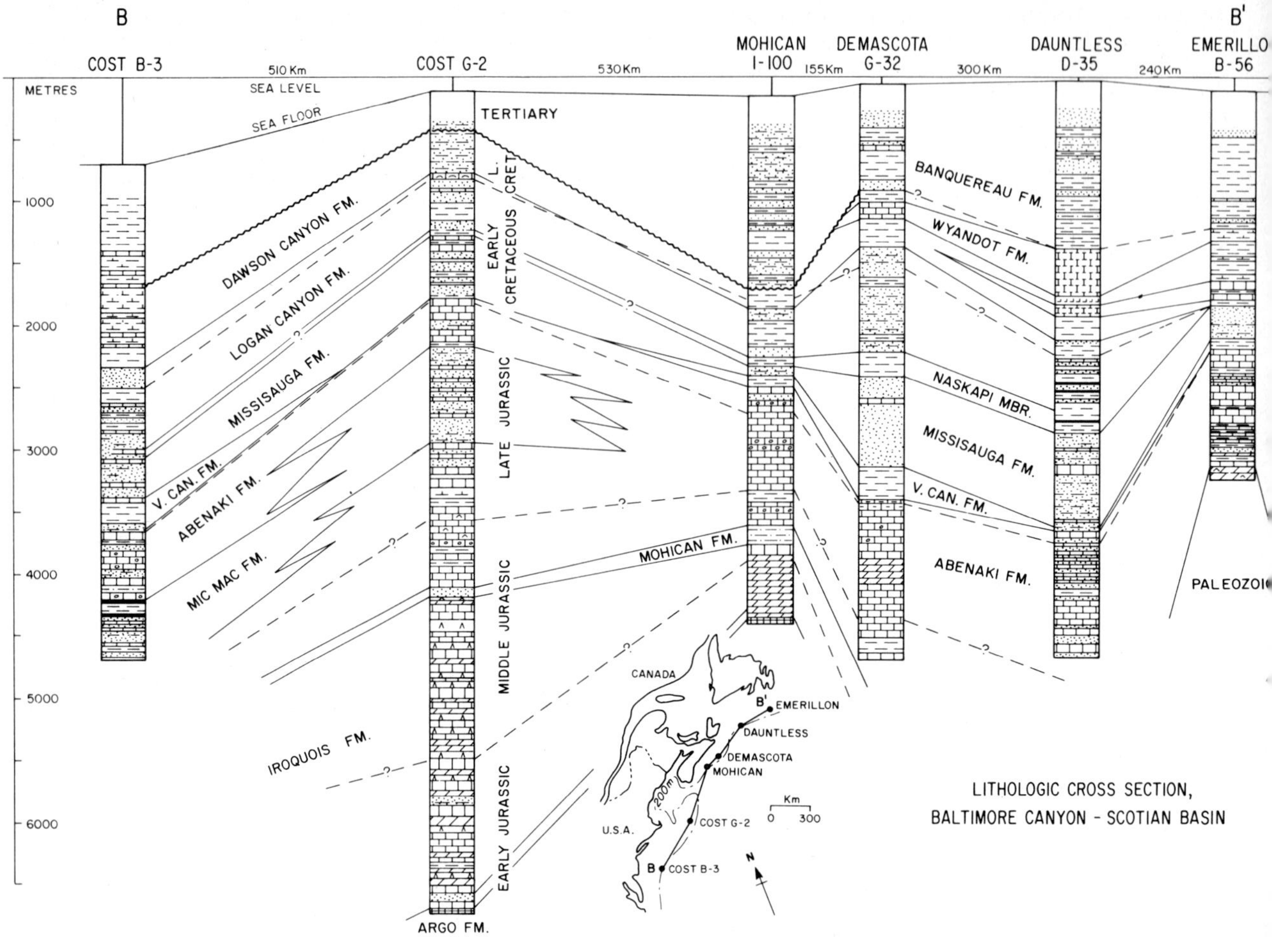

Fig. 22. Stratigraphic cross-section through wells located near the shelf edge showing extensive development of carbonate platforms during the Jurassic along the margin and the lateral similarity of the lithofacies development

carbonate platforms were described from the Scotian Basin (Jansa and Wade, 1975), and similar environmental conditions were recognized in the Georges Bank Basin (Schlee et al., 1977) and the Baltimore Canyon Trough (Scholle, 1977). The Verril Canyon Formation (Figs. 15, 19) of Scotian Basin and Baltimore Canyon Trough can be regarded as equivalent to the Hauterivian prodelta muds of Site 397.

In the onshore basin the mid-Cretaceous transgression is dated as middle Albian (Fig. 20). It may be associated with a disconformity (Martinis and Visintin, 1966; Querol, 1966), possibly related to the Austrian tectonics. Marine sedimentation continues with an erosional hiatus of Upper Santonian age (Subhercynian tectonics?) into the Campanian. This marine Aguidir Formation has, in the Tarfaya Basin, an average thickness of 650 m and shows a number of peculiarities. It should be subdivided into a number of members. The lower Tassagdelt Member consists of 250-600 m of greenish silty marls and shell beds ranging from late Albian to early Cenomanian. Benthics and cosmopolitan relationships prevail in the macro- and microfaunas. The Tassagdelt Member was deposited at the near-shore inner shelf. At the top of this sequence there is, as in the Essaouira-Agadir Basin, an abrupt change towards deposition of black shales, alternating with micritic to sparitic limestones and chert layers (Fig. 20). This Oued

Ouaâr Member which is, therefore, comparable with the Imi Mekki Formation north of Agadir has a particular fauna showing north temperate relationships and obvious predominance of pelagic and planctonic faunal elements. The environment is interpreted to correspond with the outer shelf. The faunal characteristics and the stratigraphically young age of these black shales (if compared with the Atlantic mid-Cretaceous black shale event) is to be related to coastal upwelling (Einsele and Wiedmann, 1975, this Vol.; Wiedmann, 1976). The thickness of the Oued Ouaâr Member does not exceed 100 m.

During late Coniacian and early Santonian time the organic carbon content of the marly facies decreases, while again shell beds occur and chert layers continue (Fig. 20). This Houiselgua Member attains an average thickness of 50 m, it has been deposited under mid- to inner-shelf conditions. The Campanian (and Maastrichtian) may be separated as Sebkha Tah Member. It unconformably follows the lower Santonian and is represented by a sequence of 100-180 m of silty marls with high contents of benthic faunas and subordinated oyster shell beds. In the SPANSAH shelf well the Maastrichtian is also documented in the facies of marly limestones and nanno ooze. In the Aaiun area, however, Paleocene phosphatic sands as well as Eocene phosphatic sands and clays with calcareous and chert layers unconformably follow the Campanian, whereas in the Tarfaya area the Campanian is truncated by the Miocene Aaiun Formation (Querol, 1966; Choubert et al., 1967).

The onshore disconformities can be related with Subhercynian (pre-Campanian), Laramid (pre-Eocene) and Pyrennean (post-Eocene) tectonic activities, the latter producing the main uplift of the onshore basin. Very similar conditions were recognized at the SPANSAH shelf well, where the marine Upper Cretaceous - having more than 1000 m of thickness - is separated by a distinct (Austrian) disconformity from the deltaic Lower Cretaceous and by an erosional Laramid hiatus from the incomplete Paleogene. The Eocene consists in the shelf area of marls and limestones with porcelanites. Further erosional unconformities separate the last marine incursion of Miocene age from the Eocene below and the continental Plio-Pleistocene above.

Pronounced differences, however, exist towards the slope, continental rise and the abyssal plain. In DSDP Site 369 on the slope, the marly and open-marine mid- and Upper Cretaceous is interrupted by a Cenomanian-Turonian hiatus pointing to the installation of a new deep-sea circulation pattern with contour currents at the continental slope, possibly related to the first coastal upwelling at the northwestern African margin (see Einsele and Wiedmann, this Vol.). Two more erosional gaps were recognized in the Paleogene, at which time also upwelling conditions were postulated off Northwest Africa. The mid-Eocene porcelanite bearing limestones are unconformably separated from the Maastrichtian below and the Oligocene-Miocene marly pelagic ooze with slump structures above (see frontcover figure). Strong contour currents did remove the complete Cretaceous-Paleogene sequence at Site 397, upper rise, from Barremian to Oligocene. Reworked pebbles at the base of Miocene debris flows permit recognition of pre-existing Aptian, Turonian and Oligocene marls, Albian limestones and Eocene porcelanites. The sedimentary sequence thins considerably towards the abyssal Sites 137 and 138 (Fig. 20), where late Albian was found to be the oldest sediment on oceanic crust. At Site 137, nearly 150 m of Albian-Cenomanian marls and Turonian porcelanite bearing beds were deposited above CCD, and about 250 m of Late Cretaceous-Tertiary zeolitic clays below CCD. These observations permit the conclusions that during the mid-Cretaceous the bathymetry and current system of the modern Atlantic possibly were formed.

On the Northeast American margin, the Georges Bank Basin Cretaceous exhibits intimate similarities to the Scotian Basin except for a pronounced thinning of the late Cretaceous Dawson Canyon Formation (Fig. 25). But, moreover, the Cretaceous of the basinal COST G-2 well (Fig. 19) has considerable relationships to the Tarfaya Basin. The Early Cretaceous Missisauga Formation starts (Valangian to Hauterivian) with a similar development of Wealden-type clastics, later (Barremian) followed by oolithic shallow-water limestones; the early Aptian Naskapi Formation is clastic again. The mid-Cretaceous ingression is similarly realized at the late Aptian. The equivalent Logan Canyon Formation which includes the Albian was deposited under inner shelf to nearshore conditions with the facies of marls, sandstones and shell beds. The upper Cretaceous Dawson Formation is reduced to 300 m of silty marls and ooze, and is of Cenomanian to Santonian age. The upper part of the Cretaceous is absent again, probably due to Early Tertiary erosion. The upper Cretaceous black shales are completely missing there. This asymmetry of Turonian black shale distribution in the North Atlantic is a further support of the assumption of local coastal upwelling off Morocco (Einsele and Wiedmann, 1975, this Vol.).

As mentioned above, the onshore Paleogene is characterized by pre- and post-Eocene uplift and erosion. Eocene (and Paleocene remnants) are only of local importance and restricted to the central part of the Aaiun Basin. Up to 1000 m of clastics were deposited onshore, grading upward and towards the shelf into calcareous and siliceous shales with cherts (Fig. 20). These sediments are known as well from the shelf as also from DSDP Sites 369 and 397 (reworked). The Paleogene is rapidly thinning towards the abyssal zeolitic clays of Site 137.

At the continental rise, Oligocene bottom currents cut into the basal slope, eroding the Cretaceous to Paleogene sequence. Oversteepening of the slope was apparently the cause of considerable Miocene mass flows, drilled at Site 397 (Arthur et al., 1979; von Rad, Ryan et al., 1979). At this time the depocenter shifted from its Mesozoic position below the shelf and slope to the upper rise. A last Neogene sea level rise led to the deposition of Miocene marls on the shelf and in the coastal area of the northern Tarfaya Basin (Aaiun Formation). During the late Miocene sedimentation rates increased again due to high carbonate and siliceous productivity. This environmental change definitely developed under upwelling conditions (Arthur et al., 1979). Terrigenous input was considerably reduced or eliminated from the sedimentary record of late Neogene.

As in the northern transect, also in the central transect of the North Atlantic the Tertiary evolution of the contingent basins became unrelated. In the Georges Bank Basin subsidence slowed down. In the COST G-2 well less than 500 m of clastic Tertiary sediments are unconformably resting on Upper Cretaceous (Fig. 19). In the Baltimore Canyon Trough (Fig. 21), however, more than 1500 m of Tertiary sediments were deposited. In this basin, calcareous and open marine Eocene is likewise separated by a (Laramid?) disconformity from the Maastrichtian below, the Oligocene is characterized by a regressive pulse and, finally, the Miocene by an increasing rise in sea level and the deposition of prograding wedges of clay on the shelf.

Laramid tectonics and the associated common uplift, the Oligocene regression and the final Miocene sea level rise are, thus, the very general Tertiary congruencies of these basins with the conjugate Tarfaya-Aaiun Basin.

4.3. Southern Transect

4.3.1. Blake Plateau Basin (Fig.23)

The Blake Plateau Basin is the most southerly basin discussed here. In this basin one COST well (COST GE-1, Amato and Bebout, 1978) was drilled on the shelf and two Deep Sea Drilling Project holes (Sites 390 and 392) were located seaward on the nose of the Blake Plateau (Benson, Sheridan et al., 1978; Fig. 3).

The COST GE-1 well bottomed in Devonian volcanics and phyllitic rocks (Simonis, 1979), which are overlain by lower Cretaceous clastic rocks comprising of nonmarine variegated reddish-brown and grey shale, fine- to coarse-grained sandstone, coal and anhydrite (Rhodehamel,1979; Fig. 23). The sequence is about 835m thick, and was dated as Barremian-Aptian, with the lower part of the sequence undated for lack of fossils, but probably Lower Cretaceous in age (Poag and Hall, 1979). Lack of calcareous microfossils and the lithology suggest continental deposition with an occasional influence of restricted-marine conditions, as

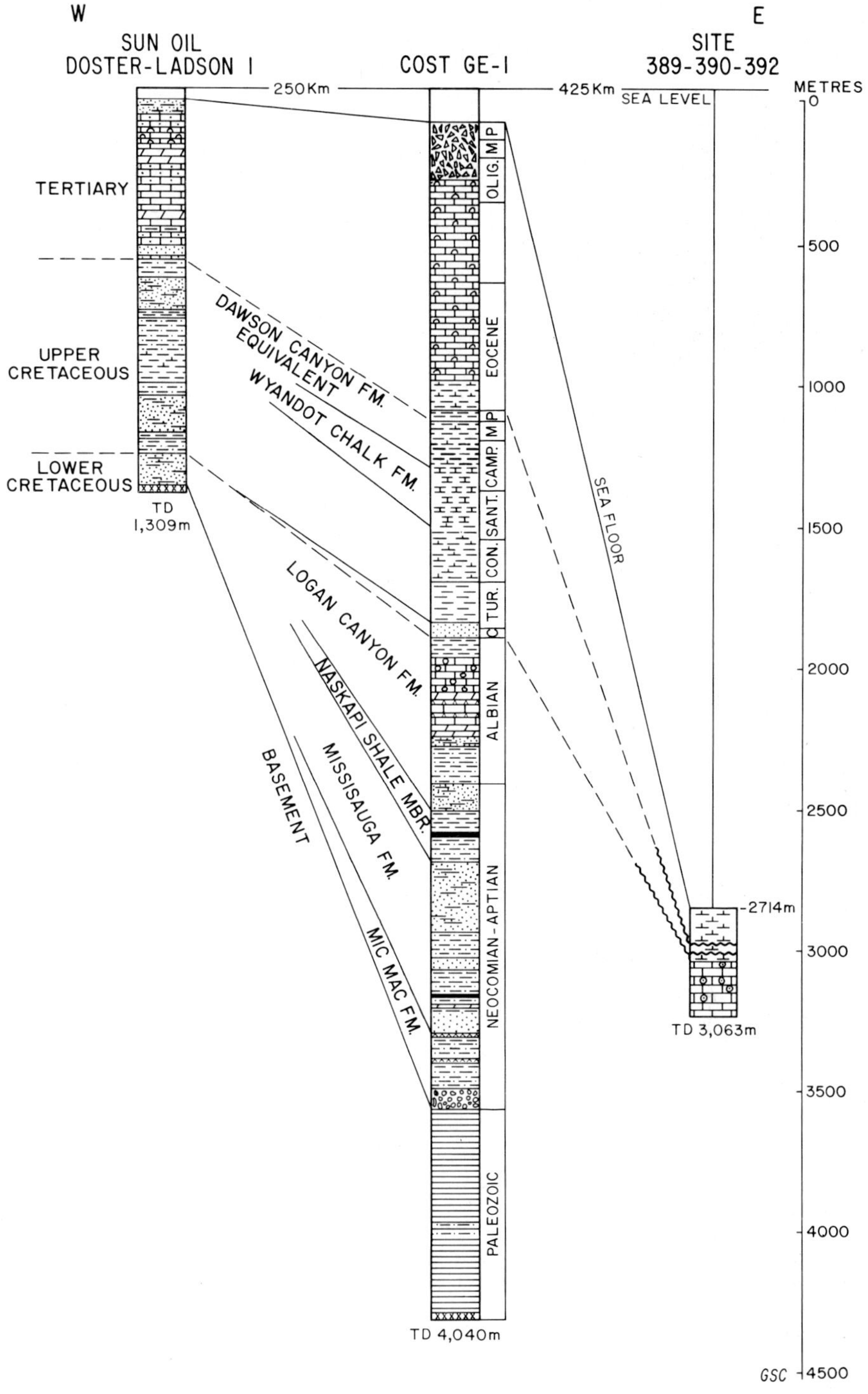

Fig. 23. Lithologic cross-section through the Blake Plateau Basin. SUN OIL Doster-Ladson-1 well was drilled on land. The interpreted correlation with the Scotian Basin shows that the major development even in such a distant basin remains similar (see Fig. 5 for location). The predominance of clastic deposits in the Cretaceous of the COST GE-1 well suggests, that more clastics are present in the basin than are indicated by the reflection seismic section (see Fig. 12)

indicated by the presence of the anhydrite. These continental clastics are synchronous and are, except for the presence of the anhydrite, similar to the Missisauga Formation on the Scotian Shelf.

There is no direct information regarding the nature of the Lower Cretaceous and older rocks seaward beneath the Blake Plateau. From seismic evidence (Shipley et al., 1978) it is very probable that most of the Blake Plateau is underlain by a Jurassic-Lower Cretaceous carbonate platform similar to the Bahama Banks (Fig. 12). Shallow-water carbonate bank deposits, represented by fenestral and oolitic limestones of Barremian and older age, were sampled on the Blake Nose (Benson, Sheridan et al., 1978). The oxidized and leached surface of the shallow carbonate bank suggests a period of subaerial exposure at the end of the Barremian, which concurs with the general development of the Upper Cretaceous on the eastern North American margin. This surface corresponds to the culmination of the Lower Cretaceous regression on the east coast margin.

The shallow-water carbonate bank deposits on the Blake Nose are overlain by variegated yellow, pale brown and pink upper Aptian-Albian nannofossil ooze (Benson, Sheridan et al., 1978), deposited during the upcoming transgression. The influence of this transgression is recognizable in the updip part of the basin (COST GE-1 well), where at this time clastic sediments became intercalated with beds of oolitic, pelleted and other shallow-water limestones (Scholle, 1979). This 778 m thick mid-Cretaceous sequence is separated by the unconformity from the Turonian-Maastrichtian deposits (Steinkraus, 1978) which are dominated by mudstones and clays. Light grey and white micrite and biomicrite were deposited during the Santonian-Campanian in the basin because of progressive deepening of the shelf environment during the late Cretaceous. The upper continental slope environment was established in the updip part of the basin during Santonian time (Poag and Hall, 1979). The chalk unit deposited here is synchronous and correlates lithologically with the Wyandot Formation in the Scotian Basin. At the outer part of the Blake Plateau Basin the mid-Albian variegated pelagic ooze is separated by the hiatus from the upper Campanian-Maastrichtian nannofossil ooze (Benson, Sheridan et al., 1978). The hiatus on the Blake Nose occurred at the same time, when upper slope depositional conditions were established in the 'nearshore' part of the basin. Therefore, we suggest here that the 'hiatus' is related to the nondeposition and development of the condensed sequence. Another minor hiatus at the Blake Nose separates upper Maastrichtian nannofossil ooze from the Paleocene clayey nannofossil ooze. Deposition of the latter ooze continued into Eocene, which is covered by a thin veneer of Quaternary foraminiferal sand and manganese nodules (Benson, Sheridan et al., 1978). The post Eocene hiatus is the result of pre-Miocene erosion, which most dramatically influenced deposition on the continental slope. The developing contour currents deeply scoured various parts of the continental slope of eastern North America (Jansa et al., 1979), which retreated locally for as much as 5 to 30 km (Schlee et al., 1979). In the onshore basin oolitic platform carbonates developed during most of the Paleogene, having an approximate thickness of 800 m.

4.3.2. Senegal Basin (Fig. 24)

The stratigraphic information about the Senegal Basin is mainly based on Spengler et al. (1966) (see also Wissmann, this Vol.). Also the sedimentary succession of Maio, Cape Verde Islands, has been reinvestigated (Stahlecker, 1935; Serralheiro, 1968, 1970; Robertson and Bernoulli, this Vol.).

The continental basement of the Mauritania-Senegal Basin is formed also by Hercynian structures truncated in the onshore part of the basin by marine Cretaceous. As we know from seismic evidence (Uchupi et al., 1976; Lehner and Ruiter, 1977: Fig. 12) the present continental shelf is built over a Jurassic to Lower Cretaceous carbonate platform, and finally buried by Upper Cretaceous to Tertiary shales. Moreover, the existence of an Early Jurassic (?) evaporite basin can be concluded from the

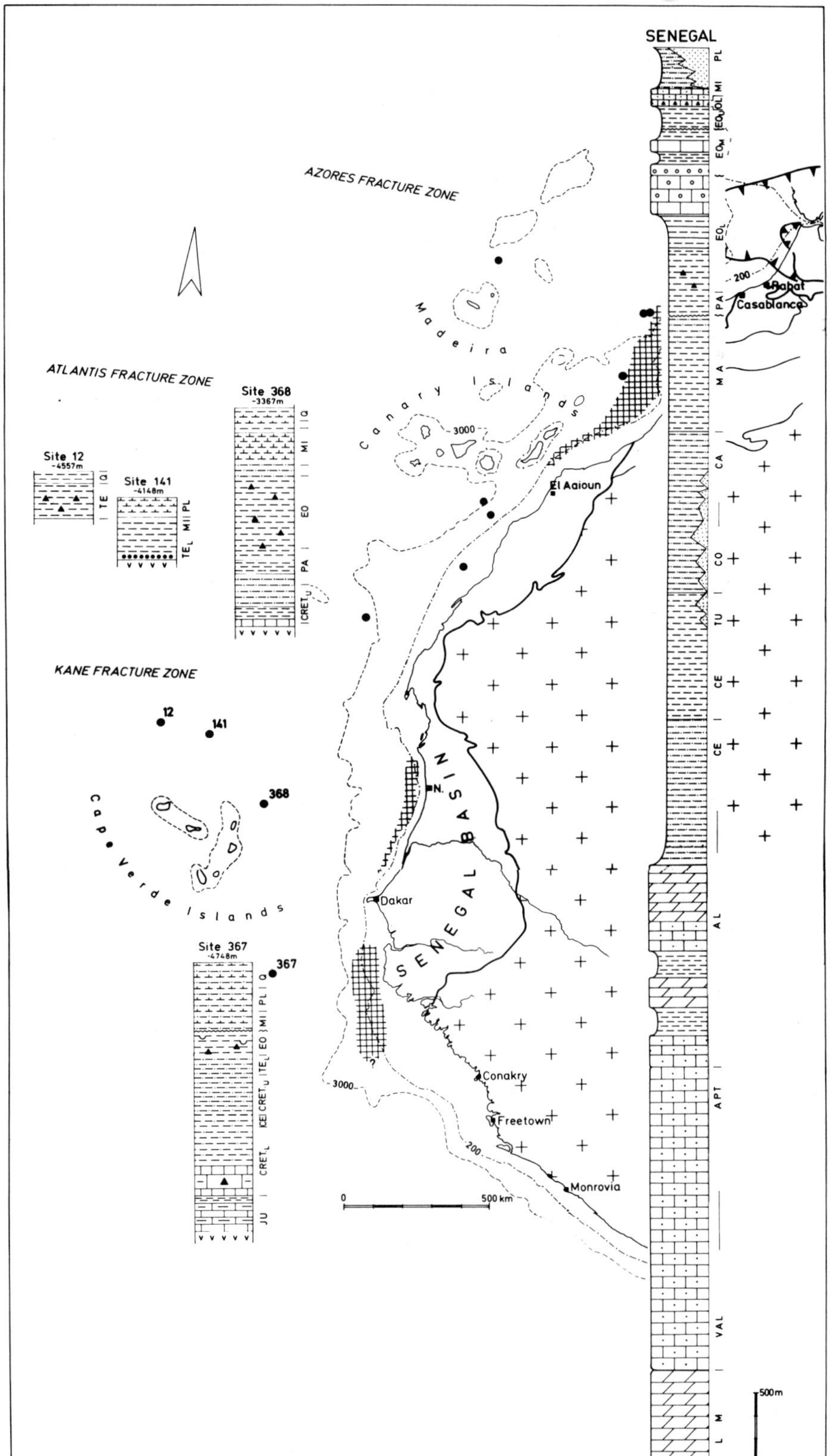

Fig. 24. Onshore-offshore correlation between Senegal onshore basin and DSDP Sites 367, 368, 141, and 12

diapiric zones in the area of the actual shelf and slope (Fig. 14; backcover figure).

In the Senegal onshore basin (Fig. 24) Upper Jurassic carbonates form the base of the stratigraphic column. The N'Diass well east of Dakar bottomed in Upper Jurassic dolomitic and oolitic limestones with a thickness of more than 760 m. The fossil content (algae, bryozoans, echinoderms) point to a shallow-water inner shelf environment. Terrigenous influx is completely absent. Parallel to the more northerly basins, the input of silicoclastics from the east sets in and rapidly increases from the Jurassic-Cretaceous boundary.

During Early Cretaceous (Valanginian to Aptian) 500-1300 m of calcarenites, quartzites, dolomitic limestones and clays were deposited in the western part of the basin, in contrast to red clays and sandstones in the east. The association of benthic foraminifers and ostracodes in the carbonate facies is again characteristic for inner shelf environments. Passing into the mid-Cretaceous, the same situation continues into the lower Albian, represented in the coastal area by calcarenites with orbitolinid assemblages. In the same area, during Late Albian to Early Cenomanian time, the sedimentary regime changes to the deposition of quartzites, sandstones and dolomites at the base passing into clays and silty clays with planktonic faunas in the upper portion of a series of 1500 m thickness. The maximum of subsidence is also found in Late Cenomanian and Turonian, comparable to the basins further north. About 450 to 700 m of clays with planktonic microfaunas were deposited at that time in the coastal area of the Senegal Basin.

The Late Cretaceous exhibits varying environments. In the Coniacian and Santonian the terrigenous influx increases again, while the Campanian represents - as in the northern basins - a further maximum of subsidence in accordance with the drastic appearance of planktonics. The trend to outer shelf environmental conditions continues into the Maastrichtian. In the coastal zone near Dakar, this stage is documented by a series of 100 to 2000 m of clays and silty clays with *Afrobolivina* appearing at first and a mainly planktonic microfauna later on. Towards the east of the basin the marine coastal Cretaceous passes into coarse red sands, sandstones and clays, named the "Sables Aquifères". At the present-day coast line, the Maastrichtian is truncated by pre-Tertiary erosion due to a general regression at the end of the Cretaceous.

Throughout the basin the Paleogene disconformably covers the Cretaceous (Fig. 24). At the actual coast line the Paleocene is developed as a series of *Globorotalia* bearing marls ("Marnes des Madeleines") with varying thicknesses of 50 to 300 m. Towards the central and eastern portion of the basin these marls become replaced by 50 to 100 m of limestones and marls, mostly mollusc shell beds. The lower Eocene, however, consists of bright marls and clays with cherts, phosphatic and glauconitic layers showing a representative fauna of planktonics at first, continuing into marls with ostracodes and limestones towards the top of the almost 500 m thick sequence. It is replaced by red sandstones and white kaolinitic clays towards the east, and seems to be separated from the underlying Paleocene by an erosional gap. The middle Eocene is represented by approximately 200 m of clays and limestones with pelagic foraminifers in the west or by 100 m of *Nummulites* bearing limestones, again with cherts and phosphatic layers in the east. It may be separated as well by an unconformity from the Lower Eocene. It seems to be eroded at the actual coastline, since Upper Eocene immediately follows the lowermost Eocene in the well Dakar-2. Here, in the Dakar area, during Upper Eocene time about 100 m of marls and clays with *Uvigerina* and *Siphonodosaria* were deposited ("Argiles de Yoff"); towards the east they grade into white and black siltic clays, including some lignitic levels. In the innermost part of the basin, however, continental conditions were installed and continued through the Neogene. Oligocene is represented by limestones with *Lepidocyclina* and is of reduced thickness in the west, while it is unknown in the east. Here, in the furthermost east, 150 m of continental and marine sands and multicoloured clays were deposited during the Miocene and possibly also during the Pliocene (the so-called "Continental Terminal"). It is replaced towards the west by marine limestones and clays with Heterosteginas, alternating with sands in some places, including lignitic layers. This sequence cannot be traced towards the actual coastline where the Dakar volcanics extruded during the Miocene. After a short depositional break at Pliocene time during which lateritic decomposition took place, similar conditions became reinstalled during the Pleistocene: volcanics near Dakar, marine siltic clays and limestones at the lower Senegal River area, and continental dunes in the east.

The observation in the Senegal Basin can be supplemented by the sedimentary succession of Maio, Cape Verde Islands. This succession is not too well known despite a series of updated stratigraphic papers. The most important dating of the Maio succession is due to Stahlecker (1935), followed by Serralheiro (1968, 1970), Grunau et al. (1975) and Robertson and Bernoulli (this Vol.).

The basal portion of this succession is formed by the "Shale-limestone eruptive complex" of Serralheiro (1970) consisting of an alternation of shales, greywackes and chert bearing limestones with alkaline volcanics. Zeolitic shales point to deposition in greater water-depth. A middle Jurassic to Oxfordian age of this complex can be assumed. It grades into the following "Limestone with chert unit" (Serralheiro, 1970), which can, at least, be partially correlated with Stahlecker's horizons 1-10. Pillow lavas and tuff layers are also intercalated in this unit. Rigassi (1974) described a variety of radiolarians, calpionellids and saccocomas from this unit, attributed to Kimmeridgian to Lower Neocomian ages while Stahlecker's ammonite faunas are corresponding with a Hauterivian to Aptian age.

Above follows a series of fine-laminated shales, light marls and marly to siliceous limestones with few intercalations of basaltic lavas. This unit may correlate with Stahlecker's horizons 10-14 and, thus, correspond with an Aptian to Albian (?) age. So far, no common agreement has yet been achieved on the continuation of the section. Grunau et al. (1975) described an upper complex of greywackes, siliceous shales, tuffs and pillow lavas attributed to a mid-Cretaceous to Eocene age and to deep-water environments of deposition. Serralheiro (1970), however, separated a further unit of light-coloured massive limestones with chert layers, similar to the Late Jurassic-Early Cretaceous sequence, but above the "Shale unit". This would point to a possibly Upper Cretaceous age of this unit which cannot be regarded as substantiated. Finally, a Miocene age is attributed to the "Pedro Vaz Conglomerates" (Serralheiro, 1970) being mainly derived from igneous rocks, but laterally passing into sands, shales and clays with foraminifers.

An important observation has been made by de Paepe et al. (1974) when recognizing that the Jurassic pillow lavas which are conformably followed by the oldest sediments of Middle Jurassic or younger ages, have tholeiitic patterns characteristic for mid-ocean ridges.

Important additional information was made available by DSDP Leg 41 (Lancelot, Seibold et al., 1978). Site 367 (Fig. 24) was placed in the Cape Verde Basin within the Jurassic magnetic quiet zone to provide information about the initial stages of sedimentation. It bottomed in oceanic crust overlain by Oxfordian-Kimmeridgian red argillaceous limestones. The Tithonian to Hauterivian sequence can be characterized by an alternation of Aptychi bearing cherty limestones and gray marls. Variegated Barremian claystones grade upwards into Aptian to Cenomanian black shales which are related with the North Atlantic black shale event (Fischer and Arthur, 1977). Multicoloured silty clays continuously grade from Late Cretaceous into Paleocene and are overlain by Eocene zeolitic clays with chert and porcelanite. Eocene is followed by Late Miocene through Pleistocene nanno marls. Thus, the widespread Oligocene hiatus can be traced into the Cape Verde Basin.

These data are to some extent supplemented by Site 368 (Fig. 24) which was placed 750 km further north on the Cape Verde Rise. This site bottomed in the mid-Cretaceous black shales with the interstratification of Miocene diabasic sills, probably related to the peak of volcanic activity on the Cape Verde Islands. Upper Cretaceous and Paleocene (?) are represented by silty clays and claystones. These terrigenous sediments were deposited below CCD and partially by turbidity currents. The terrigenous input continues on into the thick sequence of silty clays with chert of Eocene age, while pelagic carbonates are concentrated in the Miocene to Pleistocene sequence of nanno marls and oozes. Uplift of the Cape Verde Rise is comparatively young, possibly of Miocene age (Lancelot, Seibold et al., 1978).

Not too much information can be added from Sites 12 and 141 which were drilled 200 km north of the Cape Verde Islands and located in a zone of abyssal hills. Both sites (Fig. 24) exhibit an extremely reduced Tertiary sequence which seems to be complete and consists of variegated clays at the base and calcareous ooze in the upper part. Site 141 was drilled on a "diapiric" structure, but no salt has been discovered and, thus, the piercement structure was interpreted as igneous intrusion (Pimm and Hayes, 1972).

When comparing the stratigraphic sections of the two conjugate basins of the southern transect of the North Atlantic (Figs. 23, 24) as well as the interpreted seismic profiles (Figs. 12-14), the congruency of the evolution of the contingent basins is even more obvious. It remains, however, uncertain, whether an early Triassic graben stage existed or not. Isolated continental Triassic occurrences are known from the proximal portion of the Blake Plateau (Fig. 12), where they are overlaying metamorphic basement. Salt is known to occur below the platform carbonates at the Mauritania-Senegal shelf and slope and is piercing the Meso-Cenozoic sedimentary cover (Fig.14). This salt could, however, be younger than Triassic. The first marine sediments occurring simultaneously in the conjugate basins are platform carbonates of Upper Jurassic age. Within the coastal basins, these carbonates generally truncate the Paleozoic basement, in the oceanic basin they are placed on oceanic crust. In both basins the Cretaceous is integrated in the development of these carbonate platforms. Towards the inner border of the coastal basins, these limestones grade into thick clastic sequences, similar to the general facies change at the Jurassic-Cretaceous boundary in the basins further north. At this time, deposition of mid-Cretaceous black shales is not restricted to the northwestern African margin, but is similarly developed at the Blake Bahama Ridge (Site 101A) and the Hatteras Abyssal Plain. Also the Late Cretaceous claystone unit can be traced through both basins as well as the facies of variegated zeolitic clays of the top Cretaceous and Paleogene, in the latter associated with porcelanites and chert. Strong Miocene volcanism is restricted to the Northwest African margin.

At the same time there are obvious relationships to the basins further north, especially on the northeastern American margin. These relationships permit the compilation of a synthetic stratigraphic section (Fig. 25) at least of the North American Atlantic basins. Main features of common evolution of Blake Plateau and Senegal Basins are, thus, the comparatively young age of these basins due to the progressive southward extension of the North Atlantic, and the prevalence of stable platform conditions during most of their Mesozoic history.

5. Summary on the Evolution of the North Atlantic Conjugate Margins

5.1 Fracture Zones, Proto-Atlantic

In summary, the sedimentary fill of the Mesozoic-Cenozoic marginal basins on the eastern North American and the northwestern African continental margin shows a very similar depositional history, especially during their Triassic, Jurassic and Cretaceous evolution. This is thought to reflect the uniform subsidence along these margins as well as the influence of global and regional sea level changes and the regional character of tectonics, all documented in the sedimentary and fossil record.
The common origin of the presently separated basins on both sides of the North Atlantic has been postulated by the continental fit. The new continental plate reconstruction bas-

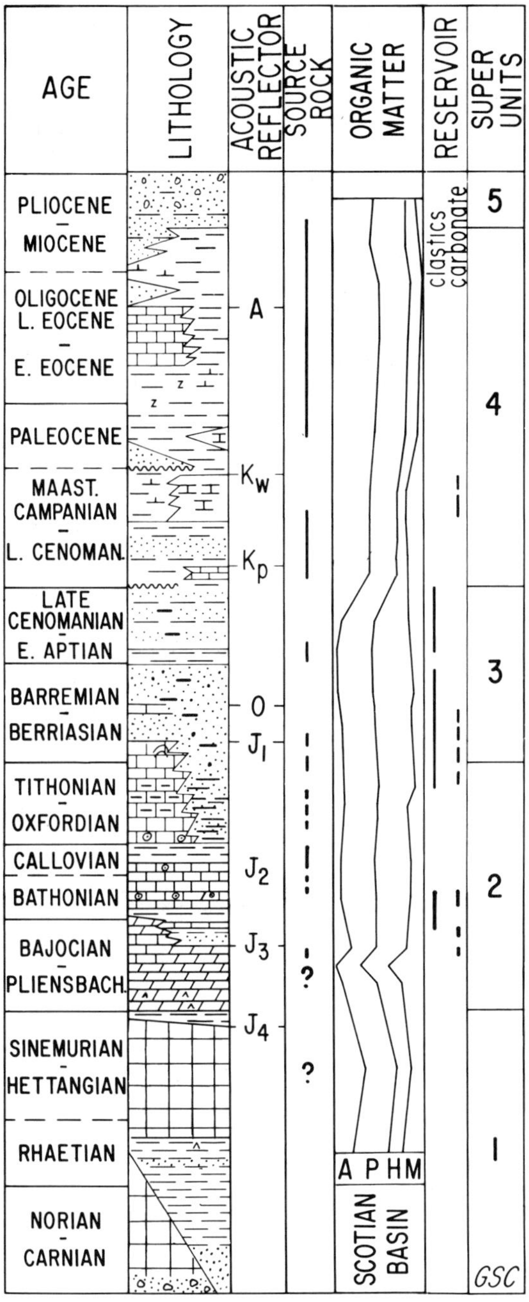

Fig. 25. Idealized stratigraphic column for the eastern North American shelf, and indicating the position of major acoustic reflectors and boundaries of lithostratigraphic super-units. Reproduced are the main parameters for the evaluation of hydrocarbon potential of the margin, such as stratigraphic position of the source rocks, reservoir rocks and organic matter types: A - amorphous, P - herbaceous, H - woody, M - coaly (after Bujak et al., 1977)

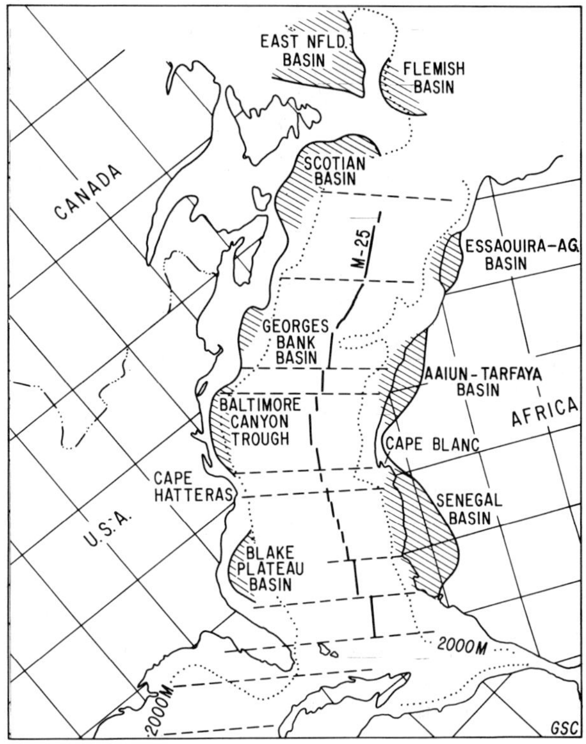

Fig. 26. Reconstructed position of the conjugate North Atlantic margins at the time of magnetic lineation M-25 (approximately Oxfordian-Kimmeridgian boundary). Reconstruction is based on identification of major transform faults in the North Atlantic (Schouten and Klitgord, 1980). Note that major Mesozoic-Cenozoic sedimentary basins on one margin are situated against basement highs on the opposite margin, except the southern transect

ed on tracing major fracture zones across the North Atlantic provided further base for the comparison of the conjugate sedimentary basins (Fig. 26). The reconstruction shows that the marginal basins on one continent are situated against basement promontories and viceversa. An exception is the southern transect. But scarcity of geological data in this region does not allow a conclusive solution of this discrepancy. The reconstruction demonstrates that each of the marginal sedimentary basins should show certain uniqueness of its development in a general context of similarity in the basin development as related to the North Atlantic basin evolution.

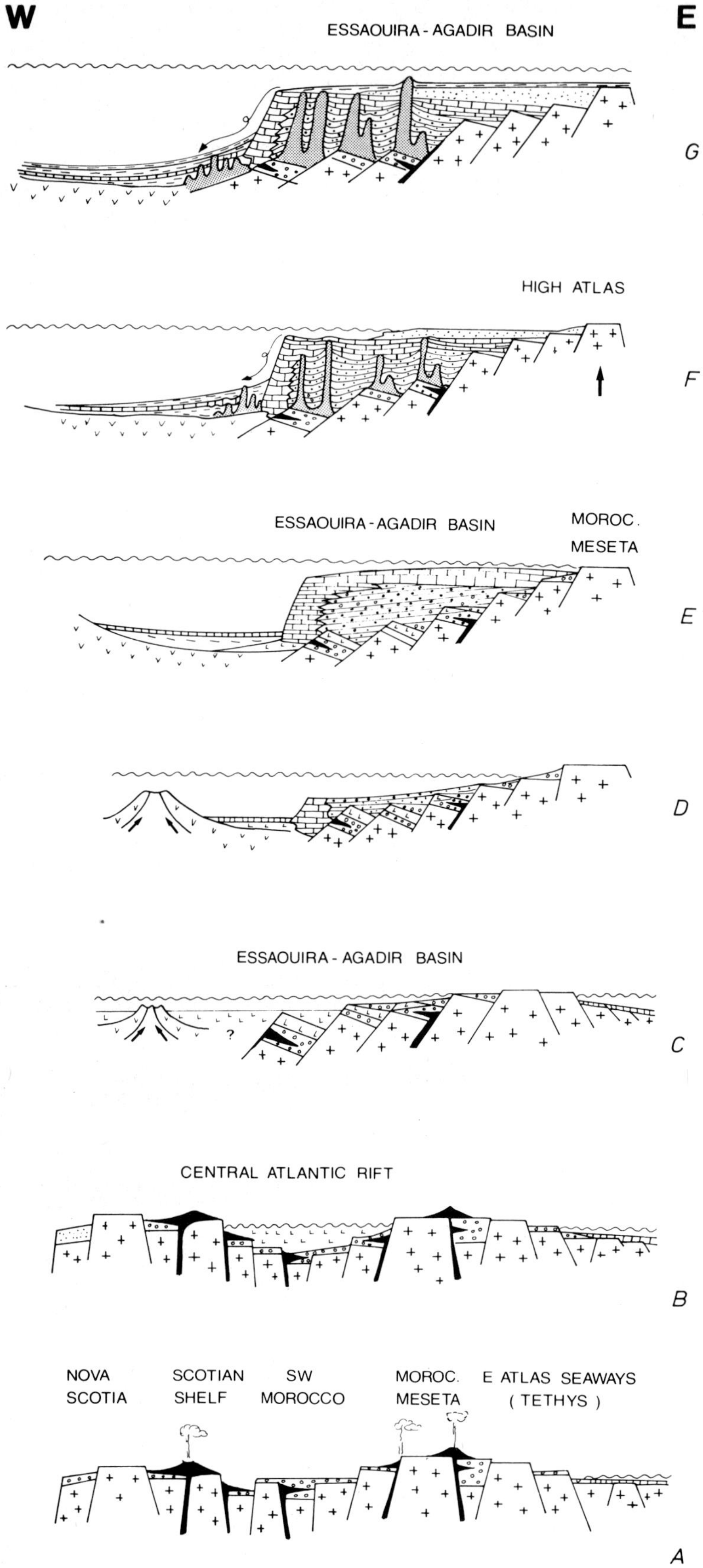

Fig. 27. Evolution of the passive continental margin at the Essaouira-Agadir Basin (modified after Lancelot and Winterer, 1980).
A Late Triassic graben stage; B Early Liassic early rifting; C Early Dogger early drifting; D Late Dogger shelf initiation; E Early Malm carbonate platform; F Early Cretaceous platform drowning and delta plains; G Mid-Cretaceous maximum transgression

One pronounced difference between both regions is the position of the North American basins on the outer shelf, while parts of the Northwest African basins are accessible onshore (Fig.26). This underlines the importance of the northwestern African margin. From a combination of onshore based studies and DSDP offshore activities the steps of marginal evolution can actually be depicted. They are schematically reproduced on Fig.27 which is for the major part adopted from Lancelot and Winterer (1980).

Not too much is known, however, about the earliest stage of graben evolution. It is interesting to note, that this first step of evolution towards the future North Atlantic parallels previous Paleozoic structures, i.e. the North American Acadian-Alleghenian orogen and the Moroccan Hercynian fold belt (proto-Atlantic).

5.2 Permo-Triassic

In comparing the stratigraphic columns of the Essaouira-Agadir and the Scotian Basins (Figs.17, 18), it becomes obvious that graben evolution and deposition of thick continental redbeds started at the eastern margin of the rift valley during the Permian (Duffaud et al., 1966; Uchupi et al., 1976). After an episode of tectonic activity at the Permian-Triassic boundary, the graben stage continued into the Triassic with the accumulation of continental redbeds. The late rifting stage, common to both, the northeastern American and the northwestern African margin, can be placed within the Late Triassic (Fig. 27 A) and can be related to Early Cimmerian tectonic activity. This is the time of beginning extrusion of basaltic and doleritic sills, which continued into the Early Jurassic and is followed by the deposition of marine evaporites (Fig. 27 B). At this time, moreover, the central rift graben shifted towards the west.

The extension of salt deposits is visible from Figs. 1 and 6. Salt has recently been drilled at the Mazagan Plateau (Site 546; Hinz, Winterer et al., this Vol.), but is proved by seismic data to extend to the north below the Rif-Betic olistostromes and to the south into the Tarfaya shelf (Hinz et al., this Vol.; see backcover figure). It reappears again at the Mauritanian slope (Wissmann, this Vol.), but it is unknown whether the salt is younger as the Tarfaya salt or of the same age. In the northwestern Atlantic salt deposits were penetrated on the Grand Bank, Scotian Basin and Georges Bank Basin, and seismic data indicate that these evaporites may extend to the Baltimore Canyon Trough (Grow et al., 1979) and even as far south as to the Carolina Trough, where a small diapiric structure was discovered in about 2000 m of water depth (Dillon et al., 1979).

5.3 Jurassic

As the margin subsided, the Early Jurassic marine sediments onlapped either on the Paleozoic basement or across the early rift unconformity which near the hingeline was cut into Triassic deposits. Carbonate rocks with minor anhydrite were deposited in a semi-restricted environment with broad tidal flats on the subsiding margin. Clastics were accumulated only around the margins of the developing sedimentary basins.

During Late Liassic the sedimentary environment on the margins deepened as a result of progressing transgressions and became open-marine, but probably not deeper than 100-200 m. Comparatively stable marine conditions were installed and are documented by the coeval deposition of reef limestones at the bases of the Amsittene Formation in the Essaouira-Agadir Basin and of the Iroquois Formation in the Scotian Basin.

With the late Early Jurassic - early Middle Jurassic initial continental drift and formation of the incipient oceanic crust began in the central North Atlantic (Jansa and Wade, 1975; Van Houten, 1977; Lancelot and Winterer, 1980; Fig. 27 C). Tectonic activity associated with this event can be concluded by a widespread mid-Jurassic regression as documented by depositional gaps and a general increase in clastics in the Moroccan Ameskhoud and the Scotian Mohican Formations. The presumed superposition of evaporitic sediments on oceanic crust at the Moroccan shelf (Lancelot and Winterer, 1980; Fig.27 C) was related to this Middle Jurassic regression. This theory was disproved by Recent drilling at the deep continental margin off the Mazagan Plateau during Leg 79 (Hinz, Winterer et al., this Vol.), which indicates that the evaporites are older than Pliensbachian and were deposited on continental crust.

The Middle Jurassic wedge of clastics formed at the same time the initiation of a marginal shelf on the subsiding marginal fault blocks (Fig. 27 D). During the Late Jurassic subsidence continued. Construction of a carbonate shelf is the main feature of the northwestern African Upper Jurassic (Fig. 27 E). This continuous and widespread evolution is interrupted by the "Atlantic" transgression of Callovian age which is the first one common to all basins surrounding the North Atlantic and can, thus, be regarded as "maturation" event of the young ocean (Lancelot and Winterer, 1980). In the deep sea (Site 416) this may correspond with a short episode of pelagic sedimentation which is to correlate with the "blue" reflector (Lancelot and Winterer, 1980).

As has been proved by Sites 370/416, the deposition of allochthonous turbiditic series at the foot of the Moroccan slope started at least during Late Jurassic if not earlier. Terrigenous clastics played an increasingly more important role by slowly filling the basins, building the coastal plains seaward, and displacing the carbonate platforms and banks in offshore direction (Fig. 21). Late Jurassic coastal plain prograded most in the Baltimore Canyon area, where the wedge is located near the present shelf edge.

5.4 Cretaceous

The regression at the Jurassic-Cretaceous boundary and coeval tectonic activity are the reasons for a drastic change in sedimentation on both sides of the North Atlantic. In the Essaouira-Agadir Basin this is the change from the final episode of platform carbonates of the Imouzzer Formation (with reef limestones, evaporites and sabkha sediments) to silicoclastic deltaic and inner shelf sediments of the Early Cretaceous Aghroud Formation. The boundary level itself is marked by a distinct hardground (Wiedmann et al., 1978a). Tectonics are not only restricted to first documented salt motion (Lancelot and Winterer, 1980), but viceversa diapirism is most likely initiated by Late Cimmerian tectonic movements, leading to the first uplift of the Atlas mountain chain (Fig.27 F). Thick turbiditic sequences were accumulated at the continental foot during Early Cretaceous and led to stimulating discussions on platform drowning and source areas or passage ways by Schlager (1980) and Price (1980). On the shelf and in the onshore basins deltaic Wealden-type sediments were deposited during the complete Lower Cretaceous or at least parts of it. This sedimentation is especially well developed in the Tarfaya-Aaiun Basin (Ranke et al., this Vol.; Einsele and von Rad, 1979) and in the Senegal Basin (Castelain, 1966). In the onshore Essaouira-Agadir Basin the Early Cretaceous is a time of competition between marine incursions and continental influx leading to an extreme degree of coastline variation. Transgressive trends of the Valanginian, Barremian and Upper Aptian are separated by redbed incursions during Hauterivian and Lower Aptian times (Wiedmann et al., 1978a; this Vol.).

A similar but more continuous environmental change can be observed at the boundary level on the Canadian margin. Here, the Baccaro platform carbonates of Oxfordian-Kimmeridgian age grade into the alternation of Abenaki platform and biohermal limestones, shales and sandstones, ranging from Tithonian to Valanginian and marking in the Scotian Basin the regressive peak. Minor tectonic activity seems to be observable at least at the shelf edge area during earliest Cretaceous (Jansa et al., 1980). Locally the top of the carbonate sequence is marked by the development of hardgrounds similar to the Essaouira-Agadir Basin.

In the more southerly troughs the Early Cretaceous covered the carbonate banks. Only in the distal parts of the basins and on the intrabasinal highs (Mazagan Plateau) the carbonate deposition continued into Early or even Late Cretaceous. The Cretaceous delta plains were built far into the basins during this time, but the deposition of clastics was interrupted by minor ingressions during Hauterivian, Aptian and ?Albian times.

The following transgressive sedimentary cycle started in the middle Albian and ended with the maximum transgression of Turonian age. The Aptian to Cenomanian Addouz Formation starts with a condensed layer of middle/upper Aptian age, probably as a consequence of eustatic oscillations. The mainly pelitic mid-shelf sedimentation of the Addouz Formation is interrupted and finally closed up by thick oyster shell beds or even oyster bioherms of latest Albian and Cenomanian age. Coeval with these high-energy sediments is the marine ingression into the Tarfaya-Aaiun Basin, leading to similar environmental conditions.

It needs to be mentioned that no black shale deposition can be recognized at this time in the Essaouira-Agadir Basin. According to Lancelot and Winterer (1980) marine organic matter has been diluted as a consequence of considerable terrigenous input.

The Turonian is the time of maximum transgression in the Essaouira-Agadir Basin (Fig.27 G). The depositional environment has completely changed; the Imi Mekki Formation consists of limestones and chert layers alternating with bituminous marls. These marls can be regarded as northern extension of the Tarfaya-Aaiun black shales which have been related to a late Cenomanian-early Coniacian upwelling event (Einsele and Wiedmann, 1975; this Vol.). The Turonian transgression covers large areas of the North African plate, opens the trans-Saharan seaway towards the east and south, and even permits recognition of the chert-black shale event in the Rif-Betic Turonian ("phtanitic" event, Wiedmann et al., in prep.). It is remarkable that clastic input at this time is scarce or absent.

The overlying Taghazout Formation, however, indicates the immediately following regression which is possibly due to first tectonic activity foreshadowing the final uplift of the Atlas mountain chain. During this (Coniacian-Santonian) time, in the Essaouira-Agadir Basin *Astarte* coquinas and cross bedded sandstones were deposited suggesting a near-shore environment; repeated depositional breaks are documented by a sequence of hardgrounds. Thereafter a last period of deepening leads to sedimentation of marly and chalky sediments of Campanian-Maastrichtian age (Aouine Formation). This period coincides with the Campanian global sea level rise stated by Sliter (1976).

Not too much is known about the offshore Upper Cretaceous. At Sites 370/416 Paleocene was found to overly poorly known Albian-Cenomanian nanno marls. This Upper Cretaceous hiatus is most probably related to erosion by bottom currents and may have happened - in contrast to the hiatus in the western North Atlantic - in a series of erosive phases ranging from Upper Cretaceous to Early Tertiary (Lancelot and Winterer, 1980; von Rad and Wissmann, this Vol.). Site 415, placed in the seaward prolongation of the Agadir Canyon is likewise unreliable to contribute to the knowledge of the offshore Upper Cretaceous, since the thick sequence of Cenomanian marl- and mudstones was found to be (1) tectonically repeated, (2) overlain by poorly dated "Albian-Coniacian" nanno marls and (3) the latter again separated from the Paleocene by a long-ranging hiatus (Lancelot, Winterer et al., 1980).

Comparison with the Upper Cretaceous of Site 137, situated on an abyssal hill about 1000 km west of Cape Blanc, may be of interest (Hayes, Pimm et al., 1972; Wiedmann et al., 1978a; this Vol.). Here, nanno marls of Albian to Cenomanian age were deposited directly on oceanic crust and well above the CCD. Upper Cenomanian and Turonian nanno marls and cherts may have been deposited near the CCD, while zeolitic clays of Upper Turonian and Senonian ages with reduced thicknesses and an impoverished microfauna point to a rapidly increasing subsidence to bathyal depth and sedimentation below CCD.

At this time, congruency of evolution at both North Atlantic margins is less obvious, but still to be observed. In most of the northeastern American basins a significant change in sediment composition occurred during the late Cenomanian. Within the shaly background sedimentation of the Dawson Canyon Upper Cretaceous in the Scotian Basin, the two global sea level highstands can be recognized again, i.e. by limestone deposition in the Turonian and by chalk sedimentation of late Santonian to Maastrichtian age. But none of these pulses is recognizable in all basins bordering the Northeast American continent. Also in the western North Atlantic the deeper basins were effected by bottom current erosion, here presumably related to one single event between late Eocene-earliest Miocene (Jansa et al., 1971; Tucholke and Vogt, 1979). This means that bottom current systems on both sides of the North Atlantic were governed by differing oceanographic regimes, as they are today.

5.5 Cenozoic

The worldwide regression at the end of the Cretaceous is not well marked in the Essaouira-Agadir Basin, where pelagic to hemipelagic sedimentation continued into the Paleocene-Eocene Imi n'Tanout Formation. Only in this basin subsidence continued, while in the Tarfaya-Aaiun Basin and on the Moroccan Meseta near-shore conditions were again established leading to the accumulation of phosphates, possibly influenced by coastal upwelling.

The uplift of the Atlas mountain chain is documented (Figs. 16, 18) by pronounced Pyrennean and Savic disconformities below and above the continental clastics of the Upper Oligocene Agadir Formation. A last and random marine incursion took place during the early Pliocene with the deposition of coastal oyster sands and marls. Not too much information is available from the starved onshore development of the Tertiary.

From Sites 370/416 and 415 some more, but also incomplete data are available. Paleocene calcareous clays and marls are of interest in relation to their low rate of sedimentation and deposition above CCD. This means that the CCD must have lowered in this part of the North Atlantic during the Late Cretaceous. A complete sedimentary record of this time would therefore be highly desirable. Strong input of coarse clastics in the Upper Paleocene of Site 415 might be explained with the first activity of the Agadir Canyon system(Lancelot and Winterer, 1980).

The Eocene is again a time of turbiditic accumulation at both sites. The pelagic sedimentation is characterized by an increase of chert accumulation as a result of increasing productivity in surface waters. This is a widespread phenomenon at the northwestern African margin and might again be related with deep oceanic circulation and coastal upwelling. This circulation may also be the reason for the late Eocene-early Miocene hiatus at Site 415 (Vincent et al., 1980). Site 370/416, however, is one of the few places of the whole Atlantic, where Oligocene has been located, i.e. as a mere continuation of the Eocene turbidites. At both sites, the Miocene is represented by an alternation of diatom oozes, marly oozes and turbidites which finally pass into the sedimentation of marly oozes of Pliocene time.

Despite the fact that considerable hiatuses may have reduced the original thickness of offshore Tertiary sediments, and despite the progressive uplift of the Atlas mountain range at this time, the starved sedimentation also of the offshore Tertiary is of interest, as well as its sedimentation above CCD.

On the northeastern American margin, the stratigraphy of the Cretaceous-Tertiary boundary levels is poorly established, but hiatuses and lowering of the sea level are noticeable at several locations in the basins. The shelf basins most deeply subsided during Eocene time, when shelf areas were submerged into water depth corresponding to the present continental slope environment. Oligocene extensive erosion of the slope by contour currents resulted in a major deep-sea unconformity (Jansa et al., 1979) which cut back the shelf edge up to 30 km (Schlee et al., 1979) and formed channels as much as one kilometer deep (Dillon et al., 1979; Fig. 17). The late Miocene-Pliocene regression resulted in renewed clastic deposition on the shelf and subaerial emergence of parts of the shelf. The rapidly outbuilding wedges of predominantly argillaceous sediments during the Neogene constructed the present continental shelf. In this way, during the Tertiary the complete separation of eastern and western North Atlantic margins became manifest.

Table 1 facilitates comparison of the main evolutionary events in the North Atlantic marginal basins treated above.

5.6 Volcanic activity

In the general opinion volcanic activity seems to be concentrated on the northwestern African margin. But this general impression is misleading and needs to be restricted to the Neogene volcanics. Oil exploration on the eastern North American margin provided data which indicate that this margin was neither volcanically nor tectonically inactive as previously thought, even though both factors played only minor roles in the passive margin development.

The first volcanic event is known from nearly all the circum-North Atlantic basins and occurred during Late Triassic-Early Jurassic. Basalt flows and sills intruded the redbed deposits of the onshore Triassic basins from northern Virginia to Nova Scotia (Van Houten, 1977). In the Fundy Basin, Nova Scotia and New Jersey, the multiple basalt flows attain thicknesses of 400 and 860 m, respectively (Smitheringale, 1973; Olsen, 1976). In addition to the basalt flows, the lower part of the Newark deposits in northern Virginia contains an unique 200 to 250 m thick sequence of predominantly andesitic and trachytic tuffs interbedded with thin basalt flows (Toewe, 1966). Dolerite sills intruded the Newark series from southern Massachusetts to Florida; their composition indicates that these are derived from an olivine tholeiite, trapped underneath the continental crust (Manspeizer et al., 1976). Radiometric analysis of the sills and dykes has yielded a rather wide range of ages, with an average of 190 m.y. (Dallmeyer, 1975; Van Houten, 1977).

From the onshore Essaouira-Agadir Basin basaltic extrusions and doleritic sills are known of Late Triassic age. A thick diabasic layer has been drilled (Chebeika-1) at the Triassic-Jurassic boundary in the Tarfaya-Aaiun Basin.

A second, minor period of volcanicity is recorded by the occurrence of tuffs and volcanic clasts within the Middle Jurassic carbonates and clastics in some of the northeastern American basins. In the Scotian Basin basalt-derived clastic grains were found in the Jurassic sequence of the Mohawk B-93 well (Bhat et al., 1975), but the authors did not specify the exact stratigraphic position. Numerous volcanic grains and altered volcanic glass were found in two more wells of the Scotian Basin in different levels of the Mohican and Iroquois Formations. Similar discoveries were made in the lower Abenaki Formation of the Georges Bank Basin (Simonis, 1980). The age of these volcanics might be Callovian-?Lower Oxfordian. The tuffs form up to 5% of the sample volume in the Georges Bank Basin and, therefore, the geographic location of the volcanic event was most probably proximal to Georges Bank. The source could have been the New England White Mountain volcanics. In this area one of the volcanic periods was dated with 160 m.y. (Foland and Faul, 1977). This event seems to be concentrated on the northeastern American margin.

But Jurassic volcanics were found to occur also on the Canary and Cape Verde Islands. They are placed at the base of the sedimentary sequences of Fuerteventura and Maio and can be dated by the subsequent sediments. At the base of the Fuerteventura sediments volcanic clasts of trachybasaltic origin were described (Rothe, 1968; Robertson and Bernoulli, this Vol.) and may be ascribed to a Late Jurassic age. The (?Late) Jurassic "eruptive complex" of Maio which forms the base of the sedimentary sequence was found to belong to the oceanic crust (De Paepe et al., 1974; Robertson and Bernoulli, this Vol.).

A third regionally widespread volcanic episode occurred during the Lower Cretaceous. At the northeastern American margin the most prominent occurrence is the mafic intrusion in the Baltimore Canyon Trough which is more than 1 km across the top of the volcanic plug (USGS seismic line No. 2, see Schlee, 1978). The emplacement of the volcanic plug produced a local unconformity with up to 400 m of sediment removed at the unconformity (Mattick, 1977); this dates the last vertical movement of the intrusion as late Lower Cretaceous. Volcanic sills also occur at different wells in the Scotian Basin which are situated along the Canso Ridge (Jansa and Wade, 1975). They are alcaline basaltic to andesitic in composition and intrude Lower Cretaceous sediments. No absolute age data are available for these rocks being presumably of Lower Cretaceous age.

In the sedimentary sequence of Maio alcaline pillow lavas and tuffs are repeatedly intercalated from Oxfordian through Eocene. Episodes of main activity most probably are the Oxfordian (mid-Cimmerian?), the Tithonian-Berriasian (Late Cimmerian) and the Albian-Cenomanian (Austrian).

The initial phase of the young Canary volcanism (Schmincke, this Vol.) seems to be related to Laramid tectonic activity at the Cretaceous-Tertiary boundary. First submarine pillow lavas and sills were dated as Eocene-Oligocene. Subaerial extrusions began with the middle Miocene, the initial age of Dakar volcanics and the Cape Verde volcanic activity. For the time being nothing is known about Tertiary volcanism at the northeastern American margin.

Several hypotheses have been advanced to explain the intraplate volcanic activity along the eastern and western North Atlantic margins (Van Houten, 1977; Foland and Faul, 1977). The new data from both margins suggest that the timing of the volcanism roughly coincides with the periods of extensional or compressional stress as recorded near the outer edges of the African and North American continental plates by periods of tectonism.

5.7 Tectonism

It is of extreme interest to note, how perfectly the times of tectonic activity at both margins correlate with some of Stille's (1924) tectonic phases. The first angular unconformity which is recognized in the Essaouira-Agadir Basin at the Permian-Triassic boundary can be correlated with the Palatinate phase of Central and Eastern Europe. It is related to the initiation of the early graben evolution.

The major late Triassic-early Jurassic tectonic activity corresponds with the Early Cimmerian tectonics and can be related to the early rifting. It is documented by angular unconformities in the Scotian Basin, the Essaouira-Agadir Basin, the Baltimore Canyon Trough, and has been accompanied by volcanism in the coastal basins. This volcanism may be related with the initiation of the broad downwarping of the margin and the development of deep-seated extentional faults.

The mid-Cimmerian tectonic phase is of local importance and seems to be restricted to the Scotian Basin. It is associated with minor volcanism on the margin and caused rejuvenation of the hinterland as seen from the deposition of coarse clastics in the Scotian, Atlas and Georges Bank Basins. The episode of tilting and warping of the coastal Newark Basin along the entire length of the belt occurred in post-Liassic time (Cornet, 1975), with the transverse extensional fractures filled by mafic dykes. This is most probably another expression of the same tectonic event.

Similar environmental conditions were installed at the Jurassic-Cretaceous boundary and are to be correlated with the Late Cimmerian tectonics. Drowning of carbonate platforms and extinction of reef biotopes which become buried by large delta plains, are the main features of that time. It can be recognized through the basins at both margins. Together with a new period of rejuvenated hinterland as source area of major clastic influx, the Late Cimmerian tectonics most probably initiated salt motion in the future diapiric zones. Volcanic activity is only of minor importance.

The next episode of major tectonic activity is best documented at the northeastern American margin. It is the time of the uplift of the central portion of the Grand Banks and the initial rise of the Avalon Uplift (Jansa and Wade, 1975), documented by an angular unconformity at the base of the Albian. These Austrian tectonics can be recognized in the onshore Tarfaya-Aayun Basin. They are again associated with major volcanic activity, especially on the northeastern American margin.

Minor unconformities have been recognized at the base of the Turonian in the Blake Plateau Basin (pre-Gosavian tectonics) and between Santonian and Campanian in the Tarfaya-Aaiun Basin (Subhercynian phase).This is at the same time the initiation of the uplift of the northwestern African coastal basins.

This uplift has been realized by Laramid tectonism and compressional tectonics in the Senegal and Tarfaya-Aaiun Basins, and finally by Pyrennean and Savic movements in the Tarfaya-Aaiun and the Essaouira-Agadir Basins. This tectonic event is associated with salt motion and led to the initiation of the Canary volcanism. Laramid(?) unconformities were possibly also recognized in the Georges Bank Basin and the Baltimore Canyon Trough. The final fate of the subsiding Northeast American basins on one side, and of basinal uplift associated with young Neogene volcanism at the Northwest African margin, has been predetermined by their dissimilar bearing during the Laramid and Pyrennean tectonic activities.

Acknowledgements

The authors are much indebted to U. von Rad for stimulation, constant interest and critical review of this paper. We gratefully acknowledge the substantial help of K.Hinz (BGR, Hannover) and J.Schlee (USGS Woods Hole, Mass.) providing us either with seismic profiles of the northwestern African margin or with several illustrations on structure and stratigraphy of the northeastern American margin. J.Bujak (Dartmouth) and G.Wissmann (Hannover) kindly reviewed the paper. The colleagues of the Atlantic Geoscience Center and the U.S.Geological Survey contributed with valuable discussions. W.Schwentke was helpful in the linguistic revise. Thanks are due to M.Cook, G.Grant, H.D.Bergner and H.Vollmer for drafting the illustrations. A long-lasting Canadian post strike made final redaction of this paper difficult.
The second author appreciates financial support of his field work by the Deutsche Forschungsgemeinschaft.

REFERENCES

Adinolfi F, Jacobson SS (1979) Geologic correlation with other wells. In: Amato RV, Simonis KEK (eds) Geologic and operational summary, COST No B-3 well, Baltimore Canyon Trough area, Mid-Atlantic OCS, US Geol Surv Open-File Rep 79-1159: 32-39

Ager DV (1974) The western High Atlas of Morocco and their significance in the history of the North Atlantic. Proc Geol Assoc 85: 23-41

Amato RV, Bebout JW (eds) (1978) Geological and operational summary, COST No GE-1 well, Southeast Georgia Embayment area, South Atlantic OCS. US Geol Surv Open-File Rep 78-668: 1-122

Amato RV, Bebout JW (eds) (1980) Geologic and operational summary, COST No G-1 well, Georges Bank area, North Atlantic OCS. US Geol Surv Open-File Rep 80-268: 112 pp

Amato RV, Simonis EK (eds) (1979) Geological and operational summary, COST No B-3 well, Baltimore Canyon Trough area, Mid-Atlantic OCS. US Geol Surv Open-File Rep 79-1159: 118 pp

Amato RV, Simonis EK (eds) (1980) Geologic and operational summary, COST No G-2 well, Georges Bank area, North Atlantic OCS. US Geol Surv Open-File Rep 80-269: 116 pp

Ambroggi R (1963) Etude géologique du versant méridional du Haut Atlas occidental et de la Plaine du Souss. Notes et Mém Serv géol Maroc 157: 321 pp

Arthur MA, von Rad U, Cornford Ch, McCoy FW & Sarnthein M (1979) Evolution and sedimentary history of the Cape Bojador continental margin, northwestern Africa. In:von Rad U, Ryan WBF et al, Init Repts Deep Sea Drilling Project 47/1: 773-816

Ascoli P (1976) Foraminiferal and ostracod biostratigraphy of the Mesozoic-Cenozoic, Scotian shelf, Atlantic Canada. Marit Sediments Spec Publ 1,(B, Paleoecology Biostratigraphy): 653-741

AUXINI (Departamento de Investigaciones Petrolíferas de Auxini) (1969) Correlación estratigráfica de los sondeos perforados en el Sahara español. Bol Geol Minero 83: 235-251

Barss MS, Bujak JP, Williams GL (1979) Palynological zonation and correlation of sixty-seven wells, Eastern Canada. Geol Surv Canada Paper 78-24: 118 pp

Bebout JW (1980) Biostratigraphy. In: Amato RV, Simonis EK (eds) Geologic and operational summary, COST No G-2 well, Georges Bank area, North Atlantic OCS. US Geol Surv Open-File Rep 80-269: 20-28

Behrens M, Krumsiek K, Meyer DE, Schäfer A, Siehl A, Stets J, Thein J, Wurster P (1978) Sedimentationsabläufe im Atlas-Golf (Kreide, Küstenbecken Marokko). Geol Rdsch 67: 424-453

Benson W, Sheridan RE et al (1978) Initial Reports of the Deep Sea Drilling Project.Vol 44, Washington (US Governm Print Off),1005 pp

Bhat H, McMillan NJ, Aubert J, Porthault B, Surin M (1975) North American and African drift - The record in Mesozoic coastal plain rocks, Nova Scotia and Morocco. In: Yorath CJ, Parker ER, Glass DJ (eds) Canada's Continental Margins and Offshore Petroleum Exploration, Canad Soc Petrol Geol Mem 4: 375-389

Bujak JP, Barss MS, Williams GL (1977) Offshore East Canada's organic type and color and hydrocarbon potential. Oil Gas J 75 (14): 198-202

Bujak JP, Barss MS, Williams GL (1977) Offshore eastern Canada; Part II, Organic type and color and hydrocarbon potential. Oil Gas J 75 (15): 96-98

Bullard EC, Everett JE, Smith AG (1965) The fit of the continents around the Atlantic. In: Blackett PMS, Bullard EC, Runcorn SK (eds) A Symposium on Continental Drift, R Soc London phil Trans (A) 258: 41-51

Castelain J (1966) Aperçu stratigraphique sur la micropaléontologie et la palynologie des sédiments secondaires et tertiaires des bassins de l'Ouest africain. In: Reyre D (ed) Bassins sédimentaires du Littoral africain, 1 Littoral atlantique, Assoc Serv géol afr, 40-51

Choubert B (1935) Recherches sur la genèse des chaînes paléozoiques et antécambriennes. Rev Géogr phys Géol dynam 8: 5-50

Choubert G, Faure-Muret A, Hottinger I (1967) Aperçu géologique du Bassin Côtier de Tarfaya. Notes Mém Serv géol Maroc 175/I: 7-106

Choubert G, Faure-Muret A, Hottinger I (1972) La série stratigraphique de Tarfaya (Maroc sud-occidental) et le problème de la naissance de l'océan Atlantique. Notes Serv géol Maroc 31: 29-40

Cornet B, Traverse A (1975) Palynological contributions to chronology and stratigraphy of the Hartford basin in Connecticut and Massachusetts. Geoscience and Man 11: 1-33

Dallmeyer RD (1975) The Palisade sill: A Jurassic intrusion? Geology 3: 243-245

De Paepe P, Klerkx J, Hertogen J et al. (1974) Oceanic tholeiites on the Cape Verde Islands: Petrochemical and geochemical evidence. Earth Planet Sci Lett 22: 347-354

Dillon RS, Sougy JHA (1974) Geology of West Africa and Canary and Cape Verde Islands. In: Nairn AEM, Stehli FG (eds) The Ocean Basins and Margins, 2 The North Atlantic: 315-390

Dillon WP, Paull CK, Buffler RT, Fail J-P (1979) Structure and development of the Southeast Georgia Embayment and northern Blake Plateau: Preliminary analysis. In: Watkins JS, Montadert L, Dickerson PW (eds) Geological and Geophysical Investigations of Continental Margins, Amer Ass Petrol Geol Mem 29: 27-41

Duffaud F, Brun L, Plauchut B (for Société Chérifienne des Pétroles) (1966) Le Bassin du Sud-Ouest Marocain. In: Reyre D (ed) Bassin sédimentaires du Littoral africain, 1 Littoral atlantique, Assoc Serv géol afr, 5-12

Eardley AJ (1951) Structural geology of North America. Harper New York, 624 pp

Einsele G, von Rad U (1979) Facies and paleoenvironment of Lower Cretaceous sediments at DSDP Site 397 and in the Aaiun Basin (Northwest Africa). In: von Rad U, Ryan WBF et al, Init Repts Deep Sea Drilling Project 47/1: 559-571

Einsele G, Wiedmann J (1975) Faunal and sedimentological evidence for upwelling in the Upper Cretaceous coastal basin of Tarfaya, Morocco. IX Congr Int Sédimentol, Nice, Thème 1: 67-74

Eliuk LS (1978) The Abenaki Formation, Nova Scotia Shelf, Canada - A depositional and diagenetic model for a Mesozoic carbonate platform. Canad Petrol Geol Bull 26: 424-514

Evans R (1978) Origin and significance of evaporites in basins around Atlantic margin. Amer Assoc Petrol Geol Bull 62: 223-234

Fischer AG, Arthur MA (1977) Secular variations in the pelagic realm. Soc Econ Paleont Miner Spec Publ 25: 19-50

Foland KA, Faul H (1977) Ages of the White Mountain intrusives - New Hampshire, Vermont and Maine, USA. Amer J Sci 277: 888-904

Folger DW, Dillon WP, Grow JA, Klitgord KD, Schlee JS (1979) Evolution of the Atlantic continental margin of the United States. In: Talwani M, Hay W, Ryan WBF (eds) Deep Drilling Results in the Atlantic Ocean: Continental Margins and Paleoenvironment, Maurice Ewing Ser 3, Amer Geophys Union, Washington DC, 87-108

Fúster JM, Muñoz M, Sagredo J, Yébenes A, Bravo T, Hernández-Pacheco A (1980) Islas Canárias. Bol Geol Minero 91/2: 351-390

Given MM (1977) Mesozoic and Early Cenozoic geology of offshore Nova Scotia. Canad Petrol Geol Bull 25: 63-91

Grow JA, Mattick RE, Schlee JS (1979) Multichannel seismic depth sections and interval velocities over outer continental shelf and upper continental slope between Cape Hatteras and Cape Cod. In: Watkins JS, Montadert L, Dickerson PW (eds) Geological and Geophysical Investigations of Continental Margins, Amer Assoc Petrol Geol Mem 29: 65-83

Grunau HR, Lehner P, Cleintuar MR, Allenbach P, Bakker G (1975) New radiometric ages and seismic data from Fuerteventura (Canary Islands), Maio (Cape Verde Islands), and Sao Tomé (Gulf of Guinea). In: Borradaile GJ et al (eds), Progress in Geodynamics, North-Holland Amsterdam-New York, 90-118

Hayes DE, Pimm AC et al (1972) Initial Reports of Deep Sea Drilling Project Vol 14,Washington (US Governm Print Off), 975 pp

Hinz K (1979) Seismic sequences of Cape Bojador, Northwest Africa. In: von Rad U, Ryan WBF et al, Init Reps Deep Sea Drilling Project 47/1: 485-489

Hinz K, Seibold E, Wissmann G (1974) Continental slope anticline and unconformities of West Africa. "Meteor" Forsch-Ergebn (C) 17: 67-73

Hollister CD, Ewing JI et al (1972) Initial Reports of the Deep Sea Drilling Project. Vol 11, Washington (US Governm Print Off) 1077 pp

Houtz E, Ewing J (1976) Upper crustal structure as a function of plate age. J geophys Res 81: 2490-2496

Jansa LF (in press) Mesozoic carbonate platforms and banks of the eastern North American margin. Marine Geol

Jansa LF, Bujak JP, Williams GL (1980) Upper Triassic salt deposits of the western North Atlantic. Canad J Earth Sci 17: 547-559

Jansa LF, Gardner JV, Dean WE (1978) Mesozoic sequences of the Central North Atlantic. In: Lancelot Y, Seibold E et al, Init Reps Deep Sea Drilling Project 41: 991-1031

Jansa LF, Enos P, Tucholke BE, Gradstein FM, Sheridan RE (1979) Mesozoic-Cenozoic sedimentary formations of the North American Basin; western North Atlantic. In: Talwani M, Hay W, Ryan WBF (eds) Deep Drilling Results in the Atlantic Ocean: Continental Margins and Paleoenvironment, Maurice Ewing Ser 3, Washington DC, Amer Geophys Union, 1-57

Jansa LF, Wade JA (1975) Paleogeography and sedimentation in the Mesozoic and Cenozoic, Southeastern Canada. In: Yorath CJ, Parker ER, Glass DJ (eds) Canada's Continental Margins and Offshore Petroleum Exploration, Canad Soc Petrol Geol Mem 4: 79-102

Jansa LF, Wade JA (1975) Geology of the continental margin off Nova Scotia and Newfoundland. In: Van der Linden WJM, Wade JA (eds) Offshore Geology of Eastern Canada, 2 Regional Geology, Canada Geol Surv Paper 74-30: 51-106

King LH (1975) Geosynclinal development on the continental margin south of Nova Scotia and Newfoundland. In: Van der Linden WJM, Wade JA (eds) Offshore Geology of Eastern Canada, 2 - Regional Geology, Canada Geol Surv Paper 74-30: 199-206

Klitgord KD, Behrendt JC (1979) Basin structure of the US Atlantic margin. In: Watkins JS, Montadert L, Dickerson PW (eds)Geological and Geophysical Investigations of Continental Margins, Amer Assoc Petrol Geol Mem 29: 85-112

Klitgord KD, Schouten H (1977) The onset of sea-floor spreading from magnetic anomalies. In: Symposium on the Development of the New York Bight, Palisades, N.Y., Lamont-Doherty Geological Observatory, 12-13

Lancelot Y, Seibold E (1978) The evolution of the central northeastern Atlantic: Summary of results of DSDP Leg 41. In: Lancelot Y, Seibold E et al, Init Reps Deep Sea Drilling Project 41: 1215-1245

Lancelot Y, Seibold E et al (1978) Initial Reports of Deep Sea Drilling Project, Vol 41, Washington (US Governm Print Off), 1259 pp

Lancelot Y, Winterer EL (1980) Evolution of the Moroccan oceanic basin and adjacent continental margin - A synthesis. In: Lancelot Y, Winterer EL et al. Init Repts Deep Sea Drilling Project 50: 801-821

Lancelot Y, Winterer EL et al (1980) Initial Reports of the Deep Sea Drilling Project, Vol 50, Washington (US Governm Print Off), 868 pp

Lehner P, De Ruiter PAC (1977) Structural history of Atlantic Margin of Africa. Amer Assoc Petrol Geol Bull 61: 961-981

Le Pichon X, Fox PJ (1971) Marginal offset, fracture zones, and the early opening of the North Atlantic. J geophys Res 76: 6294-6308

Le Pichon X, Sibuet J-C, Francheteau J (1977) The fit of the continents around the North Atlantic Ocean. Tectonophysics 38: 169-209

Makris J, Demnati A (in press) Deep seismic soundings in Morocco and a crust and upper mantle model deduced from seismic and gravity data.

Manspeizer W, Puffer JH, Cousminer HL (1976) Subduction, rifting, and sea floor spreading: A volcanic record in Morocco and eastern North America (Abst). Geol Soc Amer Abstracts with Programms 8/2: 224-225

Manspeizer W, Puffer JH, Cousminer HL (1978) Separation of Morocco and eastern North America, A Triassic-Liassic stratigraphic record. Geol Soc Amer Bull 89: 901-920

Martinis B, Visintin V (1966) Données géologiques sur le bassin sédimentaire côtier de Tarfaya (Maroc méridional). In: Reyre D (ed) Bassins sédimentaires du Littoral africain. Littoral atlantique, Assoc Serv géol afr, 13-26

Mattick RE (1977) Geologic setting. In: Scholle PA (ed) Geological Studies on the COST B-2 well, US Mid-Atlantic Outer Continental Shelf Area, US Geol Surv Circ 750: 4-7

McIver NL (1972) Cenozoic and Mesozoic stratigraphy of the Nova Scotia shelf. Canad J Earth Sci 9: 54-70

McMaster RL, Lachance TP (1968) Seismic reflectivity studies on north western African shelf: Strait of Gibraltar to Mauretania. Amer Assoc Petrol Geol Bull 52: 2387-2395

Olsen PE (1976) Stratigraphy and stratigraphic nomenclature of the Newark Supergroup of New Jersey and Pennsylvania. Unpublished manuscript.

Parsons MG (1975) The geology of the Laurentian fan and the Scotia rise. In: Yorath CJ, Parker ER, Glass DJ (eds) Canada's Continental Margins and Offshore Petroleum Exploration, Canad Soc Petrol Geol Mem 4: 155-167

Pimm AC, Hayes DE (1972) General synthesis. In: Hayes DE, Pimm AC et al, Init Reps Deep Sea Drilling Project 14: 955-975

Pitman III WC, Talwani M (1972) Seafloor spreading in the North Atlantic. Geol Soc Amer Bull 83: 619-646

Poag WC, Hall RE (1979) Foraminiferal biostratigraphy, paleoecology, and sediment accumulation rates. In: Scholle PA (ed) Geological Studies of the COST GE-1 well, US South Atlantic Outer Continental Shelf Area; US Geol Surv Circ 800: 49-63

Price I (1980) Provenance of the Jurassic-Cretaceous flysch, Deep Sea Drilling Project Sites 370 and 416. In: Lancelot Y, Winterer EL et al, Init Reps Deep Sea Drilling Project 50: 751-757

Purser, BH (ed) (1973) The Persian Gulf, Holocene carbonate sedimentation and diagenesis in a shallow epicontinental sea. Springer Berlin-Heidelberg-New York, 471 pp

Querol R (1966) Regional Geology of the Spanish Sahara. In: Reyre D (ed) Bassins sédimentaires du Littoral africain, 1 Littoral atlantique,27-39

Ratschiller LK (1971) Lithostratigraphy of the Northern Spanish Sahara. Mem Mus tridentino Sci Nat 18/1: 1-78

Reyre D (ed) (1966) Bassins sédimentaires du Littoral africain. 1ère Partie. Littoral Atlantique. Assoc Serv géol afr, Paris, VIII + 304 pp

Rhodehamel EC (1979) Lithologic descriptions. In: Scholle PA (ed) Geological Studies of the COST GE-1 well, US South Atlantic Outer Continental Shelf Area, US Geol Surv Circ 800: 24-36

Rona PA (1970) Comparison of continental margins of eastern North America at Cape Hatteras and Northwestern Africa at Cap Blanc. Amer Assoc Petrol Geol Bull 54: 129-157

Rona PA (1973) Relations between rates of sediment accumulation on continental shelves, sea-floor spreading, and eustasy inferred from the North Atlantic. Geol Soc Amer Bull 84: 2851-2872

Schlager W (1980) Mesozoic calciturbidites in Deep Sea Drilling Project hole 416A. Recognition of a drowned carbonate platform. In: Lancelot Y, Winterer EL et al, Init Reps Deep Sea Drilling Project 50: 733-749

Schlee JS (1978) Acoustic stratigraphy of Georges Bank (Abst). Geol Soc Amer Abstracts with Programs 10/2: 84

Schlee JS (1980) A comparison of two Atlantic-type continental margins. US Geol Surv Prof Paper 1167: 21 pp

Schlee JS (1981) Seismic stratigraphy of Baltimore Canyon Trough. Amer Assoc Petrol Geol Bull 65: 26-53

Schlee J, Behrend JC, Grow JA, Robb JM, Mattik RE, Taylor PT, Lawson BJ (1976) Regional geologic framework off Northeastern United States. Amer Assoc Petrol Geol Bull 60: 926-951

Schlee JS, Dillon WP, Grow JA (1979) Structure of the continental slope off the Eastern United States. Soc Econ Paleont Mineral, Spec Paper 27: 95-117

Schlee JS, Jansa LF (in press) The paleoenvironment and development of the eastern North American continental margin. Ocean Acta, Proceed Internat Geol Congress Paris, Geology of continental margins Symposium.

Schmincke HU (1976) The geology of the Canary Islands. In: Kunkel G (ed) Biogeography and Ecology in the Canary Islands, Junk The Hague, 67-184

Schmincke HU, von Rad U (1979) Neogene evolution of Canary Island volcanism inferred from ash layers and volcaniclastic sandstones of DSDP Site 397 (Leg 47A). In: von Rad U, Ryan WBF, et al, Init Reps Deep Sea Drilling Project 47/1: 703-725

Scholle PA (ed) (1977) Geological studies on the COST No. B-2 well, US Mid-Atlantic outer continental shelf area. US Geol Surv Circ 750: 71 pp

Scholle PA(ed) (1979) Geological Studies of the COST GE-1 well, US South Atlantic continental shelf area. US Geol Surv Circ 800: 114 pp

Schouten H, Klitgord KD (1977) Mesozoic magnetic anomalies, western North Atlantic, Map MF-915. US Geol Surv misc Field Stud

Schouten H, Klitgord KD (1980) The evolution of the central North Atlantic: Evidence for the continuity of the fracture zones. Congrès géologique intern Paris 1980, 3: 1371

Sclater JG, Hellinger S, Tapscott C (1977) The paleobathymetry of the Atlantic Ocean from Jurassic to the present. J Geol 85: 509-552

Seibold E, Hinz K (1974) Continental slope construction and destruction, West Africa. In: Burk CA, Drake CL (eds) The Geology of Continental Margins, Springer Berlin-Heidelberg-New York, 179-196

Serralheiro A (1968) Formações sedimentares do Arquipélago de Cabo Verde. Junta Invest Ultramar, Lisboa, 22 pp

Serralheiro A (1970) Geología da ilha de Maio. Junta Invest Ultramar, Lisboa, 103 pp

Shell Canada Ltd (1970) Well History Report, Oneida O-25. Report on file Dept Energy Mines Resources Ottawa, 50 pp

Sheridan RE (1976) Sedimentary basins of the Atlantic margin of North America. Tectonophysics 36: 113-132

Sheridan RE, Enos P (1979) Stratigraphic evolution of the Blake Plateau after a decade of scientific drilling. In: Talwani M, Hay W, Ryan WBF (eds) Deep Drilling Results in the Atlantic Ocean: Continental Margins and Paleoenvironment, Maurice Ewing Ser 3, Amer Geophys Union, Washington DC, 109-122

Sheridan RE, Gradstein FM et al (in press) Early History of the Atlantic Ocean and Gas-Hydrates in the Blake Outer Ridge - Results of the Deep Sea Drilling Project Leg 76. Geol Soc Amer Bull

Sherwin DF (1975) Scotia shelf and Grand Banks. In: McCrossan RG (ed) The Future Petroleum Provinces of Canada - Their Geology and Potential, Canad Soc Petrol Geol Mem 1: 519-559

Shipley TH, Buffler RT, Watkins JS (1978) Seismic stratigraphy and geologic history of Blake Plateau and adjacent Western Atlantic continental margin. Amer Assoc Petrol Geol Bull 62: 792-812

Simonis EK (1979) Radiometric age determinations. In: Scholle PA (ed), Geological Studies of the COST GE-1 well, US South Atlantic Outer Continental Shelf area, US Geol Surv Circ 800: 71 pp

Simonis EK (1980) Lithologic description. In: Amato RV, Simonis EK (eds) Geologic and operational summary, COST No. G-2 well, Georges Bank area, North Atlantic OCS, US Geol Surv Open-File Rep 80-269: 14-19

Sliter WV (1976) Cretaceous foraminifers from the southwestern Atlantic Ocean, Leg 36, Deep Sea Drilling Project. In: Barker PF, Dalziel IWD et al.Init Repts Deep Sea Drilling Project 36: 519-537

Smith MA, Amato RV, Furbush MA, Pert DM, Nelson ME, Hendrix JS, Tamm LC, Wood Jr G, Shaw DR(1976) Geological and operational summary, COST No. B-2 well, Baltimore Canyon Trough area, Mid-Atlantic OCS. US Geol Surv Open-file Rep 76-774: 79 pp

Smitheringale WG (1973) Geology of parts of Digby, Bridgetown, and Gaspereau Lake Map-areas, Nova Scotia. Geol Surv Canada Mem 375: 78 pp

Spengler A de, Castelain J, Couvin J, Leroy M (1966) Le bassin secondaire-tertiaire du Sénégal. In: Reyre D (ed) Bassins sédimentaires du Littoral africain. 1 - Littoral atlantique, Assoc Serv géol afr, 80-94

Stahlecker R (1934) Neocom auf der Kapverden-Insel Maio. N Jb Mineral etc BB 73: 265-301

Steinkraus W (1978) Biostratigraphy. In: Amato RV, Bebout JW (eds) Geological and operational summary, COST No. GE-1 well, Southeast Georgia Embayment area, South Atlantic OCS, US Geol Surv Open-File Rep 78-668: 29-41

Stille H (1924) Grundfragen der vergleichenden Tektonik. Borntraeger Berlin, 443 pp

Summerhayes CP, Nutter AH, Tooms JS (1971) Geological structures and development of the continental margin of northwestern Africa. Tectonophysics 11: 1-25

Talwani M, Eldholm O (1972) Continental margin off Norway: A geophysical study. Geol Soc Amer Bull 83: 3575-3606

Thiede J (1979) History of the North Atlantic Ocean: Evolution of an asymmetric zonal paleo-environment in a latitudinal ocean basin. In: Talwani M, Hay W, Ryan WBF (eds) Deep Drilling Results in the Atlantic Ocean: Continental Margins and Paleoenvironment, Maurice Ewing Ser 3,Am Geophys Union, Washington DC, 275-296

Toewe EC (1966) Geology of the Leesburg quadrangle, Virginia. Virginia Div Miner Resources Rep Invest 11: 52 pp

Tucholke BE, Houtz RE, Ludwig WJ (in press) Isopach map of sediments in the western Atlantic Ocean. Amer Assoc Petrol Geol Map Ser

Tucholke BE, Mountain GS (1979) Seismic stratigraphy, lithostratigraphy and paleosedimentation patterns in the North Atlantic Basins. In: Talwani M, Hay W, Ryan WBF (eds) Deep Drilling Results in the Atlantic Ocean: Continental Margins and Paleoenvironment Maurice Ewing Ser 3, Amer Geophys Union,Washington DC, 58-86

Tucholke BE, Vogt PR (1979) Western North Atlantic: Sedimentary evolution and aspects of tectonic history. In: Tucholke BE, Vogt PR, et al, Init Reps Deep Sea Drilling Project 43: 991-825

Uchupi E, Emery KO, Bowin CO, Phillips JD (1976) Continental margin off Western Africa: Senegal to Portugal. Amer Assoc Petrol Geol Bull 60: 809-878

Vail PR, Mitchum Jr RM, Thompson III S (1977) Global cycles of relative changes of sea level. In: Vail PR et al (eds) Seismic Stratigraphy and Global Changes of Sea Level, Amer Assoc Petrol Geol Mem 26: 83-98

Van Hinte JE (1976) A Jurassic time scale. Amer Assoc Petrol Geol Bull 60: 489-497

Van Hinte JE (1976) A Cretaceous time scale. Amer Assoc Petrol Geol Bull 60: 498-516

Van Houten FB (1977) Triassic-Liassic deposits of Morocco and Eastern North America. Amer Assoc Petrol Geol Bull 61/I: 79-99

Vincent E, Lehmann R, Sliter WV, Westberg MJ (1980) Calpionellids from the Upper Jurassic and Neocomian of Deep Sea Drilling Project Site 416, Moroccan Basin, Eastern North Atlantic. In: Lancelot Y, Winterer EL et al, Init Reps Deep Sea Drilling Project 50: 439-465

Viotti C (1967) Résultats stratigraphiques du sondage Puerto Cansado I du Bassin Côtier de Tarfaya. Notes Mém Serv Géol Maroc 175/I: 223-253

von Rad U, Arthur MA (1979) Geodynamic, sedimentary and volcanic evolution of the Cape Bojador continental margin (NW Africa). In: Talwani M, Ryan WBF (eds) Deep drilling results in the Atlantic Ocean: continental margins and paleoenvironment. Maurice Ewing Ser 3, Am Geophys Union, Washington DC, pp 187-203

von Rad U, Einsele G (1980) Mesozoic-Cenozoic paleobathymetry of the northwest African continental margin (Aaiun Basin to DSDP Site 397). R Soc London phil Trans 294: 37-50

von Rad U, Ryan WBF et al (1979) Initial Reports of the Deep Sea Drilling Project, Vol 47 pt 1, Washington (US Governm Print Off) 835 pp

Wade JA (1978) The Mesozoic-Cenozoic history of the north-eastern margin of North America. 10th Offshore Technology Conference, Houston, 1849-1855

Watkins JS, Hoppe KW (1979) Seismic reflection reconnaissance of the Atlantic margin of Morocco. In: Talwani M, Hay W, Ryan WBF (eds) Deep Drilling Results in the Atlantic Ocean: Continental Margins and Paleoenvironment,Maurice Ewing Ser 3, Amer Geophys Union, Washington DC, 205-217

Wiedmann J (1976) Geo- und hydrodynamische Prozesse im Schelfbereich in ihrer Auswirkung auf mesozoische Fossil-Vergesellschaftungen. Zentralbl Geol Palaeontol II 1976: 424-439

Wiedmann J (1979) Early Cretaceous Mollusks from DSDP Hole 397A off northwest Africa. In: von Rad U, Ryan WBF et al, Init Repts Deep Sea Drilling Project 47/1: 283-286

Wiedmann J (in press) Subsidenzgeschichte und Faunenverschiebungen an einem passiven Kontinentalrand: Das nordspanische Kreidebecken. N Jb Geol Paläont Abh

Wiedmann J, Butt A, Einsele G (1978a) Vergleich von marokkanischen Kreide-Küstenaufschlüssen und Tiefseebohrungen (DSDP) Stratigraphie, Paläoenvironment und Subsidenz an einem passiven Kontinentalrand. Geol Rundsch 67: 454-508

Wiedmann J, Einsele G, Immel H (1978b) Evidence faunistique et sédimentologique pour un upwelling dans le Bassin Côtier de Tarfaya/Maroc dans le Crêtacé supérieur. Actes 6ème Colloque Africain de Micropaléontologie, Tunis 1974. Ann Mines Géol 28/II: 415-441

Williams GL (1975) Dinoflagellate and spore stratigraphy of the Mesozoic-Cenozoic, offshore eastern Canada. In: Van der Linden WJM, Wade JA (eds)Offshore Geology of Eastern Canada, 2 - Regional Geology. Geol Surv Canada Paper 74-30: 107-162

Williams H (1978) Tectonic lithofacies map of the Appalachians. Memorial Univ Newfoundland, Map No 1, 2 sheets, scale I/I,000,000

Wissmann G (1979) Cape Bojador slope, an example for potential pitfalls in seismic interpretation of outer margin drilling. In: Von Rad U, Ryan WBF et al, Init Reps Deep Sea Drilling Project, 47/1: 491-499

Evolution of Volcanism

12 Volcanic and Chemical Evolution of the Canary Islands

Hans-Ulrich Schmincke[1]

ABSTRACT

The Canary Islands, a group of seven major volcanic islands, extends for almost 500 km roughly east-west 100 km off Northwest Africa. The islands formed chiefly during the last 20 Ma, although volcanic activity started during the Oligocene and possibly Eocene in the eastern island of Fuerteventura. Ages of the rapidly formed subaerial shields decrease irregularly from east to west but melting anomalies in the sub-Canarian mantle are presently active across the entire belt. Total volumes of individual islands are about 10 to 20 x 10^6 km^3 of which the subaerial part generally makes up less than 10%.

The dominant, shield-building magma type is saturated to moderately undersaturated alkali basalt with local tholeiite. Low pressure fractionation of olivine, clinopyroxene, and plagioclase was generally moderate, owing to rapid replenishment of the fast upward growth of the shield volcanoes and their magma chambers. Highly differentiated magma columns developed chiefly during the waning stages resulting in minor (quartz)-trachyte in the eastern and phonolitic plugs in the central and western islands. Major differentiated magma reservoirs on Gran Canaria and Tenerife culminated in large caldera-forming ash flow eruptions. Surface eruption of basalt magmas was generally inhibited during evolution and periodic partial emptying of such large differentiated zoned magma columns. Late stage basanites to nephelinites are locally nodule-bearing, of small volume, and are only slightly fractionated. High Ca/Al ratios and variable K-contents of these primitive magmas suggest garnet and phlogopite as residual phases during very low degrees of partial melting.

Multiphase episodic magmatic evolution consisting of two or more magmatic phases is characteristic of most Canary Islands and is best developed on Gran Canaria where two major multiphase cycles are distinguished. Multiphase magmatic evolution is common on other islands in the Central North Atlantic with alkali basalt shield magmas being broadly similar. It is less well developed on smaller islands and those close to the Mid-Atlantic Ridge. Highly alkalic, mafic, undersaturated magmas appear to be restricted to (large volume?) islands on thicker lithosphere (Canaries and Cape Verde Islands), presumably due to low heat flow and thus small degrees of partial melting at greater depth.

Intra-archipelago differences in melting conditions and mantle composition are reflected by consistently higher alkalinity and different trace element ratios between the western and central islands contrasted with Lanzarote and Fuerteventura to the east. Canary Island magmas on the whole are richer in Ti, Fe, and Zr and lower in Al than Azorean and Madeira magmas. Canary Island magmas may be derived from garnet-bearing mantle leaving residual garnet. The mantle beneath the Canaries is not very radiogenic with respect to $^{87}Sr/^{86}Sr$ as is characteristic for the eastern central Atlantic encompassing the Cape Verde Islands and Madeira. The mantle area south of about 30° to 35° N may be distinct from, and less heterogeneous than the mantle farther north.

There is no geological or geochemical evidence for the existence of continental crust beneath any of the Canary Islands. The origin of the Canary Island melting domain is not adequately explained by (a) an oceanic fracture zone, (b) extension of the South Atlas fault, (c) mantle plume and (d) propagating fracture zone. Unspecified mantle instabilities along the boundary between oceanic and continental lithosphere may have been instrumental in generating the unusually long-lived mantle anomaly with west to east translation of the lithosphere leading

1 Institut für Mineralogie, Ruhr Universität Bochum, 4630 Bochum 1, FRG

to an irregular non-linear age progression. Age data presently available for island volcanism in the Eastern Central North Atlantic suggest episodes of high activity between about 18 and 10 Ma and 5 Ma to the present, separated by a period of lesser magmatic productivity.

1. Introduction

"Passive margins" are distinguished from "active margins" by the absence of trenches, thick accretionary wedges of deformed sediments, high seismicity, and volcanism which are general features related to subduction processes. The passive continental margin off NW Africa conforms to such a scheme, except that it shows abundant volcanism in the form of island groups and seamounts on the continental rise. The Canary Islands are the major group of such volcanoes (Figs. 1,2).

Study of volcanic activity along a passive continental margin contributes to several problems of wide-ranging geological interest. For example, one of the major questions discussed in several chapters in this book is the location of the boundary between oceanic and continental lithosphere along the African continental margin. Is it possible by major or trace elements or isotopic tracers of volcanic rocks from the islands to differentiate between two types of lithosphere which were the source of the magmas or through which the magmas passed, and thus help locate the suture zone?

Also, a group of islands, rising several 1000 m from the seafloor and having basal diameters of 50 to 100 km form major morphological barriers to currents along and down the continental slope and thus can exert a strong influence on sedimentation patterns. The influx of volcaniclastic detritus generated by pyroclastic and epiclastic processes on the islands into the non-volcanic continent-derived sedimentary budget forms a significant part of the sedimentary columns to distances of 100 km or more away from an island. Ash layers derived from explosive eruptions on the islands provide excellent stratigraphic marker beds in the sediment column. The heat transfer from the subvolcanic magma chambers to the surrounding sedimentary pile must greatly accelerate maturation of hydrocarbons and other diagenetic processes.

Eustatic changes in sea level can be studied and dated more precisely on the islands than in the sedimentary column. Tectonic displacements, exceeding 1000 m in vertical throw, some of rather young age which occur on several islands, are difficult to document by any other method than by studying the islands. They are also clear evidence that dynamic processes off Northwest Africa, one of which is the generation of the volcanic islands, are still operative today, a conclusion also reached from the frequent and voluminous volcanic eruptions in historic times.

The Canarian archipelago consists of seven major islands: Lanzarote and Fuerteventura to the east, Gran Canaria, Tenerife and Gomera in the central part, and La Palma and Hierro to the west (Figs.1,2). Three islands (Gran Canaria, Tenerife, La Palma) are 2000 m or higher and Tenerife, dominated by the 3700 m high Pico de Teide, is the third largest oceanic volcano on earth, after Mauna Kea and Mauna Loa. The archipelago extends for almost 500 km from 115 km west of the African coast westwards and for 100 - 200 km north-south, between 27°38' - 29°25' latitude and 13°20' and 18°9' longitude (see front cover figure). Total surface area of the islands is ca. 7500 km^2. The very small volcanic island remnants of the Selvagens Islands, Conception Bank to the north and several seamounts including Echo Bank to the southwest are regarded as part of the Canary Island volcanic province.

The Canary Islands stand out among the major oceanic island groups in the world by
1) their long magmatic history - spanning 20 Ma (Ma = million years) or more on some islands; and
2) their unusually wide spectrum of volcanic and plutonic igneous rocks, ranging from highly undersaturated melilite nephelinites to peralkaline rhyolitic welded tuffs.

Outcrops are magnificent on the dry leeward side of the highly eroded older islands, allowing an unusually complete reconstruction of ocean island evolution. No wonder, therefore, that the islands have attracted geologists' attention for over 150 years. During the last 15 years, geological and petrological work on the islands has been carried out chiefly by Fuster and collaborators (Fuster et al., 1968 a-d) and my own group, and many studies are presently in progress.

Recent detailed reviews are given by Mitchell-Thome (1976), discussing much of the older literature, and Schmincke (1976), emphasizing volcanic, petrologic and structural evolution of the islands. In this report, I will not attempt to comprehensively review the geology of the Canary Islands nor give an up to date account of the recent literature. Instead, I will concentrate on three selected aspects, of which two are documented by new data:
1) structural and temporal history;
2) chemical evolution;
3) theories on the origin of the Canaries.

The temporal and chemical evolution of the Canaries is compared with other island

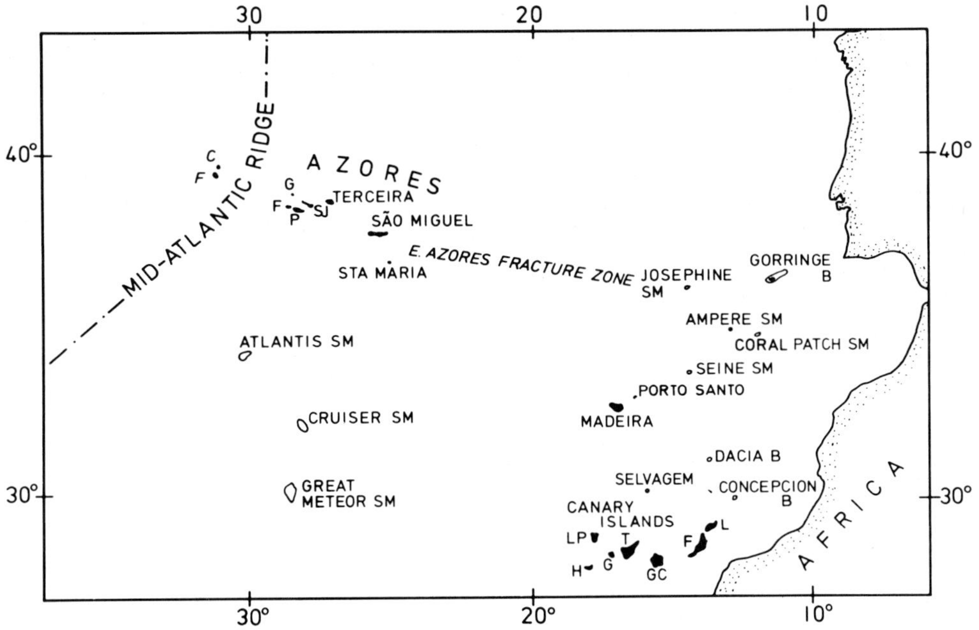

Fig. 1 Eastern Central North Atlantic showing location of Canary Islands, Madeira and Azores Islands. The African plate is bounded to the west by the Mid-Atlantic Ridge and to the North by the East Azores Fracture zone between the Azores and Gibraltar. For abbreviations of Canary Islands see Figure 2. Azores: SJ = São Jorge; P = Pico; G = Graciosa; F = Fayal; C = Corvo; F = Flores

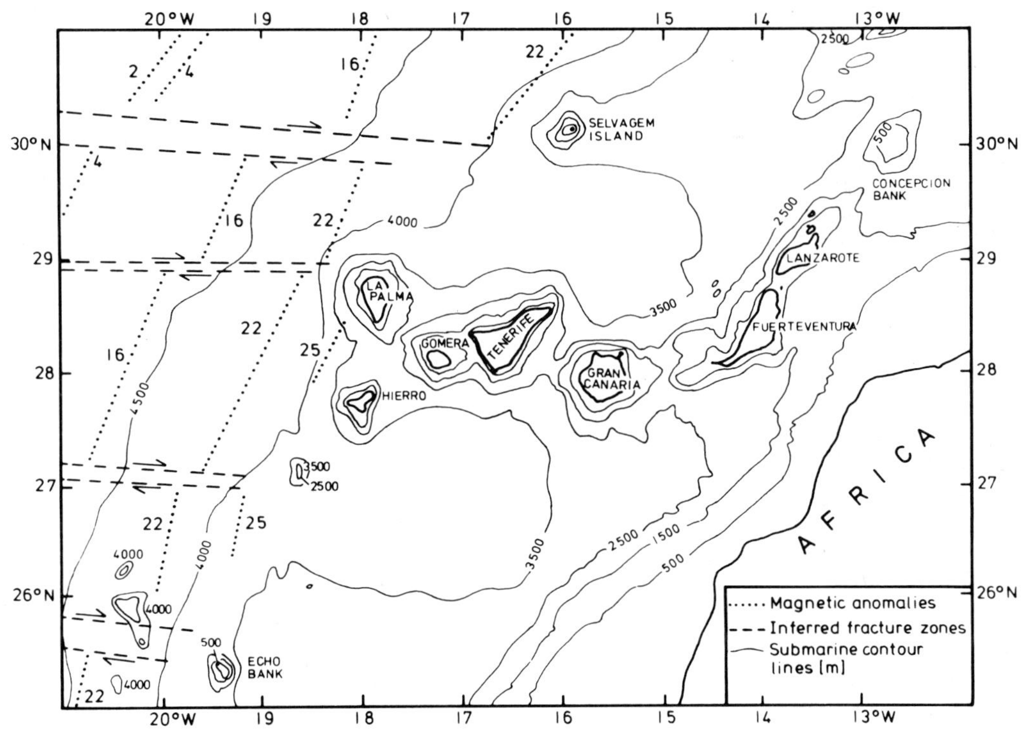

Fig.2 Bathymetry and magnetic anomalies (simplified from Rabinowitz and Hayes, 1977) in the vicinity of the Canary Islands. Offsets between magnetic anomalies are interpreted as fracture zones. Modified from Staudigel (1981)

groups in the Eastern Central North Atlantic and the intraplate volcanic Eifel province. However, comparisons are greatly hampered by a lack of age, chemical and structural data. Generalizations derived from the analogies thus remain highly speculative.

2. Structural Evolution

2.1 Growth Stages

The structure, composition and evolution of the submarine part of the islands is largely unknown except for La Palma, where a continuous, 3.5 km-thick section of the uplifted submarine part is exposed (Schmincke and Staudigel, 1976; Staudigel and Schmincke, 1981; Staudigel, 1981). This ophiolite-like section can be subdivided into four parts:

a) a central plutonic complex, dominantly alkalic gabbros;

b) a sheeted dike swarm with minor keratophyre to alkali basalt screens but with abundant sills;

c) an extrusive lower section, ca. 650 m thick, made up almost exclusively of pillow lavas and

d) an upper extrusive section, ca. 1150 m thick, composed to more than 50% of pillow fragment breccias and several types of hyaloclastites. Rocks from the uppermost part were probably erupted in very shallow water.

Another approach to monitor the submarine development of the Canaries is the study of volcaniclastic sediments encountered in drill holes, such as those found in DSDP Site 397, about 100 km SSE of Gran Canaria (Schmincke and von Rad, 1979) (Fig.3). In this hole, hyaloclastite debris flows (V-3 and V-4), 6 m thick (ca. 17 Ma old), are interpreted as representing the submarine stage of an island, possibly Fuerteventura, transported for more than 100 km to the south and southwest. Similar near-source hyaloclastite flow deposits occur in the uplifted seamount part of La Palma (Staudigel, 1981). They are overlain by volcaniclastic debris flows containing abundant tachylite and a wide variety of subaerial volcanic and subvolcanic (?) rock fragments, ranging from microgabbro to trachyte but are dominated by alkali basalt. These debris flows form excellent seismic reflectors (Wissmann in Schmincke and von Rad, 1979), and the volume of the flows probably exceeds 50 km^3 as based on seismic data. The debris flows are thought to reflect the rapid growth of an island during the shield stage, an especially large amount of clastic debris being generated during the transition period seamount/island by phreatomagmatic and magmatic explosive activity as well as erosion of freshly formed clastic deposits. Younger ash layers of Sites 369 and 397 reflect later, differentiated subaerial stages of an island, such as the 13-14 Ma old rhyolite ashes from Gran Canaria or the phonolitic Pliocene/Pleistocene ash layers, most of which are probably the result of Plinian eruptions on Tenerife (Rothe and Koch, 1978; Schmincke and von Rad, 1979) (Figs.3,4).

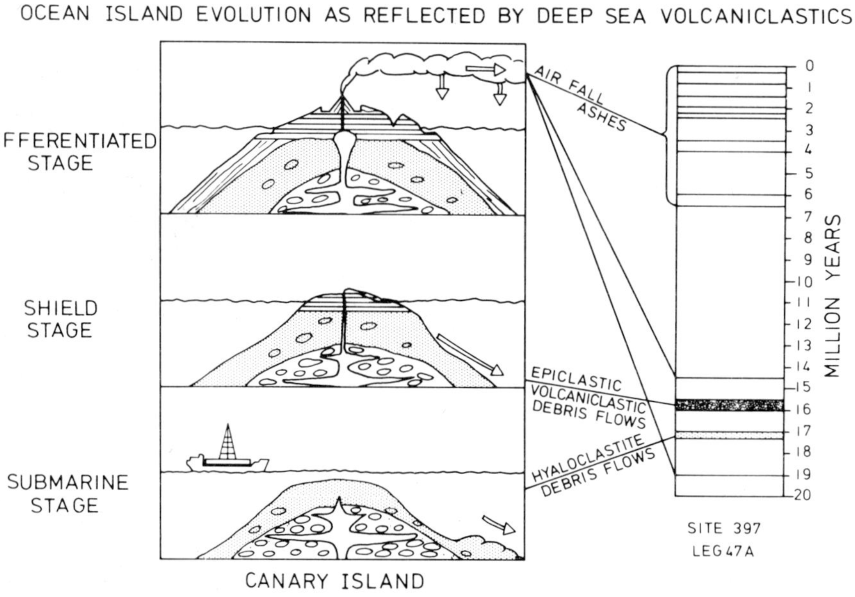

Fig.3 Three schematic growth stages of an oceanic island as reflected in three main types of submarine volcaniclastic rocks drilled south of the Canary Islands (See text for discussion)

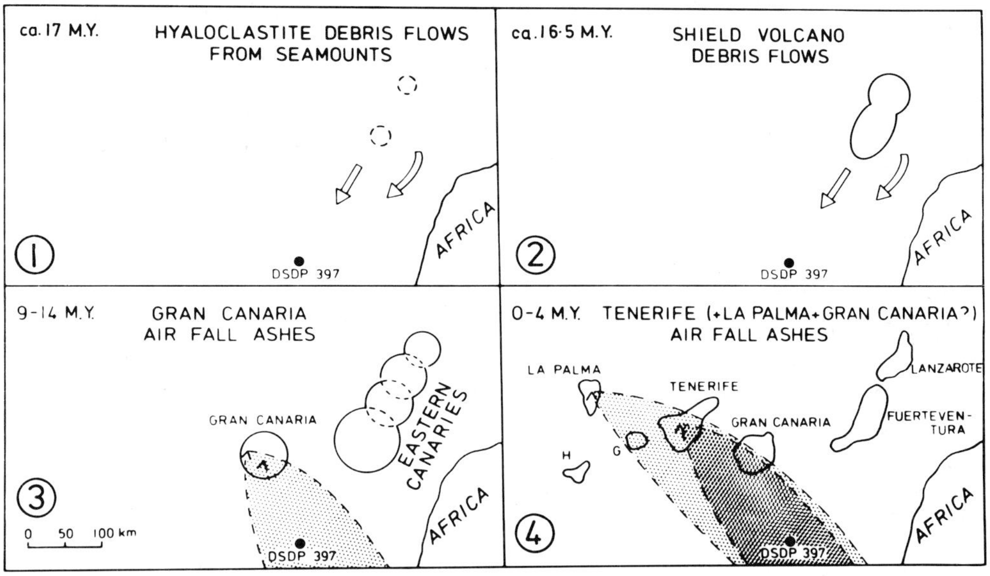

Fig.4 Four main stages of development of the Canary Islands as deduced from submarine volcaniclastic rocks drilled at DSDP site 397 (See also Fig.3). Stages 1 and 2 show the major shield phase on the eastern Canaries, neglecting the core of the island. Stage 3 shows the main shield phase of Gran Canaria, including eruption of peralkaline rhyolitic ashes around 13 - 14 Ma. Stage 4 shows the appearance of the western islands during which air fall ashes were deposited at site 397. Shape and direction of ash lobes are centered on drill site and not on - unknown - depositional axis

Other methods to study the thickness and lateral extent of the volcaniclastic apron around an island are seismic methods (Wissmann and Schmincke, in prep.).Seismic refraction data indicate a 1.5 to 2.0 km thick low velocity (2-3 km/s^{-1}) upper section beneath some islands such as Tenerife (Dash and Bosshard, 1968). This is similar to the submarine La Palma section where low velocities of this magnitude, reflecting the dominance of hyaloclastites and pillow breccias, would be expected to characterize the upper 1 km or so of a seamount.

Most of the subaerial part of the islands is made of basaltic shields, consisting of thin lava flows, generally less than a few hundred meters in lateral extent, and local scoria cones. These shield series are only up to about 1000 m thick in the more highly eroded eastern central islands but may have exceeded 1500 m or even 2000 m judging from the youthful westernmost islands Hierro and La Palma, Tenerife showing intermediate thicknesses. All islands include a wide spectrum of evolved rock series including highly differentiated rocks such as trachytes, rhyolites, and phonolites. Younger volcanic episodes occurred on all islands, as discussed below, typically consisting of scoria cones and associated lava flows, although in Tenerife and Gran Canaria central volcanoes with highly differentiated magmas developed during later stages.

2.2 Volumes and Eruptive Rates

Data on volumes of the submarine and subaerial portions of volcanic islands are important to calculate the total volume of magma erupted and intruded and, in conjunction with age data, eruptive rates. From this information, past and present dynamic states of the mantle melting domains beneath the archipelago can be assessed. Volumes of the submarine edifices were estimated by planimetry of the submarine contour lines, assuming a horizontal surface beneath the most distant contour line reflecting the shape of the submarine part of an island. Bathymetric maps of Lamont-Doherty Geological Observatory (D.Hayes, unpubl.) were used for the area off the Northwest African margin and that of Laughton et al. (1977, Deep-Sea Research, 22:791-810) for the Madeira-Azores area.

These estimates include much of the submarine apron, but not the core of an island that extends through the sediments to greater depths. The volume so determined includes non-volcanic sedimentary material, overlying and interbedded with the volcanogenic detritus of the apron, which is discussed more fully elsewhere (Wissmann and Schmincke, in prep.). Boundaries between islands were

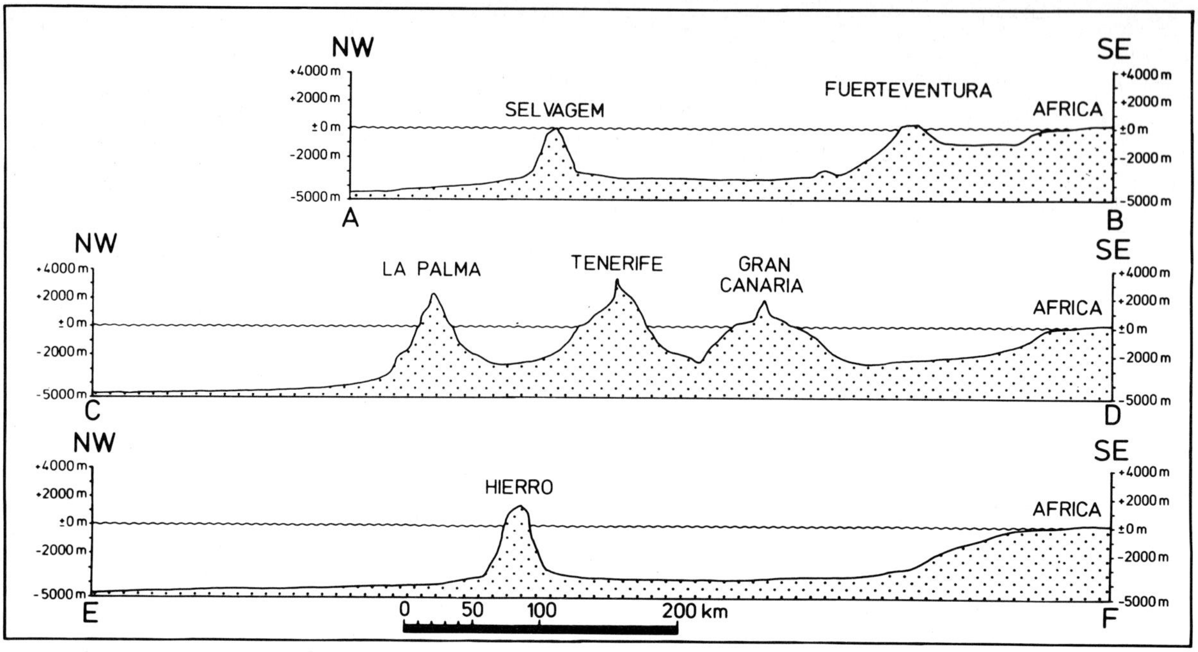

Fig.5 Three southeast-northwest cross-sections through the southern central and northeastern part of the Canary Island archipelago, including the Selvagens Islands

drawn about halfway, except where the shape of the submarine contours indicated an asymmetric distribution, which leads to an overestimate of the younger island. Estimates are believed to be least reliable in the eastern islands, because they are older, because of the higher non-volcanic sedimentation rate, and, in the case of Fuerteventura because of an unspecified portion of uplifted non-volcanic crust. Estimates are best around the young islands of Hierro and La Palma. I emphasize that these measurements are still quite subjective. They should be redone more precisely once more detailed bathymetric maps are available. More important still are seismic data needed to estimate the extent and thickness of the volcaniclastic aprons and the amount the crust has been depressed by the volcanic island. Estimates are believed to be realistic within a factor of about 2. Volumes of the subaerial part were measured using topographic maps at the scale 1 : 100.000. The data are shown in Table 1 and Figures 5 and 6.

Total volumes of islands range from 5.3×10^3 km^3 (Hierro) to 23.8×10^3 km^3 (Gran Canaria), excluding Fuerteventura whose slightly higher volume ($30.6 \times 10^3 km^3$) is difficult to evaluate for the reasons discussed above. The average total volume for one island is 19.24×10^3 km^3. This average volume is remarkably similar to that of 15 shields of the Hawaiian Chain ($19.7 \times 10^3 km^3$, as calculated from the data of Bargar and Jackson, 1974). These authors used slightly different methods in that they subdivided most islands into two or more shields. On the other hand, this is compensated, at least in part, by extending the contour lines farther than in the present approach. The almost exact agreement in the average total volume is hardly significant, but it does appear that volumes on the order of about 20×10^3 km^3 are optimum values for magma

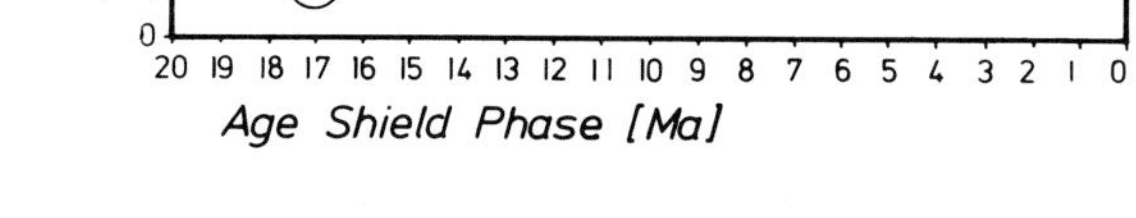

Fig.6 Percentage of subaerial to total volume of individual Canary Islands (F = Fuerteventura, L = Lanzarote, GC = Gran Canaria, Go = Gomera, Te = Tenerife, LP = La Palma, Hi = Hierro) versus the approximate age of the main shield building phase

Table 1 Submarine and subaerial volume of islands and seamounts in the Eastern Central Atlantic

	Island	Depth m (contour line)	Volume submarine ($10^3 km^3$)	subaerial	total	Percentage subaerial volume	Age of major subaerial shield phase (Ma)
Canary Islands	Canary Islands						
	Fuerteventura	3000	30.2	0.4	30.6	1.2	~ 17
	Lanzarote	3000	23.3	0.3	23.6	1.1	~ 15-17
	Gran Canaria	3000	23.0	0.8	23.8	3.5	13.5-14.0
	Tenerife	3500	21.6	2.0	23.6	8.4	~ 5
	Gomera	3000	10.5	0.5	11.0	4.5	~ 12
	La Palma	4000	16.0	0.8	16.8	4.9	~ 2-3
	Hierro	3500	5.1	0.2	5.3	3.2	~ 2-3
	Madeira	3000	8.8	0.4	9.1	4.2	2-3
	Porto Santo	3600	5.0	0.006	5.0	0.1	12-13
Azores	Santa Maria	1800	1.50	0.017	1.52	1.1	4-5
	São Miguel	1800	5.0	0.22	5.2	4.3	~ 4
	Terceira	900	1.63	0.11	1.74	6.1	< 1
	Graciosa	1800	1.89	0.097	1.89	0.5	?
	Pico	1600	1.7	0.2	1.8	10.2	< 1?
	Fayal	1250	0.8	0.05	0.8	6.2	< 1?
	São Jorge	1450	1.8	0.08	1.9	4.3	< 1?
Seamounts	Concepcion	2000	55.8	-	-	-	?
	Ampere	4700	11.0	-	-	-	?
	Gettysburg	4850	16.0	-	-	-	?
	Ormonde	4850	15.5	-	-	-	?
	Coral Patch	4700	13.1	-	-	-	?
	Echo	3000	14.6	-	-	-	?
	Selvagen	3600	48.5				

generation and rise into the crust to form oceanic islands in two of the major groups of oceanic islands in the ocean basins. Volumes of larger seamounts in this part of the Atlantic are of the same order of magnitude while the Azores islands have much smaller volumes (Table 1). For both areas, values must be corrected for porosity etc. in order to estimate real magma volumes, but seismic reflection data needed for such calculations are still scarce.

The percentage of the subaerial to the total volume ranges from around 1 in the eastern to 3 to 8% in the central and western islands (Fig.6). In conjunction with data on the east-to-west age progression of the archipelago discussed below, these are tentatively interpreted as follows: the subaerial portion of an island may not exceed about 10%, a maximum value reached when the constructive activity surpasses erosional removal. Thus, Tenerife appears to be still at peak activity while the ratio decreases roughly with increasing age in the islands Gomera, Gran Canaria, Fuerteventura, and Lanzarote. The young islands Hierro and La Palma, on the other hand, are still on the "ascending limb" of their growth curve.

Among 11 Hawaiian shields with percentage of subaerial rocks of 1% or more, the average subaerial percentage is 5.27%, again remarkably similar to the Canarian average of 3 to 8%. In the Hawaiian islands, the subaerial percentages are extreme for the mature volcanoes Mauna Kea (13%) and Mauna Loa (21%!) while growing Kilauea volcano has reached 4%.

The total volume of magma risen above the sediment surface in the Canary Islands (about 1.5 x 10^5 km^3) results in a production rate of 7.5 x 10^3 $km^3/10^6$a or 7.5 km^3 in 10^3a if the age of the archipelago is taken as 20 Ma. Historic eruptions have been recorded for only about 500 years, but exact volume estimates of historic flows were not found in the literature. The total area covered by historic flows (the past 500 years) of La Palma and Lanzarote is estimated to be about 250 km^2. Assuming an average thickness of 10 m, volumes are about 0.4 km^3 for La Palma and 2.1 km^3 for Lanzarote. An unknown amount of lava flowed into the sea, especially in La Palma where slopes are quite steep. Together with the numerous scoria cones on both islands and the volume of magma erupted in the several historic eruptions on Tenerife, the sum of historic eruptions is estimated to be about 8 km^3/500 a or 15 km^3 per 10^3a. This value, again, is remarkably similar to the average eruptive rate along the Hawaiian Chain, calculated as 18 $km^3/10^3$a by Bargar and Jackson (1974). It is difficult to interpret the higher present rate of about 15 $km^3/10^3$a with the average rate of 7.5 $km^3/10^3$a discussed above, owing to large uncertainties in the ages of the shields and in the crude methods of volume calculations. Nevertheless, eruptive rates in the entire archipelago appear to be higher at present judging from the abundant historic to prehistoric eruptions across the entire chain, a distance of some 450 km. Interestingly, present eruptive rates for Kilauea volcano are about 110 $km^3/10^3$a (Swanson, 1972), much higher than the average rate for the Hawaiian Chain (Bargar and Jackson, 1974). The problem of of global correlation of periods of high rates of magma eruption will be addressed again at the end of this report.

2.3 Age relationships

The main problem areas addressed by age dating studies are:

1) Beginning of subaerial volcanism in the area compared to that of the surrounding volcanic sea floor (ocean crust);
2) Duration of magmatic phases on single islands;
3) Migration of volcanism in the archipelago in time and space; and
4) Time relationship to structural movements in Africa, sea floor spreading rates, change in sedimentation pattern along the slope and growth of other islands.

Information about these questions is necessary to develop models about magnitude, duration and cause of melting events in the mantle.

The problem of when subaerial volcanism began in the Canary Islands has been the subject of a lively debate for more than a century. The fossiliferous calcarenites of the famous Las Palmas terrace, once thought to be early to middle Miocene (Rothpletz and Simonelli, 1890), but now known to be late Miocene to Pliocene in age (Lietz and Schmincke, 1975), are the only known major sedimentary unit interbedded with subaerial volcanics among the "older rocks" of the islands containing datable fossil assemblages. It is remarkable that, despite the pioneering work of Rothpletz and Simonelli, speculations on the age of the islands continue to be published with ages as old as early Mesozoic and late Paleozoic.

About 150 K/Ar age determinations of rocks from the subaerial part of the islands have been published, of which almost 90 are from Gran Canaria (Rona and Nalwalk, 1970; Abdel Monem et al., 1971, 1972; Lietz and Schmincke, 1975; Grunau et al., 1975; McDougall and Schmincke, 1976; Feraud et al., 1981) (Fig.7).

The oldest known submarine volcanics overly uplifted Cretaceous rocks in the Be-

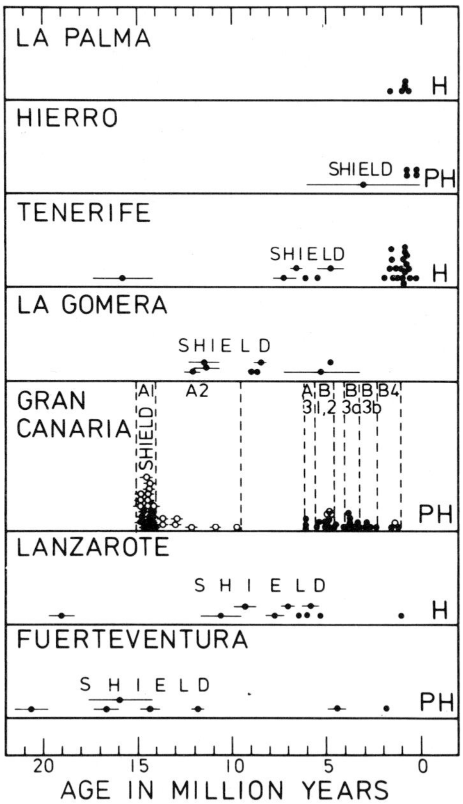

Fig.7 Potassium-argon ages from rocks of all Canarian Islands. Modified from Schmincke (1976) and McDougall and Schmincke (1976); including new data from Feraud et al. (1981).
H = historic, PH = prehistoric eruptions.
For explanations of Gran Canaria magmatic cycles and phases see text and Fig.14

tancuria Massif of Fuerteventura and are thought to be of Oligocene age based on paleontological evidence from interbedded calcarenites (Robertson and Bernoulli, this Vol.). Intrusive rocks from the same area also show Oligocene ages (Rona and Nalwalk, 1970; Abdel Monem et al., 1971; Grunau et al., 1975). Volcanic components in DSDP-sites 397 and 369, 100-150 km south of the area, are practically absent in rocks older than Miocene, but the upper Early Cretaceous to Paleogene record is missing in 397 and incomplete in 369 and holes closer to the area are not available (von Rad and Arthur, 1979). Seismic evidence north of the Canaries and east of Dacia Bank has been interpreted as indicating submarine volcaniclastic aprons of Eocene or slightly pre-Eocene age (Watkins and Hoppe, 1979).

Leaving aside the unresolved details of early submarine volcanism in the eastern part of the archipelago, there is also little known about the submarine phase of volcanism of the central and western islands. This is a major problem since these constitute 90% or more by volume of an island. The extrusive section (about 2000 m thick) of the uplifted submarine part of La Palma appears to have been constructed rapidly, perhaps during much less than 1 Ma (Staudigel, 1981). The base of the submarine apron of Gran Canaria and Tenerife may be around 18 Ma old, as suggested by seismic correlation to DSDP drill sites (Wissmann and Schmincke, in prep.). This would imply a period of 4 Ma (Gran Canaria) or more (Tenerife) for the growth of the submarine part of these islands. Since the 1000 m high subaerial part of Gran Canaria's shield formed during less than 0.5 Ma (McDougall and Schmincke, 1976), this would indicate increasing growth rates. This is difficult to accept. Clearly, the dating of the submarine phase of a volcano by seismic methods, while promising, should be checked by independent methods such as drilling through such an apron.

The subaerial part of Gran Canaria is dated in detail and is not older than Mid-Miocene (14 Ma) as shown by

1) abundant K/Ar-age data (McDougall and Schmincke, 1976); and
2) correlations of peralkaline rhyolites overlying the shield basalts with submarine ashes of the same composition, dated on biostratigraphic grounds (Schmincke and von Rad, 1979).

The hypothesis of Storetvedt et al. (1979) that the subaerial shield phases of Gran Canaria and Tenerife are Mesozoic or even early Mesozoic in age based on paleomagnetic data is thus at odds with all other geological data (Schmincke, 1979). Even though the very scant K/Ar data available for the shield phases of all other islands show a spread of several Ma (Abdel Monem et al., 1971, 1972), there is no evidence yet for significant subaerial (island) volcanism in the Canaries prior to the Miocene.

In a very general way, there is a decrease in age of subaerial shield volcanism from east to west. The main shields on the eastern islands of Fuerteventura and Lanzarote may have formed between 17 and 20 Ma, while the shield phases on the westernmost islands of La Palma and Hierro appear to be younger than about 4 Ma. This age decrease has been interpreted in terms of sea floor spreading by analogy to the more pronounced age progression observed in Pacific intraplate island chains (Schmincke, 1967; Abdel Monem et al., 1971; 1972). The exceptions to this pattern are discussed elsewhere (Schmincke,

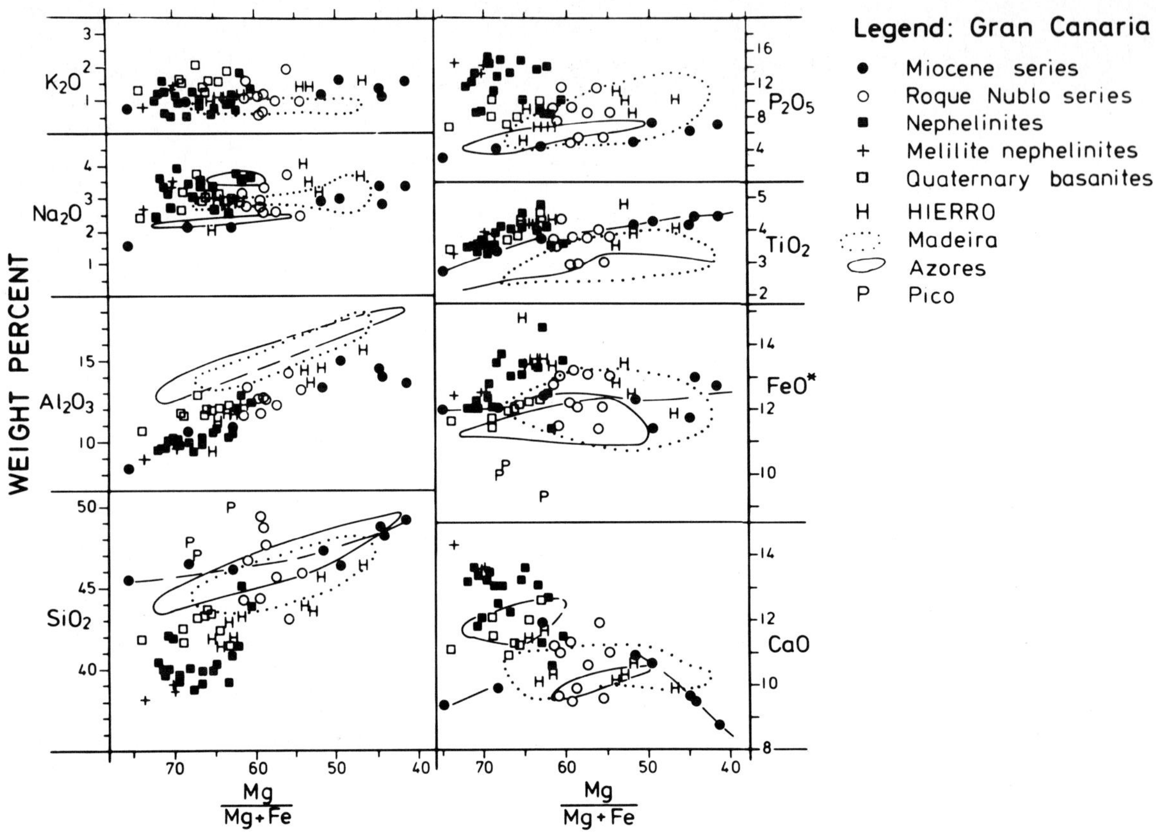

Fig.8 Major element concentrations versus Mg-number (Fe_2O_3 = 1.5% of total Fe) with emphasis on the mafic younger lavas of Gran Canaria. The Azores data are mostly from Santa Maria (Feraud et al., 1981) but include those from some other islands (Schmincke et al., in prep.). A high (younger) and a low (basement) group is present on Santa Maria, Pico magmas (P) contain less Fe and more SiO_2 than those from the other Azores islands at equivalent Mg-numbers. Data from the Canaries: Table 2; Schmincke, 1976; Lietz and Schmincke, 1975 and unpublished. Single line shows trend for Gran Canaria Miocene shield series except for TiO_2 where lower line represents very tightly grouped data from Azores islands

1976). Much more extensive dating of the shield lavas and submarine aprons of all islands is required, however, to enable a more precise evaluation of the age progression. The problem is discussed in more detail in section 3 of this report.

3. Chemical evolution

3.1 Magma types

Even though the Canary Islands are regarded as one of the petrologically and chemically best studied oceanic volcanic areas (Chayes, 1979), knowledge about the source, origin and differentiation of the magmas is sparse at best, because tightly constrained integrated field and laboratory studies are still lacking. The igneous rocks of the islands probably encompass the broadest spectrum of chemical compositions and mineralogy known in any intraplate oceanic setting (Table 2, appendix).

The rocks range from

1) hy-normative tholeiitic basalts to highly undersaturated melilite nephelinites among the mafic lavas;
2) oversaturated groundmass-quartz-bearing low silica rhyolites to extremely silica-undersaturated hauyne phonolites among the differentiated felsic rocks; and
3) layered ultramafic rocks and gabbros to alkali gabbros and syenites among the plutonic rocks (Figs.8,9).

They encompass one major ophiolite-like complex and several smaller pillow series as well as ash flow fields ranging from unwelded to extremely welded and crystallized ones.

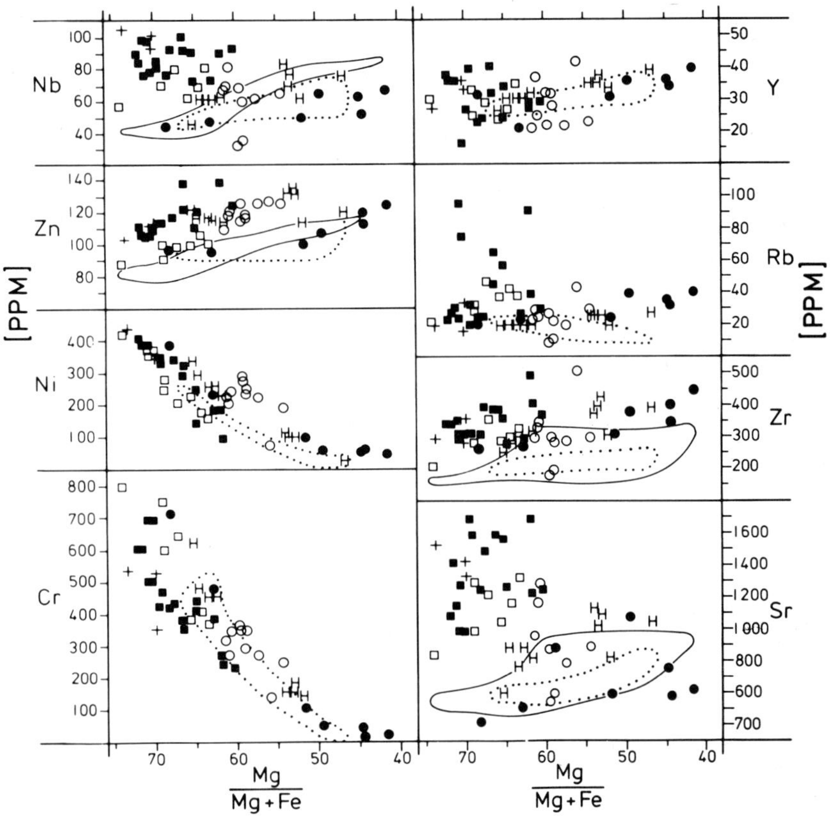

Fig.9 Trace element concentrations versus Mg-number for Canary Island magmas compared with those from Madeira and Azores. Symbols and source of data as in Fig.8

Three main types of magmatic evolution are distinguished:

1) Primitive magmas which, as discussed below, are characteristic of the late stage of a multiphase magmatic evolution and have risen to the surface without forming a crustal magma reservoir. They encompass basanites and olivine and melilite nephelinites.
2) Alkali basaltic magmas differing in degree of silica saturation form the shield and thus the dominant magma type on all islands. The paucity of primitive magmas (> 8% MgO), except for cumulitic rocks, as well as the occurrence of moderately evolved rocks (hawaiites and rare mugearites) indicate the existence of larger magma reservoirs. These are believed to be episodically recharged by more primitive magmas from the mantle which kept the system alive but prevented more extreme fractionation. Clinopyroxene and olivine are fractionated in these systems, followed by plagioclase. Typical late stage lavas are plagioclase-phyric hawaiites to mugearites.
3) At some point during upward growth of a volcano and magma chamber, highly differentiated magmas developed, commonly starting a period of emplacement of strongly evolved magmas, lacking eruptions of basalt. This is the cause for the so-called "Bunsen-Daly gap" in composition, for which Gran Canaria is probably the most clear cut example among oceanic islands (Schmincke, 1969).

Two types of differentiated magmatism are distinguished. Most common are the local endogenous and exogenous domes which are either saturated to oversaturated trachytes-syenites as on Fuerteventura (also Porto Santo, Madeira, Fayal) or undersaturated phonolites as on La Palma, Gomera and in part Tenerife and Gran Canaria (Roque Nublo formation). More voluminous and explosive volcanism generating ash flows and calderas (10-20 km in diameter) developed in the large central magma chamber systems of Gran Canaria (Miocene and, in part, Roque Nublo) and Tenerife, comparable to, but larger than, the caldera systems of western and central São Miguel and Terceira (Schmincke and Weibel, 1972; Self, 1976). These magma chamber systems must have been sustained by continued supply of basaltic magmas to their base. They developed strong mineralogical and chemical

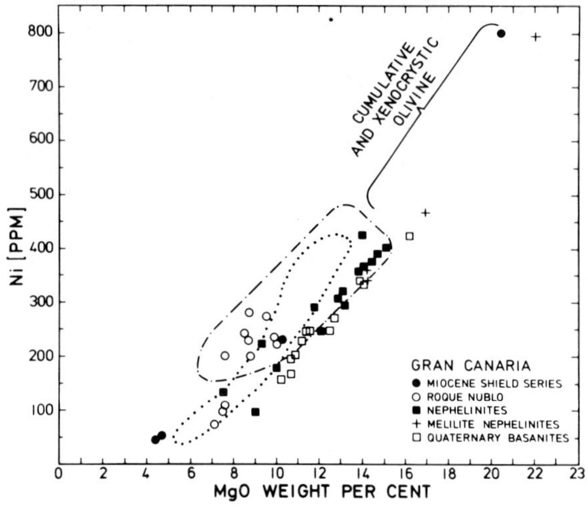

Fig.10 Ni versus MgO for mafic (Mg-number > 60) magmas from Gran Canaria, emphasizing posterosional lavas. Dotted field encloses basalts and differentiates from 5 other volcanic series, dashed lines are basalts containing high-pressure xenoliths (from Hart and Allegre, 1980)

gradients, as reflected in zoned ash flow sheets (Schmincke, 1969, 1973, 1976). The most spectacular of these is a composite rhyolite-basalt cooling unit several hundred km^2 in areal extent, starting the Miocene differentiated sequence on Gran Canaria. The low alumina content of the Canarian basaltic magmas in general (see below) and the fractionation of plagioclase during the waning shield stages in particular resulted in highly peralkaline felsic compositions.

In order to facilitate comparison among island groups and to enable discussion of mantle heterogeneity, I will now briefly discuss the most primitive magmas in the Canaries which have been neglected in the previous literature.

3.2 Primitive magmas

Shield basalt magmas are generally less mafic than posterosional mafic magmas on Gran Canaria and other islands. Basalts of the Miocene Gran Canaria shield, e.g. as well as the Roque Nublo shield, have MgO-contents almost exclusively < 10% (Mg-number < 65), except for apparently accumulative rocks (Figs. 8,10). In contrast, posterosional nephelinites, melilite nephelinites and basanites erupted on Gran Canaria have MgO-contents mostly > 10%, nearly all lying between 10 and 15% (Mg-numbers 60-72) (Figs. 8,10). Clearly accumulative rocks such as those melilite nephelinites containing desintegrated peridotite nodule material (xenocrystic olivine and orthopyroxene) have MgO-contents ranging from 16-22%. Posterosional magmas did not accumulate in high level magma reservoirs but were erupted directly to the surface from mantle depths. The near absence of more differentiated magmas in these series (Fig.11), the fact that dikes of these primitive lavas extend to the surface in many parts of Gran Canaria, and the presence of mantle derived peridotitic xenoliths (Table 3) in some lavas further support this interpretation.

The Ni-MgO data for Gran Canaria (Figs.8-10) indicate that primitive magmas of the posterosional nephelinite-basanite series had probably not suffered major olivine fractionation (Sato, 1978; Hart and Davis, 1978;

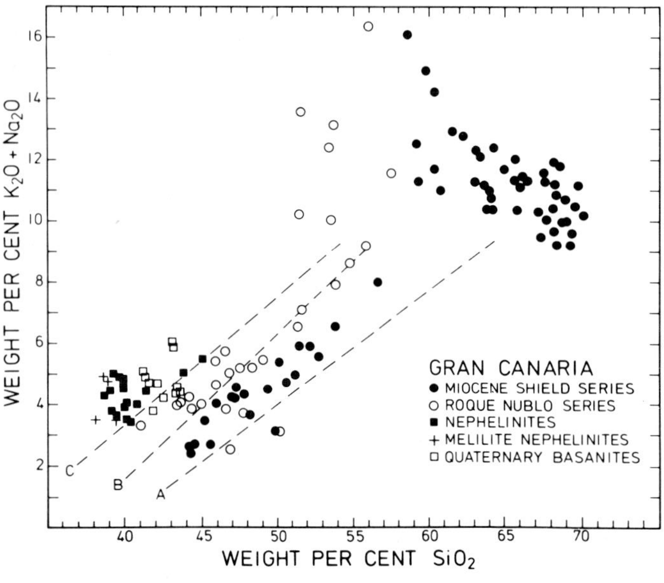

Fig.11 Total alkalis versus silica diagram for the Miocene and Pliocene (Roque Nublo) shield series and posterosional nephelinites, melilite nephelinites and basanites. Dashed lines separate tholeiites from alkali basalts (A) in Hawaii (McDonald and Katsura, 1964); alkali basalts and basanites (B) on Moheli Islands (Indian Ocean) (Strong, 1972); and basanites and nephelinites (C) in Eastern Africa (Saggerson and Williams, 1968). (Modified from Schmincke, 1976, Fig.29, including new data, Table 2 and unpubl.)

Table 3 Magmatic cycles and phases on Gran Canaria and on the Hawaiian Islands

	Phase	Parent magma	Duration of eruption (Ma)	Duration of non-eruptive interval to preceeding phase (Ma)	Volume (km^3)	Relative volume of total cycle (%)	Most differentiated derivative magma	Relative volume to parent magma (%)
Gran Canaria	Main shield (A 1)	Alkali to transitional basalt	< 1	-	$20x10^3$	> 95	Trachyte Comendite Pantellerite	< 10
	Minor shield (B 1)	Transitional to tholeiitic basalt	0,1?	-	1	< 1	-	-
Hawaiian Islands	Shield	Tholeiite	< 0,5	-	$20x10^3$	> 99	Rhyodacite	<< 1
Gran Canaria	A 2	Alkali basalt? (inferred)	-	< 1?	?	?	Trachyphonolite	< 10
	B 2	Alkali basalt, basanite	< 1	< 1	< 10	< 90	Ne-phonolite Hauyne phonolite	< 1
Hawaiian Islands	Alkalic	Alkali basalt	?	-	-	1	Trachyte	< 1
Gran Canaria	A 3	Nephelinite	0,1	4-6	<< 1	<< 1	-	
	B 3a	Nephelinite	0,5	0,5-0,8	< 10	< 10	-	
	B 3b	Melilite nephelinite	0,1	0,5-0,8	< 1	< 1	Pegmatoid	Mantle peridotite nodules
	B 4	Basanite	< 1,5	< 0,5		1	-	
Hawaiian Islands (Oahu)	Posterosional	Alkalic basalt Basanite Nephelinite Melilite nephelinite	< 0,5	2-4	< 10	<< 1	Pegmatoid	Mantel peridotite nodules

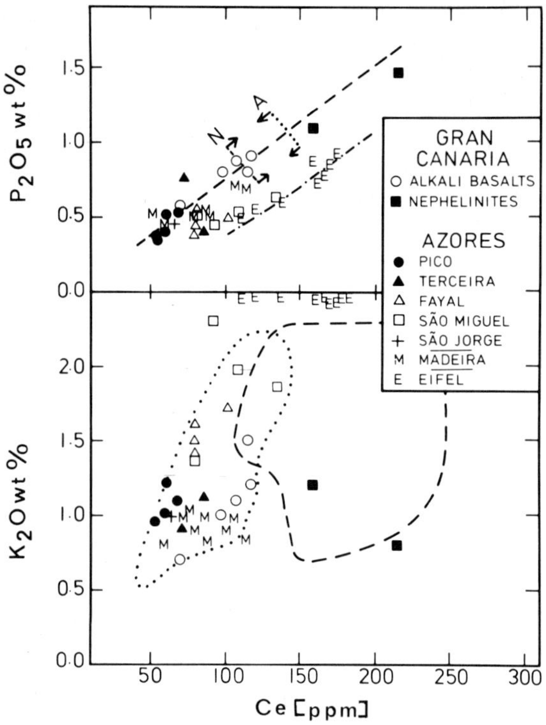

Fig.12 P_2O_5 and K_2O versus Ce-concentrations of mafic lavas from Gran Canaria, Azores and Eifel. Dotted and dashed lines show fields of alkali basalts (A) and nephelinites (N) from Sun and Hanson (1978). Data for Azores from Flower et al. (1976) and White et al (1979), Gran Canaria from Schmincke and Bowman (unpubl.) and Eifel from Schmincke and Gijbels (unpubl.). Note that São Miguel data follow the trend of the potassic Eifel magmas with higher Ce at equivalent P-concentrations

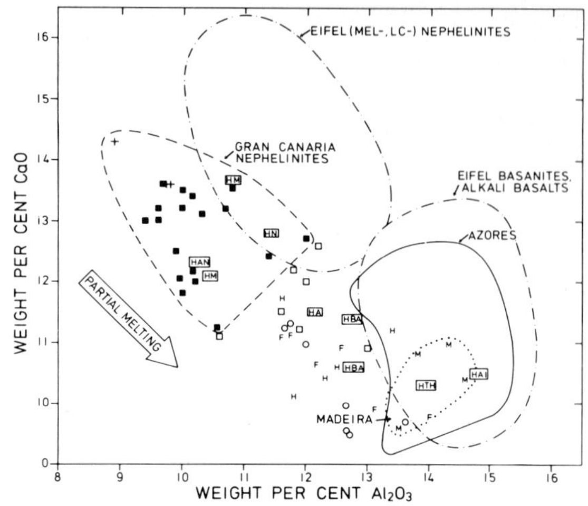

Fig.13 CaO versus Al_2O_3 of mafic lavas (Mg-number > 60) for Gran Canaria (symbols as in Fig.8), Fuerteventura (F) and Hierro (H). For comparison are shown fields of Azorean and Madeira lavas, Quaternary and Tertiary Eifel lavas and average compositions from the Hawaiian Islands (HM melilite nephelinite; HN nephelinite; HAN ankaratrite; HM mimosite; HA ankaramite; HBA basanite and basanitoid; HTH tholeiite; HAI alkali olivine basalt; data from McDonald, 1968, Table 8). Eifel data from Duda and Schmincke (1978), Mertes and Schmincke (in prep.) and unpubl. Arrow indicates direction of increasing partial melting, towards high Al and low Ca magmas with lower LIL-element concentrations

Hart and Allegre, 1980). Compared to other suites, the Gran Canaria posterosional magma series show high MgO-contents for equivalent Ni-values. K shows no positive linear relationship with either Ce or P for the highly undersaturated mafic magmas from Gran Canaria (Schmincke and Flower, 1974) (Fig.12) the colinear increase of both being interpreted as reflecting decreasing degrees of partial melting (Sun and Hanson, 1975). Residual phlogopite is still the most convincing explanation for the low K-content of the nephelinites (Schmincke and Flower, 1974; Flower et al., 1975; Sun and Hanson, 1975). The high CaO/Al_2O_3 ratio of the nephelinite and melilite nephelinite magmas (Fig. 13), which are much higher than chondritic, is commonly interpreted as reflecting residual garnet in the mantle (Frey et al., 1979; Brey, 1978). This interpretation is further strengthened by a low concentration of HREE in the melilite nephelinite (Fig.15), a correlation also characteristic of continental intraplate nephelinites and melilite nephelinites (Schmincke, 1979 b).

3.3 Magmatic Cycles and Phases

In terms of the dynamic history of the mantle-magma-volcano system, one of the most fundamental aspects of the magmatic evolution of the Canary Islands is a succession of magmatic phases (or stages) on single islands. The correspondence of eruptive growth stages to the pattern of magmatic evolution is clearly important evidence for interpreting mantle melting and evolution of the overall volcanic system. This pattern is best developed on Gran Canaria, where two major cycles are distinguished (Schmincke, 1973, 1976) (Figs.11-15; Table 3).

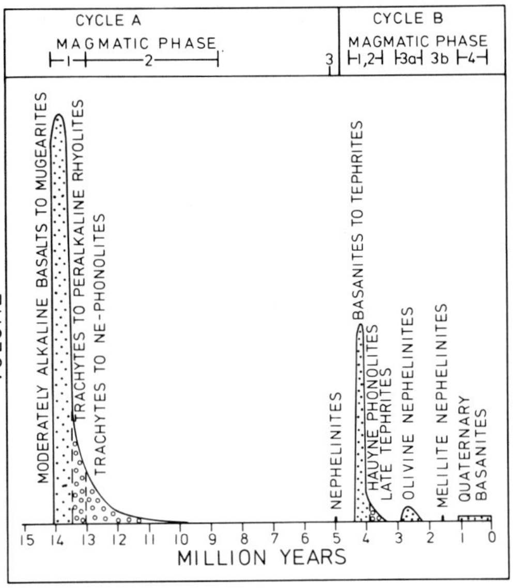

Fig.14 Volume versus age of major magmatic phases on Gran Canaria, interpreted to occur in two magmatic cycles (A and B). See text for discussion. (Modified from McDougall and Schmincke, 1976)

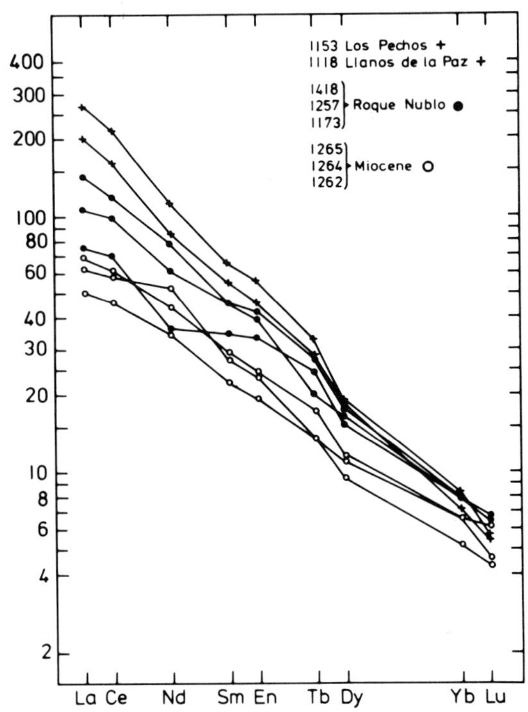

Fig.15 Representative chondrite-normalized REE compositions of Gran Canaria magmas (Miocene shield lavas, Roque Nublo lavas, and posterosional nephelinite (1118) and melilite nephelinite (1153). (Data from Bowman and Schmincke, unpubl.)

Cycle A

A I Lavas of this cycle comprise the initial or shield phase, during which 90% or more of the island is produced. The lavas are dominantly alkalic to mildly alkalic slightly undersaturated basalts. Subaerial effusion takes place rapidly, within less than 0.5 Ma (McDougall and Schmincke, 1976). Analysis of the uplifted submarine part of La Palma suggests that the early, submarine stage of island formation is also rapid (Staudigel, 1981). Derivative magmas erupted subsequently range from trachytes to peralkaline low silica rhyolites and these are also produced during less than 0.5 Ma.

A II These oversaturated derivative magmas are followed by large volumes of mildly undersaturated trachyphonolites and nepheline phonolites whose eruption - and intrusion of chemically similar magmas - may have lasted from about 12.5 to 8.5 Ma (McDougall and Schmincke, 1976), although age data are insufficient to fully assess the temporal evolution of this formation. These magmas may represent the decaying of phase I. Alternatively (Schmincke, 1976) they represent a new stage characterized by influx of more alkalic magmas (A II). These basaltic parent magmas remained at depth possibly due to development of a low density trachyte to rhyolite core of the island.

A IIIVery small volumes of nephelinites were erupted 5 Ma ago in several parts of the island, following the most extensive erosional interval in the island's history. Chemically, they resemble the younger nephelinites of phase B III.

Cycle B

B I,II The second major period of "basaltic volcanism", lasting from about 4.4 to 3.8 Ma ("Roque Nublo Group") was formerly considered to represent phase III (alkalic phase) in the island's evolution. I now prefer to interpret this phase as a new cycle. This cycle begins with a thin unit of hy-normative basalts that differ in trace elements from the basanitic-tephritic suite which is representative of the Roque Nublo volcanics (Brey and Schmincke, 1980). The basal hy-normative basalts are considered "very low volume phase I" and the bulk of the formation can be interpreted as "large volume phase II", although such definitions may force the issue. The final magmas of these phases are highly undersaturated hauyne phonolites, erupted about 3.7 to 3.5 Ma ago.

B III Following a second larger, non-volcanic-erosional interval, lasting from about 3.5 to 2.8 Ma, nephelinite and olivine nephelinite magmas were erupted in the center and eastern half of the island, chiefly between 2.8 and 2.6 Ma, but lasting to about 2.2 Ma. They were followed, after another erosion interval at 1.5 Ma by melilite nephelinites carrying abundant peridotite nodules. Local eruption of basanitic and, more rarely, nephelinitic lavas occurred from about 1.2 Ma (age of the "oldest" basanite of this phase) to the present.

Magmatic cycles broadly similar to those on Gran Canaria occur on other Canary Islands (Schmincke, 1973, 1976; Schmincke and Flower, 1974; Fuster, 1975) and have been described from Hawaii (Stearns, 1940; McDonald and Katsura, 1964; McDonald, 1968), the Indian Ocean (Flower, 1973) and from some continental intraplate settings (e.g. Duda and Schmincke, 1978).

The development of such cycles which represent one of the fundamental evolutionary processes of intraplate mantle/magma systems is little understood, but boundary conditions can be brought into focus by comparing the magmatic evolution on Hawaii and Gran Canaria (Table 3). Clearly, a "primary melting episode" (Schmincke, 1976) occurs, during which the bulk of an island is produced. This is followed by one or more "secondary" melting episodes, generating much smaller volumes of more alkalic magmas including, and often dominated by, nephelinite magmas. The terms "primary" and "secondary" were chosen to emphasize that the one or more magmatic stages following the shield building phase are a consequence of the initial, primary melting episode. In other words, the cause of the primary melting episode triggers a chain of events that follows basically similar patterns in many intraplate settings, which may last for several million years, must effect the entire lithosphere above and around the melting area and must, therefore, be explainable in the framework of plate tectonics.

Distances between major basaltic shields in the Canary Islands are about 80-110 km and degrees of melting may be around 15%, lower than beneath Hawaii (Jackson and Wright, 1970) as indicated by the more alkalic composition and the higher degree of enrichment in low-K_D elements in the Canary Island shield magmas, assuming broadly similar mantle starting compositions (Frey et al., 1978). Thus, for volumes of magma of 20×10^3 km^3 and 15% partial melting, a mantle volume of 135×10^3 km^3 (e.g. a cylinder 50 km in diameter and 70 km in height) will be involved in the partial melting process. This melting - or at least the withdrawal of the melt - must proceed very rapidly. For Gran Canaria, the shield phase may have been constructed during as little as 0.5 to as much as about 3 Ma (see above).

The nephelinitic or second or third stage (posterosional) magmas are similar in Gran Canaria and Oahu suggesting similar melting conditions. Green (1971) invoked deep fractures reaching the "enriched" asthenosphere to explain generation of nephelinite magmas beneath the Hawaiian chain. The pattern of posterosional Honolulu series vents on Oahu are at right angles to that of the tholeiitic shield and were interpreted as possibly due to uplift caused by erosional unloading of the island (Jackson and Wright, 1970). A similar change in major dike orientation associated with nephelinite volcanism (ca. 2.8 to 2.2 Ma) on Gran Canaria (Lietz and Schmincke, 1975) might also reflect a major change in stress regime during magmatic evolution. Cooling of the lithosphere following a primary melting episode may also result in deep fractures that could tap the asthenosphere and generate posterosional volcanism as proposed for Hawaii (Nakamura, pers. comm., 1979).

3.4 Multiphase Magmatic Evolution on Islands in the Eastern Central North Atlantic

Multiphase magmatic evolution with the more alkalic and silica undersaturated, generally more mafic magmas erupted during later stages in an island's evolution are now well documented from other Canary Islands (e.g. Fuster, 1975; Pellicer, 1979). In the Azores, slight chemical differences between the shield and the less saturated magma composition of later stages are described from Santa Maria and São Miguel (Feraud et al., 1981; Fernandez, 1980).

The Cape Verde Islands were traditionally regarded as very alkaline (Mitchell-Thome, 1976; Klerkx and de Paepe, 1976). More recent work, however, shows this interpretation to be an artifact of biassed sampling, nephelinites and melilite nephelinites only forming a thin cover above older and more voluminous (shield?) series of alkali basalt (de Paepe and Klerkx, 1976), thus resembling the trend exemplified by Gran Canaria.

Neither Lanzarote (Ibarrola, 1969) nor Madeira (Aires-Barros et al., 1974) and Porto Santo or some of the Azores Islands show evidence for multiphase magmatic evolution leading to more alkalic composition. However, such products may have been removed from the highly eroded island of Porto Santo and some of the Azores are indeed very young islands.

Several theories have been advanced to explain regional differences in the chemical composition of ocean island basalt magmas in the Central Eastern Atlantic ocean basin. McBirney and Gass (1967) postulated a decrease in silica saturation in volcanic rocks from islands close to the Mid-Atlantic Ridge, such as the Azores, to those close to continents such as the Canary Islands which they attributed to increased depth of magma generation and thus smaller degrees of partial melting away from the Mid-Atlantic Ridge.

Schmincke (1969, 1973, 1976) showed, however, that the Canary Islands and even single islands such as Gran Canaria span the whole range of quartz Niggli values found in volcanic rocks from the entire Atlantic (Fig.40 in Schmincke, 1969). He argued that, in view of the difference in chemical composition of multiphase magmatic evolution, regional comparisons should be made on equivalent evolutionary stages and that magmatic evolution on the Azores, a group of mostly young islands in their shield stage, might not have progressed to more alkalic and silica undersaturated stages, such as found on older Canary Islands.

Nevertheless, the question remains why multiphase magmatic evolution is very well developed on some islands like Gran Canaria but barely detectable on others like Santa Maria. An answer to this question may lie in a combination of the models of McBirney and Gass (1967) and Schmincke (1973).

The decrease in heat flow values from the MAR to the continents upon which McBirney and Gass (1967) based their theory is now explained by cooling of the lithosphere while it moves away from a mid-ocean ridge. A steady value and roughly constant lithosphere thickness are attained after some 60-80 Ma (Parsons and Sclater, 1977; Forsyth, 1977; Oxburgh, 1980). The shield basalt magmas in the Azores, Madeira, Gran Canaria and the eastern Canaries are broadly similar, silica-saturated or mildly undersaturated alkali basalts and their silicic derivative magmas are correspondingly similar, even though they were generated in areas of regionally thin lithosphere (20-30 km around the Azores) as well as "mature" oceanic lithosphere (ca. 70-80 km around Madeira and Canaries). As a first approximation, the similarity in the chemical composition of the shield magmas is interpreted as reflecting broadly similar degrees of partial melting. If pressure is singled out as a major factor controlling degree of melting, for simplicity disregarding factors such as CO_2/H_2O-ratios etc., and taking the base of the lithosphere as the main site of partial melting, two alternative scenarios can be defined: The lower boundary of the lithosphere (the wet peridotite solidus?) may have risen appreciably during partial melting in the shield stage of islands sitting on thicker, mature lithosphere such as the Canary Islands and Madeira. This may have resulted in a temporary thickness of the lithosphere that was broadly similar beneath all three island groups. Alternatively, the major melting areas beneath the Azores may reside well below the base of the lithosphere in that area.

On the other hand, lithosphere thickness cannot be increased in the Azores to allow high pressure undersaturated magmas to be formed. Viewed in this context, magmatic evolution in the Azores may never lead to strongly silica undersaturated basanitic-nephelinitic magma types characteristic of the posterosional magmatism on Oahu and Gran Canaria. Available chemical data of rocks from the oldest island (Santa Maria, Azores) show that nephelinites and strongly undersaturated basanites have not so far been found. Iceland, e.g. which geodynamically and geochemically combines features of both oceanic islands and mid-ocean ridges, while having erupted alkalic basalts locally, does not contain highly undersaturated basanitic or nephelinitic magmas, even though the island is some 20 Ma old (Jakobsson, 1972).

The reason why islands such as Lanzarote and the Madeira group having developed on thick lithosphere apparently have not erupted highly alkalic posterosional magma types as chemically distinct as those e.g. of Gran Canaria is not clear. Lithosphere thickness alone does not appear to be sufficient to explain the occurrence or absence of multiphase magmatic evolution of strongly contrasting compositions. Part of the reason may be sought in the size of the thermal pertubation in the mantle and the depth extent of the zone of partial melting. If the volume of the primary melting area is a major factor - as suggested by some but not all data on oceanic islands situated on thick lithosphere - cooling contraction, following the shield phase, will be of sufficient magnitude only in large volume systems, thus allowing lithospheric fractures to be formed through which posterosional mafic nephelinitic to basanitic magmas can rise. This reasoning does not apply to systems along or near mid-ocean ridges.

In summary, shield basalt magmas are broadly similar in the eastern North Atlantic including the island groups Azores, Madeira, Canaries, and possibly Cape Verde Islands, disregarding smaller differences in degree of melting and mantle heterogeneity as discussed below. Distance from the Mid-Atlantic Ridge and thus dependance on lithosphere thickness and geotherm does not appear to be the overriding factor controlling the chemical composition of the shield magmas, the reason for the similarity in melting conditions remaining unknown.

The magma suite: (basanite)-olivine nephelinite-melilite nephelinite does, however, appear to be restricted to islands developed on relatively thick lithosphere. It appears petrologically reasonable to view these magmas as part of the spectrum with MORB-magma generated at low pressures (ca. 20-30 km) beneath mid-ocean ridges characterized by very shallow geotherms at one end (e.g. Green, 1971) and kimberlite magmas generated in cratons with thick lithosphere at great depth characterized by steep geotherms (low heat flow) (Dawson, 1980).

3.5 Chemical Differences between Islands, Island Groups and Mantle Heterogeneity

Chemical differences between island magmas from the central eastern North Atlantic have been inferred traditionally from major elements, especially from the difference in Na/K-ratio (Le Maitre, 1962; Schmincke and Weibel, 1972; Schmincke, 1973). Extremes are the low K-Madeira magmas on one and the high K São Miguel magmas on the other hand (Schmincke and Weibel, 1972; Schmincke, 1973). Although differences in major elements such as Na and K can be due to variable melting conditions and residual mineralogy, it is nevertheless remarkable that $^{87}Sr/^{86}Sr$-ratios in ocean island basalt magmas between 20° and 36° N are highest on São Miguel and lowest on Madeira (Hawkesworth et al., 1979; Staudigel and Schmincke, unpubl.), indicating real mantle heterogeneity. Differences in mantle composition can be recognized in radiogenic isotopes and trace and major elements which are briefly reviewed.

Oversby et al. (1971) postulated a heterogeneous mantle with respect to Pb-isotopes beneath the Canaries, with a western domain encompassing La Palma and Tenerife and a less radiogenic eastern domain beneath Fuerteventura and Lanzarote. Based on much more extensive and more precise data, Sun (1980, p. 416) found no correlation of Pb-isotopic composition of lavas "with geographic position, age or rock types". Inspection of his data shows, however, that $^{206}Pb/^{204}Pb$, $^{207}Pb/^{204}Pb$, and $^{208}Pb/^{204}Pb$ for Lanzarote are mostly lower than those of other islands. The two samples (LZ 3 and 105) overlapping with the field of other islands are Miocene basalts, while one Miocene trachybasalt (LZ 106) is similar to the Quaternary and Recent basalt compositions. Moreover, there is a tendency on all islands for younger rocks to show lower Pb isotope ratios and for highly differentiated rocks (phonolite, trachyte) to have very radiogenic Pb isotopic compositions (as also shown by data from Gough and the Azores, Sun, 1980, Table 5). Although it appears that islands such as Hierro and La Gomera may occupy distinct fields, the number of data is too small for definite conclusions. Thus, I do not share Sun's skepticism and suspect that despite some overlap, correlations can be found between Pb-isotopes of Canarian magmas and age, rock type and geography. Major and trace element data are commonly not presented together with isotope compositions, preventing a more detailed assessment of the suspected correlation, in some islands, of degree of magma fractionation with isotope composition. It appears likely, that restudy of a large island group such as the Canaries using only primitive basalts, basanites and nephelinites would yield detailed information on the composition and evolution of different melting domains in the sub-Canarian mantle. Lead isotope data from the Azores and Cape Verde Islands presented by Sun (1980) are similar to, and partly overlap, with those of the Canaries, with the Azores showing generally high and the Cape Verde Islands low Pb-isotope ratios.

$^{87}Sr/^{86}Sr$-data on rocks from the Canaries (O'Nions and Schmincke, unpubl.; Staudigel, unpubl.; and Sun, 1980, Fig. 19) are between 0.7o29 and 0.7035, with most clustering between 0.7030 and 0.7033. Nephelinites on Gran Canaria have lower Sr-isotope ratios than the alkali basalts and basanites from the shield and other phases. Sr-isotope values from the Cape Verde Islands (Klerkx et al., 1974) are similar to those from the Canaries while Madeira shows exceptionally low values (0.7028 - 0.7032) (Staudigel, unpubl.). Generally higher values (0.7033 - 0.7035) are reported from the Azores with Fayal and Pico values ranging up to 0.7039 and São Miguel showing exceptionally high values up to 0.7053 (White et al., 1979; Hawkesworth et al., 1979).

A broad range of trace elements has recently been determined for lavas from the Azores (Flower et al., 1976; White et al., 1979; Feraud et al., 1981; Schmincke, unpubl.), Madeira (Hughes and Brown, 1972; Feraud et al., 1981; unpubl. data) and Canary Islands (Table 2; unpubl. data). During the last few years, it has been found that ratios of trace element pairs having nearly identical distribution coefficients such as Zr-Hf and Ta-Nb have identical ratios in all

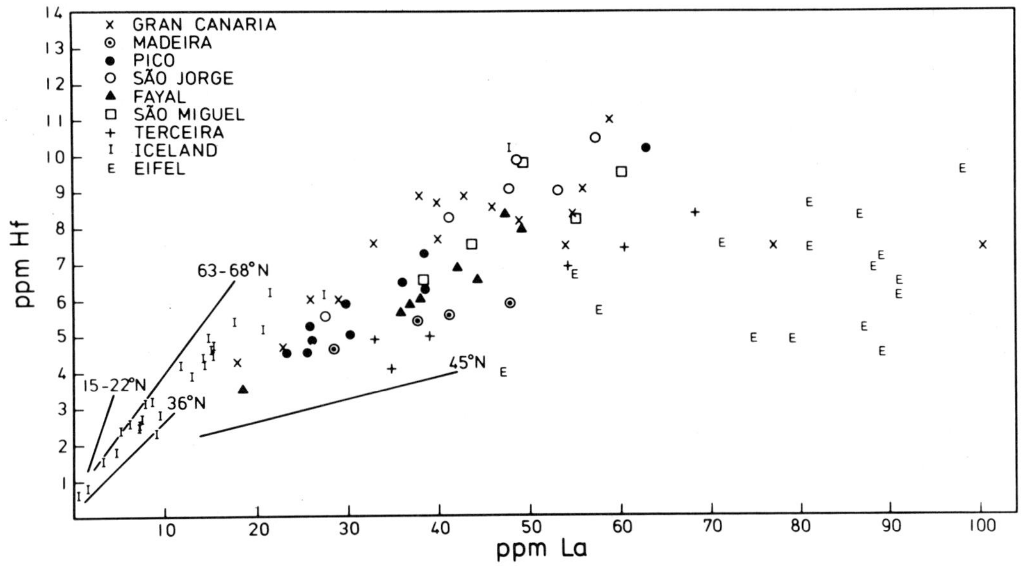

Fig.16 Comparison of Hf/La-ratios of MORB lavas from the North-Atlantic (Bougault et al., 1979; Tarney et al., 1979) and eastern central North Atlantic islands and Eifel (data from Flower et al., 1976; White et al., 1979 and Schmincke, Bowman, Gijbels and Wänke, unpubl.)

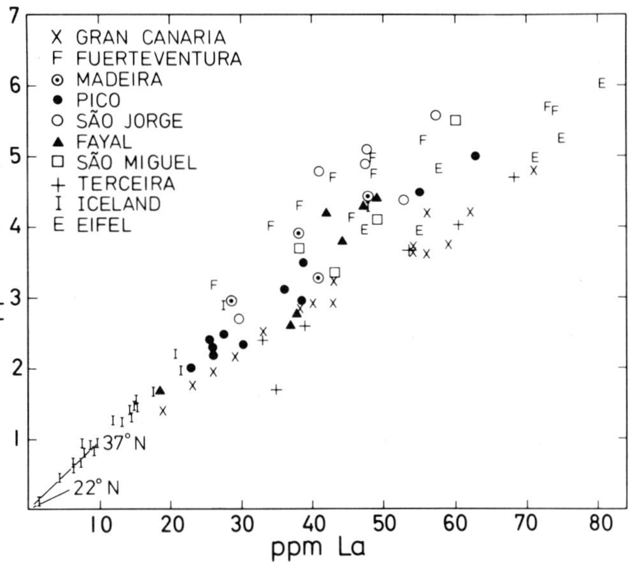

Fig.17 Comparison of Ta/La-ratios of MORB lavas from North Atlantic MORB and OIB lavas and Eifel. Source of data as in Fig.16

basalts, contrasting with others such as Hf/Ta, Ta/Th and Nb/Zr that show distinct differences in different parts of the North Atlantic which are maintained along mantle flow lines (Wood et al., 1979; Bougault et al., 1979) (Figs. 16-18). Apparently, mantle domains of various sizes exist that are chemically distinct. The data for OIB basalts from the three areas under discussion fall broadly between the MORB data for 36° and 45° N (high Nb/Zr-ratios) and 63° N (low Nb/Zr-ratios) (Fig. 18) as defined by Wood et al. (1979). They are much higher than those for MORB data from 22° N. However, distinct inter-island differences exist, indicating mantle heterogeneities on the scale of 10-100 km. The Hf/La-ratios are also intermediate between the two MORB-groups listed, are different from 22° N MORB and show similar distinct inter-island differences. Among the trace elements, Zr is unusually high in Canarian magmas except for the two tholeiitic Roque Nublo lavas (Figs.9,18). Higher concentrations in Zn, Nb, Rb and Sr are typical of the posterosional magmas of Gran Canaria, such magmas being absent in the Azores and Madeira group.

Major element differences between primitive basaltic magmas may largely reflect variable degrees of melting and fractionation rather than real differences in mantle composition. Nevertheless, they can reflect mantle heterogeneity if supported by other criteria, as was discussed in the case of major differences in K-contents such as between Sao Miguel and Madeira. Inspection of Fig. 8 representing a rather small sample of analyses done in one laboratory shows that at least some Canary Island magmas are clearly lower in alumina than those from the Azores and Madeira and this holds for different phases on a single island (and thus

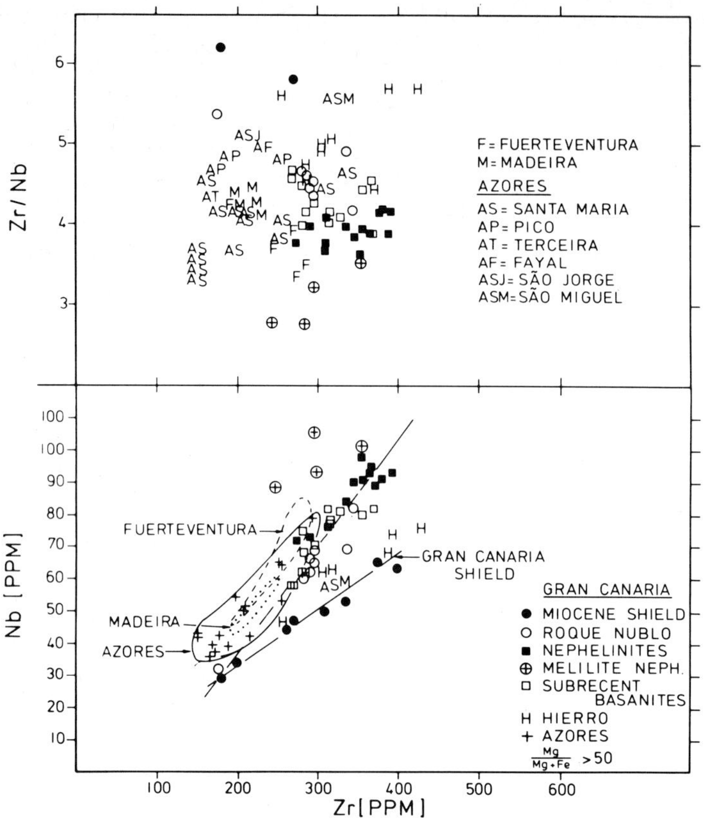

Fig.18 Zr versus Nb and Zr/Nb for Gran Canaria magmas (emphasis on posterosional mafic lavas) compared to the Canarian Islands Fuerteventura and Hierro as well as Azores and Madeira. Basement lavas from Santa Maria have higher Zr/Nb than younger series. Left line in lower figure connects younger Gran Canaria lavas. Sources of data as in Fig.8

presumably different degrees of partial melting) as well as between islands. In fact, Canarian magmas appear to be lower in Al than most other ocean island magmas compared on equivalent Mg-number and degree of low K_D element enrichment values (Fig.8). Canarian magmas are also higher in Ti and, in part, Fe than those from the other island groups, Pico (Azores) being especially low in Fe. The other differences in Fig.8 are chiefly between the Miocene shield lavas of Gran Canaria and Madeira and Azores lavas on one and the more alkalic and undersaturated magmas of Gran Canaria on the other hand, with Roque Nublo (Gran Canaria) and Hierro magmas being intermediate.

Some of these differences may be largely or entirely due to differences in melting conditions. For example, the low alumina content of the Canarian basalt magmas may indicate derivation from a garnet bearing source with garnet as a residual phase as also suggested by the sparse REE-data (see above). The higher Fe-content of the Gran Canaria nephelinites compared to the basanites and alkali basalts (Fig.19) can also be explained by lower degrees of partial melting (Langmuir and Hanson, 1980). However, there is no evidence from concentrations of P and LIL elements that Canary Island (especially Hierro) alkali basalt to basanite magmas which are very iron-rich represent much lower degrees of partial melting than alkali basalts from the Azores and Madeira. The differences might indicate a sub-Canarian mantle significantly richer in Fe than that beneath the Azores and Madeira, although large temperature differences could be another explanation (Langmuir and Hanson, 1980).

The high Ti, in part Na and low Al-concentrations of the Canary Island magmas are regionally matched by the distinction of a southern (south of 29° N) province of MORB basalts being richer in Na, Ti and lower in K and Al than MORB basalts north of 29° N (Melson and O'Hearn, 1979; Dmitriev et al., 1979). Island magmas on the African plate (Madeira, Canaries, Cape Verde Islands) have relatively low $^{87}Sr/^{86}Sr$-ratios corresponding to low Sr-isotopes of MORB basalts in this area, although trace element ratios of island basalts do not match those of MORB from 22° N (see above). Much work is presently being done to study the composition of MORB magmas in the central North Atlantic and forthcoming data will thus allow more detailed comparisons in the near future. It is here speculated that the tentative subdivision of the North Atlantic oceanic crust and mantle into a more heterogeneous northern and a more homogeneous southern part is reflected in the island magmas as well. If

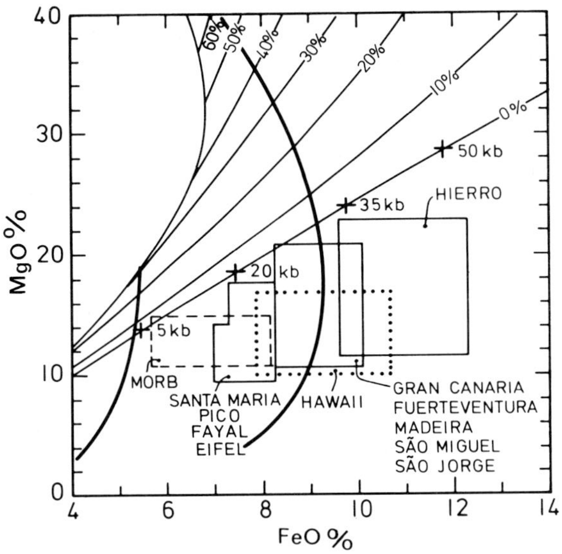

Fig.19 Cation mole percentage of MgO versus FeO (after Langmuir and Hanson, 1980) for primitive (Mg-number 60) basalts from the Canaries, Madeira, Azores, Hawaii and MORB. Percentages are degree of melting; heavy lines olivine fractionation curves; MgO and FeO of melt at solidus increase with increasing pressure (5 to 50 kb) (Langmuir and Hanson, 1980, Fig.3). Data for Canaries, Madeira, Azores, Eifel see Figs.8 and 16; for MORB and Hawaii see Langmuir and Hanson.

supported by future data, this would be a major argument against the popular idea that MORB and OIB magmas are derived from very different mantle sources (e.g. Sun and Hanson, 1975). Instead, it would indicate similar long term mantle histories, with some of the heterogeneities being developed during earlier phases of ocean crust formation and subsequent subduction. A comparison of Paleozoic proto Atlantic ophiolite geochemistry and that of present Atlantic basalts may shed more light on this problem. Recently, Chase (1981) has suggested that ancient subduction processes may have lead to secondary enrichment in the mantle reservoir, thus explaining the more radiogenic ocean island Pb-isotopes.

4. Origin of the Canary Islands

4.1 Continental or Oceanic Crust/Lithosphere beneath the Canary Islands?

The presence or absence of continental crust beneath the Canary Islands is one of the oldest and most controversial problems concerning the origin of the islands. Opposing views as to the nature of the crust beneath the archipelago are found in the most recent literature. The problem has been addressed through geological (tectonic, lithologic, stratigraphic and physical age data) and geophysical approaches (especially seismic refraction data). Hausen, who published monographs of each of the islands between 1956 and 1973 held the traditional view that the Canarian archipelago represented a microcontinent, underlain by sialic crust, later faulted, with the islands representing unsunken blocks. He based his idea chiefly on the tectonic interpretation of the steep western flanks of many islands and the misinterpretation of felsic plutonic rocks as continental granites. Schmincke (1967) suggested that at least the central and western Canaries are oceanic islands. Rothe and Schmincke (1968) presented further arguments for a subdivision of the island group into an eastern part (Lanzarote and Fuerteventura) possibly underlain by continental crust, and the larger oceanic central and western part of the group. A similar interpretation resulted from the first modern marine geophysical studies in the area (Dash and Bosshard, 1968; Bosshard and MacFarlane, 1970). Dietz and Sproll (1970), emphasizing this view, postulated that the eastern Canaries represent part of the African continental lithosphere that had been separated from the African continent. Sauer and Rothe (1972) stress the presence of continental crust beneath the eastern Canaries while Schmincke (1973, 1976) holds that the occurrence of quartz-bearing xenoliths in some lavas of the eastern Canaries does not require the presence of continental crust but merely quartz-bearing sedimentary rocks of the type cropping out on Fuerteventura. Similarly, the former presence of landbridges between the eastern Canaries and Africa postulated to account for the presence of ostrich eggs on Lanzarote (Rothe, 1968; Sauer and Rothe, 1972) meets serious geological problems (Schmincke, 1976): if such landbridges existed, they must have disappeared prior to deposition of the Mesozoic and Tertiary sediments that are known to exceed 10.000 m in thickness, filling sedimentary basins between the eastern Canaries and the African shelf (Roeser et al., 1971). This would also require that Lanzarote would have been in existence and above sea level for some 180 Ma - instead of not more than about 20 Ma as appears plausible from data available. Moreover, no continent-derived detritus has been found beneath, or interlayered with, the basal shield or submarine lavas. The only alternative to a very long-lived landbridge is the temporary existence of volcanic islands (or horst blocks) between the eastern Canaries and Africa, or a short term drastic lowering of sea level.

The major faults that were believed by Hausen and earlier authors to bound the westward sides of the Canary Islands were shown to be non-existent (Schmincke, 1968, 1976). On the other hand, major uplifts exceeding 1000 m must have occurred on both Fuerteventura and La Palma as shown by the Cretaceous sediments and submarine volcanics on the former (Rothe, 1968; Robertson and Stillman, 1979) and submarine intrusive and extrusive series on La Palma (Schmincke and Staudigel, 1976; Staudigel, 1981). Moreover, there is convincing seismic evidence for offshore faulting as in the area west of Fuerteventura (Wissmann and Schmincke, in prep.). Thus, "when all geological evidence presently available is critically evaluated, there is no compelling proof for the existence of continental crust beneath the eastern Canaries" (Schmincke, 1976, p.80).

The idea that the Canary Islands are underlain by continental crust has recently been revived (Storedtvedt et al., 1979) but the arguments listed above and others are not in accord with this hypothesis (Schmincke, 1979a).

Thus, while the weight of presently available geological, geochemical and geophysical evidence would indicate the boundary zone to be situated between the eastern Canaries and Africa, the geological and geochemical idiosyncracy of the eastern Canaries is just sufficient to allow - but not to require - that they have a different deep basement composition than the other Canaries. Evidence is more persuasive, however, for a different past and present dynamic history (tectonic uplift, degree of partial melting etc.) beneath the eastern than the central Canary Islands than for a different composition. The key to interpreting the initiation of melting beneath the Canary Islands may lie beneath the eastern rather than the western islands as discussed below.

4.2 Origin of the Canary Island melting anomaly

Wilson (1963) in his classic paper explaining age progression of oceanic volcanic chains by migration of what is now called lithosphere over hot spots also used a purported age progression in the Canary Islands as evidence for such migration. The age data used by Wilson are wrong but an age progression can nevertheless be recognized. This was postulated by von Fritsch (1867) who based this on decreasing degrees of erosion from east to west, just as Dana (1849) had noted increasing degrees of erosion - and therefore older age - from southeast to northwest in the Hawaiian Islands. The interpretation of this "erosional" age pattern in terms of sea floor spreading (Schmincke, 1967) was put on a more quantitative base by the K/Ar-age determinations of Abdel-Monem et al. (1971, 1972). Volcanic age progressions calculated from these data are not in good agreement with known sea floor spreading rates (Schmincke, 1973). These discrepancies led Anguita and Hernan (1975) to invoke propagating fracture zones to explain Canary Island volcanism in time and space. More detailed work showed that the ages for the shield lavas of Gran Canaria (and possibly other islands as well) were probably in error by several million years reducing some of the discrepancies noted by the above authors (McDougall and Schmincke, 1976). Nevertheless, detailed age studies of all islands are required before the question of age progression can be discussed in more precise terms.

The Canary Islands are considered to be the site of a mantle plume by some authors, i.e. mantle material upwelling from great depth which partially melts during rise by decompression. Morgan believes he can detect the former trace of this plume as far back as 120 - 180 Ma where it was located between Labrador and Greenland, "pre-weakening" the split that became the Labrador Sea (pers. comm. 1976). Perhaps the two main problems with the plume or hot spot model, if regarded as fixed in the mantle and geographically relatively well defined are:

a zone of seamounts and islands extends for about 1000 km both north and south of the Canaries and up to 500 km west of the African coast. The Madeira group to the north and the Cape Verde Islands to the south emphasize the existence of such a "volcanic belt". Although they occur farther west of the continental rise, Tertiary volcanics occur just west of the boundary between continental and oceanic crust west of Dakar, i.e. due east of the Cape Verde Islands (Fig.4 in Lehner and de Ruiter, 1977). A multitude of "plumes" of various sizes would seem to be required to explain the large number of these volcanic centers.

Although multiphase magmatic evolution similar to that in the Hawaiian chain characterizes many of the Canary Islands, there are many exceptions. For example, within the last 5000 years, all of the Canary Islands with the exception of Gomera have erupted lavas; the chemical composition and time pattern of some phases such as the historic tholeiite eruptions of Lanzarote do not fit easily into the simple pattern of multistage magmatic history discussed above. Even though the young island of La Palma, the westernmost island, is most active at present (as might be expected from its position at the western end of the "chain") it is not known whether the voluminous basaltic eruptions on Lanzarote in the 18th and smaller eruption in the 19th century might not herald a new major cycle, resembling the Roque Nublo volcanic cycle on Gran Canaria 4.4 Ma ago. The Canaries can thus be regarded as an east-west volcanic belt about 450 km long, underlain by several presently active melting areas. It should be possible to detect this major elongate zone of partial melting in the mantle by travel time delay studies of teleseisms. The area of partial melting should be especially large and shallow at present beneath the Eastern Canaries, because the composition of lavas erupted during recent and subrecent times suggests larger degrees of partial melting possibly at shallower depth.

Perhaps the most traditional and most common view in the literature on the origin of the Canary Islands is a connection with tectonic features on Africa (such as the South Atlas fault) or, more recently, fracture zones in the oceanic lithosphere (see review in Schmincke, 1976). However, no fracture zone is known to intersect with the Canarian archipelago (Fig.2) nor is there any evidence for a fracture zone tentatively postulated to explain the volcanism of the Cape Verde Islands (de Paepe and Klerkx, 1976) (see Fig.4 in Lehner and de Ruiter, 1977). Moreover, recent seismic data do not show the South Atlas fault to continue to the Canary Islands (Hinz et al., and Stets and Wurster, both this volume). This does not exclude the possibility of synchronous deformational events in Northern Africa and phases of magmatic activity in the Canaries as discussed by Anguita and Hernan (1975). Both however may be independent expressions of the same cause.

In the framework of plate tectonics, oceanic islands are commonly regarded to exemplify "intraplate" volcanism. However, there may be few examples of "real" intraplate volcanism such as Hawaii. Some, like Iceland, seem to have formed on or close to a divergent plate boundary, others, like the Azores, are on or close to leaky transform faults (Fig.20). The Canary Islands may owe (part) of their origin to mantle instabilities along a lithosphere boundary not commonly thought of as "fertile" (Schmincke and Staudigel, 1976):

The "volcanic belt" encompassing the Canaries and seamounts north and south of the archipelago off the coast of NW-Africa extends along an old suture between continental and oceanic lithosphere, formed about 180 Ma ago, and lies just west of a subsiding belt of thick sediment accumulations. This boundary might constitute a deep-reaching zone of weakness, along which diapiric uprise of peridotite or other mechanisms might induce melting processes in the mantle. The dynamic state of this zone is also reflected in the unusually high Jurassic to Early Cretaceous rates of subsidence off Cape Bojador which cannot be explained adequately by a combination of lithospheric cooling contraction plus sediment loading (von Rad and Arthur, 1979; von Rad and Einsele, 1980). Whatever triggered the initial melting in the mantle beneath the Canaries, an east to west progression was superimposed on this melting anomaly, possibly due to plate translation, and a string of melting anomalies developed which now underly the entire volcanic chain, generating different types of magma at different sites and at different times.

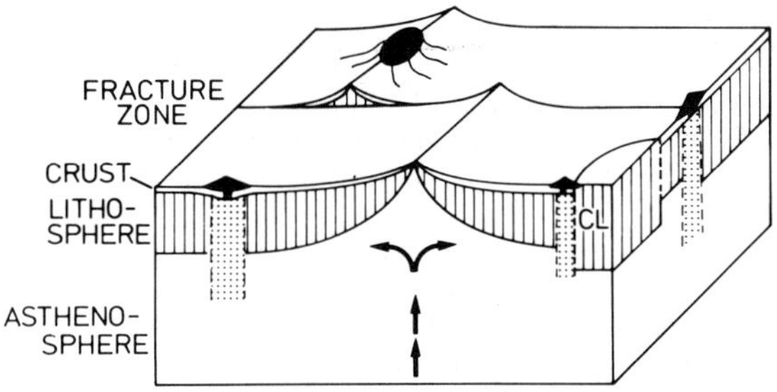

Fig.20 Schematic tectonic setting of four main types of oceanic volcanic islands. Clockwise from upper left: island on or close to mid oceanic ridge; island on leaky transform fault; island along boundary oceanic-continental (CL) lithosphere; island within plate

Despite different tectonic settings of the island groups (with Madeira and the Cape Verde Islands being difficult to place into the pigeon holes of figure 20), volcanism and thus melting in the Eastern Central Atlantic does not seem to have occurred entirely independently at different sites.

4.3 Temporal evolution of island volcanism in the Eastern Central North Atlantic

The shield phases and especially the basal complex of Fuerteventura probably represent the oldest preserved islands, with the possible exception of the Selvagens Islands. The Cretaceous complex on Maio (Cape Verde Islands) represents uplifted ocean crust (Klerkx and de Paepe, 1976), while three K/Ar ages from the central igneous complex of Maio range from 12.2 to 21.1 Ma (Grunau et al., 1975). Porto Santo, however, the small island about 50 km east of Madeira, consists dominantly of submarine rocks, about 12 - 13 Ma old (Feraud et al., 1981), slightly younger than the shield building phase of Gran Canaria. In contrast to the latter, however, younger volcanism does not seem to have occurred on Porto Santo. During the middle to late Miocene (ca. 12 - 5 Ma), subaerial volcanism apparently has been restricted to the Canary Islands. The subaerial part of the oldest island in the Azores (Santa Maria) was formed mainly between ca. 5.5 and 3.2 Ma (Feraud et al., 1981). All other Azores islands appear to be Pliocene and Pleistocene in age (Feraud et al., 1980) with ages apparently decreasing from Santa Maria in the East to Pico/Fayal in the West in the Azores Islands east of the Mid-Atlantic Ridge, reminiscent of the east to west age decrease in the Canaries discussed earlier. In the Madeira group, the old island Porto Santo also lies east of Madeira of which all 15 K/Ar age data published so far (Watkins and Abdel-Monem, 1971; Feraud et al., 1981) are between 0.7 and 3.05 Ma old, although large parts of the island are poorly known.

Judging from data presently available, volcanic activity in the Central Atlantic island groups was not uniform. A major peak apparently occurred in the middle Miocene between about 10 and 17 Ma, with major uncertainties concerning the age of the subaerial shield series on Lanzarote, Fuerteventura and Tenerife. Major seamounts in the area such as Great Meteor and Josephine also were active at this time (Wendt et al., 1976). Between about 5 and 10 Ma, activity was generally low, with the possible exception of the islands of Gomera and Tenerife for which few age data are available. Activity was high during the last 5 Ma and especially the last 2 Ma, with major younger phases on the older large islands of Gran Canaria and Tenerife and the formation of several larger islands such as Hierro, La Palma, Madeira and most of the Azores Islands. Data are not yet detailed enough to attempt to regionally analyze eruptive rates in the island groups during the past 5 Ma, except for Gran Canaria.

During the past 15 years, the temporal evolution of oceanic islands has been considered chiefly in terms of age decrease as a function of plate movement (e.g. Wilson, 1963, 1965; McDougall, 1964; McDougall and Schmincke, 1976), although some workers also discussed ocean island volcanism in the framework of episodic, regionally synchronous volcanic activity (Vogt, 1972, 1979; Shaw, 1973; Jackson, 1976; Schmincke, 1976). In the Eastern Pacific, major pulses of Neogene volcanic activity occurred during the Middle Miocene (t = 14 to 16 Ma) and Quaternary (t = ca. 2 Ma to present) (Kennett et al., 1977), and Vogt (1972, 1979) has noted broad, globally synchronous magmatic episodes around 10 to 20 and 7 to 0 Ma (Fig.21). The cause for such globally synchronous volcanism is not clear, although there appears to be some evidence for correlation with changes in spreading rate (Rea and Scheidegger, 1979; Vogt, 1979). Peaks of volcanic activity on Gran Canaria and in the Tertiary to late Quaternary volcanic fields of Central Europe seem to coincide with those in the Pacific (Fig.21), further suggesting that the cause for volcanism in these areas is part of a global pattern.

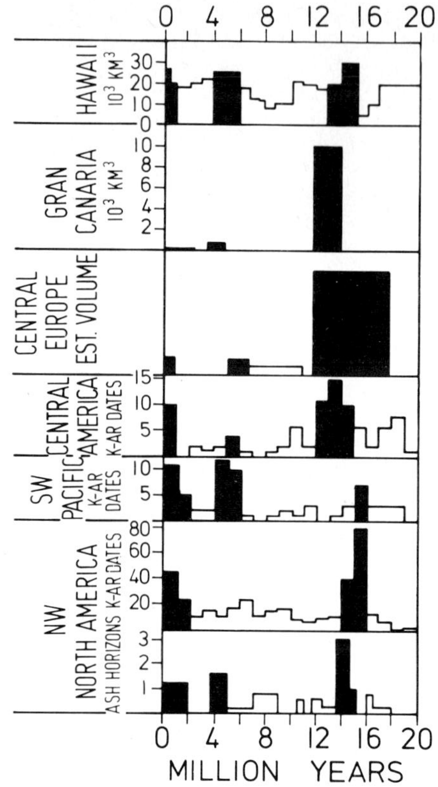

Fig.21 Globally synchronous volcanic episodes in oceanic islands (Hawaii and Gran Canaria), continental intraplate setting (Central Europe) and Pacific subduction zone setting. Gran Canaria data see Fig.7, Central Europe literature in Schmincke, in prep. Other data from Vogt (1978) and Kennett et al. (1977)

Acknowledgments

Research on Atlantic islands and sea floor basalts has been generously supported by the Deutsche Forschungsgemeinschaft. U.Bednarz calculated the island volumes. G.Pritchard and H.Niephaus did the chemical analyses; figures were prepared by K.Bauszus and A.Fischer, and the manuscript was typed by V.Helmke. G.Brey, M.J.F.Flower, G.Hanson, K.Hinz, H. Roeser, U.von Rad, and G.Wissmann provided many helpful comments on the manuscript.

References

Abdel-Monem A, Watkins ND, Gast PW (1971) Potassium-argon ages, volcanic stratigraphy, and geomagnetic polarity history of the Canary Islands: Lanzarote, Fuerteventura, Gran Canaria, and La Gomera. Am J Sci 271: 490-521

Abdel-Monem A, Watkins ND, Gast PW (1972) Potassium-argon ages, volcanic stratigraphy, and geomagnetic polarity history of the Canary Islands: Tenerife, La Palma, and Hierro. Am J Sci 272: 805-825

Abdel-Monem A, Fernandez CA, Boone GB (1975) K-Ar ages from the eastern Azores group (Santa Maria, São Miguel, and the Formigas Islands). Lithos 8: 247-254

Aires-Barros L, Matias M-J, Miranda AM (1974) Preliminary note on the petrology of Madeira Island. Bol Mus Lab Min Geol Fac Cienc Lisboa 14: 5-27

Anguita F, Hernan F (1975) A propagating fracture model versus a hot spot origin for the Canary Islands. Earth Planet Sci Lett 27: 11-19

Bargar KE, Jackson ED (1974) Calculated volumes of individual shield volcanoes along the Hawaiian-Emperor chain. J Res US Geol Surv 2: 545-550

Bosshard E, MacFarlane DJ (1970) Crustal structure of the western Canary Islands from seismic refraction and gravity data. J Geophys Res 75: 4901-4918

Bougault H, Joron J, Treuil M (1979) Alteration, fractional crystallization, partial melting, mantle properties from trace elements in basalts recovered in the North Atlantic. In: Talwani M, Harrison CG, Hayes DE (ed) Deep Drilling results in the Atlantic ocean: Ocean Crust. M Ewing Ser 2 Am Geophys Union Washington: 352-368

Bourcart J, Jeremine E (1937) La Grande Canarie. Bull Volcanol Ser II 2: 1-77

Brey G (1978) Origin of olivine melilitites-chemical and experimental constraints. J Volcanol Geotherm Res 3: 61-88

Brey G, Schmincke H-U (1980) Origin and diagenesis of the Roque Nublo breccia, Gran Canaria (Canary Islands) - Petrology of Roque Nublo Volcanics II. Bull Volcanol 43: 15-33

Chase CG (1981) Oceanic Island Pb: two stage histories and mantle evolution. Earth Planet Sci Lett 52: 277-284

Chayes F (1977) The oceanic basalt-trachyte relation in general and in the Canary Islands. Am Mineral 62: 666-671

Dana JD (1849) Geology. Volume 10 of the United States exploring expedition during the years 1838-1839, 1840, 1841, 1842. Sherman and Co, Philadelphia pp 1-756

Dash BP, Bosshard E (1968) Crustal studies around the Canary Islands. Proc Int Geol Congr 23rd Session Czech 1:249-260

Dawson JB (1980) Kimberlites and their xenoliths. Springer Verlag, Heidelberg p 1-259

De Paepe P, Klerkx J (1976) Données petrographiques et chimiques sur les laves cenozoiques des iles de Sal et Maio (archipel du Cap Vert). Mus roy Afr Cent Tervuren Rapp Ann 1975: 77-90

Dietz RS, Sproll WP (1970) East Canary Islands as a microcontinent within the Africa - North America continental drift fit. Nature 226: 1043-1045

Dmitriev LV, Sobolev AV, Suschevskaja NM (1979) The primary melt of the oceanic tholeiite and the upper mantle composition. In: Talwani M, Harrison CG, Hayes DE (ed) Deep Drilling Results in the Atlantic Ocean: Ocean Crust. M Ewing Ser 2, Am Geophys Union Washington: 302-313

Duda A, Schmincke H-U (1978) Quaternary basanites, melilites, nephelinites and tephrites from the Laacher See area (Germany). Neues Jahrb Mineral Abh 132: 1-33

Feraud G, Kaneoka I, Allegre CJ (1980) K/Ar ages and stress pattern in Azores: Geodynamic implications. Earth Planet Sci Lett 46: 275-286

Feraud G, Schmincke H-U, Lietz J, Gastaud J, Pritchard G, Bleil U (1981) New K/Ar ages, chemical analyses and magnetic data of rocks from the islands of Santa Maria (Azores), Porto Santo and Madeira (Madeira Archipelago) and Gran Canaria (Canary Islands). Bull Volcanol in press

Fernandez LA (1980) Geology and petrology of the Nordeste volcanic complex, São Miguel, Azores: Summary. Geol Soc Am Bull 91: 675-680

Flower MFJ (1973) Evolution of basaltic and differentiated lavas from Anjouan, Comores Archipelago. Chem Geol 12: 81-91

Flower MFJ, Schmincke H-U, Bowman H (1976) Rare Earth and other trace elements in historic Azorean lavas. J Volc Geotherm Res 1: 127-147

Flower MFJ, Schmincke H-U, Thompson RN (1975) Phlogopite stability and the $^{87}Sr/^{86}Sr$ step in basalts along the Reykjanes Ridge. Nature 254: 404-406

Forsyth DW (1977) The evolution of the upper mantle beneath mid-ocean ridges. Tectonophys 38: 89-118

Frey FA, Green DH, Roy SD (1978) Integrated models of basalt petrogenesis: a study of quartz tholeiites to olivine melilitites from south eastern Australia utilizing geochemical and experimental petrological data. J Petrol 19: 463-513

Fritsch K von (1867) Reisebilder von den Canarischen Inseln. Petermanns Geogr Mitt (Gotha) Ergänzungsh 5 (22): 1-44

Fuster JM (1975) Las Islas Canarias: un ejemplo de evolución espacial y temporal del vulcanismo oceánico. Estud Geol 31: 439-463

Fuster JM, Cendrero A, Gastesi P, Ibarrola E, Ruiz JL (1968a) Geology and volcanology of the Canary Islands, Fuerteventura. Inst "Lucas Mallada" Madrid Int Symp Volcanol Tenerife, 1968, Spec Publ p 1-239

Fuster JM, Santin SF, Sagredo A (1968b) Geology and volcanology of the Canary Islands, Lanzarote. Inst "Lucas Mallada" Madrid, Int Symp Volcanol, Tenerife, 1968, Spec Publ p 1-177

Fuster JM, Hernandez-Pacheco A, Muñoz M, Badiola ER, Cacho LG (1968c) Geology and volcanology of the Canary Islands, Gran Canaria. Inst "Lucas Mallada", Madrid, Int Symp Volcanol Tenerife, 1968, Spec Publ p 1-243

Fuster JM, Araña V, Brändle JL, Navarro M, Alonso U, Aparicio A (1968d) Geology and volcanology of the Canary Islands, Tenerife. Inst "Lucas Mallada" Madrid, Int Symp Volcanol Tenerife, 1968, Spec Publ p 1 -218

Gast PW (1968) Trace element fractionation and the origin of tholeiitic and alkaline magma types. Geochim Cosmochim Acta 32: 1057-1086

Green DH (1971) Composition of basaltic magmas as indicators of conditions of origin: Application to oceanic volcanism. Philos Trans R Soc London Ser A 268: 707-725

Green DH, Ringwood AE (1967) The genesis of basaltic magmas. Contrib Mineral Petrol 15: 103-190

Grunau HR, Lehner P, Cleintuar MR, Allenbach P, Bakker G (1975) New radiometric ages and seismic data from Fuerteventura (Canary Islands), Maio (Cape Verde Islands) and São Tome (Gulf of Guinea). In: Borradaille GJ et al (eds) Progress in geodynamics. K Nederl Akad Wetensch Amsterdam p 90-118

Hart SR, Allegre CJ (1980) Trace element constraints on magma genesis. In: Hargraves RB (ed) Physics of magmatic processes, Princeton Univ Press, New Jersey: 121-159

Hart SR, Davis KE (1978) Nickel partitioning between olivine and silicate melt. Earth Planet Sci Lett 40: 203-219

Hausen H (1956) Contributions to the geology of Tenerife. Soc Sci Fenn Comm Phys-Math 18: 1-247

Hausen H (1958) On the geology of Fuerteventura (Canary Islands). Soc Sci Fenn Comm Phys-Math 22: 1-211

Hausen H (1973) Outlines of the geology of Hierro (Canary Islands). Soc Sci Fenn Comm Phys-Math 43: 65-148

Hawksworth CJ, Norry MJ, Roddick JC, Vollmer R (1979) $^{143}Nd/^{144}Nd$ and $^{87}Sr/^{86}Sr$ ratios from the Azores and their significance in LIL-element enriched mantle. Nature 280: 28-31

Hughes DJ, Brown GC (1972) Basalts from Madeira. Contrib Mineral Petrol 37: 91-109

Ibarrola E (1969) Variation trends in basaltic rocks of the Canary Islands. Bull Volcanol 33: 729-777

Jackson ED (1976) Linear volcanic chains on the Pacific plate. In: The geophysics of the Pacific Ocean Basin and its margin. Geophys Monogr 19: 319-338

Jackson ED, Wright TL (1970) Xenoliths in the Honolulu volcanic series, Hawaii. J Petrol 11: 405-430

Jakobsson SP (1972) Chemistry and distribution pattern of Recent basaltic rocks in Iceland. Lithos 5: 365-86

Kennett JP, McBirney AR, Thunell RC (1977) Episodes of Cenozoic volcanism in the circum-Pacific region. J Volcanol Geotherm Res 2: 145-163

Klerkx J, Deutsch S, de Paepe P (1974) Rubidium, strontium content and strontium isotopic composition of strongly alkalic basaltic rocks from the Cape Verde Islands. Contrib Mineral Petrol 45: 107-118

Klerkx J, de Paepe P (1976) The main characteristics of the magmatism of the Cape Verde Islands. Ann Soc géol Belg 99: 347-357

Langmuir CH, Hanson GN (1980) An evaluation of major element heterogeneity in the mantle sources of basalts. Phil Trans R Soc Lond A 297: 383-407

Laughton AS, Whitmarsh RB (1975) The Azores-Gibraltar plate boundary. In Kristiansson L (ed) Geodynamics of Iceland and the North Atlantic area. Reidel, Dordrecht: 63-81

Lehner P, de Ruiter PAC (1977) Structural history of Atlantic margin of Africa. Am Ass Petrol Geol Bull 61: 961-981

Lietz J, Schmincke H-U (1975) Miocene-Pliocene sea level changes and volcanic episodes on Gran Canaria (Canary Islands) in the light of new K/Ar ages. Palaeogeogr Palaeoclimatol Palaeoecol 18: 213-239

MacDonald GA (1968) Composition and origin of the Hawaiian lavas. Geol Soc Am Mem 116: 477-522

MacDonald GA, Katsura T (1964) Chemical composition of Hawaiian lavas. J Petrol 5: 82-133

McBirney AR, Gass IG (1967) Relations of oceanic volcanic rocks to mid-oceanic rises and heat flow. Earth Planet Sci Lett 2: 265-276

McDougall I (1964) Potassium-argon ages from lavas of the Hawaiian Islands. Geol Soc Am Bull 75: 107-128

McDougall I, Schmincke H-U (1976) Geochronology of Gran Canaria, Canary Islands: Age of shield building volcanism and other magmatic phases. Bull Volcanol 40: 1-21

Melson WG, O'Hearn T (1979) Basaltic glass erupted along the Mid-Atlantic Ridge between 0-37°N: Relationships between composition and latitude. In: Talwani M, Harrison CG, Hayes DE (eds) Deep Drilling Results in the Atlantic Ocean: Ocean Crust. M Ewing Ser 2. Am Geophys Union, Washington, pp 249-261

Mitchell-Thomé RC (1976) Geology of the Middle Atlantic Islands. Gebr Bornträger, Berlin, Beitr Geol Erde 12: 1-382

Morgan WJ (1972) Plate motions and deep mantle convection. In: Studies in earth and space sciences: Harry H Hess Vol Shagam R et al (eds) Geol Soc Am Mem 132: 7-22

O'Hara MJ (1968) The bearing of phase equilibrium studies on the origin and evolution of basic and ultrabasic rocks. Earth Sci Rev 4: 69-133

Oversby VM, Lancelot J, Gast PW (1971) Isotopic composition of lead in volcanic rocks from Tenerife, Canary Islands. J Geophys Res 76: 3402-3413

Oxburgh ER (1980) Heat flow and magma genesis. In: Hargraves RB (ed) Physics of magmatic processes. Princeton, New Jersey, pp 161-199

Parsons B, Sclater JG (1977) An analysis of the variation of ocean floor bathymetry and heat flow with age. J Geophys Res 82: 803-827

Pellicer JM (1979) Estudio geochimico del vulcanismo de la isla de Hierro Archipiélago Canario. Estud geol 35: 15-29

Rea DK, Scheidegger KF (1979) Eastern Pacific spreading rate fluctuation and its relation to Pacific area volcanic episodes. J Volcanol Geotherm Res 5: 135-148

Robertson AHF, Stillmann CJ (1979) Submarine volcanic and associated sedimentary rocks of the Fuerteventura Basal Complex, Canary Islands. Geol Mag 116: 203-214

Roeser HA, Hinz K, Plaumann S (1971) Continental margin structure in the Canaries Inst Geol Sci Rep 70/16: 27-36

Rona PA, Nalwalk AJ (1970) Post-early Pliocene unconformity on Fuerteventura Canary Islands. Geol Soc Am Bull 81: 2117-2122

Rothe P (1968)Mesozoische Flyschablagerungen auf der Kanareninsel Fuerteventura. Geol Rundsch 58: 314-322

Rothe P, Koch R (1978) Miocene volcanic glass from DSDP sites 368, 369, and 370. In: Lancelot Y and Seibold E et al. Initial Repts Deep Sea Drill Proj. 41: 1061-1064

Rothe P, Schmincke H-U (1968) Contrasting origins of the eastern and western islands of the Canarian Archipelago. Nature 210: 1152-1154

Rothpletz A, Simonelli T (1890) Die marinen Ablagerungen auf Gran Canaria.Z.Deutsch Geol Ges 42: 667-736

Sato H (1977) Nickel content of basaltic magmas: identification of primary magmas and a measure of the degree of olivine fractionation. Lithos 10: 113-120

Sauer EGF, Rothe P (1972) Ratite eggshells from Lanzarote, Canary Islands. Science 176: 43-45

Schmincke H-U (1967) Cone sheet swarm, resurgence of Tejeda Caldera and the early geologic history of Gran Canaria. Bull Volcanol 31: 153-162

Schmincke H-U (1968) Faulting versus erosion and the reconstruction of the Mid-Miocene shield volcano of Gran Canaria. Geol Mitt 8: 23-50

Schmincke H-U (1969) Petrologie der phonolitischen bis rhyolitischen Vulkanite auf Gran Canaria, Kanarische Inseln. Habilitationsschrift. Univ Heidelberg, pp 1-151

Schmincke H-U (1973) Magmatic evolution and tectonic regime in the Canary, Madeira and Azores Island Groups. Geol Soc Am Bull 84: 633-648

Schmincke H-U (1976) The geology of the Canary Islands. In: Kunkel G (ed) Biogeography and ecology of the Canary Islands. Junk, The Hague. pp 67-184

Schmincke H-U (1979a) Age and crustal structure of the Canary Islands. J Geophys 46: 217-224

Schmincke H-U (1979b) The late Quaternary Eifel volcanic fields. In Illies et al: The Quaternary uplift the Rhenish shield in Germany. Tectonophysics 61: 97-125

Schmincke H-U, Flower MFJ (1974) Magmenevolution auf Atlantischen Vulkaninseln. Naturwiss 61: 288-292

Schmincke H-U, Staudigel H (1976) Pillow lavas on central and eastern Atlantic Islands. Bull Soc Geol Fr, p 871-883

Schmincke H-U, von Rad U (1979) Neogene evolution of Canary Island volcanism inferred from ash layers and volcaniclastic sandstones of DSDP site 397 (Leg 47 A) In: von Rad U, Ryan WFB et al (eds), Init Rep Deep Sea Drilling Project 47: 703-725

Schmincke H-U, Weibel M (1972) Chemical study of rocks from Madeira, Porto Santo and São Miguel, Terceira (Azores). Neues Jahrb Mineral, Abh 117: 253-281

Searle RC (1980) Tectonic pattern of the Azores spreading centre and triple junction. Earth Planet Sci Lett 51: 415-434

Self S (1976) The recent volcanology of Terceira, Azores. J Geol Soc London 132: 645-666

Shaw HR (1973) Mantle convection and volcanic periodicity in the Pacific: evidence from Hawaii. Geol Soc Am Bull 84: 1505-1526

Staudigel H (1981) Der basale Komplex von La Palma. Ph D thesis, Univ Bochum

Staudigel H, Schmincke H-U (1981) Effects of contrasting tectonic environments on the structure of the oceanic crust: The uplifted seamount of La Palma and DSDP/ IPOD sites 417 and 418. Chapman Conference, Arlington (Va) Abs.

Stearns HT (1940) Supplement to the geology and ground-water resources of the Island of Oahu, Hawaii. Hawaii Div Hydrograph Bull 9: 3-55

Storedtvedt UM, Sralestad S, Thomassen K, Langlie AA, Nergård A, Gidskehaug A: Magnetic discordance in Gran Canaria/ Tenerife and its possible relevance to the formation of the northwest African continental margin. J Geophys 44: 317-332

Sun SS (1980) Lead isotopic study of young volcanic rocks from mid-ocean ridges, ocean islands and island arcs. Phil Trans R Soc London A 297: 409-445

Sun SS, Hanson GN (1975) Evolution of the mantle: Geochemical evidence from alkali basalt. Geology 3: 297-302

Swanson DA (1972) Magma supply rate at Kilauea volcano 1952-1971. Science 175: 169-170

Tarney J, Wood DA, Varet J, Saunders AD, Cann JR (1979) Nature of mantle heterogeneity in the North Atlantic: Evidence from Leg 49 basalts. In: Talwani M, Harrison CG, Hayes DE (ed) Deep Drilling Results in the Atlantic Ocean: Ocean Crust. M Ewing Ser 2, Am Geophys Union Washington: 285-301

Vogt PR (1972) Evidence for global synchronism in mantle plume convection and possible significance for geology. Nature 240: 338-342

Vogt PR (1974) Volcano spacing, fractures and thickness of the lithosphere. Earth Planet Sci Lett 21: 235-252

Vogt PR (1979) Global magmatic episodes: New evidence and implications for the steady-state mid-ocean ridge. Geology 7: 93-98

von Rad U, Arthur MA (1979) Geodynamic, sedimentary and volcanic evolution of the Cape Bojador continental margin (Northwest Africa) In: Talwani M, Hay W, Ryan WBF (eds) Deep Drilling results in the Atlantic Ocean: Continental margins and palaeo environment. M Ewing Ser 3. Am Geophys Union, Washington, pp 187-203

von Rad U, Einsele GC (1980) Mesozoic-Cenozoic subsidence history and paleobathymetry of the Northwest African continental margin (Aaiun Basin to DSDP Site 397). Phil Trans R Soc London A 294: 37-50

Watkins ND, Abdel-Monem A (1971) Detection of the Gilsa Geomagnetic Polarity Event on the Island of Madeira. Geol Soc Am Bull 82: 191-198

Watkins JS, Hoppe KW (1978) Reconnaissance geology of Atlantic margin of NW Africa, 25°-34°N. Second Maurice Ewing Memorial Symp "Implications of Deep Sea Drilling Results in the North Atlantic". New York, pp 22-25

Watkins JS, Hoppe KW (1979) Seismic reflection reconnaissance of the Atlantic margin of Morocco. In: Talwani M, Hay W, Ryan WBF (eds) Deep Drilling Results in the Atlantic Ocean: Continental margins and palaeoenvironment. M Ewing Ser 3 Am Geophys Union, Washington, pp 205-217

Wendt I, Kreuzer H, Müller P, von Rad U, Raschka H (1976) K/Ar ages of basalts from Great Meteor and Josephine seamounts (eastern North Atlantic). Deep Sea Res 23: 849-862

White WM, Schilling J-G (1978) The nature and origin of geochemical variation in Mid-Atlantic ridge basalts from the Central North Atlantic. Geochim Cosmochim Acta 42: 1501-1516

White WM, Tapia DMM, Schilling J-G (1979) The petrology and geochemistry of the Azores Islands Contrib Mineral Petrol 69: 201-213

Wilson JT (1963) A possible origin of the Hawaiian Islands. Can J Phys 41: 863-870

Wilson JT (1965) Evidence from oceanic islands suggesting movement in the earth Phil Trans R Soc London Ser A 258: 145-167

Wissmann G (1979) The Cape Bojador slope, an example for potential pitfalls in seismic interpretation without the information of outer margin drilling. In: von Rad U, Ryan WFB et al (eds) Init Rep Deep Sea Drilling Project 47: 491-499

Wissmann G, von Rad U (1979) Seismic structure, continental basement, and Mesozoic sediments from the Mazagan Plateau of Morocco. "Meteor" Forsch Ergebn R C 31: 1-20

Wood DA, Joron JL, Treuil M, Norry M, Tarney J (1979) Elemental and Sr isotope variations in basic lavas from Iceland and the surrounding ocean floor. Contrib Mineral Petrol 70: 319-339

Table 2

Selected major and trace element chemical analyses of primitive basalt lavas from Gran Canaria, Hierro and La Palma

SAMPLE	GC 1378	GC 1262	GC 1268	GC 1383	GC 719	GC 1391	GC 1174	GC 1257
SiO_2	45.6	44.6	47.8	48.5	45.6	40.3	48.7	45.0
TiO_2	3.68	2.48	4.10	4.35	4.18	4.73	2.93	3.70
Al_2O_3	10.7	7.5	14.3	13.5	14.8	10.4	12.6	12.1
Fe_2O_3	4.48	4.66	5.66	7.52	8.79	5.09	5.15	7.13
FeO	8.24	8.09	6.46	5.71	3.30	9.89	7.40	6.55
MnO	0.19	0.18	0.16	0.17	0.19	0.19	0.16	0.17
MgO	10.20	20.10	4.56	4.31	5.37	11.90	8.61	8.59
CaO	11.90	8.81	9.55	8.95	10.70	11.40	9.48	10.80
Na_2O	2.14	1.48	3.33	3.37	3.00	2.88	2.53	2.62
K_2O	0.99	0.63	1.40	1.59	1.61	1.11	0.62	1.04
P_2O_5	0.43	0.27	0.62	0.69	0.72	0.86	0.47	0.84
H_2O^+	0.50	1.30	1.70	0.80	1.96	1.20	0.60	1.48
CO_2	0.10	0.17	0.06	0.09	0.19	0.20	0.12	0.28
Total	99.15	100.23	99.70	99.55	100.41	100.15	99.37	100.30
Mg No	62.7	76.8	44.4	41.1	49.2	62.5	59.1	57.3
Cr	473	1217	43	26	50	381	363	269
Co	73	118	80	54	53	90	71	80
Ni	230	809	53	44	67	182	281	225
Cu	111	76	136	178	111	47	94	86
Zn	95	86	111	123	108	105	113	124
Rb	23	17	34	39	39	26	8	19
Sr	505	281	730	603	1070	847	536	777
Y	21	22	35	38	36	19	32	22
Zr	266	178	391	436	376	287	174	280
Nb	46	29	62	66	65	72	32	61
Ba	118	108	319	179	340	324	286	305

Table 2 (continued)

SAMPLE	GC 1136	GC 1142	GC 1424	GC 1426	GC 1501	GC 1415	GC 1410	GC 1591
SiO_2	40.1	38.5	38.1	39.9	38.6	37.2	44.6	40.8
TiO_2	4.02	3.94	4.01	3.33	3.66	3.77	3.58	3.55
Al_2O_3	11.7	10.1	9.2	10.1	9.7	9.3	12.7	11.4
Fe_2O_3	6.44	9.92	9.71	4.57	6.18	4.52	6.70	8.26
FeO	6.30	4.20	4.62	7.82	6.70	7.96	5.23	4.06
MnO	0.22	0.21	0.20	0.19	0.20	0.19	0.17	0.18
MgO	9.67	12.90	13.70	14.10	14.10	13.70	8.88	12.40
CaO	12.50	13.00	13.10	13.60	13.60	14.40	10.60	12.20
Na_2O	3.62	2.66	3.03	3.34	3.37	3.40	3.62	3.19
K_2O	0.70	1.08	1.21	1.31	1.34	1.40	1.80	1.54
P_2O_5	1.38	1.40	1.49	1.35	1.32	1.39	0.85	1.02
H_2O^+	2.50	1.10	0.70	0.53	0.20	0.90	0.60	0.32
CO_2	0.14	0.10	0.28	0.21	0.05	1.05	0.19	0.67
Total	99.29	99.11	99.38	100.35	99.05	99.18	99.52	99.59
Mg No	61.8	66.5	67.5	70.5	69.9	69.7	61.5	68.6
Cr	262	379	429	500	527	351	241	591
Co	72	87	78	93	79	67	69	89
Ni	179	290	341	367	359	341	97	272
Cu	82	87	79	77	66	95	32	71
Zn	139	135	115	111	109	106	118	89
Rb	88	44	24	23	15	32	39	31
Sr	1888	1564	1466	1283	1404	1281	1199	1271
Y	27	40	24	37	37	32	29	25
Zr	556	381	385	310	296	341	398	276
Nb	149	98	92	97	92	97	89	69
Ba	789	587	988	941	886	589	577	484

Table 2 (continued)

SAMPLE	GC 1499	GC 1431	GC 1369	GC 1475	HI 1	HI 3	HI 4	HI 5
SiO_2	41.9	40.9	41.7	42.7	43.8	41.9	41.6	46.0
TiO_2	3.70	4.48	3.35	3.63	4.81	4.23	4.25	3.99
Al_2O_3	11.6	12.1	10.5	12.8	13.7	11.6	9.3	15.7
Fe_2O_3	3.51	4.89	3.15	4.03	4.22	3.61	4.76	6.08
FeO	8.10	7.64	8.80	7.10	9.60	10.30	10.50	5.89
MnO	0.18	0.19	0.18	0.18	0.20	0.19	0.19	0.20
MgO	12.40	10.10	16.20	10.60	7.51	11.30	13.80	4.96
CaO	12.10	12.60	11.20	11.10	10.40	11.70	11.20	9.80
Na_2O	2.63	2.91	2.47	3.85	3.52	2.97	2.05	3.66
K_2O	1.59	1.88	1.33	2.11	1.41	1.11	0.93	1.58
P_2O_5	0.78	0.99	0.67	0.72	1.02	0.89	0.51	1.05
H_2O^+	0.60	0.70	0.10	0.10	0.14	0.23	0.13	0.47
CO_2	0.06	0.16	0.05	0.14	0.03	0.03	0.05	0.04
Total	99.15	99.54	99.70	99.06	100.36	100.06	99.32	99.42
Mg No	69.0	62.9	73.8	66.7	53.1	62.8	65.4	46.9
Cr	744	366	808	595	183	468	623	0
Co	65	96	84	88	59	68	88	41
Ni	245	156	424	196	109	265	338	24
Cu	87	86	68	68	65	100	68	38
Zn	100	101	88	98	136	115	120	121
Rb	30	37	21	49	26	18	18	27
Sr	974	1306	830	1242	1082	881	594	1030
Y	33	35	30	35	37	30	26	38
Zr	279	323	212	364	429	283	257	393
Nb	67	80	57	81	76	60	46	74
Ba	594	743	509	816	441	423	331	415

Table 2 (continued)

SAMPLE	HI 6	HI 7	HI 8	HI 9	124793	124794
SiO_2	45.3	42.9	42.9	43.2	45.6	44.6
TiO_2	3.86	4.26	4.88	3.51	3.31	3.26
Al_2O_3	14.5	12.1	13.5	14.2	16.0	13.7
Fe_2O_3	5.82	4.62	3.73	6.09	4.16	2.80
FeO	7.19	9.17	9.89	7.18	6.72	9.65
MnO	0.19	0.19	0.19	0.21	0.21	0.20
MgO	6.66	10.60	7.48	7.38	5.18	9.10
CaO	10.70	10.30	10.70	10.20	9.17	10.60
Na_2O	3.14	2.89	3.46	4.08	5.10	3.68
K_2O	1.13	1.17	1.42	1.38	2.29	1.27
P_2O_5	0.84	0.68	0.86	1.13	0.87	0.65
H_2O^+	0.58	0.17	0.07	0.53	0.36	0.11
CO_2	0.03	0.02	0.04	0.10	0.03	0.03
Total	99.94	99.07	99.12	99.19	99.00	99.65
Mg No	52.0	61.7	53.3	54.1	50.0	60.2
Cr	143	414	148	150	73	340
Co	49	76	62	56	40	68
Ni	93	223	103	114	54	159
Cu	66	51	76	62	68	110
Zn	113	114	131	130	116	111
Rb	19	20	27	25	57	28
Sr	825	804	997	1115	1404	889
Y	33	32	35	35	36	30
Zr	307	318	3870	366	412	261
Nb	62	63	680	82	107	61
Ba	365	353	4210	535	787	432

Sample locations

Gran Canaria

Guigui and Hogarzales Formations

GC 1378 Alkali basalt. GC 7 in McDougall and Schmincke, 1976
GC 1262 Picrite. Re-analysis, see Schmincke, 1976
GC 1268 Hawaiite. Re-analysis, see Schmincke, 1976
GC 1383 Hawaiite. See GC 14 in McDougall and Schmincke
GC 719 Alkali basalt separating lower and upper Mogan Formations, see Schmincke 1976, GC 697

El Tablero Formation

GC 1391 Nephelinite. See GC 24 in McDougall and Schmincke, 1976

Roque Nublo Group

GC 1174 Tholeiite. Basal Roque Nublo flow.Fortaleza Grande
GC 1257 Alkali basalt. Roque Nublo Group, Mesa de Junquillo, 6oo m above sea level

Los Llanos Formation

GC 1136 Hauyne-bearing nephelinite. Basal Los Llanos flow. Rincon-El Mirador section
GC 1142 Olivine nephelinite. Top flow Rincon section See GC 79, McDougall and Schmincke 1976
GC 1424 Olivine nephelinite, see GC 71, McDougall and Schmincke
GC 1426 Olivine nephelinite, see GC 73, McDougall and Schmincke

Los Patchos Formation

GC 15o1 Melilite nephelinite, Pico de Viento, 72o m above sea level
GC 1415 Melilite nephelinite, see GC 61, McDougall and Schmincke

La Calderilla Formation

GC 141o Basanite, see GC 56 McDougall and Schmincke
GC 1591 Basanite, subrecent, Hoya del Gallego, Jinamar
GC 1499 Basanite, subrecent, Agaete intracanyon flow
GC 1431 Basanite, subrecent, Tejeda, 13oo m above sea level
GC 1369 Basanite, subrecent, Bandama Crater
GC 1475 Basanite, subrecent, Volcan de Las Arenas, road Cruz de Tejeda - Artenara

Hierro

HI 1 Alkali basalt. Intermediate Series. 2oo m south Mirador de las Playas
HI 3 Alkali basalt. Subrecent. 5oo m north of Restinga

HI 4 Ankaramite. Basal Series. Miradoro de la Penha

HI 5 Hawaiite. Basal Series. Road junction Parador Harbor

HI 6 Alkali basalt. 3o m above HI 5

HI 7 Alkali basalt. Intermediate Series. Airport Road. 3oo m from junction with Harbor Road

HI 8 Alkali basalt. Subrecent. Tamaduste

HI 9 Alkali basalt. Intermediate Series. 5o m above sea level, 5oo m south of Tamaduste

La Palma

124793 Hawaiite, 1949 flow, early aa-facies, La Palma, La Mancha

124794 Alkali basalt, 1949 flow, late olivine-phyric pahoehoe facies, La Mancha

Cretaceous Stratigraphy, Sedimentation and Paleoenvironment

13 Cretaceous Foraminiferal and Nannoplankton Biostratigraphy and Paleoecology Along the West African Continental Margin

Uwe Pflaumann[1] and Pavel Čepek[2]

Contents

ABSTRACT

The paper synthesizes the stratigraphic results of six legs of the Deep Sea Drilling Project in the tropical and subtropical eastern North Atlantic Ocean off West Africa as reflected by the calcareous nannoplankton and the planktonic foraminifera. The zonations adopted from various sources are partly emended. The Cretaceous epoch is divided into 21 nanno-zones and/or 30 foram-zones, however, not all zones could be distinguished in the study area. A special section deals with the distribution of the planktonic foraminifera in the Upper Cretaceous sediments of Site 369A, not yet discussed in the Initial Reports of the DSDP.

The zonations derived by nannofossils and foraminifers are correlated with each other and compared with the foraminiferal and ammonite record onshore in the Atlas basins to ascertain and, if possible, to refine the age determinations. Though the age assignments by the various fossil groups generally agree well, some questions remain unsolved especially within the Early Cretaceous.

By means of the established stratigraphical framework the paleoecology and preservation history of the region is outlined briefly, followed by a discussion of hiatuses and missing intervals. Finally the paleogeography is discussed regarding the motion of the African plate relative to the paleolatitudes as far as they may have influenced the evolution of the pelagic fossil assemblages.

1. Introduction

Within the last 12 years six legs of the Deep Sea Drilling Project covered the Eastern Atlantic continental margin off Northwest Africa. "Glomar Challenger" drilled 24 holes at 38 sites, and 13 of them reached and two of them passed through the Jurassic. (The results of IPOD Leg 79 to the Mazagan Plateau which was finisched in May 1981, are not covered in this paper). Consequently the Cretaceous is the oldest period which has been completely penetrated by Deep Sea Drilling in the study region.

Biostratigraphic studies on various fossil groups independently

1 Geologisch-Paläontologisches Institut der Universität Kiel, Olshausenstraße 40/60, 2300 Kiel, FRG

2 Bundesanstalt für Geowissenschaften und Rohstoffe, Stilleweg 2, 3000 Hannover 51, FRG

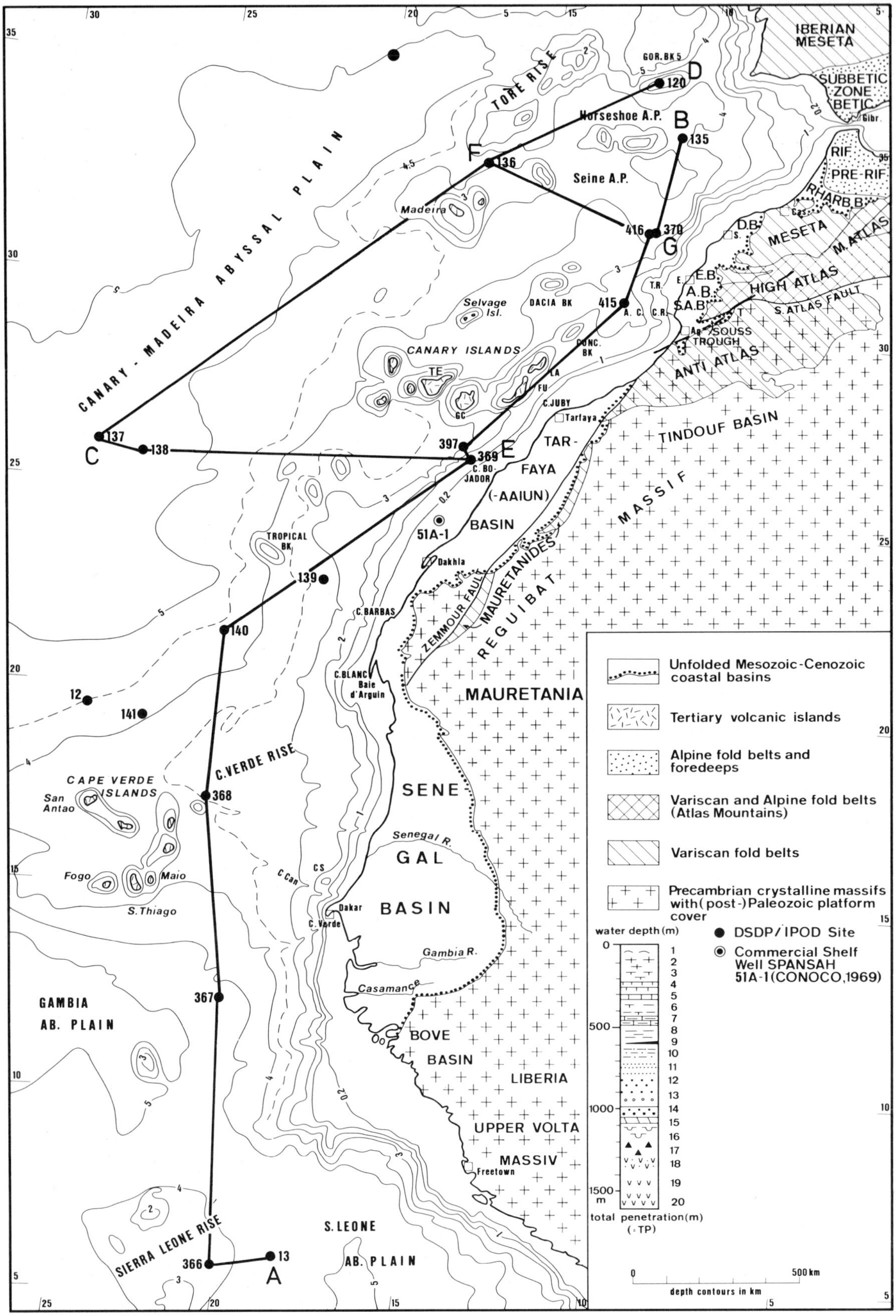
IBERIAN MESETA
SUBBETIC ZONE BETIC
RIF
PRE-RIF
RHARB B.
MESETA
M. ATLAS
HIGH ATLAS
S. ATLAS FAULT
SOUSS TROUGH
ANTI ATLAS
TINDOUF BASIN
TARFAYA (-AAIUN) BASIN
MASSIF
MAURETANIDES
ZEMMOUR FAULT
REGUIBAT
MAURETANIA
SENEGAL BASIN
BOVE BASIN
LIBERIA
UPPER VOLTA
MASSIV
CANARY - MADEIRA ABYSSAL PLAIN
TORE RISE
Horseshoe A.P.
Seine A.P.
Madeira
Selvage Isl.
DACIA BK
CANARY ISLANDS
TROPICAL BK
C. VERDE RISE
CAPE VERDE ISLANDS
San Antao
Fogo
Maio
S. Thiago
GAMBIA AB. PLAIN
SIERRA LEONE RISE
S. LEONE AB. PLAIN
Unfolded Mesozoic-Cenozoic coastal basins
Tertiary volcanic islands
Alpine fold belts and foredeeps
Variscan and Alpine fold belts (Atlas Mountains)
Variscan fold belts
Precambrian crystalline massifs with (post-) Paleozoic platform cover
DSDP/IPOD Site
Commercial Shelf Well SPANSAH 51A-1 (CONOCO, 1969)
water depth (m)
total penetration (m) (= TP)
500 km
depth contours in km

Table 1. Summary of data of the DSDP/IPOD sites with Mesozoic sections in the eastern North Atlantic off West Africa.

Leg	Site	Lat	Long	Water depth in m	Total penetr. in m	Location
3	13	6°02.4'N	18°13.71'W	4588	463	Sierra Leone Rise
13	120	36°41.39'N	11°25.94'W	1711	253.4	On the north slope of Gorringe Bank and north of the Azores-Gibraltar fracture zone
14	135	35°20.80'N	10°25.46'W	4152	689	35 km southeast of the Horsehoe Abyssal Plain
14	136	34°10.15'N	16°18.19'W	4169	313	160 km north of Madeira and 900 km southwest of Gibraltar in an area of abyssal hills
14	137	25°55.53'N	27°03.64'W	5361	401	1000 km west of Cap Blanc Africa in an area abyssal hills close to the foot of the continental rise
14	138	25°55.37'N	25°33.79'W	5288	422	850 km west of Cap Blanc and 130 km east of Site 137 at the foot of the continental rise
14	140	21°44.97'N	21°47.52'W	4483	651	About 450 km west of Cap Blanc at the foot of the continental rise
41	366	5°40. 7'N	19°51. 1'W	2870	850.5	Sierra Leone Rise
41	367	12°29. 2'N	20°02. 8'W	4748	1153	Cape Verde Basin
41	368	17°30. 4'N	21°21. 2'W	3366	984.5	Cape Verde Rise
41	369	26°35. 5'N	14°59. 'W	1752	488.5	Continental slope off Cape Bojador, Spanish Sahara
41	370	32°50. 2'N	10°46. 6'W	4214	1176.5	Deep Basin off Morocco
47	397	26°50. 7'N	15°10. 8'W	2900	1453	Uppermost Rise off Cape Bojador
50	415	31°01.72'N	11°39.11'W	2794	1079.5	Agadir Canyon
50	416	32°50.18'N	10°48.06'W	4191	1624	Morocco Basin

tried to resolve the Cretaceous history of the Atlantic Ocean and its margins.

Deep Sea Drilling sites are generally believed to offer ideal sections for setting standards in stratigraphy because tectonic diturbances and small-scale regional peculiarities common on the continents can be avoided by careful selection of drill sites. However, the ideal stratigraphic note-book contains some empty pages, as a result of following: at sites which have subsided below the carbonate compensation depth (CCD) the calcareous tests of the fossils are dissolved, the continental margin sent erosion debris far into the adjacent marine environment; bottom currents have prevented the deposition of sceletons of faunas and floras; technical handicaps or bad weather conditions during drilling may be responsible for a poor recovery; and finally time restrictions in coring, in sampling, and in researching. To reduce these handicaps we collected data from separate sets of samples, treated by independent methods and described and discussed by various groups of scientists.

To avoid multiple reference to the same authors we indicate that - unless distinctly noted - the follow-

Fig. 1. Location of the sites drilled during DSDP Legs 3, 13, 14, 41, 47, and 50 (modified from front cover figure). Lines connecting the sites refer to the correlations at Fig. 9.
Lithological symbols: 1 = shells (coquina), 2 = foraminiferal nanno ooze, 3 = nanno ooze, 4 = nanno chalk, 5 = limestone, 6 = marl, 7 = marlstone, 8 = clay, shale, 9 = coal, lignite, 10 = silty/sandy clay, 11 = silt/fine sand, 12 = sand(stone), 13 = coarse sand (stone)/conglomerate, 14 = turbidite, 15 = dolomite, 16 = siliceous (radiolarian, diatom) ooze, 17 = porcellanite, chert, 18 = volcaniclastic sandstone/conglomerate, 19 = basalt (oceanic basement), 20 = basalt (silt).
Stratigraphical abbreviations: Q = Quaternary, Tp = Pliocene, Tm = Miocene (Tmm, Tml = upper, middle, lower Miocene), TO = Oligocene, Te = Eocene, Tpa = Paleocene, Tl = lower Tertiary, Ju = upper Jurassic (Kimmeridgian/ Tithonian). Ma = Maastrichtian, Ca = Campanian, Sa = Santonian, Co = Coniacian, Tu = Turonian, Ce = Cenomanian, Al = Albian, Ap = Aptian, Ba = Barremian, Ha = Hauterivian, Va = Valanginian, Be = Berriasian

ing results are compiled from the Initial Reports of the Deep Sea Drilling Project and that responsible authors are listed in the headings of Fig. 3 and 5. Emended zonal and or stage attribution are announced separately. In Tables 2 and 3 we list all taxa of calcareous nannofossils and planktonic foraminifera referred in the Initial Reports cited, irrespective of synonymy, because the discussion of this extensive problem would exceed the framework of this paper. A similar synthesis of the Cretaceous radiolarian biostratigraphy is given by Wolfart (this Vol.).

2. Calcareous Nannofossil Biostratigraphy and Zonation

A number of calcareous nannofossil zonations have been proposed for the Cretaceous system. Those dealing specifically with the Early Cretaceous include: Worsley (1971), Thierstein (1971, 1973, and 1974), Roth (1973), Bukry (1974), Sissingh (1977), and Verbeek (1977), and partly Manivit (1971 and 1972). For the Late Cretaceous we refer to Čepek and Hay (1969 and 1970), Bukry (1969, 1973, 1974), Bukry and Bramlette (1970), Manivit (1971, 1972), Perch-Nielsen (1972 and 1977), Roth and Thierstein (1972), Risatti (1973), Roth (1973), Thierstein (1974), Sissingh (1977) and Verbeek (1977).

We selected the zonation of Perch-Nielsen (1977) with some modifications as a framework for the zonation of Late Cretaceous used in this paper. There are, however, some differences in the range of some zones and in the use of zonal names. For the Early Cretaceous the zonation scheme of Thierstein (1973) was applied along with van Hinte's (1976) geochronologic scale. In Fig. 3 the Cretaceous nannofossil zonation used in this report is correlated with the zonations applied in the six legs of "Glomar Challenger" under discussion. The recovery of Cretaceous sediments differs from voyage to voyage dependent not only on the geological conditions but also on the type of coring. Due to spot coring, especially during legs 2-14, long intervals are missing and it is impossible to assess the thickness of zone intervals or hiatuses and equally impossible to compute sedimentation rates. This difficulty is probably the reason for the relatively poor record of Upper Cretaceous sediments in comparison to that of the Lower Cretaceous. For example, Campanian *Tetralithus aculeus* and *T. gothicus* Zones are present in none of the 15 sites, however, all the Lower Cretaceous zones are recorded.

In sections, where the nannofossil index species are missing, other calcareous nannofossils were used to determine the age of the sediment sections.

The sequence of the zones from these sites is mentioned in the remarks about each zone. Some intervals are not zoned due to the absence of indicator species. Such intervals are not discussed in our remarks, but in the comments about the particular stage. Though not determined for our study area, some zones are listed with both the definition and range, because they may be needed for the later correlations of this area with other ones.

Nannoconus colomi Zone

Author: Worsley (1971, modified by Thierstein (1971)

Definition: Interval from the first occurrence of *Nannoconus colomi* (de Lapparent) to the first occurrence of *Cretarhabdus angustiforatus* (Black).

Age: Berriasian.

Important common species: *Nannoconus colomi* (de Lapparent), *Parhabdolithus asper* (Stradner) and *Polycostella senaria* Thierstein.

Remarks: The *Nannoconus colomi* Zone was encountered at Site 367, but the top of this zone is unclear, because it is impossible to identify any zone in the younger sediments. At Site 416, this zone was difficult to identify, because the upper and lower boundaries are questionable. The first occurrence of *Cretarhabdus angustiforatus* (Black) does not coincide with the upper boundary of this zone. To define the base of the Berriasian Čepek et al. (1980) used the first occurrence of *Parhabdolithus asper* (Stradner) and *Polycostella senaria* Thierstein. *Nannoconus colomi* (de Lapparent) occurs 6 meters above the lower limit of this horizon. The top of the Berriasian at Site 416 is defined by the overlying Valanginian, recorded by the presence of *Bipodorhabdus colligatus* (Black), *Calcicalathina oblongata* (Worsley) and *Tubodiscus verenae* Thierstein. *Cretarhabdus angustiforatus* occurs retarded 10 m above this boundary.

Cretarhabdus angustiforatus Zone

Author: Thierstein (1971).
Definition: Interval from the first occurrence of *Cretarhabdus angustiforatus* (Black) to the first occurrence of *Calcicalathina oblongata* (Worsley).
Age: Late Berriasian to early Valanginian.
Important common species: *Cretarhabdus angustiforatus* (Black).
Remarks: This zone was determined only in the basal section of Site 370. At Site 120 according to the species lists of Stradner (1973) and Gartner (1973), the *Cretarhabdus angustiforatus* Zone probably is present. Gartner (1973) attributed this section to the interval late Valanginian to early Hauterivian, however, according to Stradner (1973) the same section is dated as Barremian to Aptian. In the present paper this interval is referred to the *Cretarhabdus angustiforatus* to *Lithraphidites bollii* Zones, late Berriasian to early Barremian, accordingly.

Calcicalathina oblongata Zone

Author: Thierstein (1971).
Definition: Interval from the first occurrence of *Calcicalathina oblongata* (Worsley) to the first occurrence of *Lithraphidites bollii* (Thierstein).
Age: Late Valanginian to early Hauterivian.
Important common species: *Calcicalathina oblongata* (Worsley).
Remarks: The *Calcicalathina oblongata* Zone was determined in Site 370. The exact top of this zone at this site cannot be identified, because the species *Lithraphidites bollii* (Thierstein) was not observed.

Lithraphidites bollii Zone

Author: Thierstein (1971).
Definition: Interval from the first occurrence of *Lithraphidites bollii* (Thierstein) to the last occurrence of *Calcicalathina oblongata* (Worsley).
Age: Early Hauterivian to early Barremian.
Important common species: *Lithraphidites bollii* (Thierstein), *Calcicalathina oblongata* (Worsley), and *Cruciellipsis cuvillieri* (Manivit).
Remarks: In the study area this zone, as it is defined by Thierstein (1971) cannot be singled out. The *Lithraphidites bollii* Zone is represented only as the upper part of the undifferentiated interval of the *Calcicalathina oblongata* to *Lithraphidites bollii* Zone. The last interval seems to be present also at Site 120. More about these intervals is given in Wind and Čepek (1979) for Site 397 and Čepek et al. (1980) for Site 416.

Micrantholithus obtusus Zone

Author: Thierstein (1971), modified by Thierstein (1973).
Definition: Interval from the last occurrence of *Calcicalathina oblongata* (Worsley) to the last occurrence of *Nannoconus colomi* (de Lapparent) (in the sense of Thierstein, 1973), which should be equivalent to the first occurrence of *Chiastozygus litterarius*.
Age: Barremian.
Important common species: *Nannoconus colomi* (de Lapparent) and *Micrantholithus obtusus* Stradner.
Remarks: This zone was encountered at Site 367 and 370. At both sites the *Lithraphidites bollii* Zone is missing, therefore, it is difficult to determine the lower boundary of this zone. Čepek (1978) used the extinction of the species of *Cruciellipsis cuvillieri* (Manivit) to ascertain the base of the zone in discussion. At Site 367 the top of this zone was placed in Sample 367-25-4 (135-136 cm) (Čepek, 1978). After reexamination of the Cores 23 to 25 of this site we assign this interval now to the *Chiastozygus litterarius* Zone and the lower part of the *Parhabdolithus angustus* Zone (see remarks within the discussions of these zones).

Chiastozygus litterarius Zone

Author: Thierstein (1971), modified by Thierstein (1973).
Definition: Interval from the last occurrence of *Nannoconus colomi* (de Lapparent) in the sense of Thierstein (1973) (equivalent to first occurrence of *Chiastozygus litterarius* (Gorka) to the first occurrence of *Parhabdolithus angustus* (Stradner) being contemporaneous with the first appearance of *Lithastrinus floralis* Stradner.
Age: Early Aptian.
Important common species: *Chiastozygus litterarius* (Gorka), *Conusphaera mexicana* Trejo and *Micrantholithus obtusus* Stradner.

Remarks: The Chiastozygus litterarius Zone was recognized at Sites 135, 367 and 416. According to Čepek (1978) this zone ranges from Sample 367-25-1 (16-17 cm) to 367-25-3 (38-39 cm) (see remarks within the discussion of the Micrantholithus obtusus Zone). At Sites 367 and 370 the base of this zone was determined by the last occurrence of Nannoconus colomi (de Lapparent) (in the sense of Thierstein, 1973). The presence of this species in Sample 370-32-4 (100-101 cm) is considered due to contamination.

Parhabdolithus angustus Zone

Author: Manivit (1971), modified by Thierstein (1973).
Definition: Interval from the first occurrence of Parhabdolithus angustus (Stradner) to the first occurrence of Prediscosphaera cretacea (Arkhangelsky).
Age: Late Aptian to early Albian.
Important common species: Parhabdolithus angustus and Lithastrinus floralis Stradner.
Remarks: The Parhabdolithus angustus Zone is present at Sites 120, 136, 367 and 370. At Sites 367 Čepek (1978) recorded the base of this zone in Sample 367-25-1 (6-7 cm) (see remarks of the Micrantholithus obtusus Zone).

Prediscosphaera cretacea Zone

Author: Thierstein (1971), modified by Thierstein (1973).
Definition: Interval from the first occurrence of Prediscosphaera cretacea (Arkhangelsky) to the first occurrence of Eiffellithus turriseiffeli (Deflandre).
Age: Early to middle Albian.
Important common species: Prediscosphaera cretacea (Arkhangelsky), Podorhabdus albianus Black, Parhabdolithus angustus (Stradner) and Corollithion achylosum (Stover).
Remarks: This zone was encountered at Sites 369 and 416.

Eiffellithus turriseiffeli Zone

Author: Thierstein (1971), emended Roth (1973).
Definition: Interval from the first occurrence of Eiffellithus turriseiffeli (Deflandre) to the first occurrence of Lithraphidites alatus Thierstein.
Age: Late Albian.
Important common species: Eiffellithus turriseiffeli (Deflandre), Podorhabdus albianus Black, and Parhabdolithus asper (Stradner).
Remarks: This zone was recovered at Sites 137 and 370. At Sites 367 and 368 the top of this zone cannot be defined as the marker species of the following zones are missing.

Lithraphidites alatus Zone

Author: Roth (1973), emended Čepek (in press).
Definition: Interval from the first occurrence of Lithraphidites alatus Thierstein to the last occurrence of Lithraphidites acutum Verbeek and Manivit.
Age: Latest Albian to Cenomanian.
Important common species: Lithraphidites alatus Thierstein, Lithraphidites acutum Verbeek and Manivit, Parhabdolithus asper (Stradner), Corollithion exiguum Stradner.
Remarks: The Lithraphidites alatus Zone was encountered at Sites 137 and 415. The poor presence of Lithraphidites alatus Thierstein at Site 415 virtually restricts this zone only to one sample.

Corollithion exiguum Zone

Author: Roth (1973), emended in this paper.
Definition: Interval from the last occurrence of Lithraphidites acutum Verbeek and Manivit to the first occurrence of Micula staurophora (Gardet).
Age: Early Turonian.
Important common species: Corollithion exiguum Stradner and Parhabdolithus asper (Stradner).
Remarks: Roth (1973) has defined the range of this zone as the interval from the first occurrence of Corollithion exiguum Stradner to the first occurrence of Micula staurophora (Gardet). This is changed with respect to the range of the Lithraphidites acutum defining the top of Lithraphidites alatus Zone which is younger than the first occurrence of Corollithion exiguum Stradner (Verbeek 1977, Sissingh 1977, and Perch-Nielsen 1979). The Corollithion exiguum Zone seems to be present at the Site 137.

Micula staurophora Zone

Author: Roth (1973).
Definition: Interval from the first occurrence of Micula staurophora (Gardet) to the first occurrence of Marthasterites furcatus (Deflandre).
Age: Middle to late Turonian.
Important common species: Micula staurophora (Gardet), Gartnerago obliquum (Stradner) and Eiffellithus eximius (Stover).
Remarks: This zone seems to be present at Sites 136 and 368.

Marthasterites furcatus Zone

Author: Čepek and Hay (1969), emended Perch-Nielsen (1977).
Definition: Interval from the first occurrence of Marthasterites furcatus (Deflandre) to the first occurrence of Broinsonia parca (Stradner).
Age: Coniacian to Santonian.
Important common species: Marthasterites furcatus (Deflandre) and Eiffellithus eximius (Stover).
Remarks: This zone was determined only at Site 369.

Eiffellithus eximius Zone

Authors: Bukry and Bramlette (1970), emended Perch-Nielsen (1977).
Definition: Interval from the first occurrence of Broinsonia parca (Stradner) to the last occurrence of Marthasterites furcatus (Deflandre).
Age: Early Campanian.
Important species: Broinsonia parca (Stradner), Marthasterithes furcatus (Deflandre), Eiffellithus eximius (Stover).
Remarks: According to Perch-Nielsen (1977) this zone is attributed to the interval of late Santonian to early Campanian. According to Thierstein (1976) and Sissingh (1977) the first occurrence of Broinsonia parca (Stradner) lies at the Santonian/Campanian boundary, however, Perch-Nielsen (1979) placed it within the early Campanian. We restrict the range of the Eiffellithus eximius Zone to the lower Campanian.

This zone is perhaps represented at Site 13, but for a definitive decision the information in the description of Site 13 (Maxwell et al., 1970), are not conclusive enough.

Broinsonia parca Zone

Author: Perch-Nielsen (1977).
Definition: Interval from the last occurrence of Marthasterites furcatus (Deflandre) to the first occurrence of Tetralithus aculeus (Stradner).
Age: Early Campanian.
Remarks: This zone was recognized at Site 369 as Eiffellithus eximius Zone.

Tetralithus aculeus Zone

Authors: Čepek and Hay (1969), emended Perch-Nielsen (1977).
Definition: Interval from the first occurrence of Tetralithus aculeus (Stradner) to the first occurrence of Tetralithus gothicus Deflandre.
Age: Early Campanian.
Remarks: This zone was not recognized in the cores of the study region.

Tetralithus gothicus Zone

Author: Martini (1976).
Definition: Interval from the first occurrence of Tetralithus gothicus Deflandre to the first occurrence of Tetralithus trifidus (Stradner).
Age: Middle Campanian.
Remarks: This zone was not recognized in the cores of the study area.

Tetralithus trifidus Zone

Authors: Bukry and Bramlette (1970).
Definition: Interval of the total range of Tetralithus trifidus (Stradner).
Age: Late Campanian to early Maastrichtian.
Important common species: Tetralithus trifidus (Stradner), Tetralithus gothicus Deflandre, Broinsonia parca (Stradner), and Eiffellithus eximius (Stover).
Remarks: The Tetralithus trifidus Zone was encountered at Holes 13A and 369A. The identification of this zone at Site 135 is questionable.

Arkhangelskiella cymbiformis Zone

Author: Perch-Nielsen (1972), emended Martini (1976).
Definition: Interval from the last occurrence of Tetralithus trifidus

(Stradner) to the first occurrence of Lithraphidites quadratus Bramlette and Martini.

Age: Early Maastrichtian to ?early middle Maastrichtian.

Important species: Arkhangelskiella cymbiformis Vekshina.

Remarks: This zone is attributed to different ages according to different authors: Perch-Nielsen (1972) - early to middle Maastrichtian, Martini (1976), Sissingh (1977) - late Maastrichtian (early part), Čepek (in press) - early Maastrichtian, Verbeek (1977) did not use this zone. These different ages of the Arkhangelskiella cymbiformis Zone possibly result from the various ways of dividing the Maastrichtian. Sissingh (1977) divided the Maastrichtian stage into two parts only, the early and the late Maastrichtian. His Arkhangelskiella cymbiformis Zone has a range corresponding to the interval from the upper part of the Arkhangelskiella cymbiformis Zone to the Lithraphidites quadratus Zone as used in this paper. On the other side his late Maastrichtian includes also middle Maastrichtian in the sense of a tripartite Maastrichtian. According to Martini (1976) the mid-Maastrichtian age of the zone is based on the interpretation of the type Maastrichtian locality. He determined this zone in the Gulpen Chalk C to E. However, according to Čepek and Moorkens (1979) the Gulpen Chalk E (Lixhe Chalk 2) has to be attributed at least to the Lithraphidites quadratus Zone (older samples were not available). Accordingly, the Arkhangelskiella cymbiformis Zone in the sense of this paper is older and in relation to other authors it is attributed to the interval early Maastrichtian to early middle Maastrichtian. This zone is present at Site 369.

Lithraphidites quadratus Zone

Authors: Čepek and Hay (1969), emended Perch-Nielsen (1972).

Definition: Interval from the first occurrence of Lithraphidites quadratus Bramlette and Martini to the first occurrence of Micula mura (Martini).

Age: Middle to late Maastrichtian.

Important common species: Lithraphidites quadratus Bramlette and Martini and Arkhangelskiella cymbiformis Vekshina.

Remarks: This zone is present at Sites 366 (sample 55,CC). In Hole 369A, this zone was recognized by Čepek (1978) and was defined as the range from the last occurrence of Tetralithus trifidus Stradner to the first occurrence of Micula mura (Martini). This interval now belongs to two zones as defined in the present paper. The species Lithraphidites quadratus Bramlette and Martini is present only in the upper part of this section (Čepek, 1978). This part (Sample 369A-33, CC) is therefore now assigned to the Lithraphidites quadratus Zone while the lower part of the section represents the Arkhangelskiella cymbiformis Zone.

Micula mura Zone

Author: Martini (1969).

Definition: Interval of the total range of Micula mura (Martini).

Age: Late Maastrichtian.

Important common species: Micula mura (Martini), Lithraphidites quadratus Bramlette and Martini, and Arkhangelskiella cymbiformis Vekshina.

Remarks: In Fig. 3 this zone is mentioned as the Micula mura-/Nephrolithus frequens Zone. This title should express the range of the interval, which at other localities is divided into two zones: the Nephrolithus frequens Zone (the older one) and the Micula mura Zone (the younger one). Both species are mutually exclusive at all uppermost Maastrichtian localities. Following Worsley and Martini (1970) it seems, that Micula mura (Martini) and Nephrolithus frequens (Gorka) have climatically restricted occurrences. The first species is a more low latitude inhabitant while Nephrolithus frequens is more indicative for high latitudes. This zone was determined at Site 369, the only one in the eastern tropical and subtropical Atlantic Ocean in which uppermost Maastrichtian sediments were recovered.

3. Planktonic Foraminiferal Biostratigraphy and Zonation.

The composite planktonic zonation used in this paper is based on recent schemes by Bolli (1966), Pessagno (1967), Premoli-Silva and Bolli (1973), Longoria (1974), Sigal (1977, 1979), and Drushtchitz and Gorbatschik (1979).

The Cretaceous time was subdivided into 30 zones though not all of them could be defined in the study area. Fig. 5 shows the correlation of this proposed zonal scheme with the zonations applied in the Initial Reports of the six legs of the Deep-Sea Drilling Project under discussion. The geochronologic scale was adopted from van Hinte (1976).

The strategy for the zonal definitions was to favour the use of first appearance data of well defined species. Only in few cases the zones are defined as intervals between the last occurrence of one species and the first occurrence of another one. In those cases the zone was named after a common species which appears slightly below this interval.

Globuligerina gulekhensis (informal) Zone

Preliminary definition: Interval from the first occurrence of the index species to the first occurrence of Globuligerina hoterivica (Subbotina).

Age: Berriasian to early Hauterivian.

Important species: Globuligerina gulekhensis (Gorbatschik and Poroshina), G. caucasica (Gorbatschik and Poroshina).

Remarks: The zonal markers have been recorded from the Berriasian and Valanginian of Azerbeidzhan. Up to now in the North Atlantic Ocean there are no records of these species, therefore this time interval could not be indicated by planktonic foraminifera in this area.

Globuligerina hoterivica (informal) Zone

Definition: Interval from the first occurrence of Globuligerina hoterivica (Subbotina) to the first occurrence of Hedbergella tushepensis (Antonova).

Age: Late Hauterivian.

Important species: Globuligerina hoterivica (Subbotina), Gubkinella graysonensis (Tappan), Hedbergella occulta (Longoria), H. sigali (Moullade).

Remarks: Due to missing or not recorded Hedbergella tushepensis (Antonova) the top of the zone is not defined at Site 370. At Site 397 Globuligerina hoterivica was recorded (Butt, 1979). According to Rösler et al. (1979) it co-occurs with Favusella stiftia (Rösler, Lutze and Pflaumann). After Grigjalis and Gorbatschik (1980) the last species is a junior synonym of Favusella tardita (Antonova), which starts at its type locality as late as in the Late Barremian. Onshore, in the Agadir Basin G. hoterivica starts within the range of Crioceratites duvali. According to Sigal (1977), G. hoterivica and H. sigali should not co-occur, the former species being restricted to late Hauterivian the latter starting with the Barremian Nicklesia pulchella Zone. The co-occurrence of both species in DSDP-samples, attributed by nannofossils to an age of no younger than the late Hauterivian is still open for discussion.

Hedbergella tushepensis (informal) Zone

Author: Sigal 1979

Preliminary definition: Interval from the first occurrence of the index species to the first occurrence of Hedbergella aptica (Agalarova).

Age: Early Barremian.

Accompanying species: Clavihedbergella eocretacea Neagu, Hedbergella globigerinelloides (Subbotina), H. sigali Moullade, Globigerinelloides ultramicrus (Subbotina).

Remarks: As the index species has not been recorded in the study area the zone could not be singled out, but may be present at Sites 370 and 416. According to Cita (1973) early Barremian is also represented at Site 120, but without planktonic foraminifera.

According to Sigal (1977) the first occurrence of Clavihedbergella eocretacea lies within the range of Nicklesia pulchella and Emericeras emerici, attributed by Sigal to the early Barremian, while fide Wiedmann et al. (1978) this interval should be assigned to the middle and late Barremian. According to Sigal (1979) C. eocretacea starts later than Hedbergella tushepensis together with H. sigali. There we encounter the same problems as mentioned in the remarks to the previous zone..

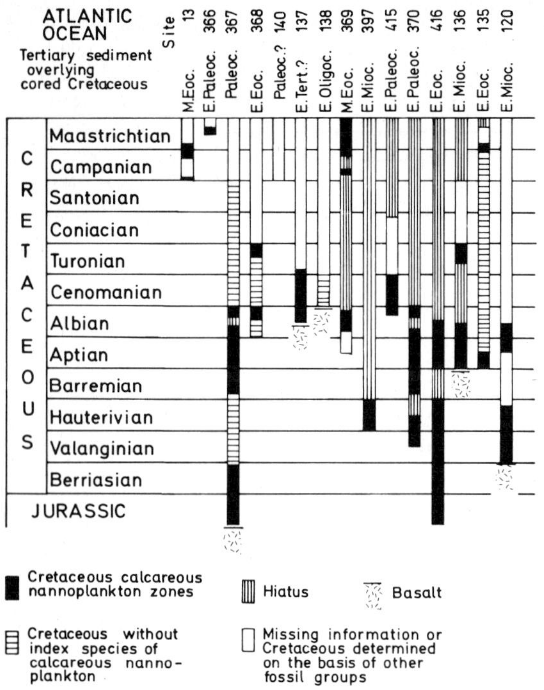

Fig. 2. Distribution of the Cretaceous calcareous nannoplankton in DSDP Sites of the eastern North Atlantic Ocean

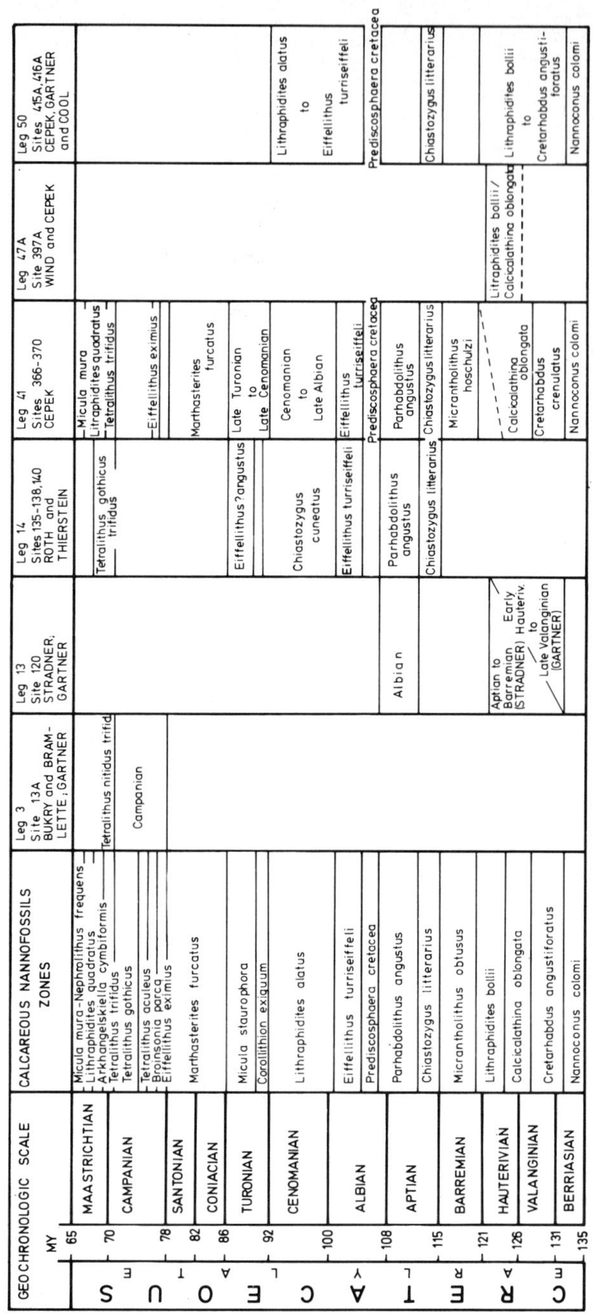

Fig. 3. Correlation of Cretaceous nannoplankton zones used in this paper with the geochronologic scale of van Hinte (1976) and with the nannoplankton zonation applied on DSDP Legs 3, 13, 14, 41, 47, and 50

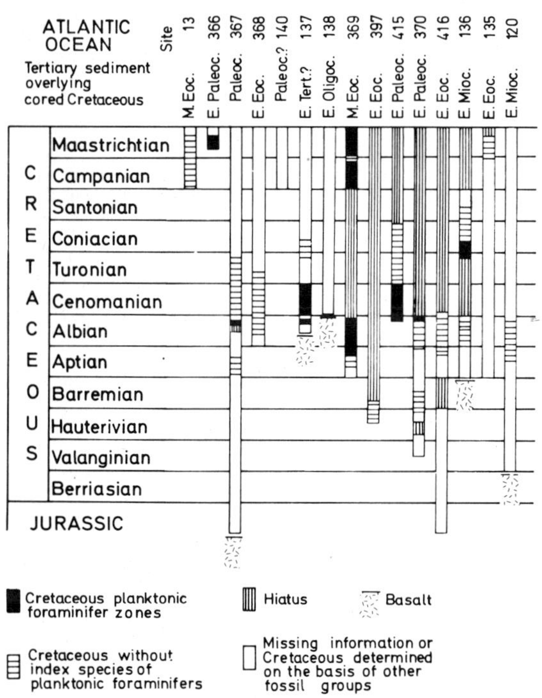

Fig. 4. Distribution of the Cretaceous planktonic foraminifera in DSDP Sites of the eastern North Atlantic Ocean

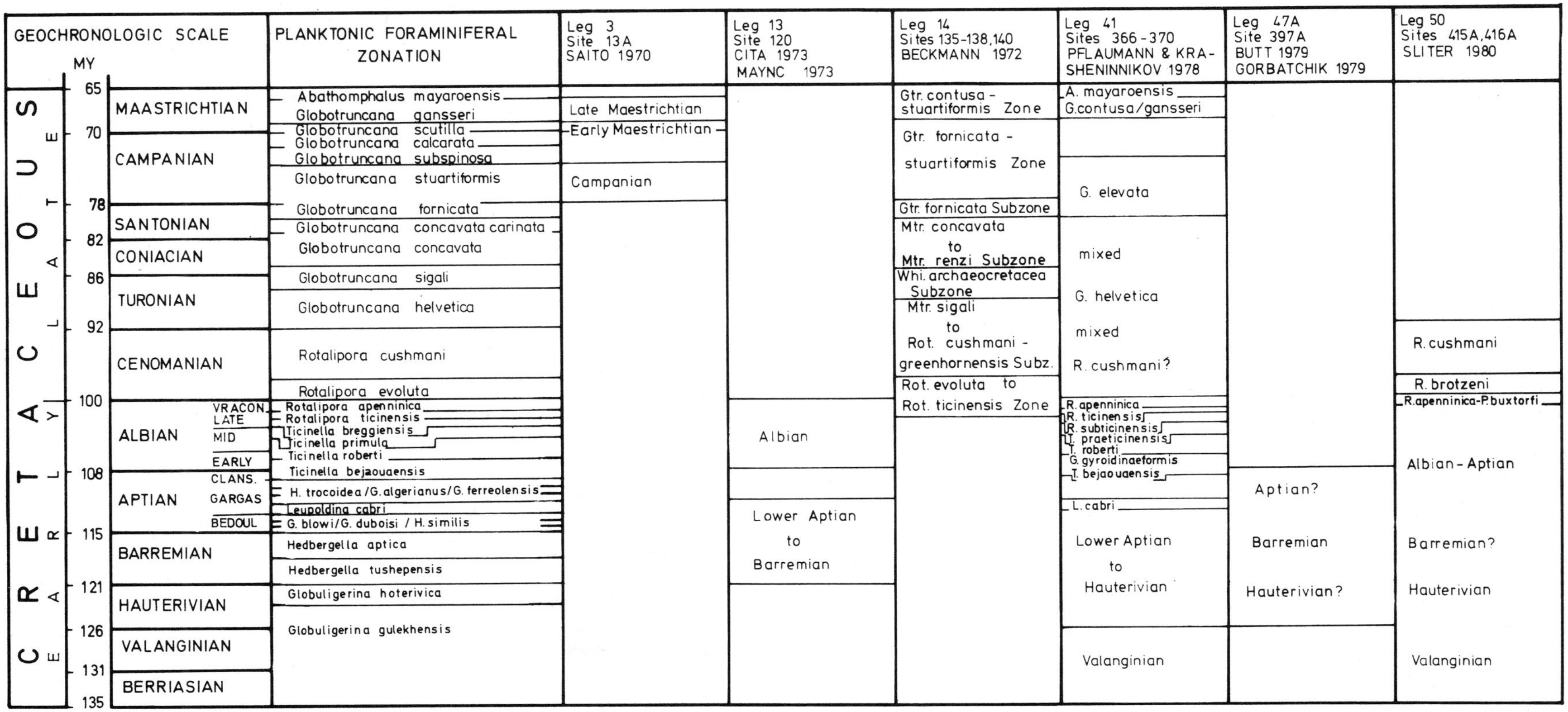

Fig. 5. Correlation of the Cretaceous planktonic foraminiferal zones used in this paper with the geochronologic scale of van Hinte (1976) and with the foraminiferal zonations applied on DSDP Legs 3, 13, 14, 41, 47, and 50

Hedbergella aptica (informal) Zone

Author: Sigal (1979)
Preliminary definition: Interval from the first occurrence of the index species to the first appearance of Hedbergella similis Longoria.
Age: Late Barremian.
Associated species: Hedbergella tardita (Antonova), H. gr. similis Longoria.
Remarks: Barremian sediments are dated only by the nannofossil record. The planktonic foraminifera of this age are Globigerinelloides ultramicrus (Subbotina), Hedbergella gorbachikae Longoria, and H. globigerinelloides (Subbotina) at Site 370. H. aptica was reported by Sigal (1979) from Site 398 as first occurring within the range of H. tardita, that is before the first occurrence of Hedbergella gr. similis. The latter species marks the base of the Deshayesites forbesi Zone, which represents the base of the Aptian stage.

Hedbergella similis Zone

Author: Sigal 1977
Definition: Interval from the first occurrence of Hedbergella similis Longoria to the first occurrence of Globigerinelloides duboisi (Chevalier).
Age: Earliest Aptian.
Accompanying species: Hedbergella quadricamerata (Antonova), H. tardita (Antonova), H. cf. planispira (Tappan), and H. sigali Moullade.
Remarks: The zone has not been recorded from the North Atlantic Ocean.

Globigerinelloides duboisi Zone

Author: Sigal 1977
Definition: Interval from the first occurrence of the index species to the first occurrence of Globigerinelloides blowi (Bolli).
Age: Early Aptian.
Accompanying species: Globigerinelloides gottisi (Chevalier), Hedbergella semielongata Longoria, Globuligerina hoterivica (Subbotina).
Remarks: The zone has not been recorded in the study area.

Globigerinelloides blowi Zone

Author: Sigal 1977
Definition: Interval from the first appearance of Globigerinelloides blowi (Bolli) to the first appearance of Leupoldina cabri (Sigal).
Age: Upper part of early Aptian.
Accompanying species: Globigerinelloides maridalensis (Bolli), Globulogerina hoterivica (Subbotina), Globigerinelloides duboisi (Chevalier), Hedbergella excelsa Longoria and H. planispira (Tappan).
Remarks: The species defining the upper limit of this zone are missing in our region, so this zone could only indirectly be recorded at Site 369A (47,CC). Gorbatschik (in Basow et al. 1979) recorded "Blowiella" blowi from Site 397A, attributing this section to "most probable" Barremian to Early Aptian, while Butt (1979) favours a Barremian age for the same section.

Leupoldina cabri Zone

Author: Sigal 1977
Definition: Range of the zonal marker.
Age: Lower part of the Gargasian, late Aptian.
Associated species: Globigerinelloides gottisi (Chevalier), G. saundersi (Bolli), Leupoldina pustulans (Bolli), L. reicheli (Bolli).
Rem The absence of the index species prevents the differentiation of the zone in the study area.

Globigerinelloides ferreolensis Zone

Author: Sigal 1977
Definition: Interval from the first occurrence of Globigerinelloides ferreolensis (Moullade) to the first occurrence of Globigerinelloides algerianus Cushman and Ten Dam.
Age: Gargasian, late Aptian.
Accompanying species: Hedbergella delrioensis (Carsey), H. infracretacea (Glaessner), Globigerinelloides gottisi (Chevalier), G. blowi (Bolli).
Remarks: This zone is recorded from Site 369 (Core 46).

Globigerinelloides algerianus Zone

Author: Sigal 1977
Definition: Interval from the first occurrence of Globigerinelloides algerianus Cushman and Ten Dam to the first appearance of Hedbergella trocoidea (Gandolfi).
Age: Middle part of the late Gargasian, late Aptian.

Important species: Globigerinelloides ferreolensis (Moullade), G. barri (Bolli, Loeblich and Tappan), G. blowi (Bolli), Hedbergella infracretacea (Glaessner), H. delrioensis (Carsey).

Hedbergella trocoidea Zone

Author: Sigal 1977
Definition: Interval from the first occurrence of Hedbergella trocoidea (Gandolfi) to the first occurrence of Ticinella bejaouaensis Sigal.
Age: Uppermost part of the Gargasian, late Aptian.
Accompanying species: Planomalina chenouriensis (Sigal), Globigerinelloides ferreolensis (Moullade), G. algerianus Cushman and Ten Dam, Hedbergella infracretacea (Glaessner).
Remarks: Core 45 at Site 369 is attributed to this zone.

Ticinella bejaouaensis Zone

Author: Sigal 1977
Definition: Interval from the first occurrence of the zonal marker of the first occurrence of Ticinella roberti (Gandolfi).
Age: Latest Aptian (Clansayesian) to lower part of early Albian.
Accompanying species: Hedbergella trocoidea (Gandolfi), H. infracretacea (Glaessner), Globigerinelloides gyroidinaeformis Moullade.
Remarks: This zone was recorded from Site 369A (44-2 to CC). At Site 370 Globigerinelloides gyroidinaeformis may indicate both this zone and/or the following one.

Ticinella roberti Zone

Author: Dalbiez (1955) emended.
Definition: Interval from the first occurrence of the zonal marker of the first occurrence of Ticinella primula Luterbacher.
Age: Upper part of the early Albian to lower part of middle Albian.
Accompanying species: Ticinella bejaouaensis Sigal, Hedbergella trocoidea (Gandolfi), H. infracretacea (Glaessner), H. planispira (Tappan), Globigerinelloides gyroidinaeformis Moullade.
Remarks: This zone is recorded from Site 369A (43,CC to 44-1). The presence of Globigerinelloides gyroidinaeformis at Site 370 may include it within the distribution area of the zone.

Ticinella primula Zone

Author: Sigal 1977
Definition: Interval from the first occurrence of Ticinella primula Luterbacher to the first occurrence of Ticinella breggiensis (Gandolfi).
Age: Upper part of middle Albian.
Associated species: Ticinella praeticinensis Sigal, T. bejaouaensis Sigal, Globigerinelloides gyroidinaeformis Moullade, Hedbergella planispira (Tappan).
Remarks: This zone is recorded at Site 369A (43-1 to 4).

Ticinella breggiensis Zone

Author: Sigal 1977
Definition: Interval from the first occurrence of Ticinella breggiensis (Gandolfi) to the first occurrence of Rotalipora ticinensis (Gandolfi).
Age: Lower part of late Albian.
Accompanying species: Ticinella praeticinensis Sigal, R. raynaudi Sigal, Praeglobotruncana delrioensis (Plummer), Hedbergella simplicissima (Mangé and Sigal).
Remarks: At Site 369A, Sample 42,CC is attributed to this zone.

Rotalipora ticinensis Zone

Author: Dalbiez (1955), renamed by Bolli (1957).
Definition: Interval from the first occurrence of Rotalipora ticinensis (Gandolfi) to the first occurrence of Rotalipora apenninica (Renz).
Age: Upper part of late Albian.
Accompanying species: Schackoina cenomana (Schacko), Ticinella subticinensis (Gandolfi), Globigerinelloides bentonensis (Morrow), G. caseyi (Bolli, Loeblich and Tappan).
Remarks: This zone was recorded from Sites 137, 367, 369 and 370.

Rotalipora apenninica Zone

Author: Dalbiez (1955), renamed by Bolli (1957), emended.
Definition: Interval from the first occurrence of Rotalipora apenninica

(Renz) to the first appearance of Rotalipora evoluta Sigal.
Age: Uppermost Albian, Vraconian.
Important species: Planomalina buxtorfi (Gandolfi), Praeglobotruncana delrioensis (Plummer), Hedbergella simplicissima (Mangé and Sigal), H. delrioensis (Carsey).
Remarks: This zone was recorded from Sites 370 and 415A.

Rotalipora evoluta Zone

Author: Pessagno (1967).
Definition: Interval from the first occurrence of the zonal marker to the first appearance of Rotalipora cushmani (Morrow).
Age: Early Cenomanian.
Accompanying species: Rotalipora globotruncanoides Sigal, R. brotzeni Sigal, R. reicheli (Mornod), R. gandolfii Luterbacher and Premoli Silva, R. greenhornensis (Morrow), Praeglobotruncana stephani (Gandolfi).
Remarks: Sliter (1980) attributed Core 13 at Site 415 to this zone. At Site 137 Cores 12 to 15 have the same age.

Rotalipora cushmani Zone

Author: Borsetti (1962), here emended.
Definition: Range of the zonal marker, from its first appearance to the first occurrence of Globotruncana helvetica Bolli.
Age: Middle and late Cenomanian.
Important species: Rotalipora greenhornensis (Morrow), R. reicheli (Mornod), R. brotzeni (Sigal), R. globotruncanoides (Sigal), R. montsalvensis Mornod, Hedbergella delrioensis (Carsey), H. simplicissima (Mangé and Sigal), Clavihedbergella simplex (Morrow), Globotruncana algeriana Caron.
Remarks: This zone was noted from Sites 137 (Core 8-11) and 415.

Globotruncana helvetica Zone

Author: Sigal (1955)
Definition: Range of the zonal marker.
Age: Early and middle Turonian.
Accompanying species: Praeglobotruncana stephani (Gandolfi), Hedbergella brittonensis Loeblich and Tappan, Globotruncana algeriana Caron, G. imbricata Mornoed, G. marginata (Reuss), G. sigali Reichel and G. schneegansi Sigal have their first occurrences within this zone.
Remarks: Species of this zone are present at Site 369A, but mixed with representatives of older and younger zones.

Globotruncana sigali Zone

Author: Sigal (1977)
Definition: Interval with the index species from the last occurrence of Globotruncana helvetica Bolli to the first occurrence of Globotruncana concavata (Brotzen).
Age: Late Turonian to early Coniacian.
Important species: Globotruncana schneegansi Sigal, G. indica Jacob and Sastry, G. marginata (Reuss), G. renzi Gandolfi, G. pseudolinneiana (Pessagno), Heterohelix moremani (Cushman).
Remarks: According to Beckmann (1972) sections of Site 136 (Cores 5 to 7) and Site 137 (Core 7) are attributed to this zone. At Site 369A this zone could not be differentiated due to mixing of older with younger faunas.

Globotruncana concavata Zone

Author: Sigal (1955), emended.
Definition: Interval from the first appearance of Globotruncana concavata (Brotzen) to the first appearance of Globotruncana concavata carinata Dalbiez.
Age: Early Coniacian to early Santonian.
Accompanying species: Globotruncana sigali Reichel, G. schneegansi Sigal, G. angusticarinata Gandolfi, G. canaliculata (Reuss), Hedbergella simplicissima Mangé and Sigal, Clavihedbergella simplex (Morrow), Globigerinelloides asper (Ehrenberg), and Stensioeina exsculpta (Reuss).
Remarks: The index fossil of this zone was recorded at Site 369A, but mixed with older faunas.

Globotruncana concavata carinata Zone

Author: Casati and Tomai (1969) fide Premoli Silva and Bolli (1973).

Definition: Range of the zonal marker.
Age: Early and lower parts of late Santonian.
Important species: Globotruncana concavata (Brotzen), Sigalia deflaense (Sigal).
Remarks: This zone could not be distinguished in the study area.

Globotruncana fornicata Zone

Author: Bolli (1957), emended.
Definition: Interval with Globotruncana fornicata from the last occurrence of Globotruncana concavata carinata Dalbiez to the first appearance of the Globotruncana stuartiformis Dalbiez.
Age: Late Santonian.
Accompanying species: Globotruncana ventricosa White, G. arca (Cushman), Planoglobulina glabrata (Cushman), and the benthic species Bolivinoides strigillatus (Chapman).
Remarks: Interval is not recorded in the study area.

Globotruncana stuartiformis Zone

Author: Sigal (1977)
Definition: Interval from the first appearance of Globotruncana stuartiformis Dalbiez to the first appearance of Globotruncana subspinosa Pessagno.
Age: Early and lower part of Campanian.
Important species: Globotruncana stuarti (de Lapparent), G. caliciformis (de Lapparent), G. flexuosa Van der Sluis, Pseudotextularia elegans (Rzehak), Pseudoguembelina costulata (Cushman), Globigerinelloides volutus (White), and the benthic species Bolivinoides decoratus (Jones).
Remarks: At Site 369A the interval 39-2 to 39,CC is attributed to this zone as well as Site 13A Core 4, which contains an early Campanian microfauna with Rugoglobigerina tradinghousensis Pessagno (Saito 1970).

Globotruncana subspinosa Zone

Author: Van Hinte (1976)
Definition: Interval from the first appearance of Globotruncana subspinosa Pessagno to the first appearance of Globotruncana calcarata Cushman.
Age: Middle part of the late Campanian.
Remarks: According to the compilation of van Hinte (1976) the base of this zone coincides with the first appearance of the benthic foraminifer Bolivinoides draco miliaris Hiltermann and Koch. At Site 369A this zone was recorded between Samples 39-1 and 38,CC; however, Bolivinoides draco miliaris Hiltermann and Koch appears later (in Sample 38-4).

Globotruncana calcarata Zone

Author: Herm (1962)
Definition: The range of the zonal marker.
Age: Latest Late Campanian.
Accompanying species: Rugoglobigerina rugosa (Plummer), Globotruncanella havanensis (Voorwijk), Globotruncana linneiana (d'Orbigny), G. fornicata Plummer, G. subcircumnodifer Gandolfi, G. ventricosa White, G. gagnebini Tilev.
Remarks: At Site 13A and 369 sediments of this zone may be present, however, the index species was not recovered.

Globotruncana scutilla Zone

Author: van Hinte (1976)
Definition: Interval from the last occurrence of Globotruncana calcarata Cushman to the first occurrence of Globotruncana gansseri Bolli.
Age: Early Maastrichtian.
Accompanying species: Globotruncana scutilla Gandolfi, G. subcircumnodifer Gandolfi, G. subpetaloidea Gandolfi, G. stuarti (de Lapparent), G. arca (Cushman), G. fornicata Plummer, G. rosetta (Carsey), G. ventricosa White, Pseudoguembelina costulata (Cushman), Pseudotextularia elegans (Rzehak).
Remarks: According to van Hinte (1976) Globotruncana scutilla Gandolfi evolves from G. fornicata Plummer at the base of the zone, and Globotruncanella havanensis (Voorwijk) has fully evolved. By the presence of G. havanensis and G. petaloidea (Gandolfi) Core 2 at Site 13A is attributed to this zone. At Site 369A the section between 37-4 and 38-5 are assigned to this zone as Pseudoguembelina costulata (Cushman), Globotruncana gagnebini Tilev, G. petaloidea Gandolfi

and G. elevata (Brotzen) co-occur before the first appearance of G. gansseri Bolli and G. subspinosa Pessagno.

Globotruncana gansseri Zone

Author: Bolli (1957)

Definition: Interval from the first appearance of the index species to the first appearance of Abathomphalus mayaroensis Bolli.

Age: Upper part of early Maastrichtian to lower part of late Maastrichtian.

Accompanying species: Globotruncana caliciformis (de Lapparent), G. subpennyi Gandolfi, G. stuarti (de Lapparent), and G. falsostuarti Sigal.

Remarks: At Site 369A the first appearance levels within this zone are in following order: Globotruncana contusa Cushman, G. gagnebini Tilev, G. stuarti (de Lapparent), G. falsostuarti Sigal, G. conica White, Abathomphalus intermedius (Bolli). In the higher part of the zone Racemiguembelina fructicosa (Egger) appears together with Pseudoguembelina excolata (Cushman). By the recognition of P. excolata at Site 13 (Saito,1970) a late Maastrichtian age without zonal marker is attributed to this sample, which was taken from the center of the bit. According to Beckmann (1972) this zone is documented at Site 135 by the presence of Racemiguembelina fructicosa (Egger).

Abathomphalus mayaroensis Zone

Author: Bolli (1957)

Definition: Interval of the range of the index species.

Age: Uppermost Maastrichtian.

Accompanying species: Racemiguembelina fructicosa (Egger), Abathomphalus intermedius (Bolli), Globotruncana contusa (Cushman).

Remarks: This time interval, the last foraminiferal zone of the Mesozoic era was definitely recorded by the presence of the index species at Site 369A only.

4. Upper Cretaceous Stratigraphy by Planktonic Foraminifera of Site 369A, drilled on the Continental Slope off Cape Bojador

Upper Cretaceous sediments of Site 369 were continuously cored but recovered only moderately. They are classified as argillaceous nanno-limestones with minor porcellanite and chert of mostly gray to greenish colors, nannobearing gray marls and chalk. Slumps within this section are composed of a variety of convoluted, varicolored marls, clays, chalks, and limestones. The whole section is about 50 m thick, and is limited by gaps: at the lower boundary the time interval between the uppermost Albian and Santonian is missing, and on the top the highest Maastrichtian zone is covered by sediments of the middle Eocene.

Within the autochthonous sediments the foraminiferal fauna is dominated by planktonic species, the benthic component is represented by various calcareous forms as buliminids, rotaliids and lagenids.

Using the planktonic foraminifers the Upper Cretaceous sediments can be attributed to five time-intervals (Figs. 6, 7.4):

1. The Globotruncana stuartiformis Zone between Core 39 section 2 and CC with Globotruncana stuartiformis, G. rosetta, G. ventricosa, G. obliqua, G. bulloides, G. elevata, G. stephensoni, Pseudotextularia elegans, Rugoglobigerina rugosa, Heterohelix punctulata;

2. The undifferentiated Globotruncana subspinosa to calcarata Zone with Globotruncana subspinosa, G. arca, Globotruncanella havanensis, G. petaloidea, Globigerinelloides subcarinatus, G. messinae messinae, Biglobigerinella multispina. Globotruncana calcarata could not be identified, therefore it was impossible to divide this section into two zones.

3. The Globotruncana scutilla Zone of the early Maastrichtian is recorded by the first occurrence of Heterohelix lata, H. ultimatumida, Pseudoguembelina costulata, P. palpebra, Gublerina robusta, Globotruncana morozovae, Rugoglobigerina hexacamerata, R. subcircumnodifer, and R. reicheli (Cores 37-6 and 38).

4. The Globotruncana gansseri Zone is marked by the first occurrence of the index species together with Globotruncana contusa, G. gagnebini, G.

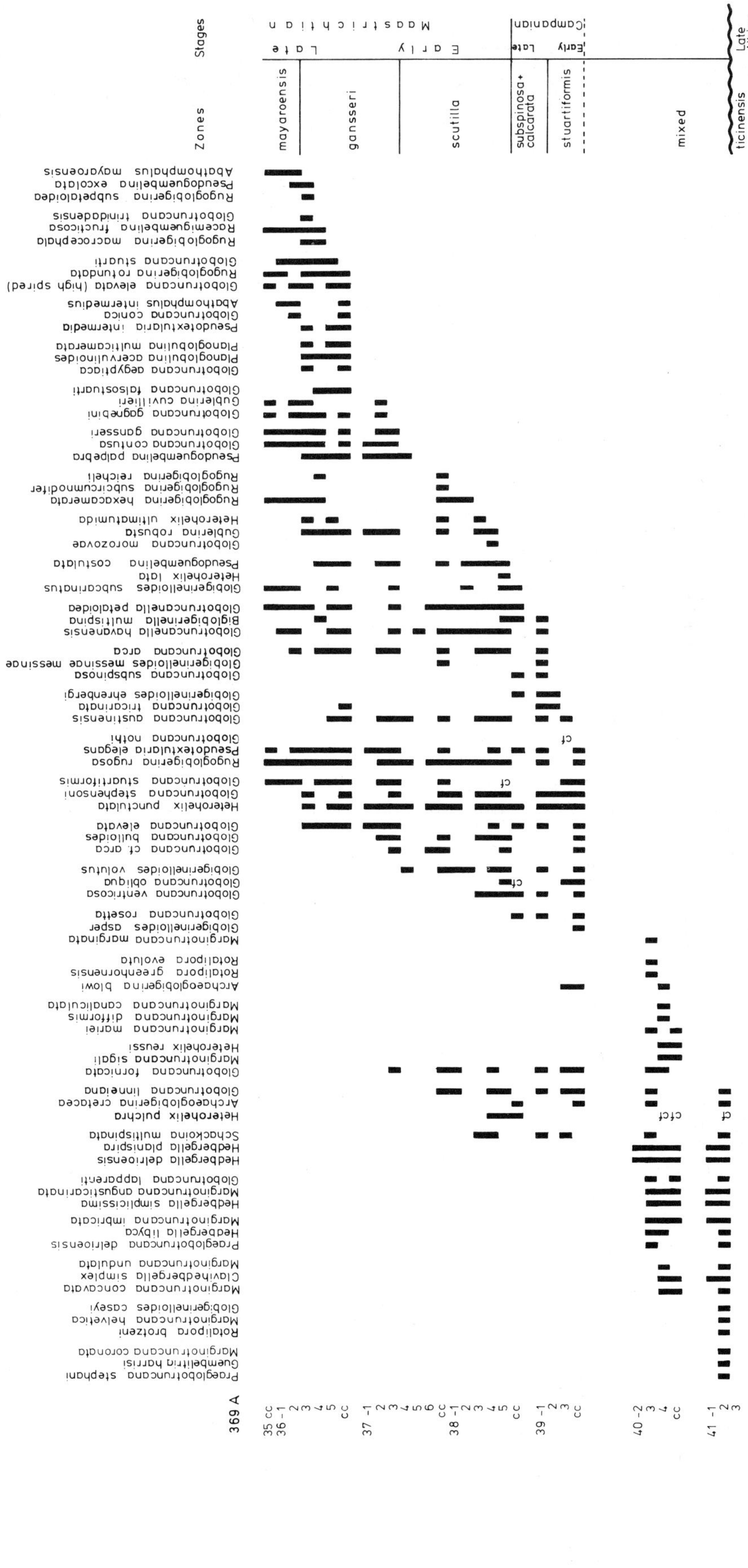

Fig. 6. Range chart of planktonic foraminifera in Late Cretaceous sediments from Site 369, drilled on the continental slope off Cape Bojador

falsostuarti, G. aegyptiaca, G. conica, G. stuarti, G. trinidadensis, Gublerina cuvillieri, Planoglobulina acervulinoides, P. multicamerata, Pseudotextularia intermedia, Abathomphalus intermedius, Rugoglobigerina rotundata, R. macrocephala and Racemiguembelina fructicosa (Core 37-4 to 36,CC).

5. Possibly separated by a small slump the Abathomphalus mayaroensis Zone covers the sediments of the last mentioned zone. It is indicated by the presence of the index species associated with Globotruncana contusa, G. gansseri, G. gagnebini, G. conica, G. stuarti, Abathomphalus intermedius, Racemiguembelina fructicosa, and Pseudotextularia elegans (Core 35,CC to 36-3).

As indicated in Fig. 6 the base of the Late Cretaceous is represented by a slump, which contains mixed faunas with species indicative for nearly all foraminiferal zones of the Cenomanian to Santonian: like Rotalipora evoluta, R. brotzeni, R. greenhornensis, Praeglobotruncana delrioensis, P. stephani, Globotruncana (Marginotruncana) helvetica, G. sigali, G. coronata. The accompanied benthic fauna is not different to that of the overlying sediments, indicating the derivation of the slumped sediments from nearby regions of the same ecologic environment. The youngest species within the slump are attributed to the Globotruncana concavata Zone, so the age of the slump may be within the time-interval Coniacian to Campanian.

5. Correlation of the floral and faunal Record

Pre-Cretaceous

The oldest sediments of the tropical and subtropical Eastern North Atlantic Ocean recovered at Sites 367 and 416A are of a late Jurassic age.
However, the Mazagan sites of Leg 79 are not covered.

At Site 367 (Cape Verde Basin) the oldest sediments contain generally a rare and poorly preserved calcareous nannoplankton flora dated as Oxfordian to Kimmeridgian (Čepek, 1978). Above these sediments the youngest Jurassic coccolith zone, the Parhabdolithus embergeri Zone, was recognized. This zone has been determinated also at Site 416 (Deep Basin off Morocco, but with a more abundant and better preserved calcareous nannoplankton.

The base of the Parhabdolithus embergeri Zone is defined by the first occurrence of the index species. The top of the zone, coincident with the Jurassic/Cretaceous boundary is usually defined by the first occurrence of Nannoconus colomi (de Lapparent). However, at Site 416 this species occurs later, within the Berriasian, therefore Čepek et al. (1980) used the first occurrence of Parhabdolithus asper (Stradner) for the determination of the base of the Cretaceous a level coindiding with the first common occurrence of Polycostella senaria Thierstein.

Berriasian

Berriasian is represented by the Nannoconus colomi Zone and by the early parts of the Cretarhabdus angustiforatus Zone. At Site 367 (Cape Verde Basin) the coccoliths of the Nannoconus colomi Zone are generally more abundant and better preserved than at Site 416A (Deep Basin off Morocco). At Site 416A Nannoconus colomi becomes common as late as within the latest parts of the zone. Its absence in the earlier parts may be due to its restriction to the tropical and subtropical paleolatitudes during the Jurassic (Thierstein, 1976) as Site 416A lies about 14 degrees farther north than Site 367. During the Berriasian the climatic regime at Site 416 may have improved.

At Site 367 a Berriasian age could not be proven by the foraminifers, while at Site 416A a narrow interval of reddish-brown mudstone and greenish-gray, micritic limestone contains an overlap of species whose range includes the Late Jurassic and Early Cretaceous and species restricted to the Lower Cretaceous. This section is tentatively attributed to the Berriasian (Sliter, 1980).

Valanginian to Hauterivian

Marine sediments of this time interval were deposited in three areas, the Cape Verde Basin (Site 367), including the island of Maio, the basin off Morocco (Sites 370, 416A), including the Gorringe Bank (Site 120), Fuerteventura (Robertson and Bernoulli, this Vol.), and the coastal basin of Agadir as part of the South Atlas Basin, and the uppermost continental rise off Cape Bojador (Site 397A). The southern regions contain limestones and marlstones, while the deep basin off Morocco contains flysch-like sediments deposited with silty to sandy turbidites and pelagic intercalations. In the coastal basin of Agadir the Valanginian is represented by silty clays with a central dolomite. In the deep-sea sediments the age can be determined more readily by the calcareous nannoplankton, while onshore an ammonite zonation can be applied.

The Valanginian is composed of parts of two nannofossil zones: the later part of the *Cretarhabdus angustiforatus* Zone and the early parts of the *Calcicalathina oblongata* Zone. At Site 370 the section below the first occurrence of *Calcicalathina oblongata* formerly were assigned to the *Cretarhabdus crenulatus* Zone (Čepek, 1978), now they are understood belonging into the *C. angustiforatus* Zone. The Hauterivian is represented by the later parts of the *Calcicalathina oblongata* Zone and the lower parts of the *Lithraphidites bollii* Zone. The index species of the last mentioned zone were recovered at Site 397A only, consequently at Sites 120 and 370 and 416A Valanginian and Hauterivian could not be discerned from each other.

Onshore, Wiedmann et al. (1978) divided the Valanginian marine section of the Agadir basin into three ammonite faunal zones and the Hauterivian into 6 ammonite zones. According to the same authors the first appearances of the benthic foraminifers *Lenticulina ouachensis* and *Epistomina caracolla* were in the Valanginian time, while *Dorothia hauterivian* starts in the Hauterivian. At Site 416A Sliter (1980) placed the Valanginian/Hauterivian boundary between the last occurrence of *Lenticulina busnardoi* and the first appearance of *Lenticulina ouachensis multicella*, *Lagena hauteriviana cylindracea* and *Dorothia hauteriviana*. The first planktonic foraminifer in the onshore samples, *Globuligerina hoterivica* occurs in horizons dated in other sections of the Agadir Basin by the late Hauterivian ammonite *Crioceratites duvali*. In the oceanic sediments the earliest planktonic foraminifera have been recovered at the Sites 370 and 397A, in sediments dated as Hauterivian by nannofossils. *Globuligerina hôterivica* is recorded from both Sites, accompanied at Site 370 by *Hedbergella occulta*, *H.* aff. *similis*, and *H.* aff. *sigali*, while at Site 397A *Clavihedbergella stiftia* or *F. tardita* are present. The known stratigraphic ranges of these species comprise an interval from late Hauterivian to early Aptian, the co-occurrence in one sample may indicate either expanded ranges of the species or mixing during drilling. At Site 120 (Gorringe Bank) there are also some problems in the age determination by benthic foraminifera and nannoplankton. On account of the presence of *Gavelinella* aff. *barremiana* and *Haplophragmium aequale* in Core 7, Maync (1973) attributed these sediments to the Barremian, in spite of the presence of *Bigenerina* cf. *antiquissima* in Core 7 and *Lingulina biformis* in Core 3, both species hitherto known only from sediments no younger than Valanginian. Contrary to Stradner (1973) we believe the nannofossil age of Core 7 is no younger than early Barremian.

On the Cape Verde Island Maio Stahlecker (1934) attributed a sequence of about 200 m aptychi-bearing bedded limestones to the Hauterivian. At Site 367 pale gray limestones and compact marls of Cores 26-29 contain rare benthic foraminifera, attributed to a Valanginian age by Kuznetsova and Seibold (1978). A similar, more diverse fauna was recovered by the same authors at Site 370 in Cores 38-49.

Barremian

Barremian sediments are reported from the Cape Verde Basin (Site 367) and from the Morocco Basin (Site 370). They contain nannofossils of the undifferentiated *Lithraphidites bollii* and *Micrantholithus obtusus* Zones. The absence of the index species *L. bollii* at both sites prevents a more exact zonation. In the southern basin marlstones and limestones contain poor benthic foraminiferal faunas whereas in the northern basin siltstones also include planktonic species. Core 370-34 contains a faunula of tiny hedbergellas (*H. sigali*, *H. planispira*, *H. delrioensis*, *H.* aff. *gorbachikae*, *H.* aff. *infracretacea*), and *Globigerinelloides* (*G. aptiense*, *G. ultramicrus*,

G. gottisi). The common to abundant, well preserved nannoplankton dates this core no younger than Barremian.

Onshore, there are records of Barremian from Maio (Stahlecker, 1934) in the facies of grayish limestones with pelitic intercalations. According to the ammonitic fauna at least 20 m of this section can be attributed to the Barremian, the next 20 m above this section are classified as transitional Barremian to Aptian.

The onshore basin of Agadir contains silty clays, sandy or dolomitic limestones and light calcilutites. Three ammonite zones are recorded by Wiedmann et al. (1978). The first appearance date of Hedbergella infracretacea is assigned to the early Barremian, while Hedbergella sigali occurs together with Heteroceras emericianum, which is assigned to a late Barremian age (contrary to Sigal 1976, who attributes this species to the early Barremian).

Aptian to Albian

This time interval can be divided into five nanno-zones and/or 13 foraminiferal zones. The Chiastozygus litterarius Zone and the lower parts of the Parhabdolithus angustus Zone are ascribed to an Aptian age, while the Albian comprises the later parts of the P. angustus Zone, the Prediscosphaera cretacea Zone, the Eiffellithus turriseiffeli Zone and the lowermost parts of the Lithraphidites alatus Zone. According to Sigal (1977) the early Aptian starts with the Hedbergella similis Zone, the Aptian-Albian boundary falls within the range of the Ticinella bejaouaensis Zone and the Albian ends with the upper limit of the Rotalipora apenninica Zone.

Aptian and Albian marine sediments have been recorded from ten DSDP-sites, from Maio, Fuerteventura, and from the onshore basins of Tarfaya, Agadir and Essaouira. Among the deep-sea sites the Aptian-Albian interval was completely investigated at Sites 367 and 370, while at the other sites only parts of it were recovered, mainly due to spot coring. At Sites 367 and 370 only the Prediscosphaera cretacea Zone and the Lithraphidites alatus Zone were not recovered, however, at Site 416 both zones were referred to on account of the presence of the index species.

According to our present zonation the assemblages of Cores 2 and 3 of Site 120 (with Parhabdolithus angustus and without Prediscosphaera cretacea) are now attributed to the P. angustus Zone, replacing the Albian age (Stradner, 1973) by "late Aptian to early Albian" .

For a section on the Cape Verde island Maio Stahlecker (1934) reported variegated clayish marls with an Aptian or younger ammonitic fauna and with increasing amounts of Inoceramus shells towards the top. A similar associated change in color of the sediments can be observed at Site 368, Core 24 contains variegated bioturbated claystones with Hedbergella infracretacea, H. globigerinelloides and Globigerinelloides sp., formerly attributed to the Barremian and lower Aptian, now assigned to the Chiastozygus litterarius Zone and restricted to an early Aptian age.

At Site 135 (Horseshoe Abyssal Plain) the early Aptian Chiastozygus litterarius Zone was reached in Core 9. The presence of Micrantholithus obtusus in Sample 135-9-1 is striking as at all the other sites in the study area this species becomes extinct within the underlying zone bearing its name. Comparison with the foraminiferal stratigraphy is impossible due to the absence of this fossil group in this section. Onshore, in the Agadir Basin the early Aptian is represented by dolomitic limestones and psammites with synsedimentary slumps containing Bedoulian ammonites and a charophyte flora, considered by Wiedmann et al. (1978) to be a regressive facies.

At Site 370, drilled in the adjacent offshore Deep Basin off Morocco, Core 32 contains with grayish brown shales and slump-material a very poor benthic microfauna with agglutinated tubes, some fish debris, and calcisphaerulids. The nannoflora is classified as Chiastozygus litterarius Zone of early Aptian age.

The Gargasian part of the Aptian contains four foraminiferal zones. The oldest one, the Leupoldina cabri Zone was indirectly proven at Site 369A, where in Core 46 Globigerinelloides ferreolensis occurs with Hedbergella sigali. The present zonation suggests an attribution to the somewhat younger Globigerinelloides ferreolensis Zone with an admixture of species from older strata. However, the nannofossils indicate an age of early Albian *Prediscosphaera cretacea* Zone.

Onshore, the Gargasian stage is condensed in the Agadir Basin displaying a diverse ammonitic fauna with a delayed first occurrence of the genus *Globigerinelloides*. Aptian marine sediments in the Basin of Essaouira transgressed during the middle Aptian *Cheloniceras martini* Zone. They in turn are covered by late Aptian silty marls which are dated by the presence of *Hypacanthoplites nolani* and *Globigerinelloides algerianus*.

The nannofossil *Parhabdolithus angustus* Zone is assigned to an age of late Aptian to early Albian. At Site 120, drilled on the Gorringe Bank the assemblages of Cores 2 and 3 (with the index species and without *Prediscosphaera cretacea*) are now attributed to the *P. angustus* Zone replacing the Albian age (Stradner 1973) by the longer range of late Aptian to early Albian. The presence of *Hedbergella* cf. *infracretacea* in Core 120-2-1 does not allow a more precise age determination. The sediments of Site 136 Core 8 have the same nannozonal assignment. These are dark nannofossil marls with small specimens of *Hedbergella planispira* and *Globigerinelloides tururensis*, accompanied by calcareous benthic foraminifera.

At Site 369A, Core 44-3 and 44-CC contain olive gray nannochalk at the level of the first appearance of the genus *Ticinella* and the last occurrence of *Globigerinelloides ferreolensis*. We attribute this section to the *Ticinella bejaouaensis* Zone, which is assigned to the uppermost Aptian and lowermost Albian. However, the nanno-record assigns this section to the *Prediscosphaera cretacea* Zone of the middle Albian, a disagreement between foraminiferal and nannofossil zonations, which needs further investigations and discussions.

Nevertheless, in the onshore Basin of Agadir the *Ticinella bejaouaensis* Zone is documented by the first occurrences of the index species and *Globigerinelloides algerianus* in clays with the ammonites *Acanthoplites seunesi* and *Hypacanthoplites nolani* of Late Aptian age (Wiedmann et al., 1978). In the Basin of Essaouira the Late Aptian is indicated by the presence of *Cheloniceras martini* and *Hypacanthoplites nolani* together with *Globigerinelloides algerianus* in silty marls, while in the Basin of Tarfaya a marine ingression of the Clansayesian could be recognized only locally (Wiedmann et al., 1978). Olive black claystones in Core 30 at Site 370 contain *Globigerinelloides gyroidinaeformis*, indicating the early Albian part of the *Ticinella bejaouaensis* Zone, and the nanno-age *Parhabdolithus angustus* Zone agrees well with these findings.

Albian

According to Wiedmann et al. (1978) the early Albian of the onshore Essaouira Basin contains a rich and diverse microfauna, the lithologic unit continues up to the *Ticinella breggiensis* Zone, and the ammonites allow a crude separation of early and tentatively middle Albian ages. In the Agadir Basin early and middle Albian is composed of greenish silty marls with impoverished megafossil record but with a relatively rich fauna of planktonic foraminifera. During the upper Albian and Vraconian more and more intercalations of limestones occur, ultimately forming a debris limestone with a reduced content of planktonic species, all ammonite zones, however, could be recovered (Wiedmann et al., 1978). The *Stoliczkaia dispar* Zone, coincident with the *Rotalipora apenninica* Zone also encompasses parts of the overlying lithologic unit with green silty marls which are barren in ammonites.

In the Basin of Essaouira the lithology changed during the late Albian from silty marls with limestone intercalations to a thick limestone sequence similar to the Basin of Agadir. The microfauna contains distinctly reduced pelagic components.

The general marine ingression in the Basin of Tarfaya starts in the late middle Albian, depositing clays and marls with *Ticinella roberti*. This marine sedimentation continues up into the *Rotalipora apenninica* Zone. At Site 370 Cores 27 and 29 contain pale green silty claystones, carbonaceous olive black silty clays and shales with *Hedbergella globigerinelloides*, *H. trocoidea*, *H. infracretacea* and *H. delrioensis*. The age of this section is bracketed by the underlying *Ticinella bejaouaensis* Zone and the overlying dark greenish gray clays in the Cores 25 to 26 containing *Praeglobotruncana*, a genus

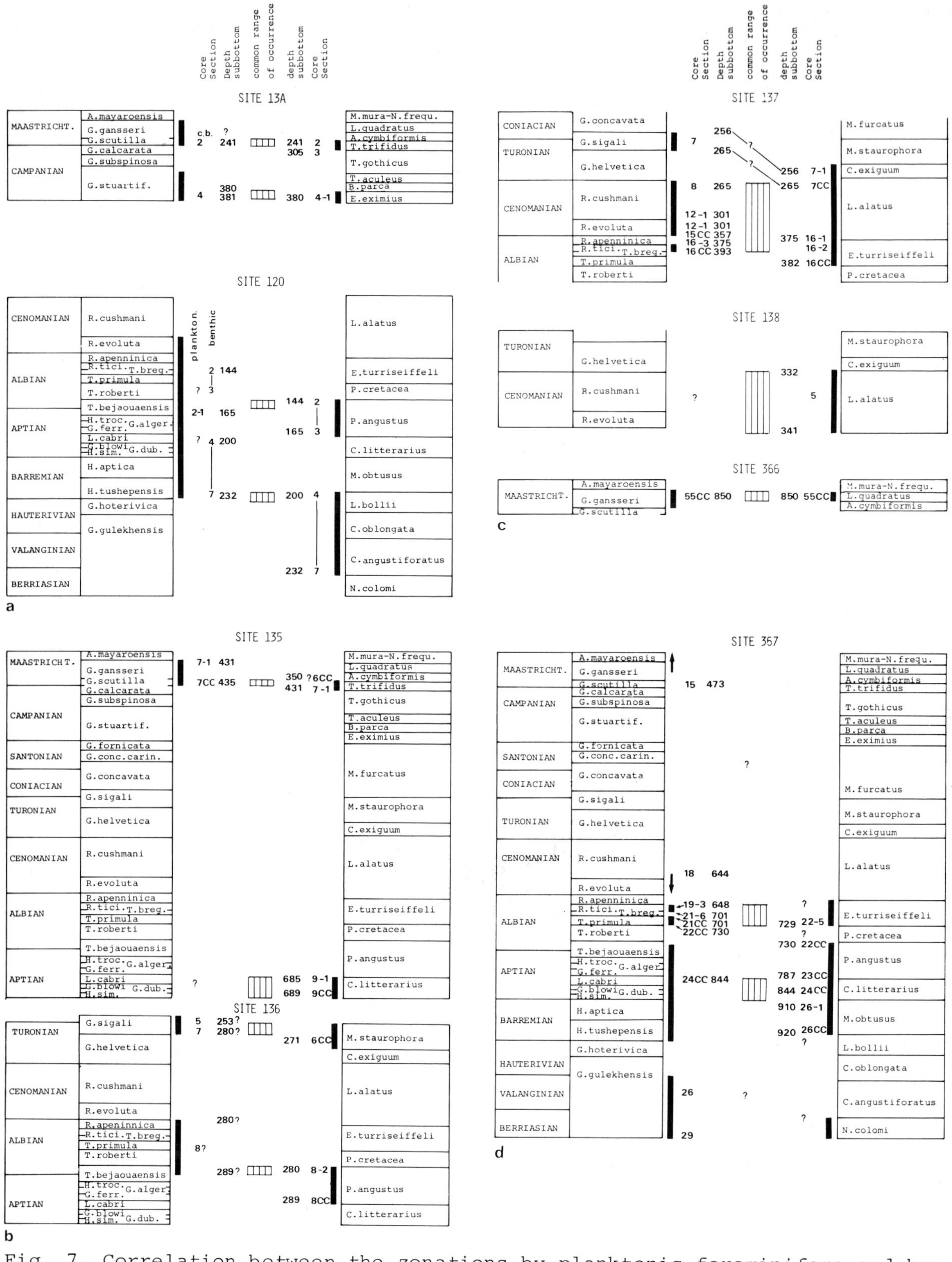

Fig. 7. Correlation between the zonations by planktonic foraminifera and by ► the calcareous nannoplankton for the Cretaceous sections of the DSDP Sites off West Africa. Hatched areas mark congruent age assignments by both groups

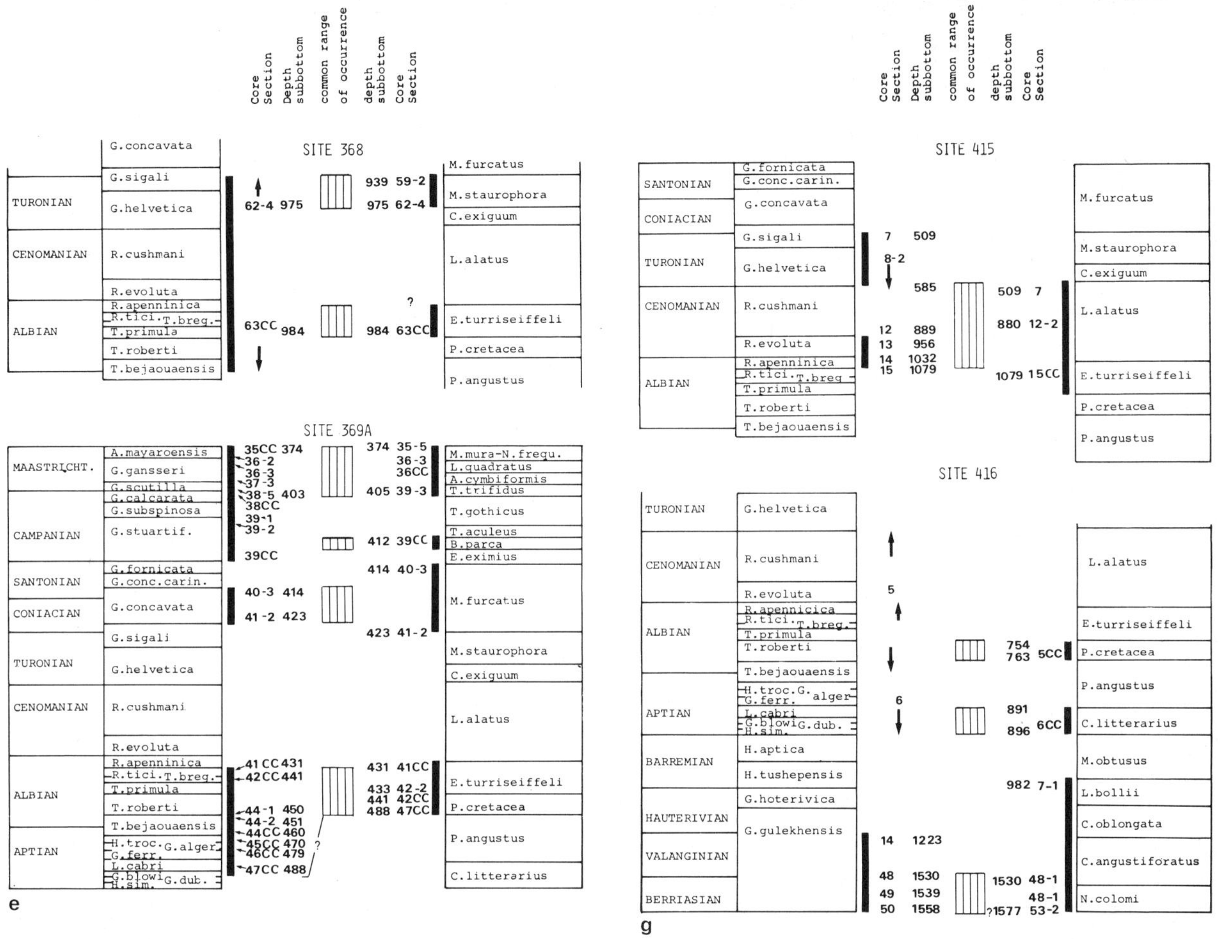

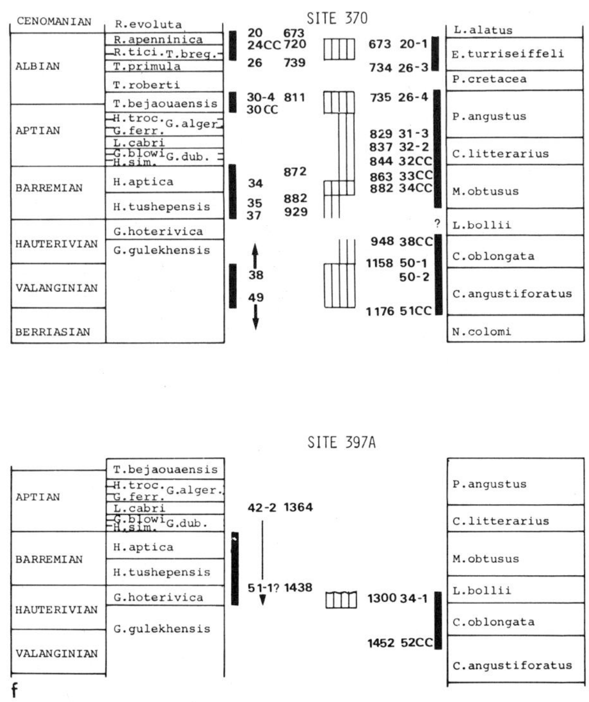

which evolves during the late Albian. So an early to middle Albian age is assigned to the interval Site 370 Cores 27 to 29. The nannofossils of this section indicate the Parhabdolithus angustus Zone and suggest an age restricted to early Albian. Prediscosphaera cretacea indicating the mid Albian was not encountered, thus the upper limit of the Parhabdolithus angustus Zone is not fixed. At the adjacent Site 416 the Prediscosphaera cretacea Zone was recognized at a level parallel to Core 28 of Site 370, where uncored sections and poor recovery results in spotty knowledge of the stratigraphy.

At Site 369, olive black nanno-bearing marls and olive gray silts and chalks were recovered in Samples 43,CC and 44-1. They belong to the Ticinella roberti Zone as indicated by the presence of the index species. The uppermost part of Core 43 is attributed to the Ticinella primula Zone as the index species was recorded. In Sample 369-42,CC the first true Ticinella breggiensis occur, defining the base of the Ticinella breggiensis Zone. The upper part of the Albian is marked by the co-occurrence of Rotalipora ticinensis and Ticinella breggiensis. At Site 369 the section 41,CC to 42-3 is attributed to the Rotalipora ticinensis Zone, recovered in olive black nannobearing marls and argillaceous chalk and carbonate layers. New species that evolved at this time are Schackoina cenomana and Praeglobotruncana delrioensis. The base of the nannofossil Eiffellithus turriseiffeli Zone lies within the last mentioned section.

At Site 367 the Cores 21,CC and 22 contain Ticinella primula and T. raynaudi, and the coccoliths indicate Eiffellithus turriseiffeli Zone. Therefore a composite age of Late Albian and Vraconian is concluded for this sequence of black shales.

From Site 137 (Abyssal Hills west off Cape Blanc) Beckmann (1972) reported from samples 16-3 and SW 1 greenish grey to dark greenish grey nannomarls and oozes with a fauna containing Rotalipora ticinensis and Planomalina buxtorfi. He attributed them to an Early Vranconian age, which accords with the nannofossil age determination of Eiffellithus turriseiffeli Zone.

At Site 368 the section between Cores 62-4 and 62,CC is composed of black shales drilled from beneath a diabase sill. They contain only a limited number of planktonic foraminiferal species. Heterohelix sp., Hedbergella delrioensis, H. planispira, H. simplicissima, and Guembelitria sp. allow only the wide-ranged age determination of Albian to Turonian. The nannofossil record is similar: Core 63 belongs to the Eiffellithus turriseiffeli Zone, Sample 62-4 to the Micula staurophora Zone.

The higher parts of the Vranconian are indicated by the foraminiferal Rotalipora apenninica Zone without Ticinella. This time interval was recovered at Site 370 with the zonal marker and Praeglobotruncana delrioensis, Schackoina cenomana, various hedbergellas and heterohelicids in claystones and shales in the section from Core 20 to 24.

At Site 415 (Agadir Canyon) Cores 14 and 15 revealed a similar fauna of the same age in dark greenish grey calcareous claystones and shales. The calcareous nannoplankton is somewhat more diverse: Eiffellithus turriseiffeli Zone to Lithraphidites alatus Zone, upper Albian to Cenomanian resp.. However, the L. alatus Zone was recovered in one sample only.

Cenomanian

The best record of Cenomanian in the deep-sea cores of the study area is reported from Site 137. Greenish to brownish nanno-marl and chalk oozes contain a rich fauna of planktonic foraminifera and an excellent nannofossil record. The Rotalipora evoluta Zone is the oldest foraminiferal zone of the Upper Cretaceous and starts in Core 16, and overlies a section of about 20 m of sediment partly uncored and partly barren in planktonic foraminifera, where the Rotalipora apenninica Zone of the Vranconian may be hidden. Lithraphidites alatus, the index species of the latest Albian and Cenomanian first occurs within the range of Rotalipora apenninica. Middle and late Cenomanian are represented by the presence of Rotalipora cushmani, R. reicheli, R. greenhornensis, Praeglobotruncana stephani and Hedbergella brittonensis in Cores 8 to 12 in similar lithology (Beckmann, 1972, Berger and von Rad, 1972).

At Site 415, the Cenomanian is represented by a sequence of more than 400 m thick dark greenish grey calcareous clays, marlstones and calcarenites partly of very dark colours. However, due to spot coring and restricted recovery the information on stratigraphy is relatively meager. The nannofossil content is strongly reduced, the Lithraphidites alatus Zone could be recorded from one sample only. The planktonic foraminifera indicate Rotalipora evoluta Zone in Core 13 (equivalent to R. brotzeni-montsalvensis Zone of Sliter, 1980) and Rotalipora cushmani Zone between Cores 12 and 8. Turbidites occur in the middle of the Rotalipora cushmani Zone indicating enhanced accumulation rates during the middle and upper Cenomanian. At Site 367 black shales of Cores 19 to 21 contain no zonal marker, as keeled forms among the planktonic foraminifera are missing. The presence of Guembelitria harrisi, Heterohelix moremani and Hedbergella brittonensis allow to assign this section to the unzoned Cenomanian only. A similar impoverished record is noted in the nannoplankton, where no zonal assignment is possible.

At Site 368 the age determination of the black shales is less precise. Hedbergella delrioensis, H. planispira, H. simplicissima, Heterohelix sp. sp., Guembelitria sp. indicate an age between Albian and Turonian. The nannofossils are poorly preserved and rare, thus not permitting a more precise age determination. Onshore, the late Albian and Cenomanian are represented by 80 m of greenish clays and marls, siltstones, dolomitic limestones and locally very thick lumachelle beds. According to Wiedmann et al. (1978) the base of the early Cenomanian in the Tarfaya Basin is coincident with the first occurrence of Rotalipora apenninica, but ammonites that may have confirmed that date are absent. We believe that the earliest occurrence of R. apenninica in that region is not caused by evolution but is dependent on the ecology. In the Agadir Basin the first occurrence of R. apenninica and Planomalina buxtorfi falls within the ammonitic Stoliczkaja dispar Zone, thus confirming the uppermost Albian age of the Rotalipora apenninica Zone.

The Albian-Cenomanian boundary falls within a section of green silty marls where no ammonites were recovered. The age determination is based on planktonic foraminifera only, however, they are rich and diverse, allowing the division into the Rotalipora evoluta Zone and Rotalipora cushmani Zone.

The Cenomanian of the Essaouira Basin starts with carbonates originating in very low water-depths. Some rotaliporas occur in the middle Cenomanian and only Rotalipora cushmani indicates the presence of later Cenomanian. The last species was recorded in thin sections only (Butt, personal communication).

Turonian

At most deep-sea sites of the study area the Turonian is not recorded due to hiatuses. Additional problems arise in the nannoplankton stratigraphy due to the absence of Lithraphidites alatus and L. acutum in the Cenomanian, so the base of the Turonian remains undefined.

At Site 137 Sample 7-1 (90-92 cm) may be attributed to the early Turonian Corollithion exiguum Zone, as Lithraphidites alatus could not be detected in spite of the abundant and well preserved flora, and it is predent in the samples below this one from Sample 8-2 downward. The planktonic foraminifera represented by globotruncanas support this age assignment, however, the "confer"-determinations allow only tentative zonal attribution to the Marginotruncana sigali Zone (late Turonian).

At Site 136 (Core 6), late Turonian was recorded by the presence of Eiffellithus eximus (=E. angustus Bukry) but without Marthasterites furcatus. The accompanying planktonic foraminifer fauna contains the globotruncanas: G. indica, G. pseudolinneiana, G. renzi in Core 6 indicating the presence of Globotruncana sigali Zone, while in Core 7 some hedbergellas and Heterohelix globulosa allow only a crude age determination of Turonian to Santonian.

At Site 369 Globotruncana (Marginotruncana) helvetica and G. imbricata were recorded, indicating the presence of early Turonian sediments. These sediments are intercalated within sediments containing mixed faunas of older and younger zones. These mixed sediments rest on late Albian sediments and contain Rotalipora evoluta, R. greenhornensis, Globotruncana concavata in addition to the aforementioned species. Therefore nearly all zones from the early Cenomanian to the Santonian are mixed in this sequence, lithologically described as "slumps". The Turonian of Site 368 has been mentioned in our discussion of the Cenomanian section.

Onshore, Wiedmann et al. (1978) recorded the Turonian of the Essaouira Basin with Globotruncana cf. helvetica, Heterohelix moremani and H. globulosa, attributed to the transition of early to middle Turonian, whereas late Turonian is indicated by the ammonite Romaniceras develianum. In the Agadir Basin, laminated nanno-marls rich in organic carbon, limestones and silica beds show a pelagic community of foraminifera and ammonites. Turonian was divided into lower early Turonian with Praeglobotruncana lehmanni and P. praehelvetica and into upper early Turonian with the first occurrence of Globotruncana helvetica, middle Turonian with the first occurrence of Globotruncana sigali and upper Turonian with the first occurrence of Globotruncana schneegansi. The age assignments are confirmed by the ammonites Spathitoides pioti, Selwynoceras sp. and Romaniceras deverianum. Finely laminated bituminous marls, limestones and cherts of the Tarfaya Basin contain the following associations: Selwynoceras reymenti, Watinoceras sp. with Globotruncana helvetica (early Turonian), Mammites nodosoides with Globotruncana sigali (mid Turonian) and Glebosoceras glebosum with Globotruncana schneegansi (late Turonian).

Coniacian

In deep-sea cores this stage was recorded within the nannofossil zone Marthasterites furcatus, which was recognized at Site 369 only within the mixed sediment mentioned above. At Site 136 in Cores 5 to 7 the presence of Marginotruncana renzi may indicate a Coniacian age, however, missing Globotruncana helvetica and G. concavata do not allow a definite zonal assignment.

Oneshore, in the Agadir Basin silty marls and lumachelle limestones contain Parapuzosia linderi and Forresteria sp., which are assigned to the early parts of the Coniacian stage. Co-occuring are the planktonic foraminifera Globotruncana lapparenti and Heterohelix moremani. In the Basin of Tarfaya, Otoscaphites bladenensis and Globotruncana angusticarinata are indicative of the early Coniacian, Prionocycloceras guyabanum and Gauthiericeras margae define the middle Coniacian, Globotruncana concavata appears in the late Coniacian. The lithological description states finely laminated bituminous marls, limestones and cherts. In the Essaouira Basin, silty marls, limestone-concretion with shell debris and dolomitic limestones may be attributed partly to the Coniacian, but no zonal assignment is possible.

Santonian

No Santonian age sediments have been recorded in deep-sea cores of this study area, however, onshore there are both ammonite and planktonic foraminifera faunas indicative for this stage. In the Tarfaya Basin, Texanites texanum co-occurs with the first Globotruncana coronata in silty clay, marls and lumachelles. In the Basin of Agadir, Texanites texanum occurs with Globotruncana concavata concavata and G. concavata carinata in sandstones and overlying silty marls and lumachelle limestones. These sediments are attributed to the early parts of the Santonian, while the later part is characterized by the first occurrence of Globotruncana fornicata and G. coronata, associated with the benthic species Neoflabellina praerugosa and N. ovalis. The Basin of Essaouira contains a condensed section of silty marls, debris limestones and dolomitic limestones with Globotruncana concavata carinata.

Campanian

With respect to previous stages, Campanian sediments are recovered more frequently in the study area. Site 13 (Sierra Leone Rise) contains sediments with a microfauna and nannoflora indicative for Campanian. The Tetralithus trifidus Zone intergrades between late Campanian and early Maastrichtian. Tentatively, sediments of Site 135 are attributed to the last mentioned nannofossil zone. The planktonic foraminifera, if present, attribute these sediments to the Maastrichtian exclusively. The best record for Campanian in the deep-sea cores is at Site 369, where nearly all planktonic foraminiferal zones, except the Globotruncana calcarata Zone, could be determined. Within the nannoplankton record only the Eiffellithus eximius Zone could not be proven directly.

Onshore, the Campanian is defined both by foraminifera and by ammonites. In the Agadir Basin silty marls, sands, dolomites and dolomitic sandstones contain the early Campanian species Bevahites subquadratus, Eulophoceras bererense and Globotruncana elevata, however, Globotruncana stuartiformis appears somewhat later. Late Campanian is indicated by the presence of Globotruncana subspinosa and G. cf. obliqua. G. calcarata was not determined, perhaps due to ecologic restrictions.

The Campanian in the Essaouira Basin is indirectly recorded by the presence of Globotruncana cretacea, G. coronata and the benthic species Palmula cushmani. In the Basin of Tarfaya, Globotruncana stuartiformis, G. arca, Neoflabellina rugosa, and Bolivina incrassata in lumachelles and silty marls and clays mark the

Campanian which discordantly overlies the Santonian. Globotruncana arca from sediments of the Moroccan Meseta indicate Santonian to Campanian age.

Maastrichtian

The most complete section of the Maastrichtian stage in the study area is recorded from Site 369. It is represented by an abundant and very well preserved calcareous nannoplankton of the Tetralithus trifidus Zone, the Arkhangelskiella cymbiformis Zone, the Lithraphidites quadratus Zone and the Micula mura Zone.

The planktonic foraminifera contains diverse faunas of the Globotruncana scutilla, G. gansseri and Abathomphalus mayaroensis Zones. The absence of Globotruncana calcarata in the samples of this site obscures the history of the Campanian/Maastrichtian boundary.

Additional occurrences of Maastrichtian sediments are reported from Sites 13 and 366 on the Sierra Leone Rise and from Site 135. These sediments contain microfaunas of the Globotruncana gansseri Zone, which at the equatorial sites is equivalent to the Lithraphidites quadratus nannofossil Zone, and off Madeira was tentatively assigned to the Tetralithus trifidus Zone.

In the Agadir Basin, the presence of Globotruncana caliciformis and Globotruncanella havanensis may assign a Maastrichtian age to the sediments without strictly zonal determination. The other basins of Morocco contain only very poor records of the last Cretaceous stage. Planktonic foraminifera are missing in the Basin of Essaouira and on the Meseta, allowing only a very rough age determination (Maastrichtian to early Eocene). Due to a gap in sedimentation the Maastrichtian is not recorded in sediments from the Basin of Tarfaya.

The Cretaceous/Tertiary boundary could not be studied in any of the 19 sites in the tropical and subtropical eastern North Atlantic Ocean. In most cases its absence may be due to major hiatuses correlative with time intervals of non-deposition or erosion. One of the most promising regions for the study of an undisturbed, continuous transition from the Mesozoic to the Cainozoic was drilled on the Sierra Leone Rise (Site 366), however, due to technical difficulties, there remained a gap of about 20 m between the early Paleocene Cruciplacolithus tenuis Zone and the late Maastrichtian Lithraphidites quadratus Zone. That gap is corresponding to the interval of the foraminiferal Globorotalia trinidadensis Zone to the Globotruncana gansseri Zone, thus causing the non-recovery of the Abathomphalus mayaroensis Zone and the earliest Paleogene zones.

On the Continental Slope off Cape Bojador (Site 369A) the last Cretaceous foraminiferal zone, Abathomphalus mayaroensis Zone, is recorded as well as the last Mesozoic nannozone Micula mura Zone, but it is covered discordantly by Eocene sediments of the Globorotalia lehneri Zone, the Discoaster bifax Zone resp., with only a slight facies change.

6. Paleoecology and Preservation

Berriasian

Few foraminifers have been recorded from this time-interval at Site 367, but radiolarians are more numerous than in the underlying layers (see Wolfart, this Vol.). At Site 416, Sliter (1980) recorded an alternation in the dominance of autochthonous deep-sea and displaced foraminifers and fishdebris, the latter being present in nearly every sample. A comparison of the nannoplankton content of the two sites shows that Site 367 is characterized by a better preservation and by higher abundances (Dean et al. 1978).

Valanginian to Hauterivian

According to Kuznetsova and Seibold (1978) the composition of the benthic foraminiferal faunas at Site 367 and 370 indicate that they developed in a warm water basin at a level above the carbonate compensation depth. In the Valanginian section of Site 416 the greatest diversity of benthic foraminifera was observed in rapidly deposited cycles of brownish terrigenous beds alternating with green-grey, carbonate-rich layers (Sliter, 1980). High abundances of the nannofossil Micrantholithus obtusus in the middle parts of the combined Valanginian to Hauterivian section at Site 416 point to nearshore conditions during the time of their deposition (Thierstein, 1976) and Butt (in press). According to Sliter (1980), however, the sediments of this Site were deposited in a quiet abyssal environment below the CCD. Opposite opinions are evidently due to the influx of terrigenous and redeposited shallow-water material reaching a maximum during the late Valanginian. At Site 397A the calcareous nannoplankton of the Calcicalathina oblongata to Lithraphidites bollii Zones are well preserved and abundant, indicating a deposition above the CCD. In the Cape Verde Basin the Hauterivian record is very poor. Core 25 of Site 367 contains light gray nannolimestones with a rare and poorly preserved benthic microfauna. Gavelinella, Lenticulina and some other calcareous benthic species co-occur with lots of radiolarians indicating a paleodepth still above the CCD.

Sites 397 and 370 contain very small and poorly preserved planktonic foraminifera, the nannofossil floras assign these samples to the late Hauterivian. The benthic foraminifera are composed of small lagenids, spirillinids and Dorothia. Basov et al. (1979) interpreted the benthic fauna of Site 397 as typical for populations of a partly closed basin with a limited exchange of water. The absence of bioturbation may be explained by oxygen-depleted conditions in the bottom water. The relatively good preservation of the foraminifera, the presence of coccoliths and thin shelled pelecypods suggests moderate water depths. According to Wind and Čepek (1979) the nannoplankton shows a heterogenous mode of preservation. The coexistence of well-preserved and poorly preserved specimens in the same samples and the occurrence of siderite crystals in samples with totally unmodified specimens suggest that much of the section recovered at Hole 397A consists of allochthonous sediments, deposited primarily in fairly shallow water (cf. also Einsele and von Rad, 1979; Butt, 1979).

In the onshore Agadir Basin, Wiedmann et al. (1978) and Butt (in press) observed a highly diversed Valanginian macrofauna and microfauna, indicating warm, oxygen-rich water in a shelf environment. The diversity of the planktonic and benthic faunal components is strongly reduced during the early Hauterivian, the mid-Hauterivian regression results in nearly littoral conditions whereas during the late Hauterivian the fauna becomes again more diverse, and the first planktonic foraminifera appear. Wiedmann et al. (1978) point to the similiarity of the Agadir Basin fauna with that of the North Atlantic faunal province during the Hauterivian in contrast to the Mediterranean and/or cosmopolitan character of the associations of the other Early Cretaceous stages.

Barremian

Highly diverse planktonic foraminiferal microfaunas in the Barremian of the Agadir Basin are indicative for an open circulation of the surface-waters under oceanic conditions.

In the deep basin off Morocco (Site 370) higher quantities and better preservation of nannofossils are recorded. The planktonic foraminifera evolved to a fairly diverse microfauna, while the benthic foraminifers are composed of calcareous forms: Lagenids, Gyroidina, Cibicides and spirillinids accompanied by some primitive agglutinated species, indicating a mixture of abyssal and slope assemblages.

At Site 367 Barremian sediments contain a calcareous nannoplankton of distinctly lower diversity and of poorer preservation than in the North. The benthic component of the foraminiferal fauna includes rare agglutinated species of poor preservation indicating a deep-water paleoenvironment. Planktonic foraminifera are missing.

In Maio, Stahlecker (1934) noted the missing of the ammonite genera Silesites and Holcodiscus and the abundance of the pulchellids. Therefore he attributed the fauna to the Mediterranean bathyal province. The benthic megafossil component is strongly reduced indicating free pelagic conditions at this time. Missing terrigenous components point to relatively large distances from the continent.

At Site 120, Maync (1973) recorded a diverse but low abundant benthic foraminiferal fauna in laminated bathyal oozes deposited far away from land. Though this region may have been a relatively deep basin, the presence of a diverse calcareous benthos indicates a deposition above the CCD. No similarities exist between the fauna of this Site and the adjacent Portugese shelf and platform facies onshore.

Aptian - Albian

At Site 120 the bathyal character of the benthic fauna persisted. At Site 136 an unusual faunal composition with gavelinellids and lagenids is believed by Beckmann (1972) to represent at least partly autochthonous deposits of the neritic to upper bathyal realm.

Excellent planktonic foraminiferal faunas accompanied by calcareous benthic species and by a rich and diverse nannoplankton are recorded from the Albian of Site 137, indicating that these sediments which cover the oceanic basalt were deposited well above the CCD.

At Site 369 *Gavelinella* and *Neobulimina* are the most common benthic species indicating moderate water depths on the slope. The coccoliths are well preserved and abundant, and the planktonic species dominate among the foraminifera. At Site 415 the benthic foraminiferal fauna is a mixture of bathyal agglutinated species with reworked neritic ones, indicating a gravitative sedimentation along this passive continental margin. The nannoplankton record is generally poorly preserved, higher frequencies occur only sporadically.

At Site 367, benthic foraminifera are only poorly recorded in the Aptian to Turonian black shale facies. Very rare agglutinated species of normal size are found in some interbeds within sediments with a small-sized but abundant planktonic component of the foraminiferal fauna, which is almost completely lacking in benthic forms. A strongly reduced oxygen concentration in the depositional depth may be responsible for the composition of those thanatocoenoses (see Pflaumann and Krashenninikov, 1978, Dean et al. 1978). Bottom currents with higher oxygen content or times of reduced input of organic carbon resulting in short-period improvements interrupted the stagnant bottom-water realm only occasionally. At Site 368 the black shales contain a similar heterogenous faunal record, poor agglutinated assemblages are replaced by planktonic or calcareous benthic ones. The nannoplankton is also poorly preserved and of low abundance. During the same time the benthic fauna at Site 370 contains the genus *Recurvoides* and associates. A slightly enhanced diversity and the presence of agglutinated species with a more complicated test morphology point to better life conditions with a more intensive vertical and horizontal mixing of the water masses than in the south. As indicated at Site 416 the redeposition and shallow water influence was greatly reduced, leading to a regime of hemipelagic sedimentation and less input by distal turbidites during Albian and Cenomanian.

Cenomanian - Turonian

In the South, at Site 367 and 368 as well as at Site 138, primitive agglutinated foraminifera of low diversity and reduced abundances indicate an abyssal depositional depth of the sediments below the CCD. The agglutinated faunas are slightly more diverse at Site 137, but the poor nannofossil content indicates a deposition close to the CCD. At Site 415 *Lithraphidites alatus* was recognized only in one sample. Its absence in the other samples may be due to the fact that this species prefered more oceanic conditions, whereas the Site was situated too close to the coast during the earliest Late Cretaceous. Layers with high abundances of *Neobulimina* and with a reduced input of neritic sediment indicate either non-deposition of terrigenous material on the continental margin or an enhanced circulation along the slope, which prevented a sediment transport to the locations of the site. The last interpretation may be supported by the coeval sedimentary hiatus at Site 397.

At Site 369 there are no changes in the ocean-bottom realm as indicated by the benthic foraminifera but the record of the planktonic ones points to a mixing of older with younger faunas, reported as "slumps" and slide-structures in the lithologic core descriptions.

Site 137, far offshore, contains only a few benthic foraminifera with agglutinated and calcareous species, the planktonic species dominating the faunas. The calcareous nannoplankton is rich, diverse and well preserved indicating a deposition well above the CCD.

At Site 136 the change from a mixture of lagenids with gavelinellas to poor primitive agglutinated faunas may indicate that the Site has crossed the CCD at this time.

The sediments of the Tarfaya Basin indicate an increase of water depths ranging down to marginal ones of the outer shelf and upper slope. The Turonian sediments of the Agadir Basin were deposited on the quiet outer shelf, the nearly total absence of benthic species within the foraminiferal fauna indicates a pelagic paleoenvironment (Wiedmann et al. 1978).

Coniacian - Santonian

With the exception of Site 369, all Deep-Sea Drilling Sites with a sedimentary record of Coniacian to Santonian are attributed to abyssal paleodepths below the CCD, as inferred by their primitive agglutinated faunas and the poor preservation and abundance of the nannofossils. Site 369 Cores 40 and 41 contain a mixture of mainly calcareous benthic foraminifers of the slope. The calcareous nannoplankton is well preserved and abundant indicating the position of the Site above the CCD. According to the faunal compositions the Tarfaya Basin reached its greatest water-depth during the Coniacian, but already during the Santonian it was shoaling. Beginning in the early Coniacian, a regression occurred in the Agadir Basin, which ultimately led to an inner-shelf or sublittoral environment.

Campanian - Maastrichtian

Most of the sites in the study area contain sediments of this time interval, and most of these sediments are showing poor, agglutinated foraminiferal faunas, indicating abyssal depths and low terrigenous influence.

Sites 13 and 366 contain benthic foraminiferal faunas indicative of water depths not principally different from present depths at these sites. The calcareous nannoplankton at Site 366 is relatively rare and poorly preserved in the Maastrichtian stage.

At Site 135, larger foraminifera mixed with abyssal faunas indicate shallow-water faunas displaced into the deep-sea by an enhanced input from the continental margin. At Site 369, the benthic foraminifera are represented mainly by buliminids and Stensioeina attributed to water depths of the upper slope, which is in line with high abundance and excellent preservation of the calcareous nannoplankton. Some redeposition may be indicated by the faunal record, interrupted by hiatuses during the Campanian. The presence of the nannofossil Micula mura and the absence of Nephrolithus frequens at Site 369 indicate a tropical environment (Worsley and Martini, 1970).

In the Basin of Agadir diverse and rich microfaunas of the early Campanian indicate outer-shelf conditions with bottom waters rich in nutrients and oxygen. The Late Cretaceous regression is indicated for the time from late Campanian onward by a more and more reduced plankton content of the foraminiferal faunas, by the increasing number of intercalations of coquinas in the sedimentary column and by the crossbedded deltaic sands of late Maastrichtian age that complete this cycle of sedimentation.

7. Hiatuses and Missing Intervals

Within the study area none of the 15 sites off northwest Africa contained a complete section of the Cretaceous sediments. This incompleteness arises from spot coring or hiatuses. Spot coring was preferred at many sites especially in the first phase of the Deep Sea Drilling Project it is difficult to decide whether a missing interval is a hiatus or merely a time of low sedimentation rates.

Berriasian to Barremian

The oldest defined hiatus in the early Cretaceous in this area is between the Hauterivian and the Barremian. This hiatus was recognized at Site 370 (Čepek, 1978 and present study). The assemblage of coccoliths in Core 35 contains Calcicalathina oblongata (Worsley) and Cruciellipsis cuvillieri (Manivit). According to Thierstein (1976) and Perch-Nielsen (1979) the second species has its last occurrence in the late Hauterivian and Calcicalathina oblongata (Worsley) in the early Barremian. Both species are missing in Core 34. Lithraphidites

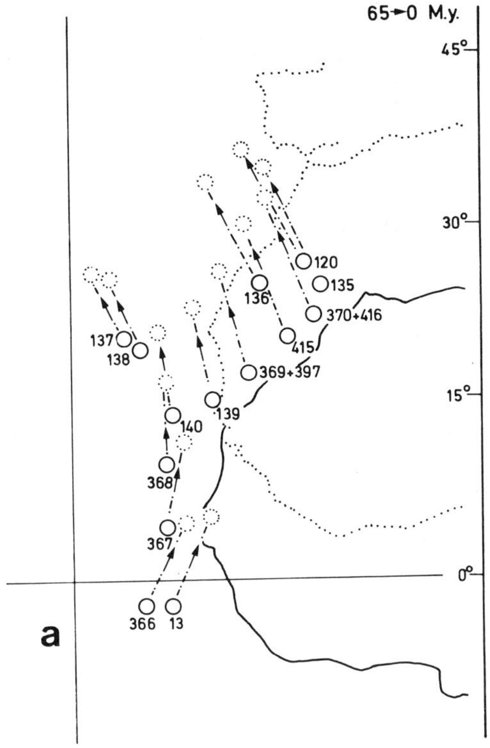

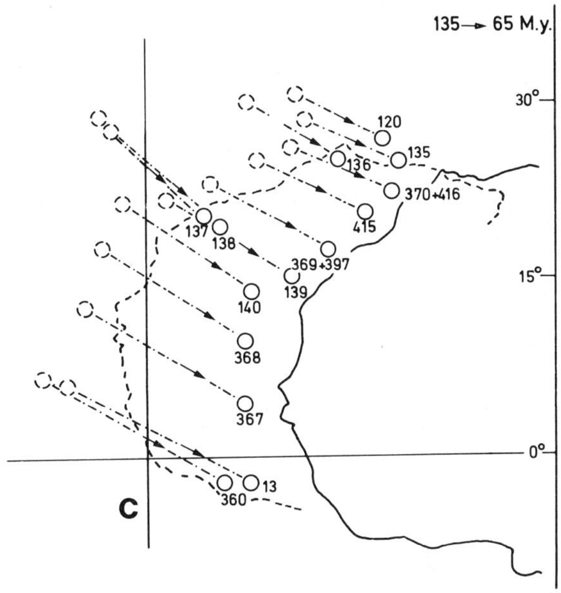

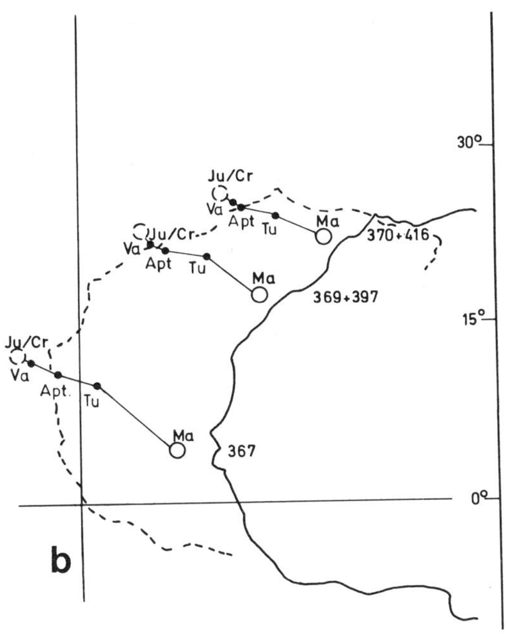

Fig. 8a-c. Motion of the African continent during the Cretaceous and up to the present period. a: General motions of the position of the sites during the Cainozoic era. b: Motion of the positions of Sites 367, 369 and 397, 370 and 416 during the Cretaceous in five time sections. Ju/Cr = Jurassic/Cretaceous boundary, around 135 million years ago, Va = latest Valanginian, about 126 million years ago, Apt = latest Aptian, about 115 million years ago, Tu = latest Turonian, about 86 million years ago, Ma = latest Maastrichtian, about 65 million years ago. c: General motion of the positions of the Sites during the Cretaceous period. Positions of the African Continent (Sclater et al., 1977) and of DSDP Sites in the eastern North Atlantic Ocean: a) dashed symbols: Jurassic/Cretaceous boundary, b) solid symbols: Maastrichtian, c) dotted symbols: present time

bollii (Thierstein), by the first occurrence of which early Hauterivian would be determined, is missing in Core 35. Thus Core 35 should be no younger than early Hauterivian. Sample 34,CC belongs to the Micrantholithus obtusus Zone which has its base in the early Barremian, indicating that the duration of this hiatus is at least from the early Hauterivian to the early Barremian comprising about six million years. If Lithraphidites bollii (Thierstein) is missing in this hole due to ecological reasons, the hiatus may be restricted to the late Hauterivian. The same

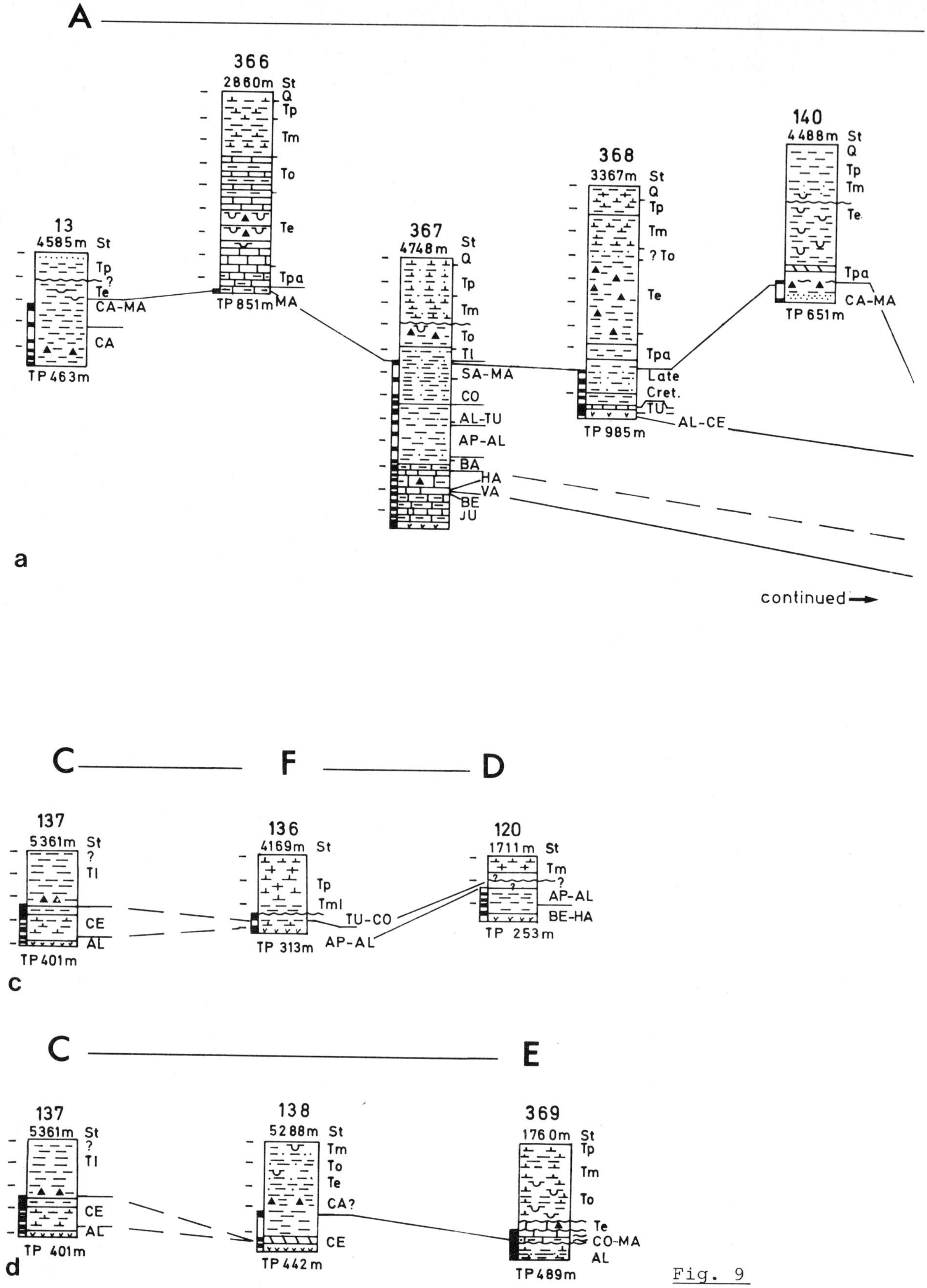
A
13
4585m St
Tp
?
Te
CA-MA
CA
TP 463m
366
2860m St
Q
Tp
Tm
To
Te
Tpa
TP 851m
MA
367
4748m St
Q
Tp
Tm
To
Tl
SA-MA
CO
AL-TU
AP-AL
BA
HA
VA
BE
JU
368
3367m St
Q
Tp
Tm
? To
Te
Tpa
Late
Cret.
TU
AL-CE
TP 985m
140
4488m St
Q
Tp
Tm
Te
Tpa
CA-MA
TP 651m
a
continued
C
F
D
137
5361m St
?
Tl
CE
AL
TP 401m
136
4169m St
Tp
Tml
TU-CO
AP-AL
TP 313m
120
1711m St
Tm
?
AP-AL
BE-HA
TP 253m
c
C
E
137
5361m St
?
Tl
CE
AL
TP 401m
d
138
5288m St
Tm
To
Te
CA?
CE
TP 442m
369
1760m St
Tp
Tm
To
Te
CO-MA
AL
TP 489m

Fig. 9

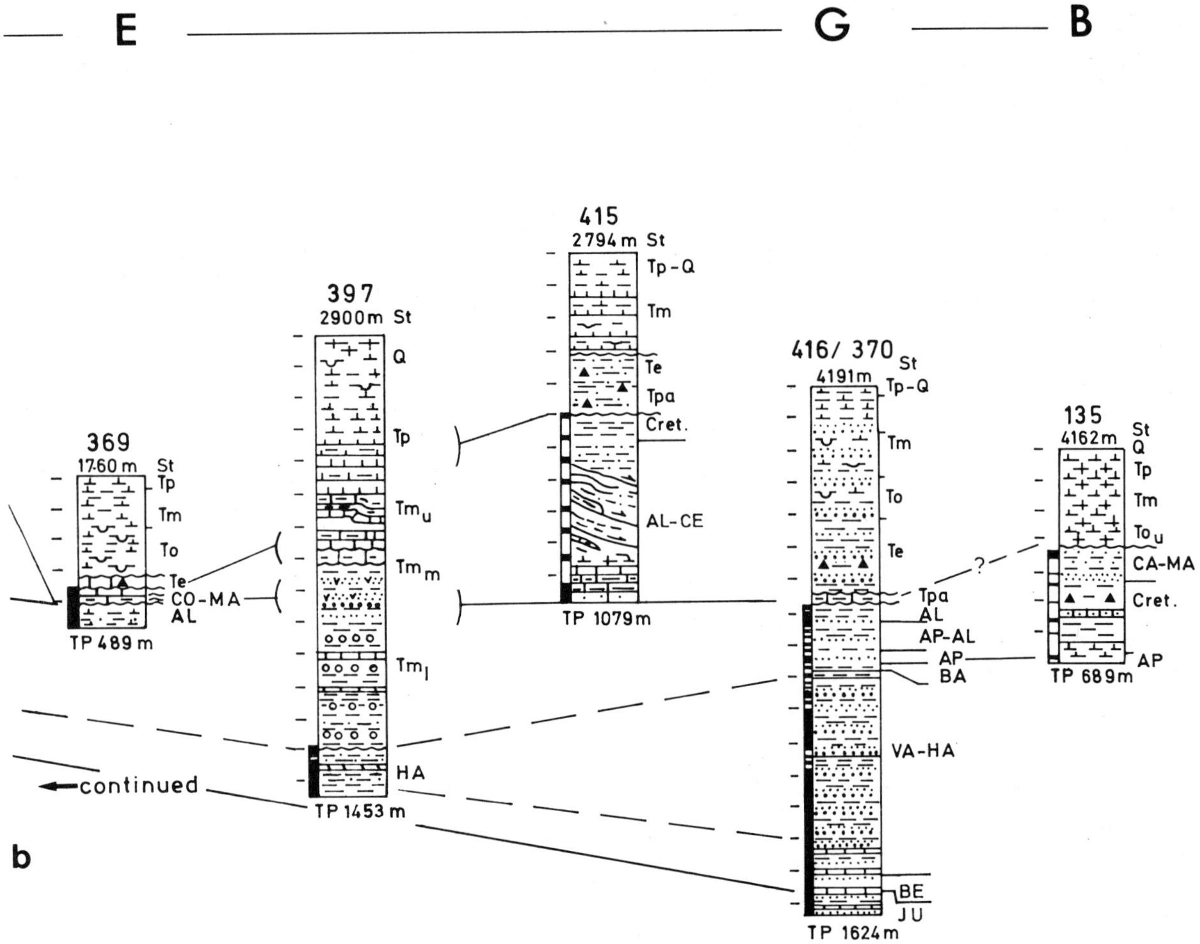

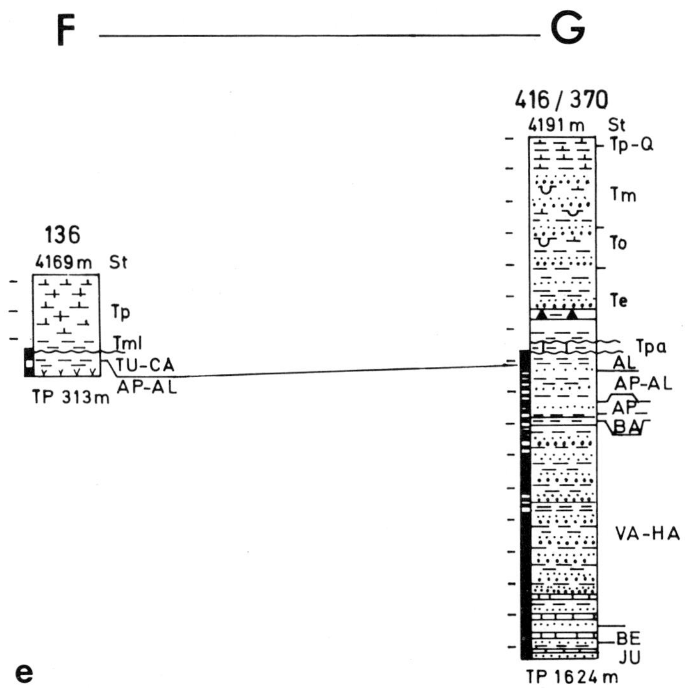

Fig. 9. Correlation of Cretaceous deposits of DSDP sites in the North Atlantic Ocean off West Africa (modified from von Rad, Front cover figure, this Vol.). For location of the lines and explanation of the symbols see Fig. 1

hiatus is probably also present at Site 367 (Čepek, 1978). At Site 12o (Ryan and Hsü, 1973) the determination of this hiatus is difficult due to spot coring. At Site 416 this hiatus is probably between late Hauterivian and the base of the Aptian where the top of Lithraphidites bollii Zone and the whole Micrantholithus obtusus Zone are missing. At Site 397 Wind and Čepek (1979) determined a large gap representing non-deposition or erosion during 100 million years between the top of the Hauterivian and the earliest Miocene.

Aptian to early Albian

The results of drilling at Sites 120, 135, 136, 137, 367 and 370 demonstrate that intervals are missing within the Aptian/Albian period. It is difficult to decide if the missing intervals represent real erosional events because this period was drilled at these sites by spot coring. The absence of the Prediscophaera cretacea Zone (middle Albian) at Sites 367 and 370 (Čepek, 1978) represents a short hiatus of about 2.5 m.y.. At both Sites Prediscosphaera cretacea (Arkhangelsky) occurs together with Eiffellithus turriseiffeli (Deflandre). This hiatus is confirmed indirectly by the planktonic foraminifera as the middle Albian zones could not be recognized completely. However, at the adjacent Site 416, the Prediscosphaera cretacea Zone was proven, thus reducing the length of the gap.

Late Albian to Cenomanian

The missing of the interval between the Lithraphidites alatus Zone and the Broinsonia parca Zone as well as the absence of autochthonous assemblages of the Rotalipora apenninica Zone to the Globotruncana fornicata Zone determines a hiatus extending from the latest Albian to the early Campanian, encompassing a time span of about 22 m.y..

At Site 416 (Čepek et al., 1980), a long hiatus representing 70 m.y. is present between the late Albian and the early Eocene. At Site 136 (Roth and Thierstein, 1972) Core 7 is barren of calcareous nannofossils. In Core 8 Parhabdolithus angustus Zone and in Core 6 Eiffellithus ? angustus Zone were determined, which corresponds to the late Turonian, probably the Micula staurophora Zone. So, three nanno-zones are missing here, corresponding to a gap of about 1.6 million years. The youngest Cretaceous sediments determined at Site 120 (Stradner, 1973) are of lower Albian age. Because the interval between this recovered level and the next one (early Miocene) is spot-cored, any statement about hiatuses is impossible. At Site 370 (Čepek et al., 1978), a hiatus representing about 37 million years may exist between the latest Albian and early Paleocene. Perhaps the hiatus at this site is slightly shorter, as between the proven upper Albian and lower Paleocene sediments an interval of 9.5 m thickness is missing due to spot coring.

Turonian to Santonian

This period is mostly missing at Sites 369, 370, 397, 415 and 416 due to hiatuses. Spot coring, the absence of coccoliths or specific index species, bad preservation, and low abundance allow only approximate stratigraphic assignments with little possibility of determining hiatuses (Sites 120, 135, 136, 137, 138, 367, and 368). At Site 136, however, Roth and Thierstein (1972) identified a hiatus between Turonian and Miocene representing about 54 million years. At Site 369 (40-2 to 41-2) the slumps contain planktonic foraminifera of the Globotruncana concavata Zone as well as nannofossils of the contemporary Marthasterites furcatus Zone, however, this section is believed to be redeposited.

Campanian to Maastrichtian

This interval was recovered only at Sites 13, 135, 366, and 369, but only at Sites 135 and 369, we were able to determine hiatuses for this time interval. At Site 135, a hiatus of about 16 m.y. is present spanning the late Maastrichtian to middle early Eocene (Roth and Thierstein, 1972). At Site 369, this time interval contains two hiatuses: the first one just mentioned is extending up to the Eiffellithus eximius Zone (base of the Campanian) and the second one is corresponding to the Tetralithus aculeus and T.

gothicus Zones. This is in agreement with the record of the planktonic foraminifera, the zonation of which is somewhat unprecise, due to the absence of Globotruncana calcarata.

8. The Motion of Africa during the Cretaceous and Paleogeographic Remarks

According to Sclater et al. (1977) all sites within the study area have higher latitudinal positions now than in Cretaceous time, especially the sites of the northern part of the area. The motion of the African plate between Cretaceous and present time has generally a northward component (Fig.8a). During the Cretaceous the sites moved in an opposite direction - generally to the south-east (Fig. 8b), the motion during the interval between the Jurassic/Cretaceous boundary time and the end of the Maastrichtian is drawn in a more detail for Sites 367, 369/397, and 370/416 in Fig. 8c. At the end of the Turonian (about 86 m.y.) the NW African sites were located about 2 to 3 degrees more southward than during the Jurassic/Cretaceous boundary time. During the last 20 m.y. of the Cretaceous period the speed of southeastward motion was greater than at any other time before. By this eastward motion during the Cretaceous the latitude of the sites was changed southward only slightly. At Site 367 in the southern part of the study region the difference in latitude between the base and the top of the Cretaceous is about 8 degrees, at Sites 370 and 416 (in the northern part of the study area) only about half of it (about 4 degrees). This difference may be explained by the counter-clockwise rotation of the African continent during the Cretaceous. In general, the equatorward change of paleolatitude is relatively minor and it cannot be expected that the variations of coccolith assemblages and planktonic foraminiferal faunas during the Cretaceous are due to latitudinal differences.

9. Acknowledgements

We are indepted to the Deep Sea Drilling Project for provide us with the opportunity to participate in Legs 41, 47A, and 50. Ms. B. Domeyer did the artwork, Mrs. C. Schmidtmann typed the manuscript and the camera-ready copy. The English text was reviewed by Dr. W. Coulbourn, and Dr. H. Stradner, Wien and Dr. A. Butt, Tübingen critically reviewed the manuscript. Financial support came from the Deutsche Forschungsgemeinschaft, Bonn, which also funded the establishment of a Cretaceous biostratigraphic data bank at the Bundesanstalt für Geowissenschaften und Rohstoffe, Hannover. Into this data bank all the nannoplankton data used in this report have been incorporated. We wish to express our sincere thanks for all support.

10. REFERENCES

Basov VA, Lopatin BG, Gramberg IS, Danyushevskaya AI, Kaban'kov VY, Lazurkin VM, and Patrunov DK (1979) Lower Cretaceous Lithostratigraphy of the Continental Rise off the western Sahara. In: von Rad U, Ryan WBF et al Initial Repts DSDP 47: 579-594

Beckmann JP (1972) The foraminifera and some associated microfossils of Sites 135 to 144. In: Hayes DE, Pimm AC et al Initial Repts DSDP 14: 389-420

Berger WH, von Rad U (1972) Cretaceous and Cenozoic sediments from the Atlantic Ocean. In: Hayes DE, Pimm AC et al Initial Repts DSDP 14: 787-954

Blow WH (1970) Deep Sea Drilling Project, Leg 3 Foraminifera from selected samples. In: Maxwell AE, von Herzen RP et al Initial Repts DSDP 3: 629-661

Bolli HM (1957) The genera Praeglobotruncana, Rotalipora, Globotruncana, and Abathomphalus in the Upper Cretaceous of Trinidad. B.W.I. US Nat Mus Bull 215: 51-60

Bolli HM (1966) Zonation of Cretaceous to Pliocene marine sediments based on planktonic foraminifera. Bol Inform As Venezol Geol Min Petrol 9: 3-32

Borsetti AM (1962) Foraminiferi planctonici di una serie Cretacea dei dintorni de Piobbico (Prov. di Pesara). Gior Geol Ann Mus Geol Bologna ser 2a 29: 19-75

Bukry D (1969) Upper Cretaceous coccoliths from Texas and Europe. Univ Kansas Paleont Contr Protista 2 art 51 79 pp

Bukry D (1973) Low-latitude coccolith biostratigraphic zonation. In: Edgar NT, Saunders JB et al Initial Repts DSDP 15: 685-703

Bukry D (1974) Cretaceous and Paleocene coccolith stratigraphy, Deep Sea Drilling Project, Leg 26. In: Davies TA, Luyendyk BP et al Initial Repts DSDP 26: 669-673

Bukry D, Bramlette MN (1970) Coccolith age determinations Leg 3, Deep Sea Drilling Project. In: Maxwell AE, von Herzen RP et al Initial Repts DSDP 3: 589-611

Butt A (1979) Lower Cretaceous foraminiferal biostratigraphy, Paleoecology, and depositional environment at DSDP Site 397, Leg 47A. In: von Rad U, Ryan WBF et al Initial Repts DSDP 47: 257-271

Butt A (in press) Micropaleontological bathymetry of the Cretaceous of Western Morocco. Cushman Found Spec Publ 20

Čepek P (1978) Mesozoic calcareous nannoplankton of the Eastern North Atlantic, Leg 41. In: Lancelot Y, Seibold E et al Initial Repts DSDP 41: 677-687

Čepek P (in press) Mesozoic calcareous - nannoplankton stratigraphy of the Central North Pacific (Mid-Pacific Mountains and Hess Rise), Deep Sea Drilling Project Leg 62. In: Thiede J, Vallier T et al Initial Repts DSDP 62

Čepek P, Gartner S, Cool TH (1980) Mesozoic calcareous nannofossils, Deep Sea Drilling Project Sites 415 and 416, Moroccan Basin. In: Lancelot Y, Winterer EL et al Initial Repts DSDP 50: 345-351

Čepek P, Hay WW (1969) Calcareous nannoplankton and biostratigraphic subdivision of the Upper Cretaceous. Trans Gulf Coast Assoc Geol Soc 19: 323-336

Čepek P, Hay WW (1970) Zonation of the Upper Cretaceous using calcareous nannoplankton. XXIII Internat Geol Congr (Prague). Proc Paleobot Sect B 3: 333-340

Čepek P, Johnson D, Krasheninnikov V, Pflaumann U (1978) Synthesis of the Leg 41 biostratigraphy and paleontology, Deep Sea Drilling Project. In: Lancelot Y, Seibold E et al Initial Repts DSDP 41: 1181-1198

Čepek P, Moorkens TH (1979) Cretaceous/Tertiary boundary and Maastrichtian - Danian biostratigraphy (Coccoliths and Foraminifera in the Maastrichtian type area. Cretaceous-Tertiary boundary events Symposium Copenhagen 1979. II Proc: 137-142

Cita MB (1973) as member of the Shipboard Scientific Party Gorringe Bank Site 120. In: Ryan WBF, Hsü KJ et al Initial Repts DSDP 13/1: 19-41

Dalbiez F (1955) The genus Globotruncana in Tunisia. Micropaleont 1: 161-171

Dean WE, Gardner JV, Jansa LF, Čepek P, Seibold E (1978) Cyclic sedimentation along the continental margin of Northwest Africa. In: Lancelot Y, Seibold E et al Initial Repts DSDP 41: 965-986

Drushtchitz VV, Gorbatschik TN (1979) Zonengliederung der Unteren Kreide der südlichen UdSSR nach Ammoniten und Foraminiferen. In: Aspekte der Kreide Europas IUGS Series A 6: 107-116

Einsele G, von Rad U (1979) Facies and paleoenvironment of Lower Cretaceous sediments at DSDP Site 397 and in the Aaiun Basin (Northwest Africa). In: von Rad U, Ryan WBF et al Initial Repts DSDP 47: 559-577

Gartner S (1970) Coccolith age determinations Leg 3 Deep Sea Drilling Project. In: Maxwell AE, von Herzen RP et al Initial Repts DSDP 3: 613-627

Gartner S (1973) Calcareous nannofossil age determinations, Deep Sea Drilling Project, Leg 13. In: Ryan WBF, Hsü KJ et al Initial Repts 13: 823-827

Gradstein FM (1978) Foraminifera from DSDP Site 370, Leg 41, Eastern North Atlantic Ocean. In: Lancelot Y, Seibold E et al Initial Repts DSDP 41: 779-782

Grigjalis AA, Gorbatschik TN (1980) K sistematike jurskich i rannemelovich Globigerinacea. Akad Nauk SSSR. Paleont Jurnal 1980-1: 20-30

Herm D (1962) Stratigraphische und mikropaläontologische Untersuchungen der Oberkreide im Lattengebirge und Nierental. Bayer Akad Wiss Math naturw Kl Abh NF 104: 1-119

Krasheninnikov VA (1978) Importance of Oceanic Deposits for Elaboration of the Mesozoic and Cenozoic Stratigraphic Scale (Pacific and Atlantic Oceans). Voprosy mikropaleont 21: 42-161 (in Russian)

Krasheninnikov VA, Pflaumann U (1978) Cretaceous agglutinated foraminifera of the Atlantic Ocean off West Africa, Leg 41, Deep Sea Drilling Project. In: Lancelot Y, Seibold E et al Initial Repts DSDP 41: 565-580

Kuznetsova KI, Seibold I (1978) Foraminifers from the Upper Jurassic and Lower Cretaceous of the Eastern Atlantic (DSDP Leg 41, Sites 367 and 370). In: Lancelot Y, Seibold E et al Initial Repts DSDP 41:515-537

Longoria JF (1974) Stratigraphic, morphologic and taxonomic studies of Aptian planktonic foraminifera. Rev Esp Micropaleont num extraord: 1-107

Manivit H (1971) Nannofossiles calcaires du Crétacé francais de l'Aptien au Danien. Essai de biozonation, appuyee sur les stratotypes. Thèse Fac Science d'Orsay 187 pp

Manivit H (1972) Les nannofossiles du Crétacé (Aptien-Turonien inclus). B R G M France No 77: 165-169

Martini E (1969) Nannoplankton aus dem Latdorf (locus typicus) und weltweite Parallelisierungen im oberen Eozän und unteren Oligozän. Senckenbergiana lethaea 50: 117-159

Martini E (1976) Cretaceous to Recent calcareous nannoplankton from the Central Pacific Ocean (DSDP Leg 33). In: Schlanger SO, Jackson ED et al Initial Repts 33: 383-423

Maxwell AE, von Herzen RP et al (1970) Site 13. In: Maxwell AE, von Herzen RP et al Initial Repts DSDP 3: 27-70

Maync W (1973) Lower Cretaceous foraminiferal fauna from Gorringe Bank, Eastern North Atlantic. In: Ryan WBF, Hsü KJ et al Initial Repts DSDP 13: 1075-1135

Perch-Nielsen K (1972) Remarks on Late Cretaceous to Pleistocene coccoliths from the North Atlantic. In: Laughton SA, Berggren WW et al Initial Repts DSDP 12: 1003-1069

Perch-Nielsen K (1977) Albian to Pleistocene calcareous nannofossils from the Western South Atlantic, DSDP Leg 39. In: Supko PR, Perch-Nielsen K et al Initial Repts DSDP 39: 699-825

Perch-Nielsen K (1979) Calcareous nannofossils from the Cretaceous between the North Sea and the Mediterranean. Aspekte der Kreide Europas. IUGS Series A, No 6: 223-272

Pessagno EA (1967) Upper Cretaceous Planktonic Foraminifera from the Western Gulf Goastal Plain. Palaeontographica Americana 5: 245-441

Pflaumann U, Krasheninnikov VA (1978) Early Cretaceous planktonic foraminifers from Eastern North Atlantic, DSDP Leg 41. In: Lancelot Y, Seibold E et al Initial Repts DSDP 41: 539-564

Premoli Silva I, Bolli HM (1973) Late Cretaceous to Eocene planktonic foraminifera and stratigraphy of Leg 15 Sites in the Caribbean Sea. In: Edgar NT, Saunders JB et al Initial Repts DSDP 15: 499-547

Risatti JB (1973) Nannoplankton biostratigraphy of the Upper Bluffport Marl - Lower Prairie Bluff Chalk interval (Upper Cretaceous), in Mississippi. Proc Symp Calc Nannofossils Gulf Coast Sect Soc Econ Paleont Miner: 8-57

Rösler W, Lutze GF, Pflaumann U (1979) Some Cretaceous planktonic foraminifers (Favusella) of DSDP Site 397 (Eastern North Atlantic). In: von Rad U, Ryan WBF et al Initial Repts DSDP 47: 273-281

Roth PH (1973) Calcareous nannofossils - Leg 17, Deep Sea Drilling Project. In: Winterer EL, Ewing JL et al Initial Repts DSDP 17: 695-793

Roth PH, Thierstein HR (1972) Calcareous nannoplankton: Leg 14 of the Deep Sea Drilling Project. In: Hayes DE, Pimm AC et al Initial Repts DSDP 14: 421-485

Ryan WBF, Hsü KJ et al (1973) Gorringe Bank - Site 120. In: Ryan WBF, Hsü KJ et al Initial Repts DSDP 13: 19-41

Saito T (1970) as member of the Shipboard Scientific Party Site 13. In: Maxwell AE, von Herzen RP Initial Repts DSDP 3: 27-70

Sclater IG, Hellinger B, Tapscott CH (1977) The paleobathymetry of the Atlantic Ocean from the Jurassic to the Present. J Geol 85: 509-552

Sigal J (1955) Notes micropaléontologiques nordafricaines 1. Du Cenomanien au Santonian: Zones et limites en facies pelagique. Soc Geol France CRSS 1955: 157-160

Sigal J (1977) Essai de zonation du Crétacé mediterranéen à l'aide des foraminifères planctoniques. Géol Mediterran 4 no 2: 99-108

Sigal J (1979) Chronostratigraphy and Ecostratigraphy of Cretaceous formations recovered on DSDP Leg 47B, Site 398. In: Sibuet JC, Ryan WBF et al Initial Repts DSDP 47/2: 287-326

Sissingh W (1977) Biostratigraphy of Cretaceous calcareous nannoplankton. Geologie en Mijnbouw 56: 37-65

Sliter WV (1980) Mesozoic foraminifers and deep-sea benthic environments from the Deep Sea Drilling Project Sites 415 and 416, Eastern North Atlantic. In: Lancelot Y, Winterer EL et al Initial Repts DSDP 50: 353-427

Stahlecker R (1934) Neocom auf der Kapverden-Insel Maio. N Jb Mineral Beil Bd 73 B: 265-301

Stradner H (1973) Nannofossils in Gorringe Bank-Site 120. In: Ryan WBF, Hsü KJ et al Initial Repts DSDP 13/1: 25-27 and 1187-1190

Thierstein HR (1971) Tentative Lower Cretaceous calcareous nannoplankton zonation. Eclogae Geol Helv 64: 459-488

Thierstein HR (1973) Lower Cretaceous calcareous nannoplankton biostratigraphy. Abh geol Bundesanst Wien 29: 1-52

Thierstein HR (1974) Calcareous nannoplankton-Leg 26 of the Deep Sea Drilling Project. In: Davies TA, Luyendyk BP et al Initial Repts DSDP 26: 619-667

Thierstein HR (1976) Mesozoic calcareous nannoplankton biostratigraphy of marine sediments. Mar Micropaleont 1: 325-362

Van Hinte JE (1976) A Cretaceous time scale. Amer Assoc Petrol Geol Bull 60: 498-516

Verbeek JW (1977) Calcareous nannoplankton biostratigraphy of Middle and Upper Cretaceous deposits in Tunisia, Southern Spain and France. Utrecht Micropaleont Bull 16:1-157

Wind FH, Čepek P (1979) Lower Cretaceous calcareous nannoplankton from DSDP Hole 397 A (Northwest African Margin). In: von Rad U, Ryan WBF et al Initial Repts DSDP 47, Part 1: 221-255

Wiedmann J, Butt A, Einsele G (1978) Vergleich von marokkanischen Kreide-Küstenaufschlüssen und Tiefseebohrungen (DSDP): Stratigraphie, Paläoenvironment und Subsidenz an einem passiven Kontinentalrand. Geol Rundsch 67: 454-508

Worsley TR (1971) Calcareous nannofossils and Lower Cretaceous sediments from the western Atlantic. Proc. Second Planktonic Conf Rome, 1970, 2: 1301-1322

Worsley TR, Martini E (1970) Late Maastrichtian Nannoplankton Provinces. Nature 225 (5239): 1242-1243

Table 2. Cretaceous nannofossil species considered in the study area listed alphabetically according to the species epithets (retrieved from Cretaceous biostratigraphic data bank, BRG Hannover)

	Legs					
	3	13	14	41	47A	50
Corollithion achylosum (Stover)			+	+		+
Zygolithus achylosus (Stover)		+				
Tetralithus aculeus (Stradner)				+		
Corollithion acutum Thierstein			+			
Braarudosphaera africana Stradner	+		+	+		
Lithraphidites alatus Thierstein		+				+
Podorhabdus albianus Black						+
Hayesites albiensis Manivit			+	+		
Chiastocygus amphipons (Bramlette & Martini)				+		
Cretarhabdus angustiforatus (Black)						+
Parhabdolithus angustus (Stradner)		+	+	+		+
Parhabdolithus asper (Stradner)			+	+	+	+
Rhagodiscus asper (Stradner)		+				
Eiffellithus augustus Bukry			+			
Masovella barnesae (Black)	+	+				
Watznaueria barnesae (Black)		+	+	+	+	+
Polycostella beckmannii Thierstein						+
Microrhabdulus belgicus Hay & Towe				+		
Broinsonia bevieri Bukry			+			+
Flabellites biforaminis Thierstein				+		+
Watznaueria biporta Bukry			+	+	+	+
Biradiatus discorhabdus (Worsley)					+	
Biscutum sp.					+	+
Lithraphidites bollii (Thierstein)					+	
Braarudosphaera sp.						+
Watznaueria britannica (Stradner)		+	+		+	+
Broinsonia sp.						+
Nannoconus bucheri Brönnimann			+	+	+	
Hayesites bulbus Thierstein			+			
Lithraphidites carniolensis Deflandre	+	+	+	+	+	+
Lucianorhabdus cayeuxii Deflandre			+			+
Cruciellipsis chiastia (Worsley)			+			+
Markalius circumradiatus (Stover)		+	+			+
Cyclococcolithus circumradiatus (Stover)		+				
Bipodorhabdus colligatus (Black)					+	+
Nannoconus colomi (de Lapparent)		+	+	+	+	+
Watznaueria communis Reinhardt			+		+	+
Staurolithites compactus (Bukry)			+			
Cretarhabdus conicus Bramlette & Martini	+		+	+	+	+
Biscutum constans (Gorka)			+	+		+
Microstaurus conus (Worsley)					+	
Corollithion sp.						+
Cretarhabdus coronadventis Reinhardt			+	+		+
Cylindralithus coronatus Bukry		+				
Cretarhabdus crenulatus Bramlette & Martini		+	+	+	+	
Prediscosphaera cretacea cretacea Arkhangelsky			+			
Prediscosphaera cretacea ponticula Bukry			+			
Prediscospharea cretacea (Arkhangelsky)	+			+		+
Cretarhabdus sp.		+				+
Staurolithites crux (Deflandre)			+			
Zygolithus crux (Deflandre & Fert)		+				
Chiastozygos cuneatus (Lyuleva)			+	+		+

Table 2 (continued)

	Legs					
	3	13	14	41	47A	50
Cruciellipsis cuvillieri (Manivit)		+	+	+	+	+
Cruciplacolithus cuvillierri (Manivit)		+				
Cyclolithella sp.		+				
Axopodorhabdus cylindratus (Noel)					+	
Arkhangelskiella cymbiformis Vekshina	+		+	+		
Microrhabdulus decoratus Deflandre	+		+	+		
Tetrapodorhabdus decoratus (Deflandre)					+	
Podorhabdus decorus (Deflandre)			+			
Micula decussata Vekshina	+			+		
Octopodorhabdus decussatus (Manivit)				+		
Cyclagelosphaera deflandrei (Manivit)					+	
Broinsonia dentata Bukry			+			
Podorhabdus dietzmanni (Reinhardt)			+	+		+
Zygodiscus diplogrammus (Deflandre)				+	+	+
Zygolithus diplogrammus (Deflandre)		+				
Glaucolithus diplogrammus (Deflandre)			+			
Gartnerago diversum Thierstein			+			
Laguncula dorotheae Black					+	
Cribrosphaerella ehrenbergi (Arkhangelsky)	+		+	+		+
Eiffellithalid sp.						+
Eiffellithus sp.					+	
Glaucolithus elegans (Gartner)			+			
Zygodiscus elegans Gartner	+			+		+
Corollithion ellipticum Bukry			+		+	
Parhabdolithus embergeri (Noël)		+	+	+	+	+
Broinsonia enormis (Shumenko)				+		+
Zygolithus erectus Deflandre		+				
Zygodiscus erectus (Deflandre)				+		+
Arkhangelskiella erratica Stover			+			
Corollithion exiguum Stradner		+				
Tranolithus exiguus Stover			+	+		
Eiffellithus eximius (Stover)				+		
Watznaueria fasciata Wind & Čepek					+	
Reinhardites fenestratus (Worsley)			+		+	+
Lithastrinus floralis Stradner		+	+	+		+
Scapholithus fossilis Deflandre & Fert				+		+
Marthasterites furcatus (Deflandre)				+		
Tranolithus gabalus Stover			+	+		+
Ethmorhabdus gallicus Noël					+	
Nannoconus globulus Brönnimann					+	
Lithastrinus grilli Stradner		+	+			
Helenea sp.					+	
Sollasites horticus (Stradner, Adamiker & Maresch)			+	+	+	+
Micrantholithus hoschulzi (Reinhardt)		+	+	+		
Discorhabdus ignotus (Gorka)				+		
Discolithus incohatus Stover				+		
Mitosia infinitae Worsley						+
Parhabdolithus infinitus (Worsley)			+	+	+	
Rucinolithus irregularis Thierstein		+	+	+		
Tubodiscus jurapelagicus (Worsley)					+	
Nannoconus kamptneri Brönnimann		+			+	
Stephanolithion laffittei Noël		+	+	+	+	+
Broinsonia lata (Noel)			+	+		
Diazomatholithus lehmani Noël			+	+	+	+

Table 2 (continued)

	Legs					
	3	13	14	41	47A	50
Chiastozygus litterarius (Gorka)			+	+		+
Zygolithus litterarius (Gorka)		+				
Cretarhabdus loriei Gartner			+			+
Kamptnerius magnificus Deflandre				+		
Tetralithus malticus Worsley						+
Cyclagelosphaera margereli Noël		+	+	+	+	+
Watznaueria martelae Noël				+		+
Vagalapilla matalosa (Stover)				+		+
Staurolithites matalosus (STOVER)			+			
Grantarhabdus meddii Black					+	
Metadoga mercurius Wind & Čepek					+	
Conusphaera mexicana Trejo					+	+
Micrantholithus sp.						+
Microstaurus sp.					+	
Micula sp.	+					
Micula mura (Martini)				+		
Nannoconus sp.					+	+
Tetralithus nitidus nitidus Martini	+					
Tetralithus nitidus trifidus (Stradner)	+					
Gartnerago obliquum (Stradner)				+		
Calcicalathina oblongata (Worsley)		+		+	+	+
Tetralithus obscurus Deflandre				+		
Micrantholithus obtusus Stradner		+	+	+	+	+
Octopodorhabdus sp.					+	
Ahmuellerella octoradiata (Gorka)			+	+		
Podorhabdus orbiculofenestrus (Gartner)			+	+		
Tranolithus orionatus (Reinhardt)			+			+
Broinsonia orthocancellata Bukry			+			
Rhabdolekiskus parallelus Wind & Čepek					+	
Broinsonia parka (Stradner)				+		
Reticulofenestra parvidendata (Deflandre)		+				
Manivitella pectem Black					+	
Manivitella pemmatoidea (Deflandre ex Manivit)			+	+	+	+
Cricolithus pemmatoides		+				
Perissocyclas sp.					+	
Tranolithus phacelosus Stover		+		+		
Lucianorhabdus phlaskus Wind & Čepek					+	
Podorhabdus sp.						+
Tetralithus pyramidus Gardet				+		
Lithraphidites quadratus Bram. & Martini				+		
Diadorhombus rectus Worsley			+	+	+	+
Watznaueria reinhardti Rood, Hay & Barnard					+	
Corollithion rhombicum (Stradner & Adamiker)			+			
Bidiscus rotatorius Bukry			+			
Discorhabdus rotatorius (Bukry)				+	+	
Cretaturbella rothii Thierstein = C. mexicana Trejo		+	+	+		
Rhagodiscus rugosus (Noël)		+				
Trapezopentus sarmatus Wind & Čepek					+	
Cretarhabdus schizobrachiatus (Gartner)				+		
Polycostella senaria Thierstein						+
Broinsonia signata (Noël)			+			
Corollithion signum Stradner		+	+	+		+
Prediscosphaera spinosa (Bramlette & Martini)			+			
Parhabdolithus splendens (Deflandre)			+	+	+	+

Table 2 (continued)

	Legs					
	3	13	14	41	47A	50
Rhabdolithina splendens (Deflandre)		+				
Staurolithites sp. -		+				
Micula staurophora (Gardet)				+		
Micula staurophorus (Gardet)				+		
Nannoconus steinmanni Kamptner		+	+			
Microrhabdulus stradneri Bramlette & Martini				+		
Tegumentum stradneri Thierstein			+			
Vagalapilla stradneri (Rood, Hay & Barnard)				+		+
Vekshinella stradneri (Rood, Hay & Barnard)					+	
Arkhangelskiella striata Stradner		+				
Ericsonia subpertusa Hay & Mohler				+		
Biscutum supracretaceum (Reinhardt)				+		
Cretarhabdus surirellus (Deflandre)			+	+	+	+
Parhabdolithus swinnertonii Black					+	
Biscutum testudinarium Black		+				
Eiffellithus trabeculatus (Gorka)			+	+		
Tetralithus trifidus (Stradner)				+		
Nannoconus truitti Brönnimann		+	+		+	
Eiffellithus turriseiffeli (Deflandre)	+		+	+		+
Vekshinella sp. -						+
Tubodiscus verenae Thierstein				+	+	+
Nannoconus wassalli Brönnimann					+	
Watznaueria sp. -						+
Rucinolithus wisei Thierstein				+		+
Zeugrhabdotus sp. -						+
Zygodiscus sp. -				+		

Table 3. Cretaceous planktonic foraminifera described from the DSDP Legs 3, 13, 14, 41, 47A, 50 and onshore (= Mo) by Wiedmann et al. (1978)

	Legs						
	3	13	14	41	47A	50	Mo
acervulinoides (Egger), Planoglobulina				x			
aegyptiaca Nakkady, Globotruncana				x			
algerianus Cushman and TenDam, Globigerinelloides							x
amabilis (Loeblich and Tappan), Hedbergella			x				
angusticarinata Gandolfi, Globotruncana				x			x
apenninica (Renz), Rotalipora			x	x		x	x
aperta Sigal, Ticinella raynaudi				x			
aptiensis Longoria, Globigerinelloides				x			
arca Cushman, Globotruncana			x	x			x
asper (Ehrenberg), Globigerinelloides	x			x			
atlantica , Globotruncana							x
austinensis Gandolfi, Globotruncana				x			
balernaensis Gandolfi, Rotalipora			x				
bejaouaensis Sigal, Ticinella				x			x
bentonensis (Morrow), Globigerinelloides			x	x		x	
blowi (Bolli), Archaeoglobigerina				x			
blowi (Bolli), Globigerinelloides				x	x		
bosquensis (Pessagno), cf., Hedbergella			x				
breggiensis (Gandolfi), Ticinella			x	x			x
brittonensis Loeblich and Tappan, Hedbergella			x	x			
bronnimanni Pessagno, Praeglobotruncana			x				
brotzeni (Sigal), Rotalipora			x	x			x
bulbosa Belford, Hedbergella							x
bulloides (Vogler), Globotruncana				x			
buxtorfi (Gandolfi), Planomalina			x		x		x
caliciformis (De Lapparent), Globotruncana							x
canaliculata (Reuss), Globotruncana			x				x
carinata Dalbiez, Globotruncana concavata							x
casey (Bolli, Loeblich and Tappan), Globigerinelloides			x	x		x	
cenomana Cushman, Guembelitria						x	
cenomana (Schacko), Schackoina			x	x		x	
citae (Bolli), Globotruncanella	x						
concavata (Brotzen), Globotruncana				x			x
conica White, Globotruncana				x			
contusa (Cushman), Globotruncana			x	x			x
convexa Longoria, aff. Loeblichella				x			
coronata (Bolli), Globotruncana				x			x
costulata (Cushman), Pseudoguembelina	x			x			
cretacea (d'Orbigny), Archaeoglobigerina				x			
cretacea (d'Orbigny), Globotruncana							x
cushmani (Morrow), Rotalipora			x			x	
cuvillieri Kikoine, Gublerina				x			
delrioensis (Carsey), Hedbergella				x		x	x
delrioensis (Plummer), Praeglobotruncana			x	x		x	x
difformis Gandolfi, cf., Globotruncana			x	x			
digitalis Sigal, Ticinella raynaudi			x	x			x
eggeri Cushman, Ventrilabrella							x
ehrenbergi (Barr), Globigerinelloides				x			
elegans (Rzehak), Pseudotextularia				x			
elevata (Brotzen), Globotruncana				x			x
eocretacea Neagu, Clavihedbergella				x		x	
evoluta Sigal, Rotalipora			x	x		x	x
excolata (Cushman), Pseudoguembelina	x		x	x			
falsostuarti Sigal, Globotruncana				x			x
ferreolensis (Moullade), Globogerinelloides				x			x

Table 3 (continued)

	Legs						
	3	13	14	41	47A	50	Mo
fornicata Plummer, Globotruncana			x	x			x
fructicosa (Egger), Racemiguemberina			x	x			
gagnebini Tilev, Globotruncana	x			x			
gandolfii Luterb. and Premoli Silva, Rotalipora							x
gansseri Bolli, Globotruncana				x			
gautierensis (Bronnimann), Hedbergella			x				
globigerinelloides (Subbotina), Hedbergella				x			
globulosa (Ehrenberg), Heterohelix			x				x
gorbachikae Longoria, Hedbergella				x			
gottisi (Chevalier), Globigerinelloides				x			
graysonensis (Tappan), Gubkinella				x		x	
greenhornensis (Morrow), Rotalipora			x	x			x
gyroidinaeformis Moullade, "Globigerinelloides"				x			
hagni (Scheibnerova), Globotruncana							x
harrisi Tappan, Guembelitria				x			
havanensis (Voorwijk), Globotruncanella	x			x			x
helvetica Bolli, Globotruncana				x			x
hexacamerata Bronnimann, Rugoglobigerina				x			
hilli Pessagno, Globotruncana				x			
hoelzli (Hagn and Zeil), Rugoglobigerina				x			
hoterivica(Subbotina),Hedbergella,Caucasella,Globuligerina				x	x		x
imbricata Mornod, Globotruncana			x	x			
indica, Jacob and Sastry, Globotruncana			x				
infracretacea (Glaessner), Hedbergella		x		x		x	
intermedia de Klasz, Pseudotextularia				x			
intermedius (Bolli), Abathomphalus				x			
kugleri (Bolli), Hedbergella				x	x		
labocaensis Longoria, aff., Hedbergella				x			
lapparenti Brotzen, Globotruncana				x			x
lata (Egger), Heterohelix				x			
lehmanni, Praeglobotruncana							x
libyca Barr, Hedbergella				x			
linneiana (d'Orbigny), Globotruncana			x	x			
macrocephala Bronnimann, Rugoglobigerina				x			
marginata (Reuss), Globotruncana				x			
mariei (Gandolfi), Marginotruncana				x			
maslakovae Longoria, Hedbergella				x			
mayaroensis Bolli, Abathomphalus				x			
messinae messinae (Bronnim.),Globigerinelloides				x			
moremani (Cushman), Heterohelix			x	x		x	x
morozovae , Globotruncana				x			
moulladei Longoria, Loeblichella				x			
multicamerata (De Klasz), Ventilabrella				x			
multiloculata Morrow, Ticinella							x
multispina (Lalicker), Biglobigerinella				x			
multispinata (Cushm. and Wickenden),Schackoina				x		x	
nothi (Bronnim. and Brown), Globotruncana				x			
obliqua Herm, Globotruncana				x			x
occulta Longoria, Hedbergella				x			
palpebra Brönnim. and Brown, Pseudoguembelina				x			
paradubia Sigal, Hedbergella							x
petaloidea (Gandolfi), Globotruncanella	x			x			
planispira (Tappan), Hedbergella			x	x		x	x
portsdownensis (Williams-Mitchell), Hedbergella				x			
praehelvetica Trujillo, Praeglobotruncana							x
praeticinensis Sigal, Ticinella				x			x

Table 3 (continued)

	Legs						
	3	13	14	41	47A	50	Mo
primula Luterbacher, Ticinella				x			x
pseudolinnieana (Pessagno), Globotruncana			x				
pulchra (Brotzen), Heterohelix	x		x	x			
punctulata (Cushman), Heterohelix				x			
pustulata Bronnimann, Rugoglobigerina	x						
raynaudi Sigal, Ticinella				x			
reicheli (Mornod), Rotalipora			x				x
reicheli Bronnimann, Rugoglobigerina	x			x			
renzi Gandolfi, Globotruncana			x				x
reussi (Cushman), Heterohelix	x			x			
roberti (Gandolfi), Ticinella				x			x
robusta De Klasz, Gublerina				x			
rosetta (Carsey), Globotruncana	x			x			
rotundata Bronnimann, Rugoglobigerina				x			
rugosa (Plummer), Rugoglobigerina	x			x			
schneegansi Sigal, Globotruncana							x
semielongata Longoria, Hedbergella				x			
sigali (Reichel), Globotruncana			x				x
sigali (Moullade), Hedbergella				x			x
similis Longoria, Hedbergella				x			
simplex Morrow, Hedbergella				x	x	x	x
simplicissima (Magne and Sigal), Hedbergella			x	x		x	x
stephani (Gandolfi), Praeglobotruncana			x			x	
stephensoni Pessagno, Globotruncana				x			
stiftia Rösler, Lutze, Pflaumann, Favusella					x		
striata (Ehrenberg), Pseudoguembelina	x		x				
stuarti (de Lapparent), Globotruncana			x	x			
stuartiformis Dalbiez, Globotruncana			x	x			x
subcarinatus (Bronnimann), Globigerinelloides	x			x			
subcircumnodifer (Gandolfi), Rugoglobigerina				x			
subcretacea (Tappan), aff., Hedbergella				x			
subdigitata (Carman), Clavihedbergella				x			
subpetaloidea (Gandolfi), Rugoglobigerina				x			
subspinosa Pessagno, Globotruncana				x			x
subticinensis (Gandolfi), Rotalipora				x			
tardita Antonova, "Globigerina"					x		
ticinensis (Gandolfi), Rotalipora			x	x			x
tradinghousensis Pessagno, Rugoglobigerina	x						
transitoria Longoria, Ticinella				x			
triangulata Antonova, "Globigerina"					x		
tricarinata (Quereau), Globotruncana				x			x
trinidadensis Gandolfi, Globotruncana				x			
trocoidea (Gandolfi), Hedbergella			x	x			
tururensis , Globigerinelloides			x				
ultimatumida (White), Heterohelix			x	x			
ultramicrus (Subbotina), Globigerinelloides				x			
undulata Lehmann, Marginotruncana				x			
ventricosa White, Globotruncana				x			
volutus (White), Globigerinelloides	x			x			
washitensis (Carsey), Favusella							x

14 Cretaceous Radiolaria from the Northwest African Continental Margin

Reinhard Wolfart[1]

It has long been known that Radiolaria in the Atlantic Basin are not as abundant or as well preserved as in the Pacific Basin (Murray, 1876; Foreman, 1977). Johnson (1976) suggested that the poor preservation of siliceous skeletons may be due to the relatively small amount of silica being deposited in the Atlantic Basin. Examinations of the Mesozoic DSDP cores from the Northwest African continental margin again confirm this observation. With few exceptions, paucity and poor preservation of radiolarian faunas are apparently the rule in this region of the Atlantic, too.

Localities

The boreholes of the Deep Sea Drilling Project which penetrated Mesozoic deposits off Northwest Africa are indicated by circles on Fig. 1. The black circles represent sections of Mesozoic sediments with occasionally well-preserved Radiolaria; the white circles indicate Mesozoic sediments without (well preserved) Radiolaria. From some sites (135, 366, 370, 397, 416) radiolarians are either not reported or simply not present, from others (368, parts of 369, 415) only very poorly preserved radiolarians were mentioned. Results of the DSDP Legs 3 (Site 13), 13 (Site 120), 14 (Sites 135, 136, 137, 138, 140), 41 (Sites 366-370), 47 (Site 397), and 50 (Sites 415, 416) are included in this report.

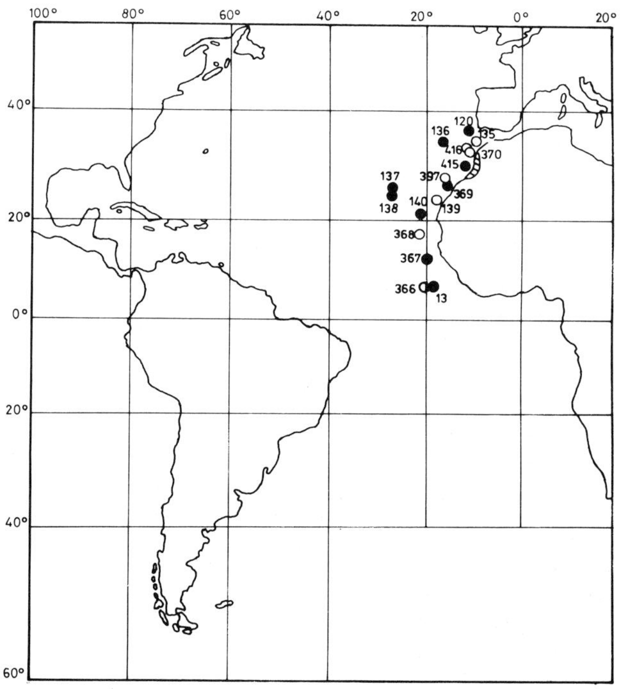

Fig.1. DSDP sites off northwest Africa where Cretaceous deposits have been penetrated.

● with Cretaceous radiolarians

○ without Cretaceous radiolarians

Agadir and Tarfaya Basin

Site 13 : Radiolaria from the Campanian and early Maestrichtian
Site 120 : " " " early Aptian to middle Albian
Site 136 : " " " Aptian Albian and Senonian (Post-Turonian)
Site 137 : " " " ? Albian, Cenomanian, early Campanian Maestrichtian
Site 138 : " " " Cenomanian, early Campanian
Site 140 : " " " Maestrichtian
Site 367 : " " " Tithonian (or older), Berriasian, Valanginian, Hauterivian, and Aptian
Site 369A : " " " late Campanian, Maestrichtian
Site 415A : " " " Cenomanian and late Albian

Berriasian to Late Jurassic localities generally yielded poorly preserved radiolarians with rare (to common) abundances (Figs. 2, 3). Neocomian localities (exclusive of Berriasian) also contain generally

1 Bundesanstalt für Geowissenschaften und Rohstoffe, Stilleweg 2
3000 Hannover 51, FRG

Fig. 2. Radiolaria bearing Cretaceous rocks off Northwest Africa according to DSDP results

Radiolaria possibly present

Basaltic basement

Possible hiatuses

B Final depth

N No Radiolaria present or reported

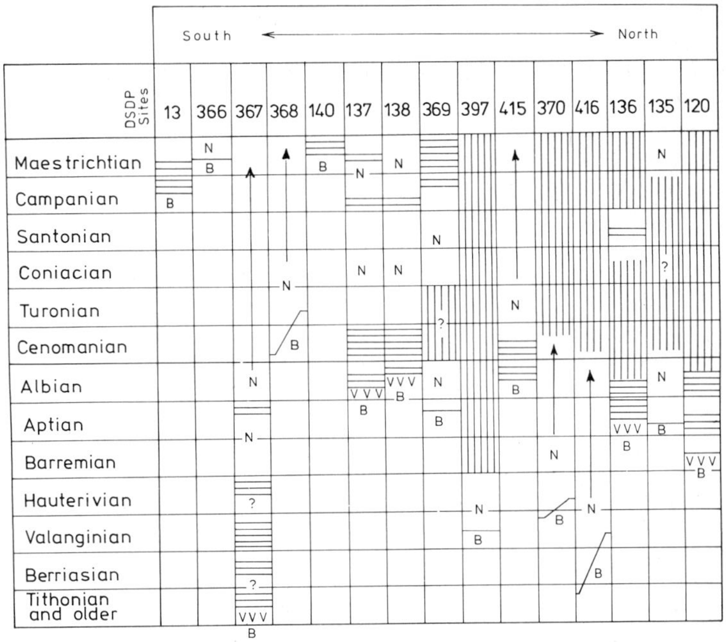

rare and poorly preserved radiolarians though the Valanginian Sethocapsa trachyostraca Zone could be recognized at Site 367. The latter hole penetrated a sequence of Upper Jurassic deposits (Tithonian to Oxfordian) also containing some recognizable radiolarians (Fig. 3), before it stopped in the basaltic basement. Sections of the Cenomanian and the late Lower Cretaceous are very similar to those of the Neocomian, with faunas generally poor and sparse. Upper Cretaceous localities, Campanian, Maestrichtian, yielded by far the most abundant faunas which are in general much better preserved than in the Upper Jurassic/Lower Cretaceous deposits. No long sequences are known.

Foreman (1977) stated that in the Atlantic Basin Radiolaria are rather consistently present in sediments of (1) late Late Cretaceous age, (2) Cenomanian/Albian age and of the late Neocomian (minor occurrences), and (3) early Neocomian/Late Jurassic (rather consistently again). The radiolarian communities of the first and the third group can be divided into two subgroups each (first group = early Campanian and late Campanian/Maestrichtian; third group = Hauterivian/Valanginian and Berriasian/Late Jurassic). In the region of the Northwest African continental margin, the radiolarian distribution coincides largely with the pattern of the Atlantic Basin outlined by Foreman. This pattern is in contrast to the Pacific where Radiolaria are generally lacking in the late Late Cretaceous or, when present poorly preserved, are completely missing in the Maestrichtian. In the Pacific, fairly common but poor early Late and late Early Cretaceous Radiolaria and rich sequences of Aptian and Neocomian radiolarian communities are known.

Radiolarian Biostratigraphic Zonation

The radiolarian zonation of the Cretaceous (Figs. 3, 4) is still very much in an uncertain stage (Foreman, 1977). Five major communities of Cretaceous radiolarian fossils were described by Pessagno (1963), Aliev (1965), Foreman (1968), Dumitrica (1970), Kozlova (1972), and Foreman (1975), and others which can also be recognized off Northwest Africa: (1) Maestrichtian/late Campanian; (2) early Campanian; (3) Cenomanian/Albian; (4) Barremian/Hauterivian or Valanginian; (5) early Neocomian/Late Jurassic.

The first coarse radiolarian zonation of the late Late Jurassic and the Cretaceous was given by Moore (1973). It was based on the Pacific

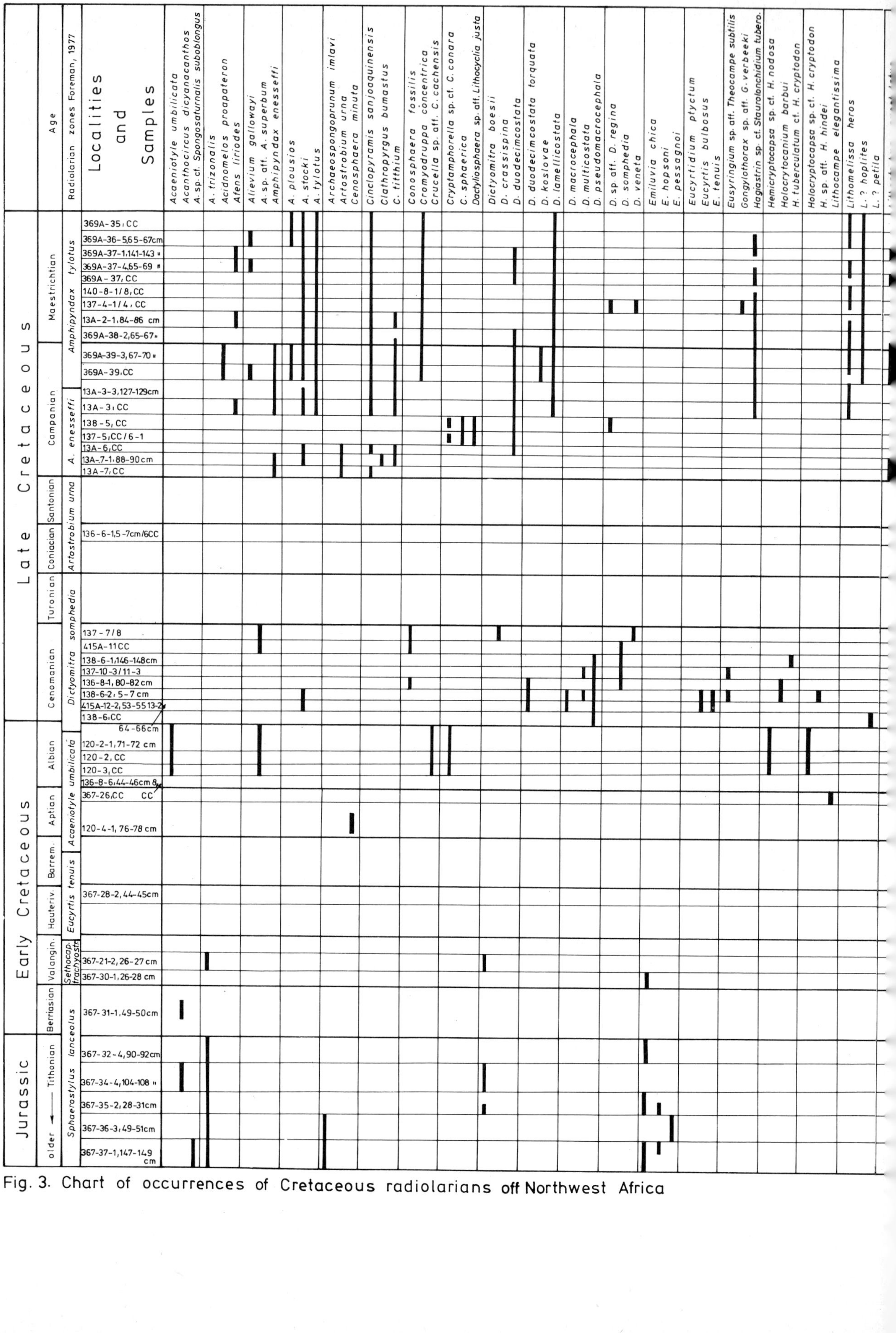

Fig. 3. Chart of occurrences of Cretaceous radiolarians off Northwest Africa

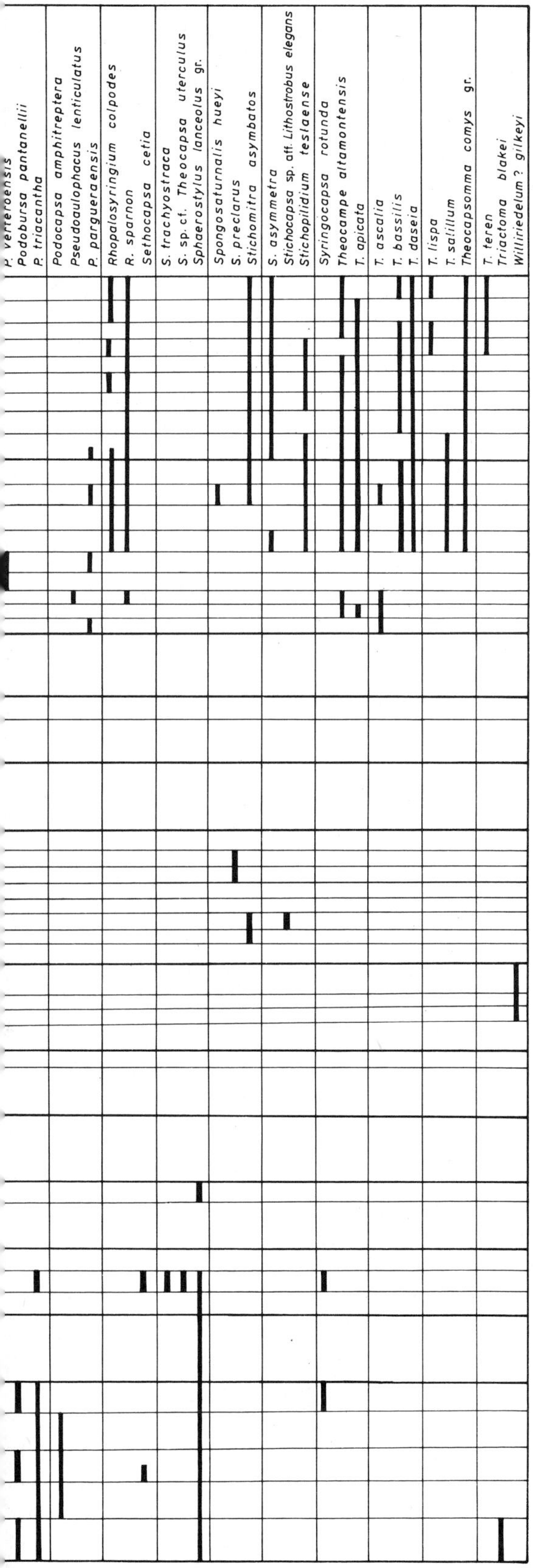

Leg 17 results and consisted of seven zones. Poor preservation, lack of comparative faunas and inadequate calcareous control made the Cretaceous zones of Moore difficult to apply. Riedel and Sanfilippo (1974) introduced another coarse zonation for the same period of time as Moore, which is also composed of seven zones. Some of these zones were widely applicable but they again had very imprecise calcareous control. This zonation was modified by Foreman (1975, 1977) to make it applicable to the Cretaceous deposits of the Atlantic Basin. Based on Pacific Leg 62 material, Schaaf (1980) proposed three new zones to substitute the *Eucyrtis tenuis* and part of the *Acaeniotyle umbilicata* zone of Foreman (1975) by the *Dibolachras tytthopora* Zone, the *Crolanium pythiae* Zone, and the *Archicapsa similis* Zone. The top of each zone is defined by the base of the overlying zone. Since the new zones are of Barremian and early Aptian age, they could not be traced in the region off Northwest Africa. All of these zonations are largely based on material from the Deep Sea Drilling Project.

Detailed zonations based on well-preserved material with good calcareous control from the Cretaceous of California and the Cenomanian of Romania were published by Pessagno (1976, 1977) and Dumitrica (1975), respectively. While Dumitrica's zones are open-ended, the zones of Pessagno could, for the most part, not be recognized in the Atlantic Basin.

Foreman's (1975, 1977) radiolarian zones of the Cretaceous within the Atlantic Basin, which are also applied in the present paper, are defined as follows.

Sphaerostylus lanceolus Zone Riedel and Sanfilippo (1974), emend. Foreman (1975). The base of the *S. lanceolus* Zone is defined by the first evolutionary appearance of *S. lanceolus*; the top is marked by the first appearance of *Sethocapsa trachyostraca*. So far, the upper limit of the *S. lanceolus* Zone cannot be precisely traced within the Northwest African continental margin because the zone fossil (*S. trachyostraca*) has only been found in one sample (367-29-2, 26-27 cm). Correspondingly this applies to the ma-

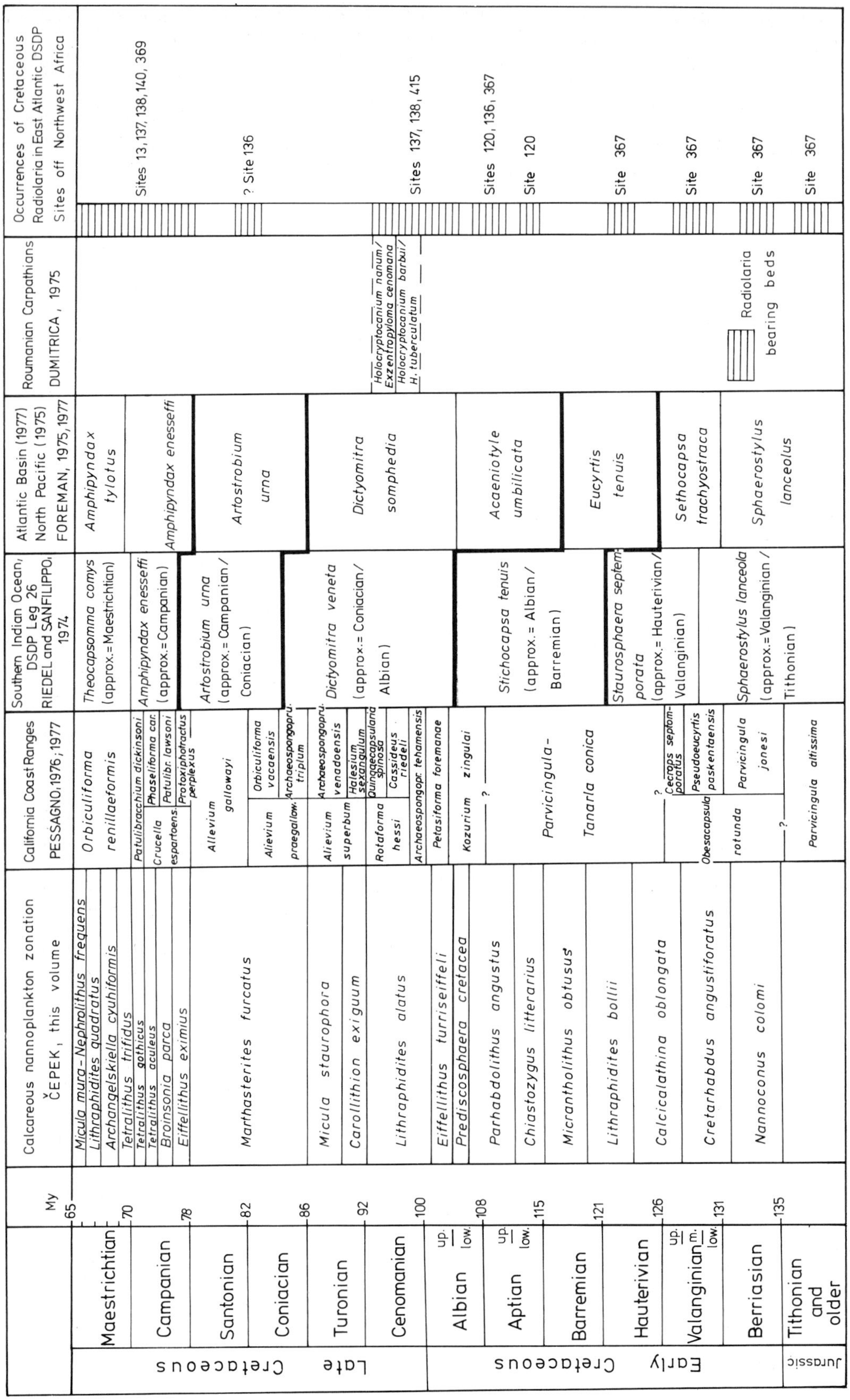

Fig. 4. Radiolarian zonations of the Cretaceous.

jority of the Cretaceous radiolarian zones off Northwest Africa.

Sethocapsa trachyostraca Zone Foreman (1975). The base of this zone is defined by the first appearance of S. trachyostraca, the upper limit by the first occurrences of Dibolachras tytthopora, Staurosphaera septemporata, and Sethocapsa ? orca and by the last occurrence of Dictyomitra cosmoconica. The base of the zone can not be exactly located, nor are the latter four radiolarian species known in the region.

Eucyrtis tenuis Zone Riedel and Sanfilippo (1974), emend. Foreman (1975). Foreman defined the base of the zone by the first appearance of Dibolachras tytthopora, which is believed to be approximately synchronous with or slightly earlier than the first appearance of E. tenuis. The upper limit is defined by the last occurrence of members of the Sphaerostylus lanceolus group which may be approximately synchronous with the last occurrences of D. tytthopora. Specific radiolarians of the E. tenuis Zone are completely unknown in the region.

Acaeniotyle umbilicata Zone Foreman (1975). The base is defined by the extinction of the Sphaerostylus lanceolus group, the upper limit by the first morphological appearance of Dictyomitra somphedia and by the last occurrence of A. umbilicata. Only the Albian part of the A. umbilicata Zone is represented by several radiolarian species. Lower and upper limit of the zone cannot be exactly drawn in the region.

Dictyomitra somphedia Zone Foreman (1975). The base is defined by the first morphological appearance of D. somphedia; in more southern latitudes this may be approximately synchronous with the first appearance of Dictyomitra veneta. The upper limit is characterized by the first morphotypic appearance of Artostrobium urna. Included are the last occurrences of Dictyomitra pseudomacrocephala, D. somphedia, and Lithomelissa ? petila and the first appearance of Spongosaturnalis ? preclarus and others. Only the lower portion of the zone, which is Cenomanian in age, contains radiolarians such as D. somphedia, L. ? petila and others. S. preclarus appears in the Cenomanian already and cannot be used in the region, therefore, to draw the upper limit of the zone.

Artostrobium urna Zone Riedel and Sanfilippo (1974). The base is defined by the first morphotypic appearance of A. urna, the upper limit by the last occurrence of A. urna and by the first morphological occurrence of Amphipyndax enesseffi, which may be approximately coincident with the first appearance of Lithostrobus punctulatus. Radiolarians of the Coniacian/Santonian A. urna Zone are not known in the region. A. urna is not restricted to the A. urna Zone since it is still present in the lower part of the A. enesseffi Zone.

Amphipyndax enesseffi Zone Riedel and Sanfilippo (1974), emend. Foreman (1977). The base is defined by the first morphological appearance of A. enesseffi and Lithostrobus punctulatus. Near the base Clathropyrgus titthium, Theocampe apicata, Theocampe daseia, Afens liriodes, and Theocampe bassilis make their first appearance. The top is defined by the base of the A. tylotus Zone. T. daseia, A. liriodes, and T. bassilis make their first appearance in the region not before the uppermost part of the zone or even the lower part of the A. tylotus Zone.

Amphipyndax tylotus Zone Foreman (1977). The base is defined by the first evolutionary appearance of A. tylotus. Near its base, Lithomelissa ? hoplites and Lophophaena polycyrtis make their first and Clathropyrgus titthium and Afens liriodes their last appearance. The top is defined as the Maestrichtian/Danian boundary recognized by the last occurrences of many radiolarian species of the Cretaceous. This is the only radiolarian zone of the Cretaceous which can be defined in the region off Northwest Africa itself. C. titthium and A. liriodes range far into the A. tylotus zone.

Late Jurassic-Early Neocomian

Off Northwest Africa, the biostratigraphic distribution of radiolarian faunas is very variable (Fig. 3). The best DSDP Atlantic sequence so far of Late Jurassic/early Neocomian Radiolaria was encountered at Site 367 (Leg 41), Cores 28 through 37. The Radiolaria are common but poorly preserved in most cases. In the nonsilicified sediments of Cores 15 through 28, Mesozoic Radiolaria are either altered to pyrite or absent. Siliceous Radiolaria from Cores 29-32 are too poor, or too sparse, to indicate anything but a broad Neocomian-Late Jurassic age with the exception of Sample 367-29-2, 26-27 cm which contains Sethocapsa cetia and S. trachyostraca and must be considered, therefore, to be low in the S. trachyostraca Zone, the base of which is defined by the first appearance of S. trachyostraca. Cores 34-37 contain more abundant and slightly better preserved Radiolaria than the former cores. Since Late Jurassic Radiolaria are not well enough known (Foreman, 1978), the age assignments must be based on foraminiferal and nannofossil data. Nevertheless the samples of Cores 34-37 yielded a diversified radiolarian assemblage of about 20 species including Sphaerostylus lanceolus; consequently they are considered to belong to the S. lanceolus Zone. It was not before the end of the Hauterivian that the zone fossil S. lanceolus dissappeared in the region off Northwest Africa.

Early Cretaceous

Radiolarian faunas of the Barremian have not yet been reported from the region off Northwest Africa, while Aptian radiolarians have only been recovered at Site 120-4-1, 76-78 cm and at Site 367-26, CC. Albian radiolarians have been particularly found in 120-2-1, 71-72 cm through 120-3, CC, besides a find of ?early Albian Radiolaria in 136-8-6, 44-46 cm and 6, CC. Dumitrica (1973) described the Albian radiolarians from Site 120 as an abundant and diversified assemblage with pyritized shells, while a poor assemblage of the early Aptian (120-4-1) is preserved as silica and consists only of rare specimens of Cenosphaera minuta. The late Aptian Sample 367-26, CC yielded Conosphaera fossilis and Lithocampe elegantissima, which do not allow any zonal specification. The Albian radiolarian assemblage of Site 220 is composed of seven species with the zone fossil Acaeniotyle umbilicata. Since the known range of A. umbilicata is from Late Jurassic through the Albian (Foreman, 1975), and since the remaining six radiolarian species are not precisely defined, the presence of A. umbilicata does not indicate any exact zonal assignment.

Late Cretaceous

Rather abundant radiolarian assemblages of Cenomanian age were recovered from the Sites 136-138 and 415A (Fig. 3). They correspond to the lower part of the Dictyomitra somphedia Zone; from its upper part, which is Turonian in age, radiolarians were not reported so far. All of the sites yielded Dictyomitra somphedia and D. pseudomacrocephala. Spongosaturnalis preclarus makes its first appearance in the Cenomanian and cannot be used, therefore, to draw the upper boundary of the D. somphedia Zone as suggested by Foreman (1975). Dumitrica's Cenomanian zone fossils Holocryptocanium barbui and H. tuberculatum have been recovered from Holes 136 and 138, while the late Cenomanian zone fossils Holocryptocanium nanum and Exantropyloma cenomana are obviously missing.

Very sparse radiolarian faunas, only Lithelidae in the Samples 136-6-1, 5-7 cm through 6 CC, have been found in deposits of Coniacian and Santonian age which were determined by means of foraminifers and coccoliths. Thus, the large time span between Turonian and Santonian (14 MA) which corresponds to the upper part of D. somphedia and to the entire A. urna Zones cannot be traced by means of Radiolaria in the region off Northwest Africa.

The late Upper Cretaceous deposits of the Campanian and the Maestrichtian contain much more diversified radiolarian assemblages than the preceding beds. There are two radiolarian zones, the Amphipyndax enesseffi Zone, which is approximately restricted to the lower part of the Campanian, and the A. tylotus Zone, comprising the upper part of the Campanian and the Maestrichtian. Amphipyndax enesseffi and Lithostrobus punctulatus are contemporaneously beginning at the base of the A. enesseffi Zone which coincides more or less with the base of the Campanian. These are probably not the true first appearances of A. enesseffi and L. punctulatus, since no radiolarians are preserved in the underlying rocks. Clathropyrgus titthium and Theocampe apicata make their first appearances near the base of the A. enesseffi Zone, while Theocampe bassilis, T. daseia, and Afens liriodes do not appear before the uppermost part of that zone.

The upper Campanian/Maestrichtian Amphipyndax tylotus Zone has been established in the Atlantic Basin by Foreman (1977). It is characterized by occurrences of the zone fossil itself, Lithomelissa ? hoplites, Lophophaena polycyrtis, Clathropyrgus titthium, and Afens liriodes. The top of the A. tylotus Zone coincides with the Maestrichtian/Danian boundary and can easily be recognized by the last occurrences of many Late Cretaceous species.

In the region off Northwest Africa Cretaceous radiolarian assemblages of six zones have been recognized:

1. Late Jurassic/Berriasian Sphaerostylus lanceolus Zone
2. Valanginian Sethocapsa trachyostraca Zone
3. Albian part of the Acaeniotyle umbilicata Zone
4. Cenomanian part of the Dictyomitra somphedia Zone
5. Campanian Amphipyndax enesseffi Zone
6. Campanian/Maestrichtian Amphipyndax tylotus Zone.

Comparison of Radiolaria in the Moroccan Coastal Basins and in the Region of the Northwest African Continental Margin

According to Wiedmann et al. (1978), there are neither radiolarians nor planktonic Foraminifera in the Lower Cretaceous deposits of the Moroccan coastal basins. In the Tarfaya Basin Upper Cenomanian, Turonian, and Coniacian are represented by finely laminated, bituminous marl, limestone and siliceous limestone with layers of siliceous concretions. Wiedmann et al. considered these sedimentary rocks to be deposited in the upper bathyal and/or outer shelf region, and they assumed that in the Moroccan coastal basins during the Upper Cenomanian/Coniacian upwelling conditions developed for the first time. Due to the upwelling the bioproductivity was strongly increased (planktonic foraminifers, coccoliths, and radiolarians). Radiolarians have not yet been studied in the Moroccan coastal basins. They are only mentioned as a group of fossils with siliceous skeletons which might have served as a source for the siliceous concretions in the Upper Cenomanian/Coniacian deposits of the Tarfaya Basin.

Further occurrences of radiolarians are reported by Wiedmann et al. from Turonian and Santonian/Campanian of the Agadir Basin. Again the occurrences of radiolarians are traced back to upwelling conditions and increased bioproductivity which resulted in the formation of dark, bituminous marl and shale containing microfossils of the pelagic plankton group (foraminifers, coccoliths, radiolarians, fish remains).

Arthur and Natland (1979) outlined that in the central North Atlantic, sediments of virtually all paleodepths reflect conditions of intermittent anoxia and generally low oxygen content in deep water masses from the Hauterivian (or earlier) through the Cenomanian. Cenomanian and Turonian mudstones contain exceptionally high amounts of marine organic matter in the eastern central North Atlantic reflecting upwelling conditions.

The Cretaceous remainder of the post-Cenomanian/Turonian central North Atlantic apparently was dominated by well-oxygenated deep waters.

In the Moroccan coastal basins (Wiedmann et al. 1978), Radiolaria are mainly restricted to bituminous sedimentary rocks which have been deposited under conditions of upwelling (Agadir Basin: Turonian, Campanian; Tarfaya Basin: Turonian). In the region of the DSDP Sites, however, the presence of radiolarian fossils certainly is not correlated with upwelling indices such as dark, bituminous shale. Diversified radiolarians also occur in the Campanian and Maestrichtian rocks which were generally formed under well-oxygenated conditions. In contrast to the coastal basins, no radiolarians are known from the Turonian, Coniacian and Santonian of the Northwest African continental margin.

Acknowledgements. The writer wishes to thank W.R. Riedel for his constructive review of the manuscript. The Deutsche Forschungsgemeinschaft provided some financial support for the Cretaceous data base.

SPECIES LIST

Acaeniotyle umbilicata (Rüst, 1898): 7, pl. 1, fig. 9. -- Foreman, 1975: 607, pl. 2E, figs. 14-17.
Acanthocircus dicranacanthos (Squinabol, 1914). -- Foreman, 1973b: 260, pl. 4, figs. 4-5.
Acanthocircus sp. cf. Spongosaturnalis suboblongus Yao, 1972: pl. 3, figs. 1-6.
Acanthocircus trizonalis (Rüst, 1898). -- Foreman, 1975: 610, pl. 20, figs. 1-4.
Acidnomelos proapteron Foreman, 1978: 749, pl. 5, fig. 13.
Afens liriodes Riedel and Sanfilippo, 1974: 775, pl. 11, fig. 11; pl. 13, figs. 14-16.
Alievium gallowayi (White, 1928). -- Pessagno, 1972: 299, pl. 25, figs. 4-6; pl. 26, fig. 5; pl. 31, figs. 2, 3.
Alievium sp. aff. A. superbum (Squinabol, 1914). -- Pessagno, 1972: 302, pl. 24, figs. 5-6; pl. 25, fig. 1.
Amphipyndax enesseffi Foreman, 1966: 356, text-figs. 7-11.
Amphipyndax plousios Foreman, 1968: 79, pl. 8, fig. 11.
Amphipyndax stocki (Campbell and Clark, 1944). -- Foreman, 1968: 78, pl. 8, figs. 12a-c.
Amphipyndax tylotus Foreman, 1977: 315, pl. 1, fig. 1.
Archaeospongoprunum imlayi Pessagno (in press). -- Foreman, 1978: 745, pl. 1, fig. 12.
Artostrobium urna Foreman, 1971: pl. 4, figs. 1-2.
Cenosphaera minuta Pantanelli in Rüst 1888. -- Dumitrica, 1973: 832, pl. 18, fig. 1.
Cinclopyramis sanjoaquinensis (Campbell and Clark, 1944): 22, pl. 7, fig. 2.
Clathropyrgus bumastus Riedel and Sanfilippo, 1974: 775, pl. 12, figs. 6-8.
Clathropyrgus titthium Riedel and Sanfilippo, 1974: 775, pl. 3, fig. 12; pl. 12, figs. 9-12.
Conosphaera fossilis Parona 1890. -- Petrushevskaya and Kozlova, 1972: 521, pl. 1, fig. 8.
Cromyodruppa concentrica Lipman, 1952: 29, pl. 1, figs. 8-9.
Crucella sp. aff. C. cachensis Pessagno, 1971: 53, pl. 9, figs. 1-3.
Cryptamphorella sp. cf. C. conara (Foreman, 1968): 35, pl. 4, figs. 11a-b.
Cryptamphorella sphaerica (White, 1928). -- Dumitrica, 1970: 82, pl. 12, fig. 73.
Dactyliosphaera sp. aff. Lithocyclia justa Rüst, 1888. -- Petrushevskaya and Kozlova, 1972: 528, pl. 3, fig. 16.
Dictyomitra boesii Parona, 1890: 170, pl. 6, fig. 9.
Dictyomitra crassispina (Squinabol, 1903). -- Moore, 1973: 829, pl. 9, fig. 6.
Dictyomitra duodecimcostata (Squinabol, 1903). -- Foreman, 1975: 614, pl. 1G, fig. 5; pl. 7, fig. 10.
Dictyomitra duodecimcostata torquata (Foreman, 1971): 1676, pl. 3, fig. 4.
Dictyomitra koslovae Foreman, 1975: 614, pl. 7, fig. 4.
Dictyomitra lamellicostata Foreman, 1968: 65, pl. 7, figs. 8a-b.
Dictyomitra macrocephala Squinabol, 1904: 230, pl. 9, fig. 10.
Dictyomitra multicostata Zittel, 1876. -- Foreman, 1971: 1692, pl. 5, fig. 16.
Dictyomitra pseudomacrocephala (Squinabol, 1903). -- Foreman, 1975: 614, pl. 7, fig. 10.
Dictyomitra sp. aff. D. regina (Campbell and Clark, 1944). -- Foreman, 1968: 68, pl. 8, figs. 5a-c.
Dictyomitra somphedia Foreman, 1973b: 264, pl. 14, fig. 18.
Dictyomitra veneta (Squinabol, 1903). -- Foreman, 1975: 614, pl. 1G, fig. 4.
Emiluvia chica Foreman, 1973b: 262, pl. 8, fig. 7.
Emiluvia hopsoni Pessagno (in Press). -- Foreman, 1978: 744, pl. 1, fig. 3.
Emiluvia pessagnoi Foreman, 1973b: 262, pl. 8, fig. 6.
Eucyrtidium ptyctum Riedel and Sanfilippo, 1974: 778, pl. 5, fig. 7; pl. 12, figs. 14-15.
Eucyrtis bulbosus Renz, 1974: pl. 7, figs. 26-29; pl. 12, figs. 15a-b.
Eucyrtis tenuis (Rüst, 1885). -- Foreman, 1975: 615, pl. 21, figs. 7-9.
Eusyringium sp. aff. Theocampe subtilis (Squinabol, 1904). -- Petrushevskaya and Kozlova, 1972: 549, pl. 1, fig. 1.
Gongylothorax sp. aff. G. verbeeki (Tan Sin Hok, 1927). -- Foreman, 1973a: 429, pl. 13, fig. 4.
Hagiastrin sp. cf. Staurolonchidium tuberosum Rüst, 1898. -- Riedel

and Sanfilippo, 1974: 779, pl. 14, figs. 5-8.
Hemicryptocapsa sp. cf. H. nodosa (Tan Sin Hok, 1927). -- Dumitrica, 1973: 841, pl. 1, fig. 6; pl. 2, figs. 1-2.
Holocryptocanium barbui Dumitrica, 1970: 76, pl. 17, figs. 105-108.
Holocryptocanium tuberculatum Dumitrica, 1970: 75, pl. 16, figs. 102, 103a-c.
Holocryptocapsa sp. cf. H. cryptodon (Dumitrica, 1970). -- Dumitrica, 1973: 841, pl. 2, figs. 3-6.
Holocryptocapsa sp. aff. H. hindei Tan Sin Hok, 1927: 53, pl. 10, fig. 75. -- Dumitrica, 1970: 74, pl. 15, figs. 100a-d.
Lithocampe elegantissima Cita, 1964: 148, pl. 12, figs. 2-3.
Lithomelissa heros Campbell and Clark, 1944. -- Foreman, 1968: 25, pl. 3, figs. 5a, b.
Lithomelissa ? hoplites Foreman, 1968: 26, pl. 3, figs. 2a-c.
Lithomelissa ? petilla Foreman, 1975: 616, pl. 1G, figs. 2-3; pl. 6, fig. 3.
Lithostrobus punctulatus Pessagno, 1963: 210, pl. 1, fig. 1.
Lophophaena ? polycyrtis Foreman, 1968: 23, pl. 3, figs. 3a-c.
Mirifusus mediodilatatus Rüst, 1885. -- Foreman, 1975: 616, pl. 2K, fig. 2; pl. 6, fig. 17.
Myllocercion minima Petrushevskaya and Kozlova, 1972: 535, pl. 2, figs. 19, 20.
Paronaella ? diamphidia Foreman, 1973b: 262, pl. 8, figs. 3-4.
Patellula planoconvexa (Pessagno, 1963): 199, pl. 3, figs. 4-6.
Patellula verteroensis (Pessagno, 1963): 199, pl. 3, figs. 1-3; pl. 6, figs. 2, 3; pl. 7, figs. 3-6.
Podobursa pantanellii (Parona, 1890). -- Riedel and Sanfilippo, 1974: 780, pl. 8, fig. 5; pl. 13, fig. 6.
Podobursa triacantha (Fischli, 1916). -- Foreman, 1973b: 266, pl. 13, figs. 1-7.
Podocapsa amphitreptera Foreman, 1973b: 267, pl. 13, fig. 11.
Pseudoaulophacus lenticulatus (White, 1928). -- Pessagno, 1963: 202, pl. 2, figs. 8-9.
Pseudoaulophacus pargueraensis Pessagno, 1963: 204, pl. 2, figs. 4-7; pl. 6, figs. 4-5.
Rhopalosyringium colpodes Foreman, 1968: 57, pl. 6, fig. 6.
Rhopalosyringium sparnon Foreman, 1968: 56, pl. 6, fig. 5.
Sethocapsa cetia Foreman, 1973b: 267, pl. 12, fig. 1; pl. 16, fig. 19.
Sethocapsa trachyostraca Foreman, 1973b: 268, pl. 9, figs. 5-7.
Sethocapsa sp. cf. Theocapsa uterculus Parona, 1890. -- Foreman, 1975: 617, pl. 2J, figs. 21, 22.
Sphaerostylus lanceolus gr. (Parona, 1890). -- Foreman, 1975: 609, pl. 2E, figs. 3-6.
Spongosaturnalis hueyi (Pessagno, in press). -- Foreman, 1975: 641, pl. 1A, fig. 6; pl. 4, fig. 10.
Spongosaturnalis ? preclarus Foreman, 1975: 611, pl. 1A, figs. 4-5; pl. 4, fig. 8.
Stichomitra asymbatos Foreman, 1968: 73, pl. 8, figs. 10a-c.
Stichomitra asymmetra Foreman, 1978: 748, pl. 5, fig. 2.
Stichocapsa sp. aff. Lithostrobus elegans Squinabol, 1903. -- Petrushevskaya and Kozlova, 1972: 546, pl. 2, fig. 7.
Stichopilidium teslaense Campbell and Clark, 1944. -- Foreman, 1968: pl. 8, fig. 13.
Syringocapsa rotunda (Hinde, 1900). -- Foreman, 1978: 749, pl. 2, fig. 2.
Theocampe altamontensis (Campbell and Clark, 1944). -- Foreman, 1968: 53, pl. 6, figs. 14a, b.
Theocampe apicata Foreman, 1971: 1679, pl. 4, fig. 6.
Theocampe ascalia Foreman, 1971: 1678, pl. 4, fig. 4.
Theocampe bassilis Foreman, 1968: 50, pl. 6, fig. 10.
Theocampe daseia Foreman, 1968: 48, pl. 6, figs. 9a, b.
Theocampe lispa Foreman, 1968, pl. 6, fig. 11.
Theocampe salillum Foreman, 1971: 1678, pl. 4, fig. 5.
Theocampe comys gr. Foreman, 1968: 29, pl. 4, figs. 2a-c.
Theocampe teren Foreman, 1968: 32, pl. 4, fig. 4.
Triactoma blakei (Pessagno, in press). -- Foreman, 1978: 743, pl. 1, fig. 15.
Williriedelum ? gilkeyi Dumitrica, 1973: 841, pl. 3, figs. 4, 6; pl. 4, figs. 1-2.

REFERENCES

Aliev KS (1965) Radiolarians of the Lower Cretaceous deposits of northeastern Azerbaidzhan and their stratigraphic significance. Akad Nauk Azerbaidzhan SSR Inst Geol, pp 1-123

Arthur MA, Natland JH (1979) Carbonaceous sediments in the North and South Atlantic: The role of salinity in stable stratification of Early Cretaceous basins. In: Talwani M, Hay W, Ryan WBF (eds) Deep drilling results in the Atlantic Ocean: Continental margins and paleoenvironment. Am Geophys Union, Washington, pp 375-401

Campbell AS, Clark BL (1944) Radiolaria from Upper Cretaceous of Middle California. Spec Pap Geol Soc Am 57: 1-61

Cita MBS (1964) Ricerche micropaleontologiche e stratigrafiche sui sedimenti pelagici del Giurassico superiore e del Cretaceo inferiore nella catena del Monte Baldo. Mem Riv Ital paleontol Stratigr 10: 1-182

Dumitrica P (1970) Cryptocephali and cryptothoracic Nassellaria in some Mesozoic deposits of Romania. Rev Roum Geol Geophys Geogr Ser Geol 14: 45-124

Dumitrica P (1973) Cretaceous and Quaternary Radiolaria in deep sea sediments from the northwest Atlantic Ocean and Mediterranean Sea. Init Rep DSDP 13, 2: 829-901

Dumitrica P (1975) Cenomanian Radiolaria at Podul Dimbovitei (Excursion B). In: Micropaleontological guide to the Mesozoic and Tertiary of the Romanian Carpathians. 14th Eur Micropaleontol Colloq, pp 87-89

Fischli H (1916) Beitrag zur Kenntnis der fossilen Radiolarien in der Riginagelfluh. Mitt Naturwiss Ges Winterthur 11: 44-47

Foreman HP (1966) Two Cretaceous radiolarian genera. Micropaleontology 12: 355-359

Foreman HP (1968) Upper Maestrichtian Radiolaria of California. Palaeontol Assoc Spec Pap Palaeontol 3: 1-82

Foreman HP (1971) Cretaceous Radiolaria, Leg 7, DSDP. Init Rep Deep Sea Drilling Project 7, 2: 1673-1693

Foreman HP (1973a) Radiolaria of Leg 10 with systematics and ranges for the families Amphipyndacidae, Artostrobiidae, and Theoperidae. Init Rep Deep Sea Drilling Project 10: 407-474

Foreman HP (1973b) Radiolaria from DSDP Leg 20. Init Rep Deep Sea Drilling Project 20: 249-305

Foreman HP (1975) Radiolaria from the North Pacific DSDP Leg 32. Init Rep Deep Sea Drilling Project 32: 579-676

Foreman HP (1977) Mesozoic Radiolaria from the Atlantic Basin and its borderlands. Stratigr Micropalaeontol Atlantic Basin and Borderlands, pp 305-320

Foreman HP (1978) Mesozoic Radiolaria in the Atlantic Ocean off the Northwest Coast of Africa DSDP Leg 41. Init Rep Deep Sea Drilling Project 41: 739-761

Hinde GJ (1900) Description of fossil Radiolaria from the rocks of Central Borneo. In: Molengraaff GAF (ed) Borneo-Expedite, Geologische Verkenningstochten in Central-Borneo (1893-1894), Appendix 1, pp 1-51

Johnson TC (1976) Controls on the preservation of biogenic opal in sediments of the eastern tropical Pacific. Science 192: 887-890

Kozlova GE (1972) (See Petrushevskaya and Kozlova 1972)

Lipman RKh (1952) Materialy morphographicheskomu izuchenyu radiolyarii verkhnemelovykh otlozhenii russkoi platformy. Paleontol Stratigr Trudy VSEGEI 1952: 24-51.

Moore TC jr (1973) Radiolaria from Leg 17 of the DSDP. Init Rep Deep Sea Drilling Project 17: 797-869

Murray J (1876) Preliminary reports to Professor Wyville Thomson, F.R.S., Director of the civilian staff, on work done on board the 'Challenger'. Proc R Soc London 24: 471-544

Parona DF (1890) Radiolarie nei noduli selciosi del calcare giurese di Cittiglio presso Laveno. Boll Soc Geol Ital 9, 1: 1-46

Pessagno EA jr (1963) Upper Cretaceous Radiolaria from Puerto Rico. Micropaleontology 9, 2: 197-214

Pessagno EA jr (1971) Jurassic and Cretaceous Hagiastridae from the Blake-Bahama Basin (Site 5A, JOIDES Leg 1) and the Great Valley Sequence, California Coast Ranges. Bull Am Paleontol 60, 264: 5-81

Pessagno EA jr (1972) Cretaceous Radiolaria. 1. The Phaseliformidae, new family, and other Spongodiscacea from the Upper Cretaceous portion of the Great Valley Sequence. 2. Pseudoaulophacidae Riedel from the Cretaceous of California and the Blake-Bahama Basin (DSDP Leg 1). Bull Am Paleontolol 61, 270: 269-328

Pessagno EA jr (1976) Radiolarian zonation and stratigraphy of the Upper Cretaceous portion of the Great Valley Sequence, California Coast Ranges. Micropaleontol Spec Pap 2: 1-95

Pessagno EA jr (1977) Lower Cretaceous radiolarian biostratigraphy of the Great Valley Sequence and Franciscan Complex, California Coast Ranges. Cushman Found Foraminiferal Res Spec Publ 15: 1-87

Petrushevskaya MG, Kozlova GE (1972) Radiolaria Leg 14 DSDP. Init Rep Deep Sea Drilling Project 14: 495-648

Renz GW (1974) Radiolaria from Leg 27 of the DSDP. Init Rep Deep Sea Drilling Project 27: 769-841

Riedel WR, Sanfilippo A (1974) Radiolaria from the southern Indian Ocean, DSDP Leg 26. Init Rep Deep Sea Drilling Project 26: 771-813

Rüst D (1885) Beiträge zur Kenntnis der fossilen Radiolarien aus Gesteinen des Jura. Palaeontographica 31: 273-321

Rüst D (1888) Beiträge zur Kenntnis der fossilen Radiolarien aus Gesteinen der Kreide. Palaeontographica 34: 181-213

Rüst D (1898) Neue Beiträge zur Kenntnis der fossilen Radiolarien aus Gesteinen des Jura und der Kreide. Palaeontographica 45: 1-67

Schaaf A (1980) Late Early Cretaceous Radiolaria from Leg 62 of the DSDP. Init Rep Deep Sea Drilling Project 62 (manuscript)

Squinabol S (1903) Le Radiolarie dei soduli selciosi nella Scaglia degli Euganei. 1 Riv Ital Paleontol 9: 105-144

Squinabol S (1904) Radiolarie cretacee degli Euganei. Atti Mem R Accad Sci Lett Arti Padova 20: 171-244

Squinabol S (1914) Contributo alla conoscenza dei Radiolarii fossili del Veneta. Appendice: Di un genere di Radiolari caratteristico del Secondario. Mem Ist R Univ Padova 2: 249-306

Tan Sin Hok (1927) Over de samenstelling en het onstaan van krijten mergelgesteenten van de Mulukken. Jaab Mijnw Nederlandsch-Indie 1926 Verh 3: 5-165

White MP (1928) Some index Foraminifera of the Tampico Embayment area of Mexico. J Paleontol 2, 4: 280-317

Wiedmann J, Butt A, Einsele G (1978) Vergleich von marokkanischen Kreide-Küstenaufschlüssen und Tiefseebohrungen (DSDP): Stratigraphie, Paläoenvironment und Subsidenz an einem passiven Kontinentalrand. Geol Rundsch 67: 454-508

Yao A (1972) Radiolarian fauna from the Mino Belt in the northern part of the Inuyama Area, Central Japan. J Geosci 15: 21-65

Zittel KA (1876) Über einige fossile Radiolarien aus der norddeutschen Kreide. Z deutsch geol Ges 28: 75-86

15 Cretaceous Stratigraphy, Environment, and Subsidence History at the Moroccan Continental Margin

Jost Wiedmann, Arif Butt, and Gerhard Einsele[1]

Abstract

In contrast to other regions around the North Atlantic, good exposures in the Moroccan coastal basins offer an excellent opportunity to study the Mesozoic development of a passive continental margin including the relationship between oceanic and coastal sediments and datum levels of the pelagic fossils. From south to north, the Cretaceous sediments of the coastal basins of Tarfaya, Agadir, Essaouira, and at the margin of the Meseta are described and compared regarding macro- and microfauna, sedimentology, and paleoenvironment. For the mainly marine 2500 m or 1700 m thick Cretaceous sequences of Agadir and Essaouira, respectively, we propose a correlation of the ammonite and foraminiferal zones. Probably both sections formed in one basin (the "Atlas Basin", see front cover), but certain facies differences were caused by different water depths since Middle Cretaceous times.

Most of the Early Cretaceous sediments of the Tarfaya region in the south and of the Meseta in the north are of continental origin. Late Cretaceous sediments of these regions reveal, however, principal differences, especially in view of the macrofauna.

On the Meseta, the macrofauna is typical of the Mediterranean faunal province, whereas the Tarfaya fauna is characterized by north-"Boreal" elements. This fauna in combination with a special black shale facies is explained by the influence of upwelling in connection with the initiation of an oceanic deep-sea circulation, which can be traced into the Agadir section. Generally, during the Cretaceous period, we observe a tendency toward decreasing sedimentation rates and a relative increase of the percentage of pelagic components, such as planktonic foraminifers. This indicates a sharp decrease of the terrigeneous influx and a landward transgression of the oceanic water mass.

If the global eustatic curve for the Cretaceous oceans is compared with the evolution of paleodepth of the Moroccan coastal basins, one can distinguish between local phenomena and global events (Turonian and Campanian transgressions). After comparing the Cretaceous sections of DSDP Sites 370/416 and 137 with the onshore Agadir sequence, we infer the development of the marginal North Atlantic as influenced by an increase of the paleodepth and CCD variations.

The thick and rather complete shallow-marine Cretaceous sequences of the coastal basins permit to infer the interrelationship between the rates of sedimentation and subsidence in some detail. To find a "true" subsidence curve, the influence of global sea-level changes has to be ruled out. In the Tarfaya and Agadir region, the coastal basins became deeper only during Turonian/Coniacian times.

1. Introduction

On continuing the Deep Sea Drilling Project - International Phase of

1 Institut für Geologie und Paläontologie, Universität Tübingen, Sigwartstraße 10, 7400 Tübingen, FRG

Ocean Drilling - it proved to be increasingly desirable to include also coastal basins into the passive margin investigations. The purpose is two-fold: (1) to arrive at a stratigraphical correlation of datum levels in offshore and onshore sections, and (2), more important, to provide a set of regional data for the reconstruction of the evolution of passive margins and their paleoenvironment. Finally, the onshore data can supplement and calibrate the geophysical pre-site data and facilitate the choice of new drilling sites.

A look at the circum-North Atlantic Mesozoic basins (Fig.1) shows that such investigations are first of all possible along the eastern side of the North Atlantic and, in particular, along its south-eastern margin. Mesozoic depositional basins encroach onto the inner shelf and the continent of Morocco and the Iberian Peninsula, whereas comparable basins of the western North Atlantic (Georges, Scotian, E-Newfoundland and Flemish Basin) lie offshore below the present shelf and continental slope and do not present any coastal exposures. The Moroccan area was chosen not only because of its excellent outcrops, but also because it is an area of special German activities in marine geology and DSDP-related investigations (Berger and von Rad, 1972; Diester-Haass et al., 1974; Hinz et al., 1974; Lancelot, Seibold et al., 1978; von Rad, Ryan et al., 1979). First onshore investigations had already been effected by a small team from Tübingen (Einsele and Wiedmann, 1975; Wiedmann, 1976; Wiedmann et al., 1978a, 1978b).

In 1974 a team from Bonn University also started investigations with similar aims in the Cretaceous of the Moroccan "Atlas Gulf" (Wurster and Stets, Stets and Wurster, Behrens and Siehl, Stamm and Thein, all this Vol.). The investigations of the "Tübingen group" are focused upon:

(1) the stratigraphy and lithofacies of coastal standard sections in the Tarfaya, Agadir and Essaouira Basins, and on the Moroccan Meseta (see Fig.2),
(2) the onshore-offshore correlation between coastal basin sections and DSDP sites close to the margin, and
(3) an investigation of the relations between subsidence, sedimentation, and paleo-bathymetry at various locations of a passive margin.

This paper is a shortened and somewhat updated English version of a

Fig. 1
Circum-North Atlantic Mesozoic coastal basins, plotted into the presumed Middle Cretaceous paleogeography, with the actual 200 and 3000 m isobaths. (Modified after Jansa and Wade, 1975 and Berggren and Hollister, 1974)

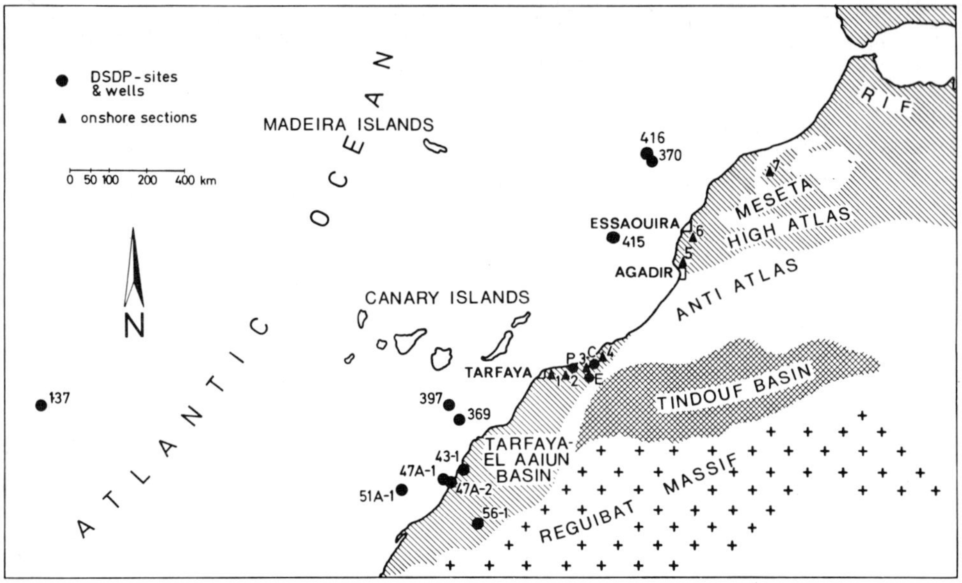

Fig. 2. Location map of onshore sections and wells, shelf wells, and DSDP sites
P Puerto Cansado; E El Amra; C Oued Chebeika; 1 Sebkha Tah; 2 Sebkha Houiselgua; 3 Oued Ouaâr; 4 Oued Chebeika; 5 Agadir; 6 Smimou; 7 Settat

previous paper (Wiedmann et al., 1978a). More detailed sedimentological and paleoenvironmental aspects of the "mid-Cretaceous" are covered by two papers by Einsele (this Vol.) and Einsele and Wiedmann (this Vol.). Ranke et al. (this Vol.) attempt a synthesis of the structure, geology, paleoenvironment and subsidence history of the Tarfaya Basin and its extension onto the West Saharan marginal basin.

2. Discussion of the Standard Sections

2.1. Tarfaya - El Aaiun Basin

The evolution of this basin is studied in detail by Ranke et al., this Vol. and by Jansa and Wiedmann, this Vol..

At the well Puerto Cansado 1 the Lower Cretaceous sediments of continental origin are 1300 m thick and finely clastic, being overlain by a 600 m thick series of light-coloured partly greenish marine clays and marls which pass into bituminous marls in the uppermost part of the sequence. This latter series, yielding *Ticinella roberti* and *Rotalipora appenninica*, represents the time span from Middle Albian to Early Cenomanian and, due to its considerably greater thickness and the absence of thick shell beds (see below), it can be distinguished very clearly from the corresponding onshore section of the northeastern part of the basin.

There is an exception to this general Middle Albian transgression in the farthest north of the Tarfaya Basin, near the Draa River mouth. Here, marine intercalations of the uppermost Aptian (Clansayesian) with *Hypacanthoplites (Nolaniceras) nolani* (Seunes) occur within the upper part of the continental sediments.

In the onshore sections of the northeastern basin the marine Cretaceous transgression typically starts in the late Middle Albian (*Dipoloceras cristatum* Zone). The Upper Albian and Lower Cenomanian are here reduced in thickness (80 m). For this period intercalations of greenish clays and marls, siltstones, dolomitic limestones, and shell beds are characteristic; the latter locally reach great thickness and are built up by trigoniids and oyster shells. Figure 3 roughly reproduces the Cretaceous sequence of the Tarfaya Basin; a more detailed description is given by

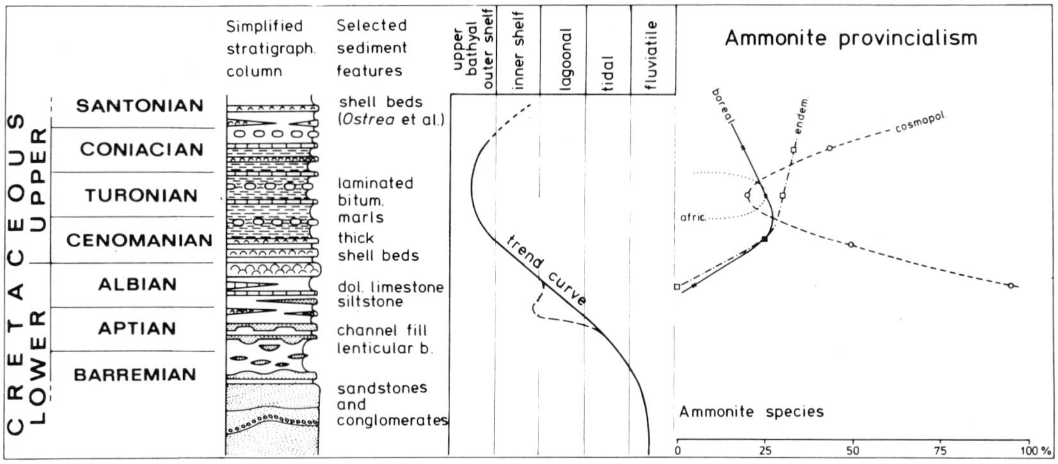

Fig. 3. The Cretaceous of the Tarfaya Basin. Idealized lithology, bathymetry, and ammonite provincialism. (After Wiedmann, 1976). Not to scale

Einsele and Wiedmann (this Vol.). The most important data are briefly summarized as follows:

(1) The marine sedimentation is continuous from the Middle Albian to the Lower Santonian what is shown by many well-documented ammonite zones and coeval foraminiferal and inoceramid faunas. So far, only the zones from the uppermost Albian to the lowermost Cenomanian and the topmost Coniacian are poorly supported by faunal evidence. The Lower Santonian is unconformably overlain by the Campanian and this, in turn, by locally preserved Miocene or by Mogrebian (Pliocene-Pleistocene) beds respectively. The thickness of the marine Cretaceous varies between 820 m in the west and 300 m in the east.

(2) The Upper Cenomanian, Turonian and Coniacian are represented by laminated, bituminous marls, limestones and siliceous limestones with intercalations of siliceous concretions up-section, whereas the Santonian and Campanian part of the sequence is again composed of an interstratification of oyster shell beds with silty marls and clays.

(3) The mainly Turonian "black shale" period of the Tarfaya Basin is described by Einsele and Wiedmann (1975 and this Vol.). The deve-

The development of the Cretaceous depositional environment may be briefly summarized as follows:

During the earliest Cretaceous the coastal basin was kept filled to above sea level by fluvial deposits indicating an enormous terrigenous influx. In the Aptian and Early Albian, intertidal and lagoonal conditions were prevailing. From Middle Albian to Early Cenomanian littoral and inner shelf conditions were established, as documented by biofacies and sediments, possibly due to a global sea-level rise. Ongoing subsidence and decreasing terrigenous influx led to outer shelf conditions from Middle Cenomanian to Coniacian. In Santonian time, the retrograde process evolved which closed up with a pre-Campanian emersion phase. During the Campanian subsidence increased again, leading to inner shelf sediments. Because of pre-Miocene emersion and denudation, the following development cannot be studied. However, it is known from the southern part of the Aaiun-Tarfaya Basin (Western Sahara) that this second depositional period continued beyond the Cretaceous-Tertiary boundary, at which phosphatic deposits were formed (Querol, 1966).

In the northern Tarfaya-Aaiun Basin, after a long hiatus, Miocene Sediments onlap onto the Cretaceous sediments, whereas older Tertiary deposits are restricted to the southern Tarfaya-Aaiun Basin.

There is more and more support for the suggestion. (Einsele and Wiedmann, 1975) that the Turonian black shale facies was deposited under coastal upwelling conditions (Einsele and

Wiedmann, this Vol.). This interpretation is mainly based on faunistic evidence, the high percentage of northern, "Boreal" faunas (Fig.3), the absence of typical Mediterranean biotas, the extreme decrease of benthic forms in macro- and microfaunas and high faunal density and mass mortality.

2.2. Agadir Basin

Despite extensive regional monographs on the Mesozoic Sequence of the High Atlas Ranges (Roch, 1930; Ambroggi, 1963), there were hardly any data available which referred to the mid-Cretaceous evolution of this area. The few faunal data available for this period (e.g., the citation of the ammonite genera Hemitissotia and Plesiotissotia by Ambroggi, 1963: p.207) favor a connection between the High Atlas, the Moroccan Meseta and the Tethys and thus a clear separation from the Aaiun-Tarfaya Basin to the South. At the same time it was necessary to examine the question whether the coastal western part of the actual High Atlas was part of one single Cretaceous basin or whether it separated two distinct basins: a northern Essaouira Basin and a southern Agadir (or Haha) Basin (Choubert and Faure-Muret, 1962; Ambroggi, 1963; Duffaud et al., 1966; Behrens et al., 1978; Stets and Wurster, this Vol.).

For a better understanding of this problem, both areas are treated here separately.

2.2.1. Stratigraphical Review

The pre-Cretaceous development of the Agadir Basin is summarized by Wiedmann et al. (1978a). In contrast to the more uniformly developed Jurassic of Tarfaya, the Jurassic of the Atlas Basin can be subdivided into numerous sedimentary cycles. This may reveal a greater mobility of the Atlas Basin. This aspect continues into the Cretaceous. The Tithonian starts with a thick sequence of anhydrite and ends with well-bedded dolomites and limestones, just as thick as the anhydrites; Nerinea and algae are locally abundant. Duffaud et al. (1966), however, included the entire Berriasian in the Tithonian sequence, whereas according to Ambroggi (1963) the following cycle starts with the Berriasian. In reality, however, only the Middle and Upper Berriasian are to be included into the marly base of the Cretaceous, while the lower Berriasian is still developed in the limestone and dolomite facies of the underlying Tithonian (Fig.4). The final regression of the Jurassic coincides with moderate (late Cimmerian) movements, and, in consequence, a first rise of diapirs of the evaporitic Triassic. In the area studied, the contact between Jurassic and Cretaceous seems to be conformable; however, the final Jurassic regression is locally documented by a distinct hardground at the top of the Tithonian-Lower Berriasian limestones. Up to this time a rough similarity does exist to the development of the likewise continental Triassic and epicontinental Jurassic of Tarfaya, however, some differences are already visible. This is especially true of the evaporitic Triassic, which has its maximum thickness offshore. Salt motion possibly influenced the subsidence of the Atlas Basin; and the later structure of the High Atlas as well as that of the shelf and the slope were influenced. A similar evolution is known from the Vascocantabric coastal basin in Northern Spain (Lotze, 1955; Wiedmann et al., in press).

The standard section of the Agadir Cretaceous was taken along a coastal strip approximately 5 km wide between Agroud, Tarhazoute and Aourir, north of Agadir. The total thickness of the almost exclusively marine Cretaceous attains 2550 m. In Figure 4 the standard lithology of the Agadir section is reproduced in relation with the most important faunistic data. The occurences of macrofaunas and microfaunas are quantified, the plankton-benthos ratio is calculated, and a standard zonation of foraminifers, ammonites and inoceramids is introduced. The paleo-bathymetric curve is the result of all available paleontological and sedimentological (Fig.5) data. For details of the Cretaceous sequence and stratigraphy we refer to Wiedmann et al. (1978a).

There is no marker horizon at the Cretaceous-Tertiary boundary. The so-called "boundary sandstone" was found to be of a lower Upper Maastrichtian age. Since Bolivina incrassata (Upper Maastrichtian) was recognized in the lower part of the marls following up-section and yielding Globigerina triloculinoides in their upper portion, this boundary must be placed somewhere within these marls.

The limestone-marl sedimentation continues from the Paleocene right into the Eocene. Once again intercalated chert layers and phosphate-bearing sandstones are frequent. Toward the end of the Middle Eocene an overall regression occurs, associated with initial orogenic movements which resulted in the uplift of the Agadir basin. In the southermost part of the investigated area, Middle Eocene and older strata are unconformably overlain by Oligocene conglomerates and sandstones, which indicate a restriction of the marine influence to the coastal region of Agadir.

2.2.2. Paleoenvironment

The Jurassic deposits consist of alternating continental and inner shelf sediments. The greatest depositional depths were probably achieved during deposition of the cephalopod facies of the Callovian and Oxfordian. The thick-bedded shallow-water carbonates of the Upper Jurassic and Lower Berriasian are indicative of littoral conditions, especially in their upper parts, with their fauna consisting of Nerinea and oysters. The limestone-marl alternation of the Upper Berriasian, bearing a diversified macrofauna (cephalopods, bivalves, echinoderms, brachiopods, and many more) and an abundant micro-benthos (nodosariids, textulariids, lituolids), points to an increase in water depth which continues up into the Valanginian. A high faunal diversity is preserved as is to be seen in Fig.4. From the Valanginian onwords the microbenthos includes also rotaliids (Epistomina). The fauna as a whole suggests warm, oxygen-rich water of the outer to inner shelf; similar conditions were described by Bettenstaedt (1973) for the Cretaceous near Helgoland (Germany). Faunal diversity and cephalopod content decrease rapidly toward the Lower Hauterivian, foreshadowing the Middle Hauterivian regression, during which almost littoral conditions were achieved. However, already in Late Hauterivian time this trend was reversed and one more a high diversity was attained with the first appearance of planktonic foraminifers. In the Hauterivian a clear North Atlantic-NW-German component in the cephalopod fauna - in particular for the Crioceratites - is noticed which is in contrast to the more Mediterranean-cosmopolitan ammonite faunas in the rest of the Lower Cretaceous. In spite of these similarities in faunal composition the stratigraphic ranges of individual species are different in both areas. In the Barremian, water depth increased even more as reflected by a rapid increase of planktonic foraminifers. The appearance of an abundant Hedbergella fauna suggests for the first time an open oceanic water circulation in the Agadir Basin. The next regressive phase is found in the Lower Aptian (Fig.4). Sporadic intercalations of agglutinating foraminifers (Ammobaculites, Haplophragmoides) and few drifted-in ammonites reveal ingressions of short duration during deposition of the dominating littoral nearshore facies.

The following ingression during the Middle Aptian led to the deposition of a condensed horizon, which shows some reworking of faunas. According to Bergner (1979) there is, locally, a condensation of the entire Middle Aptian. From the Upper Aptian onwards, the sedimentary facies and the plankton/benthos ratio suggest a renewed increase in water depth which, in the Lower Albian, may be equivalent to a position on the middle shelf or thereabouts. The contents of planktonic foraminifers (Hedbergella washitensis, Ticinella bejaouaensis and other species) vary between 5% and 15%; benthonic foraminifers show a slightly higher diversity and contain nodosariid, textulariid, and rotaliid species (Epistomina colomi, Gyroidinoides gracillima, Gavelinella spp.). This faunal assemblage is indicative of a strong oceanic circulation. Although the absence of macrofossils - especially of ammonites - is striking, there is no support for the interpretation that the thick monotonous, olive-green and silty marls of the Albian and Cenomanian of the coastal basins were deposited under stagnant conditions as were the black shales of the deeper oceanic basins (Schlanger and Jenkyns, 1976). The few ammonites found are confined to the scarce limestone beds of this sequence. In the late Middle Albian and especially in the Upper Albian,

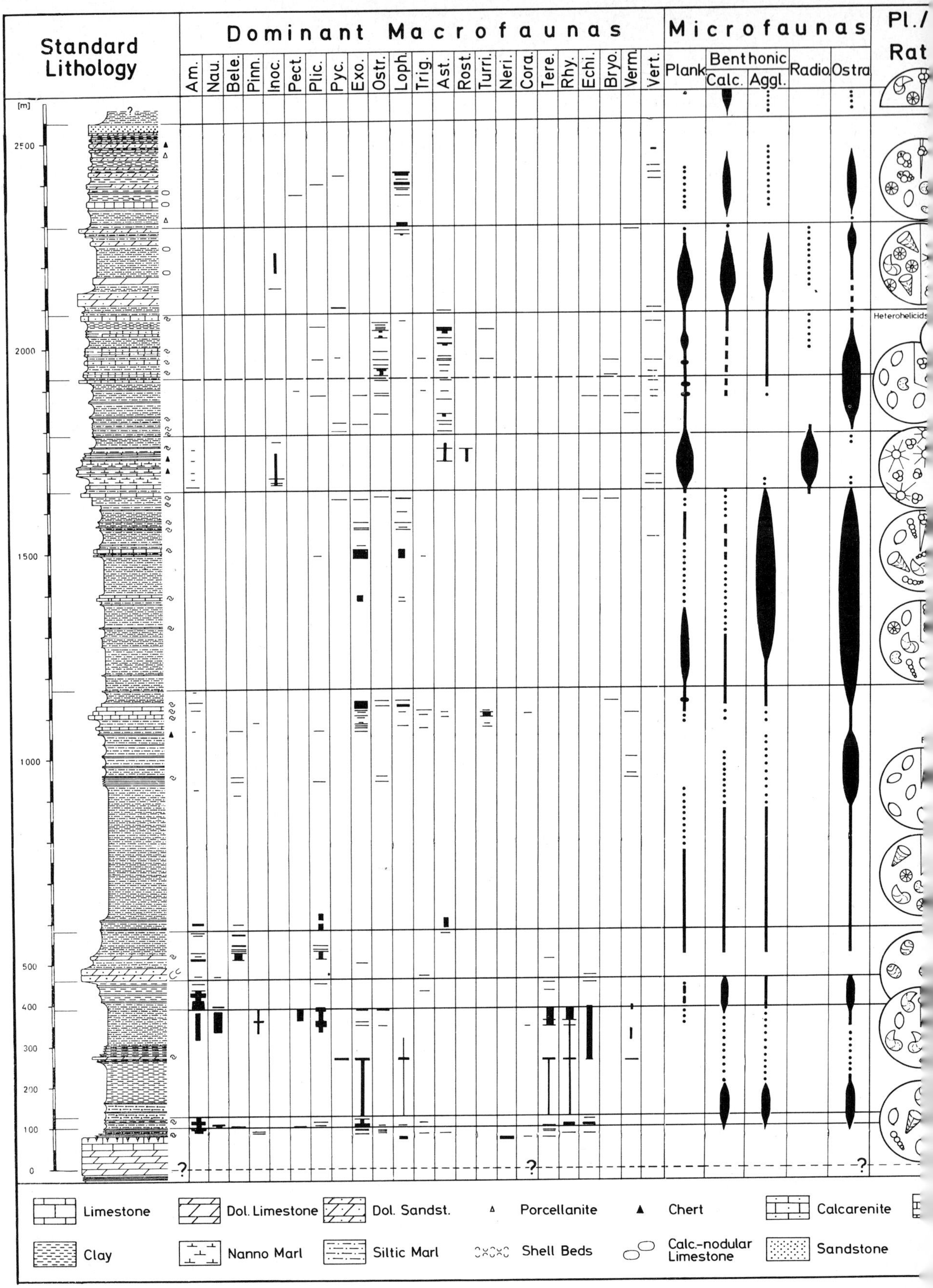

Fig. 4. Standard section of the Agadir Cretaceous: lithology, selected macrofau biostratigraphy

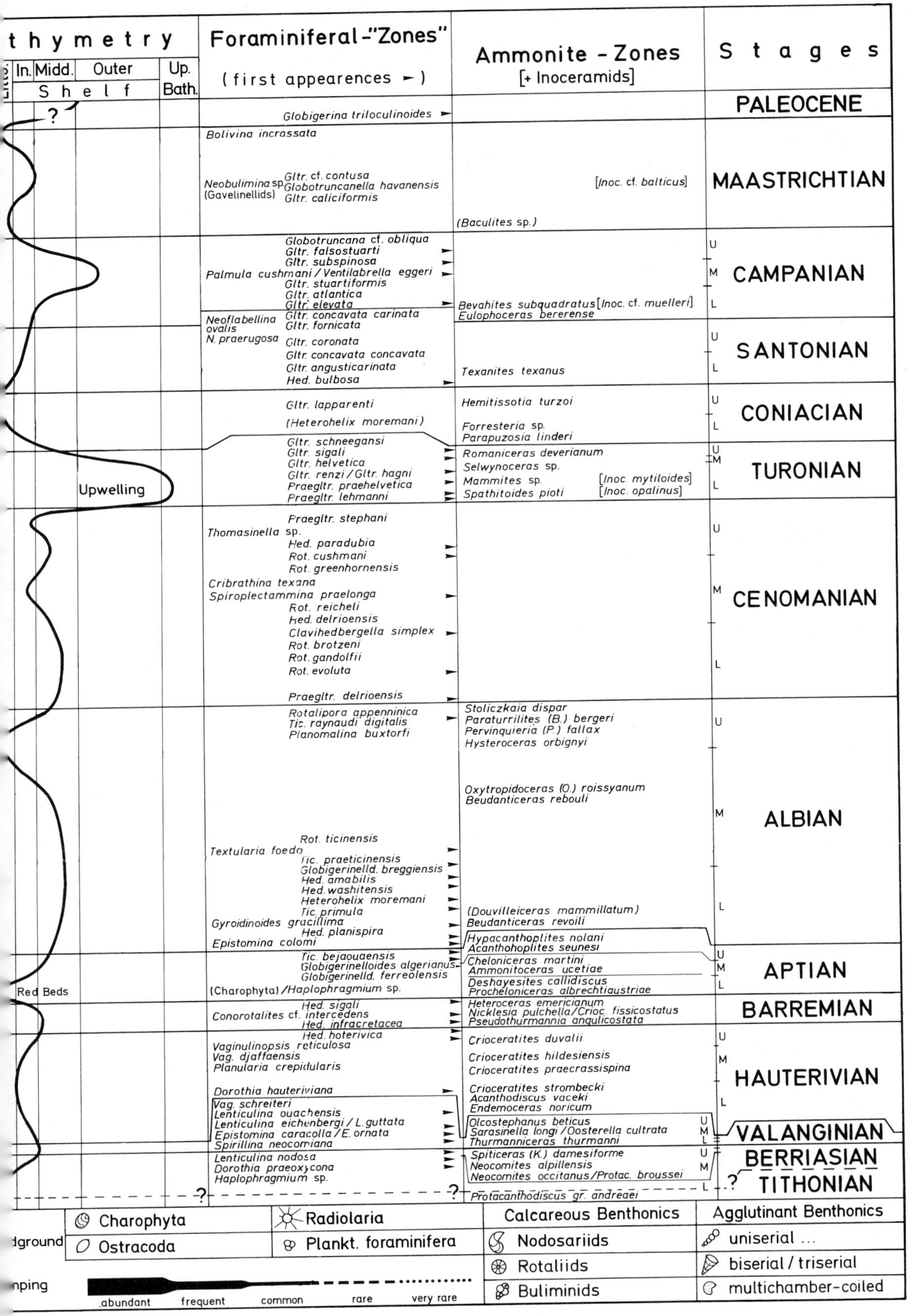

microfauna and -flora, plankton/benthos-ratio, bathymetry, index fossils, and

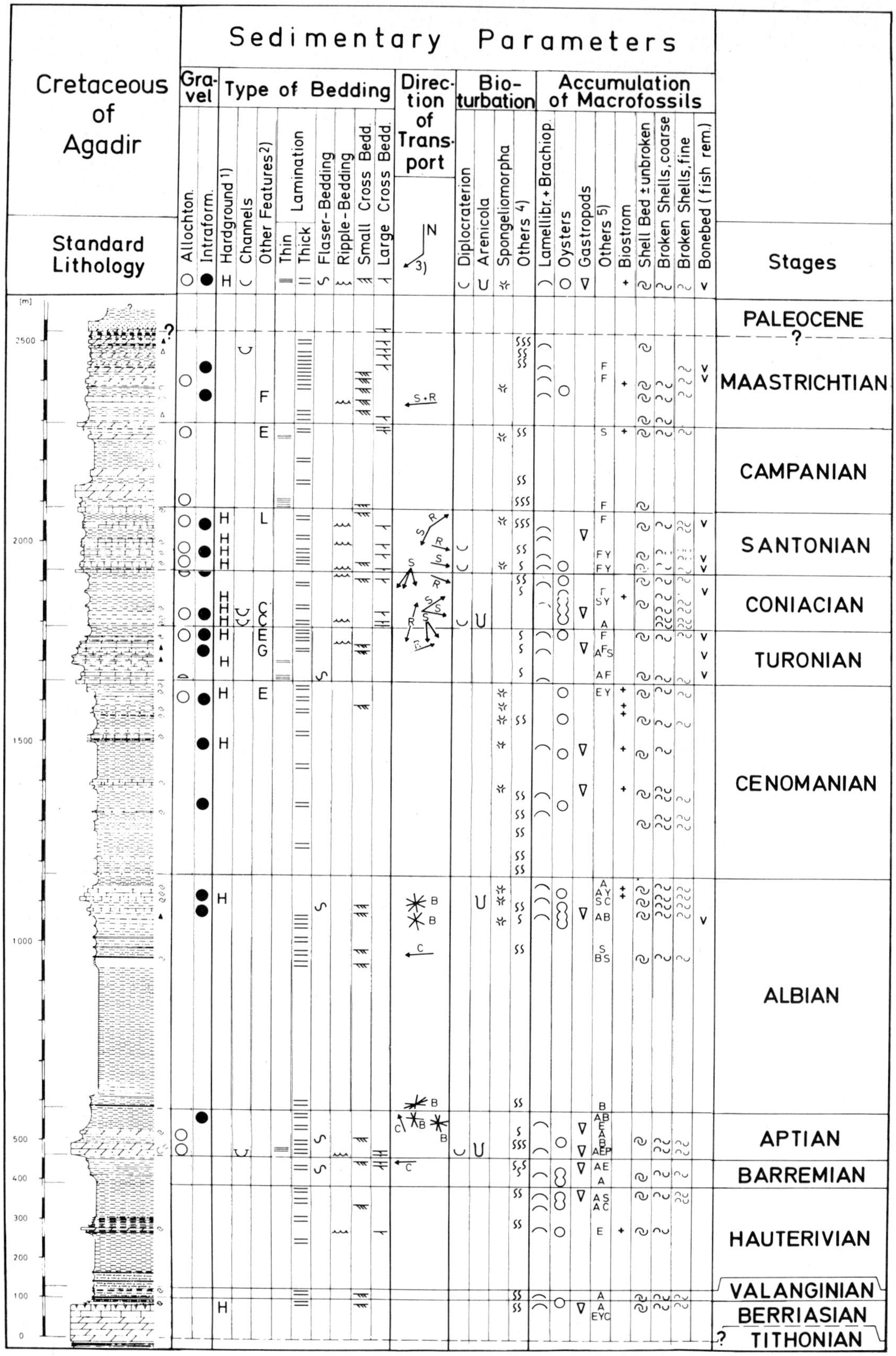
Cretaceous of Agadir
Standard Lithology
Sedimentary Parameters
Gravel
Allochton.
Intraform.
Type of Bedding
Hardground 1)
Channels
Other Features 2)
Lamination
Thin
Thick
Flaser-Bedding
Ripple-Bedding
Small Cross Bedd.
Large Cross Bedd.
Direction of Transport
N
3)
Bioturbation
Diplocraterion
Arenicola
Spongeliomorpha
Others 4)
Accumulation of Macrofossils
Lamellibr. + Brachiop.
Oysters
Gastropods
Others 5)
Biostrom
Shell Bed ± unbroken
Broken Shells, coarse
Broken Shells, fine
Bonebed (fish rem.)
Stages
[m]
2500
2000
1500
1000
500
400
300
200
100
0
PALEOCENE
MAASTRICHTIAN
CAMPANIAN
SANTONIAN
CONIACIAN
TURONIAN
CENOMANIAN
ALBIAN
APTIAN
BARREMIAN
HAUTERIVIAN
VALANGINIAN
BERRIASIAN
TITHONIAN

the coastal basin became shallow again. This is documented by thick oyster shell beds and a rich microbenthos, where ostracods are abundant. In the Lower Cenomanian, conditions of the middle shelf reappeared. Between the Lower and Upper Cenomanian the plankton/benthos ratio fluctuated from 20:80 up to 5:95. The benthonic species belong to nodosariids and gavellinids, whereas the planktonic ones belong to Rotalipora, Hedbergella, and Praeglobotruncana. While ostracods occur during the whole Cenomanian, we notice a drastic decrease in the content of calcareous foraminifers. They are completely replaced by agglutinating forms (Spiroplectammina, Cribratina, Thomasinella and others) and ostracods. These observations and the renewed occurrence of thick oyster shell beds point to a seaward shifting of the coastline in the course of which inner shelf conditions may have developed.

The Turonian is differently developed. Turbidites and small slump sheets are locally intercalated with thinly laminated bituminous marls alternating with well-bedded calcilutites, carbonate concretions and chert layers. The pelagic calcilutites as well as the bituminous marls, consist to a large extent of coccoliths, pelagic foraminifers, radiolarians, and fish remains. The character and the vast distribution of these sediments further inland suggest calm depositional conditions on a deeper outer shelf, or, according to the microfauna (with a plankton content of 100%), an upper slope or upper bathyal environment. The ammonite fauna shows a mixture of Mediterranean (Vascoceras) and Boreal species (Selwynoceras). For an environmental interpretation of these findings see Einsele and Wiedmann (this Vol.).

The Senonian is also characterized by a fluctuation in water depth. A regression with its maximum in the lowermost Santonian is announced by repeated intercalations of shell beds in the lower Coniacian which are in this case and in the Santonian built up by astartids and rostellariids. In addition to shell beds cross-bedded sandstones and hardgrounds characterize the depositional area which may have oscillated between sublittoral and inner shelf depth. This is also proved by a benthic microfauna consisting of ostracods and lituolids. Dwarfed heterohelicids point to a productive epipelagic environment. After multiple fluctuations during Santonian times there was once more, in the Middle Campanian, an increase in water depth leading to outer shelf conditions. The plankton/benthos ratio increases again, as also the number of globotruncanids. An increase in size, density, and diversity of benthics (Neoflabellina, Palmula, Gavelinella, Gaudryina and others) point to oxygen-rich and eutrophic waters at the sea floor.

With the Late Campanian the final Cretaceous regression starts, also in the Agadir Basin; it is only interrupted by a short-term climax of planktonics in the Middle Maastrichtian. Buliminids are dominant in the microfauna of the Maastrichtian; gavellinids and ostracods are less abundant. Haplophragmoides is rare, planktonic foraminifers are very rare and often corroded. Fish remains increase again. At the same time oysters become abundant again. Bioturbation increases also during this period. The buliminid associations of the Maastrichtian point to a probably independent basin within the margin. The end of this depositional period might be seen in cross-bedded deltaic sands of the Upper Maastrichtian, but it does not coincide with the Maastrichtian-Peleocene boundary. The overlying marls, until now allocated to the Paleocene contain Bolivina incrassata in their basal part and Globigerina triloculinoides above; this means that the Cretaceous-Tertiary boundary has to be placed

Fig. 5. Standard section of the Agadir Cretaceous: sedimentary and biostratinomic criteria
1 Frequent borings; 2 E erosional disconformity; C cellular dolomite; F fissures with red sediments; G graded sands or shell beds; L load casts; S slump structures; SB shell beds; 3 Direction of transport from: B long axis of belemnites; R current ripples; S cross bedding; C channels; I imbrication of gravel; 4 S rare; SS common; SSS abundant; 5 A ammonites; B belemnites; C corals; F fish teeth; P plant remains; S serpulids; Y bryozoa

within the lower portion of these marls. After the deposition of these marls the Agadir coastal basin emerged.

The most important sedimentological observations, as well as some biostratinomic data for the different Cretaceous stages of the Agadir area, are shown in Fig.5.

The persisting influence of the nearby continent (see Behrens and Siehl, this Vol.) can be deduced from the occurrence of allochthonous gravel in small quantities, but in many horizons. Some measurements of the paleocurrent direction from cross-bedding, ripples or sole marks show transport directions towards the west, northwest, or southwest (Aptian/ Albian, Maastrichtian). Most of the current measurements, however, revealed a random orientation which is characteristic for shallow-marine environments. Large and small-scale cross-bedding of sandy layers is present in nearly all formations, particularly in the Barremian and Aptian and in the Coniacian to the Maastrichtian. Thin laminated silt and sand layers with burrows and some other sole marks, alternate with silty shales (for instance in the Albian sequence), and may represent storm layers. Shallowing of the sea leading to intertidal conditions is documented by horizons with flaser bedding, burrowing organisms of the groups Diplocraterion and Arenicola, as well as by sand-filled channels in Aptian/ Albian and Maastrichtian sediments. On the other hand, the thinly and, parallel laminated bituminous marls of Turonian age which locally contain some thin graded beds, certainly have been deposited below wave base. However, such conditions were achieved only for a relatively short period. Finally, it should be noted that even the Cretaceous sequence of Agadir, which can be regarded as biostratigraphically complete, contains many hardgrounds, intraformational pebbles and erosional unconformities, particularly in its upper part.

These observations confirm and substantiate our assumptions for the bathymetric evolution of the Agadir Basin shown in Fig.4. Most of the time, the environmental conditions remained shallow-marine with some excursions into an intertidal regime or to depositional conditions somewhat below wave base.

2.3. Essaouira Basin

Southeast of Essaouira a near-shore cross-section, composed of three individual sections, has been investigated. The sequence from the Berriasian to Albian has been examined in the area between the northern flank of the Djebel Amsittene and the peak of the Djebel Amardma, east of Smimou. The Albian-Turonian series has been investigated along the Oued Taitoust, west of Ou Soulil and the Turonian-Maastrichtian between Dar Caid Neknafa and Zaouit ou Hassane.

2.3.1. Stratigraphical Review

According to the meager drilling data available and the paleogeographical summary given by Choubert et al. (1962), the Triassic-Jurassic history of the basin is largely parallel to that of the Agadir Basin; the formational thicknesses, however, are generally smaller. Basically, this applies also to the Cretaceous, where the total thickness can be estimated at 1700 m. The standard lithology, fossil record and bathymetry of the Cretaceous are compiled in Fig.6, analogous to Fig.5. For detailed information we refer, again, to Wiedmann et al. (1978a).

The evolution of the Tertiary corresponds to that in the Agadir Basin.

2.3.2. Paleoenvironment

While the Lower Berriasian is still to be ascribed to the Late Jurassic shallow-water sedimentation (with dolomicritic, dolosparitic and oolitic limestones), there is an increase in marly sediments, pelagic microfauna and micro-benthos corresponding with a first deepening of the coastal basin during the Late Berriasian. The maximum of this development was achieved in the Valanginian, and its microfauna suggests a deposition on the outer portion of the inner shelf. During the Early Hauterivian a rapid change took place from sedimentation in littoral environments (with coral bioherms) toward continental red bed series in the Middle Hauterivian. Calcareous foraminifers are completely absent. Few agglutinating forms of shallow and brackish water (Ammobaculites, Haplophragmoides), small-sized gastropods and bivalves indicate fluctuations between lagoonal and continental environments. During the Late Hauterivian and Barremian water depth increased again and rapidly passed from inner shelf, with calcareous benthos (nodosariids) and agglutinating forms (Dorothia), to an open middle shelf position in which the first planktonics evolved.

After the subsequent Early Aptian regression, a further phase of deeper environment can be observed during the Late Aptian and Early Albian. This is reflected by the deposition of silty and argillaceous marls in which occasionally limestones and dolomites are intercalated. The amount of planktonic foraminifers increased again and their relation to the calcareous benthics (nodosariids, gavellinids) points to living conditions of the middle shelf.

The establishment of a certain platform during the Late Albian resulted in the deposition of cross-bedded shell beds and dolomites intercalated with hardgrounds. Littoral shallow-water limestones containing miliolids, cuneolines, textularians and prealveolines characterize the Upper Albian-Cenomanian. Rotalipora and charophytes occur sporadically and this indicates changing oceanic and continental influxes.

According to the evidence of foraminiferal fauna, there was a deeper basin in Turonian times again, probably installing water depth of the middle shelf. The influence of upwelling, as observed in the Tarfaya and Agadir Basin,can no longer be detected here. Ammonites referring to the Early and Middle Turonian are unknown and, therefore, it is not possible to answer the question whether this area belongs to the Vascoceras Province of the Meseta or to the Selwynoceras Province of Tarfaya. The coquina-rich sediments of Coniacian/Santonian age suggest a turbulent and high-energetic environment of the littoral zone. Again, the shell beds are mainly composed of astartids. The marly sequence of this period bear a microfauna consisting of heterohelicids, lituolids, and ostracods. Accoording to Sliter (1972), an increase of heterohelicids points to an epipelagic fertility in coastal water masses (see above). Intercalation of shell beds decreases in the course of the Middle Campanian; the yellowish marls of this period exhibit an abundant microfauna, predominantly that of calcareous benthics (Lenticulina, Nodosaria, Palmula, Frondicularia, Neoflabellina) which shows both a high diversity and a distinct increase in size. This could make one think of an efficient aeration of the relatively warm bottom waters even in the Essaouira region. Also agglutinating forms (Gaudryina, Lituola) are common and globotruncanids dominate among the plantonics. Fauna and facies suggest a middle-shelf environment. In Maastrichtian time calcareous concretions (up to 1 m Ø) intercalate within the lower part of alternating limestones, dolomites, and marls, whereas brownish, more or less massive chert layers occur in the upper part. The microfauna of these marls is rich in buliminids and less abundant in gavellinids and ostracods; planktonics are completely missing. Locally, fish remains are highly concentrated. Here too, the predominance of buliminids suggests an isolated basin with a largely reduced interaction with the open ocean. The formation and accumulation of phosphates may have started in the upper part of this stage, but was intensified mainly during the Paleocene/Eocene period, during which sedimentation ceased for most of this area.

2.4. The Moroccan Meseta

Although a comparable standard section of this region does not yet exist, the development of the Cretaceous in the Meseta near Settat may be worth mentioning here for comparison with the mid-Cretaceous development in Tarfaya.

2.4.1. Stratigraphical Review

Descriptions of the largely reduced Mesozoic sequence in the central Meseta of Settat are given by Gigout (1951) and Michard (1976). According to these authors the basement, consolidated during the Hercynian orogeny, became re-integrated in the area of sedimentation since the Triassic. Sequences of argillaceous and sandy red beds are followed by a basal conglomerate, continental/fluviatile sandstones and reddish clays, into which

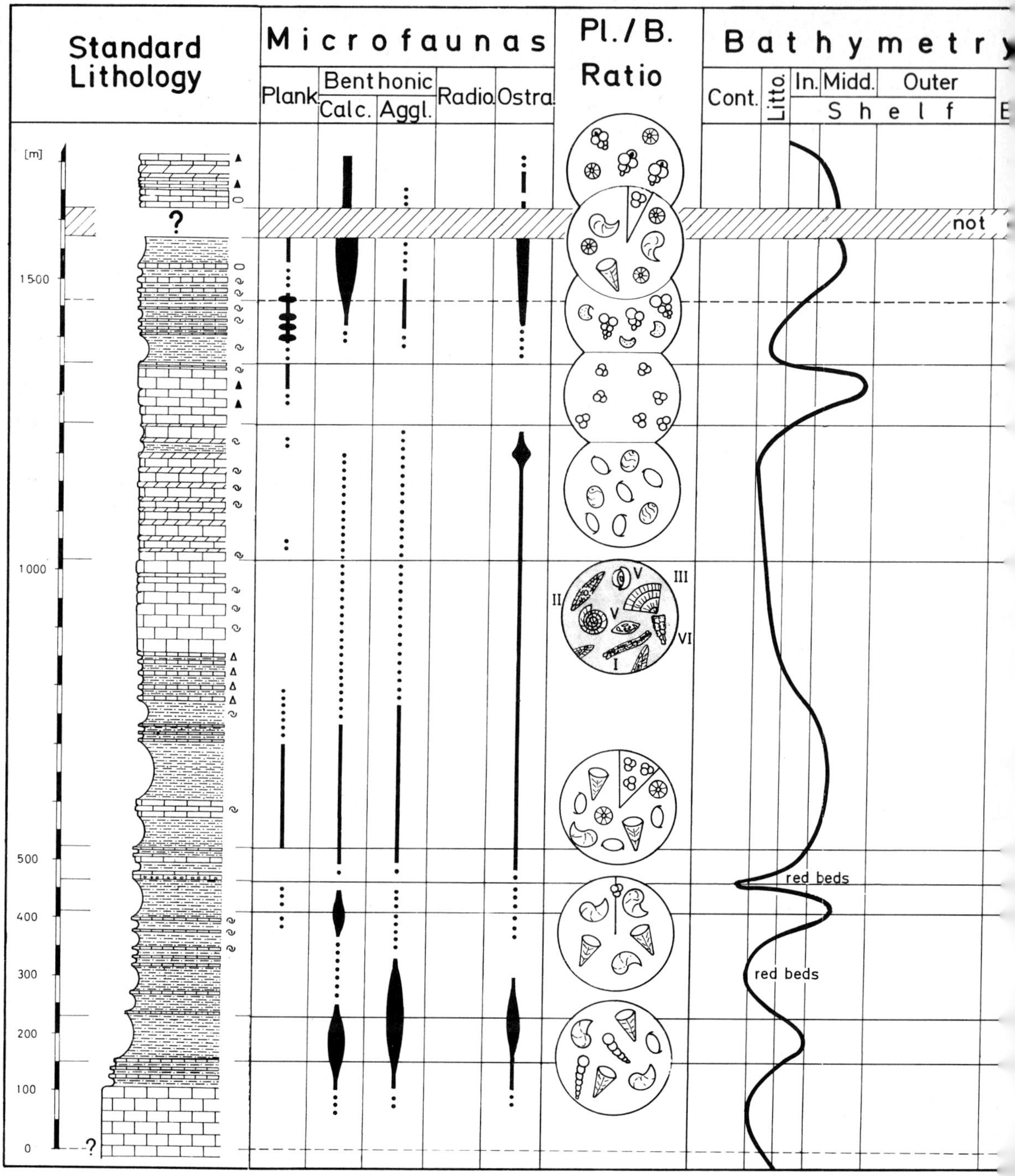

Fig. 6. The Essaouira Cretaceous standard section: Lithology, microfaunas, plankton/benthos-ratio, bathymetry, index-fossils, biostratigraphy. (Symbols as in Fig.5)

raminiferal-"Zones" (irst appearences ►)		Ammonite - Zones	Stages
ulimina sp. inellids		*Baculites* sp.	MAASTRICHTIAN
			?
ula *nani*	*Globotruncana fornicata* *Globotruncana cretacea* *Globotruncana coronata*		CAMPANIAN
abellina *ugosa*	*Gltr. concavata carinata* *Hedbergella bulbosa*		SANTONIAN CONIACIAN
	Globotruncana cf. *helvetica* ► *Heterohelix globulosa* *Heterohelix moremani*	*Romaniceras deverianum*	TURONIAN
ina sp. (I) *lveolina* sp. (II) ►			CENOMANIAN
lina sp. (III) *aria* sp. (IV) ds (V) *mina colomi*	*Heterohelix moremani* *Globigerinelld. breggiensis* *Hedbergella washitensis* ► *Ticinella bejaouaensis* *Hedbergella planispira* ►	*Pervinquieria (P.) fallax* *Oxytropidoceras (O.) roissyanum* *Hypengonoceras* sp. *Beudanticeras rebouli* *Leymeriella* cf. *tardefurcata* *Protanisoceras* sp.	ALBIAN
ina sp.	*Globigerinelld. algerianus*	*Hypacanthoplites nolani* *Cheloniceras martini*	APTIAN
	Hedbergella sigali ► *Hed. infracretacea*	*Leptoceras puzosianum* *Nicklesia compressissima* *Hamulina subcylindrica*	BARREMIAN
aria crepidularis *ulina ouachensis* icro-gastropods/Pelecypods		*Acanthodiscus vaceki*	HAUTERIVIAN
inopsis schreiteri *nina caracolla/ Vaginulina kochi* *na neocomiana* ►		*Distoloceras curelense* *Kilianella superba* *Thurmanniceras thurmanni*	VALANGINIAN
lina nodosa ► *a* sp. *hragmium* sp.		*Berriasella boissieri*	BERRIASIAN
?		?	? TITHONIAN

a sequence of marine limestones and dolomites is intercalated. These were ascribed to the Valanginian by Gigout (1951). The clastic sequence might be equivalent with the Lower Cretaceous, since the overlying formation has to be assigned to the Cenomanian. The Cenomanian is developed as an alternation of silty marls and oyster shell beds (Exogyra). Above this, a massive dolomitic limestone follows, locally with a rich ammonite fauna proving a Lower to Middle Turonian age: Nigericeras lamberti, N. gignouxi, Vascoceras depressum, V. triangulare, V. cf. tavense, Paravascoceras evolutum, Gombeoceras cf. gongilense, Schindewolfites sp., Pseudaspidoceras footeanum and others. A detailed description of this interesting fauna is in preparation.

The light-colored dolomitic limestones of the Lower Turonian are followed by silty marls which, so far, have been termed "Senonian". Only the Campanian, however, has been dated in these marls, by the foraminifers _Globotruncana arca_ and _Frondicularia cordata_ cited by Gigout (1951: p.117). The top layer of this sequence, finally, is composed of partly chalky, partly siliceous microcrystalline and oolithic limestones and marls enriched in phosphate, and attributed to the Maastrichtian to Lower Eocene. The phosphatic sequence is superposed by partly silicified gastropod-bearing limestones ("Calcaires à Thersitées") of the Lutetian; they represent the end of the sedimentation and enable us to date the emergence of vast areas of the Meseta. The total thickness of the Cretaceous in this area is probably less than 300 m.

2.4.2. Paleoenvironment

The reduced thickness and the facies development make it obvious that the Cretaceous of the Moroccan Meseta was formed under more stable conditions than in the coastal basin farther south. The Lower Cretaceous consists of a Wealden-like continental facies, being divided by a single marine ingression which may be referred to the Valanginian. The ammonite, bivalve, and brachiopod fauna described by Gigout (1951) suggests inner shelf conditions. On the other hand, the typical "African" oyster facies of the transgressive Cenomanian indicates an environment within the littoral zone, whereas the platform limestones of the Turonian may be interpreted as having formed on the middle shelf. So far, the Senonian has only been incompletely documented, but it surely initiated with a regressive pulse, while the plankton-rich Campanian might correspond with the last transgressive sea level rise superposed onto the Meseta, episodically creating again inner shelf (?) conditions. One can presumably accept a Senonian "large Interatlas-Gulf" as outlined by Choubert and Salvan (1949), even though this may have been mainly a very shallow depositional area. The phosphate series of Maastrichtian-Eocene age are considered as coastal deposits of a regressive cycle, even if they seem to onlap here and there transgressively upon older Cretaceous sediments (Gigout 1951: p.122). The still marine and thersiteas-bearing limestones of Lutetian age pass into lagoonal formations further eastward and mark the final emergence of this "Interatlas-Gulf".

3. Comparison between the Coastal Basins

3.1. Facies Evolution and Bathymetry of the Coastal Basins

It is interesting to see that the Cretaceous of Tarfaya and of the Meseta show considerable similarities in their subsidence and sedimentation history, although this has taken place at quite different water depths reflected by differing faunal assemblages. On the other hand, both parts of the Atlas Basin, the Agadir and the Essaouira Basins, are so similar in their evolution, that there is no reason to assume a former separation of the two basins.

The most important stages of the Cretaceous evolution in the described basins are compiled along a North-South cross-section extending approximately from Settat-Meseta to Tarfaya (Fig.7).

On the Meseta, the continental-littoral development continued during the whole Early Cretaceous. In Tarfaya, this development persisted into the Middle Albian. A short-term ingression (ammonite facies) can be noticed in both areas though, however, at slightly different times: on the Central Meseta during the Valanginian and in the North Tarfaya Basin in the latest Aptian. Both ingressions correspond with transgressive trends in the Atlas Basin (cephalopod marls) which had been occupied by a marine platform already during the Jurassic. The intermittent regressions in this latter area (red bed facies with charophytes) are of Middle Hauterivian and Lower Aptian age. For this reason, we note a short time concurrence of the depositional factors within the Atlas and Tarfaya onshore basins during the Early Aptian.

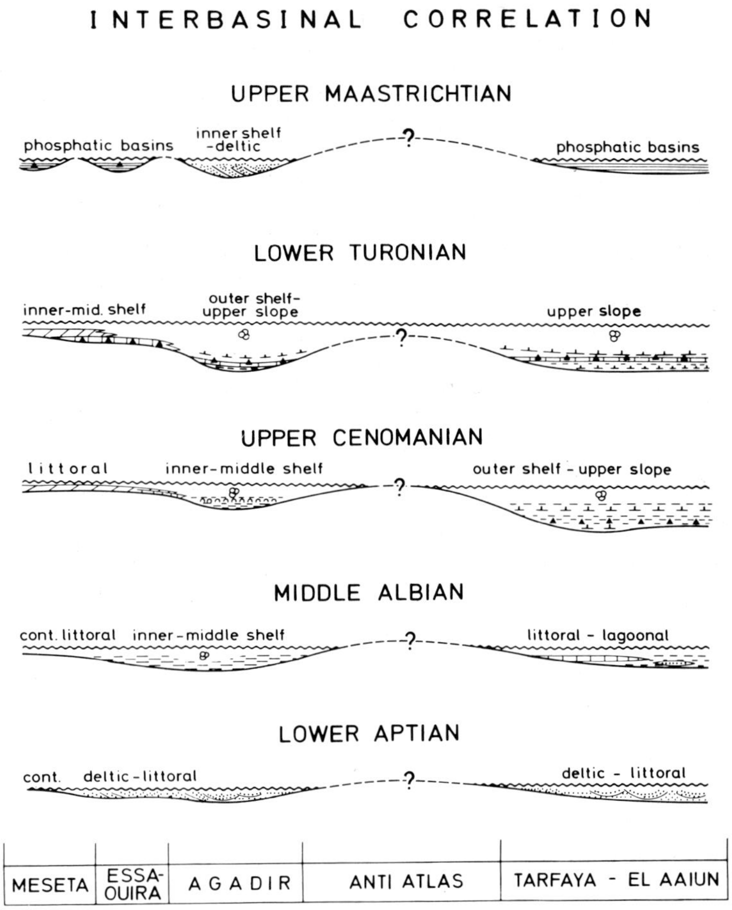

Fig. 7. Simplified schematic correlation of the Cretaceous sediments of the Essaouira, Agadir, and Tarfaya Basins and Meseta

If, indeed, the Essaouira and Agadir Basins were not separated during their Cretaceous history (compare Stets and Wurster, this Vol.), then the question remains why such strong differences in the composition of the index ammonite faunas, as expressed in Figs. 4 and 6, and particularly pronounced during the Berriasian to Barremian period, exist. Only very few of the index-species are identical in both areas. In this connection it is also surprising that species of the Northern German genus Crioceratites occur in the Hauterivian of Agadir, whereas the bulk of lowermost Cretaceous ammonites belongs to the Mediterranean faunal province. The associated benthic foraminifers, however, enable us not only to correlate both subbasins, but they also permit onshore/offshore correlations and, at the same time, a comparison between Mediterranean and Boreal faunal provinces. As expected, the evolution of planktonic foraminifers starts near the Hauterivian-Barremian boundary. These faunas remain an indispensable tool for both paleoecological interpretation (plankton-benthos-ratio) and stratigraphical correlation. But in some cases, divergences in first appearances occur which are partly controlled by facies (e.g., late appearances of Globotruncana angusticarinata and concavata).

The evolution of the Atlas and Tarfaya Basins reflects a parallel deepening of the basins from Early to Late Albian and diverges again afterwards. On the Meseta, continental conditions still dominated at this time (Fig.7). It seems that the large deltaic fans which were formed on the Meseta, and in particular during the Lower Cretaceous at Tarfaya, were an important source to the deep sea, in which prodelta and turbiditic series were accumulated (Sites 397, 370, 415 and 416; Einsele and von Rad, 1979; von Rad and Arthur, 1979; Price, 1980; Ranke et al., this Vol.).

After deposition of the thick oyster shell beds of the Middle and Upper Albian, the Essaouira Basin followed an individual evolution with the installation of a carbonate platform composed of thick miliolid limestones.

In contrast to their precursors, the ammonite faunas of the Aptian and Albian correlate perfectly across the basins. While the deposition of platform carbonates continued in the Essaouira Basin, a pronounced facies change to a deeper environment can be observed in all other basins during the Cenomanian and Turonian. The maximum of this trend is reached in the Early Turonian. Boreal faunal elements predominated at this time in the Tarfaya Basin; in the Agadir Basin Boreal faunas are mixed with Mediterranean faunas. Mediterranean faunas and inner-middle shelf conditions characterize, however, the Essaouira Basin and the Meseta.

Evolutionary trends were largely parallel within the Moroccan coastal basins during the following Late Cretaceous. The Coniacian and Santonian show clear regressive tendencies which are documented by numerous sedimentological and faunal observations. The climax of this development coincides with a pre-Campanian emersion in the Tarfaya Basin. The Campanian transgression seized all basins, including the Meseta. This phase is documented by hemipelagic sediments and a rich globotruncanid fauna.

As a result of the Maastrichtian regression, the direct connection between the basins was interrupted. In the Agadir region, delta fans characterize the peak of the regression. However, pelagic sedimentation returned during the next and last transgressive cycle in the Paleocene and Eocene before final emergence. In the Essaouira basin, on the Meseta and in the El Aaiun-Tarfaya Basin sedimentation likewise continued across the Cretaceous-Tertiary boundary into the Eocene.

3.2. Global Sea Level Fluctuation - Local Changes in Depositional Depth

It seems to be of general interest to compare the bathymetric evolution in the Moroccan coastal basins with the cumulative curve of extension of the Cretaceous world oceans compiled by Sliter (1976) and with summaries on global sea level changes (e.g. Vail and Mitchum, 1979). This enables us to distinguish local phenomena from global events (Fig.8) and to reconsider the question of worldwide transgressions and regressions presently being discussed (Janšin, 1973; Wiedmann, 1973; Matsumoto, 1977; Vail, Mitchum and Thompson, 1977; Pitman, 1978; Vail and Mitchum, 1979).

A comparison of the bathymetric curves of the four areas discussed with Sliter's cumulative curve immediately reveals that global fluctuations of sea level were effective in the Cretaceous of Morocco. It can be seen, that the transgressive peaks of the Moroccan Turonian and Middle Campanian can be correlated with global transgressions, just as the regressive peaks at the Jurassic-Cretaceous and Cretaceous-Tertiary boundaries with global regressions. The remaining local phenomena are confined to the apparently less stable Atlas Basin. They are of low magnitude (continental to middle shelf at the most) and, therefore, hardly to be linked with major tectonic movements, but rather with salt motion or changes in sedimentary supply and the displacement of estuaries and coast lines (see also Chap.5).

It may be worth mentioning that the distribution of transgressions and regressions in Northern Spain (Vascogotic Basin), Northwestern Europe and Columbia during Cretaceous times (Wiedmann, 1969: figs.21, 22), shows an even more striking correspondence with the Sliter curve. The Cretaceous in Northern Spain, in particular, is affected by several synchronous transgressions (in the Early Cenomanian, Early Turonian, Early Coniacian and Campanian). Comparable to the situation in Morocco (Fig.8), in the southern Celtiberic Basin, i.e. towards the Iberian Meseta, these peaks merge into one Early Turonian transgression (Wiedmann, 1975: fig.7).

Part of the Cretaceous and especially the post-Cretaceous smaller transgressions and regressions may be solely caused by changes of the rate of sedimentation and a long-term sea-level rise or fall (Pitman, 1978; see also Chapt.5).

4. Onshore-Offshore Correlation

Referring to model investigations by Bernoulli (1972) and Bernoulli and

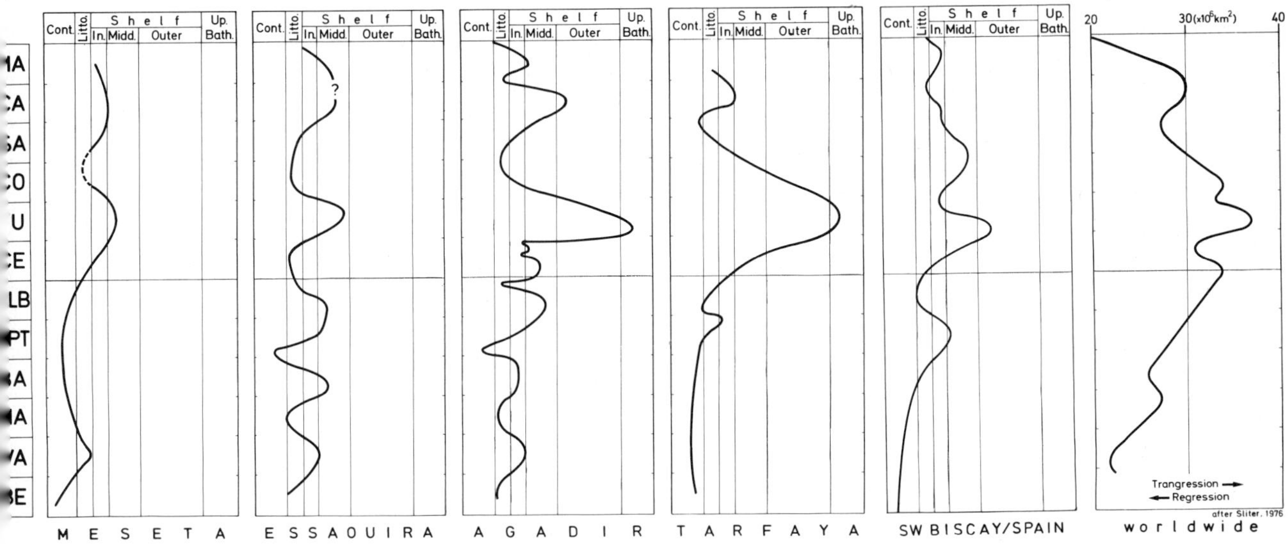

Fig. 8. Comparison of the Cretaceous bathymetric curves of the Meseta, the Essaouira, Agadir, and Tarfaya Basins of Morocco and the Bascocantabric Basin of Spain with Sliter's (1976) curve of the extension of the world oceans

Jenkyns (1974), this present report finally tries to correlate the results obtained in the Moroccan coastal basins with those of the deep sea cores off Northwest Africa. First of all, a biostratigraphical correlation is attempted (Fig.9), followed by an analysis of the paleoenvironments. Hence, this study will approach the problems of paleobathymetry, CCD, sea level fluctuations and shifting of coast lines, spreading events, subsidence and finally of paleooceanography in general.

The standard sections of Agadir and Essaouira and Sites 370/416 (Legs 41/50, Moroccan basin) and 137 (Leg 14) have been chosen for correlation. Site 137 does not exactly fit into the direct prolongation of the section line, but furnishes comparatively good and interesting data for comparison. In addition to the previously published results of these sites (Hayes, Pimm et al., 1972, Berger and von Rad, 1972; Lancelot, Seibold et al., 1978; Lancelot, Winterer et al., 1980) we studied 200 additional samples obtained from the Cretaceous parts in the cores of Sites 370 and 137.

4.1. Sites 370/416

At the foot of the continental rise off Cape Santin (Morocco) in a water-depth of about 4200 m, 1624 m of sediments were penetrated. The Cretaceous part of the drilling record exceeds 850 m in thickness and comprises the Valanginian to Cenomanian interval (Lancelot, Seibold et al., 1978; Lancelot, Winterer et al., 1980). The sediments of the Valanginian-Aptian (Fig.9) are approximately 800 m thick and show a remarkable terrigenous influx. Silty clays, siltstones, and occasionally conglomerates, transported by turbidity currents, are intercalated with nannofossil marls. A Turonian-Maastrichtian hiatus suggests submarine erosion, possibly by bottom currents.

The marly intercalations in the Lower Cretaceous contain a relatively rich foraminiferal and a meager ostracod fauna. The planktonics mainly consist of dwarfed hedgergellids (H. planispira, H. infracretacea, H. sigali). In comparison, there is quite a diversity among the benthics which can be attributed to nodosariids (Lenticulina ouachensis, Marginulina, Dentalina), textulariids (Dorothia hauteriviana) and primitive agglutinating forms such as Ammodiscus, Glomospira, Reophax and others (Fig.10A).

The thickness of the turbidites of Site 416 (800 m), considerably exceeds that of equivalent sediments in the coastal basins of Agadir (500 m), and Essaouira (400 m) (Fig.9). It seems that both shallow-marine sediments and to a lesser extent also the

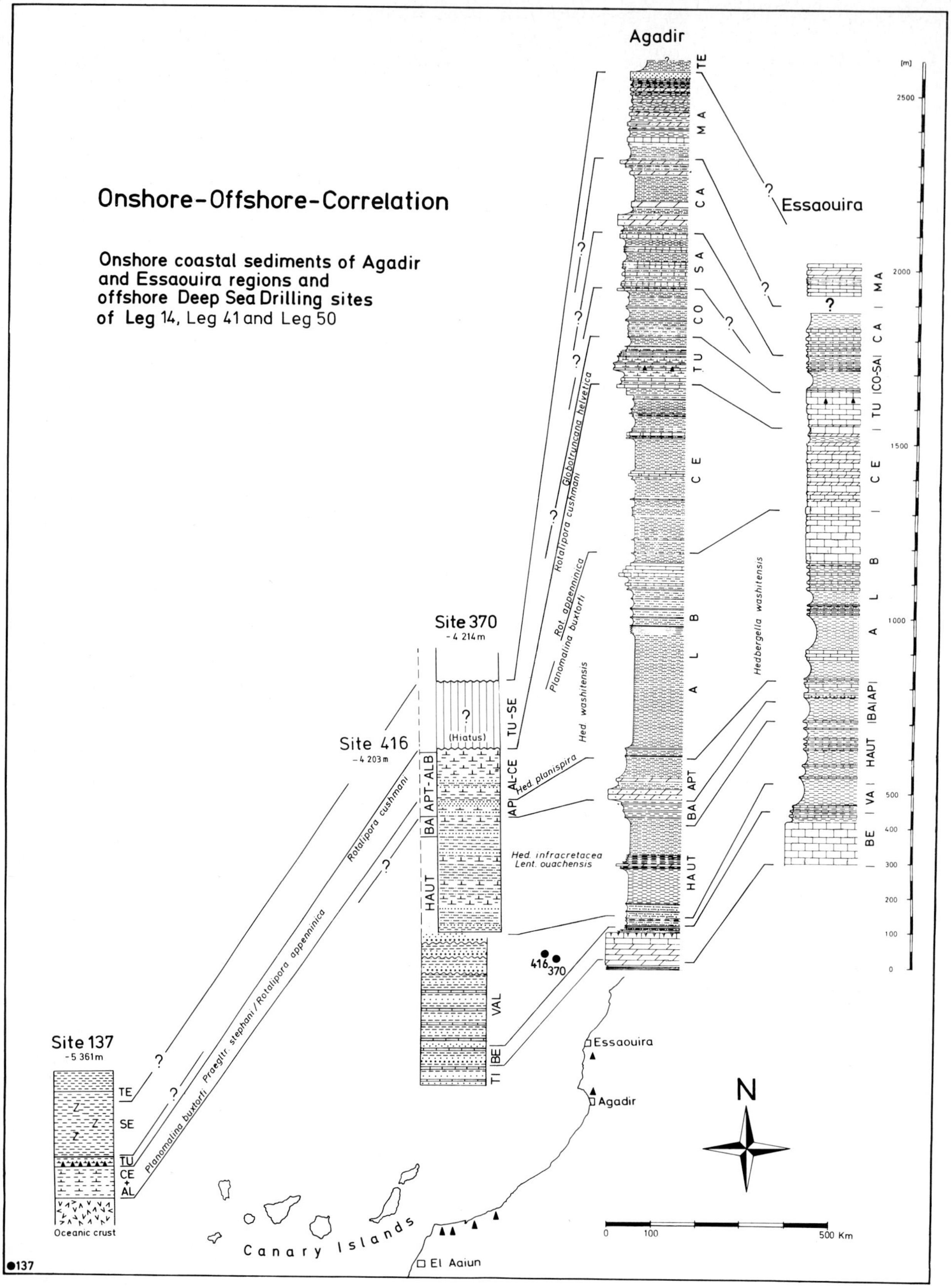

Fig. 9. Onshore/offshore correlation of the Cretaceous sections of Agadir and Essaouira with DSDP Sites 370/416 and 137

important continental clastics have fed the offshore basins. The foraminiferal faunas of the coastal basins and of the paleobathymetrically deeper Sites 370/416 are very similar, as can be seen from the comparision of Figs. 10A and B. The association of calcareous benthics with primitive agglutinating forms might point to an upper bathyal environment, but the occurrence of planktonics suggests that the sediments of Barremian-Aptian age were not deposited below the CCD. On the other hand, there is an increase of rotaliids (particularly Gyroidinoides and Gavelinella) during the Late Aptian-Early Albian, indicating also a deepening of the offshore basin. The relatively low rate of planktonics, however, may be explained by a high terrigenous influx. Therefore, it can be assumed, that the Early Cretaceous North Atlantic sediments were deposited above CCD.

These observations are in contrast to the stimulating recent discussion by Schlager (1980) and Price (1980) on nature, depositional depth and source of Early Cretaceous turbidites at Site 416. Here deposition below CCD is presumed.

At Site 370 about 100 m of Albian-Cenomanian sediments are equivalent to more than 1000 m near Agadir and approximately 700 m near Essaouira. Unlike the sedimentation of the Valanginian-Aptian (and possibly Lower Albian) interval, the sediments of the Albian-Cenomanian transition (Cores 26-24) consist of nanno-plankton-bearing greenish shales with silty intercalations. The rich microfauna (Fig.10C) consists mainly of primitive agglutinating foraminifers (Rhizammina, Reophax, Ammodiscus, Glomospira, Pelosina, Haplophragmoides, Recurvoides) which show a great resemblance to the faunas of Alpine flysch basins below CCD (Hesse and Butt, 1976). Thickness and facies of the Albian-Cenomanian sediments and faunal assemblages suggest subsidence of the offshore basin to greater depth below the foraminiferal lysocline and, probably, below CCD. It should be called to mind once more, that there is an excellent correlation between the Albian-Cenomanian transition zone of Site 370 and the Agadir standard section, although in the latter section there is an abundant plankton.

Kuznetsova and Seibold (1977) documented in detail the Jurassic to Early Cretaceous benthic foraminifers from Sites 367 and 370, Cape Verde and Moroccan Basin. At Sites 416 and 415 off southern Morocco Sliter (1980) reported foraminifers from the flysch-like terrigenous facies. He distinguished two types of faunas: in the laminated layers calcareous foraminifers and in the interbedded claystones deep-sea agglutinated benthics. Sliter believed that the faunas were deposited below the CCD at a paleodepth of 2000 m. Interestingly, this conclusion of Sliter is in agreement with the interpretation of Hesse and Butt (1976) in the Cretaceous Alpine turbidite basins.

In contrast, the uppermost Cenomanian consists of nannoplankton-bearing clays (Cores 24-20) with a high content of planktonics (Rotalipora, Praeglobotruncana, Hedbergella) and calcareous benthics (Gyroidinoides, Osangularia), but with only few agglutinating forms (Spiroplectammina). This fauna was definitely deposited above CCD. A deepening or fluctuation of CCD during the Cenomanian might be connected with an increase of the plankton productivity in the Late Cenomanian (Pflaumann and Krasheninnikov, 1978).

The following Upper Cretaceous (Turonian-Maastrichtian) interval is missing what is possibly related to deep currents. This perfectly matches with the indications for Turonian upwelling which clearly influenced the transgressive eupelagic series in the sections of Agadir and Tarfaya.

4.2. Site 137

DSDP Site 137, about 1000 km off Cape Blanc, was drilled on an abyssal hill at a depth of 5361 m. The first sediments overlying oceanic crust are of Albian (Cenomanian) age (Hayes, Pimm et al., 1972). There seems to be a continuous sequence up into the Neogene which reflects highly pelagic and oceanic conditions. After formation of the crust before the Late Albian, sedimentation began with nannoplankton-bearing marls and chalk sediments containing abundant planktonics of Late Albian age (Globigerinelloides breggiensis, Planomalina buxtorfi and Rotalipora ticinensis) and Cenomanian (Rotalipora appenninica, R. evoluta, R. reicheli, and Praeglobotruncana delrioensis). Density and preservation of this fauna (Fig.10D) proves that Site 137 was situated

Fig. 10. Comparison of onshore and offshore foraminiferal assemblages. A B Upper Valanginian to Lower Hauterivian assemblage; abundance of calcareous benthics, especially nodosariids, such as Lenticulina ouachensis (arrow), L. guttata, Dentalina, Marginulina, Vaginulina. A offshore assemblage of Site 370/416, Core 40-1, 101-103 cm. B onshore assemblage, sample 6518, Agroud, N of Agadir. C D Assemblages of the offshore Mid-Cretaceous. C Primitive agglutinant foraminifers of the Albian to Cenomanian (Rhizammina, Reophax, Glomospira, Pelosina, Recurvoides), Site 370/416, Core 25-1, below CCD. D Well-preserved planktonic assemblage (especially Rotalipora and Praeglobotruncana) of Lower to Middle Cenomanian, Site 137, Core 11-5, above CCD. Scale 1 mm

above CCD at this time. Further subsidence due to the contraction of the cooling oceanic crust took place during the Upper Cenomanian and Turonian, lowering the site beyond the paleo-CCD as documented by noncalcareous clays intercalated with nanno and chert layers. Final subsidence below CCD is documented by zeolitic deep-sea clays of the Late Turonian and Senonian, showing a selectively impoverished microfauna consisting of agglutinating foraminifers and fish teeth. The total thickness of the Cretaceous does not exceed 300 m.

4.3. Comparison

Conclusions on this onshore-offshore correlation are limited by many gaps of observation: There is a large distance between the Sites 370/416 and 137, and there are hardly any comparative data from the Moroccan shelf, except the available geophysical informations (Hinz et al., this Vol.). Only such data could help to reduce inconsistencies between the standard section of Agadir and Sites 370/416.

Therefore it is surprising, that facies and especially microfaunal similarities are quite evident, especially prior to the Aptian and, in particular, the concurrence of benthic faunas compared in Figs.10A and B is asthonishing. Even if redeposition of these faunas in the deep sea has to be considered, deposition below CCD can be excluded. If the Neocomian fauna of Sites 370/416 must be interpreted as upper bathyal, this surely defines the lower limit of depositional depth.

The large accumulation of sediments in the coastal basins during the Albian-Cenomanian period seems to be connected with pronounced subsidence along the continental margin. Evidence for greater waterdepths is found at Site 370 which was found to have reached CCD between Middle Albian to Middle Cenomanian times. It is of interest that Site 137 still was situated at the flank of the Mid-Atlantic Ridge and hence above CCD at this time and thus shows stronger faunal relations to the coastal basins. A further indication for the drastic subsidence of the Atlantic in Albian times is given by the opening of the South Atlantic towards the North Atlantic as recently proved by Bolli, Ryan et al. (1978) and Wiedmann and Neugebauer (1978).

The next distinct change seems to have occurred in the Middle Cenomanian. Subsidence below CCD is noted in the area of Site 137 during the Upper Cenomanian. This area still remained at great oceanic depth during its subsequent evolution, while sediments similar to those of the coastal basins were deposited. In contrast, Site 370 was now placed above CCD for a short time. We agree with the explanation for these fluctuations in CCD by Pflaumann and Krashennikov (1978) as being due to an increase in plankton production. Moreover, this does excellently match with the indications of upwelling in the Tarfaya Basin (Upper Cenomanian-Ccniacian) and also in the Agadir subbasin (Turonian).

A synchronous break in the evolution of microfaunas has also been observed in the Middle Cenomanian of the northern North Atlantic (Hart and Tarling, 1974) and might be related to the installation of a modern North Atlantic current system (Wiedmann, 1976: p.432). A corresponding deep contour current, of probably younger age, is the cause for the break of sedimentation at Site 370. In this way our observations (Einsele and Wiedmann, 1975; Wiedmann, 1976; Wiedmann et al., 1978a, 1978b) may modify the recent reconstructions fo the current systems in the North Atlantic during the Cretaceous (Gordon, 1973; Bartlett, 1973; Berggren and Hollister, 1974).

5. Relationship between Sedimentation Rate, Paleodepth, and Subsidence in the Moroccan Coastal Basin

5.1. General Remarks

As described above, biostratigraphic boundaries as well as paleodepth are well established for the majority of the coastal and shallow water (shelf) sediments of Tarfaya, Agadir, and Essaouira which are best exposed along the present coast line.

In order to determine rates of sedimentation and subsidence, an absolute time scale is necessary. Unfortunately, there are great uncertainties in this time scale, when larger periods are subdivided into shorter intervals (e.g. biostratigraphic zones). Therefore, also the rates of sedimentation and subsidence as determined for single stages or zones of the stratigraphic system may become somewhat problematic. Nevertheless, an attempt was made to refine this method in contrast to the common rough estimates of average rates for long periods.

In Fig.11 three different absolute time scales published recently for the Cretaceous period are plotted. Scales II and III are similar, but show large deviations from scale I, especially for the Mid-Cretaceous stage boundaries. As a result, the rates of sedimentation and subsidence differ, if different scales are used (indicated in Fig.11a and d). The principal tendency of the subsidence curves or the sedimentation rates, however, is little affected by the use of different time scales. To get comparable data, the following calculations are based on scale I (van Hinte, 1976).

5.2. Sedimentation Rates of Different Sediment Types

In relation to their position between the coastline and the lower continental slope, the sickness and thus also the sedimentation rate of the continental margin deposits vary considerably for certain time intervals.

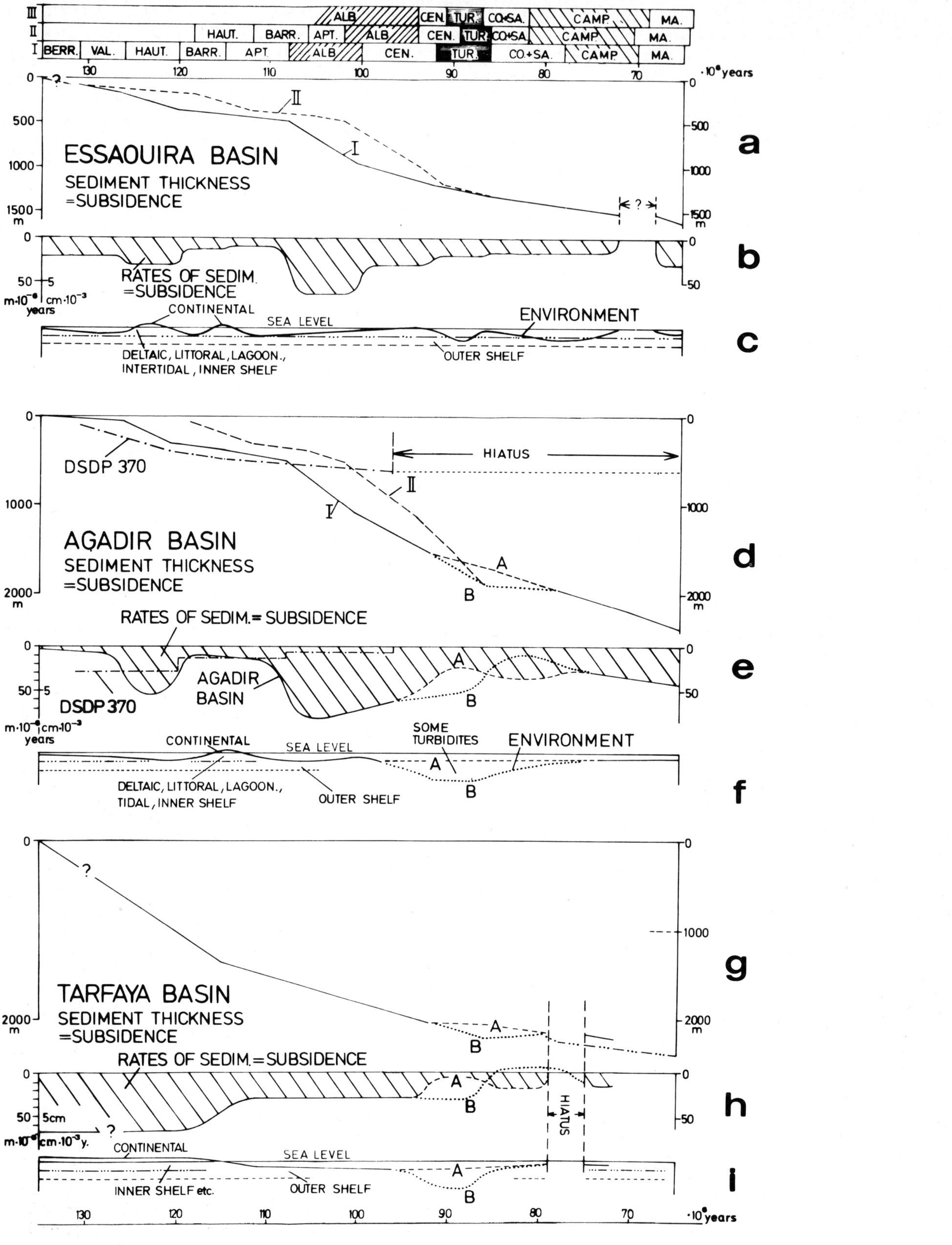
III ALB CEN. TUR. CO+SA. CAMP MA.
II HAUT. BARR. APT. ALB CEN. TUR. CO+SA. CAMP MA.
I BERR. VAL. HAUT. BARR. APT. ALB CEN. TUR. CO+SA. CAMP MA.
130 120 110 100 90 80 70 ·10⁶ years
ESSAOUIRA BASIN
SEDIMENT THICKNESS =SUBSIDENCE
a
RATES OF SEDIM. =SUBSIDENCE
b
CONTINENTAL
SEA LEVEL
ENVIRONMENT
DELTAIC, LITTORAL, LAGOON., INTERTIDAL, INNER SHELF
OUTER SHELF
c
HIATUS
DSDP 370
AGADIR BASIN
SEDIMENT THICKNESS =SUBSIDENCE
d
RATES OF SEDIM.= SUBSIDENCE
AGADIR BASIN
DSDP 370
e
SOME TURBIDITES
DELTAIC, LITTORAL, LAGOON., TIDAL, INNER SHELF
f
TARFAYA BASIN
SEDIMENT THICKNESS =SUBSIDENCE
g
h
HIATUS
INNER SHELF etc.
i

Over large areas, where sediment supply can always compensate subsidence, the sedimentation rates chiefly depend on the rate of subsidence. Nevertheless, certain sediment types may have characteristic average values, some of which are listed in Table 1.

With the possible exception of the Turonian bituminous marls, at all locations mentioned in Table 1, the sedimentation rates were sufficient to keep pace with subsidence. The high rates for the Jurassic shelf carbonates and the Early Cretaceous prodelta deposits are not unrealistic, if they are compared to Recent sediments in corresponding environments (e.g. Gould, 1970; Sarnthein, 1973; Milliman, 1974).

During shorter time intervals, i.e. the single stages of the Cretaceous, the sedimentation rates varied considerably. In the Agadir Basin, the highest rates were found for the Albian-Cenomanian interval, when silty clays were dominating. The intercalated thick and widely extending oyster reefs probably grew much faster than the normal sediments, but where they occur, often omission planes or hardgrounds can be observed. In contrast, the sedimentation rate of the Turonian "black shales", here interpreted by upwelling, is relatively low. It corresponds with the rates of pelagic calcareous deep sea muds or chalky sediments with a certain proportion of opaline silica.

5.3. Subsidence, Sea Level Changes, and Paleodepth

The interrelationship between subsidence, sedimentation rate and paleodepth in a section across the continental margin at Cape Bojador was already described in a more general way by Einsele and von Rad (1979); von Rad and Einsele (1980) (cf. also Ranke et al., this Vol.).

At all locations from coastal drill holes to DSDP Site 397 at the present lower continental slope, the rate of subsidence was found to slow down more or less exponentially from Jurassic to Neogene time. Along the present coast and at some distance farther inland, even some young uplift must have occurred. In the same cross-section, paleodepth was determined, but as soon as the sedimentary environment has to be assumed to have been deeper than shallowmarine and shelf, all interpretations become very uncertain.

On the basis of our sedimentological and biostratigraphic studies of the thick and more or less complete Cretaceous coastal profiles of Tarfaya, Agadir, and Essaouira, we can present a much more accurate history of subsidence and paleodepth for these areas than that derived for the Cape Bojador margin. In the coastal basins, water depth hardly exceeded the limit beyond which the interpretation of environment from sediment and faunal characteristics becomes problematic. However, for a more accurate treatment of this problem, the additional influence of sea level changes has to be taken into ccount (Watts and Steckler, (1979).

In a first step, we neglect this influence and assume that most of the time the water depth of the coastal basins fluctuated only between littoral and inner to middle shelf conditions. Then, the resulting subsidence curves (Fig.11a, d, and g) represent subsidence in relation to the (changing) sea level. Note that the Early Cretaceous part of curve g for the Tarfaya Basin is not as well established as the other curves, because in the Tarfaya region continental deposits were prevailing during that time.

As mentioned above, the sediment fill of the coastal basin was adjusted to a changing sea level. To determine true subsidence as a result of endogenic processes, these changes must be known. Unfortunately, our knowledge on global sea level changes

Fig. 11. Subsidence in relation to the changing sea level, sedimentation rates, and bathymetry in the coastal basins during the Cretaceous. a,b,c Essaouira Basin (near the present coast); d,e,f Agadir Basin (coastal exposures); g,h,i Tarfaya Basin (near present coast, Southern Morocco) Biostratigraphic stages over absolute time scale after: I van Hinte (1976), II Ernst and Kreuzer (1976); III Obradovich and Cobban (1975). The diagrams a and d are based on scales I and II, all others on I. Subsidence is not corrected for compaction, i.e. present sediment thickness represents subsidence

Table 1. Average Accumulation rates of some Sediment Types at the West African Continental Margin (related to the present stage of compaction)

Sediment type	Location	Age	Sedim. Rate m/m.y.	Remarks
Shelf carbonates	Slope off Cape Bojador	Jurassic	ca.100	From seismic records (Ranke et al., this Vol.)
Prodelta silts and clays	Continental slope and Rise off Cape Bojador	Early Cretaceous	up to 140	" "
Chiefly clastic sediments (2,400 m)	Agadir and Tarfaya coastal Basin (near present coast)	Cretaceous	35	
Chiefly clastic sediments (1,600 m)	Essaouira Basin (near present coast)	Cretaceous	23	
Chiefly silty claystones	Agadir Basin (along the present coast)	Albian to Cenomanian	50-80	Including thick oyster reefs
Bituminous laminated marls	" "	Turonian	10-20	

during Cretaceous time is still very limited and far from being in accordance, if the work of different authors is compared (e.g. Sliter, 1976; Hancock and Kauffman, 1979; Sleep, 1976; Vail and Mitchum, 1979; Watts and Steckler, 1979; this paper). Especially quantitative data is very uncertain and problematic.

In spite of these shortcomings, in Fig.12 an attempt is made to rule out the influence of sea level changes on one of the subsidence curves of Fig.11 (Agadir Basin, curve d). It is interesting to note that the resulting "true" subsidence curve shows a more regular gradient with breaks less pronounced than the curve related to a fluctuating sea level. The general trend of the curve, however, is not substantially changed. Therefore, at least at two of the three localities studied the Cretaceous history of subsidence appears to show the same common tendency. After a period of slow subsidence the rate of subsidence is increasing during the interval from the Albian to the Cenomanian Stage. This "Mid-Cretaceous Event" has also been observed along the concomitant Atlantic continental margin of the United States (Whitten, 1976) and Canada (Gradstein et al., 1976). Therefore, we assume that the increased Albian and Cenomanian subsidence is not only a local phenomenon. In the outer part of the coastal basins (along the present coast line) it enabled, in combination with the sea level rise, the deposition of 700 to 1000 m sediments. The high sediment supply from the nearby continent always maintained shallow-marine conditions until the end of Cenomanian time.

During the Turonian and the following stages of the Late Cretaceous, subsidence somewhat decreased and, at the same time, the input of terrigenous material, as well as the total rate of sedimentation slowed down. Because of uncertainties in the bathymetric interpretation of the Turonian bituminous marls of the Agadir and Tarfaya Basin, we have to discuss two possibilities for the corresponding subsidence curves of these basins (Fig.11d,g):

Curve A is found under the assumption that the depositional environment remained shallow as before. As a consequence, the rate of subsidence is equal to the sedimentation rate, and subsidence becomes slower already at the transition from the Cenomanian to the Turonian.

Curve B (Fig.11d and g, Fig.12) represents the alternative explanation assuming the persistence of strong subsidence, whereas accumulation of sediment was decreasing. Now, the rates of subsidence and sedimentation differ (Fig.11e, h) and as a result, the coastal basins became deeper (Fig. 11f, i). Since we could not observe indications of current and wave action or reworking of sediment, but instead of these a few turbidites in the Turonian section of Agadir, we favour the second hypothesis. In the case of Fig.11d and g (subsidence related to sea level) this would imply that at the end of the Turonian the subsidence curves display a striking bend, and near Tarfaya, curve B would indicate a period with some uplift. Such an uplift is, however, rather unlikely in an area of long-lasting subsidence. On the other hand, in the Tarfaya Basin the Upper Santonian and Lower Campanian are missing (Choubert et al., 1967), but it is not known at which water depth the marine sediments were eroded. The sediments above the hiatus are interpreted as shallow-marine, because they contain sands and shell beds of oysters.

The "sea level-corrected" subsidence curve (Fig.12) does not show this marked bend at the transition from the Turonian to the Coniacian, if the rate of subsidence during the Cenomanian is assumed to continue during the Turonian (compare curve B of Fig.11d). Therefore this curve is thought to be the best approximation possible at present, which also explains the deepening of the coastal basins during the Turonian. However, it is obvious that better quantitative data is necessary for different aspects (absolute time scale, determination of paleo-waterdepth and stratigraphic boundaries, rise and fall of sea level) in order to refine subsidence curves for a relatively short period, i.e. the Cretaceous.

The amount of total subsidence during Cretaceous time varied between approximately 1600 and 2500 m. This is not surprising, if the different position of the coastal exposures within a large sediment wedge between the Cretaceous coastline and the paleoslope is taken into consideration (see also von Rad and Einsele, 1980).

In conclusion, we want to stress the following two facts:

(1) During Cretaceous time subsidence of the coastal basins was much greater than that of the adjacent Atlantic oceanic crust below moderate sediment cover. There,

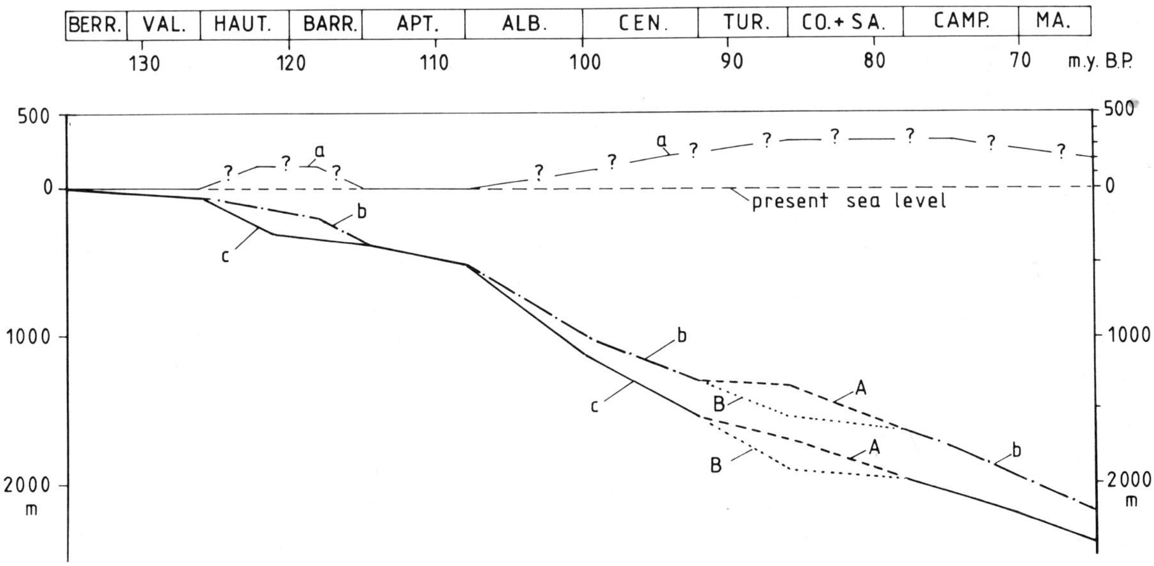

Fig. 12. Comparison of two subsidence curves for the Agadir Basin. a Tentatively assumed sea level changes, b Subsidence curve related to (changing) sea level (compare Fig.11d, A and B see text), c "true" subsidence curve, corrected for sea level changes. Time scale after van Hinte /1976)

according to Sclater et al. (1977), subsidence must have been on the order of 800 m. The higher subsidence rate below the coastal basins appears to have been caused not only by the higher sediment load, but also by a thinning of the underlying continental or transitional crust. Possibly, further processes such as metamorphism may have also contributed (see e.g. Watts and Ryan, 1976; Watts and Steckler, 1979).

(2) The Cretaceous sequences of the coastal basins are not only surprisingly thick, but in contrast to the Deep Sea Drilling Sites of the Atlantic (see e.g. Arthur, 1979), rather continuous and complete, especially those of Agadir and Essaouira. This is in part a consequence of (1). Furthermore, we can infer that the sediments of the coastal basins were protected against the erosive forces of deep currents which affected most of the slope and basinal sediments during Mid- and Late Cretaceous times.

Acknowledgements

Field work was generously financed by the Deutsche Forschungsgemeinschaft within the scope of the Deep Sea Drilling Project. The Moroccan Geological Survey, especially Drs. M. Bensaid, A. Boudda, E.A. Hilali, and H. Salvan, supported this work in various manners. Dr. H.G. Dietrich took an active part in the field work and the analysis of sedimentary structures and composition of the Agadir section. Further valuable help in the field was rendered by Drs. P. Brenner and H. Immel; Dipl.-Geol. H.D. Bergner provided information on a special study in the Agadir region and drafted most of the figures. We appreciate the help of Mrs. M. Simson in the preparation of the microfaunas. Moreover, we gratefully acknowledge the critical review of this paper by Dr.U.von Rad (Hannover) and Prof.Dr.D.Bernoulli (Basel).

Finally, we thank the National Science Foundation (Washington, D.C.) for providing samples from NE Atlantic DSDP drill sites.

REFERENCES

Ambroggi R (1963) Etude géologique du versant méridional du Haut Atlas occidental et de la Plaine du Souss. Notes et Mém Serv géol Maroc 157: 321 pp

Arthur MA (1979) Paleoceanographic events - Recognition, resolution, and reconsideration. Rev Geophys Space Phys 17: 1474-1494

Bartlett G (1973) The Canadian Atlantic continental margin: Paleogeography, paleoclimatology and seafloor spreading. Geol Surv Canada Paper 71-23: 43-72

Beckmann JP (1972) The foraminifera and some associated microfossils of sites 135 to 144. In: Hayes DE, Pimm AC et al, Init Repts Deep Sea Drilling Project 14, 389-420, 11 Tab, Washington

Behrens M, Krumsiek K, Meyer DE, Schäfer A, Siehl A, Stets J, Thein J, Wurster P (1978) Sedimentationsabläufe im Atlas-Golf (Kreide, Küstenbecken Marokko). Geol Rdsch 67: 424-453

Berger WH, von Rad U (1972) Cretaceous and Cenozoic sediments from the Atlantic Ocean. In: Hayes DE, Pimm AC et al, Init Repts Deep Sea Drilling Project 14: 787-954

Berggren WA, Hollister CD (1974) Paleogeography, paleobiogeography and the history of circulation in the Atlantic Ocean. In: Hay WW (ed), Studies in Paleo-Oceanography Soc econ Paleont Mineralog spec Publ 20: 126-186

Bergner HD (1979) Die Kreide im westlichsten Hohen Atlas (Marokko) unter besonderer Berücksichtigung von Kondensationshorizonten im Aptium. Diplomarb. Tübingen, 321 pp.

Bernoulli D (1972) North Atlantic and Mediterranean Mesozoic facies: A comparison. In: Hollister CD, Ewing JI et al, Init Repts Deep Sea Drilling Project 11: 801-822

Bernoulli D, Jenkyns HC (1974) Alpine, Mediterranean, and Central Atlantic Mesozoic facies in relation to the early evolution of the Tethys. In: Dott RH, Shaver RH (eds), Modern and Ancient Geosynkl Sedimentation Soc econ Paleont Mineralog spec Publ 19, 129-160

Bettenstaedt F (1973) Zur Phylogenie und Palökologie einiger Foraminiferen und Ostrakoden aus den Ober-

Hauterive und Mittel-Barrême von Helgoland. Senckenbergiana lethaea 54: 265-279

Bolli HM, Ryan WBF et al (1978) Initial Reports of the Deep Sea Drilling Project. Vol.40, XXV, 1079 pp. Washington (US Governm Print Off)

Burke K (1976) Development of graben associated with the initial ruptures of the Atlantic Ocean. Tectonophysics 36: 93-112 Amsterdam

Butt A (1949) Lower Cretaceous foraminiferal biostratigraphy, paleoecology, and depositional environment at DSDP Site 397, Leg 47A. In: von Rad U, Ryan WBF et al, Init Repts Deep Sea Drilling Project 47/1: 257-262

Choubert G, Faure-Muret A (1962) Evolution du domaine atlasique marocain depuis les temps paléozoiques. Livre Mém Paul Fallot 1: 447-514 Paris

Choubert G, Faure-Muret A, Hottinger I (1967) Aperçu géologique du Bassin Côtier de Tarfaya. Notes Mém Serv géol Maroc 175/I: 7-106

Choubert G, Salvan H (1949) Essai paléogéographique du Sénonien au Maroc. Notes Mém Serv géol Maroc 74: 13-50

Diester-Haass L, Schrader HJ, Pflaumann U, Rühl N, Thiede J (1974) Auftriebseinfluß in Sedimenten. Nachr dt geol Ges 9: 39-44

Duffaud F, Brun L, Plauchut B (for Société Chérifienne des Pétroles) (1966) Le Bassin du Sud-Ouest marocain. In: Reyre D (ed) Bassins sédimentaires du Littoral africain 1: 5-12 Paris

Einsele G, von Rad U (1979) Facies and paleoenvironment of Lower Cretaceous sediments at DSDP site 397 and in the Aaiun Basin (Northwest Africa). In: von Rad U, Ryan WBF et al, Init Repts Deep Sea Drilling Project 47/1: 559-571

Einsele G, and Wiedmann J (1975) Faunal and sedimentological evidence for upwelling in the Upper Cretaceous coastal basin of Tarfaya, Morocco. 9th intern Congr Sedim Nice 1975, Thème 1, Indicateurs sédimentologiques, 67-72, Nice

Ernst G, Kreuzer H (1976) Calibration of stratigraphic methods in the Boreal Cretaceous. Unpublished annual report, I.G.C.P.-Key Project

Gigout M (1951) Etudes géologiques sur la Méséta marocaine occidentale (arrière-pays de Casablanca, Mazagan et Safi). Notes Mém Serv géol Maroc 86: 507 pp

Gordon WA (1973) Marine life and ocean surface currents in the Cretaceous. J. Geol. 81: 269-284

Gould HR (1970) The Mississippi delta complex. In: Morgan JP (ed), Deltaic sedimentation. Modern and ancient, Soc.econ.Paleont.Mineralog. spec.Publ.15: 3-30

Gradstein FM, Williams GL, Jenkins WAM, Ascoli P (1975) Mesozoic and Cenozoic stratigraphy of the Atlantic continental margin, eastern Canada. In: Canad Continental Margins, Offshore Petrol Symp St Johns Newfoundland, 1974, Canad Soc Petrol Geol 103-131, Calgary Alberta

Hancock JM and Kauffman EG (1979) The great transgressions of the Late Cretaceous. J Geol Soc London, 136: 175-186

Hart MB, Tarling DH (1974) Cenomanian paleography of the North Atlantic and possible mid-Cenomanian eustatic movements and their implications. Palaeogeogr Palaeoclimat Palaeoecol 15: 95-108

Hayes DE, Pimm AC et al (1972) Initial Reports of Deep Sea Drilling Project, v. 14, Washington DC (US Governm Print Off), 975 pp

Hesse R, Butt A (1976) Paleobathymetry of Cretaceous turbidite basins of the Eastern Alps relative to the calcite compensation level..J Geol 84: 505-533

Hinte van JE (1976) A Cretaceous time scale. Amer Assoc Petrol Geol Bull 60: 498-516

Hinz K, Seibold E, Wissmann G (1974) Continental slope anticline and unconformities off West Africa. "Meteor" Forsch Ergebn 17: 67-73

Jansa LF, Wade JA (1975) Geology of the continental margin off Nova Scotia and Newfoundland. Offshore Geology of Eastern Canada. Geol Survey Canada Paper 74-30/2: 258 pp

Janšin AL (1973) (Über die sogenannten weltweiten Transgressionen und Regressionen). Bjull moskovsk obsc ispytalej prirody N S otd geol 48/2: 9-44

Kuznetsova KI, Seibold I (1978) Foraminifera from the Upper Jurassic and Lower Cretaceous of the Eastern Atlantic (DSDP Leg 41, Sites 367 and 370). In: Lancelot Y, Seibold E et al, Init Repts Deep Sea Drilling Project 41: 515-537

Lancelot Y, Seibold E et al (1978) Initial Reports of Deep Sea Drilling Project, v.41, Washington DC (US Governm Print Off), 1259 pp

Lancelot Y, Winterer EL (1980) Evolution of the Moroccan oceanic basin and adjacent continental margin - A synthesis. In: Lancelot Y, Winterer EL et al, Init Repts Deep Sea Drilling Project 50: 801-821 (Washington DC)

Lancelot Y, Winterer EL et al (1980) Initial Reports of the Deep Sea Drilling Project, v.50, Washington DC (US Governm Print Off), 868 pp

Lotze F (1955) Salzdiapirismus im nördlichen Spanien. Z. dt geol Ges 105: 814-822

Matsumoto T (1977) On the so-called Cretaceous transgressions. Palaeont Soc Japan spec Papers 21: 75-84

Michard A (1976) Eléments de géologie marocaine. Notes Mém Serv géol 252, 408 pp

Milliman JD (1974) Marine Carbonates. Springer, Berlin Heidelberg New York, 375 pp

Obradovich JD, Cobban WA (1975) A time-scale for the Late Cretaceous of the Western Interior of North America. In: Caldwell, WGE (ed), The Cretaceous System in the Western Interior of North America. Spec Paper Geolog Assoc Can 13: 31-54

Pflaumann U, Krasheninnikov VA (1978) Early Cretaceous planktonic foraminifers from eastern North Atlantic DSDP Leg 41. In: Lancelot Y, Seibold E et al, Init Repts Deep Sea Drilling Project 41: 539-550

Pitman III WC (1978) Relationship between eustacy and stratigraphic sequences of passive margins. Bull Geol Soc Amer 89: 1389-1403

Price I (1980) Provenance of the Jurassic-Cretaceous flysch, Deep Sea Drilling Project sites 370 and 416. In: Lancelot Y, Winterer EL et al, Init Repts Deep Sea Drilling Project 50: 751-757, Washington (US Governm Print Off)

Querol R (1966) Regional Geology of the Spanish Sahara. In: Reyre D (ed) Bassins sédimentaires du Littoral africain, 1, Littoral atlantique 27-39

Rad U von and Arthur M (1979) Geodynamic, sedimentary and volcanic evolution of the Cape Bojador continental margin (NW Africa). In: Talwany M; Hay W & Ryan WBF (eds) Deep Drilling Results in the Atlantic Ocean: Continental Margin and Palaeoenvironment, Amer Geophys Union, Maurice Ewing Ser 3: 187-203, Washington DC

Rad U von, Einsele G (1980) Mesozoic-Cainozoic subsidence history and paleobathymetry of the northwest African continental margin (Aaiun Basin to DSDP Site 397). Phil Trans Royal Soc London 294: 37-50

Rad U von, Ryan WBF et al (1979) Initial Reports of the Deep Sea Drilling Project, v.47, pt.1, Washington DC (US Governm Print Off), 710 pp

Reyre D (ed) (1966) Bassins sédimentaires du Littoral africain. 1ère Partie. Littoral Atlantique. Assoc Serv géol afr, Paris, VIII + 304 pp

Roch E (1930) Etudes géologiques dans la région méridional du Maroc occidental. Notes Mém Serv Carte géol Maroc, 542 pp

Sarnthein M (1973) Quantitative Daten über benthische Karbonatsedimentation in mittleren Breiten. Veröff Univ Inssbruck 84 (Festschr Heissel): 267-279 Innsbruck

Schlager W (1980) Mesozoic calciturbidites in Deep Sea Drilling Project hole 416A. Recognition of a drowned carbonate platform. In: Lancelot Y, Winterer EL et al, Init Repts Deep Sea Drilling Project 50: 733-749, Washington (US Governm Print Off)

Schlanger SO, Jenkyns HC (1976) Cretaceous oceanic anoxic events: causes and consequences. Geologie en Mijnbouw 55: 179-184 Amsterdam

Sclater IG, Hellinger S, Tapscott C (1977) The paleobathymetry of the Atlantic Ocean from Jurassic to the present. J. Geol 85: 509-552

Sleep NH (1976) Platform subsidence mechanism and "eustatic" sea-level changes. Tectonophysics: 36, 45-56

Sliter WV (1972) Cretaceous foraminifers - Depth habitats and their origin. Nature 239 no 5374: 514-515

Sliter WV (1976) Cretaceous foraminifers from the southwestern Atlantic Ocean, Leg 36, Deep Sea Drilling Project. In: Barker PF, Dalziel IWD et al, Init Repts Deep Sea Drilling Project 36: 519-537

Sliter WV (1980) Mesozoic foraminifers and deep-sea benthic environments from Deep Sea Drilling Project Sites 415 and 416, eastern North Atlantic. In: Lancelot Y, Winterer EL et al, Init Repts Deep Sea Drilling Project 50: 353-427

Vail PR , Mitchum Jr RM (1979) Global cycles of relative changes of sea level from seismic stratigraphy. In: Watkins JS, Montadert L, Wood Dickerson P (eds), Geological and Geophysical Investigations of Continental Margins, Amer Assoc Petrol Geol Mem 29: 469-472, Washington DC

Vail PR, Mitchum Jr RM, Thompson III S (1977) Global cycles of relative changes of sea level. In: Vail PR et al (eds). Seismic Stratigraphy and Global Changes of Sea Level, Amer Assoc Petrol Geol Mem 26: 83-98, Washington DC

Viotti C (1967) Résultats stratigraphiques du sondage Puerto Cansado I du Bassin Côtier de Tarfaya. Notes Mém Serv géol Maroc 175/I: 223-253

Watts AB, Ryan WBF (1976) Flexure of the lithosphere and continental margin basins. Tectonophysics 36: 25-44

Watts AB, Steckler MS (1979) Subsidence and eustasy at the continental margin of eastern North America. In: Talwani M, Hay W, Ryan WBF (eds). Deep Drilling results in the Atlantic Ocean: continental margins and paleoenvironment. Am Geophys Union, Maurice Ewing Series 3: 218-234

Whitten EHT (1976) Geodynamic significance of spasmodic, Cretaceous, rapid subsidence rates, continental shelf, USA. Tectonophysics 36: 133-142

Wiedmann J (1969) The heteromorphs and ammonoid extinction. Biol Rev 44: 563-602

Wiedmann J (1973) Evolution or revolution of ammonoids at Mesozoic System boundaries. Biol Rev 48: 159-194

Wiedmann J (1975) Subdivisiones y precisiones bio-estratigráficas en el Cretácico superior de las Cadenas Celtibéricas. Actas ler Symposium Cretácico Cordillera Ibérica. Cuenca 1974: 135-153 Cuenca

Wiedmann J (1976) Geo- und hydrodynamische Prozesse im Schelfbereich in ihrer Auswirkung auf mesozoische Fossil-Vergesellschaftungen. Zbl Geol Paläont II 1976 (5/6): 424-439 Stuttgart

Wiedmann J (1979a) Early Cretaceous mollusks from DSDP hole 397A off northwest Africa. In: von Rad U, Ryan WBF et al, Init Repts Deep Sea Drilling Project 47/1: 283-286

Wiedmann J (in press) Subsidenzgeschichte und Faunenverschiebungen an einem passiven Kontinentalrand: Das nordspanische Kreidebecken. N Jb Geol Paläont Abh

Wiedmann J, Butt A, Einsele G (1978a) Vergleich von marokkanischen Kreide-Küstenaufschlüssen und Tiefseebohrungen (DSDP): Stratigraphie, Paläoenvironment und Subsidenz an einem passiven Kontinentalrand. Geol Rdsch 67: 454-508

Wiedmann J, Einsele G, Immel H (1978b) Évidence faunistique et sédimentologique pour un upwelling dans le Crétacé supérieur. Actes 6ème Colloque afr Micropaléont Tunis 1974, Ann Mines Géol 28/II: 415-427

Wiedmann J, Neugebauer J (1978) Lower Cretaceous Ammonites from the South Atlantic Leg 40 (DSDP), their stratigraphic value and sedimentologic properties. In: Bolli HM, Ryan WBF et al, Init Repts DSDP 40: 709-734

Wiedmann J, Reitner J, Engeser Th, Schwentke W (in press) Subsidenzgeschichte und Faunenrespons im nordspanischen Kontinentalrandbecken während der Kreide

16 Turonian Black Shales in the Moroccan Coastal Basins: First Upwelling in the Atlantic Ocean?

Gerhard Einsele and Jost Wiedmann[1]

Abstract

In comparison with those of the deep Atlantic, the Upper Cretaceous black shales, chiefly Turonian bituminous marls of the Morrocan coastal basins take an exceptional position. They were deposited in a water depth of 200 to 300 m during a period when the deep Atlantic water circulation lead to oxygenated sea-floor conditions and incomplete sediment sequences. The Turonian shallow water "black shales" are characterized by carbonaceous laminated marls with intercalated pelagic limestones and cherts. Their occurrence in marginal basins coincides with the peak of the Cretaceous marine transgression and decreasing terrigenous sediment supply from the continent. Their sedimentation rate ranges from 5 to 20 m/m.y. and their fauna contains elements of Boreal assemblages of the Temperate zones in northern Europe and North America. They are interpreted as an indicator of the onset of coastal upwelling along the NW African continental margin, although *long-distance* cold bottom currents probably did not yet exist. The lack of a middle and upper Cretaceous black shale facies in the conjugate North American shelf basins, however, may be explained by the already existing influence of a trade wind system on coastal upwelling in the eastern North Atlantic and the onset of oceanic circulation.

1 Institut für Geologie und Paläontologie, Universität Tübingen, Sigwartstraße 10, 7400 Tübingen, FRG

1. Introduction

During Mesozoic and Tertiary times the West African continental margin was characterized by several coastal basins (Reyre, 1966; Lehner and De Ruiter, 1976; Lancelot and Winterer, 1980) which developed on subsiding continental or transitional crust (see, e.g., Weigel et al.; Hinz et al., this Vol.). These basins were filled up with continental and marine sediments of locally varying thickness (see, e.g., von Rad and Arthur, 1979; Stets and Wurster, this Vol.). Part of their sediment fill can be studied in present-day coastal exposures and farther inland in southern Morocco. Along the present coastline, the Jurassic and Lower Cretaceous sequences reach thicknesses in the order of 2000 to 4000 m. Later on, subsidence of the continental margin slowed down and, partly due to sea level rise, input of terrestrial detritus from the continent decreased. As a result, the younger deposits became thinner and incomplete (see, e.g., Wiedmann et al., this Vol.; Ranke et al., this Vol.).

This paper deals with the occurrence of Upper Cretaceous, mainly Turonian black shales. This term here is used for laminated bituminous marls with intercalated beds and nodules of limestone and chert. These sediments were deposited in the Aaiun-Tarfaya coastal basin and in the Agadir basin just about during the transition from rapid to low margin subsidence. Simultaneously, the mid-Cretaceous transgression reached its maximum (Wiedmann et al., this Vol.), and the sea intruded far inland, particularly east of Agadir into the Cretaceous Atlas Gulf (Behrens and Siehl; Wurster and

Stets; Stamm and Thein, all this Vol.). For an interpretation of this oxygen-deficient depositional environment, it appears necessary to describe briefly the Cretaceous sequence of the Tarfaya Basin and to concentrate on the properties of the bituminous deposits themselves. The Cretaceous Agadir section is described in Wiedmann et al. (this Vol.).

The Turonian black shales of these coastal basins deviate from the black shales of the deep Atlantic in two ways: They are younger and deposited in relatively shallow water depth. We therefore think that they can throw some light on the paleoenvironmental development of the Atlantic Ocean itself.

2. The Cretaceous Sequence of the Aaiun-Tarfaya Basin

Our study was restricted to the northern part of the Aaiun-Tarfaya basin in the province of Tarfaya, southern Morocco (Fig.1). A comprehensive description of this area was given by Martinis and Visintin (1966); Choubert et al. (1967, 1972); Wiedmann et al. (1978b); and Ranke et al. (this Vol.). In our context the following observations may be of interest (Fig.2):

According to the evaluation of exposures on land and some exploration wells, the Lower Cretaceous reaches a thickness of approximately 1700 m, whereas the Upper Cretaceous marine deposits make up only 600 m. The transition from terrestrial to limnic and shallow-marine facies can be observed at the northeastern rim of the basin along the present coast (Fig.1, Exp.1). Our paleocurrent measurements show that the fluvial sands (Wealden facies) came from the east, northeast, or southeast. The first lagoonal and intertidal deposits are represented by mudstones with intercalated laminated dolomitic beds, gypsum pseudomorphs, bioturbation, and the occurrence of shell beds, lenticular sands (flaser-bedding) as well as sand-filled channels (see also Einsele, this Vol.) which are probably of Aptian to Albian age. Along the coastal cliffs farther southwest (Fig.1, Exp.2) the Albian sequence is built up by dark or olive-grey silty clays and marls which contain extensive shell beds of shallow-marine origin (Fig.1, Exp.3 and Fig. 3A). The nearby Cretaceous paleo-coastline is documented by repeated intercalations of sands, as well as by some dolomitic beds deposited in a lagoonal environment.

The transition zone from Albian to Cenomanian is well exposed 10 to 15 km from the coast (Fig.1, Exp.3, Tassagdelt). It is characterized by marls, shallow-marine carbonates, and thick shell beds (oysters, Trigoniidae, Figs.2A and 3A). Some carbonate-cemented sandstones indicate that the Cretaceous coastline cannot have been very far away inland. Farther west and in the Upper Cenomanian section, marls and nodular limestones, probably caused by burrowing organisms, prevail. In the uppermost Cenomanian (Fig.1, Exp.4 and 5), these beds gradually become more bituminous and regularly laminated (Fig.3C). This is the beginning of the Turonian "euxinic event" which is described in more detail in the next chapter. The Turonian, as well as the younger Cretaceous beds, are preserved in the western part of the Tarfaya province (Fig.1, Exp.6 through 9). Similarly as the Upper Cenomanian and Turonian, at least part of the Coniacian sequence is also represented by laminated bituminous marls, limestones, and chert layers (Fig.2A). During Late Coniacian, Santonian and Campanian times, however, again silty clays and marls with oyster beds predominate (Fig.1, Exp.10). At the transition from the Santonian to the Campanian stage, a short hiatus has to be assumed (Fig.2A, Choubert et al., 1972). Campanian sandy marls, limestones, and oyster beds are known only from the neighborhood of the town of Tarfaya (Fig.1, Exp.12). They are truncated by the Miocene transgression.

Figure 2 includes a strongly simplified environmental interpretation of the Cretaceous sequences. It is based on sedimentological, as well as faunistic criteria which are partly listed in Table 1, but cannot be described here in detail. The bathymetric curve (Fig.2B) derived from these criteria shows that the maximum water depth of the coastal basin was reached during Turonian time. This coincides with the peak of the mid-Cretaceous marine transgression (see, e.g., Wiedmann et al., this Vol.) which certainly also had some influence on the formation of the bituminous marls of this period as will be discussed.

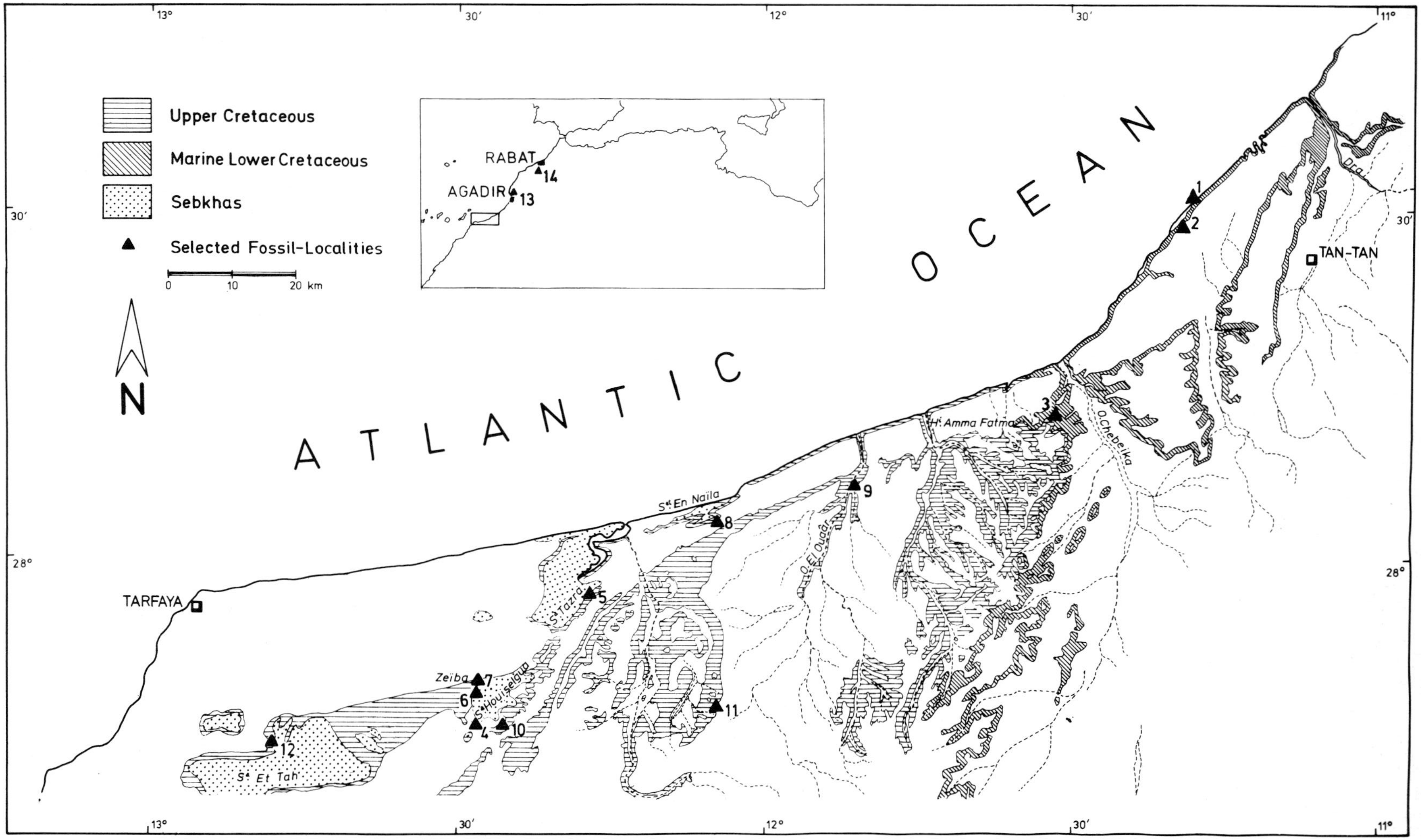

Fig. 1
Simplified geological map of the marine Cretaceous in the Tarfaya Province (after Choubert et al., 1967) including the location of exposures: 1 Shallow-marine to limnic early Albian N of Tantan Plage; 2 Marine early Albian at Tantan Plage; 3 Marine Upper Albian and Cenomanian, Tassagdelt; 4 Upper Cenomanian black shales, Sebkha Houiselgua S; 5 the same, Sebkha Tazra; 6 Early Turonian black shales, Sebkha Houiselgua W; 7 the same, Zeiba; 8 the same, Sebket Eh Naila; 9 the same, Oued El Ouaâr; 10 Coniacian-early Santonian Sebkha Houiselgua E; 11 Early Santonian, Hammada Er Rhouiba SW; 12 Early Santonian, Campanian Sebkha Tah; 13 Turonian N of Agadir; 14 Turonian of Sidi Hajaj, near Settat

Table 1. Biostratigraphy of the marine Cretaceous of Tarfaya based on Collignon (1967), Lehmann (1967), and our own observations (Wiedmann et al., 1978a)

		Ammonite Zones	Inoceramids	Foraminifera (first appearances)
CAMPANIAN		— — — — — —		*Globotruncana stuartiformis* *Gl. arca* + *Neoflabellina rugosa* + *Bolivina incrassata*
SANTONIAN	U.	—		
		Texanites texanus + *Pseudoschloenbachia*		*Gl. coronata* + *Neoflabellina deltoidea*
	L.	spp.		
CONIACIAN	U.		*I. sp. indet.*	*Gl. concavata*
	M.	*Gauthiericeras margae*	*I. chouberti*	
		Prionocycloceras guyabanum		*Gl. tricarinata*
	L.	*Otoscaphites bladenensis*		*Gl. angusticarinata*
TURONIAN	U.	*Glebosoceras glebosum*	*I. dresdenensis* *I. winkholdioides*	*Gl. schneegansi* *Gl. renzi*
	M.	*Mammites nodosoides*	*I. subhercynicus*	*Gl. sigali*
	L.	*Selwynoceras reymenti* *Watinoceras* n. sp. cf. *cobbani*	*I. mytiloides* *I. opalensis*	*Gl. helvetica*
CENOMANIAN	U.	*Acompsoceras viottii* + *Pachydesmoceras maroccanum*	*I. reachensis* + *I. pictus*	*Praeglobotruncana lehmanni* *Rotalipora cushmani*
	M.	*Calycoceras newboldi* + *Acanthoceras rhotomagense* + *Tarrantoceras*		*R. brotzeni*
		Euturrilites scheuchzerianus		
	L.	—		*R. appenninica*
ALBIAN	U.	*Salaziceras* n. sp. + *Pervinquieria fallax*		
		Anisoceras saussureanum		*R. ticinensis*
		Dipoloceras cristatum		
	M.	—		*Ticinella roberti*
	L.	—		
APTIAN	U.	*Hypacanthoplites nolani*		

3. The Turonian Bituminous Laminated Marls

The stratigraphic range of the Turonian black shales has not yet been worked out precisely. Near the Tarfaya coast, they start in the Upper Cenomanian and persist into Coniacian time. Farther inland the Turonian is eroded, but the black shales may have been restricted to a shorter time interval and finally have been replaced by shallow-marine carbonates and coastal sediments. For the Agadir Basin or Atlas Gulf, this facies change is described in some detail by Stamm and Thein (this Vol.).

3.1 Sedimentological Criteria

Figure 2C displays an example of a Turonian/Coniacian sequence of the Tarfaya province. It consists of more or less thinly laminated bituminous marls alternating with micritic limestones or nodules of limestones concentrated in layers (Figs.3C, 4A). Most of these beds also display parallel lamination, but occasionally small-scale low-angle cross-bedding and/or loss of stratification by bio-

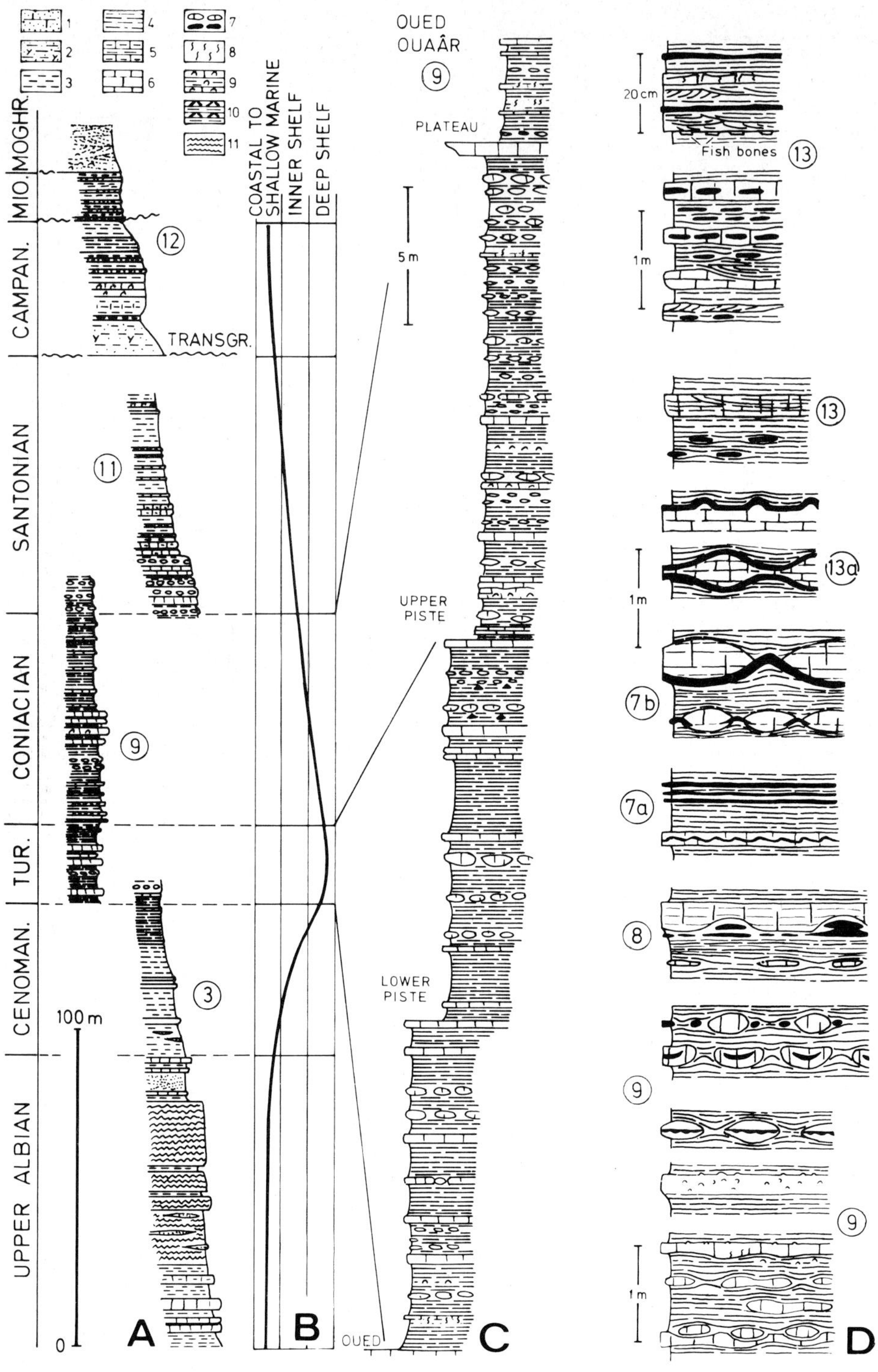

Fig. 2
A Simplified profiles of the marine Upper Cretaceous of Tarfaya (redrawn from Choubert et al., 1967, numbers signify selected outcrops, Fig.1).
B Paleobathymetric interpretation of *A*.
C Turonian to Lower Coniacian sequence of Oued Ouaâr (Exp.9, Fig.1) consisting of bituminous laminated marls alternating with micritic limestones or nodules of limestones and some chert. For further details see text.
D Selected short sections of several exposures (numbers see Fig.1) showing

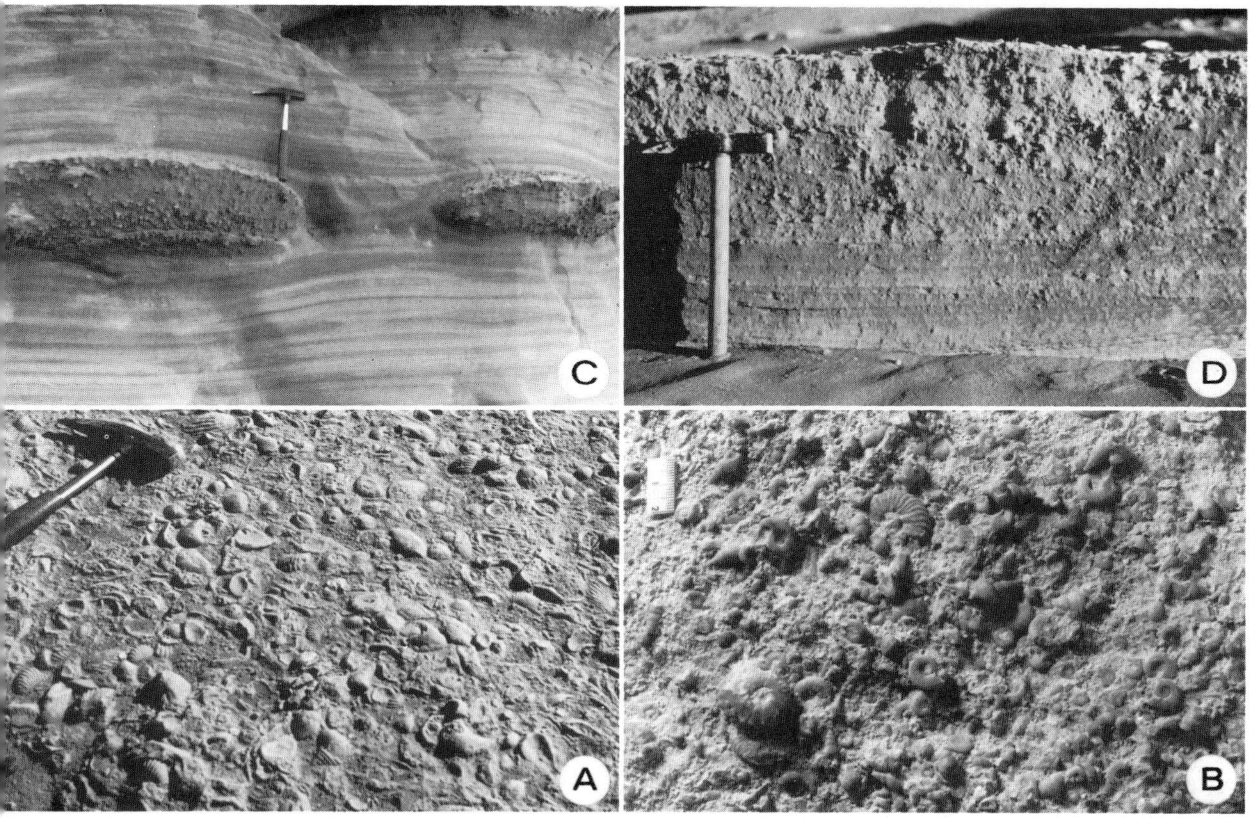

Fig. 3. Some characteristic lithologies of the Tarfaya marine Cretaceous
A Sublittoral shell bed (with Ptychomya, Trigonia, oysters) of Middle Albian age, 20 km NE of locality 3.
B Ammonite abundance and preservation, Lower Turonian, with Benuites, Watinoceras and Rostellarias, locality 6.
C Laminated bituminous marls and shell-bearing silicified limestone nodule, Upper Cenomanian, locality 4.
D Shell-bearing calcareous layer with benthic molluscs (Astarte, Rostellaria), Lower Turonian, locality 6

turbation can be observed. Some limestone beds are rich in small delicate shells of benthic molluscs (e.g., Astarte, Rostellaria) or convex-upward shells which settled vertically rather than being transported by currents (Fig.3D). Thin sections of the bituminous marls show alternating and often lenticular laminae of microfossils (chiefly foraminifers) and a brown bituminous clay matrix (Fig.4B). This micro-flaser bedding is characteristic of many oil shales (Bitterli, 1963). In the micritic limestones, microfossils, particularly "calcispheres", are abundant (Fig.4C). The average time period of one marl-limestone rhythm may be on the order of 200,000 years. This figure is, however, very uncertain, because the number of limestone beds within the Turonian section is difficult to as-

different features of bedding, particularly those indicating the relationship of limestone nodules and chert, as well as sections with erosion and small-scale cross bedding. 13 Exposure north of Agadir.
Lithology: 1 sandstone; 2 sandy marl with gypsum (y): silty clay and marl; 4 bituminous marl, well-laminated; 5 marl and marly limestone; 6 limestone; 7 nodules of limestone and chert; 8 bioturbation; 9 shell-bearing limestones and marls; 10 shell beds; 11 oyster reefs

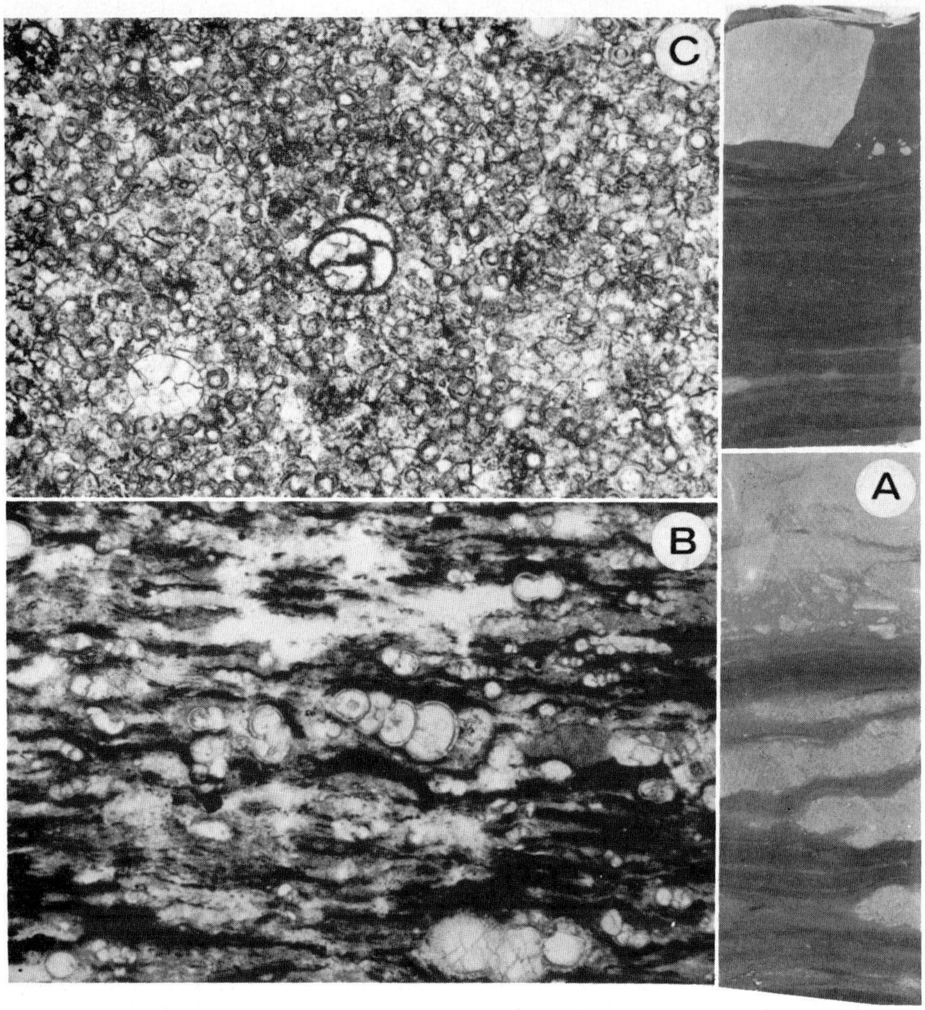

Fig. 4. Details of black shale lithology
A Two cores from a BRPM well, Oued El Amra 15 km NE locality 9, Lower Turonian, showing nodular limestones alternating with bituminous marls, part of which is bioturbated (below). Note broken limestone bed (above) indicating penecontemporaneous tectonics or mass movements.
B Thin section of Upper Cenomanian black shales with typical micro-flaser bedding and intercalated layers of foraminifera, locality 4. 90/1.
C Thin section of Lower Turonian limestone, composed of calcispheres, locality 6. 90/1

certain due to the replacement of limestone beds by more or less distinct layers of nodules. The average rate of sedimentation is on the order of 5 m/m.y. (in the Agadir basin 10 - 20 m/m.y., see Wiedmann et al., this Vol.) and therefore lower than in all the other stages of the Cretaceous. This is certainly caused by the low imput of terrigenous material, but also production (or at least preservation) of carbonate cannot have been high. Thus the hemipelagic to pelagic Turonian sequence appears to be the result of a starved marginal basins (see also Kauffman, 1977).

3.2 Limestone Nodules and Chert

The pelagic appearance is underlined by the occurrence of chert which is particularly abundant in the Upper Turonian and Lower Coniacian. It forms nodules or thin bands which are frequently associated with limestones (Fig.2D). From the interrelationship of these two constituents it can be concluded that diagenetic migration and precipitation of silica is often somewhat older than that of calcium carbonate. This is the case when chert nodules are found in the center of limestone concretions or when the shape of concretionary limestones is influenced by pre-existing bands or nodules of chert. However, the opposite sequence of diagenetic events can also be observed. Then the shape of chert nodules is dictated by the pre-existing limestone concretions (Fig.2D, 13a). The same sequence can be found in cavities of fossils which are filled by SiO_2 and $CaCO_3$ cement. From these contradictory observations we conclude that migration and precipitation of both constituents are more or less penecontemporaneous. Since the primary parallel lamination is strongly influenced by both types of concretions, the porosity of the sediments must have been much higher during the formation of the nodules than at present. Some features, such as the upward or downward bending of cherty lenses, indicate differential compaction due to the presence of solidified nodules.

For detailed discussions on silica-diagenesis and the formation of chert layers and nodules see, e.g., Lancelot (1973), Riech and von Rad (1979). We only emphasize that, according to our observations, the presence of concretions of carbonate or chert obviously hampers the molecular diffusion of the other constituent which is still in solution within the sediment. Therefore, the second constituent is often forced to precipitate directly beside or at least in the neighborhood of the first constituent already filling the pores of the sediment.

The $CaCO_3$ of the limestone nodules is probably derived from aragonitic shells which are dissolved within the sediment (see, e.g., Bathurst, 1975). The source of silica for the cherts appears to have been chiefly radiolarians, some of which have been detected in thin sections in poor preservation.

3.3 Faunal Characteristics

The unusual nature and succession of the Tarfayan ammonite fauna was detected by Collignon (1967), and has been confirmed by further detailed research (Wiedmann et al., 1978b). Whereas the Upper Albian ammonites are still totally Cosmopolitan, the Upper Cenomanian is characterized by abundant representatives of the North American genus Tarrantoceras, the early Turonian by the appearance of the Boreal[1)] genus Selwynoceras and, finally, the Coniacian by the invasion of the American Prionocycloceras. Even more remarkable is the total absence of typical Tethyan (Mediterranean, Central American, North African) groups like the Neolobitids in the Cenomanian, the Vascoceratids (abundant on the Moroccan Meseta) in the Lower Turonian, or the Tissotiids in the Coniacian. The Santonian ammonite fauna, however, appears to be again Cosmopolitan.

Fig.5A exhibits the divergence of contemporaneous ammonite associations of the Tarfaya and the neighbouring Atlas-Meseta Basin. The data for a statistical treatment could be gained by counting large numbers of ammonites on well exposed bedding planes (Fig. 3B) and subdividing them into different morphotypes. Up to 1780 specimens were found on one bedding plane! (Fig.5). The results are significant: At time equivalent levels, e.g. in the Turonian II, the Atlas-Meseta association is represented by two morphotypes (the B-type maximum with smooth evolute forms and a D-type minimum with tubercled forms). The B-type fauna corresponds to the group of Vascoceratids ("V" in Fig.5B) being representative for the Eur-African and American Mediterranean warm water realm. The equivalent Tarfaya assemblage, however, has a pronounced C-type maximum which is mainly composed of North Temperate and especially North American Selwynoceratids ("S" in Fig. 5B) and North African endemics; A-, D- and E-type minima constitute minor fractions of Cosmopolitan forms.

Our assumption of a relationship between ammonite morphotypes and environment is supported by the fact that during different times, e.g. Cenomanian V and Turonian II, a similar morphotype distribution can be observed despite the differing systematic position of these faunas. Typical warm water bivalves, such as oysters and rudists, are completely absent in the Tarfaya Upper Cretaceous,

1 The common term "Boreal" is also used here, but it can be substituted by Temperate

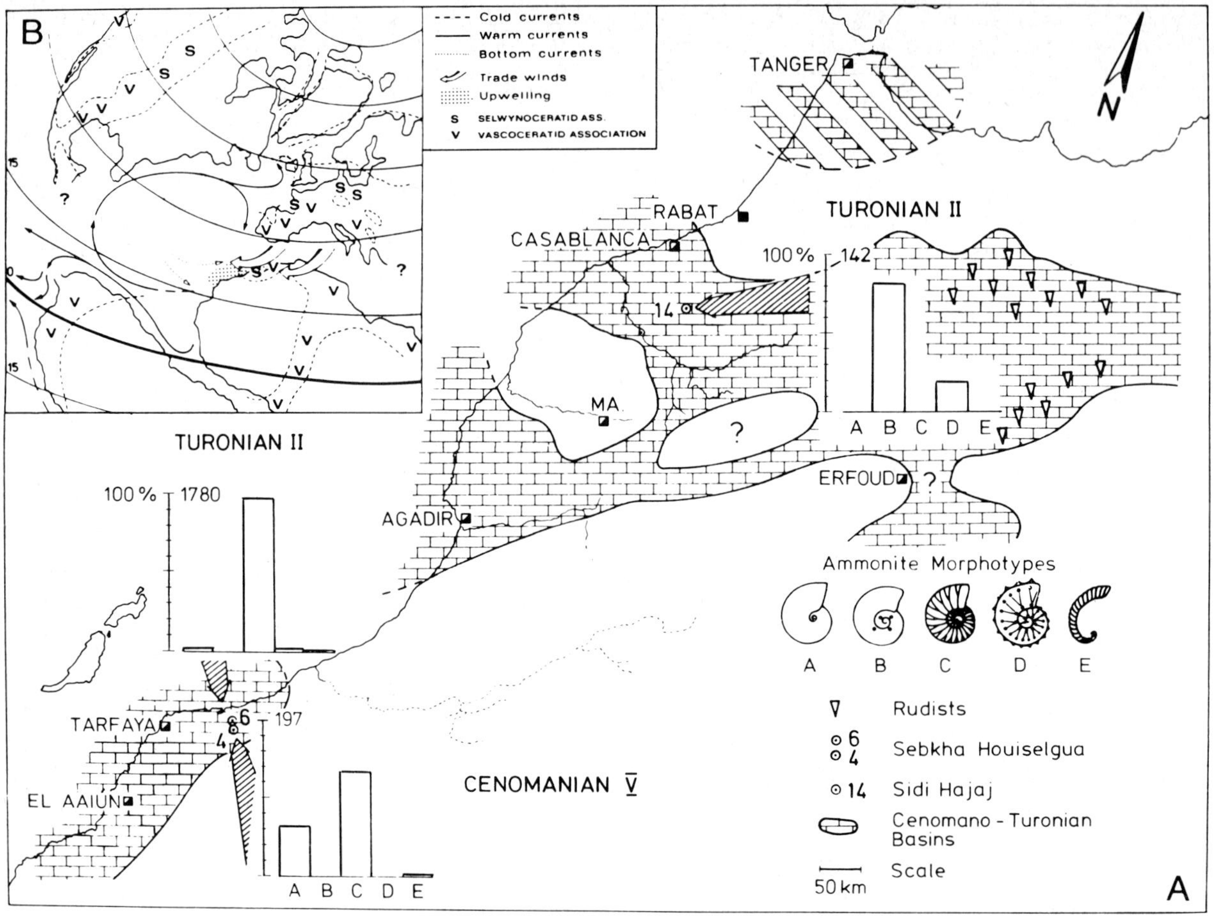

Fig. 5. Cenomanian-Turonian Coastal basins of Northwest Africa and the divergence of ammonite associations in the Tarfaya Basin from those on the Moroccan Meseta, demonstrated in quantitative spectra of ammonite morphotypes (*A*), and paleooceanographical implications (*B*)

while they are abundant in the Atlas-Meseta area. On the other hand, inoceramids which during the Upper Cretaceous seem to be adapted to cooler waters (Wiedmann, 1976), are frequent at Tarfaya.

A further hint at relatively cold water conditions may be the comparatively strong sculpturing on shells of ostracods and globotruncanids. Benthic microfaunas, prevailing in the Upper Albian (Fig.6A) or Santonian (Fig.6D), are reduced to a minimum during the black shale event (Fig.6B, C). Here, the Temperate globigerinoid forms (Fig.6B) as well as calcispheres (Fig.4C) and calcareous nannofossils (Figs.7A-F) are abundant. Globotruncanids typical of tropical waters are scarce, their faunistic relationships to the warm water environments of the Tunisian Turonian, for example, are restricted (J.Salaj, pers.commun.). The adult shell size of planktonic foraminifers is rather small and diversity is extremely low (Fig.6B).

Radiolarians have been found in many samples though often only poorly preserved. We assume that they lived in large quantities in the Late Cretaceous Tarfaya Basin, since chert layers in the bituminous shales are very common. Also fish remains such as teeth, vertebrae, and otolites have been found in remarkable quantities (Fig.6C), whereas benthic fossils (pelecypods, gastropods etc.) are rare or completely absent in these deposits.

Finally it has to be emphasized that the shales contain an extensive degraded biomass, indicating a high density of biotic elements. Moreover, it has

Fig. 6. Microfaunas of Tarfaya black shale facies and accompanying sediments. Scale 1 mm.

A Upper Albian microfauna of inner shelf environment, with planktonic foraminifers (Hedbergella washitensis, Ticinella raynaudi, Globigerinelloides breggiensis) and benthics (Trochammina, Lenticulina, Nodosaria, Bairdia, Cytherella). Locality 3.

B Late Cenomanian microfauna of black shale facies of deepening shelf environment, with planktonic foraminifers (Rotalipora cushmani, Hedbergella delrioensis, H.paradubia, Heterohelix globulosa) and fish remains. Locality 4.

C Fish remains in the Upper Turonian black shales with planktonics (Globotruncana schneegansi, Hedbergella planispira, Heterohelix globulosa) and reduced number of benthics (Veenia ?). Locality 10.

D Lower Santonian microfauna of inner to mid-shelf environment with planktonic foraminifers (Globotruncana concavata, Gl.angusticarinata) and benthics (Palmula cushmani, Veenia ?). Locality 12

to be stressed that, although some of these observations could be explained otherwise, the total range of biotic factors (Table 2) is best explained by upwelling conditions.

3.4 Geochemical Characteristics

In general the bituminous marls contain from 6% to 10% organic carbon and from 35% to more than 70% calcium carbonate. In contrast to the Early Cretaceous, dolomitic beds are absent. The abundant remains of fish mentioned above are responsible for relatively high contents of phosphate (0.1% to 0.4%).

The heavy metal concentrations in the bituminous deposits are at least partly higher than those known from normal clayey shelf deposits (Wedepohl, 1964; Förstner and Wittmann, 1979; Brumsack, 1980). If they are calculated for dried sediment free of car-

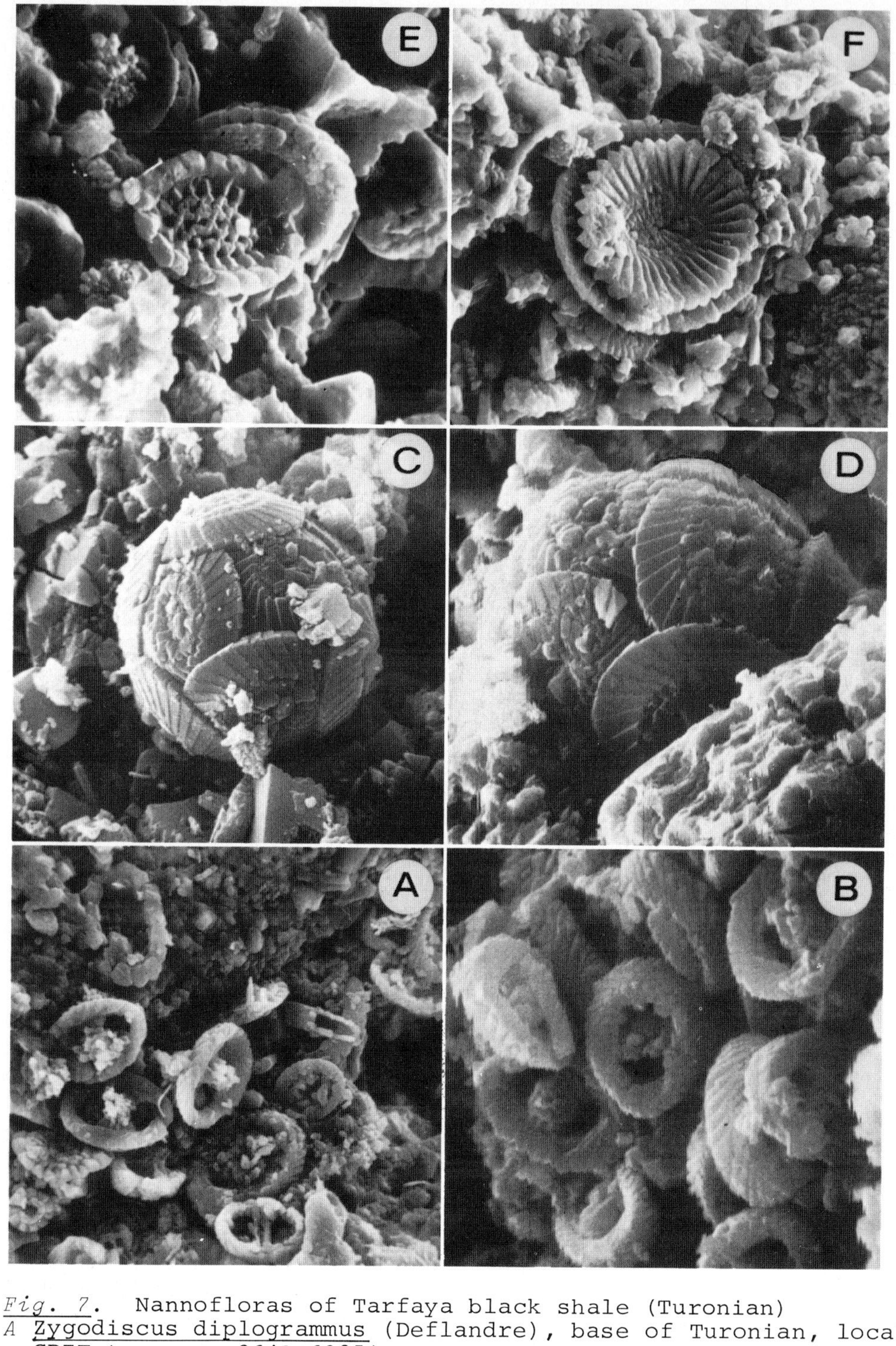

Fig. 7. Nannofloras of Tarfaya black shale (Turonian)

A Zygodiscus diplogrammus (Deflandre), base of Turonian, locality 7. 4000/1. GPIT type no. 3640-63851.

B Z. cf. diplogrammus (Deflandre), Lower Turonian, locality 9. 7000/1. GPIT type no. G483-77085.

C Watznaueria barnesae (Black), Lower Turonian, locality 9. 3500/1. GPIT type no. G483-77150.

D Watznaueria biporta Bukry, Middle Turonian, locality 7. 5000/1. GPIT type no. G482-77155.

E Cribrosphaerella ehrenbergii (Arkhangelsky), Middle Turonian, locality 7. 5000/1. GPIT type no. 3650-64056.

F Watznaueria cf. biporta Bukry, Middle Turonian, locality 7. 4000/1. GPIT type no. 3650-64057

Table 2. List of faunistic arguments for upwelling

Boreal species of ammonites and bivalves
Absence of Tethyan faunas
Predominance of planktonics (coccolithids, calcispheres, radiolarians, foraminifers)
Scarcity of benthics and bioturbation
Increase in endemics
Extreme faunal density
Low species diversity
Small sized molluscs (Crassatella) and foraminifers (Heterohelix, Globotruncana)
Predominance of globigerinoid planktonics (Hedbergella, Globigerinoides, Rugoglobigerina), scarcity of Globotruncana
Cool water sculptures of some foraminifers (Praeglobotruncana) and ostracodes (Veenia)
Abundance of fish remains
Relationship of Tarfaya foraminifers to North American groups, but less relationship to the foraminiferal assemblage of the Eastern Atlas Mountains

bonate, the mean values for zinc are approximately 400 ppm, for nickel 200 ppm, for chromium 120 ppm. Copper and lead reach values of only 20 - 30 ppm. Molybdenum was found in a range between 5 and 70 ppm (mean 17 ppm).

The significance of these heavy metal concentrations in relation to the depositional environment of the bituminous marls is not clear. Several authors (e.g., Brongersma-Sanders, 1966; Calvert and Price, 1971; Diester-Haass, 1978) have noticed that sediments of upwelling areas are particularly rich in heavy metals. In our case, a high proportion of these constituents could be correlated to the organic matter. On the other hand, Wedepohl (1964), Brumsack (1980), and Brumsack and Lew (this Vol.) have shown that oxygen depletion in sea water is sufficient to enable the precipitation of relatively high concentrations of heavy metal sulfides without the aid of organisms or adsorption by clay minerals. Therefore, high contents of heavy metals may provide just another criterion for oxygen minimum conditions which are not restricted to areas of upwelling.

4. Environmental Interpretation of the Turonian Black Shales

4.1 Paleowater Depth and Shifting of Coastline

As mentioned above, the Turonian black shales of the Tarfaya province were deposited in a coastal basin on top of subsiding continental crust. The Cretaceous sediments below the Turonian are doubtless of continental or shallow marine origin, and the younger beds also show characteristic features of shallow marine conditions. From these observations we have to conclude that the black shales of Tarfaya cannot have been accumulated in deep water. As discussed by Wiedmann et al. (this Vol.) in more detail, a global sea-level rise in combination with the reduction of terrigenous input has led to a deepening of the coastal basin, although the rate of subsidence was slowing down.

From these facts we conclude that the Turonian bituminous marls were deposited in a deep shelf sea, where waterdepth was controlled chiefly by the rise of sea level and the difference between accumulation rate and subsidence. Since the Albian and even the Lower Cenomanian sediments were deposited in shallow water, only the sea-level rise between the Albian-Cenomanian boundary and the Turonian peak has to be taken into account. One method to determine this particular proportion of Cretaceous sea level rise would be to compare the present height of the Albian-Cenomanian paleocoastline with the maximum elevation of the Turonian coastline farther inland, provided there was no vertical displacement between these two datum lines since Turonian time. Unfortunately, the Turonian coastline has been eroded in Southern Morocco, and on the other hand we cannot rule out younger tilting of the continental margin. In the Agadir-Essaouira coastal basin, the situation is still worse, because this area was strongly affected by younger tectonic movements in combination with the development of the western High Atlas Mountains (Stets and Wurster, this Vol.). We therefore use the following method to estimate the paleodepth of the Turonian shelf sea of Tarfaya:

1. From global sea level curves (Hancock and Kauffman, 1979; Vail and Mitchum, 1979) the sea level rose at least 100 m (in the time interval mentioned above).

2. The continental crust below the coastal basin subsided by approximately 200 m (Einsele and von Rad, 1979; von Rad and Einsele, 1980; Wiedmann et al., this Vol.), whereas the Cenomanian and Turonian sediments reached a total thickness of only 80 m. Consequently, a combination of both mechanisms must have caused a Turonian paleo-water depth on the order of 200 to 300 m. This rough estimation is, of course, only valid for the central part of the former coastal basin which coincides with the occurrence of the black shales

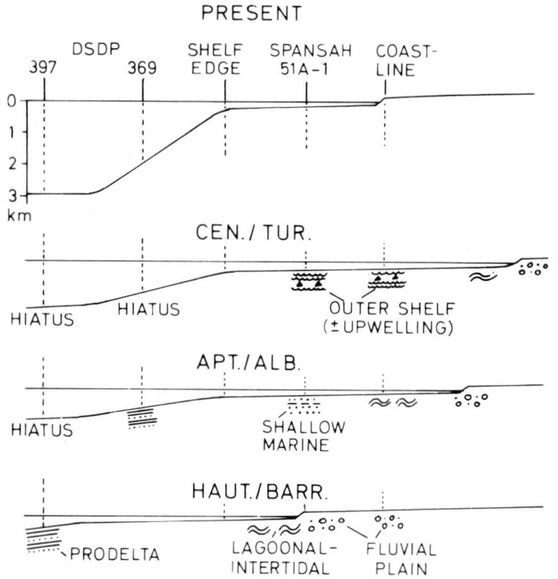

Fig. 8. Strongly simplified hypothetical sections across the continental margin at Cape Bojador. At the top present morphology and situation of drilling sites. Other sections from base to top landward facies migration, shifting of the coast line, and evolution of the shelf edge and slope during Cretaceous times

near the present coastline. The resulting marine transgression must have reached its maximum during Turonian time. The coastline of the basin migrated far inland (Fig.8) were, near the hinge line the subsidence of the marginal continental crust approached zero values.

4.2 Marginal Black Shales Versus Atlantic Deep Sea Sediments

A strongly simplified scheme of the Cretaceous facies development at the continental margin near Cape Bojador, south of Tarfaya is shown in Fig.8. From Sites 369 and 397 of the Deep Sea Drilling Project (Lancelot and Seibold et al. 1978; von Rad and Ryan et al. 1979) we know that Cenomanian/ Turonian (and partly also younger sediments) on the slope and upper continental rise are missing. Thus, the nature of Turonian deposits in this part of the deep Cretaceous Atlantic ocean is unknown. The same is true for most of the drill holes in the total central North Atlantic.

Figure 9 displays the configuration of this area during mid-Cretaceous time (after Arthur and Natland, 1979; Tissot et al., 1979). Furthermore, the most important findings of these authors are combined and shown in such a way that western and eastern North Atlantic data are separated. Secondly, the Cretaceous sediments recovered from the DSDP-sites are subdivided into three time intervals, the Barremian to Albian, the Cenomanian, and the Turonian to Santonian, in order to be able to plot the paleowater depth for the corresponding sediment (according to Arthur and Natland, 1979). Thus, the reader can easily see, where, in which time interval, and at which water depth black shales or "normal" pelagic sediments have been deposited. We again notice that the Turonian sediments (and even more so the younger Cretaceous beds up to the Campanian) in the deep sea are often partly or entirely missing. In the deep sea, Turonian black shales have been found only at DSDP-site 144 in the southernmost part of the western Central North Atlantic Basin (Fig.9 paleo-water depth 1.8 km, also see Tissot et al., 1979).

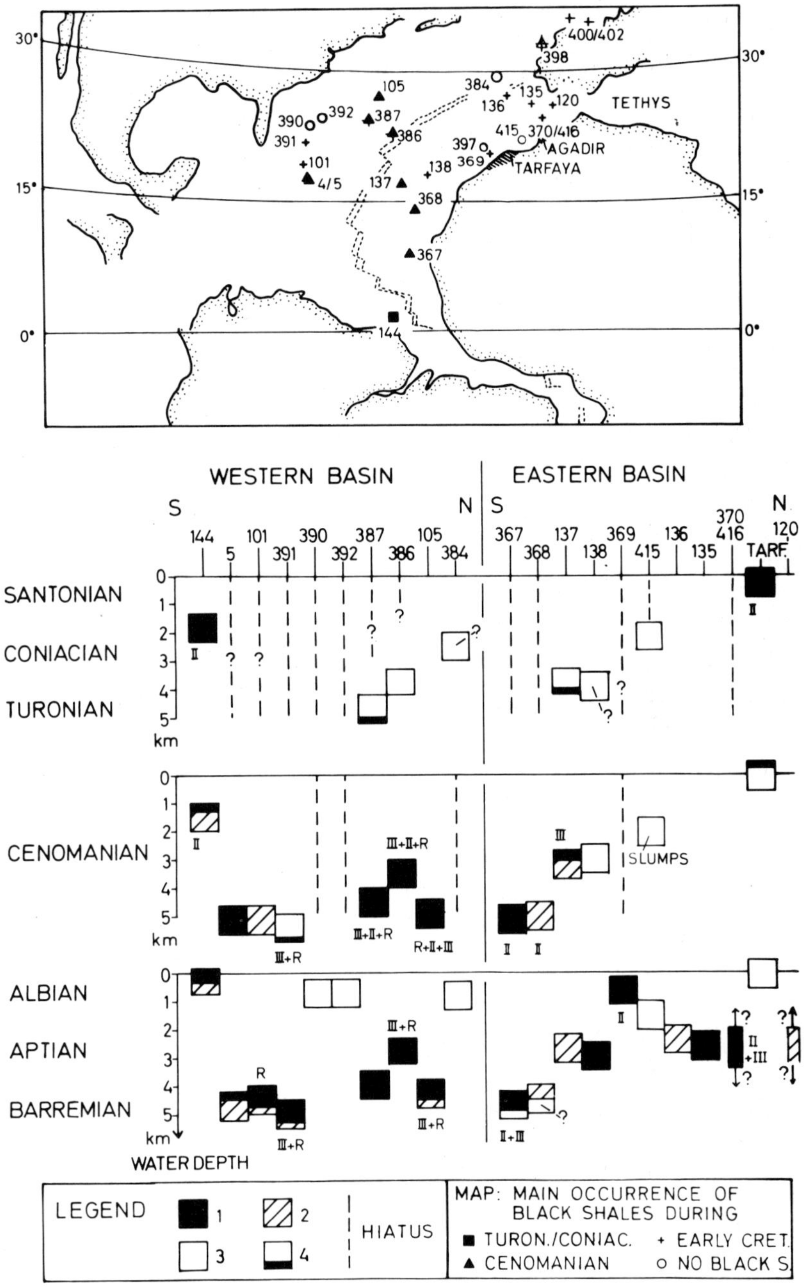

Fig. 9
Occurrence, age, and paleodepth of black shales in the central North Atlantic (data from Arthur, 1979; Arthur and Natland, 1979; classification of organic matter after Tissot et al., 1979). Note the unique position of the Turonian bituminous marls of the Agadir-Tarfaya coastal basin in comparison to the other occurrences.
1 dark-colored, relatively organic-rich sediment showing evidence of anoxic or very low oxygen conditions. Main types of organic matter: II planktonic; III terrestrial (moderately degradated); R residual
2 dark-colored organic-rich sediment showing evidence of low oxygen to oxygenated conditions
3 sediment evidencing well-oxygenated conditions
4 lowermost part corresponds to (1), higher sections to (3)
For further explanations see text. Site 397 showed a hiatus from Barremian to Oligocene time and Site 416 from Late Albian to Eocene time

4.3 Paleoceanography and the Onset of Upwelling

The hiatuses or condensed and well-oxygenated sediments of the deep central Atlantic ocean are probably the result of the destruction of stable stratification and increased circulation of deep waters (Lancelot and Seibold, 1978; Arthur and Natland, 1979; Tissot et al., 1979; Jansa et al., 1979). In addition, the rising sea level caused the entrapment of terrigenous material in marginal parts of the coastal basins, which prior to Cenomanian-Turonian time, were not inundated.

This was the situation when the Turonian bituminous marls were deposited in the coastal basins of Tarfaya and Agadir. Therefore we assume that the onset of deep ocean circulation in combination with the mineralization of organic matter enabled upwelling of nutrient-rich waters (Fig.10). Because of the different configuration of the central North Atlantic with a wide seaway to the Pacific and narrow connection with the northern polar seas (see Kauffman, 1973; Sclater et al., 1977; Tissot et al., 1979; Thiede, 1979) the current system of the Late Cretaceous ocean probably was not the same as today. According to Sclater et al. (1977) Tarfaya lay, however, only 10° farther north from the equator than at present. We therefore assume that the water masses of the coastal basin were affected by a trade wind zone similar to that of today (Fig.5B). During offshore blowing winds, upwelling water masses must have entered the relatively deep Turonian shelf area generating a high production of phyto- and zooplankton feeding other organisms described above. As a result, oxygen was depleted near the sea bottom and a relatively high proportion of organic matter was preserved. It is, however, characteristic of such upwelling zones that the oxygen minimum at the sea floor does not remain constant over longer periods, but is subject to changes in space and time (Tomczak, 1970; Weichart, 1970). This may be the reason for the intercalation of beds with bioturbation and some small-scale cross-lamination, indicating time intervals with slow bottom currents.

The type of organic matter in the Turonian bituminous marls has not yet been determined by kerogen analyses. The scarcety of plant debris in thin sections and the fauna described above indicate, however, that the organic matter of these shales is of marine origin deposited in a reducing environment. Under the microscope, plant debris was not observed. The predominance of planktonic organic matter in the black shales of the Tarfaya coastal basin is in good agreement with the general tendency described by Tissot et al. (1979) for the eastern basin of the central North Atlantic (Fig.9). Whereas in Aptian-Albian time terrestrial and residual organic matter plays a great part in all drill holes, the contribution of planktonic marine organisms considerably increases during the Cenomanian stage. The reasons for this increase are a reduced input of terrestrial plant material due to the ongoing marine transgression and possibly also climatic changes (Chamley et al., 1980). The changing oceanographic conditions described above have, however, also been of great importance.

The high carbonate content of the bituminous marls is due to zooplankton and phytoplankton. In shallow water well above the CCD these constituents are not dissolved or only partially. Preservation of carbonate is probably enhanced by sulfate reduction starting directly at the surface of the laminated sediments and thus increasing the alkalinity of pore waters (Berger,

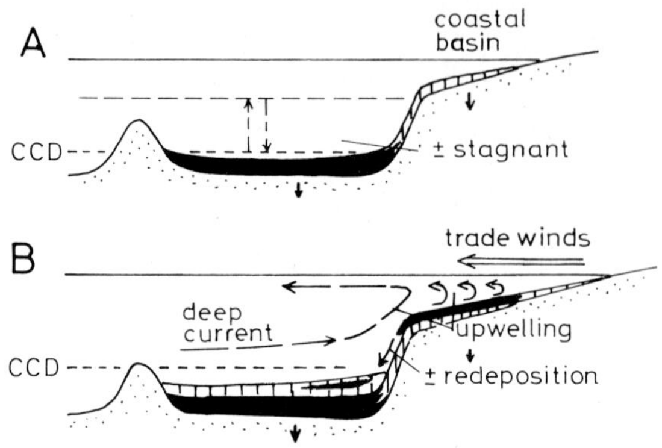

Fig. 10. Simplified model showing the change from (*A*) mid-Cretaceous ± stagnant bottom water masses and the formation of deep-sea black shales to (*B*) Upper Cretaceous deep ocean circulation and upwelling with deposition of bituminous marls on the shelf and slope

1979). Similarly, the preservation of opaline silica for the diagenetic formation of chert is enabled by a high input of radiolarian skeletons, only a small part of which can be dissolved during settling in relatively shallow water.

Farther inland in the Tarfaya basin, the Turonian sediments have been eroded, but in the Agadir coastal basin the shallow-water and coastal facies of the bituminous marls are partially preserved. Here the black shales grade into calcareous and dolomitic micrites (Stamm and Thein, this Vol.).

4.4. The Boreal Fauna of the Turonian Black Shales

A difficult problem is the evaluation of the temperature of the upwelling deeper waters and the occurrence of a Boreal fauna in the bituminous shales. At present the surface water temperature in these areas may drop by 9°C below the value which is normal for the corresponding latitude. For the Cretaceous ocean, however, many authors (e.g., Fischer and Arthur, 1977; Berger, 1979) assume reduced temperature gradients, vertically as well as between low and high latitudes. Therefore we should stress that the terms Boreal (north Temperate realm, e.g., Kauffman, 1973) or Tethyan do not necessarily signify faunal provinces of a large difference in water temperature. Other factors controlling the development of different Cretaceous faunal provinces are discussed by Kauffman (1973).

Nevertheless, the onset of an oceanwide bottom current system comparable to the present situation in the Atlantic Ocean (Einsele and Wiedmann, 1975; Wiedmann, 1976; Kauffman, 1977) cannot entirely be ruled out. Tissot et al. (1979) point out that, after Cenomanian time, the connection of the central North Atlantic with the Tethys and Mesogea became progressively wider and deeper. Consequently, high-latitude cool waters could enter and initiate a convective current system in the major part of this ocean. Anoxic conditions in deep water persisted only at the southern end of the eastern Atlantic basin (Cape Verde Basin, Demerara Rise).

5. Conclusions

In summary we conclude that the Turonian black shales of the Aaiun-Tarfaya basin are the result of the first appearance of coastal upwelling in the central North Atlantic (Fig.10). Upwelling or the overturn of oceanic water masses was, in contrast to the present situation, facilitated by low vertical temperature gradients. It also affected the sediments of coastal basins the water depth of which had increased by the mid-Cretaceous global sea-level rise as well as by decreasing accumulation rates which could not compensate the subsidence rates. Whereas the Turonian slope and deep sea sediments were removed at many localities or reworked by erosion and slumping, the shelf sediments are fully preserved and are therefore an important record for the environmental interpretation of this time period.

Like Thiede et al. (1981), we do not believe in several global "oceanic anoxic events" which are limited to a relatively short time interval (Jenkyns, 1980). At least during Late Cretaceous time, when overturn of deep water masses of unstable stratification was accomplished easily by the prevailing wind regimes or slight latitudinal differences in water density, the conception of regional circulation systems with local zones of upwelling appears to be more realistic. This would sufficiently explain the occurrence of black shales on continental margins and epeiric seas at many localities around the Atlantic and the Tethys (summarized by Jenkyns, 1980). More or less independently circulating regional water masses would lead to "local black shale events" of differing duration within a longer period of a general tendency toward the formation of black shales.

The relationship of the Turonian Tarfaya fauna to the Boreal fauna of North America and Northern Europe remains a mystery. Long-distance transport by cool bottom currents as assumed earlier (Einsele and Wiedmann, 1975; Kauffman, 1977) already requires a rather stable stratification of the Atlantic water masses, a situation which cannot easily be combined with the concept of low vertical and horizontal temperature gradients of the Cretaceous Atlantic Ocean (Berger, 1979).

Somewhat later, during Senonian to Eocene times, the widespread accumulation of phosphates and cherts in condensed deposits of the coastal basins can be regarded as evidence that upwelling along the Northwest African continental margin was going on (Lucas and Prévot, 1975).

It may be interesting to compare the observations of the NW African continental margin with the conjugate NE American shelf basins, e.g., the Scotian and Georges Bank Basins. In the central Scotian Basin, e.g., in the Mohican I-100 well (see Jansa and Wiedmann, this Vol.) there is no proof of a Cretaceous black shale facies, but instead evidence for well-oxygenated environments. In the COST G-2 well, in the central portion of the Georges Bank Basin, as well as in the southernmost Blake Plateau Basin, no signs of a black shale facies have been recognized.

The asymmetry of the NW African and NE American marginal Upper Cretaceous sediments strongly supports the conclusions of this paper to assume a low-latitudinal trade wind zone which has initiated upwelling along the NW Atlantic margin. Possibly, the modern anticyclonic wind system started during mid-Cretaceous time and gave rise to a surface circulation pattern of the North Atlantic similar to that of today.

Acknowledgements

Drs. J.Thiede and U.von Rad critically read the manuscript and made helpful suggestions. Dr. P.Čepek determined the coccoliths. The German Research Foundation (DFG) granted financial support.

REFERENCES

Arthur MA (1979) North Atlantic black shales: the record at Site 398 and a brief comparison with other occurrences. In: Sibuet JC, Ryan WBF et al (eds) Init Rep Deep Sea Drilling Project 47 (2): 719-738

Arthur MA, Natland JH (1979) Carbonaceous sediments in the North and South Atlantic: The role of salinity in stable stratification of Early Cretaceous basins. In: Talwani M, Hay W, and Ryan WBF (eds) Deep drilling results in the Atlantic Ocean: Continental margins and paleoenvironment. Maurice Ewing Ser 3. Am Geophys Union, Washington DC, pp 375-401

Bathurst RGC (1975) Carbonate sediments and their diagenesis, 2nd edn. Dev Sedimentol 12. Elsevier, Amsterdam, 658 p

Berger WH (1979) Impact of deep-sea drilling on paleoceanography. In: Talwani M, Hay W, Ryan WBF (eds) Deep drilling results in the Atlantic Ocean: continental margins and Paleoenvironment. Maurice Ewing Ser 3. Am Geophys Union, Washington DC, pp 297-314

Bitterli P (1963) Aspects of the genesis of bituminous rock sequences. Geol Mijnbouw 42: 6

Brongersma-Sanders M (1966) Origin of trace metal enrichment in bituminous shales. In: Hobson GD, Speers GC (eds) Advances in organic geochemistry. Pergamon, London, pp 231-236

Brumsack HJ (1980) Geochemistry of Cretaceous black shales from the Atlantic Ocean (DSDP Legs 11, 14, 36 and 41). Chem Geol 31: 1-25

Calvert SE, Price NB (1971) Recent sediments of the South African shelf. ICSU/SCOR Working Party 31. In: Delany FM (ed) Symposium Cambridge 1970, The Geology of the East Atlantic Continental Margin, 4, Africa, Rep 70/16. Inst Geol Sci, pp 175-185

Chamley H, Girond d'Argoud G, Robert C (1980) Clay mineralogy of Cretaceous and Cenozoic sediments off the Moroccan margin, DSDP Sites 415 and 416. In: Lancelot Y, Winterer EL et al (eds) Init Rep Deep Sea Drilling Project 50: 715-721

Choubert G, Faure-Muret A, Hottinger L (1967) Aperçu géologique du Bassin Côtier de Tarfaya. Notes Mém Serv Géol Maroc 175 (1): 7-106

Choubert G, Faure-Muret A, Hottinger L (1972) La série stratigraphique de Tarfaya Maroc sud-occidental et le problème de la naissance de l'océan Atlantique. Notes Mém Serv Géol Maroc 31: 29-40

Collignon M (1967) Les Céphalopodes crétacés du bassin côtier de Tarfaya. Notes Mém Serv Géol Maroc 175 (2): 9-78

Diester-Haass L (1978) Sediments as indicators of upwelling. In: Boje R, Tomczak M (eds) Upwelling ecosystems, Springer, Berlin Heidelberg New York, pp 261-281

Einsele G, von Rad U (1979) Facies and paleoenvironment of Lower Cretaceous sediments at DSDP Site 397 and in the Aaiun Basin (Northwest Africa). In: von Rad U, Ryan WBF et al (eds) Init Rep Deep Sea Drilling Project 47 (1): 559-577

Einsele G, Wiedmann J (1975) Faunal and sedimentological evidence for upwelling in the Upper Cretaceous coastal basin of Tarfaya, Morocco. IX Congr Int Sédimentol, Nice, Thème 1, pp 67-74

Fischer AG, Arthur MA (1977) Secular variations in the pelagic realm. In: Cook HE, Enos P (eds) Deep water carbonate environments. Soc Econ Paleontol Mineral Spec Publ 25: 19-50

Förstner U, Wittmann GTW (1979) Metal pollution in the aquatic environment. Springer, Berlin, 486 p

Hancock JM, Kauffman EG (1979) The great transgressions of the Late Cretaceous. J Geol Soc London 136: 175-186

Jansa LF, Enos P, Tucholke BE, Gradstein FM (1979) Mesozoic-Cenozoic sedimentary formations of the North American Basin; Western North Atlantic. In: Talwani M, Hay W, Ryan WBF (eds) Deep drilling results in the Atlantic Ocean: Continental margins and paleoenvironment. Maurice Ewing Ser 3. Am Geophys Union, Washington DC: pp 1-57

Jenkyns HC (1980) Cretaceous anoxic events: from continents to oceans. J Geol Soc London 137: 171-188

Kauffman EG (1973) Cretaceous Bivalvia. In: Hallam A (ed) Atlas of palaeobiogeography. Elsevier, Amsterdam London New York, pp 353-383

Kauffman EG (1977) Geological and biological overview: Western Interior Cretaceous Basin. Mt Geol 14: 3-4, 75-99

Lancelot Y (1973) Chert and silica diagenesis in sediments from the central Pacific. In: Winterer EL, Ewing JI et al (eds) Init Rep Deep Sea Drilling Project 17: 377-405

Lancelot Y, Seibold E (1978) The evolution of the central northeastern Atlantic - summary of results of DSDP Leg 41. In: Lancelot Y, Seibold E et al (eds) Init Rep Deep Sea Drilling Project 41: 1215-1245

Lancelot Y, Seibold E et al (1978) Init Rep Deep Sea Drilling Project 41, Washington (US Governm Print Office)

Lancelot Y, Winterer EL (1980) Evolution of the Moroccan oceanic basin and adjacent continental margin - a synthesis. In: Lancelot Y, Winterer EL et al (eds) Init Rep Deep Sea Drilling Project 50: 801-833

Lehmann R (1963) Etude des Globotruncanidés du Crétacé supérieur de la province de Tarfaya (Maroc occidental). Notes Mém Serv Géol Maroc 21: 133-159, 3 figs, 10 pls, Rabat

Lehmann R (1967) Les Foraminifères pélagiques du Crétacé du Bassin Côtier de Tarfaya. Notes et Mém Serv géol Maroc 175 (2): 151-176

Lehner P, De Ruiter PAD (1976) Africa's Atlantic margin typified by a string of basins. Oil Gas J Nov: 252-265

Lucas J, Prévot L (1975) Les marges continentales pièges géochimiques; L'exemple de la marge atlantique de l'Afrique à la limite Crétacé-Tertiaire. Bull Soc Géol Fr 17,4 (7): 496-501

Martinis B, Visintin V (1966) Données géologiques sur le bassin sédimentaire côtier de Tarfaya (Maroc méridional). In: Reyre D (ed) Bassins sédimentaires du Littoral africain 1: 13-26, 11 figs. Paris

Rad von U, Arthur MA (1979) Geodynamic, sedimentary and volcanic evolution of the Cape Bojador continental margin (NW Africa). In: Talwani M, Ryan WBF (eds) Deep drilling results in the Atlantic Ocean: continental margins and paleoenvironment. Maurice Ewing Ser 3. Am Geophys Union, Washington DC, pp 187-203

Rad von U, Ryan WBF et al (1979) Init Rep Deep Sea Drilling Project 47: part 1

Reyment RA (1969) Ammonite biostratigraphy, continental drift and oscillatory transgressions. Nature 224: 137-140

Reyre D ed (1966) Bassins sédimentaires du Littoral africain 1 (Littoral Atlantique), Assoc Serv Géol Afr, Paris

Riech V, von Rad U (1979) Silica diagenesis in the Atlantic Ocean: diagenetic potential and transformations. In: Talwani M, Hay W, Ryan WBF (eds) Deep drilling results in the Atlantic Ocean: Continental margins and paleoenvironment. Maurice Ewing Ser 3. Am Geophys Union, Washington DC, pp 315-340

Sclater JG, Hellinger S, Tapscott C (1977) The paleobathymetry of the Atlantic Ocean from the Jurassic to the present. J Geol 85: 509-552

Thiede J (1979) History of the North Atlantic Ocean: Evolution of an asymmetric zonal paleo-environment in a latitudinal ocean basin. In: Talwani M, Hay W, Ryan WBF (eds) Deep drilling results in the Atlantic Ocean: continental margins and paleoenvironment. Maurice Ewing Ser 3. Am Geophys Union, Washington DC, pp 275-296

Thiede J, Dean WE, Claypool GE (1981) Oxygen deficient depositional paleoenvironments in the Mid-Cretaceous tropical and subtropical central Pacific Ocean. Cretaceous Research, IGC Paris Black Shale Symposium (in press)

Tissot B, Deroo G, Herbin JP (1979) Organic matter in Cretaceous sediments of the North Atlantic: Contribution to sedimentology and paleogeography. In: Talwani M, Hay W, Ryan WBF (eds) Deep drilling results in the Atlantic Ocean: Continental margin and paleoenvironment. Maurice Ewing Ser 3. Am Geophys Union, Washington DC, pp 362-374

Tomczak M (1970) Schwankungen von Schichtung und Strömung im westafrikanischen Auftriebsgebiet während der "Deutschen Nordatlantischen Expedition 1937". "Meteor" Forschungsergebnisse A 7: 1-109

Turekian KK (1965) Some aspects of the geochemistry of marine sediments. In: Riley JP, Skirrow G (eds) Chemical oceanography, vol 2. Academic Press London New York, pp 81-126

Vail PR, Mitchum RM jr (1979) Global cycles of relative changes of sea level from seismic stratigraphy. In: Watkins JS, Montadert L, Dickerson PW (eds) Geological and geophysical investigations of continental margins. Am Assoc Petrol Geol Mem 29: 469-472

Viotti C (1967) Résultats stratigraphiques du sondage Puerto Cansado I du Bassin Côtier de Tarfaya. Notes Mém Serv Géol Maroc 175/I: 223-253, 6 pls, Rabat

Wedepohl KH (1964) Untersuchungen am Kupferschiefer in Nordwestdeutschland: Ein Beitrag zur Deutung der Genese bituminöser Sedimente. Geochim Cosmochim Acta 28: 305-364

Weichart G (1970) Kontinuierliche Registrierung der Temperatur und der Phosphatkonzentration im Oberflächenwasser des nordwestafrikanischen Auftriebsgebietes. Dtsch Hydrogr Z 23: 49-60

Wiedmann J (1976) Geo- und hydrodynamische Prozesse im Schelfbereich in ihrer Auswirkung auf mesozoische Fossil-Vergesellschaftungen. Zentralbl Geol Palaeontol II 1976: 424-439

Wiedmann J, Butt A, Einsele G (1978a) Vergleich von marokkanischen Kreide-Küstenaufschlüssen und Tiefseebohrungen (DSDP) Stratigraphie, Paläoenvironment und Subsidenz an einem passiven Kontinentalrand. Geol Rundsch 67: 454-508

Wiedmann J, Einsele G, Immel H (1978b) Évidence faunistique et sédimentologique pour un upwelling dans le Bassin Côtier de Tarfaya/Maroc dans le Crêtacé supérieur. Actes VI-ème Colloque Africain de Micropaléontologie, Tunis 1974. Ann Mines Geól 28, II: 415-441

17 Cretaceous Coastline-Connected Mass Movements, Southern Morocco

Gerhard Einsele[1]

Abstract

The migrating coastline of a marginal environment causes a variety of sedimentary facies including coastal mass movements. During Barremian to Albian time central parts of the pre-existing Moroccan coastal basins became very shallow or emerged. The interrelationship between sediment supply, subsidence, and shallow-marine to coastal and fluvial environment can be studied in good exposures along the present coastline. Large-scale load casts, ball and pillow structures, boulder beds and sandy mass flows deposited at the foot of ancient cliffs reflect the shallowest period within a rapidly changing environment. Some features require early-diagenetic carbonate cementation as known from beach rocks. The observed types of mass movements may be of diagnostic value for the recognition of ancient shorelines.

1. Introduction

In general, continental margins are the site of frequent mass movements of different nature and volume (e.g., Walker, 1973; Embley and Jacobi, 1977; Nardin et al., 1979; Dingle, 1980; Schwarz, in press). During the early stages of block-faulting, rifting, sea floor spreading, and rapid subsidence, youthful passive margins are usually characterized by a depositional environment with strong submarine relief. The slope between the shoreline and deep troughs often is subdivided into several steps of intermediate height (see e.g. Curray, 1980), and instability of slope sediments may occur at different localities and depths (Fig.1a). During a later stage of development when a thick wedge of sediments has been piled up at the continental margin, only one continuous continental slope is left (Fig.1b). On this type of slope, especially along its lower and intermediate part, occasional huge mass movements

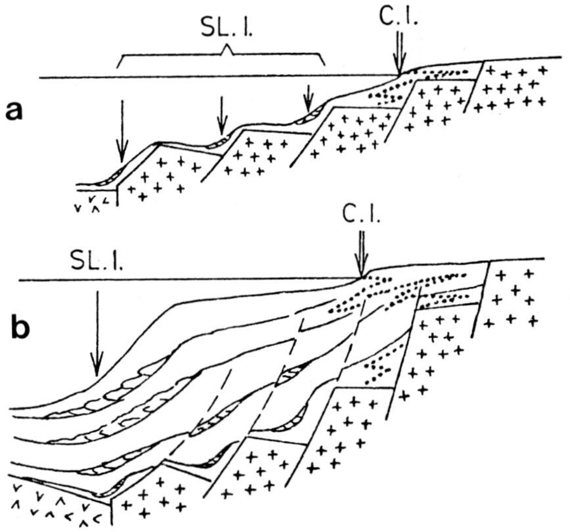

Fig. 1. Scheme to demonstrate the occurrence of slope instabilities (SL.I) and coastal instabilities (C.I.) at passive continental margins during *a* a youthful and *b* later stage of development

1 Institut für Geologie und Paläontologie, Universität Tübingen, Sigwartstraße 10, 7400 Tübingen, FRG

take place (Hinz et al., von Rad and Wissmann, this Vol.). Thus in the course of time, one can expect a development from numerous small and medium-sized mass movements to occasional events on a much larger scale.

There is, however, another process which frequently causes mass movements at continental margins. This process is related to coastal erosion, but may be also linked with contemporaneous faulting of the underlying basement (Fig.1) and thus be characteristic for a young passive margin. Studies on the stratigraphy and environment of the Cretaceous sediments deposited in Moroccan coastal basins (Wiedmann et al., 1978 and this Vol.) have revealed well-exposed examples of this second type of mass movements.

2. Occurrence and Special Features of the Coastal Mass Movements

2.1. General Setting and Stratigraphic Position

The three localitites where Early Cretaceous coastal mass movements were studied in some detail are shown in Fig.2. By chance, these localities coincide with the present coastline, providing cliffs 10 to 20 m high. There are more, but usually less well-exposed examples of contorted Cretaceous deposits or related features farther inland (Bergner, 1979; Behrens and Siehl, Wurster and Stets, this Vol.), indicating that the phenomena described here are of general importance for the understanding of the sedimentary processes in coastal basins. The coastline of the West African basins migrated back and forth several times due to sea level changes and non-equilibrium conditions between subsidence and sediment accumulation (Wiedmann et al., this Vol.). Special mass movements are, however, also restricted to certain sediment types and bed sequences.

At the localities studied the transition from the Barremian to the Aptian stage is characterized by sediments and faunas indicating a development from shelf to coastal and continental facies and back to marine environment. Shallow marine

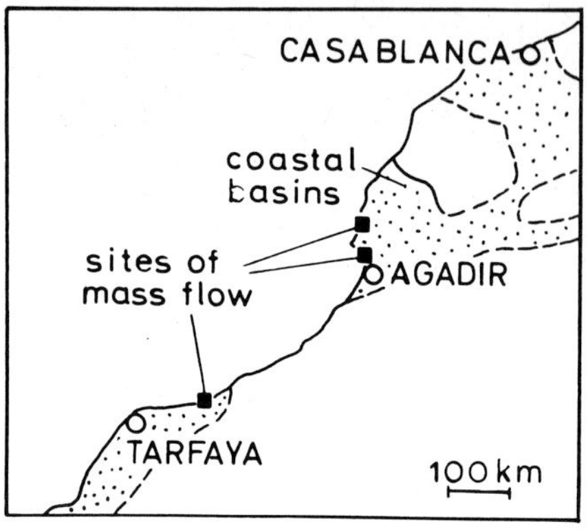

Fig. 2. Location map of sites with boulder beds and mass flows in sediments of Cretaceous coastal basins (stippled areas) of Southern Morocco. I to III see Fig.3

and coastal deposits are represented by silty clays, marls, sandy muds and sands, whereas the continental sediments are chiefly represented by sands and pebbly sands as well as intercalated clayey dark and red beds of a fluvial plain. The occurrence of dolomitic layers and rare nodules or thin bands of primary gypsum probably reflect the local development of small lagoons. Later, in higher Aptian and Albian time, marine marls reappear which alternate with some limestones and/or thin sand beds.

2.2. Section of Tan-Tan Plage

The sedimentary sequence of the three exposures demonstrating a Barremian-Aptian emergence, as well as mass flow phenomena are shown in Fig.3. The sequence with the least influence by continental sediments and only one special feature of penecontemporaneous mass movement is that of Tan-Tan Plage in the province of Tarfaya, Southwestern Morocco (Fig.3I). In the lower part of the profile, a sequence of mudstones with thin layers and lenses of sand and small sand-filled channels is observed. Wave ripples are abundant, and in some of the channels shells of pelecypods are concentrated. Similar, but considerably larger channels also occur in the upper part of the profile (beds C through D). Some of them cut deeply into the underlying beds and are filled with sand and marl. One channel has eroded more than 2.5 m pre-existing sediment of bed C which contains large cobbles and boulders of sandstone. The sand layer A and B show large-scale cross-bedding and contain, besides shells, some quartz pebbles which indicate that at least part of their material is derived from nearby land sources. Whereas the other beds described above clearly can be interpreted as characteristic layers within the intertidal realm (see e.g., Reineck and Singh, 1973; Ginsburg, 1975), the position of beds A and B is not quite clear. Either they result from the filling of wide shallow channels cut by intertidal streams near the coast, or they represent intertidal flat sand ridges or shoals (Newton and Werner, 1970; de Vries Klein, 1970). Similar beds in ancient sediments have been described by Raaf et al. (1977).

At first sight bed (C) was thought to represent a slump mass, but there is no clear evidence for slumping or mass flow in the sequence of this conspicuous layer. As indicated on Figs.3I and 4, the rounded, up to 2 m large sand bodies generally are in contact with each other. Their displacement is evident from the random orientation of their primary sedimentary structures (Fig.4), but a certain pattern of imbrication could not be ascertained. For an interpretation of this phenomenon, we have to take into account the possible influence of diagenesis. Today, most of the sands are cemented by carbonate, including dolomite. From the occurrence of primary gypsum and early diagenetic dolomite in the exposures north of Agadir we can conclude that climatic conditions favored the early formation of beach rock (see e.g., Bathurst, 1975). This would allow the following hypothesis: Beach and intertidal sands bordering the coast or emerging sandy islands are rapidly cemented by carbonate. Shortly after, the sandstones are undercut by wave erosion, and the displaced and rounded boulders are swept together to form local accumulations of blocky debris. Afterward, with continuing build-up of tidal and coastal sediments, the interstices between the sandstone boulders are filled with loose sand, and still somewhat later, the boulder bed can be scoured again by tidal channels eroding from a higher accumulation level. According to this model, the boulder bed represents the remnant of cliff erosion during a short event of emergence.

2.3. Agadir Section

The second example north of Agadir comprises more sediment types and a wider range of sedimentary structures (Fig.3II). Dark green (?) shallow-marine marls with intercalated thin marlstones are overlain by tidal flat deposits with channels similar to those of Tan-Tan Plage (E). The channels chiefly contain sand with some pebbles of quartz and schist as well as rare wood fragments. The abundant ichnofauna includes Diplocraterion and Arenicolites indicating a very shallow marine environment (see e.g., Frey, 1975).

The tidal flat deposits are overlain by dolomitic beds (F to G) which partly wedge out to be replaced by nodules or bands of gypsum and thin red beds (G). The primary origin of the gypsum is indicated by large solution cavities in one of the fine-grained dolomitic layers which, according to chemical analyses, contain approximately twice as much $CaCO_3$ as $MgCO_3$. This unit presumably represents lagoonal deposits.

The first horizon (H) with conspicuously contorted beds immediately follows on top of these lagoonal deposits. It comprises 3 m thick fine- to medium-grained sandstones with some minor mud intercalations. Large-scale trough cross-bedding and the lack of any fauna or bioturbation may signify a fluvial environment. The lower part of these sands is affected by load casting and the underlying muds show a kind of flame structure, but completely isolated "balls and pillows" are an exception.

In the following part of the sequence (beds J through K) a tidal or lagoonal regime prevails with carbonate ($CaCO_3/MgCO_3$ ratio approximately 2:1) as the most important sediment component. A second fine- to medium-grained series of contorted sandstones (L) on top of these carbonates starts with two horizons of large-scale balls and pillows which originate from a primarily coherent sand bed with megaripples. The upper part of these beds has probably been truncated by erosion.

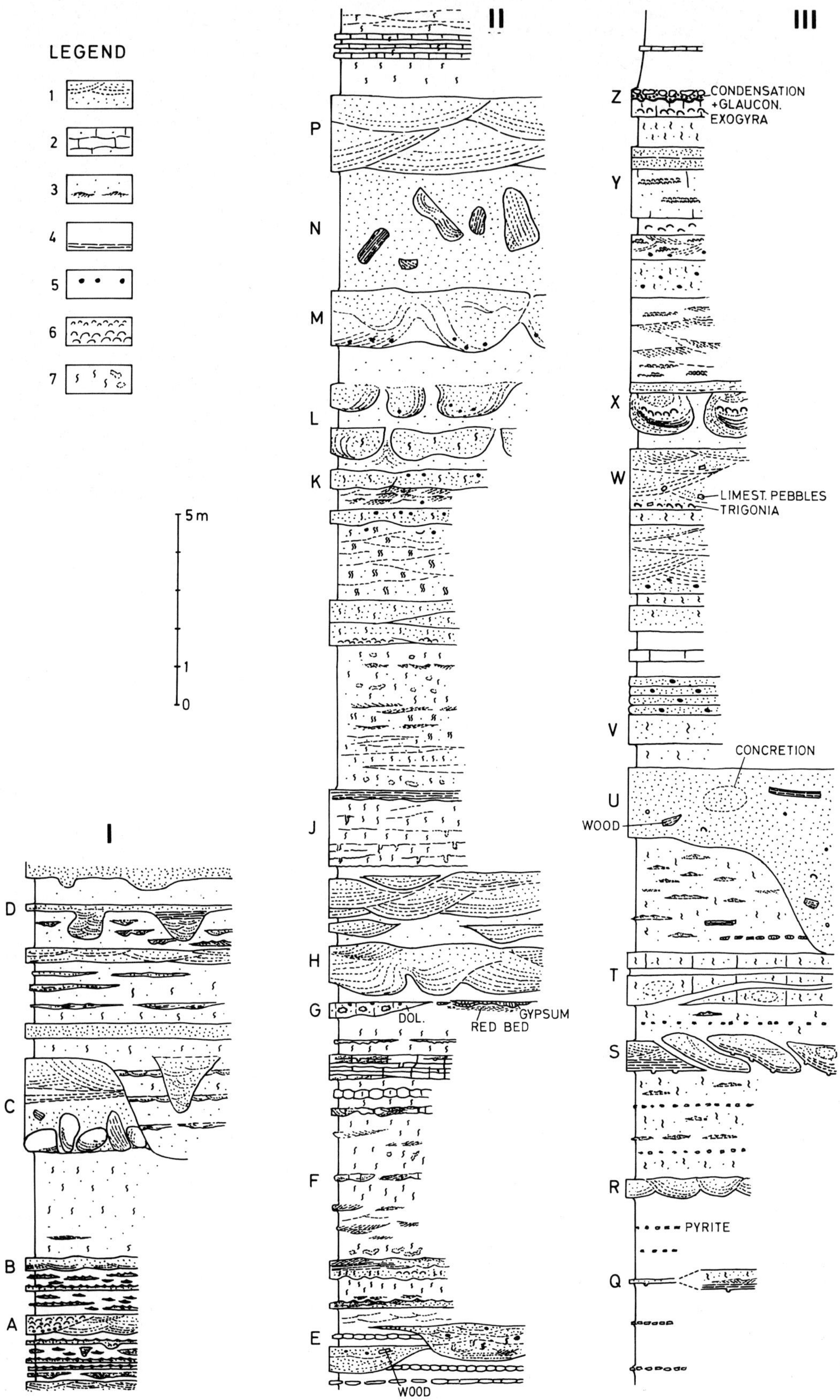
LEGEND
1
2
3
4
5
6
7
5m
1
0
I
D
C
B
A
II
P
N
M
L
K
J
H
G
DOL.
RED BED
GYPSUM
F
E
WOOD
III
Z
CONDENSATION
+GLAUCON.
EXOGYRA
Y
X
W
LIMEST. PEBBLES
TRIGONIA
V
CONCRETION
U
WOOD
T
S
R
PYRITE
Q

Fig. 4. Barremian/Aptian boulder bed exposed at a recent cliff, undercut by wave action, Tan-Tan Plage (exposure I of Figs.2 and 3). Note the different orientation and composition of boulders which usually are in contact with each other

A third, 1.5 m thick sand bed (M) resembles horizon (H) and is only affected by some load casting as described above. The next 3- to 4-m thick layer (N), however, represents a genuine mass flow (Fig.6). Its matrix consists of structureless fine to medium sand cemented by carbonate (Fig.7), and in the matrix there float large chunks and boulders of layered sandstones and mudstones in random orientation. The mass flow deposit is overlain by another sequence of cross-bedded coastal or fluvial sands (P), lagoonal, and tidal sediments.

In contrast to the Tan-Tan section, we have here clear evidence of a mass flow. The cause of this mass flow, however, may also have been cliff erosion which cut into nonindurated or partly solidified sands and thus created the necessary relief.

2.4. Section of Imsouane

In the third example, an excellent cliff esposure at the present coast south of Imsouane, the phenomena already described are repeated, but somewhat modified and amplified (Fig.3III). A few special sedimentary structures are difficult to interpret, but in general the situation may be explained as follows:

The lowermost section (Q) consists of dark marly shales with chiefly thin (1 - 3 cm) beds of carbonate-cemented fine sands and layers rich

Fig. 3. Upper Barremian to Lower Aptian bed sequences with load casting and mass flow phenomena. I Tan-Tan Plage; II coast 20 km north of Agadir; III coast south of Imsouane, 25 km north of Cap Rhir. Further explanations see text. 1 sandstone, often cross-bedded, generally carbonate-cemented, partly also containing detrital carbonate (total carbonate content varying between 10 and 70%); 2 dolomitic beds, often rather sandy; 3 muddy sands or sandy muds, often with lenses of sand, indurated; 4 clayey siltstone or silty claystone and marlstone, often containing plant debris and pyrite, partly well laminated; 5 pebbles of quartz or schist; 6 coarse or fine-grained shell bed; 7 bioturbation, with small and large burrows (including Diplocraterion and Arenicola)

in pyrite concretions. The sandstone beds (e.g., bed Q) show sedimentary structures in a sequence from parallel lamination at the base to small-scale cross-bedding and fine-grained broken shells with bioturbation at the top. These beds may represent storm layers in a shallow marine, possibly somewhat restricted environment (see e.g., Reineck and Singh, 1973; Goldring and Bridges, 1973). Bed R is a large-scale cross-bedded sandstone which is partly affected by load casting. The sand is somewhat bioturbated and appears to have been transported chiefly from a western source. The bed may be interpreted as near-shore sand migrating toward or along the coastline farther east. It is overlain by tidal deposits consisting of dark strongly bioturbated marls with small lenticular sands (flaserbedding).

In contrast to bed R, the fine to medium sand of the following bed S has been transported from east to west (N 220° - 260°), i.e., from the land toward the sea. This is documented by flute casts and large-scale cross bedding on top of the parallel laminated basal part of the layer. Burrows were observed only on the lower bedding plane. The origin of this bed is not quite clear. Either it has been formed at the shore face, or as a washover fan on top of bioturbated intertidal muds (Blatt et al., 1980). In the eastern part of this outcrop, bed S is replaced by channel deposits contorted by load casting and some mass flow (Figs.5 and 8). At the transition from bed S (in place) to the mass flow, parts of the same bed appear to have been slightly stacked up in a western or southwestern direction (Fig.3). Such a process is also indicated by slide marks at the sole of these blocks. It signifies that the sands must have been somewhat indurated prior to the mass movement. Later, the cementing carbonate was concentrated in the displaced, but still unconsolidated sandstone boulders and locally formed large concretions of dense hard sandstone. A similar concretionary cementation was also observed in an overlying dolomitic sandstone (T) which follows on top of dark silty-sandy and marly shales. This layer consists of several large lenses which show parallel lamination or low-angle cross-bedding as well as bioturbation. It may re-

Fig. 5. Overview of cliff exposure S Imsouane, site III (Figs.2 and 3). On the left-hand side, mass flow U wedges out and is underlain by tidal or lagoonal muds. On the right-hand side, part of the boulders of layer U appears to be affected only by load casting and slight lateral displacement. Layer S also shows some lateral displacement, particularly further east (right-hand side, Fig.8) which is not seen at the overlying horizon T

Fig. 6. Floating blocks and chunks of sandstones and marlstones in structure-less sandy matrix, exposure II north of Agadir, layer N (cp. Figs.2 and 3)

Fig. 7. Thin-section of fine- to medium-grained sandy matrix of layer N, exposure II (Fig.3 II and Fig.5). The grain size distribution is typical for all the beds affected by mass movement (see also Dalrymple, 1979). Note the presence of some finer grained silt and clay and the lack of distinct particle orientation. (Vertical to bedding plane)

present beach sands which truncated the flow mass of horizon S and cover it now uniformly.

After a further interval of tidal and possibly also lagoonal sedimentation (dark pyritic silty marls, flaser-bedded sands, ? primary gypsum, wood fragments), a second mass flow (bed U) has affected the sequence. In the western part of the exposure, some primary sedimentary structures are preserved at the base of the bed, whereas the higher part consists of massive structureless sands with some large lumps of angular marl and single quartz pebbles. Farther east, the flow mass deeply cuts into the underlying tidal sediments and contains many big boulders of sandstone, layered marls and large oysters floating in the sandy matrix in random orientation (Fig.8). At some distance away, however, the boulders resemble pillows originating from load casting which are only slightly displaced by lateral movement but not turned around. Hence, the exposure demonstrates the most active center of a mass flow as well as its marginal parts, which have moved slower and have exerted less influence on the underlying beds and on the orientation of displaced lumps.

The overlying sequence (V to W) is characterized by thick calcareous to dolomitic sandstones, part of which are burrowed and contain oysters. A 3 m thick section (W) shows trough cross-bedding and contains coarse limestone pebbles. This bed probably represents a second fluvial or foreshore event after the deposition of tidal sands.

Finally, an impressive load cast horizon (X) is developed, before the

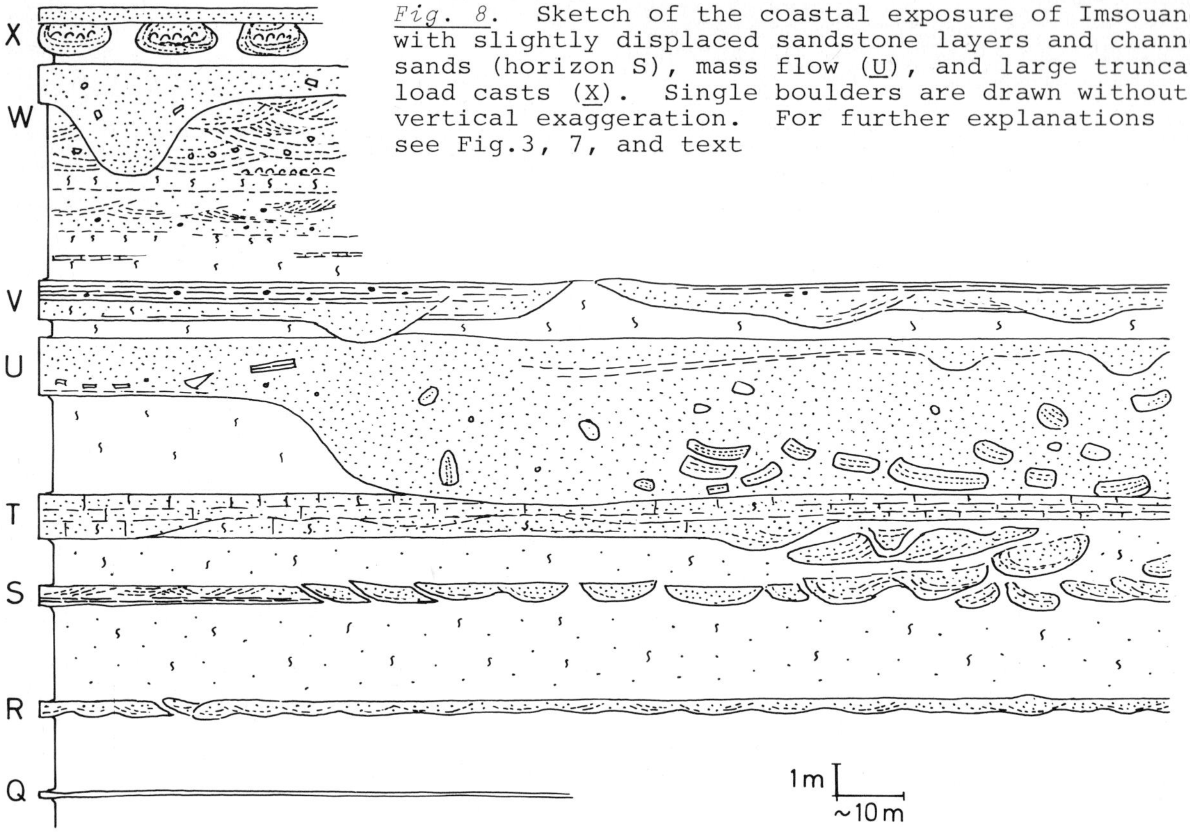

Fig. 8. Sketch of the coastal exposure of Imsouane with slightly displaced sandstone layers and channel sands (horizon S), mass flow (U), and large truncated load casts (X). Single boulders are drawn without vertical exaggeration. For further explanations see Fig.3, 7, and text

environment shifts again to fully marine conditions indicated by the recurrence of ammonites and glauconite in a condensed bed (Z). Then marls and limestones of Middle Aptian time follow. The large isolated pillows of horizon X have long axes in NE-SW direction corresponding with the paleocurrent direction of the large-scale cross-bedded sands. Layers of shells in the upper half of the pillows reflect marine influence.

3. Discussion and Conclusions

In the coastal basins of Morocco, contorted, reworked, or mass flow deposits of Barremian to Aptian age show the following features (Figs.3 and 9):

- Load casting of coherent trough cross-bedded sand layers,
- Isolated balls and pillows caused by load casting,
- Accumulations of sandstone boulders,
- Sandy mass flows with isolated lumps and boulders of sandstone and marls.

An analysis of the environmental conditions reveals that these phenomena were generated during a period of shallowing sea and short emergence (Fig.9). In all three exposures studied the facies changes from shallow marine and tidal to beach or fluvial environment, in two of the exposures repeatedly. In this respect, it is important to note that the different mass movements took place during the "shallowest" environmental interval. This does not necessarily imply an erosional phase for the generation of all the phenomena described, but the boulder bed and the sandy mass flows can hardly be explained without an emergence and some erosion. In these cases probably a slight uplift of the eastern landward part of the coastal basin has enabled the approaching sea to cut wave terraces and cliffs, and thus to produce the relief necessary for mass flows. Prior to cliff erosion and displacement, part of the emerged sands was sufficiently in-

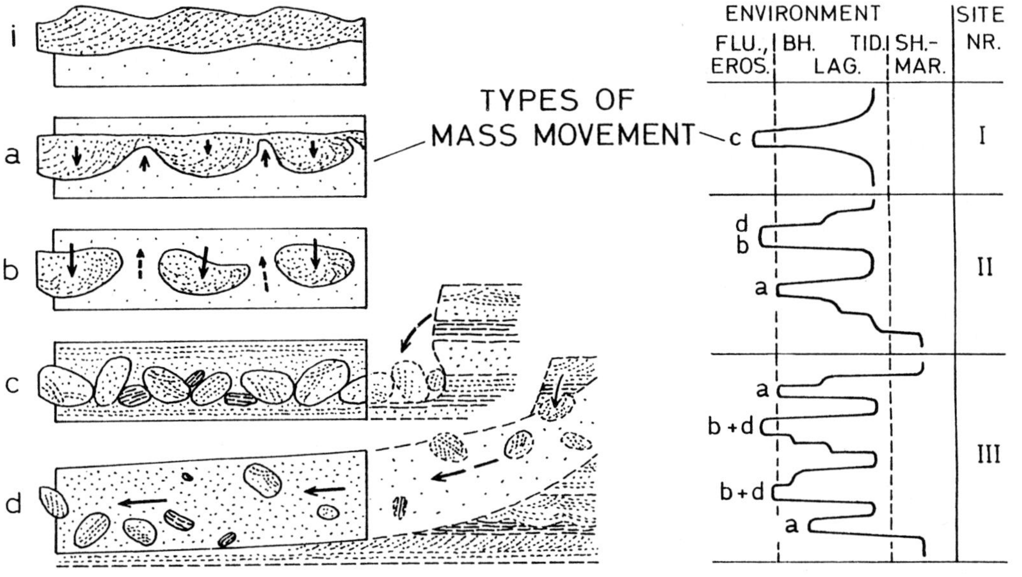

Fig. 9. Different types of mass movements in sandy coastal environment (left) and their occurrence and environmental interpretation at sites I to III (right, compare Fig.3). a load casting of thick cross-bedded sand layer (i), top truncated by erosion; b ball and pillow structures produced by load casting of thick sand bed; c boulder bed due to cliff erosion; d floating boulders of sandstone and mudstone in sandy structureless matrix (mass flow). (FLU. fluvial; EROS. coastal erosion; BH beach; LAG. lagoonal; TID. tidal; SH.-MAR. shallow-marine)

durated to form coherent boulders when the steepwalled cliffs failed. Marls or shales were solidified by desiccation.

The mechanisms of load casting, the formation of ball-and-pillow structures, as well as of sandy mass flows, have been described by many authors (e.g., Dżulyński and Walton, 1965; Middleton and Hampton, 1976; Lowe, 1979; Blatt, Middleton & Murray, 1980). Therefore, only a few remarks to the initiation of mass flow are added: Although mass flows can also occur below sea level and appear to be rather frequent within the range of subaqueous wave and current action (Lowe, 1976; Rust, 1977; Dalrymple, 1979), the examples described here are probably best explained by emergence. Relief and rapid drop of the ground water level on top of rather loosely packed sand or mud provide the most powerful mechanism for load casting and slope instabilities. Due to the lowering of the water table, effective pressure in an underlying bed is increased and initiates the transformation of the original fabric into a more densely packed structure. The excess pore water cannot escape quickly enough to prevent the loss of bearing capacity or the onset of mass flow (liquefaction). Once a sliding process has started, failure progrades backward from the neighborhood of a cliff or steep-walled fluvial channel into flat-lying emerged sediments and may affect large areas. At first, blocks and chunks from the uppermost layers ride on the flowing mass (matrix-supported flow), but are later incorporated into the matrix and sometimes concentrated at their base. Already the presence of a small proportion of "clay" may prevent the interlocking of grains of a poorly sorted sand (Fig.7) and thus substantially reduce the strength of a grain flow (Rodine and Johnson, 1976).

In a simplified way, the special episode of coastal basin sedimentation during Barremian-Aptian time is combined by Fig.10 in several stages of development. From stages I to III subsidence is decreasing or, probably more realistic, is surpassed by a relatively high rate of sedimentation. Thus, stage III is dominated by coastal and fluvial sediments. Individual sand beds of this environment are deposited so rapidly on top of soft muds that load casting takes

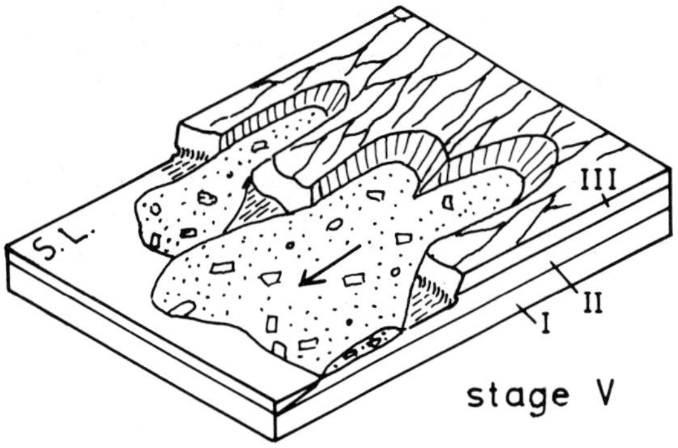

Fig. 10. Environmental interpretation of the Barremian/Aptian bed sequences in six stages of development. Stage VI again corresponds to stages I and/or II. Note that emergence can be also caused by a prograding deltaic sequence over a constantly subsiding area

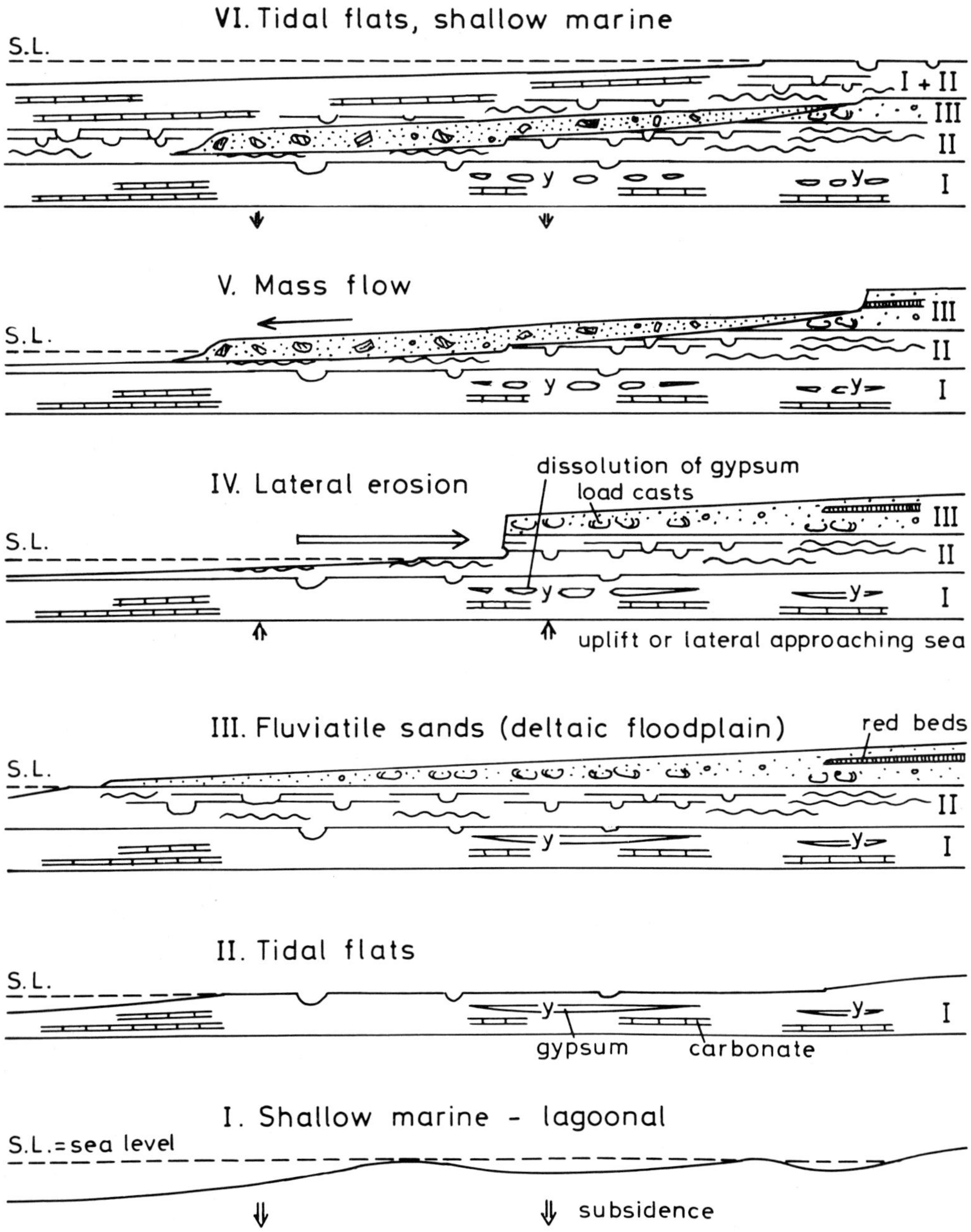

place. The returning sea starts to rework the emerged sediments and locally creates sufficient relief for the initiation of mass flows (stages IV and V). Finally, continuing subsidence and landward migration of coastline, storm flood erosion, shifting of tidal channels and estuaries again restore a tidal flat and shallow marine environment (stage VI).

Acknowledgements

The exposures mentioned in this study were visited during two joint field trips with J. Wiedmann and some other colleagues of the Geological Institute of Tübingen in 1973 and 1975. The German Research Foundation (DFG) granted financial support. Drs. U. von Rad and F. Werner critically read the manuscript and made helpful suggestions.

REFERENCES

Bathurst RGC (1975) Carbonate sediments and their diagenesis, 2nd edn.Dev Sedimentol 12: 658

Bergner HD (1978) Die Kreide im westlichsten Hohen Atlas (Marokko) unter besonderer Berücksichtigung eines Kondensationshorizonts im Aptum. Diplomarb Geowiss Fak Univ Tübingen, 321 p

Blatt M, Middleton G, Murray R (1980) Origin of sedimentary rocks, 2nd edn. Prentice-Hall, New Jersey, 782 p

Curray JR (1980) The IPOD programme on passive continental margins. In: The evolution of passive continental margins in the light of recent deep drilling results. Philos Trans Roy Soc London Ser A 294: 5-16

Dalrymple RW (1979) Wave-induced liquefaction: a modern example from the Bay of Fundy. Sedimentology 26: 835-844

Dingle RV (1980) Large allochthonous sediment masses and their role in the construction of the continental slope and rise off southwestern Africa. Mar Geol 37: 33-354

Dżulyński S, Walton EK (1965) Sedimentary features of flysch and greywackes. Dev Sedimentol 7: 274

Embley RW, Jacobi RD (1977) Distribution and morphology of large submarine sediment slides and slumps on Atlantic continental margins. In: Richards AF (ed) Marine slope stability. Mar Geotechn 2: 205-228

Frey RW (ed) (1975) The study of trace fossils. Springer, Berlin Heidelberg New York

Ginsburg RN (ed) (1975) Tidal deposits, a casebook of recent examples and fossil counterparts. Springer, Berlin Heidelberg New York

Goldring R, Bridges P (1973) Sublittoral sheet sandstones, J Sediment Petrol 43: 736-747

Lowe DR (1976) Subaqueous liquefied and fluidized sediment flows and their deposits. Sedimentology 23: 285-308

Lowe DR (1979) Sediment gravity flows: their classification and some problems of application to natural flows and deposits. Soc Econ Paleontol Mineral Spec Publ 27: 75-82

Middleton GV, Hampton MA (1976) Subaqueous sediment transport and deposition by sediment gravity flows. In: Stanley DJ, Swift DJP (eds) Marine sediment transport and environmental management. Wiley, New York, pp 197-218

Nardin TR, Hein FJ, Gorsline DS, Edwards, BD (1979) A review of mass movement processes, sediment and acoustic characteristic, and contrasts in slope and base-of-slope systems versus canyon-fan-basin floor systems. Soc Econ Paleontol Mineral Spec Publ 27: 61-73

Newton RS, Werner F (1970) Form und Schichtungsgefüge periodischer Sandkörper im Strömungsfeld des Außenelbewatts. Geol Rundsch 60: 321-330

Raaf JFM de, Boersma JR, Gelder A van (1977) Wave-generated structures and sequences from a shallow marine succession, Lower Carboniferous, Country Cork, Ireland. Sedimentology 24: 451-483

Reineck HE, Singh JB (1973) Depositional sedimentary environments. Springer, Berlin Heidelberg New York

Rodine JD, Johnson AM (1976) The ability of debris, heavily freighted with coarse clastic materials, to flow on gentle slopes. Sedimentology 23: 213-234

Rust, BR (1977) Mass flow deposits in a Quaternary succession near Ottawa, Canada: diagnostic criteria for subaqueous outwash. Can J Earth Sci 14: 175-184

Schwarz HU (in press) Modern subaqueous slope failures - overview and occurrences. J Sediment Petrol

Vries Klein de G (1970) Depositional and dispersal dynamics of intertidal sand bars. J Sediment Petrol 40: 1095-1127

Walker RG (1973) Mopping up the turbidite mess. In: Ginsburg RN (ed) Evolving concepts in sedimentology. John Hopkins Univ Press, Baltimore London, pp 1-37

Wiedmann J, Butt A, Einsele G (1978) Vergleich von marokkanischen Kreide-Küstenaufschlüssen und Tiefseebohrungen (DSDP): Stratigraphie, Paläoenvironment und Subsidenz an einem passiven Kontinentalrand. Geol Rundsch 67 (2): 454-508

18 Sedimentation in the Atlas Gulf I: Lower Cretaceous Clastics

Manfred Behrens and Agemar Siehl[1]

ABSTRACT

The transgressions and regressions in the Atlas Gulf control the facies of the clastic deposits during the Early Cretaceous. As the coastline oscillates, the succession of varied environments migrates systematically landward and seaward. Large terrigenous supply is characteristic of regressive periods (Upper Hauterivian, Lower Aptian) and small supply of transgressive periods (Valanginian, Barremian, Upper Aptian). The distributive province of the terrigenous detritus in the eastern hinterland of the gulf (e.g., the Precambrian crystalline massifs of the Central High Atlas) remains constant. The clastics, transported westward through the Atlas Basin and decreasing in grain size, reach the continental slope and are deposited as turbidites in the contemporaneous offshore sediments in DSDP Site 370/416.

The following sedimentological and petrographical parameters were used for the analysis of the paleogeographical and paleoenvironmental development: The directional distribution of cross bedding and sole marks, the feldspar content, the regional distribution of the main constituents (insoluble residue, calcite, and dolomite), the calcium content of the dolomites, the change of the sediment thickness, and the occurrence of corals and redbeds.

1. Introduction

At the end of the Jurassic period a broad westward dipping carbonate platform under the present sea level was developed off Northwest Africa (cf. Hinz et al., this Vol.). Landwards, it extends to the longitude of Chichaoa in the region of the western Atlas Mountains. Its marine to lagoonal limestones, dolomites, and marls of Bathonian to Portlandian age already distinctly shape the coastal basins along the passive continental margin. During the Jurassic/Cretaceous transition, the Atlantic withdrew westward: only between Essaouira and Agadir along the present coast are continuously marine sections found. In the Meseta and Atlas regions continental redbeds were deposited on marine Upper Jurassic. In Valanginian times, the sea once again advanced toward the east. The Moroccan Meseta was very extensively flooded east of the Safi – El Jadida line and hence experienced the first marine transgression since the Hercynian orogeny. In the western Atlas an extended marine sedimentation area developed during the Early Cretaceous, reaching continuously from the coastal basins toward the east and bordered in the south by the uplands of the Anti-Atlas: The Atlas Gulf (Stets and Wurster, 1977; Behrens et al., 1978; Wurster and Stets, this Vol.; Stamm and Thein, this Vol.; cf. also Roch, 1930; Choubert and Faure-Muret, 1962). Here the Lower Cretaceous series show a systematic facies change from well-stratified marine deposits in the west to fluvial, undifferentiated accumulations of sediments in the east. The coastline repeatedly oscillated due to the contrasting and discontinuously acting processes of sea-level changes, subsidence, and sedimentation (Fig. 1).

The flooding which began with the Valanginian continued until the close of the Early Hauterivian. It ended with the formation of a broad-faced coral reef parallel to the coast. During the Late Hauterivian the

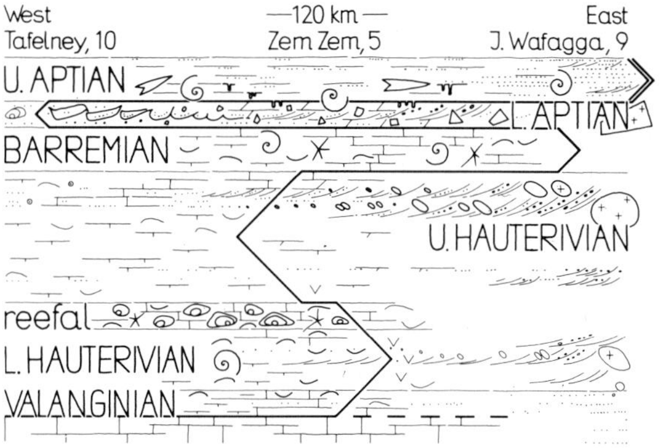

Fig. 1. Facies diagram of the early Lower Cretaceous in the western Atlas Gulf. Symbols of lithology and sedimentary structures explained in Fig. 13

1 Geologisches Institut der Universität Bonn, Nussallee 8, D–5300 Bonn 1, FRG

Atlantic retreated from the wedge of sand and fine gravel shed from the east. The Barremian sea invaded north of the High Atlas toward the east beyond the meridian of Chichaoua–Taroudannt (Roch, 1930) and produced a thin, but extensive and uniform carpet of fossiliferous marls and limestones. In the Early Aptian, a conspicuous regression ensued once again, during which limnic conditions occurred even in the persistent coastal basins (Wiedmann et al., 1978). Finally during the Aptian, the Atlantic flooded the Atlas Gulf particularly far toward the east, as proven by ammonite discoveries near Ait Ourir and at the plateau du Kik near Azni (Moret, 1931; Choubert and Faure-Muret, 1962). In the research area this event is marked by a marine layer 30–50 m thick, which is distinguished by frequent hardgrounds, as well as by the stratigraphically condensed ammonite fauna rich in species and individuals (Wiedmann et al., 1978). The characteristic succession of the Aptian presents an excellent key horizon for correlation of the sections.

The main features of this facies zonation were already described by Roch in 1930; he dated the sequence by means of macrofauna. Duffaud supplemented the subdivision in 1960 with the aid of microfaunas in the western Atlas area. Ambroggi (1963) described and classified the Cretaceous strata near Agadir. Finally, as part of the Deep Sea Drilling Project during the 1970's, several teams from the Geological Institutes of the Universities in Bonn and Tübingen studied the Moroccan Cretaceous, in part biostratigraphically (Wiedmann et al., 1978), in part sedimentologically, emphasizing the paleogeography (Behrens et al., 1978). In the following we will analyze and interpret the facies development in the Atlas Basin during the Early Cretaceous (Valanginian – Aptian).

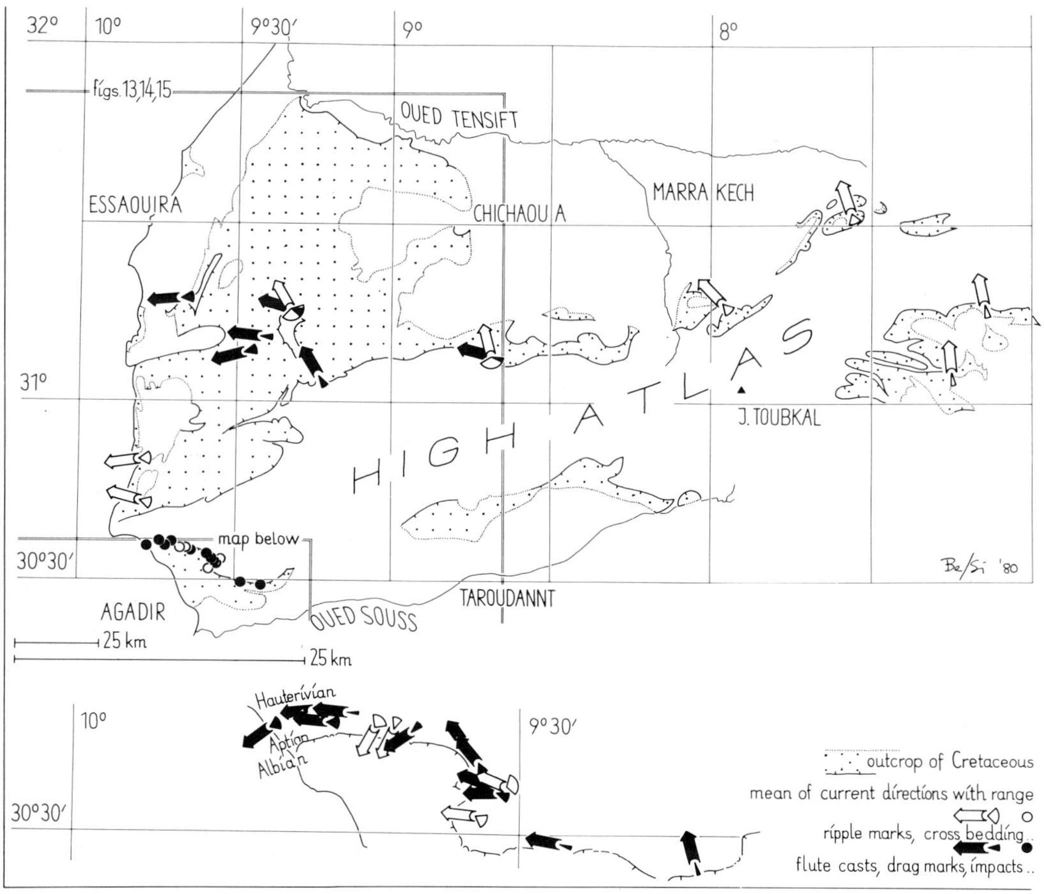

Fig. 2. Transport directions in the Lower Cretaceous north and south of the High Atlas Mountains. Cross-bedding, abundant in the continental series of the eastern part, indicates the general direction of braided rivers transporting the material from the source area toward the north and west. Sole marks prevail in the marine sediments of the western part; they indicate westward-directed paleocurrents following a plain slightly inclined toward the continental slope

2. Transport Directions

Each current direction indicated in Fig. 2 applies to the entire stratigraphic column of the Lower Cretaceous. In the sections southeast of Marrakech the dating is not quite certain: the samples are from the upper third of the redbeds under the marine Aptian-Albian successions, or, where this is missing, from the middle part of the redbeds below the marine Turonian. Cross-bedding and transverse ripple marks on the one hand, as well as sole marks on the other, were evaluated and treated separately. These two groups display greater directional divergence from each other than the measurements of each of these groups in one stratigraphic sequence. The current direction vectors are based on about 400 cross-bedding planes and about 100 ripple marks (open arrows) in addition to about 500 sole marks (black arrows).

Cross-bedding and ripple marks dominate in the continental redbeds of the eastern area. Here the transport directions point northwest- and westward, partly with a definite angle to the chain of the present High Atlas. They indicate the main direction of the drainage system from the highlands which presumably existed in the outcrop area of the Precambrian crystalline basement of the Toubkal-Ouzellarh massif in the High Atlas as well as in the northern part of the central Anti-Atlas. In the marine environment to the west, the transport directions generally follow the main angle of inclination down to the continental slope. In the area north of Agadir which we worked out in detail, we noticed a considerable local meandering of the transport direction, which may also be assumed elsewhere in the Atlas Gulf.

3. Feldspar Distribution

Being clasts which are sensitive to transport, feldspar grains can illustrate the dispersal pattern more clearly than, for example, transport-resistant heavy minerals. Qualitative and quantitative analyses of the feldspar spectra of sand and sandstone have repeatedly furnished concise paleogeographical arguments (Füchtbauer, 1967; Heim, 1966; Schnitzer, 1957; Valeton, 1953).

Figure 3 shows the total feldspar content as well as the Na+Ca-feldspar / K-feldspar ratios for certain localities, treating separately the sections of the Upper Hauterivian (lower semicircle) and that of the Lower Aptian (upper semicircle). In the east and south, where this stratigraphic differentiation was not possible, only the upper or the middle section of the redbeds was examined. In total, 550 diffractometrical analyses of ground rock were evaluated and then calculated sectionwise as average rank values. These range from 0 (= no feldspar) to 5 (= largest proportion of feldspar). The sealing of the pores due to the mainly dolomitic carbonate cement and the intensive neomorphic sparitization protected the feldspars from the influence of weathering and intrastratal solution, so that they generally appear fresh in thin section. Only detrital or even overgrown feldspars were observed, quite frequently microcline (see Figs. 6, 7). The proportion of feldspar decreases in general from east to west and northwest (Fig. 3). At the same time the grain size of the clastics from middle or fine gravel is reduced to fine sand and silt. This distribution is presumably due to a transport segregation, with the original grain size of quartz being smaller than feldspar, which would be in harmony, for instance, with a granite. Such source rocks occur immediately south of Sections 26, 25, 23 and 22 in the ancient crystalline massifs of the Central High Atlas, as well as in the Anti-Atlas south of Sections 24 and 27. While these areas are most likely the source areas during the Early Cretaceous, the Hercynian massif of the Tichka granite, lying 50 km westward between Sections 9 and 28, was not reached by erosion until the early Tertiary (Brix, 1981).

In the *Upper Hauterivian* the decrease of the feldspar content from east to west confirms the transport directions already discussed. This distribution is true even for several areas in the marine regime (e.g., Taroudannt, Agadir, Kap Ghir). Repeated reworking of siliclastic material together with bioclasts took place in the coastal fringe. The pebbles are well rounded (Figs. 4, 5, 6). Polymictic breccias and conglomerates, characteristic of the Lower Aptian, are missing in the Upper Hauterivian.

The coarse sand fractions (2.0 – 0.2 mm) contain large amounts of feldspar compared with the fine fractions (0.2 – 0.06 mm). Low average quartz/feldspar ratios of 3.1 (coarse) and 6.5 (fine) can be observed in Sections 9 and 26 north of the central High Atlas and high quartz proportion (22.8, coarse / 52.4, fine) in the westernmost sections (17). It is remarkable that the feldspars evidently were not much affected by transport. Otherwise they would have moved into finer fractions. The good preservation of the feldspars is an argument against their recycling from Triassic clastics in the hinterland. In the *Lower Aptian* the ratio of Na+Ca-feldspar to potassium feldspar is similar to that in the Upper Hauterivian. This points to the continuous influence of one single source. In many cases the proportion of feldspar is larger in the Aptian than in the Hauterivian. The same applies to the grain size. For a coastal section (17), the average quartz/feldspar ratios are 4.8 (coarse) and 17.9 (fine). In the Lower Aptian, the feldspars are accompanied by immature pebbles of angular to subrounded Precambrian rhyolites, quartzites and dolomites up to 10 cm in size (Figs. 7, 8, 9). The high feldspar content and the nature of the pebbles document considerable uplift and erosion of a crystalline mountain range to the east, probably not very far away. The fluvial to limnic environment occupied the Atlas Gulf far to the west. Charophytes indicate freshwater conditions, even near the recent Atlantic coast (Wiedmann et al., 1978).

A systematic evaluation of the heavy mineral content was not carried out because of the dolomite cement which can only be dissolved by an intensive acid treatment which falsifies the heavy mineral spectrum. Only a small number of samples (about 70) from the Upper Hauterivian yielded good separations with cold acetic acid. In all samples studied a stable spectrum dominates with large proportions of zircon, besides tourmaline,

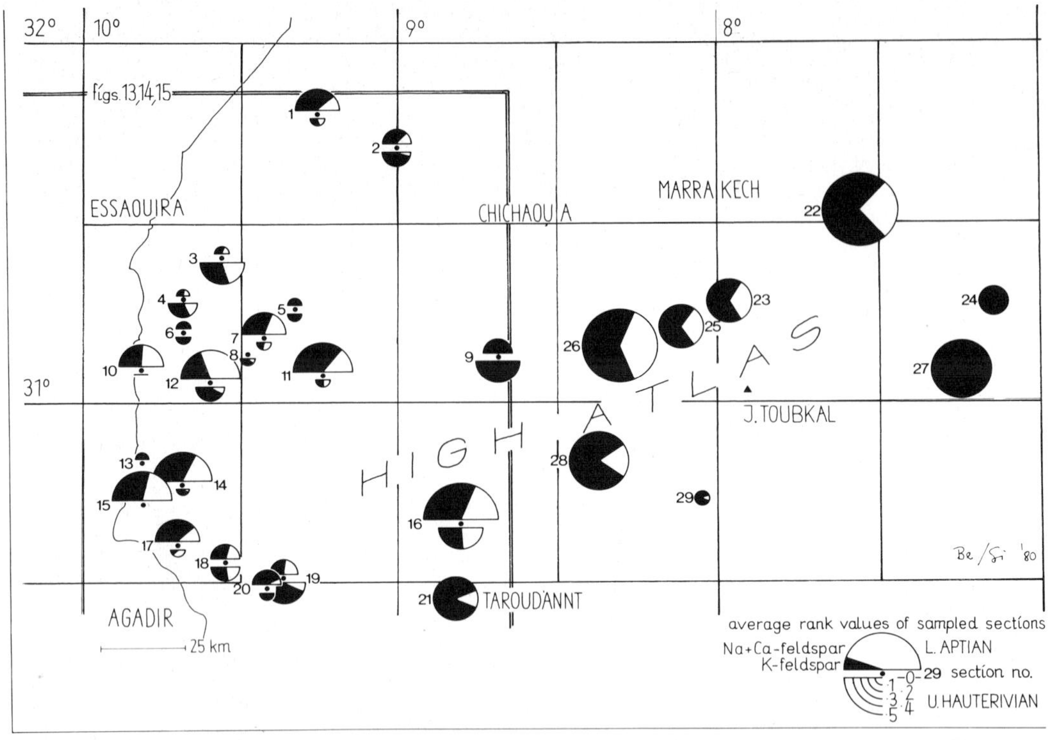

Fig. 3. Regional pattern of feldspar content in the clastics of the Upper Hauterivian and Lower Aptian estimated from diffractograms of total rock samples as rank values and averaged section-wise. Average rank values correspond to diameters of circles, scale is given in figure. High quantities of feldspar are correlated with coarse clastics in the surroundings of the Central High Atlas and disclose the proximity of the distributive province. Sections sampled by the authors are numbered as follows: **1** J. Skiat (31°46.0'N, 9°15.6'W); **2** J. Jefra (31°41.3'N, 9°1.2'W); **3** J. Tamergate (31°24.5'N, 9°33.5'W); **4** O. Tidzi (31°17.2'N, 9°41.2'W); **5** Zem Zem (31°14.8'N, 9°19.8'W); **6** Smimou (31°12.0'N, 9°40.3'W); **7** Takoucht (31°11.8'N, 9°26.2'W); **8** Imin Tlet (31°9.7'N, 9°30.3'W); **9** J. Wafagga (31°8.6'N, 8°42.7'W); **10** C. Tafelney (31°5.6'N, 9°49.0'W); **11** Talezza (31°5.3'N, 9°14.4'W); **12** Ougouelloul (31°2.5'N, 9°35.7'W); **13** Imsouane (30°51.3'N, 9°48.8'W); **14** Ait Messaoud (30°45.3'N, 9°40.8'W); **15** Tamri (30°42.7'N, 9°49.0'W); **16** Afensou (30°41.5'N, 8°47.8'W); **17** Tagourzamt (30°36.6'N, 9°42.8'W); **18** Tamraght (30°33.6'N, 9°33.2'W); **19** Ameskhoud (30°31.4'N, 9°21.7'W); **20** Bouyaya (30° 29,4'N, 9°24.6'W); **21** J. Moughal (30°27.4'N, 8°49.8'W); **22** O. Zat (31°32.6'N, 7°34.2'W); **23** Moulay Brahim (31°17.2'N, 7°58.1'W); **24** Animiter (31°16.6'N, 7°8.3'W); **25** Aouzzer (31°12.6'N, 8°6.6'W); **26** Azegour (31°8.2'N, 8°19.6'W); **27** Amezgane (31°6.2'N, 7°15.8'W); **28** Tizin Test (30°50.2'N, 8°24.6'W); **29** Aoufour (30°44.6'N, 8°2.0'W)

rutile, and brookite. Among the instable minerals apatite is always present; it is especially abundant in samples with a high feldspar content. Garnet is relatively rare; however, it also occurs in the coastal sections, while pyroxene and amphibole are only present in the landward sections. All this indicates a spectrum which is already relatively poor primarily and originates from acidic magmatites, eroded in the eastern margins of the Atlas Basin. This is in full agreement with the conclusion drawn from the regional distribution of the feldspars. Authigenic barite is quite frequent, anatase is also present.

4. Stratigraphical Sections

The biostratigraphical dating of the Lower Cretaceous sequence in the High Atlas proceeds from the full marine facies developed in the present coastal region (Roch, 1930; Ambroggi, 1963; Wiedmann et al., 1978). The excellent outcrops and the lithologically very clearly subdivided strata permit a reliable assignment to the different stages of the Cretaceous, even in the area where the beds alternate with those more strongly influenced by the continent. This holds true for the western Atlas Gulf approximately as far east as the line Chichaoua–Taroudannt, where the Cretaceous beds form a continuous cuesta landscape. The detailed fauna and strata descriptions by Roch (1930) are helpful, as is the excellent Geological Map 1 : 100 000 (Sheet El Khemis des Meskala, 1970, Tamanar, 1970 and Tarha-

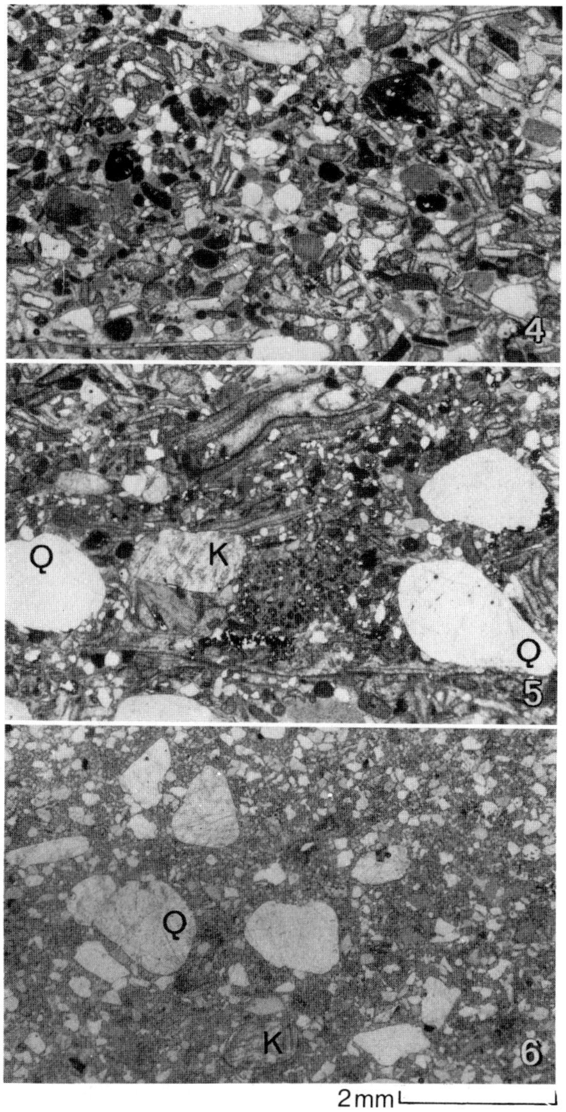

Figs. 4, 5, 6. Facies succession of the Upper Hauterivian from west to east in the south of the High Atlas near Agadir. **4** Grapestone with sparitic cement, micritic envelopes around bioclasts, and a small amount of fine-grained terrestrial debris; deposited in a high energetic, fully marine offshore environment (Tagourzamt, 17). **5** Wackestone with bioclasts, coarse grains of quartz (Q) and potassium feldspar (K) deposited in marine environment in the range of the coarse grained terrigenous influx (Tamraght, 18). **6** Early diagenetic dolomite with small xenotopic to hypidiotopic anhedral grains between 5 μ and 30 μ and large clasts of quartz (Q) and feldspar (K) without discernible bioclasts. Region of nearshore sebkha sedimentation (Ameskhoud, 19)

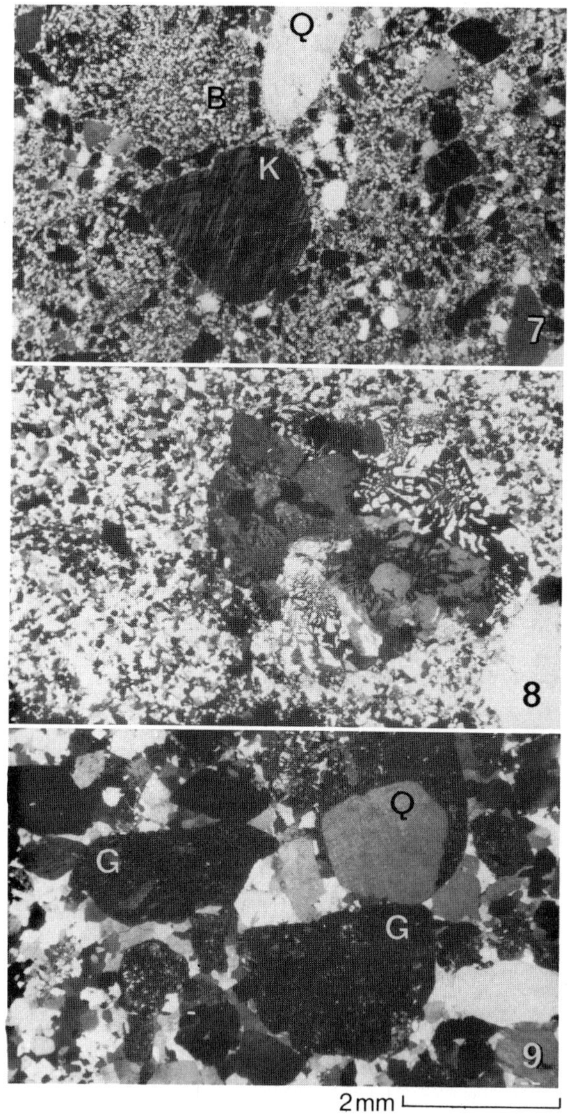

Figs. 7, 8, 9. Marine conglomerates of early Aptian age suggesting the advance of clastic marine deltaic fans toward the north and west in the Atlas basin with large feldspar grains and igneous rock fragments, cemented by dolomite. **7** Potassium feldspar (K), quartz (Q), bioclast (B) (Talezza, 11). **8** Clast of ? Precambrian rhyolite with characteristic graphophyric structure (Ougouelloul, 12). **9** Fragments of rhyolite (G) and quartz porphyry (Q), derived from the Precambrian basement (Tagourzamt, 17)

zoute, 1971). The compilation of all available data for the western Atlas Gulf results in ten cross sections (Figs. 13, 14) which represent the basis for our paleogeographical conclusions. As an example of the stratigraphic subdivision of the 59 sections surveyed by the authors the Section Zem Zem from the central area of the western Atlas Gulf will be described here, showing the interfingering of marine and terrestrial sediments (Fig. 10).

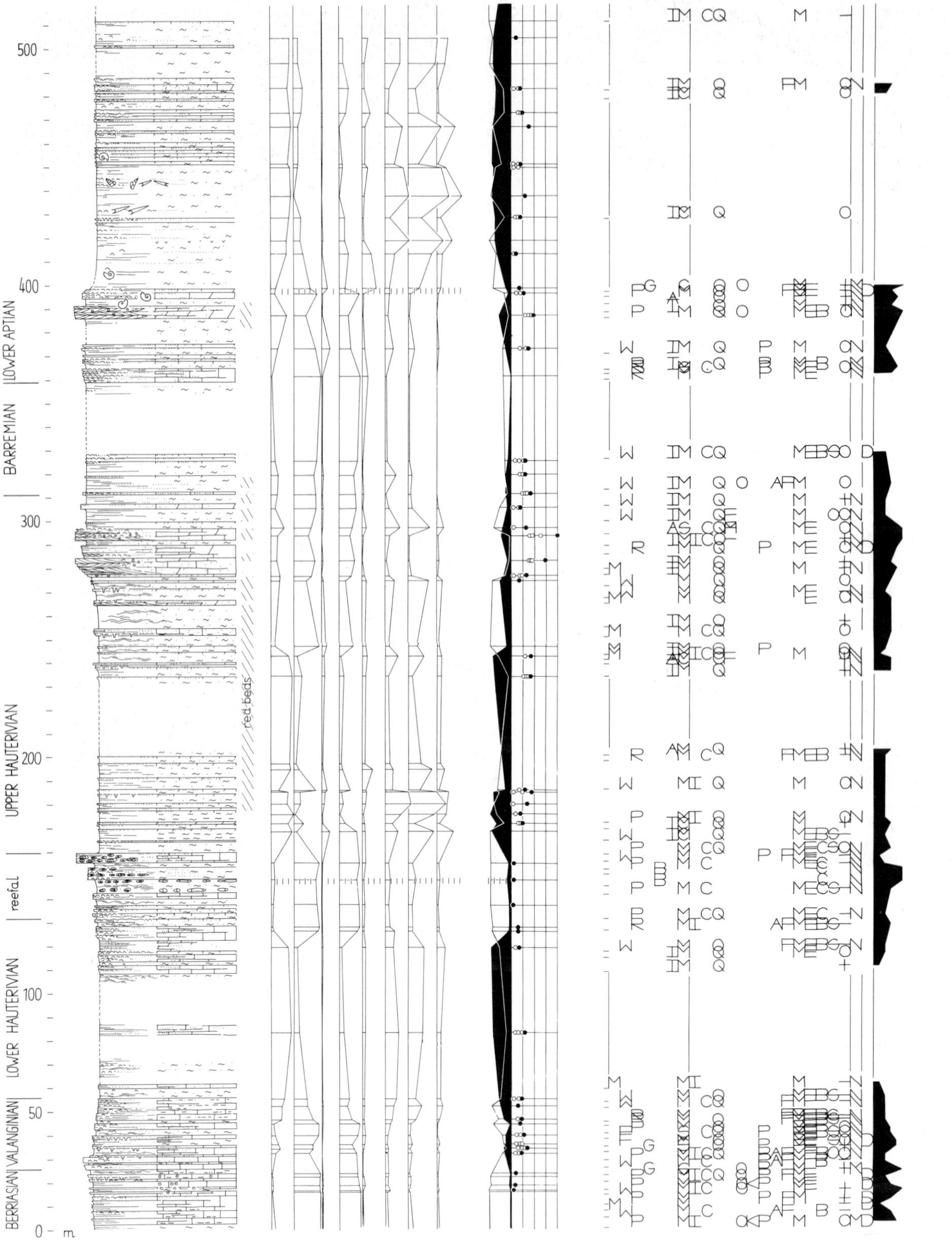
ZEM ZEM
Be/Si '80
LOWER APTIAN
BARREMIAN
UPPER HAUTERIVIAN
reefal
LOWER HAUTERIVIAN
BERRIASIAN VALANGINIAN
red beds
500
400
300
200
100
50
0 m
40 20 R CO_2
10 20 Mg Ca
.4 P
2 Fe
1 Al
4% H_2O
100% R
c f m c f
Si Sand Gravel
1 2 3 4 5 6 7 8 9 0 1 2 3 4 5 6 7 8 9 0 1 2 3
l m
Energy

Berriasian (about 25 m)

Pelmicrites, oolites, wackestones, and packstones with bioturbation, erosion, and resedimentation are quite common, i.e. shallow water sediments: the dying out of the uniform Upper Jurassic carbonate sedimentation.

Valanginian (30 m)

Beginning with a well-banked lumachelle (floatstones and rudstones). The increasing percentage of quartz, silt, and clay indicate the onset of a terrigenous influx of fine clastics.

Lower Hauterivian (105 m)

In the lower half a monotonous alternation of argillaceous micrites, grey marls, and siltstones followed by rudstones and packstones, then isolated coral colonies on a marly substratum, and finally solid banks of coralline limestone. These form a very characteristic key horizon near the entire western Atlas Gulf.

Upper Hauterivian (105 m)

At the base grey-green siltstones and sandy limestones, as well as yellow dolomitic wackestones and packstones. Directly above this a rapid increase in the grain coarseness of the terrigenous components, together with a color change to red and purple. The fossil content decreases sharply. The early diagenetic dolomite crusts contain cracks and intraclasts (Fig. 11). In the uppermost quarter coarse clastic deposits with an increased feldspar content and polymictic conglomerates occur together with lumachelles, indicating the marine influence.

Barremian (50 m)

Alternating strata of grey marls and light-colored limestones with a rich marine fauna which may be traced up to the region of Chichaoua and Imin Tanaoute (Roch, 1930).

Lower Aptian (40 m)

At the base lumachelle banks with an increasing terrigenous content which signifies the renewed withdrawal of the sea. These are overlain by red, cross-bedded,

Fig. 11. Cross-bedded wackestones and rudstones with quartz pebbles of Upper Hauterivian age in the transitional zone between marine and terrestrial sedimentation. In the lower part of the picture cracks in a dolomitic layer are filled by terrigenous material (Section Zem Zem at 284 m), see Fig. 10

Fig. 12. Large scale ball-and-pillow structure typical for Lower Aptian sandstones in the coastal region. Outcrop at the road between Agadir and Essaouira near Assaka

pebbly, dolomite-cemented sandstones and conglomerates with coarse and heterogeneous material. Toward the west yellow, well-sorted and cross-bedded fine sandstones and siltstones prevail, alternating cyclically. The saturation of the sediments by interstitial water in an intertidal environment lead to a collapse of the not yet consolidated fabric and produced widespread, large-scale convolute bedding, slumping, as well as sliding of cross-bedded sandstone lenses on flat shear plaines (Fig. 12; see also Einsele, this Vol.). The instability of the stratification was possibly due to a rapid accumulation. In contrast, immediately north of the axis of the High Atlas (Ait Messaoud, 16; Cross section D in Fig. 13) coral reefs develop in an area with a smaller supply of terrigenous material and perhaps with less subsidence.

Upper Aptian (about 30 m)

Packstones and grainstones, oolites, as well as the termination of the dolomitization indicate the abrupt end

◄

Fig. 10. Section Zem Zem (31°14'45''N, 9°19'47''W), north of the High Atlas. Symbols of sedimentary structures and lithology explained in Fig. 13. Stratigraphic column, geochemical and sedimentological data: **R** insoluble residue in $HCl+HNO_3$; CO_3, **Ca**, **Mg** proportional to the carbonate content; **P** maximum value up to 4,5 %; **Fe**, **Al**, H_2O proportional to the content of clay minerals; **R** *black curve* insoluble residue in HCl, *white curve* terrigenous material less than 63 μ; silt (coarse), sand (fine, medium, coarse), gravel (fine) *open dots* Q_3, M, Q_1 (Trask); *black dots* 10 percentile; 1 mudstone M; 2 wackestone W, floatstone F; 3 packstone P, rudstone R; 4 grainstone G, rudstone R; 5 boundstone B; 6 siltstone I, sandstone A; 7 micrite M, sparite S, clay C; 8 intraclasts I; 9 carbonate lithoclasts C; 0 quartz Q; $\underline{1}$ feldspar F, magmatites M; $\underline{2}$ oolites O; $\underline{3}$ oncoids K; $\underline{4}$ pellets P, peloids P; $\underline{5}$ algae A; $\underline{6}$ foraminifera F; $\underline{7}$ molluscs M, brachiopods B; $\underline{8}$ echinoderms E; $\underline{9}$ bryozoans B, corals C; $\underline{0}$ ostracods O, serpulids S; $\overline{1}$ sorting good +, medium o, poor –; $\overline{2}$ micritization M, neomorphism N; $\overline{3}$ erosional discontinuities D; *last column* relative energy of sedimentation low l, medium m

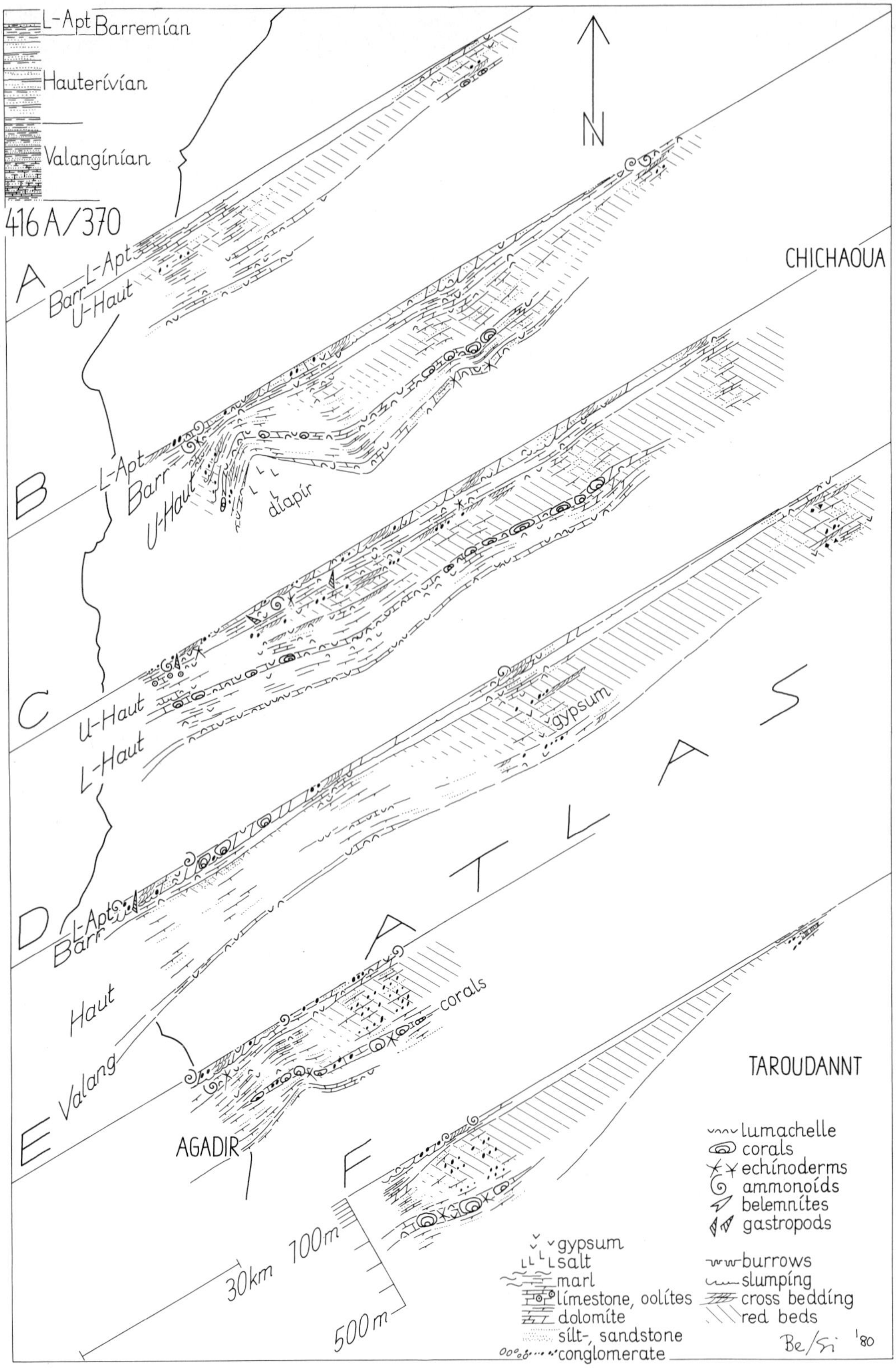

Fig. 13

Figs. 13, 14. Cross sections in NE – SW directions (A–F) and N – S directions (I–IV) through the western part of the Atlas Gulf. Note the decrease of marine influence from west to east (e.g., redbeds) and toward the High Atlas Mountains (e.g., gypsum, lack of coral beds). Diapirs of Triassic salt in the coastal region of Essaouira have marked, but local effect on the facies

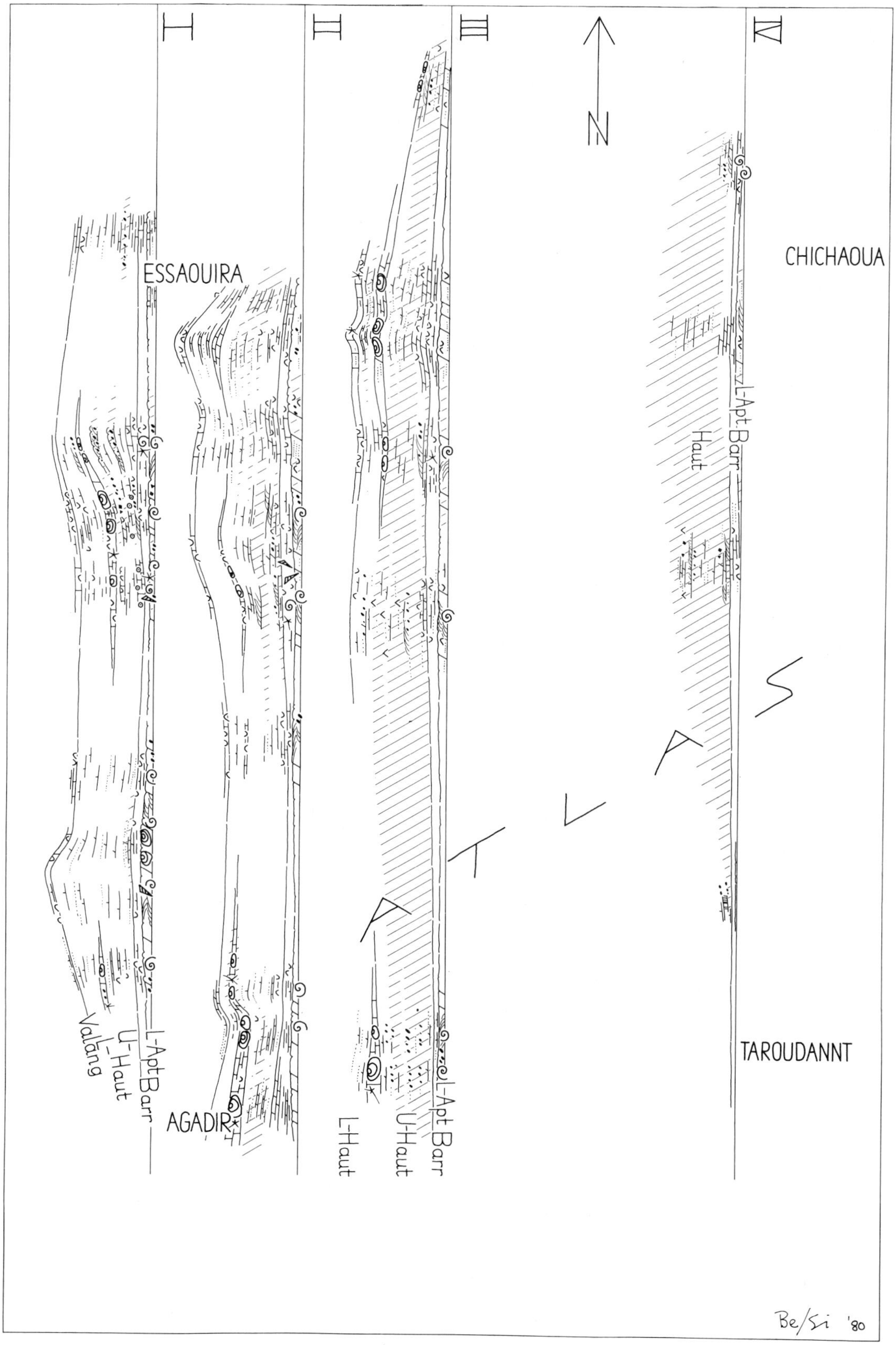

I
II
III
IV
N
ESSAOUIRA
AGADIR
CHICHAOUA
TAROUDANNT
ATLAS
L-Apt Barr
U-Haut
L-Haut
Valang
L-Apt Barr
U-Haut
L-Haut
L-Apt Barr
Haut
Be/Si '80

of the short-lived early Aptian regression. The renewed submergence displaces the coastline far toward the east. The clastic sediments temporarily no longer reach the western Atlas Gulf, where a condensation horizon with ammonitic lumachelle and hardgrounds is formed, caused by extremely low sedimentation rates. This layer is overlain by a cyclic marine sequence with alternating blue-green marls and yellowish siltstones. The marls occasionally turn to black, bituminous layers of clay (black shale facies). In the western sections a dwarfed, in part pyritic fauna of ammonites, belemnites, and gastropods appears. Corresponding to these in the coeval strata of the eastern Atlas Gulf, sedimentary cycles occur with sandstones and siltstones containing a normal ammonite fauna (Roch, 1930). In the early Aptian the proximal fine sand and silt facies of the detrital fan was deposited near the coast; during the late Aptian this facies zone shifted more than 100 km to the east to the High Atlas.

5. Facies Pattern

Near the early/late Hauterivian boundary, a wide barrier coral reef developed in the western Atlas Gulf stretching from north to south. This reef is, however, interrupted in the western extension of the Atlas Mountains and marks the areas with normal marine conditions, optimal water depths of some ten meters, and a lack of considerable terrigenous supply. East of the reefal facies lies an extended coastal plain with precipitation of gypsum in a sebkha environment and abundant accumulation of clastic sediments in front of the river mouths. The western reef front is partially discernible near Essaouira. There, in particular, a confined facies change dominates, due to the local differences in subsidence affected by salt tectonics. Deeply sinking basins between the rising diapirs of Triassic salt absorb large thicknesses of sediments (over 600 m) and do not offer any favorable conditions for reefal growth. Apparently large water depths with higher rates of sedimentation also prevailed in the western extension of the Atlas Mountains near Kap Ghir, while the western Atlas itself probably represented a subareal to lagoonal region of relatively high altitude and slow sedimentation under sebkha conditions, as verified by the lack of corals and by the presence of gypsum alternating with red siltstones (cf. Section D in Fig. 13 and III in Fig. 14).

Thick sequences of the Upper Hauterivian consist of brown to dark purple argillaceous sediments as well as yellow-red, sandy, gravelly sediments. These red beds end in the west at a facies boundary running north–south and parallel to the recent coastline (Fig. 15). The interfingering with the grey and yellow marine sediments indicates that the coastline oscillated.

The terrigenous content of the rocks, averaged for each section, ranges between 25 % and 50 % in the Upper Hauterivian (lower semicircle) of the central basin (Fig. 15). A certain increase is observed toward the east and north. The carbonate phase (calcite/dolomite ratio) reflects a more differentiated distribution pattern. The generally high proportions of calcite in the west are increasingly replaced by higher portions of dolomite toward the east and the Atlas range. Beyond this, the high Ca content of the dolomites in the west decreases toward the Atlas with a decreasing calcite/dolomite ratio, until almost stoichiometrical dolomites prevail. This trend is especially pronounced on the southern edge of the Atlas Mountains between Tagourzamt (56.5 % Ca, 17) through Tamraght (55.3 %, 18) and Bou Yaya (51.0 %, 20) as far as Ameskhoud (50.0 %, 19). According to Füchtbauer and Goldschmidt (1965), the negative correlation of the Ca excess with the total dolomite content might be indicative of an early diagenetic dolomitization, which in our case increases toward the east and is concentrically arranged around the Atlas Mountains. The dolomitization in the Upper Hauterivian is restricted to individual beds.

While shallow water conditions with a rich littoral fauna generally dominated in the western Atlas Gulf, redbeds, dolomitic crusts fractured in situ, erosion channels, and intraclasts, together with a severely impoverished fauna, indicate lagoonal conditions in the eastern Atlas Gulf. The greater Ca surplus of dolomite in the west occurs in the region of normal marine sedimentary conditions, while the dolomites with low Ca content in the east were deposited in a more highly saline environment, probably under arid conditions. The facies distribution on the northern flank of the Atlas Gulf fits into the overall picture of a paleogeographically controlled epigenetic dolomitization. In the section of the Talezza (11) stoichiometrically balanced dolomites alternate in layers with gypsum strata and red siltstones; at the same time, corals grow only 30 km further north and west from here.

After the advance and, in turn, the following recession of the Barremian sea, comparable conditions prevail once again in the early Aptian (upper semicircle in Fig. 15). However, the facies zonation is more strongly accentuated around the present High Atlas. The northwestern Essaouira Basin only contains calcite in the carbonate phase. Dolomite appears toward the east, southeast, and south, and finally dominates there over the proportion of calcite. At the same time, the Ca content of the dolomites decreases. In a comparison of the Ca excess and the dolomite content of the Lower Aptian with that of the Upper Hauterivian it becomes evident that during the early Aptian a particularly far-reaching and intensive epigenetic dolomitization took place. The data of the two easternmost sections do not contradict this general picture. Near Jebel Wafagga (9) in the north and near Afensou (16) in the south no Lower Aptian sediments could be recognized with certainty. On top of the clastic redbeds we found marls and siltstones with marine pelecypods and gastropods, which belong either to the Barremian or to the Upper Aptian and from which the examined samples originate.

6. Discussion

According to the transport vectors and the mineral distribution, the source rocks of the Lower Cretaceous clastics in the Atlas Gulf are mainly the Precambrian

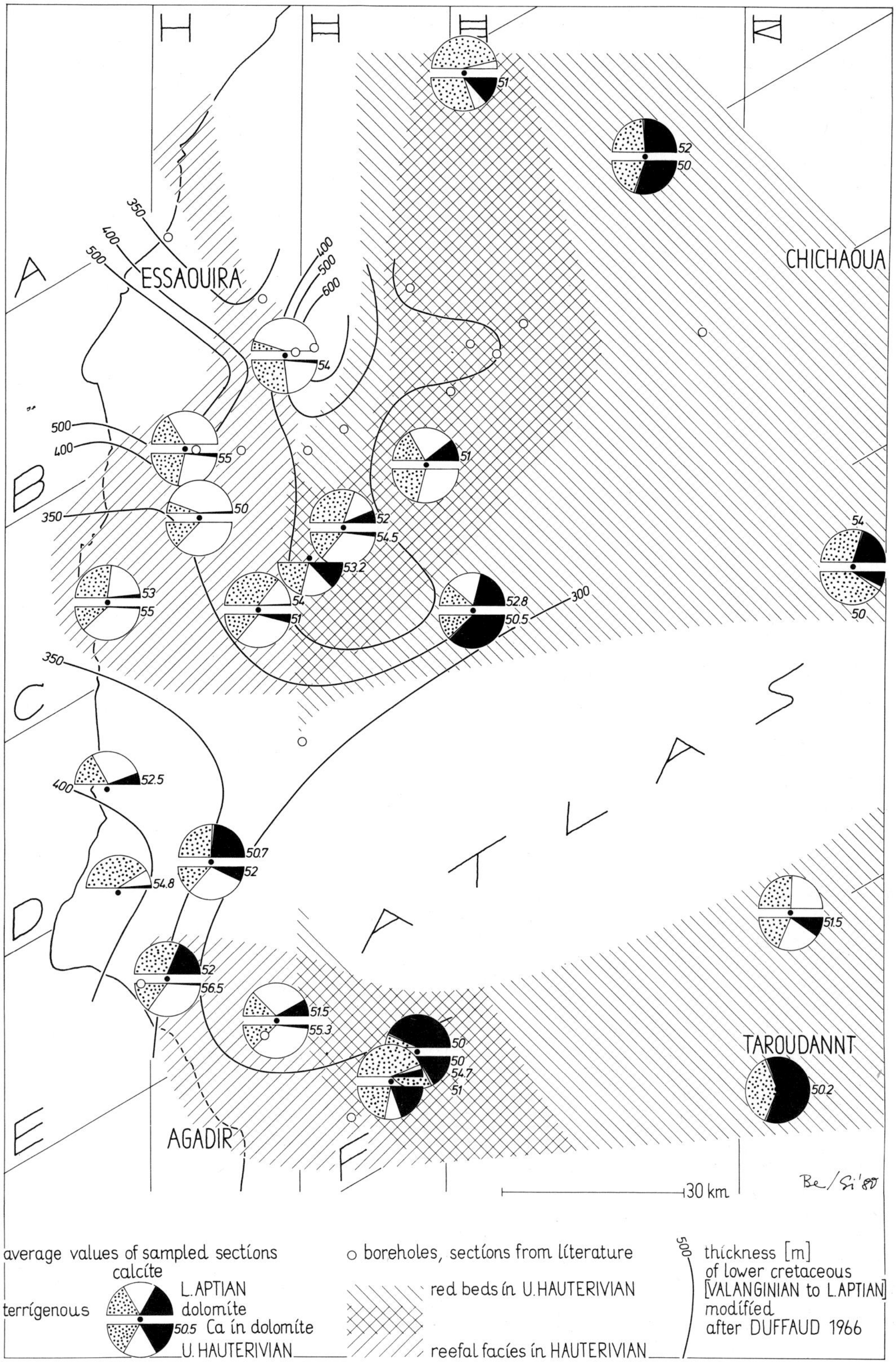

Fig. 15. Average ratios of terrigenous material, calcite, dolomite and percentage of Ca in dolomite in sampled sections showing decreased Ca excess with higher dolomite content. Distribution of redbeds and reefal facies

acidic magmatites of the Toubkal-Ouzellarh Massif (cf. also Price, 1980; Lancelot and Winterer, 1980). The Moroccan Meseta in the north is less likely to have supplied detrital material to the Atlas Gulf (Price, 1980); the Tichka massif in the western High Atlas was not cut by erosion during Early Cretaceous times (Brix, 1981).

On the coast the fluvial sediment was enriched with bioclastic material. Its finest clasts flow downslope over the edge of the shelf platform as turbidity currents. In the lower Valanginian at DSDP Site 370/416 the Bouma cycles contain a high proportion of hardened micrite particles, calcarenites and ooids, the erosive products of the Jurassic carbonate platform along the continental margin. In the upper Valanginian a change occurs from green and grey calcitic turbidites to brown siliciclastic turbidites, which build up the Hauterivian sequence up to 500 m thick, with fine sandstone, siltstone and calcareous and quartzose mudstone (Schlager, 1980). We interpret this material as being derived from the Atlas Gulf based on all petrographical and paleogeographical evidence, as well as on the contemporaneous increase of siliciclastic components in both regions. The question still remains whether the intensified erosion in the hinterland was due to the en bloc uplift of the High Atlas or to a lowered sea level. It is certain, however, that the erosion cannot be regarded as the result of an orogeny in the Atlas, since there is no indication of any Mesozoic folding. The turbidites in Site 370/416 may therefore well be defined as being flysch-like with regard to their sedimentary structures (Price, 1980), but the tectonic aspect of flysch as a sediment deposited in connection with intense crustal deformation is missing.

Acknowledgments. Dr. R. Stamm gave valuable comments on microfacies of our carbonate thin sections, Dr. A. Butt analyzed several samples from the Zem Zem section with respect to foraminifera, the results of which will be published later.

For the opportunity of examining unpublished borehole sections we are indebted to the Service Géologique du Maroc, Rabat. The study was supported by the Deutsche Forschungsgemeinschaft.

We thank Prof. Dr. H. Füchtbauer and Dr. U. von Rad for critical review of the manuscript and helpful suggestions.

REFERENCES

Ambroggi R (1963) Etude géologique du versant méridional du Haut Atlas Occidental et de la plaine du Souss. Notes Mém Serv Géol Maroc 157: 321 p

Behrens M, Krumsiek K, Meyer DE, Schäfer A, Siehl A, Stets J, Thein J, Wurster P (1978) Sedimentationsabläufe im Atlas-Golf (Kreide Küstenbecken Marokko). Geol Rundsch 67: 424–453

Brix M (1981) Schwermineralanalyse und andere sedimentologische Untersuchungen als Beitrag zur Rekonstruktion der strukturellen Entwicklung des westlichen Hohen Atlas/Marokko. Inaug Diss Univ Bonn: 247 p

Choubert G, Faure-Muret A (1962) Evolution du domaine atlasique marocain depuis le temps paléozoique. Livre à la mémoire du professeur Paul Fallot ... Soc Géol Fr 1: 447–527

Choubert G, Faure-Muret A, Hottinger L (1966) Aperçu Géologique du bassin côtier de Tarfaya. Notes Mém Serv Géol Maroc 175: 7–219

Duffaud F (1960) Contribution à l'étude stratigraphique du bassin sécondaire du Haut Atlas occidental (Sud-Ouest marocain). Bull Soc Géol Fr 2 (7): 728–734

Duffaud F (1966) Le bassin du sud-ouest marocain. In Reyre D (ed), Bassins sédimentaires du litoral africain. Assoc Serv Géol Afr 1: Paris 5–12

Füchtbauer H (1967) Der Einfluß des Ablagerungsmilieus auf die Sandstein-Diagenese im mittleren Buntsandstein. Sediment Geol 1: 159–179

Füchtbauer H, Goldschmidt H (1965) Beziehungen zwischen Calciumgehalt und Bildungsbedingungen der Dolomite. Geol Rundsch 55: 29–40

Heim D (1966) Petrographische Beiträge zur Paläogeographie des Buntsandsteins. Notizbl Hess Landesamt Bodenforsch Wiesbaden 94: 235–258

Lancelot Y, Winterer EL (1980) Evolution of the Moroccan oceanic basin and adjacent continental margin – a synthesis. Init Rep Deep Sea Drilling Project 50: 801–821

Meyer DE (1978) Microfacies and microfabrics of early Middle Cretaceous sediments selected from site 370 DSDP Leg 41 (Deep basin of Morocco). Init Rep Deep Sea Drilling Project 41: 497–498

Moret L (1931) Recherches géologiques dans l'Atlas de Marrakech. Notes Mém Serv Géol Maroc 18: 262 p

Price I (1980) Provenance of the Jurassic-Cretaceous flysch, Deep Sea Drilling Project sites 370 and 416. Init Rep Deep Sea Drilling Project 50: 751–757

Roch E (1930) Etudes géologiques dans la région méridional du Maroc occidental. Notes Mém Serv Géol Maroc 9: 542 p

Schlager W (1980) Mesozoic calciturbidites in Deep Sea Drilling Project Hole 416 A. Recognition of a drowned carbonate platform. Init Rep Deep Sea Drilling Project 50: 733–749

Schnitzer WA (1957) Die Lithologie und Paläogeographie des Buntsandsteins am Westrand der Böhmischen Masse. Erlanger Geol Abh 24: 130 p

Stets J, Wurster P (1977) Sedimentationsmodell Atlas-Golf (Kreide Küstenbecken Marokko). Nachr Dtsch Geol Ges 17: p 75

Valeton I (1953) Petrographie des süddeutschen Hauptbuntsandsteins. Heidelb Beitr Mineral Petrogr 3: 335–379

Wiedmann J, Butt A, Einsele G (1978) Vergleich von marokkanischen Kreide-Küstenaufschlüssen und Tiefseebohrungen (DSDP): Stratigraphie, Paleoenvironment und Subsidenz an einem passiven Kontinentalrand. Geol Rundsch 67: 454–508

19 Sedimentation in the Atlas Gulf II: Mid-Cretaceous Events

Paul Wurster and Johannes Stets[1]

ABSTRACT

According to paleogeographic research in the Cretaceous sequence of western Morocco, a funnel-like Atlas Gulf existed between Agadir and Safi, stretching farther toward the east in the present High Atlas region. Bordered by structural highs of the Anti-Atlas in the south, and the Moroccan Meseta in the north, the Atlas depression subsided slowly and continuously.

Within the Atlas Gulf transgressions and regressions of the Cretaceous Atlantic can be recorded by an almost synchronous migration of the littoral zones. Distribution of shallow-marine, littoral, lagoonal, deltaic, coastal plain and fluvial facies are pointed out. Regimes, processes and kinematics of sedimentation are discussed by a sedimentation model.

Significant events in Barremian, Aptian, and Turonian time are explained by an accelerated rise of the sea level. Our sedimentation model proposes eustasy as a major cause for the Mid-Cretaceous events, documented in the sedimentary record. Relations to similar rock sequences in the Sahara Basin, at the opposite margin of the Atlantic Ocean, and in the deep-sea basins, are discussed with regard to the Atlas Gulf sedimentation model.

1. Preliminary Remarks

The description and reconstruction of the environments of the past are the objectives of *paleogeography*. This aim implies that the geological time interval under discussion is well documented and well defined in the geological record. The Deep Sea Drilling Project (DSDP) helps us to approach this final goal of paleogeography. Increasing penetration and closer spacing of deep-sea boreholes meanwhile provide us with a more or less complete stratigraphic record of the oceans. By means of growing geological and geophysical data the geodynamic evolution of the deep ocean basins since the Mesozoic becomes more and more evident. Due to our recent knowledge on the evolution of the Atlantic Ocean, an immense gap in the paleogeography of the Earth begins to close. As far as worldwide Cretaceous paleogeography is concerned, the position and rough contours of continents and oceans can now be traced: During the Early Cretaceous, the central part of the Atlantic widened, and became a broad and deep ocean basin (Sclater et al., 1977), connected to the Pacific by the Caribbean. During Late Cretaceous first connections toward the south to the later South Atlantic opened. The northern part of the Atlantic originated at the end of Cretaceous and in Tertiary time. Connections already existed to the Arctic and Tethys seas at that time.

We are able to compare Cretaceous conditions to recent ones, if we take into account the position of the poles and of the equator, and the differences of polar icecaps, climate, shelf conditions, continental margin position, ocean currents, etc. Indeed, one of the objectives of IPOD/DSDP is the global synthesis of the Cretaceous paleoenvironment.

One of the objectives of our investigations is the search for the conspicuous *Cretaceous events* in paleogeographic evolution, which led to worldwide transgressions and to striking chemical and biological changes in the areas of sedimentation. One of the most important prerequisites to approach this aim is the investigation of the fossil environment (deep sea, shelf, shallow sea, lagoonal, or continental conditions). Concerning IPOD/DSDP data on the Atlantic Ocean, it therefore seems necessary to explore also neighboring marginal and continental areas, and especially those onshore sediments which are equivalent to those represented in the deep sea.

The Cretaceous of the Moroccan coastal basins is well suited for such an analysis, as good outcrops along the coast and in the High Atlas region provide a close network of cross-sections. The sedimentary sequences of this region contain changes in facies, ranging from shallow marine to continental. Last but not least, structural deformation during the Alpine orogeny did not destroy the primary sedimentary pattern, as happened in other mountain ranges.

One of the most important features we had to look for in our analysis of Cretaceous paleogeography was the fossil *coastline* and its migration in the course of time. Since the systematic sedimentological research of Recent coasts has reached a rather high standard

1 Geologisches Institut der Universität Bonn, Nussallee 8, D–5300 Bonn 1, FRG

(Reineck and Singh, 1973), its application to fossil coasts seems necessary; but fossil coasts and coastlines and their geological inventory are dubious and enigmatic. Also in the present case, a reliable identification of the Cretaceous coastline was difficult to establish.

The construction of paleogeographic maps, of facies diagrams, and of sedimentation models from the sedimentary record is a good way to solve these problems. Nevertheless, the sea level and its intersection with the continent is the predominant reference horizon for geodesy, geography, and also for paleogeographic reconstructions. It is also the best reference level for a sedimentation model, because along the coastline lithosphere, hydrosphere, and atmosphere meet. Characterized by numerous sedimentary pecularities, a specific complex is formed, the *littoral* zone. This facies complex can more easily be traced in the sedimentary record than the exact position of the shoreline.

The best approach to a sedimentation model is to delineate the hypsographic curve of the time interval under discussion. Deviations from the standard are mainly caused by the *relief* represented onshore and offshore. These deviations influence the development of the littoral facies. Reconstruction of a sedimentation model therefore requires the knowledge of the relief, which we inferred roughly by structural analysis of the coastal basin (Stets and Wurster, this Vol.) and of its sediments. If the relief remains stable for the time interval under discussion, we are able to look for the littoral zone in the facies diagram, constructed from a series of columnar profiles. On the other hand, if the relief is still unknown, statements on the other factors controlling the model are doubtful.

Climatic conditions and the sediment *supply* from the adjacent continents are important factors controlling the littoral zone and also the sedimentation model. Especially evaporites, originating in lagoons and sabkhas near the coastal plain, indicate the position of the littoral zone. Supply of coarse-grained sediments gives rise to conglomerates or to beach ridges and plains. Decreasing supply of this material from the hinterland results in mud plains and tidal flats. If even pelitic material is kept back by a rising sea level, shallow-water carbonates originate under favorable conditions.

Animal and *plant* remains, as well as *sedimentary structures,* give additional clues to the conditions of life and to the paleobathymetry of the ancient shelf.

Taking into account all these different factors, we tried to establish a sedimentation model of the Mid-Cretaceous (Aptian to Cenomanian) near the Atlantic coast of Morocco. Our main problem is to explain in a simple way the final paleogeographic result of an open system which contains so many unknown factors. Fortunately, the reconstruction of a sedimentation model underlies clear rules and is not only a result of arbitrary assumptions: Some of the controlling factors exclude each other, such as deep-sea sediments and continental sediments. Delineation of a sedimentation model therefore has to be handled as a cautious adaptation to actualistic sedimentological patterns. We start from a limited number of well-defined sedimentation areas of regular arrangement and distribution, and take into consideration the lateral development of the facies pattern known until now.

2. Structural Framework

The *structural evolution* of the subsurface is another factor controlling basin development and sedimentation processes; but the influence of crustal movements can only be calculated if the structural setting and the reaction of the deeper crust during the geological evolution are known. Concerning the deeper crust and upper mantle of Morocco, only a few data and details are known (Michard, 1976; Makris and Demnati, unpubl.). On the other hand, the geological and structural setting is quite clear. According to that, Cretaceous sediments in Morocco cover quite different structural units. Their contrasting behavior allows calculation of crustal movements at that time. For a further understanding of these problems, it seems necessary to give a short review (see Fig. 1). For more details the reader is referred to the geological maps of Morocco at the scale 1 : 500 000 or to the elements of Moroccan geology by Michard (1976).

The deepest structural level, containing the oldest rock units in Morocco, is found in domal uplifts of the *Anti-Atlas.* Precambrian crystalline rocks, covered by platform sediments of Late Precambrian to Early Paleozoic age build up this huge anteclise at the northern edge of the African craton. Crustal thickness here reaches about 32 km. This region was not influenced by the Variscan orogeny, as it is situated outside the ancient Variscan mobile belts. During the whole Mesozoic and Cainozoic the Anti-Atlas remained a rather high stable block with reference to the position of the sea level. It has been the main source area of sedimentary supply throughout that time until today.

Paleozoic rock series predominate in the *Moroccan Meseta,* which is part of the Variscan mobile belt. Crustal thickness also amounts to 30 km. In the deep subsurface this region consists of basement complexes equivalent to those of the Anti-Atlas. During the Variscan orogeny the Paleozoic series were consolidated. Since then, the Meseta has also reacted as a stable block. Erosion and peneplanation prevailed. During the Mesozoic and Early Tertiary until today this region has remained in a rather high structural position. In contrast to the Anti-Atlas, however, it was covered by thin platform sediments. This means that the structural position with reference to the sea level was not as high as that of the Anti-Atlas. The Morocco-Oranian Meseta, which is situated farther toward the east, has a similar position. It is a part of the adjacent High Plateaus of Algeria.

On the contrary, the *High* and *Middle Atlas* are narrow mobile belts. The High Atlas reaches a width of only 50 to 120 km. It consists mostly of rhomboidal blocks, tilted against each other. Also here crustal thickness amounts up to 30 km (J. Makris, pers. comm.). The genesis of the deeper stockwerks in the western High Atlas is the same as that of the Meseta. Farther toward the east and in the central High Atlas it is similar to that of the Anti-Atlas.

Covering the ancient basement, thick Mesozoic sedimentary series are found throughout the narrow belts. This gave rise to the idea that a deep furrow of high subsidence existed between the Mesetas to the north

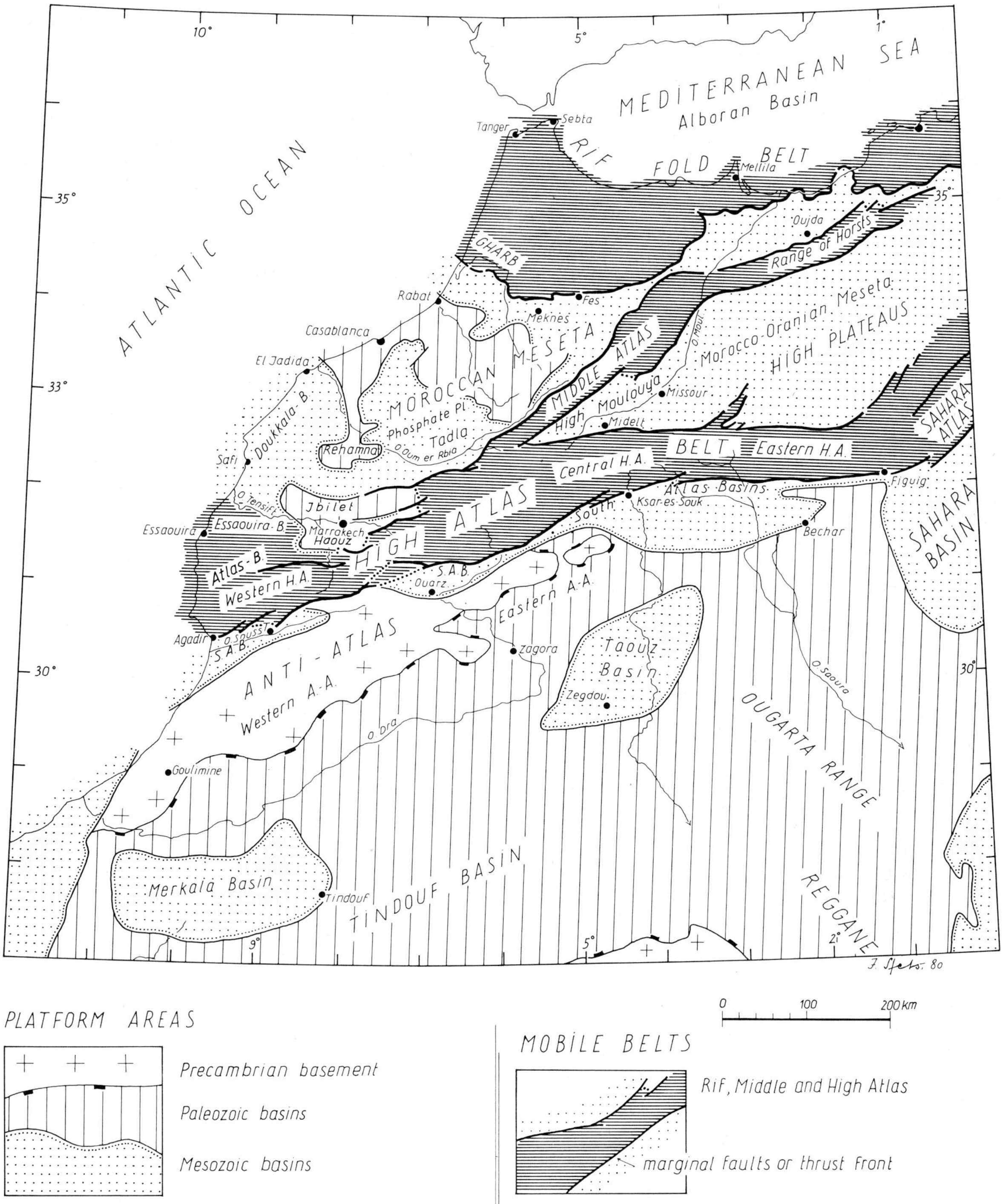

Fig. 1. Relationship of Mesozoic mobile belts and platform areas, containing Precambrian basement, Paleozoic and Mesozoic cover, and epicontinental basins in Morocco. *S.A.B.* South Atlas Basins

and the old African craton to the south. Its shape during Triassic to Mid-Cretaceous times resembles that of a complicated rift or graben structure (Atlas Rift, sensu Stets and Wurster, this Vol.). During the Late Cretaceous, this belt began to rise by structural inversion, and became a high mountain range which now reaches heights up to 4000 m above sea level.

The *Atlantic shelf* off Morocco covers heterogeneous geological units such as the Anti-Atlas, High Atlas, and Moroccan Meseta. In this area the crust has been extensively fractured and thinned during the genesis of the passive continental margin in a complicated process since the Triassic. A sedimentary wedge originated during the Cretaceous in the area where the Atlas mobile belt met the Atlantic margin. This area, to be discussed later in more detail, is called the Atlas Gulf. Structural relations between Atlas and Atlantic and their geological evolution since the Triassic were already outlined in a previous paper (Stets and Wurster, this Vol.).

The well-defined crustal units and their lateral relationship make it possible to reconstruct the relative structural movements against each other. The adjacent African craton and its position to the sea level of the Atlantic Ocean gives a rather reliable, perhaps even an absolute, reference level to calculate crustal movements.

3. The Atlas Gulf in Mid-Cretaceous Times

3.1 Atlas Gulf

A well-definable and uniform sedimentation area resulted (Fig. 2) from compilation of the main Lower and Mid-Cretaceous profiles, studied in the High Atlas region and in the coastal basins at distances of about 50 km. On the structural highs of the Moroccan Meseta in the north and northeast, as well as in the Anti-Atlas in the southeast, thin sedimentary series of a few meters occur only in marginal position (Profile 11). In the interior of the high areas no sedimentation is found. Toward the Atlantic Ocean in the west, the basin opens like a funnel. Thickness increases in that direction. Special attention has to be paid to the complete fossiliferous profiles near to the Atlantic coast north of Agadir, containing mainly open-marine facies (Profile Agadir: Wiedmann et al., 1978; Profile 02: Behrens and Siehl, unpubl.), or east of Essaouira (Profile 15: Behrens, Meyer and Siehl, unpubl.; Profile Smimou: Wiedmann et al., 1978; and this Vol.).

Toward the east, the thickness of the sequence decreases gradually. Red series, intercalated into the sequence, are increasingly found in this direction. The direct relationship between all these single profiles is evident, especially if one compares those from the southern margin (Profiles 07, 10) with those in the north (Profiles 20, 22). According to thickness and facies development, a uniform sedimentation area, which we call *Atlas Gulf,* can be traced opening toward the Atlantic Ocean. Its position roughly coincides with the Mesozoic Atlas Rift, which here intersects with the Atlantic continental margin, causing an area of continuous subsidence. Local thickness variations may be due to movements of tilted blocks in the former rift area or to halokinetic processes in the nearshore salt diapir province (Hinz et al., this Vol.; Stets and Wurster, this Vol., Fig. 1).

In contrast to the opinion cited above, Ambroggi (1963), Choubert and Faure-Muret (1962) and Duffaud et al. (1966) postulate two separate main basins: the Agadir Basin and the Essaouira Basin, stretching along the southern and northern flank of the High Atlas. A structural high is assumed in the region of the western High Atlas (massif ancien), where no sediments are found today. This high is said to separate the two basins in the east. This contradiction is not easy to solve, as the sedimentary record only shows shallow-sea to marginal conditions along the High Atlas mountain range. We will discuss this problem later in detail (3.4).

3.2 Apto-Albian Sequence

Within the Atlas Gulf different provinces can be distinguished during the Apto-Albian according to facies development. They are characterized by four main columnar sections, each of which is typical for one of the provinces. As fossils are rare or even lacking in the interior and near the margins of the gulf, we were forced to distinguish rock units within the sequence. According to geological mapping and step-by-step correlation, it is possible to distinguish these units even in areas where strong variations of facies occur. Only in the western part of the gulf did we succeed in stratigraphical correlation with ammonites and microfossils (Wiedmann and Butt, pers. comm.). Rock units, stratigraphic correlation, and facies provinces are shown in Fig. 3.

Profile 04: Askouti
(position: 30° 30' N, 9° 27,8' W, see Fig. 2)
Unit 1 starts at the base with a conglomerate, consisting mostly of quartz pebbles. It represents the main transgressive event of the Aptian. It is overlain by a well-bedded sequence of grey marls and micritic limestones, containing shells of molluscs (b), relics of echinoids (d), and cephalopods (e). Marls and limestones are mostly rich in organic matter. This rock unit reaches a thickness of 32 m.

Unit 2 consists of green to olive, partly bituminous marls. Intercalation of yellow layers of calcareous sandstone are frequent. Trace fossils (a), thin shells of lamellibranchs (b), serpulids (c) and echinoid fragments (d) are found. Small pyritized cephalopods (e) are observed only in the lower part. As a fault cuts through this unit, the thickness of this profile is not complete. According to adjacent profiles the total thickness in this region averages about 400 m.

The most striking characteristics of unit 3 are rock sequences with a thickness of 15–20 m, consisting of reddish calcareous sandstones, grey and white limestones and dolomites, intercalated with grey marls. Trace fossils (a) are rare, whereas oyster beds (maximum diameter of shells up to 12 cm) and coquina beds are very abundant. The total thickness reaches 235 m in this profile.

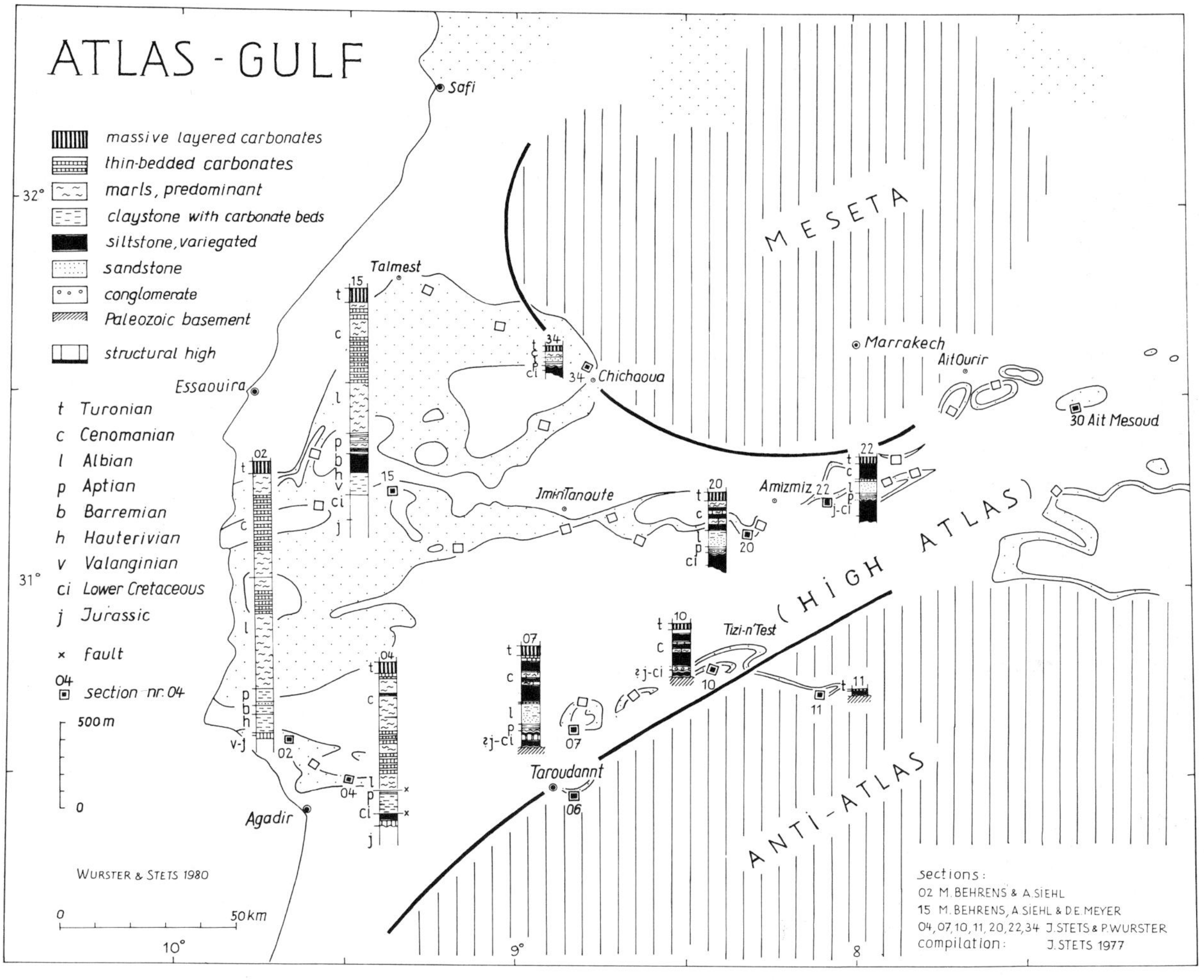

Fig. 2. General configuration of the Cretaceous Atlas Gulf; thick marine rock series (*W*) interfinger with thin continental sequences (*E*); stable high areas of the Meseta (*N*) and Anti-Atlas (*S*) without sediments

The stratigraphic position is given left of the column (Fig. 3). By means of ammonites only Late and Terminal Aptian (Gargasian, Clansaysian) was determined in unit 1. Unit 2 according to ammonites and foraminifera belongs to Latest Aptian and Early to Middle Albian (determined according to J. Wiedmann and A. Butt, pers. comm.). Foraminifera and macrofaunal assemblages in unit 3 indicate a Late Albian (Vraconian) age. The Profile Askouti characterizes the open-marine but shallow-sea facies in the outer part of the Atlas Gulf. It is similar to the Agadir Profile (Wiedmann et al., 1978). The total thickness of the Apto-Albian of about 650 – 700 m indicates high subsidence in this part of the gulf, which was mostly compensated by terrigenous supply. No deep-sea facies was observed within this profile.

Profile 22: Aouzer
(position: 31° 12,9' N, 8° 6,7' W, see Fig. 2)
Unit 1 here starts with a transgressive series of grey to yellowish calcareous sandstones overlying a series of continental redbeds. Well-bedded sandy marls and sandstones, which are rich in organic matter, follow on top of it. Often allochthonous plant remains (f) are observed. This unit is rich in trace fossils (a). Shells of molluscs (b) are rare. The thickness amounts to 30 m.

Unit 2 consists mostly of yellowish to yellow calcareous and dolomitic sandstones, locally showing cross-stratification. They contain layers of conglomerates (mostly quartz pebbles) and shell beds. Intercalations of grey to greenish marls occur, increasing toward the top of the unit. Trace fossils (a) and mollusc shells (b) are often observed. Cephalopods (e) were found only in the upper part, plant remains (f) in the lower part of the section. The thickness reaches 80 m.

Unit 3 is 10 m thick and represented by only one rock sequence of grey to white dolomites and sandy dolomites, containing layers of quartz pebbles and shell beds.

Ammonites in unit 2 belong to Late Aptian/Albian as in the Profile Askouti (determined by J. Wiedmann, pers. comm.). The Profile Aouzer, representing a sandy,

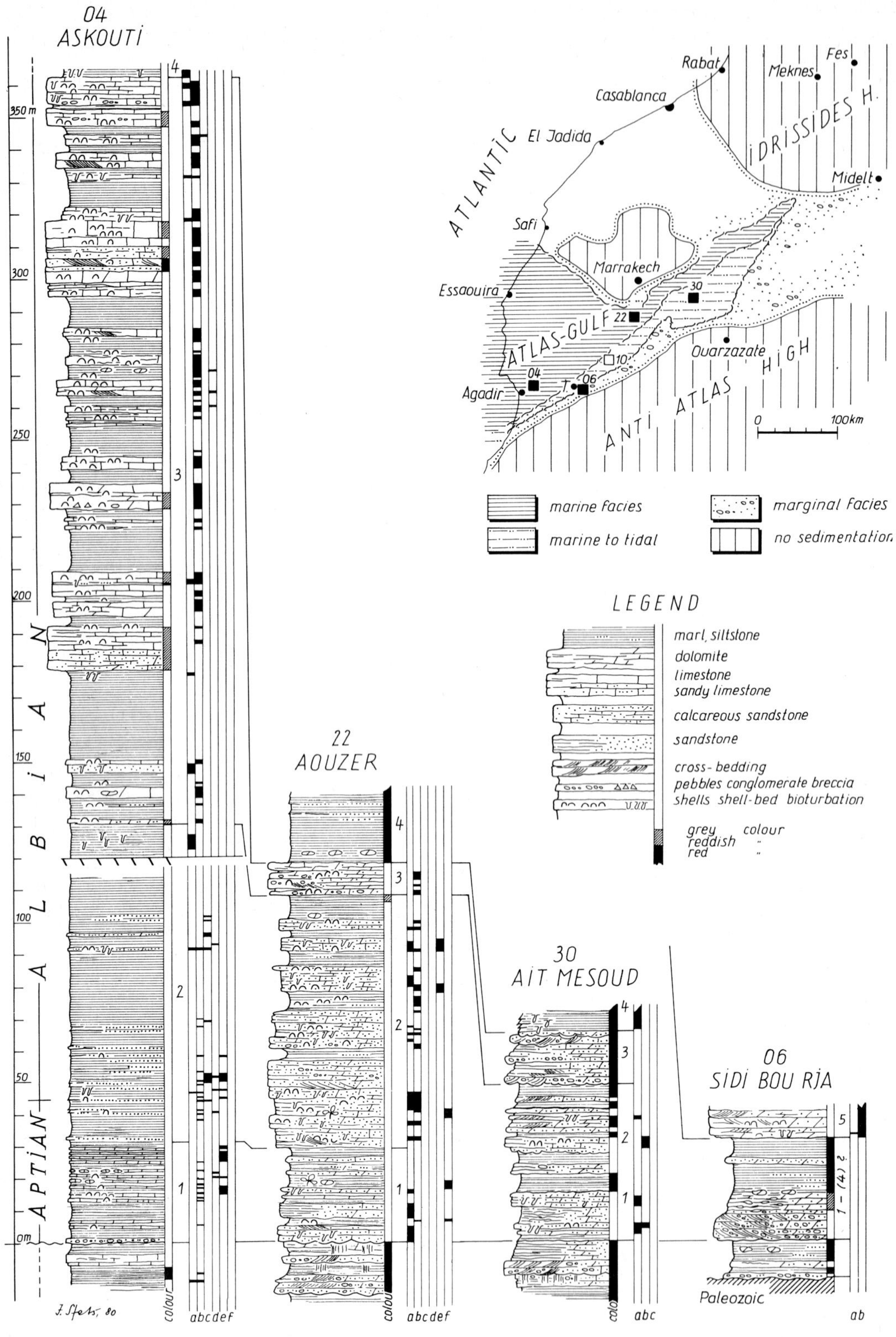

Fig. 3. Apto-Albian sequence: four typical profiles and a paleogeographic map show facies distribution in the Atlas Gulf; rock units 1 to 4 and fossils are indicated; in the *right columns*: *a* trace fossils; *b* lamellibranchiata; *c* serpulids; *d* echinoids; *e* cephalopods; *f* plant remains

open-marine shallow-sea environment, characterizes the inner part of the Atlas Gulf during the Apto-Albian. The total thickness amounts to 120 m, indicating that the subsidence was less in the inner gulf than in the outer part.

Profile 30: Ait-Mesoud
(position: 31° 28,6' N, 7° 18,2' W)
In the Ait-Mesoud Profile rock units 1 and 2 can no longer be separated. They consist of grey calcareous to dolomitic sandstones at the base, rich in trace fossils (a). This sequence ends with a hardground (hd), followed by red marls. This alternation is repeated several times within this unit. Shell beds (b) are rare.

Unit 3 is built up of conglomerates and sandstones in several cycles. The pebbles within the conglomerates are derived from the Precambrian of the Anti-Atlas (black quartzites, rhyolites). Locally pebbles of red sandstones indicate reworking of earlier Mesozoic rocks. The total thickness of Apto-Albian amounts to 65 m, showing that no high subsidence occurred here. We suggest a shallow-sea to tidal-flat environment for this province.

Profile 06: Sidi-bou-Rja
(position: 30° 27,5' N, 8° 49,8' W, see Fig. 2)
The Sidi-bou-Rja Profile shows a condensed rock sequence. We suppose that the 15 m-thick grey to reddish conglomerate and nodular carbonate series may belong to the Apto-Albian sequence. Up to now, a reliable correlation is not possible. This profile characterizes the southern marginal facies of the gulf toward the Anti-Atlas.

3.3 Cenomano-Turonian Sequence

Another facies pattern can be stated for the Cenomanian and Turonian. With respect to details of the Turonian facies the reader is referred to Stamm and Thein (this Vol.). Since no exact stratigraphic boundary can be given by means of micropaleontological data in the main profiles, we distinguished another rock unit (no. 4) reaching from the top of the Vraconian unit 3 to the base of a massive carbonate series, mainly belonging to the Turonian (unit 5, "barre calcaire turonienne"). This Turonian carbonate rock unit can be traced throughout the whole Meseta and Atlas region ("domaine atlasique", Choubert and Faure-Muret, 1962; Michard, 1976). It represents a distinct marker horizon in the field. Rock sequences, stratigraphic correlation, and facies provinces of Cenomano-Turonian are shown in Fig. 4.

Profile 04: Askouti
(position: 30° 30' N, 9° 27,8' W, see Fig. 2)
Unit 4 consists mainly of grey marls in the Askouti region. There are intercalations of limestones and calcarenites containing shell layers and coquina beds, partly showing cross-stratification. Rich assemblages of oysters, reaching shell diameters up to 15 cm, suggest autochthonous oyster reefs in places. On the other hand we observed dolomites and marls. Trace fossils (a), molluscs (b) and relics of echinoids (d) indicate rich benthonic life. The thickness of this unit reaches 345 m, suggesting a high subsidence rate. According to this evidence the sediments accumulated under shallow-marine to lagoonal conditions. Only open-marine shelf conditions are found in the neighboring Agadir region (Wiedmann et al., 1978). The Profile Askouti is located just in the transition zone between the open-marine province to the west and the adjacent province characterized by predominating lagoonal conditions.

Toward the top of the Askouti Profile, open-marine conditions prevail in this region, thus indicating the transgressive event typical for the Latest Cenomanian and for Turonian time.

Profile 10: Tafinegoult
(position: 30° 47,5' N, 8° 24,0' W, see Fig. 2)
The first dolomitic arenite sequence, containing trace fossils (a) and molluscs (b), is taken as base of unit 4 in this profile. This boundary is not quite certain, as no rock sequences of unit 3 type are found. Within unit 4 red marls and siltstones predominate. Three sequences containing layers of gypsum and dolomite and four intercalations of grey to yellowish mostly unfossiliferous dolomites and dolomitic arenites can be distinguished. The thickness amounts to about 145 m. The paleoenvironment is lagoonal with a few ingressions of the open sea. The transgression by the uppermost Cenomanian to Turonian open sea begins near the base of carbonate unit 5.

Profile 22: Aouzer
(position: 31° 12,9' N, 8° 6,7' W, see Fig. 2)
In unit 4 fine-grained red siltstones prevail. Occasional dolomitic sandstones are intercalated. Trace fossils (a) and remnants of molluscs (b) are rare. In the vicinity of Aouzer, one thin layer of gypsum was found within this sequence, indicating a lagoonal or sabkha environment. Unit 4 measures about 85 m in thickness. Subsidence is not very pronounced in this interior part of the gulf. The transgression during latest Cenomanian time is indicated by dolomitic limestones, grey marls and oolite-bearing limestones, situated a few meters below the massive carbonate series of unit 5.

Profile 30: Ait-Mesoud
(position: 31° 28,6' N, 7° 18,2' W)
Within unit 4 red marls and siltstones predominate, indicating the environment of an extensive mud plain. Locally, medium to fine-grained red sandstones occur, showing cross-stratification. Trace fossils (a) are abundant in the lower part of this unit. No mollusc shells or plant remains were observed in this sequence. This sort of facies characterizes most of the areas farther to the east. The abrupt occurrence of Turonian carbonates of unit 5 indicates the transgression of the open sea into the interior of the gulf.

3.4 Transverse Cross-Section (N–S)

The Mesozoic filling of the gulf and the basin development during the Cretaceous is more impressively documented by a cross-section, compiled from a number

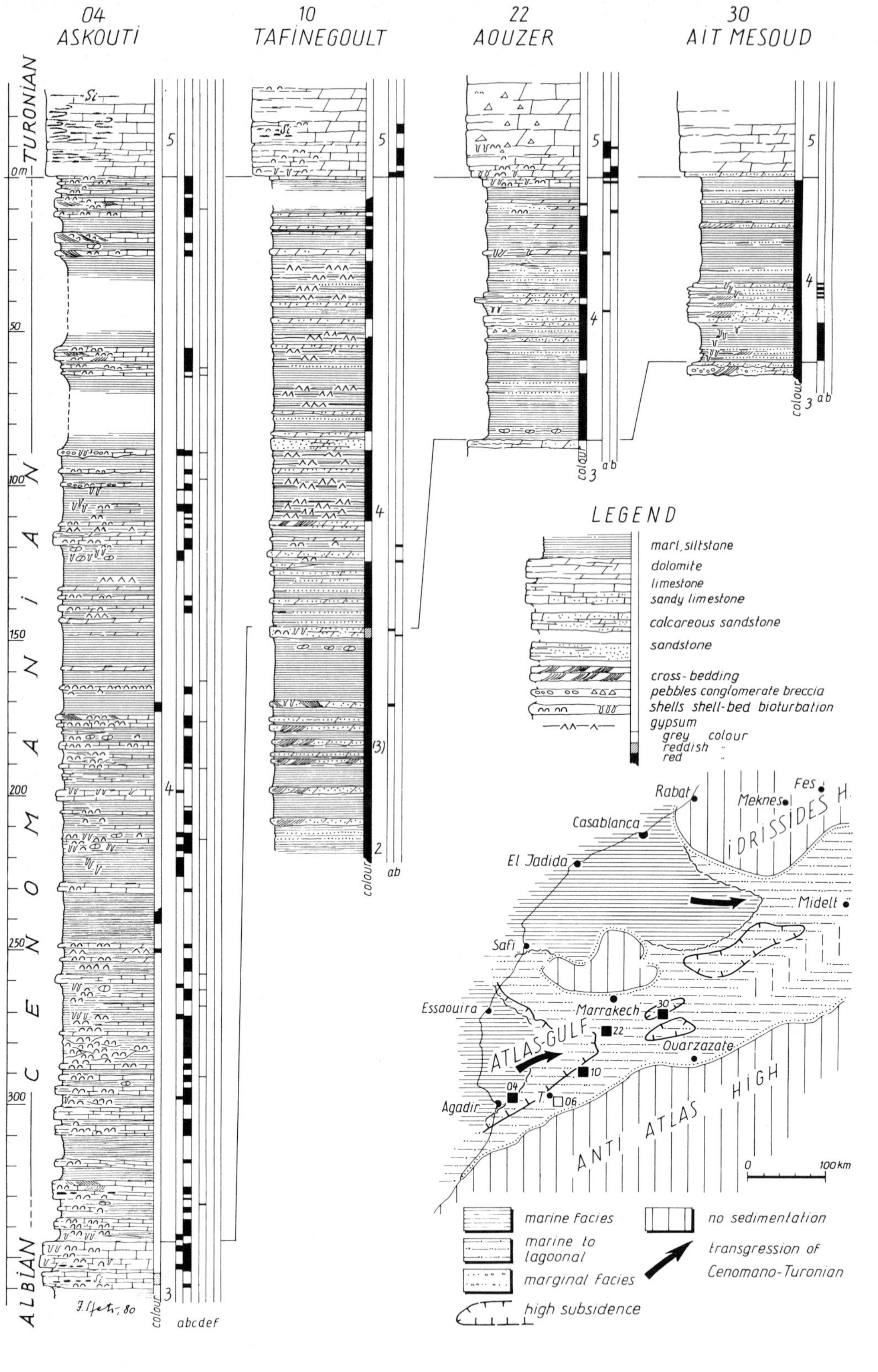

Fig. 4. Cenomano-Turonian sequence; four typical profiles and a paleogeographic map show facies distribution in the Atlas Gulf; areas of higher subsidence are indicated; rock units and fossils as in Fig. 3

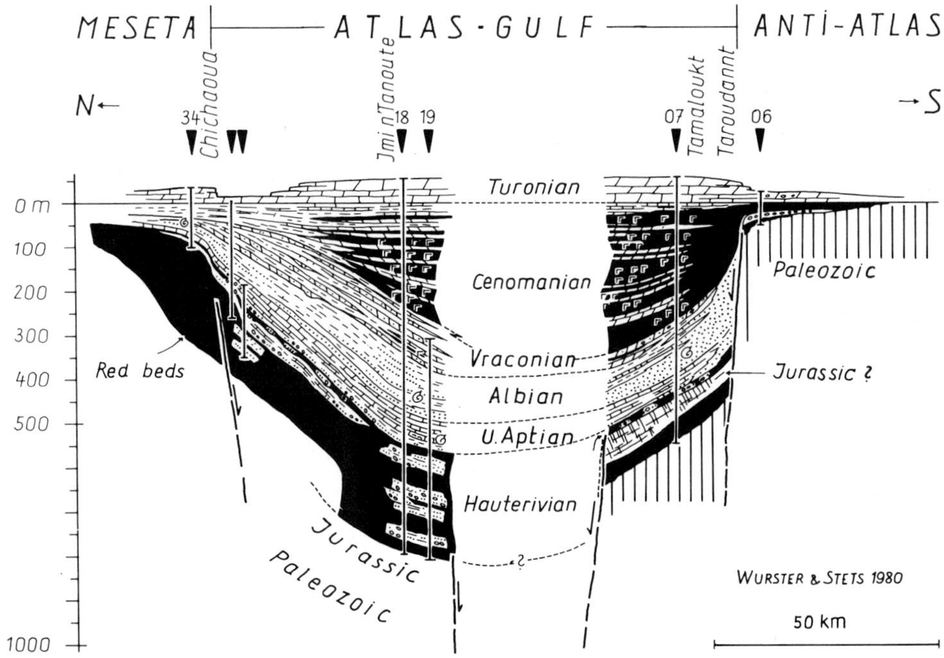

Fig. 5. North – south section across the Cretaceous Atlas Gulf; location of Profiles 06, 07, 18, 19, 34 see Fig. 2; legend see Fig. 3 and 4

of columnar sections. Syngenetic normal faults, active since the Early Mesozoic, are clearly shown, as well as the configuration of the Atlas Gulf in relation to the Atlas Rift (Fig. 5). This section crosses the Atlas Gulf between Chichaoua in the north and Taroudannt in the south, about 100 km east of the present Atlantic coast. The uniform Turonian carbonate unit (unit 5) is taken as reference line for the construction of the diagram.

The stable platform areas of the Meseta to the north and of the Anti-Atlas to the south are characterized by a thin Cretaceous sedimentary cover of less than 100 m (Profiles 06, 34). Toward the center, the thickness increases to more than 800 m in the north, and more than 600 m in the south. Correlation of equivalent rock units on both sides of the High Atlas is possible. It proves that the Atlas Rift persisted throughout the Early and Mid-Cretaceous. Abrupt changes in thickness from the platform areas toward the interior, especially in the south, show the syngenetic activity of marginal normal faults. Smaller changes in the interior are due to block movements along syngenetic normal faults. The mobile belt character during Early and Mid-Cretaceous times is also shown by differential subsidence of individual blocks at different times: In the south a condensed sequence of different facies (Profile 06, Sidi-bou-Rja, Fig. 3) covers the Paleozoic basement. North of Taroudannt, a very thin Neocomian sequence overlies Jurassic redbeds and limestones. The Mid-Cretaceous facies on top of it resembles that of the gulf. In the interior, near Imi-n'Tanoute, Neocomian redbeds, about 200 m thick, cover gypsiferous rock sequences of Jurassic age.

In our opinion, this N – S cross-section gives a convincing answer to the question, whether a uniform Atlas Gulf existed, as we believe, or a central Atlas high during Early and Mid-Cretaceous times. No isolated narrow channel-like basins can be observed along the northern and southern flanks of the High Atlas. Neither thickness nor facies nor sedimentology gives any hints of a central swell or high. The described rock units of Apto-Albian and Cenomano-Turonian can be correlated without any doubt from the northern flank to the south. All these facts are good arguments for the existence of a uniform Atlas Gulf.

3.5 Longitudinal Cross-Sections (E – W)

East to west sections illustrate the facies development in the direction of the gulf axis. They supply another basis for the construction of the sedimentation model. In the longitudinal direction the changes in the position of the coastline and of the littoral zone can be traced more easily than in the transverse cross-section.

The longitudinal section along the *southern* flank of the High Atlas was compiled from the data of five main profiles (Fig. 6). Supplementary profiles in the vicinity are not shown in the diagram. The southern section runs somewhat oblique to the main axis of the gulf. The most striking feature along this line is a sedimentary wedge of about 200 km in length, reaching a maximum thickness of more than 1000 m near the Atlantic Ocean. This wedge is caused by continuous subsidence, reaching its maximum rates near the continental margin. In the westernmost part of the gulf (Profiles 02, 04), grey to greenish marls and siltstones, containing intercalations of sandstones, arenites, and limestones, predominate. According to lithology, se-

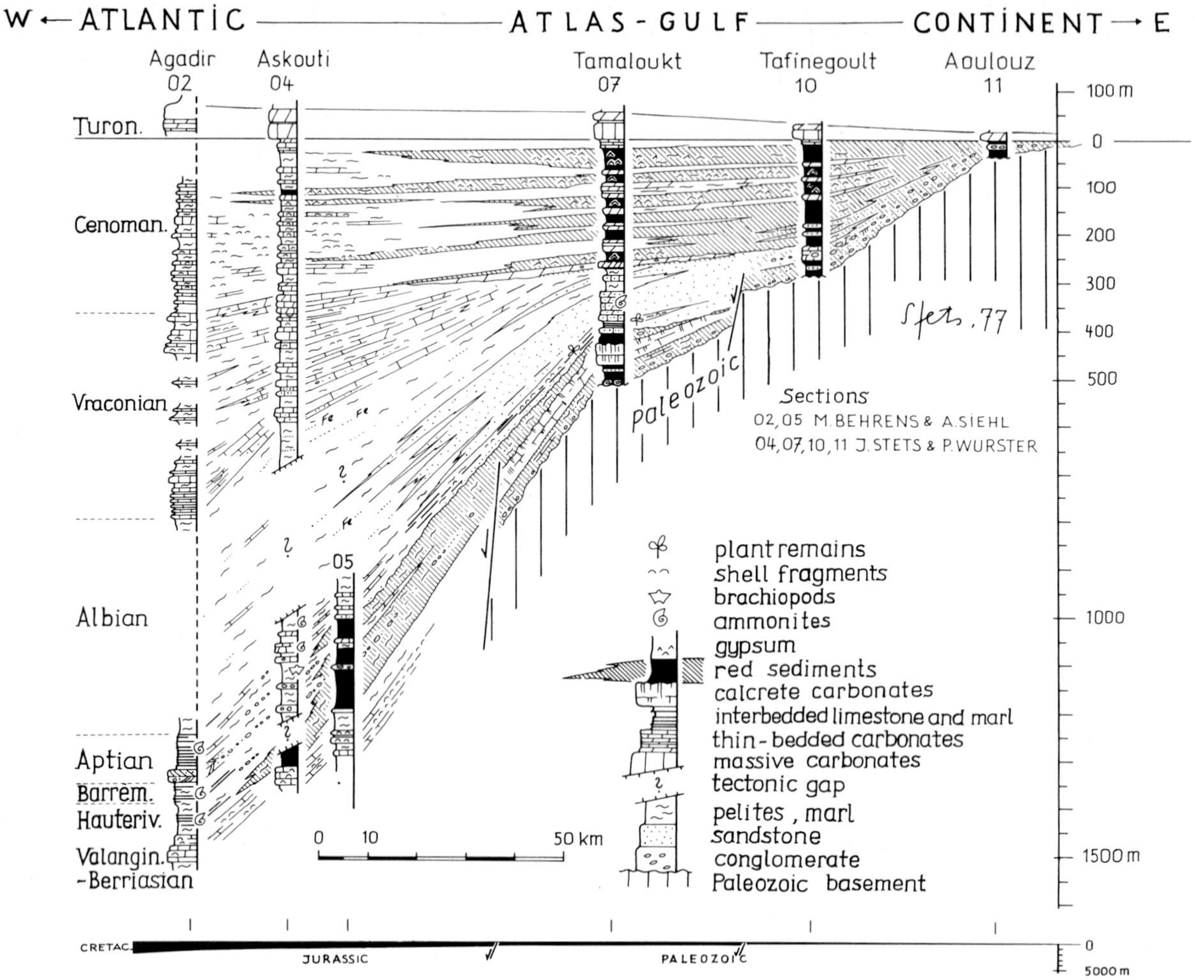

Fig. 6. Longitudinal cross-section along the southern flank of the High Atlas; note extensive migration of the littoral zone, generalized in Fig. 9; location of Profiles 02, 04, 07, 10, 11 see Fig. 2

dimentary structure, and fauna, the sediments show conditions of a continuous shallow-marine, open-sea environment. High-energy conditions and reworking can be inferred in places, especially during Vraconian time. In the easternmost area (Profile 11), marginal conditions are indicated by continental redbeds and strongly decreasing thicknesses. Between these two facies provinces, i.e., in the interior of the gulf (Profiles 04, 07, 10), open-marine, littoral, and continental sediments interfinger. Shifting of the coastline and of the littoral zone are evident: Beginning in the earliest *Neocomian*, the sea retired after a short period of transgression. During the *Barremian*, the sea tried to invade the gulf, but with little effect. In the *Aptian*, a striking transgression took place, and the coastline shifted rather quickly toward the east. During the *Albian*, the sea remained within the gulf. Within the *Cenomanian*, the littoral zone oscillated to a large extent, as the different peaks in Fig. 6 show. In the Latest Cenomanian and in the *Turonian* no littoral zone is visible in the longitudinal cross-section, as the transgression moved farther toward the east.

According to this cross-section the coastline and the littoral zone had no fixed position during the Early and Mid-Cretaceous. Within a distance of 100 km, repeated interfingering of facies occurs, obscuring the littoral zone, which we need for the construction of the sedimentation model. As a reference, we arbitrarily use the abrupt change from red to grey or greenish rock colors, taking into account all the difficulties mentioned above. In the longitudinal cross-section (Fig. 6) two significant transgressions can be observed, one of Aptian and another of latest Cenomanian to Turonian age. Changes in subsidence rates and of sedimentary supply from the adjacent continents caused only small-scale variations within the mobile belt. In our opinion, the two extreme and sudden shifts of the coastline toward the east are due to global events, e.g., sea-level changes.

Another longitudinal cross-section, which was constructed along the *northern flank* of the High Atlas, allows also a detailed analysis of the Cretaceous paleoenvironment (Fig. 7). It shows an almost similar evolution. The *Neocomian* series is composed of red coarse

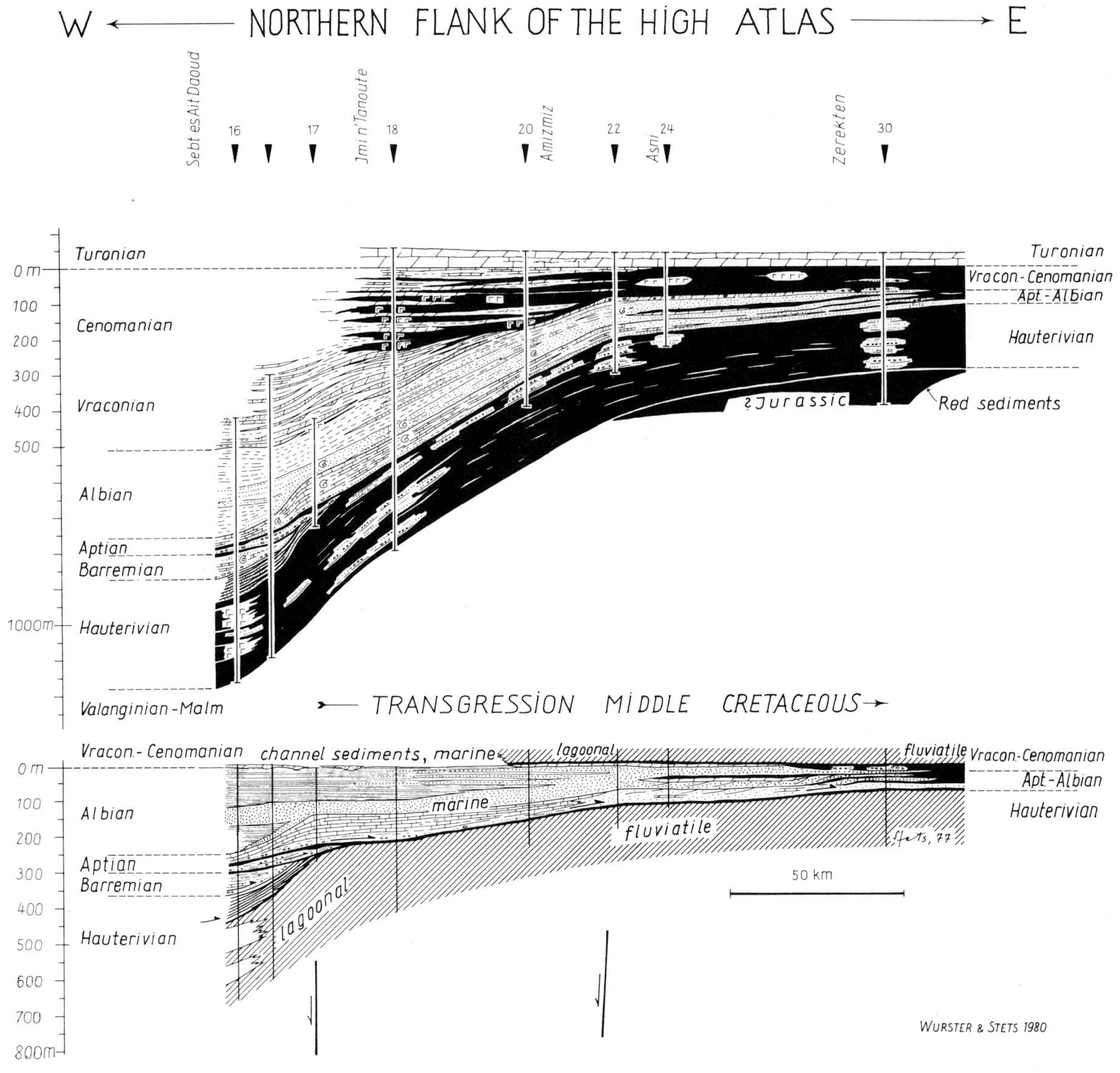

Fig. 7. Longitudinal cross-section along the northern flank of the High Atlas; the lower diagrammatic section illustrates the Aptian transgression as a significant Mid-Cretaceous event; location of Profiles 16, 17, 18, 20, 22, 24, 30 see Fig. 2

conglomerates, cross-bedded sandstones and siltstones. We have to assume a fluvial plain environment with channels and fans. Relics of subsoil calcification and of diagenetic encrustation suggest an occasionally submerged mud plain. Farther toward the west, evaporites, bioturbation, and changing rock colors indicate that no distinct high-energy littoral zone existed. Lagoonal and tidal-flat environments invaded the interior of the gulf on the vast fluvial mud plain. During the *Barremian*, a short ingression of the open-sea environment is documented by ammonites in the area west of Imi-n'Tanoute.

The *Aptian* transgressive event began with clastic sediments at the base, consisting of grey to yellowish sandstones and local conglomerates. In our opinion, the Aptian transgression started rapidly, shifting the littoral scenery toward the east at a high-energy level. At that time the sea level rose quickly. Consequently the following sedimentation occurred below wave base. Bituminous marls and limestones indicate preservation of organic matter. On the other hand, ammonites show that aerobic conditions dominated, whereas anaerobic conditions must have been restricted to the bottom of the gulf.

During the *Albian*, terrigenous detrital sediments such as well-sorted fine-grained sandstones and grey to olive siltstones filled the gulf from the east. We suggest that they are deposits of a shallow-shelf environment in front of a stationary or slowly westward-prograding coast. Hardgrounds, conglomerates and bio-

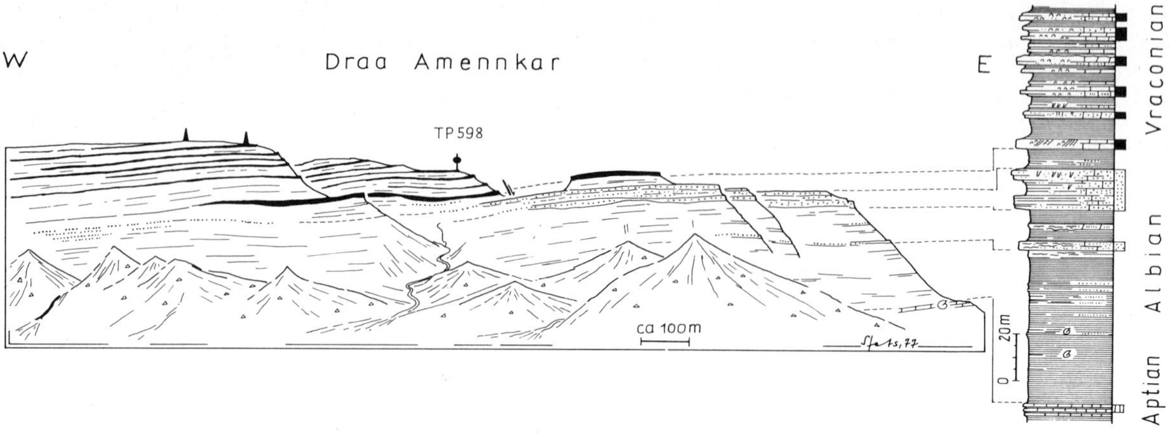

Fig. 8. Field section along the northern flank of the Jebl Ibail anticlinal structure SW of Chichaoua shows interfingering of Apto-Albian rock units; location see Fig. 2

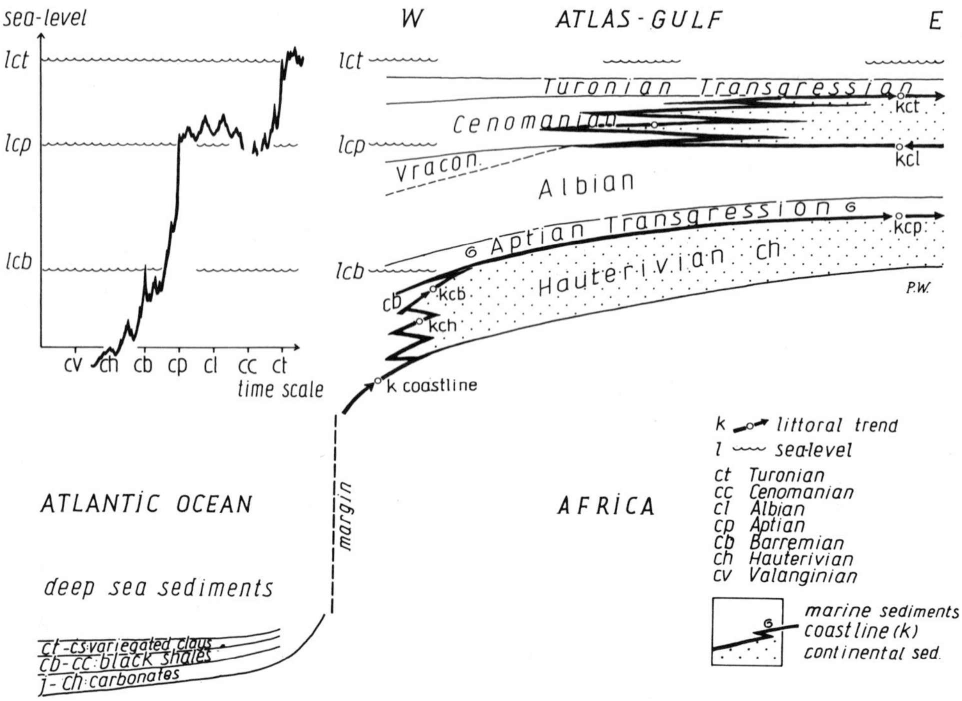

Fig. 9. Sedimentation model: interpretative sketch elucidates the migration of the littoral zone and the facies distribution, caused by three steps of a rising sea level

clastics of the Vraconian mark the littoral to shallow-marine environment retreating toward the Atlantic. At that time the deep-water event of Aptian time was already compensated by accumulation of sediments.

During the *Cenomanian*, grey siltstones, dolomites, calcareous dolomites and coquina beds represent a shallow-marine environment. Red siltstones alternating with dolomites, gypsiferous in places, and gypsum layers characterize lagoonal or sabkha environments. Coming from the east, a fluvial to lacustrine mud plain grades into the open sea environment in the west. Minor variations in the amount of subsidence or oscillations of the sea level within the gulf caused major ingressions and regressions of the sea on a flat plain of red swamps and lagoons.

In *Turonian* time, another rise of the sea level, maybe less than that of Aptian time, caused the final inundation of the Cenomanian mud plain and a transgression onto the bordering flat areas of the gulf.

The field section of Fig. 8 along the northern flank of the Jbel Ibail anticline southwest of Chichaoua shows the filling of the gulf during Albian time in a larger scale. Aptian limestone series, belonging to unit 1, are followed by greenish to olive marls and yellow calcareous sandstones of unit 2. They thin toward the west and finally wedge out, interfingering with the mudstone facies. The Vraconian rock sequences of unit 3, consisting of dolomitic calcarenites and coquina beds, show the same tendency toward the outer part of the gulf.

4. Sedimentation Model of the Atlas Gulf

Summarizing all the data obtained until now, we are able to establish the sedimentation model. This model is based on the following assumptions:

- The crustal movements were quite conservative. Continuous subsidence occurred, controlled by the structural framework. Local tilting of separate blocks modified the subsidence rates, but did not change the facies distribution.
- The relief in the continental area bordering the gulf remained nearly unchanged throughout the Early and Mid-Cretaceous.
- There are no signs that during Early to Mid-Cretaceous times major climatic changes took place.

In our opinion, these assumptions allow us to explain the geological processes in a rather simple and convincing way. Arbitrary interpretation is now restricted to one or two factors only. Considering these facts we prefer distinct oscillations of the sea level to explain the sedimentation processes in the Atlas Gulf.

The sketch of Fig. 9 shows the curve of the rising sea level (left figure), with regard to the position of the littoral zone (right figure). Because of a very low sea level during the Early Neocomian, the coastline (k) lies near the outer part of the gulf or closer to the continental slope. During the Hauterivian the coastline oscillates in a rather stationary position, dependent on the supply from the continent and shallow-marine sedimentation. During Barremian time (cb), a marginal ingression occurred, because of a distinct rise of the sea level. The coastline (k) shifted landward into the gulf for about 50 km to the east. During the Aptian (cp), the sea level rose quickly, reaching level l_{cp} and causing an extensive transgression.

Inundation covered the whole gulf for a distance of about 250 km. During the Albian (cl) the rise of the sea level slowed down or remained nearly in the previous position. The delay of further sea level rise caused the coastline of the Late Albian (Vraconian) to begin to migrate back toward the west, but still to remain within the gulf. As the gulf was nearly filled with sediments, no relief in the sedimentation area existed during Cenomanian (cc) time. Therefore small-scale oscillations of the sea level caused already major inundations and regressions within the inner part of the gulf. Another rise of the sea level during Late Cenomanian to Turonian times (l_{ct}) was sufficient to flood the whole gulf and the adjacent areas.

Fig. 10 illustrates these processes by a series of block diagrams. Each block diagram shows in a schematical manner the geological features. They are in agreement with the sedimentary record, found in the main pro-

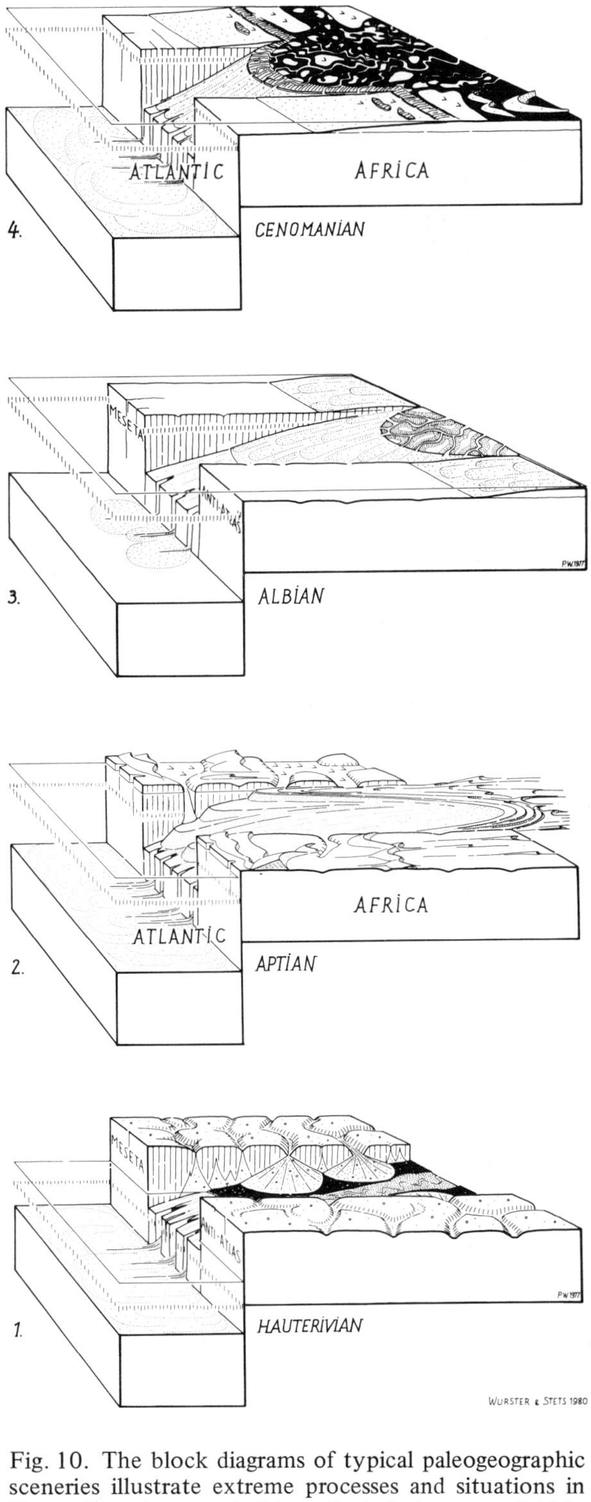

Fig. 10. The block diagrams of typical paleogeographic sceneries illustrate extreme processes and situations in the sedimentary evolution; selected stages are highly exaggerated and simplified

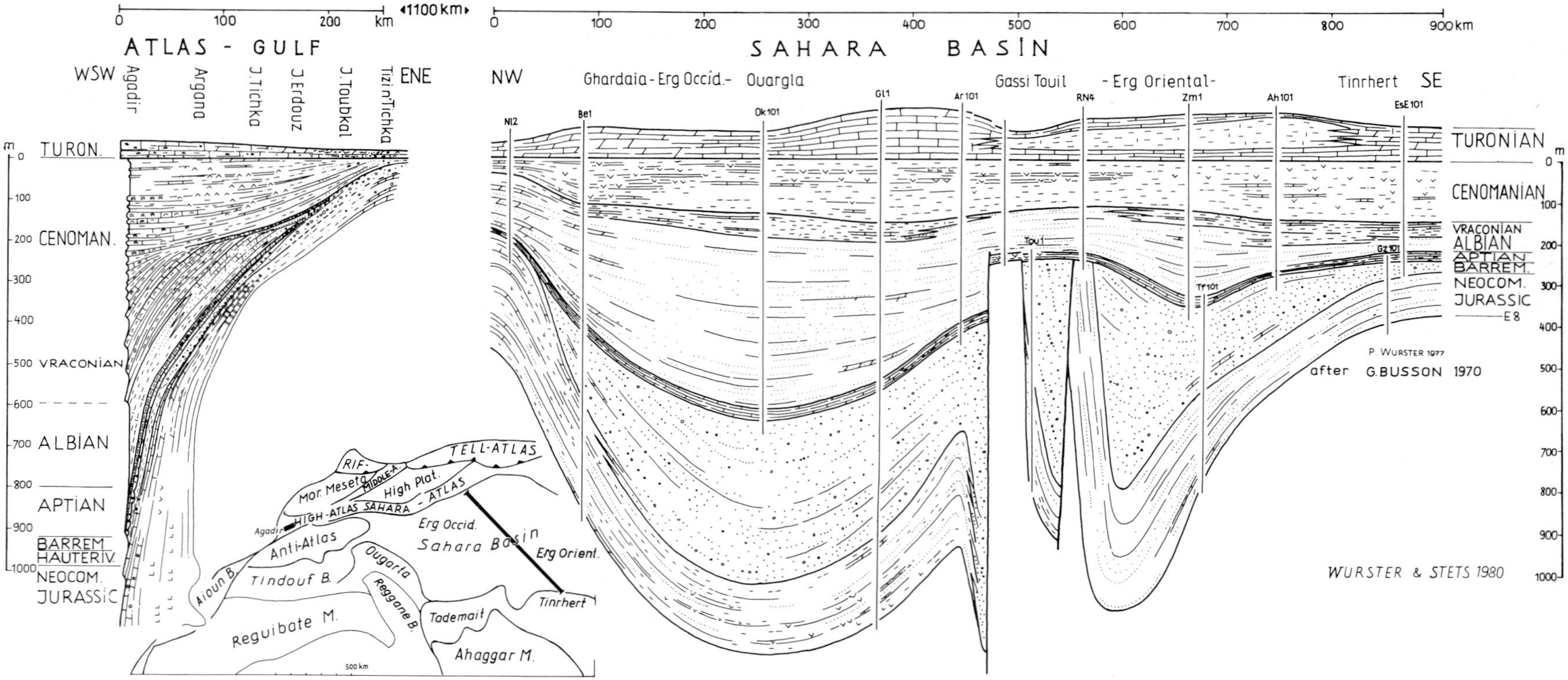

Fig. 11. Synoptic diagram of the Atlas Gulf (*left side*) and Sahara Basin (*right side*, after Busson); similarities of facies distribution and synchronous events indicate changes of sea level; legend see Figs. 6 and 7

files and within the longitudinal cross-sections. We know that the geological processes of one time interval are much more complicated than can be drawn in the graphs. Therefore we take the extreme positions of the sea level for the illustration. The Atlas Rift and the Atlas Gulf were geomorphologically not so pronounced as drawn in the block diagrams; but we had to exaggerate the relief, in order to better illustrate the processes within the gulf and in the adjacent ocean basin (see also Fig. 13).

1. During the Neocomian, especially during the Late *Hauterivian*, the very low sea level allowed the continental detritus to prograde toward the outer part of the gulf. Red sediments are found near the Recent Atlantic coast (Behrens and Siehl, this Vol.). Much detritus accumulated in the shelf area and was transported further down the continental slope toward the deep ocean basins.

2. In contrast to the Hauterivian, the *Aptian* block diagram shows the consequence of a quickly raised sea level. According to its high position the coastline and littoral zone migrated far into the gulf. Marginal areas of the neighboring high platforms were flooded and open marine conditions predominated within the gulf. Because of the high sea level, coarse-grained detritus was retained in the inner part of the gulf. Only fine-grained pelitic material reached the outer part of the gulf, the shelf, and went down toward the ocean floor. In the Late Aptian sequences of limestones and marls originated below the wave base in the shallow sea.

3. During *Albian* time, the sea level remained rather stable. The continental detritus, derived from the adjacent high areas, was reworked in the littoral zone. Proceeding from the inner part, the gulf was filled in by shallow-marine or deltaic sediments. Well-sorted fine-grained material prograded slowly toward the outer gulf. At the beginning, not much material reached the open ocean, as the littoral zone was situated far in the inner gulf. Anaerobic conditions prevailed at the bottom of the outer gulf, as indicated by the fauna.

4. During the *Cenomanian*, the growth of the sedimentary pile within the gulf reached the sea level. Coastal swamps, sabkha plains, and lagoons originated near the shore, whereas oyster reefs and oyster banks were built up in a shallow-marine environment. A great deal of detrital material was transported to the shelf and further down to the ocean floor.

The most striking event within this paleogeographic scenery of the Atlas Gulf is that of the Aptian transgression. We suggest that it is of global extent. This interpretation will be analyzed and discussed in the following excursions.

5. Excursion to the Sahara Basin

The Atlas Gulf can be regarded as a water-gauge station of the huge Atlantic Ocean, able to register and to control sea-level changes during the Cretaceous. If the transgressions and regressions postulated in the previous chapters for the Atlas Gulf are of global extent, they should also be observed in areas outside the gulf. As there are no possibilities of investigating such processes within the ocean by conventional geological methods, we have to look for them first on the adjacent North African Platform. The example nearest at hand is the vast Sahara Basin in Algeria, recently studied in detail by Busson (1970, 1972). In nearly every respect this sedimentary basin shows properties similar to those of southwest Morocco, concerning crustal structure, relief, and climatic conditions. It measures about 1000 km in the N–S direction, and about 2000 km in the E–W direction, i.e., it is about 100 times larger than the Atlas Gulf. Continuous subsidence within this large basin is documented by a 5000 m thick sedimentary pile, containing the whole sedimentary record from the Triassic to the Cretaceous. Shallow-marine sediments of the Sahara Atlas in the north, detrital and evaporitic sediments of an extensively oscillating littoral zone, and continental sediments near the Hoggar Massif in the south interfinger in the basin within a conformable and complete sedimentary sequence. The Cretaceous sedimentary rock sequences of the Sahara Basin resemble those of the Atlas Gulf (Fig. 11).

In Early *Neocomian* times marine sediments originated. With the reservation to different stratigraphic correlations (Hauterivian of the Atlas Gulf, Barremian of the Sahara Basin), continental rock series follow on

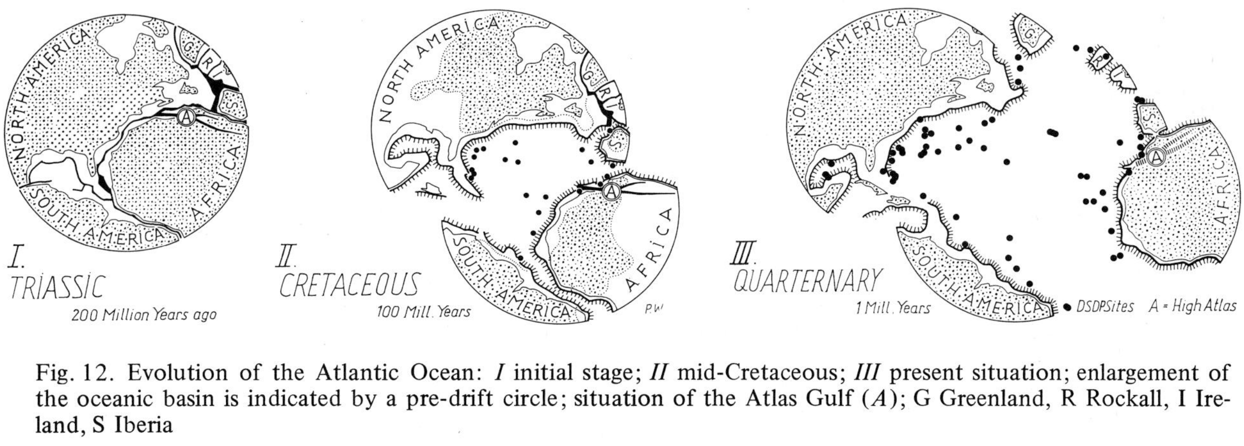

Fig. 12. Evolution of the Atlantic Ocean: *I* initial stage; *II* mid-Cretaceous; *III* present situation; enlargement of the oceanic basin is indicated by a pre-drift circle; situation of the Atlas Gulf (*A*); G Greenland, R Rockall, I Ireland, S Iberia

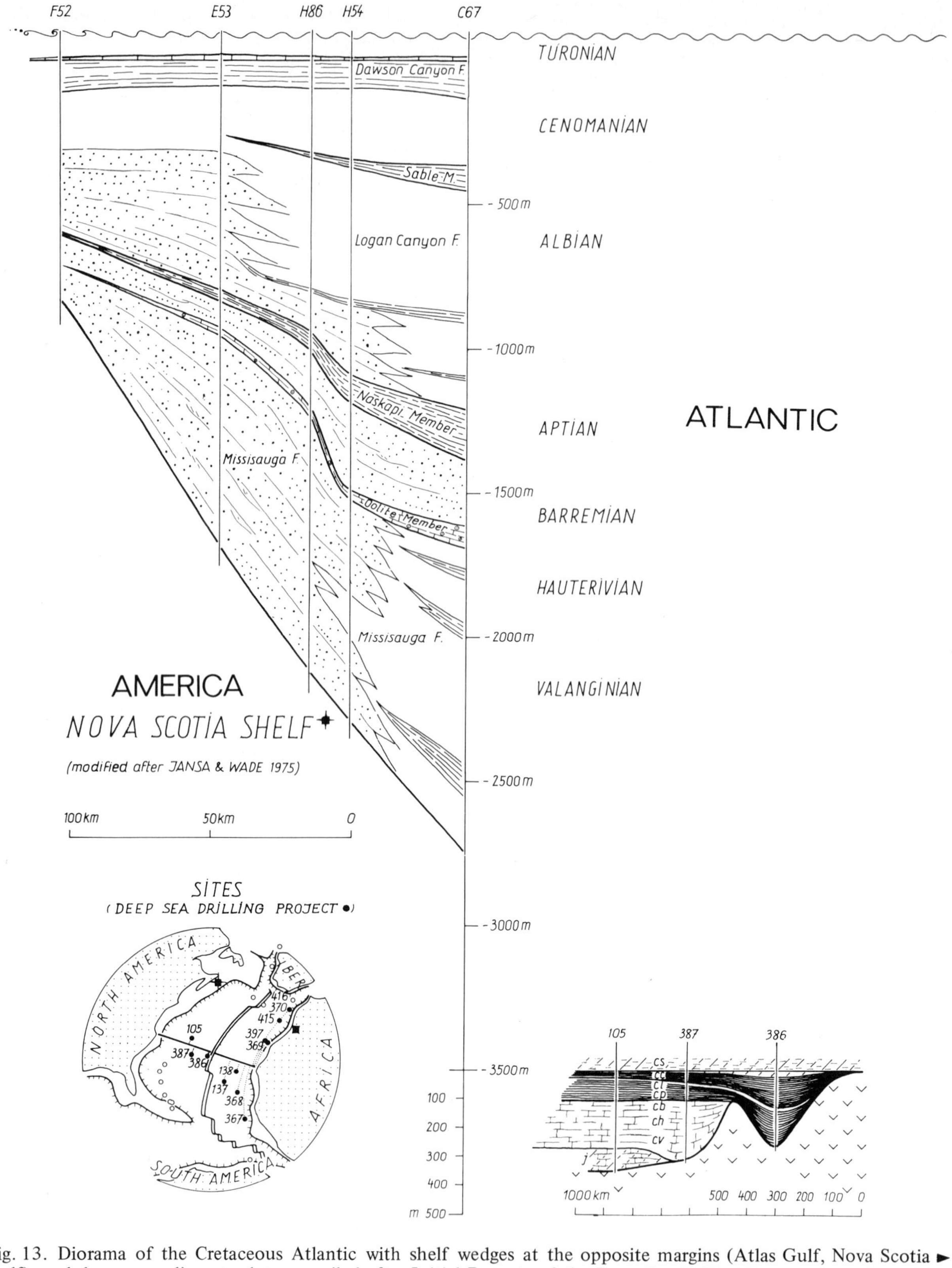

Fig. 13. Diorama of the Cretaceous Atlantic with shelf wedges at the opposite margins (Atlas Gulf, Nova Scotia ► shelf), and deep-sea sediments; data compiled after Initial Reports of the Deep Sea Drilling Project and references cited in the text; the diorama elucidates conspicuous relations and consequences of the Mid-Cretaceous events; Note: vertical scale is the same for deep-sea and shelf wedges, horizontal scale is 10 times reduced for the deep-sea area

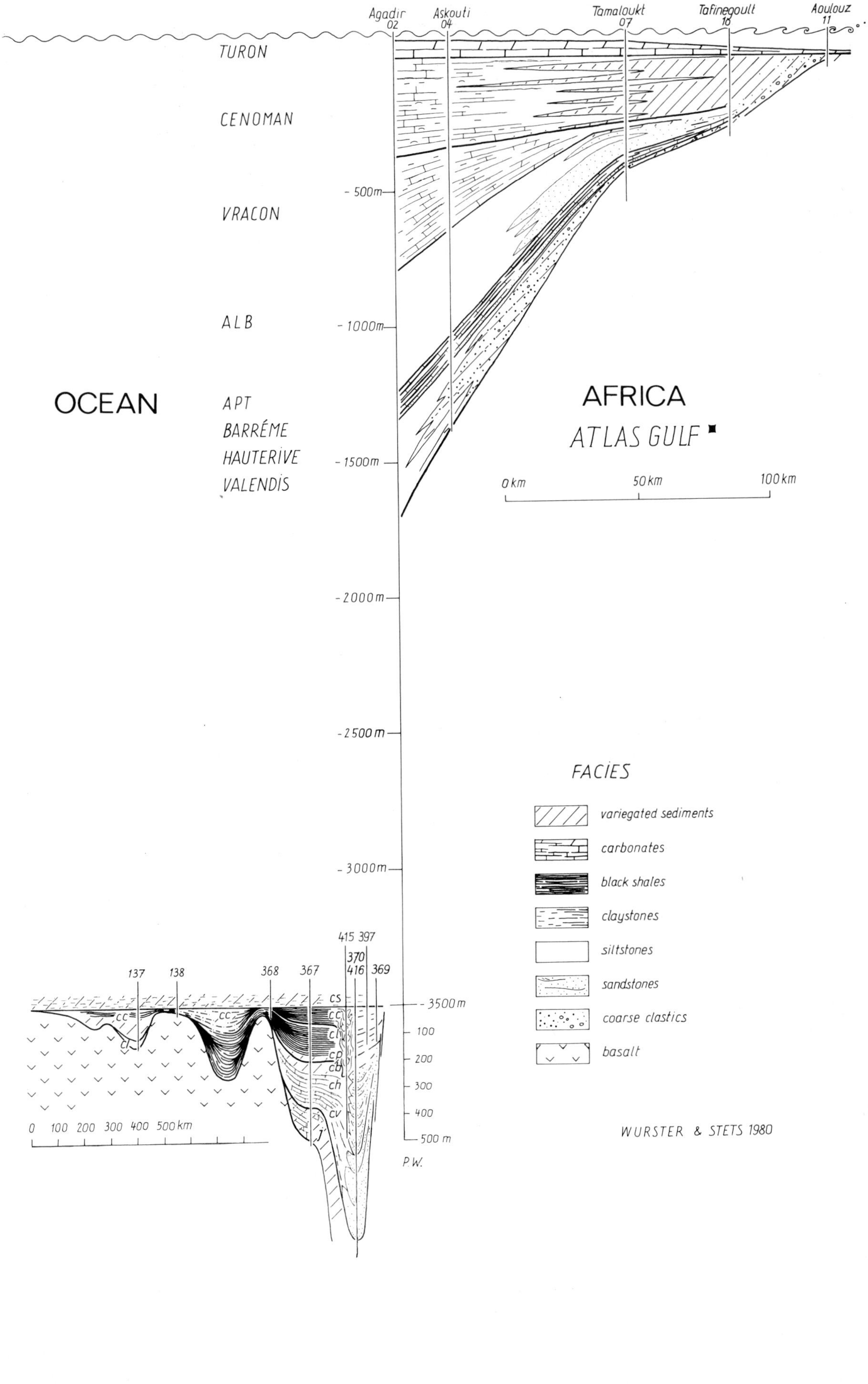
Agadir 02
Askouti 04
Tamaloukt 07
Tafinegoult 10
Aoulouz 11
TURON
CENOMAN
VRACON
ALB
APT
BARRÉME
HAUTERIVE
VALENDIS
OCEAN
AFRICA
ATLAS GULF
0 km
50 km
100 km
- 500m
- 1000m
- 1500m
- 2000m
- 2500m
- 3000m
- 3500m
FACIES
variegated sediments
carbonates
black shales
claystones
siltstones
sandstones
coarse clastics
basalt
137
138
368
367
415 397
370
416
369
cs
cc
cl
cp
cb
ch
cv
j
100
200
300
400
500 m
0 100 200 300 400 500 km
P.W.
WURSTER & STETS 1980

top of the lowermost Neocomian sequence. This "continental intercalaire" reaches a thickness up to 500 m, consisting of red conglomerates, sandstones, siltstones, and marls. With respect to the sediments and to facies distribution, Busson infers a fluvio-deltaic environment for this time interval. The correlation of this environment with that of the Atlas Gulf indicates an extremely low sea level of the adjacent ocean. Lowering of the sea level revived the erosional processes and accelerated the reworking on the African continental platform. The sedimentary detritus filled the continental Sahara Basin and went down to the deep ocean basins of the Tethys.

During the *Aptian*, the correlation of the sedimentary sequence of both areas became better, as the so-called "barre aptienne" covered the whole Sahara Basin. This series consists of a rather thin carbonate rock sequence. The environment is said to be open-marine to lagoonal. The lithology of the sedimentary sequence of the Atlas Gulf and of the Sahara Basin correspond well, although no marine faunas of Aptian age were found within the Sahara Basin until today. The facies changes from the underlying red series to the similar calcareous "barre aptienne". This means that the transgression postulated in the Atlas Gulf was wider and even reached the Hoggar massif in the south. The northwestern part of the African craton was flooded at this time. In our opinion there is no evidence for intense crustal movements.

From the *Albian* onward, a corresponding facies development took place: Fine-grained sandstones with intercalations of marls and siltstones predominate. Coarser material is found only in marginal positions near the Ougarta and Hoggar highs. Prograding delta fans and reworked littoral sediments interfinger with shallow-marine marls in the southwestern Sahara Basin, as well as in the outer part of the Atlas Gulf.

Sediments of Late Albian (Vraconian) age are also found in the Sahara Basin. Diachronous interfingering of different lithological facies can be stated: according to Busson, the open marine calcareous sequence prograded toward the south. This interpretation is in contrast to our opinion about the Vraconian of the Atlas Gulf. Here a regressive tendency was postulated. Further research is necessary to solve this problem. We suggest that a highly oscillating sea level has to be taken into consideration.

During the *Cenomanian*, gypsiferous redbeds of nearly equal thickness covered the Atlas Gulf and the Sahara Basin. Coming from the Atlantic Ocean and from the Tethys sea, fossiliferous marine series prograded temporarily into both basins. The relief was flat within the basins and smooth on the nearby platform areas. The model of a vast mud plain inferred for the Atlas Gulf can be extended to the Sahara Basin.

The *Turonian* transgression easily flooded both basins and the neighboring African platform. A major change in the sedimentary conditions at that time is reflected by the change of red to white and grey rock colors.

Taking into account all the facts and contradictions mentioned above, we suppose that the Atlas Gulf sedimentation model is also valid for the Sahara Basin. Concerning the genetic interpretation, no major crustal movements, either local or regional, might have had decisive influence on the facies development. This seems plausible because of the extent of the stable basement complex of the North African craton. An alternation of synchronous subsidence and uplift seems to be improbable, if one tries to explain the Mesozoic sedimentary pattern. The facies variations can more easily be interpreted by oscillations of the sea level. Correlation work should be extended to other North African sedimentary basins step by step, as to the Ghadames, Murzuk, Syrte, Kufra, and Western Desert basins (Klitzsch, 1970; Van Houten, 1980).

Eustatic sea level changes may be of different origin. In any case the cause of the events leading to the transgressions in Northwest Africa lies outside of the areas discussed above.

6. Excursion to the Atlantic Ocean

Synchronism of the Mid-Cretaceous events in the Atlas Gulf and in the Sahara Basin indicates that similar reasons are responsible for them. To check the model anew, it seems indispensable to include also the paleogeographic setting of the opening Atlantic Ocean (Fig. 12; see also Jansa and Wiedmann, this Vol.).

Proceeding from the assumption that during the Triassic the Atlantic furrow began to break up (Fig. 12, I), and that during the Jurassic the central North Atlantic began to open by drifting, we have to postulate a deep ocean basin of intermediate size already during Mid-Cretaceous time (Fig. 12, II). It was bordered by the North American continent, by Western Europe, Northwest Africa, and South America, and was open only toward the west by the Caribbean.

Shallow and narrow graben-like seaways existed only in Northwest Europe, in the Rif area, and between Guiana and Guinea, which granted rather restricted connections with the Tethys, the Arctic and Antarctic seas. The influx of cold bottom water was therefore largely choked. Calculations concerning the position of the carbonate compensation depth (CCD), the origin of carbonates, or upwelling and circulation of water masses within this Early to Mid-Cretaceous Atlantic basin, which are based upon modern oceanographic conditions, are therefore of hypothetical value.

In order to prove the sedimentation model of the Atlas Gulf, we first look at the Cretaceous sedimentary wedge, which originated on the opposite shelf of Canada at nearly the same time (Jansa and Wade, 1975). It stretches along the western North Atlantic, hidden by younger sediments of the coastal plain and of the shelf area. No onshore outcrops are available. Therefore, the research on the Mesozoic in this region is based only on well data and geophysics. Although the results on both sides of the Atlantic were gained by quite different methods, facies development and distribution are comparable (Fig. 13). Rapid Mesozoic subsidence rates can be observed at the two opposite margins. At the North American margin, much more detritus was deposited during Early and Mid-Cretaceous time than at the Moroccan margin. Even if we are not able to delineate the coastline in detail as was done in the Atlas Gulf, striking breaks in facies development are evident. There are the "Oolite Member" in the Hauterivian/Barremian similar to the Barremian event in the Atlas Gulf, the "Naskapi Member" equivalent to the Aptian transgression, the "Sable Member" corresponding to the Vraconian, and, last but not least, the "Dawson Canyon Formation" equivalent to the Turonian transgression. The distribution of fine-grained clastics shows also certain analogies.

Symmetrical arrangement and structure of the sedimentary wedges on both sides of the Atlantic of course do not prove our sedimentation model of the Atlas Gulf. Unequivocal hints are, however, evident, that continuous subsidence at the passive continental margins was superimposed by discontinuous events. The significant variations in facies development and distribution, also visible in the sedimentary record of the Canadian shelf wedge, obviously indicate major transgressions, caused by eustatic sea level changes (Vail et al., 1978).

It is also interesting to study the effects of these events on the sedimentary pattern of the deep-sea sediments in the Cretaceous Atlantic. A diorama or diagrammatic cross-section through the deep sea connecting the continental margins is shown in Fig. 13. It contains drill-site data of *Glomar Challenger* cruises on both sides of a line, crossing the Atlantic Ocean between Cape Hatteras and Cap Blanc. Borehole data were projected onto the cross-section, following the outer shape of the continental margin. Simplified sedimentary phenomena are traced at the same vertical scale used for the shelf wedges. The deep-sea cross-section was added at a standardized water depth of 3500 m according to Chenet and Francheteau (1980). This sketch, roughly schematized and standardized, gives a general idea of the conditions at the shelf and in the deep sea (Arthur, 1979; Tucholke and Vogt, 1979; Lancelot and Winterer, 1980).

The first striking difference is indicated by a strongly reduced thickness of the deep-sea sediments. Furthermore, differences of facies development are clearly visible. On the other hand, an amazing relation exists between the facies pattern in the deep sea and that of the shelf wedges:

Carbonate bearing and silty Lower Cretaceous rock series in the deep ocean correspond to a low sea level in the Atlas Gulf. The most striking phenomenon is that the black shale sedimentation in the deep sea coincides with the Mid-Cretaceous rise of the sea level in the Atlas Gulf. Did the circulation of deep water masses in the Atlantic stagnate at that time? Variegated deep-sea sediments were deposited in Turonian and post-Turonian time, when the sea level rose anew and reached its highest level.

All the findings and results mentioned above may have to be modified, if more details become known by drilling and geophysics. However, the present observations are sufficient to support the arguments stated in our sedimentation model. The Cretaceous events in the Atlas Gulf, in the Sahara Basin, on the Canadian and Moroccan margin, as well as in the deep basins of the widening Atlantic, have to be traced back to one impetus. They are all controlled by synchronous but different processes. We do not know the cause of these events, nor do we have any preconceived ideas or speculations on it. In our opinion, however, climatic changes or glacial processes must be excluded. We suppose that crustal movements of global extent, as for example break-up and opening of new seaways toward the South Atlantic, changes in the volume of spreading centers, acceleration of crustal spreading, or subduction processes far from our area of investigation should be taken into consideration.

Acknowledgments. We gratefully acknowledge much assistance that enabled the performance of this study. Field work, done in 1975–1978, was largely supported by the German Science Foundation (DFG, grants No. Wu 32/15-19). The Moroccan authorities, especially the staff of the Geological Survey at Rabat, kindly facilitated the investigations. Dr. Winfried Zimmerle, Celle, and Dr. Ulrich von Rad, Hannover, reviewed the English manuscript and contributed many helpful and critical comments. Thanks are due to all of them.

REFERENCES

Ambroggi R (1963) Etude géologique du versant méridional du Haut Atlas occidental et de la plaine du Souss. Notes Mém Serv Géol Maroc 157: 321 pp

Arthur MA (1979) North Atlantic Cretaceous black shales: the record at site 398 and a brief comparison with other occurrences. Init Rep Deep Sea Drilling Project 47: 719–751

Behrens M, Krumsiek K, Meyer DE, Schäfer A, Siehl A, Stets J, Wurster P (1978) Sedimentationsabläufe im Atlas-Golf (Kreide Küstenbecken Marokko). Geol Rundsch 67: 424–453

Busson G (1970) Le Mésozoique saharien, II, Essai de synthèse des données des sondages algéro-tunesiens. ed CNRS, 811 pp, Paris

Busson G (1972) Principes, méthodes et résultats d'une étude stratigraphique du Mésozoic saharien. Mém Mus Hist Nat C 26: 441 pp

Chenet PY, Francheteau J (1980) Bathymetric reconstruction method: application to the central Atlantic basin between 10° N and 40° N. Init Rep Deep Sea Drilling Project 51, 52, 53: 1501–1513

Choubert G, Faure-Muret A (1962) Evolution du domaine Atlasique Marocain dépuis les temps Paléozoiques. In: Durand-Delga M (ed) Livre à la mémoire P Fallot. Mem H-S Soc Géol Fr 1: 447–527

Duffaud MF (red) et al (1966) Le bassin du Sud-ouest Marocain. In: Reyre D (ed) Bassins sédimentaires du litoral Africain. Assoc Serv Géol Afr, pp 5–12

Jansa LF, Wade JA (1975) Geology of the continental margin off Nova Scotia and Newfoundland. Geol Surv Can Pap 74–30: 51–105

Klitzsch E (1970) Die Strukturgeschichte der Zentralsahara. Geol Rundsch 59: 459–527

Lancelot Y, Winterer EL (1980) Evolution of the Moroccan Oceanic Basin and adjacent continental margin – a synthesis. Init Rep Deep Sea Drilling Project 50: 801–821

Makris J, Demnati A (1977) Deep seismic soundings in Morocco and a crust and upper mantle model deduced from seismic and gravity data. 15 pp (unpubl manuscript)

Michard A (1976) Eléments de géologie marocaine. Not et Mém Serv Géol Maroc 252: 408 pp

Pitman WCIII (1978) Relationship between eustasy and stratigraphic sequences of passive margins. Geol Soc Am Bull 89: 1389–1403

Reineck HE, Singh IB (1973) Depositional sedimentary environments. Springer, Berlin Heidelberg New York

Sclater JG, Hellinger S, Tapscott C (1977) The paleobathymetry of the Atlantic Ocean from the Jurassic to the present. J Geol 85: 509–552

Tucholke BE, Vogt PR (1979) Western North Atlantic: sedimentary evolution and aspects of tectonic history. Init Rep Deep Sea Drilling Project 43: 791–825

Vail PR, Mitchum RMJr, Thompson SIII (1977) Seismic stratigraphy and global changes of sea level. Am Assoc Petrol Geol Mem 26:49–212

Van Houten FB (1980) Latest Jurassic – Early Cretaceous regressive facies, northeast Africa craton. Am Assoc Petrol Geol Bull 64:857–867

Wiedmann J, Butt A, Einsele G (1978) Vergleich von marokkanischen Kreide-Küstenaufschlüssen und Tiefseebohrungen (DSDP): Stratigraphie, Paläoenvironment und Subsidenz an einem passiven Kontinentalrand. Geol Rundsch 67: 454–508

20 Sedimentation in the Atlas Gulf III: Turonian Carbonates

Roger Stamm and J. Thein[1]

ABSTRACT

The Turonian carbonates of the High Atlas region have been analyzed with respect to their lithofacies and the distribution of their chemical major elements by means of 42 stratigraphical sections. The geographical distribution of litho- and chemofacies illustrates contrasting oceanic and continental influences. Cross-sections make it possible to correlate the major stages of the environmental evolution from an outer shelf, over a separating threshold, to the western confines of the Saharan epicontinental sea. Deep sea drilling results reveal similar stratigraphic developments concerning the offshore area. A schematic model of sedimentation for two extreme evolution phases: transgression with a maximum pelagic realm and regression with a continentally influenced shallow water regime, is discussed.

1. Introduction

During the intensive deep sea drilling of the last years a typical black shale facies, assigned to middle Cretaceous, has been cored in the Atlantic (DSDP Sites 367, 368, 137). This period, however, is often represented in the ocean areas by a large hiatus covering a timespan from middle Cretaceous to Tertiary. The facies of black shales, continuously increasing in thickness (100 m), can be traced up to the shelf, where it interfingers with a broad carbonate platform. Here the stratigraphic sedimentary evolution is characterized by a significant change from hemipelagic limestones in the lower part of the sections to a dolomitic shallow-water facies in the upper part. A similar development may be seen in DSDP Site 367, where bituminous marls of early Turonian age are covered by younger siltstones which where deposited under oxidizing conditions. Even in the abyssal hills west of the Canary Plain (DSDP Site 137), the post-Cenomanian sedimentation is marked by a sudden break from pyritic and chertiferous nanno-oozes in the lower part to carbonate-free, zeolitic clays above (Berger and von Rad, 1972).

The lower calcareous strata in the Atlas Gulf could possibly be connected with a transgressive cycle which is superseded by dolomites after a rapid regression. This sharp facies change on the shelf might correspond chronologically to the "oceanic anoxic event" of Schlanger and Jenkyns (1976) at the end of the middle Cretaceous. This verifies the assumption of Fischer and Arthur (1977) and of Jenkyns (1980) that a causal connection exists between the flooding of widespread shallow water areas in the course of transgressions and the simultaneous development of euxinic environments in the bordering oceans.

Much information on the geochemistry and lithofacies of deep-sea sediments can be found in Lange (1974), Hsü and Jenkyns (1974), Cook and Enos (1977), Migdisov et al. (1980) and Renard et al. (1980). Detailed research on shallow-water carbonates was done for instance by Graf (1960), Friedman (1969), Honjo and Tabuchi (1970), Wedepohl (1970) and Veizer (1977). Comparisons of both environments, however, have not yet been accomplished. In order to comprehend such environmental transitions in time and space, we give in this paper account of the stratigraphical and regional information based on the geochemical and structural analyses of more than 1500 samples of vertical sections of Moroccan Turonian. (Detailed trace element analyses will be published later.) The Atlas Gulf paleogeography provides the basis for a model of the compact relationships of an epeiric sea and the oceanic realm.

2. Regional Geology (Fig. 1)

The carbonate sequence of the "barre turonienne", which includes after Roche (1930), Choubert (1948), Busson (1970), Schroeder (1976) and Wiedmann et al. (1978) the upper Cenomanian and the Turonian, is well accessible today in the northernmost West African coastal basin, the Cretaceous "Atlas-Gulf" (Stets and Wurster, 1977; Behrens et al., 1978; Stets and Wurster, this Vol.) between the Atlantic Ocean and the Sahara region. It is exposed in the west in coastal cliffs (Agadir)

1 Geologisches Institut der Universität Bonn, Nussallee 8, D–5300 Bonn 1, FRG

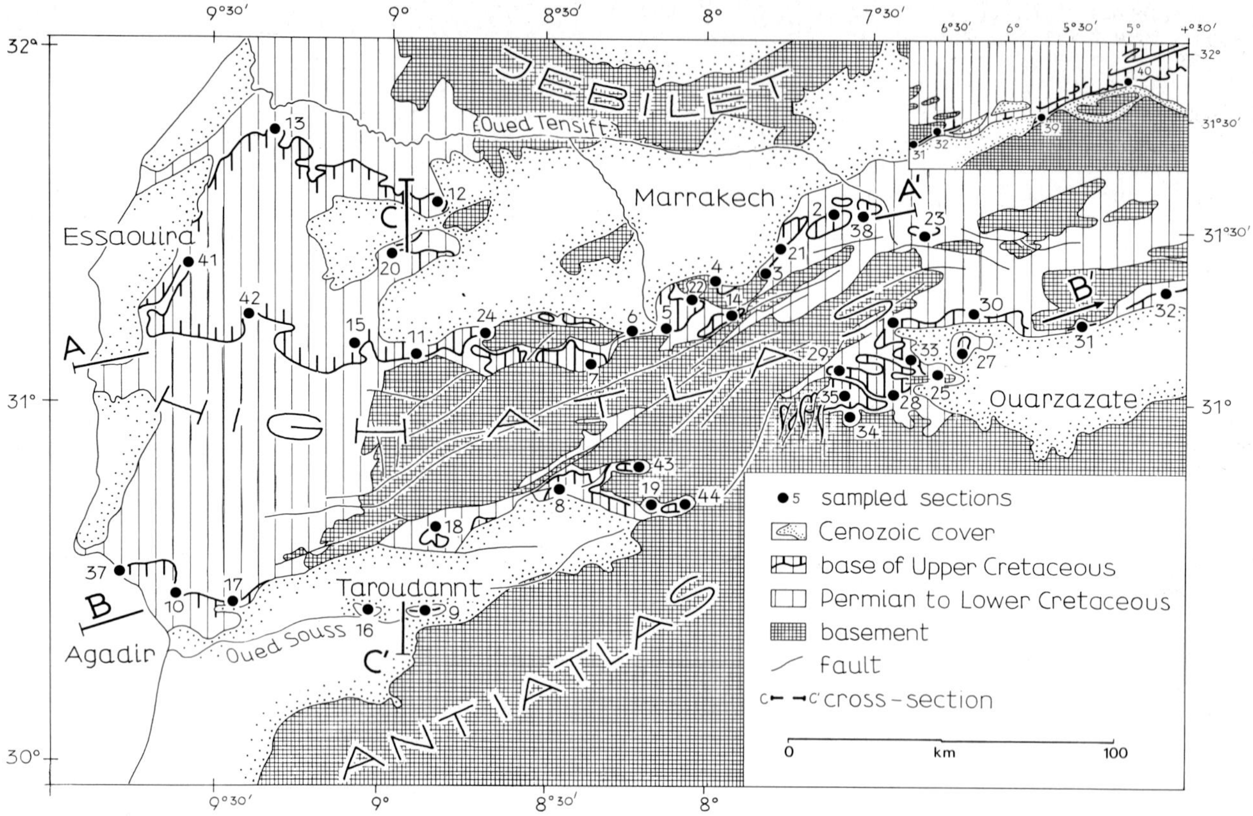

Fig. 1. Schematic geological map of the area of investigation

and salt domes (Essaouira) with a thickness of about 100 m, decreasing eastward to less than 40 m. To the north and south of the High Atlas, Cretaceous strata are tilted around the Paleozoic and Precambrian core of the Atlas orogen. They are covered by the young continental debris of the plains of Marrakech and of the Oued Souss. To the south and the east of the Souss plain, Turonian sediments reappear as a sequence, only 10 m thick, separated by a thin carpet of coarse-grained terrestrial lower Cretaceous rocks from the Precambrian basement of the Anti-Atlas, which was extensively flooded by the upper Cenomanian transgression.

To the north of the investigated area, thin Turonian sediments with strong terrigenous influence, and a continental lower Cretaceous member, overlie the Paleozoic rocks of the Jebilet. In these southern and northern marginal positions, the "barre turonienne" can easily be discerned between clastic lower and lagoonal to continental upper Cretaceous, as being the only marine horizon in typical dolomitic facies.

In the western region, the stratigraphic limitation of Turonian is more difficult. There carbonate horizons were deposited already in the lower Cretaceous, and the upper Cretaceous is almost entirely developed as carbonates. However, the latest biostratigraphic results (Wiedmann et al., 1978) permit the correlation of these sections. Moreover, in this region, the Turonian shows a very typical facies of hemipelagic, bituminous, calcareous laminites, rich in cherts. To the east, however, this facies grades rapidly into dolomites. The lower Cretaceous and the Senonian become clastic to the east and more strongly influenced by the continent.

The paleogeographic position of the High Atlas area during Cretaceous times is still problematic, because neither Turonian nor lower or higher Cretaceous rocks are preserved. Whereas older publications (Choubert and Faure-Muret, 1962; Duffaut, 1966) assume a Cretaceous land for this region, modern detailed investigations emphasize a depositional basin (Behrens et al., 1978; Wurster and Stets, this Vol.). It is unlikely that the area mentioned has supplied the Turonian carbonates with terrigenous material. On the other hand, the regional facies development suggests that it may have been a morphologic threshold since upper Turonian. The orogenetic uplift of the Atlas mountains began in Senonian time (Brix, 1980).

3. Vertical and Regional Distribution of Facies

The high lithologic diversity of the High Atlas Turonian can be reduced to a few facies types which allow a regional differentiation into facies zones, reflecting the changing influences of the Atlantic pelagic realm and the epicontinental Saharan sea (Fig. 2). In connection with the distribution pattern of the chemical major elements (Fig. 3), the facies zones represent a western basin with a major zone of subsidence along the northern edge of the High Atlas (Sections 11, 14, 15, 41, 42). The thickness of sediments decreases rapidly when approaching the northern and the southern land masses. Diminished thicknesses can be observed on an eastern threshold region between the Atlantic and the Sahara. Further to the east (Sections 1 to 40), a configuration of the facies zones, comparable with that of the western basin is indicated. In the Sahara Basin, east of the area of investigation, stronger subsidence caused again an intensified deposition of Turonian sediments with a thickness of about 100 m (Busson, 1970).

Although the terrigenous influence is minimal, distribution of land and sea, connected with paleoenvironmental parameters fundamentally affect the distribution pattern of the geochemical elements. The further described results reveal a strong dependence of Ca, Sr, C org., Fe and Mn from Eh-pH conditions. The insoluble residue "R", Al, Fe, and Mn decrease with the distance from land; Mg reacts on salinity changes.

During the Turonian epoch, the facies belts were shifting with trans- and regressions. The schematic lithofacies map of Fig. 2 depicts a time of maximum transgression.

3.1. The Zone of Calcareous Laminites (Section 37, Fig. 4)

This facies type is limited to an area near the actual coast and is characterized by a high content of organic carbon (up to 22 % in the carbonate free sediment), in addition to chert and clay. These sediments with very little benthonic fauna were deposited continuously under anaerobic to "dysaerobic" (Byers, 1977) conditions. The abundance of micro- and nannoplankton (globigerinas, radiolarians, calcispheres) and of nekton (cephalopods, fishes, saurians) emphasizes the strong pelagic influence of the Turonian "Atlantic". Some bonebeds lead to an enrichment of phosphorus up to 1.8 %, whereas the bulk of the samples show relatively low phosphorus content. Sr-values reach more than 800 ppm, limited not only to the apatitic bonebeds. The high contents of Sr are the result of the substantial mass of calcitic foraminiferal tests resistant to diagenetic alteration, and of an obstructed migration of Sr during diagenesis, due to the high bitumen and clay contents of the sediment (Wedepohl, 1970; Veizer and Demovic, 1974). Although the content of clay is relatively high, the residue/aluminum ratio of the sediment reaches maximum values (350), which is explained by the very high silica production of the radiolarians. The occurrence of pyrite in the reducing environment and the adsorption of Fe by clay minerals produce a slight increase of the Fe-values. The enrichment of Mn in bituminous carbonates, described by Hartmann (1964) and Wedepohl (1970), could not be confirmed here, as the laminites are characterized by extremely low Mn-values (25 ppm). These may result from the solution of Mn under strongly reducing conditions and the continuous transport of Mn out of the sedimentation basin.

The vertical development of Section 37 is rather uniform, only interrupted by two omission layers with subsolution and enrichment of nekton.

3.2. Calcareous Micrites (e.g., Section 41, Fig. 4)

The laminite zone grades to the east into a ca. 50 km broad belt of calcareous micrites. They are built up besides of the substantial planktonic fauna, by thin-tested benthos, occasionally accumulated in allochthonous biorudites. It forms in this zone already an important additional source for the carbonate formation. The lower part of the sections is still affected by the pelagic, anaerobic realm of the calcareous laminites of Section 37, manifested in the deposition of chertiferous, thin-bedded limestones. After a sudden change to dolomites in the upper part, the environment is characterized by an aerated, agitated shallow-water regime with its typical, rapidly changing petrology. While the pelagic lower part is chemically comparable with Section 37 (Ca, Mg, Sr), the terrigenous influence increases in the upper strata, verified by higher values of lithogeneous elements (R, Al, Fe, Mn). The bioturbation by benthonic fauna and the content of bioclasts increase continuously to the east, and finally, matrix supported, ruditic bioclasts are the major petrographic component of a dominant lumachelle facies between Essaouira and Marrakech.

3.3. Lumachelle Facies (e.g., Section 24, Fig. 5)

The mainly benthonic fauna (pelecypods, gastropods, echinoderms, bryozoans) in ruditic, and, further to the east in more arenitic grain size, indicates an aerobic environment with frequent redeposition. Strong water agitation periodically created grain-supported oolitic bioarenites and brecciated erosion layers. Only the southern sections are still definitely subdivided into a lower calcareous and an upper dolomitic sequence, whereas the northern profiles are already entirely dolomitic (except for two thin calcareous layers in Section 11). This dolomitic development in an easterly direction can be verified by the distribution pattern of the Ca/Mg ratios and by the content of Sr. Phosphorus distinguishes the northern area with low values from the south with higher concentrations. The enrichment of the terrigenous elements Fe, Al, R in the lower and the upper third of the sections indicates a vertical tripartition, which is characteristic of a zone of dolomitic micrites, adjacent to the east.

3.4. Zone of Dolomitic Micrites (e.g., Section 5, Fig. 5)

In this area, between Marrakech and Ouarzazate, within the swell position which separates the ancient Atlantic from the Saharan sea during Turonian times, dolomitic micrites are formed, mainly of completely deteriorated and homogenized benthonic tests. Bioclastic rocks are

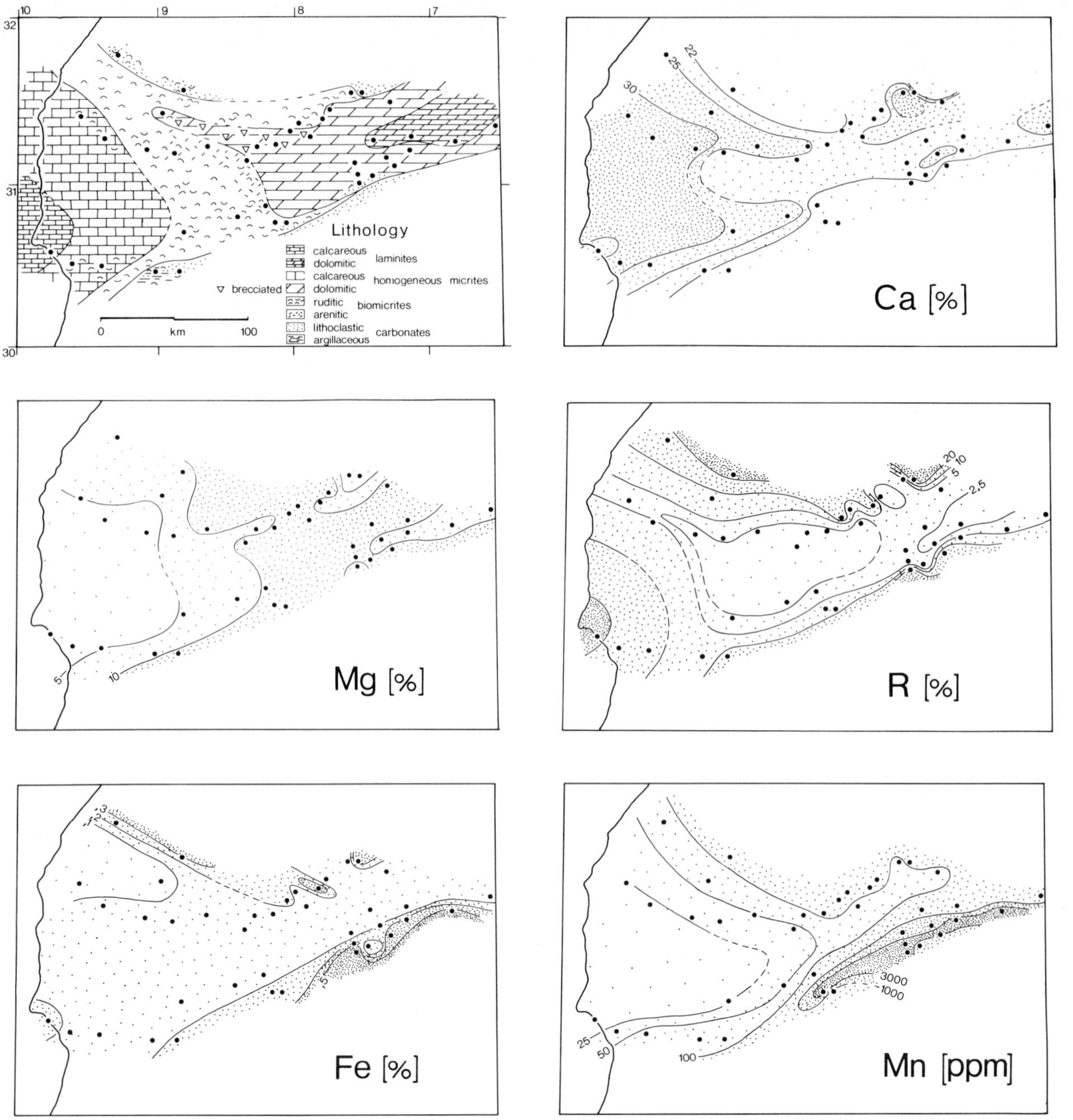

Fig. 2. Regional distribution pattern of the lithological composition and of major chemical elements (*Ca, Mg, Fe, Mn* and *R* = insoluble residue = SiO_2). Investigated sections are marked by black dots

rare and limited to the base and top of the sections, which are generally covered by thin algal laminites of an agitated intertidal zone.

The early diagenetic dolomitization caused very low Ca/Mg ratios in this region, where the swell position is determined by very high Fe/Al ratios, especially in the middle Turonian. The latter reflect low terrigenous influence, low accumulation rates and oxidizing conditions with numerous interruptions in sedimentation. Minimal values in R/Al ratios demonstrate the low planktonic silica production in this region, and the dominant argillaceous character of the overall very low quantities of terrigenous detritus. The coarse-clastic sand fraction is held back in the wide northern and southern coastal-near fringes, together with coarser bioclasts and autochthonous phosphatic ooliths (Section 4). Whereas in the north the terrigenous influence is documented only by silt, sand, iron, and possibly by phosphorus, the southern margin (Sections 26, 19) is characterized by substantial concentrations of Mn

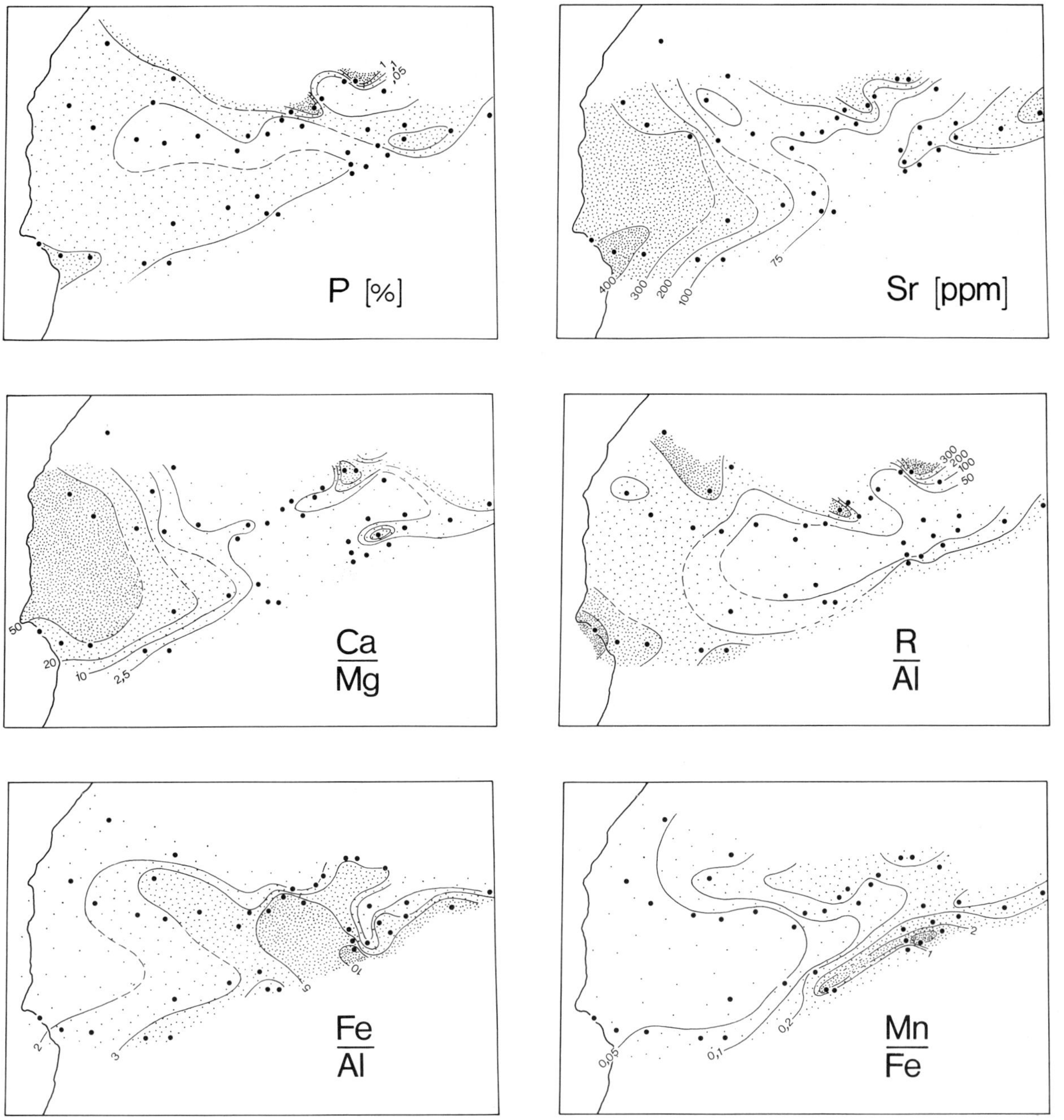

Fig. 3. Regional distribution pattern of *Sr* and *P* and of selected element ratios, showing marine basins (*Ca/Mg*) and the degree of terrigenous influence (*R/Al*)

(as shown in the distribution maps of Mn and of the Mn/Fe ratios). This enormous accumulation of Mn leads to the formation of the economic Mn-ore deposits of Imini (Section 26; Lesavre, 1976) and Tasdremt (Section 19; Bouladon and Jouravsky, 1952).

3.5. Dolomitic Laminites (e.g., Section 30, Fig. 6)

An euxinic lagoonal facies with dolomitic laminites represents the westernmost realm of the ancient epeiric Saharan sea. This facies, however, is essentially different from the pelagically influenced laminites of the Atlantic coastal region: the facies is purely dolomitic, partly even extremely saline, as it is reflected by the occasional precipitation of sulfates. However, frequent chert layers and nodules, as well as dubious relics of

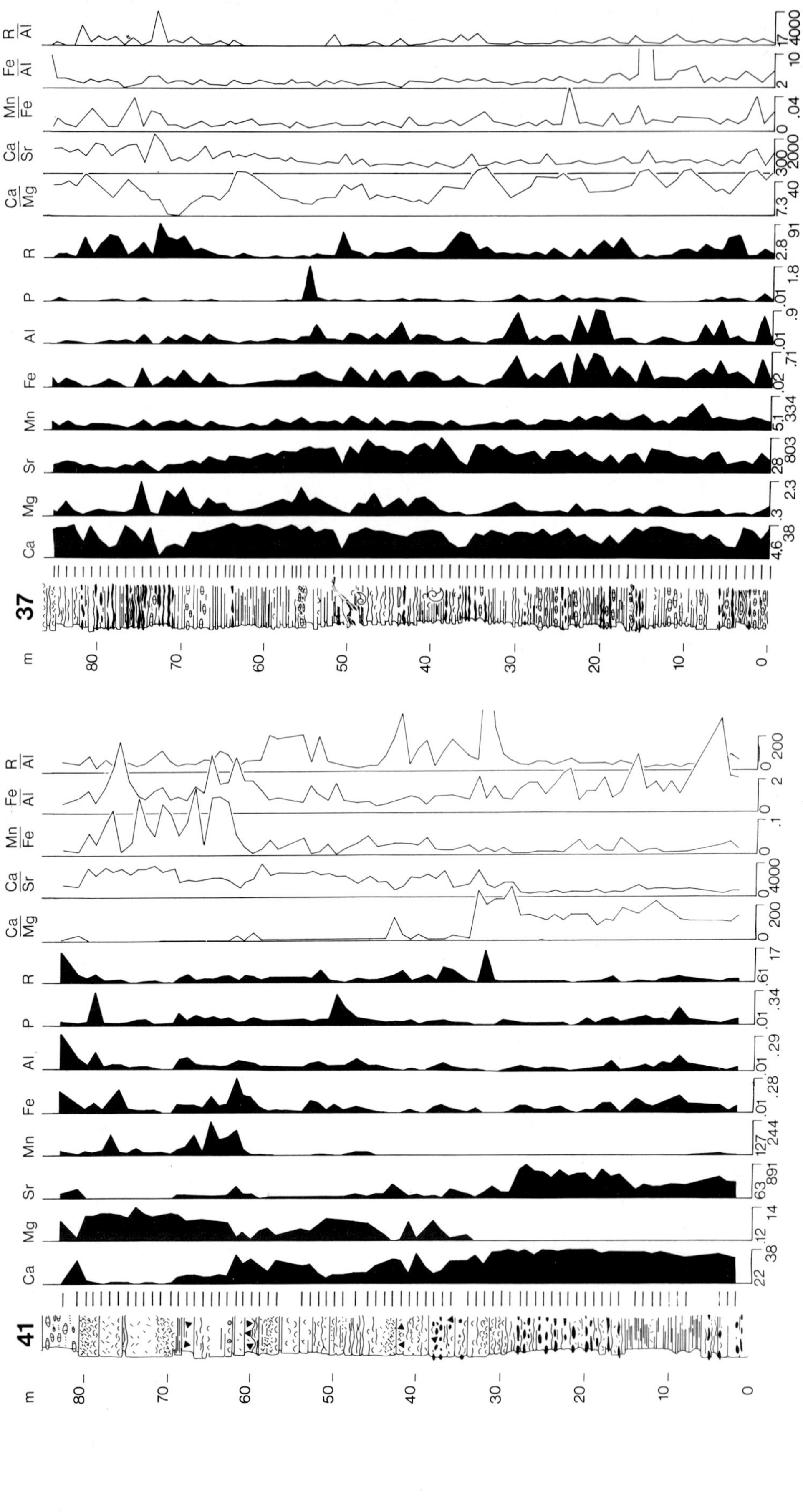

Fig. 4. Vertical distribution of selected chemical elements and element-ratios, compared with lithology, for the western zones of calcareous laminites (Section 37) and micrites (Section 41). Changes of *Ca/Mg*, *Mn/Fe* and *Fe/Al* ratios are most important for stratigraphic correlation. Section 37 represents only the middle and upper part of the Turonian. (Explanation of lithologic column is given in Fig. 5.)

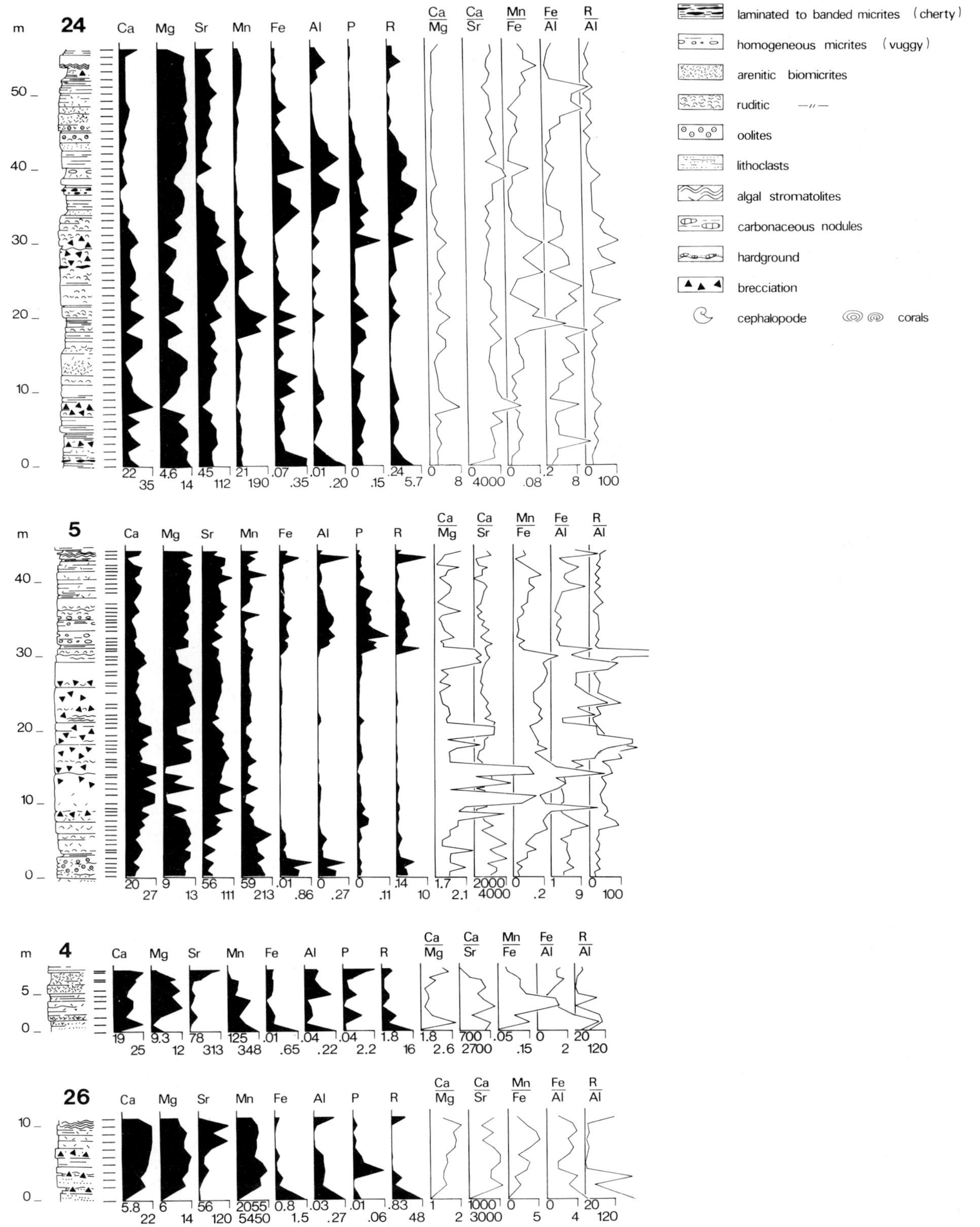

Fig. 5. Element pattern in sections of the lumachelle belt (24) of the dolomitic micrite zone (5) and of the swell region (4, 26). Note high *Mn/Fe* and *Fe/Al* values in the middle part of Section 5. The high *Ca/Mg* ratios in the brecciated part of Section 5 are the result of recalcitization, subsequent to anhydrite-solution. (The explanation of the lithologic columns are valid for all figures.)

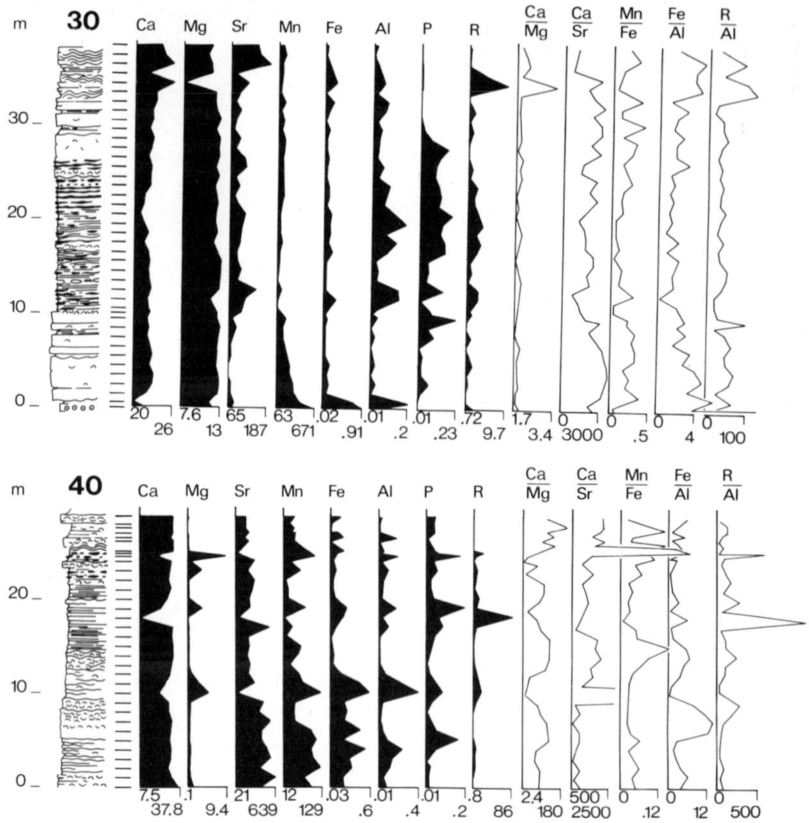

Fig. 6. Vertical element pattern in the dolomitic laminites (30) and micrites (40) of the eastern epeiric Saharan sea

foraminifers indicate some kind of pelagic regime also for this area. On the other hand, plant debris (Plate 1, D) may have played an important role in the deposition of bituminous material, which unfortunately is no longer directly demonstrable because of the recent alteration. Numerous indications, however, as missing benthos, pseudomorphs of limonite after pyrite and marcasite, silicification, and very high concentrations of typical trace elements, such as Mo (24 ppm), may be regarded as remnants of a bituminous facies.

Also in this zone, the same stratigraphic tripartition as in the swell region can be observed, but here the terrigenous elements show high values in the middle part of the sections, in which also phosphorus is concentrated because of a high content of fish debris in the sediment. Generally we observe a shift to a calcareous facies toward the east (Section 40), which also differs from that of the Atlantic realm by a rapid change of petrology; oyster banks, biomicrites, micrites, laminites, etc.

4. Correlation

The chronologic coexistence of different facies types (Fig. 2) makes a lithostratigraphic correlation of the individual sections difficult. A better criterion is given by geochemistry, which discloses significant trends and incisions in all the sections. Thus, a comparison of *Ca/Mg* mean-values of each single section with the total mean of all samples (Figs. 7, 8, 9) confines the occurrence of limestones to the basins and indicates a progressing dolomitization when approaching the swell and the coasts. In addition to these regional trends, a drastic change from calcareous to dolomitic facies in the middle part of the sections is very important for correlation. This isochronous event, separating a lower phase of euxinic limestone sedimentation from an upper dolomite formation in an aerobic environment of shallow water, may be equalized with the change from anoxic to oxic sedimentation conditions which can be observed in the oceans during the middle Turonian (Schlanger and Jenkyns, 1976; Jenkyns, 1980).

Parallel to the dolomitization in the swell region, we observe increasing *Mn/Fe* and *Fe/Al* ratios caused by stronger oxidation and a much lower rate of accumulation for this area. The beginning of the dolomitic facies coincides with a drastic rise of the Mn/Fe ratio, especially in the southern sections. This can be explained by the change in the redox situation: Fe is fixed in pyrite, in an euxinic environment, Mn coprecipitates preferentially with dolomite under aerated conditions. The same trend can be observed in the eastern sections, where for example in Section 30, an increase of the Mn/Fe ratios coincides with the change from euxinic dolomitic laminites to oxigenated micrites and stromatholites. In the western basin Mn/Fe ratios decrease at the point where the Senonian transgression, observed at the top of the sections, begins. In the dolo-

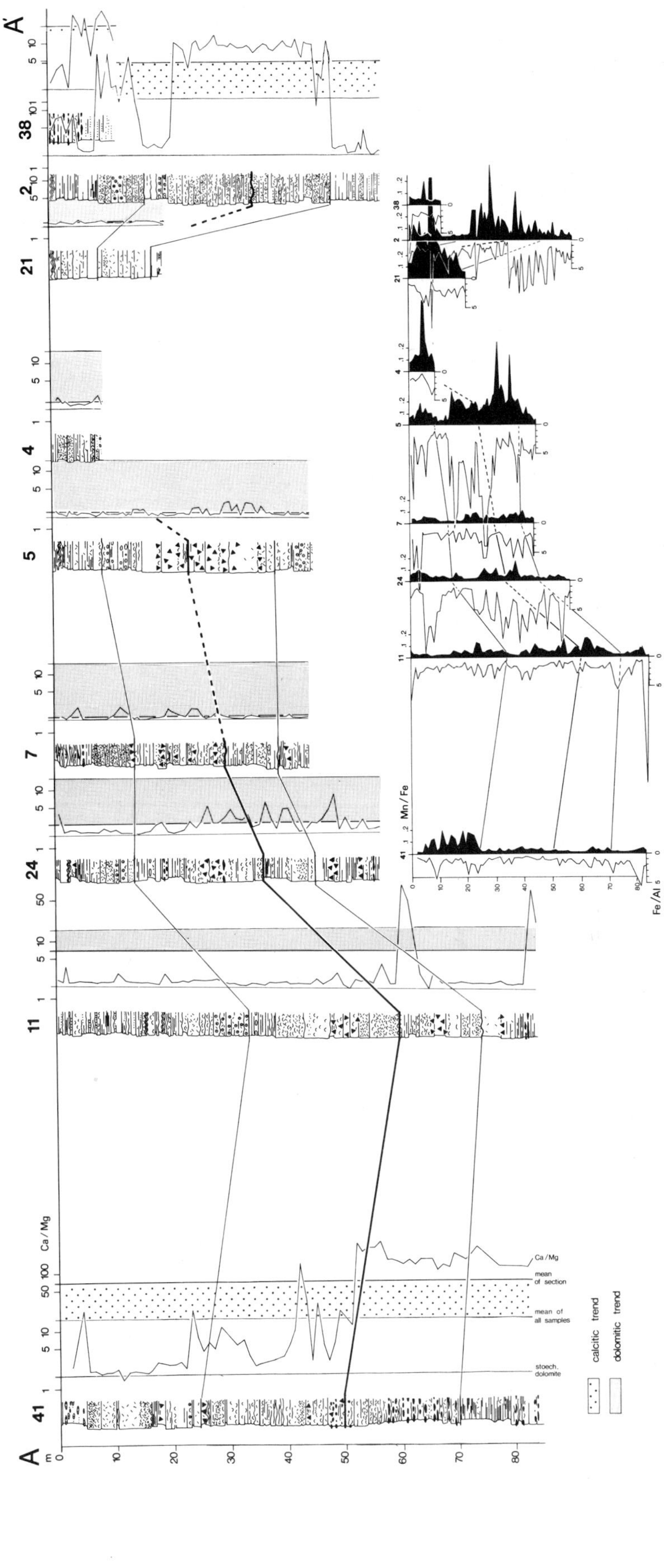

Fig. 7. Cross section north of the High Atlas. Note strongly decreasing *Ca/Mg* ratios when approaching the thin dolomitic swell sections. Correlation is done by means of the *Ca/Mg* ratios, separating a lower calcareous from an upper dolomitic part. *Mn/Fe* and *Fe/Al* ratios allow a further subdivision

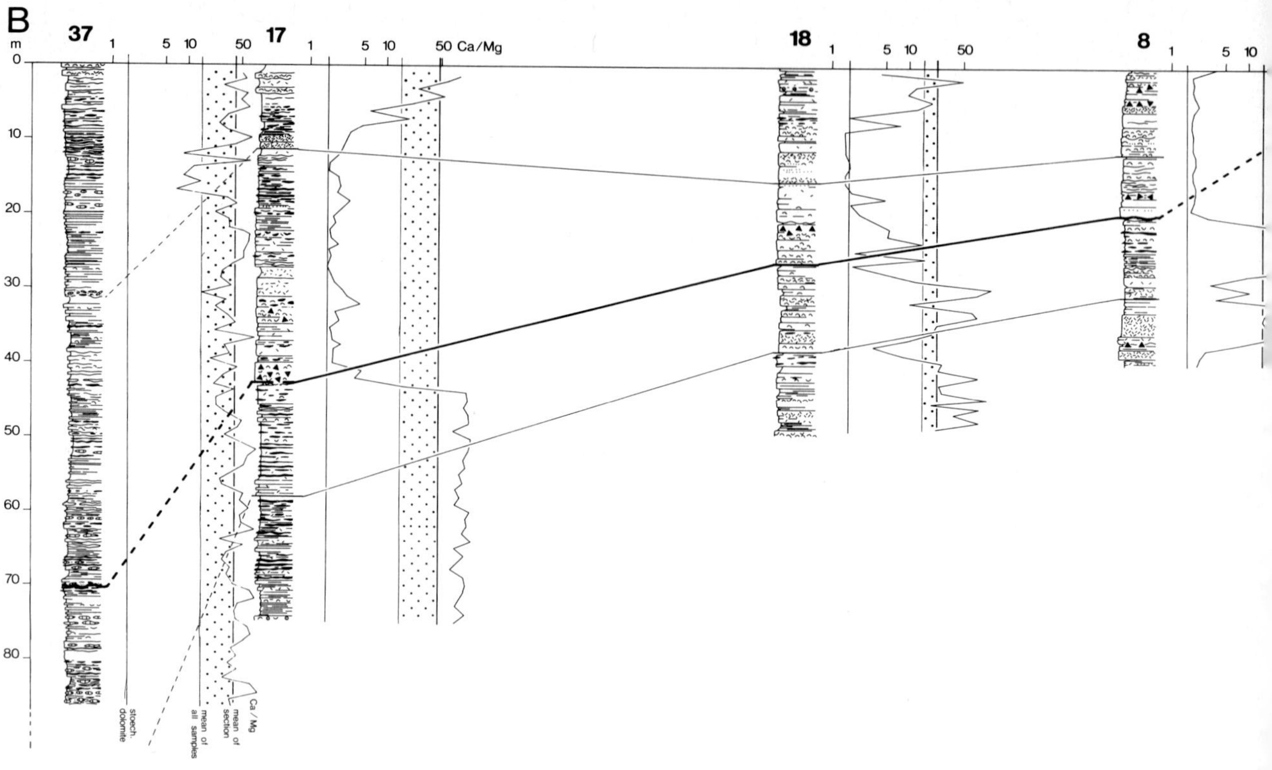

Fig. 8. Cross section along the southern Atlas margin. The calcareous facies dominates the sections and penetrates ► far to the east, on the Precambrian basement of the Anti-Atlas. The Mn-ore deposits of the southern coastal fringe of the Atlas Gulf are indicated by the extremely high *Mn/Fe* ratios in the Sections 19 and 26

mitic swell region, with continuously oxidizing environment, mainly terrigenous iron defines the variations of Mn/Fe. Periods of very high Mn/Fe values can be correlated with maximum transgression, when the supply of clastic terrigenous detritus is minimal, and hydrogeneous Mn-precipitation is high.

Fe/Al maxima correspond to Mn/Fe minima. Generally high values on the top and at the base of all sections are the consequence of high Fe-supply from detritic terrigenous compounds and, at the base, from the fixation of iron as pyrite in the lower Turonian euxinic facies. In the middle part of the swell sections, however, hydrogeneous precipitation of Fe and missing detritic supply of Al implicate very high Fe/Al ratios. A corresponding regional distribution is given in the Fe/Al map (Fig. 3). In the euxinic laminite facies of Section 30, the Fe/Al ratio remains low, despite the occurrence of pyrite. This is a consequence of a relatively high accumulation of argillaceous material.

The correlation of Section 37 with the adjacent eastern sections is only possible by using the variations of the Fe/Al ratios and of R, whereas the Ca/Mg values are distributed rather at random.

In addition to the stratigraphical separation of a transgressive and a regressive unit by means of the Ca/Mg ratios, the distribution pattern of the Fe/Al and Mn/Fe ratios, in connection with changes in lithology, allows a further subdivision of most of the sections into three distinct stages of development. The first and lowest part corresponds with the beginning of the Cenomano-Turonian transgression. In the westernmost sections the pelagic character is merely reinforced (higher plankton and bitumen content). Eastwards, the contrast between continental lower Cretaceous and the marine Cenomano-Turonian is more obvious. Continental redbeds are overlain by bioarenites and oolites of the transgressing sea. In the condensed swell sections, dolomite-cemented conglomerates and breccias with manganese ores are deposited on the Precambrian basement.

The second part is characterized, after a short regression which is noticeable by low Ca/Mg ratios in some sections (8, 11, 18, 41), by the highest sea level rise with maximum expansion of pelagic micrites to the east. It caused the development of bituminous laminites even in the basin of Ouarzazate. The transgressive phase is followed by a rapid regression, indicated by the change to low Ca/Mg ratios. It is not clear, whether the swell area fell dry already at that time. In the northern sections the environmental conditions changed to high salinity with the precipitation of anhydrite. During this regressive phase, the greatest regional differentiation of the lithofacies existed: laminites in Section 37, biomicrites in 11 and 24, hypersaline micrites in Sections 5 and 7, and finally laminite and micrite sedimentation in separate basins of the epeiric Saharan sea. At the end of this period a uniform lithofacies is developed with widespread vuggy micrites (dissolved sulfates?).

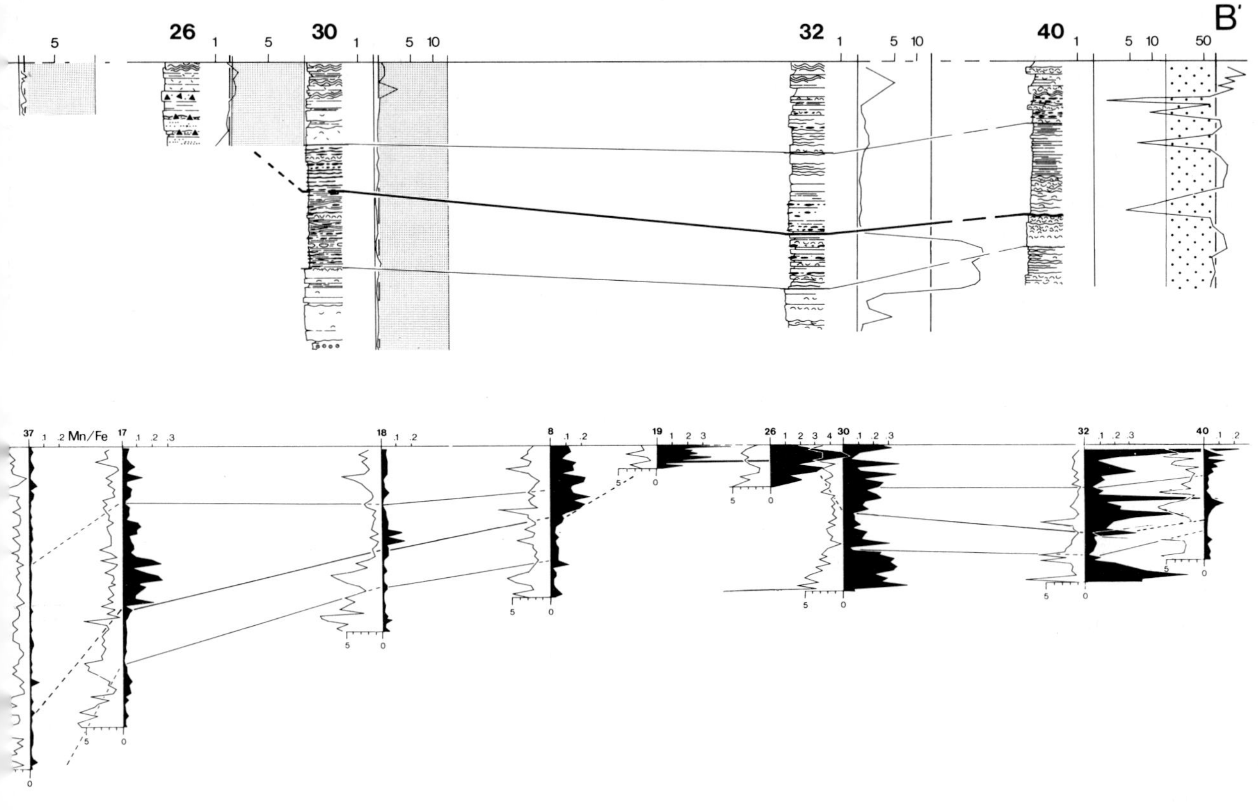

The beginning of the third part coincides with a new rise of sea level, which leaves a completely different lithofacies, compared to the first transgressions of the lower Turonian: dolomitic, bioarenitic oolites and biosparites in the north, calcareous, chertiferous micrites in the south and dolomitic biomicrites in the east. At the end of this third period, the development of the western basin is separated from that of the Saharan sea. The latter becomes continental, after a short time of intertidal conditions at the beginning of the Senonian. In the southwestern area the transgressive trend is reinforced.

5. Conclusions

Two different stages of development can be distinguished in the Turonian history of the High Atlas: transgression and regression (Fig. 10).

The "Cenomano-Turonian" transgression differs from the other Cretaceous transgressive events observed in the High Atlas region by a wide extension of the sea and by the deposition of highly pelagic carbonates with extremely low terrigenous influence. The sea invades an almost completely flattened relief (Wurster and Stets, this Vol.) which limits the coarse-grained terrigenous detritus to a narrow coastal fringe. The marine realm with its parameters of water depth, chemo-physical conditions and salinity determines the distribution pattern of litho- and chemofacies, especially during the early Turonian. In the late Turonian, the facies development is influenced not only by a major regression event, but also by the beginning uplift of the High Atlas mountains (Brix, 1980; Wurster and Stets, this Vol.).

Times of maximum sea extension are characterized by the deposition of pelagic laminites and micrites in the marginal sea, as well as in special depressions of the epicontinental sea, here developed as dolomitic facies with anhydrite. Biogenic silica, organic carbon and strontium now reach their highest concentrations. Intercalated bonebeds lead to a high accumulation of phosphates. Oxidizing conditions in areas of rather shallow water favor a rich benthonic life, producing biorudites, or, after reworking in agitated water, bioarenites. In the coastal and swell regions, phosphorus is precipitated in oolites and pellets and accumulates together with bioclasts and lithoclastic debris. Near the southern margin, manganese is concentrated in ore deposits at the boundary between reducing lagoons and well oxigenated shallow water areas. Magnesium is limited to the aerobic zones of the swell and to the epicontinental sea, where partly hypersaline conditions occur in lagoons.

During regressive times, the pelagic facies is pushed back far to the west, toward the open Atlantic. In the

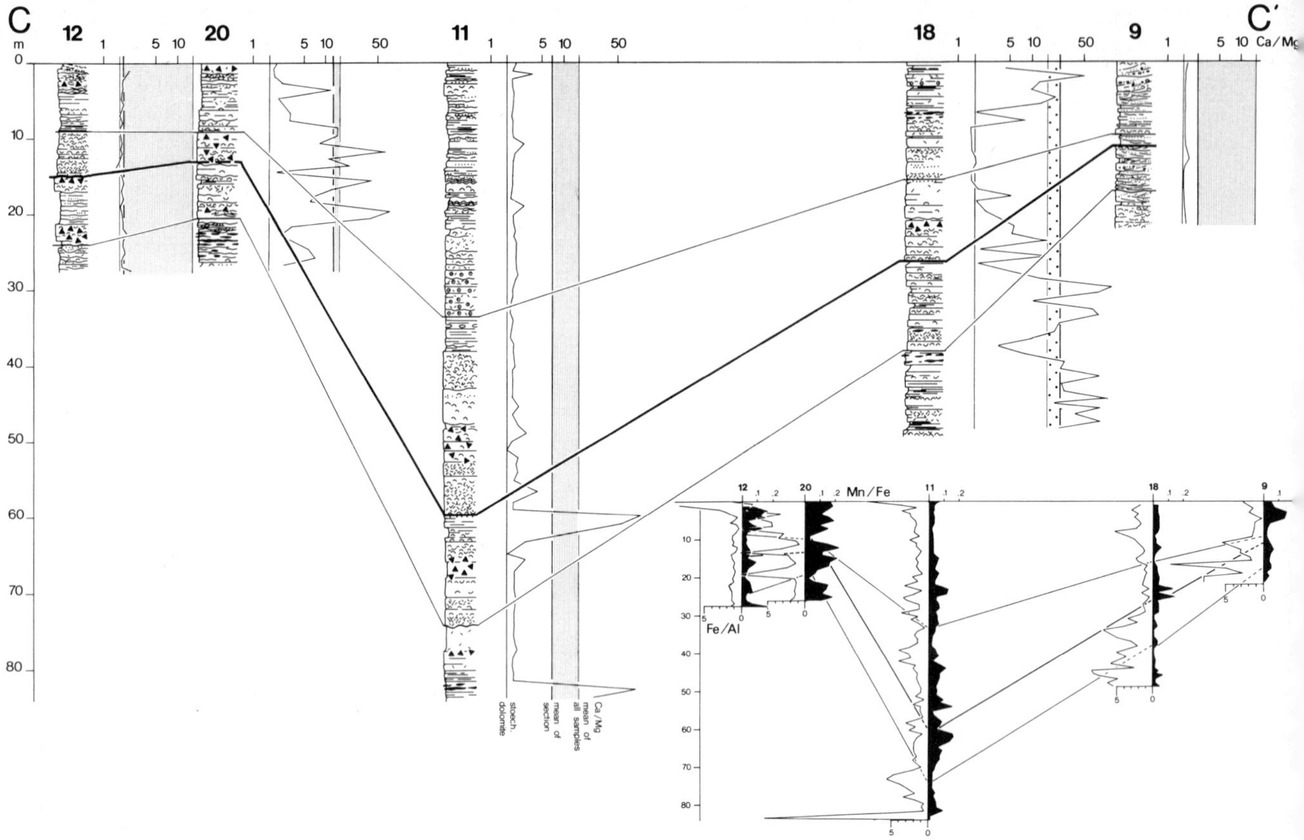

Fig. 9. Cross section in north-south direction showing a possible correlation of northern and southern profiles. Correlation is possible for the thick basin sections but remains very uncertain for the marginal regions, strongly influenced by terrigenous input

shallower shelf areas coquinas develop on the ocean side, passing landwards into wide hypersaline mudflats. On the swells, which are now falling dry, the manganese deposits are reworked to breccias. Wide-spread algal mats in the eastern epicontinental sea indicate intertidal conditions. The dolomite facies predominates in almost the whole sedimentation area.

During the transgressive phase, wide areas of the Atlas Gulf are influenced by anaerobic to dysaerobic bottom water conditions, overlain by an oxygen-rich water sheet in the photic zone with a proliferating planktonic and nektonic life. In the euxinic zone, certain elements are mobilized (C, N, H, P, Mn) and partly revolved in the nutrient cycle of the photic zone. Other elements, however, are concentrated in the sediment. They are supplied either biologically (Ca, Si, Sr, P, C) or are directly adsorbed and precipitated in a favorable anoxic environment (Fe, trace elements). The continuous concentration of these elements in the sediment necessitates an external source for them, either from the open sea by upwelling (Wiedmann et al., 1978) or derivation from the continent. Besides the terrigenous elements, we suggest that also phosphorus is derived from land, as it is always deposited together with the coarse-grained terrigenous detritus near the continent (compare also Bushinski, 1964; Slansky, 1980).

In this phase other conditions characterize the shallow lagoons of the epicontinental sea, where Sr-poor dolomite is formed, due to high salinity, together with sulfates. Here, a large supply of quartz and clay plays a certain role in the formation of chert, as is verified by frequent corrosion of the quartz grains. The principal source of silica, however, is the dissolution of biogenic opal (von Rad et al., 1978).

The manganese, which is concentrated in the southern coastal area in ore deposits, is primarily derived from the Anti-Atlas. The regional distribution of these deposits is closely linked to the Precambrian Mn ore province of the Anti-Atlas south of Ouarzazate. Mn-rich weathering products, formed during a long continental period, were reworked in the basal sediments of the transgressing Turonian sea. Under the locally reducing conditions of euxinic basins or lagoons, Mn was remobilized, also shown by the low Mn values of the middle part of Section 30. It precipitated as Mn ores at the boundary of anaerobic to oxigenated water along the margins of the lagoons.

In most cases, a comparison with offshore sediments is not possible, because Turonian strata are missing in the offshore drillholes. The nearest comparable Turonian strata are those of the Tarfaya Basin, which Wiedmann et al. (1978a, 1978b and this Vol.) describe as

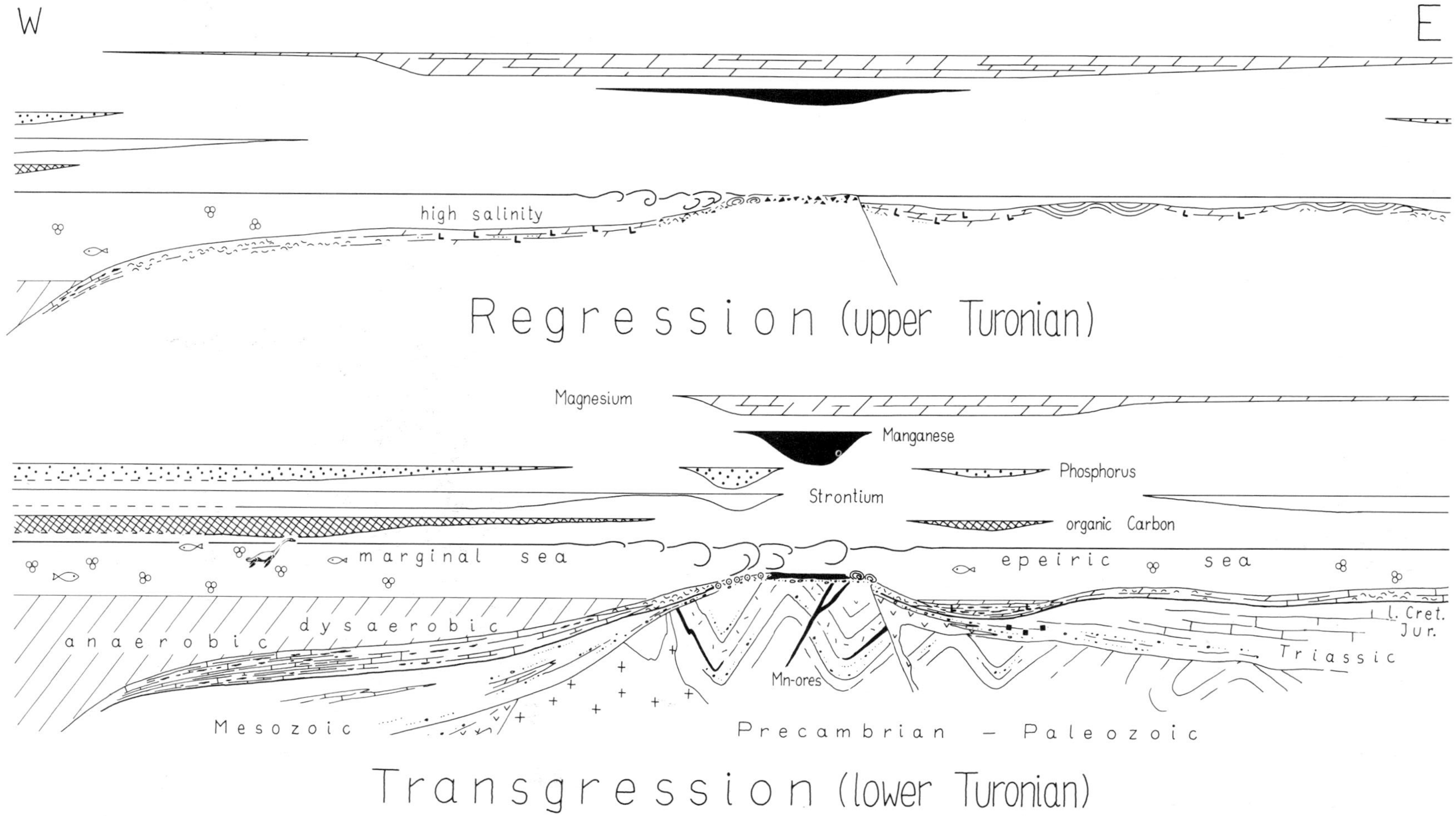

Fig. 10. Schematic model of the Turonian facies evolution, depicting two extreme environments: 1 a period of high sea level with largely spread reducing conditions and the enrichment of typical elements (C org., Sr, P). Mn accumulates in shallow water regions. 2 the regressive phase of the upper Turonian with highly saline conditions. The "oceanic" elements are pushed back far to the west and dolomites cover the shelf

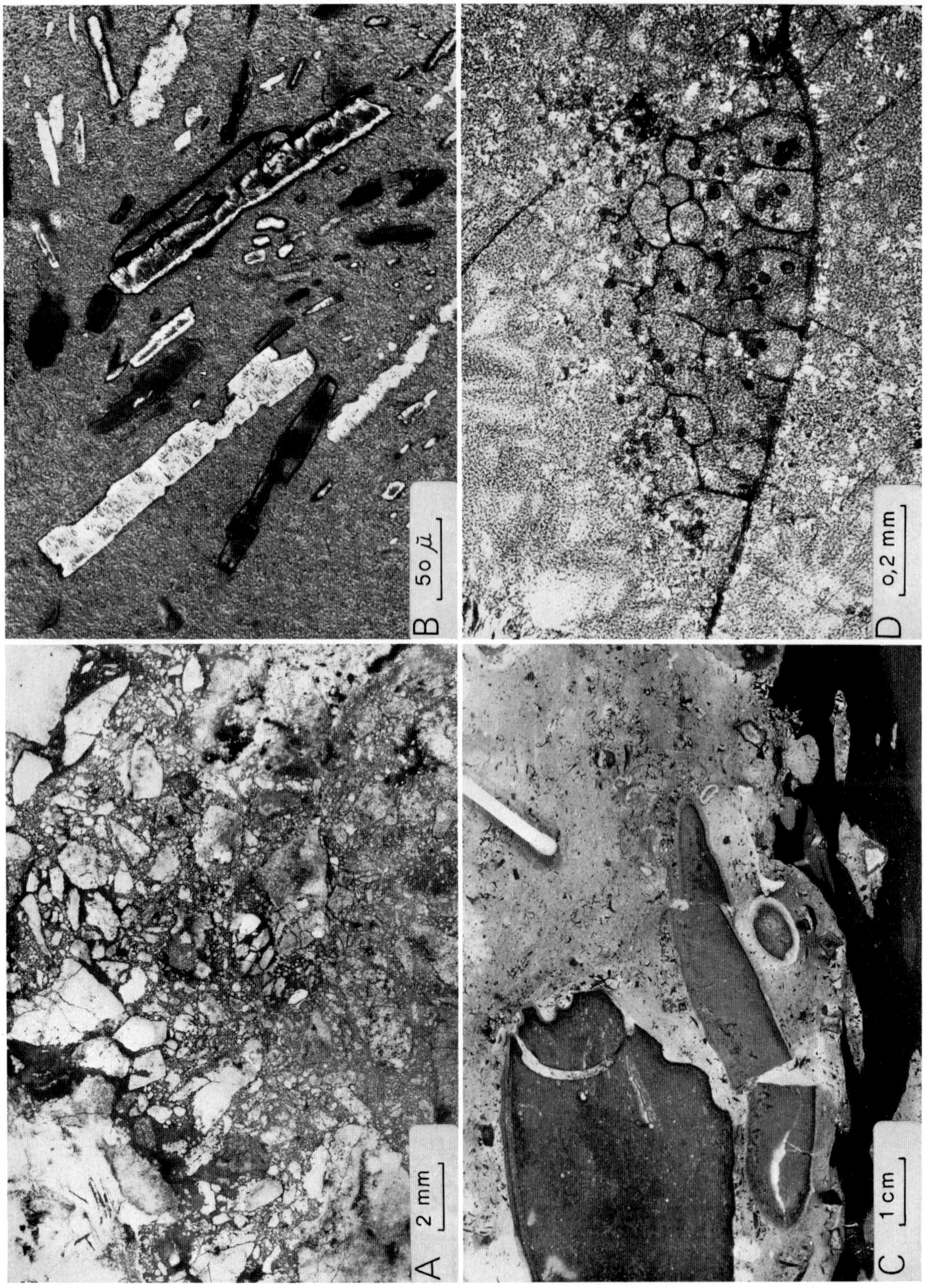
A 2 mm
B 5o µ
C 1 cm
D o,2 mm

bituminous sediments of about 100 m in thickness. This thick black shale facies can be traced oceanwards to the south, along the northern West African continental margin, to the Cape Verde Islands, where it was drilled in DSDP-Sites 367 and 368 (Lancelot, Seibold et al., 1978). In the deep sea realm west of the outer shelf, however, only decimeters to one meter of bituminous marls are present. In most cases Tertiary sediments cover lower Cretaceous with a major hiatus. This interruption of sedimentation during Turonian time may have been caused by a high CCD as described by Wiedmann et al. (1978a). A more probable explanation for the missing upper Cretaceous strata is given by later erosion (Lancelot and Winterer, 1980; von Rad and Einsele, 1980). Two marked horizons of omission and erosion in Section 37 can be correlated with significant changes of geochemical trends in the other sections. They coincide with the transgressive and regressive events described above. This agrees with the observations of Vail et al. (1980), who make rapid sea level changes responsible for the major unconformities along continental margins.

Acknowledgments. We are indebted to the Deutsche Forschungsgemeinschaft for having supported our investigations by financial grant. Without the friendly and helpful collaboration of the Moroccan Geological Survey, field work in Morocco would not have been possible. We thank Prof. K.H. Wedepohl, Prof. J. Wiedmann and Dr. U. von Rad for review and critical comments.

Plate 1. A Section 14, brecciated middle part. Completely unsorted dolomitic clasts in coarse-grained calcitic matrix. Larger clasts are extensively fractured. Recalcitization starts from fractures and from the edges of clasts
B Section 1, laminite facies. Anhydrite crystals in authigenic quartz of a chert nodule (polarized light).
C Section 37, meter 56,5. Eroded dark bituminous marls, overlain by a light-gray bone bed with flattened micritic nodules, up to 20 cm wide. The nodules are partly burrowed and their marginal zones are leached and oxidized. Finer-grained sediments with well preserved bones are contorted around the large pebbles which may be fractioned by diagenetic compaction of the surrounding sediment
D Section 30, laminite facies. Remains of leaf cuticulae with different cell nuclei, embedded in length-slow chalcedony of a chert nodule

REFERENCES

Behrens M, Krumsiek K, Meyer DE, Schäfer A, Siehl A, Stets J, Thein J, Wurster P (1978) Sedimentationsabläufe im Atlas-Golf (Kreide Küstenbecken Marokko). Geol Rundsch 67: 424–453

Berger WH and von Rad U (1972) Cretaceous and Cenozoic sediments from the Atlantic Ocean. In: Hayes DE, Pimm AC et al (eds) Initial Reports of the Deep Sea Drilling Project 14: 787–954

Brix M (1980) Heavy minerals reflecting the tectonic development of the Western High Atlas during Cretaceous and Tertiary times. IAS 1st Europ Meeting Abstr: 104–107

Bouladon J, Jouravsky G (1952) Géologie des gîtes minéraux marocains: Manganèse. XIX. Congr Géol Int Alger 3 sér 1: 44–80

Bushinski GI (1964) On shallow water origin of phosphorite sediments. In: van Straaten LMIU (ed) Developments in sedimentology, 1: Deltaic and shallow water marine deposits. Amsterdam, p 62–70

Busson G (1970) Le Mésozoique saharien, II: Essai de synthèse des données de sondages algéro-tunisiens. CNRS (ed), Paris: 811 pp

Byers CW (1977) Biofacies patterns in euxinic basins: a general model. Soc Econ Paleont Min Spec Publ 25: 5–17

Choubert G (1948) Essai sur la paléogéographie du Mésocrétacé marocain. Vol jub Soc Sc Nat Maroc: 307–329

Choubert G, Faure-Muret A (1962) Evolution du domaine atlasique marocain depuis les temps paléozoiques. In: Livre Mém Paul Fallot, Soc Géol France 1: 447–527

Cook HE, Enos P (eds) (1977) Deep water carbonate environments, Soc Econ Paleont Min Spec Publ 25: 336pp

Duffaud F (1966) Le bassin du sud-ouest marocain. In: Bassins sédimentaires du littoral africain. Assoc Serv Géol Afric: 5–12

Fischer AG and Arthur MA (1977) Secular variations in the pelagic realm. In: Cook HE, Enos P (eds) Deep water carbonate environments. Soc Econ Paleont Min Spec Publ 25: 19–50

Friedman GM (ed) (1969) Depositional environments in carbonate rocks. Soc Econ Paleont Min Spec Publ 14: 208pp

Graf DL (1960) Geochemistry of carbonate sediments and sedimentary carbonate rocks, Part III: Minor elements distribution, Part IV: Bibliography. Ill State Geol Surv Circ 301, 309: 51pp

Hartmann M (1964) Zur Geochemie von Mangan und Eisen in der Ostsee. Meyniana 14: 3–21

Honjo S, Tabuchi H (1970) Distribution of some minor elements in carbonate rocks, 1. Pac Geol 2: 41–79

Hsü KJ, Jenkyns HC (eds) (1974) Pelagic sediments: on land and under the sea. Intern Assoc Sed Spec Publ 1: 447pp

Jenkyns HC (1980) Cretaceous anoxic events: from continents to oceans. J Geol Soc London 137: 171–188

Lancelot Y, Winterer EL (1980) Evolution of the Moroccan Oceanic Basin and adjacent continental margin – a synthesis. In: Lancelot Y, Winterer EL et al Initial Reports of the Deep Sea Drilling Project 50: 801–821

Lancelot Y, Seibold E et al (1978) Initial Reports of the Deep Sea Drilling Project 41, Washington: 1259pp

Lange J (1974) Geochemische Untersuchungen an pelagischen Sedimenten des atlantischen und pazifischen Ozean (DSDP, LEG I–VII). Thesis, Göttingen

Lesavre A (1975) Le gisement de manganèse d'Imini (Maroc). Thesis, Paris: 101 pp

Migdisov AA, Girin PI et al (1980) Major and minor elements and sulfur isotops of the mesozoic and cenozoic sediments at sites 415 and 416, LEG 60, Deep Sea Drilling Project. In: Lancelot Y, Winterer EL et al Initial Reports of the Deep Sea Drilling Project 50: 675–689

Renard M, Letolle R, Bourbon M, Richebois G (1978) Some trace elements in the carbonate samples recovered from holes 390, 390A, 391C and 392A of DSDP LEG 44. In: Benson WE, Sheridan RE et al (eds) Initial Reports of the Deep Sea Drilling Project 44: 557–566

Roch E (1930) Etudes géologiques dans la région méridionale du Maroc occidental. Notes Mém Serv Géol Maroc 9: 542pp

Schlanger SO, Jenkyns HC (1976) Cretaceous oceanic anoxic events: causes and consequences. Geol & Mijnb 55: 179–184

Schroeder R (1976) Mikropaläontologische Datierung der "Barre Turonienne" in Süd-Tunesien. Geol Palaeontol Inst Univ Frankfurt Forschber: 42

Slansky M (1980) Géologie des phosphates sédimentaires. Mém BRGM 114

Stamm R (1980) Die dolomitische Fazies des Turon im Hohen Atlas zwischen Atlantik und Sahara (Marokko). Thesis, Bonn: 192 pp

Stets J, Wurster P (1977) Sedimentationsmodell Atlas-Golf (Kreide Küstenbecken Marokko). Nachr Dtsch Geol Ges 17: 75

Vail PR, Mitchum RM jr, Shipley TH, Buffler RT (1980) Unconformities of the North Atlantic. In: The evolution of passive continental margins. Philos Trans R Soc London Ser A 294: 137–155

Vail PR et al (1977) Seismic Stratigraphy and Global Changes of Sea Level. In: Payton CE (ed) Seismic Stratigraphy-application to Hydrocarbon Exploration. Amer Assoc Petrol Geol Mem 26: 49–212

Veizer J (1977) Geochemistry of lithographic limestones and dark marls from the Jurassic of Southern Germany. N Jahrb Geol Palaeontol Abh 153: 129–146

Veizer J and Demovic R (1974) Strontium as a tool in facies analysis. J Sediment Petrol 44: 93–115

von Rad U, Einsele G (1980) Mesozoic-Cainozoic subsidence history and paleobathymetry of the northwest African continental margin (Aaiun Basin to DSDP Site 397). In: The evolution of passive continental margins. Philos Trans R Soc London Ser A 294: 37–50

von Rad U, Riech U, Rösch H (1978) Silica diagenesis in continental margin sediments off north-west Africa. In: Lancelot Y, Seibold E et al Initial Reports of the Deep Sea Drilling Project 41: 879–905

Wedepohl KH (1970) Geochemische Daten von sedimentären Karbonaten, Karbonatgesteinen und ihr fazieller und petrogenetischer Aussagewert. Verh Geol Bundesanst Wien 4: 692–705

Wiedmann J, Butt A, Einsele G (1978a) Vergleich von marokkanischen Kreide-Küstenaufschlüssen und Tiefseebohrungen (DSDP): Stratigraphie, Paläoenvironment und Subsidenz an einem passiven Kontinentalrand. Geol Rundsch 67: 454–508

Wiedmann J, Einsele G, Immel H (1978b) Evidence faunistique et sédimentologique pour un upwelling dans le Bassin Côtier de Tarfaya/Maroc dans le Crétacé supérieur. 6ème Colloq Afr Micropaléont, Tunis 1974

21 Cretaceous Magnetic Stratigraphy of Southwest Morocco

Klaus Krumsiek[1]

ABSTRACT

From 1975 to 1979 two sections near the Atlantic Ocean north and south of the High Atlas Mountains, built up by predominantly marine sediments ranging from Upper Tithonian to Turonian, were sampled for the calibration of expected magnetic polarity zones. Two sections more landward to the east with predominantly terrestrial, poorly dated sediments had to be correlated magnetostratigraphically by using the magnetic polarity sequence of the western sections.

Most of the sediments showed multi-component remanent magnetization. The components were separated and the respective magnetic minerals identified. This was done by alternating field (af), thermal- and combined demagnetization with respect to the specific blocking temperatures and coercivity forces of the minerals.

Components linked to goethite and hematite pigment are found to be parallel or antiparallel to the direction of the Recent Earth's magnetic field (F_{rec}). Hematite pigment, magnetite, and specularite-borne components are aligned with the normal or reversed direction of the paleofield. Magnetite and specularite are assumed to be the carriers of remanences which originated syndepositionally or shortly after sedimentation (depositional detrital remanent magnetization DRM, postdepositional detrital remanent magnetization PDRM). As regards hematite pigment, however, components are aligned with a paleofield, this chemical remanent magnetization (CRM) was probably acquired before the folding of the High Atlas Mountains (Late Eocene).

The Cretaceous magnetic polarity sequence, based on results of the western sections, was also found in the eastern sections and supported or corrected the stratigraphic positions assumed with respect to the development of the Atlas Gulf (Wurster and Stets, this Vol.). The Cretaceous magnetic polarity sequence of SW Morocco considerably agrees with the results of other authors. Due to higher sedimentation rates the short reversals of Upper Albian and Cenomanian, until now known only from the Southern Alps (Vandenberg and Wonders, 1980) are more distinct in Moroccan sequences.

One of the Turonian polarity changes is accompanied by regional changes in lithology and geochemistry and therefore does not reflect the behavior of the Earth's magnetic field during the time of deposition. However, the direct relation between polarity changes and climatic variations has become quite firmly established (Doake, 1978; Cordell, 1980; Fairbridge, 1977). In conclusion, a magnetic triggering of geochemical events cannot be excluded from discussion.

The virtual paleomagnetic pole position (VPMP) obtained, 154 W, 76 N, confirms the Cretaceous mean position of Hailwood (1975) at 153 W, 75 N.

1. Introduction

North and south of the High Atlas Mountains predominantly marine, biostratigraphically well-determined Cretaceous sediments in the west interfinger with predominantly continental sediments of largely unknown age in the east. The region thus offers a unique possibility to apply and check magnetostratigraphic methods.

The sections in the west, Berriasian to Turonian in age, allow the biostratigraphic calibration of polarity sequences. One of my objectives was to test the possibility of applying magnetostratigraphic correlation in the eastern sections. Moreover, additional information concerning the behavior of the Earth's magnetic field during the Cretaceous was expected from the Moroccan results.

For calibration purposes we sampled the sections 100 km east of Essaouira at the northern flank of the High Atlas Mountains and at Tamrakt and Ascouti 36 km north-east of Agadir at the southern flank. On the other hand we correlated the sections of Aouzzer 80 km south of Marrakech at the northern flank and at Afensou 30 km north of Taroudannt at the southern flank by magnetostratigraphic methods.

The field work at Zem Zem was carried out in the spring of 1975 and that at the other sections in the autumn of 1976. Five hundred and nine cores, 2.5 cm in diameter and sometimes more than 20 cm in length

1 Geologisches Institut der Universität Bonn, Nussallee 8, D–5300 Bonn 1, FRG

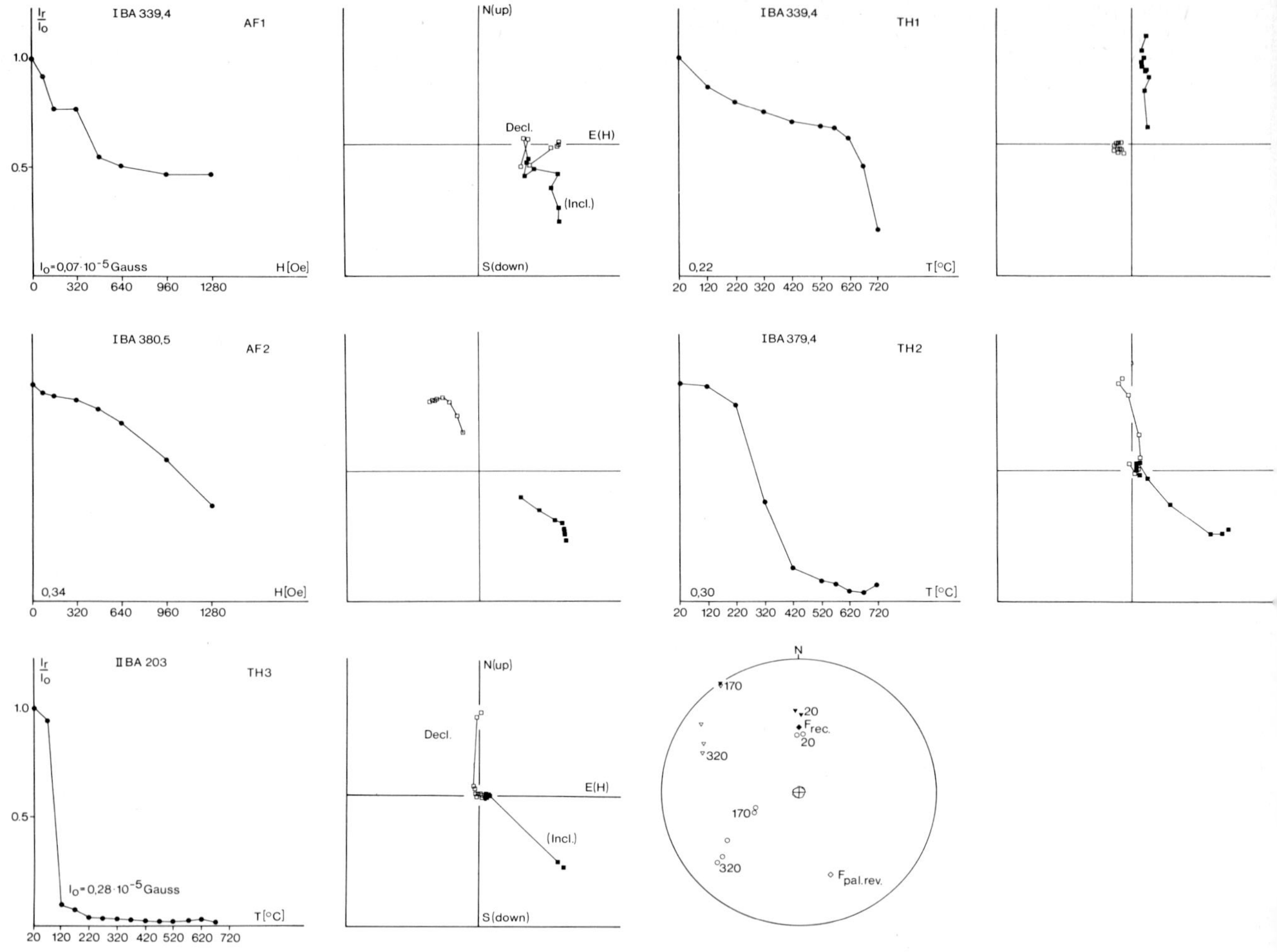

Fig. 1. Normalized af- and thermal demagnetization curves and diagrams of specimen IBA 339,4 and IBA 380,5 and of IBA 339,4 and IIBA 203, respectively. The behavior of normalized intensity (I_r/I_o) during demagnetization is shown by demagnetization curves.
The behavior of normalized intensity (I_r/I_o) and direction (D/I) of RM-vector with respect to the geographical coordinates is shown by demagnetization diagrams. Equal area projections yield higher resolution of direction behavior. The **points** represent successive positions of RM-vector during progressive demagnetization.
Symbols in drawings: **Open squares** end points of changing RM-vector projected in a horizontal plane; **closed squares** end points of changing RM-vector projected in a vertical plane.
Symbols in equal area projection: **triangles** RM-directions before tectonic correction; **squares** RM-directions after tectonic correction; **closed symbols** RM in lower hemisphere; **open symbols** RM in upper hemisphere.
For conversion in SI: 10^{-5} Gauss = 10 mA/m; 100 Oe = 8 kA/m.
Curves of IBA 339,4 specimens demonstrate the relation between af1-type and TH1-type and those of IBA 380,5 and 379,4 specimens the relations between af2-type and TH2-type. The curve of IBA 203 is typical for TH3-type demagnetization behavior. The RM-direction of specimen IBA 203 changes from the magnetic field existing in Recent times (F_{rec}) toward the Earth's magnetic field existing in pre-folding times

were drilled from 417 rock units at the four section sites. In total we measured the NRM of 3159 specimens with an UGF 4 Spinner Magnetometer (Geofisyka, Brno, CSSR) in the Laboratory of the Institute of Geology at Bonn University.

2. Demagnetization Experiments and Magneto-Mineralogy

From each core one or more pilot specimens were thermally demagnetized in 10 or 14 steps up to 720° C using a TSD-1 Thermal Specimen Demagnetizer (Schonstedt Instrument Company, Virginia). During the cooling process the Earth's magnetic field was shielded

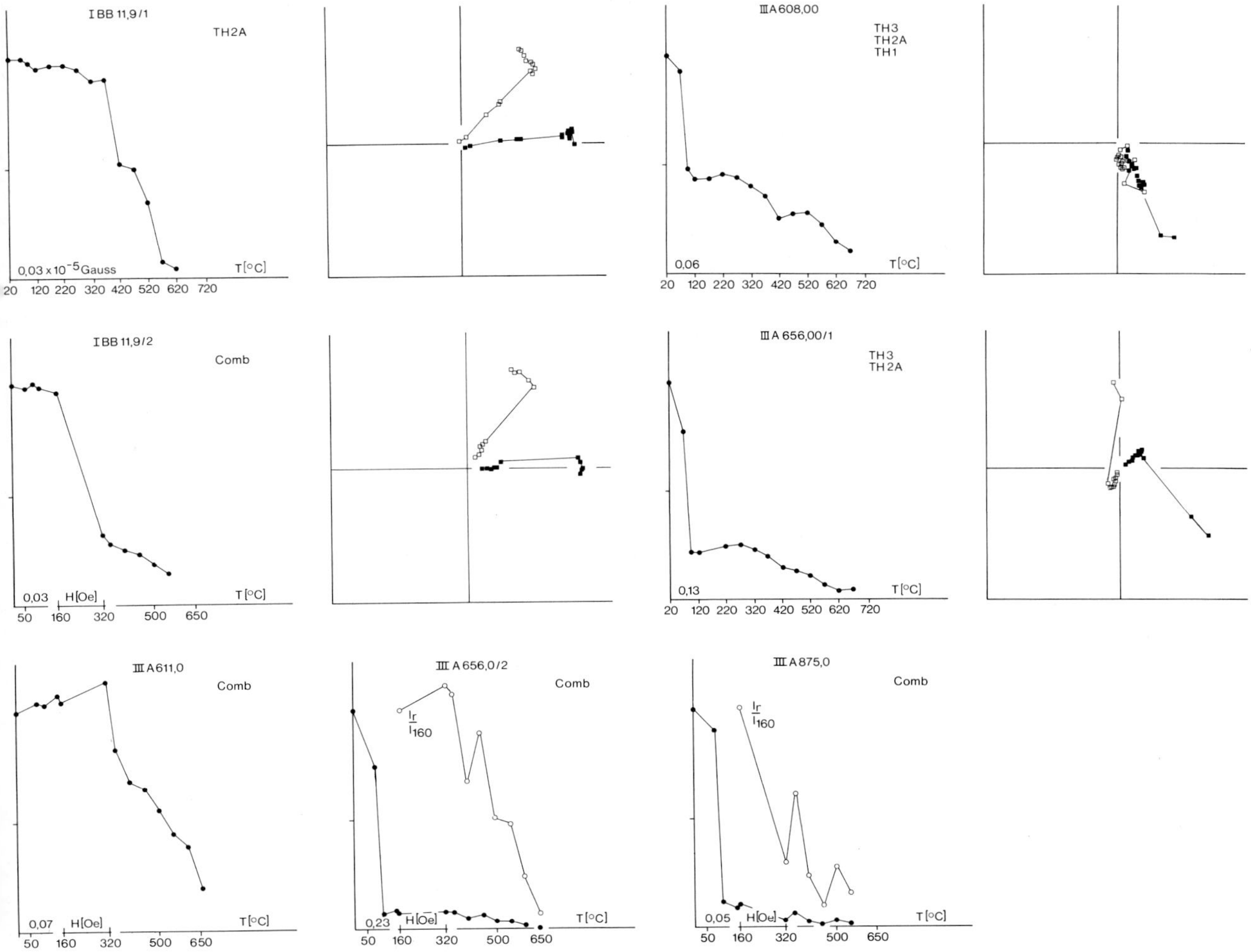

Fig. 2. Normalized curves and diagrams of specimens IBB 11,9/1, IBB 11,9/2, IIIA 608, and IIIA 656/1 and curves of IIIA 611, IIIA 656/2, and IIIA 875.
For legend see Fig. 1.
Curve of specimen IBB 11,9/1 shows TH2A-type behavior. Curves of IIIA 608 and IIIA 656/1 show a combination of different TH-types. During combined-demagnetization a decrease indicates magnetite (IBB 11,9/2, IIIA 875); stability indicates hematite. In additional presence of TH3-type normalization to I_{160} amplifies the effect (open symbols IIIA 656/2, IIIA 875)

down to less than 5×10^{-5} Gauss (10^{-5} Gauss = 1nT in SI).

From every core of the Zem Zem Section an additional specimen was af-demagnetized in seven steps up to 1280 Oe (100 Oe = 8 kA/m in SI units). The magnetic behavior was plotted automatically by means of demagnetization curves (I_r/I_o) and -diagrams (Zijderfeld, 1967; Krumsiek and Siehl, 1976). The evaluation of 586 thermal demagnetization and 162 af-demagnetization experiments shows that the alteration of the vector of remanent magnetization (RM) is described by a few basic types of curves.

Concerning the intensity, the RM-vector of af-type 1 (af 1) decreases initially and then remains stable at the higher field peak values (Fig. 1, IBA 339,4). The af 2-vector remains stable at the first three or four steps of demagnetization and then drops slowly (Fig. 1, IBA 380,5). Af 3 shows a stable behavior of RM up to the highest values of 1280 Oe. As regards the RM-direction, the af 1-vector is unstable, while the vectors of af 2 and af 3 usually remain parallel to the Recent Earth's magnetic field (F_{rec}).

The initial RM-decrease of af 1 can be attributed to magnetite, the residual stability indicates the presence of a second magnetic mineral with coercivity forces higher than 1280 Oe, which is typical for all kinds of hematite and goethite. The RM-decrease of af 2 is well known from hematite pigment particles which have grown in the recent field and slightly exceed their critical diameters of spontaneous magnetization (Krumsiek, 1980). Af 3 characterizes specimens with hematite in single domain configuration or goethite with components of hematite or goethite which obscure eventually existing magnetite components.

TH-type 1 (TH 1)-demagnetization diagrams show RM stability up to 570° C and RM decrease between 570° C (Fig. 1, IBA 339,4). According to Collinson (1974) this behavior is typical for specularite as carrier of remanence. TH 1 usually corresponds with af 1. As regards TH 2, demagnetization up to 420° C diminishes a component parallel to F_{rec} (Fig. 1, IBA 379,4) which therefore must be carried by hematite pigment. In this diagram the RM-vector moves toward the reversed paleofield direction, probably conserved by magnetite or specularite. TH 2 corresponds with af 2.

During type TH 3 demagnetization the component parallel to F_{rec} drops rapidly at 120° C (Fig. 1, IIBA 203). The remaining component is parallel to the normal or reversed paleofield direction as shown in Fig. 1, IIBA 203 for example. Usually TH 3 and af 3 occur in specimens from the same core. The high coercivity forces and low blocking temperatures shown by these types characterize goethite (Hedley, 1971; Heller, 1978). According to Heller (1978) the blocking temperatures of goethite range from 50° to 95° C. In some specimens the goethite carries more than 98 % of NRM.

Type TH 2A demagnetization shows similarities with both – type TH 2 and type TH 1 (Fig. 2, IBB 11,9/1). The RM-intensities remain stable up to about 345° (± 25° C) and diminish to zero between 570° and 620° C. The RM-directions are aligned with the normal or reversed paleofield, as well as with the normal or reversed Recent or Subrecent field directions. Corresponding blocking temperatures are well known for magnetite and hematite pigment (Collinson, 1974). Because isothermal remanent magnetization (IRM) components were not found by af-experiments, magnetite must be excluded as a carrier of post-folding directions. Furthermore, a generation of magnetite under weathering conditions is impossible (Garrels, 1960). Therefore it must be assumed that hematite pigment is the carrier of the magnetization components aligned parallel or antiparallel with the Recent field.

This pigment is generated from goethite by dehydration and the goethite itself is generated from Fe--minerals like pyrite or siderite during weathering.

The time of spontaneous magnetization cannot be determined exactly, because the lapse of time between the precipitation of the Fe hydroxide and the transition of the critical diameter of the hematite pigment particles is not known. The time TH 2A-specimens with directions normal or reversed to the paleofield acquired their magnetization ranges from short after sedimentation to the time of the folding of the High Atlas at the end of Eocene.

Of course, the polarity of the Earth's magnetic field that existed during the time of deposition is conserved in 2A-specimens containing magnetite as the carrier of remanence.

Multicomponent magnetization is common and easily recognizable by two or three different types of curves occurring during demagnetization. TH 1, TH 2, and TH 3 are combined in specimen IIIA 608 (Fig. 2) demagnetization curve indicating goethite, magnetite, or hematite pigment and specularite as magnetic minerals. Specimen 656/1 (Fig. 2) shows a combination of TH 3 and TH 2A with a goethite component following F_{rec} and a stable component aligned with the reversed direction of F_{rec}, which therefore must be carried by hematite pigment.

To distinguish between components of hematite pigment and magnetite components the combined demagnetization technique was used. According to this method, experiments with 60 specimens of TH 2a- and TH 3-type were carried out under the supervision of Dr. F. Heller at ETH Zürich laboratory where this procedure had been worked out originally. Meanwhile the method has been adapted to the laboratory and computer equipment at Bonn University. The following procedure yields reliable results: One starts with thermal demagnetization in the first three steps up to 150° C, which eliminates possibly existing goethite-bearing components. Then two af steps of 160 Oe and 320 Oe are applied. Stability of remanence indicates hematite; dropping intensity, however, is a criterion for magnetite. In the case of hematite the thermal-demagnetization between 350° and 720° C under favorable conditions separates pigment from specularite by their different blocking temperatures (Collinson, 1974). In all more than 210 Moroccan specimens were combined demagnetized.

Specimen IIIA 611 (Fig. 2) shows somewhat increasing remanence during the af-demagnetization steps, which indicates hematite; the second thermal demagnetization diminishes the remanence at 350° and 600° C, which indicates hematite pigment and specularite, respectively. The remaining stable remanence is reversed after tectonic correction. The remanence of specimen IBB 11,9/2 (Fig. 2) drops rapidly during af-cleaning. In this specimen magnetite conserves the reversed paleofield direction. In specimens containing goethite components of high intensity the effect of combined demagnetization is optimally demonstrated by normalization to I_{160} (Fig. 2, IIIA 656/2, Fig. 2, IIIA 875).

The steady decrease of remanence during thermal demagnetization is not common (Fig. 3, IIBA 194; Fig. 4, IIIFA 357) and can be sufficiently explained by multidomain configurations within the magnetic particles. A comparison of the demagnetization types and the rock material suggests that TH 1 is related to sandstone, TH 2A to dolomite, and TH 3 to limestone. Besides the Ca/Mg ratio the porosity of rocks seems to play an important role, because 2A-type demagnetization was found in porous limestones, and that of type TH 3 in dense dolomite.

Thus the most suitable sediment for paleomagnetic investigation is sandstone with specularite and/or magnetite aligned with the ancient field during or shortly after deposition. The most unreliable rock is dolomite formed under oxic conditions that alterated detritic magnetite into Fe hydroxide. Furthermore Fe hydroxides also precipitate from interstitial solutions. The time between precipitation and dehydratization to hematite pigment and between precipitation and the growth of the pigment particles up to their critical diameter can be relatively short under certain conditions, especially in a warm climate; in other cases it lasted at least several million years (Walker, 1967). During the sedimentation of limestones the conditions are slightly reducing. Particles of detrital magnetite and hematite remain unaffected and Fe-ions usually precipitate as pyrite. If there was no later oxidation which

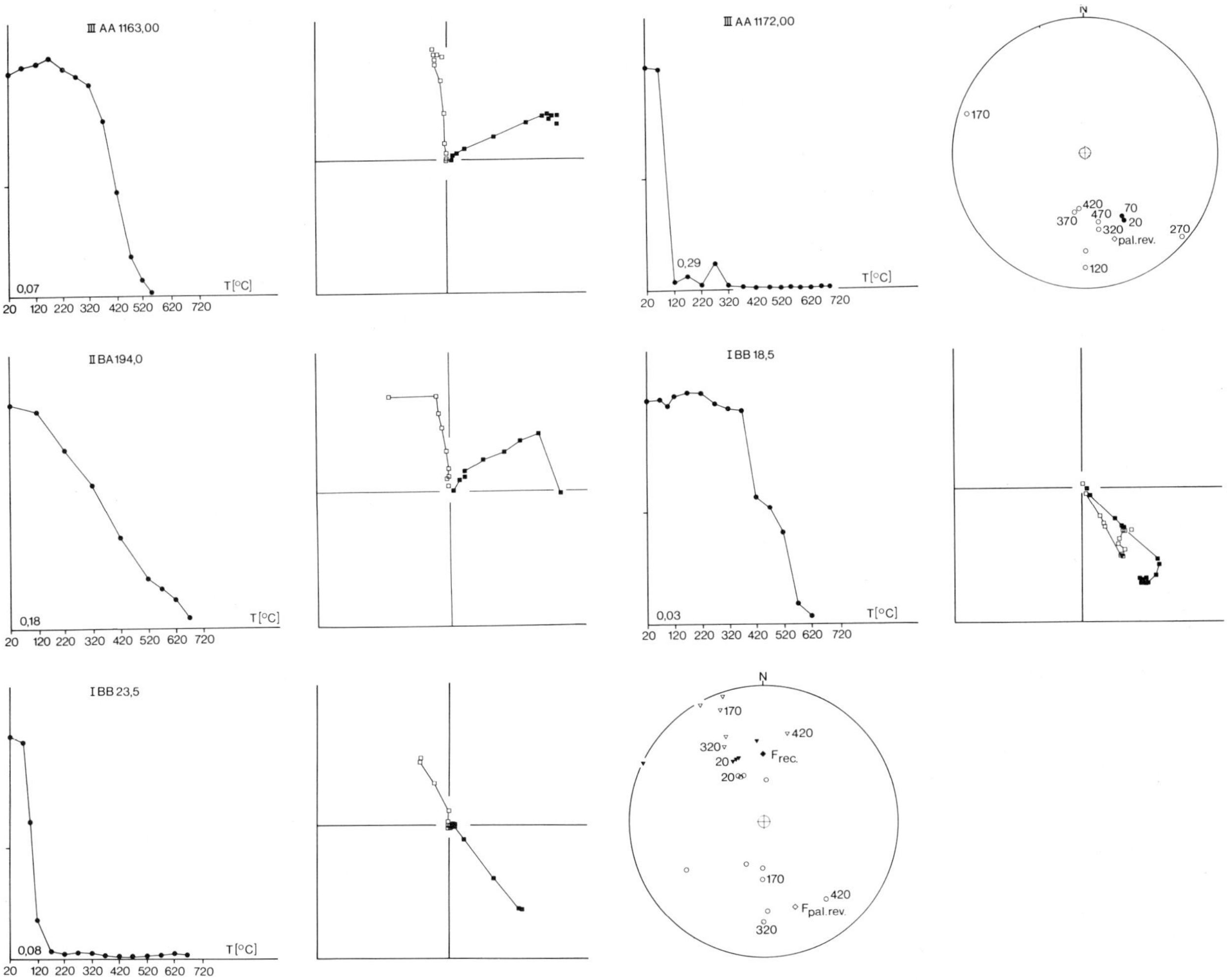

Fig. 3. Normalized curves and diagrams of pilot specimens IIIA 1163, IIIA 1172, IIBA 194,0, IBB 18,5, IBB 23,5 drilled at the Tamrakt-Ascouti and Zem Zem Sections.
For legend see Fig. 1.
See text for further information

altered the pyrite to goethite, the paleofield direction, conserved by detrital magnetite or specularite, can be separated by demagnetization techniques. Earlier contact with free oxygen can be excluded for specimens containing organic carbon. During thermal demagnetization these specimens are easily recognizable. As described by Heller (1978) the C-combustion produces darker colors during heating at 400° C, and reduces hematite to magnetite, which is indicated by a simultaneous increase of remanence. Sometimes goethite is already dehydrated to hematite pigment, well illustrated by 2A-type demagnetization of components parallel or antiparallel to the recent field direction (Fig. 2, IIIA 656). The same reaction supposedly took place, if the sediment after deposition was affected by oxidation, perhaps during short regressive periods. Therefore synsedimentary acquisition of remanence cannot be unambiguously assumed for specimens which contain neither organic carbon nor detrital magnetite or specularite. The significance of the constancy test should be considered with scepticism if the change from anoxic to oxic conditions is a regional event as for example well demonstrated for the Turonian of SW Morocco by Stamm and Thein (this Vol.; Schlanger and Jenkyns, 1976; Jenkyns, 1980).

As regards the natural remanent magnetization (NRM) intensities the highest values (9×10^{-6} Gauss; 10^{-5} Gauss = 10 mA/m in SI units) were measured from red sandstones of the Upper Hauterivian. Limestones and dolomites usually ranged within the order of 10^{-7} Gauss, but goethite-bearing limestones (TH 3-type) yielded NRM-intensities up to 2.8×10^{-6} Gauss. Values smaller than 10^{-7} Gauss occurred sparsely from NRM-measurements, but were easily reached during demagnetization experiments, which were therefore sometimes limited by the resolution of the UGF 4 Spinner Magnetometer (2×10^{-8} Gauss).

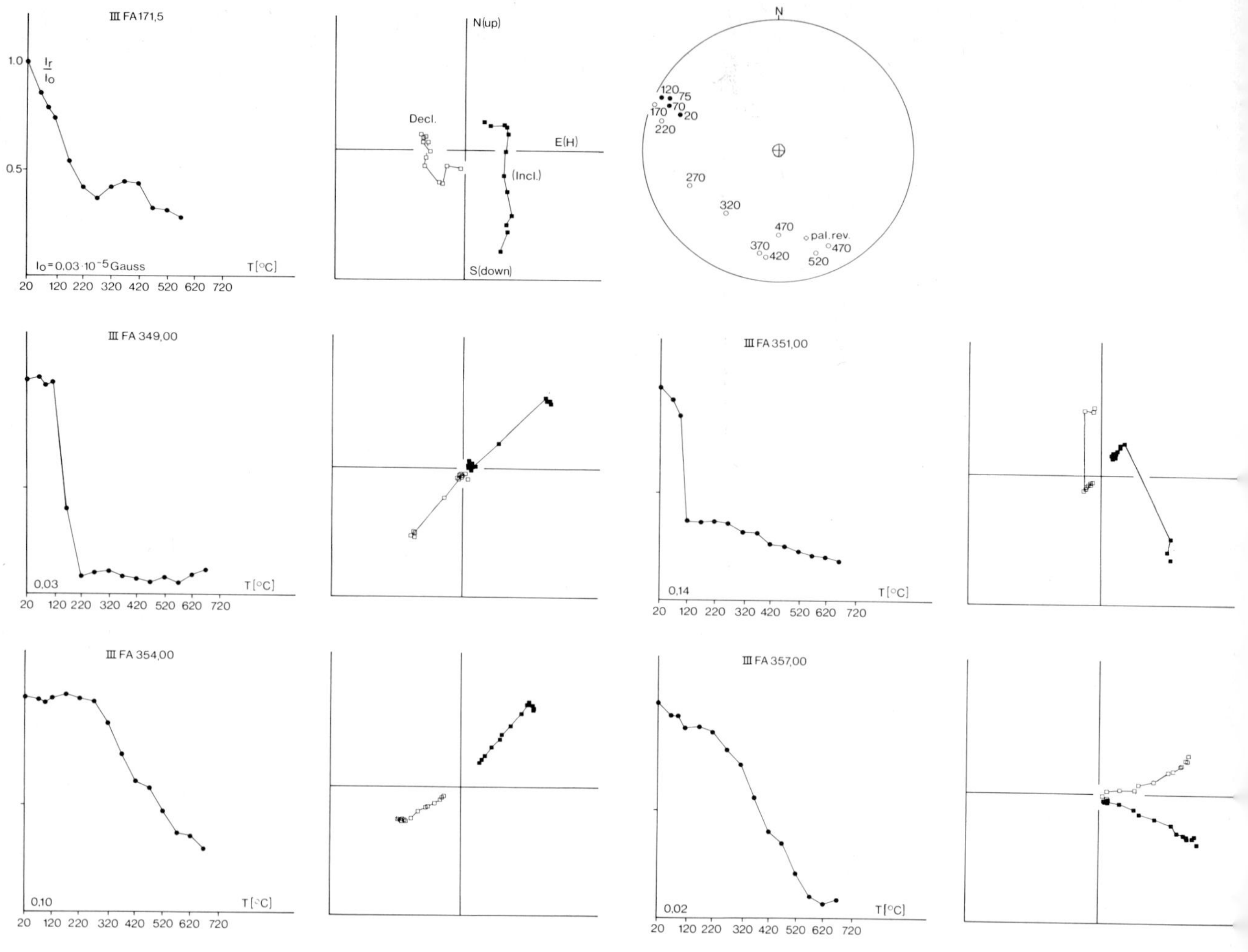

Fig. 4. Normalized curves and diagrams of pilot specimens IIIFA 171,5, IIIFA 349,0, IIIFA 354,0, and IIIFA 357,0.
For legend see Fig. 1.
See text for further information

3. Description of Sections and Discussion of Magnetostratigraphic Data

After demagnetization experiments 17 % of the cores remained fully remagnetized in F_{rec} or showed instability of remanences both in intensity and direction and therefore were rejected from further evalutation. In the case of unsatisfying demagnetization results NRM-data were used if more than two specimens of one core pointed toward reasonable paleofield direction.

The declination and inclination (DR/IR) for CARM (characteristic remanent magnetization) was plotted separately versus stratigraphic position of sections. Furthermore the NRM and CARM are presented in equal area projection demonstrating the concentration due to demagnetization experiments and tectonic corrections, respectively (Figs. 5–8).

The demagnetization behavior of a few specimens can only be presented here as demagnetization curves and -diagrams. Therefore the demagnetization parameters of the remaining specimens referenced in this paper are listed in Table 1.

3.1 Tamrakt-Ascouti Section (Fig. 5)

North of Agadir sediments ranging from Triassic to Paleocene are exposed nearly completely in fresh road-, stream- and beach cuts.

The western and (paleo) near-ocean sections (Wiedmann et al., 1978, this Vol.) are dominated by marls unsuitable for paleomagnetic sampling. So a compromise had to be found between westerly sections with continuous sedimentation but sampling gaps and easterly and more continental sections with drillable limestones, dolomites, and sandstones but sometimes

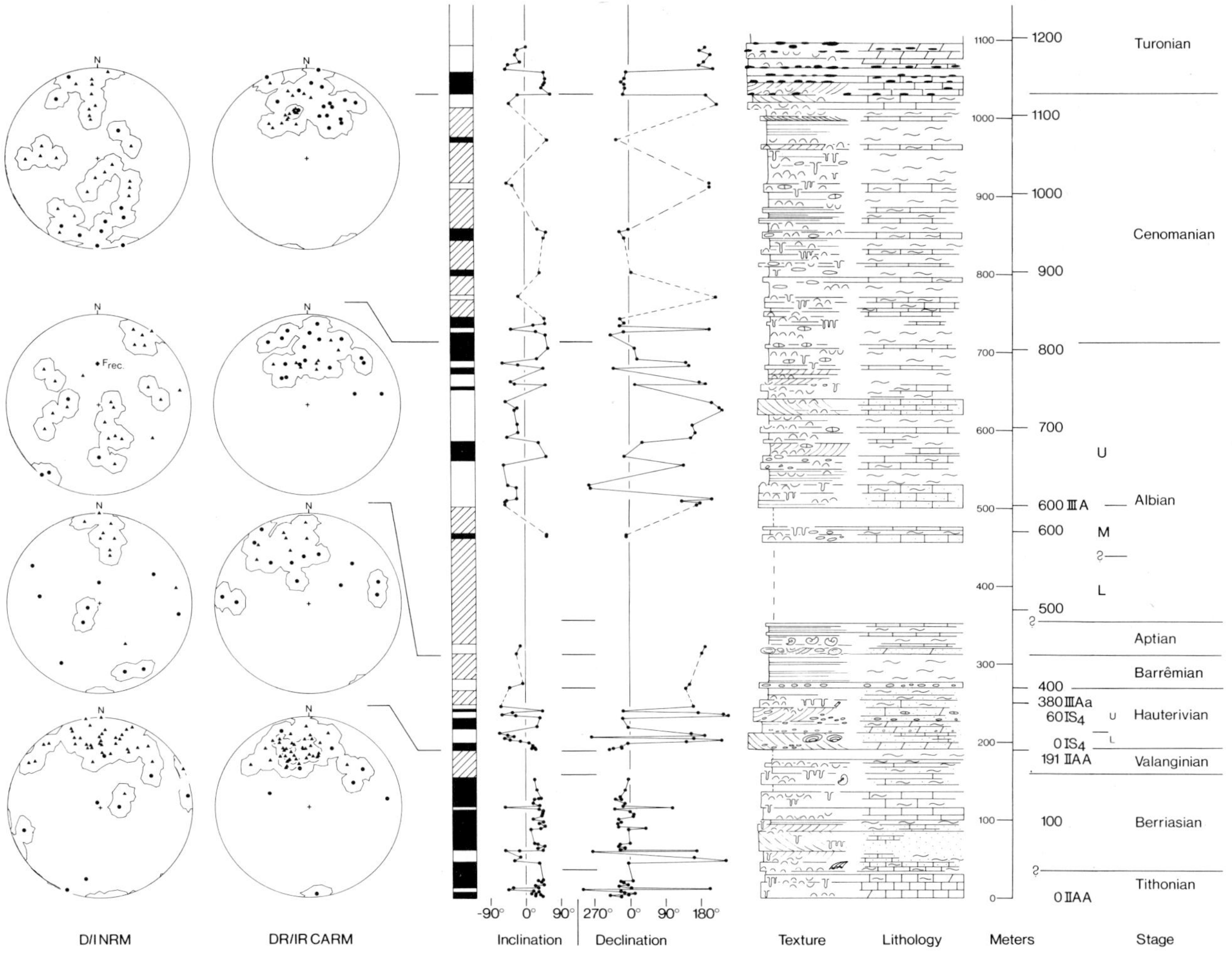

Fig. 5. Magnetostratigraphy, NRM and CARM of the Tamrakt-Ascouti Section. Declination and inclination of CARM is plotted separately versus stratigraphic position (CARM = characteristic remanent magnetization).
Black normal polarity; **white** reversed polarity; **hatched** sampling gaps.
Equal area projections show the NRM before (D/I) and the CARM after (DR/IR) tectonic correction.
Triangles positive directions in lower hemisphere; **circles** negative directions mirrored in lower hemisphere, allowing the separation of areas with different point density.
Distance of isolines: 8 %

discontinuous sedimentation. Therefore an "Agadir Section" was combined from the Assif (river) Tamrakt Section east of forrester's house ranging from Berriasian to Hauterivian, and of the Assif Tourza Section (Ambroggi, 1963) south of Ascouti ranging from Hauterivian to Turonian (hereafter called Ascouti Section; for location see Fig. 1, section 17 in Stamm and Thein, this Vol.). Both sections are well described (Ambroggi, 1963; Behrens et al., 1978; Behrens and Siehl, this Vol.; Stamm and Thein, this Vol.; Wurster and Stets, this Vol.) and their faunal elements sufficiently determined to be suitable for magnetostratigraphic calibration. The geological history of this area is presented in the papers referenced above.

The idea of Ambroggi (1963), that the sedimentation of Tithonian platform type limestones continuously passes the Triassic-Jurassic boundary, was confirmed by Wiedmann et al. (1978) on account of fossiliferous marls immediately above the limestones in the beach section indicating Middle and Upper Berriasian. Because of the eastward-trending transgression younger parts of Berriasian must be deposited on the limestones at the Tamrakt Section.

The marls occurring at IIAA 36 contain faunal elements typical for the Berriasian (*Alectrionia rectangulata* at IIAA 39,6 and IIAA 42,1, cf. Ambroggi, 1963). How much of the underlying platform limestones and dolomites was deposited in the Tithonian and which in the Berriasian cannot be decided. Intercalations of light greyish and brownish sandstones set in at IIAA 45. They become more abundant and dominate the marls between IIAA 60 and IIAA 90. This sequence is overlain at IIAA 95 by light greyish marls and limestones with sparse beds of greenish marls. Between IIAA 160 and IIAA 191 grey-greenish marls are intercalated sparsely by thin beds of limestone. According to Ambroggi (1963) the lower part can be recognized as Valanginian, due to its typical faunal elements (for example *Nautilus* sp. at IIAA 161). Nevertheless, as the Upper Hauterivian follows directly at IIAA 191 the upper part of the marly sequence represents the Lower Hauterivian as well.

The first reversal of the Agadir Section is well established by cores drilled in the dolomite at IIAA 11,2 and in the limestone at IIAA 12,5.

The Upper Berriasian sandstones were almost totally remagnetized in Recent times. However, thermal demagnetization yielded normal and reversed paleofield directions. The reversal between IIAA 48 and IIAA 60 is supported by cores taken at IIAA 48, 52, 59, and 60,5. The reversal at IIAA 116 is questionable because its stable component is not sufficiently aligned with the reversed paleofield direction.

The marls representing Valanginian and Lower Hauterivian are sampling gaps. 191 of the IIAA part is correlated with the beginning of IS4 part in which the paleomagnetic sampling was continued. The base of this sequence consists of oolithic limestone which is overlain by shell hash and fine-grained conglomerates. All these units are cross-bedded. The coral bearing unit between IS4 16,4 and IS4 21,5 is a marker horizon, well known from numerous sections of the Atlas Mountains (forrester's house horizon, Ambroggi, 1963; Behrens and Siehl, this Vol.). Terrestrial influences are indicated by red and violet colors between IS4 38 and IS4 46. These clastics are overlain by marls which still represent the Upper Hauterivian.

The basis of IS4 is magnetized parallel to the normal paleofield direction, but the shell hash between IS4 10,8 and IS4 13,4 and at IS4 15 shows reversed polarity. Also the stable components of all pilot specimens sampled at forrester's house horizon (IS4 16,5 – 21,5) are aligned with the reversed paleofield direction. At IS4 30 normal remanence was recognized after NRM-measurements, and by thermal- and combined demagnetization as well.

NRM-intensities, usually ranging in order of 10^{-7} Gauss, exceed 5×10^{-6} Gauss in specimens of Hauterivian clastics. These are the highest NRM-values measured in specimens of the Ascouti Section.

The next reliable reversal is documented by cores drilled from the coquina and gravel-bearing marl at IS4 44,7, from the overlying coarse-grained coquina limestone at IS4 45,2, and from the sandy yellow dolomite at IS4 48.

Normally magnetized dolomitic limestone at IS4 50 is succeeded by sandy and dolomitic limestone which, at IS4 56, shows reversed directions. Unfortunately this bed is overlain by greenish marls of Upper Hauterivian which are not suitable for sampling.

The base of the Barremian of the Tamrakt Section is characterized by green marls intercalated by yellow limy sandstones. More to the east the clastic portion of this horizon increases, forming an important escarpment in the Ascouti Section. This horizon was used for the correlation of the Tamrakt and Ascouti Sections (380 IIIAa = 60 IS4).

Due to complicated faulting tectonics and because of the exceptional stratigraphic situation at Ascouti, where Lower Hauterivian transgressed directly over the Tithonian carbonate platform (Ambroggi, 1963), sediments older than Barremian were not sampled.

The stable components of IIIAa 401 and IIIAa 405 pilot specimens point toward reversed paleofield directions.

The next core was drilled at IIIAa 455 from a sandy dolomite that represents the basis of the Aptian. This marker horizon, 4 m thick in the Ascouti Section, is easily identified by its typical slumping features (Wiedmann et al., 1978; Behrens and Siehl, this Vol.). Demagnetization brought about a stable reversed magnetization. The slumping layer is overlain by marls which contain faunal elements that indicate the Upper Aptian transgression. Up to IIIAa 600 the section then is dominated by marls and therefore not suitable for sampling with the exception of a marl bed at 590 IIIAa which is magnetized in normal paleofield direction and probably Middle Albian in age.

Between 604 IIIA and 800 IIIA the sequence is characterized by grey arenites and limestones that bear a considerable content of quartzsand and are intercalated with green marls. Most of these beds are distinctly cross-bedded. Argillaceous shell hash with oysters occurs above 670 IIIA. This interfingering of marine and terrigenous beds cuts off the pelagic sedimentation of Lower and Middle Albian. In general, the ocean regressed westward oscillatingly, but the coastline did not yet reach the Ascouti area. The extent of regression therefore is less important than that of Upper Hauterivian time (Ambroggi, 1963; Behrens et al., 1978). This sequence is more resistant to weathering than the marls and thus builds an escarpment of considerable size.

Some of the geologists who work in the Atlas region use the term "Vraconian" for the sequence described (Roch, 1930; Behrens et al., 1978). Although this may be an advantage for mapping and is also in full agreement with the geological timetable of the Royal Dutch Shell (1975) which refers Vraconian to the Upper Albian as a whole, it leads to misunderstanding with respect to the Cretaceous timetable of Van Hinte (1975). This timetable, which meanwhile is well established in magnetostratigraphy, restricts the term Vraconian to the uppermost part of Upper Albian. Furthermore, Ambroggi (1963) in describing lithology and faunal elements of numerous Upper Albian sections of southwest Morocco used the term in the same sense, and pointed out that the sequence is divided into five cycles, each one outlined morphologically by specific weathering. Applying these results to the Ascouti Section, the marls that overlay the predominantly arenitic escarpment between IIIA 730 and IIIA 749 are of Lower Upper Albian age due to their faunal elements (*Hystoceras orbigny*; Albian superieur ante Vraconien sensu Ambroggi, 1963). The Vraconian therefore has a minimum thickness of 60 m located between IIIA 749 and the facies boundary at IIIA 809, where the clastics are again cut off by marls. The biostratigraphic boundary in the Ascouti Section perhaps is masked within these marls which are usually mapped as Lower Cenomanian (Ambroggi, 1963; Schmitz, pers. comm.). The boundary was determined biostratigraphically within the marls of the beach section near Agadir by Wiedmann et al. (1978; this Vol.). A boundary position within the coarser clastics cannot be rejected generally because *Hystoceras orbigny* indicates the uppermost beds of Lower Upper Albian. However, this assumption would condensate the Vraconian to only a few meters thickness.

Thermal demagnetization of arenite specimens taken between IIIA 605,5 and 619 indicates one or two superimposed components while the stable components are aligned with the reversed paleofield direction (IIIA 608, Fig. 2). Between IIIA 630 and IIIA 633,5 an arenitic bed near the top of the first cycle renders reversed directions. The overlaying marls, producing a sampling gap, are succeeded between IIIA 656 and IIIA 670 by red-colored dolomitic breccias and grey, red-patched limestones, arenites, and shell hashes. Thermal demagnetization moved the RM-vector of the IIIA 656 specimen antiparallel to F_{rec}. Because the NRM of more than two specimens are aligned with the reversed paleofield, a weak primary magnetization from depositional or post-depositional magnetite is assumed. The top series, however, yield normal paleofield directions.

The next series suitable for sampling is composed of shell hashes, argillaceous shell hash with oysters, and marls. It is characterized by two beds extremely resistant to weathering (IIIA 680) and near the top (IIIA 704–710) and therefore morphologically easily recognizable as twin escarpment. The sandy marls at IIIA 688 are of normal magnetization. Reversed remanent directions appear at IIIA 694, 700,5, and 710, respectively.

The resistance against weathering of grey arenites and reddish quartz sandstones between IIIA 726 and 746 is shown by cascades of Assif Tourza. Thermal demagnetization of pilot specimens drilled at IIIA 730, 731,2, and 742 confirmed the reversed remanence which was already assumed on account of NRM-measurements.

The next part which was sampled is exposed between IIIA 762 and IIIA 790. The base consists of cross-bedded quartz sandstones, which are 3 m thick and have a grey-yellowish color. Limy sandstones 7 m thick are at the top of the escarpment. This escarpment is of higher stratigraphic position than the marls bearing *Hystoceras orbigny* and hence may be Vraconian in age according to the discussion given at the beginning. From core IIIA 762 normal directions were measured. On the other hand cores from IIIA 763 and 765 yielded reversed NRM-directions. As described above with respect to specimen IIIA 656 (Fig. 2) components antiparallel to F_{rec} were proved stable after elimination of components aligned with the reverse paleofield direction. Therefore the NRM-data were used. Pilot specimens of core IIIA 783, 787, and 790 indicated stable components aligned with normal paleofield directions at IIIA 783 and with reversed paleofield directions at IIIA 787 and IIIA 790. The top of the Upper Albian sampled at IIIA 795 and IIIA 809 is of normal magnetization.

As outlined earlier, the Albian-Cenomanian boundary cannot be fixed exactly by biostratigraphic methods and must be postulated within the marls above (IIIA 810). With the beginning of the Cenomanian the Atlantic transgressed toward the east but did not reach the extent it had in the Upper Aptian. Therefore, in the Atlas Gulf region the Cenomanian transgression does not have the significance which is usually attributed to it with respect to its global importance. Moreover, the ocean soon began to regress again, as is documented by gypsiferous marls and claystones.

The Cenomanian part of the Ascouti Section measures 326 m in thickness. It is dominated by marls with intercalation of limestones and oyster-beds, both suitable for sampling. At IIIA 1016 and 1023 colored gypsiferous clays within two beds of limestones with oysters indicate the maximum of the Cenomanian transgression in the Agadir Section.

The uppermost oyster-bed at IIIA 1135,5 was taken as top of the Cenomanian, because the overlying limestones and chert-bearing dolomites (IIIA 1136) have recently been placed into the Turonian (Stamm and Thein, this Vol.).

The excarpment at IIIA 826 built up by limestones with oyster-beds shows normal magnetization. The light grey limestones of the next escarpment contain both directions: normal at the base and reversed near the top (IIIA 830 and 833). Normal directions were found within the fine-grained arenite at the base (IIIA 838) and at the top (IIIA 840) as well. Examination of a laminated limestone at IIIA 847 had the same result.

Within the succeeding marls and claystones a rather compact, grey-yellowish limestone forms an escarpment between IIIA 874 and 877. Reversed paleofield directions resulted after removing superimposed components (Fig. 2, IIIA 875). The stable component of an arenite sampled at IIIA 906 points toward the normal paleofield direction. The marly, brown-yellowish limestones with oyster-beds at IIIA 950, 958, and 961 show normal paleofield directions again.

The lowermost gypsiferous marlstones at IIIA 1006 are followed by olive-green claystones with gypsum at IIIA 1020,5. This sequence yielded samples only from IIIA 1018 and from IIIA 1020 which both show reversed remanence. The overlying marlstones are intercalated by grey-brownish-colored, dolomitic arenites and shell-hash layers at IIIA 1076. After thermal demagnetization normal paleofield directions were well established,

Reverse magnetization prevails within the marly shell-hash layers at IIIA 1123. NRM of limestone specimens from IIIA 1133 again indicates reversed paleofield directions, but, due to the weak NRM-intensities demagnetization experiments hardly gave significant results. The uppermost Cenomanian core drilled from the shell-hash layers at IIIA 1153,3 yielded normal paleofield directions.

During Turonian time the Atlas Gulf extended to the east and was linked with the Sahara Basin by a marine street between the High Atlas and the Anti-Atlas (Busson, 1970; Stamm and Thein, this Vol.). The Turonian of the Ascouti Section, in total 64 m thick, is built up by light greyish limestones in its lower part, while the upper series from IIIA 1170 onward are composed of fine- to medium-grained dolomites of grey-yellowish color. The Turonian is characterized by layers of chert linked to distinct bedding planes. Spaced in intervals between 5 and 20 cm near the basis, these chert-containing layers become increasingly rare toward the top of the Turonian. Intercalations of shell-hash layers with oysters show a diametrically opposed trend: absent in the lower portion they become more frequent within the dolomites, thus documenting the shifting of the coastline toward the west. In contrast to the terrigenous conditions in the Northern Atlas region, marine conditions related to a small marine inlet still existed at the southern boundary of High Atlas, even during Senonian time (Stamm and Thein, this Vol.).

Initially the normal polarity within the Upper Cenomanian continues and is well established by combined demagnetization of pilot specimens drilled from limestones at IIIA 1143, 1147,6, 1150, 1155,5, and 1163 (Fig. 2). The dolomite series starting at IIIA 1166 were sampled at IIIA 1167, 1172, 1176, 1184, 1191, 1195, and 1202. The originally reversed NRM-directions of core IIIA 1167,1 were not confirmed by demagnetization experiments. Combined demagnetization yielded hematite pigment to be the carrier of the stable components in pilot specimens 1172 (Fig. 3), 1191, and 1202 and magnetite respectively in pilot specimens of IIIA 1176, 1185, and 1195. In pilot specimens of 1176 and 1185 reversed directions are disclosed after elimination of the magnetite-borne components. Consequently hematite pigment of post-depositional origin but of more stable remanence must be assumed. In spite of the unsecure age of the spontaneous magnetization of hematite pigment the magnetite-borne reversed paleofield directions of specimen 1195 remain reliable.

It may be of special interest for future investigations that both the change in magnetomineralogy from magnetite to hematite pigment and that of the magnetic polarization from normal to reversed takes place between IIIA 1164 and 1172. As will be discussed later, this simultaneous behavior may be correlated with the change of the sedimentary facies from limestone to dolomite at IIIA 1166 (Stamm and Thein, this Vol.).

3.2. Zem Zem Section (Fig. 6)

The Zem Zem Section is located at the northern boundary of the High Atlas, 46 km SE of Essaouira (see Fig. 1, Section 42 in Stamm and Thein, this Vol.). The paleogeographical situation in general is more oceanward than that of the Ascouti Section. The bedding is steeply inclined to the north and sometimes overturned. Special information about lithology, geochemistry, microfaunal elements, and biostratigraphical and magnetostratigraphical correlation data of the Zem Zem Section should be expected from Butt et al. (in prep.). The description of the Zem Zem Section therefore is restricted to a short review in which the biostratigraphical boundaries taken from the official Geological Map have to be regarded preliminary.

The Jurassic/Cretaceous boundary at Zem Zem up to now is uncertain and must be assumed within the limestones at the base of the sampled section. It can be limited between gypsiferous beds at IBA 20 and the marls that directly overlay the limestones at IBA 50, for which Upper Berriasian age is well established (Wiedmann et al., 1978 and this Vol.).

Reversals occur at IBA 86 and IBA 111–115 within the Upper Berriasian and at IBA 131–132 within the Valanginian.

The forrester's house bed at the base of the Upper Hauterivian (IBA 255), and easily determinable by its content of corals (Behrens and Siehl, this Vol.), shows normal paleofield directions. This is in contrast to the reverse directions measured in the Tamrakt Section and may be well explained with asynchronous sedimentation. Due to the regressive trend at that time and the more oceanward position a younger age at the Zem Zem region seems to be quite reasonable.

The reddish-brown sandstone sampled at IBA 339 (Fig. 1) shows reversed magnetization. The dolomite at IBA 348 is reversed, too, but synsedimentary origin is not guaranteed.

The next cores were sampled at IBA 372 from dolomite, at IBA 378 and IBA 379 (Fig. 1) from a sandy limestone, at IBA 380 from conglomerates and at IBA 393 and 396 from sandy limestone again. All specimens yield reversed paleofield directions. The sandstones of Upper Hauterivian at the Zem Zem Section are characterized by the highest NRM-intensities ever measured within Moroccan sediments (higher than 9×10^{-6} Gauss).

The parts within the marly Barremian that were suitable for sampling, and the shell-hash layers of Lower Aptian age show normal polarity of the paleofield. Although indicating reversed paleofield directions in the Ascouti Section the uppermost Lower Aptian beds, well characterized by snowball structures and cross-bedded layers, show normal magnetization at the Zem Zem Section. Again this fact may reasonably be explained by asynchronous sedimentation due to the more oceanward position of Zem Zem, but asynchronous acquisition of remanence cannot be excluded because of insufficient demagnetization experiments.

These beds are directly overlain by marls with amonites, which represent the Upper Aptian transgression and Lower Albian in an environment of middleshelf position (Wiedmann et al., 1978). Unfortunately, the material could not be sampled with our drilling technique.

During the Upper Albian and its uppermost interval, the Vraconian, mainly cross-bedded limestones, dolomites, and shell-hash layers were deposited. As at the Ascouti Section, this littoral environment was rhythmically influenced by terrigenous and pelagic facies, shifting to and fro. In contrast to the Ascouti Section, in the Zem Zem Section this sequence is especially characterized by fluvial intercalations. The Upper Albian begins at 00 IIBA (= 788,5 IBA) and is 230 m thick.

During thermal cleaning of IIBA 55, 83, and 84,4 pilot specimens the intensity was lowered to less than 1.3×10^{-8} Gauss. Therefore, the related directions could not be resolved sufficiently and the resulting reversed remanence is somewhat questionable. In contrast, the reversals at IIIBA 159, IIIBA 177,8 are well established by af-demagnetization. Shortly interrupted by a normally magnetized limestone from IIBA 181,5, the reversed remanence directions continue within the sandstone at 194,3 (Fig. 3) and the limestones at 201,5, 203, and 216. NRM-measurements of all specimens of the core taken at IIBA 216 show reversed paleofield

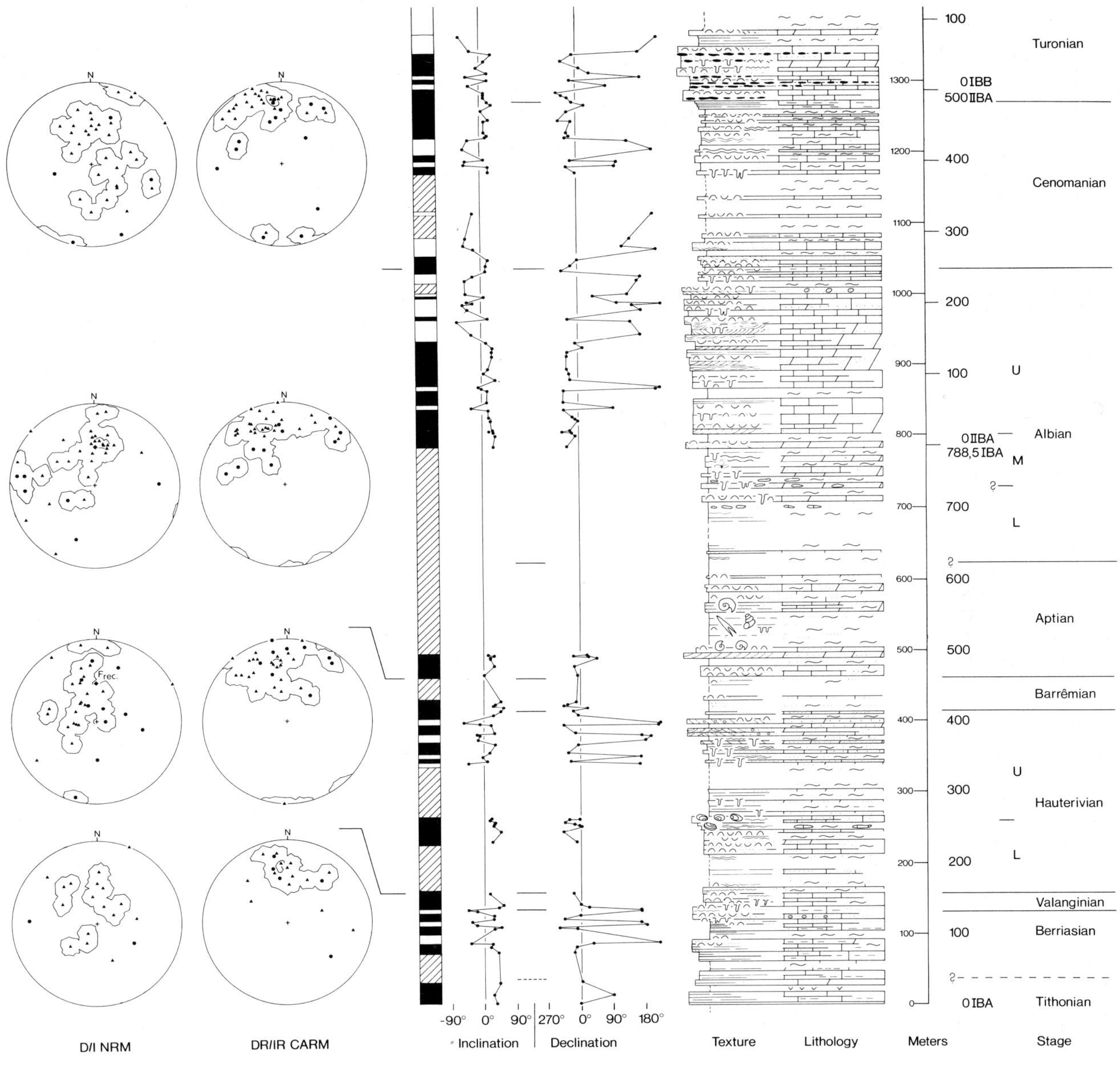

Fig. 6. Magnetostratigraphy, NRM and CARM of the Zem Zem Section. For legend see Fig. 5

directions, but the remanence vector of pilot specimens was shifted out of this direction by thermal demagnetization as described earlier with respect to IIIA 656 (Fig. 2).

The series of limestones and arenites are abruptly cut off by marls. The far off visible escarpment, modeled by extremely different weathering qualities, is commonly mapped as the Albian/Cenomanian boundary. As mentioned with respect to the Ascouti Section, the biostratigraphic boundary has to be assumed within the marls lower than IIBA 250 for which already Cenomanian age is highly probable (Butt, pers. comm.). Therefore, a correlation of the basis of the normal interval that starts at IIBA 249, and the Albian/Cenomanian boundary seems to be advantageous. In this case the reversal at IIBA 235 becomes part of the Vraconian.

Reversed magnetization of the dolomite at IIBA 279, already supposed after NRM-measurements, was proved true by combined demagnetization. After thermal cleaning the remaining components of the dolomites at IIBA 283 and 291 could not be resolved sufficiently. However, the reversal probably continues, because more than two specimens of each core yielded NRM-directions aligned with the reversed paleofield.

The stable component of specimens taken from a dolomitic intercalation at IIBA 330 within the Cenomanian marls points toward the reversed paleofield direction.

In the uppermost Cenomanian, intercalations of limestones and dolomites become more frequent. The basis of the Turonian is documented by the appearance of cherts at IIBA 482 (Stamm and Thein, this Vol.).

The reversal at IIBA 398 and IIBA 403 is somewhat questionable because the components which remained in both pilot specimens were smaller than 10^{-8} Gauss, scattering considerably around the reversed paleofield direction. As to the limestones at IIBA 421 and 433, questionable reversed remanence is assumed after NRM-measurements.

Dolomites were sampled at IBB 11,9 and IBB 18,5 whereby IBB 00 is identical with 500 IIBA. Although the curves of thermal-demagnetization (TH 2A) are congruent (Fig. 2, 11,9/1; Fig. 3, 18,5), indicating the same decay of intensity, the directions differ by nearly 180° with reversed remanence at IBB 11,9 and normal remanence at IBB 18,5. Magnetite was well established for both pilot specimens by combined demagnetization (Fig. 2, 11,9/2). For the specimens drilled from a limestone at IBB 23,5, however, the reversed direction that occurred after extinction of a goethite-borne component could not be related unambiguously to the syndepositional Earth's magnetic field, because hematite is the carrier of remanence (Fig. 3, 23,5).

The succeeding interval is characterized by normal polarity, conserved by magnetite, but well-established reversed directions occur at IBB 59,8. Specimens drilled at IBB 80,3 also show reversed magnetization, after NRM-measurements, but after demagnetization the weak remaining component could not be determined sufficiently with respect to direction.

3.3. Afensou Section (Fig. 7)

In sections north of Taroudannt, about 70 km east of Agadir, marine and terrigenous influences are balanced within a sequence reaching from Berriasian to Turonian (for location see Fig. 1, Section 18, in Stamm and Thein, this Vol.). Faunal elements suitable for biostratigraphic calibration are lacking. Therefore the stratigraphic positions have to be determined by lithological comparison with the sections at Agadir. For this purpose marker beds were outlined toward the east by a file of densely spaced sections (Behrens et al., 1978; Behrens and Siehl; Wurster and Stets; Stets and Wurster, all this Vol.), thus documenting the rapid alteration of lithofacies in the east–west direction. The sequence of Mesozoic and Cenozoic rocks between the High Atlas in the north and the Souss Plain at the south is well exposed by the Assif Ait Haj in a natural profile that cross-cuts the axis of the anticline of Amzaourou. This anticline seemed to be suitable for magnetostratigraphic purposes due to excellent geological information (Stets and Wurster, this Vol.; Schmitz, 1977; Wellpott, 1977). Schmitz (1977) presented a Geological Map of this area (1 : 25 000, sheet Tamaloukt East) and numerous delicately analyzed sections as well. He also distinguished 18 stratigraphic members advantageous for mapping.

The Mesozoic sequence starts with the Lower Red Formation, 30 m in thickness, that unconformly overlies the Paleozoic basement. On top of the Lower Red Formation follows the Pipe Limestone Formation. For both formations Cretaceous age may be assumed, but according to Schmitz (1977), a Jurassic origin of the Pipe Limestone Formation seems to be more likely. Therefore also a Triassic age of the Lower Red Formation may be possible. A sedimentation gap between Pipe Limestones and Middle Red Formation is assumed (Schmitz, 1977). After 20 m in thickness the red clastics are cut off by the Grey Clastic Formation, that together with the succeeding Limestone Marl Formation forms an excellent mapping-unit (Schmitz, 1977).

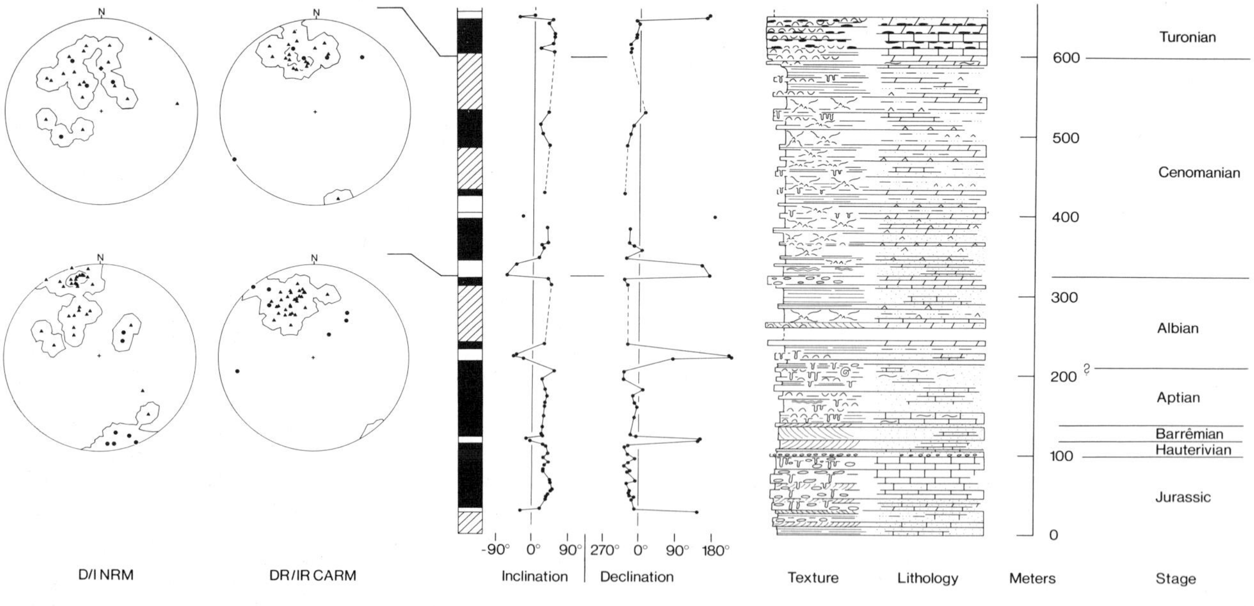

Fig. 7. Magnetostratigraphy, NRM and CARM of the Afensou Section. For legend see Fig. 5

Schmitz (1977) assumed that the Middle Red Formation represents the Hauterivian and Barremian, and that the Grey Clastic Formation with the Limestone Marl Formation are Aptian in age. However, according to Behrens and Siehl (this Vol.) Hauterivian age for the Middle Red Formation and Barremian age for the Grey Clastic Formation seemed to be more likely. This opinion is favored by the magnetostratigraphic results discussed later.

The Limestone Marl Formation is correlated with the transgression of the Aptian but biostratigraphically established by only one fossil. It follows the Carbonate Sandstone Formation beginning with oyster-containing sandstones overlain by dolomites which are more resistant to weathering and are intercalated with marly sandstones.

A dolomitic bed, 6 m in thickness, with increasing content of quartz gravels and solution features indicating regional emergences, forms the top of the formation. This dolomite is a useful marker bed in sections along the South- and North Atlas. Comparison of parallel sections supports Upper Albian age (Behrens et al., 1978; Schmitz, 1977). Regarding the development of facies during Atlas Gulf history (Wurster and Stets, this Vol.) Middle Upper Albian is most likely (Stets, pers. comm.).

The term Upper Red Formation is used for the red siltstone and claystone series which dominate the succeeding 200 m of the Afensou Section (Schmitz, 1977). Due to intercalations of dolomites and gypsum beds they are well explained as deposits of lagoon environment interrupted cyclically by terrigenous influences (lagoon sebkha cycles). Cenomanian age is assumed (Schmitz, 1977), however, this opinion is based on lithology only without any biostratigraphic control up to now.

The Upper Carbonate Formation, 60 m in thickness, represents the Turonian (Schmitz, 1977; Stamm and Thein, this Vol.). The sequence is built up by limestones and dolomites. Intercalations of shell-hash layers are common, but in comparison with the Agadir and Zem Zem Sections cherts are rare. Lithology and faunal elements indicate that the ocean was extended far to the east of Afensou. The Turonian/Coniacian boundary is established lithologically with the beginning of red clastics (Schmitz, 1977).

The lower part of the section was sampled within the southern limb (4A) of the mentioned anticline, easily accessible near Tamaloukt north of Taroudannt. Due to indeterminable fault dislocations it was continued at the northern limb south of Afensou (SW1).

The NRM-directions of all specimens from the Lower Red Formation, drilled at 4A 28,5, are aligned with F_{rec}. However, thermal demagnetization indicates reversed polarity. Although sampled in dense space intervals, no evidence of reversals was found within the Pipe Limestone Formation. This contradicts the assumption that the Pipe Limestone in toto represents the Jurassic. Instead a sedimentation during intervals of prevailing normal polarity, for example in Upper Jurassic or during the Berriasian/Valanginian will better fit the results.

The basis of the Middle Red Formation shows normal polarity at 4A 100, 110, and 112. The reversed polarity of the uppermost part of Hauterivian at SW1 119, assumed after NRM-measurements, is well established by thermal demagnetization. The same polarity is found in specimens that were sampled after a distance of 0.2 m at the base of the Lower Grey Formation. Normal polarity again occurs at SW1 123 and 124. Specimens drilled from the Limestone Marl Formation (Aptian) at SW1 148 and 159, and from Carbonate Sandstone Formation at SW1 165, 173, 181, 194, and 204 all indicate normal polarity. Reversed directions, however, were yielded from specimens taken at SW1 218, 220, and 223. Difficulties occur with respect to the age of this reversal which may be Upper Aptian or Lower Albian.

The succeeding bed suitable for sampling is the dolomitic top bed of the Albian. Cores were drilled at SW1 312, 320, and 323. Pilot specimens from SW1 312 and 320 show normal polarity. However, the stable components of specimen from SW1 323 are aligned with the reversed paleofield directions. A core taken 15 m in top from the lower part of the Upper Red Formation also shows reversed magnetization. As this position commonly is assumed as to be Cenomanian to Albian – the latter being more probable due to the eastern position of Afensou – the accord with the Ascouti and Zem Zem Sections is evident.

Normal polarity was found between SW1 346 and 384 but the reversed paleofield direction was established for SW1 388. Because of the unconsistancy of red clay and siltstone, cores could be drilled only at SW1 428, 488, and 530 all samples indicating normal polarity for the respective Cenomanian interval.

Also the compact limestones and dolomites of the Turonian still show normal polarity between SW1 605 and SW1 645. Reversed polarity, however, starts again at SW1 648. As already mentioned from the Ascouti Section, evidently simultaneous geomagnetic and geochemical changes took place within the same horizon (Stamm and Thein, this Vol.).

3.4. Aouzzer Section (Fig. 8)

This section, in the most paleo-continent-ward position, is located 45 km south of Marrakech at the northern flank of the High Atlas (see Fig. 1, Section 5, in Stamm and Thein, this Vol.). The magnetostratigraphic sampling was carried out along the banks of the Qed N'fis. Corresponding with the eastward position, terrestrial deposits by far exceed the marine deposits. With respect to the Atlas Gulf evolution, the marine intercalations must be regarded as Aptian/Albian and Turonian. Furthermore, the top bed of the Upper Albian described in connection with the Afensou Section, is again easily recognizable. The Cenomanian offers another chance for correlation which is not only bounded by the two well-known marine horizons, but also shows the typical lagoon sebkha cycles mentioned from the Afensou Section. A successful biostratigraphic determination is given from 3F 196,5, indicating Upper Aptian. The Hauterivian/Barremian, and the Barremian/Aptian boundaries which up to now are open to subjective speculation probably now could be determined better with the help of magnetostratigraphy.

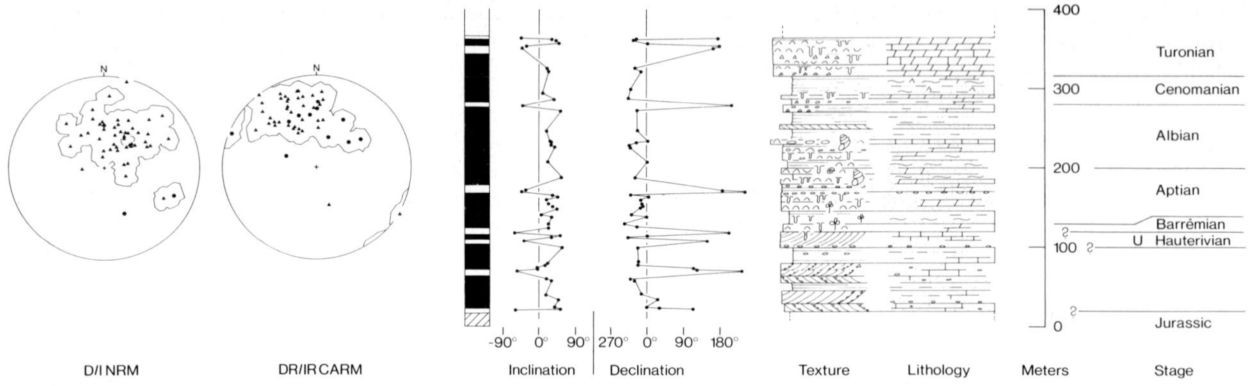

Fig. 8. Magnetostratigraphy, NRM and CARM of the Aouzzer Section. For legend see Fig. 5

The matrix of red conglomerates sampled near the base of the section at 3F 33,1 is reversely magnetized. The cycles of red-colored conglomerates, sandstones, and siltstones between 3F 23,5 and 116 usually show normal paleofield directions. Reversals occur only at 3F 69–75 and at 3F 108, respectively.

The overlying sequence between 3F 116 and 278 is built up dominantly by greyish-colored sandstones, marls, and claystones. The sequence is interrupted by a red-colored terrestrial arenite between 3F 165 and 172,5. Specimens taken at 3F 119 yielded reversed directions. The succeeding samples drilled from the base of red intercalations at 3F 165 yielded normal directions only. The top of red intercalations then is reversely magnetized again, as is documented by pilot specimens from 3F 169 and 171,5 (Fig. 4). After a sequence of normal polarization, reversed directions were yielded again from specimens drilled at 3F 278.

Stratigraphical correlation of this sequence now seems to be possible by magnetostratigraphic correlation with the other Moroccan sections. These show predominantly reversed polarization in the Upper Hauterivian and the Lower Barremian. In addition, this boundary in the Afensou Section is marked by changing colors from red to grey. The same relations were found also in the Aouzzer Section. Below and above the boundary at 3F 116, reversed polarity was established at 3F 108 and 119 as well. Consequently, the beginning of Upper Hauterivian should be assumed lower than 3F 105, and the Hauterivian/Barremian boundary at 3F 116. The basis of the Lower Aptian then may be correlated with the beginning of siltstone and claystone sedimentation at 3F 130. However, because of the well-established Lower Aptian reversed polarity (Lowrie et al., 1980), that was also well confirmed in the Ascouti Section, an Aptian age of the grey clastics that start at 3F 116, supposed by Stets and Wurster (pers. comm.), cannot be rejected. Further uncertainties still appear regarding the stratigraphic position of the red clastics up to 3F 108 and of the reversal at 3F 65 for which an Upper Jurassic to Lower Hauterivian age should be most probable.

At 3F 198 cephalopods of Upper Aptian age document the climax of transgression within this region. According to Schmitz (pers. comm.) the Aptian/Albian boundary may be assumed at 3F 200. Upper Albian age of the dolomite bed at 3F 278 is supported by reverse polarity, well known from the corresponding series at the Afensou Section.

Specimens sampled at 3F 287 and 297 yielded normal paleofield directions for the Cenomanian. Dolomites occurring at 3F 318 are regarded as deposits of the Turonian. The highly brecciated portion of this sequence between 3F 326,5 and 348 was not sampled. Specimens drilled at 3F 322,5 and 326 again yielded normal paleofield directions, but the Turonian series is reversely magnetized at 3F 349, 351, and 354. In contrast to the common TH 3 behavior, shown during demagnetization of 3F 351 specimen (Fig. 4), in which the goethite component is aligned with F_{rec}, the goethite component of specimen 3F 349 (Fig. 4) points toward the reversed paleofield direction. After normal polarity at 3F 357 (Fig. 4) which remained stable during demagnetization, reversed directions occur again in specimens taken from 3F 363. The anoxic-oxic transiton seemingly correlated with a change of magnetic polarity until now (Stamm and Thein, this Vol.) could not be identified exactly in the Aouzzer Section and may be assumed within the brecciated series.

4. Magneto-Stratigraphic Results and Their Relation to the Geomagnetic Time Scale of Other Authors

Because of its well-established biostratigraphic record the Tamrakt-Ascouti Section is of major importance for magnetostratigraphic calibration. Therefore the correlation of all other profiles is based on this section.

The lowermost reversal of the Tamrakt-Ascouti Section is within the Upper Jurassic. The lowermost reversal of the Aouzzer Section perhaps has the same stratigraphic position. The correlation of the reversal at the base of the Afensou Section is still uncertain. A Jurassic or even Triassic age is discussed by Schmitz (1977). In the Zem Zem Section perhaps the sampling intervals are too great, so that the reversed interval may have been missed.

Two reversals were found within the Upper Berriasian of the Tamrakt-Ascouti and of the Zem Zem Section as well. Although in different positions with respect to the lithological profile, they may be of synchronous origin, especially if one considers that the Berriasian sandstones of the Tamrakt-Ascouti Section show rather high sedimentation rates. The next reversal at the base of the Valanginian was identified only in the Zem Zem Section, but was probably missed in the Tamrakt-Ascouti Section because the corresponding marls could not be sampled. The Berriasian or Valanginian age of the second and third reversals at Aouzzer is still subject of speculation.

The Lower Hauterivian at the Ascouti-Tamrakt and the Zem Zem Sections shows normal polarity. At Tamrakt-Ascouti the Upper Hauterivian and the base of Barremian are characterized by four intervals of reversed polarity, whereby the Hauterivian/Barremian boundary lies within the uppermost reversal. This is in full agreement with the Afensou Section that shows reversed magnetization at the top of the Middle Red Formation (Hauterivian) and at the base of the Lower Grey Formation (Barremian). The similarities in the lithological and magnetostratigraphical sequences between Afensou and Aouzzer allows the determination of the Hauterivian/Barremian boundary also at the Aouzzer Section.

The Upper Hauterivian and the Barremian of the Zem Zem Section show the same polarity pattern as was found in the Tamrakt-Ascouti Section. Only the uppermost reversal occurs within a horizon that usually is considered as Upper Hauterivian (Behrens and Siehl, this Vol.). However, the correlation of the Hauterivian/Barremian border with the uppermost reversal would hardly affect the rough biostratigraphic record, as it only lowers the boundary for a few meters, and thus yields full agreement of all Moroccan sections.

Because of magnetite as the carrier of remanence, the reversed magentization of the Aptian base at the Tamrakt-Ascouti Section reflects the polarity of the Earth's magnetic field at the time of deposition. On the other hand, the corresponding part of the Zem Zem Section shows normal polarity. Perhaps the reversal is masked within the unsampled marls of the bottom series: Compared with the Tamrakt-Ascouti Section, the Zem Zem region is located more oceanward, so the regressive shifting of the lithofacies possibly arrived there with some delay. However, as hematite pigment was determined as the carrier of the normal remanence, a later acquisition of magnetization may also be possible.

A reversal of later Aptian age is found in the Aouzzer Section. Perhaps it may be correlated with the reversal assumed as Lower Albian in the Afensou Section, because the Aptian-Albian boundary is not well defined. In the Zem Zem and Tamrakt-Ascouti Sections the corresponding series lie within sampling gaps.

Excellent agreement occurs within the Upper Albian. It shows four reversals in the Tamrakt-Ascouti and probably five in the Zem Zem Section. The different number and the different thickness of the reversals can be explained by discontinuous sedimentation and varying sedimentation rates which is typical for intercalations of sandstones and marls.

In Afensou and Aouzzer the reversal of the Albian top bed apparently can be correlated with one of the central reversals of the corresponding series at the Zem Zem and Tamrakt-Ascouti Sections.

Biostratigraphic uncertainties on the one hand and sampling gaps of considerable extension on the other do not allow distinct correlations between the reversals within the Cenomanian. Therefore, all attempts to correlate the reversals of the Lower Cenomanian at the Ascouti, Zem Zem, and Afensou Sections must be regarded with scepticism. The results within the Upper Cenomanian directly below the Cenomanian/Turonian boundary are contradictory; they show reversed magnetization only at the Tamrakt-Ascouti Section.

The syndepositional origin of the Turonian reversals is still questionable. In the Tamrakt-Ascouti Section the polarity change from normal to reversed occurs in the middle of the Turonian at 1166 IIIA. Magnetite was found to be the carrier of stable remanence below IIIA 1166. Above IIIA 1166 hematite pigment is the carrier of stable remanence. Therefore, the polarity change was rather controlled by geochemical processes than by the behavior of the Earth's magnetic field at the time of deposition. This assumption is supported by the fact that a lithological boundary exists between limestone and dolomite which lies within the same horizon as the polarity transition. Moreover, Stamm and Thein (this Vol.) found that such a synchronous geochemical event took place within all Turonian sections of the area. Due to the geochemical record, this boundary must be placed directly below IBB 23,5 of the Zem Zem Section. It is accompanied by normal polarity, conserved by magnetite at IBB 18,9, indicating the Earth's magnetic field direction during deposition, and reversed remanence carried by hematite pigment at IBB 23,5 which acquired its remanence during an undeterminable later reversed polarity interval. In the Afensou Section the corresponding sequence was not sufficiently sampled because of its brecciation; but the specimens below the assumed boundary show normal, and those above, reversed remanence. An additional reversal may be assumed within the limestones at IBB 11,9 of the Zem Zem Section. Although magnetite is the carrier of remanence, the directions differ by more than 60° from the true paleofield direction.

At all sections the uppermost part of the Turonian shows reversed magnetization, carried by magnetite, and therefore this indicates the Earth's magnetic field direction during the time of deposition. The normal remanence found at the Zem Zem Section is perhaps related to a later normal polarity interval of the Earth's magnetic field because hematite pigment acts as carrier of the stable direction. Further information concerning the correlation of geochemical events and polarity changes can be expected from geochemical and demagnetization experiments carried out at Bonn (Krumsiek, Stamm and Thein, in prep.).

Larson and Hilde (1975) calibrated the ocean-floor polarity sequence with the help of two biostratigraphical determinations of Oxfordian and Hauterivian age. The corresponding absolute ages were taken from the Phanerozoic time scale (PTS, Casey, 1965) and used for extrapolation by assuming constant spreading rates between these time markers. This M stratigraphy extends from 109 m.y. (M-0) backward to 153 m.y. (M-25).

However, the absolute ages of the M polarity sequence must be corrected when using the time scale of Van Hinte (1976) which nowadays is usually preferred in magnetostratigraphy. According to this recalibration, the base of M-0 is of Upper Albian age (111 m.y) and the base of M-25 of Middle Oxfordian age (135 m.y.). The total time interval of M sequence is therewith condensed from 44 m.y. (PTS Casey, 1965) to 35 m.y.

The uncertainties in the age determination of ocean-floor lineations indirectly recorded as marine magnetic anomalies not only result from the problematical extrapolation mentioned above, but also from the extrapolation of the unknown time span (hiatus?) between the origin of the ocean-floor basalts and the deposition of the oldest sedimentary layer that can be determined biostratigraphically. More reliable therefore are biostratigraphic age determinations of reversals, measured directly from the same rock specimens.

Up to now the best type section for the late Cretaceous-Paleocene geomagnetic sequence was worked out at Gubbio by Lowrie and Alvarez (1977). This polarity sequence meanwhile is extended by Lowrie et al. (1980), including the Hauterivian by using sections of the Southern Alps. Furthermore, reversals of the Upper Jurassic have been found in southwest Germany (Heller, 1977) and in North America (Steiner, 1980).

According to the results of Lowrie et al. (1980) an additional correction of the absolute age of the M-anomalies is necessary, so that M-0 marks the base of the Aptian. M-1 then belongs to the Barremian. The reversed polarity of the Earth's magnetic field documented by anomaly M-3 (aged 122 m.y. at the base) lasted for the time from the Late Hauterivian to the earliest Barremian. The Jurassic/Cretaceous boundary is congruent with the base of M-18, 135 m.y. in age (Steiner, 1980). These data were taken into account for the construction of Fig. 9, in which M-0 to M-4 are shown according to Lowrie et al. (1980) and M-18 to M-20 according to Steiner (1980). The M-anomalies in between, however, are recalibrated by extrapolation, using the base of M-3 (122 m.y.) and the base of M-18 (135 m.y.) as time markers.

Lowrie et al. (1980) apparently still maintain their assumption (Lowrie and Alvarez, 1977) of a Cretaceous long normal polarity interval that existed from the Lower Aptian (114 m.y.) to the basis of the Camapanian (78 m.y.) where the anomaly 34 put an end to this period of constancy. Nearly the same view was supported by numerous authors, such as Helsley and Steiner (1970), Larson and Pitman (1972), and Pechersky and Khramov (1973), the latter restricting the time span to the Aptian-Albian and the Turonian-Cenomanian boundaries.

In contrast to this opinion are the results of DSDP site 263 (Green and Brecher, 1974; Jarrard, 1974), that yielded three short reversed polarity intervals within the lower part of the Upper Albian. Moreover, former results of the authors from SW Morocco showed numerous reversals within the Albian, Cenomanian, and Turonian (Krumsiek, 1977; Behrens et al., 1978) and, last but not least, Vandenberg and Wonders (1980) found reversed polarity intervals at the bottom and top of the *breggiensis* zone (Lower Upper Albian), and also in parts of the *Cushmani* zone (Upper Middle Cenomanian to Upper Cenomanian). The latter are meanwhile known as Valle del Mis (VM) and Quiro Schievenin (QS) reversals (Lowrie et al., 1980). Perhaps changes in inclinations, found by Hailwood (1980) at DSDP site 402A, are of comparable age. The Isea reversal, presented by Vandenberg et al. (1978), is of Upper Albian age and perhaps identical with the Gatan reversal of DSDP site 260 (Jarrard, 1974).

A comparison of the reversals deduced from the Moroccan sections with those of other authors shows particular agreements within the Upper Hauterivian and the Upper Albian. The reversals of the Upper Hauterivian and Lower Barremian correspond with M-5 and M-3, and the Upper Albian reversals with the VM reversals. Evidence for M-0 was found at the base of the Aptian at the Ascouti Section.

It is still uncertain to what extent the further Moroccan reversals can be correlated with the geomagnetic polarity sequences of the other authors. Thus the Upper Tithonian reversal near the base of the Tamrakt Section may correspond with M-19 and, though more doubtful, the reversal at the base of the Aouzzer Section may be of the same age. The Upper Berriasian reversal at Tamrakt and Zem Zem may correspond with M-15 or M-16. M-14 perhaps is represented near the base of the Valanginian in the Zem Zem Section. Eventually the Isea reversal (Upper Aptian) was found in the Aouzzer Section, but in the other sections the corresponding parts are sampling gaps.

The Lower Cenomanian reversal found in the Tamrakt-Ascouti and Afensou Sections has not yet been published by other authors. The succeeding reversal within the Cenomanian perhaps corresponds with the Qs reversal presented by Vandenberg and Wonders (1980). Reversed polarity from Upper Cenomanian to Lower Turonian was merely reported by Pechersky and Khramov (1973).

The Turonian polarity change at IBB 23,5 within the Zem Zem Section and at IIIA 1166 within the Tamrakt-Ascouti Section is probably related to a regional anoxic-oxic event (Stamm and Thein, this Vol.; Schlanger and Jenkyns, 1976; Jenkyns, 1980) and does not reflect the behavior the Earth's magnetic field during time of deposition. The short reversal found at IBB 11,9 is somewhat unreliable due to its peculiar direction. More reliability is given to the reversal within the uppermost Turonian. However, reversed polarity data from the Upper Turonian are not yet reported from any area outside Morocco.

Figure 9 shows that reversed polarity intervals younger than Late Barremian are very rare and of short duration. High sedimentation rates of 65 m/m.y., as they are calculated for the Zem Zem and Ascouti Sections are better suited to record short polarity changes than the comparatively smaller rates of 6.6 m/m.y. as for example given for the Gubbio Section (Arthur and Fischer, 1977). Moreover, the process of magnetization of ocean-floor basalts is very complex (Bleil and Petersen, 1977) and polarity intervals of short duration should not be expected to be conserved as magnetic lineations of the ocean floor.

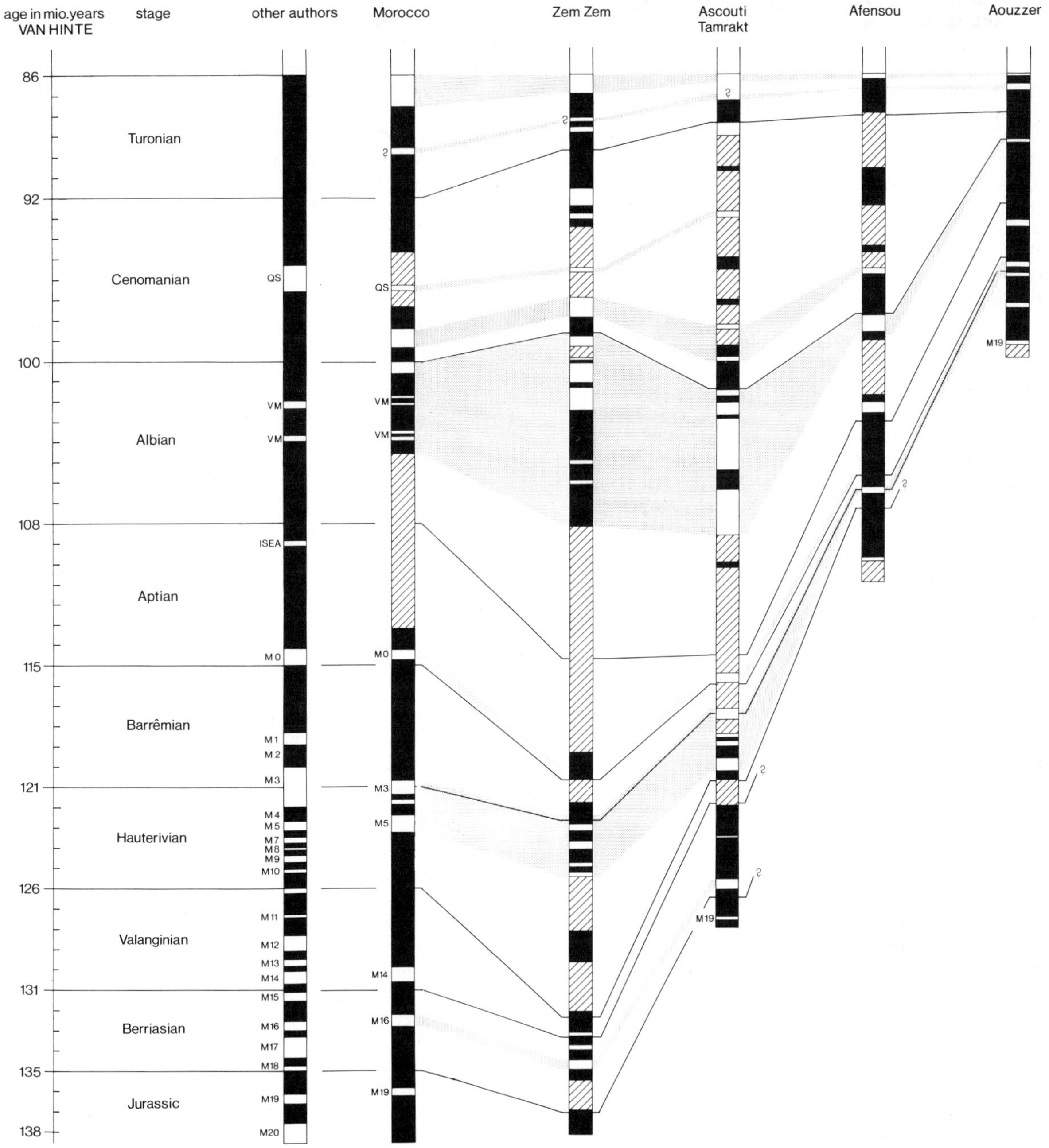

Fig. 9. Compilation of SW Morocco magnetic polarity zones (**column 2**) from Moroccan sections presented in this paper (**columns 3–6**) and comparison with Cretaceous geomagnetic polarity sequence of other authors (**column 1**). **M-0** to **M-20** according to Larson and Hilde (1975), but recalibrated to the time scale of Van Hinte (1975) using the age determinations for **M-0** to **M-4** given by Lowrie et al. (1980) and for **M-18** to **M-20** given by Steiner (1980). Isea, Valle del Mis (**VM**) and Quero-Schievenin (**QS**) reversals are presented by Vandenberg et al. (1978) and Vandenberg and Wonders (1980)

5. Paleo-Pole Positions

Discussing the pole positions corresponding with the diagrams of Figs. 1–4, it must be taken into account that the investigations were carried out for magnetostratigraphic purposes. Therefore numerous points plotted in the diagrams represent very weak rest components sometimes measurable only with little accuracy by the UGF 4 Spinner Magnetometer. Paleo-pole positions, however, should rather be reconstructed by data from specimens with a rather strong remanent magnetization.

Therefore, we can expect reliable results only from the paleo-pole positions of the predominantly clastic parts of the Moroccan sections, as for example the Berriasian-Valanginian of the Tamrakt-Ascouti Section (Position 8, Table 2; Fig. 10), the Hauterivian-Albian of the Afensou Section (Position 10, Table 2; Fig. 10), and the whole Aouzzer Section (Position 11, Table 2, Fig. 10). Within these sections the syndepositional magnetic components are in excellent agreement with the well-established Cretaceous mean direction of Morocco (143 W, 75 N), presented in 1975 by Hailwood. Moreover, even the mean paleo-pole positions reconstructed by the maximum point density method from all 11 Moroccan pole positions (at 144 W, 76 N, as listed in Table 2) verify the results of Hailwood.

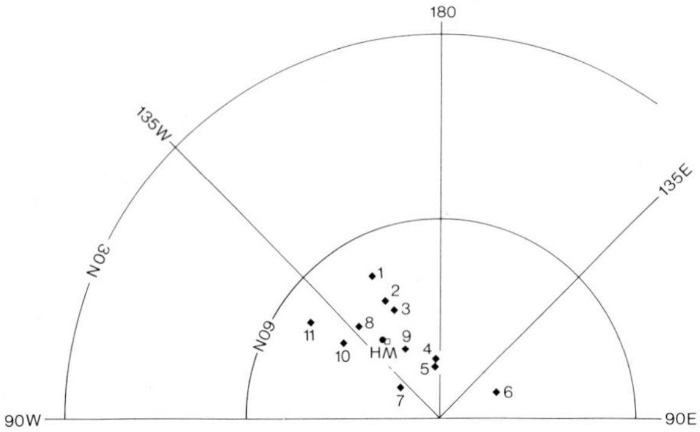

Fig. 10. The paleo-pole positions computed with the CARM of Figs. 5–8. **Numbers** refer to Table 2 containing the related paleomagnetic and statistic parameters. **M** Moroccan mean-pole of this paper reconstructed by the point density method. **H** Moroccan-Cretaceous mean-pole of Hailwood (1975)

Acknowledgments. S. Arbab, G. Dresen, R. Ollig, H. Orzekowsky, S. Schneider helped in field work. G. Schulte, supported by S. Müller and H. Orzekowsky, carried out laboratory work. G. Hahn prepared the drawings, typed the manuscript and carried out demagnetization experiments at Bonn and at ETH Zürich laboratories, kindly supervised by Dr. F. Heller. I. Nagel smoothened the author's basic English.

Dr. Pucher, Dr. v. Rad and Dr. Roeser, Hanover, reviewed the manuscript and delivered helpful remarks. The colleagues and post-graduates of the Bonn working group put their special section records at my disposal and, together with colleagues of Tübingen working group, gave fruitful comments in numerous discussions.

The Service Géologique du Maroc represented by Dr. M. Bensaid, guaranteed the success of field work delivering Geological Maps and inevitable permits. The work was supported by a grant of the Deutsche Forschungsgemeinschaft.

REFERENCES

Ambroggi R (1963) Etude géologique du versant méridional du Haut Atlas occidental et de la Plaine du Souss. Notes Mém Serv Géol Maroc 157: 321

Arthur MA, Fischer AG (1977) Upper Cretaceous-Paleocene magnetic stratigraphy at Gubbio, Italy I. Lithostratigraphy and sedimentology. Bull Geol Soc Am 88: 367–371

Behrens M, Krumsiek K, Meyer DE, Schäfer A, Siehl A, Stets J, Thein J, Wurster P (1978) Sedimentationsabläufe im Atlas-Golf (Kreide Küstenbecken Marokko). Geol Rundsch 67: 424–453

Bleil U, Petersen N (1977) Magnetic properties of basement rocks. Leg 37, site 332. Init Rep Deep Sea Drilling Project 37: 446–456

Busson G (1970) Sedimentation, transgression et paléogéographique sur les grandes plates-formes du Mésozoique l'exemple du Cenomano-Turonien du Nord-Est de la plate-form Saharienne et de Berberie. Bull Soc Géol Fr XI (7): 6

Casey R (1964) The Cretaceous period. In: The Phanerozoic Time scale. Geol Soc London Quart Journ 120 supp: 193–202

Collinson DW (1974) The role of pigment and specularite in the remanent magnetism of red sandstones. Geophys J R Astron Soc 38: 253–264

Cordell BM (1980) Martian climatic change: A magnetic trigger? Geophys Res Lett 12: 1065–1068

Doake CSM (1978) Climatic change and geomagnetic field reversals: A statistical correlation. Earth Planet Sci Lett 38: 313–318

Fairbridge RW (1977) Global climate change during the 13, 500-b p. Gothenburg geomagnetic excursion. Nature (London) 265: 430–431

Garrels RM (1960) Mineral equilibria at low temperature and pressure. Harper & Brothers, New York, pp 1–254

Green KE, Brecher A (1974) Preliminary paleomagnetic results for sediments from site 263, leg 27. Init Rep Deep Sea Drilling Project 27: 405–413

Hailwood EA (1975) The paleomagnetism of Triassic and Cretaceous rocks from Morocco. Geophys J R Astron Soc 41: 219–235

Hailwood EA (1980) Paleomagnetism of late Mesozoic to Holocene sediments from the Bay of Biscay and Rockall Plateau, Drilled on Leg 48 site 402A. Init Rep Deep Sea Drilling Project 48: 305–332

Hedley JG (1971) The weak Ferromagnetism of Goethite (alpha FeOOH). Z Geophys 37: 409–420

Heller F (1977) Paleomagnetism of Upper Jurassic limestones from South Germany. J Geophys 42: 475–488

Heller F (1978) Rockmagnetic studies of Upper Jurassic limestones from Southern Germany. J Geophys 44: 525–543

Helsley CE, Steiner MB (1970) Evidence for long intervals of normal polarity during the Cretaceous period. Earth Planet Sci Lett 5: 325–332

Jarrard RD (1974) Paleomagnetism of some leg 27 sediment cores. Init Rep Deep Sea Drilling Project 27: 415–423

Jenkyns HC (1980) Cretaceous anoxic events: from continents to oceans. J Geol Soc London 137: 171–188

Krumsiek K (1977) Cretaceous magnetostratigraphy of SW-Morocco. Trans Am Geophys Union 58 (9): 898

Krumsiek K (1980) Zur plattentektonischen Entwicklung des Indo-Iranischen Raumes (Resultate paläomagnetischer Untersuchungen in Afghanistan. Geotekt Forsch 60: 1–223

Krumsiek K, Siehl A (1976) PALM – ein Rechenprogramm für paläomagnetische Zwecke. Z Dtsch Geol Ges 127: 309–321

Larson RL, Pitman WC (1972) World wide correlation of Mesozoic magnetic anomalies and its implications. Geol Soc Am Bull 83: 3645–3662

Larson RL, Hilde WC (1975) A revised time scale of magnetic reversals for the Early Cretaceous and the Late Jurassic. J Geophys Res 80: 2586–2594

Lowrie W, Alvarez W (1977) Upper Cretaceous-Paleocene magnetic stratigraphy at Gubbio, Italy – III Upper Cretaceous magnetic stratigraphy. Bull Geol Soc Am 88: 367–389

Lowrie W, Channel JET, Alvarez W (1980) A review of magnetic stratigraphy investigations in Cretaceous pelagic carbonate rocks. J Geophys Res 85: 3597–3605

Pechersky DM, Khramov AC (1973) Mesozoic paleomagnetic scale of the USSR. Nature (London) 244: 499–501

Roch E (1930) Etudes géologiques dans la région méridionale du Maroc occidental. Notes Mém Serv Géol Maroc 9: pp 1–440

Schlanger SO, Jenkyns HC (1976) Cretaceous oceanic anoxic events: causes and consequences. Geol Mijnbouw 55: 179–184

Schmitz J (1977) Geologische Aufnahmen am Südrand des Hohen Atlas bei Taroudannt ergänzt durch karbonatpetrographische Untersuchungen. Unpubl Dipl-Arb, Bonn Univ

Steiner MB (1980) Investigation of the geomagnetic field polarity during the Jurassic. J Geophys Res 85: 3572–3586

Stets J, Wurster P (1977) Sedimentationsmodell Atlas-Golf (Kreide Küstenbecken Marokko). Nachr Dtsch Geol Ges 17: p 75

Vandenberg J, Klootwijk CT, Wonders AAH (1978) The Late Mesozoic movements of the Italian peninsula: Further paleomagnetic data from the Umbrian sequence. Geol Soc Am Bull 89: 133–150

Vandenberg J, Wonders AAH (1980) Paleomagnetism of Late Mesozoic pelagic limestones from the southern Alps. J Geophys Res 85: 3623–3627

Van Hinte JE (1976) A Cretaceous time scale. Bull Am Assoc Petrol Geol 60: 498–516

Walker TR (1967) Formation of red beds in modern and ancient deserts. Bull Geol Soc Am 78: 363–368

Wellpott A (1977) Geologische Aufnahmen am Südrand des Hohen Atlas bei Taroudannt (Marokko) ergänzt durch tonmineralogische Untersuchungen an ausgewählten Proben. Unpubl Dipl-Arb, Bonn Univ

Wiedmann J, Butt A, Einsele G (1978) Vergleich von marokkanischen Kreide-Küstenaufschlüssen und Tiefseebohrungen (DSDP): Stratigraphie, Paläoenvironment und Subsidenz an einem passiven Kontinentalrand. Geol Rundsch 67: 454–508

Zijderfeld JDA (1967) Ac demagnetization of rocks: Analysis of results. In: Collinson DW, Creer KM, Runcorn SK (eds) Methods in paleomagnetism. Elsevier Sci Publ Comp, Amsterdam, pp 254–286

Table 1. Paleomagnetic data

1	2	3	4	5	6	7	8	9	10	11	12
Tamrakt-Ascouti Section											
IIAA 11	d	TH2A,TH3	g,h	?	241/-46	320°	.73	.38	xo	R	Tithon
IIAA 12	l	TH3	g	?	202/-33	70°	2.1	.56	xo	R	Tithon
IIAA 48	s	TH3A,TH1	h	sp	235/-31	470°	2.5	.83	xx	R	Berr
IIAA 116	l	TH2A		m	110/-53	420°	7.3	2.2	xo	R	Berr
IS4 1	l	TH2A,COM		m	311/ 27	370°	.76	.26	xx	N	U.Haut
IS4 11	l	TH3,COM	g	m	139/-13	NRM	2.8		xo	R	U.Haut
IS4 13	l	TH3,TH2A,COM	g	m	227/-37	960	2.3	1.4	xx	R	U.Haut
IS4 15	l	TH3,COM	g	m	159/-51	960	1.7	.89	xx	R	U.Haut
IS4 17	s	TH3,COM	g	m	263/-28	160	3.3	.70	xo	R	U.Haut
IS4 19	s	TH2A,COM		h	191/-45	320	.56	.54	oo	R	U.Haut
IS4 21	s	TH2A,COM		m	153/-67	NRM	.35		xx	R	U.Haut
IS4 30	s	TH2A,COM		m	358/ 30	NRM	2.0		xx	N	U.Haut
IS4 45	l	TH2A,COM		m	252/-24	160	4.9	.51	xx	R	U.Haut
IS4 48	d	TH2A,COM		m	174/-28	1280	2.0	1.1	xx	R	U.Haut
IS4 50	d	TH3,TH2A,COM	g	m	344/ 44	160	3.0	1.2	xx	N	U.Haut
IS4 56	s	TH2A,COM		m	163/-60	1280	2.9	1.2	xx	R	U.Haut
IIIAa401	s	TH3,COM	g	m	139/-40	120°	87.	5.3	xx	R	Barr
IIIAa405	d	TH2A		m	148/-06	620°	8.2	1.5	xx	R	Barr
IIIAa455	d	TH3,TH2A	g	m	188/-13	170°	2.1	.84	xx	R	L.Apt
IIIA 608!	s	TH3,TH2A,TH1	g	m,sp	169/-51	620°	62.	.92	xx	R	U.Alb
IIIA 611!	l	TH3,TH2A,COM	g	h	178/-53	570°	4.8	1.0	oo	R	U.Alb
IIIA 615	d	TH3,TH2A,COM	g	h	207/-22	70°	1.2	.76	oo	R	U.Alb
IIIA 633	l	TH2A,TH1	h	sp	258/-47	NRM	.72		xo	R	U.Alb
IIIA 656!	l	TH3,TH2A,COM	g,h	m	137/-57	NRM	1.4		xx	R	U.Alb
IIIA 688	l	TH3A,COM	g	?	30/ 33	470°	2.2	.41	oo	N	U.Alb
IIIA 694	d	TH3,COM	g	m	153/-48	120°	.52	.05	xo	R	U.Alb
IIIA 700	l	TH3	g	?	165/-19	370°	1.6	.19	oo	R	U.Alb
IIIA 710	l	TH3	g	?	159/-23	520°	13.	1.2	xo	R	U.Alb
IIIA 730	l	TH3,TH2A	g	?	234/-27	620°	9.0	.50	xo	R	U.Alb
IIIA 731	s	TH3,TH2A	g	?	229/-25	NRM	3.2		xo	R	U.Alb
IIIA 740	s	TH3	g	?	212/-52	120°	5.2	2.3	xo	R	U.Alb
IIIA 762	s	TH3A	h	?	13/ 50	320°	1.0	.18	xo	N	U.Alb
IIIA 763	s	TH3A,COM	h	m	190/-29	NRM	.52		xx	R	U.Alb
IIIA 765	l	TH2,COM	h	?	177/-38	NRM	1.6		xo	R	U.Alb
IIIA 783	d	TH2A		?	322/ 42	370°	2.3	.61	xo	R	U.Alb
IIIA 787	d	TH3A,COM	h	m	149/-21	NRM	.99		xx	R	U.Alb
IIIA 790	d	TH3A,COM	g	?	142/-60	170°	6.4	.48	xo	R	U.Alb
IIIA 795	l	TH3,TH2A	g	?	21/ 27	370°	1.4	.51	xo	N	U.Alb
IIIA 809	s	TH3	g	?	14/ 57	620°	12.	.35	xo	N	U.Alb
IIIA 826	l	TH2A,COM		m	314/ 50	470°	.97	.20	xx	N	Cenom
IIIA 830	l	TH3,TH2A,COM	g	m	345/ 25	320°	3.1	1.0	xx	N	Cenom
IIIA 833	l	TH3,TH2A,COM	g,h	?	138/-41	NRM	6.2		xo	R	Cenom
IIIA 838	l	TH3	g	?	336/ 10	320°	27.	1.0	xo	N	Cenom
IIIA 840	l	TH3	g	?	349/ 48	320°	7.7	2.1	xo	N	Cenom
IIIA 847	l	TH2A	h	?	335/ 46	470°	2.9	.73	xo	N	Cenom
IIIA 875!	l	TH3,TH2A,COM	g	m	221/-21	120°	6.8	.38	xx	R	Cenom
IIIA 906	l	TH3,TH2A,COM	g	m	5/ 34	470°	.94	.12	xx	N	Cenom
IIIA 950	l	TH3,TH2A,COM	g	h	348/ 45	520°	20.	.24	oo	N	Cenom
IIIA 958	l	TH3	g	?	336/ 50	370°	.15	.09	oo	N	Cenom
IIIA 961	l	TH3	g	?	358/ 29	170°	8.1	1.4	xo	N	Cenom
IIIA1018	l	TH2A,COM		m	204/-36	525°	2.5	.31	xx	R	Cenom
IIIA1020	l	TH3,TH2A,COM	g	m	197/-50	70°	1.7	1.7	xx	R	Cenom
IIIA1076	d	TH3	g	?	327/ 52	370°	29.	.54	xo	N	Cenom
IIIA1123	l	TH2A,COM		m	224/-46	420°	3.0	1.9	xx	R	Cenom
IIIA1133	l	?		?	193/-22	NRM	.45		oo	R	Cenom
IIIA1135	l	TH2A		?	344/ 59	370°	3.6	1.7	xo	N	Cenom

Table 1. continued

1	2	3	4	5	6	7	8	9	10	11	12
IIIA1143	l	TH2A		?	348/ 38	NRM	2.3		xo	N	Turon
IIIA1150	l	TH2A,COM		m	340/ 46	490°	5.4	1.1	xx	N	Turon
IIIA1163!	l	TH2A,COM		m	349/ 44	490°	7.2	.53	xx	N	Turon
IIIA1167	d	TH3,TH2A,COM	g,h	?	216/-57	NRM	.52		--	R	Turon
IIIA1172!	d	TH3,COM	g	h	171/-47	470°	29.	.27	--	R	Turon
IIIA1176	d	TH2A,COM		h	185/-18	670°	3.1	.03	--	R	Turon
IIIA1185	d	TH3,TH2A,COM	g	h	107/-38	470°	1.4	.22	--	R	Turon
IIIA1191	d	TH3,TH2A,COM	g	h	174/-24	270°	1.8	.55	--	R	Turon
IIIA1195	d	TH3,COM	g	m	188/-04	320°	1.3	.31	xx	R	Turon
IIIA1202	d	TH3,COM	g	h	214/-25	520°	.72	.13	oo	R	Turon
Zem Zem Section											
IBA 86	l	TH3,TH2A,COM	g	h	220/-37	320°	1.2	.30	oo	R	Berr
IBA 111	d	TH3,COM	g	sp	182/-24	620°	9.1	.10	xx	R	Berr
IBA 115	d	TH3,TH2A,COM	g	m	167/-34	570°	49.	.20	xx	R	Berr
IBA 131	l	TH3,TH2A,COM	g	m	169/-20	720°	16.	.18	xx	R	Valang
IBA 132	l	TH3,TH2A,COM	g	m	171/-47	570°	40.	.16	xx	R	Valang
IBA 136	d	af1,TH3,TH1	g	sp	24/ 43	670°	3.3	1.3	xx	N	Valang
IBA 339!	s	af1,TH1		m,sp	164/-41	670°	35.	1.6	xx	R	U.Haut
IBA 348	d	af1,TH2A,COM		h	171/-03	370°	33.	.85	oo	R	U.Haut
IBA 372	d	af3,TH3,COM	g	m	180/-15	620°	120	.65	xx	R	U.Haut
IBA 378	l	af3,TH3,COM	g	?	194/-10	NRM	18.		xo	R	U.Haut
IBA 379!	s	TH2,COM	h	sp	169/-17	520°	17.	3.1	xx	R	U.Haut
IBA 393	s	TH3,TH2A,COM	g	m	214/-11	520°	9.3	3.4	xx	R	U.Haut
IBA 396	s	TH3,TH2A,COM	g	m	221/-56	620°	24.	1.7	xx	R	U.Haut
IIBA 55	l	TH3,TH3A,COM	g	?	102/-30	620°	8.2	.13	oo	R	U.Alb
IIBA 83	l	af3,TH3	g	?	222/-04	320°	1.1	.11	oo	R	U.Alb
IIBA 84	l	af3,TH3	g	?	227/-12	NRM	.43		oo	R	U.Alb
IIBA 159	s	af1,TH2A,COM		m	174/-29	420°	3.4	1.7	xx	R	U.Alb
IIBA 178	l	af1,TH2A,COM		m	147/-68	220°	2.0	1.8	xx	R	U.Alb
IIBA 181	d	TH3,TH2A,COM	g	h	332/ 18	420°	5.4	.17	oo	N	U.Alb
IIBA 194!	s	af3,TH1,COM		sp	194/-36	620°	18.	2.4	xx	R	U.Alb
IIBA 201	l	af3,TH3,COM	g	?	150/-50	520°	1.1	.18	oo	R	U.Alb
IIBA 203!	l	af3,TH3,COM	g	h	230/-24	320°	28.	.95	oo	R	U.Alb
IIBA 216	l	TH3,TH2A,COM	g	m	140/-43	NRN	1.3		xx	R	U.Alb
IIBA 235	l	TH3,TH2A,COM	g	m	165/-47	320°	1.0	.54	xx	R	U.Alb
IIBA 249	d	TH3,TH2A,COM	g	?	317/ 13	520°	.44	.10	oo	N	Cenom
IIBA 279	d	af3,TH3,COM	g	m	218/-20	NRM	1.8		xx	R	Cenom
IIBA 283	d	af3,TH3,COM	g	m	127/-48	NRM	4.5		xx	R	Cenom
IIBA 330	d	af3,TH3,COM	g	m	207/-22	470°	28.	.24	xx	R	Cenom
IIBA 398	d	af3,TH3,COM	g	?	109/-43	120°	9.7	.03	oo	R	Cenom
IIBA 403	d	af3,TH3	g	?	118/-43	120°	9.7	.03	oo	R	Cenom
IIBA 421	d	af3,TH3	g	?	174/-27	NRM	.48		oo	R	Cenom
IIBA 433	d	af3,TH3,TH2A	g	?	145/-34	NRM	3.2		xo	R	Cenom
IBB 12!	d	af1,TH2A,COM		m	86/-26	520°	3.2	1.1	xo	R	Turon
IBB 18!	d	af1,TH2A,COM		m	325/ 25	420°	2.8	1.6	xx	N	Turon
IBB 23!	l	af3,TH3,COM	g	h	179/-36	220°	8.2	.20	--	R	Turon
IBB 55	d	af1,TH2A,COM		m	353/ 34	520°	1.9	.55	xx	N	Turon
IBB 60	d	af2,TH2A,COM		m	174/-28	470°	4.1	1.7	xx	R	Turon
IBB 80	l	af3,TH3,COM	g	m	223/-56	370°	4.8	.12	xx	R	Turon
Afensou Section											
4A 28	s	TH2A,TH1	h	sp	142/-28	670°	17.	2.2	xx	R	?
4A 100	s	TH2A,TH1	h	sp	331/ 40	670°	22.	3.7	xx	N	Haut
4A 110	s	TH2A,TH1	h	sp	334/ 24	NRM	15.		xx	N	Haut

Table 1. continued

1	2	3	4	5	6	7	8	9	10	11	12
4A 112	s	TH2A,TH1	h	sp	334/ 23	NRM	8.0		xx	N	Haut
SW1 119	s	TH1		sp	122/-03	NRM	24.		xo	R	U.Haut
SW1 119,2	s	TH1		sp	147/-13	320°	14.	13.	xx	R	Barr
SW1 123	s	TH1		sp	353/ 30	670°	25.	20.	xx	N	Barr
SW1 124	s	TH1		sp	324/ 27	670°	25.	11.	xx	N	Barr
SW1 148	l	TH2A,COM		m	351/ 28	520°	50.	5.6	xx	N	Apt
SW1 204	l	TH2,TH2A	h	?	325/ 56	520°	2.7	.49	xo	N	?Apt
SW1 218	l	TH2,TH1	h	sp	81/-21	420°	1.6	.53	xo	R	?Alb
SW1 220	l	TH3A,TH1	h	sp	221/-47	NRM	.99		xx	R	?Alb
SW1 223	l	TH3A,TH1	h	sp	217/-40	NRM	1.4		xx	R	?Alb
SW1 312	d	TH3A	h	?	332/ 46	NRM	2.3		xo	N	U.Alb
SW1 320	d	TH3,TH3A	g,h	?	322/ 39	250°	1.8	.88	xo	N	U.Alb
SW1 323	d	TH3,TH3A,COM	g,h	m	212/-67	320°	4.4	.60	xx	R	U.Alb
SW1 338	d	TH2A		m	192/-41	NRM	1.4		xx	R	U.Alb
SW1 346	d	TH3,TH3A,COM	g,h	m	325/ 16	320°	4.8	.80	xx	N	Cenom
SW1 384	d	TH2A,COM		m	336/ 34	NRM	6.8		xx	N	Cenom
SW1 388	d	TH3,TH3A,TH1	g,h	sp	221/-25	270°	.96	.20	xx	R	Cenom
SW1 488	d	TH3A,TH2A	h	?	331/ 36	420°	5.5	1.7	xo	N	Cenom
SW1 530	d	TH2	h	?	12/ 37	470°	13.	2.2	oo	N	Cenom
SW1 605	l	TH2A,COM		m	336/ 49	420°	1.2	.14	xx	N	Turon
SW1 640	l	TH3,TH2A,COM	g	h	359/ 33	220°	4.1	.43	oo	N	Turon
SW1 648	d	TH3,TH2,COM	g	h	160/-30	420°	1.3	.13	--	R	Turon
SW1 650	d	TH3,COM	g	h	165/ 04	370°	2.8	.05	--	R	Turon
Aouzzer Section											
3F 22	s	TH2A,TH1,COM		m,sp	111/-61	620°	56.	1.7	xo	R	?
3F 69	s	TH1,COM		sp	231/-56	620°	7.5	5.2	xx	R	?
3F 75	s	TH1,COM		sp	120/-01	620°	11.	7.7	xo	R	?
3F 108	s	TH1,COM		sp	149/-43	NRM	2.4		xx	R	U.Haut
3F 119	d	TH2A,COM		m	200/-62	570°	.67	.28	xx	R	Barr
3F 165	l	TH2A		?	9/ 48	120°	8.2	6.4	xo	N	?Apt
3F 169	s	TH1		sp	238/-46	620°	2.3	1.8	xx	R	?Apt
3F 171	s	TH3A,TH1	h	sp	181/-37	470°	3.3	1.0	xx	R	Apt
3F 278	d	TH3A,TH2A	h	?	210/-42	370°	4.5	.87	xo	R	U.Alb
3F 287	d	TH2A		?	315/ 15	320°	8.8	1.0	xo	N	Cenom
3F 297	d	TH3,TH2A	g	?	321/ 11	170°	1.1	.51	xo	N	Cenom
3F 322	d	TH3	g	?	345/ 25	520°	6.9	.09	xo	N	Turon
3F 326	d	TH2A		?	333/ 19	570°	1.1	.90	xo	N	Turon
3F 349!	d	TH3,TH2A	g	?	180/-21	170°	3.3	1.9	xo	R	Turon
3F 351!	d	TH3,TH2A	g	?	162/-41	670°	13.	2.5	xo	R	Turon
3F 354!	d	TH2A		?	182/-31	670°	10.	3.0	xo	R	Turon
3F 357!	d	TH2A		?	3/ 49	570°	2.5	.26	xo	N	Turon
3F 363	d	TH3	g	?	173/-44	270°	1.9	.68	xo	R	Turon

◄ References to Table 1

1: specimen number (see Figs. 5 - 8); 2: rock type (s: sandstone; l: limestone; d: dolomite); 3: demagnetization types (! behind specimen number indicates related curves and diagrammes in Figs. 1 - 4); 4: carriers of secondary components (g: goethite; h: hematite pigment); 5: carriers of stable components (h: hematite pigment; m: magnetite; sp: specularite); 6: direction of stable component; 7: peak alternating field (OE) or temperature of demagnetization related to column 6; 8: RM intensity before demagnetization in 10^{-7} Gauss; 9: RM-intensity after demagnetisation in 10^{-7} Gauss related to column 7; 10: reliability of syndepositional origin of stable component (xx: reliable; xo: ambiguous due to unknown carrier of remanence or to direction not sufficiently aligned with the paleofield; oo: unreliable due to hematite pigment, indeterminable demagnetization type or intensities smaller than 10^{-8}Gauss; --: unreliable due to hematite pigment and correlation between change of polarity and geochemistry; 11: polarity; 12: age.

Table 2. Geophysical, stratigraphical, paleomagnetical and statistical parameters of Morrocan Sections

Section	Stages	E	Dm	Im	$\alpha 95$	k	Northpole	Fig.10
Tamrakt-Ascouti	Cen-Turon	35	358	41	6.9	11.84	17 W 82 N	5
(30°31'N, 9°27'W)	Apt-Alb	26	8	43	9.7	8.03	117 E 81 N	6
(30°34'N, 9°28'W)	Haut-Barr	20	352	45	20.2	2.40	128 W 82 N	7
	Berr-Val	47	344	32	10.7	3.68	139 W 71 N	8
Zem Zem	Cen-Turon	36	347	21	11.6	4.02	154 W 66 N	1
(31°30'N, 9°21'W)	Apt-Alb	35	348	28	8.8	7.61	155 W 70 N	2
	Haut-Barr	22	350	29	8.3	12.81	157 W 72 N	3
	U.Tith-Val	16	358	40	14.3	5.97	175 W 81 N	4
Afensou	Cen-Turon	22	352	37	12.1	6.13	153 W 78 N	9
(30°39'N, 8°49'W)	Haut-Alb	37	343	36	6.9	11.06	129 W 71 N	10
Aouzzer (33°13'N, 8°06'W)	?Tith-Turon	50	337	35	8.4	5.59	126 W 65 N	11

E: number of units; Dm: mean of declination; Im: mean of inclination; $\alpha 95$: semiangle of 95% cone of confidence; k: precision parameter

22 Stratigraphy, Facies, and Significance of Late Mesozoic and Early Tertiary Sedimentary Rocks of Fuerteventura (Canary Islands) and Maio (Cape Verde Islands)

Alastair H. F. Robertson[1] and Daniel Bernoulli[2]

ABSTRACT

Mesozoic deep water sedimentary rocks uplifted and exposed in basement complexes on the islands of Fuerteventura (Canary Islands) and Maio (Cape Verdes) help document the early Atlantic ocean and the volcanic history of these islands.

On Fuerteventura ca. 1.5 km of terrigenous quartzose clastics, black shales, and subordinate redeposited limestones of early Cretaceous age are overlain first by Albian-Cenomanian marls and clastics, showing extensive soft sediment slumping, then after a hiatus by Senonian nannofossil chalks with chert nodules. After a further break alkalic submarine volcanic rocks, largely hyaloclastics, were erupted, then overlain by redeposited bioclastic limestones and volcaniclastics of Oligocene age. A dyke complex was intruded, then the basement complex was uplifted, peneplaned, and overlain by Neogene plateau lavas.

On Maio the basement complex comprises mafic pillow lavas of mid-ocean ridge character overlain by up to 400 m of deep water sedimentary rocks. Upper Jurassic to lower Cretaceous radiolarian-rich pelagic limestones are succeeded by middle and upper Cretaceous sequences of mudstones, redeposited limestones, chalks, tuffs, and volcaniclastics which document the onset of volcanism. During a sedimentary hiatus plutonic rocks were intruded, uplifted, then eroded to produce conglomerates of late Miocene age. These are overlain by several series of gently dipping subaerial alkaline basalt flows.

Comparisons with D.S.D.P. sites, oil wells and on-land geology confirm that the lower Cretaceous Fuerteventura quartzose clastics formed part of a deep water fan complex located close to the Atlantic continent-ocean boundary. Comparable flysch is known in the Moroccan basin, in the Betic-Maghrebide system and the feeding Tan-Tan-Cape Bojador delta system. The inferred disconformity between Cenomanian and Senonian is part of a widespread depositional hiatus associated with slumping on the adjacent west African continental margin. On Maio pelagic limestones comparable with the Tethyan Maiolica facies were deposited on a subsiding ?middle/late Jurassic ocean ridge before uplift related to onset of volcanism probably in the late Cretaceous.

1. Introduction

Mesozoic sedimentary rocks of deep-water facies crop out extensively along with Tertiary and some Mesozoic igneous rocks to form the islands of Fuerteventura in the Canary Islands and Maio in the Cape Verde Islands (Figs. 1, 2). These rocks have a twofold significance: first, they formed part of the oldest At-

1 Grant Institute of Geology, University of Edinburgh, W. Mains Street, Edinburgh, Great Britain

2 Geologisches Institut der Universität Basel, Bernoullistr. 32, CH-4056 Basel, Switzerland

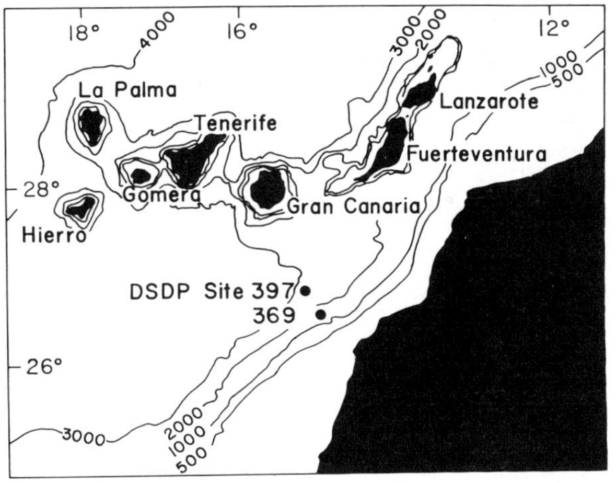

Fig. 1. Location map of Fuerteventura, Canary Islands

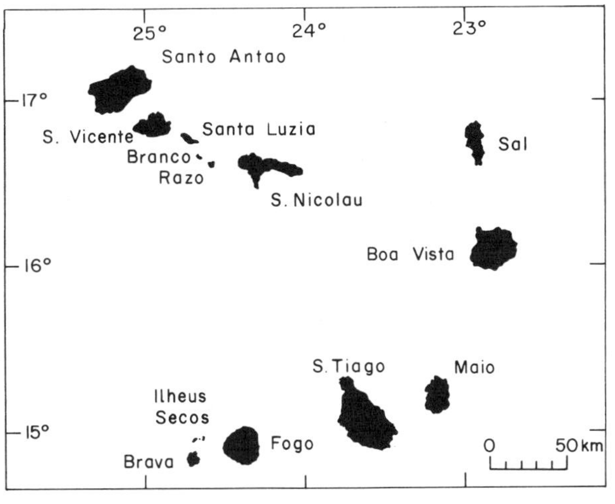

Fig. 2. Location map of Maio, Cape Verde Islands

lantic Ocean which connected the Caribbean area with the Tethys and thus supplement data on the eastern Atlantic continental margin and adjacent oceanic crust; secondly, they document the early volcanic history and uplift of these islands. Particularly, the position of Fuerteventura in the southwestern projection of the Alpine belt of North Africa means that both Alpine and Atlantic tectonics may be involved. In this paper we review available data on the Mesozoic rocks of Fuerteventura and Maio in the light of information from the Deep Sea Drilling Project (D.S.D.P.), off-shore petroleum wells, and the on-land geology. Mesozoic deep-water sedimentary rocks may also be present on Gomera (Canary Islands, Cendrero, 1971) and Sao Tomé (Gulf of Guinea, Grunau et al., 1975), but little data are available.

2. Geological Setting and History of Research

2.1 Fuerteventura

Knowledge of the presence of sedimentary rocks of Fuerteventura (Fig. 1) dates back to the mid-19th century. On Fuerteventura, von Fritsch (1867) and Gagel (1910) proposed a Paleozoic age for black shales, siltstones, and brownish crystalline limestones intercalated with basic igneous rocks. The interpretation dating from Gagel (1910), subsequently developed by Hausen (1958) and Rothe (1968), was that Fuerteventura, like the other Canary Islands, originated as a volcanic pile constructed on the edge of the African continent. This margin later disintegrated as a result of tectonic fracturing and igneous activity to produce the present islands.

Following the suggestion by Gastesi (1973) that the oldest rocks of Fuerteventura on the west coast (Fig. 3) might be an uplifted portion of Atlantic ocean floor, the pre-Miocene rocks have been extensively reinvestigated (Bennel-Baker et al., 1974; Stillman et al., 1975; Stillman and Robertson, 1977; Robertson and Stillman, 1979a,b). It is now clear that the thick pile of subhorizontal Neogene plateau basalts (Fuster and Aguilar, 1965; Fuster et al., 1968) unconformably overlies a lithologically diverse assemblage, termed the Fuerteventura Basal Complex (Fuster et al., 1968; Bennel-Baker et al., 1974). The Basal Complex is exposed extensively on the western side of the island in a number of inliers of which the largest, the Betancuria Massif, has a total outcrop area of ca. 700 km^2 (Fig. 3). The Basal Complex is dominated by an extensive dyke-in-dyke swarm in which the majority of dykes trend

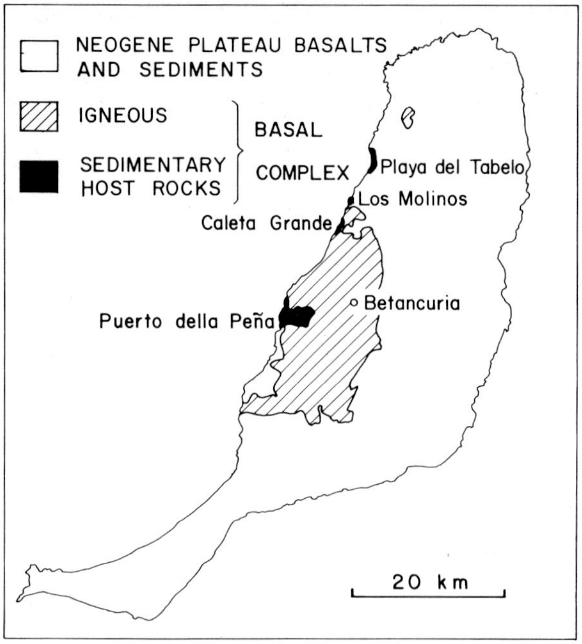

Fig. 3. Outline geological map of Fuerteventura to show location of the Mesozoic and Tertiary sedimentary outcrops

NNE–SSW; dyke intensity ranges from locally 30 % to 90 % of the total outcrop. Where visible, the host rocks to the dykes comprise Mesozoic and minor Tertiary sedimentary and submarine volcanic rocks (Fuster et al., 1968; Robertson and Stillman, 1979a,b) which were affected by thermal metamorphism of low greenschist grade (chlorite-epidote-quartz-albite).

The main dyke phase was closely associated with the emplacement of ultrabasic and alkaline plutons which largely obscure the host rocks. A variety of lines of evidence, particularly available age data (Rona and Nalwalk, 1970; Abdel-Monem et al., 1971; Grunau et al., 1975; Stillman et al., 1975), suggest that the main phase of intrusion took place over a period of ca. 22 m.y. ranging in age from late Eocene to early Miocene. Whereas deformation in the sedimentary host rock indicates major compression, possibly before the Oligocene, the dykes imply subsequent rifting by at least 10 km and possibly more (Laubscher, 1975).

Mesozoic rocks on Fuerteventura were not confirmed until Lehmann (in Rothe, 1968) identified planktonic foraminifera of Albian age (see below). The first detailed account of sedimentary features was given by Rothe (1968). Subsequently, Robertson and Stillman (1979a,b) discussed the stratigraphy, sedimentology, and structure of the Fuerteventura Mesozoic rocks in their regional context. New biostratigraphic data were obtained by Bernoulli et al. (in preparation), these and additional observations are also given here.

2.2 Maio

According to Serralheiro (1968,1970), Klerkx and de Paepe (1971), de Paepe et al. (1974) and Grunau et al. (1975) the central part of Maio (Fig. 2) comprises an "eruptive complex" composed of lava breccias and flows, cut by dykes and small intrusive bodies, including essexites and associated syenites (Fig. 4). More detailed new work by Furnes et al. (1981) shows that the "eruptive complex" is a basement complex, like that of Fuerteventura, containing bedded volcanic and sedimentary rocks. The oldest of these rocks are mafic pillow lavas with a mid-ocean ridge chemical character and are overlain by pelagic chert-bearing limestones of late Jurassic and early Cretaceous age. Volcanogenic sediments which stratigraphically overlie document subsequent stages of volcanism and uplift. Later intensive intrusion of mafic alkaline sills and plutonics was associated with strong uplift and possibly consequent folding, gravity tectonics, and erosion. The Tertiary volcanics, which are mostly subaerial, show an alkaline basic trend, and include strongly undersaturated lavas and carbonatite dykes. Serralheiro (1968, 1970) stated that there is transitional contact between his "central eruptive complex" and the stratigraphically lowest of the adjacent Mesozoic sedimentary rocks (Shale-limestone-eruptive complex). In contrast, the new work by Furnes et al. (1981) shows that the sediments and associated volca-

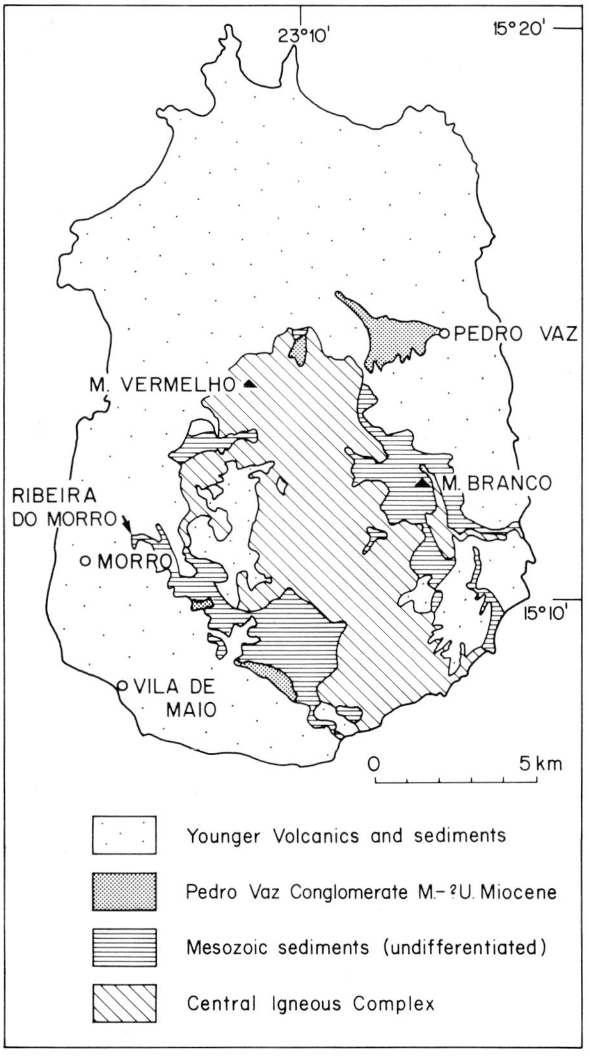

Fig. 4. Simplified geological map of Maio, modified from Serralheiro (1970)

nics entirely predate the intrusive rocks of the "central eruptive complex" and that these latter formed by later intrusion of dykes, sills, and plutonic bodies. The Mesozoic sedimentary rocks dip outward from the center of the island and are then overlain by igneous-derived conglomerates of late Miocene age (Rigassi, 1975).

The Mesozoic rocks on Maio were first recognized by Friedländer (1912, 1913) and Hennig (1913) by the discovery of aptychi in the Ribeira do Morro in the W of the island. Since then a protracted debate has taken place on the precise age and stratigraphic significance of the fossils from this locality (Mitchell-Thomé, 1974, 1976): Stahlecker (1934) divided the Mesozoic limestones of the Ribeira do Morro based on ammonite faunas into 14 horizons ranging from Hauterivian to Albian. This early Cretaceous age was confirmed by Trauth (1936, 1938) who determined Neocomian aptychi from the same section. In contrast to this, a late Jurassic age was claimed for the Maio limestones or part of them by Soares (1944-47, 1952, 1953), Serralheiro (1968), by Furon and by Saldanha (both in Mitchell-Thomé, 1979). Their age allocations, however, were based mainly on new forms or generic determinations and their arguments are therefore not convincing. Serralheiro (1970) found additional outcrops of comparable Mesozoic rocks on both the east and west flank of his "central eruptive complex" and scattered in other parts of the island. Despite the relatively well-studied state of the Ribeiro do Morro beds in the northwest, he chose to map these in an "Undifferentiated Mesozoic" (upper Jurassic and lower Cretaceous) category, while apparently similar outcrops in the eastern part of the island were regarded as upper Jurassic (Portlandian). "Eocretaceous" units (shales, marls, marly limestones, and light colored limestones with flints) were recognized by Serralheiro (1970) in both the major eastern and western outcrops. Meanwhile, a middle to late Jurassic age has been proposed for part of the Monte Branco section in the east by Rigassi (1975 and pers. comm., 1980) by characteristic microfossil assemblages. The data, however, need to be reassessed in the light of recent discoveries of reverse faults (thrusts) which duplicate some of the sedimentary sequence adjacent to Monte Branco (A. Robertson, unpubl. data).

The postulated occurrence of upper Jurassic (Maync, 1959), Senonian (Bebiano and Soares, 1951) or middle Eocene (Soares, 1948) strata on the island of S. Nicolau (Fig. 2) has not been confirmed. The inferred Senonian or middle Eocene formation

(Monte Focinho) appears to be middle Miocene in age (Rigassi, 1975), whereas the occurrence of a late Jurassic benthonic shallow-water foraminifer in a crystalline limestone (Ponta do Escorregadeiro, Maync, 1959) is, if not redeposited, at odds with the general paleogeographic setting. Alternatively this sample might be derived from a ballast pebble from Brazil, Africa, or Portugal, as suggested by Rigassi (1975).

3. The Sedimentary Succession of Fuerteventura

The pre-volcanic sedimentary rocks of Maio and Fuerteventura have not been classified in terms of modern stratigraphical usage largely because the density of intrusion complicates detailed measurements and correlation of outcrops.

Based on an investigation of the exposures shown in Fig. 5, located between Puerto de la Peña and Caleta de la Peña Vieja on the west coast of Fuerteventura, Rothe (1968) recognized the following "series": (1) Dark shale and marl with sandy beds, mostly thin-bedded; (2) Limestone-marl, parallel-bedded; (3) Dark shale with thin beds of red and gray fine-grained limestone; (4) Sand and siltstone with clay intercalations;

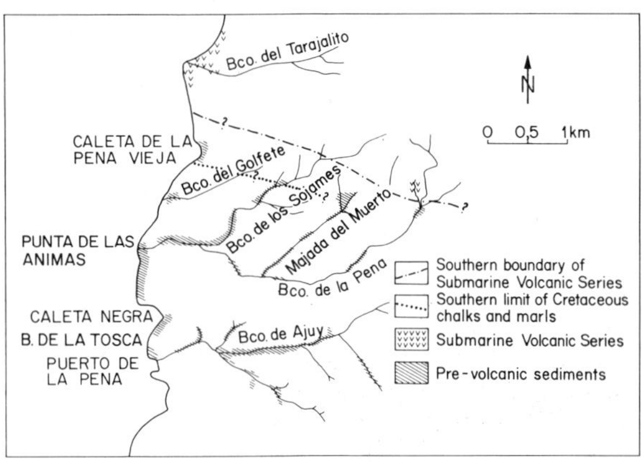

Fig. 5. Map of the Cretaceous sedimentary sequences of the southern coastal exposures, Fuerteventura (from Robertson and Stillman, 1979a)

(5) Very fine-grained limestone with quartz silt and chert nodules.

Rothe (1968) stated that his series 1–3, ca. 500–700 m thick were disposed in a major east–west-trending isoclinal overturned syncline. Sequences in the Majada del Muerto and Majada de Perra were thus correlated (Fig. 5). The evidence for folding rests on a marked swing in strike in the fold axial region. In the course of re-mapping the area, Robertson and Stillman (1979a), while confirming that the beds were predominantly overturned, also discovered that the succession becomes younger uniformly to the north. This was based on sedimentary structures in re-deposited sediments, including grading, small-scale cross-lamination and ripples. Robertson and Stillman (1979a) recognized a succession of seven units, described below, roughly corresponding to Rothe's (1968), but without involving repetitions of the lower series. Any repetition of the sequence could only be by pre-volcanic thrusting which has not been recognized.

Grunau et al. (1975), for their part, did not recognize the overturned nature of the sequence between Puerto de la Peña and Caleta de la Peña Vieja. They therefore allocated a post-Albian, late Cretaceous age to the predominantly terrigenous part of the sequence south of Caleta de la Peña Vieja (their Fig. 15). This is clearly contradicted not only by the sedimentological evidence mentioned above, but also by the discovery of an early Cretaceous ammonite in the Barranco de la Peña (Bernoulli et al., in preparation).

North of the Caleta de la Peña Vieja where Rothe's (1968) map ends, Robertson and Stillman (1979b) identified a submarine volcanic sequence composed of lava breccia and mafic pillow lavas interbedded with volcaniclastic arenites, lutites and redeposited bioclastic limestones and marls near the top of the sequence.

Robertson and Stillman (1979a) also described a second important occurrence of pre-volcanic rocks further north along the west coast of Fuerteventura, ca. 25 km north of Puerto de la Peña (Fig. 3), which comprise arenaceous lithologies of coarser grain than any seen in the southern sequences. These beds become younger southward, but when traced southward, they are unfortunately lost in a dyke swarm of 100 % intensity, and it is not possible to relate the beds stratigraphically in the two sequences.

3.1 Southern Sequences

Rothe (1968) gives a detailed description of the individual exposures, which is supplemented by Robertson and Stillman (1979a) and by new data here. The generalized sedimentary succession is shown in Fig. 6.

A) Basal Unit. No base of the pre-volcanic sedimentary sequence is known from Fuerteventura (cf. Maio, see below). The lowest beds exposed are epidotized black shales, gray siltstones, sandstones, and marbles which flank the northern margin of the major basic plutons of the Betancuria massif (Fig. 5).

B) Calcarenite-Sandstone-Shale Unit. Moving northward, up through the sequence, there is an incoming of relatively thick limestone beds within the siliciclastic-black shale-calcilutite sequence. Rothe (1968) recognized these limestones in the Barranco de Ajuy (Fig. 5), but they are best exposed slightly up-section along the coast from Bajas de la Tosca to Caleta Negra (Fig. 5). Rothe (1968) reported "impressions" of Posidonia in Barranco de Ajuy. Robertson and Stillman (1979a) describe the calcarenite beds as being amalgamated, composed of several sub-units up to 0.3 m thick. These contain matrix-supported clasts of feldspar-phyric volcanic rocks, up to 0.2 m in diameter, which are well rounded. In thin section, although the calcarenites are largely recrystallized, echinoid plates, shells, algae, and benthonic foraminifera can still be recognized. In addition, poorly preserved ooids, pisoliths, bryozoa and limestone intraclasts are also noted. The volcanic rock is trachybasalt, thus the oldest igneous rock known on Fuerteventura.

By strong contrast, the amalgamated calcarenites are inter-bedded with fine- to medium-grained yellow to pale gray quartzarenites and quartzlutites. They show the Bouma B-C-D-E, C-D-E and D-E divisions of classic turbidites (Bouma, 1962).

C) Calcilutite-Siltstone-Shale Unit. Unit B containing the thick bedded calcarenites passes transitionally upwards into several hundred meters of dark-gray to black silty shales (Fig. 6) interbedded with massive or weakly parallel or cross-laminated thin bedded calcilutites. Occasionally Zoophycos-type burrows occur. In thin section these limestones consist mostly of microspar recrystallized from micrite, with scattered unrecognizable shell fragments.

D) Sandstone-Siltstone-Shale Unit. Traced up-section along the coast, the limestone interbeds gradually thin, disappear and give way to ca. 1000 m of siliciclastic arenites and rudites interbedded with black shales and silts. Rothe's (1968) map implies the presence of an unconformity between his "Dark shales with layers of red and gray fine-grained limestone series" and his overlying "Sand and siltstone series". By contrast Robertson and Stillman (1979a) mapped a swing in strike south of Punta de las Animas (Fig. 5), implying that the E–W juxtaposition of Rothe's series is tectonic, not sedimentary. Similar siliciclastic arenites and lutites are present both above and below the "unconformity" which could not be confirmed.

A particular feature of the lower part of the Sandstone-siltstone-shale unit, remarked upon by both Rothe (1968) and Robertson and Stillman (1979a), is the presence of intercalations of massive or graded laterally continuous orthoquartzites up to 1.1 m thick. The graded quartzarenites show well-developed Bouma A-B-C-D-E divisions and combinations. Bottom structures include well-developed flute- and grove-casts. After rotation to horizontal, assuming simple deformation as in cylindrical folds, the paleocurrents indicate derivation from the east and north-east. A pro-

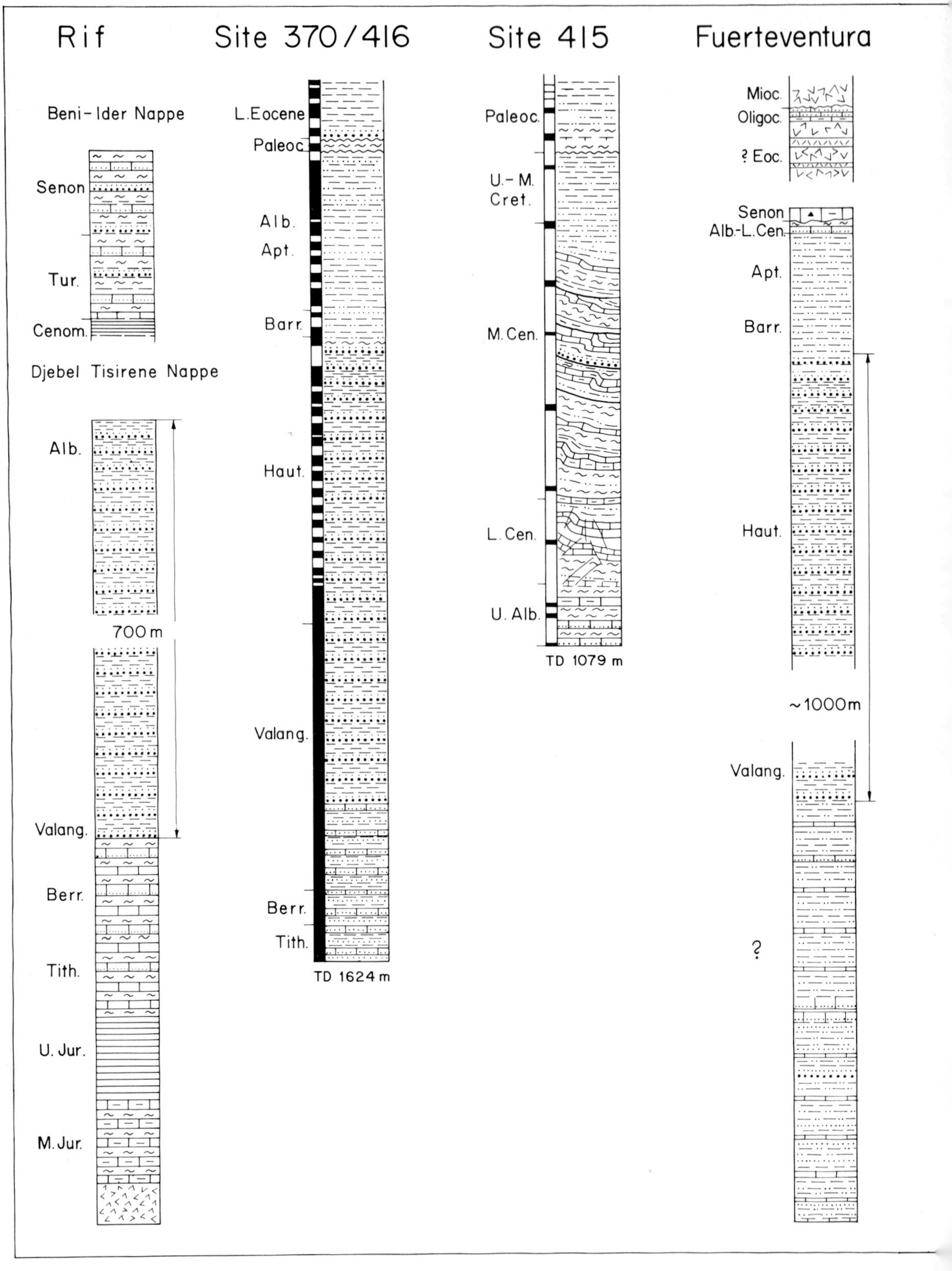

Fig. 6. Columnar sections of Mesozoic sequences along the eastern Atlantic continental margin. Sources: Andrieux (1971); Furnes et al. (1981); Lancelot,

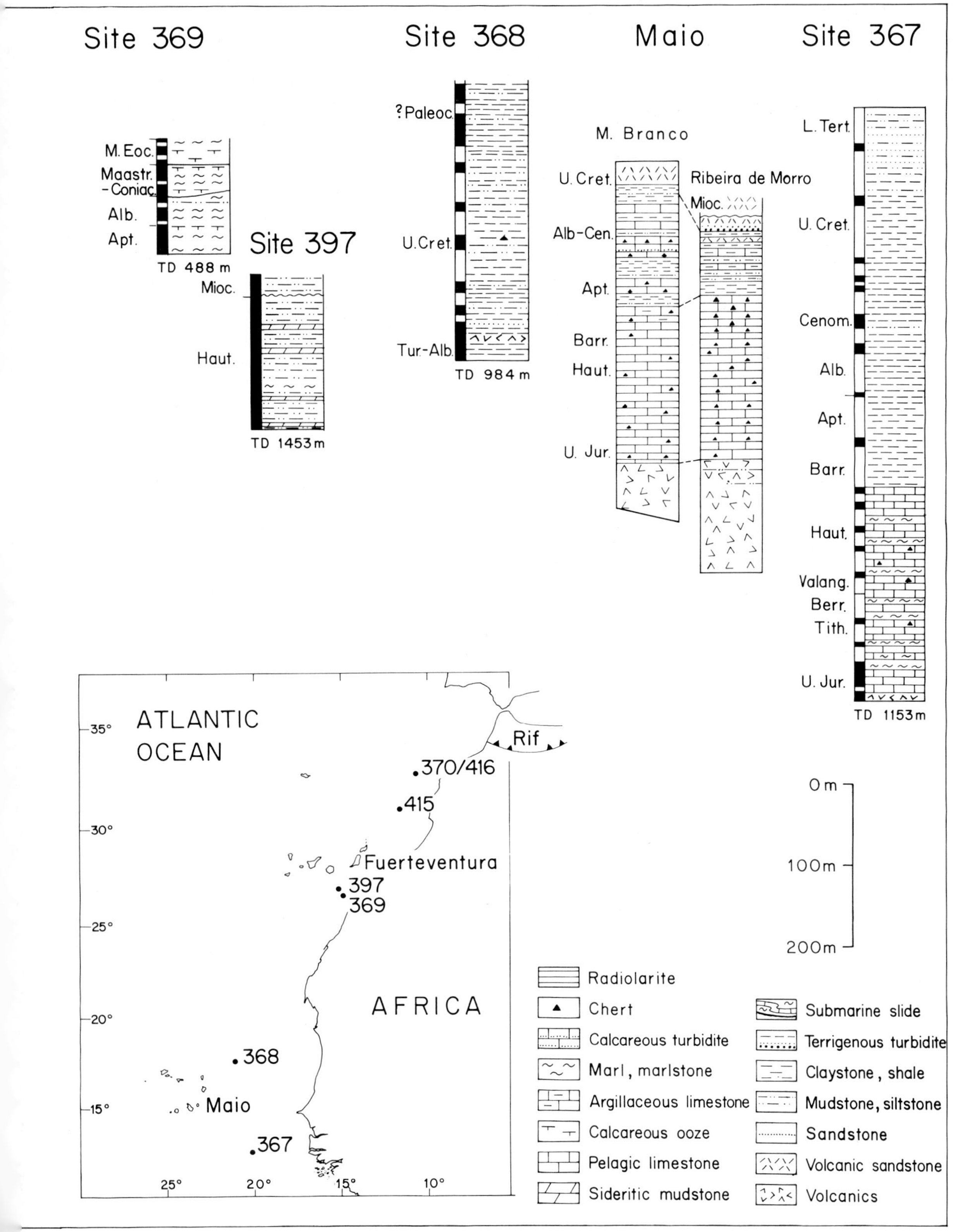

Seibold et al. (1978); von Rad, Ryan et al. (1979); Lancelot, Winterer et al. (1980); Robertson and Stillman (1979a). Black bars left of column represent cored intervals

Fig. 7. Lower Cretaceous deep-water clastics, Fuerteventura and Moroccan Basin: alternations of cross-laminated to parallel-laminated arenites with intercalated gray mudstones. A and C Unit D, Valanginian, Barranco della Peña, Fuerteventura. B Hauterivian, DSDP-41-370-32-4, 128–136 cm, Morrocan Basin. D Flute casts at the base of a thin-bedded turbidite, Unit D, Valanginian, Barranco della Peña, Fuerteventura. E Valanginian, DSDP-50-416A-15-3, 84–97 cm, Moroccan Basin. Diameter of coin and scale bars 2 cm

nounced feature of the thick Sandstone-siltstone-shale unit is the presence of repeated thickening- and coarsening-upward cycles up to 25–30 m thick. In the lower parts of individual cycles the beds average 0.18 m thick, then thicken upward to average 0.4–0.6 m. Near the base of the cycles individual quartzarenites show predominantly C-D-E and D-E divisions, while in the upper part of the cycles the Bouma A-B-C-D-E, and A-B-C divisions are also present. The thinner-bedded parts of the cycles typically show evidence of extensive reworking by bottom currents, as indicated by abundant small-scale low-angle cross-lamination, ripples, small-scale channeling on scales of a few cm (Fig. 7).

Petrographically, the quartzarenites of both the northern and southern outcrop area are mostly lithic graywacke and feldspathic wacke as defined by Pettijohn (1975). Typically quartz predominates (70 %–90 %) with subordinate alkali feldspar, perthite, and plagioclase in the andesine-labradorite range. There are also grains of microcline, muscovite, epidote, tourmaline, zircon, and other heavy minerals. Intra-formational clasts, mostly of siliciclastic rocks, are noted. Texturally, the sandstones are mostly immature to sub-mature, often with a bimodal grain-size distribution. The matrix is generally calcareous, ranging up to 35 % in the impure argillaceous quartzlutites. The thicker-bedded quartzarenites often show well-defined quartz overgrowths over iron-stained rounded quartz grains. In thicker beds the matrix is typically finely crystalline chalcedonic quartz.

In the lower part of Unit D (Barranco della Peña), an ammonite has been found which could be determined as Neocomites sp. with a Valanginian affinity (O. Renz in Bernoulli et al., in prep.). A general Valanginian to Hauterivian age of Unit D is supported by lithological correlation with the early Cretaceous sequence at Site 370/416 of the Deep Sea Drilling Project (Lancelot, Winterer et al., 1980, our Fig. 6).

Rothe (1968) described his equivalent to the Sandstone-siltstone-shale unit (his sand- and siltstone series) as extending as far as Barranco del Golfete (Fig. 5). Systematic description in the shore sections north of this is hindered by the sporadic nature of the exposures, which can be reached only by boat.

E) Homogenous Siltstone Unit. This stratigraphically higher unit is well exposed inland in the higher reaches of the Barranco de los Sojames (Fig. 5) and can also be seen in small outcrops approximately on strike in the Majado del Muerto to the east. High degrees of dyke density, together with limited exposure, preclude any attempt to recognize significant lateral variations. Vertical variation can, however, be observed in the Barranco de los Sojames where the quartzarenites become finer and thinner upward, giving way to finely laminated quartzlutites, black shales, and siltstones. The highest exposed levels consist of virtually homogenous cherty noncalcareous quartzlutites, showing minute convolute laminations and nodules of quartzose replacement chert, up to 0.2 m in diameter. A Barremian to Aptian age is tentatively assigned to Unit E (Fig. 6).

F), G) Marl-Calcarenite Unit and Thicker Bedded Marl and Chert Unit. The transition from the underlying siliciclastics to the overlying Albian to upper Cretaceous pelagic carbonates is best exposed along the almost inaccessible coast south of Caleta de la Peña Vieja (Fig. 5) (series 5 of Rothe, 1968). The lower part of the pelagic sequences comprises ca. 30 m of white to pale gray chalks and marls with subordinate inter-bedded calcarenites and occasional calcirudites. In contrast to the underlying siliciclastic succession, the Albian rocks are highly bioturbated, with numerous small (to 0.02 cm long) vertical burrows which are often pyritized. The calcarenites are graded and show basal concentrations of black shale intraclasts, phosphatic fragments, shell fragments, and occasional fish teeth. Red siltstones up to 0.03 m thick are also present as partings. Several bedding planes contain anastomozing clumps of ferruginous concretions, each up to several centimeters in diameter. Up-section the calcarenites increase in thickness and grain-size, with intercalations of rudites composed mostly of subrounded grains of black chert up to several centimeters in diameter. Typically the rudites form the basal 5–15 cm of beds which grade up into gray calcarenites then impure calcilutites. Some beds consist of densely packed intraclasts of various chalk lithologies, as seen in the underlying sequence. Many horizons show partial Bouma sequences, convolute lamination, and evidence of slumping. Many beds are distinctly lumpy; intraclasts often show diffuse "fuzzy" margins and phacoidal deformation (Voigt, 1962) suggestive of redeposition while still plastic (Fig. 8).

Up-section there is a relatively abrupt change to ca. 30 m of much paler white and pale gray marls and chalks which are more homogeneous than their counterparts below (Unit G). Coarser-grained material is confined to several centimeter-thick partings at the base of relatively thick marl and chalk units up to 2.3 m thick. The lower beds are typically mottled with small scattered intraclasts of calcilutite and quartzose siltstone up to 3 cm in diameter, also other features again indicative of slumping. The highest beds exposed at Caleta de la Peña Vieja are mostly white or pale pink chalks and marls with scattered no-

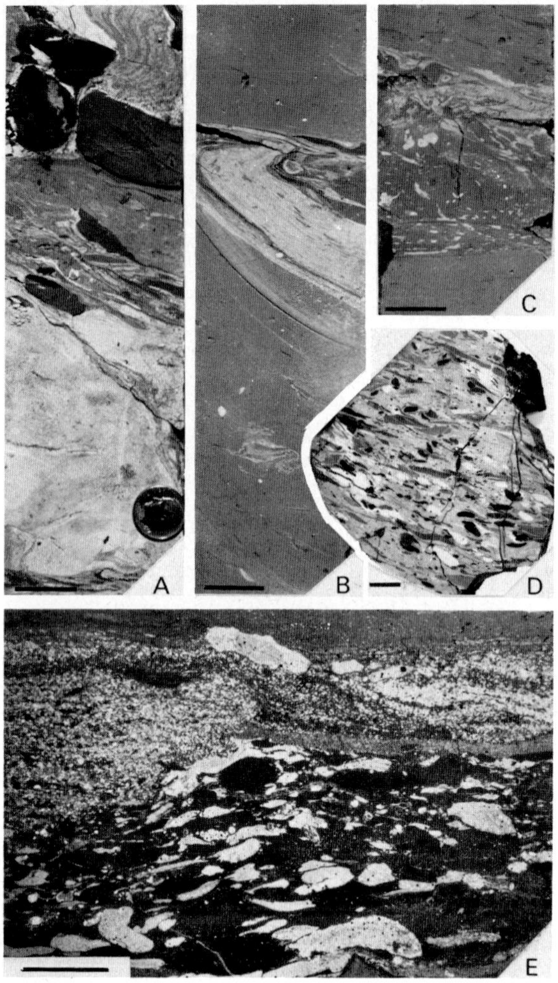

◄Fig. 8. Upper Cretaceous facies on Fuerteventura and at Site 369, D.S. D.P. continental slope off Cape Bojador. A Submarine slump composed of various clasts of chalk, nannofossil bearing marls and clays in a fluidally deformed marly matrix. Coniacian to Santonian. The slumped strata illustrated in A to C directly overlie dark-colored, laminated nanno-bearing marls of Albian age. DSDP-41-369A-40-2, 105-125 cm. Scale bar 2 cm. B Slumped marls and chalks. Coniacian to Santonian. DSDP-41-369A-41-2, 2-22 cm. Scale bar 2 cm. C Submarine slump composed of chalks, nannofossil marls, and clays. Note deformed chalk pebbles in fluidally deformed marly matrix. Coniacian to Santonian. DSDP-41-369A-41-2, 80-90 cm. Scale bar 2 cm. D Lithic rudite with siltstone, mudstone and pelagic limestone pebbles showing plastic deformation by submarine slumping, in a matrix of pelagic limestone. Upper Cretaceous. Caleta de la Peña Vieja, Fuerteventura. Polished surface scale bar 2 cm. E Lithic rudite with plastically deformed siltstone, mudstone and limestone pebbles overlain by cross-laminated quartz-rich arenite and pelagic limestone. Upper Cretaceous, Caleta de la Peña Vieja, Fuerteventura. Thin-section, scale bar 5 mm

dules of gray chalcedonic chert formed by in situ replacement.

Petrographically, the Albian to upper Cretaceous chalks are planktonic foraminiferal radiolarian micrites with occasional benthonic foraminifera and scattered shell fragments. In the matrix, recrystallized heavily overgrown undeterminable coccoliths can be recognized. In these rocks the silicic intraclasts consist of lithologies found in the underlying pre-volcanic sequence (Fig. 8).

Clearly, in units F and G two different faunal associations can be distinguished. As noted above, Lehmann (in Mitchell-Thomé, 1976) and Grunau et al. (1975) identified *Globigerina* sp., *Hedbergella* sp., *Globigerinelloides* sp., *Ticinella* sp. and *Pithonella* sp., suggesting an Albian age. More specifically, a late Albian to early Cenomanian age can be assigned to the lower part of the sequence by the presence of ammonites and planktonic foraminifera: Bernoulli et al. (in prep.) determined the following association: *Partschiceras* cf. *whiteavesi* (Kossmatt), *Schackoina gandolfii* Reichel, *Rotalipora* sp., *Hedbergella* sp. and *Gabonella* sp. A pre-Maastrichtian, Senonian age of the upper part of the sequence is documented by double-keeled globotruncanids, rare *Heterohelicidae*, that are associated with benthonic foraminifera of the genera *Stensiöina*, *Gavelinella*, *Polymorphina* and *Reussella*. The relations between the sediments containing these different associations are obscured by the dyke intrusions, however, it is notable that no Turonian sediments have been found thus far. A hiatus between the Albian-Cenomanian and the Senonian could be suggested by analogy with D.S.D.P. Site 369 (Lancelot, Seibold et al., 1977) where a hiatus between Albian marls and chalks and Senonian (Coniacian) chalks is associated with slumping and soft sediment deformation similar to that observed in Unit F (Fig. 8).

H) Submarine Volcanic Rocks and Bioclastic Calcarenites. In the upper reaches of the Barrancos de los Sojames and de la Peña, the upper Cretaceous sedimentary rocks give way upstream, apparently conformably, to submarine volcanic rocks. These are well exposed at the mouth of the Barranco del Tarajalito (Fig. 5). There, as reported by Robertson and Stillman (1979b), the volcanic breccias and basaltic pillow lavas are intercalated with volcaniclastic siltstones and sandstones. The volcaniclastic rocks contain occasional fragments of quartz, quartzite, and marble, presumably derived from the stratigraphically older pre-volcanic sequence. Recrystallized limestone which may have been originally pelagic is present within the pillow lava interstices. Robertson and Stillman (1979b) argue that the highly vesicular, mostly fragmental, nature of the submarine volcanic rocks signifies eruption in shallow water (i.e., less than 100 m). Once again the succession is interrupted by dyke densities of up to 100 % and it is not possible to establish the presence or absence of breaks in the sequence. However, within the highest levels of the host sequence of the Basal Complex, fragmentary sequences of graded and laminated bioclastic calcirudites to arenites interbedded with silty calcilutites are found (Figs. 6 and 9). Up-section volcaniclastic sandstones and siltstones occur. The bioclastic limestones contain numerous well-preserved larger benthonic foraminifera, echinoid fragments, algal rhodolites, bryozoa, occasional coral fragments, and glauconite grains, fragments of mafic igneous rocks and of older sediments. These sediments have been interpreted as of Miocene age by Fuster et al., however, an early to middle Oligocene age can be assigned to them from the association of Eulepidina dilatata (Michelotti), Nephrolepidina praetournieri H. Douvillé, Heterostegina sp., Cycloclypeus sp., Spiroclypeus, Pararotalia viennoti (Greig) (Hottinger in Bernoulli et al., in prep.). This fauna also indicates resedimentation of the shallow-water material from a source situated near the lower limit of the photic zone. Planktonic foraminifera occurring in the intercalated calcilutites and volcanic siltstones are undeterminable.

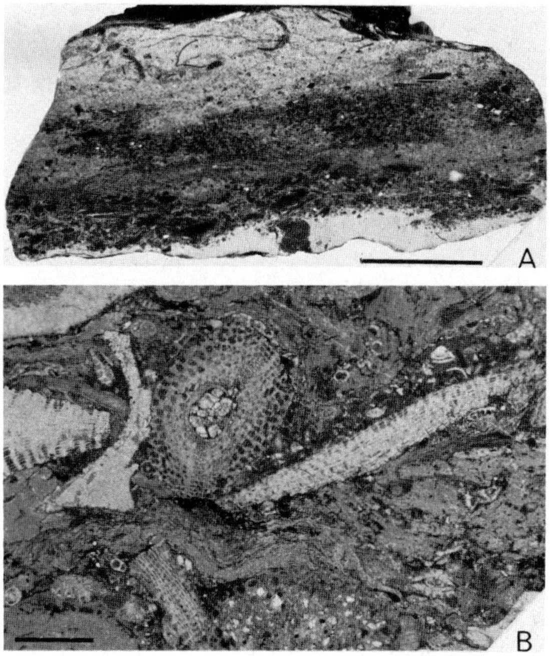

Fig. 9. Redeposited bioclastic limestones of Unit H, lower to middle Oligocene, Caleta Grande, Fuerteventura. A Redeposited limestone with abundant larger foraminifera, echinoderm fragments, bryozoans, red algae, reworked planktonic foraminifera and volcanic rock fragments. Polished surface, scale bar 2 cm. B Microfacies of a redeposited bioclastic limestone, showing closely packed larger benthic foraminifera, bryozoans, red algae, and fragments of reworked volcanic rocks and pelagic sediments. Thin-section, scale bar 2 mm

3.2 Northern outcrops

Pre-volcanic Mesozoic sedimentary rocks are exposed along the coast within 1 km north and south of Barranco del Tabelo (Fig. 3). The dyke density exceeds 70 %. Beds are steeply dipping and become younger to the south. Extensive impregnation by iron has taken place due to hydrothermal activity associated with later igneous activity. The sequence consists of alternations of quartz-arenites, quartzlutites, and indurated black shales. Subordinate rudites occur in beds up to 1.8 m thick toward the base of the exposed succession. The clasts comprise well-rounded pebbles of quartz and metamorphic rocks up to 10 cm in diameter. Such coarse-grained rocks have not been observed in the southern sequences and a detailed correlation cannot be established; there is, however, little doubt that the northern outcrops are a lateral equivalent of the lower Cretaceous terrigenous sequence, possibly of Unit D, in the south.

4. The Sedimentary Succession of Maio

Although, as mentioned earlier, there has been considerable discussion concerning the paleontology and age of the sedimentary rocks, much less published information is currently available on the sedimentology and detailed stratigraphy. A recent compilation by Grunau et al. (1975) merely suggests that volcanic rocks of probably late Jurassic age are followed by lower to middle Cretaceous basinal pelagic cherty limestones (according to Grunau et al. upper Berriasian to Barremian and Albian-Cenomanian), which in turn grade into limestones, shales, sandstones, and volcanics, locally dated as Eocene. Grunau et al. (1975) allocate a thickness of 1500 to 2500 m to the whole sequence, but this is strongly questioned by the new discovery of thrust faults in the east of the island (A. Robertson and others, unpubl.). Comparable with Fuerteventura the sedimentary sequence is intruded by essexitic and basanitic dykes, sills, and plutonic masses for which radiometric dating suggests an age mostly between 21 and 12 m.y. (early to middle Miocene; Grunau et al., 1975). The sedimentary succession dips radially away from the central eruptive complex having been tilted before being overlain by late Tertiary marine clastics and younger eruptives.

A most important point is that petrographic and geochemical studies of the Jurassic volcanics have indicated that the pillow-lavas have trace element and rare earth element patterns characteristic for ocean floor tholeiites generated at a mid-ocean ridge (Klerkx and de Paepe, 1971, 1976; de Paepe et al., 1974). Such truly ocean floor rocks have not been recognized on Fuerteventura, nor are they seen again in the Maio succession where subsequent volcanism is alkaline. De Paepe et al. (1974) make the important point that the oceanic pillow lavas conformably underlie the oldest limestones of early Cretaceous or possibly even older age. Earlier they postulated (Klerkx and de Paepe, 1971) that the Maio central igneous complex could be a segment of a Mesozoic ocean ridge, but new work by Furnes et al. (1981) indicates that the intrusive rocks of the central igneous complex entirely postdate the Mesozoic sedimentary sequence.

The only author so far to publish more details of the Mesozoic sedimentary sequence is Serralheiro (1968 and especially 1970), whose detailed geological map at 1:30.000 shows the series of lithologic units described below. His description is supplemented here by biostratigraphic data from Stahlecker (1934) and Rigassi (1975, and pers. comm., 1980).

A) Shale-Limestone-Eruptives Complex. The lowest unit of Serralheiro (1970), the "Shale-limestone-eruptive" complex consists of shales, graywackes, and chert-bearing limestones intercalated with mafic igneous rocks and crops out extensively west of Monte Branco. From this area, Rigassi (1975) describes siliceous shales and graywackes yielding small coccoliths of reportedly pre-Oxfordian aspect. Up-section intercalated cherty limestones occur with radiolarians, Saccocoma sp. and Protoglobigerina sp. and are possibly Oxfordian in age. The contact with the overlying "limestone with chert unit" is obscured by igneous rocks. The complex is also seen in the west of the island and in the northwest near Monte Vermelho where limestones are recrystallized in direct contact with granular intrusives, including essexite.

B) Limestone with Chert Unit. Serralheiro (1970) maps this unit extensively on both the east and west flanks and as isolated exposures in the south and north of the island. Particularly, in the Ribeira do Morro, these limestones are reported to be well stratified, fine-grained limestones in beds up to 1 m thick (Fig. 10). Laminations are ubiquitous, often spaced at ca. 10 cm intervals. Noumerous chert nodules are present typically as intercalations 1–15 cm thick occurring on average every 0.3–0.5 m.

In the southeastern part of the island (Monte Branco) Rigassi (1975) has recognized different microfossil assemblages in the "limestone with chert unit" that are in line with a late Jurassic to early Cretaceous age of the unit. Near the bottom of the formation an association of radiolarians and Saccocoma, late Jurassic, possibly Kimmeridgian in age, is followed by radiolarians and Saccocoma associated with calpionellids of the late Tithonian (Crassicollaria brevis Remane). Near their top, the Tithonian limestones are intercalated with a thin sequence of pillow lavas which in turn are covered by limestones of latest Tithonian to Berriasian [Calpionella alpina (Lorenz), Tintinnopsella carpathica (Murgeanu and Filipescu), Nannoconus steinmanni Kamptner, N. colomi (de Lapparent)] and Neocomian age [Tintinnopsella longa (Colom), Nannoconus colomi and N. bucheri Brönnimann]. At Monte Branco the limestones with chert are, according to Rigassi (1975), unconformably overlain by Albian to upper Cretaceous graywackes, siliceous shales, and tuffs.

While he made a detailed stratigraphic division in the Monte Branco region, Serralheiro (1970) has not differentiated the Mesozoic strata of the Ribeira do Morro succession on the northwest flank, although this is in fact by far the best stratigraphically documented section on the island. It was here that, following collection of aptychi by Friedländer (1912, 1913), described by Hennig (1913), Stahlecker (1934) made finds allowing him to subdivide the section into 14 units ranging from Hauterivian to Aptian. According to Rigassi (1975), the lowest 50 m of the section contain no recognizable macro-fossils and only undiagnostic microfossils. The upper portions of the section (Stahleckers Horizons 10–14) belong to Unit C and are of Aptian age. The "limestone with chert unit" therefore ranges here approximately up to the Barremian-Aptian boundary.

Light-colored compact limestones intercalated with chert closely similar to the "limestone with chert unit" were distinguished by Serralheiro (1970). According to him, these limestones overlie the shales, marls, and marly limestones of Unit C (below) and thus would be of late Cretaceous age. However, new observations suggest (A. Robertson, unpubl.) that these limestones are a tectonic repetition of the "limestone with chert unit".

C) Shale, Marl, and Marly Limestone Unit. Serralheiro (197o) noted that in a number of localities, principally in the Ribeira do Morro, the "limestone with chert unit" passes up into a sequence of shales and marls and marly limestones. A high density of inter-leaved basaltic sheets complicates the study of these rocks. The limestones in this unit are often white, gray, or yellow and are markedly thinner-bedded than the "limestone with chert unit". Many beds are siliceous. In addition numerous intercalations of thinly bedded, finely laminated gray and white soft-weathering shales are present. This sequence corresponds to horizons 10 to 14 in the Ribeira do Morro section (Aptian to ?Albian, Stahlecker, 1934).

Fig. 10. Early Cretaceous Maiolica facies, central Atlantic and Tethys. A Lower Cretaceous limestones, Ribeira do Morro, Maio. Note rhythmically bedded laterally continuous beds with stringers of chert nodules. B Unconsolidated nannoplankton ooze with nodule of hard, chalcedonic quartz chert. The carbonate sediment close to the chert has been lithified: in thin-section it shows essentially the same depositional texture as the chert, which is of replacement origin. Blake-Bahama Formation. Upper Tithonian-Valanginian. Cat Gap, DSDP-11-105-1-4, 99-113 cm. Scale bar 2 cm. C Lower Cretaceous pelagic limestone, with nodule of chalcedonic quartz chert of replacement origin and ubiquitous stylolites. Monte Branco, Maio. D Typical aspect of Maiolica facies: white, well-bedded calcilutites (biomicrites with calcitized radiolaria) with lenses and bands of black replacement chert. Vigla Limestone (Maiolica), upper Tithonian-lower Cretaceous. Ionian Zone, western Greece; between the villages of Khanopoulon and Gribovon, north of Arta

D) Graywackes, Siliceous Shales, Tuffs, and Volcanics. Grunau et al. (1975) and Rigassi (1975) mention above the Aptian marls and shales a sequence of graywackes, "tuffs", and siliceous shales to which they allocate an Albian/Cenomanian to Eocene age. Agglutinating foraminifera of the so-called *Rhabdammina*-fauna point to a deep-water environment of deposition.

Serralheiro (1970) describes the tilted Mesozoic rocks of Maio as being overlain by the Pedro Vaz Conglomerates, which are best exposed in the northeast of the island. This unit dips more gently than the Mesozoic rocks, from 30°–36°. Lithologically, the Pedro Vaz comprises conglomerates grading upward into sands, clays, and shales, with poorly preserved foraminifera of middle to late Miocene age (Rigassi, 1975). These coarse-igneous derived conglomerates obviously post-date the relatively deep-water marine phase of the island in the Mesozoic and earliest Tertiary.

Immediately prior to writing this review, the whole outcrop area of the Mesozoic sedimentary rocks and the central complex has been remapped by a team comprising M. Le Bas, H. Furnes, A. Robertson, C. Stillman, and J. Zielonka. Early results (Furnes et al., 1981; A. Robertson, unpubl. data) show that the existing stratigraphy in the southeast of the island is in need of major revision. In this area the mapping shows that much of the sequence is repeated by major thrust faults. In particular, the upper Jurassic pillow lavas mentioned by Rigassi (1975) on Monte Branco are likely to be a thrust intercalation of the ocean floor basement rocks identified by de Paepe et al. (1974) elsewhere on the island. Consequently, much of Rigassi's section west of these basaltic pillow lavas is probably a repeat of the overlying upper Jurassic and early Cretaceous limestone sequence. The new work shows that Serralheiro's (1970) lowest "shale-limestone complex" in fact comprises parts of the Jurassic limestone (not necessarily basal) heavily intruded by dykes and sills close to the central complex. Indeed where basal mafic pillow lavas have not been observed (e.g., west of Monte Branco), the host limestones (without extrusions) gradually give way to 100 % of intrusive rocks. Furthermore, from the evidence of thrusting it is clear that Serralheiro's (1970) "higher" light-colored limestone intercalated with chert is simply a repeat of the lower "limestone with chert unit". With these important exceptions the rest of the Mesozoic sequence more or less corresponds to that reported by Serralheiro (1970) and Rigassi (1975). A new section measured east of Monte Branco is shown in Fig. 6. An account of the revised stratigraphy and sedimentology of the Maio Mesozoic and Tertiary sedimentary rocks is in preparation by A. Robertson and others.

5. Interpretation and Comparison with D.S.D.P. and Onshore Data

The D.S.D.P. cores of Legs 14 (Berger and von Rad, 1972), 41 (Lancelot, Seibold et al., 1977), 47A (von Rad, Ryan et al., 1979) and 50 (Lancelot, Winterer et al., 1980) combined with dredge samples, piston cores (von Rad et al., 1979), published oil well logs (AUXINI, 1969; CONOCO, 1969 in von Rad and Einsele, 1980) and on-land information (Querol, 1966; Wiedmann et al., 1978) allow the Fuerteventura Mesozoic sediments to be viewed in the broader context of the eastern Atlantic continental margin. At present the sediments on Maio can only be considered in general terms, but useful comparisons can be made with the cores of D.S.D.P. Sites 367 and 368 (cf. Lancelot, Seibold et al., 1978).

5.1 Fuerteventura

Late Jurassic to Early Cretaceous: Deep Water Submarine Fan Deposition: Rothe (1968) followed earlier workers (Gagel, 1910; Hausen, 1958) in regarding the Fuerteventura Mesozoic sedimentary rocks as being deposited on the drowned West African continental margin. His belief that the lower part of the island sequence was upper Jurassic and possibly older rested heavily on a correlation with the on-land Moroccan stratigraphy (Querol, 1966; Dillon and Sougy, 1974; Wiedmann et al., 1978). Specifically, he correlated his series a-c with the Lias, Dogger and Malm sequences of the Aaiun (Tarfaya) Basin, particularly limestones and dolomites, his series d to the widespread "Sables de Tantan", and his series e to the Cenomanian to Turonian limestones and marls of Aaiun. Mitchell-Thomé (1976) pointed out that this correlation is tenuous since the Jurassic age on Fuerteventura depends on the possible occurrence of Posidonia.

Robertson and Stillman (1979a) confirmed suggestions by Fuster et al. (1968) and Grunau et al. (1975) that the Fuerteventura siliciclastics and limestones are redeposited sediments (turbidites s.l.), which thus cannot be directly compared with in situ shelf facies of the Aaiun (Tarfaya) Basin. In any case new biostratigraphic evidence (Bernoulli et al., in prep.) shows that the Fuerteventura siliciclastics are partly Valanginian in age. On the other hand, the Sables de Tantan do represent part of the widespread "Wealden" siliciclastic facies attributed to uplift and accelerated erosion of the internal crystalline massifs which took place in early Cretaceous times (Dillon and Sougy, 1974; Martinis and Visintin, 1966; von Rad and Arthur, 1979; von Rad and Einsele, 1980). The early Cretaceous siliciclastics of the Aaiun (Tarfaya) Basin may thus represent the shallow marine and continental equivalents of the deep-water facies on Fuerteventura. The pre-Valanginian (late Jurassic to Berriasian?) redeposited carbonates of our Unit B may have been redeposited from a late Jurassic carbonate platform underlying the early Cretaceous Wealden deltas (Hinz et al., this Vol.; von Rad and Einsele, 1980; cf. Price, 1980a; Schlager, 1980).

The bulk of the early Cretaceous Fuerteventura siliciclastics were laid down as classical turbidites (e.g., Bouma, 1962). The vertical organization of the turbidite beds is comparable with that of many ancient deep sea fans (e.g., Walker, 1978; Middleton and Hampton, 1973; Mutti and Ricci Lucchi, 1972, 1975). Specifically the classical turbidites of Units A–D are closely comparable to deposits of the smooth outer part of the mid-fan (suprafan) or the outer fan. In general, the gradual progressive thickening of the siliciclastics through Units A–D, followed by thinning and fining upward sequences through the upper part of Unit D into Unit E, suggests progressive progradation of the mid-fan over the outer fan, then retreat of the mid-fan (suprafan). Individual fan lobes are recorded by the numerous thickening-upward and coarsening-upward cycles of Unit D.

The relatively thick-bedded massive sandstones of the lower part of Unit D probably represent the maximum progradational phase in which mid-fan channels fed more distal fan-lobes. Unfortunately, the outcrops do not allow lateral relationships to be assessed. The thin-bedded silt turbidites of the upper part of Unit D and of Unit E can be attributed to a later transition from a lower fan to a basin plain depositional area. By contrast, the northern siliciclastic sequences contain graded intercalations of rudites deposited by mass-flow and turbidity currents. These deposits stand comparison with channelized pebbly mudstone typical of the inner mid-fan. This may suggest that the northern exposures were located close to a major supply area (inner fan and submarine canyon).

The existence of extensive deep sea fans along the northwest African margin during the late Jurassic and early Cretaceous is confirmed by the sediments recovered during the Deep Sea Drilling Project (Fig. 6). At Site 397, situated on the uppermost continental rise off Cape Bojador, a late Hauterivian sequence comprises laminated silty claystones with siderite layers (von Rad, Ryan et al., 1979; Fig. 6). These are interpreted as prodelta sediments rapidly deposited in water depth of at least 500 to 1000 m (von Rad, Ryan et al., 1979; Einsele and von Rad, 1979). Above, there are early Miocene mass-flow deposits following a major 100 million years erosional hiatus (Arthur and von Rad, 1979). At Site 369, situated on the continental slope off Cape Bojador, upper Aptian to (lower?) Albian silty nannofossil-bearing marls are unconformably overlain by slumped upper Cretaceous light-colored nannofossil marls and chalks.

The early Cretaceous Fuerteventura rocks differ from those of Site 397 by being more quartzose, coarser-grained and more thickly bedded. Whereas the early Cretaceous sediments of Sites 369 and 397 were deposited on a relatively deep water continental slope, the Fuerteventura redeposited siliciclastics were laid down on the middle to outer portions of submarine fan systems near the foot of the continental rise. Much of the relatively coarse material on Fuerteventura derived from the early Cretaceous Wealden deltas presumably by-passed the continental shelf and slope in canyons.

Further north, at Site 370/416, located in the Moroccan Basin, the early Cretaceous interval comprises uninterrupted distal turbidite deposition (Lancelot, Winterer et al., 1980; Price, 1980a). This turbidite sequence of thinly bedded sandstones, siltstones, mudstones, and Tithonian to lower Valanginian calcarenites, calcisiltites, and micrites is very similar to the turbidite sequence of Fuerteventura and to the late Jurassic to early Cretaceous flysch deposits in the Mauretanian-Massylian flysch trough of the Maghrebide chain, particularly of the Rif (Laubscher and Bernoulli, 1977; our Figs. 6 and 7). Price (1980a) points out that petrographically the turbidites at Site 416 contain a greater volume of mica, schist, and arenite fragments relative to Fuerteventura, which contains more plutonic lithic fragments including perthite, characteristic of the ancient Precambrian massifs south of the South Atlas fault belt.

Price (1980a, his Fig. 6) considered a provenance of the siliciclastic detrital material at Site 370/416 from either prograding Wealden delta fronts in the Aaiun Basin, by longitudinal transport from the Rif-Betic system or from the hinterland of the Essaouira Basin, the Moroccan Meseta, which at that time was emerged (Faure-Muret and Choubert, 1971). Of these sources an origin from the Moroccan Meseta and by-passing of the shelf and slope seems most probable to us. In fact, Wealden-type deltas are very widespread around the central Atlantic-southwest Tethys area (von Rad and Arthur, 1979, our Fig. 11) and the occurrence of widespread deep-water sands in this area may also shed light on the significance of early, upper Jurassic-lower Cretaceous flysch deposits in the Betic-Maghrebide system (Durand-Delga, 1960; Andrieux,1971; Didon et al., 1973; Lespinasse, 1975). These flysch sediments occur all along the deep (transform) furrow connecting the central Atlantic with the Ligurian segment of the Tethys ocean in late Jurassic-early Cretaceous times (Laubscher and Bernoulli, 1977). They also are associated with Middle Jurassic deep-water marls and limestones and upper Jurassic radiolarites. Along their base, slivers of ophiolites (Andrieux, 1971) possibly represent relics of oceanic crust. These flysch deposits are therefore

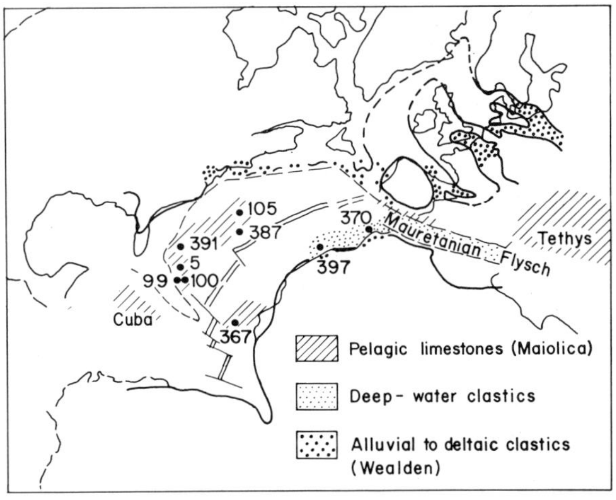

Fig. 11. Early Cretaceous (Valanginian, 125 million years) paleogeography of the central Atlantic area (von Rad and Arthur, 1979) and distribution of pelagic limestones (Maiolica) and deep water clastics

likely to be the lateral equivalents of the upper Jurassic-lower Cretaceous deep sea sands of the Moroccan Basin and Fuerteventura which are related to relative uplift and erosion along the northwest African margin rather than to incipient Alpine-type orogeny.

The black shales and siltstones which characterize the lower Cretaceous Fuerteventura sequences can be attributed to the Barremian-Albian interval of black shale deposition which affected a whole range of oceanic, continental margin, and epeiric environments, both on land and at sea (data summarized by Jenkyns, 1980). Although the explanation for these black shale episodes is still enigmatic, it is widely attributed to global depletion of oxygen in the oceans related to enhanced production of marine and terrestrial organic carbon during periods of marked marine transgression (Arthur and Natland, 1979; Arthur and Schlanger, 1979). On Fuerteventura an abundance of plant material points to a predominance of terrestrial organic matter in the black shales and siltstones. This Aptian to (early?) Albian anoxic event has also been noted in the Aptian to Albian dark-colored organic-rich nannofossil marls and shales along the Cape Bojador slope (Lancelot, Seibold et al., 1977; von Rad et al., 1979).

Late Cretaceous: Slope Instability: By (?late) Albian time, following a renewed transgression in the Aaiun (Tarfaya) Basin, sedimentation of dark-colored shale and mudstone gave way to deposition of light-colored pelagic limestones. These limestones were deposited in a more oxygenated environment under still high fertility conditions (radiolarians, fish teeth, phosphatized debris). The sediments show evidenc of extreme gravitational instability with mass flow, turbidity current deposition, reworking of older deposits etc. (Fig. 8). By analogy with Site 369 where slumped pelagic nannofossil marls and limestones of Senonian age unconformably overlie dark-colored upper Aptian to (?lower) Albian nannofossil marls (Lancelot, Seibold et al., 1977), an unconformity is inferred between the dated upper Albian to lower Cenomanian and the overlying Senonian (pre-Maastrichtian) limestones. Such an unconformity has not been observed in the field where observation is limited by outcrop conditions and the density of the dyke swarm, however, Turonian sediments are conspicuously missing both at Site 369 and on Fuerteventura. They occur occasionally as limestone fragments in early Miocene debris flow deposits at Site 397 (von Rad, Ryan et al., 1979). Turonian carbonate sediments occur, however, on the shelf off the western Sahara (Spansah 51A-1; in Einsele and von Rad, 1979).

Gravitational instability during the late Cretaceous is also documented by the sequence drilled at Site 415, in the Agadir canyon off southern Morocco (Lancelot, Winterer et al., 1980; Price, 1980b). There, lithologic repetitions and deformation of the cored sediments indicate a reportedly allochthonous unit composed of several imbricate sheets of Cenomanian shales and carbonates above a decollement level of Albian to Cenomanian shales. As revealed on multichannel seismic profiles (Price, 1980b), the drilled sequence is only the distal part of a large allochthonous mass beneath the lower slope, derived from the upper slope where a prominent angular unconformity is present. Gravitational movements and deformation predate deposition of the overlying Tertiary turbidites. Price (1980b) interprets the allochthon as a belt of gravity-driven thrust masses emplaced in connection with uplift of the western Atlas that is thought to be related to transcurrent faulting along the South Atlas fault and its oceanic prolongations during the late Cretaceous. However, slope rejuvenation, large-scale erosion and mass-wasting is known to have taken place repeatedly along the northwestern African margin during the Cenozoic (Uchupi et al., 1976; Sites 369, Lancelot, Seibold et al., 1977, and 397, Arthur et al., 1979; Embley and Jacobi, 1977). Consequently, the slide sheets reported may not be related to regional tectonic events: similarly, it now seems unlikely that the gravitational instability shown by the Fuerteventura Albian to Upper Cretaceous limestones was the result of uplift related to the onset of volcanism as proposed by Robertson and Stillman (1979a,b).

Early Tertiary: Volcanic Activity and Uplift: Whereas up to the late Cretaceous the Fuerteventura rocks reflect the general sedimentary evolution of the NW African continental margin, the early Tertiary volcanic and sedimentary rocks document an individual geological history for the (eastern) Canary Islands. Seismic evidence suggests that general uplift of the Canary Arch began in early Eocene (Watkins and Hoppe, 1979), but earlier localized submarine volcanism is probable. After a hiatus in deposition, the earliest submarine volcanic rocks on Fuerteventura were erupted possibly in latest Cretaceous, but more probably in early Tertiary times, into waters that were already relatively shallow (Robertson and Stillman, 1979b). Volcaniclastic sediments accumulated particularly during brief periods in eruption and lava extrusion led to construction of a thick volcanic pile. The lowest volcaniclastic sediments are steeply dipping in general conformity with the stratigraphically underlying pre-volcanic sedimentary rocks to the S; in contrast the lower to middle Oligocene sediments are only gently inclined, implying that the basal volcanic and the underlying pre-volcanic rocks were deformed during the period of extrusion of the submarine volcanic sequence, although no unconformity could be detected in the fragmentary relics of the sequences. There is evidence that the earlier dykes were tilted prior to the later part of submarine eruption, then tilted again prior to eruption of the Miocene plateau lavas (Stillman, in Robertson and Stillman, 1979b).

The lower to middle Oligocene bioclastic and volcaniclastic sediments intercalated in the highest exposed levels of the submarine volcanic sequence were emplaced by mass flow in an open marine environment. The predominance of algal rhodolites, red algae, bryozoans, and larger orbitoid foraminifera with only a minor fraction of coral debris in the displaced carbonate sediments points to a shallow-water source near the base of the photic zone, possibly on a seamount or an offshore shoal fringing an emergent volcanic edifice. Admixed pyroclastic material and clasts of pre-volcanic sedimentary and of intrusive and extrusive igneous rocks show that areas of pre-volcanic sediments and younger plutonic rocks of the Basal Complex were undergoing erosion at that time.

Both volcanic debris and bioclastic fragments point to local, shallow-water and even emerged sediment sources within the (eastern) Canary Islands, for the lower to middle Oligocene sediments. No Oligocene sediments are preserved in the Aaiun basin (AUXINI, 1969), whereas the Oligocene-Miocene sediments of the slope and rise, where recognized, consist of quartz- and phosphate-rich biomicrites and -sparites and quartzose sandstones, probably derived from the upper slope (von Rad et al., 1979). There are also hemipelagic siliceous clays, marls, and chalks deposited under upwelling conditions (von Rad et al., 1979; Site 369, Lancelot, Seibold et al., 1977); in the Moroccan Basin these are associated with terrigenous turbidites (Site 340/470, Lancelot, Winterer et al., 1980). On the other hand, Oligocene to lower Miocene sediments similar in composition and dominated by red algae, rhodolites, and other, bryozoans and larger benthonic foraminifera, occur in the limestone caps of other Tertiary volcanic build-ups (Selvagem Grande, Jérémine, 1951) and as redeposited material in volcaniclastic debris flows derived from the Canary Islands, probably Fuerteventura and/or Lanzarote (early to middle Miocene, Site 397, Schmincke and von Rad, 1979). With the construction of the volcanic ridges former deep water areas became isolated from adjacent continental margin sedimentation and turned into source areas for redeposition of volcaniclastic sediments.

The lower to middle Oligocene sediments of Fuerteventura are highly important for the reconstruction of the early volcanic and tectonic history of the island. Soon after the onset of pre-Oligocene submarine volcanism, Fuerteventura was affected by major folding along E–W-axes. After deposition of the younger submarine volcanics and of the Oligocene sediments that were only slightly tilted, Fuerteventura underwent a phase of crustal extension with the intrusion of a swarm of sheeted dykes. Seismic records west of Fuerteventura (Grunau et al., 1975) indicate the presence of a thick volcanic apron, probably dating from early Miocene uplift and erosion of the Basal Complex soon after sheeted dyke intrusion. This uplift may well have been associated with the shallow emplacement of plutons, including the syenitic-trachytic ring complexes. Intrusions, greenschist metamorphism in the pre-volcanic sediments and in the dyke-complex (Fuster et al., 1968; Stillman et al., 1975) predate a major hiatus and erosion followed by possibly first submarine (Schmincke and von Rad, 1979), then later subaerial up-building of an early to middle Miocene volcanic shield (18 to 11 m.y. B.P., Abdel-Monem et al., 1971; Schmincke, 1976; Schmicke and von Rad, 1979; von Rad and Arthur, 1979; Schmincke, this Vol.).

5.2 Maio: Pelagic and Hemipelagic Deposition

The Maio limestone sequence rests on tholeiitic ocean ridge-generated lavas (de Paepe et al., 1974; Furnes et al., 1981) of late Jurassic or even older age (Rigassi, 1975). The overlying limestones with chert containing planktonic crinoids, radiolarians, calpionellids, ammonite phragmocones and aptychi, and occasional bibalves are open marine pelagic deposits. Stahlecker (1934) and Krejci-Graf (1965) suggest water depth of 2000 to 3000 m. Interpretation of the Maio sediments as a sequence deposited on subsiding oceanic crust would suggest its interpretation as a deepening sequence laid down over a range of water depth in the order of 2500 to about 4500 m. The overlying shales, marls, and marly limestones would then mark the passage to the overlying siliceous shales, graywackes, and tuffs of late Cretaceous age. The recorded occurrence of fish remains in the middle Cretaceous shales (Stahlecker, 1934) may tentatively be correlated with the middle Cretaceous anoxic events.

There are insufficient data to adequately interpret the upper Cretaceous to lower Tertiary deposits (Grunau et al., 1975; Rigassi, 1975). The occurrence of deep-water agglutinating foraminifera in the sequence of siliceous shales, graywackes, tuffs, and lavas points to a deep depositional environment, however, with the addition of local volcanic sediment sources. This sequence is reported to range up to the Eocene (Grunau et al., 1975), and may record the early development of a submarine volcanic pile exposed above sea level during the Miocene.

Two D.S.D.P. holes relevant to the stratigraphy of Maio have been drilled in the Cape Verde Basin (Site 367) and on the Cape Verde Rise (Site 368, Lancelot, Seibold et al., 1977; our Fig. 6). Site 367 bottomed in basalt, overlain by upper Jurassic to lower Cretaceous limestones. These consist of Oxfordian to lower Tithonian red argillaceous limestones, comparable to the Rosso ad Aptici facies of the Jurassic Tethys, overlain by upper Tithonian to Neocomian white and gray limestones with darker marl interbeds comparable to the Tethyan Maiolica facies (Bernoulli, 1972; Bernoulli and Jenkyns, 1974; Jansa et al., 1977). The "Mediterranean aspect" of the limestones of Maio had been noticed already by Colom (1954, 1955) and been taken by this author as an indication for the existence of a Mesozoic E–W-trending ocean connecting the Caribbean (Cuba, cf. Brönnimann, 1953, 1955; Colom, 1965) with the Mediterranean area (cf. Bernoulli and Jenkyns, 1974; Bernoulli and Lemoine, 1980; our Fig. 11). In this context, the limestones with ammonites and aptychi of the Maio sequence appear to generally correspond with the Tithonian-Neocomian at Site 367, while the pre-Tithonian limestones with Saccocoma and radiolarians (Colom, 1955; Rigassi, 1975) are similar to the Oxfordian-Kimmeridgian interval at this site.

At Site 367, the Neocomian limestones pass upward into Barremian red and green claystones grading up-section into Aptian-Cenomanian black shales, then into upper Cretaceous deep sea clays deposited below calcite compensation depth. Site 368 on the Cape Verde Rise terminated in Albian-Turonian black shales, overlain by upper Cretaceous to Paleogene deep sea clays with chert and porcellanite and then Miocene nannofossil marls and oozes. This and seismic evidence (Uchupi et al., 1976) suggest that uplift of the Cape Verde Rise took place in post-Eocene times, contemporaneous with Miocene intrusion at Site 367 (Duncan and Jackson, 1977) and with the development of the central igneous complex on Maio (Grunau et al., 1975). Certainly by Miocene times the island had been uplifted and tilted to produce the coarse igneous-derived rudites of the Pedro Vaz Formation (Serralheiro, 1970; Rigassi, 1975; Grunau et al., 1975).

6. Summary of Tectonic Setting

Despite the paucity of published data, the general tectonic setting of the Mesozoic rocks of Maio is clear. Geophysical data show that Maio and the Cape Verde Islands rest on oceanic crust west of the magnetic quiet zone (Hayes and Rabinowitz, 1975). Initial Atlantic rifting on the adjacent African margin is thought to have taken place in the late Triassic to early Jurassic. At least 500 km of oceanic crust would have been created prior to deposition of the Maio limestones close to the Mid-Atlantic ridge sometimes during the late Jurassic to early Cretaceous interval. If the Maio sediments followed a normal ocean floor subsidence pattern, an upward transition to terrigenous deep-sea deposits as in the Cape Verde Basin would be expected (Lancelot, Seibold et al., 1977). However, upper Cretaceous sediments with terrigenous material derived from a continental source have not been reported. Onset of volcanicity in the island of Maio during the late Cretaceous (?) to early Tertiary apparently initiated an individual tectonic evolution with uplift and subaerial volcanism by Miocene times followed by later eruptions of alkaline lava flows.

The tectonic setting of Fuerteventura is complicated by its position near the Moroccan continental margin. Geophysical data (gravity, magnetics, seismics, see Weigel et al.; Hinz et al.; Roeser; all this Vol.) are unable to distinguish between subsided continental crust, oceanic crust with a thick sediment cover, or possibly the existence of transitional crust loaded with mafic igneous rocks. In this regard the sedimentological model proposed above suggests that Fuerteventura was located on marginal oceanic crust ca. 30–50 km oceanward of the continental shelf edge (Fig. 12).

One remaining puzzling feature is the nature of the deformational regime responsible for the relatively steep dips of the sediments on both Maio and Fuerteventura. On Maio where sediments dip symmetrically away from the central igneous complex (Serralheiro, 1970), the tilting can be plausibly related to uplift of plutons in the center of the island. However, on Fuerteventura the Mesozoic sediments are almost invariably vertical to overturned, without obvious relationship to major intrusive massifs. The southern sequences young consistently to the north, while those in the north of the island young in opposite direction (Robertson and Stillman, 1979a). Robertson and Stillman (1979a) analized the cleavage-bedding relationship and minor fold data of the southern sequences in terms of an inverted limb of a major N–E facing reclined fold. The cleavage orientation was found not to be coincident with the statistical axial plane of the fold. Such a fold could have been generated by dextral motion along a shear zone orientated almost N–S.

To explain the data, Robertson and Stillman (1979a) followed other workers (Grunau et al., 1975; Leh-

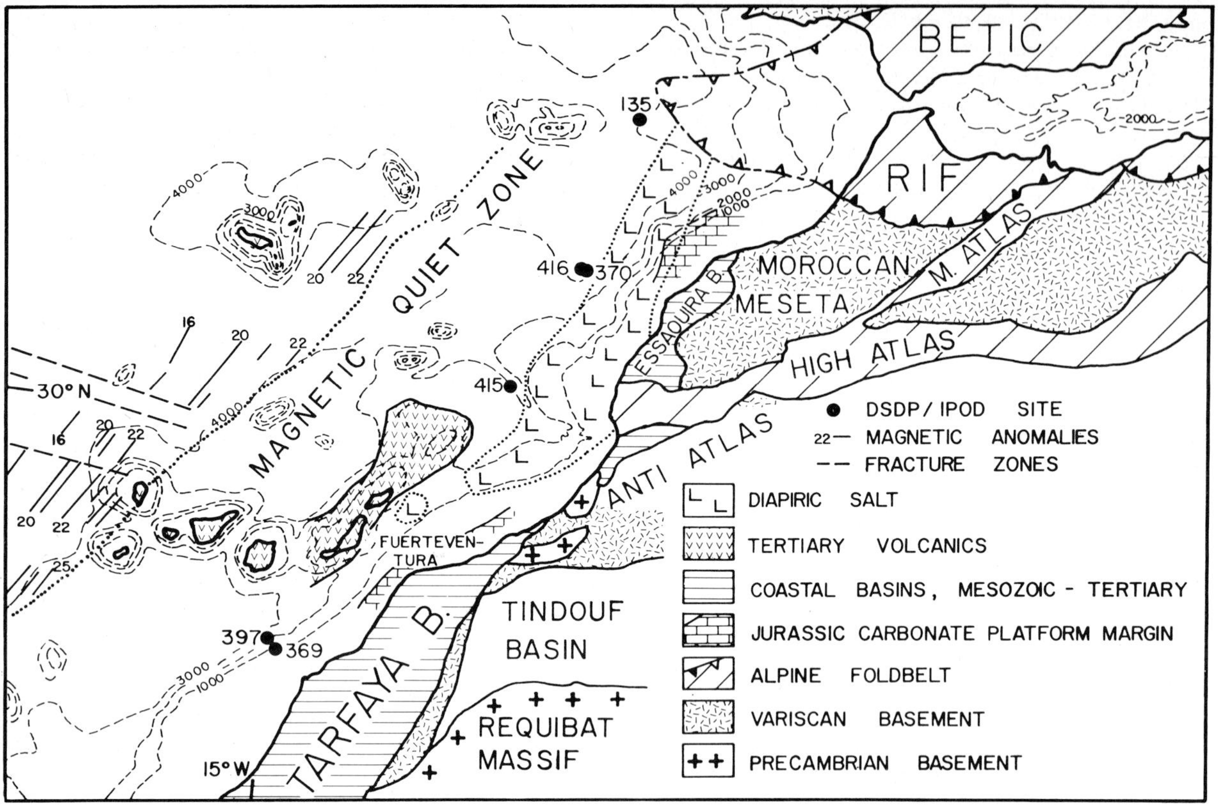

Fig. 12. Tectonic setting of Fuerteventura (Data from Bhat et al., 1975; Choubert and Faure-Muret, 1968; Hayes and Rabinowitz, 1975; see also back cover figure)

ner and de Ruiter, 1977; Anguita and Hernan, 1975) in looking for a regional solution and suggesting a possible link with the South Atlas Fault Belt. When projected seaward, the South Atlas Fault Belt could pass close to Lanzarote and might then be expected to intersect oceanic crust in the vicinity of Fuerteventura. Tectonic movements along the South Atlas could thus have affected a narrow linear zone on Fuerteventura. However, recent seismic studies (Watkins and Hoppe, 1979; Weigel et al.; Stets and Wurster; Hinz et al., all this Vol.) have not confirmed a structural link between Fuerteventura and the South Atlas. An alternative and local explanation for the tectonics of Fuerteventura might involve compression of sea-floor strata adjacent to forcefully emplaced Tertiary plutons. Repeated pulses of emplacement could have upbended the sediments, overturned them, then impressed a slightly oblique cleavage giving observed cleavage-bedding and minor fold features and together with the dyke intrusions overprinted the sediments by thermal metamorphism of greenschist grade. The overturning would explain the northward younging directions of the southern sequences adjacent to the Betancuria massif plutons; the southernly dips and younging of the northern sequence toward the S may be due to the emplacement of the Montanas Milocho and Blanca plutons (Stillman, pers. comm., 1980).

Acknowledgements. The authors were supported during this study by the British Natural Environmental Research Council (A.R.) and the Swiss National Science Foundation (D.B.). We both thank J.M. Fuster and his colleagues for their kind advice and assistance. A. Robertson thanks J. Bennel-Baker, M. Le Bas, H. Furnes, J. Smewing, C. Stillman and J. Zelonka for making field data available. D. Bernoulli thanks Y. Lancelot, E. Seibold and the Deep Sea Drilling Project for the opportunity to study cores of D.S.D.P. Leg 41 and H. Laubscher and L. Hottinger for cooperation in the field. For helpful discussions and use of additional, partly unpublished data we also thank P. Lehner, I. Price, D. Rigassi and U. von Rad. L. Hottinger and O. Renz determined various fossils. H. Laubscher, C.J. Stillman, U. von Rad and L. Hottinger commented on drafts of the manuscript.

REFERENCES

Abdel-Monem A, Watkins ND, Gast PW (1971) Potassium-argon ages, volcanic stratigraphy, and geomagnetic polarity history of the Canary Islands: Lanzarote, Fuerteventura, Gran Canaria, and La Gomera. Am J Sci 271: 490-521

Andrieux J (1971) La structure du Rif central. Notes Mém Serv Géol Maroc 235: 1-155

Anguita F, Hernan F (1975) A propagating fracture model versus a hot spot origin for the Canary Islands. Earth Planet Sci Lett 27: 11-19

Arthur MA, Natland JH (1979) Carbonaceous sediments in the North and South Atlantic: the role of salinity in stable stratification of Early Cretaceous basins. In: Talwani M, Hay W, Ryan WBF (eds) Deep drilling results in the Atlantic Ocean: continental margins and paleoenvironment. Maurice Ewing Ser 3, Am Geophys Union, pp 375-401

Arthur MA, von Rad U (1979) Early Neogene base-of-slope sedimentation at Site 397: sequential evolution of gravitative mass transport processes and redeposition along the north-west African passive margin. In: von Rad U, Ryan WBF et al. Init Rep Deep Sea Drilling Project 47 (1): 603-639

Arthur MA, Schlanger SO (1979) Cretaceous "oceanic anoxic events" as causal factors in development of reef-reservoired giant oil fields. Am Assoc Petrol Geol Bull 63: 870-885

Arthur MA, von Rad U, Cornford C, McCoy FW, and Sarnthein M (1979) Evolution and sedimentary history of the Cape Bojador continental margin, NW Africa. In: von Rad U, Ryan WBF et al. Init Rep Deep Sea Drilling Project 47(1): 773-816

AUXINI (Departamento de Investigaciones Petroliferas de AUXINI) (1969) Correlacion estratigraphica de los sondeos perforados en el Sahara espanol. Bol Geol Minero Madrid 83: 235-251

Bebiano JB, Soares JMP (1951) Note on some supposed Senonian fossils from Sao Nicolau island. 18th Int Geol Congr 14: 186-189

Bennel-Baker M, Smewing JD, Stillman CJ (1974) The Basal Complex of Fuerteventura, Canary Islands. J Geol Soc London 131: 330

Berger WH, von Rad U (1972) Cretaceous and Cenozoic sediments from the Atlantic Ocean. In: Hayes DE, Pimm AC et al. Init Rep Deep Sea Drilling Project 14: 787-954

Bernoulli D (1972) North Atlantic and Mediterranean Mesozoic facies: a comparison. In: Hollister CD, Ewing JI et al Init Rep Deep Sea Drilling Project 11: 801-871

Bernoulli D, Jenkyns HC (1974) Alpine, Mediterranean and central Atlantic Mesozoic facies in relation to the early evolution of the Tethys. In: Dott RH, Shaver RH (eds) Modern and ancient geosynclinal sedimentation. Soc Econ Paleontol Mineral Spec Publ 19: 129-160

Bernoulli D, Lemoine M (1980) Birth and early evolution of the Mesozoic Tethys: the overall situation. Mem Bur Rech Géol Min 115: 168-179

Bhat H, McMillan NJ, Aubert J, Porthault B, Surin M (1975) North American and African drift - the record in Mesozoic coastal plain rocks, Nova Scotia and Morocco. In: Yorath CJ et al (eds) Canadian continental margins and offshore petroleum potential. Mem Can Soc Petrol Geol 4: 375-389

Bouma AH (1962) Sedimentology of some flysch deposits: a graphic approach to facies interpretation. Elsevier, Amsterdam, 168 pp

Brönnimann P (1953) On the occurrence of Calpionella in Cuba. Eclogae Geol Helv 46: 263-268

Brönnimann P (1955) Microfossils incertae sedis from the Upper Jurassic and Lower Cretaceous of Cuba. Micropaleontology 1: 28-51

Cendrero A (1971) Estudio geologico y petrologico del Complejo basal de la isla de La Gomera (Canarias). Estud Geol 27: 3-74

Choubert G, Faure-Muret A (eds) (1968) Carte tectonique internationale de l'Afrique. 1:5'000'000. Assoc Afric Geol Surv, Unesco, Paris

Colom G (1954) La sedimentacion pelagica de la Isla de Maio y sus equivalentes mediterraneos. R Soc Esp Hist Nat Vol homen Prof Hernandez-Pacheco, pp 179-192

Colom G (1955) Jurassic-Cretaceous pelagic sediments of the western Mediterranean zone and the Atlantic area. Micropaleontology 1: 109-214

Colom G (1965) Essais sur la biologie, la distribution géographique et stratigraphique des Tintinnoidiens fossiles. Eclogae Geol Helv 58: 319-334

CONOCO (1969) Spansah 51A-1, offshore Spanish Sahara: Final Report by Continental Oil Company of Spain, Spanish Gulf Oil Company, Compania Espanol de Petroleas S.A. Exploration Departement, unpublished report and logs

Didon J, Durand-Delga M, Kornprobst J (1973) Homologies géologiques entre les deux rives du détroit de Gibraltar. Bull Soc Géol Fr (7) 15: 77-105

Dillon WP, Sougy JM (1974) Geology of West Africa and Canary and Cape Verde Islands. In: Nairn AEM, Stehli FG (eds) The Ocean basins and margins, vol 2. Plenum Press, New York, pp 315-390

Duncan RA, Jackson ED (1977) Geochronology of basaltic rocks recovered by D.S.D.P. Leg 41, Eastern Atlantic Ocean. In: Lancelot Y, Seibold E et al (eds) Init Rep Deep Sea Drilling Project 41: 1113-1118

Durand-Delga M (1960) Le sillon géosynclinal du flysch tithoniquenéocomien en Méditerranée occidentale. Atti Accad Lincei Cl Sci Fis Mat Nat Rend 29: 579-585

Einsele G, von Rad U (1979) Facies and paleoenvironment of Lower Cretaceous sediments at DSDP Site 397 and in the Aaiun Basin (NW Africa). In: von Rad U, Ryan WBF et al. Init Rep Deep Sea Drilling Project 47(1): 559-577

Embley RW, Jacobi R (1977) Distribution and morphology of large submarine sediment slides and slumps on Atlantic continental margins. In: Richards AF (ed) Marine slope stability. Mar Geotechn 2: 205-228

Faure-Muret A, Choubert G (1971) Domaine rifain et atlasique. In: Tectonique de l'Afrique. Unesco, Paris, pp 17-46

Friedländer I (1912) Über das Vorkommen älterer Gesteine, darunter Kalkstein mit Aptychen auf den Kapverden. Anz Akad Wiss Wien Math-natw Kl 49

Friedländer I (1913) Beiträge zur Kenntnis der Kapverdischen Inseln. Dietrich Reimer, Berlin, 109 pp

Fritsch von K (1867) Reisebilder von den Kanarischen Inseln. Petermanns Mitt Ergh 5:22: 1-44

Furnes H, Le Bas HJ, Robertson AHF, Stillman CJ, Zielonka J (1981) The Mesozoic and Tertiary history of Maio, Cape Verde Islands. Geol Soc London Newsl 10: 11

Fuster JM, Aguilar MJ (1965) Nota previa sobre la geologia del macizio de Betancuria, Fuerteventura (Islas Canarias). Estud Geol Inst Invest Geol Lucas Mallada 21: 181-197

Fuster JM, Cendrero A, Gastesi P, Ibarrola E, Lopez Ruiz J (1968) Geologia y volcanologia de los Islas Canarias - Fuerteventura. Inst Lucas Mallada Consejo Sup Invest Cient, Madrid, 239 pp

Gagel C (1910) Die mittelatlantischen Vulkaninseln. Handb Reg Geol Heidelberg 7, 10: 1-32

Gastesi P (1973) Is the Betancuria Massif, Fuerteventura, Canary Islands, an uplifted piece of oceanic crust? Nature (London) Phys Sci 246: 102-104

Grunau HR, Lehner P, Cleintuar MR, Allenbach P, Bakker G (1975) New radiometric ages and seismic data from Fuerteventura (Canary Islands), Maio (Cape Verde Islands) and Sao Tomé (Gulf of Guinea). In: Progress in geodynamics. R Soc Neth Acad Arts Sci, Amsterdam, pp 89-118

Hausen H (1958) On the geology of Fuerteventura. Soc Sci Fenn Comm Phys-Math 22, 1: 1-211

Hayes DE, Rabinowitz PD (1975) Mesozoic magnetic lineations and the magnetic quiet zone off northwest Africa. Earth Planet Sci Lett 28: 105-115

Hennig E (1913) Aptychen von den Kapverdischen Inseln. Z Dtsch geol Ges 65: 151-158

Jansa L, Gardner JV, Dean WE (1977) Mesozoic sequences of the central North Atlantic. In: Lancelot Y, Seibold E et al. Init Rep Deep Sea Drilling Project 41: 991-1031

Jenkyns HC (1980) Cretaceous anoxic events: from continents to oceans. J Geol Soc London 137: 171-188

Jérémine E (1951) Contribution à la connaissance lithologique de la Grande Selvagem. Rev Fac Cien Univ Lisboa 2: 5-20

Krejci-Graf K (1965) Die mittelatlantischen Vulkaninseln. Mitt Geol Ges Wien 57, 2: 401-431

Klerkx J, Paepe de P (1971) Cape Verde Islands: Evidence for a Mesozoic Oceanic Ridge. Nature (London) Phys Sci 233, 41: 117-118

Klerkx J, Paepe de P (1976) The main characteristics of the magmatism of the Cape Verde Islands. Ann Soc Geol Belg 99: 347-357

Lancelot Y, Seibold E et al (1978) Init Rep Deep Sea Drilling Project 41: 1259

Lancelot Y, Winterer EL et al (1980) Init Rep Deep Sea Drilling Project 50: 868

Laubscher HP (1975) Report on Excursion to Fuerteventura, Canary Islands, March 1975. Geol Newslett 2: 134-135

Laubscher HP, Bernoulli D (1977) Mediterranean and Tethys. In: Nairn AEM, Kanes WH, Stehli FG (eds) The Ocean basins and margins, vol 4A. Plenum Press, New York, pp 1-28

Lehner P, Ruiter de PAC (1977) Structural history of Atlantic margin of Africa. Am Assoc Petrol Geol Bull 61: 961-981

Lespinasse P (1975) Géologie des zones externes et des flyschs entre Chaouen et Zoumi (Centre de la chaîne Rifaine, Maroc). Thèse, Univ Paris VI, 248 pp

Martinis B, Visintin V (1966) Données géologiques sur le bassin sédimentaire côtier de Tarfaya (Maroc méridional). In: Reyre D (ed) Bassins sédimentaires du littoral africain, 1ère partie: littoral Atlantique. Assoc Serv Geol Afr, Paris, pp 13-26

Maync W (1959) The foraminiferal genera Spirocyclina and Iberina. Micropaleontology 5: 33-68

Middleton GV, Hampton MA (1973) Sediment gravity flows: Mechanics of flows and deposition. In: Middleton GV, Bouma AH (co-chairmen) Turbidites and deep water sedimentation. Pac Sect Soc Econ Paleontol Mineral, Short Course, Anaheim, pp 119-158

Mitchell-Thomé RC (1974) The sedimentary rocks of Macaronesia. Geol Rundsch 62: 1179-1216

Mitchell-Thomé RC (1976) Geology of the Middle Atlantic Islands. Gebrüder Borntraeger, Berlin, 382 pp

Mitchell-Thomé RC (1979) The Pre-Quarternary stratigraphy of the Atlantic Islands. Geol Rundsch 68: 495-522

Mutti E, Ricci Lucchi F (1972) Le torbiditi dell'Appennino settentrionale: introduzione all'analisi di facies. Mem Soc Geol Ital 9: 161-199

Mutti E, Ricci Lucchi F (1975) Turbidite facies and facies associations. In: Mutti E, Parea GC, Ricci Lucchi F, Sagri M, Zanzucchi G, Ghibaudo G, Iaccarino S (eds) Examples of turbidite facies and facies associations from selected formations in the Northern Appennines. IX Int Congr Sediment, Nice, 1975, Field Trip A 11: 21-36

Paepe de P, Klerkx J, Hertogen J, Plinke P (1974) Oceanic tholeiites on the Cape Verde Islands:

Petrographical and geochemical evidence. Earth Planet Sci Lett 22: 347-354

Pettijohn FJ (1975) Sedimentary rocks, 3rd edn. Harper-Row, London, 628 pp

Price I (1980a) Provenance of the Jurassic/Cretaceous flysch, Deep Sea Drilling Project Sites 370 and 416. In: Lancelot Y, Winterer EL et al. Init Rep Deep Sea Drilling Project Leg 50: 751-757

Price I (1980b) Gravity tectonics on a passive margin: Deep Sea Drilling Project Site 415 in the light of regional seismic data. In: Lancelot Y, Winterer EL et al. Init Rep Deep Sea Drilling Project Leg 50: 579-771

Querol R (1966) Regional geology of the Spanish Sahara. In: Reyre D (ed) Sedimentary basins of the African coasts, 1st part: Atlantic Coast. Assoc Serv Geol Afr, Paris, pp 27-39

Rigassi D (1975) Micropaleontological investigations in Cabo Verde Arcipelago. Coll Afr Geol, Leeds

Robertson AHF, Stillman CJ (1979a) Late Mesozoic sedimentary rocks of Fuerteventura, Canary Islands: implications for West African continental margin evolution. J Geol Soc London 136: 47-60

Robertson AHF, Stillman CJ (1979b) Submarine volcanic and associated sedimentary rocks of the Fuerteventura Basal Complex, Canary Islands. Geol Mag 116: 203-214

Rona P, Nalwalk AJ (1970) Post-early Pliocene unconformity on Fuerteventura, Canary Islands. Geol Soc Am Bull 81: 2117-2122

Rothe P (1968) Mesozoische Flysch-Ablagerungen auf der Kanareninsel Fuerteventura. Geol Rundsch 58: 314-332

Schlager W (1980) Mesozoic calciturbidites in Deep Sea Drilling Project Hole 416 A. Recognition of a drowned carbonate platform. In: Lancelot Y, Winterer EL et al. Init Rep Deep Sea Drilling Project 50: 733-749

Schmincke HU (1976) The geology of the Canary Islands. In: Kunkel G (ed) Biogeography and ecology in the Canary Islands. W Junk, The Hague, pp 67-184

Schmincke HU, von Rad U (1979) Neogene evolution of Canary Islands volcanism inferred from ash layers and volcaniclastic sandstones of DSDP Site 397 (Leg 47A). In: Rad von U, Ryan WFB et al. Init Rep Deep Sea Drilling Project 47 (1): 703-725

Serralheiro A (1968) Formaçoes sedimentares do Arquipelago de Cabo Verde. Junta Invest Ultramar: 7-22

Serralheiro A (1970) Geologia da Ilha de Maio (Cabo Verde). Junta Invest Ultramar: 103 pp

Soares JMP (1944-1974) A Proposito dos "Aptychi" da Ilha de Maio (Archipelago de Cabo Verde). Bol Soc Port Cien Nat 15, 1-4: 101-108

Soares JMP (1948) Observations géologiques sur les îles du Cap Vert. Bull Soc Géol Fr (5) 18: 383-389

Soares JMP (1952) A proposito da stratigrafia da Ilha de Maio. An Fac Cien Univ Porto 36: 26-35

Soares JMP (1953) A proposito dos "Aptychi" da Ilha de Maio (Arquipelago de Cabo Verde). Junta Miss Geogr Invest Ultramar: 111 pp

Stahlecker R (1934) Neocom auf der Kapverden-Insel Maio. Neues Jahrb Mineral Geol Palaeont Beil Bd Abt B 73: 265-301

Stillman CJ, Robertson AHF (1977) The dyke swarm of the Fuerteventura Basal Complex, Canary Islands. Abstr Newslett Geol Soc London 6: 8

Stillman CJ, Fuster JM, Bennel-Baker MJ, Munoz M, Smewing JD, Sagredo J (1975) Basal Complex of Fuerteventura is an oceanic intrusive complex with rift-system affinities. Nature (London) Phys Sci 257: 469-471

Trauth F (1936) Die Laevilamellaptychi des Oberjura und der Unterkreide. Ann Nat Hist Mus Wien 47: 127-145

Trauth F (1938) Die Lamellaptychen des Oberjura und der Unterkreide. Palaeontographica 88: 115-229

Uchupi E, Emery KO, Bowin CO, Phillips JO (1976) Continental margin off Western Africa. Am Assoc Petrol Geol Bull 60: 809-878

Voigt E (1962) Frühdiagenetische Deformation der turonen Plänerkalke bei Halle/Westf. als Folge einer Grossgleitung unter besonderer Berücksichtigung des Phacoid-

Problems. Mitt Geol Staatsinst Hamburg 31: 146-275

von Rad U, Arthur MA (1979) Geodynamic and volcanic evolution of the Cape Bojador continental margin (NW Africa). In: Talwani M, Hay W, Ryan WBF (eds) Deep drilling results in the Atlantic Ocean: Continental margins and paleoenvironments. Maurice Ewing Ser 3, Am Geophys Union, pp 187-203

von Rad U, Einsele G (1980) Mesozoic - Cainozoic subsidence history and palaeobathymetry of the Northwest African continental margin (Aaiun Basin to DSDP Site 397). Philos Trans R Soc London Ser A, 294: 37-50

von Rad U, Čepek P, Stackelberg von U, Wissmann G, Zobel B (1979) Cretaceous and Tertiary sediments from the Northwest African slope (dredges and cores supplementing DSDP results). Mar Geol 29: 273-312

von Rad U, Ryan WBF et al (1979) Init Rep Deep Sea Drilling Project 47 (1): 835

Walker RG (1978) Deep-water sandstone facies and ancient submarine fans: Models for stratigraphic traps. Am Assoc Petrol Geol Bull 62: 932-966

Watkins JS, Hoppe KW (1979) Seismic reflection reconnaissance of the Atlantic margin of Morocco. In: Talwani M, Hay W, Ryan WBF (eds) Deep drilling results in the Atlantic Ocean: Continental margins and paleoenvironments. Maurice Ewing Ser 3, Am Geophys Union, pp 205-217

Wiedmann J, Butt A, Einsele G (1978) Vergleich von marokkanischen Kreide-Küstenaufschlüssen und Tiefseebohrungen (DSDP): Stratigraphie, Paläoenvironment und Subsidenz an einem passiven Kontinentalrand. Geol Rundsch 67: 454-508

Cenozoic Stratigraphy, Sedimentation and Paleoenvironment

23 Oligocene and Post-Oligocene History of Sedimentation and Climate off Northwest Africa (DSDP Site 369)

Lieselotte Diester-Haass[1] and Hervé Chamley[2]

ABSTRACT

The combined study of coarse and clay fractions of Oligocene sediments recorded in DSDP Site 369, compared to results on surrounding Neogene and Quaternary sediments, leads to the following paleoenvironmental conclusions :

The whole Quaternary period off Northwest Africa was marked by alternations between humid and dry stages, linked to the shifting of the arid zone, which is presently the Sahara. The extension of the shifting can be defined by comparing data from cores located between 12° and 28° N. Arid intervals appear to be longer in the early Pleistocene than later on. Upwelling activity is most intense during periods with humid climate in the Sahara ; these are generally glacial periods (high $\delta^{18}O$ values). Upwelling is more intense in the Late Quaternary.

During the Neogene, arid stages in the northern coastal Sahara probably occurred in Early and Middle Miocene and in Early Pliocene times. Upwelling influence was locally strong during the arid Early Miocene and since the Late Pliocene beginning of the Arctic glaciation.

Compared to this Late Cenozoic history, the Oligocene sediments, marked by high opal and benthos amounts, cannot be interpreted by increased fertility due to upwelling. Oceanographic conditions different from the subsequent ones were responsible for a decrease in production of planktonic foraminifers, leading to a relative enrichment in biogenous opal and benthos. Oligocene accumulation rates of planktonic foraminifers are by a factor of about 250 lower than Quaternary ones, with similar values of carbonate dissolution in both periods. Sand-sized biogenous opal, however, has similar accumulation rates in the Late Quaternary and Oligocene.

Most of the Oligocene sedimentary changes are interpreted in terms of climatic events, sea-level changes and tectonic activity (Fig. 6). Four successive zones have been found, with a dominant influence of arid climate in the Early Oligocene, of more humid climate and numerous sea-level changes in middle-Late Oligocene, and of tectonic instability in the Latest Oligocene. Data recorded suggest that the western Sahara did not exist as a desert during the Oligocene.

1. Introduction

Coring and drilling on the continental slope provides us with pertinent information on processes in the marine realm, as well as processes on near-by land masses such as climatic changes and shifting of the shoreline. Off NW Africa, numerous sedi-

1 Geographisches Institut, Universität des Saarlandes, 6600 Saarbrücken, FRG

2 Université de Lille, E.R.A. 764 C.N.R.S., U.E.R. des Sciences de la Terre, 59655 Villeneuve d'Ascq Cedex, France

ment cores and dredges from METEOR and VALDIVIA expeditions (Seibold, 1972 ; Seibold and Hinz, 1976) and material obtained from DSDP drilling (Leg 14, Hayes et al., 1972 ; and Leg 41, Lancelot et al., 1978) allow the reconstruction of the sedimentary and climatic history during the Quaternary and Neogene. The Cretaceous is quite well studied (see Pflaumann and Čepek, Wiedmann et al., this vol.).

The Paleogene, however, and especially the Oligocene, was only rarely drilled (Sites 366, 369), cored, or dredged, and only sparse information is available on the Oligocene paleoenvironment and sedimentation (see Lancelot et al., 1978 ; von Rad et al., 1979). The aim of our investigation was to try to help fill this gap by studying the rather complete Oligocene sequence in Site 369A (26° 35,5'N, 14°59'W, 1752 m water depth) from DSDP Leg 41 (Fig. 1). Oligocene sedimentation history can be better understood if it is compared to Quaternary and Neogene sedimentation processes in the same area. The important changes in paleoenvironment, which seem to have occurred at the end of the Oligocene, require a brief review on the younger history, especially that of climate and upwelling.

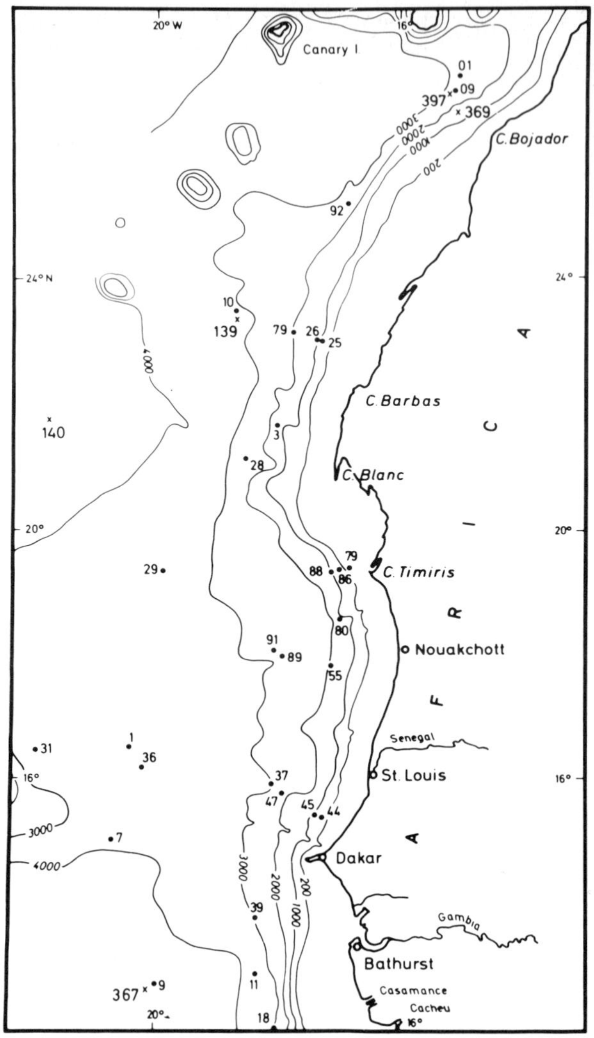

Fig. 1 . Position of investigated METEOR- and VALDIVIA-cores off NW and W Africa (dots) and of DSDP sites (Leg 14 and 41) (crosses)

2. Short Review on the Neogene and Quaternary History of Climate and Upwelling off NW Africa

2.1. Methods for Investigation of Climate and Upwelling

Our methods and results on paleoclimatic reconstructions off NW Africa have been described in several publications. Table 1 shows the general approach to the translation of sediment data into climate indicators in a highly simplified manner. In practice, of course, a great number of sedimentologic variables must be considered simultaneously. The difficulties which arise from the different weighing of the evidence, and also from differences of opinion about the environmental meaning of sedimentary parameters, are illustrated by the fact that Sarnthein and Koopmann (1980) come to conclusions regarding climatic change in the Late Quaternary which differ from ours in several ways (see Sarnthein, 1979 ; Diester-Haass, 1979a). A scheme such as that in Table 1 cannot be followed rigidly, therefore. No absolute values (e. g. x % of y component = humid or arid climate) can be given. For each specific case, possible exceptions and special conditions have to be considered. For example, the Holocene reworking of red-stained peak glacial dune sand and silt by the Senegal river leads to rather high abundances of red-stained quartz in a sediment with abundant clay and coarse terrigenous

Table 1 . Climatic interpretation of marine sediments from the NW and W African continental margin. (After Diester-Haass)

METHOD	PARAMETER	INTERPRETATION a)
1. Visual sediment observation	Sediment colors	
	brown or gray or red	Arid
	green	Humid
2. Coarse fraction analysis1)	Red-stained quartz	
	Abundant	Arid
	Rare	Humid
	Mica	
	Rare	Eolian input
	Abundant	Fluviatile input
	Wood fibers present	Fluviatile input
3. Grain-size distribution of sand-sized terrigenous matter2)	> 80 μm abundant	Fluviatile input
	> 80 μm rare or absent	Eolian input
4. Amount of < 6 μm fraction3)	High	Fluviatile input
	Low	Eolian input
5. $CaCO_3$-content	High	Low terrigenous input
	Low	High terrigenous input
6. Clay mineral analysis4) b)	Chlorite/illite ratio	
	High	Arid
	Low	Humid
	Mixed layers	
	Rare	Arid
	Abundant	Humid

a) This is a very rough guide, subject to modification, as all available evidence is considered (for example: sediment colors are also influenced by sedimentation rates; grain-size distribution can be modified by winnowing processes; $CaCO_3$-content can be modified by dissolution, etc.).

b) Applied until now only to sediments off the northern Sahara.

References : 1)Diester-Haass, 1975 ; 1976a ; 1978b; 1979a; 1979b ; Diester-Haass et al., 1973 ; Chamley et al., 1977 ; Diester-Haass and Chamley, 1978. 2) Diester-Haass, 1979b ; Diester-Haass and Chamley, 1980. 3) Lange, 1975 and unpublished results. 4) Chamley et al., 1977; Chamley and Diester-Haass, 1979 ; Chamley and Giroud d'Argoud, 1979.

fraction, plant debris, etc. (Diester-Haass, 1975). Obviously, in this case it would be quite wrong to apply the correlation red-stained quartz with arid climate.

The parameters for the recognition of upwelling influence in sediments are described in detail in Diester-Haass (1978a) and are critically reviewed in Seibold (1978) and Diester-Haass and Müller (1979), where special attention is given to the influence of river load on the fertility in the sea and the difficulty to separate it from upwelling influence.

Table 2 . Recognition of upwelling influence in marine sediments off the northern Sahara. a)

METHOD	PARAMETER b)	UPWELLING
1. Coarse fraction analysis[1])	Biogenous opal content	
	High	+
	Low	−
	Radiolarian/planktonic Foraminiferal ratio	
	High	+
	Low	−
	Benthos accumulation rate	
	High	+
	Low	−
	Plankton accumulation rate	
	High	+
	Low	−
	Benthos/plankton ratio of foraminifers	
	High	+
	Low	−
2. Micropaleontology[2])	Cool water assemblages of planktonic foraminifers	+
3. Geochemistry[3])	Opal content	
	High	+
	Low	−
	Organic carbon accumulation rate	
	High	+
	Low	−
4. Isotopes[4])	$\delta^{13}C$	
	Low	+
	High	−

a) In areas with tropical weathering, river input contains high amounts of dissolved load, leading to similar consequences for the organic production in the sea, as does upwelling in areas without river input or with river input from areas with less chemical weathering (as in Mediterranean areas). (See Diester-Haass and Müller, 1979).

b) The parameters give hints as to upwelling by comparison of upwelling-influenced and upwelling-free sediments. No absolute values can be given (saying x % of y particle means upwelling). In other upwelling areas, as, e. g., off SW Africa and Peru, other parameters exist additionally, as PO_4-, U-increase, O_2- depletion and thus absence of benthos, etc. (See Diester-Haass, 1978a).

References : [1]) Diester-Haass, 1978a ; Seibold, 1978. [2]) Thiede, 1975 ; 1977 ; Pflaumann, 1975. [3]) Müller, 1975 ; Müller and Suess, 1979 ; Suess et al., 1978 ; Diester-Haass and Müller, 1979. [4]) Berger et al., 1978 ; Killingley and Berger, 1979.

Table 2 summarizes in a schematic way the main methods of investigation and the parameters that allow an interpretation. Again, the above-emphasized caution applies in using this type of correlation list.

2.2. Late Quaternary History of Climate and Upwelling off NW and W Africa (METEOR- and VALDIVIA-cores)

Figure 2 shows schematically the principal climatic and upwelling interpretations for the Late Quaternary off NW and W Africa, based on the methods described in Tables 1 and 2. Upwelling influence in the sediments could be reconstructed only for the area off the northern Sahara (27° to 20° N). In the belt between 20° and 19° N only two cores are available. One is too far from the continent (Core 29), the other (Core 88) shows special conditions (strong quartz sand and silt input), leading to very different chemical conditions in the sediments, which do not allow a comparison with the northern cores. Furthermore this core, situated in 1000 m water depth, did not perhaps experience increased glacial upwelling. Off the northern Sahara it could be shown that the depth of strong glacial upwelling influence on the sediments was between 2000 and 3000 m depth.

South of 17° N rivers from areas with intense chemical weathering provide an important chemical load. This fertilizes the sea as does upwelling. Furthermore, the fine-grained river load creates chemical conditions favoring the preservation of organic carbon and perhaps of biogenous opal. To the present this river influence cannot be separated from that of upwelling phenomena, although it is possible to state that river influence was important in one place and not in another.

In the NW and W African area between 27° and 12° N, we distinguish four belts with different climatic evolution in the Late Quaternary. The following summary is based on a number of studies (Diester-Haass, 1975 ; 1976a ; 1978a ; 1979a ; 1980a ; 1981 ; Diester-Haass et al., 1973 ; Diester-Haass and Müller, 1979 ; Chamley et al., 1977), where details on data and interpretation may be found.

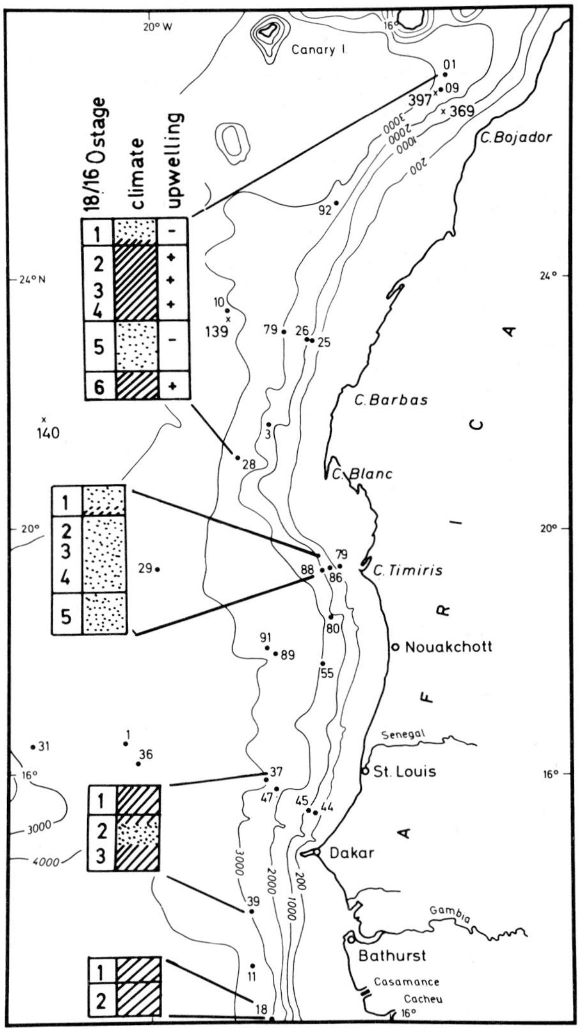

Fig. 2 . Late Quaternary sediments off NW and W Africa, climate and upwelling influence. Dotted eolian input, arid climate ; hatched fluviatile input, humid climate

1. 27° to 20° N. The interglacial Stages 1 and 5 are strongly influenced by eolian supply. The earliest part of Holocene Stage 1, however, is characterized by fluviatile supply as is the case for the entire glacial period (Stages 2, 3, 4, and 6). Upwelling influence increases in the glacial periods and is reduced in the interglacials. The best explanation for this observation is probably the increased velocity of the Canary Current (Diester-Haass, 1980a).

2. 19° to 20° N. The two cores available in this area suggest an arid

climate in glacial as well as in interglacial times. An Early Holocene humid interval could not be clearly demonstrated, although it may exist.

3. 16° to 14° N. In this area the opposite climatic trend to 20° to 27° N is found : fluviatile sediment supply to the continental slope in the Holocene Stage 1 and uppermost Stage 2 (until about 15,000y. B. P.) is preceded by eolian quartz supply (about 15,000 to 22,000 y. B. P.). Before this arid interval, river supply occurred.

4. A core near 12° N revealed river input in the interglacial Stage 1 and in the glacial Stage 2 (Diester-Haass, 1981). Due to the lack of appropriate core material between 12° and 14° N (problem of slumping on the continental slope), the southern boundary of the Sahara during Stage 2 could not be located more precisely.

Thus, the influence of the Sahara during the Holocene could be found between 27° and 19° N. In the earliest Holocene, however, the arid belt almost disappeared. The influence of the Sahara during the glacial maximum (about 15 to 22,000 y. B. P.) could be found between 19° and 14° N, showing a southward shifting of the desert belt of about 5° to 8°, that is about 600 to 900 km. At 12° N humid conditions prevailed during Holocene and late Glacial periods.

2.3. Quaternary History of Climate and Upwelling (DSDP Site 397)

DSDP Site 397 yielded a 140 m long section of Quaternary sediments (von Rad et al., 1979). The application of the approach outlined in Table 1 (especially 2 and 6) and Section 2.1 leads to the following conclusions (Diester-Haass, 1979b ; Chamley and Diester-Haass, 1979 ; Chamley and Giroud d'Argoud, 1979). In the upper part of the Pleistocene, characterized by strong variations in oxygen-isotope composition of foraminifers, at least five periods with a more humid climate have been found. These correspond to the glacial extremes of Shackleton and Cita (1979). They show increased influence of upwelling. Between these humid periods arid intervals occur with warmer water temperatures and less upwelling influence in the sediments of site 397.
In the lower Pleistocene, where oxygen isotope composition shows less intense variations, the arid intervals become longer and they contain indications of upwelling influence, which was probably less intense than in the humid Quaternary intervals.

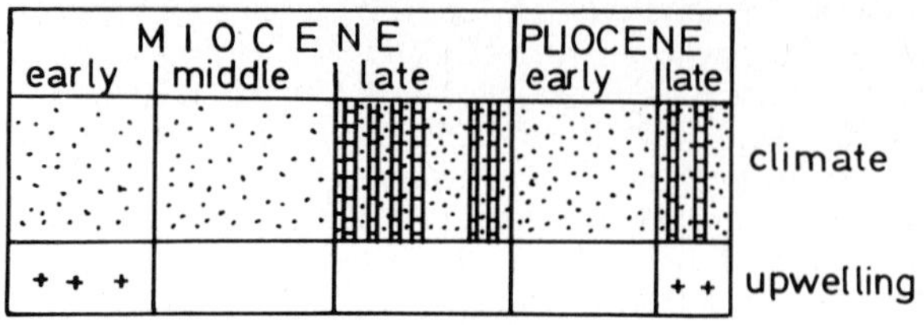

Fig. 3 . Neogene sediments off NW Africa (DSDP Sites 139, 140, 141, 369 and 397) : climate and upwelling influence. Dotted eolian input, arid climate ; hatched fluviatile input, humid climate

2.4 Neogene History of Climate and Upwelling off NW Africa (DSDP Sites 139, 140, 141, 369 and 397)

Figure 3 summarizes in a schematic graph the Neogene history of climate and upwelling off NW Africa from results described in Diester-Haass and Chamley, (1978), Chamley and Giroud d'Argoud, (1979), Sarnthein and Diester-Haass, (1977), Diester-Haass, (1978b,1979b),Chamley et Diester-Haass (1979).The application of the parameters described in tables 1 and 2 to these DSDP samples led us to conclude that in the northern coastal Sahara an arid climate existed during the early and middle Miocene and early Pliocene. In the upper Miocene several periods with river influence occurred in Sites 369 and 397, which, in the latter site, correlate to heavier oxygen isotope values (Shackleton and Cita, 1979). In the Pliocene of Site 397 two humid intervals, corresponding to higher δ^{18} O values have been found (Shackleton and Cita, 1979).

Our interpretations regarding the evolution of Neogene climate are in contradiction to those of Arthur et al. (1979), wich are based partly on poorly dated land-evidence cited from the literature. These data commonly originate from mountain areas, which may not be representative for the coastal areas (Arthur et al., 1979, p. 802). Moreover the interpretation of Arthur et al. (1979) is based on the results of Chamley and Diester-Haass (1979). Arthur et al. (1979, p. 803) mention : (1) a correlation between silty, highly stained quartz input and high mixed-layer clay and kaolinite input, (2) fine sand, poorly stained quartz input associated with high chlorite and palygorskite contents. Chamley and Diester-Haass (1979) did not report such correlations (cf. also the interpretation of Sarnthein et al., this Vol.).

Upwelling influence was strong during the arid early Miocene on the continental slope in the area of Sites 139 and 140 and of Site 369. During the Late Pliocene, starting at 3 m. y., upwelling has been found in the area of Site 397, that is with the beginning of Arctic glaciation.

3. Oligocene History of Sedimentation and Climate (DSDP Site 369)

The methods and concepts tested in Neogene and Quaternary sediments are now applied to the Oligocene sediments from DSDP Site 369, situated off the northern Sahara (Fig. 1).

The main questions posed are :

1. Was there upwelling already in the Oligocene off the Sahara, and what can be learned about Oligocene oceanography ?

2. Did the Sahara exist as a desert already in Oligocene times ?

Original and detailed data concerning these DSDP samples are published in Diester-Haass (1980b) and Diester-Haass and Chamley (1980).

3.1 Upwelling in the Oligocene ?

High opal contents as well as benthos/plankton ratios of foraminifers, which are too high for this water depth of 1752 m, have been found to be possible indicators of upwelling in areas without supply from strongly chemically weathered areas (cf. sect. 2.1, and Table 2). The high radiolarian content (Fig. 4c), as well as the high radiolarian/planktonic foraminiferal ratios (Fig. 4f) and the benthos/plankton ratios of foraminifers (Fig. 4g), which are normally about 5 in this water depth, suggest that very high fertility existed in the Oligocene at Site 369.

A schematic comparison of Oligocene percentages of planktonic foraminifers, radiolarians, benthos and benthos/plankton ratios of foraminifers with values from the Neogene (Fig. 5) shows the strong differences in sand fraction composition between both periods : smaller, but highly variable amounts of planktonic foraminifers, and much higher radiolarian and benthos amounts in the Oligocene than in the Neogene of the same Site.

The uppermost Oligocene/lowermost Miocene represents a special unit with important shallow water material (Fig. 5), and is interpreted as a period with lowered sea level and strong upwelling (Diester-Haass and Chamley, 1980).

Among the four possible reasons for the high benthos/plankton ratios of foraminifers, three are not supported by the facts :

1. A shallowing in water depth can be excluded (Lancelot et al., 1978).

2. Increased supply of benthos from shallow water. Comparing the amount of shallow water-derived glauconite and phosphorite (Fig. 4i, j) with the benthos plankton curve, we note no correlation between them, excluding a common provenance for these particles. In the Neogene of this site, however, there is a perfect correlation between both curves, suggesting that benthos and glauconite/phosphorite were supplied together from shallow water.

3. Dissolution of planktonic foraminifers, leading to an enrichment of benthonic ones. The fragmentation curve of planktonic foraminifers (Fig. 4k) can be considered as an indicator of foraminiferal dissolution (Thunell, 1976). Fragmentation values and thus dissolution in the Oligocene were not stronger than in the Neogene (Fig. 5). Furthermore, the comparison

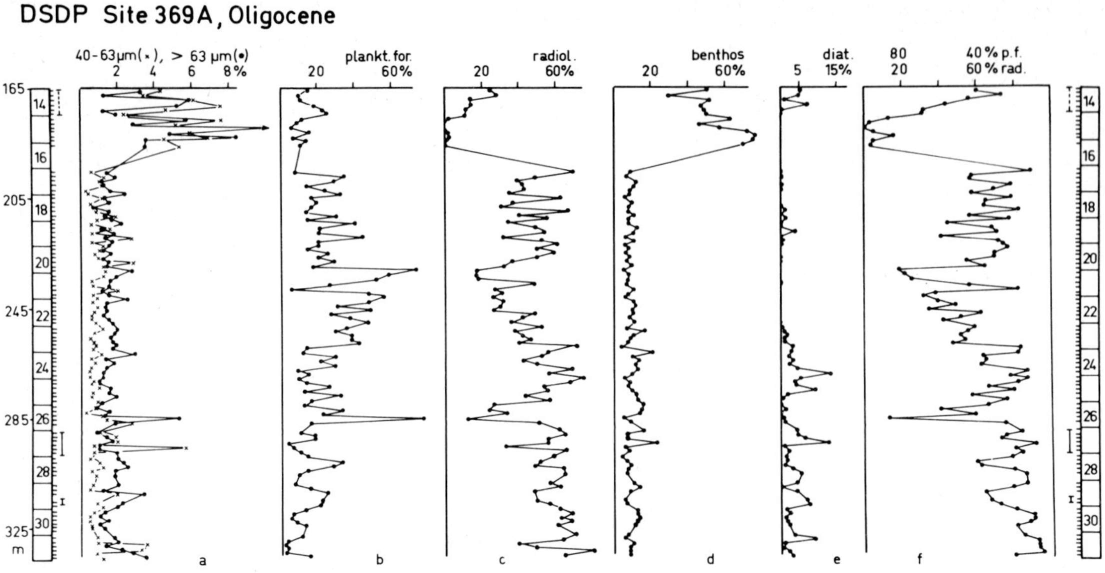

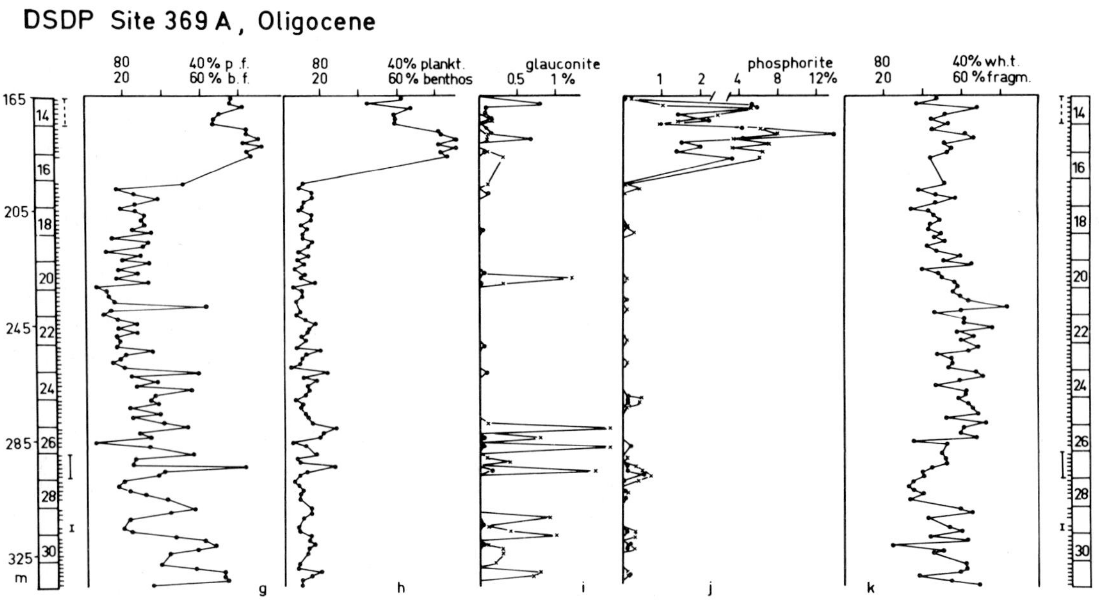

Fig. 4 . Composition of the biogenous sand fraction and amount of glauconite and phosphorite in Oligocene sediments of Site 369A.
Left and right side meters below sea floor, core numbers. Small horizontal dashes investigated samples. Vertical bars stippled supposed slump (core 14) ; solid line certain slump (core 27, 29-5).
a. % Sand (> 63 µm) fraction of total sediment ; b. % planktonic foraminifers of sand fraction ; c. % radiolarians of sand fraction ; d. % total benthos of sand fraction ; e. % diatoms of sand fraction ; f. radiolarian/planktonic foraminiferal ratio calculated as (r./r.+pl.f) x 100 ; g. benthos/plankton ratio of foraminifers, calculated as (b./b.+pl.) x 100 ; h. ratio of total benthos versus total plankton (radiolarians plus planktonic foraminifers) calculated as (b./b.+pl.) x 100. i. % glauconite of sand (dots) and 40 to 63 µm fraction (crosses) ; j. % phosphorite grains of sand (dots) and 40 to 63 µm fraction (crosses); k. ratio of fragments of planktonic foraminifers versus whole tests, calculated as (f./f.+w.t.) x 100

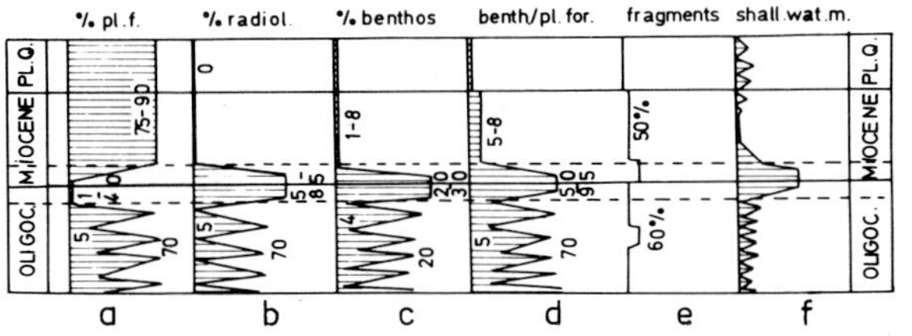

Fig. 5 . Schematic graph showing changes in biogenous sand fraction composition in the Oligocene and Neogene of Site 369/369A. Benthos/Plankton ratio of foraminifers calculated as (b./b.+pl.) x 100. Fragmentation of planktonic foraminifers calculated as (f./f.+w.t) x 100. Shallow water material in the sand fraction. Stippled line upper and lower boundary of sediment unit, which is strongly influenced by shallow water supply (Core 9 to 16, early Miocene-late Oligocene)

of the fragmentation curve and the benthos/plankton ratios of foraminifers does not show any correlation. If varying dissolution was responsible for varying benthos/plankton ratios, the curves should co-variate as they do in the Neogene of this site (Diester-Haass, 1978b).

4. Variations in productivity are, in our opinion, the best explanation for the abnormally high benthic/ planktonic foraminiferal ratio.

The highly increased radiolarian/ planktonic foraminiferal ratios in the Oligocene compared to the Neogene and Quaternary from the same site and area cannot be explained by increased foraminiferal dissolution (see above). Enhanced preservation or high production of radiolarian opal might be a reason.

To test these hypotheses, the Oligocene accumulation rates of various biogenous components were calculated after the method described in Diester-Haass (1976b), and compared with the Quaternary rates (for critical discussion see Diester-Haass, 1980b).

The method is based on two assumptions :

1. When a component constitutes 10 weight % of the sediment, it also constitutes 10 % of the length of the sediment sample. An example : 5 % of a sample consists of planktonic foraminifers. The investigated sample is 10 cm long. Then 0,5 cm of the length of the investigated sediment column consist of planktonic foraminifers.

2. For the unsampled intervals the same sediment composition has been assumed as for the sample taken above the unsampled interval.

The lengths (in cm) formed by each component in each sample have been calculated for the sampled and the unsampled intervals and summed up for the whole Oligocene. By division by the age, accumulation rates of individual biogenous components have been obtained.

The result is shown in Table 3, where accumulation rates of the total Oligocene (Cores 14 to 31), including the allochthonous (Cores 14 and 15) and the autochthonous parts (see the following paragraph), are compared to Quaternary ones (from Diester-Haass, 1976b). We distinguish various components and calculate the factor by which Quaternary accumulation rates are higher than Oligocene ones.

The main results are the following :

Accumulation rates of total sediment are about 120 times lower than in the Quaternary.

Biogenous carbonate particles have much lower accumulation rates in the Oligocene than in the Quaternary. For benthos particles the differences are not as important as for planktonic foraminifers, which were accumulated by a factor of 250 less than in the Quaternary. Since dissolution can be excluded, a strong decrease in production must be the reason.

Radiolarian skeletons were accumulated in the same order of magnitude as in the Quaternary.

These results explain the highly increased benthos/plankton ratios of foraminifers and radiolarian/planktonic foraminiferal ratios, by much reduced dilution by plankton foraminifers compared to the Quaternary. Thus the very benthos- and opal-rich Oligocene sediments are not due to increased benthos and opal produc-

Table 3 . Accumulation rates in mm/10^3 y = m/10^6 y

	OLIGOCENE Site 369A 16-2-31-6	: 14-1-31-6	QUATERNARY OFF NW AFRICA	FACTOR a) 16-2-31-6	: 14-1-31-6
Total sediment	11.0	10.1	1300	118	129
sand-sized part.:					
Planktonic foram.	0.040	0.047	10.0	250	213
Radiolarians	0.084	0.087	0.03-0.3 b)	0.36-3.6	0.34-3
Total benthos	0.015	0.053	0.41	27	8
Benth. foram.	0.014	0.051	0.30	21	6
Echinoids	0.0005	0.0008	0.03	60	38
Ostracods	0.0001	0.0001	0.008	80	80

a) Factor by which Quaternary accumulation rates are higher than Oligocene ones.
b) Holocene (small fertility) and Late Glacial (increased upwelling) values.

tion and high fertility, but to strongly reduced dilution by biogenous carbonate, especially planktonic foraminifers.

Increased amounts of benthonic foraminifers compared to planktonic ones in the Oligocene were also noted, but not explained in DSDP Site 144 (Leg 14, Hayes et al., 1972) and in Site 366 (Leg 41, Lancelot et al., 1978).

These findings exclude the assumption of upwelling in the Oligocene at Site 369.

The presented results of low planktonic and benthonic foraminiferal production rates, of a deep CCD (demonstrated by the rather low fragmentation values of planktonic foraminifers) and of a mixed siliceous-carbonate sediment facies, appear to be an accurate illustration of the "Sillen-Ocean" (Berger, 1979 ; see also Berger, 1970), which is a low-fertility ocean with near-saturation of nutrients and thus preservation of the precipitated matter. Combined low CCD and low carbonate accumulation rates (Berger, 1979) can be explained by the low foraminiferal production rates. This can be interpreted when considering the observation of Davies et al. (1977) of a cool Oligocene tropical ocean. Moreover Berger (1979) believes that a mid-ocean overturn due to low stability of the water column excludes the formation of a warm surface layer. This vertical water movement might reduce the production of planktonic foraminifers.

A fractionation of carbonate from silica, typical in the highly fertile "Harvey Ocean" (Berger, 1979), where undersaturation with nutrients leads to redissolution and biological reprecipitation, does not occur in the "Sillen Ocean". Here carbonate and siliceous materials deposit on the same place, as observed in the Oligocene of Site 369A.

3.2 Climate, Tectonics, and Eustacy in the Oligocene

The same samples that have been investigated for the question of possible upwelling influence have been studied by means of a clay and coarse fraction analysis in order to try to shed some light on the climatic history during the Oligocene (for methods see Diester-Haass and Chamley, 1980). It proved that the terrigenous particles of the clay and coarse fraction do not only contain information on the climate, but also on sea-level changes and tectonic instability, and that often the latter factors hide the paleoclimatic information.

The sedimentologic investigation led to a distinction of four zones, which are summarized schematically in Fig.

6 (original data are given in Diester-Haass and Chamley, 1980). For the interpretation of these data some remarks are necessary on the origin of the clay minerals :

1. The clay minerals do not show any sign of diagenetic influence and are considered to originate from the African continent (Mélières, 1978 ; Timofeev et al., 1979 ; Chamley and Giroud d'Argoud, 1979), except for zeolites and opal-CT (Riech and von Rad, 1979).

2. The provenance of the various clay minerals found in Site 369 is interpreted under the aspect of climate, tectonic, and eustatic movements corresponding to the opinion of Millot (1964), Pédro (1968), Paquet (1979), Chamley (1979) (see Table 4).

The coarse fraction analysis results can contribute to the climatic interpretation by two parameters : the amount and grain size of coarse terrigenous particles (see Table 1). The influence of sea-level changes can be detected by means of quantities of material supplied from shallow water, since lowering of sea level reduces the transport path of shelf particles to the slope and thus increases these particles on the slope. Red-stained quartz amounts cannot be climatically interpretated in these Oligocene sediments, since reduction of iron (pyrite formation) occurred.

The good correlation between terrigenous clay and coarse fractions allows the following interpretation of the four Oligocene sediment zones (Fig. 6, Table 4), which is summed up here.

In zone I the high smectite content and rather abundant palygorskite, as well as low kaolinite amounts, suggest a rather weak erosion on the continent, which influences only the coastal area. Upstream areas, where kaolinite is mainly formed, are not attacked by erosion. The very small amounts of coarse terrigenous particles (> 63 μm) are also in favor of weak erosion. The reason for the weak continental erosion might be either a high sea-level stand and no tectonic movement or a rather arid climate. The absence of shallow-water material on the continental slope is in favor of a high sea-level stand. The low kaolinite amount indicates a rather arid climate, as does the grain size distribution of terrigenous particles of > 63 μm, which suggests a mainly eolian input by the absence of particles of > 100

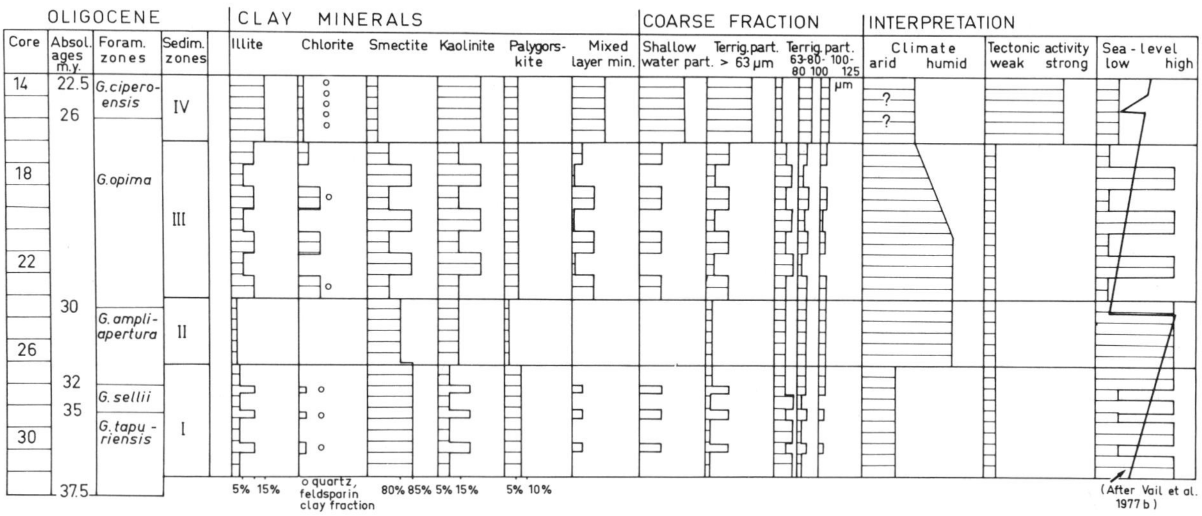

Fig. 6 . Summary graph, showing core no., absolute ages (after Berggren and Van Couvering, 1974), foraminiferal zones (Lancelot et al., 1978), our sediment zones. The variations in clay mineral composition, coarse fraction particles, and in grain size are shown schematically (for original data see Diester-Haass and Chamley, 1980). Our interpretation of the results concerns climate, tectonic activity, and sea-level changes

Table 4 . Provenance of clay minerals in Site 369. (See text for explanations)

MINERAL	AREA	PROVENANCE SOIL	CLIMATE	PRESENT OUTCROP	MAIN PERIOD OF FORMATION
Illite Chlorite	Old rocky substrates				
Fibrous clays	Marginal confined areas	Calcareous marly rocks	Arid	Coastal continental areas	Early Eocene
Smectite	Badly drained areas	Gray soils	Warm, short wet, long dry seasons	Downstream surficial soils	Paleogene
Kaolinite	Well drained areas		Hot, humid	Upstream sloping surficial areas	
Irregular mixed layers (degradations of illite, chlorite, smectite)		At the base of pedologic profiles			

μm (Diester-Haass, 1979b ; Diester-Haass and Chamley, 1980).

The three layers within zone I characterized by drastic changes (see Fig. 6) have probably been deposited during short periods of lowered sea level. This hypothesis is based on the shallow-water particles which have been provided to the slope by shelf erosion and bottom transport, and on the increase in illite and mixed layers, which is due to increased erosion attaining the deeper layers in coastal areas and also the upstream kaolinite area. Climate probably has not changed markedly during the deposition of the three layers compared to the "normal facies" of zone I.

The clay mineral association of zone II (Cores 26-4 to 23-4) suggests a rather strong continental erosion, which attacked the more distant well-drained slope areas with the kaolinite. This erosion cannot be attributed to a deepening of the erosion basis by a lowered sea level, since shallow water particles are absent. The increased erosion is attributed to increased rainfall and river erosion. This interpretation is supported by the presence of coarse terrigenous particles (80-100 and 100-125 μm size) (Diester-Haass, 1979b).

Zone III (Cores 23-3 to 17) contains two types of sediments, which alternate several times. The first type indicates continental erosion which is probably due to a marine regression. Clay minerals such as rather abundant illite and mixed layers, chlorite, and primary minerals in the clay fraction, as well as reduced kaolinite amounts, suggest erosion in the near shore basins during the regressive periods. The second type of sediment was deposited during a high sea-level stand (no shallow water particles) and less erosion, which attacked only surficial soil layers and rocks. Climatic deductions cannot be made from the "regression sediments", because the erosional influence hides the other information. The clay mineral association and the grain size spectra of the coarse terrigenous material in the "transgression sediments" point to a more humid climate similar to that of zone II and to an aridification in the upper zone III, where kaolinite decreases and smectite increases.

Zone IV (Cores 14 to 16) indicates strong erosion by the presence of abundant illite, mixed-layer minerals and quartz and feldspar in the clay fraction, and by rather high kaolinite amounts. The simultaneous increase in kaolinite and illite suggests a strong erosion in up and down stream areas. The high kaolinite percentages cannot be supplied by fluviatile erosion alone. A rejuvenation of the relief seems necessary. The shallow water particles found on the continental slope reveal a wide spectrum of grains and provenance, such as ooids from near the coastline, dolomite, as well as serpulids and other shallow-water biogenous particles. All these data suggest a regression and/or tectonic movements, affecting coastal and upstream areas. A strong tectonic activity in the uppermost Oligocene/lowest Miocene has also been suggested by Dillon and Sougy (1974), Sarnthein and Diester-Haass (1977), Diester-Haass and Chamley (1978), Diester-Haass (1978a). According to von Rad and Wissmann (this Vol.), however, no tectonic deformation of the Tertiary sediments can be noticed on seismic records.

The results concerning sea-level variations resemble the trend of the Vail et al. (1977) sea-level curve for the Oligocene (Fig. 6). Possibly the general trend of sea-level variations is due to tectonics, whereas the rapid variations in sea level might mirror Antarctic glaciations and deglaciations.

The climatic results suggest that the Sahara did not exist as a desert in Oligocene times. This assertion correlates with finding of Siesser (1978) in the Namib desert, the south-African "counterpart" of the Sahara, where humid conditions prevailed similarly in the Oligocene. The paleogeographic reconstructions of Berger and von Rad (1972) are also in agreement with this assertion, since after these authors Site 369 was in about 15° N in the Late Eocene and thus has "traversed" the subtropical semi-humid region in the Oligocene (if climatic zonality of the Tertiary can be compared to the present one). The Sahara as a desert did not develop before Early Miocene times (Sarnthein and Diester-Haass, 1977 ; Diester-Haass, 1978b ; Sarnthein, 1978 ; Diester-Haass and Chamley, 1978 ; Sarnthein et al., this Vol.).

Acknowledgments

The samples of Oligocene sediments have been provided to us by the Deep Sea Drilling Project financed by the National Science Foundation of the U. S. A. We thank W. H. Berger, U. Pflaumann, U. von Rad and E. Seibold for helpful remarks and discussions. Technical assistance of M. Acquaviva, C. F. Froget, F. Dujardin and M. Draga is gratefully acknowledged. We thank the Centre National pour l'Exploitation des Océans (France) and the Deutsche Forschungsgemeinschaft for financial support.

REFERENCES

Arthur MA, von Rad U, Cornford C, McCoy F, Sarnthein M (1979) Evolution and sedimentary history of the Cape Bojador continental margin Northwestern Africa. In : von Rad U, Ryan WBF et al. Initial Reports Deep Sea Drilling Project 47(1) : 773-816

Berger WH (1970) Biogenous deep-sea sediments : production, preservation and interpretation. In : Riley JP, Chester R (eds). Chemical oceanography, vol. 5, 2nd ed., Academic Press, London, New-York, pp 265-382

Berger WH (1979) Impact of deep sea drilling on paleoceanography. In : Talwani M, Hay W, Ryan WBF (eds). Deep Drill Res Atlantic Ocean : Continental margins and paleoenvironment M Ewing Ser 3 : 297-314

Berger WH, von Rad U (1972) Cretaceous and Cenozoic sediments from the Atlantic Ocean. In : Hayes DE, Pimm HC et al. Initial Reports Deep Sea Drilling Project 14 : 787-954

Berger WH, Diester-Haass L, Killingley JS (1978) Upwelling off North-West Africa in the Holocene: decrease as seen in carbon isotopes and sedimentological indications. Oceanol Acta 1(1) : 3-7

Berggren WA (1972) Late Pliocene-Pleistocene glaciation. In : Laughton AS, Berggren WA et al. (eds). Initial Report Deep Sea Drilling Project 12 : 953-963

Chamley H (1979 North Atlantic clay sedimentation and paleoenvironment since the late Jurassic. In : Talwani M, Hay W, Ryan WBF (eds). Deep Drill Res Atlantic Ocean : Continental margins and paleoenvironment M Ewing Ser 3 : 341-361

Chamley H, Diester-Haass L (1979) Upper Miocene to Pleistocene climates in Northwest Africa deduced from terrigenous components of site 397 sediments (DSDP Leg 47A). In : von Rad U, Ryan WBF et al. Initial Reports Deep Sea Drilling Project 47(1) : 641-646

Chamley H, Giroud-D'Argoud G (1979) Clay mineralogy of site 397, South of Canary Islands (DSDP Leg 47A). In : von Rad U, Ryan WBF et al. Initial Report Deep Sea Drilling Project 47(1) : 595-601

Chamley H, Diester-Haass L, Lange H (1977) Terrigenous material in East Atlantic sediment cores as an indicator of NW African climates. "Meteor" Forsch Ergebn 26 : 44-59

Davies TA, Hay WW, Southam JR, Worsley TR (1977) Estimates of Cenozoic oceanic sedimentation rates. Science 197 : 53-55

Diester-Haass L (1975) Sedimentation and climate in the Late Quaternary between Senegal and the Cape Verde Islands. "Meteor" Forsch Ergeb 20 : 1-32

Diester-Haass L (1976a) Late Quaternary climatic variations in North-West Africa deduced from East Atlantic sediment cores. Q Res 6 : 299-314

Diester-Haass L (1976b) Quaternary accumulation rates of biogenous and terrigenous components on the East Atlantic continental slope off NW Africa. Mar Geol 21 : 1-24

Diester-Haass L (1978a) Sediments as an indicator of upwelling. In : Boje R, Tomczak M (eds). Upwelling ecosystems.New York, pp 261-281

Diester-Haass L (1978b) Influence of carbonate dissolution,climate, sea-level changes and volcanism on Neogene sediments off Northwest Africa (Leg 41). In : Lancelot Y, Seibold E et al. Initial Reports Deep Sea Drilling Project 41 : 1 033 - 1 047

Diester-Haass L (1979a) Indicators of continental climate in marine sediments. A reply. "Meteor" Forsch Ergebn 31 : 53-58

Diester-Haass L (1979b) DSDP Site 397 : Climatological, sedimentological and oceanographic changes in the Neogene autochthonous sequence. In : von Rad U, Ryan WBF et al. Initial Reports Deep Sea Drilling Project 47(1) : 647-670

Diester-Haass L (1980a) Upwelling and climate off Northwest Africa during the Late Quaternary. In : Van Zinderen Bakker Sr EM, Coetzee JA (eds). Palaeoecol Afr 12 : 229-238

Diester-Haass L (1980b) Oligocene at Site 369A (DSDP Leg 41) : Reduced carbonate accumulation rates lead to high opal content. Mar Geol 36 : 35-48

Diester-Haass L (1981) Late Quaternary sedimentation processes on the West African continental margin and climatic history. "Meteor" Forsch Ergeb ; (in prep)

Diester-Haass L, Chamley H (1978) Neogene paleoenvironment off NW Africa based on sediments from DSDP Leg 14. J Sediment Petrol 48 : 879-896

Diester-Haass L, Chamley H (1980) Oligocene climatic, tectonic and eustatic history off NW Africa (DSDP Leg 41, Site 369). Oceanol Acta 3 : 115-126

Diester-Haass L, Müller PJ(1979) Processes influencing sand fraction composition and organic matter content in surface sediments off W Africa (12-19°N). "Meteor" Forsch Ergeb 31 : 21-47

Diester-Haass L, Schrader HJ, Thiede J (1973) Sedimentological and paleoclimatological investigations of two pelagic ooze cores off Cape Barbas, North-West Africa. "Meteor" Forsch Ergeb 16 :19-66

Dillon WP, Sougy JMA (1974) Geology of West Africa and Canary and

Cape Verde Islands. In : Nairn EM, Stehli FC (eds). The Ocean basins and margins, vol. 2, The North Atlantic. Plenum Press, New-York, London, pp 315-390

Hayes DE, Pimm AC et al. (1972) Initial Reports of the Deep Sea Drilling Project 14 : 975 p

Killingley JS, BERGER WH (1979) Stable isotopes in a mollusc shell : detection of upwelling events. Science 205, 4402 : 186-188

Lancelot Y, Seibold E et al. (1978) Initial Reports of the Deep Sea Drilling Project 41 : 1259 p

Lange H (1975) Herkunft und Verteilung von Oberflächensedimenten des west-afrikanischen Shelfs und Kontinentalhanges. "Meteor" Forsch Ergeb 22 : 61-85

Mélières F (1978) X-Ray mineralogy studies, Leg 41, Deep-Sea Drilling Project, Eastern North Atlantic Ocean. In : Lancelot Y, Seibold E et al. Initial Reports Deep Sea Drilling Project 41 : 1065-1086

Millot G (1964) Géologie des argiles. Masson, Paris, 449 p

Müller PJ (1975) Diagenese stickstoffhaltiger organisher Substanzen in oxischen und anoxischen marinen Sedimenten. "Meteor" Forsch Ergeb 22 : 1-60

Müller PJ, Suess E (1979) Productivity, sedimentation rate and sedimentary organic matter in the oceans. I - Organic carbon preservation Deep Sea Research 26A/12 : 1347-1362

Paquet H (1979) Evolution géochimique des minéraux argileux dans les altérations et les sols des climats méditerranéens et tropicaux à saisons contrastées. Mém Serv Carte Géol Alsace-Lorraine Strasbourg 20 : 210

Pedro G (1968) Distribution des principaux types d'altération chimique à la surface du globe. Présentation d'une esquisse géographique. Rev Géogr Phys Géol Dyn 10 : 457-470

Pflaumann U (1975) Late Quaternary stratigraphy based on planktonic foraminifera off Senegal. "Meteor" Forsch Ergeb 23 :1-16

Riech W, von Rad U,(1979) Silica diagenesis in the North Atlantic basin : diagenetic potential and transformations. In : Talwani M, Hay W, Ryan WBF (eds). Deep Drill Res Atlantic Ocean : Continental margins and paleoenvironment M Ewing Ser 3 : 315-340

Sarnthein M (1978) Sand deserts during glacial maximum and climatic optimum. Nature, London, 272, 5649 : 43-46

Sarnthein M (1979) Indicators of continental climates in marine sediments. A discussion. "Meteor" Forsch Ergeb 31 : 49-51

Sarnthein M, Diester-Haass L (1977) Eolian sand turbidites. J Sediment Petr 47, 2 : 868-890

Sarnthein M, Koopmann B (1980) Late Quaternary deep-sea record on North-west African dust supply and wind circulation. In : Van Zinderen Bakker Sr EM, Coetzee JA (eds). Palaeoecol Afr 12 : 239-253

Seibold E (1972) Cruise 25/1971 of R. V. "Meteor" : Continental margin of West-Africa. General report and preliminary results. "Meteor" Forsch Ergebn 10 : 17-38

Seibold E (1978) Sediments in upwelling areas, particularly off Northwest Africa. Contrib to Rapports et Procès verbaux, Ser ICES, Copenhagen, p 81

Seibold E, Hinz K (1976) German cruises to the continental margin of North West Africa in 1975. General reports and preliminary results from "Valdivia" 10 and "Meteor" 39. "Meteor" Forsch Ergebn 25 : 47-80

Shackleton NJ, Cita MB (1979) Oxygen and carbon isotope stratigraphy of benthic Foraminifers at Site 397 : detailed history of climatic change during the Late Neogene. In : von Rad U, Ryan WBF et al. Initial Reports Deep Sea Drilling Project 47(1) : 433-445

Siesser WG (1978) Aridification of the Namib desert : Evidence from oceanic cores. In : Van Zinderen-Bakker EM (eds). Antarctic glacial history and world paleoenvironments. Balkema, Rotterdam, pp 105-113

Suess E, Thiede J, Müller PJ (1978) Accumulation patterns of biogenous and terrigenous sediment constituents off NW Africa (Abstr). 10 th Int Congr Sedimentol, Jerusalem, July 9-14, II : 650-651

Thiede J (1975) Shell and skeleton producing plankton and nekton in the eastern North Atlantic Ocean. "Meteor" Forsch Ergebn 20 : 33-79

Thiede J (1977) Aspects of the variability of the glacial and interglacial North Atlantic eastern boundary current (last 150 000 years). "Meteor" Forsch Ergebn 28 : 1-36

Thunell R (1976) Optimum indices of calcium carbonate dissolution in deep-sea sediments. Geology 4 : 525-528

Timofeev PP, Eremeev VV, Rateev MN (1978) Palygorskite, sepiolite, and other clay minerals in Leg 41 oceanic sediments. Mineralogy, facies and genesis. In : Lancelot Y, Seibold E et al. Initial Reports Deep Sea Drilling Project 41 : 1087-1101

Vail PR, Mitchum Jr RM, Thompson S (1977) Seismic stratigraphy and global changes of sea-level, part 4 : global cycles of relative changes of sea-level. In : Payton CE (ed). Seismic stratigraphy-applications to hydrocarbon exploration. Am Assoc Petrol Geol Tulsa Mem 26 : 83-97

von Rad U, Ryan WBF et al. (1979) Initial Reports of the Deep Sea Drilling Project 47(1) : 835 p

von Rad U, Čepek P, Stackelberg U von, Wissmann G, Zobel B (1979) Cretaceous and Tertiary sediments from the northwest african slope (dredges and cores supplementing DSDP results). Mar Geol 29 : 273-312

24 Atmospheric and Oceanic Circulation Patterns off Northwest Africa During the Past 25 Million Years

Michael Sarnthein[1], Jörn Thiede[2], Uwe Pflaumann[1], H. Erlenkeuser[1], Dieter Fütterer[1], Bernhard Koopmann[1], Heinz Lange[1], and Eugen Seibold[1]

CONTENTS

1 Geologisch-Paläontologisches Institut und Museum, Christian-Albrechts-Universität, Olshausenstraße 40/60, 2300 Kiel, FRG

2 Institutt for geologi, Universitetet i Oslo, Postboks 1047, Blindern, Oslo 3, Norway

ABSTRACT

The sediments of the eastern Atlantic contain excellent historical records of the patterns of oceanic and atmospheric circulation in the subtropics. This is particularly the case at the low relief northwest African continental margin which favors unrestrained interaction of the land-sea climatic system and which forms a schematic, vertical cross section through the ocean from the equator to Mediterranean latitudes. Our synthesis of the Neogene and Quaternary evolution of paleoenvironments along this margin tries to show that oceans and land respond in a complementary way to global climatic events and mechanisms.

Composition and fabrics of surface sediments produce a detailed record on (1) the stratification of modern deep-water circulation (Antarctic Bottom Water, North Atlantic Deep Water, various currents along the slope such as the South Atlantic Central Water, oxygen-minimum layer), (2) the geometric patterns of ecologic surface-water masses (for example, the North Atlantic eastern boundary current, north equatorial water, upwelling water, freshwater discharge), (3) their temperature distribution in general, and (4) the major structures of climate and wind circulation (continental wetness and aridity, zonal and meridional wind systems, and the respective wind speeds, summer position of the Intertropic Convergence Zone). This record basically contributes to the understanding of the Quaternary and Neogene history which, however, shows a number of additional environmental patterns which have no modern counterparts.

Estimates from the planktonic foraminiferal record suggest that during the last interglacial climatic optimum at 6000 years B.P., surface-water temperatures of the Canary Current were higher during summer and only slightly higher during winter as compared to today. This increase of seasonality can signal a summer breakdown of coastal upwelling and of meridional Trade Winds which correlates with the reduced supply of northwest Saharan dust. The Harmattan dust outbreaks and the ITCZ occupied approximately the same summer position as today. Continental wetness spread over most parts of the Sahara and resulted in an extended discharge of fluvial muds.

During the last glacial mode at 18,000 years B.P., the estimates from planktonic foraminiferal counts show that the North Atlantic polar front had shifted to lat. 42°-45°N and bordered almost directly upon the subtropical gyre, whose geometry and temperatures differed only slightly from today. The intensified Canary Current cooled down by 6°-8°C and expanded some 500 km further south. The productivity of coastal upwelling was high as a result of both a subsurface expansion of nutrient-rich SACW and strenghtened and perennial Trade Winds.

Distribution and composition of aeolomarine dust and the upwelling record suggest that during glacial stages the ITCZ was essentially in its interglacial position with seasonal migrations comparable to that of today. Meridional winds intensified and zonal winds decreased by about 50 %. As a result, continental aridity and dune fields spread over large parts of north Africa between lat. 10° and 28°N, river load disappeared, even in front of the Niger delta.

In the deep-water circulation, the record of benthic foraminifers suggests a replacement of the well-oxygenated North Atlantic Deep Water by less oxygenated water masses. The last (short) transition from glacial to interglacial modes shows time-transgressive pulses in the occurrence of planktonic groups advancing from south to north, and a slightly delayed onset of continental wetness near 12,000 years B.P.

The Neogene oxygen-isotopic record of the Deep-Sea Drilling Sites 366 and 397 suggests six major glacial-style intervals (24-20 m.y., 18-14 m.y., 13-?9.5 m.y., 7.5-5.3 m.y., 3.2-1.9 m.y., and since 0.73 m.y.) marking a stepwise general climatic deterioration leading to the Present. Each interval correlated with enhanced oceanic circulation and deep-sea erosion, glacial-style meridional winds, and aridity spreading over north Africa. The intervening phases amounted to an equal range of time (20-18 m.y., 14-13 m.y., ?9.5-7.5 m.y., 6.8-6.0 m.y., 5.3-3.2 m.y., and 1.9-0.73 m.y.) and led to climatic stability or amelioration. They correlated with high sea level, calm oceanic and prevailing zonal wind circulation, and with continental wetness. During two time intervals, during the Early Miocene until 14 m. y. B.P.and during the last 3 m.y., sediments record enhanced fertility at the continental margin which was induced by upwelling as well as by intense river discharge. The Neogene northward drift of Africa resulted in a gradual southward shift of the position of wind systems and the correlated climatic belts measured relative to the African plate.

1. INTRODUCTION

The wind regime of the Wheel Round Latitudes (WRL) largely determines the climatic zonation of the subtropics in North Africa and at the NW African continental margin (Fig. 1). It controls both the Saharan aridity and the surface circulation in the eastern Atlantic, particularly the Canary Current and the coastal upwelling. Further north, the westerlies during summer and polar troughs during winter characterize the Mediterranean climate belt as well as the eastward-directed surface-current system in the Atlantic. Toward the south, the zone of tropical disturbances resulting in a hot and humid tropical climate is separated by a major climatic boundary, the Intertropic Convergence Zone (ITCZ). Most of these climatic and oceanographic patterns shift north and south seasonally, and vary in intensity. But are these patterns documented in the marine sedimentary record? How did they change from Interglacial to Glacial times, from pre-Quaternary to Quaternary times? What was the response to equally strong paleoenvironmental events which were documented during Pliocene and Miocene times (Kennett, 1977)? How are changes in the oceanic surface circulation interrelated with fluctuations of the intermediate and deep-water circulation? What is the series of oceanographic and sedimentary events during an environmental change and by which

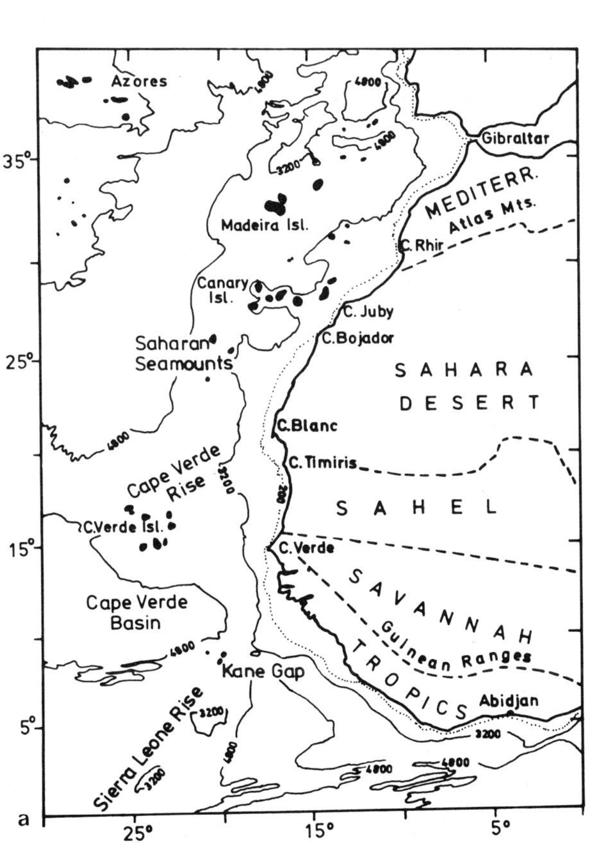

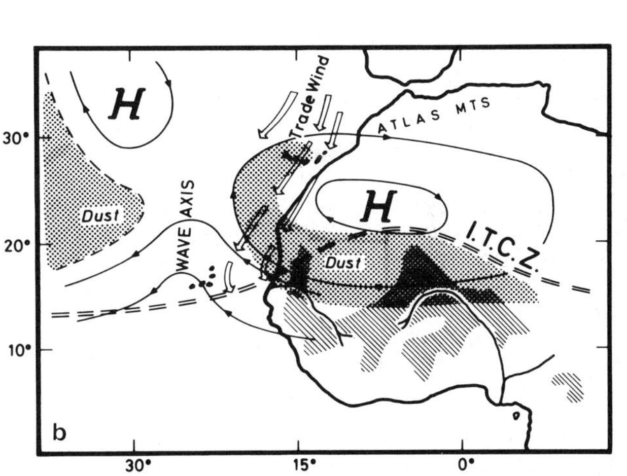

Fig. 1. Northwest African continental margin. a North African climatic zones and bathymetry of the eastern Atlantic. b Schematic representation of lower mid tropospheric flow patterns and underlying Trade Winds associated with an Easterly Wave and a Saharan air outbreak during July (Sarnthein et al., 1981). Major source areas of lateritic dust particles marked by oblique hatching. I.T.C.Z. shows the average summer position of the Intertropical Convergence Zone at the surface

major factors are the mechanisms controlled? Indeed, a number of different, partly opposed models on the paleooceanographic and paleoclimatic evolution of North Africa and the adjacent Atlantic already exist. Fragmentary evidence from continental and marine sections appears to be the major reason for the disagreement (see Ch.2.1.).

In our synthesis we try to compile the deep-sea record of paleooceanography and paleoclimate at the eastern Atlantic continental margin by comparison of modern, Quaternary and Neogene times. Our approach is based on results from almost 15 years of research, the publications of which are spread widely throughout the scientific literature.

The deep-sea sediment record is in many respects superior to that from land in addressing large-scale paleoenvironmental and climatic problems: (1) In general, the deep-sea sediments continuously store information, whereas the terrestrial records are characteristically incomplete. For example, dune fields (and/or lakes) are short-lived and rapidly shifting features in geological history, and the deposits remaining on land are permanently endangered by renewed reworking. (2) Contrary to terrestrial sediments, the stratigraphy and absolute dating of Late Cenozoic deep-sea sediment have reached a high degree of precision. (3) Finally, the fluctuations in the composition of stable-isotopes, and in calcium-carbonate contents in deep-sea sediments provide an opportunity to correlate local phenomena directly with the global evolution of climate and oceanography.

2. MAJOR CLIMATIC AND OCEANOGRAPHIC PATTERNS

2.1 Atmospheric Circulation

Briefly summarized, the following features of north African climate and, more specifically, of atmospheric circulation can be inferred from oceanic sediments in the Eastern North Atlantic Ocean:

The extent of the arid Saharan belt (with less than 100 mm annual precipitation) should be reflected by a lack of fluvial sediment supply and by the occasional deposition of dune-sand. Today this zone extends from 17° to 28°N (Schiffers, 1971; Rodier, 1975).

The southward penetration of Mediterranean winter rains should be documented by river sediments, inspite of the locally intensified river supply produced by the Atlas and Anti-Atlas mountain ranges close to the west African continental margin. The calcretes and crusts which are characteristic of the semi-arid proportions of the Mediterranean climate belt (less than 300-400 mm annual precipitation; Yaalon, 1978) also provide a specific soil-derived mineral assemblage contained in the dust contribution.

The northward penetration of monsoonal summer rains should also be recorded by the fluvial mud contribution and by characteristically lateritic soil weathering products in the dust flux.

The maximum northward penetration of the Intertropic Convergence Zone (ITCZ) during summer (today at 17°-19°N, Fig. 1) can be documented by the aeolian dust contribution from the south Saharan and Sahel zones to the Atlantik Ocean. These dust outbreaks are connected to a system of tropical disturbances (Easterly Waves) which appear south of the ITCZ at a 3- to 4-day cycle (Prospero and Carlson, 1972; Tetzlaff and Wolter, 1980). The Easterly Waves generate the dust plumes by coupling gusty surface winds with a strong ascending wind component. Once having penetrated through the north-dipping boundary layer of the ITCZ into the overlying easterly mid-tropospheric jetstream, the Harmattan or Saharan Air Layer (Semmelhack, 1934; Jaenicke and Schütz, 1978), the dust cloud moves west above the Trade Wind inversion (at 700-1500 m) with a maximum concentration around 3000 m between 17° and 21°N (Diaz et al., 1976). It passes around the upper-air high-pressure system found over the western Sahara, and finally forms a sickle-shaped trajectory over the eastern north Atlantic (Fig. 1; Tetzlaff and Wolter, 1980). This circulation pattern can also be seen on satellite imagery (Szekielda, 1978) and is corroborated by more than 30 years of weather observations in Tenerife (Abel et al., 1968; Bartels, Berlin, pers. comm.), where most Saharan dust turbidity arrives with the descending air masses above the atmospheric inversion, mainly during July and August. During these months, the winds come from the western to north-northwestern sector about two thirds of the time and from the southwest about 10 % of the time. Additional southwest Saharan dust crosses the Atlantic ocean in the wake of the Easterly Waves.

The shallow Trade-Wind layer beneath the Harmattan flow has mainly been considered responsible for the Saharan dust supply, but, due to its north-south direction, it contributes today only a limited amount of dust originating from the northwestern fringe of the Sahara to a limited offshore range, as confirmed by the suspension data of Milliman (1977) and Koopmann (1979). The Trade Winds obtain their maximum strength during late spring (peaks up to 24 m/s; Sarnthein and Walger, 1974; Hughes and Barton, 1974). They blow basically parallel to the shoreline, south of 18°N even slightly onshore toward the SSE. Accordingly, they can displace the fall-out from the dust plumes above toward the south; however, the maximum Harmattan dust outbreaks occur during June to September with some seasonal delay to the Trade Wind maximum (Tetzlaff and Wolter, 1980).

The major zones of vegetative cover (tropics, savannah, Sahel, Sahara, Mediterranean, mountain ranges) will probably be only roughly delineated with respect to their position by the marine pollen record because of large-scale mixing processes during aeolian transportation (Schulz, 1976; Agwu and Beug, 1981; Rossignol and Duzer, 1979).

Short and mid-range climatic changes (at a maximum time resolution of 200 to 2000 years) should be recorded with regard to their regional extension, their rates and amplitudes, their phase relationships and their connection to oceanic circulation and carbonate chemistry. It would be of particular interest to find further evidence to evaluate the major models which compete in interpreting the relation of subtropical to global climatic regimes during Glacial and Interglacial times. In particular, one hypothesis (Penck, 1973; Fairbridge, 1976; Rognon and Williams, 1977; Nicholson and Flohn, 1980) suggests a rather constant north-south extent of the desert belt during warm and cold stages, but distinct latitudinal shifts parallel to the expansion and shrinkage of the northern hemisphere ice caps. Another hypothesis (Newell et al., 1975; Manabe and Hahn, 1977; Sarnthein, 1978; Talbot, 1980; Maley, 1980) assumes a basically stable latitudinal position for the core of the north African arid belt, but proposed expansion during cold and contraction during warm climatic regimes.

Long-range climatic evolution (millions of years) may be related to the presence or absence of ice caps in the northern hemisphere, movements of the African plate which would lead to a change of latitudes, to the folding of the Atlas mountains, and to the formation of the oceanographic barrier of the Canary volcanic islands.

2.2 Surface Water Circulation

Surface winds exert a decisive influence upon the small and large-scale oceanic surface water circulation along the northwest African continental margin. The sluggish, wide, and shallow eastern boundary current (Wooster et al., 1976) between the Eurafrican continental margin and the central subtropical Sargasso Sea waters (Fig. 2) originates from the North Atlantic drift. It changes its name from Portugal to Canary current as it flows along the Eurafrican continental margin southward to Cape Verde. At approximately 15°N, the current diverges. The dominant fraction of this surface water leaves the continental margin and, driven by the Trade Winds, it turns southwest and west. It forms the westward-flowing North Equatorial current which associated with the equatorial upwelling zone. A minor part of the waters from the Canary current continues south and south-eastwards along the coast at least during portions of the year as Guinea current, until it joins the equatorial counter-current at approximately 5-7°N.

The broad and relatively simple pattern of the eastern boundary current is locally modified by the presence of islands (upwelling around the Canary and Cape Verde Islands), by the morphology of the shelf and the location of the shelf break, and

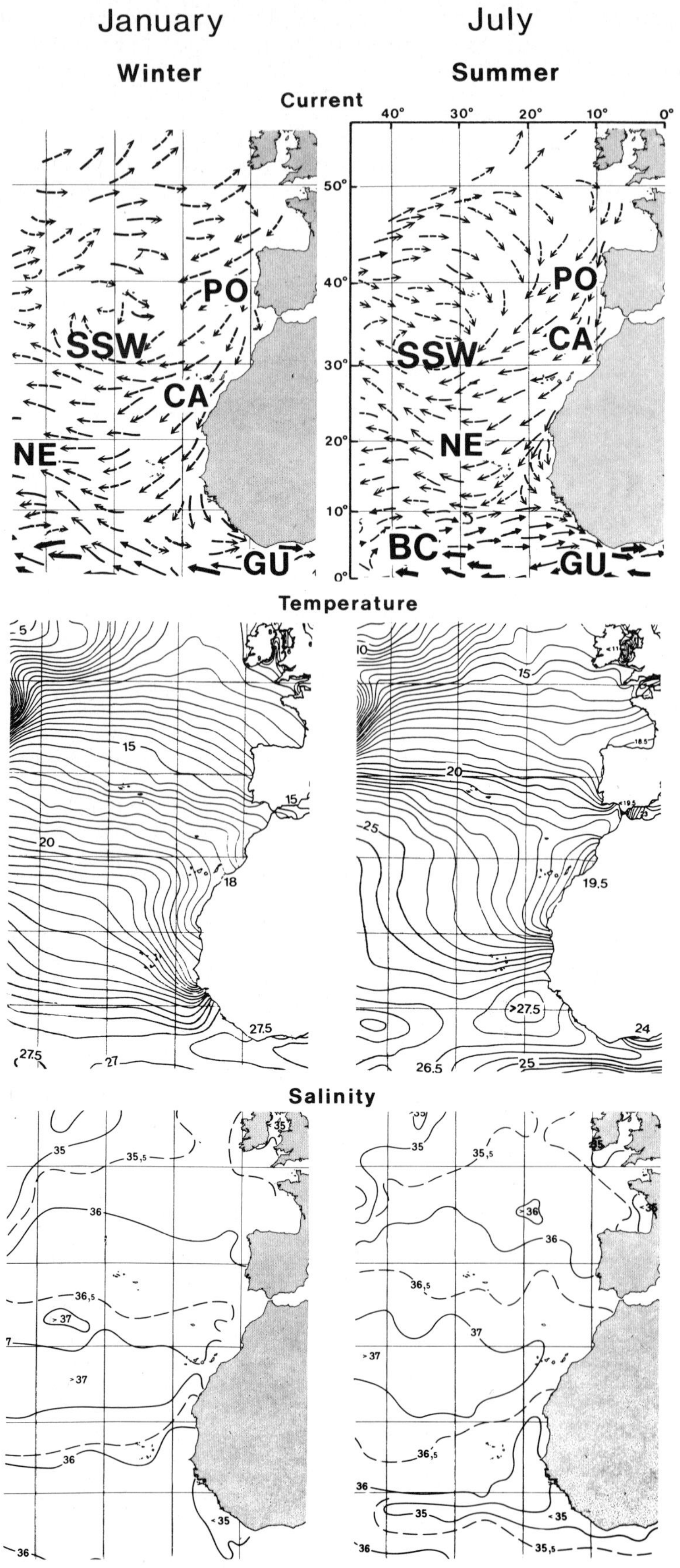

Fig. 2. Hydrography of the eastern North Atlantic Ocean off northwest Africa. (Simplified after Deutsches Hydrographisches Institut, 1967 and Robinson et al., 1979). Surface-water isotherms are average temperatures in °C; Isohalines are average salinities (salt per kg of sea water). PO Portugal Current; CA Canary Current; GU Guinea Current; NE Northequatorial Current; SSW Sargasso Sea Water; BC distal extension of Benguela Current. Winter January; Summer July

by the influence of local winds whose intensity and directions change seasonally (Schemainda et al., 1975). Monsoonal and other local winds which are blowing northward seasonally, can also produce reversals of the direction of this general flow of the surface water (Mittelstaedt et al., 1975) and they can drive warm tropical waters (Guinean Water, GW) northward along the coast as far as Cape Blanc.

The coastal upwelling of cool and nutrient-rich waters from the subsurface circulation is the most conspicuous and geologically important response of the ocean to the surface winds. Upwelling has been observed along the entire northeast Atlantic continental margin, however, it is most strongly developed throughout the year at 20°-25°N (Wooster et al., 1976). Seasonal shifts occur north and south of that area (Schemainda et al., 1975). For example, a distinct summer maximum is observed at 30°N, south of the Atlas Mountains. Upwelling is generated by a complex interaction of the Trade Winds parallel to the coast with the Canary Current and the rotation of the Earth (Ekman transport). They generate a momentum dragging the surface layer of the oceanic waters seaward and replacing it with waters from below. Contrary to common belief, coastal upwelling does not permanently affect the surface water very close to the coastline, but mainly that over the middle-outer shelf and upper continental slope. Off northwest Africa, the width of the belt of surface waters, the hydrography of which (temperature,salinity, nutrients) has been affected by upwelling, rarely exceeds 2° longitude. In the case of very narrow shelves such as during periods of lowered late Cenozoic sea levels, one can expect very steep gradients and rapidly changing hydrographic properties in the surface water. They might occur over a narrow zone along the continental margin at the transition from upwelling affected to normal waters of the Canary current.

The interaction of the Canary current, which transports cool surface waters from the north, with the coastal upwelling results in the following hydrographic properties of the surface water (Schemainda et al., 1975). As compared to the adjacent open ocean which is particularly unproductive, the upwelling waters are often as much as 10°C colder and 1 ‰ less saline. Their dissolved oxygen content is reduced, but their nutrient concentrations are as much as one order of magnitude higher, particularly where the upwelled water derives from the nutrient-rich South Atlantic Central Water (SACW), i.e., south of 22°N (Fig. 4; Schemainda et al., 1975; Codispoti and Friederich, 1978). The patchiness and irregularity of the single upwelling events which have been observed during synoptic oceanographic experiments are grossly averaged by the geological signals stored in the underlying sediments.

The interaction of coastal upwelling and the Canary current is depicted in both the biogeography of living planktonic organisms and in the biological productivity of the surface waters off northwest Africa. For example, planktonic foraminiferal species typical of temperate to cool surface water (Fig. 3 A) follow the Canary current southward along the continental margin down to very low latitudes (de Miro, 1973;, Thiede, 1975; Cifelli and Stern-Benier, 1976). However, the upwelling water masses are inhabited by plankton assemblages which are characterized by important contributions of subsurface dwellers. Accordingly, these assemblages can serve as biological indicators to outline the regional extent and the different origins and qualities of upwelling water (Fig. 3 B, C). Off northwest Africa, the upwelling is fed in the northern part from the nutrient-poorer North Atlantic Central Water (NACW) and is less productive. However, in the southern part the upwelling is fed from nutrient-rich subsurface water originating from the South Atlantic Central Water and carrying its own deep-dwelling plankton assemblages (for example, *Globorotalia menardii*, see Fig. 3 C; and certain radiolarian species). It can lead to very anomalous and complicated biogeographic patterns (Goll and Bjørklund, 1972; Thiede, 1975; Mateu, 1979; Labracherie, 1980) and to locally high concentrations of organic carbon and siliceous microfossils on the sea floor. However,these coastal-upwelling-derived sediments can only be expected in short and narrow belts parallel to the continental margin. Finally, the sediments have to pass a number of "filters" which alter the primary message (Seibold, 1980).

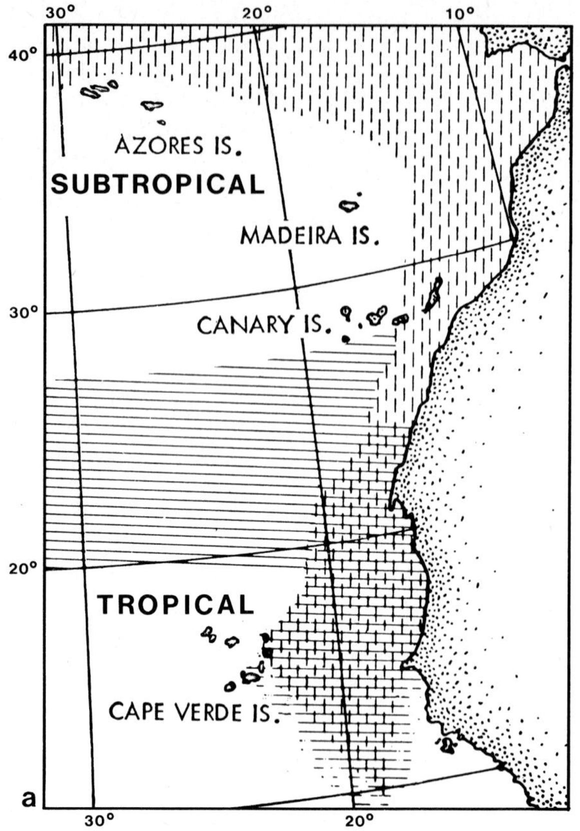

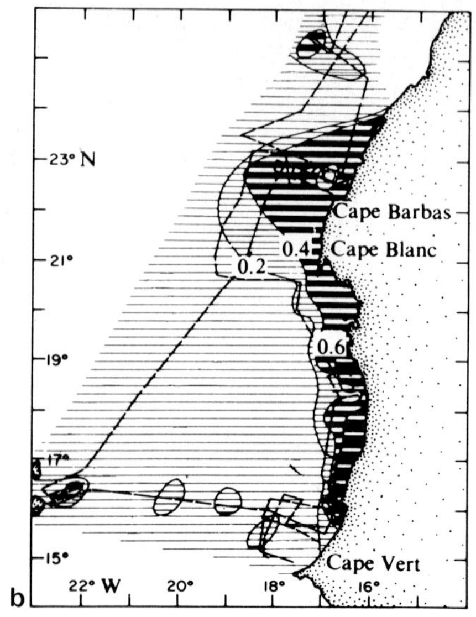

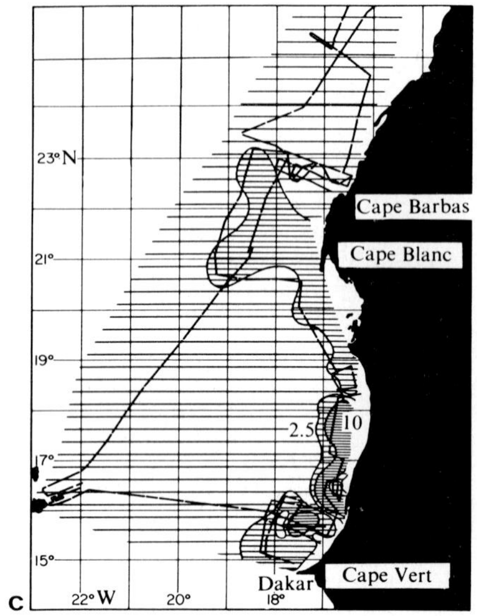

Fig. 3. Biogeographic patterns of modern planktonic foraminifers inhabiting the surface water masses off northwest Africa.
a (after Cifelli and Stern-Benier, 1976): Horizontal lines indicate a mixture of tropical and subtropical faunas, vertical lines indicate a mixture of northern and subtropical faunas, crossed lines indicate a mixture of all three faunas.
b (after Thiede): Distribution of planktonic foraminiferal assemblages typical for cool, upwelled water. Values are factor loadings of a Q-mode factor analysis.
c (after Thiede, 1975): Distribution of *Globorotalia menardii* (in percent of the total planktonic foraminiferal faunas) in surface-water plankton samples

Fig. 4. Stratification of oceanic circulation (arrows) near the sea bottom along the eastern North Atlantic continental margin. (Modified from Lutze, 1980, with supplements).
a Distribution of bottom water temperatures (°C) and oceanic currents (long time averages). AABW Antarctic Bottom Water; NADW North Atlantic Deep Water; MOW Mediterranean Outflow Water; SAIW South Atlantic Intermediate Water; EUC Equatorial Undercurrent; NACW North Atlantic Central Water; SACW South Atlantic Central Water; SW Surface Water; CC Canary Current; GW Guinea Water Mass. Derivation of data see text.
b Distribution of bottom-water oxygen (ml/l)

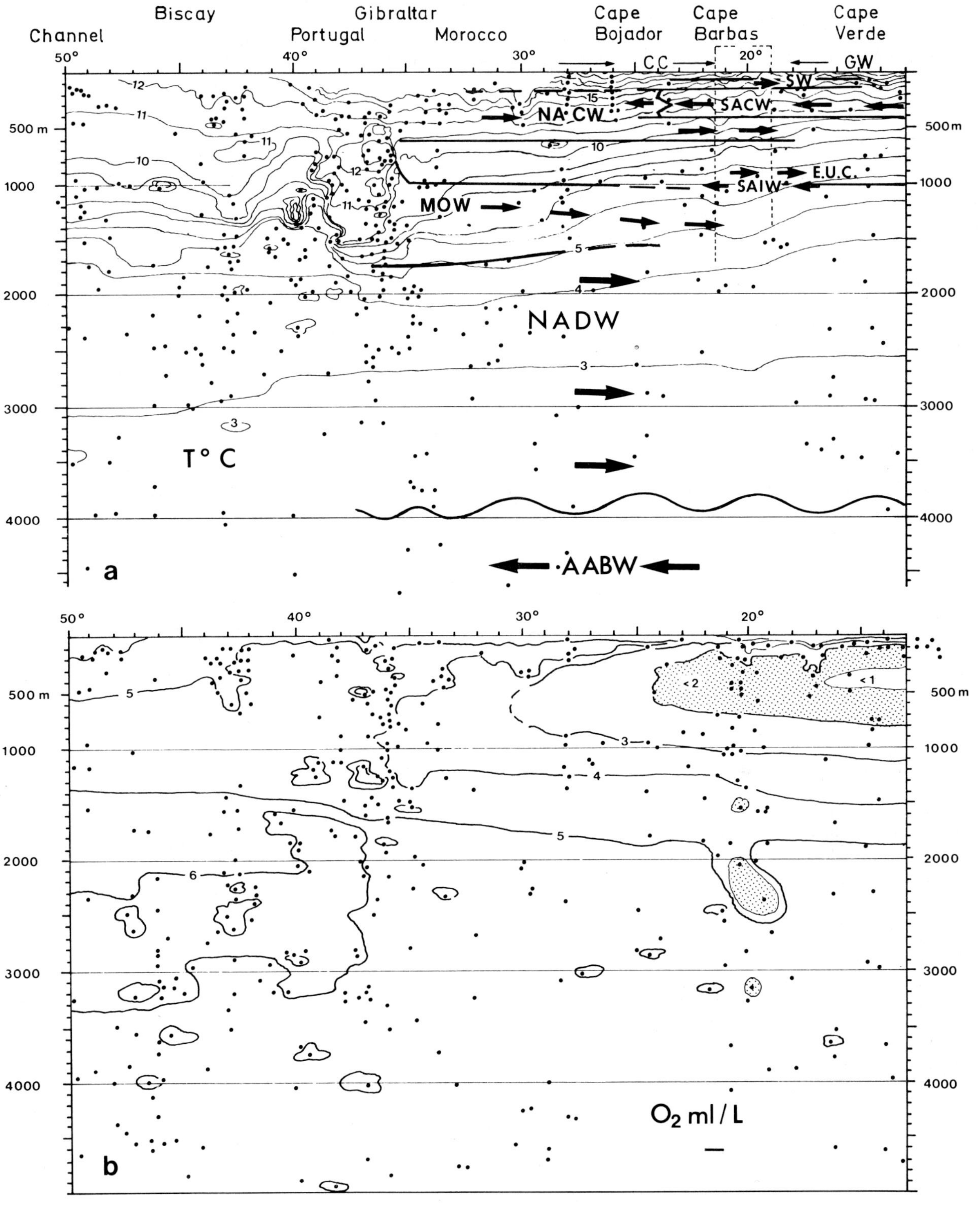
Channel
Biscay
Portugal
Gibraltar
Morocco
Cape Bojador
Cape Barbas
Cape Verde
CC
GW
SW
NA CW
SACW
E.U.C.
SAIW
MOW
NADW
AABW
T° C
a
O2 ml / L
b

The surface-water currents also carry planktonic larvae of benthonic organisms which inhabit shallow waters offshore into the open ocean (Scheltema, 1972; Thiede, 1975); at times the larval shells probably contribute significant quantities of calcium carbonate to the deep ocean.

2.3 Stratification of Deep-Water Circulation

The water-mass stratification at the NW African continental margin (Fig. 4) is substantiated by detailed data (e.g., from moored current meters) only for the uppermost 500-1500 m (e.g., Defant, 1941; Mittelstaedt, 1972; Hughes and Barton, 1974; Mittelstaedt et al., 1975; Schemainda et al., 1975; Johnson et al., 1976; Shaffer, 1976; Huber et al., 1977; Codispoti and Friederich, 1978; Fahrbach and Meincke, 1978; Tomczak and Hughes, 1980). The position of the deeper water masses can be inferred solely from sporadic historical temperature, salinity, current, and oxygen data, and by analogy to the western Atlantic Ocean (Wüst, 1933, a, b, 1936; Dietrich, 1969; Lonsdale, 1978; Lutze, 1980; Fig. 4). In summary, the following features have been recorded below the surface currents (generalizing strong seasonal and short-term variations):

Southflowing Subtropical Water (SW) north of 15°N resulting in a subsurface salinity maximum at 50-100 m depth on the shelf;

A northflowing undercurrent: It is centered between 150 and 400 m depth, up to 70 km wide, and can be traced north in form of isolated water bodies up to 25°-26°N, with velocities up to 20 cm/s. Its water originates from South Atlantic Central Water (SACW), forms the oxygen-minimum layer off West Africa, and provides a characteristically nutrient-rich source water for the southern upwelling region. The SACW shows a well-defined South Atlantic radiolarian assemblage (Labracherie, 1980).

The North Atlantic Central Water (NACW) follows the Canary Current from the north at 100-600 m depth. The NACW gradually veers out from the continental slope to the open sea south of 23°N. It is more saline, somewhat more oxygen-rich, and poorer in nutrients than the SACW which it replaces north of 22°-23°N as the major source of upwelling water.

An equator-ward undercurrent (up to 4 cm/s) has been recorded by Shaffer (1976) at 900 m, 19°N, being approximately 40 km wide and again somewhat richer in oxygen.

The Subantarctic Intermediate Water (SAIW) has been identified by Mittelstaedt (1972) offshore Cape Blanc at 980 m by lowered salinity.

The Mediterranean Outflow Water (MOW) has been traced south along the continental margin from Gibraltar to 20°N with a center of highly saline and relatively oxygen-rich water gradually sinking from 1250 m to 1450 m depth (Mittelstaedt et al., 1975; Huber et al., 1977).

The North Atlantic Deep Water (NADW) appears to be rather unstructured in terms of temperature, salinity, and a high oxygen content (4-5 ml/l) below 2000 m (except for a single oxygen minimum closely over the bottom, below the upwelling-induced maximum of surface productivity off Cape Blanc; Weichart, 1974).

The Antarctic Bottom Water (AABW) can hardly be recognized with the temperature-salinity data existing at the NW African continental margin. An approximation of the boundary is offered by Wüst's (1935) "abyssal thermocline" near 2°C potential temperature below some 3800 m water depth (Gardner, 1975; Berger, 1978). Accordingly, a rather mixed version of the AABW can be traced up to 35°N and is exerting effective control over the position of the carbonate compensation depth (CCD) because of its calcium carbonate undersaturation (Berger, 1978). Close to the surface of the AABW, Lonsdale (1978) has determined apparently an overprint of tidal currents which reach up to 17 cm/s in 4000 m water depth in channels between seamounts (Fig. 5).

In summary, it appears that the net-transport rates of suspension in north- and southward flowing currents along the continental margin might largely balance each other. More than 60 km away from the continental slope, current velocities decrease to less than 2 cm/s below 400 m (Defant, 1941; Shaffer, 1976). At the continental slope and rise, internal waves and turbulence at the various

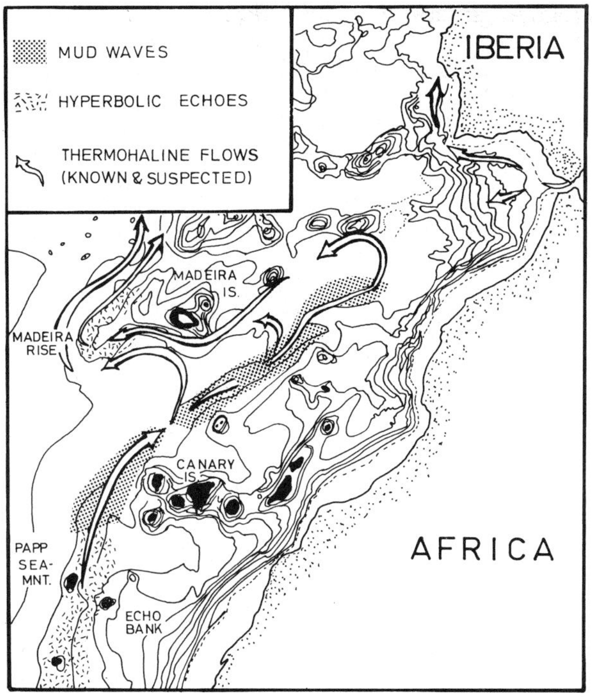

Fig. 5. Hypothetical pattern of bottom currents as inferred from the distribution of current-induced bedforms and current-meter results on the northwest African continental rise. (Simplified after Lonsdale, 1978)

water mass boundaries can effect strong sediment sorting and transport, particularly during times of vigorous glacial circulation (Jacobi et al., 1975; Bein and Fütterer, 1977; Fahrbach and Meincke, 1978; Diester-Haass and Müller, 1979). The high organic productivity in the southern sector of coastal upwelling (above the range of the SACW) forms another overlapping factor. It strongly increases calcium-carbonate dissolution along the slope because of CO_2-production due to bacterial decomposition of organic particles. This remineralization is also indicated by the position of the O_2-minimum layer at the upper continental slope (Fig. 4; Weichart, 1974; Berger, 1976; Lutze, 1980).

3. MODERN DEPOSITIONAL REGIME: ATMOSPHERIC AND OCEANIC CONTROLS

The modern sedimentary record on some major structures of modern environments is outlined in Table 1 and discussed in the following chapters.

3.1 Effects of Morphology

Both the atmospheric and oceanic circulation and their sedimentary record are clearly subject to influences from morphology of the continent and the seafloor.

A low relief characterizes most of the northwest African landscape and is the major requirement for free atmospheric circulation and associated moisture and sediment transport. Only the Guinean Ranges south of 12°N and the Atlas Mountains north of 29°N act as local obstacles and as drainage areas for rain and snow. The Atlas Mountains therefore also add to the development of aridity of the northernmost Sahara. The generally narrow shelf of northwest Africa (40-120 km, up to 200 km only off W. Guinea, shelf edge at 110-120 m) promotes sediment transfer from river mouths to the continental slope. Further, shallow-water sediment transport on a narrow shelf should follow generally the same direction during times of high and low sea level, at least as long as the effective wind directions did not change.

The continental slope has steep gradients (3°-6°) over large areas. An abundance of steep canyons north of 25°N and south of 21°N dissects up to 40 % of the slope area. Each of them can generate a significant overprint over the paleoenvironmental record of the sediments by various means (Egloff, 1972; Seibold and Hinz, 1974; Dept. of Fisheries, 1977):The grain-by-grain downslope transport, triggered by bioturbation, is greatly enhanced by steep gradients (Bein and Fütterer, 1977). Turbidites and sand flows can be generated easily (Sarnthein and Diester-Haass, 1977). Slumps are common features compared to the western Atlantic continental margin (Rona, 1970; Jacobi, 1976; Seibold and Fütterer, 1980; Haake et al.,

Table 1. Major features of the modern depositional regime

Patterns of environment	Sedimentary record
Deep-water masses	
Antarctic Bottom Water (Upper boundary)	Contourite mudwaves and erosion near 3800 to 4300 m water depth
	Calcite lysocline below 4000 m depth, reduced sedimentation rates
North Atlantic Deep Water	Middle Bathyal Fauna of benthic foraminifers
Currents along slope	Grain-by-grain downslope sediment transport
	Belt of maximum debris deposition at 1000-2000 m depth coinciding with the maximum of organic carbon in the sediments and of fragmentation of planktonic foraminifers
South Atlantic Central Water	Antarctic Radiolarian index species
	Position of aragonite compensation depth at 300-700 m
Oxygen-minimum layer (at 200-8oo m)	Maximum benthic foraminiferal numbers
	No record by the organic-carbon distribution
Surface water masses	
Major ecologic water masses (for example, Canary Current)	Planktonic foraminiferal provinces
	Transfer temperatures deduced from planktonic foraminiferal assemblages
Coastal water temperatures	Benthic skeletons requiring a specific temperature range and being reworked to the deep sea
Provinces of upwelling and resulting belts of lowered temperatures; high productivity as result of both wind-induced upwelling intensity and nutrient-rich upwelled subsurface waters	Enhanced production and sedimentation rates of biogenic opal and organic carbon and subsequently reduced $\delta^{13}C$ values
	Maximum foraminiferal numbers
	Patchy occurrence of organisms indicative of lowered surface temperatures and of cold upwelled subsurface waters; influence on seasonality record
Continental freshwater discharge	Planktonic foraminiferal index species;
	Increased sedimentation rates of biogenic opal, organic carbon, and fluvial mud; Reduced $\delta^{18}O$ and $\delta^{13}C$ values

Table 1. (continuation)

Climates / Wind systems:	
Continental wetness:	Fluvial sediments and turbidites (with plant fibers and a characteristic excess of clay); aeolomarine pollen record
Continental aridity:	Aeolian-sand turbidites; aeolomarine pollen record
Various wind systems: meridional versus zonal circulation	Dust composition and pollen supply characteristic of different source areas
	Upwelling activity in conjunction with shore-parallel meridional winds
	Position and direction of major surface-water current systems as deduced from transfer-temperature gradients
Position of the Intertropic Convergence Zone (ITCZ): (during summer)	Latitude of outbreaks of aeolian dust in the Saharan Air Layer (Harmattan Wind)
Wind speeds:	Quantitative evaluation of grain-size distribution of dust sediments

this Vol.) and result in high and steep escarpments. On the other hand, "slope walls" (Seibold and Hinz, 1974) such as off Cape Timiris (19°N) and Cape Verde (15°) can effectively act as sediment traps.

Extensive sediment slides and debris flows have frequently occurred near the upper margin of the continental rise, below 2000-2700 m water depth. They reach distances of more than 500 km offshore despite the low morphologic gradients of 0.1° (Uchupi et al., 1976; Embley, 1976; Embley et al., 1978). Major forms of sediment redistribution, i.e., a destruction of the normal hemipelagic record, have been noted by Egloff (1972), Jacobi et al. (1975), Jacobi (1976), Uchupi et al. (1976), and Embley et al. (1978) along the channels on the rise, and adjacent to and on top of seamounts, the Sierra Leone Rise, and the oceanic fracture zones. Partly upslope directed sediment waves are concentrated near the eastern side and downstream of deep-sea gaps and major island obstacles as an effect of the AABW circulation (Fig. 5). Examples occur west of Cape Verde, west and northwest of the Cape Verde Islands, between the continental slope and the North and South Banks Plateau (26°N), around the Saharan Seamounts (24°-27°N), and north to northwest of the Canary Islands, but also south of the Kane Gap (9°N). It appears that most sediment waves are presently remnants of ancient (? glacial) conditions, although their formation might be also related to modern ephemeral turbiditic events (Normark et al., 1980) and to deep-water tidal currents (Lonsdale, 1978).

3.2 Aeolian Versus Fluvial Sediment Supply

The recognition of aeolian dust presents no major problem in non-turbiditic, silt-sized, distal deep-sea sediments deposited more than some 150-200 km seaward of the continental margin: Hardly any other means of transport but aeolian supply can travel the vast distances from the con-

Table 2. Composition of dust- and river-borne deep-sea sediments

	Aeolomarine dust deposits and aerosol samples		Sahelian lowland river discharge (Senegal river)
	< 500 km offshore	> 500 km offshore	
Grain sizes (insoluble residue)			
Clay	< 20-30 %	> 50-98 %	35-50 %, (max. 90 %)
Silt modal sizes	> 25 µm	12-25 µm	11-12 µm
Sand > 63 µm	5-25 %	Traces	< 3 %
Sand > 125 µm (in % of total sand)	10-15 %	-	5-12 %
Composition			
Resistant fecal pellets	-	-	< 20 %
Plant fibres (> 63 µm)	Traces	Traces	4-50 %
Muscovite/biotite (> 63 µm)	0.1-0.5	Not determined	0.5-1.0
Limonitic silt aggregates	-	-	< 20 %
Stained quartz in % of total quartz	5-75	5-75	35-55
Irregular clay aggregates resistant to ultrasonic and peroxide treatment	Abundant	-	-

tinental sources. However, it is increasingly difficult to separate (Table 2) an aeolian from a fluvial input when approaching the upper continental rise and slope (Bein and Fütterer, 1977; Lutze et al., 1979; Sarnthein, 1979; Diester-Haass, 1979, 1980). A most reliable feature of fluvial input is the basically bimodal terrigenous grain-size distribution with an excess of clay (Fig. 6A). Poorly defined mixed-layer clay minerals have been also claimed as typically river-borne in the fossil record (Chamley et al., 1977; Chamley and Diester-Haass, 1979). However, they turned out to be irregular clay aggregates which are resistant to conventional laboratory treatments and the modern occurrence of which is confined to dust fall-out and the evaporite environment of sebkhas (Lange, 1981).

Moreover, the Saharan dust discharge as a whole is by no means uniform because the composition varies with the different source provinces. For example, the south Saharan and Sahel regions (south of 20° to 25°N) are largely depleted of carbonates and covered by diatomaceous lake deposits, lateritic nodular soils and iron crusts (Fig. 1; Conrad, 1969; Servant and Servant, 1972; Schmidt-Lorenz, 1977). Such deposits have accumulated during ?Paleogene, Neogene, and Quaternary phases of humid environment (e.g., Michel, 1973, 1980). Generally, the formation and destruction of a lateritic horizon appears to be lengthy, requiring several hundred thousands of years (Nahon and Lappartient, 1977; Nahon, 1980).

In summary, two major types of Saharan dust, a southern and a northern,

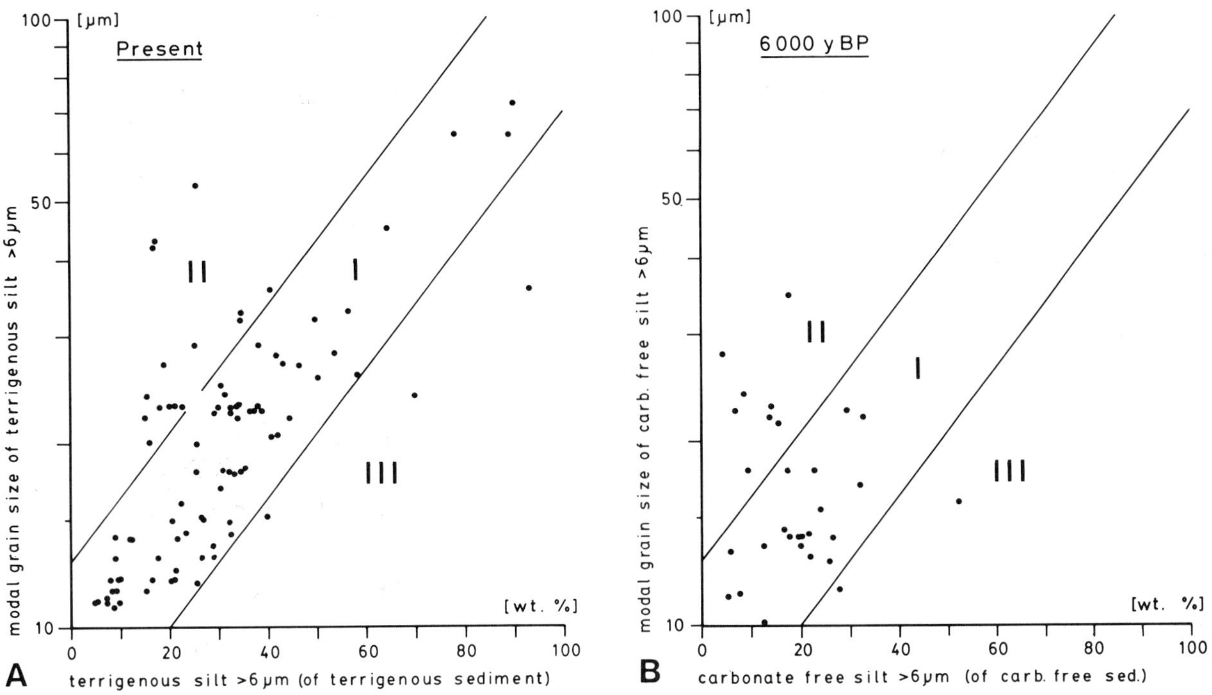

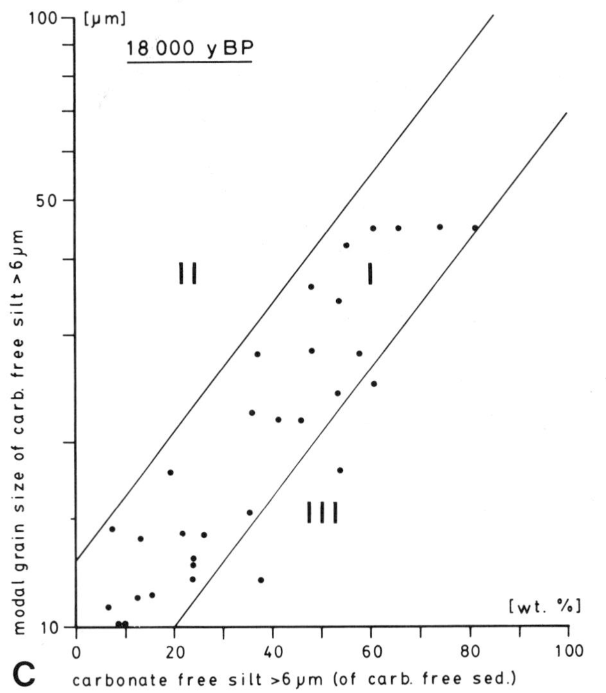

Fig. 6. Proportions of the terrigenous sediment fraction > 6 μm versus modal grain sizes of the terrigenous silt fraction > 6 μm (on a carbonate and opal free basis) (after Koopmann, 1979).- I prevailing sample group with well-sorted grain sizes; II sample group with a relative excess of fine fraction (< 6 μm) in the bimodal grain-size distributions, restricted to small-scale areas seaward off river mouths (see Fig. 7A); III sample group with a relative deficit of fine fraction (< 6 μm), restricted to current winnowed areas along the upper continental slope. A Surface samples; B 6000 yr B.P.; C 18,000 yr B.P.

and a spectrum of intermediate types (Table 3) can be expected in the deep-sea sediments at the northwest African continental margin, in addition to various kinds of river-supplied muds. The N Saharan dust (Table 3) reflects its source environment, the vast Neogene and Quaternary, marine and lacustrine carbonate deposits as well as pedogenetic calcretes and salt crusts north of a line from Air to Hoggar to Tibesti (Conrad, 1969). The different composition of aeolomarine dust deposits hardly implies differences of contemporary climates, but a different heritage

Table 3. Composition of south and north/central Saharan dusts

	South Saharan and Sahelian Dust	Central and North Saharan Dust
Carbonate	< 4-8 %	Up to 20-50 %
Dolomite	-	Common (up to > 2 %)
Quartz	Up to 20 %	< 15 %
Stained quartz (in % of total quartz)	Up to 75 %	5-15 %
Typical clay minerals	Kaolinite (20-35 %) and montmorillonite	Illite, chlorite, palygorskite, and montmorillonite
Feldspar	Abundant	Very abundant
Freshwater diatoms	Common	Common
Phytoliths	Common	Very common
Data compiled from:		
	Folger (1970), Parkin et al. (1971), Aston (1973), Parmenter et al. (1974), Emery et al. (1974), Kiper (1977), Johnson (1979), Fütterer (1980), Glaccum and Prospero (1980); Biscaye (1965), Griffin et al. (1968), Conrad (1969), Chester et al. (1972), Lange (1975 and in press), Chamley et al. (1977).	Chester et al. (1972; offshore S-Morocco), Conrad (1969; NW-Saharan sections), Koopmann et al. (1979; Baie du Levrier, 21°N), Yaalon and Ganor (1977; loess in Israel), unpubl. data on Moroccan loess and a north-south Saharan transect.

of (ancient) climate-controlled sediments in the source regions (Nahon, 1980). Finally, the generation of dust itself is not related directly to aridity. A complicated set of meteorological conditions which are not necessarily restricted to deserts but rather to semi-arid regions, favor formation and uptake of dust.

3.3 The Sedimentary Record on Modern Climates and Wind Circulation: The Terrigenous Fraction

Surface-sediment samples average the record of the last 500-2000 years, assuming a benthic mixing depth of 6-8 cm (Berger and Johnson, 1978; Wetzel, 1981). Within this order of magnitude, a number of variables document both the pattern of fluvial sediment supply and the prevailing dust trajectories in the terrigenous (opal- and carbonate-free) sediment fraction (Koopmann, 1979).

Fluvial runoff and the corresponding wetness of continental climate are recorded by lobate, small-scale areas with an excess of clay in the basically bimodal grain-size distribution (Fig. 6A) and by the occurrence of abundant fecal pellets on the upper continental rise. Such areas occur off Guinea, up to 17°N, and off the mouth of the Wadi Sous, at 30°N (Fig. 7A).

The eddy-shaped dust outbreaks which extend over more than 1200 km offshore in the Saharan Air Layer (= Harmattan), are reflected by the distribution of terrigenous suspension in the surface water (Emery et

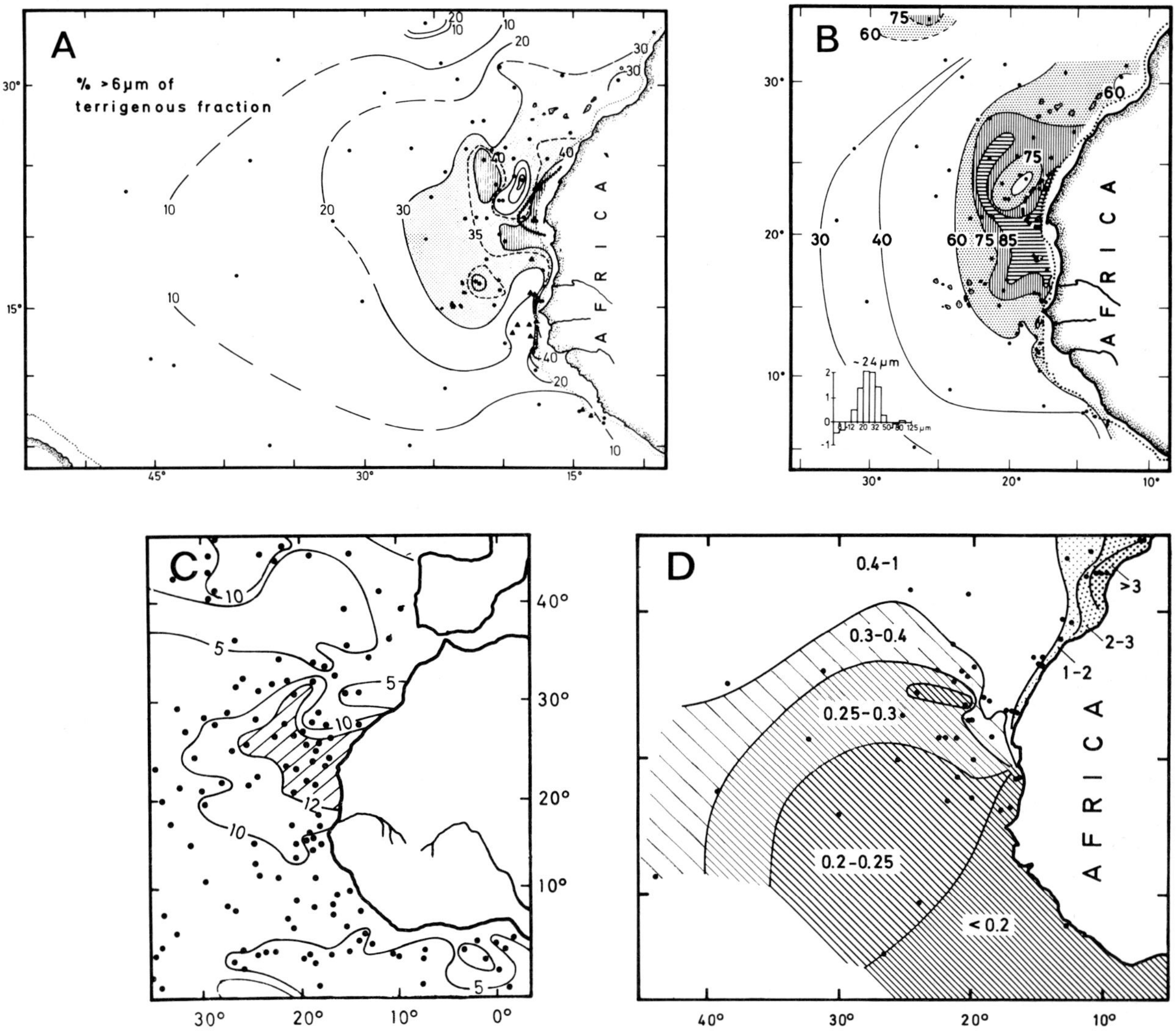

Fig. 7. Distribution maps of the terrigenous sediment fractions (on a carbonate and opal free basis; Koopmann, 1980) in northeast Atlantic surface sediments showing the dispersal pattern of proximal Harmattan dust which originates from the south Sahara and Sahel zones.
A Relative abundance (%) of the grain-size fraction > 6 µm. (After Koopmann, 1979). Samples marked by triangles show a relative excess of fine fraction (< 6 µm) in the bimodal grain-size distributions (see Fig. 6 A) and are restricted to areas seaward off river mouths.
B Concentrations (%) of quartz. (Redrawn after Kolla et al., 1979).
C Distribution of the terrigenous silt mode around 24 µm. (see histogram). This grain-size assemblage is weighted for the respective proportion of fine fraction (< 6 µm). Values are factor loadings of a Q-mode principal component analysis. (Sarnthein et al., 1981).
Area with dominant Trade Wind dust supply is surrounded by a bold dotted line in A and C.
D Chlorite to kaolinite abundance ratio (after Lange, 1981). Note: Chlorite is abundant in both river load and wind load originating from the Atlas Mountains

al., 1974; Milliman, 1977) and particularly in the sediments of the eastern Atlantic. For example (Fig. 7B) they are documented by the abundance of quartz (Kolla et al., 1979), by the grain-size fraction >6 µm, by a 24 µm-mode (Fig. 7C) (Sarnthein and Koopmann, 1980; Sarnthein et al., 1981), and more vaguely, by kaolinite (Fig. 7D) (Biscaye, 1965; Lange, 1981). Approximately the same pattern of kaolinite distribution can be inferred from the dust collections of Chester et al. (1972). The contours of sediment dispersal locate the long-term center of (Harmattan) dust outbursts from the coast near 16°-22°N and accordingly, the northernmost summer position of the ITCZ near 16°-19°N.

The modern dust carried by the Trade Winds cannot be distinguished from the Harmattan dust supply by means of grain-size analysis, but only by compositional features such as the abundance of unstained quartz, chlorite, and illite (Koopmann et al., 1979; Sarnthein, 1979; Lange, 1981). Accordingly, modern dust deposits derived from the meridional Trade Winds are confined to a small, proximal sector offshore 19°-25°N, which lies off the most arid part of the Sahara (Fig. 7). The distribution pattern of Harmattan dust is only slightly affected by north-south displacement in the surface Trade Winds, except for the fact that the near-shore maximum of clay-sized kaolinite is shifted somewhat toward south as compared to that of silt grain sizes, and even farther as compared to that of quartz.

3.4 Terrigenous Surface Sediments and Oceanic Circulation

The impact of oceanic currents on the large-scale dispersal of terrigenous particles seems to be minor in the open sea. This is largely due to a lumping of dust grains in fecal pellets near the sea surface (Smayda, 1971; Schrader, 1971) and their subsequent rapid deposition. Features made by AABW currents which are indicated by 3.5 kHz precision depth recorder registrations can be largely excluded as a factor controlling grain-size patterns because of their local restriction; within the area of modern mud waves grain sizes are not coarsened but slightly reduced (Jacobi et al., 1975; Jacobi, 1976; Uchupi et al., 1976; Embley et al., 1978; Lonsdale, 1978).

Nevertheless, wave and current transport are crucial on the shelf and upper slope (Summerhayes et al., 1976; Diester-Haass and Müller, 1979; Fütterer, 1980). Large amounts of aeolian dust fall out 50-100 km offshore (Lepple, 1975; Jaenicke and Schütz, 1978). The vast majority of this dust together with the supply of rivers is, however, washed across the shelf. Subsequently, the sediments are transported further downslope "grain by grain" by a complex interaction of resuspending bioturbation, currents, and gravity. The result of these processes is recorded by a number of tracer particles of shallow-water origin (e.g., benthic shelf foraminifers) and by grain-size parameters. Even at the base of the slope, one quarter to one third of the fine-sand fraction of non-turbiditic sediments derives from this type of downslope transport (Diester-Haass, 1975; Bein and Fütterer, 1977; Bein and Sass, 1978; Lutze et al., 1979; Lutze, 1980). Accordingly, aquatic transport dominates over the atmospheric transport in the terrigenous sediment record down to some 800-1000 m water depth. Maximum sedimentation rates are characteristic for this critical threshold of sediment transport (Diester-Haass and Müller, 1979; Fütterer, 1980; Erlenkeuser et al., in prep.). The northward undercurrent of SACW (100-600 m depth) is documented in the clay fraction even down to the upper continental rise: Lange (1975 and in prep.), Diester-Haass and Müller (1979), and Sarnthein and Koopmann (1980) report a northward transport of excess clay-grain sizes and montmorillonite which are derived from the Senegal load, over more than 250 km distance.

The combination of this south-north current transport and the two aeolian dust trajectories from different source areas results in a characteristic clay-mineral distribution north of 18° N: Illite and chlorite prevail on the upper slope, and montmorillonite and kaolinite dominate on the rise (Fig. 7D). This pattern spuriously suggests a general process of clay-mineral sorting by direct downslope transport across the slope (Chamley et al., 1977), which cannot be confirmed in the light of our recent evidence.

3.5 The Sedimentary Record on Surface-Water Circulation: Planktonic Foraminiferal Assemblages

The surface water of the Canary Current carries rich plankton assemblages which produce shells and skeletons providing a major proportion of the pelagic and hemipelagic sediments offshore northwest Africa. As in other parts of the ocean, the distribution patterns of fossils of planktonic organisms serve to outline the flux of pelagic sediment components from the different oceanic surface-water masses (Bé and Tolderlund, 1971; Kipp, 1976). For the upwelling areas off northwest Africa, foraminifers (Fig. 8) are the most thoroughly studied planktonic fossil group (Thiede, 1975; Pflaumann, 1980), whereas calcareous

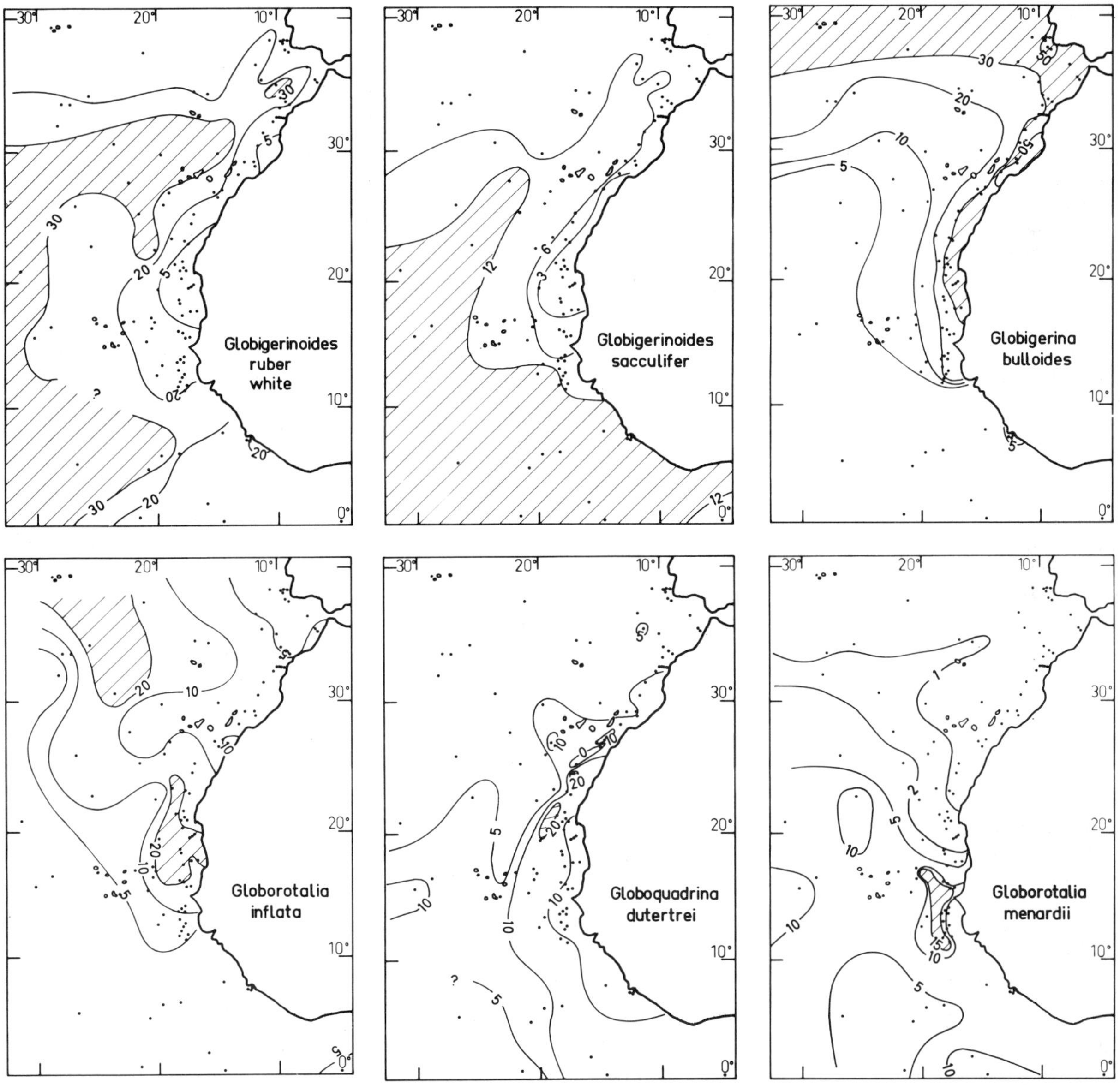

Fig. 8. Relative abundance (%) of some dominant species of planktonic foraminifers (>150 µm) in northeast Atlantic surface sediments

nannofossils (McIntyre and Bé, 1967), diatoms (Schrader, 1972; Richert, 1976), and pelagic gastropods (Diester-Haass and van der Spoel, 1978) have proven to be less useful in reconstructing the surface-water provinces. A recent study of Labracherie (1980) shows, however, that radiolarians are particularly useful to establish the influence of subsurface circulation.

The distribution patterns of planktonic foraminiferal tests in 135 surface-sediment samples document at least three major biogeographic provinces which reflect hydrographic differences in the surface water. Globigerinoides ruber and Globigerinoides sacculifer (Fig. 8A,B) are limited to the tropical warm surface waters to the south and west of the Canary Current (Fig. 2). Globigerina bulloides (Fig. 8C) is probably the most important species inhabiting the cool surface waters of the modern Canary Current (Fig. 2) and it is therefore also one of the dominant species in the underlying pelagic sediment (de Miró, 1973; Kipp, 1976). Additional conspicuous maxima concur with cold waters derived from local centers of coastal upwelling such as off southern Morocco and South Portugal. The distribution of G. bulloides suggests that the influence of the Canary Current surface waters is limited to a zone of only 100 to 200 km width at its narrowest extension south of the Canary Islands. Further to the south, the eastern boundary current gradually grows in width under the influence of coastal upwelling at Cape Blanc, particularly where it veers from the continental margin to the southwest and joins the North Equatorial Current. The latter divergence is recorded by the tropical assemblage (Fig. 8A) replacing the cool and temperate-water species within a zone up to 1000 km broad.

Surface waters affected by the coastal upwelling regime along the northwest African continental margin are populated by a highly specialized assemblage of planktonic foraminifers dwelling usually in subsurface waters (Thiede, 1975; Pflaumann, 1980). Globorotalia inflata and Globoquadrina dutertrei (Fig. 8D,E) might serve as the best examples of such species in the surface sediments. Globorotalia menardii reaches its maximum abundance in sediment-surface samples from the region off Cape Verde where we have observed a seasonally complicated pattern of current divergence and convergence (Fig. 8F) under an at least partial influence of warm Guinean Water.

A more careful examination of Fig. 8 reveals that the distribution pattern of planktonic foraminiferal species is even more complicated than outlined above. From this we assume that the distribution of major water masses is by no means as simple as outlined above, but that mixing processes and seasonal shifts disturb the oceanic fronts between the surface-water masses considerably. The fact that sediment-surface samples integrate signals over a time span of several decades or hundreds of years suggests that the hydrographic features which generate the patchy distributions of planktonic fossils (Fig. 8) have occurred repeatedly and over long periods over the same area. Thus the distribution of planktonic foraminifers in the sediments can serve as a useful tool for reconstructing past patterns of surface circulation, despite their complexity.

3.6 The Opal and Carbon Record of Surface-Water Productivity and Upwelling

Both coastal upwelling and the input from major rivers such as the Senegal River can equally enhance fertility in the surface-water masses (Diester-Haass and Müller, 1979). This fertility is reflected in the sediments particularly by the abundance of biogenic opal (Fig. 9), despite the various "filtering" processes during silica deposition and dissolution (Schrader, 1972; Seibold, 1979). Maximum opal values of more than 0.5 % - 2.0 % (20-40 mg cm^{-2} 1000 yr^{-1}), occur at 20°-27°N and south of 17° N on the upper rise and, in addition, at the windward side of the Cape Verde Islands. The highest abundance (>12 %, 800-1700 mg cm^{-2} 1000 yr^{-1}), was observed in the very shallow and protected Baie du Levrier (21°N) which traps the upwelling triggered plankton production (Koopmann et al., 1979).

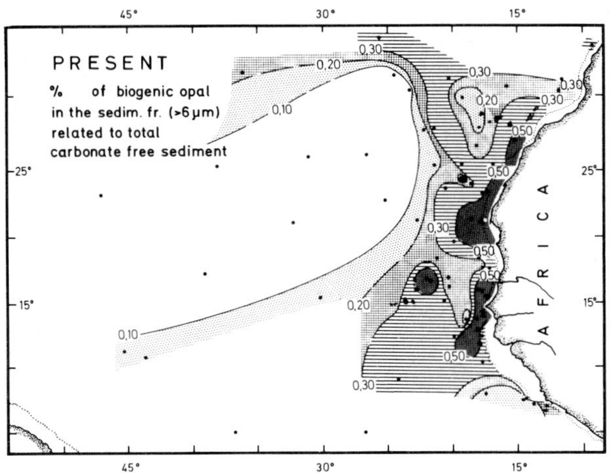

Fig. 9. Concentrations (in weight percent) of biogenic opal (> 6 µm) in the bulk carbonate free surface sediments of the subtropical northeast Atlantic (values are estimated from smear-slide counts of six grain-size classes; after Koopmann, 1980)

Abundant diatom frustules and sponge spicules in general reflect an increased fertility of surface waters. However, the abundance of radiolarian tests appears to be linked rather to Antarctic provenance areas and a high nutrient content of the subsurface waters which feed the upwelling. For example, radiolarians clearly help to distinguish the nutrient-rich SACW south of 22° N(Fig. 4) from the nutrient-poorer NACW in the northern part of the upwelling belt (Labracherie, 1980).

The organic-matter content in sediments per se can only be regarded as a rough indicator of enhanced oceanic productivity because its preservation depends on such factors as sedimentation rates and abundance of fine-grained sediment fraction (Müller and Suess, 1979; Suess, 1980). However, a simple plot of the percentages of organic carbon (Fig. 10) can also identify at least two salient regions with high abundance. One lies below the center of upwelling activity at Cape Blanc (21°N), and a second off the mouth of the Senegal River. Both lie in the critical water depths of maximum sediment deposition at 800-2000 m, but they do not coincide with the depth positions of the oxygen-minimum layer (Fig. 4).

Diversity and intensity of bioturbation are roughly correlated with the content of organic matter in the sediments, which in turn is a function of both primary production and sedimentation rates (Müller and Suess, 1979). Thus oceanic circulation(by upwelling)and continental climate (by terrigenous input) can regulate bioturbation. Wetzel (1981) describes the overall differentiation of bioturbational struc-

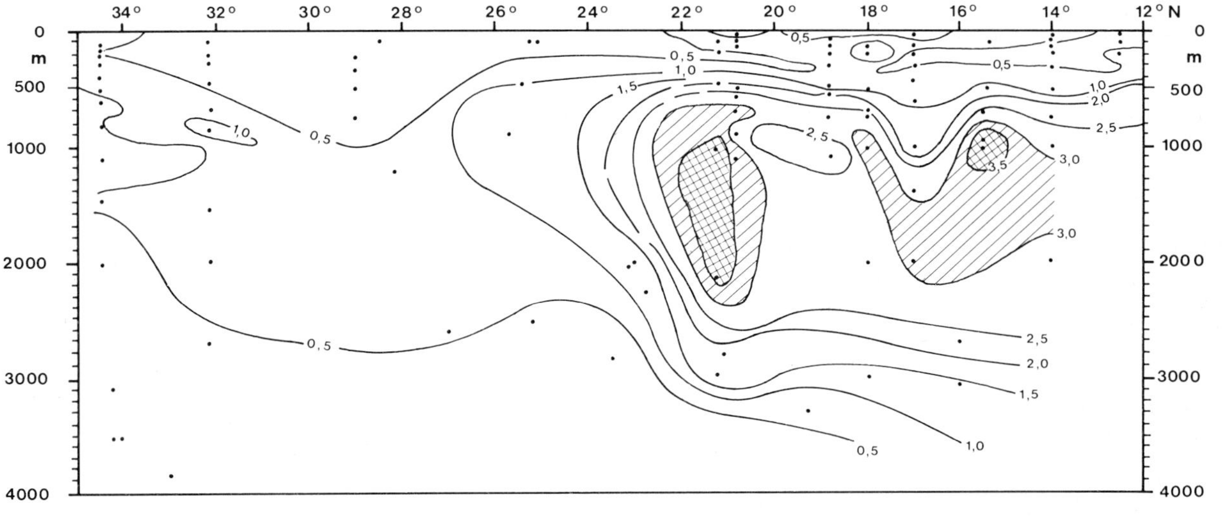

Fig. 10. Depth distribution of organic carbon concentrations (weight percent) in surface sediments along the northwest African continental margin. (Compiled from Diester-Haass and Müller, 1979; Diester-Haass, 1978; Hartmann et al., 1976)

tures between 12° and 28°N with respect to cold and warm climates. Biogenic deformation structures, which he attributes to increased organic carbon content, are limited to sediments off the river mouths of Senegal, Gambia, and Casamance where they are associated with a locally enhanced nutrient input during phases of humid climate. In the area of Cape Blanc, they dominate in the sediments of oxygen-isotope stage 2 where they might indicate the influence of increased upwelling during cold climates (Diester-Haass et al., 1973; Thiede, 1977; Labracherie, 1980). The diversity of lebensspuren culminates in sediments which have accumulated during cold stages (Wetzel, 1981), and also coincides with increased sedimentation rates and content of organic matter (Hartmann et al., 1976; Müller and Suess, 1979). The controls on the diversity of lebensspuren are not known.

3.7 The Impact of Bottom-Water Masses on Sediment Distribution

The intense stratification of the bottom water along the northwest African continental margin (Fig. 4) has an important impact on the depositional environment; bottom-water currents are particularly effective in sediment transport at the upper slope. For example, a combination of bioturbation and current activity has already been shown (Sect. 3.4) to effectively displace downslope fine-grained sediment particles from the site of initial deposition on a large scale. Another bottom-water effect is the differential dissolution of calcareous (Berger, 1971), siliceous (Schrader, 1970), and organic matter (Müller and Suess, 1979). In this way bottom water strongly modifies the flux of sediment particles produced by planktonic organisms. The spatial distribution pattern of benthic organisms, especially that of benthic foraminifers (Zobel and Ranke, 1978; Haake, 1980; Lutze, 1980), can reflect the vertical and horizontal variations of the bottom water on the continental slope with great detail and high sensitivity, although in most cases only via food control (Douglas and Woodruff, 1980).

Seiler (1975), Zobel and Ranke (1978), Lutze (1980), and Haake (1980) have established the distributional ranges of three major benthic foraminiferal species assemblages at 10°-40°N (Fig. 11). A Shelf-Fauna extends down to 100-200 m water depth. The Slope-Fauna spreads down to approximately 1000-1500 m. Its occurrence corresponds to the depth position of the intermediate, complex water masses impinging on the slope. This depth interval concontains the oxygen-minimum layer in the upper part and the MOW (Fig. 4) near the lower boundary. Finally, the habitat of the Middle-Bathyal-Fauna conforms closely to the widespread and largely uniform range of the NADW. Other patterns of species distribution (Fig. 11), of the standing stock and of foraminiferal numbers (Fig. 12) have a latitudinal extension. These ranges may reflect the regional pattern of nutrient supply from river discharge and from the productivity of coastal upwelling, because benthic foraminifers feed on organic debris (Lutze, 1980). Seaward of the mouth of the Senegal River, maximum foraminiferal numbers coincide with the oxygen-minimum layer (Fig. 4) rather than with the maximum abundance of organic matter in the sediments (Fig. 10); in the upwelling controlled area off Cape Blanc (21°N) the maxima of foraminiferal numbers are at an intermediate depth position.

The ranges of the three living benthic foraminiferal assemblages correlate well with those of empty tests in the size fraction coarser than 63 µm (Fig. 11) and particularly, in size fractions >250 µm (Lutze, 1980). We infer, therefore, that sediment of an equivalent grain size

Fig. 11. Distribution of benthic foraminifers along the northwest African continental margin. (After Lutze, 1980):
A,B Comparison between depth zonations of living and dead assemblages. Broken line 35 % isoline of Slope Fauna.
C-F Depth-distribution patterns of selected index species.
C,D Distribution patterns of living and dead small shelf-*Bolivina* tests. Differences reflect downslope transport.
E,F Examples of slope and middle bathyal species

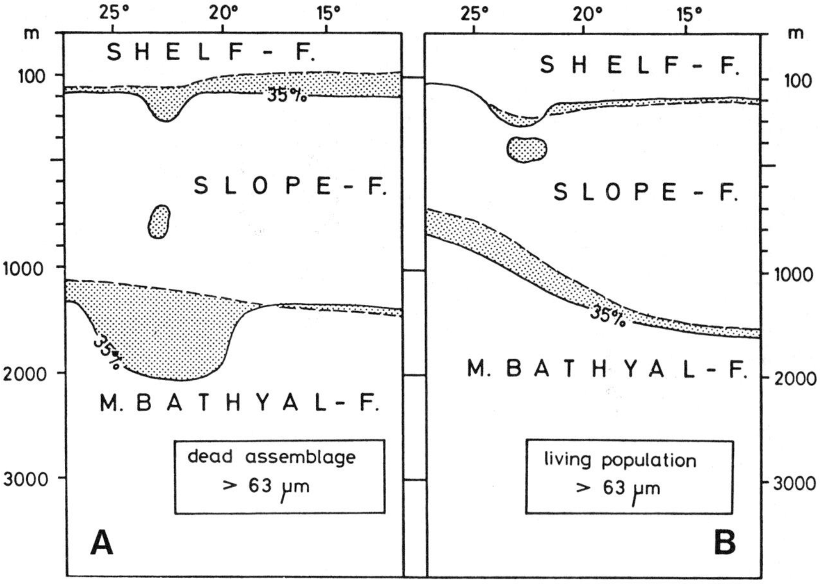

SHELF-F.
SLOPE-F.
M. BATHYAL-F.
35%
dead assemblage
> 63 μm
living population
> 63 μm
A
B

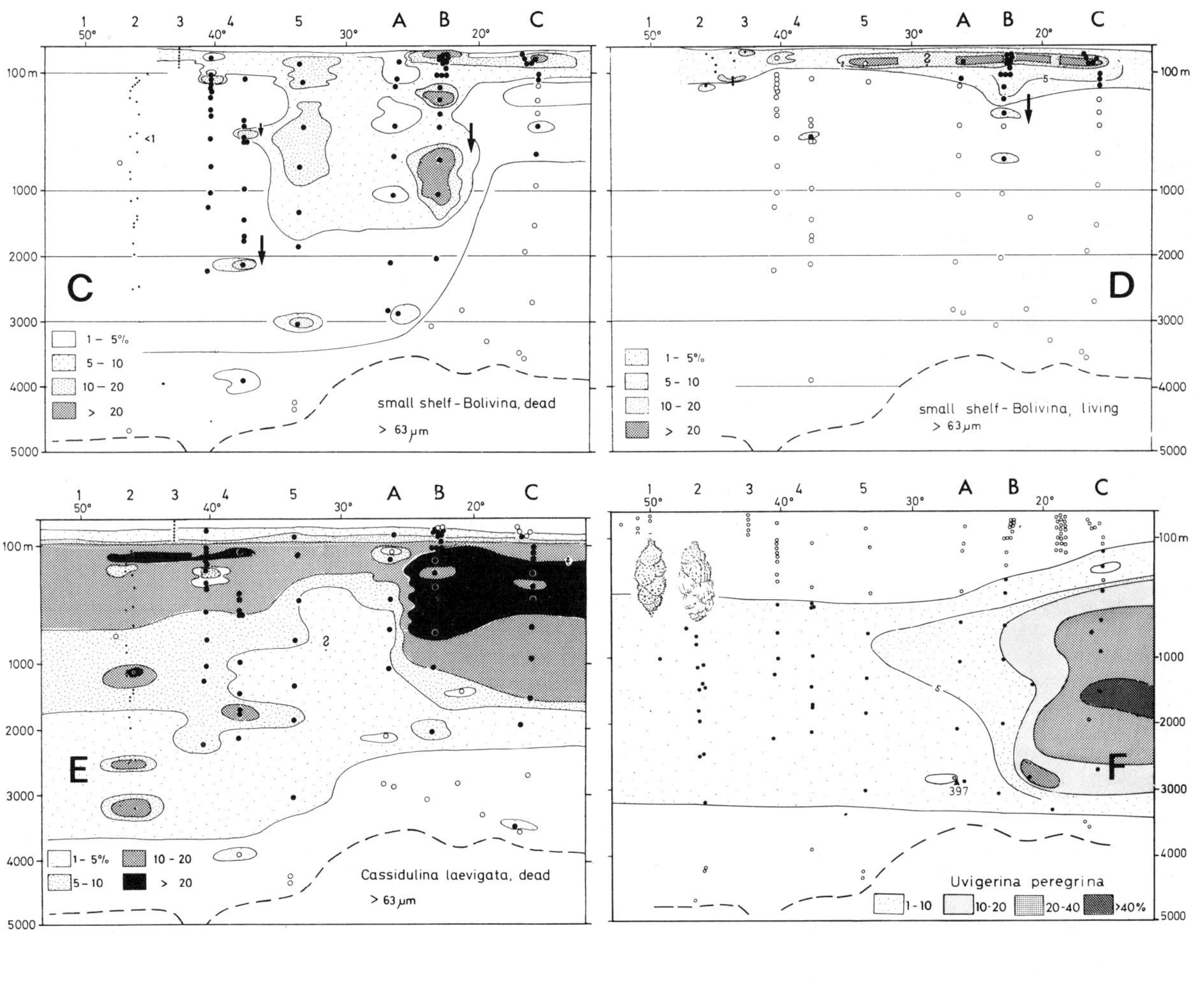

small shelf-Bolivina, dead
> 63 μm
small shelf-Bolivina, living
> 63 μm
Cassidulina laevigata, dead
> 63 μm
Uvigerina peregrina
1 - 5%
5 - 10
10 - 20
> 20
1-10
10-20
20-40
>40%
C
D
E
F

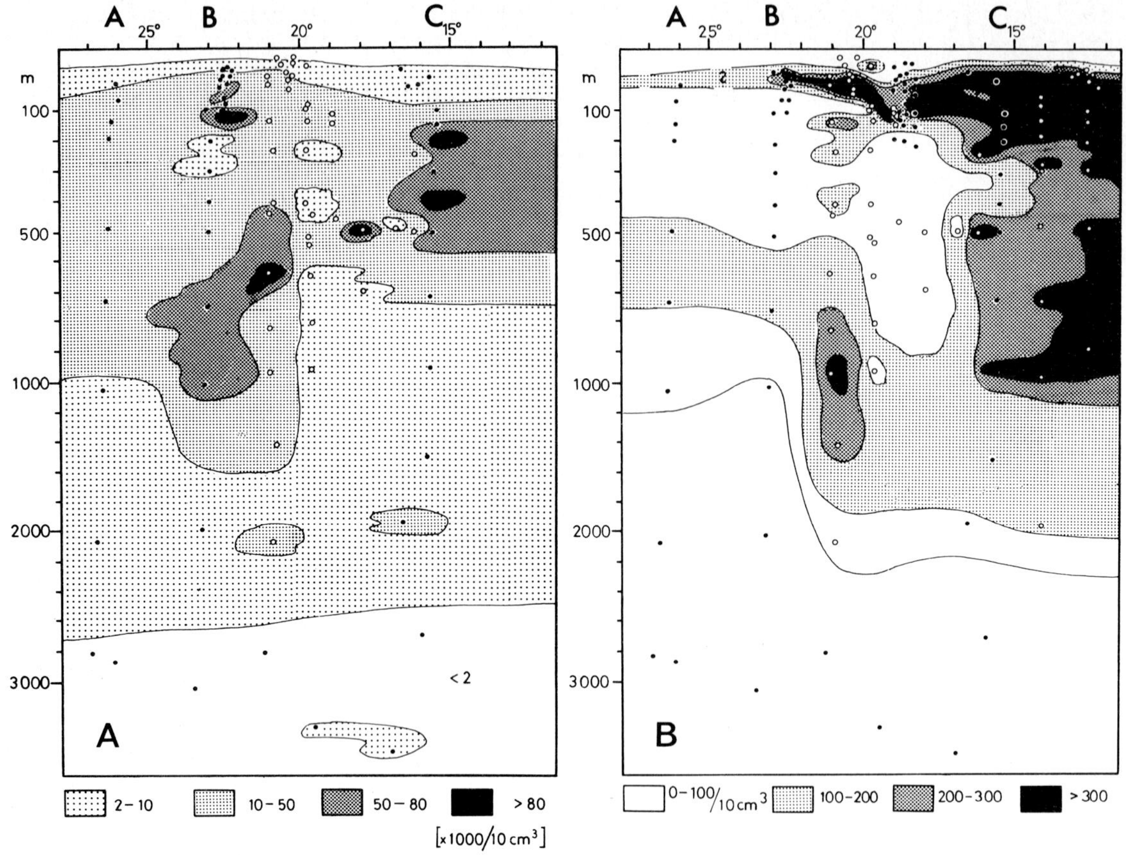

Fig. 12. A Standing stock of benthic foraminifers (stained tests per 10 c.c. wet surface sediment).
B Benthic foraminiferal numbers (empty tests per 10 c.c. wet surface sediment).
Both figures reveal the influence of nutrient supply from upwelling and tropical rivers on the foraminiferal distribution patterns

is rarely displaced as isolated grains across the deeper parts of the continental slope. The gentle, but persistent grain-by-grain transport which is bioturbation- and current-controlled only applies to the silt-sized and finest-sand fraction. It is demonstrated by high percentages (up to one quarter of the fauna) of shelf-derived Bolivina tests of the 63-125 μm fraction in otherwise autochthonous sediments from the rise and lower slope (Lutze et al., 1979).

Undersaturated bottom waters are known to result in the differential dissolution of calcareous microfossils. Dissolution affects both the surfaces of individual carbonate grains and the species composition of the calcareous grain assemblages as a whole. Off northwest Africa, the Canary and Cape Verde-Gambia Basins locally exceed 6 km water depth, but are completely covered by calcareous sediments (Biscaye et al., 1976). The Calcite Compensation Depth (CCD) is nowhere fully reached because the corrosive AABW (Fig. 4) is renewed very slowly in the northeast Atlantic (Wüst, 1939). Nevertheless, the carbonate content of sediments below some 5000 m water depth is clearly reduced to less than 50 % or 25 % of the bulk sediment, indicating that some dissolution is going on. Dissolution can also locally affect the calcareous components in shallower depth along the continental slope because of the locally high flux and decay of organic carbon. As a result, the aragonite lysocline is very shallow, at 300-700 m water depth. In addition, a band of intensive fragmentation of (calcite) planktonic foraminiferal tests is reported from 1000-2000 m water depth where it coincides with the maximum abundance of organic matter (Fig. 10; Thiede, 1973; Diester-Haass and Müller, 1979; Zobel and Ranke, 1978; Berger, 1976).

4. ENVIRONMENTAL CHANGES: HIGH-RESOLUTION STRATIGRAPHY AND GLOBAL CORRELATION

4.1 Oxygen Isotope Stratigraphy

Oxygen-isotope record (Emiliani, since 1955; Shackleton and Opdyke, 1973, 1976; Berger, 1979) has provided a tool for time-calibrating the signals of ocean and continental paleoenvironment in deep-sea cores on a 500-5000 years scale. In addition, the high sediment-accumulation rates (in general 2-20 cm/1000 yr on the continental rise) arising from both continental input and high oceanic productivity at the northwest African continental margin, result in a high time resolution which is unusual for the deep-sea sedimentary record and which permits a detailed cross-check with the record from land sections (Pastouret et al., 1979; Erlenkeuser et al., in prep.).

Long time series in the range of a half to several million years are provided by oxygen-isotope curves from the Sierra Leone Rise (Erlenkeuser et al., in prep.; Schneider, in prep.; Vergnaud-Grazzini and Rabussier-Lointier, 1980), a Vema core off Cape Blanc (Parkin and Shackleton, 1974), and from DSDP Site 397 off Cape Bojador at 27°N (Shackleton and Cita, 1979; Fig. 25). Most curves are indirectly time-calibrated by the paleomagnetic record. They show that glacial-interglacial fluctuations in the range of 20,000 to 120,000 years have characterized the Earth's climate for the past 3.4 m.y. Before this time, the middle Pliocene (4.5-3.4 m.y.) had a period of relatively stable climate which resembles Holocene and warmer interglacial conditions such as have also been reported from some Pacific and Caribbean cores (Shackleton and Opdyke, 1977; Keigwijn, 1979). Both the amplitudes of glacial extremes and the means of $\delta^{18}O$ increased at about 2.4 m.y. and especially markedly at 0.7 m.y. ago, possibly resembling the overall growth of polar ice caps.

Distinct and high frequency fluctuations of climate similar to those of the Quaternary also occurred between 6.0 to 4.5 m.y. ago (i.e., about the time of Miocene/Pliocene boundary) during and after the Messinian salinity crisis, however, some of the observed fluctuations may be due merely to increased sedimentation rates and thus improved stratigraphic resolution during this time span (Fig. 25). In comparison with Holocene isotope values, most of the measurements from Miocene and Pliocene times deviate in the "warm" (=less glaciated) direction, with single "cool" (=more glaciated) extremes which are isotopically little heavier (0.3 per mil $\delta^{18}O$) than the Holocene values. Fluctuations during the upper Miocene show a wavelength corresponding to several 100,000 years. The "cool" extremes are at Holocene $\delta^{18}O$ level, the "warm" ones are about 0.7 per mil lighter (Shackleton and Cita, 1979).

Excellent illustrations of $\delta^{18}O$ records with a detailed time resolution are given in sediment cores from about 1000 m water depth in front of the Niger and Senegal River mouths (Pastouret et al., 1978; Erlenkeuser et al., in prep.). For example, the transition from the last glacial (oxygen-isotope stage 2) to the Holocene climatic optimum, the Termination I, can be resolved into 2 or 3 very short steps in cores with high sedimentation rates such as 40 cm per 1000 yr (Fig. 13). In both the planktonic (surface water) and benthonic (bottom water) records, signals of deglaciation which last only hundreds of years occur about 14,000-13,000 yr B.P., about 12,000 yr B.P., and about 9000 yr B.P. ending at 8500 to 8700 yr B.P. (Note: The radiocarbon ages are derived from measurement on planktonic foraminiferal carbonate >125 µm which is less subject to age contamination than total carbonate or organic carbon). The bottom-water changes occur some 500-1500 years earlier than the corresponding surface-water changes (Fig. 13 and 14). From this evidence we assume a gene-

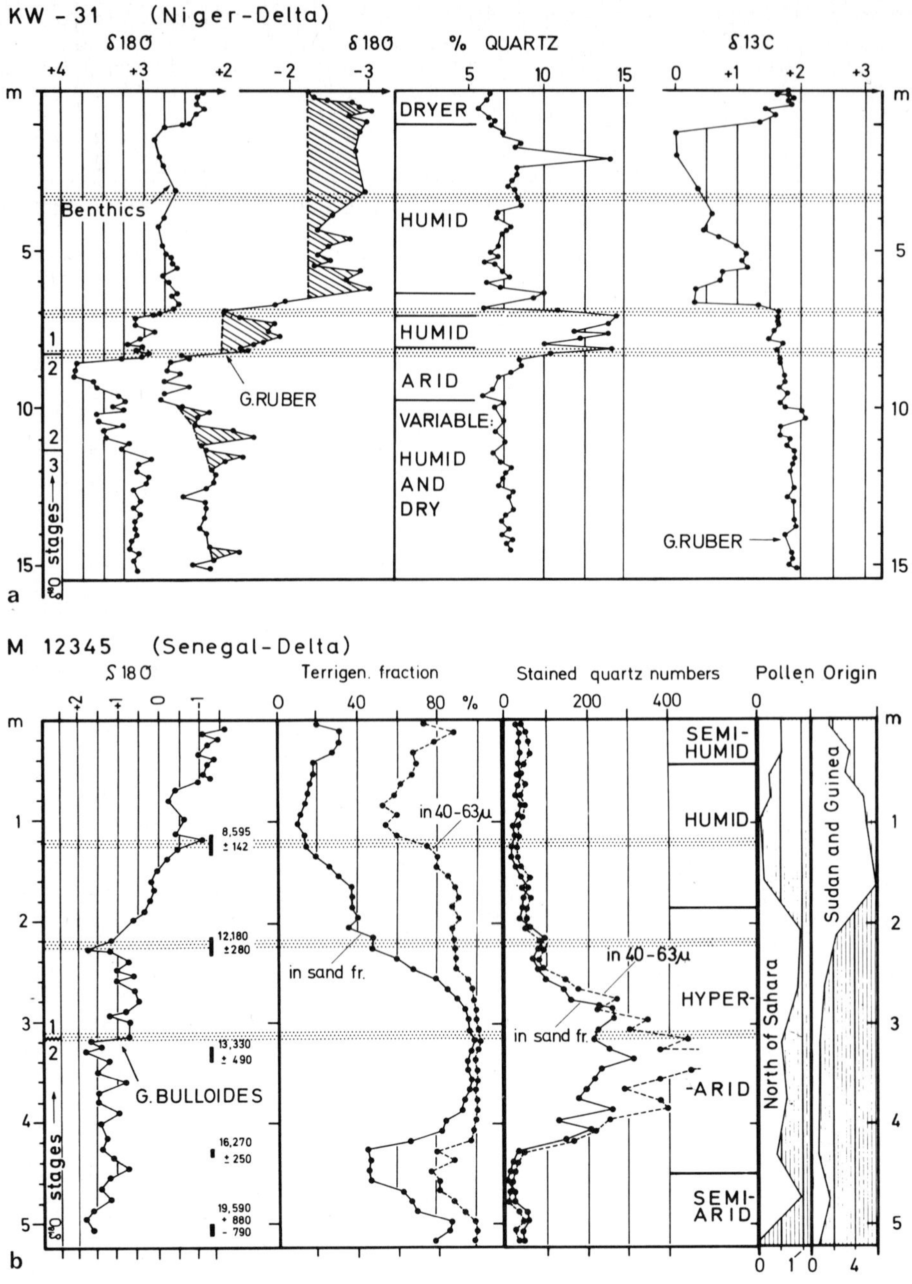

Fig. 13. High resolution sedimentary records of climatic changes in north-west Africa near the oxygen-isotope stage 1-2 boundary (Termination I). Sediment core WALDA/KW-31 is derived from the outer Niger Delta. (Redrawn from Pastouret et al., 1978). Core 12 345-5 from the outer Senegal Delta. (Modified from Diester-Haass, 1975; Rossignol-Strick and Duzer, 1979; Erlenkeuser et al., in prep.). For the exact position of core 12 345-5 see map on Fig. 14. Dotted pattern across figures indicates the 3 synchronous steps of Termination I in both cores

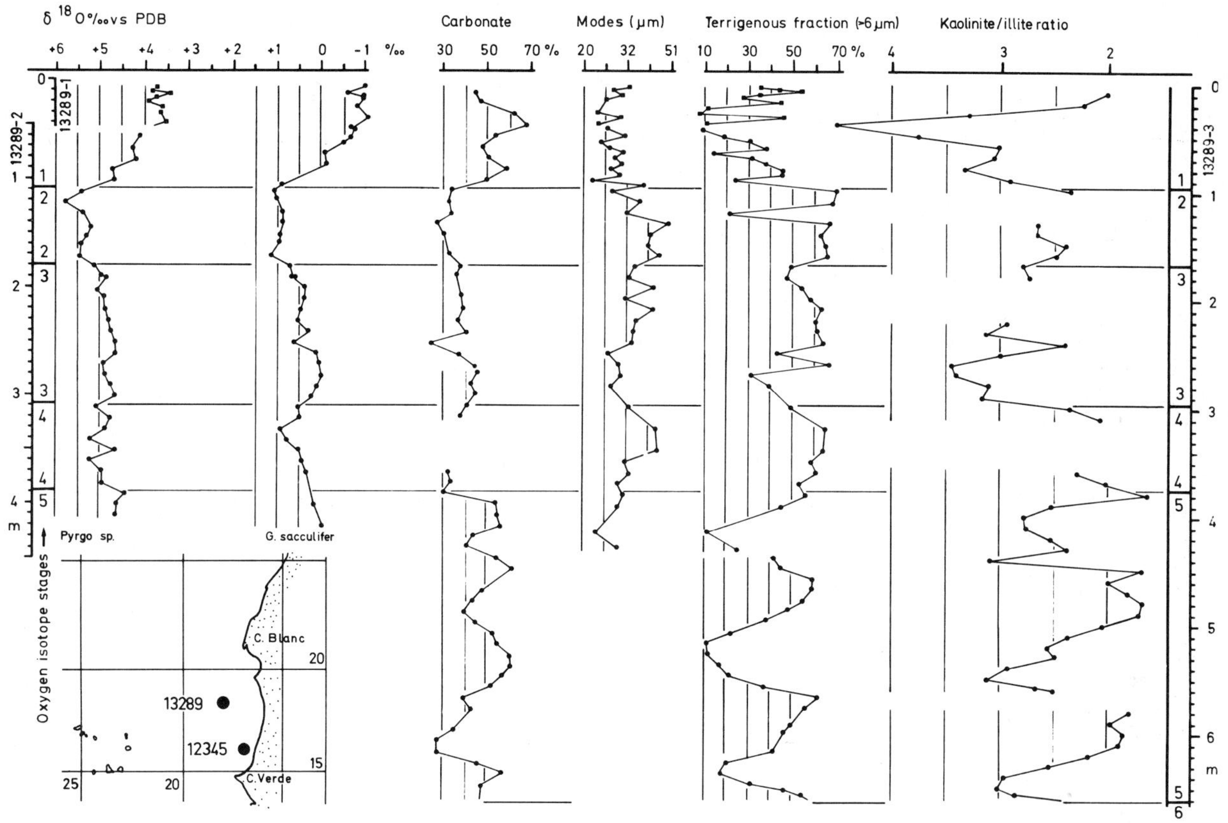

Fig. 14. Sediment cores 13 289-1,-2,-3: Late Quaternary fluctuations of carbonate concentrations and aeolian dust composition (grain-size modes and proportions of the terrigenous silt fraction > 6 µm; kaolinite to illite ratio) as compared to the oxygen isotopic stratigraphy of benthic (Pyrgo sp.) and planktonic (G. sacculifer) foraminifers. High proportions of illite mark an enhanced contribution of Trade Wind dust. Note: Many oscillations of the benthic oxygen-isotope curve show a slight "lead" as compared to the fluctuations of the other curves, furthermore, the dust composition has changed during the late oxygen-isotope stage 1, indicating an increased Trade Wind supply

ral time lag of the surface warming and a delayed decrease of upwelling at the east Atlantic continental margin as compared to polar deglaciation and global climatic change. The same time lag holds also true for cooling phases, e.g., at the base of glacial stages 2 (27,000-27,500 yr B.P.) and 4 (~75,000 yr B.P.).

Toward the end of the warming phases, enhanced freshwater discharge from the Niger River and, to a lesser extent, from the Senegal River imposes apparently anomalous peaks of light oxygen onto the local isotopic surface-water record (Fig.13). The sediment record indicates a slightly delayed post-glacial onset of precipitation and monsoonal moisture supply over the Sahelian drainage area as also suggested by the pollen record (Rossignol-Strick and Duzer, 1979; see below).

4.2 Advance and Retreat of Surface Water Masses in the Late Quaternary: Time Transgressive Features of Plankton Stratigraphy

Planktonic foraminifers provided the major tool for subdividing the Late Quaternary stratigraphically before an oxygen isotopic stratigraphy was established. Originally, ecostratigraphic zones were based on the presence and absence of globorotalids (Schott, 1935) or more specifically on the presence and absence of the

Globorotalia menardii complex (Ericson and Wollin, 1956). This species group is limited to the tropical and subtropical low-latitudes as recorded from the Atlantic water column (Bé and Tolderlund, 1971) and from the surface sediments (Fig. 8). According to Cifelli and Stern-Benier (1976), the species extends up to 25°N in endemic occurrence. These ecological limits of the Globorotalia menardii complex prevent isochronous stratigraphic boundaries from being drawn over a wide latitudinal range (Kennett and Huddlestun, 1972; Ruddiman and McIntyre, 1973; Pflaumann, 1975; Zobel and Ranke, 1978).

Figure 15 shows the time-transgressive postglacial immigration of Globoratalia menardii along a north-south transect at the West African continental margin. The species appears in minor quantities as far north as 21°N around 13,000 years ago. After disappearing from 11,000 to 10,500 years ago, it gradually spread from 5°N,approximately corresponding to the thermal equator, to 27°N where the species group did not appear until 5000 years ago. The abundance of the species decreases at 16° to 18° and again at 21° to 23°N. Strikingly, Globorotalia menardii has disappeared again during the past 1000 years between 23° and 27°N. We can interpret these trends of distribution in two ways: 1) They provide a record of the advance and retreat of warm Guinean surface waters (Figs. 2 and 4), possibly triggered by the varying intensity of upwelling contributing cold water to the shelf edge during the Holocene. 2) The variations of G. menardii reflect the SACW in the undercurrent (Fig. 4).

Other events, such as the extinction of Globorotaloides hexagonus (Bolli and Premoli-Silva, 1973; Pflaumann, 1975; Zobel and Ranke, 1978), and the changes in dominant coiling direction of various species, have been used successfully for stratigraphic correlation (Thiede, 1973; Ericson and Wollin, 1956; Ruddiman, 1972; Imbrie et al., 1973; Pflaumann, 1975). However, these events are al-

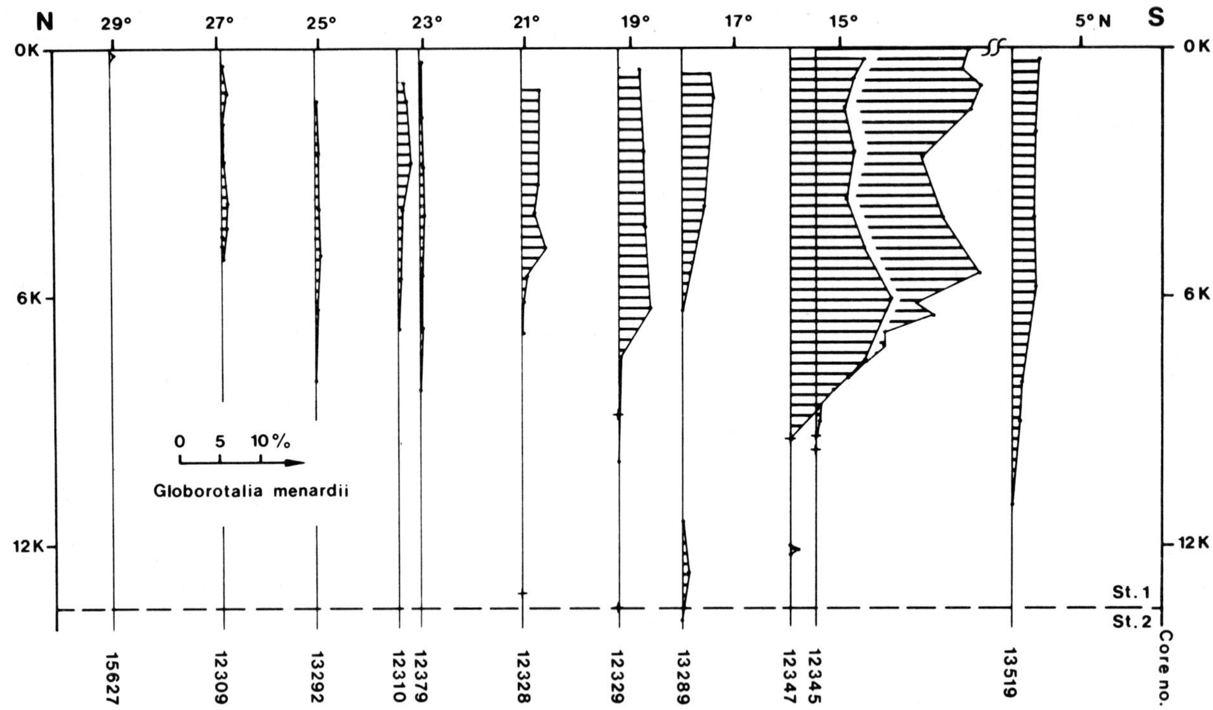

Fig. 15. Occurrence of Globorotalia menardii group during the last 14,000 yr in a north-south transect along the West African continental margin. Note the time-transgressive first appearance of the group, proceeding from south toward north. Chronostratigraphy based on radio-carbon-dated oxygen-isotope curves. (Erlenkeuser et al., in prep.)

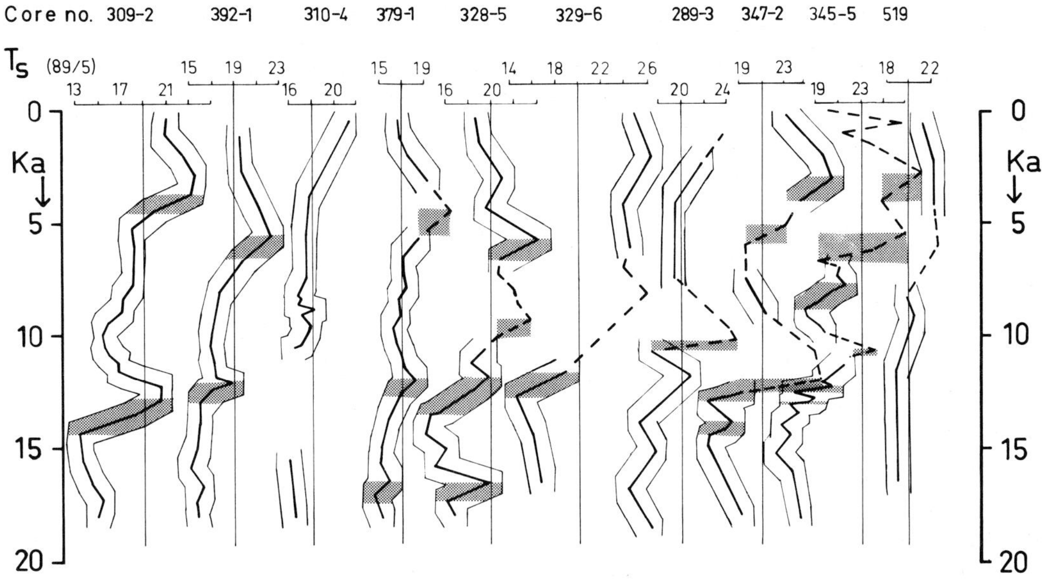

Fig. 16. North-south transect of summer surface tempteratures (°C) during the last 18,000 yr as derived by transfer function technique. Phases of pronounced temperature increase are shadowed. Range of error (± 1°C, ± 500 yr) is indicated by parallel thin lines. Broken line marks extrapolated temperature values. For chronostratigraphy see Fig. 15

so ecology-dependent and time-transgressive, when a time resolution of less than 10,000 years is considered.

Temperature gradients computed by the transfer-function technique of Imbrie and Kipp (1971) indicate that the Holocene surface temperatures have increased during several time-transgressive pulses (Fig. 16).

Layers rich in aragonitic pteropods at 1000 to 3000 m water depth are among several striking features formed during the last deglaciation period and the peak Holocene in the deep-sea cores off West Africa, Portugal, and elsewhere (Kudrass, 1973; Diester-Haass and van der Spoel, 1978). In some cores, only a single layer can be detected, but where accumulation rates are sufficiently high, several pteropod layers are distinguished (Fig. 17). Diester-Haass and van der Spoel (1978) assumed the age of the single layer as late Würmian, possibly centered around 14,000 yr B.P.. Berger (1978b) attributed the feature to a special, 1000-years lasting aragonite-preservation spike controlled by a global melt-water lid. Detailed time control has now revealed that late and post-glacial pteropod-rich layers, composed of different species at different locations off northwest Africa, recur at least two or three times, with a latitudinal shift, and parallel the previously discussed warming pulses of surface water (Figs. 16 and 17). The age of the first pteropod pulse after the 18,000 yr B.P. glacial maximum is about 17,000 yr B.P.. A second pteropod horizon is centered around 13,000 yr B.P.. The next spike starts about 11,000 yr B.P. in the south and ends around 9000 yr B.P. at 23°N; the last one occurs between 8000 and 5000 yr B.P. at 23°-27°N. This is the most prominent pteropod horizon. Off Portugal, the pteropod layers seem to belong to the first two pulses, although no sufficiently detailed dating is presently available to confirm this age assignment (Kudrass, 1973).

Under present conditions, the strong supply of upwelling-derived organic carbon induces a drastic rise of the aragonite compensation depth (ACD) from an average of 2,800 m in the central Atlantic Ocean (Berner, 1977) to about 600 m along the northwest African continental margin south of Cape Bojador (27°N) (Diester-Haass and Müller, 1979; Fütterer, 1980). North of it, the ACD lies between 2000 and 2400 m. Accordingly, our results on the distribution of pteropod layers can be interpreted in a

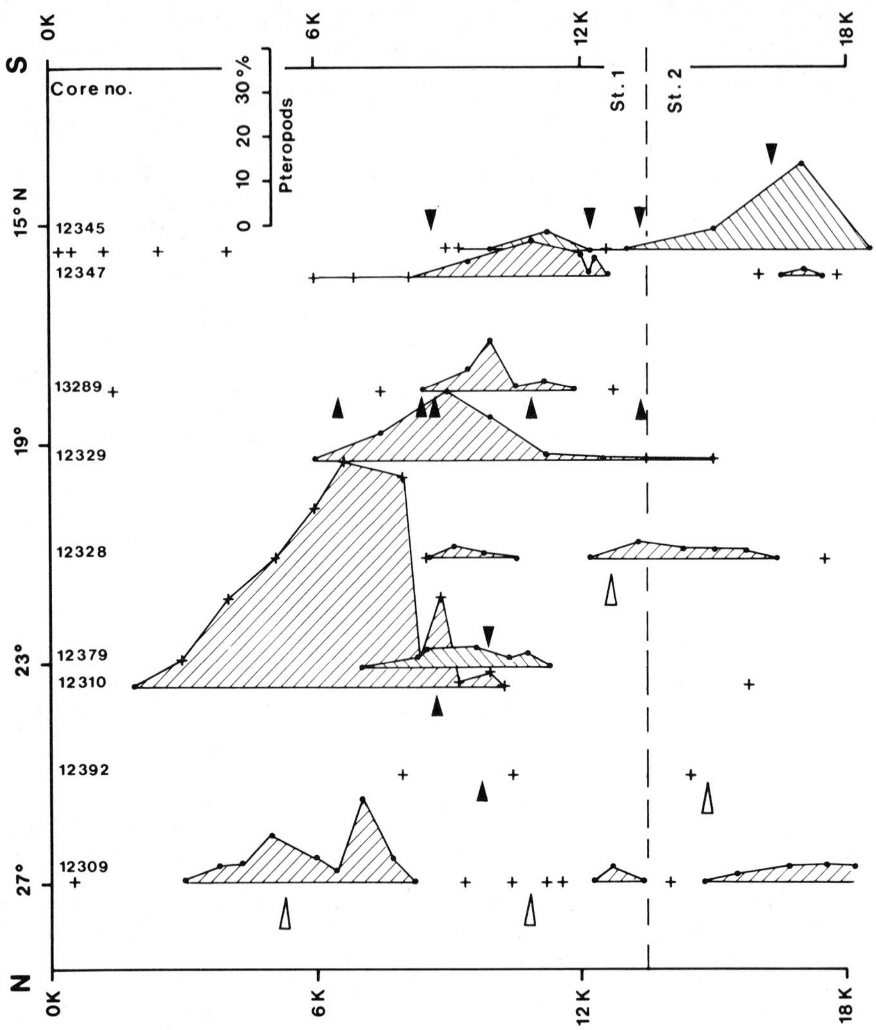

Fig. 17. North-south transect of pteropod concentrations (in the sand fraction), along the northwest African continental margin during the last 18,000 years. For chronostratigraphy see Fig. 15. Crosses indicate minor abundance. Note the irregular and multiple occurrence of high concentrations at different latitudes. Water depth of core locations: 2100-3300 m; except for Core 12 345: 945 m. Full triangles: ^{14}C ages from planktonic carbonate. Open triangles: ^{14}C ages from organic carbon (by Geyh, Hannover, in Thiede, 1977). Pteropod data partly from Diester-Haass et al. (1973) and Diester-Haass (1975)

way that local pulses of reduced coastal upwelling fertility, such as by a blocked advection of SACW, dominate the formation of pteropod accumulations in the sediments. On the other hand, global excursions of ocean chemistry and the aragonite compensation depth appear to be of subordinate influence in this region. Indeed, the time-transgressive immigration of *Globorotalia menardii*, the occurrence of which is to some degree related to tropical upwelling off northwest Africa (Fig. 3C), follows just above the pteropod layers in all cores investigated (compare Figs. 15 and 17). Apparently, the preservation of the layers might serve as indicator of a reduced upwelling fertility in most prominent upwelling regions. Such exceptional conditions of circulation should have prevailed, for example, during the early and middle oxygen-isotope stage 1.

4.3 Late Quaternary Evolution of Bottom-Water and Atmospheric Circulation: Benthic and Lithostratigraphic Record

Both the total carbonate content and the terrigenous sediment fraction of the eastern Atlantic continental margin generally vary in parallel with the oxygen-isotopic record on global climatic fluctuations. Philippi (1910), Olausson (1969), Ruddiman (1971), Hays and Peruzza (1972), and Gardner (1975) have pointed out that carbonate contents are low in the northern and equatorial Atlantic during glacial stages (Fig. 14). According to Philippi (1910), the reduced carbonate content is usually related to increased dissolution by AABW which is thought to have been delivered in greater amounts during glacial times.

An increased incursion of AABW can also explain the decreased numbers of "warm benthic foraminifers" during glacial phases at 2800 m water depth (Site 397; Figs. 18 and 24). Lutze (1979) has regularly observed such a deviation throughout Pleistocene sediments. These fluctuations do not reflect real temperature variations in the deep sea but rather depend on fluctuating oxygen contents and food supply. During the present warm stage, the site is in contact with well-oxygenated NADW which possibly has been replaced by a much less oxygenated AABW during cold stages. However, both a glacial attenuation of oxygen and increased carbonate dissolution can also be caused by a locally enhanced flux of organic matter. This flux can originate from increased upwelling und surface-water productivity, and finally, from an increased inflow of nutrient-rich SACW to the northwest African continental margin.

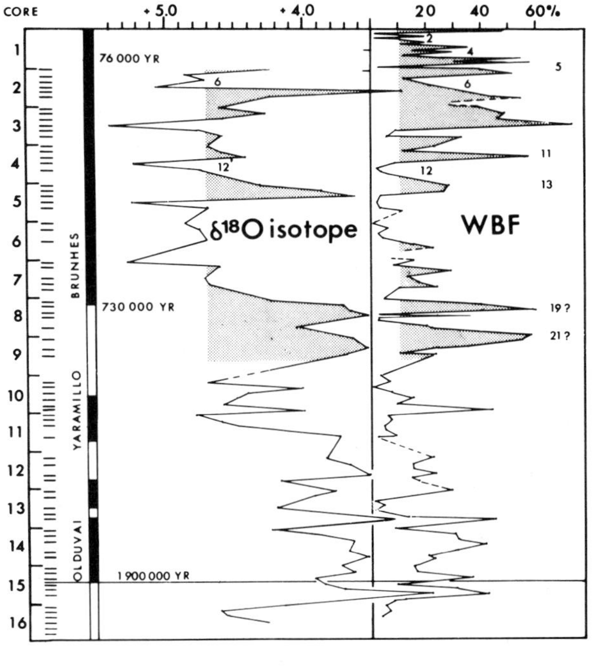

Fig. 18. Comparison of the WBF curve (abundance fluctuations of "warm" benthic foraminifers") with the oxygen-isotope curve. (Shackleton and Cita, 1979) at DSDP-Site 397. (Lutze, 1979). Differences between the curves are partly due to different sampling positions

Inconsistencies between the Pleistocene records of total carbonate content and of oxygen isotopes (Fig. 14) are of special interest, because they are most probably controlled by the masking supply of non-carbonate sediments from the African continent, i.e., by river supply and beyond the river influence, by the north African dust discharge (Hays and Peruzza, 1972); Thiede, 1977; Lutze et al., 1979). In principle, glacial stages parallel (1) an increase and coarsening of the aeolian silt-sized fraction, (2) an increase of its accumulation rates, (3) enhanced proportions of Trade Wind-derived chlorite, illite, and irregular clay aggregates diluting Harmattan-related kaolinite (Fig. 14) (Parkin and Shackleton, 1973; Chamley et al., 1977; Lutze et al., 1979; Sarnthein and Koopmann, 1980; Lange, 1981). However, when examined more closely, the Trade-Wind dust record can vary markedly from the global $\delta^{18}O$ record. For example, Late Holocene aeolomarine deposits from the last 5000-6000 yr appear to gradually approach a glacial rather than interglacial type of dust record (Figs. 14, 13). Similar "leads" of the dust signal as compared to the isotopic signal are observed at earlier transitions from warm to cold stages, for example, at the stage 4-5 boundary and at older stratigraphic boundaries in other cores. On the other hand, the dust signal can also lag behind by up to 1500 yr at transitions from cold to warm stages. The transitional phases from one regime to another can be very short, sometimes not exceeding a few hundred years (Fig. 14).

The leads and lags of atmospheric circulation intensity are confirmed by the pollen most of which are dust-borne. A good example is shown in the high time-resolution section of Meteor-core 12345 (Fig. 13): Rossignal-Strick and Duzer (1979) show a distinct continuation of the extremely arid pollen spectrum from glacial stage 2 over into early stage 1 (up to 12,000 yr B.P.). Furthermore, the onset of more semi-arid pollen accumulation following the mid-Holocene, during which humid pollen was dominant, occurred some 5500 yr ago. These observations agree with pollen records from a number of cores further north (21°-25°N) (Caratini et al., 1979; Agwu and Beug, 1981). In conclusion, it appears that the Trade Winds supplying dust and pollen from the NW Saharan fringe have become more vigorous *ahead* of polar ice-cap formation and have ceased slightly *after* the first major phase of polar deglaciation as documented by oxygen isotopes. But the dust grain size which is the parameter of wind strength and has led also the deglaciations (Fig. 14, stage 4-3 boundary and 2-1 boundary), is controlled not solely by the meridional Trade Wind but also by the zonal Harmattan wind system. The post-glacial time lag of increasing vegetational cover coincides with the oxygen isotopic message of the fluvial freshwater supply (Fig. 13), and with land data which show that the widespread dune fields at the South Saharan fringe also became dormant slightly after the end of the last glacial, about 12,000 yr ago (Servant and Servant, 1972; Michel, 1973; Sarnthein, 1978).

However, the late Quaternary dust record can reflect the evolution of continental aridity only to a limited degree. For example, dominantly coarse and abundant dust deposition during stage 3 depicts a regime of increased wind vigor (Fig. 14). This record contrasts with contemporaneously widespread humidity all over the Sahara as evidenced from land data (Rognon, 1976, and unpubl. data).

5. QUATERNARY TIME SLICES AND NO-ANALOG SITUATIONS

5.1 6000 years B.P.: the Interglacial Regime

The time slice of 6000 yr B.P. was selected for reconstruction because it represents the youngest warm climatic extreme. It occurred near the end of the Holocene climatic optimum starting near 8700 yr B.P., as indicated by the oxygen isotopic curves and radiocarbon dates, and predated a short, but marked phase of cooling, and of vigorous aridification on land near 5000 yr B.P. (Gabriel, 1978; Sarnthein, 1978; Maley, 1980).

Major features of the 6000 B.P. surface hydrography were estimated by transfer-function analysis (Kipp, 1976) of planktonic foraminiferal assemblages from 16 deep-sea cores. The 6000 B.P. surface-water temperatures were mapped for the seasonal extremes (Fig. 19). They differed from the estimates for the present (Pflaumann, 1980) by an increase of 1°-2°C during winter and 2°-3°C during summer. Along the continental margin, between 19° and 27°N, the seasonal temperature differences were more pronounced than today (Fig. 20) because the 6000 yr B.P. winter temperatures increased less than those during the summers. The enhanced seasonality off Cape Blanc may be due to strongly reduced summer upwelling of deeper waters which can retain a constant temperature range over the year. Such a conclusion is favored by the radiolarian record (Labracherie, 1980), by the evaluation of the dust components in the marine sediments (Sarnthein et al., 1981), the distribution of organic matter (Müller and Suess, 1979), and the record of the coastal mollusc faunas (Petit-Maire, 1980).

A northward retreat of the polar to temperate foraminiferal assemblage suggests that 6000 yr ago, the intensity of the Canary Current had weakened (Thiede, 1977). Both a reduced Canary Current and a seasonally reduced or absent upwelling system suggest a short, but important variation of the present asymmetry of the eastern and western North Atlantic ocean at 6000 yr B.P.; the warm surface-water provinces extended along the eastern margin farther

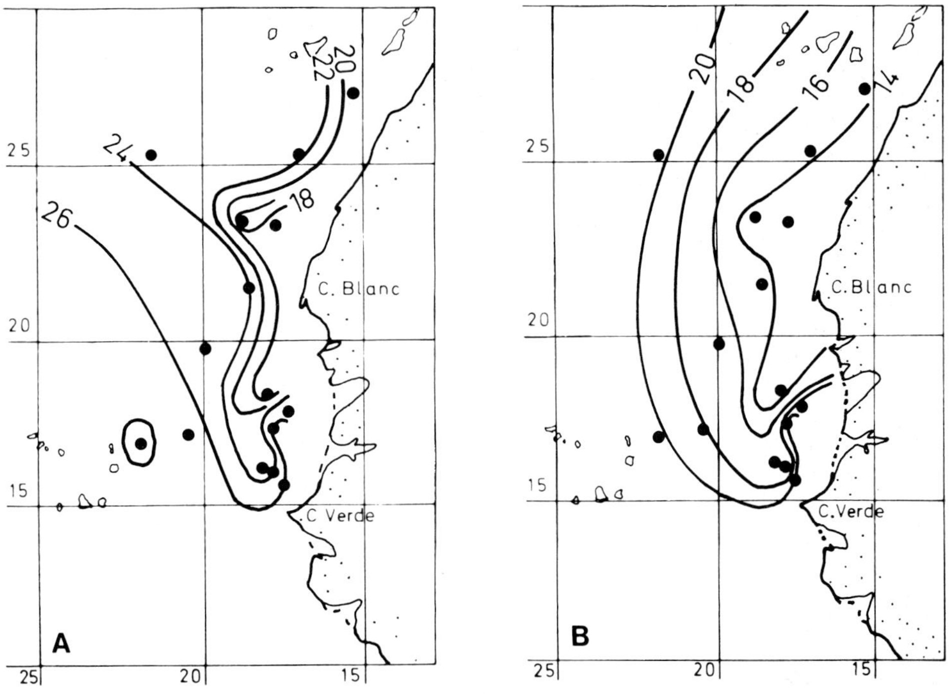

Fig. 19. Surface-water isotherm maps for the Holocene climatic optimum as derived by transfer-function technique. Contour interval is 2°C. A Map for August 6000 yr B.P. B Map for February 6000 yr B.P. (Redrawn from Pflaumann, 1980). Shoreline is modified using an elevated sea level of 2.5 m.

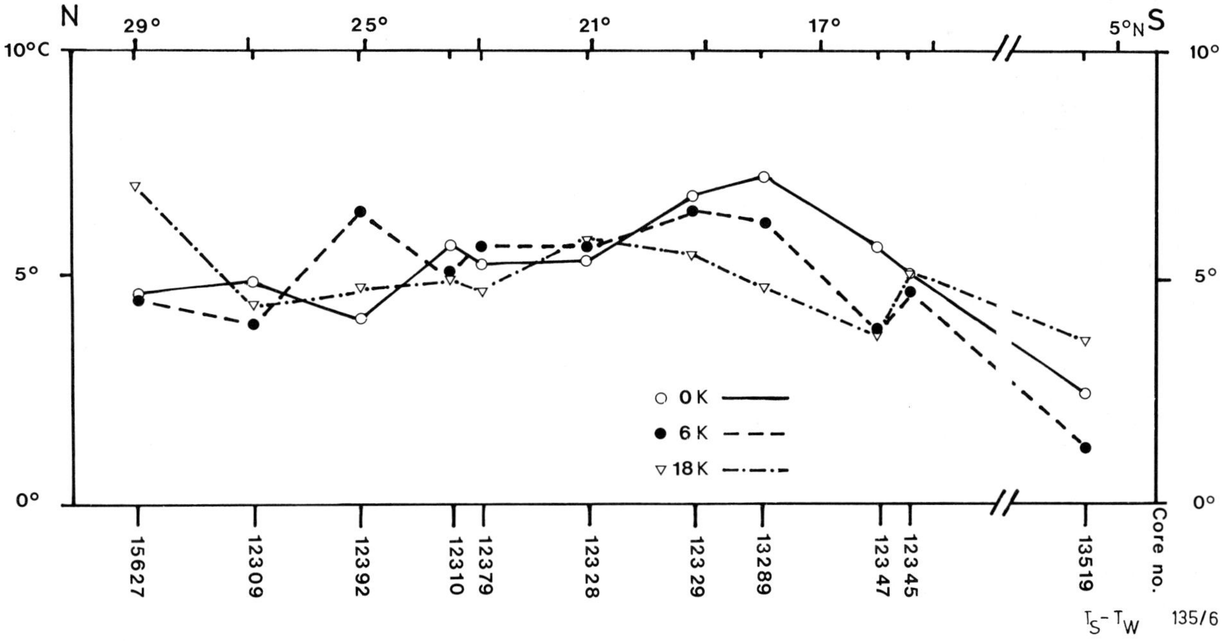

Fig. 20. North-south transect of seasonal temperature anomalies (in °C) as derived by transfer-function technique for the time slices today, 6000 yr B.P., and 18,000 yr B.P.

poleward and reduced the locally accentuated temperature gradients off northwest Africa. Not too much is known about the characteristics of the probably more sluggish 6000 yr B.P. deep-water circulation in the eastern Atlantic. The reduced upwelling indicates a decreased inflow of SACW. Benthic foraminifer distributions disclose a slight advance of the less oxygenated AABW (with the "cool" species C. wuellerstorfi) from south of 20° to some 30°N (Lutze et al., 1979) after 6000 yr B.P.

The paleo-wind regime and continental climates 6000 yr B.P. were estimated from the terrigenous fraction in 34 deep-sea cores. An excess of fine-grained silt and clay (11 µm grain-size mode) largely dominated deposition on the upper continental rise where coarser-grained proximal-dust sediments (24 µm mode) prevail today. The fine-grained deposits derived from long-distance transport and close to the continent, from extensive fluvial mud supply (Fig. 6B). The fluvial component (Fig. 21) was widespread, but typically small-scale and lobate, and reflected predominantly wet climatic conditions over most parts of the western Sahara. The peak Holocene humidity is corroborated by the (Trade Wind borne) pollen record from sediment cores between 16° and 25°N (Agwu and Beug, 1981; Caratini, 1979; Strick-Rossignol and Duzer, 1979). The contribution of Trade-Wind dust was greatly reduced during the early (peak) Holocene north of 20°, indicating a strongly weakened, perhaps seasonally ceasing surface-wind regime from NNE.

Nevertheless, the dust outbursts from the Harmattan level continued to be active, although somewhat reduced. Their latitudinal position was constant at 16°-20°N. This is shown by the distribution patterns of dust grain sizes beyond the belt of river influence (Figs. 21A and B), clay minerals, and stained quartz (Sarnthein, 1979). From this we infer an unchanged northward penetration of the ITCZ during peak Holocene summers as compared to today, although the fluviomarine sediments as well as the land record (Street and Grove, 1976; Sarnthein, 1978; Petit-Maire, 1980) show that the monsoonal rain fronts have advanced deeply northward into the Saharan realm 6000 yr B.P. ago.

5.2 18,000 years B.P.; the Last Glacial Regime

The time slice of 18,000 yr B.P. was selected to complement the study by CLIMAP (Cline and Hays, 1976). Another reason for this selection was to compare the youngest glacial maximum with the 6000 yr B.P. time slice, the opposite climatic extreme. Glacial advances and retreats have occurred repeatedly during the past 700,000 years and probably also in earlier times. Pleistocene land records and marine oxygen-isotope records indicate that the interval around 18,000 yr B.P. lies near the end of the longest and most pronounced cold phase during oxygen-isotope stage 2. The time slice coincides with a maximum extent of most ice caps in the northern hemisphere and a considerably greater extent of winter sea ice in the Southern Hemisphere (Dreimanis, 1977; CLIMAP, 1976).

Our estimates of the surface hydrography at 18,000 yr B.P. are based on the record of planktonic foraminifers from 50 cores (Fig. 22) and a compilation of data from various sources (CLIMAP, 1976; McIntyre et al., 1976; Prell et al., 1976; Gardner et al., 1976; Molina-Cruz and Thiede, 1978; Pflaumann, 1980, Pflaumann, in prep.).

A sharp oceanic front recorded in the sediments, crossed the northern part of the North Atlantic Ocean. Steep temperature gradients mark the southern limit of the polar water province at approximately lat 42°N (Fig. 22). During winters the transition between cold polar surface waters and the warm subtropical gyre was compressed to a width of only 500 km. South of this thermal barrier, the warm North Atlantic Drift flowed eastward across the ocean to impinge on the Iberian peninsula. The boundary probably also indicates the paleo-position of the Westerly wind system. Furthermore, the steep gradients in the surface-water temperatures are consistent with the assumption that strong winter storms persistently formed at this latitude under conditions similar to the modern Antarctic "Howling Fifties" (see below).

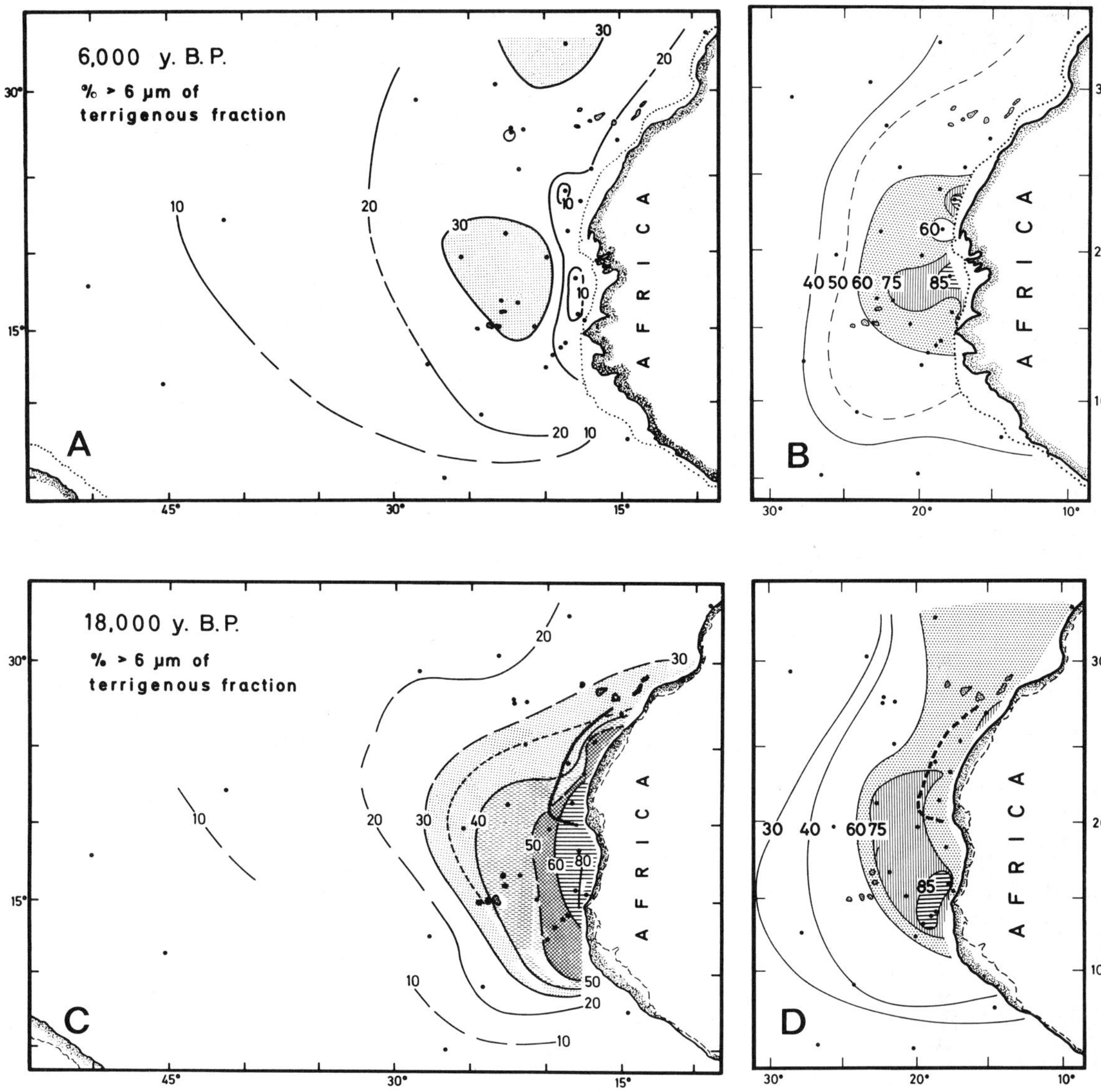

Fig. 21. Distribution of terrigenous grain-size fractions (on a carbonate and opal free basis) in Late Quaternary sediments of the northeast Atlantic. See Fig. 7 legend for explanation.
A Relative abundance (%) of the grain-size fraction > 6 µm for 6000 yr B.P. Note the numerous triangles which indicate abundant supply of fluvial mud (cf. Fig. 6 B) with an excess of fine fraction < 6 µm.
B Distribution of the terrigenous silt mode around 24 µm (see Fig. 7 C) for 6000 yr B.P.
C Relative abundance (%) of the grain-size fraction > 6 µm for 18,000 yr B.P.
D Distribution of the terrigenous silt mode around 24 µm (see Fig. 7 C) for 18,000 yr B.P.
18,000 yr B.P. shoreline corresponds to a lowered sea level of about 100 m

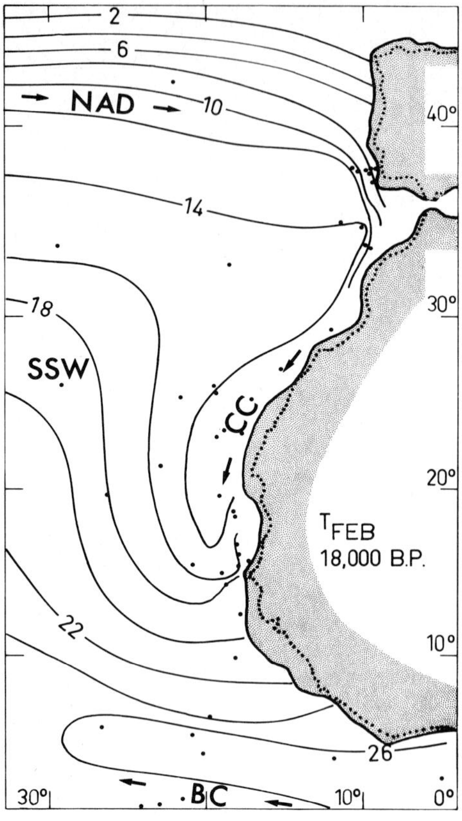

Fig. 22. Surface-water isotherm map for February 18,000 yr B.P., the last glacial maximum. Contour interval is 2°C. (Data compiled from McIntyre et al., 1976; Molina-Cruz and Thiede, 1978; Pflaumann, 1980). NAD North Atlantic Drift; SSW Subtropical Sargasso Sea Water; BC Benguela Current; CC Canary Current. Shoreline corresponds to a lowered sea level of about 100 m

During summers, however, temperate and transitional waters extended farther north in the Eastern Atlantic as far as the center of the Bay of Biscay (Molina-Cruz and Thiede, 1978), thus shifting and weakening the thermal front.

The thermal front was deflected to the southwest off southern Portugal and northwest Africa during winters and summers, thereby delineating the course of a cold eastern boundary current (Fig. 22). Both this flow pattern and its low temperature level are corroborated by ice-rafted gravel cored off southern Portugal and Morocco (Kudrass, 1973). The very sharp paleotemperature gradients north of lat 30° we interpret as indicating strongly increased current intensity, since upwelling is rather insignificant in this region of Westerly winds.

North of Cape Blanc, the surface waters of the Canary Current were as much as 8°C colder at 18,000 yr B.P. than today during winter and 6°C colder during summer. The 18,000 yr B.P. summer isotherms have much steeper gradients in this region than today and form a distinct front off Cape Verde (McIntyre et al., 1976). The cold Canary Current extended somewhat further offshore and about 500 km further south 18,000 years ago. The zone of lateral transition to tropical and subtropical waters was compressed as compared to today.

Except for nearshore cores, the 18,000 yr B.P. equatorial region differed from modern conditions only by 1°-2°C during February, but August temperature estimates were 2°-8°C cooler (Gardner et al., 1976).

Increased equatorial divergence between 0° and 6°N and the Benguela Current may have produced the distinct temperature minimum along the equator during Southern Hemisphere winters. The thermal equator remained in the same position around lat 5°N with seasonal migrations like those of today. The difference between glacial and interglacial current regimes is one of intensity: The cooler sea-surface temperatures suggest that the glacial circulation near the geographic equator was stronger during southern winters. Since this circulation is closely related to the Trade Winds, they were probably more vigorous.

In the Gulf of Guinea, the high fertility of planktonic foraminifers suggests that salinity of the Last Glacial surface waters was almost normal marine. At that time the freshwater discharge from the Niger was low (Pastouret et al., 1978).

The 18,000 yr B.P. North Atlantic Subtropical Water, the Sargasso Sea Water, as reconstructed from its salinity characteristics was compressed in the east by some degrees of latitude and was displaced to the west, and included the Caribbean (Prell et al., 1976; Balsam et al., 1980; Crowley, 1981). The temperature patterns of the surface waters, however, point to a rather unchanged extension of the subtropical water in comparison to modern conditions. In its center there were only slightly reduced temperatures during winter, while summer temperatures were at least as high as today (McIntyre et al., 1976).

Seasonality, i.e., the difference between the estimated annual temperature extremes, was about 5°C along the Northwest African margin 18,000 yr ago (Fig. 20). Slightly increased seasonality occurred at some sites further seaward off Portugal, off Morocco, and near Cape Verde. Contrary to modern patterns (Thiede, 1977), the 18,000 B.P. seasonality did not increase toward the coast at the continental margin, but it did increase offshore and reached a maximum along the seaward periphery of the Canary Current. Thus the current may have shifted in position seasonally and, owing to the sea-level drop 18,000 yr ago, was displaced toward the continental margin.

Reduced seasonal contrasts along the inner continental margin 18,000 yr ago may have resulted from a continuous upwelling regime throughout the year, because upwelled subsurface waters have temperature characteristics independent of seasons. The 18,000 yr B.P. coastal upwelling was much more productive than today and, like the Canary Current, it shifted westward over the continental slope ("Peru-model" of Hart and Currie, 1960; Diester-Haass and Schrader, 1979; Müller and Suess, 1979). The distribution of some radiolarian tracer species (Theocalyptra davisiana and Phormospyris stabilis scaphipes; Labracherie, 1980) supports the conclusion that the high fertility was a result of an oceanic circulation pattern which has no present-day analogue. The nutrient-rich subsurface SACW penetrated much further north and substituted for the NACW which ceased to form (McIntyre et al., 1976). Müller and Suess (1979) showed that the rates of primary production increased by a factor of 2 to 3 at lat 25°N, in areas where river-supplied nutrients as an alternative to those supplied by upwelling can be excluded.

Transfer-function technique derived temperatures from planktonic foraminiferal associations show that upwelling induced local temperature anomalies of about -4°C. As a further characteristic of intense upwelling, deep-water species (G. pachyderma, G. inflata) contributed up to 3/4 of the total planktonic foraminifer fraction at the slope off Senegal (Pflaumann, 1975). Furthermore, unusually high abundances of the subpolar species G. bulloides are evidence for intensive Glacial upwelling at 27° lat N also (Lutze et al., 1979). At the same latitude, the opal proportion amounted up to 15 % of the carbonate-free sediment fraction > 6 µm (Koopmann, 1980). However, it appears that the extremes of upwelling may have occurred not around 18,000 yr B.P., but earlier during oxygen-isotope stage 3 (Müller and Suess, 1979) and again, later, for short period during the early Stage 1. This timing is also corroborated by the low ratio of $^{13}C/^{12}C$ in planktonic foraminiferal shells (Berger et al., 1978; Lutze et al., 1979).

The pattern and the intensity of deep-water circulation underwent major changes between interglacial and glacial modes. We have mentioned the record of a strengthened flux of SACW replacing the NACW (Fig. 4). At the upper rise, around 3000 m water depth, 18,000 yr B.P. benthic foraminiferal assemblages show a complete change to "cool" water species (Lutze et al., 1979) (Fig. 18). They had a specific preference for a less oxygenated and more nutrient-rich environment. For explanation we can infer either a larger influx of AABW (Streeter and Shackleton, 1979; Lutze, 1979) or a locally enhanced flux of organic debris from the strengthened upwelling production. The deeper-water circulation was, with high probability, intensified in all levels along the continental slope. This is evidenced by the increased proportions of "grain-by-grain" reworked shallow-water particles at the upper continental rise. However, the exposed shelf might also have contributed to this reworking by a drastic reduction of the distance from the shoreline to the deep sea. At the continental rise, intensified contour currents within the AABW mass may have formed widespread mud-wave fields during the last glacial. However, their exact chronology is not substantiated.

From a theoretical point of view, the climatic belts on the European and African continents might be expected to basically parallel the outlined zonation of ocean temperatures and water masses in their latitudinal position and extent (McIntyre et al., 1976; Flint, 1971). Accordingly, the 18,000 yr B.P. realm of arctic-subantarctic climate spread parallel with the polar water mass as far as the southern margin of the Alps. Almost juxtaposed or with a little distance to the south, we should expect the belt of (dry) subtropical climate corresponding to the position of the subtropical water mass in the Atlantic. Thus the intervening zones of temperate and Mediterranean climates would have been compressed to a narrow band. This presently widespread and seasonally humid belt which is associated with the Westerly wind system, was perhaps only a few hundred kilometers wide or locally completely absent. Manabe and Hahn (1977) arrived at similar conclusions when simulating the ice-age climate with a mathematical model on the base of CLIMAP data (CLIMAP, 1976). However, the discussion on the glacial paleoclimate in the northwest Sahara is still controversial, as outlined in Section 2.1.

The Saharan coastal climate is influenced also by coastal upwelling. The resulting cold surface waters minimize the amount of both local evaporation and nearshore precipitation, as shown for example by Schiffers (1971), UNESCO (1977), and Fig. 22. Thus both the enhanced Trade Winds and the stronger coastal upwelling, a product of the Trade Winds, would have strenghtened the aridity during glacial times in the west Saharan coastal region, except for the Atlas Mountains which serve as a local rain-catchment.

This general paleoclimatic model was examined under the light of evidence from terrigenous deep-water sediments. They show that river discharge was absent from the whole northwest African continental margin from south of the Atlas Mountains to as far as the Gulf of Guinea (Pastouret et al., 1978) during the last glacial maximum at 18,000 yr B.P.. This conclusion is especially based on the reduced abundance both of clay fraction and within the silt fraction, of the 11 µm mode in sediments from lat 10° to beyond 28°N (Figs. 21C, 6C). These deposits also are devoid of plant remains and other constituent particles characteristic of river supply (Table 2, Fig. 13). Thus our interpretation indicates, in agreement with the model, glacial aridity for the northwest Sahara such as for the south Sahara (cf. Sarnthein, 1979; Diester-Haass, 1979; Sarnthein et al., 1981). New evidence for this interpretation is based on (1) grain-size data, (2) the so-called mixed layer clays (Chamley et al., 1977), now identified as irregular clay aggregates originating from Trade Wind dust falls (Lange, in press), (3) the reinterpretation of stained quartz grains as indicators of mainly lateritic source regions instead of being mere characteristics of aeolian transportation ("Wüstenquartz"). The model of glacial aridity is clearly confirmed by the pollen record in deep-sea cores between 15° and 25°N. It shows a maximum discharge of steppe and desert-related species of Ephedra, Artemisia, and Chenopodiacea, and a disappearance of almost all remaining species during glacial times (Caratini et al., 1979; Agwu and Beug, 1981; Rossignol-Strick and Duzer, 1979; Labracherie, 1980).

The Harmattan winds supplied abundant and coarse aeolian dust (24 µm mode, Fig. 21D) with abundant quartz (Kolla et al., 1979) to continental-margin sediments at 18,000 yr B.P. and until the end of oxygen-isotope stage 2. However, the bulk of the dust penetrated scarcely further west than it does today. Despite a reduced sample coverage, the dust appears dispersed with a simple sickle-shaped pattern over the eastern Atlantic (Fig. 21D), a pattern which parallels the modern Harmattan trajectory some 300 km further south. From this we can conclude that the trajectory of glacial Harmattan dust outbursts roughly followed the same type of anticyclonic gyre as it does today.

Northeast of the sickle pattern the 18,000 yr B.P. Trade Wind regime is documented by a fringe of even coarser grain sizes with a dominating grain-size mode of 45 µm extending up to 200 km offshore between lat 18° and 27°N and suppressing the grain-size assemblage of the 24 µm mode shown in Fig. 21D. North of some 20°-22°N, these deposits differ also in composition from the Harmattan dust discharge in the south (Chamley et al., 1977; Sarnthein,

1979). Thus the major compositional boundary of this region which does not match the grain-size boundary at lat 17°N (Fig. 21D), retained the same latitudinal position as today. North of this latitude, illite irregular clay aggregates, chlorite in the more northerly cores, and unstained quartz were dominant. By analogy with modern deposits, this sediment is also regarded as of aeolian origin (Sarnthein, 1979). The expanded distribution of coarse grain sizes indicates intensified glacial Trade Winds from north-northeast during times of exposed shelves and reduced length of dust trajectories (Koopmann, 1979). Based on the evaluation of grain-size data, the average speed of the meridional surface Trade Winds must have increased by 50 % during the last glacial as compared to today, whereas the speed of the zonal (mid-tropospheric) Harmattan dust plumes would have decreased by one third (Sarnthein et al., 1981).

The strengthened meridional surface circulation should have largely intensified the displacement of dust falls from the overlying Harmattan level toward the SSW, according to the different grain sizes involved by an amount of several hundred km (compare Figs. 21D and 7C). Based on this reasoning and particularly on the stationary boundary of dust composition near 20°N, we can conclude that the actual center of Harmattan dust outbursts from the African continent did not shift, but retained its position near 15°-20°N during glacial times just as at 6,000 yr B.P. and today. This also implies, on the long-term average, a constant summer position of the ITCZ and of the adjoining upper-air high-pressure system over the western Sahara with seasonal migrations comparable to todays's, such as have been suggested by Gardner et al. (1976). This is essential for an understanding of the interglacial-glacial structures of the Hadley cell, the form of which has remained basically constant (see Sect. 2.1).

Accordingly, not the tropical but the Mediterranean and temperate zones appear to be those climatic belts which are squeezed and expanded by the varying extension of the north polar ice caps, as expected from the distribution of oceanic water masses. Thus a mechanism of shifting subtropical climate belts (Rognon and Williams, 1977) could no longer serve to explain the contrasting widespread glacial aridity and interglacial wetness over the whole of Northwest Africa as they are recorded, for example, by fluctuating amounts of ground-water recharge (Sonntag et al., 1980), by the varying extent of dune fields on the continent (Sarnthein, 1978), or by the fluctuating fluvial sediment supply to the ocean. The prime factor for the changing wetness in the southern Sahara appears to be the variable strength of the meridional Trade Winds coupled with the variable strength of the (zonal) tropical disturbances and the varying supply of water vapor from the equatorial Atlantic (Sarnthein et al., 1981). The northern Sahara remained arid 18,000 yr ago because also the above-mentioned winter cyclones originating from the polar front did not provide much precipitation, since evaporation was strongly reduced over the North Atlantic when cold polar waters spread to 42°N.

5.3 No-Analog Processes: Formation of Aeolian-Sand Turbidites

Glacial regimes induced a rather unique coincidence of several partly interdependent environmental features at the west African continental margin between lat 14° and 20°N: (1) A hyperarid climate in the presently arid to semi-arid Sahel zone, (2) vigorous and dominantly offshore blowing (Trade) surface winds from NE which effected a strong migration of the Ogolien sëif-dune fields in the southwest Sahara (Sarnthein and Walger, 1974), and (3) a lowered sea-level which shifted the shoreline to the shelfbreak, close to steep offshore slopes and numerous canyon heads. This set of glacial conditions is recorded by well-defined aeolian-sand turbidites and dated by the bracketing pelagic sediments in the deep-sea (Sarnthein and Diester-Haass, 1977). We can assume an immense glacial discharge of dune sands in the range of hundreds of cubic km to the shoreline and further, by slumps and turbidity currents down to the continental rise up to distances of more than 1200 km offshore.

Spectacular examples of aeolian-sand turbidites were found off Senegal and Mauritania (Fig. 23) and in the Gulf of Guinea (Sarnthein and Diester-Haass, 1977) documenting glacial regimes of strong offshore-blowing Trade Winds in both the southern Sahara and the northern Congo basin, which then formed a northern extension of the Kalahari desert (de Ploey, 1965).

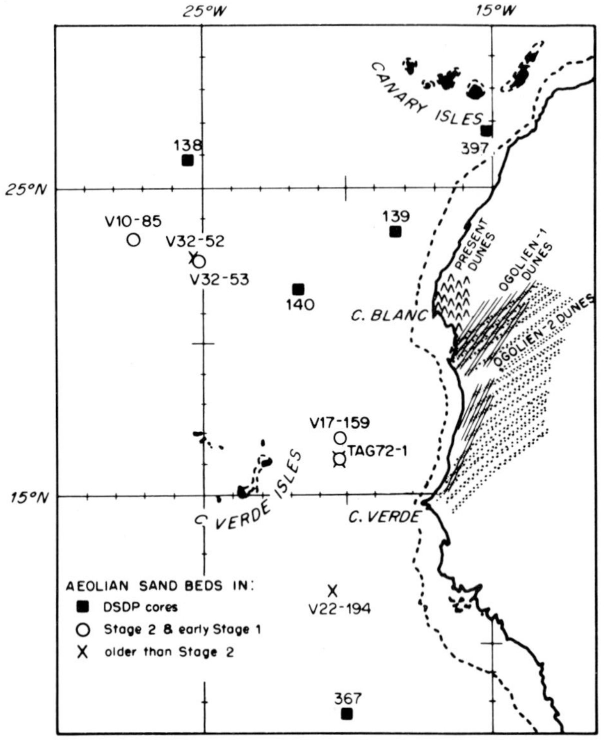

Fig. 23. Location and orientation of Pleistocene and Holocene dune systems in northwest Africa, and position of deep-sea cores and Deep Sea Drilling Sites which contain aeolian-sand turbidites. Broken line at the shelf break marks the approximate peak-glacial shoreline.

6. EVOLUTION OF NEOGENE PALEOENVIRONMENTS

The record of Neogene oceanic and atmospheric environments is interpreted from sediments cored at 15 DSDP sites along the NW African continental margin (Fig. 24). Of the 15 sites, only Site 366 and 397 and the Pliocene of Site 141 have thick pelagic and hemipelagic carbonate sediments (Fig. 25) which display an almost continuous and high-resolution stratigraphic record. At present, all three sites are located beneath the eastern Atlantic boundary current. Most remaining sites contain incomplete sections due to spot coring, however, comprehensive interpretation of the Cenozoic paleoenvironments is still possible because of the broad background information which has been compiled for modern and Late Quaternary times in the previous sections.

Following the palinspastic reconstructions of Francheteau (1970) and Sclater et al. (1977), the African continent has drifted as far as 6° in latitude to the north during the past 25 m.y.B.P. The implications for the latitudinal position of the paleoclimatic zones in this region are demonstrated in Fig. 26. Oblique lines indicate the theoretical extent by which the climatic belts were gradually displaced southward relative to the drifting continental plate during this time. The same displacement should hold true for the dune fields between lat 14° and 20° which were migrating to the Atlantic shoreline during glacial times (Fig. 23). Accordingly, we should regard this shifting latitudinal belt as a zone where aeolian-sand turbidites are to be expected during more ancient times of glacial-style climate. Finally, the position of Site 366 would have gradually moved northward across the equator toward the zone of influence of the northeastern Atlantic boundary current.

Eleven major discrete steps of paleoenvironmental evolution are distinguished in the Neogene sections, each phase having lasted on the order of 0.7 to some 4.0 million years and representing a long-term average of a paleoenvironmental record (Sarnthein, 1978b;Arthur et al., 1979; Cita and Ryan, 1979). This averaging was originally due to both the inadequate time resolution of the generally wide-spread standard sampling in DSDP cores and the reduced time scales used in Figs. 25 and 26. However, the long-term phases appear to reflect real trends or states of environmental evolution. The actual major climatic fluctuations are superimposed on the numerous small scale fluctuations suggested in the

Fig. 24. Physiographic setting of Deep-Sea Drilling sites off northwest Africa which are used for the transect of Fig. 26. (Base map redrawn from Berger and von Rad, 1972). Additional Sites no. 13, 137, 415, and 416 omitted because not used for Fig. 26

various sedimentary components in Fig. 25, in the oxygen-isotope curve (Shackleton and Cita, 1979), and in the striking marlstone-limestone cycles of Site 366 (Dean et al., 1981). Their duration is as short as 20,000 to 50,000 years and even shorter, and compares well with that of Quaternary climatic fluctuations (Hays et al., 1976).

Step 1: Earliest Miocene: Evidence of Enhanced Oceanic Circulation and Meridional Surface Winds

(Lower boundary poorly defined, near the Oligocene-Miocene boundary; upper boundary in the upper part of NN2, around 20 m.y.B.P.)

After a short, but marked warming near the end of the Oligocene, $\delta^{18}O$ curves once again indicate a number of phases of strong surface- and bottom-water cooling before temperatures gradually increased until the latest Early Miocene (Fig. 26; Shackleton

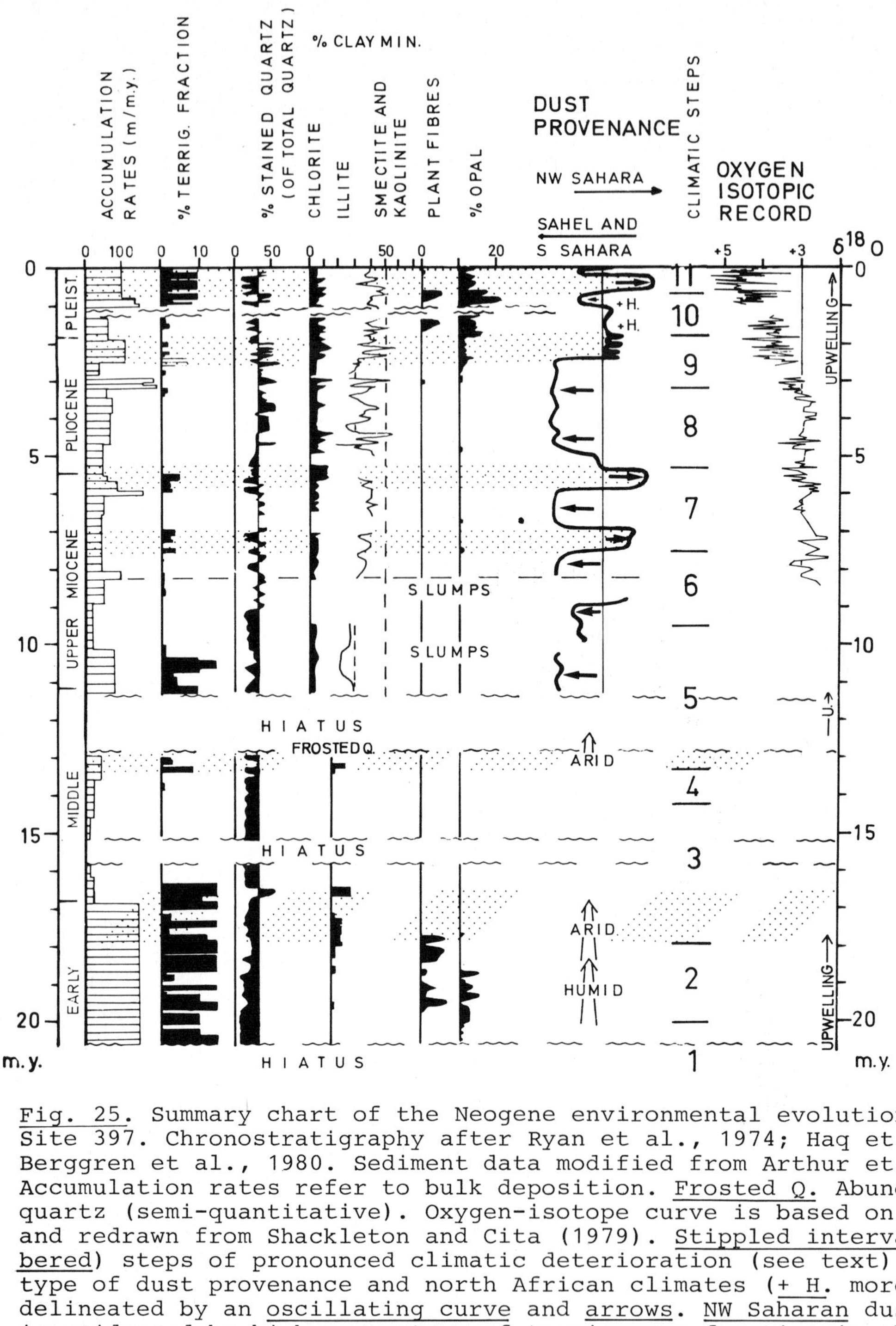

Fig. 25. Summary chart of the Neogene environmental evolution at DSDP Site 397. Chronostratigraphy after Ryan et al., 1974; Haq et al., 1980; Berggren et al., 1980. Sediment data modified from Arthur et al., 1979. Accumulation rates refer to bulk deposition. Frosted Q. Abundance of frosted quartz (semi-quantitative). Oxygen-isotope curve is based on dense sampling and redrawn from Shackleton and Cita (1979). Stippled intervals mark (numbered) steps of pronounced climatic deterioration (see text). Dominant type of dust provenance and north African climates (+ H. more humid) are delineated by an oscillating curve and arrows. NW Saharan dust provenance is evidenced by high percentage of terrigenous fraction in the sand fraction, by reduced abundance of stained quartz and kaolonite, and by abundant chlorite and illite. Southern dust provenance is inferred from abundant stained quartz and kaolinite. Fluvial sediment input/humidity is evidenced by plant debris (and high clay proportions). For the evidence of upwelling see text and abundances of biogenic opal

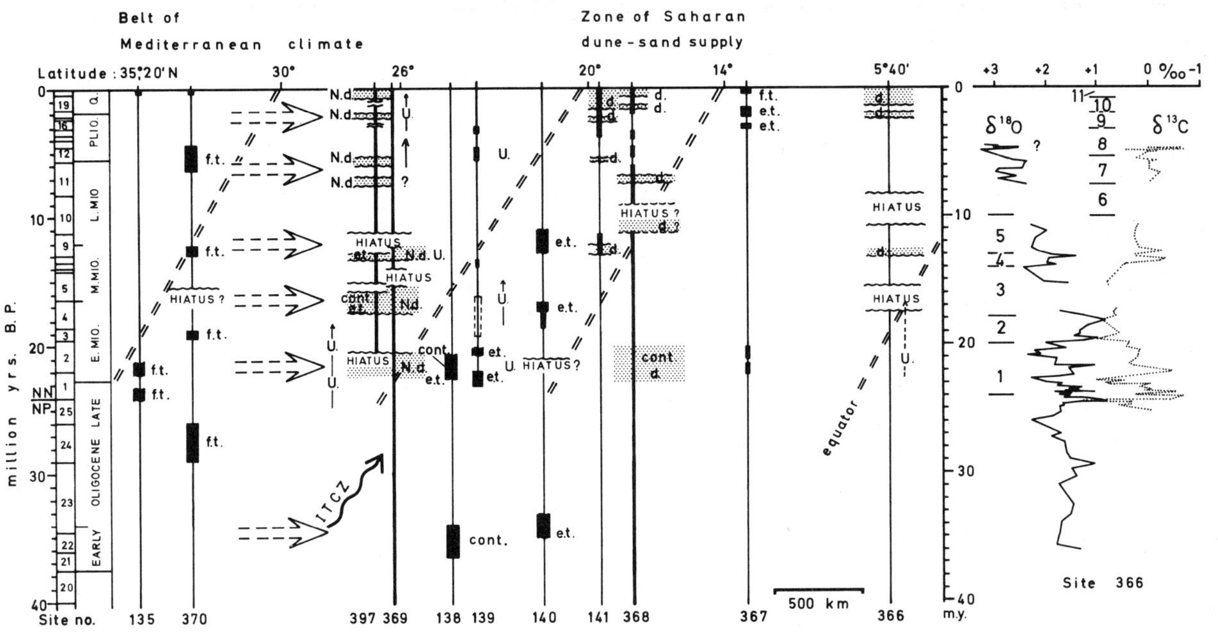

Fig. 26. Evolution of Neogene paleoenvironments along the northwest African continental margin compiled in a north-south transect (for DSDP site positions see Fig. 24; modified after Sarnthein 1978b). Arrows delineate (numbered) steps of marked climatic deterioration (cf. Fig. 25) as indicated by increased supply (stippled bands) of Harmattan dust (d.) and Trade Wind dust (N.d.), aeolian-sand turbidites (e.t.), strong upwelling (U.), the formation of contourites (cont.), and hiatuses. f.t. fluvial-sand turbidites. Hatched oblique lines delineate the time-transgressive southward drift of the north African climatic belts due to the northward shift of the African plate. ITCZ approximate summer position of Intertropical Convergence Zone. Chronostratigraphy after Ryan et al., 1974; Gartner, 1977; Hardenbol and Berggren, 1978; Haq et al., 1980; Berggren et al., 1980. Oxygen and carbon isotope curves of Site 366 are derived from benthic foraminifers, after Vergnaud-Grazzini and Rabussier-Lointier, 1980, stratigraphy emended. For characteristics of Harmattan dust and Trade Wind dust see Ch. 3.2 and 3.3, and Fig. 25

and Kennett, 1975; Boersma and Shackleton, 1978; Vergnaud-Grazzini and Rabussier-Lointier, 1980). These fluctuations accompanied a marked sea-level drop (cycle TM 1.2 of Vail et al., 1977) which was associated with abundant deposition of reworked glauconite at the continental rise (Diester-Haass, 1978; Sarnthein 1978b). During this time, a permanent steep gradient between the glaciated south polar and the tropical regions had already been establishd (Kennett, 1977; Matthews et al., 1980). An intensified bottom-water circulation, perhaps going on since the Early Oligocene (Kennett, 1977), led to widespread erosion and sculpturing of the West African upper continental rise north of Cape Blanc (Sites 140, 397, 370; Arthur et al., 1979; Hinz et al., this Vol.) and to the deposition of contourites at Sites 138 and 368 (although poorly dated). Minor fragmentation of planktonic foraminifers at the Sierra Leone Rise (Site 366; Diester-Haass, 1978) corresponds to evidence of a generally deep CCD (Berger and von Rad, 1972; van Andel, 1975; Thiede et al., 1980).

Intensified coastal and equatorial upwelling comparable to that of the glacial regimes, and a resulting strong oxygen-minimum layer on the upper slope are inferred from the biogenic opal deposits in Site 140, 366, 368, 369, and 397, from maximum rates of sediment accumulation under the equator (Site 366) (Pflaumann, in Lancelot and Seibold, 1978), from reworked shallow-water fauna characteristic of cool-temperate temperatures (Sarnthein, 1978b) and finally, from large amounts of organic-rich mudstones at Sites 369 and 397 (Shipboard Scientific Party, 1978, 1979). They derive from mass flows which originated in an oxygen-depleted zone along the upper slope (Arthur et al., 1979). Possibly, this was the time of initiation of the Canary Current when the seaway between the North Atlantic and the Tethys gradually closed (Berggren and Hollister, 1974).

The intensified surface-water circulation was linked to enhanced Trade Winds. These winds resulted in (1) the dominance of proximal dust sediments derived from the northwest Saharan fringe (Site 369) (Diester-Haass, 1978) and (2) the formation of aeolian-sand turbidites in Sites 138 and 139. These sand layers indicate both the effect of a fall of sea level and an early phase of Saharan aridity and dune activity in those paleolatitudes where it might be expected. Evidence of a similar desert event was recorded only once before in the Early Oligocene (Fig. 26). On land, the regression series forms the base of the Continental Terminal, a lateritic, detrital series which is wide-spread over Western Africa up to 23°N and represents most of the Neogene (Michel, 1973).

Step 2: Upper Early Miocene Phase of Steady Warming and Continental Wetness

(Lower boundary near the upper boundary of NN2, 20 m.y.B.P., upper boundary in the middle of NN4, near 18 m.y.B.P.)

This phase shows marked global ocean warming as reflected by the oxygen-isotope data of Vergnaud-Grazzini and Rabussier-Lointier (1980) (Site 366), and occurs at the same time as a continuous sea-level rise (Vail et al., 1977). As a result surface and bottom-water currents were weakened, permitting rather continuous sediment deposition at the NW African continental margin (Fig. 26), but also causing a distinct (cyclic) shoaling of the CCD (Site 366; Diester-Haass, 1978; Dean et al., 1981). Contourites were formed only occasionally (Site 397). Waning silica deposition may also serve as evidence for a more sluggish surface circulation and reduced upwelling because there is no indication of dilution by carbonate sediments. This model is confirmed by profilic growth of solitary corals and green algae at paleolat 22°N showing the warmest coastal waters being recorded until the Present from an interpolated age of 19 m.y. until the end of Step 2 (Site 397; Shipboard Scientific Party, 1979). The first appearance of *Globigerinoides subquadratus*, which is homeomorphous with the modern tropical form *G. ruber* (Fig. 8) (Krasheninnikov and Pflaumann, 1978), might likewise be related to the favorable environment.

During this time, fluvial deltas supplied extensive clayey mud with abundant plant debris (Fig. 25). The deposits were reworked downslope in giant pebble-mud flows, capable of cutting canyons. We believe that this episode demonstrates a wet continental paleoclimate south of the contemporaneously folded Atlas Mountains (Site 397; Shipboard Scientific Party, 1979). Deposition of aeolomarine dust then almost completely disappeared as a result of the vegetative cover evidenced by the plant debris and of the sluggish (Trade and Harmattan) winds. If dust was present at all, it was strongly diluted by fluvial mud.

Step 3: Upper Early to Middle Miocene Time of Climatic Deterioration and Increasing Submarine Erosion

(Lower boundary within NN4, near 18 m.y.B.P., upper boundary in the upper part of NN6, near 14 m.y.B.P.)

At this time, the deep-water became isotopically heavier (Fig. 26). The sea-level continued to rise simultaneously (Vail et al., 1977). Thus we infer no polar ice-volume build up but a cooling of bottom water. Calcium-carbonate dissolution continued to be high in the eastern Atlantic, also at intermediate water depths of less than 3000 m (Thiede et al., 1979). Major parts of Fuerteventura volcanic island were built (Schmincke, 1979) and may have influenced the bottom-water circulation by barring and/ or shifting contour currents off the continent. Erosion continued along the West African continental margin, as evidenced by all Sites with sufficient stratigraphic time control (Sites 366, 369, 397). Major hiatuses occurred at the beginning and the end of Step 3 (Fig. 26).

A general drop of surface-water temperatures is supported by the disappearance of benthic skeletal debris remains that are characteristic of warm coastal waters at Site 397. Upwelling-induced fertility is recorded only in sediments from Site 369 (Diester-Haass, 1978) which then lay near paleolat 21°N. Lat 21°N is the optimum latitude for present-day coastal upwelling (Fig. 26).

Coarse-grained (including coarse silt) and greyish Trade-Wind dust dominated the enhanced aeolomarine dust deposition around the Early-Middle Miocene boundary (Fig. 26; particularly at Site 369; cf. data of Diester-Haass, 1978b) and reflected a phase of enhanced meridional surface winds. Most probably, they gave rise to widespread continental aridity such as also demonstrated by aeolian-sand turbidites in Sites 140 and 397.

Step 4: Middle Miocene Climatic Optimum

(Duration: from the upper part of NN6, near 14 m.y.B.P., until upper part of NN7, near 13 m.y.B.P.)

The last large pulse of surface and bottom-water warming during the Miocene (Site 366; Vergnaud-Grazzini and Rabussier-Lointier, 1980) was linked to the most salient sea-level highstand since the Late Eocene (Vail et al., 1977). Cycles of strong calcium-carbonate dissolution led to low rates of sediment accumulation, particularly near the beginning and the end of this interval. The first occurrence of the benthic foraminiferal "cool-water" species *Cibicidoides wuellerstorfi* and *Uvigerina peregrina* s.l. (Site 397; Lutze in Shipboard Scientific Party, 1979) might actually reflect a reduced oxygen content of the deep-water, a regime like that resulting from possible incursions of AABW during the Quaternary (Lutze et al., 1979). A few contourites indicate some current activity in the bottom water near the beginning of the interval. The occurrence of reworked oöids suggests relatively warm temperatures in the subtropical coastal waters up tc 23°N paleolatitude (Shipboard Scientific Party, 1979). The evolution of the *Globorotalia fohsi* group may be associated with high surface-water temperatures in the equatorial region (Site 366; Krasheninnikov and Pflaumann, 1978).

Reddish Harmattan-derived dust dominated the terrigenous continental slope sedimentation offshore the northern Sahara (Site 369) and implies a paleoclimate with prevailing zonal winds and associated wetness such as found during the Holocene climatic optimum. A large part of the riverborne sediment may have been trapped on the continent because of the far-reaching marine transgression. Fluvial-mud turbidites are, however, fairly abundant at the upper continental rise (Site 397).

Step 5: Sudden Onset of Deep-Sea Erosion and of Higher Amplitude Climatic Cycles Near the Middle-Upper Miocene Boundary

(Duration: from within NN7, near 13 m.y.B.P., until the upper NN10 zone, near approximately 9 to 10 m.y.B.P.; no continuous, undisturbed section available)

The final closure of the western Tethyan equatorial pathways for deep-water circulation and the opening of an outlet for cold Norwegian Sea Overflow Water across the Island-Faeroe Ridge (Thiede, 1979) initiated a phase of extensive current erosion at the northwest African continental rise (hiatuses at Sites 366, ?368, 369, 397). Calcium-carbonate dissolution continued to be strong in all Sites considered (Diester-Haass, 1978; Shipboard Scientific Party, 1979).

Kennett (1977) assigned the final build-up of the Antarctic ice shield to the same interval. This process resulted in both a two-step major sea-level drop (Vail et al., 1977) which led to abundant turbidite deposition (Sites 140, 397; Arthur et al., 1979) and in a two-step enrichment in $\delta^{18}O$ in both the surface and bottom water record of Site 366 (Vergnaud-Grazzini and Rabussier-Lointier, 1980). After the Mid-Miocene regime of global warmth, the final build-up of the unipolar, Antarctic glaciation probably induced a marked planetary asymmetry of climate for the Late Miocene, i.e., a distinct displacement of the thermal equator toward the Northern Hemisphere (Flohn, 1978). The marked global environmental change may be the reason for an increased number of extinctions and first appearances of planktonic species during this time, as revealed by the detailed planktonic stratigraphy (Thunell, 1981, arrived at a similar result in the Caribbean region). But except for the oxygen-isotope data and some barnacle debris which replaced the former coral debris in turbidites, we have only minor evidence for a regime characteristic of "cold" episodes at the northwest African continental margin. This could be accounted for by a northward shift of the thermal equator. The rather general abundance of fish debris possibly suggests a slight increase of surface-water productivity, but biogenic opal was absent.

The continental climate was generally characterized by strong variability. Strong pulses of unstained terrigenous input rich in chlorite (Diester-Haass, 1978; Chamley and Giroud d'Argoud, 1979; Mélières, 1978), characterize the sedimentation at Sites 369 and 397 and can be related to dominating Trade Wind dust supply. In addition, the deposition of some aeolian-sand turbidites at Sites 140 and 397 just after 13 m.y.B.P. (Fig. 26) depicted a regime of sand-dune migration and the lowered sea level. At the more southern Sites 141 and 368, Diester-Haass and Chamley (1978) observed strong pulses of (Harmattan) dust input from the Sahel zone. From all these features, we infer a time of major aridity coupled with generally enhanced meridional surface winds which coincided with the outlined phase of drastic ocean cooling. More arid stages may have short-term alternated with wetter phases of prevailing zonal wind circulation and occasional river supply from the north Saharan realm, as shown by fluvial-mud turbidites in Site 397 (Arthur et al., 1979).

Step 6: Late Tortonian Phase of High Dissolution and Prevailing Zonal Winds

(Exact duration poorly defined due to stratigraphic gaps and incomplete sections: from the upper NN10 zone, near 9 m.y.B.P., until lower part of NN11 zone, near 7.5 m.y.B.P.)

The late Tortonian was characterized by a fairly constant sea level, in a position comparable to that during the Holocene (Vail et al., 1977). This episode is reflected in the sediments by oxygen-isotope data which were generally lighter than Holocene values (Fig. 25; Shackleton and Cita, 1979). The most obvious characteristic was the pronounced and cyclic dissolution of calcium carbonate extending to water depths of less than 3000 m (Diester-Haass, 1978; Cita and Ryan, 1979). The shoaling of the lysocline resulted in extremely low rates of pelagic sediment accumulation (Fig. 25) and possibly caused the extended hiatus in Site 366 (Vergnaud-Grazzini and Rabussier-Lointier, 1980; Fig. 26). Bottom-water circulation is poorly recorded from this time. The sites off NW Africa appear devoid of current induced sediment structures such as contourites. Quantitative estimates of planktonic foraminiferal assemblages show generally persisting warm surface-water temperatures during the late Tortonian at Site 366 (Pflaumann, in prep.) such as also shown by oxygen-isotope data, although our records do not contain sufficient time-resolution for exact correlation.

Both the absence of coastal upwelling and of north Saharan Trade Wind dust (Fig. 25) during major parts of the interval reflect an extended phase of sluggish meridional surface winds. The dust supply by zonal Harmattan winds was likewise reduced along the West African continental margin. Large proportions of the conspicuous lateritic and ferruginous crusts of the "Continental Terminal" in Senegal and Mauritania probably formed during this time under a tropical climate (Michel, 1973).

Step 7: Messinian Disturbances of Oceanic and Continental Paleoenvironment

(Duration: from lower NN11 zone, near 7.5 m.y.B.P., until the middle of NN12, around the Miocene-Pliocene boundary, near 5.3 m.y.B.P.)

A number of major changes control the eastern Atlantic paleoenvironment during Messinian time. Distinct fluctuations of oxygen isotopic ratios of bottom waters near Site 366 (Fig. 26) and 397 (Fig. 25) are recorded in benthic foraminifers (Shackleton and Cita, 1979; Vergnaud-Grazzini and Rabussier-Lointier, 1980, emended stratigraphy). A first, gradual deviation in the "cool" direction, actually a glacio-isotopic effect, culminated near 6.5 m.y.B.P. This "cooling" trend was followed at both sites by a rise toward "warm" peaks near 6.0 m.y.B.P., a date which coincides with the onset of the Mediterranean salinity crisis (Cita and Ryan, 1979). Not during, but after the onset, we observe again a number of "cool" $\delta^{18}O$ extremes, well below the Holocene level (+ 3,2 ‰ $\delta^{18}O$). They occurred through the late Messinian until near 5.3 m.y.B.P., when the Mediterranean salinity crisis terminated. The late Messinian fluctuations correlated with a short, but marked sea-level drop (Kennett, 1967; Vail et al., 1977) which is reflected in sediments from the continental rise by an increased supply of glauconite and other trace particles from the shelf.

The preceding world-wide, but little understood carbon-isotope shift at 6.3 m.y.B.P. (Haq et al., 1980) has as yet not been clearly identified in our sites. An analogous shift occurs one million years later in Site 366 (Fig. 26; Vergnaud-Grazzini and Rabussier-Lointier, 198o).

Strong calcium-carbonate dissolution continued on the continental rise through the Messinian. The extinction of some benthic foraminiferal species (Lutze in Shipboard Scientific Party, 1979) might announce a change of deep-water circulation. Cita and Ryan (1979) suggest that this event is related to the disconnection of the Mediterranean from the Atlantic, depriving its abyssal circulation of the main source of salty, warm water. However, no bottom-current-induced sediment structures were reported.

Species and oxygen-isotopic composition of planktonic foraminifers indicate distinct, but minor cooling trends at Site 366, the only site with quantitative estimates available (Pflaumann, in prep.; Vergnaud-Grazzini and Rabussier-Lointier, 1980). Neither coastal nor equatorial upwelling activity is recorded in the sediments.

At Site 366 and 397, increased relative abundance of illite and chlorite (Mélière, 1978; Chamley and Giroud d'Argond, 1979), and of terrigenous components (Diester-Haass, 1978; Arthur et al., 1979) record two phases of markedly enhanced meridional surface winds (Figs. 25 and 26), at the beginning and toward the end of this Step 7. Between the two phases, reddish Harmattan dusts indicate a continuation of the zonal wind regime from 6.9 to 6.0 m.y.B.P. The Trade Winds may have picked up the dust partly from the periodically exposed shelf and carried it southward as far as the Cape Verde Islands (Site 141; reinterpreted from Diester-Haass and Chamley, 1978). The glacial-style regimes and possible associated phases of aridity are in good agreement with records on major Messinian age glaciations in the Southern and Northern Hemispheres (Kennett, 1977; Clark, 1971).

A Messinian age phase of aridity was also observed in land sections of Senegal associated with the remains of a major regression (Michel, 1973).

Step 8: Early Pliocene Phase of Steady and Calm Circulation

(Duration from the Miocene-Pliocene boundary, near 5.3 m.y.B.P., until the top of the Lower Gauss normal magnetic interval, near 3.2 m.y.B.P.)

During the Early Pliocene, the polar ice caps again retreated (Kennett, 1977) and a high sea level was re-established (Vail et al., 1977). It correlated with a comparatively stable oxygen-isotope composition of the bottom water at Site 397 (substantiated by a relatively dense sample spacing). Most oxygen isotopic excursions went toward lighter values with respect to the Holocene ones except for a short interval around 4.7 m.y.B.P. (Shackleton and Cita, 1979). Parallel to this evolution, however, the calcium-carbonate dissolution decreased markedly at the continental rise (Berger and von Rad, 1972; Diester-Haass, 1978; Diester-Haass and Chamley, 1978; Cita and Ryan, 1979). This was particularly the case after 4.5 m.y. B.P., a time of global descent of the CCD (van Andel, 1975). Accordingly, the rates of sediment accumulation were generally higher than in the late Miocene.

During the early Pliocene, there is no evidence of bottom currents, since sediment reworking and hiatuses are absent, but the deep-water environment apparently remained well-oxygenated. (For example, ubiquitous pyrite in Site 397 is conspicuously absent in the sediments from this interval; Shipboard Scientific Party, 1979). Based on planktonic foraminiferal assemblages, Pflaumann (in prep.) observes a rise of tropical surface-water temperatures which reached a maximum from 5.0 to 3.5 m.y.B.P. at Site 366. In the whole region, signals of upwelling conditions such as biogenic opal (Fig. 25), did not occur before 3.2 m.y.B.P.

Reddish Harmattan dust dominated the terrigenous sediment supply which was extremely small, however, at the northwest African continental margin. Strongly reduced meridional winds can be inferred from the absence of Trade Wind dust. In summary, we assume a regime of relative stability in terms of a prevailing zonal atmospheric circulation. Its combination with warm sea-surface temperatures probably induced widespread wetness in the Saharan realm. During this time, the formation of lateritic crusts resumed in the regions of the present south Sahara and Sahel zone (Michel, 1973). In addition, Conrad (1969) reported widespread, undifferentiated lacustrine (carbonate and siliceous) sediments from the northwestern part of the Sahara, which corroborates the concept of humid subtropical conditions in the early Pliocene also for this region.

Step 9: Onset of Enhanced Circulation and Progressive Climatic Deterioration in the Late Pliocene

(Duration: from the top of the lower Gauss magnetic interval, near 3.2 m.y.B.P., until the Pliocene-Pleistocene boundary around 1.9 m.y.B.P)

A number of environmental changes start simultaneously in the middle Pliocene near 3.2 and reached maxima after 2.5 m.y.B.P., following the prior extended phase of relative stability and calm. The overall level and the light extremes of oxygen isotopic values of benthonic foraminifera (Fig. 25) became markedly heavier than those found in the Holocene. In addition, the amplitudes of the fluctuations became much more pronounced (Shackleton and Cita, 1979). Vail et al (1977) showed a large magnitude stepwise sea-level drop in conjunction with the first occurrence of extensive glaciations on the Northwestern Hemisphere continents (McDougall and Wensink, 1966; Worsley and Herman, 1980). The onset of turbidite deposition (Sites 367 and 397; Sarnthein, 1978b; Diester-Haass, 1979b) might correlate to the same event. Reworking of older strata and a minor erosional hiatus at Site 397 mark the simultaneous enhancement of bottom-water circulation (Cita and Ryan, 1979). For the same timespan, planktonic foraminiferal assemblages indicate a marked decrease of equatorial surface-water temperatures at Site 366 (Pflaumann, in prep.),and biogenic silica and abundant pyrite indicate that coastal upwelling conditions began at Site 397 (Fig. 26; Arthur et al., 1979). Finally, this event was associated with the first major trend toward lighter carbon isotopic composition of benthic foraminifers since the late Miocene (Shackleton and Cita, 1979). Only the dissolution of calcium carbonates continued to decrease without interruption across the Early-Late Pliocene boundary, but became more oscillatory (Cita and Ryan, 1979; Diester-Haass, 1978).

Concurrent changes of atmospheric circulation also point toward a progressive climatic deterioration and toward increased variability. In agreement with the enhanced upwelling, the distribution pattern of terrigenous sediment components suggests the beginning of numerous phases of accentuated, glacial-style meridional Trade Winds, particularly after 2.5 m.y.B.P.. Their dust load contributed as far as Site 141 (Diester-Haass and Chamley, 1978), coinciding with the deposition of conspicuous aeolian-sand turbidites at Site 367 (Fig. 26; Sarnthein, 1978b). From these deposits we can infer phases of marked desertification with dune fields being active even in the southern Sahel zone. This should not be related to climatic deterioration only, but also to the gradual northward drift of the African continent. For the same reason, Harmattan dust may have reached the position of Site 366 for the first time (Fütterer, 1978) which then had arrived at the proper latitudes. The pronounced climatic change was also documented on land, in the northwestern Sahara between the Atlas and Hoggar Mountains. Conrad (1969) reported extensive and thick freshwater series from the "Plio-Villafranchian" which finally were covered by carbonate and evaporite crusts associated with dune-sand deposits during that time.

Step 10: Early Quaternary Standstill of Climatic Deterioration

(Duration: Pliocene-Pleistocene boundary near 1.9 m.y.B.P., until the Brunhes-Matuyama boundary, at 0.73 m.y.B.P.)

Only minor changes are reported from this time interval. For example, bottom water erosion again culminated near 1.0 m.y.B.P. at the upper continental rise (Site 397, Fig. 25). As indicated by an increased deposition of biogenic opal, coastal upwelling fertility was strongly intensified and reached its absolute maximum since the Early Miocene offshore Cape Bojador (Figs. 25 and 26). This was not necessarily caused by enhanced Trade Winds only, but was possibly triggered by a further advance of nutrient-rich South Atlantic Central Water (SACW) (cf. discussions in Sects. 3 and 5). Northwest Saharan rivers contributed abundant plant remains to the sediments at Site 397 (Fig. 25). This process was favored by a low sea level placing the shoreline close to the shelf-break.

In summary, no long-term climatic change was established at the Pliocene-Pleistocene boundary as compared to the drastic variations before. The oceanic and atmospheric circulation apparently continued to fluctuate within the range obtained during the latest Pliocene (Cita and Ryan, 1979; Arthur et al., 1979). Apart from the short arid phase in the late Pliocene, climatic continuity was also found in the land record, where Villafranchian "pluvial" lakes spread again over the northwest Sahara, and tropical lateritic Fe-crusts occur up to north of the Aïr and Hoggar Mountains (lat. 25°N) during the early Quaternary (Conrad, 1969; Michel, 1973).

Step 11: The Second Strong Climatic Deterioration Since the Middle Pliocene: High-Amplitude Climatic Cycles During the Middle and Late Pleistocene

(Duration since the Brunhes-Matuyama boundary, near 0.7 m.y.B.P.)

At the start of the latest interval, all sedimentary parameters which we considered as indicative of paleoenvironment began to vary with the largest short-term amplitudes found since the earliest Miocene. The time-slices of 18,000 and 6,000 yr B.P. which were described in Sections 4 and 5 exemplify cyclic pairs of such extreme conditions which occurred approximately ten times during the Brunhes magnetic epoch.

Near 0.7 m.y.B.P., the following features of paleoenvironment display the most radical changes (Fig. 25): The build-up of major ice caps in Europe and North America, i.e., the beginning of extreme glacial conditions resulted in a marked increase of the "cold" peaks of the oxygen isotopic composition of bottom water. Concurrently, several typical Pliocene benthic foraminiferal species disappear (Lutze, 1979). Furthermore, the fertility of coastal upwelling suddenly retreated to much lower intensity offshore the northern Sahara (biogenic opal in Site 397). However, this trend is only poorly understood because at the same time bulk sediment accumulation rates slightly decreased (Fig. 25) and the supply of northwest Saharan Trade Wind dust and the accordant meridional surface-wind circulation became strongly enhanced and would imply an increase of upwelling. The change in the wind regime agrees with a simultaneous disappearance of plant debris in the sediment input and reveals a pulse of advancing desertification in North Africa also shown by the land record (Conrad, 1969; Michel, 1973). Thus, the high, early Pleistocene offshore fertility may have been rather correlated to a fertilizing effect of local river discharge.

Other characteristics of paleoenvironment show the greatest rate of change already somewhat earlier, near 0.9 m.y.B.P.. For example, we suggest a drop of equatorial sea-surface temperatures for that time as based on planktonic foraminiferal assemblages (Site 366; Pflaumann, in prep.), and a general increase of terrigenous sediment discharge (Sites 141 and 397) which correlates with an early increment of sedimentation rates (Fig. 25). Finally, the range of fluctuation of calcium-carbonate dissolution in the deep water remained almost constant throughout the Pleistocene.

In conclusion, our estimates demonstrate how the Neogene paleoenvironments along the northwest African continental margin may have evolved always in the broad context of global climatic evolution. After the deep climatic breaks during the Oligocene and the Early Miocene, a phase of general recovery of environmental conditions followed and reached almost an Eocene level. However, since the last 18 m.y.B.P. and more specifically, since the last 13 m.y.B.P., several distinct major episodes of climatic deterioration have determined the general trend of evolution leading to the Present. These steps have always correlated with enhanced oceanic circulation, glacial-style meridional wind systems and rather arid climates in northern Africa. Intervening phases of climatic amelioration and stability were bound to prevailing zonal wind circulation and continental wetness. Altogether, both alternating types of climate lasted approximately for an equal range of time. Enhanced fertility which was induced by coastal upwelling and/or by major river discharge, was basically restricted to two time intervals, to the Early and Middle Miocene until 14 m.y.B.P., and to the latest 3 m.y.B.P.. Increased productivity by equatorial upwelling already terminated near 17.5 m.y.B.P. and only moderately returned in the Middle and Late Pleistocene.

In conjunction with the Neogene northward drift of Africa, we indeed observe a gradual southward advance of the position of wind systems and the correlated climatic belts (as shown by the distribution of aeolian-sand turbidites and dust; Fig. 26) such as expected by theory.

ACKNOWLEDGEMENTS

We gratefully acknowledge the support and helpful discussions of many colleagues at Kiel University and from the CLIMAP group. We are particularly indebted to K. Romine-Loutit and M. Leinen (Kingston, R.I.) for careful peer reviews, to W. Coulbourn (Kiel) and B. Zobel (Hannover) who critically read this paper, to G.F. Lutze for discussions and to Mrs. Bornhöft who has carefully typed the camera-ready manuscript.

The studies were generously funded for many years by the Deutsche Forschungsgemeinschaft.

REFERENCES

Abel N, Jaenicke R, Junge C, Kauter P, Prieto RG, Seiler W (1969) Luftchemische Studien am Observatorium Izana (Teneriffa). Met Rundsch 22: 158-167

Agwu C, Beug HJ (1981) The history of the vegetation and climate as recorded in marine sediments off the west African coast. 'Meteor' Forsch Erg C 36 (in press)

Arthur M, von Rad U, Cornford C, McCoy FW, Sarnthein M (1979) Evolution and sedimentary history of the Cape Bojador Continental Margin, Northwest Africa. In: Ryan WBF, von Rad U et al. Init Reps Deep Sea Drilling Project 47: 773-816

Aston SR, Chester R, Johnson LR, Padgham RC (1973) Eolian dust from the lower atmosphere of the eastern Atlantic and Indian Oceans, China Sea and Sea of Japan. Mar Geol 14: 15-28

Balsam WL, Flessa KW, Kipp NG, DuBois LG (1980) Planktonic foraminiferal diversity in the interglacial and glacial North Atlantic: A test of diversity gradients as a paleooceanographic technique. Geology 8: 582-585

Bé AHW, Tolderlund DS (1971) Distribution and ecology of living planktonic foraminifera in surface waters of the Atlantic and Indian Oceans. In: Funnell BM, Riedel WR (eds) Micropaleontology of Oceans. Cambridge University Press: 105-149

Bein A, Fütterer D (1977) Texture and composition of continental shelf to rise sediments off the northwestern coast of Africa: An indication for downslope transportation. 'Meteor" Forsch Erg C 27: 46-74

Bein A, Sass E (1978) Analysis of log-probability plots of recent Atlantic sediments and its analogy with simulated mixtures. Sedim 25 (4): 575-582

Berger WH (1971) Sedimentation of planktonic foraminifera. Mar Geol 11: 325-358

Berger WH (1976) Biogenous deep sea sediments: Production, preservation and interpretation. In: Riley JP, Chester R (eds) Chemical Oceanography 5, 2nd ed. Academic Press: 265-387

Berger WH (1978) Sedimentation of deep-sea carbonate: maps and models of variations and fluctuations. J Foram Res 8 (4): 286-302

Berger WH (1978) Deep-sea carbonate: pteropod distribution and the aragonite compensation depth. Deep-Sea Res 25 (5): 447-452

Berger WH (1979) Stable isotopes in foraminifera. In: Foraminiferal Ecology and Paleoecology; SEPM Short Course No. 6, Houston, Texas: 156-198

Berger WH, Diester-Haass L, Killingley JS (1978) Upwelling off North-West Africa: the Holocene decrease as seen in carbon isotopes and sedimentological indicators. Oceanologica Acta 1 (1): 3-7

Berger WH, Johnson RF (1978) On the thickness and the 14 C age of the mixed layer in deep-sea carbonates. EPSL 41: 223-227

Berger WH, von Rad U (1972) Cretaceous and Cenozoic sediments from the Atlantic Ocean. In: Hayes DE, Pimm AC et al (eds) Init Reps Deep Sea Drilling Project 14: 787-954

Berggren WA, Burckle LH, Cita MB, Cooke HBS, Funnell BM, Gartner S, Hays JD, Kennett JP, Opdyke ND, Pastouret L, Shackleton NJ, Takayanagi Y (1980) Towards a Quaternary time scale. Quat Res 13: 277-302

Berggren WA, Hollister CD (1974) Paleogeography, paleobiogeography and the history of circulation in the Atlantic Ocean. In: SEPM Spec Publ 20: 126-186

Berggren WA, van Couvering JA (1974) The late Neogene. Dev in Palaeont Strat 2. Elsevier

Berner RA (1977) Sedimentation and dissolution of pteropods in the ocean. In: Andersen NR, Malahoff A (eds) The Fate of Fossil Fuel CO_2 in the Oceans. Plenum Press: 243-260

Biscaye PE (1965) Mineralogy and sedimentation of recent deep-sea clay in the Atlantic Ocean and adjacent seas and oceans. GSA Bull 76: 803-832

Biscaye PE, Kolla V, Turekian KK (1976) Distribution of calcium carbonate in surface sediments of the Atlantic Ocean. J Geophys Res 81: 2595-2603

Blanc PL, Rabussier D, Vergnaud-Grazzini C, Duplessy JC (1980) North Atlantic deep water formed by the later middle Miocene. Nature 283: 553-555

Boersma A, Shackleton N (1977) Oxygen and carbon isotope record through the Oligocene, Site 366 (equatorial Atlantic). In: Lancelot Y, Seibold E et al (eds) Init Reps Deep Sea Drilling Project 41: 957-962

Bolli HM, Premoli Silva I (1973) Oligocene to Recent planktonic foraminifera and stratigraphy of the Leg 15 sites in the Caribbean Sea. In: Edgar TN, Saunders JB et al. Init. Reps Deep Sea Drilling Project 15: 475-497

Caratini C, Bellet J, Tissot C (1979) Etude microscopique de la matière organique: Palynologie et palynofacies. Orgon III-Mauritanie, Senegal, îles de Cap-Vert. Ed du CEPM-CNEXO, Paris: 215-247

Chamley H, Diester-Haass L, Lange H (1977) Terrigenous material in East Atlantic sediment cores as an indicator of NW African climates. 'Meteor' Forsch Erg C 26: 44-59

Chamley H, Diester-Haass L (1979) Upper Miocene to Pleistocene climates in NW Africa deduced from terrigenous components of Site 397 sediments. In: Ryan WBF, von Rad U et al. Init Reps Deep Sea Drilling Project 47 A: 641-646

Chamley H, Giroud d'Argoud G (1979) Clay mineraloy of Site 397, south of Canary Islands. In: von Rad U, Ryan WBF et al. Init Reps Deep Sea Drilling Project 47 A: 595-602

Chester R, Elderfield H, Griffin JJ, Johnson LR, Padgham RC (1972) Eolian dust along the eastern margins of the Atlantic Ocean. Mar Geol 13/2: 91-106

Cifelli R, Stern-Benier C (1977) Planktonic foraminifera from near the west African coast and a consideration of faunal parcelling in the North Atlantic. J Foram Res 6: 258-273

Cita MB, Ryan WBF (1978) Late Neogene paleoenvironment. Part 4. Evolution of ocean paleoenvironment. Interpretation. In: Ryan WBF, von Rad U et al. Init Reps Deep Sea Drilling Project 47 A: 447-459

Clark DL (1971) Arctic Ocean ice cover and its late Cenozoic history . GSA Bull 82 (12): 3313-3324

CLIMAP Project Members (1976) The surface of ice-age earth. Science 191: 1131-1137

Cline RM, Hays JD (eds) (1976) Investigation of late Quaternary paleoceanography and paleoclimatology. GSA Mem 145, p 464

Codispoti LA, Friederich GE (1978) Local and mesoscale influences on nutrient variability in the northwest African upwelling region near Cabo Corbeiro. Deep-Sea Res 25: 751-770

Conrad G (1969) L'évolution continentale post-hercynienne du Saraha algérienne. Centre de Recherches sur les Zones Arides, CNRS, Sér géol 10, 527 p

Crowley TC (1981) Temperature and circulation changes in the eastern north Atlantic during the last 150,000 years: evidence from the planktonic foraminiferal record. Marine Micropaleont 6: 97-129

Dean WE, Gardner JV, Cepek P (1981) Tertiary carbonate-dissolution cycles on the Sierra Leone Rise, eastern equatorial Atlantic Ocean. Mar Geol 39: 81-101

Defant A (1941) Die absolute Topographie des physikalischen Meeresniveaus und der Druckflächen, sowie die Wasserbewegungen im Atlantischen Ozean. Wiss Ergebnisse der Deutschen Atlant Exp 'Meteor' 1925-27 6/1 (5): 191-260

De Miro Orell M (1973) Sedimentos recientes del margen continental de Mauritania (expedición Sahara II). Res Exp Cient B/O Cornide 2: 1-12

de Ploey J (1965) Position géomorphologique, génèse et chronologie de certains dépôts superficiels au Congo occidental. Quaternaria VII: 131-154

Dept. of Fisheries and the Environment (1977) CSS Baffin offshore survey Senegal and the Gambia 1. Minister of Supply and Services Canada, Ottawa, p 127

Diaz HF, Carlson TN, Prospero JM (1976) A study of the structure and dynamics of the Saharan air layer over the Northern Equatorial Atlantic during BOMEX. NOAA Techn Memorandum ERL WMPO-32, p 61

Diester-Haass L (1975) Late Quaternary sedimentation and climate in the East Atlantic between Senegal and Cape Verde Islands. 'Meteor' Forsch Erg C 20: 1-32

Diester-Haass L (1978a) Sediments as indicators of upwelling. In: Boje R, Tomczak M (eds) Upwelling Ecosystems. Springer: 261-280

Diester-Haass L (1978b) Influence of calcium-carbonate dissolution, climate, sea-level changes and volcanism on Neogene sediments off NW Africa. In: Lancelot Y, Seibold E et al (eds) Init Reps Deep Sea Drilling Project 41: 1033-1047

Diester-Haass L (1979a) Indicators of continental climates in marine sediments. A reply. 'Meteor' Forsch Erg C 31: 53-58

Diester-Haass L (1979b) DSDP Site 397: Climatological, sedimentological and oceanographic changes in the Neogene autochthonous sequence. In: Ryan WBF, von Rad U et al. Init Reps Deep Sea Drilling Project 47: 647-670

Diester-Haass L (1980) Upwelling and climate off Northwest Africa during the Late Quaternary. Palaeoecology of Africa 12: 229-238

Diester-Haass L, Chamley H (1978) Neogene palaeoenvironment off NW Africa based on sediments from DSDP Leg 14. J Sed Petrol 48: 879-896

Diester-Haass L, Müller PJ (1979) Processes influencing sand fraction composition and organic matter content in surface sediments off W-Africa (12-19°N). 'Meteor' Forsch Erg C 31: 21-47

Diester-Haass L, Schrader HJ (1979) Neogene coastal upwelling history of northwest and southwest Africa. Mar Geol 29 1/4: 39-54

Diester-Haass L, Schrader HJ, Thiede J (1973) Sedimentological and paleoclimatological investigations of two pelagic ooze cores off Cape Barbas, Northwest Africa. 'Meteor' Forsch Erg C 16: 19-66

Diester-Haass L, van der Spoel S (1978) Late Pleistocene pteropod-rich sediment layer in the northeast Atlantic and protoconch variation of Clio pyramidata L.1967. Paleo Paleo Paleo 24/2: 85-110

Dietrich G (1969) Atlas of hydrography of the northern North Atlantic Ocean. Copenhagen, 1-140

Douglas R, Woodruff F (1980) Deep-Sea benthic foraminifera. The Sea VII,(in press)

Dreimanis A (1977) Correlation of Wisconsin glacial events between the eastern Great Lakes and the St. Lawrence Lowlands. Géogr phys Quat 31 (1-2): 37-51

Egloff J (1972) Morphology of ocean basin seaward of northwest Africa: Canary Islands to Monrovia, Liberia. AAPG Bull 56 (4): 694-706

Embley RW (1976) New evidence for occurrence of debris flow deposits in the deep sea. Geology 4: 371-374

Embley RW, Rabinowitz PD, Jacobi RD (1978) Hyperbolic echo zones in the eastern Atlantic and the structure of the southern Madeira Rise. EPSL 41: 419-433

Emery KO, Lepple F, Toner L, Uchupi E, Rioux RH, Pople W, Hulburt EM (1974) Suspended matter and other properties of surface waters of the northeastern Atlantic Ocean. J Sed Pet 44/4: 1087-1110

Emiliani C (1955) Pleistocene temperatures. J Geol 63: 538-578

Ericson DB, Wollin G (1956) Micropaleontological and isotopic determinations of Pleistocene climates. Micropaleont 2, 3: 257-270

Erlenkeuser H, Koopmann B, Sarnthein M (1981) Details of timing the aeolian dust and fluvial mud supply at the eastern Atlantic continental margin. Oceanologica Acta (in prep)

Fahrbach E, Meincke J (1978) Hochfrequente Strömungsschwankungen in Bodennähe über dem Kontinentalabfall. 'Meteor' Forsch Erg A 20: 1-12

Fairbridge RW (1976) Effects of Holocene climatic change on some tropical geomorphic processes. Quat Res 6: 529-557

Flint RF (1971) Glacial and Quaternary geology. J. Wiley & Sons Inc, p 892

Flohn H (1978) Comparison of Antarctic and Arctic climate and its relevance to climatic evolution. In: Antarctic glacial history and world paleoenvironments. van Zinderen-Bakker EM (ed). Balkema, Rotterdam: 3-13

Folger DW (1970) Wind transport of land-derived mineral, biogenic, and industrial matter over the North Atlantic. Deep-Sea Res 17: 337-352

Francheteau J (1970) Paleomagnetism and plate tectonics. SIO Reference 70-30, p 345

Fütterer D (1978) Late Neogene silt at the Sierra Leone Rise (Leg 41 Site 366): Terrigenous and biogenous components. In: Lancelot Y, Seibold E et al. Init Reps Deep Sea Drilling Project 41: 1049-1059

Fütterer D (1980) Sedimentation am NW-afrikanischen Kontinentalrand: Quantitative Zusammensetzung und Verteilung der Siltfraktion in den Oberflächensedimenten. 'Meteor' Forsch Erg C 33: 15-60

Gabriel B (1978) Klima- und Landschaftswandel der Sahara. In: Sahara 10,000 Jahre zwischen Weide und Wüste. Museen der Stadt Köln (A. von Gagern Herausgeb.): 22-34

Gardner JV (1975) Late Pleistocene carbonate dissolution cycles in the eastern Equatorial Atlantic. In: Sliter WV, Bé AWH, Berger WH (eds) Dissolution of deep-sea carbonates. Cushman Found Foram Res Spec Publ 13: 129-141

Gardner JV, Hays JD (1976) Responses of sea-surface temperature and circulation to global climatic change during the past 200,000 years in the eastern Equatorial Atlantic Ocean. GSA Bull Mem 145: 221-246

Gartner S (1977) Nannofossils and biostratigraphy: An overview. Earth Sci Rev. 13: 227-250

Glaccum RA, Prospero JM (1980) Saharan aerosols over the tropical north Atlantic-Mineralogy. Marine Geol 37 (3/4): 295-323

Goll RM, Bjørklund KR (1971) Radiolaria in surface sediments of the North Atlantic Ocean. Micropaleont 17 (4): 434-454

Griffin JJ, Windom H, Goldberg ED (1968) The distribution of clay minerals in the world ocean. Deep-Sea Res 15: 433-459

Haake FW (1980) Benthische Foraminiferen in Oberflächen-Sedimenten und Kernen des Ostatlantiks vor Senegal/Gambia (Westafrika). 'Meteor' Forsch Erg C 32: 1-29

Haq BU, Worsley TR, Burckle LH, Douglas RG, Keigwin LD Jr, Opdyke ND, Savin SM, Sommer MA II, Vincent E, Woodruff F (1980) Late Miocene carbon-isotopic shift and synchroneity of some phytoplanktonic biostratigraphic events. Geology 8 (9): 427-431

Hardenbol J, Berggren WA (1978) A new Paleogene numerical time scale. Amer Assoc Petrol Geol, Studies in Geology 6: 213-234

Hart TJ, Currie RJ (1960) The Benguela current. Discovery Rep 31: 123-298

Hartmann M, Müller PJ, Suess E, van der Weijden CH (1976) Chemistry of Late Quaternary sediments and their interstitial waters from the NW African continental margin. 'Meteor' Forsch Erg C 24: 1-67

Hays JD, Imbrie J, Shackleton NJ (1976) Variations in the earth's orbit: Pacemaker of the ice ages. Science 194 (4270): 1121-1132

Hays JD, Peruzza A (1972) The significance of calcium carbonate oscillations in eastern Equatorial Atlantic deep-sea sediments for the end of the Holocene warm interval. Quat Res 2: 355-362

Huber K, Mittelsteadt E, Weichert G (1977) Meteor-Fahrt Nr. 30, 23.1. -2.3.1973, Seegebiet NW-Afrika. Zur Hydrographie der Gewässer vor Marokko. Physikalische und chemische Daten. Deut Hydrogr Inst, Meeresk Beob u Erg 46: 1-131

Hughes P, Barton ED (1974) Stratification and water mass structure in the upwelling area off northwest Africa in April/May 1969. Deep-Sea Res 21: 611-628

Imbrie J, Kipp NG (1971) A new micropaleontological method for quantitative paleoclimatology: Application to a late Pleistocene Caribbean core. In: Turekian KK (ed) The late Cenozoic glacial ages. Yale Univ. Press: 71-181

Imbrie J, van Donk J, Kipp NG (1973) Paleoclimatic investigation of a late Pleistocene Caribbean deep-sea core: Comparison of isotopic and faunal methods. Quater Res 3: 10-38

Jacobi RD (1976) Sediment slides on the northwestern continental margin of Africa. Mar Geol 22: 157-173

Jacobi RD, Rabinowitz PD, Embley RW (1975) Sediment waves on the Moroccan continental rise. Mar Geol 19: M61-M67

Jaenicke R, Schütz L (1978) Comprehensive study of physical and chemical properties of the surface aerosols in the Cape Verde Islands region. J Geoph Res 83 (C 7): 3585-3599

Johnson DR, Barton ED, Hughes P, Mooers CNK (1975) Circulation in the Canary current upwelling region off Cabo Bojador in August 1972. Deep-Sea Res 22: 547-558

Johnson LR (1979) Mineralogical dispersal patterns of North Atlantic deep-sea sediments with particular reference to eolian dusts. Mar Geol 29 1/4: 335-346

Keigwin LD Jr (1979) Middle Pliocene climatic change in the western Mediterranean from faunal and oxygen isotopic trends. Nature 282 (5736): 294-296

Kennett JP (1967) Recognition and correlation of the Kapitean Stage (Upper Miocene, New Zealand). New Zealand Journal of Geol and Geoph 10/4: 1051-1063

Kennett JP (1977) Cenozoic evolution of Antarctic glaciation, the circum-Antarctic ocean and their impact of global paleoceanography. J Geoph Res 82: 3843-3860

Kennett JP, Huddlestun P (1972) Abrupt climatic change at 90,000 yr BP: Faunal evidence from Gulf of Mexico cores. Quatern Res 2: 384-395

Kiper M (1977) Sedimente und ihre Umwelt im Senegaldelta. Unpubl. Dipl. Arb. Univ. Kiel: p 61

Kipp NG (1976) New transfer function for estimating past sea-surface conditions from sea-bed distribution of planktonic foraminiferal assemblages in the North Atlantic. GSA Mem 145: 3-42

Kolla V, Biscaye PE, Hanley A (1979) Distribution of quartz in Late Quaternary Atlantic sediments in relation to climate. Quater Res 11: 261-277

Koopmann B (1979) Saharastaub in Sedimenten des subtropisch-tropischen Nordatlantik während der letzten 20,000 Jahre. Thesis Univ. Kiel: p 107

Koopmann B (1980) Quantitative determination of silt sized biogenic silica in Atlantic deep-sea sediments. IAS, Abstracts 1st European Regional Meeting 1980, Bochum: 30-33

Koopmann B (1981) Sedimentation von Saharastaub im subtropischen Atlantik während der letzten 25.000 Jahre. 'Meteor' Forsch Erg C 35 (in press)

Koopmann BJ, Lees A, Piessens P, Sarnthein M (1979) Skeletal carbonate sands and wind-derived silty marls off the Saharan Coast: Baie du Levrier, Arguin Platform, Mauritania. 'Meteor' Forsch Erg C 30: 15-57

Krasheninnikov VA, Pflaumann U (1978) Zonal stratigraphy of Neogene deposits of the eastern part of the Atlantic Ocean by means of planktonic foraminifers, Leg 41 Deep Sea Drilling Project. In: Lancelot Y, Seibold E et al. Init Reps Deep Sea Drilling Project 16: 613-657

Kudrass HR (1973) Sedimentation am Kontinentalhang vor Portugal und Marokko im Spätpleistozän und Holozän. 'Meteor' Forsch Erg C 13: 1-63

Labracherie M (1980) Les radiolaires temoins de l'évolution hydrologique depuis le dernier maximum glaciaire au large de Cap Blanc (Afrique du Nord-Ouest). Palaeo Palaeo Palaeo 32 (1/2): 163-184

Lancelot Y, Seibold E (1978) The evolution of the central northeastern Atlantic - Summary of results of DSDP Leg 41. In: Lancelot Y, Seibold E et al (eds) Init Reps Deep Sea Drilling Project 41: 1215-1245

Lange H (1975) Herkunft und Verteilung von Oberflächensedimenten des westafrikanischen Schelfs und Kontinentalhanges. 'Meteor' Forsch Erg C 22: 61-84

Lange H (1981) Distribution of chlorite and kaolinite in eastern Atlantic sediments off north Africa. Sedim., 10 ms.-p (in press)

Lepple KF (1975) Eolian dust over the north Atlantic ocean. Ph D Thesis Univ. of Delaware, p 270

Lonsdale P (1978) Bedforms and the benthic boundary layer in the North Atlantic: A cruise report of indomed Leg 11. SIO Reference 78-30

Lutze GF (1979) Benthic foraminifers at Site 397: faunal fluctuations and ranges in the Quaternary. In: von Rad U, Ryan WBF (eds) Init Reps Deep Sea Drilling Project 47, 1: 419-431

Lutze GF (1980) Depth distribution of benthonic foraminifera on the continental margin off NW Africa. 'Meteor' Forsch Erg C 33: 31-80

Lutze GF, Sarnthein M, Koopmann B, Pflaumann U, Erlenkeuser H, Thiede J (1979) Meteor cores 12309: Late Pleistocene reference section for interpretation of the Neogene of Site 397. In: von Rad U, Ryan WBF et al. Init Reps Deep Sea Drilling Project 47: 727-739

McDougall J (1979) The present status of the geomagnetic polarity time scale. In: McElhinny MW (ed) The Earth: Its origin, structure, and evolution. Academic Press, London: 543-566

McDougall J, Wensink H (1966) Paleomagnetism and geochronology of Pliocene-Pleistocene lavas in Iceland. EPSL 1: 232-236

McIntyre A, Bé AWH (1967) Modern Coccolithophoridae of the Atlantic Ocean. 1. Placoliths and cyrtoliths. Deep-Sea Res 14: 561-597

McIntyre A, Kipp NG, Bé AWH, Crowley T, Kellogg T, Gardner JV, Prell W, Ruddiman WF (1976) Glacial North Atlantic 18,000 years ago: A CLIMAP reconstruction. GSA Mem 145: 43-76

Maley J (1980) Etudes palynologiques dans le Bassin du Tchad et paléoclimatologie de l'Afrique nordtropicale de 30,000 ans à l'époque actuelle. Thèse Univ. de Montpellier, France, p 600

Manabe S, Hahn DG (1977) Simulation of the tropical climate of an ice age. J Geoph Res 82 (27): 3889-3909

Mateu G (1979) Foraminiferos planctonicos des area de afloramiento del Atlantico del NW de Africa: Estructura, origen y evolucion de sus comunidades. Rev Española Micropal 11 (1): 135-157

Matthews RK, Curry WB, Lohmann KC, Sommer MA, Poore RZ (1980) Late Miocene palaeo-oceanography of the Atlantic: oxygen-isotope data on planktonic and benthic foraminifera. Nature 283: 555-557

Mélières F (1978) X-ray mineralogy studies, Leg 41, Deep Sea Drilling Project, eastern North Atlantic Ocean. In: Lancelot Y, Seibold E et al, Init Reps Deep Sea Drilling Project 41: 1065-1086
Michel P (1973) Les bassins des fleuves Sénégal et Gambie, étude géomorphologique. Mém ORSTOM 3 tomes, p 752
Michel P (1980) The southwestern Sahara margin. Sediments and climatic changes during the Recent Quaternary. Palaeoecology of Africa 12: 297-306
Milliman JD (1977) Effects of arid climate and upwelling upon the sedimentary regime off southern Spanish Sahara. Deep-Sea Res 24: 95-103
Mittelstaedt E (1972) Der Hydrographische Aufbau und die zeitliche Variabilität der Schichtung und Strömung im nordwestafrikanischen Auftriebsgebiet im Frühjahr 1968. 'Meteor' Forsch Erg A 11: 1-57
Mittelstaedt E, Pillsbury D, Smith RL (1975) Flow patterns in the Northwest African upwelling area. Dt Hydrogr Z 28: 145-167
Molina-Cruz A, Thiede J (1978) The glacial eastern boundary current along the Atlantic Eurafrican continental margin. Deep-Sea Res 25: 337-356
Müller PJ, Suess E (1979) Productivity, sedimentation rate, and sedimentary organic matter in the oceans.- I. Organic carbon preservation. Deep-Sea Res 26: 1347-1362
Nahon D (1980) Soil accumulations and climatic variations in Western Sahara. Palaeoecology of Africa 12: 63-68
Nahon D, Lappartient JR (1977) Time factor and iron crust genesis. Catena 4: 249-254
Newell RE, Herman GF, Gould-Steward S, Tanaka M (1975) Decreased global rainfall during the past Ice Age.Nature 253: 33-34
Nicholson SE, Flohn H (1980) African environmental and climatic changes and the general atmospheric circulation in Late Pleistocene and Holocene. Climatic Change 2: 313-348
Normark WR, Hess GR, Stow DAV, Bowen AJ (1980) Sediment waves on the Monterey fan levee: A preliminary physical interpretation. Mar Geol 37: 1-18
Olausson E (1969) On the Würm-Flandrian boundary in deep-sea cores. Geol en Mijnb 48, 3: 349-361
Parkin DW, Phillips DR, Sullivan RAL, Johnson LR (1970) Airborne dust collections over the North Atlantic. J Geophys Res 75: 1782-1793
Parkin DW, Shackleton NJ (1973) Trade Wind and temperature correlations down a deep-sea core off the Saharan Coast. Nature 245 (5426): 455-457
Parmenter C, Folger DW (1974) Eolian biogenic detritus in deep-sea sediments: A possible index of Equatorial Ice Age aridity. Science 185 (4152): 695-698
Pastouret L, Chamley H, Delibrias G, Duplessy JC, Thiede J (1978) Late Quaternary climatic changes in western tropical Africa deduced from deep-sea sedimentation off Niger Delta. Oceanologica A 1/2: 217-232
Penck A (1913) Die Formen der Landoberflächen und die Verschiebungen der Klimagürtel. Sitz ber Kgl Preuß Akad Wiss Halbbd I,4: 77-97
Petit-Maire N (1980) Holocene biogeographical variations along the northwestern African coast (28°-19°N). Palaeoclimatic implications. Palaeoecology of Africa 12: 365-377
Pflaumann U (1980) Variations of the surface-water temperatures at the Eastern North Atlantic continental margin (Sediment surface samples, Holocene climatic optimum, and Last Glacial maximum). Palaeoecology of Africa 12: 191-212
Pflaumann U (in prep) Neogene surface water temperatures in the tropical Eastern Atlantic (DSDP Site 366)
Philippi E (1910) Die Grundproben der Deutschen Südpolar Expedition 1901-1903. Dtsch Südpolar Exped 1901-1903, 2 (4): 411-616
Prell WP, Gardner JV, Bé AWH, Hays JD (1976) Equatorial Atlantic and Carribean foraminiferal assemblages, temperatures, and circulation: Interglacial and Glacial comparisons. GSA Mem 145: 247-266
Prospero JM, Carlson TN (1972) Vertical and aerial distribution of Saharan dust over the Western Equatorial North Atlantic Ocean. Jour Geophys Research 77: 5255-5265

Richert P (1976) Relationship between diatom biocoenosis and taphocoenosis in upwelling areas off West Africa. Abstr 4th Symposium on Recent and Fossil Diatoms, Oslo
Robinson, MK, Bauer RA, Schroeder EH Atlas of North Atlantic-Indian Ocean monthly mean temperature and mean salinity of the surface layer. Naval Oceanographic Off. (Washington D.C.) (in prep)
Rodier JA (1975) Evaluation de l'écoulement annuel dans le Sahel tropical Africain. Travaux et Docum de l'ORSTOM 46: 1-121
Rognon P (1976) Essai d'interprétation des variations climatiques au Sahara depuis 40,000 ans. Rev de géogr phys et de géol dyn II, 18 (2-3): 251-282
Rognon P, Williams MAJ (1977) Late Quaternary climatic changes in Australia and North Africa: a preliminary interpretation. Palaeo Palaeo Palaeo 21: 285-327
Rona PA (1970) Comparison of continental margins of Eastern North America at Cape Hatteras and Northwestern Africa at Cap Blanc. Bull AAPG 54/1: 129-157
Rossignol-Strick M, Duzer D (1979) Late Quaternary pollen and dinoflagellate analysis of marine cores off West Africa. 'Meteor' Forsch Erg C 30: 1-14
Ruddiman WF (1971) Pleistocene sedimentation in the equatorial Atlantic: Stratigraphy and faunal paleoclimatology. Geol Soc Amer Bull 82: 283-302
Ruddiman WF, McIntyre A (1973) Time-transgressive deglacial retreat of polar water from the North Atlantic . Quater Res 3: 117-130
Ryan WBF, Cita MB, Dreyfus Rawson M, Burckle LH, Saito T (1974) A paleomagnetic assignment of Neogene stage boundaries and the development of isochronous datum planes between the Mediterranean, the Pacific and Indian Oceans in order to investigate the response of the world ocean to the Mediterranean "Salinity Crisis". Riv Ital Paleont 80/4: 631-688
Sarnthein M (1978) Sand deserts during glacial maximum and climatic optimum. Nature 271 (5648): 43-46
Sarnthein M (1978b) Neogene sand layers off NW Africa: Composition and source environment. In: Lancelot Y, Seibold E et al (eds) Init Reps Deep Sea Drilling Project 41, suppl vol.: 939-959
Sarnthein M (1979) Indicators of continental climates in marine sediments, a discussion. 'Meteor' Forsch Erg C 31: 49-51
Sarnthein M, Diester-Haass L (1977) Eolian-sand turbidites. J Sed Petrol 47 (2): 868-890
Sarnthein M, Koopmann B (1980) Late Quaternary deep-sea record on NW African dust supply and wind circulation. Palaeoecology of Africa 12: 239-253
Sarnthein M, Tetzlaff G, Koopmann B, Wolter K, Pflaumann U (1981) Glacial and interglacial wind regimes over the eastern subtropical Atlantic and NW Africa. Nature, 13 ms-p (in press)
Sarnthein M, Walger E (1974) Der äolische Sandstrom aus der W-Sahara zur Atlantikküste. Geol Rdsch 63: 1065-1087
Scheltema RS (1972) Reproduction and dispersal of bottom dwelling deep-sea invertebrates: A speculative summary. In: Brauer RW (ed) Baro-biology and the experimental biology of the deep sea. Univ. N. Carolina, Chapel Hill, N.C.: 58-66
Schemainda R, Nehring D, Schulz S (1975) Ozeanologische Untersuchungen zum Produktionspotential der nordwestafrikanischen Wasserauftriebsregion 1970-1973. Geod. Geophys Veröff 4 (16): p 88
Schiffers H (1971) Die Sahara und ihre Randgebiete. Bd. 1. Physiogeographie. Weltforum Verlag, München, p 674
Schmidt-Lorenz R (1977) Feuchttropen-Reliktböden im Aufforstungsprojekt Ouagadougon/Obervolta (Sudan-Sahel-Klima). Mitt d Bundesforschungsanstalt für Forst- und Holzwirtschaft: 1-10
Schmincke HU (1979) Age and crustal structure of the Canary Islands. A discussion. J Geophys 46: 217-224
Schneider R (1981) 750,000 Jahre Klimageschichte und Paläoozeanographie am Sierra Leone Rücken. Unpubl. Dipl. Arb. University of Kiel
Schott W (1935) Die Foraminiferen in dem äquatorialen Teil des Atlantischen Ozeans. Wiss Ergeb Deutsch Atlant Exped 'Meteor' 1925-27 3 (3): 43-134
Schrader HJ (1971) Fecal pellets: Role in sedimentation of pelagic diatoms. Science 174: 55-57

Schrader H (1972) Kieselsäure-Skelette in Sedimenten des ibero-marokkanischen Kontinentalrandes und angrenzender Tiefsee-Ebenen. 'Meteor' Forsch Erg C 8: 10-36

Schulz E (1976) Aktueller Pollenniederschlag in der zentralen Sahara und Interpretationsmöglichkeiten quartärer Pollenspektren. In: Palaeoecology of Africa 9: 8-14

Sclater JG, Hellinger S, Tapscott C (1977) The paleobathymetry of the Atlantic Ocean from the Jurassic to the present. J Geol 85: 509-552

Seibold E (1979) Sediments in upwelling areas, particularly off Northwest Africa. Symp. on the Canary Current: Upwelling and Living Resources. Rapp Proc Verbaux Ser ICES Kopenhagen, 11 ms-p

Seibold E (1980) Climate indicators in marine sediments off Northwest Africa. Palaeoecology of Africa 12: 175-187

Seibold E, Fütterer D (1981) Sediment dynamics on the continental margin of the North West African continental margin. In: Heezen memorial volume. Wiley & Chichester (in press)

Seibold E, Hinz K (1974) Continental slope construction and destruction, West Africa. In: Burk CA, Drake CL (eds) The geology of continental margins. Springer, New York Heidelberg: 179-196

Seiler WC (1975) Tiefenverteilung benthischer Foraminiferen am portugiesischen Kontinentalhang. 'Meteor' Forsch Erg C 23: 89-101

Semmelhack W (1934) Die Staubfälle im nordwest-afrikanischen Gebiet des Atlantischen Ozeans. Ann Hydrogr Marit Meteor 62: 273-277

Servant M, Servant-Vildary S (1972) Nouvelles données pour une interpretation paléoclimatique de séries continentales du Bassin Tchadien (Pleistocene récent, Holocene). In: van Zinderen Bakker EM (ed) Palaeoecology of Africa 6: 87-92

Shackleton NJ, Cita MB (1979) Oxygen and carbon isotope stratigraphy of benthonic foraminifera in Site 397: Fine structure of climatic change in the Late Neogene. In: Ryan WBF, von Rad U et al Init Reps Deep Sea Drilling Project 47, Pt. 1: 433-445

Shackleton NJ, Kennett JP (1975) Paleotemperature history of the Cenozoic and the initiation of Antarctic glaciation: Oxygen and carbon isotope analysis in DSDP Sites 277, 279 and 281. In: Kennett JP, Houtz R et al. (eds) Init Reps Deep Sea Drilling Project 29: 743-755

Shackleton NJ, Opdyke ND (1973) Oxygen isotope and palaeomagnetic stratigraphy of equatorial Pacific core V28-238: Oxygen isotope temperatures and ice volumes on a 10^5 year scale. Quatern Res 3: 39-55

Shackleton NJ, Opdyke ND (1976) Oxygen-isotope and paleomagnetic stratigraphy of Pacific core V28-239 Late Pliocene to Latest Pleistocene. Geol Soc Amer Mem 145: 449-464

Shackleton NJ, Opdyke ND (1977) Oxygen isotope and palaeomagnetic evidence for early northern hemisphere glaciation. Nature 270 (5634): 216-219

Shaffer G (1976) A mesoscale study of coastal upwelling variability off NW-Africa. 'Meteor' Forsch Erg A 17: 33-70

Shipboard Scientific Party (1978) 5. Site 369: Continental Slope off Cape Bojador, Spanish Sahara. In: Lancelot Y, Seibold E et al, Init Reps Deep Sea Drilling Project 41: 327-420

Shipboard Scientific Party (1979) 2. Site 397. In: von Rad U, Ryan WBF et al, Init Reps Deep Sea Drilling Project 47, Pt 1: 17-217

Smayda TJ (1971) Normal and accelerated sinking of phytoplankton in the sea. Mar Geol 11: 105-122

Sonntag C, Thorweihe U, Rudolph J, Löhnert EP, Junghans C, Münnich KO, Klitzsch E, ElShazly EM, Swailem FM (1980) Isotopic identification of Saharan groundwaters, groundwater formation in the past. Palaeoecology of Africa 12: 159-171

Street FA, Grove AT (1978) Global maps of lake-level fluctuations since 30,000 BP. Quater Res 12: 83-118

Streeter SS, Shackleton NJ (1979) Paleocirculation of the deep North Atlantic: 150,000 year record of benthic foraminifera and oxygen-18. Science 203: 168-171

Suess E (1980) Particulate organic carbon flux in the oceans -

surface productivity and oxygen utilization. Nature 288: 260-263
Summerhayes CP, Milliman JD, Briggs SR, Bee AG, Hogan C (1976) Northwest African shelf sediments: Influence of climate and sedimentary processes. J Geol 84: 277-300
Szekielda KH (1978) Eolian dust in the northeast Atlantic. Oceanogr Mar Biol Ann Rev 16: 11-41
Talbot MR (1980) Environmental responses to climatic change in the West African Sahel over the past 20,000 years. In: Williams MAJ, Faure H (eds) The Sahara and the Nile. A.A. Balkema, Rotterdam: 37-62
Tetzlaff G, Wolter W (1980) Meteorological patterns and the transport of mineral dust from the North African continent. Palaeoecology of Africa 12: 31-42
Thiede J (1973) Planktonic foraminifera in hemipelagic sediments: Shell preservation off Portugal and Morocco. GSA Bull 84: 2749-2754
Thiede J (1975) Shell- and skeleton-producing plankton and nekton in the eastern North Atlantic Ocean. 'Meteor' Forsch Erg C 20: 33-79
Thiede J (1977) Aspects of variability of the glacial and interglacial North Atlantic eastern boundary current (last 150,000 years). 'Meteor' Forsch Erg C 28: 1-36
Thiede J (1979) History of the North Atlantic Ocean: Evolution of asymmetric zonal paleo-environment in a latitudinal ocean basin. In: Maurice Ewing Series 3 (Amer. Geophys. Union, Washington): 275-296
Thiede J, Agdestein T, Strand JE (1980) Depth distribution of calcareous sediments in the Mesozoic and Cenozoic North Atlantic Ocean. EPSL 47: 416-422
Thunell RC (1981) Cenozoic palaeotemperature changes and planktonic foraminiferal speciation. Nature 289: 670-672
Tomczak M Jr, Hughes P (1980) Three dimensional variability of water masses and currents in the Canary Current upwelling region. 'Meteor' Forsch Erg A 21: 1-24
Uchupi E, Emery O, Bowin CO, Phillips JD (1976) Continental margin off Western Africa: Senegal to Portugal. Am Ass Petrol Geol Bull 60: 809-878
Unesco Paris, CERCG-CNRS (1977) Repartition mondiale des regions arides. Note technique 7 du MAB, Unesco
Vail PR, Mitchum RM Jr, Thompson S (1977) Seismic stratigraphy and global sea-level changes, vol. 1. AAPG Mem 26: 63-98
Van Andel T (1975) Mesozoic/Cenozoic calcite compensation depth and the global distribution of calcareous sediments. Earth and Planet Sci Lett 26: 187-194
Vergnaud-Grazzini C, Rabussier-Lointier D (1980) Compositions isotopiques de l'oxygène et du carbone des foraminifères tertiaires en Atlantique équatorial (site 366 du DSDP). Rev géol dynam geogr phy 22: 63-74
Weichart G (1974) Meereschemische Untersuchungen im nordwestafrikanischen Auftriebsgebiet 1968. 'Meteor' Forsch Erg A 14: 33-70
Wetzel A (1981) Ökologische und stratigraphische Bedeutung biogener Gefüge in quartären Sedimenten am NW-afrikanischen Kontinentalrand. 'Meteor' Forsch Erg C 34: 1-47
Wooster WS, Bakun A, McLain DR (1976) The seasonal upwelling cycle along the eastern boundary of the North Atlantic. J Mar Res 34, 2: 131-141
Worsley TR, Herman Y (1980) Episodic ice-free Arctic Ocean in Pliocene and Pleistocene time: Calcareous nannofossil evidence. Science 210: 323-325
Wüst G (1933) Das Bodenwasser und die Gliederung der Atlantischen Tiefsee. Wiss Ergebn Deut Atlant Exped 'Meteor' 1925-1927 6 (1) Atlas Beilagen
Wüst G (1933) Bodenwasser und Bodenkonfiguration der Atlantischen Tiefsee. Zschr Ges f Erdkd 1933 (1/2): 1-18
Wüst G (1935) Die Stratosphäre. Wiss Ergebn Deut Atlant Exped 'Meteor' 1925-1927 6 (1+2): 1-288
Wüst G (1936) Die Tiefenzirkulation im Raume des Atlantischen Ozeans. Naturwissensch 24 (9): 133-141
Wüst G (1939) Die Grenzen der Ozeane. Ann Hydrogr marit Meteorol, Beih z. Maiheft
Yaalon DH (1978) "Geoderma" - continental sedimentation, calcrete, desert loess and paleosols, sand dunes and eolionites. 10th Int Congr Sedim Jerusalem 1978, Post Congress Excursion Y 5: 193-238

Yaalon DH, Ganor E (1977) East Mediterranean trajectories of dust carrying storms from the Sahara and Sinai. In: Morales CH (ed) Saharan dust: Mobilization, transport, deposition. Wiley, New York

Zobel B, Ranke U (1978) Zusammensetzung, Stratigraphie und Bildungsbedingungen der Sedimente am Kontinentalhang vor Sierra Leone (Westafrika). 'Meteor' Forsch Erg C 29: 21-74

25 Late Quaternary Fluxes of Major Sediment Components to the Sea Floor at the Northwest African Continental Slope

Jörn Thiede[1], Erwin Suess[2], and Peter J. Müller[3]

ABSTRACT

Rates of accumulation of the bulk sediment and of organic carbon, nitrogen, phosphorus, sulfur, of calcareous and opaline matter, of benthic and planktonic foraminifers and radiolarians, of quartz, mica and residual terrigenous constituents are used to assess the flux of sediments in terms of oceanic and atmospheric processes to a core location off northwest Africa during the past 130,-140,000 yrs. The time resolution between samples is in the order of 500 yrs. during glacial stages and 2000 yrs. during interglacial stages. Changes from the glacial to the interglacial mode of deposition and *vice versa* explain the dominant part of the variability of the signals. It can be shown that the atmospheric circulation, whose effect is traced by means of accumulation rates of terrigenous components, was strongest during the (glacial) oxygen isotope stages 2 and 6. It also triggered high particle fluxes from primary producers in the photic zone (marine algae) to the sea floor which in turn allowed high standing stocks of bathyal benthos to build up. Only the zooplankton, as documented by the fluxes of foraminiferal shells and radiolarian skeletons to the sediments, did not respond to the strong productivity changes in the surface waters. Climatic changes which document to have caused the rapid fluctuations of the accumulation rates of the hemipelagic sediment off northwest Africa, and the periods of transition from the glacial to the interglacial circulation regimes and *vice versa*, lasted only a few hundred to a thousand yrs. These climatic changes affected both the atmospheric and ocean surface circulation at the same time.

1 Institutt for geologi, Universitetet i Oslo, Postboks 1047, Blindern, Oslo 3, Norway

2 School of Oceanography, Oregon State University, Corvallis, OR 97 331, USA

3 Geologisch-Paläontologisches Institut und Museum, Christian-Albrechts-Universität, Olshausenstraße 40/60, 2300 Kiel, FRG

1. Introduction

Marine sediments along continental margins are the products of complex interactions between atmospheric, oceanic, biologic and diagenetic processes. The individual effects, however, are often very difficult to separate from each other. The rate at which sediments and their individual components accumulate on the sea floor as well as rate changes are expected to be important bearers of information about these processes. In this study we will attempt to interprete the fluxes of a number of important biogenic and terrigenous sediment components to the sea floor at a site of apparently continuous sediment accumulation on the continental slope off northwest Africa (Fig. 1) during the past 140,000 yrs. in terms of such interacting processes. The area off northwest Africa is particularly interesting, but also complex, because of the interaction of a continental margin, the eastern boundary current and trade wind system (Canary Current, cf. Wooster et al., 1976).

Accumulation rates and their rates of change are considered particularly useful to track the history of this regime back in time because many of the sediment components can be related to specific source functions. However, the calculation of accumulation rates requires a complete set of numerous variables which can only be collected for a few cores of exceptional qualities. We have chosen to present the data for only one core (M 12392-1) to be able to describe the history of its depositional site, but many more cores with similar qualities are available from the northwest African continental margin (Seibold, 1972); this core may be considered representative of a large region. We have chosen to use a definition of accumulation rates which expresses the fluxes of sediment components in terms of weight per unit area and time. This approach requires data for water and salt contents as well as porosities and goes back to the original paper by Koczy (1951), defining flux as sedimentation intensity.

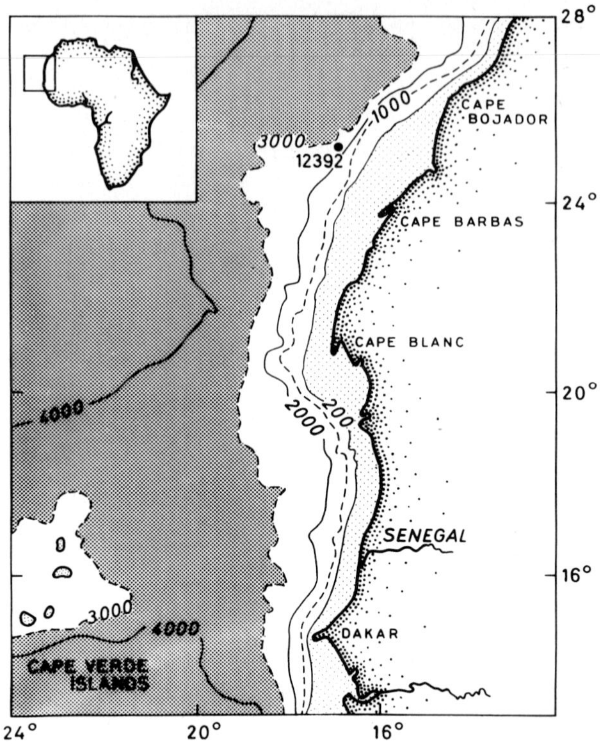

Fig. 1
Location map of site M 12392-1 and generalized bathymetry of the Atlantic continental margin off Northwest Africa

2. Methods

To generate accumulation rate data it is necessary to combine chemical, mineralogical, micropaleontological and geotechnical property measurements with the chronostratigraphic framework of a sedimentary section under study in as detailed a way as possible. Here we attempt to achieve such a complete account of the depositional budget from the site of core M 12392-1 (Fig. 1). This core is 9.6 m long and was taken by means of a long box corer (Kasten core) in 2375 m water depth at 25°10,3'N latitude and 16°50,7'W longitude off Cape Bojador. Its sediment consists of light olive green to gray, modestly bioturbated, texturally homogenous calcareous muds (Hartmann et al., 1976, Diester-Haass, 1976, Thiede, 1977). The site is situated at the northern margin of the region whose sediments are intensely affected by coastal upwelling (Diester-Haass, 1978) and upslope from a large slide scar (Embley, 1976). The detailed record and its perfect stratigraphic correlation with counterparts from the eastern equatorial Pacific (Thiede, 1977) suggest that the pelagic mode of sedimentation had not been disturbed during the past 130.-140.000 yrs.

2.1 Concentration Measurements

Concentrations of sediment components as discussed in this study were determined by a wide variety of methods. Compositional analyses of the coarse fractions > 0.15 mm (Table 1) were carried out in conjunction with micropaleontological and stratigraphic studies of the planktonic foraminiferal faunas published earlier (Thiede, 1977). The frequency concentration of dissolution resistant planktonic foraminifers is a composite of the percentage of *G. dutertrei*, *G. crassaformis*, *G. hirsuta*, *G. inflata* and *G. truncatulinoides*. The loss of $CaCO_3$ due to dissolution was then estimated according to the formula $L = (1 - \frac{R_1}{R_2})$, (L = estimated loss, R_1 = Sum of resistant species in the least dissolved sample, R_2 = Sum of resistant species per sample), cf. Berger (1971). The lowest concentrations of dissolution resistant species of planktonic foraminifers (23.6%) was observed 15 cm below the sediment surface.

Analytical procedures for chemical and mineralogical determinations of sediment constituents (Table 2) are described in detail by Hartmann et al. (1976) and were as follows: Carbonate-CO_2 was measured gas-volumetrically of CO_2 evolved by acid treatment and likewise organic carbon as CO_2 liberated by oxidation with potassium-dichromate-sulfuric acid. Total nitrogen was determined by Kjeldahl digestion of nitrogenous compounds using a selenium catalyst and subsequent ammonia destillation and titration. Organic phosphorus was taken as the analytical difference of acid soluble phosphorus before and after ignition of sediment for 2 hours at 400°C to convert organically bound phosphorus into soluble phosphate. Phosphate phosphorus was measured by the molybdate-blue method as total phosphate but after treatment of samples with concentrated hydrochloric acid. Sulfur was done by automatic titration of SO_2 and SO_3 with potassium iodate and a starch/iodide indicator after oxidative combustion in the LECO-induction furnace.

Opal and quartz concentrations were determined by quantitative X-ray diffraction techniques before and after heat conversion of opaline silica to cristobalite; alumina internal standard and calibration mixtures of different sediment matrixes with varying opal and quartz-contents were used for quantification (Goldberg, 1958, Ellis, 1972, Molina-Cruz & Price, 1977).

2.2 Geotechnical Properties of the Sediment

The calculation of the accumulation rates requires the measurement of a number of geotechnical properties of the bulk sediment: i.e. either porosity and wet bulk density or dry bulk density and water content. Both means were employed here. Most samples were taken from the wet core (Thiede, 1977). The dry weight per volume of wet sediment was determined from these by directly weighing a known volume (usually 20 cm^3) of wet sediment after drying at 105°C and subsequent correction for salt content using porosity data. Wet density, dry density and porosity (Table 2) were calculated using a scheme by which dry densities for pure quartz, calcite and clay residue were assigned to each sample according to the respective mineral proportions. These combined with the dry weight measurements allowed for sufficiently accurate determinations of wet density and porosity to be used in accumulation rate estimates. These data are in good agreement with actual measurements of water content and wet densities as obtained for a random selection of 10 samples (Table 3) which were kindly made available by F.-C. Kögler, Univ. Kiel (pers. comm.).

2.3 Dating and Correlation to Chronostratigraphy

Measurements of $\delta\ ^{18}O$ ratios in the benthic foraminfer _Uvigerina peregrina_ (Fig. 2) in the samples from core M 12392-1 (Shackleton, 1977) and the easy correlation of their depth distribution with other distant pelagic cores (Thiede, 1977) document the superb stratigraphic qualities of M 12392-1. Correlation of the stratigraphy of this core to an absolute time scale is based on a number of sample intervals which were radiometrically dated or which can be related to absolutely dated events in other cores or regions. Table 4 summarizes the dates used in this chronostratigraphic correlation; specifically: the time control (Fig. 2a) for the time span from the Present to 30,000 yrs. B.P. is based on ^{14}C datings of this core and of an outer Niger Delta core using $\delta\ ^{18}O$ ratios in _Uvigerina_-species for correlation (Pastouret et al., 1978). Sections older than 30,000 yrs. B.P. are dated by ages assigned to oxygen isotope stage boundaries or maxima and minima of the oxygen isotope distribution and calibrated by the Barbados, Bermuda and New Guinea sea level curves (Harmon et al., 1981, Steinen et al., 1973, Veeh & Chappel, 1970). In the core section > 30,000 yrs. ago, due

Table 1. Core M 12392-1: Percentages of sand-sized (> 0.063 mm) material (in weight percent of the bulk sediment), of material > 0.15 mm (in weight percent of the total sand fraction), of major biogenic coarse fraction (> 0.15 mm) components (in percent of the total grain assemblages), of dissolution resistant planktonic foraminifers (in percent of the total planktonic foraminiferal faunas) and of the estimated loss of $CaCO_3$ due to dissolution (in weight percent). Data are from Thiede, 1977

Level (cm)	% > 0.063 mm of bulk sed.	% > 0.15 mm of sand fract.	% Plankt. foram.	% Radiolar.	% Benthic foram.	% Estimated loss of $CaCO_3$	% Dissolution resistant plankt. foram.
surface	29.0	78.9	-	-	-	-	-
4.5	16.9	65.3	98.1	.1	.7	32	34.9
10.0	17.6	68.0	98.0	.3	.7	24	31.2
15.0	19.8	68.2	98.7	.1	.8	0	23.6
20.0	22.9	68.9	97.8	0	1.3	41	40.2
25.0	17.9	62.2	97.6	0	1.4	46	43.5
32.5	17.2	55.3	96.9	.1	1.3	48	45.7
42.5	13.1	51.6	96.4	.5	1.0	58	55.6
52.5	11.1	54.1	93.3	2.5	1.1	54	51.4
62.5	11.3	55.4	95.1	2.0	1.9	53	49.9
72.5	14.7	55.6	96.9	1.1	1.0	53	50.5
77.5	16.1	58.8	95.4	1.5	1.3	55	52.6
82.5	14.0	50.3	95.1	2.8	1.0	52	49.2
92.5	14.2	37.9	91.9	4.6	1.4	50	46.8
102.5	15.0	39.1	93.9	2.3	1.9	58	56.7
112.5	9.8	53.8	93.8	2.3	2.7	55	53.0
122.5	10.4	58.6	93.6	2.7	1.6	57	54.3
132.5	10.2	55.0	95.9	1.9	.8	48	45.5
142.5	9.8	55.0	93.6	2.9	2.0	56	53.3
152.5	12.3	45.5	90.0	2.5	3.0	53	50.1
162.5	11.0	18.3	88.2	2.6	4.9	51	48.1
172.5	7.7	55.8	92.4	3.3	2.7	58	56.1
182.5	8.5	57.0	95.4	2.1	1.1	50	47.1
192.5	9.0	55.0	94.3	1.8	2.5	37	37.3
202.5	9.0	53.4	93.8	2.7	1.9	41	39.7
212.5	9.9	52.2	90.2	1.8	2.7	43	41.6
220.5	11.4	59.3	94.9	1.5	1.6	43	41.2
232.5	8.3	56.3	93.7	2.9	2.4	46	43.8
242.5	9.5	53.3	90.3	5.3	3.2	49	46.0
252.5	13.6	51.9	94.3	2.0	2.2	48	45.1
262.5	11.3	52.9	92.0	3.7	1.8	49	46.0
272.5	10.2	55.6	93.5	2.6	2.2	42	40.4
282.5	7.0	64.7	88.4	8.7	1.9	54	51.3
292.5	8.9	54.9	86.9	10.3	1.6	46	44.1
302.5	8.8	55.2	92.2	4.9	.9	46	43.4
312.5	9.9	56.5	94.9	2.7	.9	47	44.3
322.5	9.9	65.0	92.8	4.6	1.7	50	46.8
332.5	7.1	55.3	88.1	7.3	3.2	47	44.2
342.5	7.1	53.1	85.9	11.0	1.6	40	39.6
352.5	7.5	62.4	79.0	18.1	2.2	59	57.3
361.0	10.9	55.5	88.5	8.8	1.9	45	43.2
375.5	12.0	54.1	93.8	2.5	2.1	54	51.1
385.5	16.7	64.4	94.3	2.4	2.0	60	58.3
395.5	9.1	62.3	95.9	1.5	1.8	57	55.4
405.0	15.5	65.3	94.0	3.3	1.1	40	58.9
415.5	17.8	58.0	96.1	2.0	.8	54	51.0
425.5	18.4	63.1	95.7	2.5	.7	60	59.2

Table 1. Cont.

Level (cm)	% > 0.063 mm of bulk sed.	% > 0.15 mm of sand fract.	% Plankt. foram.	% Radiolar.	% Benthic foram.	% Estimated loss of $CaCO_3$	% Dissolution resistant plankt. foram.
458.5	14.5	60.9	93.7	2.8	1.1	56	53.2
465.5	12.6	51.7	90.8	4.9	1.9	60	58.7
475.5	14.6	56.7	93.0	4.3	1.6	49	46.3
485.5	16.4	61.2	96.2	.9	2.2	50	47.5
495.5	18.6	40.5	96.8	1.4	.9	42	41.0
505.5	14.0	47.3	94.3	2.8	1.3	43	41.1
515.5	10.5	44.1	92.5	4.7	1.9	43	41.3
525.5	8.5	47.1	94.8	1.9	1.7	44	41.8
535.5	7.9	47.4	94.4	1.2	1.7	46	43.7
545.5	7.4	54.2	91.8	2.4	3.5	54	51.3
555.5	7.2	53.4	92.4	2.9	3.0	51	48.5
565.5	6.9	64.2	95.6	1.4	2.0	53	50.5
575.5	6.5	60.4	96.5	1.3	1.6	54	50.8
585.5	7.6	57.6	94.6	.9	2.7	53	50.0
595.0	4.2	60.7	96.3	.9	1.6	51	48.3
605.5	5.8	61.1	97.8	.2	1.3	40	39.6
615.5	12.6	61.3	97.9	0	1.6	32	34.8
625.5	16.8	61.5	99.2	0	.5	36	36.8
635.5	19.7	65.0	98.9	0	.5	43	41.6
645.5	12.3	59.2	99.0	.1	.4	48	45.4
655.5	7.8	50.1	98.7	.8	.9	47	44.4
665.5	9.1	60.1	98.1	0	1.6	47	44.7
675.5	12.9	62.0	98.3	0	1.2	53	50.4
685.5	12.0	63.0	96.9	0	1.8	55	52.3
695.5	9.8	59.7	96.8	0	1.1	51	47.8
705.0	15.1	64.2	98.2	0	.7	46	43.4
715.5	15.2	59.9	97.7	0	1.0	45	43.1
725.5	11.2	58.8	98.8	0	.4	41	40.1
735.5	19.3	66.3	98.5	0	.9	50	47.4
745.5	12.4	59.4	98.7	0	.6	45	43.3
755.5	13.8	66.1	97.5	0	1.4	55	52.6
765.5	12.4	70.3	98.8	0	.2	54	51.3
775.5	11.3	32.5	97.0	0	2.0	54	51.4
785.0	10.9	67.1	95.9	0	2.2	54	51.8
795.5	11.9	40.7	97.0	.1	.7	54	51.2
805.5	13.1	66.0	98.6	0	.7	51	48.5
815.5	17.5	67.0	98.3	.1	1.3	49	45.9
825.5	17.0	65.6	97.0	.5	1.3	53	50.4
835.5	17.3	63.7	94.9	3.4	.9	58	56.4
846.0	12.6	53.4	82.7	14.9	1.4	60	58.3
857.5	16.5	44.7	84.9	10.4	2.4	43	41.6
865.5	11.8	47.5	90.6	6.0	2.1	39	38.8
875.5	12.7	49.6	94.2	1.9	2.1	46	43.4
885.5	19.1	70.2	94.3	1.1	3.1	54	51.5
895.5	10.4	49.2	93.7	1.8	3.3	38	37.9
905.5	9.1	52.2	93.6	1.9	3.4	52	49.1
915.5	8.8	49.0	92.2	2.1	4.0	35	36.4
925.5	5.8	50.6	92.5	3.8	3.5	43	41.1
935.5	6.9	48.4	91.4	3.7	3.1	29	33.2
945.5	7.8	52.6	96.6	1.5	3.6	41	40.1
954.5	6.4	53.8	94.9	2.4	3.7	41	39.9
963.0	8.6	57.3	92.9	2.8	3.3	55	52.9

Table 2. Distribution of physical properties and major sediment components in core M 12392-1. Concentrations in (%), or in (ppm) of the bulk dry sediment

Depth (cm)	Dry Weight (g/cm^3 wet)	Quartz (%)	C-org (%)	Opal (%)	Residue (%)	N-total (ppm)	Calcite (%)	Sulfur (%)	P-org (ppm)	P-PO_4 (ppm)	Wet Density (g/cm3)	Dry Density (g/cm^3)	Porosity (vol %/100)	Accumulation Rates(g/cm2/ 1000 YR)
2- 10	.976	11.7	.35	1.1	34.75	654	52.1	.032	128	405	1.61	2.50	.609	4.59
10- 20	1.031	7.4	.34	.7	27.26	558	64.3	.040	98	399	1.65	2.54	.594	4.85
20- 29	1.026	5.4	.35	.7	24.35	523	69.2	.048	103	419	1.65	2.56	.599	4.82
30- 40	1.035	5.2	.37	.9	18.93	545	74.6	.046	109	387	1.66	2.59	.600	5.79
40- 50	.965	10.5	.89	1.6	26.31	1170	60.7	.068	120	451	1.61	2.54	.620	5.40
50- 60	.818	10.0	1.07	2.4	29.53	1392	57.0	.140	108	458	1.52	2.52	.676	4.58
60- 70	.796	8.8	1.13	3.1	24.77	1455	62.2	.240	125	453	1.51	2.54	.687	9.79
70- 80	.709	9.6	1.24	2.9	23.26	1478	63.0	.278	119	487	1.46	2.55	.722	8.72
83- 90	.778	11.6	1.11	4.5	22.29	1392	60.5	.316	106	494	1.50	2.54	.694	9.56
90-100	.754	19.7	1.27	2.1	18.03	1522	58.9	.314	87	517	1.49	2.58	.708	9.27
100-110	.792	11.7	1.23	3.2	21.67	1484	62.2	.354	98	507	1.51	2.55	.689	9.75
110-120	.587	12.5	1.83	2.2	18.97	2138	64.5	.436	94	562	1.39	2.57	.772	12.92
120-130	.589	11.8	2.33	2.0	20.87	2628	63.0	.424	133	508	1.39	2.56	.770	12.96
130-140	.556	9.1	1.87	1.8	22.43	2151	64.8	.420	141	502	1.37	2.56	.783	12.50
140-150	.575	8.2	2.26	2.6	21.44	2539	65.5	.428	140	501	1.37	2.56	.775	13.23
150-160	.624	10.5	2.01	2.3	22.19	2265	63.0	.438	158	500	1.41	2.55	.755	14.35
160-170	.878	10.6	1.85	2.8	22.65	2159	62.1	.304	139	505	1.56	2.55	.655	20.20
170-180	.606	13.2	2.27	1.7	21.13	2532	61.7	.428	164	487	1.40	2.56	.763	13.93
180-190	.558	10.8	1.62	3.4	22.98	1809	61.2	.340	134	458	1.37	2.54	.781	12.82
190-200	.628	9.6	1.91	3.2	25.49	2259	59.8	.462	157	466	1.41	2.53	.752	14.43
200-210	.611	9.7	1.72	2.9	25.88	1979	59.8	.432	143	460	1.40	2.53	.759	8.85
210-220	.629	10.7	1.85	1.5	28.75	2161	57.2	.436	149	473	1.41	2.52	.750	9.12
220-230	.563	10.5	2.38	1.5	29.42	2845	56.2	.468	193	482	1.37	2.52	.777	8.16
230-240	.562	11.1	2.43	1.7	28.97	2820	55.8	.518	192	472	1.37	2.52	.777	6.69
240-250	.493	11.9	2.05	1.6	27.85	2437	56.6	.444	154	474	1.33	2.52	.805	5.86
250-260	.629	14.4	1.91	1.7	27.99	2258	54.0	.448	145	446	1.41	2.52	.750	7.49
260-270	.665	13.0	2.10	1.4	29.30	2375	54.2	.504	168	457	1.43	2.52	.736	7.90
270-280	.644	12.0	2.42	1.7	25.18	2800	58.7	.484	158	447	1.42	2.54	.747	7.66
280-290	.562	9.5	3.03	1.6	28.67	3447	57.2	.520	215	437	1.37	2.52	.777	6.69
290-300	.561	10.5	2.39	1.7	27.51	2750	57.9	.520	176	440	1.37	2.52	.777	6.67

300-310	.611	10.4	2.24	1.4	27.06	2583	58.9	.532	180	441	1.40	2.53	.759	7.27
310-320	.663	10.7	2.02	1.4	28.38	2354	57.5	.504	159	434	1.43	2.52	.734	4.71
320-330	.578	8.8	2.71	1.9	27.69	3041	58.9	.580	191	443	1.38	2.52	.771	4.11
330-340	.611	12.8	1.75	1.9	26.45	2054	57.1	.428	101	408	1.40	2.53	.759	4.33
340-350	.680	12.2	1.95	1.5	27.85	2231	56.5	.492	157	418	1.44	2.53	.731	4.82
350-360	.477	10.5	2.49	3.7	26.51	2921	56.8	.478	185	418	1.32	2.52	.811	3.38
363-373	.661	12.4	1.59	2.0	25.41	2129	58.6	.404	153	407	1.43	2.54	.740	4.69
373-383	.642	11.2	1.82	1.5	23.98	2191	61.5	.420	142	398	1.42	2.55	.748	4.56
383-393	.605	9.2	1.88	1.9	20.72	2162	66.3	.400	153	392	1.40	2.57	.765	4.29
393-403	.674	8.9	1.85	2.6	21.85	2081	64.8	.406	154	405	1.44	2.56	.737	4.79
403-413	.639	9.1	1.52	2.3	21.48	1798	65.6	.344	112	383	1.42	2.56	.750	4.41
413-423	.690	9.7	1.20	2.2	22.20	1445	64.7	.248	104	380	1.45	2.56	.730	4.76
423-433	.707	10.6	1.40	1.4	23.00	1551	63.6	.360	117	367	1.46	2.56	.724	4.88
433-443	.727	11.3	1.31	1.2	25.69	1522	60.5	.350	139	335	1.47	2.54	.714	5.02
443-453	.728	9.5	1.78	1.1	26.32	2021	61.3	.512	131	384	1.47	2.54	.713	5.02
453-403	.755	8.7	1.94	1.0	20.86	2141	67.5	.420	107	354	1.49	2.57	.706	3.77
463-473	.674	10.0	1.52	1.2	23.58	1597	63.7	.404	98	348	1.44	2.55	.736	3.37
473-483	.814	11.3	1.24	1.3	27.46	1301	58.7	.364	93	353	1.52	1.53	.678	4.07
483-493	.743	10.8	1.09	1.2	25.01	1200	61.9	.244	87	348	1.48	2.55	.708	3.71
493-503	.844	12.3	1.13	1.2	24.57	1231	60.8	.484	79	352	1.54	2.55	.669	4.22
503-513	.866	18.4	.81	1.6	27.89	876	51.3	.220	62	384	1.55	2.53	.658	4.85
513-523	.834	16.9	.85	1.6	29.45	948	51.2	.298	64	383	1.53	2.52	.669	4.67
523-533	.835	16.3	.68	1.5	30.22	915	51.3	.344	86	382	1.53	2.52	.669	4.67
533-543	.834	13.3	.69	1.3	29.91	871	54.8	.304	80	403	1.53	2.52	.669	4.67
543-553	.797	13.2	.56	1.3	28.34	622	56.6	.144	78	412	1.51	2.53	.685	4.47
553-563	.781	12.1	.86	1.1	28.44	1033	57.5	.302	104	415	1.50	2.53	.691	4.22
563-573	.766	12.0	.86	1.3	30.44	1100	55.4	.348	94	396	1.49	2.52	.696	4.14
573-583	.753	13.5	.81	1.4	33.19	1023	51.1	.420	89	400	1.48	2.50	.699	4.07
583-591	.819	10.9	.74	1.2	32.06	962	55.1	.392	87	398	1.52	2.51	.674	4.42
590-598	.881	15.3	.63	1.5	39.17	796	43.4	.442	72	402	1.55	2.47	.643	4.76
603-613	1.011	8.6	.62	.7	24.88	710	65.2	.320	71	358	1.64	2.56	.605	5.46
613-623	.901	5.4	.50	.3	19.50	652	74.3	.208	65	394	1.58	2.59	.652	4.86
623-633	.969	5.7	.39	.3	20.31	511	73.3	.196	58	452	1.62	2.59	.626	4.85
633-643	.991	5.8	.47	.3	23.83	564	69.6	.178	65	469	1.63	2.57	.614	4.96
643-653	.892	13.5	.38	1.2	35.32	592	49.6	.272	54	445	1.56	2.49	.642	4.46
653-663	.867	11.4	.48	1.2	29.82	541	57.1	.200	73	408	1.55	2.52	.656	4.34
663-673	.862	4.6	.41	.5	27.49	549	67.0	.184	84	442	1.55	2.55	.662	4.31
673-683	.862	7.5	.50	.7	27.00	579	64.3	.226	78	445	1.55	2.55	.662	4.31
683-693	.884	9.3	.45	.9	30.15	598	59.2	.192	68	446	1.56	2.53	.651	3.05
693-703	.917	9.2	.43	.8	29.87	544	59.7	.276	59	485	1.58	2.53	.638	3.16

Table 2. Cont.

Depth (cm)	Dry Weight (g/cm3 wet)	Quartz (%)	C-org (%)	Opal (%)	Residue (%)	N-total (ppm)	Calcite (%)	Sulfur (%)	P-org (ppm)	P-PO4 (ppm)	Wet Density (g/cm3)	Dry Density (g/cm3)	Porosity (vol %/100)	Accumulation Rates (g/cm2/1000 YR)
703-713	.975	6.3	.33	.3	24.27	494	68.8	.240	61	531	1.62	2.56	.619	3.36
713-723	1.048	7.6	.41	.4	27.19	571	64.4	.192	50	562	1.66	2.55	.589	3.62
723-733	.961	6.8	.36	.5	26.04	460	66.3	.268	73	516	1.61	2.55	.623	5.53
733-743	.963	7.8	.37	.6	26.93	500	64.3	.212	48	468	1.61	2.55	.622	5.54
743-753	.816	13.3	.45	1.1	29.65	589	55.5	.352	57	461	1.52	2.52	.676	4.69
753-763	.797	9.0	.56	.7	28.84	734	60.9	.284	63	461	1.51	2.53	.685	4.58
763-773	.781	9.5	.55	.8	29.45	711	59.7	.368	71	459	1.50	2.53	.691	4.49
773-783	.887	8.9	.55	.8	32.75	664	57.0	.272	77	477	1.56	2.51	.647	5.10
783-793	.803	10.1	.43	.8	33.27	544	55.4	.296	66	480	1.51	2.51	.680	4.61
793-803	.839	10.8	.48	1.0	35.32	601	52.4	.322	56	489	1.53	2.50	.664	4.83
803-813	1.017	7.5	.42	.5	28.88	572	62.7	.396	67	510	1.64	2.54	.600	5.85
813-823	1.037	4.4	.33	.3	20.87	409	74.1	.204	57	644	1.66	2.59	.599	5.96
823-833	1.116	4.3	.40	.5	18.10	546	76.7	.344	44	540	1.71	2.60	.571	6.41
833-839	1.049	6.9	1.81	1.0	26.19	1966	64.1	.442	95	588	1.66	2.54	.587	6.03
842-848	.846	11.1	.83	1.2	26.67	982	60.2	.384	67	500	1.54	2.54	.667	4.74
848-855	.746	15.4	1.18	2.1	26.42	1274	54.9	.508	74	507	1.48	2.53	.705	4.18
855-863	.695	14.5	1.24	2.1	26.26	1288	55.9	.460	69	524	1.45	2.53	.725	3.89
863-873	.792	15.8	1.16	2.2	22.24	1214	58.6	.450	64	529	1.51	2.55	.689	4.43
873-883	.789	14.3	1.17	1.7	20.63	1207	62.2	.372	56	546	1.51	2.57	.691	11.05
883-893	.702	9.4	.60	1.3	19.50	718	69.2	.596	57	501	1.46	2.58	.728	9.83
893-903	.638	10.3	1.83	2.1	19.87	1762	65.9	.464	88	585	1.42	2.57	.752	8.94
903-913	.738	11.7	1.73	1.7	19.67	1687	65.2	.394	85	550	1.48	2.57	.713	10.33
913-923	.724	11.2	1.98	2.1	20.92	1894	63.8	.376	99	567	1.47	2.56	.717	10.13
923-933	.707	11.3	1.87	1.7	21.93	1777	63.2	.444	97	557	1.46	2.56	.724	9.90
933-943	.625	9.9	1.94	1.7	24.16	1937	62.3	.464	106	585	1.41	2.55	.755	8.75
943-953	.592	10.2	2.23	1.5	25.37	2318	60.7	.482	113	552	1.39	2.54	.767	8.29
953-963	.608	10.8	1.84	1.5	23.96	1937	61.9	.428	92	516	1.40	2.55	.762	8.51

Table 3. Comparison between measured and predicted physical properties of core M 12392-1

Depth (cm)	Water content (wt-% dry sed.)		Wet density ($g.cm^{-3}$wet sed.)	
	meas.[x]	cal-cul.	meas.[x]	cal-cul.
17- 27	68	61	1.56	1.65
110-120	125	132	1.38	1.39
330-340	119	124	1.44	1.40
393-403	116	109	1.43	1.44
505-513	94	77	1.49	1.53
575-585	106	93	1.46	1.48
685-695	85	74	1.54	1.56
753-763	94	86	1.51	1.51
803-813	70	62	1.63	1.64
953-963	122	125	1.40	1.40

[x]Measured physical properties data from Kögler (1972; unpublished)

to the large uncertainty in the assigned ages (Table 4), there are several possibilites in interpreting the age-depth relationship. We choose the one illustrated by the trend of the smoothed heavy line in Fig. 2a for sedimentation rate estimates rather than the trend emphasising more abrupt changes (dashed line in Fig. 2a). The rational is, that if the sample spacing were infinitesimally small, the sedimentation rate changes with time would be better approximated by a smoothed curve. The very detailed time control in the section younger than 30,000 yrs. shows such a trend nicely, and we have interpolated the sedimentation rates for the period > 30,000 yrs. ago accordingly. The maximal difference between the two estimates is ± 25% in the core sections between 300 - 400 cm and between 800 - 900 cm of depth. In all other sections both interpretations agree to within 10% of each other. These variable errors affect the accumulation rate estimates by the same margin. Termination I and II (Broecker & van Donk, 1970) mark the steepest gradients of the oxygen isotope curve and correspond to the very rapid deglaciation after the penultimate

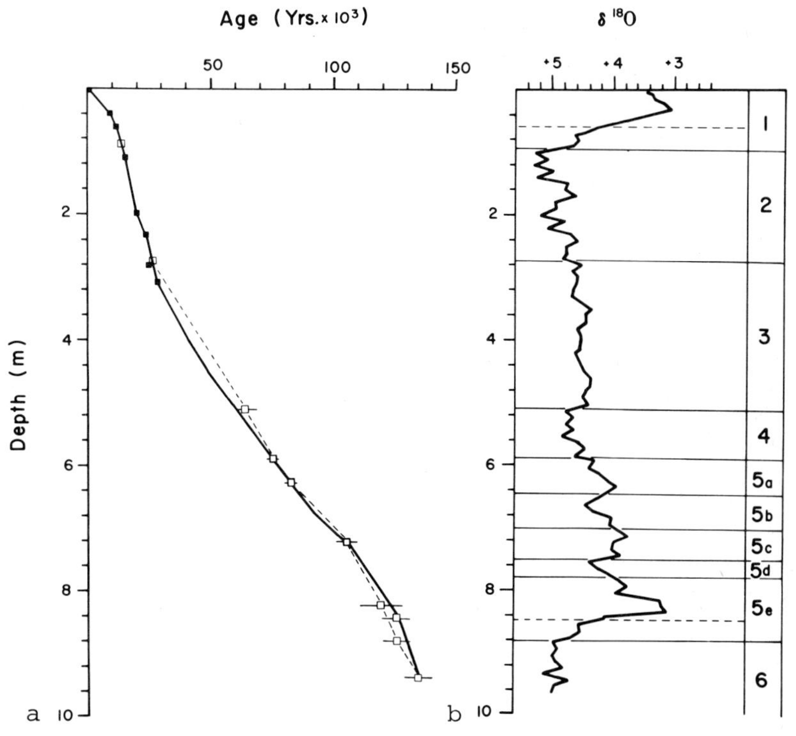

Fig. 2
Oxygen isotope ratios (a) expressed in $\delta^{18}O$ in Uvigerina peregrina after Shackleton (1977) and (b) age-depth relationship of core M 12392-1

Table 4. Radiometric dates and ages of oxygen isotope stage boundaries used in the chronostratigraphic correlation of core M 12392-1

Depth (cm)	Age (x10 yrs)	Remarks
0- 7	710±230	C-org: Geyh, 1976
30- 32	6,000	δ ^{18}O minimum
38- 42	9,760±230	$CaCO_3$: Erlenkeuser, in: Koopmann, 1979
70	11,000	Termination I: Broecker and van Donk, 1970
97	13,500	Termination I: Ruddiman & McIntyre, 1976
105-110	15,145±400	C-org: Geyh, 1976
121	16,500	δ ^{18}O maximum, in: Koopmann, 1979
195-200	19,000±1020	C-org: Geyh, 1976
222-230	22,800±500	$CaCO_3$: Erlenkeuser, unpubl.
281	24,600±2,400	C-org: Pastouret et al., 1977;δ ^{18}O correlation with Niger core
306-312	28,250±1060	C-org: Geyh, 1976
510	64,000±4,000	Oxygen isotope stage boundary 3/4: Shackleton & Opdyke, 1976, Lutze et al., 1979
590	75,000±3,000	Oxygen isotope stage boundary 4/5: Stuiver et al., 1978, Ninkovich et al., 1978
634	83,000±2,000	δ ^{18}O minimum, Barbados I: Shackleton & Matthews, 1977
714	105,000±5,000	δ ^{18}O minimum, Barbados II: Harmon et al., 1979
824	118,000±9,000	δ ^{18}O minimum, Barbados III: Harmon et al., 1979
845	126,000	Termination II: Broecker & van Donk, 1970
880	126,000	Termination II: Ruddimann & McIntyre, 1976
935	134,000±5,000	δ ^{18}O maximum: interpolated after time scale of Ruddimann & McIntyre, 1976

and ultimate glaciation with assigned ages of 11,000 yrs. and 126,000 yrs., respectively. More recently, these boundaries are placed within the upper 1/3 of the two oxygen isotope maxima (Ruddiman & McIntyre, 1976). For Termination I the age increases thereby to 13,500 yrs. but does not affect the sedimentation rate, whereas for Termination II the age increase is entirely within the uncertainty of 126,000 $\pm$ 5,000 yrs. We have in our interpretation and illustrations (except for Fig. 2, where we show the original and the new definitions for Terminations I and II) throughout only used the original one by Broecker and van Donk (1970). The stratigraphic boundaries of the oxygen isotope stage 1 to 5 and the upper boundary of stage 6 (Emiliani, 1955) as well as the substages a-e of stage 5 (Shackleton, 1969) are well defined in this core. Changes of the oxygen isotope ratios adjacent to Terminations I and II are surprisingly similar.

The detailed age-depth relationship (Fig. 2) also yields a more accurate age calibration of the oxygen isotope curve than presently in use. Significantly affected is the oxygen isotope stage boundary 2/3 of 26,000 yrs. ago at 285 cm of depth in core M 12392-1 rather than at 235 cm as used by Thiede (1977), Lutze et al. (1979) and Koopmann (1979). Accumulation rates of bulk sediment and sediment constituents are based on this age-depth relationship and sediment geotechnical property data as listed in Table 2 (columns 11, 12 and 14).

A comparison of the distribution of calcium carbonate contents throughout core M 12392-1 with a core from the western equatorial Atlantic (Fig. 3) and the similarity of sequences of carbonate maxima and minima document the excellent stratigraphic qualities of both cores. This is surprising since one core came from the western flank of the Mid-Atlantic Ridge (Bé et al., 1976) whereas the other one was taken on the continen-

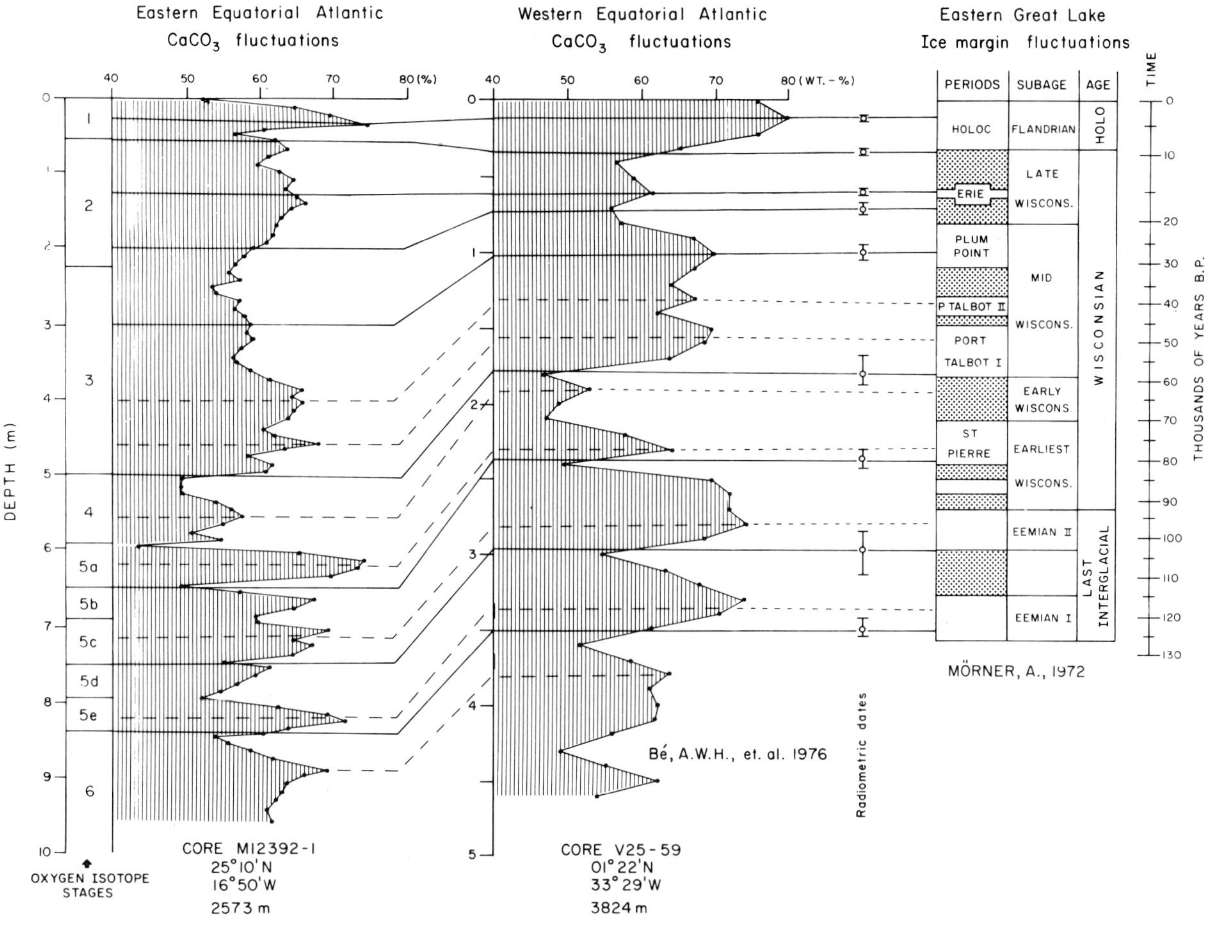

Fig. 3
Correlation of the carbonate stratigraphy of cores M 12392-1 and V25-59 from the western equatorial Atlantic (after Bé et al., 1976) to the ice margin fluctuations adjacent to the eastern Great Lakes (after Mörner, 1972)

tal slope off northwest Africa. Although it is known that continental margin sediments are often affected by reworking processes (Sarnthein et al., this Vol.) the sediments of core M 12392-1 seem to show little effect of such reworking, but rather they seem to reflect the pelagic depositional environment throughout the past 120,-140,000 yrs. of the subtropical Atlantic Ocean.

2.4 Calculation of Accumulation rates

The bulk sediment flux ($Bulk_{acc}$) is equal to the product of sedimentation rate (ω) and the bulk dry density (p_{dry}) (van Andel et al., 1975; Koczy, 1951):

$Bulk_{acc} = \omega \cdot p_{dry}$; whereby

$Bulk_{acc} = g \cdot cm^{-2} \cdot 10^{-3} \cdot y^{-1}$

$\omega = cm \cdot 10^{-3} \cdot y^{-1}$

$p_{dry} = g \cdot cm^{-3}$, dry weight per wet volume. The fluxes of the individual sediment components can then be calculated by apportioning the bulk sediment flux according to the concentration data:

$$Comp._{acc} = \omega \cdot p_{dry} \cdot \frac{\%\ Comp.}{100}$$

or

$$Comp._{acc} = \omega \cdot (p_{wet} - 1.026 \cdot \frac{\Phi}{100}) \cdot \frac{\%\ Comp.}{100}, \text{ whereby}$$

$p_{wet} = g.cm^{-3}$, wet weight per wet volume

Φ = porosity, vol-%

1.026= density of interstitial water, $g.cm^{-3}$

3. Results

The detailed age correlation obtained from this core yields an age calibration of the distribution of the stable oxygen and carbon isotopes ratios (Fig. 4) more accurate than previously used for the past 130-140,000 yrs. The accumulation rates of the bulk sediment and its individual components are based on this age-depth relationship and are plotted against age rather than depth because of a better assessment of the rates of changes with time.

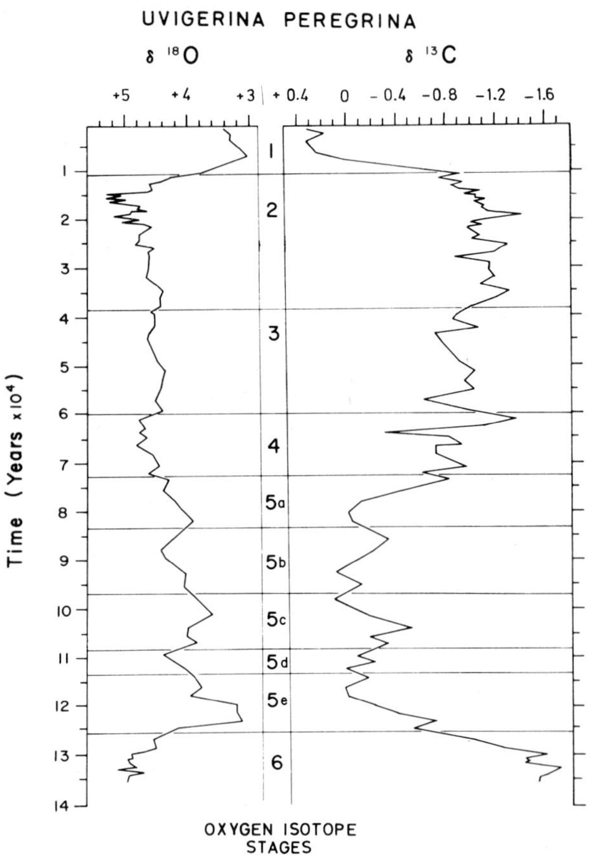

Fig. 4
Oxygen and stable carbon isotope ratios in *Uvigerina peregrina* in core M 12392-1 versus time (after Shackleton, 1977 and Thiede, 1977) and new ages listed in Table 4

3.1 Accumulation Rates of the Bulk Sediment

The bulk accumulation rates at site M 12392 (Fig. 5 and Table 2) vary from 4 to 14 g $g.cm^{-2}.10^{-3}.yr^{-1}$. These data are affected by a maximal error of ± 25% between 36 and 53.10^{3} yrs. ago and 114 and 126.10^{3} yrs. ago but only by an error of ± 5% for the other periods. Except for two short intervals which correspond to parts of oxygen isotope stages 6 and 2, the average flux of the bulk sediment was remarkably constant and low over the past 130-140.000 yrs. This is very surprising because the carbonate concentrations (Fig. 3) which fluctuate between 40 and 80 weight % of the bulk sediment, suggest that the sediment composition was highly variable. During most of the oxygen isotope stages 2 and 6 the bulk sediment accumulation rates increased 2-3 fold. The sharp pulses with maximal rates lasted only a few thousand years and correspond roughly to the maxima of the oxygen isotope values. Such a distribution reflects a very basic pattern which can be seen in many of the other sediment components with some modifications. The bulk accumulation rate is the basis for the calculation of the constituents accumulation rates.

3.2 Fluxes of Terrigenous Components

Terrigenous components of the hemipelagic deposits along the nortwest African continental margin are highly variable in composition. For this study we have quantified the concentrations of quartz in the bulk sediment and mica in the coarse size fraction (Tables 1 and 2). A fine grained residue, consisting mainly of clay minerals (Chamley et al., 1977), is defined by subtracting the calcareous, opaline, organic carbon and quartzose fractions from the bulk sediment.

The fluxes of the three different terrigenous components are highly variable (Fig. 6). Since the core is located at a distance from any major river mouth, we assume that the dominant portion of this material was transported by wind to site of deposition. Quartz accumulation rates in their basic pattern trace the accumulation rates of the bulk sediment. Three narrow and well defined maxima

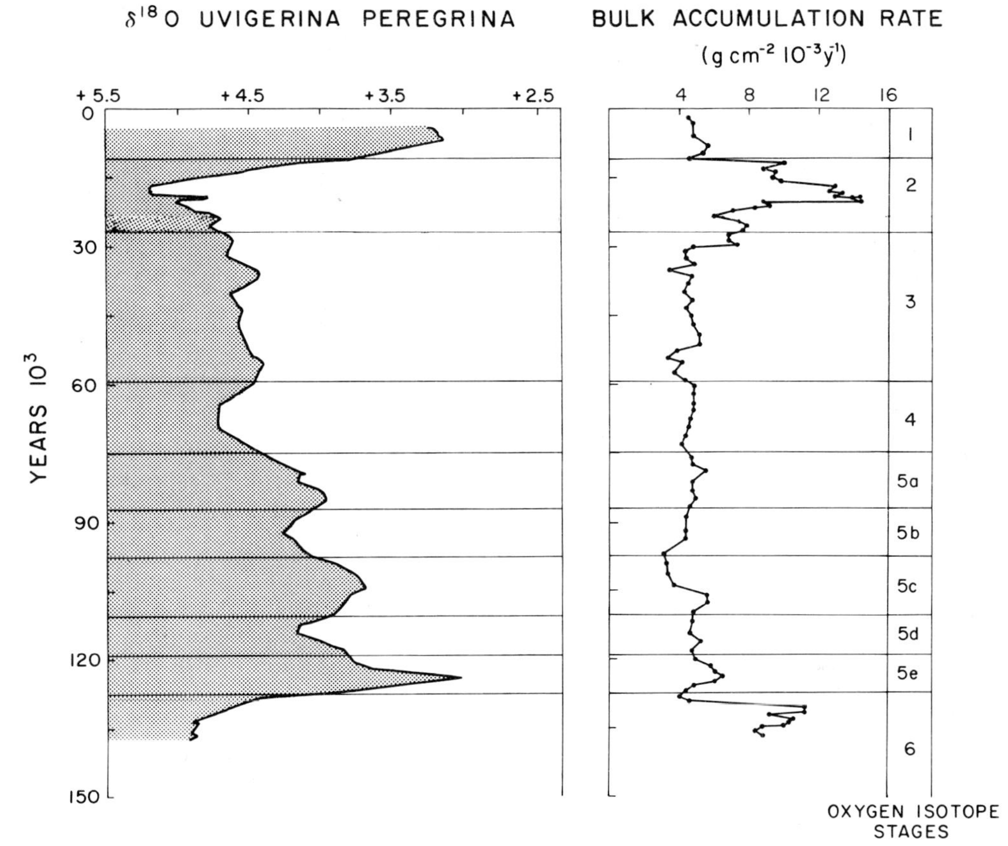

Fig. 5
Bulk accumulation rates ($g.cm^{-2}.10^{-3}.yr^{-1}$) of core M 12392-1 sediments during the past 140.000 yrs. Oxygen isotope ratios are given for correlation; for maximal and minimal errors of estimates, see text

during which quartz accumulation rates increased 2-3 fold (> 1.5 g cm^{-2} $10^{-3}yr^{-1}$) coincide with intervals of the oxygen isotope stages 6 and 2 and to a lesser degree with stage 4. During these times the source area for quartz, the Saharan Desert, is known to have been considerably drier and the atmospheric circulation to have been more vigorous than today (Sarnthein, 1978, Koopmann, 1979, CLIMAP, 1976, Parkin, 1974). There exists a remarkable correlation of the variations of the quartz accumulation rates to the oxygen isotope values, with time, which suggest an intimate relationship of the climatic changes affecting the wind regime over the Saharan Desert and the rates at which water is transferred from the ocean to the ice shields in the polar regions. The calculated accumulation rate of coarse mica grains (> 0.15 mm) is based on much less reliable and representative data (Thiede, 1977) than that of quartz or the clay residue. This might be responsible in part for the scatter of these data. However, Figure 6 also reveals clearly a correlation to the oxygen isotope curve because virtually no mica in this size fraction was transferred to the site of deposition during isotope stages 5 and 1, whereas the flux of coarse mica grains was largely confined to oxygen isotope stages 2-4 and 6. However, also during this time span, intervals occurred of a few thousand years in duration, when almost no mica reached the site of this core. Such was the case approximately 25,-30,000 yrs. B.P., 40,-45,000 yrs. B.P. and 50,-55,000 yrs. B.P.

The signal of the residue accumulation rates is more complex than the mica and quartz signals. The lowest residue accumulation rates (> 1.3 g $cm^{-2}.10^{-3}.yr^{-1}$) are observed during the late isotope stage 3 and 5 a - c.

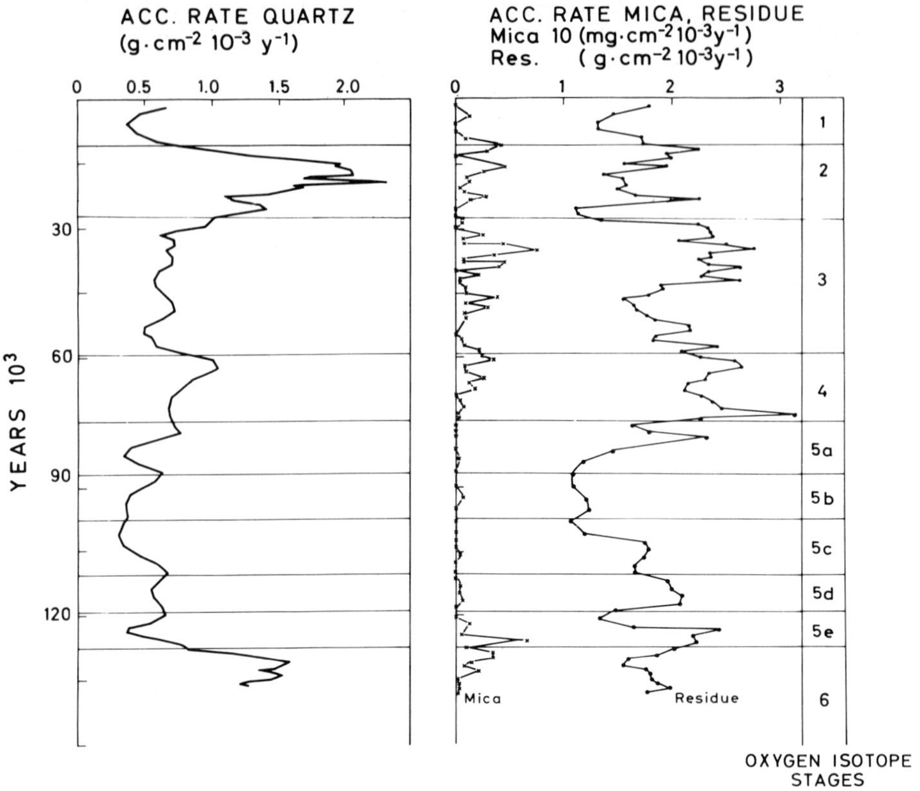

Fig. 6
Stratigraphy and accumulation rates (in g or mg.cm^{-2}.10^{-3}.yr^{-1}) of major terrigenous sediment components: Quartz and residue in bulk sediment, mica in coarse fraction > 0.15 mm

High accumulation rates of this material are confined to the middle part of stage 3 and to stage 4. Its input rates during stages 3 and 4 are considerably higher than those of stages 6 and 2, and they are henceforth not following the general trend outlined for the accumulation rates of the bulk sediment and of the quartz. The differences of the mica and residue accumulation rates from those of quartz suggest that these materials probably reached the continental margin from different source regions and by another mode of transport.

3.3 Fluxes of Biogenic Components

The fluxes of biogenic sediment components are typified by the accumulation rates of calcareous and opaline matter as well as organic carbon (Fig. 7). Although the absolute values differ considerably from one component to the other, they trace the trend of accumulation rates of the bulk sediment. The fluxes of all three components during isotope stages 2 and 6 are 2-5 fold higher than those of the intermediate time spans when the variability of the accumulation rates was very small. The sharp decrease of the accumulation rates in the course of the deglaciations after the penultimate and the ultimate glacial maxima are particularly noteworthy. Whereas the carbonate accumulation rates remained rather constant during isotope stages 5, 4, 3 and 1 (fluctuating between 3 - 5 g.cm^{-2}.10^{-3}.yr^{-1}9 the values of the opaline matter and of the organic carbon have minimal values during stages 1 and 5 but they grew from stage 4 through stage 3 at an increasing pace until they reached their peak values during stage 2.

The proportion of sand sized material is modest throughout this core (Thiede, 1977), the dominant portion of the calcareous material consist therefore of coccoliths, rather than of foraminifers. Diatoms supersede the abundance of radiola-

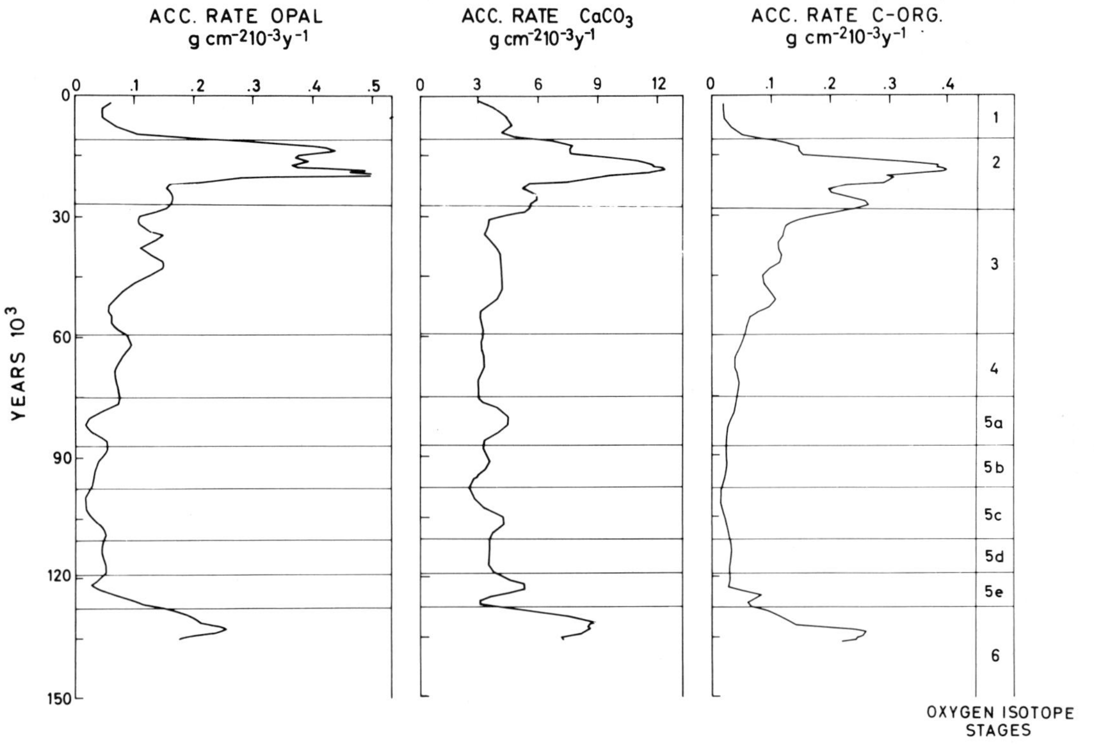

Fig. 7
Stratigraphy and accumulation rates of major biogenic sediment components (in $g.cm^{-2}.10^{-3}.yr^{-1}$): bulk calcium carbonate, opal and organic carbon

rians approximately by a factor of 10 and those of sponge spicules by a factor of 2-5 (Diester-Haass et al., 1973). It can therefore be expected that the dominant portion of the signal seen in the accumulation rates of the calcareous and opaline fractions of the bulk sediment as well as the organic carbon (Müller & Suess, 1979) reflects a message of the production and preservation of different types of marine algal material. Since algae represent basically the primary producers within the marine food chains they should be good indicators of the intensity of the fertility and primary productivity of the surface waters within the photic zone. If this assumption is reasonable, then the coincidence of large amounts of airborne quartz, which suggest aridity in its source regions and a vigorous atmospheric circulation, with phases of high fluxes of marine algal matter to the sea floor, should mark periods of high surface fertility due to increased coastal upwelling. The low sea level stands during the Glacial maxima resulted in the zone of coastal upwelling being moved parallel to the coast-line westward to an area over the present continental slope. Although we have no direct indicators for a large contribution of displaced material in this core, the general picture developed from the presence of the eolian-sand turbidites in the adjacent deep-sea basins (Sarnthein & Diester-Haass, 1977) supports the importance of the sea level fluctuations for the westward movement of the coastal upwelling zone.

To support the interpretation developed above we have attempted to estimate the fluxes of secondary producers such as protozoans (benthic and plaktonic foraminifers as well as radiolarians) and we have measured a number of important chemical components in the sediment (Tables 1 and 2; Fig. 8). The benthic calcareous and planktonic foraminiferal as well as radiolarian data yield only a rough impression of the accumulation rates of these components because the size fraction > 0.15 mm represents a very small proportion of the bulk sediment.

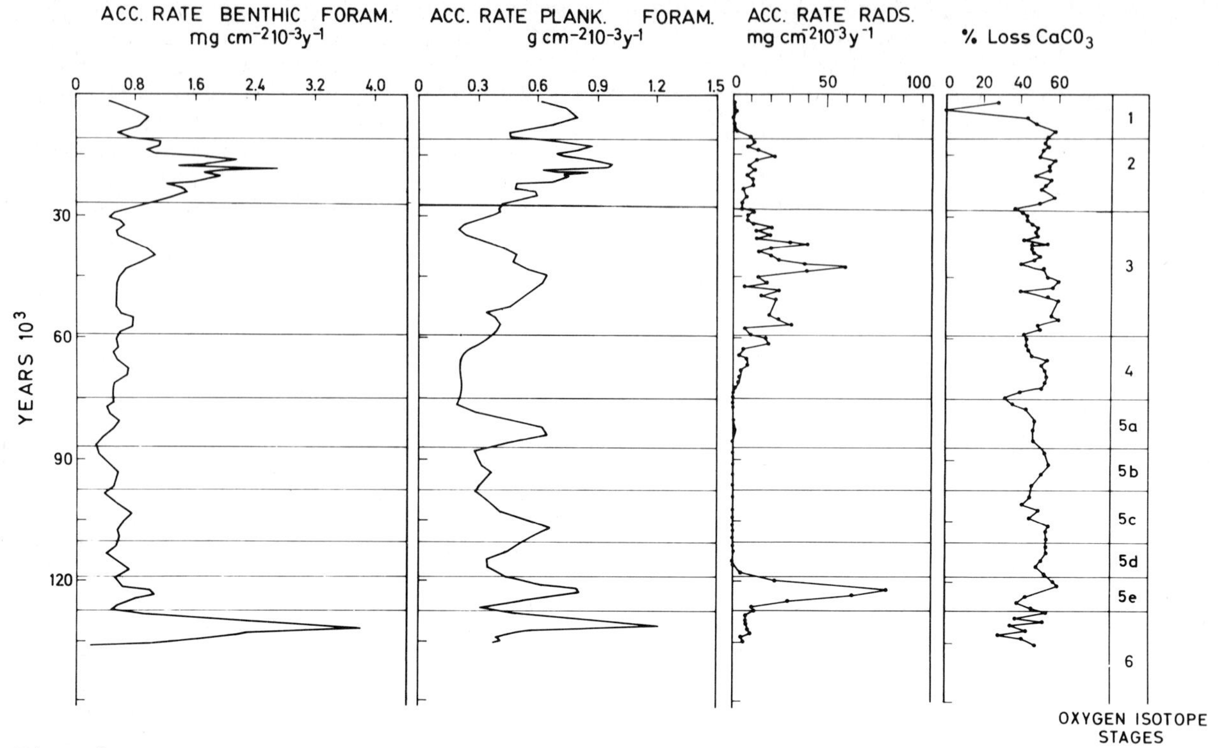

Fig. 8
Stratigraphy and accumulation rates (in $mg.cm^{-2}.10^{-3}.yr^{-1}$) of planktonic and benthic biogenic coarse fraction (> 0.15 mm) components: planktonic and benthic foraminiferal shells. The loss of $CaCO_3$ has been estimated based on planktonic foraminiferal census data (cf. Methods)

In addition, the effects of the temporal changes of composition and the shell size of assemblages in the sand fraction remain in question. It is clear that the trends of the accumulation rates of the benthic foraminifers (Fig. 8) trace those of surface water productivity. Peaks coincide with the high accumulation rates of the bulk sediment, of quartz, opal and organic carbon. From studies of modern benthic foraminiferal communities (Sarnthein et al., this Vol., Haake, 1980, Lutze, 1980) we know that the size of their standing stock is closely correlated to organic carbon concentrations in the surface sediments. The increase of benthic foraminiferal accumulation rates during isotope stages 2 and 6 reflect therefore the immediate response of the bathyal benthos to the high rate of supply of food during times of high surface water productivity.

The signal obtained from accumulation rates of plankton, foraminifers and radiolarians in the size fraction > 0.15 mm is more complicated than that of the benthos (Fig. 8) and it does not seem to trace the accumulation pattern of the primary producers. Planktonic foraminiferal fluxes were high during isotope stage 6 and 2, but the peaks are less pronounced than for the primary producers. During intervals corresponding to isotope stages 5e, 5c and 5a as well as approximately 45,000 yrs. ago and during the Holocene the planktonic foraminiferal faunas yielded higher fluxes of shell material than during the adjacent intervals. This reponse cannot be seen in the accumulation rates of some of the other biogenic components (Fig. 7), but it coincides rather with minima in the quartz accumulation rates suggesting phases of sluggish atmospheric circulation and consequently low primary productivity above this zone of the northwest African continental margin. There also seems to be a faint negative correlation between accumulation rates of the residue and of planktonic constituents, supporting the idea of reduced terrigenous input during those periods of sluggish atmospheric circulation. Radiolarian fluxes (Fig. 8) were highest 127,-121,000 yrs. B.P. Virtually no radiolarians were preserved during the time spans 120,-75,000 yrs. B.P. and from 10,000 yrs. ago to the present time. These two barren intervals bracket a phase of continuous modest to high accumulation rates of radiolarian skeletons. Maximal accumulation rates are seen approximately 40,-45,000 yrs B.P. in the middle of stage 3 and 15,000 yrs. B.P. just after the oxygen isotope maximum marking the last glacial extreme. This pattern resembles in part that of the planktonic foraminifers, but it also suggests that planktonic secondary producers respond in an individual mode to late Quaternary changes of the surface water circulation along the northwest African continental margin, whereas benthos is closely related to primary producers and processes in the photic zone.

4. Preservation of Biogenic Components

4.1 Elemental Fractionation of the Organic Matter

The dominant portion of the organic matter produced in the photic zone is oxidized in the water column. Therefore only a small fraction reaches the ocean floor. Recent sediment trap studies (Suess, 1980), suggest that about 2% of the original organic carbon production reaches the sea floor at the site of core 12392-1. At present the average annual primary production rate is about 75 g C $m^{-1} \cdot y^{-1}$ (Schemainda et al., 1975). The rate at which this material is preserved, seems to be related to the speed of sediment accumulation (Heath et al., 1977, Müller & Suess, 1979) at the individual depositional sites. In Fig. 9a are shown the C-org/N-total and the C-org/P-org weight ratios and their changes with depth. The patterns suggest a trend of accumulation of organic matter with depth and age that is increasinly N- and P-depleted. Such trends probably reflect long-term diagenetic changes of the organic matter after deposition, but it is superimposed upon important fluctuations which may be related to the primary signal of the bulk sediment input; i.e. to fluctuations of calcareous, opaline, organic carbon constituents and of benthic foraminifers (Figs. 5-7).

The source and/or reservoirs for the organic matter as defined by Suess & Müller (1980) may here be thought of as detritus and benthic biomass. Organic geochemical data of the composition of the kerogen (Debyser et al., 1979), stable carbon isotope ratios of total sedimentary organic matter (Dr. Erlenkeuser, University of Kiel, pers. comm.), C/N and C/P ratios (Fig. 9a), and the accumulation rate pattern of the primary producers in core M 12392-1 (Fig. 7) suggest that the dominant share of the organic material is detrital and pelagic in origin rather than terrigenous.

Upon descent through the water column biogenous detrital organic matter undergoes strong elemental fractionation by preferential removal of nitrogen and phosphorus containing organic compounds (e.g. Cobler & Dymond, 1980, Knauer et al., 1979, Smetacek, 1978). At the water-sediment interface, a portion of the detritus is converted into biomass by benthic organisms, which concentrate to carbon. These two processes are reflected in the elemental composition of sedimentary organic matter (Suess & Müller, 1980). Hence, C/N and C/P ratios of core M 12392-1 should have been *a priori* higher during isotope stages 2,3 and 6 since during these times the bulk sedimentation rates were high and favored the burial of detrital organic matter rather than its conversion to benthic biomass. The reverse, however, is the case during times of slow bulk sediment accumulation. In Fig. 9a we have plotted the C/N and C/P weight ratios against depth to investigate this relationship.

Both ratios actually show the predicted differentiation in that stage 1 and 5 sediments have lower C/N and C/P ratios than deposits from stages 2, 3 and 6. The true organic matter C/N ratios are higher by about 1.2 C/N units in interglacial sediments and by about 0.5 C/N units in glacial sediments than the ratios given in Fig. 9a, since the latter are based on total nitrogen determinations. They thus include some inorganic nitrogen bound to clay minerals in the

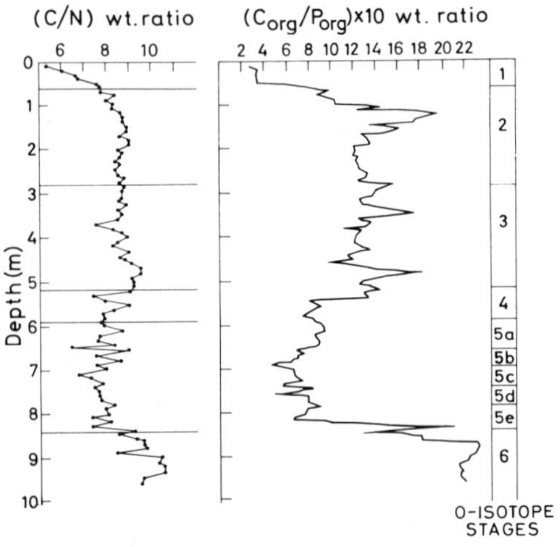

Fig. 9a
Depth distribution of C-org/N and C org/P-org weight ratios in core M 12392-1

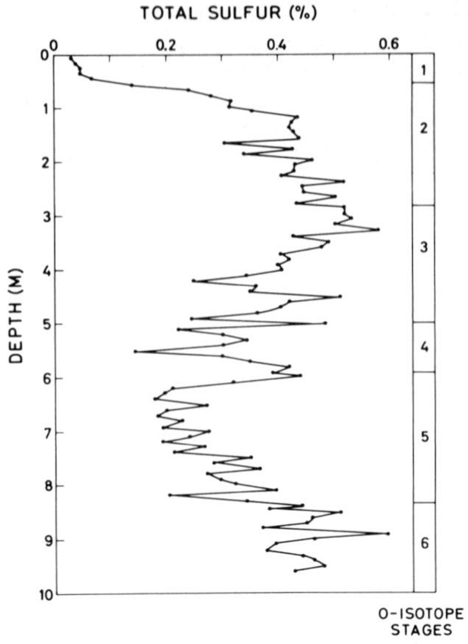

Fig. 9b
Depth distribution of total sulfur in core M 12392-1

form of exchangeable and fixed ammonium. Inorganic nitrogen concentrations fluctuate around 100 ppm in this core (Müller, 1975) and thus account for 13 - 23% of the total nitrogen in interglacial sediments and for 3 - 10% in glacial deposits. The differences of organic matter C/N ratios between interglacial and glacial sections are therefore smaller - - although still measureable - than the differences apparent from Fig.9a.

Diagenesis has to some degree wiped out the original signals as indicated by the general increase of the ratios with depth and age if interglacial and glacial core sections are compared separately. This is the result of preferential removal of nitrogen- and phosphorus-bearing organic compounds. This is indenpendently verified by amino acid analyses (Müller, unpubl. data) which show that the sum of amino acids, if normalized to total sedimentary organic carbon, decreases by about 55% during the past 140.000 yrs. in this core.

The principal decomposition reaction of organic matter in this core is by bacterial oxidation via sulfate reduction. This has been verified by interstitial water studies (Hartmann et al., 1976) and is also documented in the sediments by relatively high sulfur contents (Table 2, Fig. 9b). According to Kaplan et al. (1964) and Hartmann et al. (1976), sedimentary organic matter has a C/S weight ratio of < 0.03. The highest possible organic sulfur contribution is thus 0.02% in interglacial and 0.08% in glacial sediments. The majority of sulfur can therefore be traced back to interstitial sulfide formation from bacterial sulfate reduction and subsequent precipitation of metal sulfides or elemental sulfur.

The sulfur distribution with depth (Fig. 9b) closely mirrors the general distribution patterns of organic matter, i.e. the highest contents occur in stages 2, 3 and 6 and the lowest in stages 1 and 5. This was to be expected since higher bulk sedimentation rates and higher productivity during glacial stages resulted in higher organic carbon accumulation rates (cf. Fig. 7) creating strongly anoxic conditions in the interstitial water close to the sediment surface at these times. The present depositional regime at site M 12392-1 is different, so that oxic and suboxic diagenesis, as defined by Froelich et al. (1979), is likely to predominate in the upper few decimetres of Holocene age. Stage 5 sediments probably formed under less reducing conditions. Their present sulfur content is the result of anoxic diagenesis, once the sediments were buried deep enough to leave the zone of denitrification.

The sedimentary sulfur contents allow an estimate of the organic matter amounts remineralized during anoxic diagenesis, applying the principle model for anoxic organic matter decomposition originally proposed by Richards (1965):

$$(CH_2O)_{106}(NH_3)_{16}(H_3PO_4) + 53\ SO_4^{2-}$$
$$= 106\ CO_2 + 106\ H_2O + 16\ NH_3 + H_3PO_4 + 53\ S^{2-}$$

Hence, the formation of 0.5% sulfur by sulfate reduction which is approximately the maximum concentration in this core suggests that at least 0.37% organic carbon, 0.066% organic nitrogen, and 0.009% organic phosphorus have been remineralized by the above process. For organic carbon, this amount is equivalent to 15 - 20% of the organic matter still present in the organically-rich sections of stages 2, 3 and 6. Diagenesis was thus not strong enough to obliterate the primary organic input signals.

4.2 Intensity of Remineralization

Sediment particles, especially biogenic ones which accumulate on the ocean floor are known to be subject to remineralisation processes (Suess, 1976) which often change the composition of the grain assemblages and the appearance of the single grains. These alteration processes are intense at the sediment water interface and they affect in particular the components consisting of organic carbon, calcareous, opaline and phosphatic particles (Berger, 1971, Berner, 1979, Schrader, 1971, Suess, 1981). As shown for the organic matter in the preceeding section, it is often difficult - if not impossible - to interprete accumulation rates of these sediment components and to separate the primary signal of their production from the secondary ones of syngenetic and diagenetic alterations.

Although frequently observed in other cores, the planktonic foraminiferal shell assemblages of core M 12392-1 were generally well preserved and not significantly fragmented. We have therefore attempted to estimate the loss of calcium carbonate based on the planktonic foraminiferal census (cf. Figs. 8 and 10). This approach is not without problems, since these estimates are highly sensitive to changes of the primary input which certainly happened frequently in this part of the northeastern Atlantic boundary current regime (Berger, 1971, Thiede, 1977). The estimates based on the planktonic foraminiferal census data (Fig. 8) suggest that, except during a short interval in the Holocene, considerable amounts of $CaCO_3$ have been dissolved throughout the past 130,-140,000 yrs. when considering the accumulation of quartz, it is apparent from Fig. 10 that the carbonate accumulation rates are at their minima during the cold intervals of stage 5. Estimates for the dissolved carbonate fraction, on the other hand, reach maximum values during these intervals. In oxygen isotope 5 several events of this type of "dissolution cycles" follow each other. Their synchronous appearance with high accumulation rates of opal supports loss due to dissolution as a result of increased aggressiveness of bottom water from oxidative decomposition of organic matter. Accordingly, the carbonate maxima recorded during interglacial stage 5 might not indicate periods of increased $CaCO_3$ production but rather

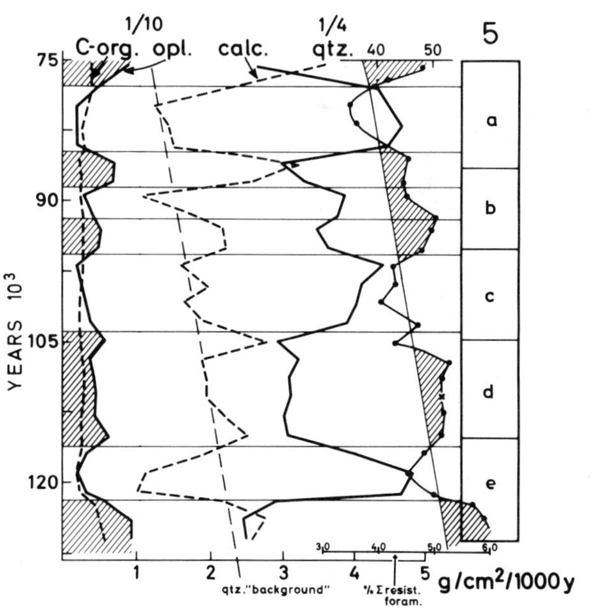

Fig. 10
Distribution of accumulation rates of the bulk sediment, opal, organic carbon, quartz and calcareous material and the estimated loss of $CaCO_3$ due to dissolution during oxygen isotope stage 5

a
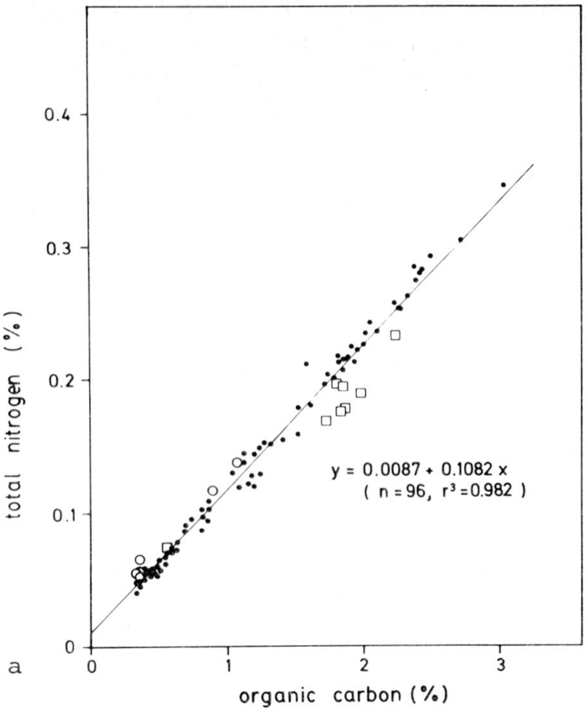

b
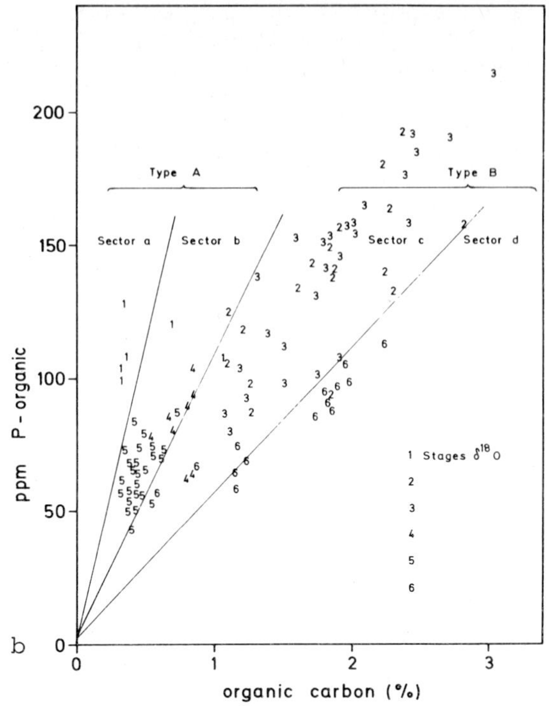

c
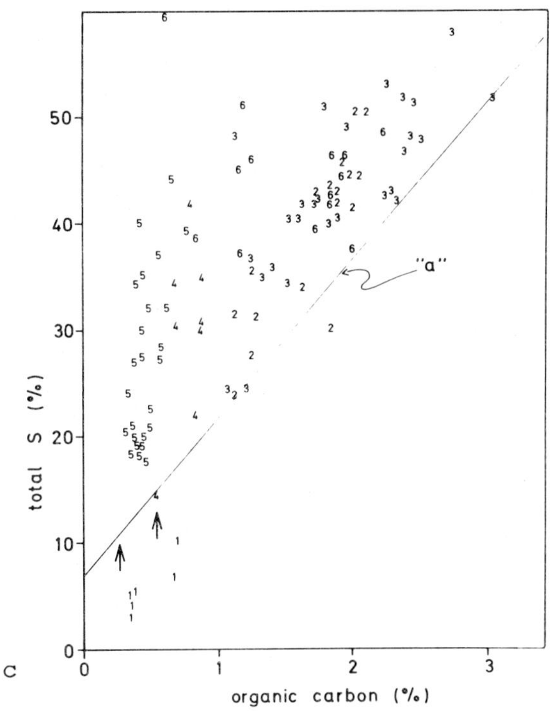

Fig. 11
Scatter plots of c-org vs. P-org (b), C-org vs. N-total (a) and C-org vs. S-total (c) for all samples of core M 12392-1. In Fig. 11a the open circles indicate samples from oxygen isotope stage 1, the open squares from stage 6, the black dots from stages 2, 3, 4 and 5 (undifferentiated); in Fig. 12b and c the data-points are grouped into isotope stages 1-6

periods of less intensive dissolution (Suess et al., 1978). Lack of appropriate data prevented a similar comparative study for the opaline particles.

Fig. 11 shows scatter plots of all the C-org and N as well as the C-org and P-org concentrations. The trend towards more N- and P-depleted organic matter when comparing data between oxygen isotope stages 1 and 5 on the one hand and 2, 3 and 6 on the other, reflects longterm diagenetic processes of preferential mineralization of N-org and P-org over C-org regardless of the source material. The source, as discussed earlier, is controlled by different proportions of benthic biomass carbon to detrital carbon and is also responsible for characteristic C-org/N-total and C-org/P-org signatures. To separate the character of the source from diagenetic effects we compared separately 2 sets of similar source materials through time (i.e. stages 1 and 5 = type A sedimentary organic matter with noticeable benthic biomass carbon; stages 2, 3 and 6 =

type B sedimentary organic matter with overwhelmingly detrital carbon. Their respective changes through time can be seen from the grading of type A material from sector a to b and of type B material from sector c to d of Fig. 11b. Isotope stage 4 contains a mixture of type A and B material; this is in agreement with the generally transitory nature of this stage in the glacial-interglacial cycles. A comparable resolution of source type and diagenesis on organic matter could not be achieved for the C-organic and nitrogen relationship. This is primarily due to the inaccurate organic nitrogen concentrations, because the analytical data contain forms of inorganic nitrogen as well. Nevertheless, Fig. 11a shows that the youngest sedimentary organic matter (isotope stage 1) contains considerably more nitrogen than the oldest matter (isotope stage 6) at constant organic carbon contents. Unfortunately, both stages also differ in source; i.e. stage 1 contains type A and stage 6 contains type B material; thus precluding further differentiation.

The sulfur content (Fig. 9b) of sediments from core M 12392-1, as a first approximation, is controlled by the organic carbon content. Sedimentary sulfur here consists of reduced SO_4-species, S^{2-} and S^0, which became fixed during the course of organic matter decomposition (Hartmann et al., 1976). The scatter plot (Fig. 11c) illustrates:
(1) that sediments from isotope stage 1 still undergo significant organic matter decomposition by SO_4-reduction and consequently their S-content will still increase with future burial (direction of arrows);
(2) that with increasing C-org contents also increasing amounts of sulfur have become incorporated (line "a" gives this trend);
(3) that in addition a characteristic amount of "excess" sulfur has become fixed in sediments; i.e. deviation above line "a";
(4) that this "excess" sulfur is higher in progressively older sediments (compare progressive deviation of samples from stage 2 through 6 in Fig. 11c).
This trend of ± continuous sulfur increase with age is unlike the C-org, N-total and P-org relationships because it solely depends on the amount of decomposable organic matter and not on its source. This is consistent with the almost imperceptible changes in C-org/N-total and C-org/P-org ratios with age of different types of sedimentary organic matter, in that both indicate continuing diagenesis throughout the 14,000 yrs. record which core M 12392-1 represents.

5. Response of the Depositional Environment off Northwest Africa to Late Quaternary Climatic Changes: Glacial vs. Interglacial

The core M 12392-1 penetrated sediments deposited during the past 135-140,000 yrs. This time span comprises two true interglacial, three glacial and a number of intermediate intervals, corresponding to the oxygen isotope stages 1-6. The trends of accumulation rate pattern during the past 30,000 yrs. of the bulk sediment, of its calcareous and terrigenous fractions with high fluxes during the glacial stage and a sharp decrease during the ensuing deglaciation and postglacial times are not restricted to this continental margin, but they have been observed over a wide region of the Atlantic Ocean (Grigoriev, 1976). Although probably modified by the proximity of the adjacent continent, the core M 12392-1 allows to deduce responses of the North Atlantic depositional regime to paleoclimatic changes during the past 140,000 yrs., rather than only responses to changes of a very local environment. The major difference between the glacial and interglacial stages as documented by the series of measurements obtained from this core are the high glacial and relatively low interglacial accumulation rates of the bulk sediment and most of its major components.

5.1 Leads and Lags

Changes of the depositional environment off northwest Africa as described here reflect a response to the late Quaternary climate fluctuations; this can be exemplified by a comparison of the oxygen isotope ratios measured in benthic foraminiferal shell material (Shackleton, 1977) and sea sur-

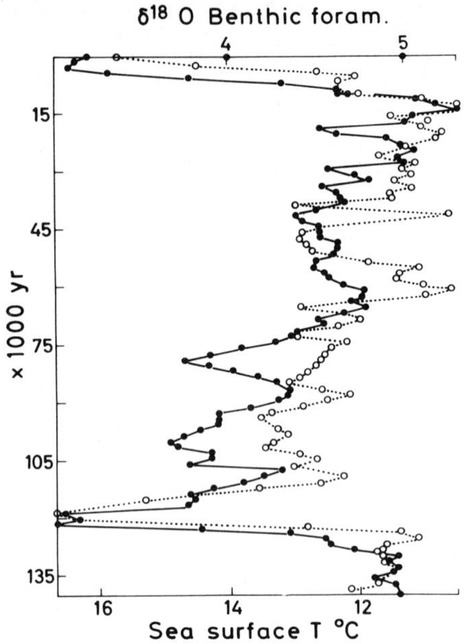

Fig. 12
Oxygen isotope ratios and estimated sea surface temperature changes in core M 12392-1 after Thiede, 1977

face temperatures (Thiede, 1977) estimated from the transfer function technique developed by Imbrie & Kipp(1971) (Fig. 12). Both signals trail each other generally, though not in their minute details. The core is situated today within the depth interval filled with North Atlantic Deep Water (cf. Sarnthein et al., this Vol.). It is therefore interesting to note that changes of the oxygen isotope ratios in benthic foraminifers preceded corresponding changes of the sea surface temperatures (Fig. 12). This relationship is best developed in the large fluctuations of both signals during the deglaciations and interglacials, whereas it is less clear during the glacial and intermediate stages when the North Atlantic Deep Water was probably replaced by a different type of glacial bottom water (Schnitker, 1980). We have tried to quantify this time difference by assuming that the lag is constant throughout the entire time span covered by this core and that the data points are equally spaced with respect to time. Both assumptions represent oversimplifications of the true data and it remains to be seen if these results can be verified for a longer time series as well. However, applying a cross correlation function of oxygen isotope values vs. sea surface temperatures revealed that the bottom water signal preceded that of the surface water by one sample point which corresponds to a time span of approximately 1000 - 2000 yrs. A lead of the benthic oxygen isotope record over the sea surface temperatures has also been observed elsewhere in the main North Atlantic Basin (Ruddiman & McIntyre, 1979) and has been used to document changes in the polar realms before the tropical-subtropical climatic zones could respond. Because of the shortness of the time series obtained from this core, we have not attempted to quantify leads and lags of other variables in the same systematic fashion.

5.2 Rates of Changes

The intervals adjacent to Terminations I and II which document the fast deglaciations after the ultimate and penultimate glacials, and the preceding glacials are those with the most important changes of the depositional regime off northwest Africa. In Fig. 13 the accumulation rates of quartz as a measure of the atmospheric circulation, of calcareous and opaline material and of organic carbon as a measure of sea surface productivity, are shown together for oxygen isotope stage 2. The accumulation rates of the three components of biogenic origin reflect probably instantaneous responses of processes in the photic zone to changes of the atmospheric circulation which is driving the sea surface currents. Small leads and lags can be observed, but they do not occur in a systematic pattern and, based on the short time series from

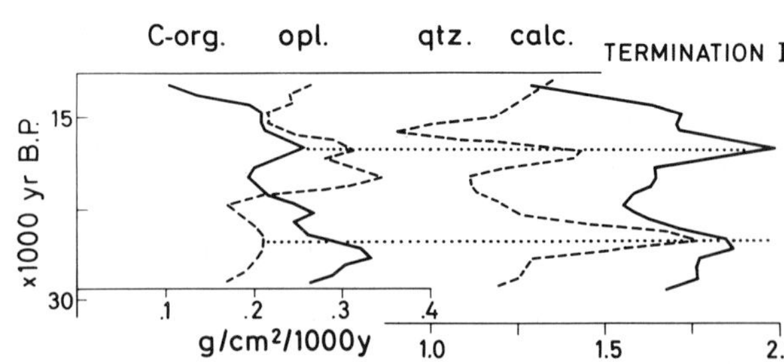

Fig. 13
Accumulation rates of organic carbon, opaline and calcareous matter and quartz during isotope stage 2

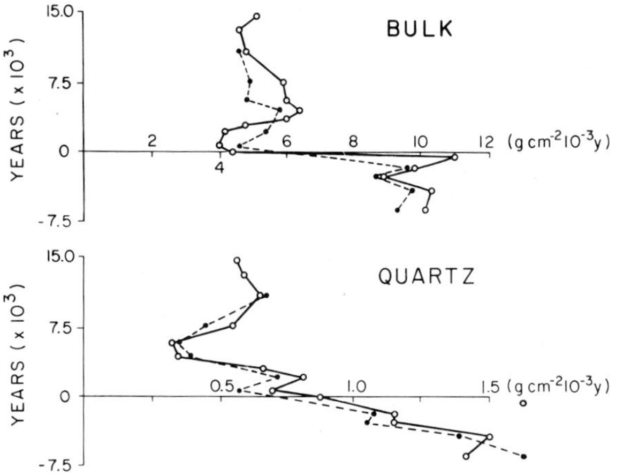

Fig. 14
Accumulation rates of the bulk sediment and of quartz during the deglaciations of Terminations I and II.

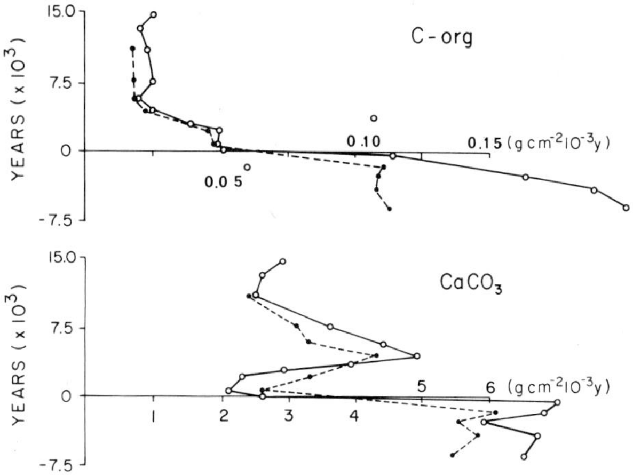

Fig. 15
Accumulation rates of calcareous matter and of organic carbon during the deglaciations of Terminations I and II. For both Figs. 14 and 15 the datum level Termination I = 11,000 yrs. B.P. and Termination II = 127,000 yrs. B.P. were used; both the quartz and organic carbon accumulation rate patterns show individual samples (three in all) which deviate considerably from the trends illustrated. These samples could perhaps represent very thin turbidite layers which would naturally not conform to the accumulation rate pattern of pelagic sedimentation.

this zone, we have to conclude that each of the events leading to high accumulation rates is documenting an individual and highly complex history of circulation changes of very short duration.

A comparison of analogous accumulation rate changes across Termination I and II, however, reveal a great deal of similarity (Figs. 14 and 15), both with respect to absolute fluxes of single components and to the temporal development of the individual signals. Accumulation rates of the bulk sediment, of quartz, of organic carbon and of calcareous material drop sharply across both terminations during intervals which lasted probably less than 1000 yrs. and when sea level rose steeply as indicated by the sharply decreasing $\delta\ ^{18}O$ values. Since the rising sea level must have resulted in an important lateral displacement of the shore-line and of the upwelling regime towards the east, we are not able to quantify and to separate local from ocean-wide causes of these changes. However, the general similarity of this core to distant Atlantic isochronous sedimentary records suggests that the important changes across Terminations I and II are representative for the entire pelagic realm of the main North Atlantic Basin. The remarkable similarity of the magnitudes of accumulation rate changes among the dominant sediment constituents in core M 12392-1 across Terminations I and II may also be used to infer trends which might develop over the next 5000 yrs. or so.

5.3 Response of the Marine Food Chain

The accumulation rate data presented here offer fragmentary information about the fluxes of biogenic particles from different levels of the marine food chain. We have argued for marine algae to dominate the signals of the organic carbon, of the bulk opaline and of the bulk carbonate accumulation rates in core M 12392-1. They represent primary producers growing in the photic zone where the dominant share of biogenic components of these sediments originate. The amount of organic matter (Rowe & Staresinic, 1979) reaching the sea floor and being available as food particles have also been shown to control the accumulation rates of fossil remains of the bathyal benthos (foraminifers). However, the planktonic secondary producers (planktonic foraminifers and radiolarians, cf. Fig. 8) deviate considerably from the basic accumulation rate pattern of bulk sediment and quartz, as well as from the biogenic components produced by algae. The reason for this discrepancy is not quite clear, but radiolarians are often brought to the continental slope off Northwest Africa by subsurface counter-currents from the south (Labracherie, 1980). Modern planktonic foraminiferal faunas are hardly found in the coastal water masses of the upwelling region itself, whereas they occur more abundantly further offshore and in the southern part of the region affected by upwelling (Thiede, 1975). This area is more frequently invaded by warm surface waters from the south. The fragmentary record of the response of the marine food chain can therefore not be expected to be fully understood through the analysis of the depositional history of only one core, but has to await a better regional coverage within the coastal upwelling regions.

6. Conclusions

The fluxes of the sediment particles to the site of core M 12392-1 represent the result of a multitude of processes controlling the input into the depositional regime and filtering the fluxes of particles through the air, through the water column and finally through the sediment water interface. In this study we have attempted to account for the accumulation rates and their changes of the bulk sediment, of its main terrigenous and biogenic components. The stratigraphic resolution available from this core allows a detailed assessment of changes of the atmospheric circulation and the response of the ocean surface circulation off Northwest Africa over the past 135-140,000 yrs. Concentrations of some of the variables measured here have been altered with time due to diagenetic changes; however, none of these alterations have been drastic enough to obscure the stratigraphy of fluctuations caused by changes of the primary input. It is obvious that sea level and atmospheric circulation which develop in the course of the glacial maxima cause the dominant portion of the changes of accumulation rates, which in general are high during glacial and low during interglacial periods.

They control also a response of the marine ecosystem, because the particle output of primary producers (Fig. 8) increased several-fold during glacial over interglacial stages. The increased availability of food on the sea floor caused in addition a dramatic rise of benthic animal standing stocks and/or turnover rates (Fig. 8). However, it remains unclear, how planktonic protozoans (foraminifers, radiolarians) and eventually higher members of the marine food chain (fish e.g.) respond to the dramatic accumulation rate increase of the primary producers during glacial stages.

The distribution of benthic foraminiferal oxygen isotope ratios and carbonate concentrations and the generally fine-grained appearance of sediments suggest that deposition at the site of this core is typical for the pelagic environment of the main North Atlantic Basin (Ruddiman & McIntyre, 1979). The speed of accumulation rate changes and the similarity of the accumulation rates close to Terminations I and II (Figs. 14 and 15) offer a vivid impression of how quickly, in the course of approximately 1000 yrs., the oceanography of the North Atlantic Basin switched from a glacial to an interglacial mode. Most of the variables discussed in this paper are related directly or indirectly to the atmospheric and the ensuing sea surface circulation; changes in the driving mechanism of both regimes appear to have affected most variables discussed here instantaneously. Only in the case of the planktonic protozoans we can see how individual members of the marine food chain with a specific position within the marine planktonic ecosystem deviate significantly from this immediate relationship.

7. Acknowledgements

The detailed set of data discussed in this paper has been produced in the laboratories of the Department of Geology, University of Aarhus (Denmark), Department of Geology,

University of Kiel (FRG), and School of Oceanography, Oregon State University (U.S.A.). The core has been collected during cruise no. 25 of R.V. "Meteor". The measured geotechnical properties have been made available by Dr. F.-C. Kögler (Department of Geology, University of Kiel, FRG). We are extending our sincere thanks to all colleagues who helped to collect the data discussed in this paper and who commented on the manuscript. This work was partially supported by Deutsche Forschungsgemeinschaft (DFG) and the U.S. Office of Naval Research grant N00014-79-C-0004, Project NR083-1026 to Oregon State University, Corvallis/Oregon-USA.

8. References

Bé, A.W.H., J.E. Damuth, L. Lott and R. Free (1976) Late Quaternary climatic record in western equatorial Atlantic sediment. Geol. Soc. Amer. Mem. 145:165-200.

Berger, W.H. (1971) Sedimentation of Planktonic foraminifera. Marine Geol. 11:325-358.

Berner, R.A. (1979) A new look at biogenous material in deep sea sediments. Ambio Spec. Rep. 6:5-10.

Broecker, W.S. and J. van Donk (1970) Insolation changes, ice volumes, and the O^{18} record in deep-sea cores. Rev. Geophys. Space Phys. 8:169-198.

Chamley, H., L. Diester-Haass and H. Lange (1977) Terrigenous material in East Atlantic sediment cores as an indicator of NW African climates. "Meteor"Forsch.-Ergebn. C26:44-59.

CLIMAP (1976) The surface of the ice-age earth. Science 191:1131-1137.

Cobler, R. and J. Dymond (1980) Sediment trap experiment on the Galapagos Spreading Center, Equatorial Pacific. Science 209:801-802.

Debyser, Y., F. Gadel, C. Leblond and J.-P. Cambon (1979) Géochimie des kérogenes dans les sédiment. Orgon 3:375-403.

Diester-Haass, L. (1976) Late Quaternary climatic variations in northwest Africa deduced from East Atlantic sediment cores. Quatern. Res. 6:299-314.

Diester-Haass, L. (1978) Sediments as indicators of upwelling. In "Upwelling ecosystems" by Boje, R. and M. Tomczak (eds.), p. 261-281, (Springer Verlag) Heidelberg.

Diester-Haass, L., J.-J. Schrader and J. Thiede (1973) Sedimentological and paleoclimatological investigations of two pelagic ooze cores off Cape Barbas, northwest Africa. "Meteor" Forsch.-Ergebn. C16:19-66.

Ellis, D.B. (1972) Holocene sediments of the Atlantic Ocean, the calcite compensation depth and concentrations of calcite, opal and quartz. Ms-thesis, School of Oceanography, Oregon State University, pp. 77.

Embley, R.W. (1976) New evidence for occurrence of debris flow deposits in the deep sea. Geology 4:371-374.

Emiliani, C. (1955) Pleistocene temperatures. J. Geol. 63:538-578.

Froelich, P.N., G.P. Klinkhammer, M.L. Bender, N.A. Luedtke, G.R. Heath, D. Cullen, P. Dauphin, D. Hammond, B. Hartman and V. Maynard (1979) Early oxidation of organic matter in pelagic sediments of the eastern Equatorial Atlantic: suboxic diagenesis. Geochim. Cosmochim. Acta 43, 1075-1090.

Geyh, M.A. (1976) ^{14}C routine dating of marine sediments. Proc. 9th Radiocarbon Conference, Los Angeles

Goldberg, E.D. (1958) Determination of opal in marine sediments. J. Marine Res. 17:178-192.

Grigoriev, D.L. (1976) Geochemistry of North Atlantic deep-sea sediments. Variations in climate and accumulation rates. Ms-thesis, University of Miami.

Haake, F.W. (1980) Benthische Foraminiferen in Oberflächen-Sedimenten und Kernen des Ostatlantiks vor Senegal/Gambia (Westafrika). "Meteor" Forsch.-Ergebn. C32:1-29.

Harmon, R.S., T.-L. Ku, R.K. Matthews and P.L. Smart (1979) Limits of U-series analysis: Phase 1 results of the Uranium-Series Intercomparison Project. Geology 7:405-409.

Harmon, R.S., L.S. Land, R.M. Mitterer, P. Garrett, H.P. Schwarcz and G.J. Larson (1980) Bermuda sea level during the last interglacial. Nature 289:481-483.

Hartmann, M., P.J. Müller, E. Suess and C.H. van der Weijden (1976) Chemistry of Late Quaternary sediments and their interstitial waters from the NW African continental margin. "Meteor" Forsch.-Ergebn. C24:1-67.

Heath, G.R., T.C. Moore and J.P. Dauphin (1977) Organic carbon in deep-sea sediments. In "The Fate of Fossil Fuel CO_2 in the Oceans" by Andersen, N.R. and A. Malahoff (eds.), Plenum Press, p. 605-635.

Imbrie, J. and N.G. Kipp (1971) A new micropaleontological method for quantitative paleoclimatology: Application to a late Pleistocene Caribbean core. In "Late Cenozoic Glacial Ages" by Turekian, K.K. (ed.), p. 71-181.

Kaplan, I.R., K.O. Emery and S.C. Rittenberg (1963) The distribution and isotope abundance of sulphur in recent sediments off Southern California. Geochim. Cosmochim. Acta 27, 297-331.

Knauer, G.A., J.H. Martin and K.W. Bruland (1979) Fluxes of particulate carbon, nitrogen, and phosphorus in the upper water column of the northeast Pacific. Deep-Sea Res. 26A, 97-108.

Koczy, F.F. (1951) Factors determining the element concentration in sediments. Geochim. Cosmochim. Acta 1:73-85.

Koopmann, B. (1979) Saharastaub in den Sedimenten des subtropisch-tropischen Nordatlantiks während der letzten 20.000 Jahre. PhD thesis, University of Kiel, pp. 107.

Labracherie, M. (1980) Les radiolaires temoins de l'évolution hydrologique depuis le dernier maximum glaciaire au large de Cap Blanc (Afrique de Nord-Ouest). Palaeogeogr., Palaeoclimat., Palaeoecol. 32:163-184.

Lutze, G.F. (1980) Depth distribution of benthic foraminifera on the continental margin off NW Africa. "Meteor" Forsch.-Ergebn. C32:31-80.

Lutze, G.F., M. Sarnthein, B. Koopmann, U. Pflaumann, H. Erlenkeuser and J. Thiede (1979) Meteor Cores 12309: Late Pleistocene reference section for interpretation of the Neogene of Site 397. In von Rad, U, Ryan, W.B.F. et al. 1979. Initial Reports of the Deep Sea Drilling Project, Volume 47, Part 1: Washington (U.S. Government Printing Office) pp. 727-739.

Mörner, N.-A. (1972) World climate during the last 130.000 yrs. 24th Intern. Geol. Congr. (Montreal) 12:72-79

Molina-Cruz, A. and P. Price (1977) Distribution of opal and quartz on the ocean floor of the subtropical southeastern Pacific. Geology 5:81-84.

Müller, P.J. (1975) Diagenese stickstoffhaltiger organischer Substanzen in oxischen und anoxischen marinen Sedimenten. "Meteor" Forsch.-Ergebn. C22:1-60.

Müller, P.J. and E. Suess (1979) Productivity, sedimentation rate, and sedimentary organic matter in the oceans. I. Organic carbon preservation. Deep-Sea Res. 26A:1347-1362.

Ninkovich, D., N.J. Shackleton, A.A. Abdel-Monem, J.D. Obradovich and G. Izett (1978) K-Ar age of the Late Pleistocene eruption of Toba, north Sumatra. Nature 276:574-577.

Parkin, D.W. (1974) Trade-winds during glacial cycles. Proc. Roy. Soc. London A337:73-100.

Pastouret, L., H. Chamley, G. Delibrias, J.C. Duplessy and J. Thiede (1978) Late Quaternary climatic changes in western tropical Africa deduced from deep-sea sedimentation off the Niger Delta. Oceanol. Acta 1:217-232.

Richards, F.A. (1965) Anoxic basins and fjords. In "Chemical Oceanography" by Riley, J.P. and G. Skirrow (eds.), vol. 1, p. 611-645.

Rowe, G.T. and N. Staresinic (1979) Sources of organic matter to the deep-sea benthos. Ambio Spec. Rep. 6:19-23.

Ruddiman, W.F. and A. McIntyre (1976) Northeast Atlantic Paleoclimatic Changes over the Past 600,000 yrs. Geol. Soc. Amer. Mem. 145:111-146.

Ruddiman, W.F. and A. McIntyre (1979) Warmth of the subpolar North Atlantic Ocean during northern hemisphere ice-sheet growth. Science 204:173-175.

Sarnthein, M. (1978) Sand deserts during glacial maximum and climatic optimum. Nature 272:43-46.

Sarnthein, M. and L. Diester-Haass (1977) Eolian-sand turbidites. J. Sediment. Petrol. 47:868-890.

Schemainda, R., D. Nehring and S. Schulz (1975) Ozeanologische Untersuchungen zum Produktionspotential der nordwest-afrikanischen Wasserauftriebsregion 1970-1973. Geodät. Geophys. Veröff. IV(16), 85 pp., (Akad. Wiss. DDR) Berlin.

Schnitker, D. (1980) Quaternary deep-sea benthic foraminifers and bottom water masses. Ann. Rev. Earth Planet. Sci. 9:343-370.

Schrader, H.-J. (1971) Ursache und Ergebnis der Auflösung von Kieselskeletten in den oberen Sedimentbereichen am Beispeil zweier Kernprofile vor Marokko und Portugal. Proc. 2nd Plankt. Conf. 2:1149-115 .

Seibold, E. (1972) Cruise 25/1971 of R.V. "Meteor": Continental margin of West-Africa - General report and preliminary results. "Meteor" Forsch.-Ergebn. C10:17-38.

Shackleton, N.J. (1969) The last interglacial in the marine and terresrial records. Proc. Roy. Soc. London B174:135-154.

Shackleton, N.J. (1977) Oxygen isotope stratigraphy of the middle Pleistocene. In "British Quaternary Studies" by F.W. Shotton (ed.), p. 1-16, (Clarendon Press) Oxford.

Shackleton, N.J. and N.D. Opdyke (1976) Oxygen-isotope and paleomagnetic Stratigraphy of Pacific core V28-239. Late pliocene to Latest pleistocene. Geol. Soc. Amer. Mem. 145, 449-464.

Shackleton, N.J. and R.K. Matthews (1977) Oxygen isotope stratigraphy of Late Pleistocene coral terraces in Barbados. Nature 268:618-620.

Smetacek, V., K. von Bröckel, B. Zeitschel and W. Zenk (1978) Sedimentation of particulate matter during a phytoplankton spring bloom in relation to the hydrographical regime. Mar. Biol. 47:211-226.

Steinen, R.O., R.S. Harrison and R.K. Matthews (1973) Eustatic low stand of sea level between 125,000 and 105,000 B.P.: evidence from the subsurface of Barbados, West-Indies. Geol. Soc. Amer. Bull. 84: 63-70.

Stuiver, M., C.J. Heusser and J.C. Vang (1978) North American glacial history extended to 75,000 yrs. ago. Science 200:16-20.

Suess, E. (1976) Porenlösungen mariner Sedimente - Ihre chemische Zusammensetzung als Ausdruck frühdiagenetischer Vorgänge. Habil.-thesis, University of Kiel. pp. 193.

Suess, E. (1980) Particulate organic carbon flux in the oceans: surface productivity and oxygen utilization. Nature 288:260-263.

Suess, E. (1981) Phosphate regeneration from sediments of the Peru continental margin by dissolution of fish debris. Geochim. Cosmochim. Acta 45:577-588.

Suess, E. and P.J. Müller (1980) Productivity, sedimentation rate and sedimentary organic matter in the oceans: II Elemental fractionation. Colloq. Intern. du C.N.R.S., No. 293:17-26, Paris (France).

Suess, E., J. Thiede and P.J. Müller (1978) Accumulation patterns of biogenous and terrigenous sediment constituents off NW Africa. 10th Intern. Congr. Sediment. (Jerusalem), Abstract. p. 652-653.

Thiede, J. (1975) Shell- and skeleton -producing plankton and nekton in the eastern North Atlantic Ocean. "Meteor" Forsch.-Ergebn. C20:33-79.

Thiede, J. (1977) Aspects of the variability of the glacial and interglacial North Atlantic eastern boundary current (last 150,000 yrs.). "Meteor" Forsch.-Ergebn. C28:1-36.

van Andel, Th.H., G.R. Heath and T.C. Moore (1975) Cenozoic history and paleoceanography of the central equatorial Pacific Ocean. Geol. Soc. Amer. Mem. 143:134 p.

Veeh, H.H. and J. Chappel (1970) Astronomical theory of climatic change: Support from New Guinea. Science 167:862-865.

Wooster, W.S., A. Bakun and D.R. Mc Lain (1976) The seasonal upwelling cycle along the eastern boundary of the North Atlantic. J. Marine Res. 34:131-141.

26 Benthic Foraminifera: Depth Distribution and Redeposition

Friedrich-Wilhelm Haake[1], William T. Coulbourn[2], and Wolfgang H. Berger[2]

ABSTRACT

This study demonstrates that benthic foraminifera can be quantitatively related to depth by an indexing method. An Apparent Depth Zone (ADZ) is computed for each major species in the taxonomic lists at-hand and the combination of those values is then used to produce a Depth ESTimate (DEST) for each sample. For W Africa, species' ADZs compare favorably in value and in sequence with published depth ranges of the same species. In four of the five study areas examined the ADZ-based sample depth predictions are systematically related to the actual sample depth. Anomalous samples are identifiable as outliers on depth-DEST plots. The method, therefore, is a promising data-editing technique that could be applied to eliminate reworked samples prior to further statistical analyses.

Introduction

The depth distribution of benthic foraminifera on the continental shelf and slope and on the abyssal sea floor has become an important tool both in paleoecologic reconstruction and in economic exploration (see Lipps et al., 1979). Essentially, this development began with the studies of Natland (1935) who assigned dominant species of the Ventura Basin to five depth zones ranging from lagoonal to 2800 m depth. All other studies from the first half of the century describe population distribution without emphasizing depth zonation (Cushman, 1910-1917; Awerinzew, 1911; Pearcy, 1914; Cushman, 1918-1931; Cushman, 1920; Cushman, 1921; Wiesner, 1931; Cushman, 1932-1934; Chapman and Parr, 1937; Schott, 1935; Chapman, 1941).

After World War II the number of studies on the depth zonation of foraminifera increased rapidly. Today, improved sampling techniques and intense oceanographic exploration generate so much information that computer-based statistical analyses are commonplace in the micropaleontological literature and are often the basis of discussion of the biogeography of planktonic and benthic foraminifera. These techniques generally involve a search for structure within the distributions at-hand and attempt to relate the structured data to some combination of external variables. For example, depth (and a combination of associated environmental measures) is traditionally chosen as a factor closely related to benthic foraminiferal distributions. In this study we assume that benthic foraminiferal assemblages are depth-related.

The literature available on depth zonations of foraminifera is extensive. In the following list the studies are grouped according to geographic region and cover the time span from 1950 to the present:

---

1 Geologisch-Paläontologisches Institut der Universität Kiel, Olshausenstraße 40/60, 2300 Kiel, FRG

2 Geological Research Division, Scripps Institution of Oceanography, La Jolla, CA 92093, USA

A. California
Bandy, 1953; Walton, 1955; Resig, 1958; Uchio, 1960a; Bandy, 1961; Phleger, 1964; Streeter, 1972.
B. Eastern Coast of the United States including the Gulf of Mexico
Phleger, 1951; Parker, 1954; Phleger, 1956; Wilcoxon, 1964; Pflum and Frerichs, 1976.
C. Western Coast of Central America
Bandy and Arnal, 1957; Smith, 1964; Golik and Phleger, 1977.
D. Western Coast of South America
Bandy and Rodolfo, 1964; Ingle et al., 1980; Coulbourn, 1980.
E. Eastern Atlantic
Caralp et al., 1970; Seiler, 1975; Zobel and Ranke, 1978; Haake, 1980; Lutze, 1980.
F. Mediterranean Sea
Parker, 1958; Chierici et al., 1962; Haake, 1977.
G. Asiatic Pacific Coasts
Polski, 1959; Saidova, 1960; Waller, 1960; Ishiwada, 1964; Chiji and Konda, 1970; Saidova, 1975, 1976; Matoba, 1976.
H. Indian Ocean
Frerichs, 1970; Subba Rao et al., 1979; Zobel, 1973; Lutze, 1974.
I. Boreal and Arctic Seas
Phleger, 1952; Green, 1960; Anderson, 1963; Vilks, 1969; Lagoe, 1977; Belanger and Streeter, 1980.
J. Antarctic Seas
Uchio, 1960b; McKnight, 1962; Bandy and Echols, 1964; Pflum, 1966; Kennett, 1968; Theyer, 1970; Echols, 1971.

The foraminifera from these studies show depth-related distributions. The actual dependency undoubtedly varies from region to region, the most commonly suggested controls being temperature, salinity, substrate, water mass, and hydrostatic pressure or some combination of these and perhaps other ignored variables. We think food supply probably plays a crucial role, that is, the fertility of overlying waters (which, of course, shows a general correlation with depth). We expect, therefore, that the depth range of species will vary between regions,and also that redeposition may locally obscure depth-related zonations (Bandy, 1953; Parker, 1958; Green, 1960; Uchio, 1960a; Andersen, 1963; Wilcoxon, 1964; Haake, 1980; Lutze, 1980). Our study explores relationships between benthic foraminiferal populations and water-depth for selected regions of the world. In the course of the analysis some samples are tagged as being unusual; we attribute their idiosyncrasy to redeposition.

Data and Method of Analysis

Among the studies available, we selected data from the following: Parker, 1958; Caralp and others, 1970; Echols, 1971; Haake, 1980; Lutze, 1980. From percentage occurrences of selected species in samples from each of these study areas, we computed three statistics that are the basis of our interpretations. The mathematical approach is not new; it has been applied in various forms in deriving dissolution rankings and solution indices, and temperature-related indices for planktonic foraminifera in deep-sea carbonate samples (Berger, 1968; Berger and Gardner, 1975; Coulbourn et al., 1980). The basic equation for paleodepth determination which we use is given in Berger (1978). In the form used here the method is based on a sequence of linear combinations of census data and sample depths as outlined in Figure 1. ADZ_a, which is calculated as shown in Step 1 (Fig. 1), may be taken as the "typical depth" for species a. In Step 2 the ADZ values derived in Step 1 for each column of the table are then applied in computing the summary values DEST for each row of the same table. Thus the method is recursive. The parameter DEST is an estimate of the depth at which a given sample was taken, based on its population of foraminifera. The result of each analysis is a comparison in the form of a regression equation relating observed and estimated depth of each sample (Step 3).

Depth Dependence of Benthic Foraminifera off W Africa (Haake Data)

We begin with the data of Haake(1980) to show the type of results our method produces. The area off Senegal and Gambia is a passive continental margin with a smooth and gently inclined

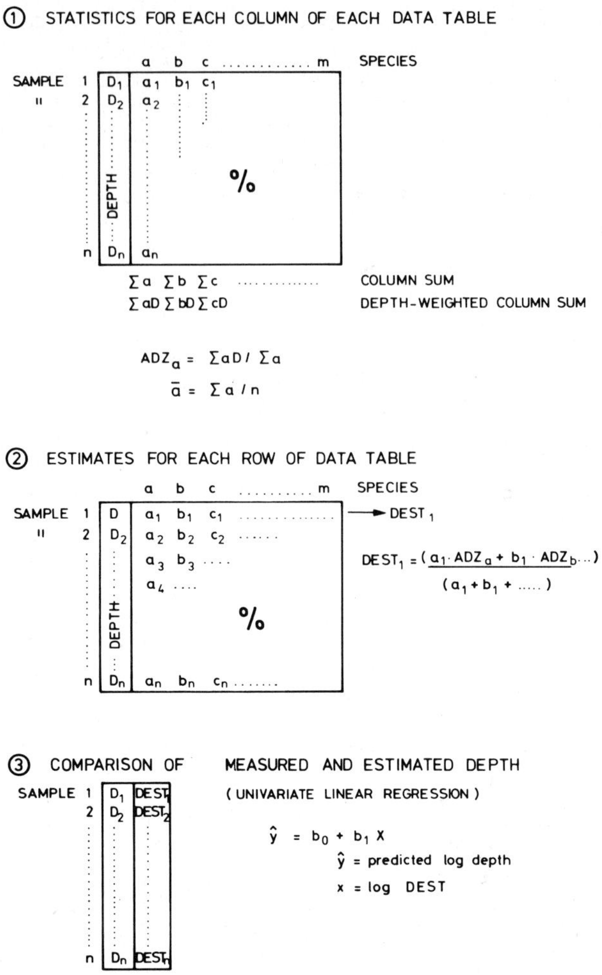

Fig. 1. Schematic diagram of the method used to compute the values ADZ, DEST, and linear regression coefficients described in the following analyses of depth distribution of benthic foraminifera

shelf. The continental slope is dissected by many small canyons and one big one, the Dakar Canyon. Bottom samples from this canyon consist of intensively bioturbated mud. Turbidite sedimentation does not seem to occur at present, but some current activity was observed by television (Seibold and Hinz, 1974). All of Haake's samples are box cores of surface sediments taken at some distance away from major bathymetric irregularities, and we assume a relatively stable substrate for the habitat of those foraminiferal assemblages. The samples span a depth range of about 2000 m.

The result of Step 1 (Fig. 1), the "Apparent Depth Zone" (ADZ) for each species, is given in Table 1. We excluded from the analyses all species with maximum percentage occurrence of less than 5%.

The indexing method works quite well. This is apparant in a comparison of the ADZ list and Haake's (1980) listing of depth ranges of the same species (Table 1 and Fig. 2). In Step 2, values for ADZ are substituted in the equations for the "Depth Estimate" (DEST), and it is the values of DEST that are plotted against sample depth in Figure 3 and in the following X-Y plots. Values for each plot are listed in Appendix 1. In all of our examples the best correspondence of DEST with the actual sample depth is for samples within the middle of the depth range. In this instance that range is roughly between 100 m and 1000 m. Here and in the following examples DEST does not accurately predict depth for samples from the extremes of the data set. This departure of derived from observed values is expected and it arises from a "no-analog" situation. There are no species that predict zero meters water depth, nor are species that are abundant at depths greater than the deepest of our samples accurately weighted by ADZ. Our predictions, therefore, tend to be biased toward the average depth of the sample set; that is, they are too great for the shallowest samples within each data set, and too little for the deepest of samples at-hand.

This tendency is corrected by fitting a regression equation to the data points (Step 3: the form of the equation is $Y = b_o + b_1X$). Y and X stand for the logarithms of water depths. It is necessary to use logarithms rather than linear depth measures because changes in assemblages become progressively less pronounced with increasing water depth. Thus, the uncertainty in assigning estima-

Fig. 2. Depth ranges of benthic foraminifera off Senegal/Gambia, W Africa (after Haake, 1980). Rank numbers are used for comparison with the sequence of ADZs assigned to the same species in Table 1

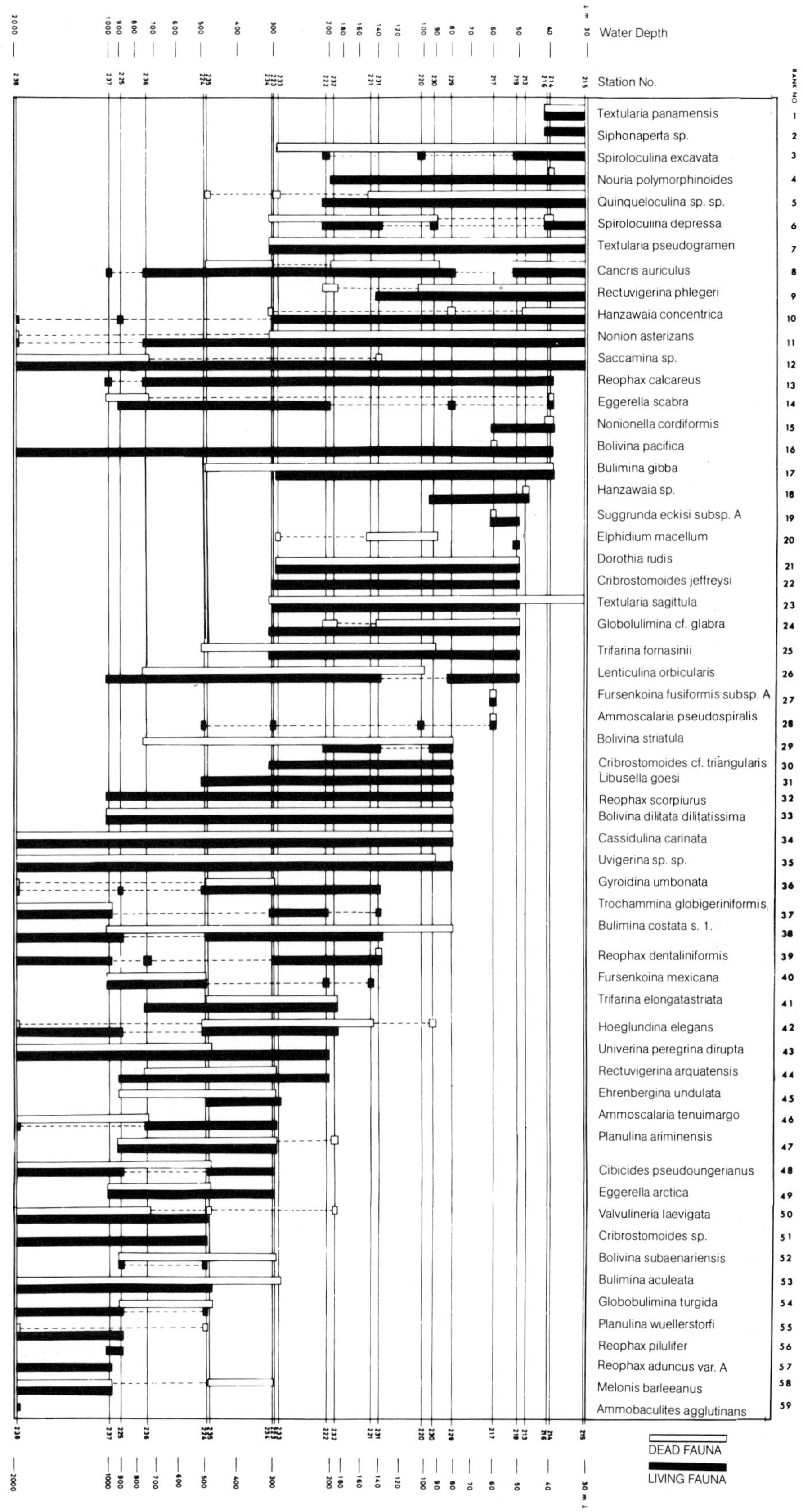
Water Depth
Station No.
RANK NO
Textularia panamensis 1
Siphonaperta sp. 2
Spiroloculina excavata 3
Nouria polymorphinoides 4
Quinqueloculina sp. sp. 5
Spiroloculina depressa 6
Textularia pseudogramen 7
Cancris auriculus 8
Rectuvigerina phlegeri 9
Hanzawaia concentrica 10
Nonion asterizans 11
Saccamina sp. 12
Reophax calcareus 13
Eggerella scabra 14
Nonionella cordiformis 15
Bolivina pacifica 16
Bulimina gibba 17
Hanzawaia sp. 18
Suggrunda eckisi subsp. A 19
Elphidium macellum 20
Dorothia rudis 21
Cribrostomoides jeffreysi 22
Textularia sagittula 23
Globolulimina cf. glabra 24
Trifarina fornasinii 25
Lenticulina orbicularis 26
Fursenkoina fusiformis subsp. A 27
Ammoscalaria pseudospiralis 28
Bolivina striatula 29
Cribrostomoides cf. triangularis 30
Libusella goesi 31
Reophax scorpiurus 32
Bolivina dilitata dilitatissima 33
Cassidulina carinata 34
Uvigerina sp. sp. 35
Gyroidina umbonata 36
Trochammina globigeriniformis 37
Bulimina costata s. 1. 38
Reophax dentaliniformis 39
Fursenkoina mexicana 40
Trifarina elongatastriata 41
Hoeglundina elegans 42
Univerina peregrina dirupta 43
Rectuvigerina arquatensis 44
Ehrenbergina undulata 45
Ammoscalaria tenuimargo 46
Planulina ariminensis 47
Cibicides pseudoungerianus 48
Eggerella arctica 49
Valvulineria laevigata 50
Cribrostomoides sp. 51
Bolivina subaenariensis 52
Bulimina aculeata 53
Globobulimina turgida 54
Planulina wuellerstorfi 55
Reophax pilulifer 56
Reophax aduncus var. A 57
Melonis barleeanus 58
Ammobaculites agglutinans 59
DEAD FAUNA
LIVING FAUNA
2000
1000
900
800
700
600
500
400
300
200
180
160
140
120
100
90
80
70
60
50
40
30 m

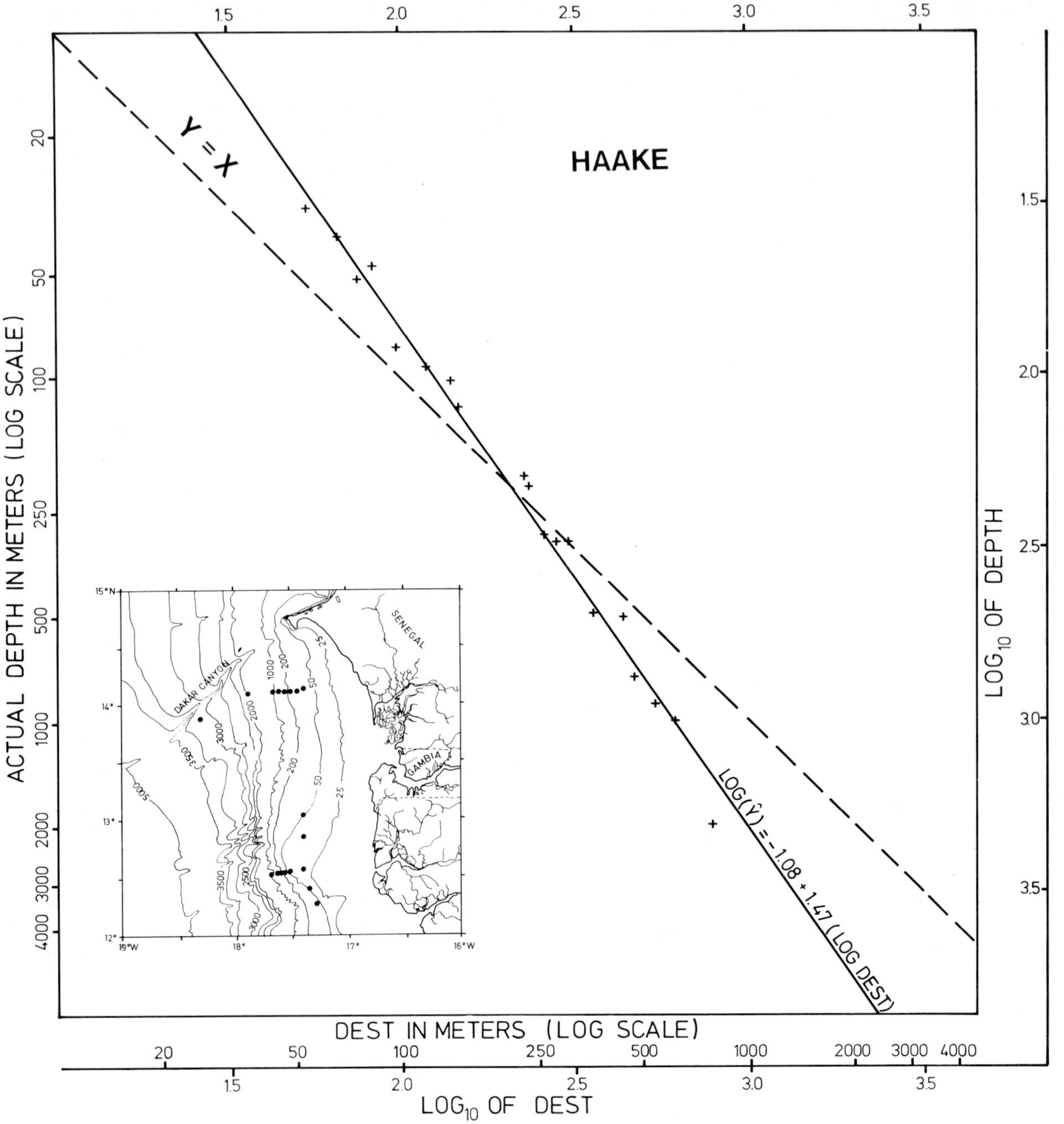

Fig. 3. Plot of depth estimate (DEST) versus depth for samples cored from W Africa (Haake, 1980)

ted sample depths increases downslope: it might be better expressed as percentage of water depth than as an amount in meters. The log-transformation is appropriate in such a set of data. The regression coefficient b_o would equal zero and the coefficient b_1 would equal one for a one-to-one correspondence between the estimated and the observed values. In our example (Fig. 3), the coefficient b_1 produces the necessary range extension in DEST which counters the tendency toward central values; and b_o defines the point of intersection between the lines x = y and the fit, that is the central value (on a log scale). In Figure 3 the fit of the regression line to the data points is excellent; the correlation coefficient is 0.993.

Other samples gathered from the W African continental margin can be assigned a DEST based on their faunal composition and on the values for ADZ already calculated. The abundance of each species in the new sample would be weighted according to its respective ADZ as listed in Table 1. Assuming that our data base is broad enough, the position of a new sample would be close to the regression line as it is plotted in Figure 3. Significant departures would indicate an unusual faunal composition with respect to depth from which the sample was taken. In the usual case, such a deviation would alert us to redeposition. The degree of departure could be used as a "redeposition index".

Depth Dependence of Foraminifera off Northwest Africa (Lutze Data)

We reexamined the depth distribution of 49 species of benthic foraminifera contained in 39 samples from the continental margin off Senegal (Table 2 and Fig. 4). Lutze (1980) collected these samples from three transects across the continental margin. The geologic setting is comparable to that off Gambia. The reports of Seibold and Hinz (1974) and of Jacobi (1976), however, present seismic reflection profiles and maps of sediment slides indicating that downslope reworking of bottom sediments is commonplace in the area of Transect C off Dakar (Fig. 4). Seismic profiles showing slumps and canyons off Cape Barbas and off Cape Bojador indicate unstable substrates for Transects A and B as well. Indeed, Lutze (1980) noted differences in the distribution of live and of empty tests of B. pseudoplicata, B. difformis, B. variablis, and B. pacifica and attributed the disparity to downslope reworking. The microfaunal population apparently does indicate some reworking but perhaps not on the scale anticipated from the data available on the geologic processes in the area. Apparently, downslope displacement of foraminifera is significant in the size-fraction less than 125 µm (Lutze et al., 1979) and of minor importance to the census counts of the fraction greater than 250 µm (Lutze, 1980). Counts from the coarse fraction produce the pattern shown in Figure 4.

Our analysis suggests a continuous succession of species depth preferences except for one possible "faunal break" registered between the species Uvigerina finisterrensis (ADZ = 851) and Sphaeroidina bulloides (ADZ = 1318). This break may be due to a combination of sampling gaps and the exclusion of rare species from the analysis.

The regression coefficient b_0 = -0.28 and b_1 = 1.11 indicate that the values of DEST need little adjustment: they are close to the actual depths. The correlation between observed and predicted values is 0.98. If downslope reworking is occurring, it is:

1. proceeding so slowly with respect to sedimentation rate that the surface samples are not recording its effects,
2. affecting all stations in all three traverses similarly (displacing all foraminifera by a certain percentage of their optimum depth),
3. not affecting the coarse fraction analyzed in Lutze's study,
4. not affecting these samples because they may have been fortuitously taken from stable areas.

Comparison between West African Study Areas

A comparison of ADZs between the Haake and Lutze study areas indicates a general agreement for species common to both areas (Table 3). In general the values for species from the Gambia study area are shallower than those recorded for species occurring to the north of Dakar. The greater than 250 µm size-fraction was used in both studies, therefore, this discrepancy probably reflects the more intense downslope reworking reported for the continental margin off NW Africa.

Depth Dependence of Foraminifera from the Eastern Mediterranean Sea (Parker Data)

Parker (1958) tabulated benthic foraminiferal population in Mediterranean surface sediments in order to produce

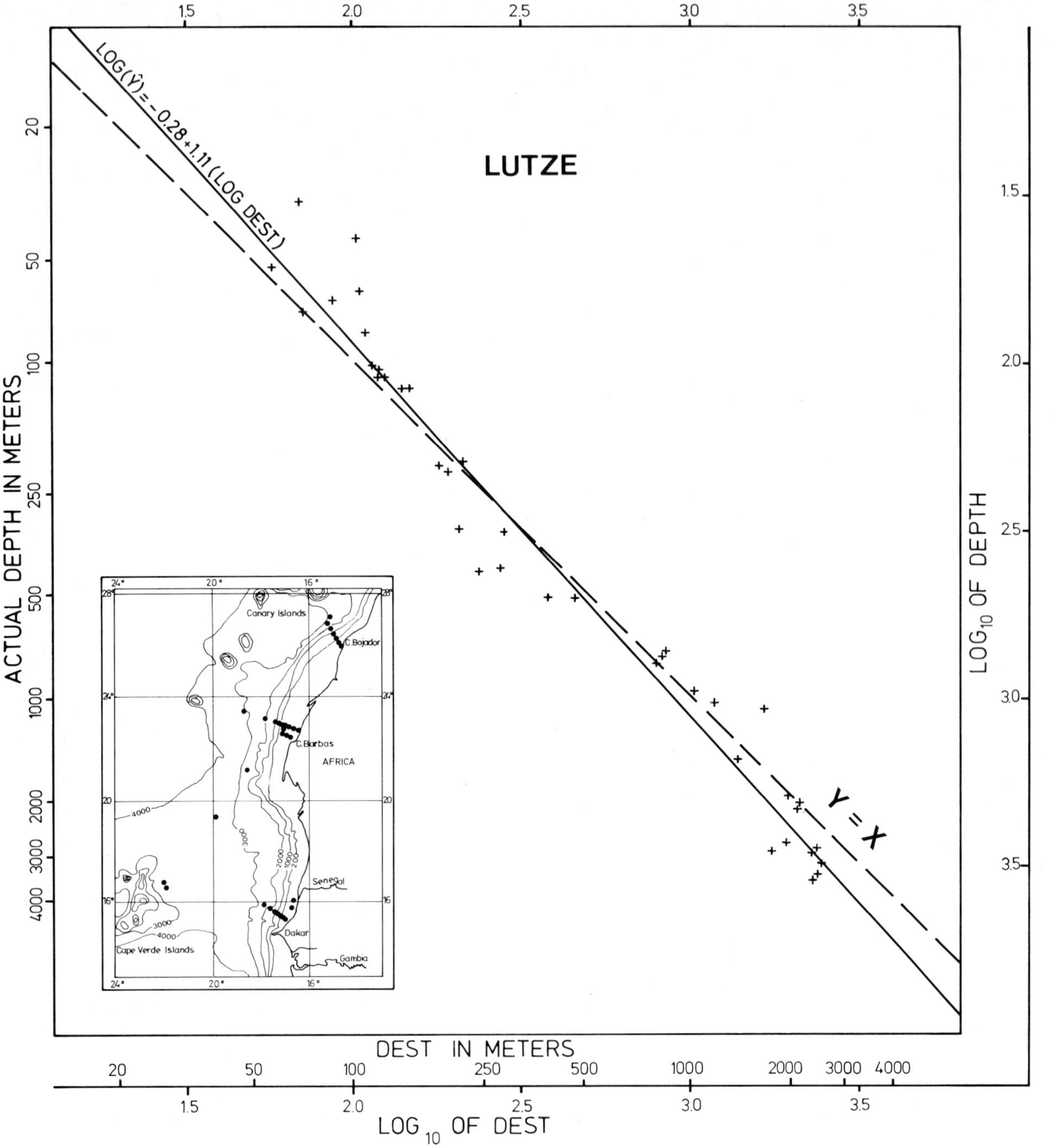

Fig. 4. Plot of depth estimate (DEST) versus depth for samples cored from W Africa (after Lutze, 1980)

criteria for interpreting fossil faunas in establishing the geologic history of the core sediments. She identified five faunal depth groups: 0 m to 25 m, 0 m to 51 m, 143 m to 205 m, 500 m to 700 m, and 1000 m to 1300 m, and noted that "many distributional changes occur between these depths but there are more at the boundaries which give the assemblages within them a relatively uniform character" (Parker, 1958, p. 227).

We used 55 of Parker's 60 samples (several samples were very poor in foraminifera). The samples ranged in water depth from about 50 m to 3500 m. Thirty species were selected from her census counts (Table 4). The samples were originally collected using a short gravity corer. They were washed over a 0.149 mm screen and we expect, therefore, that her counts include specimens smaller than were examined in the Haake (1980) and Lutze (1980) sample sets.

The ADZs we computed suggest the largest increments between _Hyalinea balthica_ (ADZ = 314) and _Bolivina marginata_ (ADZ = 457), between _Epistominella rugosa convexa_ (ADZ = 776) and _Gyroidina altiformis_ (ADZ = 1000), and between _Gyroidina_ cf. _neosoldanii_ (ADZ = 1259) and _Bolivina pseudoplicata_ (ADZ = 2188). ADZs increase gradually between these "faunal breaks" (Table 4). As mentioned before, these boundaries are somewhat arbitrary in that they are in part produced by gaps in sample depth range and are further enhanced by our selection of dominant species.

The DESTs for Parker's (1958) samples are distributed in a linear pattern on our logarithmic plot (Fig. 5). The scatter is greater than in the two previous examples and is expressed in a smaller correlation coefficient (r = 0.843). In part this must arise from the inhomogeneity of the region covered in her study. Parker identified a number of samples as containing displaced faunas; all of these were from "moderately deep water very close to the coast" (Parker, 1958, p. 232). The samples flagged by Parker are circled in Figure 5. In four out of the six cases plotted our analysis agrees with Parker's assessment. Samples marked by diamonds in Figure 5 were identified by Parker as relict on the basis of poor preservation and infilling of tests with glauconite. Surprisingly, the faunal composition of these samples is not aberrant with respect to depth. These results suggest that the depth-indexing method may register the effects of reworking, but will not necessarily be sensitive to the exposure of relict assemblages if they are in place.

Depth Distribution of Benthic Foraminifera in the Gulf of Gascogne (Caralp et al. Data)

Caralp, Lamy and Pujos (1970) studied assemblages of benthic foraminifera collected by a variety of coring methods from the Gulf of Gascogne. We analyzed the depth distribution of 25 of those species from 42 of their samples spanning a depth range of about 4000 m (Table 5 and Fig. 6). Caralp and others (1970) mapped the depth distribution of several agglutinated foraminifera and the genera (1) _Uvigerina_, (2) _Eponides_, (3) _Bulimina_, (4) _Nonion_, (5) _Cibicides_, and (6) _Planulina_. Of these groupings, species of _Cibicides_ appear to be the most closely related to depth. Considering the ensemble of groupings they recognized three depth-related assemblages:

1. Shelf: 0 m to 200 m
2. Upper Continental Slope: 200 m to 1700 m
3. Deep Water: 1700 m to 4000 m
   a) Lower Continental Slope 1700 m to 2500 m
   b) Upper Abyssal Plain 2500 m to 3000 m.

In their analysis of depth dependence Caralp and others used only 25 samples from their study area. A glance at the inset bathymetric map of the study area suggests why (Fig. 6). High relief, uneven bottom topography, and numerous submarine canyons characterize the region. The canyons are associated with secondary gullies and intensively dissected sea floor near the upper continental slope (Belderson and Kenyon, 1976).

In our analysis of the Gulf of Gascogne foraminifera we found that the largest increments in species' ADZs correspond to the faunal divisions recognized in the original study (Table 5). This is all the more remarkable in that we inadvertently included 15 samples that Caralp and others excluded from their analysis of depth dependence. Two of those samples came from submarine canyons (Fig. 6). The locations of those samples on our Depth-DEST plot are indicated by circles and diamonds (Fig. 6). Those samples are a source of much of the scatter in our DEST plot and must bias our ADZ values toward excessive depths. The trend in the DEST plot, and the scatter indicate that the DEST values are not very sensitive indicators of

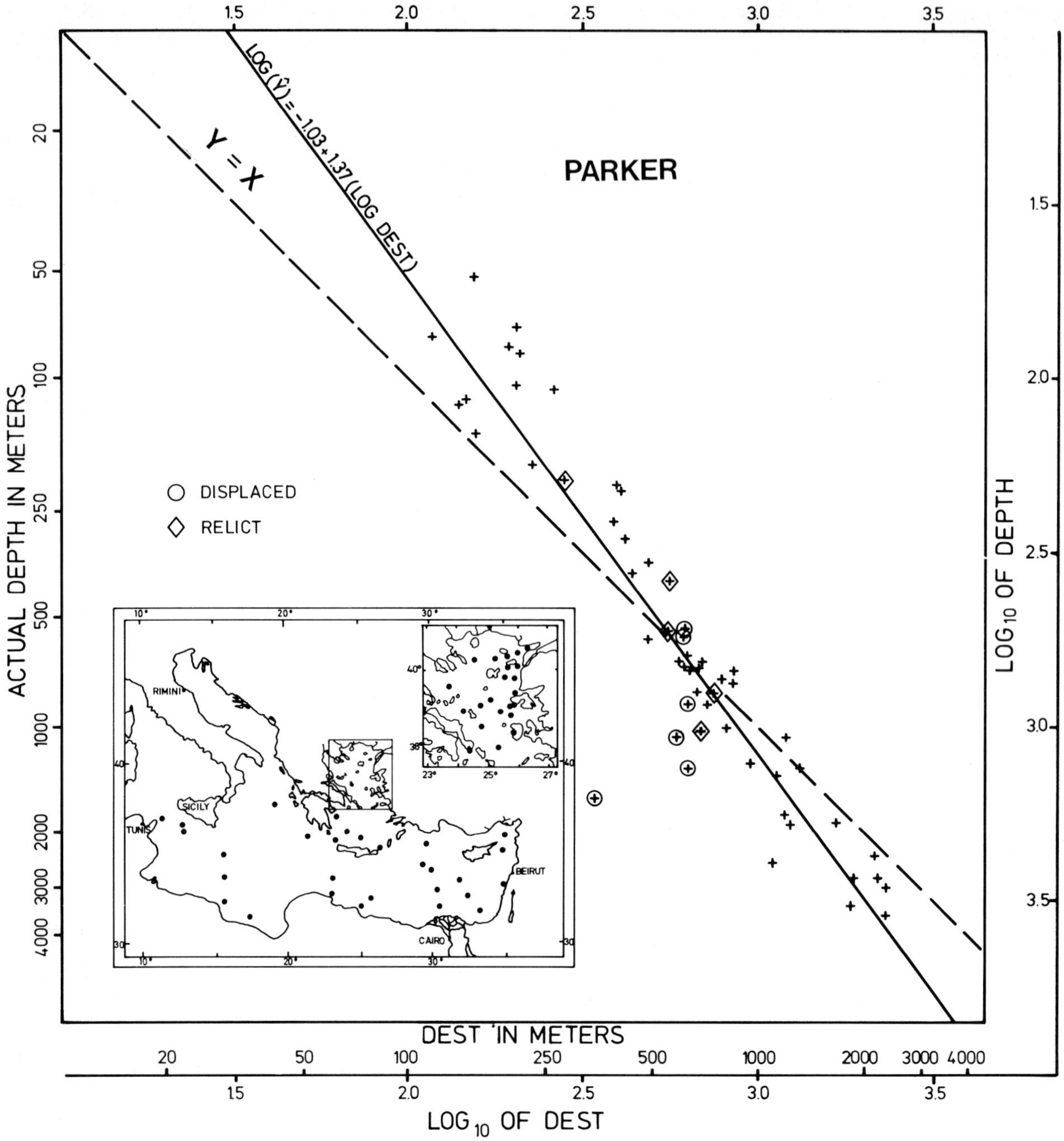

Fig. 5. Plot of depth estimate (DEST) versus depth for samples cored from the eastern Mediterranean Sea. Coordinates in circles represent reworked samples, those enclosed in diamonds represent relict samples.

actual sample depths. We dispense with a regression equation. Orientation of the trend and the scatter of values suggest that the entire sample set is dominated by mixed assemblages.

Depth Distribution of Benthic Foraminifera from the Scotia Sea (Echols Data)

Echols (1971) studied 90 samples collected by gravity corer, grab sampler,

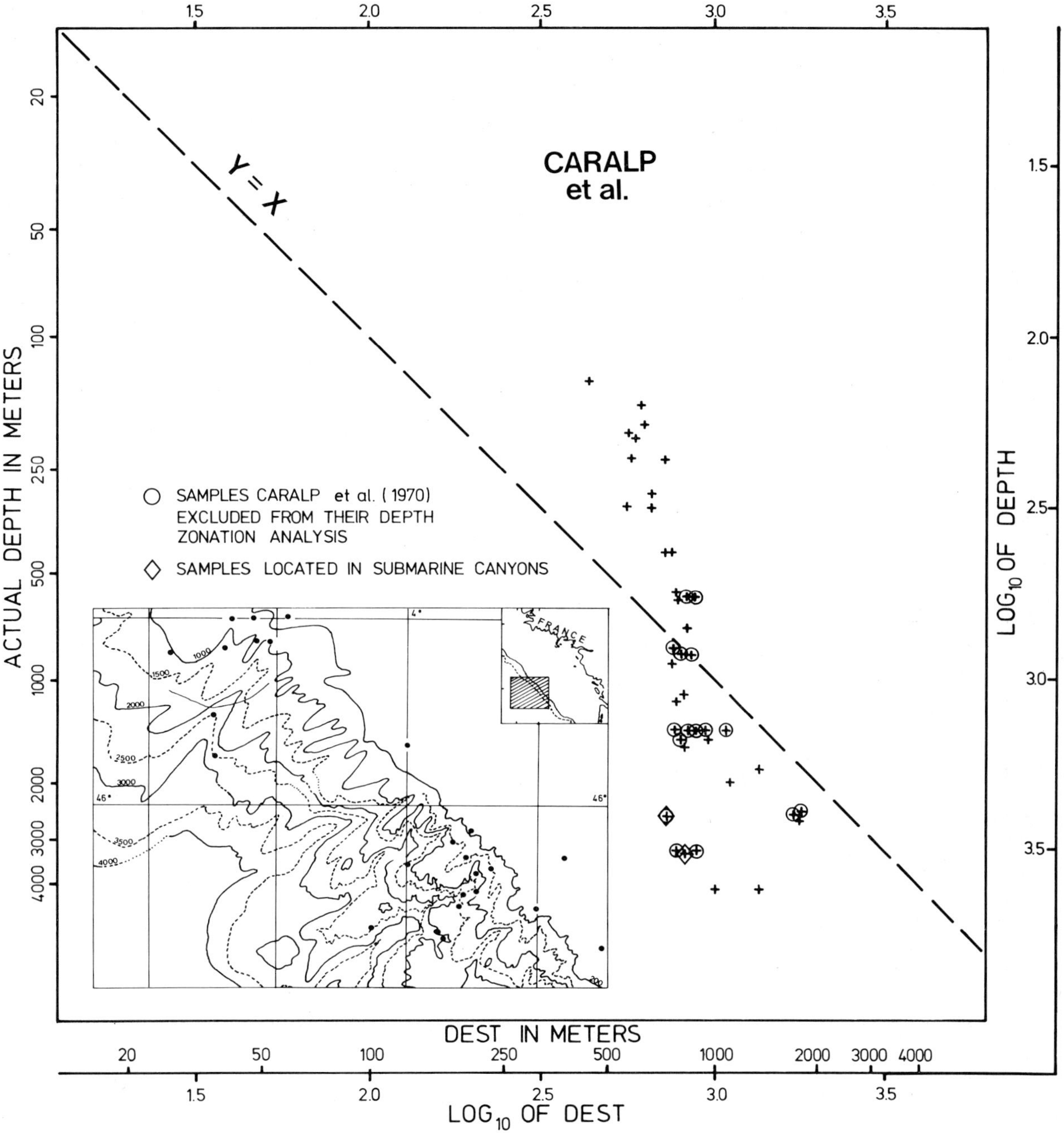

Fig. 6. Plot of depth estimate (DEST) versus depth for samples cored and dredged from the Gulf of Gascogne. Points enclosed in diamonds represent samples from submarine canyons, those enclosed in circles were not used in Caralp and others (1970) depth zonation analysis

and trawl from the southwestern Atlantic Ocean, and tabulated populations of empty foraminifera tests for 75 of those samples that were washed over a 62 micron mesh. He identified a distinctive upper bathyal fauna occupying a depth range of 330 m to 750 m. In computing ADZs for 40 species extracted from Echols' list of 200, we also recognize this shallow group (Table 6). Echols considered calcareous and agglutinated species in separate groups, whereas we combined them. Perhaps for this reason

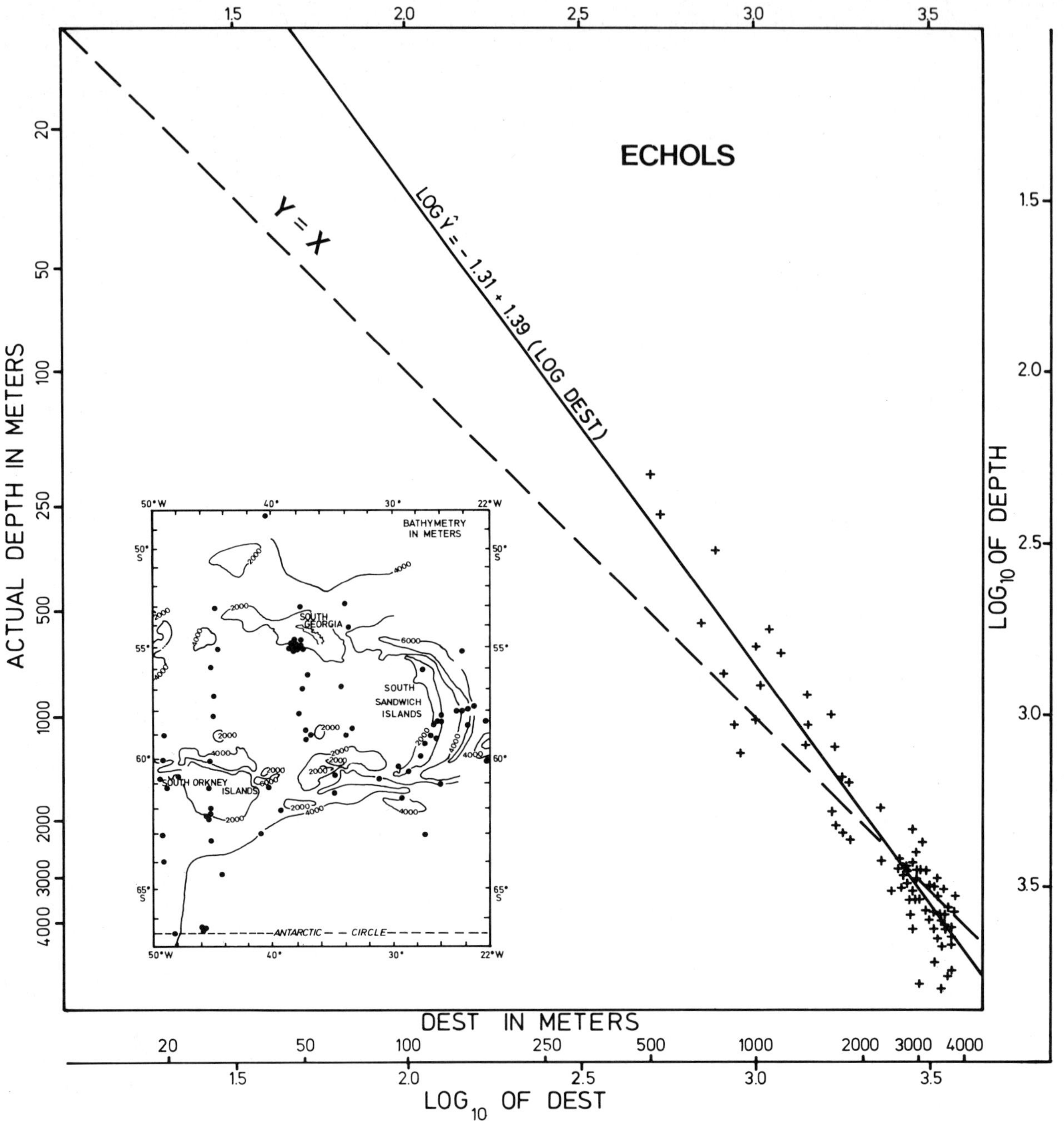

Fig. 7. Plot of depth estimate (DEST) versus depth for samples cored from the Scotia Sea (Echols, 1971)

we do not have a direct correspondence of deeper groupings. Large increments in ADZ occur above and below Pseudobolivina antarctica (ADZ = 1123), and between abyssal agglutinated species at the end of our list (Table 6).

Our DEST plot differs from the previous examples in the concentration of samples at great depth. The log-log scale compresses these into a small cluster in the lower, right corner of the graph (Fig. 7). The correlation coefficient of 0.94 indicates that DEST is a useful depth index for this sample set. Considerable redeposition may be indicated at 4000 m depth and below.

Conclusions

We have described a depth-indexing method that generates species depth zonations and sample depth estimates based on census counts of benthic foraminifera. The application of the method to published data from a variety of geologic settings shows that it provides reasonable estimates for species depth preferences, accurately predicts sample depths, and can be used to identify mixed and redeposited assemblages.

Our results indicate a strong correlation between species of benthic foraminifera and water depth, but say nothing about the environmental control behind this correlation. As a by-product of our investigation we have compiled a list of ADZs of a few key, indicator species (Table 7). Ranges for Hanzawaia spp., Planulina ariminensis, Bulimina aculeata, and Textularia sagittula are quite similar in widely separated regions of the world ocean. Finally, the apparent depth zones suggest a continuum of species habitats for the west African margin. There may be more clearly bounded depth zones in the eastern Mediterranean and Scotia Sea.

Acknowledgements

The work of F.-W. Haake was supported by the Deutsche Forschungsgemeinschaft, that of W. Coulbourn partly by National Science Foundation grants OCE 76-84029 AO2 and OCE 78-25587 (Berger, Vincent). Thanks are also due to G.F. Lutze who critically reviewed the manuscript and gave suggestions for its improvement. We are indebted to Ms. S. Mühlhan who key-punched the data cards, to Ms. B. Rippe who drafted the illustrations, and to Ms. C. Schmidtmann who typed the text.

REFERENCES

Anderson GJ (1963) Distribution patterns of recent foraminifera of the Bering Sea. Micropaleontology 9(3): 305-317

Awerinzew S (1911) Zur Foraminiferen-Fauna des Sibirischen Eismeeres. Acad Imp Sci St Petersburg Mem 8 Ser Res Sci Exped Polaire Russe 1900-1903 Sect E: Zool 2(3): 1-28

Bandy OL (1953) Ecology and Paleoecology of some California Foraminifera. Part I. The frequency distribution of recent foraminifera off California. J Paleontol 27(2): 161-182

Bandy OL (1961) Distribution of foraminifera, radiolaria and diatoms in sediments of the Golf of California. Micropaleontology 7(1): 1-26

Bandy OL, Arnal RE (1957) Distribution of recent foraminifera off West Coast of Central America. Bull Amer Ass Petrol Geol 41(9): 2037-2053

Bandy OL, Echols RJ (1964) Antarctic foraminiferal zonation. Amer Geophys Union Antarct Res Ser 1: 73-91

Bandy OL, Rodolfo KS (1964) Distribution of foraminifera and sediments, Peru-Chile Trench area. Deep Sea Res 11: 817-837

Belanger PE, Streeter SS (1980) Distribution and ecology of benthic foraminifera in the Norwegian-Green-Land Sea. Mar Micropaleontol 5: 401-428

Belderson RH, Kenyon NH (1976) Long-range sonar views of submarine Canyons. Mar Geol 22: M70-M74

Berger WH (1968) Planktonic foraminifera: Selective solution and paleoclimatic interpretation. Deep Sea Res 15: 31-43

Berger WH (1978) Paleobathymetric analysis. In: Fairbridge RW, Bourgeois J (eds) The Encyclopedia of Sedimentology. Dowdon, Hutchkinson, Ross, Stroudsburg, Penn: 524-528

Berger WH, Gardner JV (1975) On the determination of Pleistocene temperatures from planktonic foraminifera. J Foraminiferal Res 5: 102-113

Caralp M, Lamy A, Pujos M (1970) Contribution à la connaissance de la distribution bathymetrique des foraminifères dans le Golfe de Gascogne. Rev Esp Micropaleontol 2(1): 55-84

Chapman F (1941) Report on foraminiferal soundings and dredgings of the FIS "Endeavour" along the continental shelf of the south-coast of Australia. Trans Roy Soc South Australia 65(2): 145-211

Chapman F, Parr WJ (1937) Foraminifera.- Australasian Antarct Exped 1911-14. Sci Rep Ser C Zool Bot 1(2): 1-190

Chierici MA, Busi MT, Cita MB (1962) Contribution á une étude écologique des foraminifères dans la Mar Adriatique. Rev Micropaleontol 5(2): 123-142

Chiji M, Konda I (1970) Depth distribution of foraminiferal assemblages in the bottom sediments around Okushiri Island, north Japan Sea. Bull Osaka Mus Nat Hist 23: 35-70

Coulbourn WT (1980) Relationship between the distribution of foraminifera and geologic structures at the Arica Bight, South America. J Paleontol 54(4): 696-718

Coulbourn WT, Parker FL, Berger WH (1980) Faunal and solution patterns of planctonic foraminifera in surface sediments of the North Pacific. Mar Micropaleontol 5: 329-399

Cushman JA (1910-1917) A monograph of the foraminifera of the north Pacific Ocean. Bull US Nat Mus 71: pt1-6

Cushman JA (1918-1931) The foraminifera of the Atlantic Ocean. Bull US Nat Mus 104: pt 1-8

Cushman JA (1920) Foraminifera. In: Report of the Canadian Arctic Expedition 1913-18 Vol IX: Annelids, Parasitic Worms, Protozoans etc Southern Party 1913-16: 1M-12M

Cushman JA (1921) Foraminifera of the Philippine and adjacent Seas. Bull US Nat Mus 100 4: 1-608

Cushman JA (1932-1934) The foraminifera of the tropical Pacific Collection of the "Albatross", 1899-1900. Bull US Nat Mus 161: pt 1-3

Echols RJ (1971) Distribution of foraminifera in sediments of the Scotia Sea area, Antarctic waters. Antarct Oceanol I: Antarct Res Ser 15: 93-168

Frerichs WE (1970) Distribution and ecology of benthonic foraminifera in the sediments of the Andaman Sea. Contrib Cushman Found Foraminiferal Res 21(4): 123-147

Golik A, Phleger FB (1977) Benthonic foraminifera from the Gulf of Panama. J Foraminiferal Res 7: 83-99

Green KE (1960) Ecology of some arctic foraminifera. Micropaleontol 6(1): 57-78

Haake FW (1977) Living benthic foraminifera in the Adriatic Sea: influence of water depth and sediment. J Foraminiferal Res 7(1): 62-75

Haake FW (1980) Benthische Foraminiferen in Oberflächen-Sedimenten und Kernen des Ostatlantiks vor Senegal/Gambia (Westafrika). "Meteor"-Forschungsergebn Reihe C Nr 32:1-29

Ingle JC jr, Keller G, Kolpack RL (1980) Benthic foraminiferal biofacies sediments and water masses of the southern Peru-Chile Trench area, southeastern Pacific Ocean. Micropaleontol 26(2): 113-150

Ishiwada Y (1964) Benthonic foraminifera off the Pacific coast of Japan referred to biostratigraphy of the Kazusa group. Geol Surv Japan Rep 205: 1-45

Jacobi RD (1976) Sediment slides on the northwestern continental margin of Africa. Mar Geol 22: 157-173

Kennett JP (1968) Ecology and distribution of foraminifera. In: The fauna of the Ross Sea part 6. New Zealand Dep Sci Ind Res Bull 186 New Zealand Oceanogr Inst. Mem 46: 1-48

Lagoe MB (1977) Recent benthic foraminifera from the Central Arctic Ocean. J Foraminiferal Res 7(2): 106-129

Lipps JH, Berger WH, Buzas MA, Douglas RG, Ross CA (1979) Foraminiferal Ecology and Paleoecology. SEPM Short Course Nr 6. Houston. Soc Econ Paleontol Mineral 1-198

Lutze GF (1974) Benthische Foraminiferen in Oberflächen-Sedimenten des Persischen Golfes. Teil 1: Arten. "Meteor"-Forschungsergebn Reihe C 17: 1-66

Lutze GF (1980) Depth distribution of benthic foraminifera on the continental margin off NW Africa. "Meteor"-Forschungsergebn Reihe C 32: 31-80

Lutze GF, Sarnthein M, Koopmann B, Pflaumann U, Erlenkeuser H, Thiede J (1979) Meteor Cores 12309: Late Pleistocene Reference Section for Interpretation of the Neogene of Site 397. In: von Rad U, Ryan WBF (eds), Initial Repts DSDP 47, Pt 1 (US Government Printing Office) Washington DC: 727-739

Matoba Y (1976) Foraminifera from off Noshiro, Japan, and postmortem destruction of tests in the Japan Sea. Progr Micropaleontol Spec Publ: 169-189

McKnight WM jr (1962) The distribution of foraminifera off parts of the Antarctic Coast. Bull Amer Paleontol 44: 63-157

Natland ML (1935) The temperature and depth distribution of some recent and fossil foraminifera in the southern California region. Scripps Inst Oceanogr. Techn Ser 3(10): 225-230

Parker FL (1954) Distribution of the foraminifera in the northeastern Gulf of Mexico. Bull Mus Comp Zool 111(10): 453-588

Parker FL (1958) Eastern Mediterranian Foraminifera. Rep Swed Deep-Sea Exped Vol VIII Sediment Cores from the Medit Sea and the Red Sea No 4: 219-283

Pearcy FG (1914) Foraminifera of the Scottish National Antarctic Expedition. Trans Roy Soc Edinburgh 49 (4) Session 1913-14: 991-1044

Pflum CE (1966) The distribution of foraminifera in the eastern Ross Sea, Amundsen Sea, and Bellinghausen Sea, Antarctica. Bull Amer Paleontol 50 Nr 226: 151-209

Pflum CE, Frerichs WE (1976) Gulf of Mexico deep-water foraminifers. Cushman Found Foraminiferal Res Spec Publ 14: 1-124

Phleger FB (1951) Ecology of foraminifera, northwest Gulf of Mexico. Part I. Foraminifera distribution. Geol Soc Amer Mem 46: 1-88

Phleger FB (1952) Foraminifera distribution in some sediment samples from the Canadian and Greenland Arctic. Contrib Cushman Found Foraminiferal Res 3(2): 80-89

Phleger FB (1956) Significance of living foraminiferal populations along the central Texas Coast. Contrib Cushman Found Foraminiferal Res 7:106-151

Phleger FB (1964) Patterns of living benthonic foraminifera, Gulf of California. In: Andel TH van, Shor GG jr (eds) Marine geology of the Gulf of California. Amer Assoc Petrol Geol Mem 3: 377-394

Polski W (1959) Foraminiferal biofacies off the North Asiatic Coast. J Paleontol 33(4): 569-587

Resig JM (1958) Ecology of foraminifera of the Santa Cruz Basin, California. Micropaleontol 4(3):287-308

Saidova CHM (1960) Distribution of foraminifera in sediments of the Okhotsk Sea. Acad Nauk CCCP Inst Oceanogr 32: 96-157 (in Russian)

Saidova CHM (1975) Benthic foraminifera of the Pacific Ocean part III. Acad Nauk CCP Inst Oceanogr: 1-586 (in Russian)

Saidova CHM (1976) Benthic foraminifera of the world oceans (zonation and quantitative distribution). Acad Nauk CCCP Inst Oceanogr: 1-159 (in Russian)

Schott W (1935) Die Foraminiferen in dem äquatorialen Teil des Atlantischen Ozeans. Wiss Ergebn Dt Atlantische Exped Forschungs- und Vermessungsschiff "Meteor" 1925-1927 3(3): 43-134

Seibold E, Hinz K (1974) Continental slope construction and destruction. In: Burk CA, Drake CL (eds). The Geology of Continental Margins. Springer, Berlin Heidelberg New York: 179-196

Seiler WC (1975) Tiefenverteilung benthischer Foraminiferen am portugiesischen Kontinentalhang. "Meteor" Forschungsergebn Reihe C Nr 23: 47-94

Smith PB (1964) Recent foraminifera off Central America: Ecology of benthonic species. Geol Surv Prof Pap 429-B: B1-B55

Streeter SS (1972) Living benthonic foraminifera of the Gulf of California, a factor analysis of Phleger's (1964) data. Micropaleontol 18(1): 64-73

Subba Rao M, Vedantam D, Nageswara Rao J (1979) Distribution and ecology of benthonic foraminifera in the sediments of the Visakhapatnam Shelf, east coast of India. Paleogeogr Paleoclimato Paleoecol 27: 349-369

Theyer F (1970) Benthic foraminiferal studies in the Pacific basin. Antarct J 55:180-181

Uchio T (1960a) Ecology of living benthonic foraminifera from the San Diego, California, area. Contrib Cushman Found Foraminiferal Res Spec Publ 5: 1-72

Uchio T (1960b) Benthic foraminifera of the Antarctic Ocean. Seto Mar Biol Lab Biol Results Japanese Antarct Res Exped 12: 1-20

Vilks G (1969) Recent foraminifera in the Canadian Arctic. Micropaleontol 15(1): 35-60

Waller HO (1960) Foraminiferal biofacies off the south China Coast. J Paleontol 34(6): 1164-1182

Walton WR (1955) Ecology of living benthonic foraminifera, Todos Santos Bay, Baja California. J Paleontol 29(6): 952-1018

Wiesner H (1931) Die Foraminiferen der Deutschen Südpolar-Expedition 1901-1903. In: Drygalski E von: Deutsche Expedition 1901-1903, 20. Band Zoologie 12. Band: 53-165

Wilcoxon JA (1964) Distribution of foraminifera off the southern Atlantic coast of the United States. Contrib Cushman Found Foraminiferal res 15(1): 1-24

Zobel B (1973) Biostratigraphische Untersuchungen des indisch-pakistanischen Kontinentalrandes (Arabisches Meer). "Meteor"-Forschungsergebn Reihe C 12: 9-73

Zobel B, Ranke U (1978) Zusammensetzung, Stratigraphie und Bildungsbedingungen der Sedimente am Kontinentalhang vor Sierra Leone (West Afrika). "Meteor" Forschungsergebn Reihe C 29:21-74

Table 1. Apparent depth zones (ADZ) for species of benthic foraminifera off W Africa (Haake, 1980). Rank numbers compare the sequence of species with that of Haake's (1980) report reproduced herein as Figure 2

Species	Ranking numbers (Haake 1980) see Figure 2	ADZ
1. *Textularia panamensis*	1	32
2. *Siphonaperta* sp.	2	32
3. *Nonionella cordiformis*	15	44
4. *Hanzawaia concentrica*	10	45
5. *Spiroloculina excavata*	3	47
6. *Quinqueloculina* spp.	5	48
7. *Suggrunda eckisi*, subsp., A	19	48
8. *Rectuvigerina phlegeri*	9	50
9. *Cribrostomoides jeffreysii*	22	62
10. *Nouria polymorphinoides*	4	63
11. *Textularia pseudogramen*	7	63
12. *Bulimina gibba*	17	76
13. *Nonion asterizans*	11	76
14. *Hanzawaia* sp.	18	79
15. *Spiroloculina depressa*	6	81
16. *Textularia sagittula*	23	89
17. *Elphidium macellum*	20	102
18. *Globobulimina* cf. *glabra*	24	105
19. *Cancris auriculus*	8	110
20. *Dorothia rudis*	21	110
21. *Cribrostomoides* cf. *triangularis*	30	123
22. *Ammoscalaria pseudospiralis*	28	158
23. *Bolivina striatula*	29	158
24. *Bolivina pacifica*	16	158
25. *Saccammina* sp.	12	170
26. *Reophax calcareus*	13	170
27. *Trifarina fornasinii*	25	234
28. *Lenticulina orbicularis*	26	251
29. *Liebusella goësi*	31	257
30. *Reophax scorpiurus*	32	282
31. *Eggerella scabra*	14	282
32. *Cassidulina carinata*	34	282
33. *Uvigerina* spp.	35	316
34. *Bolivina dilatata dilatatissima*	33	339
35. *Trifarina elongatastriata*	41	363
36. *Fursenkoina mexicana*	40	363
37. *Rectuvigerina arquatensis*	44	398
38. *Ehrenbergina undulata*	45	398
39. *Gyroidina umbonata*	36	398
40. *Hoeglundina elegans*	42	398
41. *Bulimina costata* s.l.	38	427
42. *Planulina ariminensis*	47	479
43. *Bolivina subaenariensis*	52	550
44. *Eggerella arctica*	49	708
45. *Cibicides pseudoungerianus*	48	759
46. *Reophax dentaliniformis*	39	812
47. *Trochammina globigeriniformis*	37	832
48. *Cribrostomoides* sp.	51	832
49. *Bulimina aculeata*	53	871
50. *Uvigerina peregrina dirupta*	43	912
51. *Reophax pilulifer*	56	1000
52. *Globobulimina turgida*	54	1047
53. *Ammoscalaria tenuimargo*	46	1148
54. *Valvulineria laevigata*	50	1175
55. *Reophax aduncus*, var. A	57	1349
56. *Melonis barleeanus*	58	1585
57. *Planulina wuellerstorfi*	55	1820
58. *Ammobaculites agglutinans*	59	1995

Table 2. Apparent depth zones (ADZ) for species of benthic foraminifera off NW Africa (Lutze, 1980)

Species	ADZ
1. *Quinqueloculina pulchella*	42
2. *Ammonia beccarii*	44
3. *Textularia panamensis*	52
4. *Quinqueloculina vulgaris*	60
5. *Ammoscalaria pseudospiralis*	65
6. *Elphidium crispum*	66
7. *Nonion asterizans*	87
8. *Textularia pseudogramen*	91
9. *Planorbulina mediterranensis*	93
10. *Dorothia rudis*	107
11. *Cibicides lobatulus*	117
12. *Globobulimina* "324"/"339"	126
13. *Cancris auriculus*	135
14. *Hanzawaia concentrica*	138
15. *Amphicoryna scalaris*	195
16. *Trifarina fornasinii*	209
17. *Ehrenbergina undulata*	251
18. *Pyrgo fornasinii*	263
19. *Uvigerina peregrina* var.	282
20. *Hoeglundina elegans* subsp. "339"	331
21. *Siphotextularia caroliniana*	339
22. *Uvigerina elongatastriata*	339
23. *Liebusella goësi*	363
24. *Discanomalina vermiculata*	372
25. *Cribrostomoides triangularis*	380
26. *Hyalinea balthica*	389
27. *Planulina ariminensis*	513
28. *Cibicides pseudoungerianus*	676
29. *Eggerella arctica*	724
30. *Bulimina aculeata*	759
31. *Jaculella obtusa*	776
32. *Uvigerina finisterrensis*	851
33. *Sphaeroidina bulloides*	1318
34. *Uvigerina peregrina*	1318
35. *Marsipella elongata*	1445
36. *Melonis zaandami*	1698
37. *Oridorsalis umbonatus*	1778
38. *Globobulimina hoeglundi turgida*	1820
39. *Sigmoilopsis schlumbergeri*	1820
40. *Rhabdammina abyssorum*	1950
41. *Ceratobulimina pacifica*	1995
42. *Cyclammina pusilla*	1995
43. *Cribrostomoides subglobosum*	2042
44. *Hoeglundina elegans*	2239
45. *Cibicides kullenbergi*	2291
46. *Saccorhiza* + *Rhizammina*	2455
47. *Hormosina globulifera*	2630
48. *Cibicides wuellerstorfi*	2630
49. *Pyrgo murrhina*	2818

Table 3. Comparison of ADZ's for W African study areas. Note: Lutze's shallow water subspecies of Hoeglundina elegans was used in this comparison and in Table 7. Numbers in brackets denote the rank numbers of Lutze's species (see Table 2)

Species	ADZ (Haake 1980)	ADZ (Lutze 1980)
1. (3.) Textularia panamensis	32	52
4. (14.) Hanzawaia concentrica	45	138
11. (8.) Textularia pseudogramen	63	91
13. (7.) Nonion asterizans	76	87
19. (13.) Cancris auriculus	110	135
20. (10.) Dorothia rudis	110	107
21. (25.) Cribrostomoides triangularis	123	380
22. (5.) Ammoscalaria pseudospiralis	158	65
27. (16.) Trifarina fornasinii	234	209
29. (23.) Liebusella goësi	257	282
33. (19.) Uvigerina spp.	316	282
35. (22.) Trifarina elongatastriata	363	339
38. (17.) Ehrenbergina undulata	398	251
40. (20.) Hoeglundina elegans	398	331
42. (27.) Planulina ariminensis	479	513
44. (29.) Eggerella arctica	708	724
45. (28.) Cibicides pseudoungerianus	759	676
49. (30.) Bulimina aculeata	871	759
56. (36.) Melonis barleeanus	1585	1698
57. (48.) Planulina wuellerstorfi	1820	2630

Table 4. Apparent depth zones (ADZ) for species of benthic foraminifera from the eastern Mediterranean Sea (Parker, 1958). Dashed horizontal lines indicate suggested faunal groups

Species	ADZ
1. Asterigerina mamilla	89
2. Elphidium advenum	102
3. Textularia sagittula	135
4. Textularia calva	138
5. Rosalina obtusa	148
6. Elphidium complanatum	155
7. Cassidulina cf. oblonga	155
8. Hanzawaia rhodiensis	170
9. Bulimina gibba	182
10. Cibicides lobatulus	200
11. Cassidulina carinata	204
12. Bolivina catanensis	214
13. Bigenerina nodosaria	245
14. Hyalinea balthica	314
15. Bulimina marginata	457
16. Nonion barleeanus	537
17. Planulina ariminensis	575
18. Bulimina costata	575
19. Uvigerina mediterranea	589
20. Hoeglundina elegans	617
21. Spiropthalmidium acutimargo	708
22. Bolivina dilatata dilatatissima	724
23. Cibicides aff. floridanus	741
24. Epistominella rugosa convexa	776
25. Gyroidina altiformis	1000
26. Involutina cf. planorbis	1023
27. Glomospira gordialis	1047
28. Gyroidina cf. neosoldanii	1259
29. Bolivina pseudoplicata	2188
30. Glomospira charoides	2291

Table 5. Apparent depth zones (ADZ) for species of benthic foraminifera from the Gulf of Gascogne (Caralp et al., 1970)

Species	ADZ
1. *Textularia sagittula*	182
2. *Cibicides refulgens*	240
3. *Cibicides lobatulus*	331
4. *Quinqueloculina seminula*	372
5. *Bulimina buchiana*	479
6. *Cassidulina laevigata*	589
7. *Bulimina striata*	616
8. *Cibicides ungerianus*	661
9. *Uvigerina auberiana*	708
10. *Trifarina angulosa*	776
11. *Cibicides pseudoungerianus*	776
12. *Cassidulina crassa*	832
13. *Valvulina capreola*	832
14. *Elphidium lidoense*	977
15. *Gavellinopsis praegeri*	1047
16. *Trifarina bradyi*	1288
17. *Bolivina dilatata*	1349
18. *Uvigerina peregrina*	1445
19. *Uvigerina parvula*	1622
20. *Sigmoilina schlumbergeri*	1820
21. *Bulimina inflata*	1862
22. *Gyroidina soldanii*	1862
23. *Uvigerina dirupta*	2042
24. *Hyperammina* spp.	2138
25. *Nonion pompilioides*	2754

Table 6. Apparent depth zones (ADZ) for species of benthic foraminifera from the Scotia Sea (Echols, 1971)

Species	ADZ
1. *Cassidulina subglobosa*	245
2. *Ammobaculites agglutinans*	398
3. *Portatrochammina antarctica*	427
4. *Miliammina arenacea*	468
5. *Cassidulinoides parkerianus*	538
6. *Fursenkoina fusiformis*	692
7. *Textularia earlandi*	794
8. *Astronion echolsi*	794
9. *Angulogerina earlandi*	794
10. *Ammoflintina argentea*	871
11. *Bulimina aculeata*	891
12. *Fursenkoina earlandi*	891
13. *Pseudobolivina antarctica*	1123
14. *Cribrostomoides arenacea*	1388
15. *Nonionella iridea*	1479
16. *Haplophragmium canariensis*	1659
17. *Gyroidina subplanulata*	1820
18. *Conotrochammina bullata*	2042
19. *Trochammina glabra*	2138
20. *Epistominella exigua*	2239
21. *Haplophragmoides parkerae*	2344
22. *Trochammina multicamerata*	2344
23. *Eilohedra weddelensis*	2399
24. *Haplophragmoides quadratus*	2512
25. *Textularia wiesneri*	2754
26. *Portatrochammina eltanensis*	2818
27. *Eggerella parkerae* var. B	2951
28. *Martinottiella antarctica*	3090
29. *Cribrostomoides umbilicatulum soldanii*	3162
30. *Reophax* sp. A	3162
31. *Saccammina tubulata*	3311
32. *Adercotryma glomerata*	3311
33. *Psammosphaera fusca*	3388
34. *Glomospira charoides*	3467
35. *Tritaxis inhaerens*	3548
36. *Nuttalides umbonifera*	3548
37. *Cribrostomoides subglobosus*	3715
38. *Cyclammina pusilla*	3715
39. *Cribrostomoides sphaeriloculus*, mikrosph.	4266
40. *Conotrochammina kennetti*	4898

Table 7. Depth preferences (ADZ) for selected species of benthic foraminifera in five different study areas

Species	Haake 1980	Lutze 1980	Parker 1958	Caralp et al. 1970	Echols 1971
Melonis barleeanus	1585	1698	537	-	-
Hoeglundina elegans	398	331	617	-	-
Uvigerina peregrina s.l. (except spinose forms)	316	282	-	1445	-
Hanzawaia rhodiensis/concentrica	45	138	170	-	-
Planulina ariminensis	479	513	575	-	-
Bulimina aculeata	871	759	-	-	891
Textularia sagittula	89	-	135	182	-
Bolivina dilatata + subsp.	339	-	724	1349	-

APPENDIX

Samples located off Senegal/Gambia, W Africa (after Haake, 1980)

Stat.	Lat.	Long.	Depth	DEST
215	$12^{o}34,06'$	$17^{o}23,02'$	32	54
214	$12^{o}51,06'$	$17^{o}22,99'$	40	66
213	$13^{o}03,10'$	$17^{o}22,98'$	48	83
219	$12^{o}32,75'$	$17^{o}31,49'$	52	76
229	$14^{o}08,04'$	$17^{o}22,91'$	81	98
230	$14^{o}07,52'$	$17^{o}27,04'$	94	120
220	$12^{o}32,34'$	$17^{o}33,51'$	103	141
221	$12^{o}32,25'$	$17^{o}34,55'$	148	155
232	$14^{o}07,06'$	$17^{o}31,36'$	195	229
222	$12^{o}32,08'$	$17^{o}35,45'$	207	234
233	$14^{o}06,94'$	$17^{o}31,77'$	290	257
223	$12^{o}32,05'$	$17^{o}35,92'$	299	282
234	$14^{o}06,85'$	$17^{o}32,40'$	300	302
235	$14^{o}06,78'$	$17^{o}32,82'$	488	355
224	$12^{o}31,91'$	$17^{o}37,08'$	497	437
236	$14^{o}06,64'$	$17^{o}35,99'$	744	468
225	$12^{o}31,55'$	$17^{o}38,43'$	900	537
237	$14^{o}08,45'$	$17^{o}38,96'$	996	617
238	$14^{o}04,81'$	$17^{o}52,05'$	1983	776

Samples located off NW Africa (after Lutze, 1980)

Stat.	Lat.	Long.	Depth	DEST
390	22°40,30'	16°27,60'	33	69
311	22°22,00'	16°47,30'	43	102
350	15°42,50'	16°47,00'	51	58
312	22°26,00'	17°04,20'	62	105
302	26°00,20'	14°40,00'	64	87
352	16°16,70'	16°41,00'	70	71
313	22°28,50'	17°09,00'	82	110
338	15°19,00'	17°03,30'	103	115
315	22°42,50'	17°09,50'	107	120
314	22°30,10'	17°13,40'	109	123
316	22°55,50'	17°08,00'	109	120
303	26°11,80'	14°47,60'	151	178
339	15°20,60'	17°07,50'	151	145
340	15°21,00'	17°11,20'	201	214
304	26°13,80'	14°48,60'	202	182
317	22°51,70'	17°09.00'	214	195
322	22°54,20'	17°09,80'	316	209
341	15°23,00'	17°11,80'	324	282
342	15°23,00'	17°15,00'	405	275
323	22°54,40'	17°11,10'	418	240
305	26°17,10'	14°51,30'	501	380
343	15°23,90'	17°16,60'	506	457
344	15°25,70'	17°20,50'	725	851
306	26°19,60'	14°55,70'	741	832
325	22°59,40'	17°19,70'	768	794
345	15°28,00'	17°21,60'	966	1023
326	23°02,10'	17°24,60'	1024	1175
307	26°24,30'	14°57,70'	1070	1660
346	15°34,20'	17°32,70'	1517	1380
332	16°11,60'	22°20,70'	1948	1950
327	23°07,90'	17°44,10'	2032	2089
308	26°38,70'	15°03,20'	2090	2089
347	15°49,50'	17°50,70'	2710	1905
328	21°08,70'	18°34,40'	2822	1738
309	26°50,30'	15°06,60'	2849	2344
301	27°03,00'	15°03,10'	2896	2291
310	23°29,90'	18°43,00'	3075	2455
329	19°22,00'	19°55,80'	3314	2344
330	16°46,10'	22°01,00'	3488	2291

Samples from the E Mediterranean Sea (after Parker, 1958)

Stat.	Lat.	Long.	Depth	DEST
4662A	39°47,00'	26°00,00'	51	155
4666	40°04,00'	25°46,00'	71	204
4661	40°05,00'	26°04,00'	75	117
4658	39°58,00'	25°47,00'	82	195
4667	39°47,00'	25°41,00'	86	209
4657A	39°05,00'	25°58,00'	104	204
4649A	40°23,00'	25°25,00'	106	263
4654A	39°12,00'	25°13,00'	115	148
4668	37°54,00'	24°33,00'	117	141
4654B	38°54,00'	25°29,00'	143	158
4721A	36°49,00'	12°54,00'	179	229
4652	38°50,00'	25°52,00'	201	282
BS21A	39°21,00'	26°03,00'	205	398
BS20	39°19,00'	26°00,00'	210	407
4657B	39°04,00'	25°56,00'	256	389
BS29	38°23,00'	25°56,00'	287	417
4648	38°56,00'	24°18,00'	338	490
4647	37°09,00'	23°38,00'	366	437
4657	38°32,00'	24°54,00'	384	562
4718	35°06,00'	15°44,00'	541	550
4703	32°48,00'	23°13,00'	552	617
4694	36°02,00'	15°03,00'	567	617
4723	37°11,70'	11°33,50'	567	490
4687	31°49,00'	33°29,00'	631	631
4685	32°06,00'	30°41,00'	647	603
4664	40°28,00'	26°25,00'	658	631
4660	40°24,00'	26°02,00'	665	617
4714	32°30,00'	15°49,00'	665	692
4656	37°59,00'	25°31,00'	680	676
4713	31°37,00'	17°32,00'	684	851
BS9	36°28,00'	12°55,00'	731	794
4654	39°04,00'	24°54,00'	746	851
4686	32°06,00'	32°05,00'	799	676
4702	32°05,00'	25°16,00'	799	741
4659	40°18,00'	25°45,00'	850	631
4650	40°14,00'	25°42,00'	859	724
4670	35°49,00'	23°27,00'	996	813
4671	36°16,00'	24°18,00'	1016	692
4675	35°21,00'	26°40,00'	1064	589

Samples from the E Mediterranean Sea (after Parker, 1958) continued

4693	35°08,00'	34°56,00'	1073	1202
4649	39°35,00'	23°53,00'	1265	955
4688	32°42,00'	32°43,00'	1312	1318
4691	33°17,00'	34°59,00'	1331	631
4716	33°52,00'	15°47,00'	1378	1122
4665	40°15,00'	25°18,00'	1587	339
4705	33°44,00'	23°18,00'	1788	1202
4673	35°55,00'	25°16,00'	1844	1230
4678	35°34,00'	29°49,00'	1862	1660
4683	32°59,00'	30°28,00'	2358	2138
4689	33°32,00'	32°09,00'	2442	1096
4698	34°24,00'	29°34,00'	2738	1862
4701	32°33,00'	26°03,00'	2760	2188
4681	34°03,00'	30°12,00'	2852	2291
4710	37°49,00'	19°19,00'	3241	1820
4708	36°01,00'	21°37,00'	3499	2291

Samples from the Gulf of Gascogne (after Caralp et al., 1970)

Stat.	Depth	DEST	Stat.	Depth	DEST
6721	135	437	6595	1100	813
65119	160	617	6707	1150	776
65108	180	631	6723A	1400	832
6591	190	562	6723B	1400	871
6642	200	589	6723D	1400	1072
6710A	230	458	6723E	1400	933
6710B	230	724	6711	1420	776
6704	290	661	6709A	1480	794
6722A	315	562	6709B	1480	759
6722B	315	661	6717	1540	813
6719A	430	724	6719	1800	1349
6719B	430	741	6709	1975	1096
6592	560	776	6718A	2480	1738
6702A	580	832	6718B	2480	1698
6702B	580	776	6718C	2480	1738
6702C	580	851	6712A	2500	724
6720A	700	832	6717A	3200	891
6706A	850	794	6717B	3200	813
6706B	850	794	6717C	3200	794
6706C	850	851	6707	4175	1000
6716	900	759	6714	4200	1349

Samples from the Scotia Sea (after Echols, 1971). Depths listed are taken from Echols' (1971) Table 2.

Stat.	Lat.	Long.	Depth	DEST
673	54°07,00'	38°09,00'	201	501
497	61°10,00'	45°13,00'	256	537
675	55°44,00'	38°04,00'	329	776
674	54°44,00'	38°04,00'	531	708
652	58°20,00'	26°17,00'	558	1096
685	54°55,00'	38°06,00'	636	955
658	59°05,00'	26°39,00'	659	1202
676	54°47,00'	38°01,00'	750	813
659	59°05,00'	26°38,00'	805	1047
651	58°20,00'	26°09,00'	851	1413
503	62°14,00'	45°20,00'	993	1660
679	54°54,00'	37°06,00'	1032	1023
569	56°04,00'	27°28,00'	1043	871
654	57°51,00'	27°35,00'	1058	1778
630	59°53,00'	27°42,00'	1244	1413
650	58°19,00'	26°06,00'	1244	1698
502	62°14,00'	45°19,00'	1279	912
680	54°58,00'	37°59,00'	1500	1778
504	62°14,00'	45°22,00'	1589	1862
649	58°17,00'	25°54,00'	1841	2291
681	55°01,00'	38°00,00'	1900	1660
460	53°06,00'	44°46,00'	2105	1698
506	62°16,00'	45°25,00'	2159	2818
537	60°47,00'	48°02,00'	2189	1778
686	55°02,00'	38°10,00'	2288	1862
624	60°21,00'	29°30,00'	2345	3020
1067	59°01,00'	37°09,00'	2485	2884
544	60°02,00'	40°10,00'	2627	2291
1087	60°49,00'	49°25,00'	2644	2570
725	54°05,00'	33°40,00'	2722	2818
533	61°09,00'	48°48,00'	2753	2692
1035	60°40,00'	34°54,00'	2790	2951
1074	62°05,00'	39°15,00'	2796	2570
481	58°05,00'	44°56,00'	2800	3090
708	59°10,00'	37°08,00'	2823	2951
728	52°53,00'	33°57,00'	2864	2884
705	58°51,00'	37°18,00'	2937	2884
528	63°03,00'	49°09,00'	2968	2630
648	58°03,00'	24°46,00'	2976	3311

Samples from the Scotia Sea (after Echols, 1971), continued

736	53°02,00'	37°39,00'	3084	2692
693	56°53,00'	37°31,00'	3138	2630
712	58°43,00'	33°21,00'	3157	3236
716	56°51,00'	34°17,00'	3171	3162
617	61°59,00'	27°40,00'	3199	2455
1040	60°34,00'	28°37,00'	3206	3467
700	58°04,00'	37°47,00'	3212	2818
476	57°11,00'	44°53,00'	3382	3311
1068	59°01,00'	34°00,00'	3400	2818
524	64°05,00'	49°08,00'	3422	3715
689	55°15,00'	37°03,00'	3468	2951
467	55°03,00'	44°33,00'	3472	2884
1062	61°42,00'	34°58,00'	3644	3548
473	55°55,00'	45°32,00'	3678	3162
672	54°57,00'	38°22,00'	3697	3090
520	66°33,00'	48°11,00'	3722	3715
622	61°39,00'	29°15,00'	3737	3236
614	61°05,00'	26°02,00'	3770	2754
513	63°15,00'	44°56,00'	3784	3467
1046	60°54,00'	31°08,00'	3806	3002
548	59°00,00'	49°02,00'	3854	3162
574	55°12,00'	24°07,00'	3993	3388
1021	63°02,00'	40°55,00'	4110	3467
607	60°05,00'	22°24,00'	4130	3236
516T	66°28,00'	45°42,00'	4196	3631
517	66°26,00'	45°45,00'	4196	3467
516B	56°28,00'	45°42,00'	4196	2818
647	57°55,00'	24°06,00'	4493	3311
1016	64°30,00'	44°06,00'	4513	3631
618	63°02,00'	27°07,00'	4558	3631
604	58°27,00'	22°22,00'	4644	3388
486	60°07,00'	45°14,00'	5289	3236
638	58°35,00'	23°53,00'	5545	3631
562	48°13,00'	40°26,00'	5587	3548
646	57°52,00'	24°00,00'	6076	2951
645	57°54,00'	23°49,00'	6167	3388

Inorganic/Organic Geochemistry

27 Inorganic Geochemistry of Atlantic Ocean Sediments with Special Reference to Cretaceous Black Shales

Hans-J. Brumsack[1,2] and Myron Lew[2]

ABSTRACT

258 samples of Atlantic Ocean sediments (DSDP Sites 105, 112, 135, 137, 138, 144, 149, 328 to 330, 354, 355, 357 to 360, 362, 364, 366 to 370, 386, and 391) comprising shales, marls, and limestones from the Lower Cretaceous to the Quarternary were analyzed for the elements $C_{org.}$, S, $N_{org.}$, Fe, Mn, Zn, Cu, and their carbonate content. For a great number of samples also the sulfur isotopic composition was determined. An age-dependent trend was found for $C_{org.}$, S, and $N_{org.}$, showing a minimum during the Paleogene coinciding with the most positive $\delta^{34}S$ values. The mean values for Mn and Fe are the same on a carbonate-free basis for the different age groups. Zn and Cu are enriched in Cretaceous samples.

In 88 samples from the Upper and Lower Cretaceous, including 38 black shale samples, the trace elements Zn, Cu, Cr, Pb, Ni, Co, Mo, V, Ba, Sr, and in selected samples Ag, Cd, Tl, Bi, and B and the $\delta^{13}C_{org.}$ composition were determined. The high accumulation of heavy metals in black shales relative to average shale cannot be explained by the following processes:

1. water composition of marginal basins controlled by continental runoff ("Kupferschiefer"-type),
2. hydrothermal solutions connected with ridge volcanism,
3. metabolism of marine organisms.

The order of heavy metal enrichment relative to average shale ($Pb<Co<Cr<Ni<Cu<V<Zn<Mo<Ag$) corresponds well (except Ag) with the increasing concentrations of these elements in normal seawater. The presence of pyrite, sphalerite, and Cu-sulfides could be demonstrated by microprobe scannings. The sulfur is of bacteriogenic origin. Pyrite and metal sulfides are not in isotopic equilibrium. This is an indication of the heavy metal sulfides being formed in the water column and pyrite in the sediment. The Cretaceous proto-Atlantic can be compared to the present euxinic stage of the Black Sea. Model calculations show that a normal productivity is sufficient for the explanation of the high $C_{org.}$ content of black shales.

Atlantic Ocean sediments contain 3.0 to 4.2 x 10^{14} t S, the black shale contribution is 44% to 52% of the total amount. The bulk of sulfur is isotopically light, the mean sulfur isotopic composition of Atlantic Ocean sediments is -18.6‰ (rel. CDT).

1. Introduction

The aim of this chapter is to give some new and more representative data about the sulfur, organic carbon, and trace element contents of sediments from the Atlantic Ocean. The paper is divided into two parts. In the first part the results of up to 258 analyses of organic carbon ($C_{org.}$), total sulfur (S_{total}), organic nitrogen ($N_{org.}$), total iron

1 Geochemisches Institut der Universität Göttingen, Goldschmidtstraße 1, 3400 Göttingen, FRG

2 Present address: Scripps Institution of Oceanography, La Jolla, CA 92093, USA

(Fe_{total}), manganese, zinc, and copper are shown. Furthermore the total sulfide and sulfate sulfur isotopic composition will be discussed. In the second part a geochemical characterization of the Atlantic black shales is given by several trace elements, and the possible sources and mechanisms of heavy metal enrichment are shown.

2. Samples and Chemical Analysis

The samples were cored during Legs 11 to 41 of the DSDP (1970 to 1975) by D/V Glomar Challenger. Two hundred and fifty eight samples were chosen for chemical analysis comprising shales, marls, and limestones of different geological ages (Lower Cretaceous to Quaternary). Most of the Cretaceous black shale samples (66%) originate from off the NW African coast.

Figure 1 shows the present location of the drill sites and the DSDP site numbers. For further information see Brumsack (1979, 1980) and Lew (1980) and the respective Initial Reports.

The analyses mostly were carried out by spectrometric and coulometric methods. The relative standard deviation for duplicate analyses is between 2% and 15% (Table 1). An international standard reference material (SGR-1) was analyzed for testing the accuracy of the methods involved.

3. Chemistry of Atlantic Ocean Sediments

3.1 Organic Carbon ($C_{org.}$)

Figure 2 shows the frequency distribution diagram of $C_{org.}$. The values

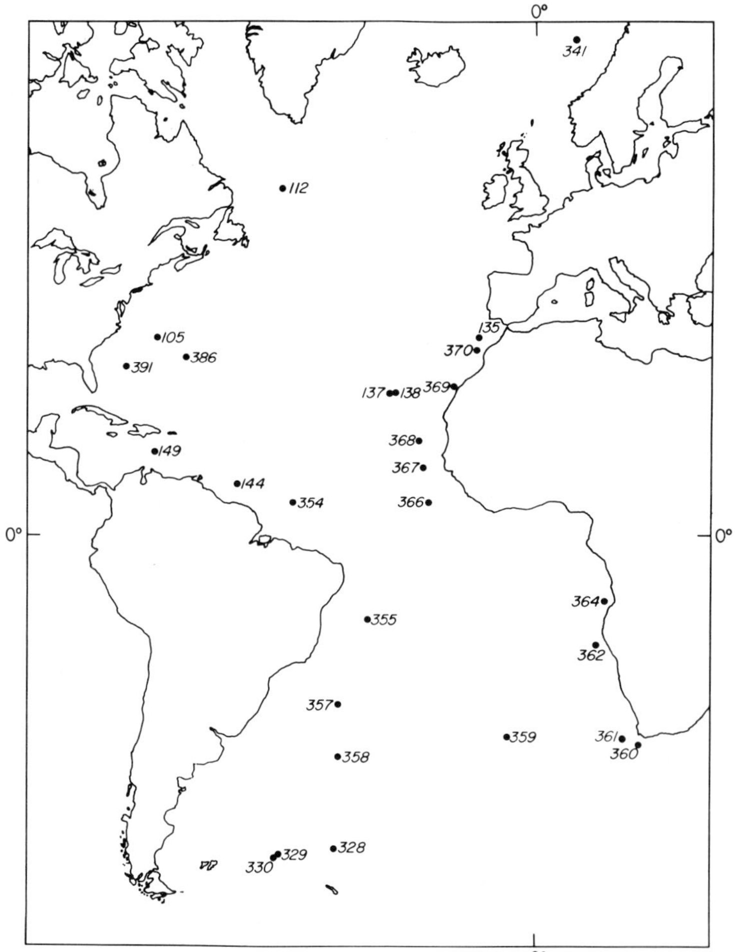

Fig. 1. Location of DSDP Drill Sites

Table 1. Analytical Methods

Element[a]	Method	References
$Fe^{(2)}$ $Mn^{(3)}$ $Zn^{(3)}$ $Cu^{(4)}$ $Ni^{(5)}$	AAS (C_2H_2/air-flame)	Herrmann (1975)
$Cu^{(7)}$ $Ni^{(5)}$ $Cr^{(3)}$ $Co^{(4)}$ $Pb^{(9)}$ $Mo^{(5)}$ $Sr^{(6)}$ $Ba^{(8)}$ $V^{(6)}$ $Ag^{(-)}$	AAS (graphite furnace) partly N_2, otherwise N_2/H_2 as carriergas	Heinrichs (1979[b])
$Cd^{(12)}$ $Tl^{(8)}$ $Bi^{(15)}$	AAS (graphite furnace) after separation and enrichment by fract. distill.	Heinrichs and Lange (1973) Heinrichs (1979[a])
$Mo^{(-)}$ $V^{(-)}$	ES, Fe as internal standard	Wedepohl (1964)
$B^{(9)}$	ES, Be as internal standard	Harder (1959)
$C^{(3)}$	coulometric titration	Herrman and Knake (1973)
$S^{(8)}$	coulometric titration	Lange and Brumsack (1977)
$\delta^{34}S$	mass-spectrometry, separation as CdS and reaction to SO_2	Ricke (1964), Kiba et al. (1955) Kiba and Kishi (1957)
$N_{total}^{(4)}$	Kjeldahl-dist. and titration following acid digestion	Bremner (1965)
$N_{fix.}^{(3)}$	Kjeldahl-dist. following destruction of $C_{org.}$ and extraction by 2m KCl	Bremner (1965) Fleige et al. (1971)
$N_{exc.}^{(14)}$	Titration following extraction	Keeney and Bremner (1966)
$N_{org.}$	$N_{org.} = N_{total} - (N_{fix.} + N_{exc.})$	

[a]relative % standard deviation in brackets
AAS = Atomic Absorption Spectrometry
ES = Emission Spectroscopy

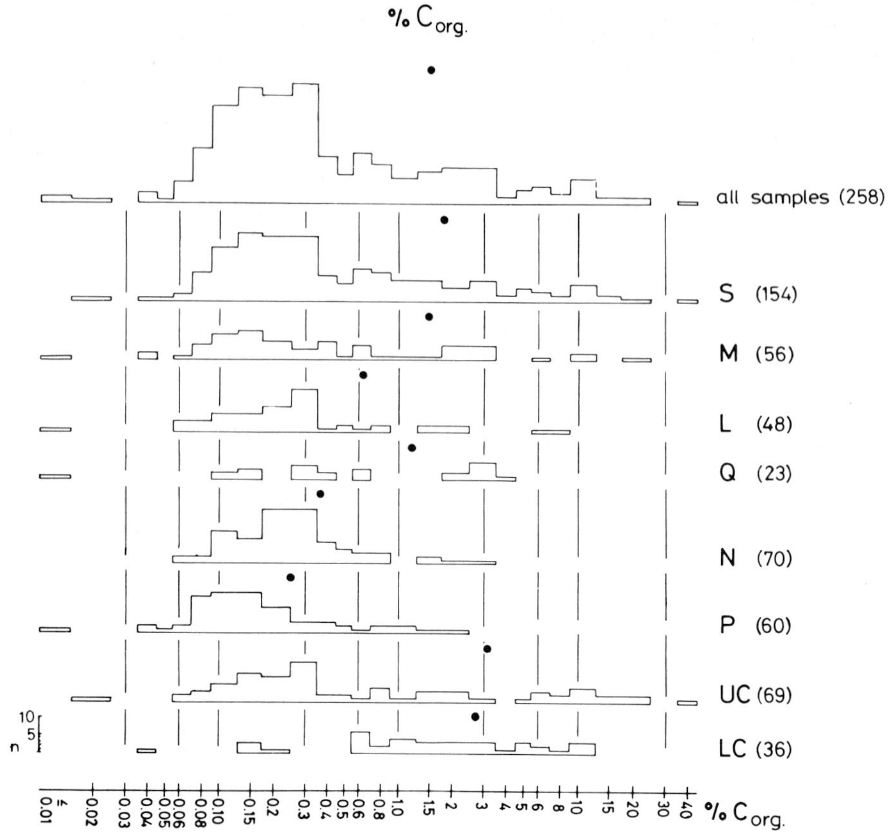

Fig. 2. Frequency distribution of organic carbon ($C_{org.}$). S = shales, M = marls, L = limestones, Q = Quaternary, N = Neogene, P = Paleogene, UC = Upper Cretaceous, LC = Lower Cretaceous, ● = mean value (see Table 2)

range from 0.01 to more than 30%, the arithmetic mean of all 258 samples is 1.49% (see also Table 2). As expected, the mean values decrease from shales (S) to limestones (L) because of dilution by carbonate material. The samples were also divided into five different age groups. Samples from the Lower and Upper Cretaceous show the highest $C_{org.}$ contents (2.76% and 3.16% respectively). The Cretaceous Atlantic Ocean is characterized by the "black shale" facies which is responsible for the high $C_{org.}$ mean values. A jump to relatively low values occurs during the Paleogene (0.25%) followed by a slight increase to the Neogene (0.36%) and Quaternary (1.18%). Whether this latter trend reflects changes in the worldwide organic primary production or is artificially produced by sample selection cannot be decided unequivocally. For comparison also the mean values of the 96 NW African samples (NWA) are listed in Table 2. The high proportion of black shales (35%) is responsible for the high $C_{org.}$. The $\delta^{13}C$ ratio of the total organic matter was determined on 24 samples from the Upper and Lower Cretaceous. The values range from -22.2% to -28.1‰ (mean value -25.6‰ rel. PDB) and support the occurrence of both marine and terrestrial sources. Similar results were obtained by Erdman and Schorno (1977) on the lipid fraction of some samples from Site 368.

3.2 Total Sulfur (S_{Total})

Figure 3 shows the frequency distribution for S_{total} in the different groups mentioned before. In this and the following figures black squares represent samples with $C_{org.}$ contents >1%, hatched squares represent samples with intermediate $C_{org.}$ concentrations <1% >0.5%, whereas samples with $C_{org.}$ <0.5% are white. The general parallelism in the dis-

Table 2. Mean values of Atlantic Ocean sediments

Element	All Samples	NWA	Q	N	P	UC	LC	S	M	L
% $C_{org.}$	1.49^{258}	2.62^{96}	1.18^{23}	0.36^{70}	0.25^{60}	3.16^{69}	2.76^{36}	1.80^{154}	1.44^{56}	0.64^{48}
% S_{total}	0.79^{258}	1.44^{95}	0.38^{23}	0.24^{70}	0.24^{60}	1.51^{69}	1.62^{36}	0.97^{154}	0.70^{56}	0.32^{48}
ppm N_{total}	676^{168}	1072^{42}	1181^{22}	461^{69}	325^{58}	2377^{12}	1206^{7}	834^{91}	704^{39}	269^{38}
ppm $N_{fix.}$	167^{168}	201^{42}	184^{22}	160^{69}	132^{58}	312^{12}	222^{7}	214^{91}	140^{39}	81^{38}
ppm $N_{org.}$	492^{168}	846^{42}	970^{22}	288^{69}	181^{58}	2021^{12}	953^{7}	598^{91}	551^{39}	177^{38}
$C_{org.}/N_{total}$*	14.3^{168}	20.1^{42}	10.4^{22}	8.0^{69}	8.0^{58}	27.7^{12}	19.4^{7}	15.3^{91}	13.2^{39}	10.0^{38}
$C_{org.}/N_{total}$	9.3^{168}	12.2^{42}	8.9^{22}	8.4^{69}	7.5^{58}	19.1^{12}	17.8^{7}	8.5^{91}	10.0^{39}	10.7^{38}
$C_{org.}/N_{org.}$*	19.7^{168}	25.4^{42}	12.7^{22}	12.8^{69}	14.4^{58}	32.5^{12}	24.6^{7}	21.4^{91}	16.9^{39}	15.2^{38}
$C_{org.}/N_{org.}$	17.4^{168}	17.1^{42}	14.4^{22}	15.8^{69}	17.4^{58}	26.8^{12}	26.5^{7}	16.1^{91}	18.3^{39}	19.6^{38}
% Fe_{total}	3.33^{257}	3.58^{95}	2.52^{23}	2.70^{70}	3.23^{60}	4.02^{69}	3.92^{35}	4.29^{153}	2.54^{56}	1.19^{48}
% Fe_{total}**	4.56^{257}	4.54^{95}	4.42^{23}	4.43^{70}	4.31^{60}	4.96^{69}	4.51^{35}	4.56^{153}	4.70^{56}	4.41^{48}
% Mn	0.12^{257}	0.11^{95}	0.07^{23}	0.10^{70}	0.12^{60}	0.13^{69}	0.14^{35}	0.10^{153}	0.21^{56}	0.07^{48}
% Mn**	0.16^{257}	0.14^{95}	0.12^{23}	0.16^{70}	0.16^{60}	0.16^{69}	0.16^{35}	0.11^{153}	0.39^{56}	0.26^{48}
% $CaCO_3$	27^{257}	21^{96}	43^{23}	39^{70}	25^{60}	19^{69}	13^{35}	6^{153}	46^{56}	73^{48}
ppm Cu	78^{256}	108^{95}	42^{23}	38^{70}	45^{59}	137^{69}	121^{35}	102^{152}	55^{56}	27^{48}
ppm Zn	217^{257}	359^{95}	104^{23}	76^{70}	97^{60}	448^{69}	326^{35}	280^{153}	140^{56}	108^{48}
$\delta^{34}S_{total}$	-15.8^{140}	-15.2^{68}	-15.2^{13}	-14.6^{30}	-7.8^{12}	-16.5^{55}	-19.0^{30}	-15.3^{83}	-19.0^{33}	-12.7^{24}
$\delta^{34}S_{total}$*	-19.1^{140}	-16.4^{68}	-23.7^{13}	-21.1^{30}	-15.0^{12}	-18.7^{55}	-18.8^{30}	-19.0^{83}	-19.2^{33}	-18.5^{24}
$\delta^{34}S_{sulfide}$	-17.4^{153}	-22.1^{42}	-29.6^{12}	-20.3^{60}	-10.3^{51}	-19.1^{22}	-19.2^{7}	-14.9^{90}	-16.8^{30}	-25.1^{32}
$\delta^{34}S_{sulfide}$*	-22.2^{20}	-16.5^{14}				-22.1^{17}	-22.5^{3}			
$\delta^{34}S_{sulfate}$	-0.8^{160}	-10.7^{44}	$+3.5^{17}$	$+0.1^{63}$	$+4.8^{51}$	-16.6^{22}	-11.4^{7}	-4.0^{89}	-1.1^{39}	$+8.3^{32}$
$\delta^{34}S_{sulfate}$*	-20.3^{20}	-19.1^{14}				-21.7^{17}	-12.3^{3}			

Number of samples in superscript
* = weighted mean value
** = calculated carbonate-free

NWA = North-West Africa, Q = Quaternary, N = Neogene, P = Paleogene, UC = Upper Cretaceous, LC = Lower Cretaceous, S = shales, M = marls, L = limestones

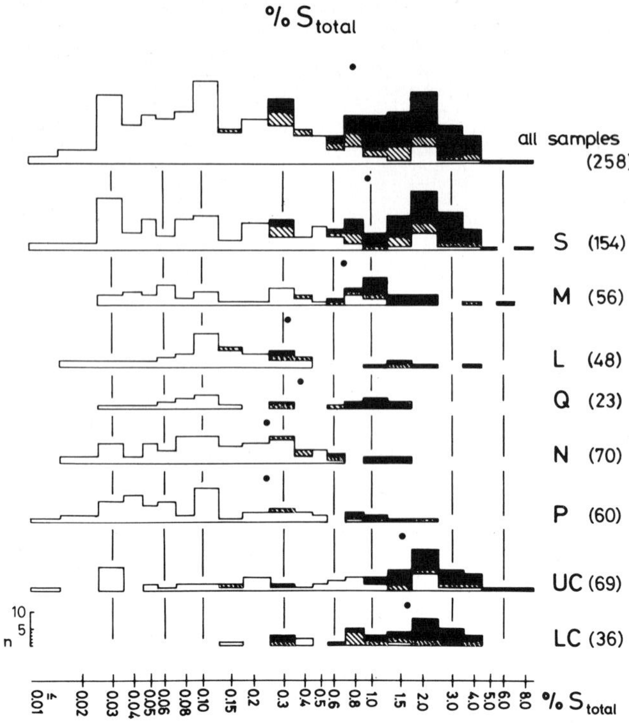

Fig. 3. Frequency distribution of total sulfur (S_{total})

tribution of $C_{org.}$ and S_{total} is obvious. The high arithmetic mean of 0.79% S for all samples (see also Table 2) is the result of the high incorporation of sulfur into the Cretaceous black shales.

3.3 Nitrogen

The main nitrogen compounds of living organisms are proteins, containing 15% to 19% N with a C/N ratio (weight) about 3 (Borodovskiy, 1965). Therefore protein-rich marine organisms show low C/N ratios (e.g., phytoplankton about 6), whereas higher terrestrial plants which contain less proteins show higher C/N ratios. In open ocean areas the deposition of terrestrial organic matter is less important (Degens and Mopper, 1976).

N-containing compounds are preferentially decomposed while settling through the water column. The C/N ratio of particulate organic matter increases from about 7.5 close to the surface to 15.1 at the sediment/seawater interface (Gordon, 1971). C/N ratios between 6 and 15 were determined in a large variety of marine sediments (Arrhenius, 1950; Emery and Rittenberg, 1952; Hartmann et al., 1971; Marchig, 1972; Stevenson and Cheng, 1972; Müller, 1975). C/N ratios of less than 6 are due to higher concentrations of fixed nitrogen in clay minerals.

The results of our investigation are shown in Figs. 4 and 5 and Table 2. As is to be expected, nitrogen is well correlated with $C_{org.}$ (Corr. Coeff. $C_{org.}$ - N_{total} = 0.93; $C_{org.}$-$N_{org.}$ = 0.93). Therefore the different age groups show a trend similar to that described for $C_{org.}$ and

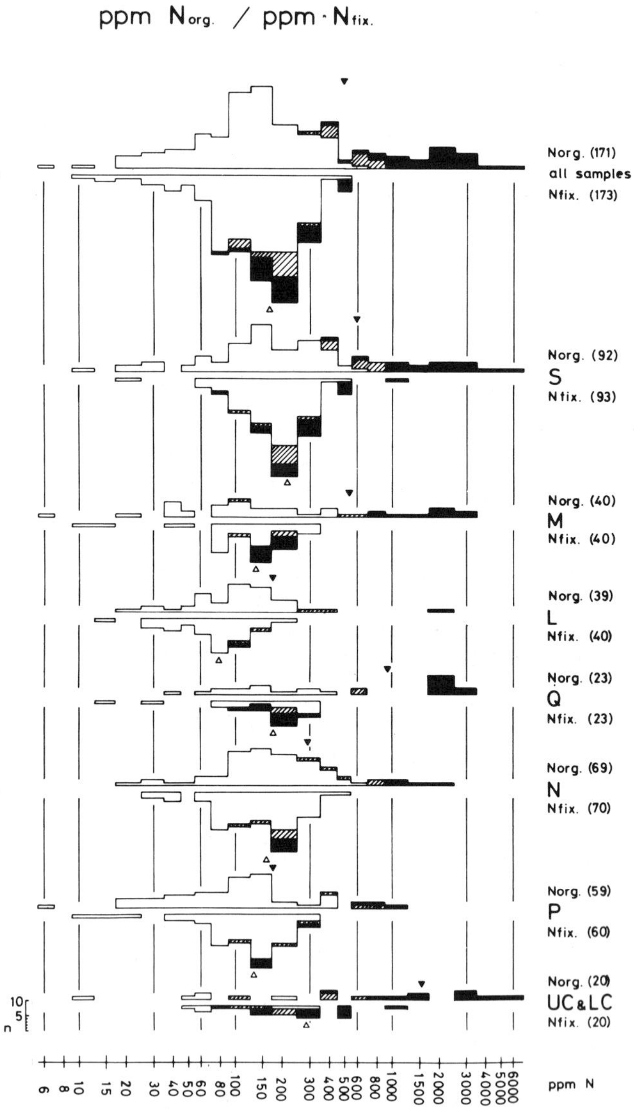

Fig. 4. Frequency distribution of organic nitrogen ($N_{org.}$) and fixed nitrogen ($N_{fix.}$)

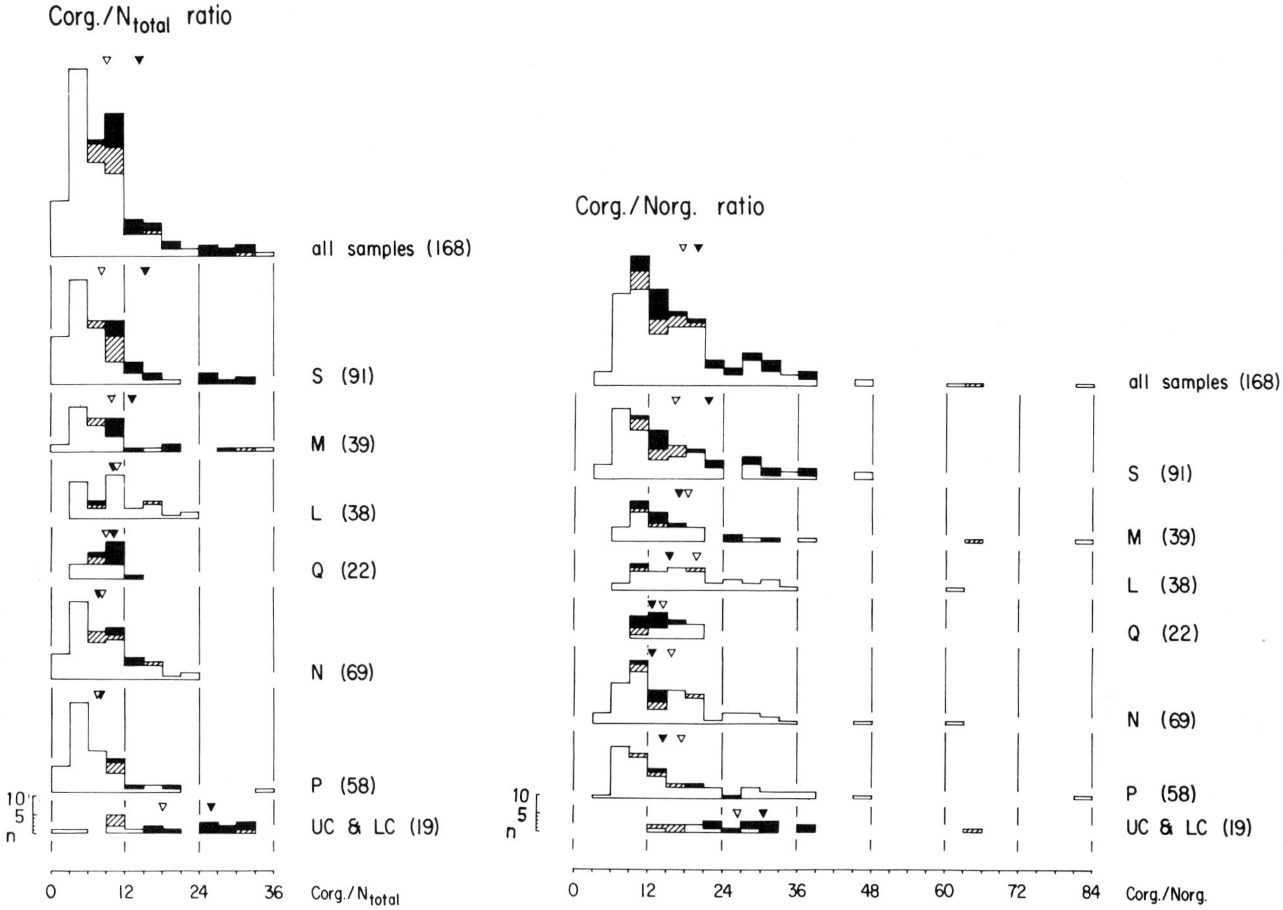

Fig. 5. Frequency distributions of the $C_{org.}/N_{total}$ and $C_{org.}/N_{org.}$ ratios

S_{total}. The $C_{org.}/N_{total}$ ratio follows this general trend, whereas the $C_{org.}/N_{org.}$ ratio increases with age and shows a maximum during the Cretaceous. The continuous decomposition of N-containing compounds corresponds with higher incorporation of nitrogen in clay minerals in $C_{org.}$-rich sediments. Shales generally show higher nitrogen values and higher C/N ratios than marls and limestones.

An extremely high C/N ratio was determined in one sample from Site 368 which was taken close to a basaltic sill. The $C_{org.}/N_{total}$ ratio of 111 and the $C_{org.}/N_{org.}$ ratio of 258 indicate higher temperatures and preferential nitrogen loss. 56% of the remaining nitrogen here is present as fixed nitrogen (459 ppm).

3.4 Fe_{Total} and Mn

The frequency distribution of Fe_{total} is shown in Fig. 6. The arithmetic mean for 257 samples is 3.33% Fe. The dilution effect due to different $CaCO_3$-contents is shown in the three lithological groups (S, M, L.). The increasing carbonate content from 13% during the Lower Cretaceous (LC) to 43% during the Quaternary (Q) is also responsible for the decreasing Fe total concentration in the five age groups. On a carbonate-free basis the mean values of all groups are very uniform (mean of all samples 4.56% Fe).

Fe belongs (in addition to Mn) to those elements which are mobilized under reducing conditions (as Fe^{2+}). This fact is not reflected by the frequency distribution of this element. Samples rich in $C_{org.}$ do not show a distinct Fe-depletion. If hydrogen sulfide is present Fe^{2+} is

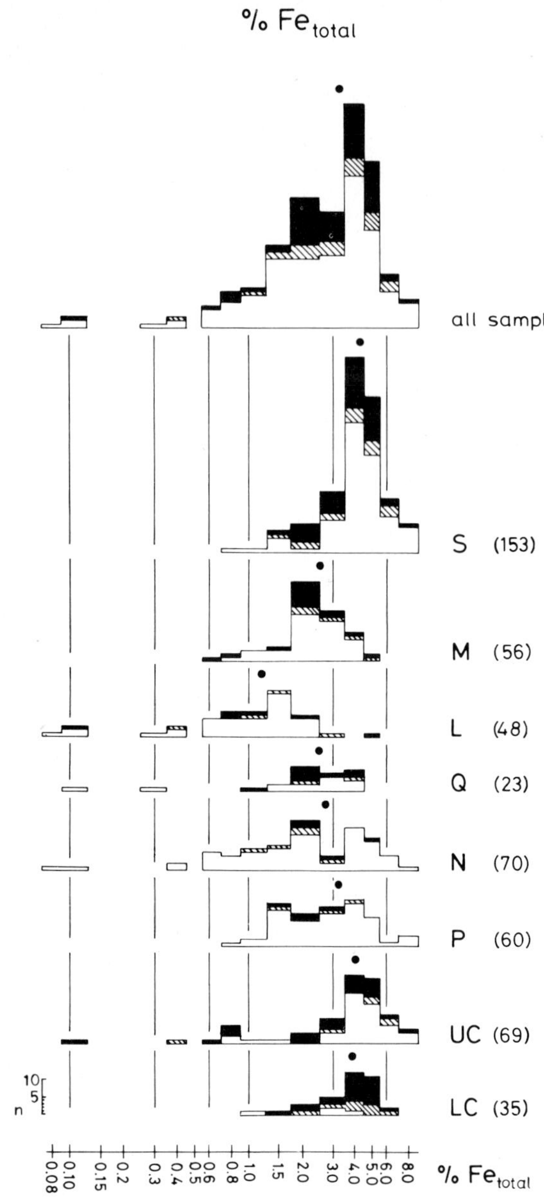

Fig. 6. Frequency distribution of total iron (Fe_{total})

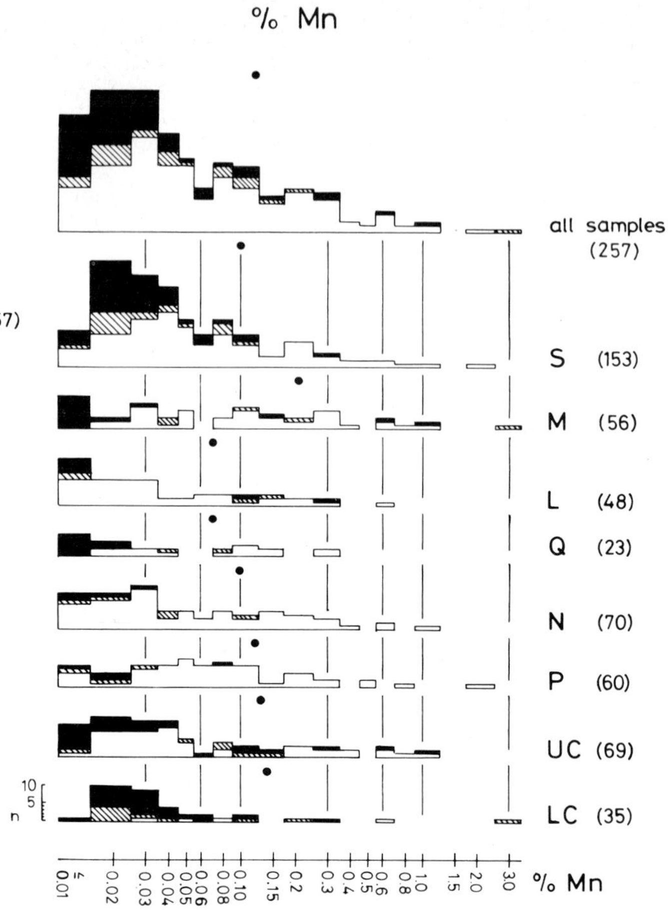

Fig. 7. Frequency distribution of Mn

fixed as iron sulfide immediately. Samples rich in $C_{org.}$ also show high S_{total} concentrations as was demonstrated earlier. Therefore this "reactive" iron must be one of the main factors controlling the pyrite concentration in black shales. If the available iron reservoir is exhausted the hydrogen sulfide produced will leave the place of its generation. Because of their low concentrations other heavy metals are of minor importance for the sulfur fixation. The uniform Fe distribution indicates that generally enough hydrogen sulfide was present in $C_{org.}$-rich sediments to fix the mobilized "reactive" iron as pyrite.

In contrast to Fe the mobilization of Mn (as Mn^{2+}) is evident, as shown in the frequency distribution of this element (Fig. 7). Samples rich in organic matter are depleted in Mn, but surprisingly Mn also shows a uniform mean value (0.16%) in the five different age groups if calculated carbonate-free. This indicates that Mn, like Fe, was fixed in the sedimentary column. But Mn^{2+} should migrate over longer distances than does Fe^{2+}, otherwise the Mn-depletion of $C_{org.}$-rich samples should not be reflected in the frequency distribution diagram. Precipitation of Mn-oxides in oxidizing horizons and incorporation of Mn in the carbonate phase must be responsible for the uniformity of the mean values.

This latter effect is clearly shown in the high mean values of marls (and limestones, if calculated $CaCO_3$-free). The presence of idiomorphic Fe/Mn-carbonate in Cretaceous marls could be demonstrated by microprobe scannings.

3.5 Cu and Zn

The trace elements Cu and Zn (Figs. 8 and 9) have been determined in almost all samples. The arithmetic mean values are 78 ppm Cu and 217 ppm Zn. In comparison to other literature values (e.g., Wedepohl, 1970) these mean values seem to be too high. Considering only samples from Paleogene to Quaternary the mean values are 41 ppm Cu and 88 ppm Zn. Therefore, the Cretaceous black shales are responsible for the much higher mean values especially from the NWA samples.

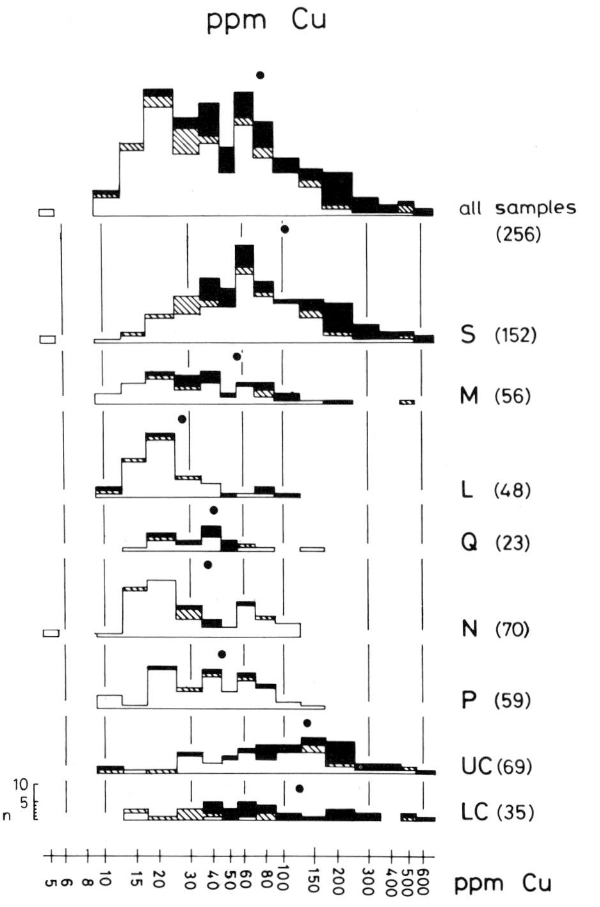

Fig. 8. Frequency distribution of Cu

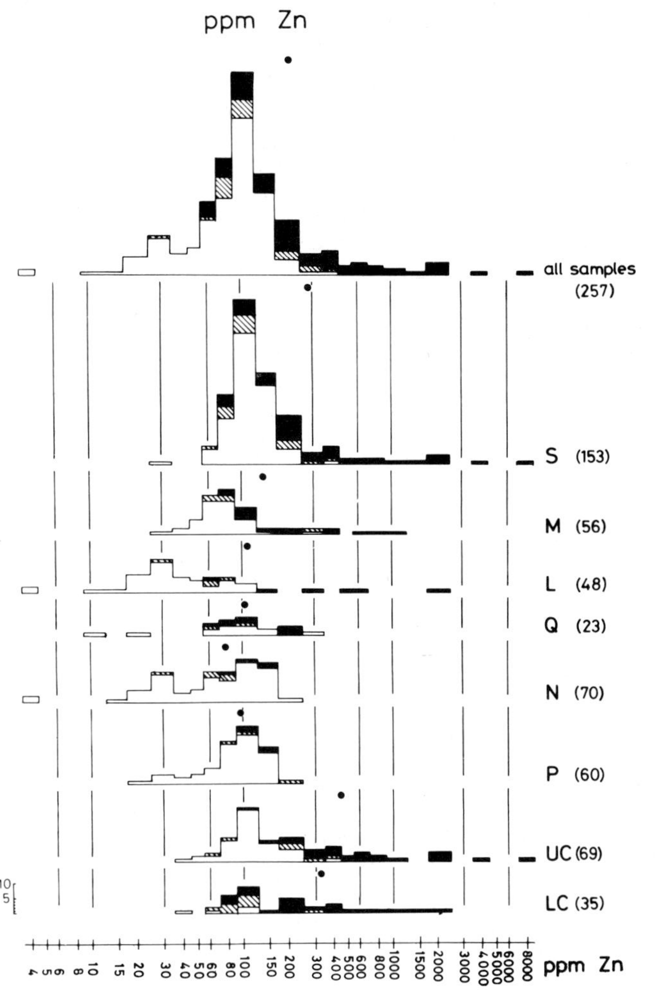

Fig. 9. Frequency distribution of Zn

As is to be expected, limestones and marls are depleted in both elements relative to shales. This dilution effect is stronger for Cu than for Zn. Zn is generally correlated with the $C_{org.}$ contents, even in the P to Q age groups, whereas Cu is mostly enriched in Cretaceous black shale samples.

The heavy metal content and genesis of the black shales will be discussed later in this publication. Generally, the NWA samples show higher values of $C_{org.}$, S, N, Cu, and Zn than the mean of all samples investigated, whereas the Fe, Mn, and carbonate contents are about the same. This reflects the relatively higher proportion of black shales sampled in this area.

3.6 Sulfur Isotopes

The main processes for the fixation of sulfur in the geological environment are the precipitation of gypsum (and anhydrite) from evaporating seawater and pyrite formation in marine sediments due to bacterial sulfate reduction.

The first process has been investigated quite well, and an age-dependent $\delta^{34}S$ curve has been established from different evaporite deposits (Holser and Kaplan, 1966; Nielsen, 1965, 1979). At the end of the Proterozoic values are around +30 ‰ (relative to the Canyon Diabolo Meteorite = CDT) and decrease to +10 ‰ in the Permian. During the Triassic a 10 ‰ jump to +20 ‰ has been found, followed by minor variations (±5 ‰). The present seawater $\delta^{34}S$ is relatively homogeneous, showing a value of about +20 ‰. These variations in the sulfur isotopic composition of seawater being documented over a period of 600 m.y. are explained by changes in fluxes between the individual sulfur reservoirs of the sedimentary cycle, e.g., the rate of river inflow, the rate of sulfate precipitation in evaporite deposits, the rate of pyrite formation, etc. In general the proportion of sulfide to sulfate is of significant importance. This proportion is closely linked to the amount of organic matter stored in the sediments and should also affect the amount of atmospheric oxygen (Holland, 1973; Schidlowski et al., 1977).

The second process has already been mentioned before. The formation of pyrite due to bacterial sulfate reduction is combined with sulfur isotope fractionation, as could be demonstrated by various authors (Hartmann and Nielsen, 1969; Kaplan et al., 1963; Thode et al., 1960). Pyrite generally shows lighter values compared to the corresponding seawater. Most sulfides in the marine environment show negative $\delta^{34}S$ ratios. According to Jørgensen (1979) and Nielsen (1979) "open" and "closed" system conditions are to be distinguished during pyrite formation. Open system conditions should prevail in the uppermost bioturbidated part of the sedimentary column. There the exchange of sulfate with the overlying water column is most effective, the sulfate reservoir keeps its constant seawater isotopic composition. In deeper sediment layers still enough organic matter may be present for the metabolism of sulfate-reducing bacteria. The diffusive supply of sulfate is limited in this "semi-closed" system. A relatively higher portion of porewater-sulfate will be reduced and the sulfide produced will become heavier. Between both systems fluent transitions exist. In both cases mostly iron sulfides (which later are transformed to pyrite) are formed (Berner, 1970). Therefore in "old" sediments the porportion of "open" to "closed" systems cannot be determined. It can only be assumed that the most negative $\delta^{34}S$ pyrite values were produced in a more "open" system.

A hint for the two processes described may be given by heavy metal sulfides. In anoxic environments they should be formed close to the sediment/seawater boundary or in the H_2S-containing water column, because a considerable diffusive transport into deeper sediment layers is very unlikely. Formed at an earlier stage of "open" system conditions the sulfur from acid-soluble sulfide should be isotopically lighter than the bulk of pyrite. At temperatures from 10° to 20°C the $\delta^{34}S$ difference in equilibrium between pyrite and sphalerite (the most abundant heavy metal sulfide besides pyrite) should be about 3 ‰ (Sakai, 1969: Ohmoto, 1972). In one black shale sample from side 367 there was enough heavy metal sulfide available to measure both types of sulfides with a $\delta^{34}S$ difference of 13 ‰. At least in this case pyrite and sphalerite have not been in equilibrium during their formation. The most simple explanation would be to expect the formation of the two sulfides at different locations. Such sulfur isotope data support the assumption of an anoxic water column in the Cretaceous Atlantic Ocean.

The results of this investigation are shown in Figs. 10 and 11. $\delta^{34}S_{total}$ was determined in 140 samples, $\delta^{34}S_{sulfide}$ in 152 samples and $\delta^{34}S_{sulfate}$ in 160 samples. As expected, the frequency distribution of the sulfur isotopic composition is very scattered, showing values ranging from -48 ‰ to +33 ‰. Samples rich in organic matter show negative $\delta^{34}S$ values in most cases.

Regarding the $\delta^{34}S_{total}$ distribution of the three lithological groups (Fig. 10), marls show the most nega-

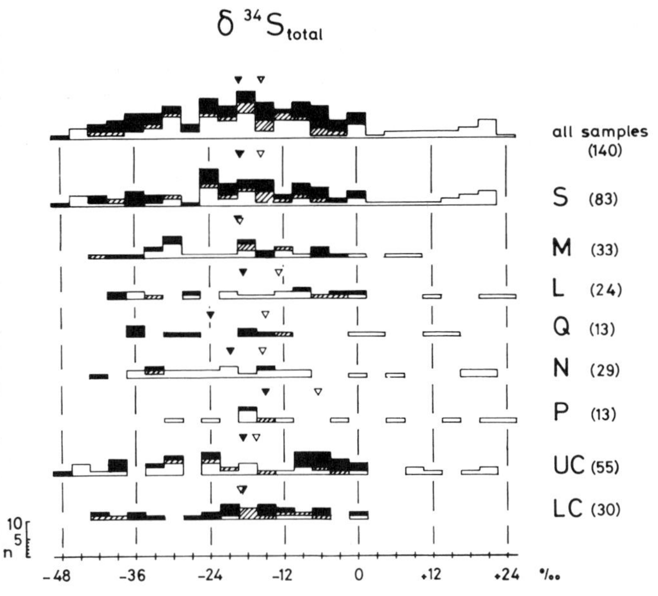

Fig. 10. $\delta^{34}S_{total}$ ratios of Atlantic Ocean sediments

tive mean sulfur isotopic composition (-19 ‰), followed by shales (-15.3 ‰) and limestones (-12.7 ‰). But a simple consideration shows that the arithmetic mean values might be erroneous. Regardless of their S_{total} content all samples have been treated equally. But 20 samples with a positive $\delta^{34}S$ ratio have a mean sulfur content of only 0.18% compared to 118 samples with negative $\delta^{34}S$ values with a sulfur mean value of 1.31%. This clearly shows that the bulk of sulfur being stored in marine sediments is light compared to seawater because it is pyrite formed by the activity of sulfate-reducing bacteria. For this reason the samples have been weighted according to their S_{total} contents. These "weighted" mean values are also shown in Fig. 10 (black triangles). Most mean values are shifted to more negative $\delta^{34}S$ ratios and, surprisingly, the difference of the mean values for the lithological units disappears (-19 ‰ for S, M, L.). On the other hand a trend is still visible for the five different age groups. Compared to the other groups the Paleogene samples (P) show the smallest fractionation (the most positive mean value). From P to Q a trend toward lighter values is recognizable. This minimum of fractionation during the P coincides with the lowest $C_{org.}$ and $N_{org.}$ mean values and low S_{total} contents.

Besides $\delta^{34}S_{total}$ also the sulfur isotopic composition of the sulfate and sulfide was determined (Fig. 11). Unfortunately the absolute amount of sulfate and sulfide has only been analyzed in a few samples. Therefore "weighted" mean values could not be calculated.

The difference of the mean $\delta^{34}S$ values for sulfide and sulfate ($\Delta\delta^{34}S$) is 16.6 ‰ for all samples. But there are still many sulfate samples which show negative $\delta^{34}S$ ratios. It seems that a portion of the sulfide is oxidized in the sediment itself or, more probably, during sample storage and treatment in the laboratory. The heavy sulfate values around +20 ‰ represent porewater but do not necessarily reflect the contemporaneous seawater isotopic composition. They may also represent the sulfate reservoir remaining after the metabolizable organic matter had been exhausted and might therefore be heavier than the seawater during that time. Another interesting trend is visible in the different lithological groups. The $\Delta\delta^{34}S$ difference increases from 10.9 ‰ in shales to 15.7 ‰ in marls and 33.4 ‰ in limestones. This is perhaps an effect created by the different availability of "reactive" iron and metabolizable organic matter.

In the five age groups the $\Delta\delta^{34}S$ is decreasing from 33.1 ‰ in the Quaternary (Q) to 2.5 ‰ in the Upper Cretaceous (UC), respectively 7.8 ‰ in the Lower Cretaceous (LC). Generally the sulfides show a trend toward more positive values with increasing depth (age), whereas the sulfate data show an opposing trend. The most plausible explanation is, that oxidation of sulfide to sulfate (probably during sample treatment, as mentioned before) is responsible for the $\delta^{34}S_{sulfate}$ shift towards lighter values. The sulfide data by contrast might indicate continuing bacterial activity in deep sedimentary layers. The sulfate supplied by diffusion (see porewater data by R.S. Miller et al., 1979), and to a less extent by the compactional flow, is completely reduced to sulfide in this "closed" system.

On the other hand it cannot be excluded that global changes in the seawater isotopic composition influence the $\delta^{34}S$ ratio of marine sediments and vice versa. As the fluxes between the individual sulfur reser-

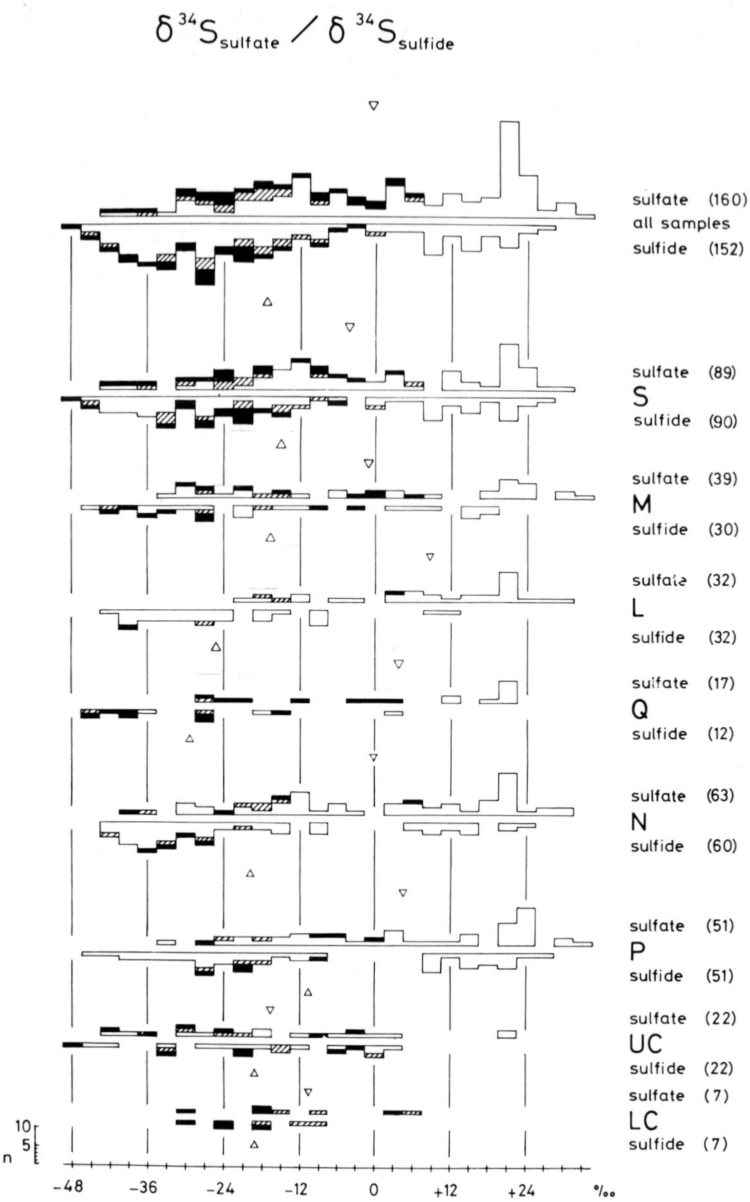

Fig. 11. $\delta^{34}S_{sulfate}$ and $\delta^{34}S_{sulfide}$ ratios of Atlantic Ocean sediments

voirs are not known for the geological past, it is not possible to calculate if the higher incorporation of reduced sulfur into Cretaceous sediments is at least partly responsible for the shift to heavier values in the evaporite age curve during that time.

4. Geochemistry of Black Shales from the Atlantic Ocean

Black shales are characterized by unusual amounts of $C_{org.}$ and S_{total}. Bituminous sediments may also contain considerable amounts of certain trace elements (Vine and Tourtelot, 1970), with various sources of the metals. Lange et al. (1977) have observed an accumulation of V, Cr, Ni, Cu, Zn, Mo, e.g., in six samples of Cretaceous black shales (Leg 41, Site 367) relative to average shales low in $C_{org.}$. Whereas the geological aspects of the genesis of balck shales have already been discussed in detail by several authors in the Initial Reports (e.g., Gardner et al., 1977; Jansa et al., 1977; Lancelot and Seibold, 1977), geochemical information on this type of sediment, covering large areas of the Atlantic, is still meager.

4.1 Minor Elements

Eighty-eight samples of Lower and Upper Cretaceous were chosen for chemical analysis. According to their $C_{org.}$ content, the samples were divided into three different groups:

Group 1 = 38 samples with $C_{org.}$ contents >1% (black squares in the following figures).

Group 2 = 11 samples with $C_{org.}$ >0.5 <1% (hatched squares).

Group 3 = 39 samples with $C_{org.}$ <0.5% (white squares).

Group 1 samples are subsequently called "black shales". A subdivision of these samples into different regional groups was not made because of the relatively small number of samples. But 24 out of 38 black shale samples originate from off NWA, therefore the mean values given in the following section should to a great extent also represent this area. Group 3 samples are chemically closely related to normal pelagic clays, marls, or carbonate oozes. Group 2 samples are neither black shales nor normal pelagic sediments.

The frequency distribution diagrams for $C_{org.}$, S_{total}, Fe_{total}, and Mn have already been shown in the first part of this publication. The mean values of the different groups are listed in Table 3. Frequency distributions of minor elements are plotted in Fig. 12. Zn is highly enriched in black shales (max. 7150 ppm) relative to normal shales (110 ppm). Cu, Cr, and Ni show a similar trend but are not accumulated to the same extent. On the other hand Pb does not show any preference for any group. The same is valid for Co, although a slight enrichment in $C_{org.}$-rich samples cannot be excluded.

V and Mo belong to those elements which are preferentially accumulated in black shales. Especially Mo (max. = 606 ppm → accumulation factor 233) shows, besides Ag (factor 329) the highest enrichment relative to normal shales. Ag was determined in 21 samples of Site 367 only, the extreme value is 23 ppm.

Sixteen samples of Site 105 were analyzed for Cd, Tl, and Bi. In contrast to Cd and Tl, Bi has not

Table 3. Mean values of Atlantic Ocean black shales (Group 1) and neighboring sediments

Element		Group 1	Group 2	Group 3
$C_{org.}$	(%)	6.18	0.69	0.20
S_{total}	(%)	2.19	1.44	0.76
$S_{sulfide}$	(%)	1.27	0.65	0.42
$S_{sulfate}$	(%)	0.90	0.80	0.35
Fe_{total}	(%)	3.54	4.15	4.43
Mn	(ppm)	910	720	1617
Zn	(ppm)	828	161	117
Cu	(ppm)	156	117	104
Co	(ppm)	27	28	25
Cr	(ppm)	202	164	134
Ni	(ppm)	186	90	98
Pb	(ppm)	16	14	19
V	(ppm)	822	272	207
Mo	(ppm)	59	8	3
Ag	(ppm)	4.5	0.7	0.8
Ba	(ppm)	530	382	401
Sr	(ppm)	278	279	164
B	(ppm)	84	73	97
B (Carb.-free)	(ppm)	97	117	107
Cd	(ppb)	923	-	933
Tl	(ppb)	1757	-	607
Bi	(ppb)	141	-	313

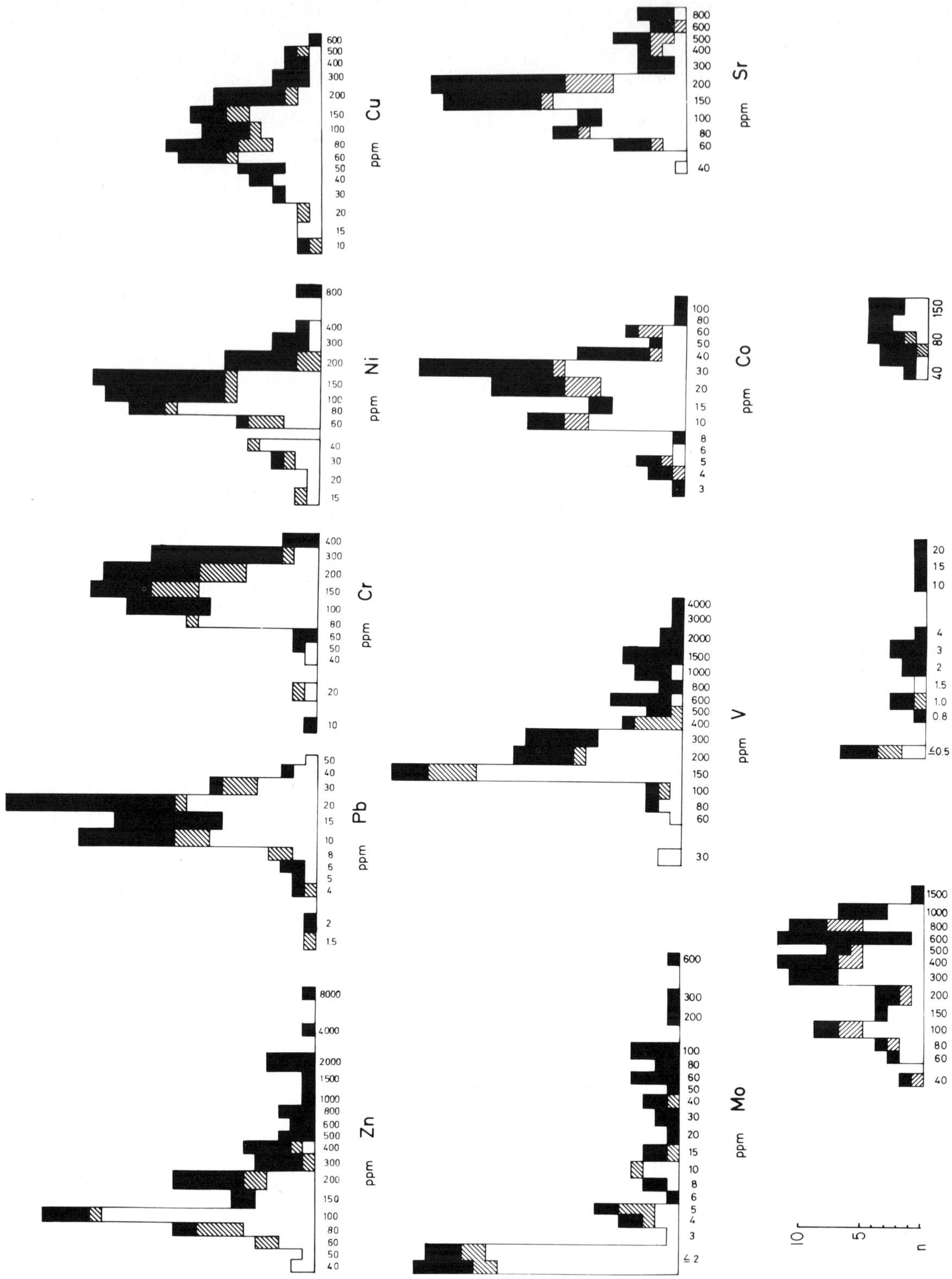

Fig. 12. Frequency distributions of minor elements in Atlantic Ocean black shales

been accumulated in black shales. For Ba, Sr, and B a definite trend for any group could not be detected. Ba shows a slight enrichment in shales relative to carbonate-rich samples. Sr is preferentially fixed in biogenic carbonates, whereas dolomite and ankerite do not show any enrichment. B (25 selected samples) is mainly fixed in the clay mineral fraction. The amount of B in clay minerals may be used as a paleosalinity indicator, if the minerals were able to equilibrate with seawater. Harder (1959) could demonstrate that clays deposited in the high salinity environment of evaporites have higher B concentrations (>1000 ppm) than those deposited in normal seawater (about 100 ppm) or brackish and freshwater (<80 ppm). The black shale samples show B concentrations about 100 ppm; there is no indication for a much higher salinity of the deep water.

4.2 Origin of Metals

As can be seen from the frequency distributions (Fig. 12) the Cretaceous black shales are characterized by a relatively high enrichment of heavy metals. With regard to the paleogeographic situation during the Cretaceous the following processes may lead to metal enrichments in sediments:

1. precipitation in marginal basins with water partly from continental runoff,
2. hydrothermal solutions in connection with ridge volcanism,
3. metabolism of marine organisms,
4. normal seawater.

1. An example for this kind of heavy metal enrichment is the Permian "Kupferschiefer", covering a large area in Central Europe. Wedepohl (1964, 1980) showed by trace-element and lead isotopic analysis that the metals were mobilized from sediments and precipitated as sulfides under stagnant water conditions. The largest amounts of sulfides occur in the former coastal areas. He showed by balance calculations that the "Kupferschiefer" metals cannot be derived from "normal" seawater. A comparison of analyses of the Atlantic black shales with "Kupferschiefer" data shows that especially the proportion of Pb, Ag, Cu and Zn is never reached in the Atlantic samples (Table 4).

2. Several metal-bearing hydrothermal deposits were recently discovered in connection with ridge volcanism (e.g., Bauer Deep). They are characterized by high concentrations of Fe, Mn, Cu, Ni, Co, and Zn (Dymond and Corliss, 1977; Bonatti et al., 1976). A comparison of the Cu, Ni, and Zn concentrations of the samples from the Bauer Deep with mean values of black shales shows the differences of both types. Black shales have

Table 4. Comparison of "Kupferschiefer", Atlantis II Deep brine, and a mock mixture of brine and average shale (SYNSED) with black shales from the Atlantic Ocean

Element	A	B	C	C/A	C/B	D	E	F
Pb	16	46	25000	1562.5	543.5	1500	23	171
Co	27	94	450	16.7	4.8	300	19	47
Cr	202	430	350	1.7	0.8	10	90	82
Ag	4.5	23.0	9000	2000.0	391.3	200	0.07	20
Ni	186	870	1000	5.4	1.1	50	68	66
Cu	156	635	25000	160.3	39.4	5000	39	535
V	822	3627	5000	6.1	1.4	100	130	127
Zn	828	7158	50000	60.4	7.0	26000	110	2699
Mo	59	606	1500	25.4	2.5	200	2.6	22

A = Mean values of 38 Atlantic black shales (Group 1).
B = Extreme values of Atlantic black shales.
C = Extreme values of "Kupferschiefer" (Wedepohl, 1964).
D = Atlantis II Deep brine (Miller et al., 1966).
E = Normal shales, low in $C_{org.}$ (Wedepohl, 1970).
F = SYNSED, consisting of 10% D and 90% E.
All values in ppm.

higher Zn and lower Cu and Ni contents.

Another comparison was made with the Atlantis II Deep brines from the Red Sea. These hydrothermal solutions are characterized by high concentrations of Zn, Cu, and Pb (Miller et al., 1966). A mock sediment (SYNSED) consisting of 10% of brine and 90% normal shale should have the composition computed in Table 4 (column F). While Co, Ag, Cu, Zn, and possibly Mo show about the same concentration, Pb is too much enriched and Cr, Ni, and V are depleted in SYNSED relative to black shales. Organic matter may accumulate V and Ni (in porphyrins), but no process is known which accumulates Cr exclusively. Also the generally low Pb concentration of the black shales should be taken into account.

Furthermore, it should be considered that the black shales cover a large area of the Atlantic Ocean and were deposited over a long geological period. No reasonable mechanism is known which supplies over a comparable time the required high amounts of hydrothermal solutions continuously over large areas. The Red Sea type deposits are formed from brines deposited in depressions close to the mounds and consist mostly of massive ore bodies.

On the other hand decreasing metal contents of the sediments should be expected with increasing distance from the ridge, equivalent to increasing metal concentrations with increasing depth in the core. This is definitely not the case in the cores investigated.

3. Few data are available on the chemical composition of $C_{org.}$-rich sediments deposited in regions of upwelling. Several authors (e.g., Brongersma-Sanders, 1966) suggest that marine organisms are responsible for the special heavy metal content of bituminous sediments. The correlations between metals and $C_{org.}$ (Table 5) should represent an indication for this process.

Table 6 lists a broad spectrum of literature data on heavy metals in marine plankton. It is difficult to obtain reliable mean values. Line A in Table 6 represents the most probable ones.

A hypothetical sediment composition was calculated for various $C_{org.}$ contents by using the data for normal shales (Table 4, column E) and the plankton data. However, by simple deposition of mixtures of unaltered organic material and shale such heavy metal concentrations as are found in the black shales are not attained. Therefore another sediment mixture was calculated on the assumption that 90% of the $C_{org.}$ is oxidized on its way through the water column, but the total load of heavy metals remains fixed in the 10% residue (10-fold enrichment). This sediment would contain enough Zn and Cu, but too much Pb and too little Ni, Cr, and V. Ni and V may be accumulated by porphyrins formed during diagenesis (Palmer and Baker, 1978), but the high Cr and low Pb concentrations of black shales cannot be explained by this mechanism.

Calvert and Price (1970) and Chester et al. (1978) have observed a direct relation between sediments and seawater particulates for the elements Cu, Pb, and Zn in the upwelling area off SW Africa. But this process of heavy metal accumulation seems to be of subordinate significance for the black shale genesis.

4. Seawater represents another possible source for the heavy metal content of black shales. Although these elements are only present in trace amounts (ppb range; Table 7, column E), the oceans represent an inexhaustable reservoir. On the other hand, these concentrations are above the solubility products of certain sulfides under seawater ionic strength. Therefore they should be precipitated under reducing conditions in water containing hydrogen sulfide (Brewer and Spencer, 1974).

Table 5. Correlation coefficients

Element	$C_{org.}$	S
Zn	0.91	0.70
Cu	0.50	0.67
Co	0.20	0.45
Ni	0.67	0.70
Cr	0.49	0.62
V	0.91	0.69
Pb	-0.03	0.24
Mo	0.63	0.56
Ag	0.96	0.80
S	0.71	-
HM	0.93	0.76

HM = Zn+Cu+Co+Ni+Cr+V

Table 6. Heavy metal contents of marine organisms

Ag	Co	Cr	Cu	Mo	Ni	Pb	V	Zn	
0.12-0.26[a]	0.8[c]	1.2[c]	(238[g])	3[g]	65[g]	83[g]	3.1[c]	2600?[h]	
0.67[b]	2[d]	1[a]	(200[h])	10[h]	36[h]	5[h]	3.1[a]	150[i]	
	1[a]	54?[e]	11[i]	0.45[i]	3[i]	8[i]	4.2[e]	150[c]	
	0.87[b]	4.9[b]	14[c]	0.3[c]	2.8[c]	2.1-31[a]	21[f]	165[d]	
		55?[f]	13[d]	5[d]	9[d]	160?[e]		180-780[a]	
			11.5-57.5[a]	2[a]	8.4-11.6[a]	11[b]		3700?[e]	
			340?[e]	56?[f]	83?[e]	64[f]		483[b]	
			39[b]	0.38[k]	8.1[b]	15[l]		880?[f]	
			280?[f]		93?[f]			150[m]	
0.4	1.2	2.4	24	0.4	7.2	12	3.6	180	A
1.0	3.0	6.0	60	1.0	18	30	9	450	B

[a]Martin and Knauer, 1973; [b]Fowler, 1977; [c]Yamamoto and Fujita, 1966; [d]Presly et al., 1972; [e]Bostrøm et al., 1974; [f]Moore and Bostrøm, 1978; [g]Nicholls et al., 1959; [h]Vinogradova et al., 1962; [i]Bowen, 1966; [k]Sugawara, 1961; [l]Wedepohl, 1974; [m]Ishibashi et al., 1965.
All values in ppm.
Column A = Mean value used for the calculation.
Column B = Mean value x 2.5; under the assumption that the dry matter of marine organisms consists of 40% $C_{org.}$.

Table 7. Comparison of the mean values of black shales from the Atlantic Ocean, normal shales, and seawater

Element	A	B	C	D	E	A/D	B/D	C/D
Pb	16	21	46	23	0.002[b]	0.70	0.91	2.00
Co	27	38	94	19	0.02[a]	1.42	2.00	4.95
Cr	202	257	430	90	0.2[d]	2.24	2.86	4.78
(Ag)	(4.5)	(4.9)	(23.0)	(0.07)	(0.28)[i]	(64.29)	(70.00)	(328.57)
Ni	186	263	870	68	0.5[k]	2.74	3.87	12.79
Cu	156	194	635	39	0.8[c,f]	4.00	4.97	16.28
V	822	1281	3627	130	1.9[h,l]	6.32	9.85	27.90
Zn	828	1207	7158	110	2.0[a,c]	7.53	10.97	65.07
Mo	59	84	606	2.6	11.0[e,g]	22.69	32.31	233.08

Order of elements in seawater: Pb < Co < Cr < (Ag) < Ni < Cu < V < Zn < Mo
Order of accumulation A/D : Pb < Co < Cr < Ni < Cu < V < Zn < Mo < (Ag)
Order of accumulation B/D : Pb < Co < Cr < Ni < Cu < V < Zn < Mo < (Ag)
Order of accumulation C/D : Pb < Cr < Co < Ni < Cu < V < Zn < Mo < (Ag)

A = Mean values of 38 Atlantic black shales; B = Atlantic black shales, carbonate-free; C = Extreme values of Atlantic black shales; D = Normal shales, low in $C_{org.}$ (Wedepohl, 1964); E = Seawater
Column A, B, C, and D in ppm, column E in ppb.

[a] Brewers et al., 1976; [b] Doe, 1974; [c] Knauer et al., 1973; [d] Loveridge et al., 1960; [e] Manheim and Landergren, 1978; [f] Moore, 1978; [g] Morris, 1975; [h] Riley and Taylor, 1972; [i] Schutz and Turekian, 1965; [k] Sclater et al., 1976; [l] Sugaware et al., 1956.

In Table 7 the mean values of the black shales, normal shales and normal seawater are listed in the order of metals with increasing seawater concentration. Comparing the order of metal contents of seawater with the order of accumulation of these elements in black shales (column A/D) a coincidence is conspicuous. Only Ag is out of order, but it shows a strong tendency to organic matter (cor. coeff. = 0.96) and accumulates in the mock sediment mixture of normal shale and decomposing organic matter.

In summary, normal seawater seems to represent the most important source of the heavy metals occurring in black shales. The biogenic portion can be neglected for several elements. It may only influence the Cu, Zn, and Ag contents of black shales. In our opinion the correlations between heavy metals and $C_{org.}$ (Table 5) are not a direct indication of the accumulation of these elements by organic matter, but show that the euxinic basins may act as a sink for both $C_{org.}$ and metals. The anoxic environment preserves organic matter from being oxidized and forms the condition for sulfate-reducing bacteria which produce the hydrogen sulfide needed for the precipitation of heavy metal sulfides. The occurrence of Fe, Zn, Ni, and Cu as sulfides could be demonstrated by microprobe scannings. Further information about the mechanism of fixation and accumulation of metals in black shales proportional to seawater concentrates is needed.

4.3 Models and Recent Examples

Two separate depositional environments, leading to anaerobic sediments, are distinguished today:

1. oxygen minimum zones in regions of high organic productivity, and
2. stagnant basins, due to restricted deep water circulation.

Recent examples exist for both and may be useful in understanding anaerobic environments.

Examples for the first case are the areas of upwelling along continental margins, like the northwestern Indian Ocean (Von Stackelberg, 1972) or off South West Africa (Calvert and Price, 1970). Highly fertile deep water reaches the surface producing large amounts of organic matter. The dissolved oxygen content of the underlying seawater approaches zero because it is consumed for the oxidation of the dead organic material settling through the water column. Anoxic sediments cover the seafloor where this mid-water oxygen minimum (at water depths ranging from 150 to 800 m) intersects the continental slope. At greater depths the oxygen content may increase again, benthic foraminiferal faunas may exist. Another example for this model are the sapropels, deposited at topographic heights of the Pacific during the Cretaceous (Ryan and Cita, 1977). Little is known about the chemical composition of such sediments. They seem to have higher concentrations of Cu, Ni, Pb, and Zn (Calvert and Price, 1970). Such sediments do not occur in the deep basin centers.

The Black Sea represents an intensively studied example for the second model (Degens and Ross, 1974). It is an enclosed basin with only a small connection to the Mediterranean Sea. The waters are stratified. Dense, saline seawater from the Mediterranean is overlaid by oxygenated, less saline surface water. A marked pycnocline exists as deep as 150 to 200 m. The water exchange between both volumes is restricted. According to Östlund (1974) the mean residence time of the deep water is 935 yr. Brewer and Spencer (1974) could show that considerable amounts of heavy metal sulfides are precipitated in the hydrogen sulfide-rich environment. In contrast to the first model, the sapropels are also deposited in the deeper central basins. Benthic faunas are missing. The following balance calculations will help to decide between the two models for the proto North Atlantic.

4.4 Calculation for the Explanation of the $C_{org.}$ Content of Black Shales

Generally there exist two mechanisms leading to sediments rich in $C_{org.}$. At any rate the oxidation of the major fraction of organic matter has to be prevented. Samples from delta areas, for example, often contain considerable amounts of organic material. The sediment accumulation rates (SR) are very high.

If the sediment accumulation rates are low (<20 m/m.y.) the environment of deposition and the primary production of organic matter (PP) is of decisive significance. Ryther (1969) distinguishes three regions according to their primary production:

1. open ocean 50 g C/m^2yr
2. nearshore areas 100 g C/m^2yr
3. upwelling areas 300 g C/m^2yr

Another important factor is the degree of preservation (PR), indicating what portion of the primary production is finally buried in the sediment. Menzel and Ryther (1970) argue that this degree is generally less than 0.1%. A simple calculation for the open ocean supports this fact:

$$\% \; Corg._{sed.} = \frac{PP \times PR}{SR \times D}$$

$$PR\;(\%) = \frac{Corg. \times SR \times D}{PP}$$

The open ocean values are: $C_{org.}$ = 0.3%, SR = 3 m/m.y., PP = 50 g C/m^2 yr, density (D) = 1.1 g/cm^3, resulting in a degree of preservation (PR) of 0.02%, which is in the same order of magnitude as the value given by Menzel and Ryther. In contrast to the open ocean this value is considerably higher (4%) for the Black Sea (Deuser, 1971). Table 8 shows the expected concentration of $C_{org.}$ if the sedimentation rate (SR), the primary production (PP), and the degree of preservation (PR) are varied. A calculation with changing SR, PP, and PR shows that high organic productivity is not necessary to obtain sediment with $C_{org.}$ in the 10% range. The rate of sediment accumulation (diluting of organic matter by detritus) and the degree of preservation (being in the 1% range in anoxic basins) are the main factors. The sediment accumulation rates of the black shales are not well determined, because of the bad preservation of microfossils; they are generally between 3 and 20 m/m.y.

Assuming a mean $C_{org.}$ concentration of 6%, a primary production of 50 to 100 g C/m^2yr, a sedimentation rate of 5 to 10 m/m.y., and a density of 1.1 g/cm^3, the resulting degree of preservation of the organic matter in Black Shales is about 1% which is similar to the Black Sea value.

5. Estimate of the Total Sulfur Content and the Mean Sulfur Isotopic Composition of Atlantic Ocean Sediments

Two very simple calculations were made, based on the following assumptions:

1. sediment mass of the Atlantic Ocean equals 5 x 10^{16} t,
2. the area of the Atlantic Ocean (106.5 x 10^6 km^2) was created at a constant spreading rate during the past 150 m.y.,
3. mean values from Table 2 are representative for the Atlantic,

4a. model No. 1: constant rate of sediment accumulation during all time intervals,

4b. constant total sediment input during all time intervals.

The calculation was made in 300 steps, each lasting 0.5 m.y. In Table 9 the sediment mass, being deposited in the five geological time intervals and the corresponding S_{total} and $\delta^{34}S$ mean values are listed for both

Table 8. Calculation for the explanation of the $C_{org.}$ content of black shales

PP	SR = 1.25	2.5	5	10	20	40	80	
50	0.73	0.36	0.18	0.09	0.05	0.02	0.01	
100	1.45	0.73	0.36	0.18	0.09	0.05	0.02	PR = 0.02%
300	4.36	2.18	1.09	0.55	0.27	0.14	0.07	
50	36.4	18.2	9.09	4.55	2.27	1.14	0.07	
100	72.7	36.4	18.2	9.09	4.55	2.27	1.14	PR = 1.00%
300	-	-	54.6	27.3	13.6	6.82	3.41	

PR = Degree of preservation (%). PP = Primary productivity (g C/m^2y).
SR = Sedimentation rate ($m/10^6y$). Density = 1.1 g/cm^3.

Table 9. Estimate of the total sulfur content and the mean sulfur isotopic composition of Atlantic Ocean sediments

Age group (m.y.)	Sediment mass x 10^{16} t		$\%S_{Total}$		S Mass x 10^{14} t		$\delta^{34}S$	$\delta^{34}S_{weighted}$
Model No. 1								
Q (1.5)	0.10		0.38		0.04		-15.2	-23.7
N (24.5)	1.48		0.24		0.36		-14.6	-21.1
P (39)	1.81		0.24		0.43		- 7.8	-15.0
		0.35[a]		2.77		0.97		
UC (35)	1.05		1.51		1.59		-16.5	-18.7
		0.70[b]		0.89		0.62		
		0.17[a]		2.00		0.34		
LC (36)	0.52		1.62		0.84		-19.0	-18.8
		0.17[c]		1.09		0.19		
		0.18[d]		0.22[e]		0.04		
All (136)	4.96				3.26	2.99	-15.8	-18.6
Model No. 2								
Q (1.5)	0.05		0.38		0.02		-15.2	-23.7
N (24.5)	0.82		0.24		0.20		-14.6	-21.1
P (39)	1.30		0.24		0.31		- 7.8	-15.0
		0.39[a]		2.77		1.08		
UC (35)	1.17		1.51		1.77		-16.5	-18.7
		0.78[b]		0.89		0.69		
		0.40[a]		2.00		0.80		
LC (36)	1.20		1.62		1.94		-19.0	-18.8
		0.40[c]		1.09		0.44		
		0.40[d]		0.22[e]		0.09		
All (136)	4.54				4.24	3.63	-16.9	-18.6

[a] 1/3 black shale; [b] 2/3 low $C_{org.}$ (<1%) samples; [c] 1/3 low $C_{org.}$ (<1%) samples; [d] 1/3 limestone; [e] mean value of 37 limestones ($C_{org.}$ <1%) from Q to P age groups.

models. The Cretaceous samples are furthermore divided into subunits, otherwise the total sulfur content would have been overestimated because of the rare sampling of LC limestones. It was assumed that 1/3 of the Cretaceous sediments comprises of black shales with $C_{org.}$ >1%. For the LC samples 1/3 of limestones was added, which S mean value was calculated from 37 samples from Quaternary to Paleogene with $C_{org.}$ <1% (mean value 0.22% S).

By using these data the total sulfur mass is 3.0 to 4.2 x 10^{14} t. The black shale contribution to the total sulfur accumulation is 44% to 52%, indicating the importance of this sediment type for the global sulfur cycle.

A more simple calculation shows similar results. Until the end of the Upper Cretaceous (65 m.y.) the Atlantic Ocean should have had an areal extension of about 6 x 10^7 km^2. Assuming a mean rate of sediment accumulation of 10 m/m.y. (density 1.1 g/cm^3) and a stagnant period of 10 m.y. (according to Jansa et al., 1978, the thickness of Cretaceous black shales often exceeds 100 m) about 6.6 x 10^{15} t of black shales were deposited. The mean S_{total} content of this sediment type is about

2%, resulting in a total sulfur mass of 1.3×10^{15} t. The results for model No. 1 (1.3×10^{15}t S) and model No. 2 (1.88×10^{14} t S) are in the same order of magnitude.

The mean $\delta^{34}S$ value is not very much affected by mass changes (see model No. 1 and 2), being about -18.6‰ (rel. CDT). Whether this mean value is the same for Pacific and Indian Ocean sediments remains an unanswered question because, for example, red clays, being an important sediment type, have not yet been sufficiently sampled and analyzed.

6. Conclusions

The 258 samples investigated show an age-dependent trend for $C_{org.}$, S_{total}, and $N_{org.}$ with general low concentrations during the Paleogene and high values during the Cretaceous. The Fe and Mn distributions reflect redox-conditions in the different sediment types. Mn is more mobile compared to Fe, Fe^{2+} is in situ fixed as iron sulfide. Cu and Zn show high accumulation in Cretaceous black shales.

Sulfur isotopes indicate the bacteriogenic origin of the sulfides. Paleogene samples show the most positive $\delta^{34}S$ mean value. This might perhaps reflect changes in productivity, water temperature, and resulting oxygen content.

The Cretaceous black shales are characterized by high amounts of $C_{org.}$ and S and show a special trace element composition. The bulk of heavy metals, being accumulated in the black shales, probably originated from normal seawater. The organic matter represents a mixture of marine and terrestrial sources.

The Cretaceous proto-North Atlantic can be compared to the present euxinic stage of the Black Sea.

Model calculations show that high organic productivity is not necessary to obtain sediments with $C_{org.}$ in the 10% range. 3.0 to 4.2×10^{14} t S are probably deposited in Atlantic Ocean sediments. The $\delta^{34}S$ mean value is -18.6‰ (rel. CDT), reflecting the bulk of sulfur being stored in reduced form as pyrite. Further information about the mechanism of fixation and accumulation of metals in black shales proportional to seawater concentrations is needed.

Acknowledgements. We would like to thank Prof. K. H. Wedepohl for helpful discussions and support of this investigation.

Profs. J. M. Gieskes, M. Kastner, K. H. Wedepohl, and J. Rullkötter kindly reviewed the manuscript.

This work was financially supported by the Deutsche Forschungsgemeinschaft to the first author.

REFERENCES

Arrhenius G (1950) Carbon and nitrogen in subaquatic sediments. Geochim Cosmochim Acta 1: 15-21

Barker PF, Dalziel IWD et al (1976) Initial reports of the DSDP, vol 36. US Government Printing Office, Washington, 1079 pp

Benson WF, Sheridan RE et al (1978) Initial reports of the DSDP, vol 44. US Government Printing Office, Washington

Berner RA (1970) Sedimentary pyrite formation. Am J Sci 268: 1-23

Bewers JM, Sundby B, Yeats PA (1976) The distribution of trace metals in the western North Atlantic off Nova Scotia. Geochim Cosmochim Acta 40: 687-696

Bolli HM, Ryan WBF et al (1978) Initial reports of the DSDP, vol 40. US Government Printing Office, Washington

Bonatti E, Honnorez-Guenstein MB, Honnorez J, Stern D (1976) Hydrothermal pyrite concretions from the Romanche Trench (Equatorial Atlantic): metallogenesis in oceanic fracture zones. Earth Planet Sci Lett 32: 1-10

Bordowskiy OK (1965) Accumulation of organic matter in bottom sediments. Mar Geol 3: 33-82

Bostrom K, Joensuu O, Brohm I (1974) Plankton: its composition and its significance as a source of pelagic sediments. Chem Geol 14: 255-271

Bowen HJM (1966) Trace elements in biochemistry. Academic Press, London New York

Bremner JM (1965) Inorganic forms of nitrogen. In: Black CA (ed) Methods of soil analysis, part 2. Agronomy 9: 1238-1255

Brewer PG, Spencer DW (1974) Distribution of some trace elements in Black Sea and their flux between dissolved and particulate phases. In: Degens ET, Ross DA (eds) The Black Sea - geology, chemistry and biology. Am Assoc Petrol Geol Mem 20: 137-143

Brongersma-Sanders M (1966) Origin of trace metal enrichment in bituminous shales. In: Hobson GD, Speers GC (eds) Advances in organic geochemistry. Pergamon Press, New York, pp 231-236

Brumsack H-J (1979) Geochemische Untersuchungen an kretazischen Atlantik-Schwarzschiefern der Legs 11, 14, 36 und 41 (DSDP). PhD Thesis, Göttingen, 57 pp

Brumsack H-J (1980) Geochemistry of Cretaceous black shales from the Atlantic Ocean (DSDP Legs 11, 14, 36 and 41). Chem Geol 31: 1-25

Calvert SE, Price NB (1970) Minor metal contents of recent organic-rich sediments off South West Africa. Nature (London) 227: 593-595

Chester R, Griffiths A, Stoner JH (1978) Minor metal contents of surface seawater particulates and organic-rich shelf sediments. Nature (London) 275: 308-309

Degens ET, Mopper K (1976) Factors controlling the distribution and early diagenesis of organic material in marine sediments. In: Riley JP, Chester R (eds) Chemical oceanography, vol 6. Academic Press, London New York, pp 59-113

Degens ET, Ross DA (1974) The Black Sea - geology, chemistry, and biology. Am Assoc Petrol Geol Mem 20: 633

Deuser WG (1971) Organic carbon budget of the Black Sea. Deep Sea Res 18: 995

Doe, pers comm (1974) Lead. In: Wedepohl KH (ed) Handbook of geochemistry, vol II/4 Springer, Berlin Heidelberg New York

Dymond J, Corliss JB (1977) History of metalliferous sedimentation at Deep Sea Drilling Site 319 in the South Eastern Pacific. Geochim Cosmochim Acta 41: 741-753

Edgar NT, Sanders JB et al (1973) Initial reports of the DSDP, vol 15. US Government Printing Office, Washington

Emery KO, Rittenberg SC (1952) Early diagenesis of California basin sediments in relation to origin of oil. AAPG Bull 36: 735-806

Erdman JG, Schorno KS (1977) Geochemistry of carbon: DSDP Leg 41. In: Lancelot Y, Seibold E et al (eds) Initial reports of the DSDP, vol 41. US Government Printing Office, Washington, pp 849-853

Fleige H, Meyer BU, Scholz H (1971) Fraktionierung des Boden-Stickstoffs für N-Haushalts-Bilanzen. Göttinger Bodenkundl Ber 18: 1-37

Fowler, SW (1977) Trace elements in zooplankton particulate products. Nature (London) 269: 51-53

Gardner JV, Dean WE, Jansa L (1977) Sediments recovered from the Northwest African continental margin, Leg 41, DSDP. Initial reports of the DSDP, vol 41. US Government Printing Office, Washington, pp 1121-1134

Gordon DC Jr (1971) Distribution of particulate organic carbon and nitrogen at an oceanic station in the central Pacific. Deep Sea Res 18: 1127-1134

Harder H (1959) Beitrag zur Geochemie des Bors, Teil II. Nachrichten der Akademie der Wissenschaften in Göttingen, II. Math-Phys Kl 6: 123-183

Hartmann M, Nielsen H (1969) $\delta^{34}S$-Werte in rezenten Meeressedimenten und ihre Deutung am Beispiel einiger Sediment-Profile aus der westlichen Ostsee. Geol Rundsch 58: 621-655

Hartmann M, Lange H, Seibold E, Walger E (1971) Oberflächen-sedimente im Persischen Golf und Golf von Oman. I. Geologisch-hydrologischer Rahmen und erste sedimentologische Ergebnisse. "Meteor" Forsch Ergebn C No 4: 1-76

Hays DE, Pimm AC et al (1972) Initial reports of the DSDP, vol 14. US Government Printing Office, Washington, 975 pp

Heinrichs H (1979a) Determination of bismuth, cadmium and thallium in 33 international reference rocks by flameless atomic absorption spectrometry. Fresenius Z Anal Chem 294: 345-351

Heinrichs H (1979b) Determination of lead in geological and biological materials by graphite furnace atomic absorption spectrometry. Fresenius Z Anal Chem 295: 355-361

Heinrichs H, Lange J (1973) Trace element analysis and microanalysis of silicate and carbonate rocks by flameless atomic absorption spectrophotometry. Fresenius Z Anal Chem 265: 256-260

Hermann AG (1975) Praktikum der Gesteinsanalyse. Springer, Berlin Heidelberg New York, 204 pp

Hermann AG, Knake D (1973) Coulometrisches Verfahren zur Bestimmung von Gesamt-, Carbonate- und Nichtcarbonat-Kohlenstoff in magmatischen, metamorphen und sedimentären Gesteinen. Fresenius Z Anal Chem 266: 196-201

Holland HD (1973) Systematics of the isotopic composition of sulfur in the oceans during the Phanerozoic and its implications for atmospheric oxygen. Geochim Cosmochim Acta 37: 2605-2616

Hollister CD, Ewing JI et al (1972) Initial reports of the DSDP, vol 11. US Government Printing Office, Washington, 1077 pp

Holser WT, Kaplan IR (1966) Isotope geochemistry of sedimentary sulfates. Chem Geol 1: 93-135

Ishibashi M, Fujinaga T, Yamamototo T, Fujita T, Watanabe K (1965) The zinc contents of seaweeds. J Chem Soc Jpn Pure Chem Sect (Nippon Kagaku Zasshi) 86: 728

Jansa L, Gardner JV, Dean WE (1977) Mesozoic sequences of the central North Atlantic. Initial reports of the DSDP, vol 41. US Government Printing Office, Washington, pp 991-1031

Jørgensen BB (1979) A theoretical model of the stable sulfur isotope distribution in marine sediments. Geochim Cosmochim Acta 43: 363-374

Kaplan IR, Emery KO, Rittenberg SC (1963) The distribution and isotopic abundance of sulfur in recent marine sediments off Southern California. Geochim Cosmochim Acta 27: 297-331

Keeney DR, Bremner JM (1966) Determination and isotope-ratio analysis of different forms of nitrogen in soils: 4. Exchangeable ammonium nitrate, and nitrite by direct-distillation methods. Soil Sci Soc Am Proc 30: 583-587

Kiba T, Kishi I (1957) Microcolorimetric determination of sulfate by reduction to hydrogen sulfide with Tin(II)-strong phosphoric acid. Chem Soc Jpn Bull 30: No 1, 44-48

Kiba T, Takagi T, Yoshimura Y, Kishi I (1955) Tin(II)-strong phosphoric acid. A new reagent for the determination of sulfate by reduction to hydrogen sulfide. Chem Soc Jpn Bull 28: no 9, 641-644

Knauer GA, Martin JH (1973) Seasonal variations of cadmium, copper, manganese, lead, and zinc in water and phytoplankton in Monterey Bay, California. Limnol Oceanogr 18: 597-604

Lancelot Y, Seibold E (1977) The evolution of the central Northeastern Atlantic - summary of results of DSDP Leg 41. Initial reports of the DSDP vol 41. US Government Printing Office, Washington, pp 1215-1245

Lancelot Y, Seibold E et al (1977) Initial reports of the DSDP vol 41. US Government Printing Office, Washington, 1259 pp

Lange J, Brumsack H-J (1977) Total sulphur analysis in geological and biological materials by coulometric titration following combustion. Fresenius Z Anal Chem 286: 361-366

Lange J, Wedepohl KH, Heinrichs H, Gohn E (1977) Notes about the chemical composition of "black shales" from Site 367 (Leg 41). Initial reports of the DSDP, vol 41. US Government Printing Office, Washington, pp 875-877

Laughton AS, Berggren WA et al (1972) Initial reports of the DSDP, vol 12. US Government Printing Office, Washington

Lew M (1980) Die Verteilung von Stickstoff, Schwefel, Schwefel-Isotopen sowie Mn, Zn, Fe und Cu in Sedimenten des atlantischen Ozeans (DSDP-Bohrkerne). Thesis, Göttingen, 44 pp

Loveridge BA, Milner GWC, Barnett GA, Thomas AM, Henry WM (1960) Determination of Cu, Cr, Pb, and Mn in seawater. At Energy Res Establ UK, r-3323

Manheim FT, Landergren S (1978) Molybdenum. In: Wedepohl KH (ed) Handbook of geochemistry, vol II. Springer, Berlin Heidelberg New York, 42-H-1 - 42-L-3

Marchig U (1972) Zur Geochemie rezenter Sedimente des Indischen Ozeans. "Meteor" Forsch Ergebn C 11: 1-104

Martin JH, Knauer GA (1973) The elemental composition of plankton. Geochim Cosmochim Acta 37: 1639-1653

Menzel DW, Ryther JH (1970) Distribution and cycling of organic matter in the oceans. In: Hood DW (ed) Organic matter in natural waters. Symp Univ Alaska Sept 2-4, 1968. Inst Mar Sci Occas Pub. 1: 31-54

Miller AR, Densmore CD, Degens E, Hathaway JC, Manheim FT, McFarlin PF, Pocklington R, Jokela A (1966) Hot brines and recent iron deposits in deeps of the Red Sea. Geochim Cosmochim Acta 30: 341-359
Moore C, Böstrom K (1978) The elemental composition of lower marine organisms. Chem Geol 23: 1-9
Moore RM (1978) The distribution of dissolved copper in the Eastern Atlantic Ocean. Earth Planet Sci Lett 41: 461-468
Morris AW (1975) Dissolved molybdenum and vanadium in the Northeast Atlantic Ocean. Deep Sea Res 22: 49-54
Müller PJ (1975) Diagenese stickstoffhaltiger organischer Substanzen in oxischen un anoxischen marinen Sedimenten. "Meteor" Forsch Ergebn C 22: 1-60
Nicholls GD, Curl HC, Bowen VT (1959) Spectrographic analysis of marine plankton. Limnol Oceanogr 4: 472
Nielsen H (1965) Schwefelisotope im marinen Kreislauf und das $\delta^{34}S$ der früheren Meere. Geol Rundsch 55: 160
Nielsen H (1979) Sulfur isotopes. In: Jäger J, Hunziker JC (eds) Lectures in isotope geology. Springer, Berlin Heidelberg New York, pp 283-312
Östlund HG (1974) Expedition "Odysseus 65": radiocarbon age of Black Sea water. In: Degens ET, Ross DA (eds) The Black Sea - geology, chemistry, and biology. Am Assoc Petrol Geol Mem 20: 127-132
Ohmoto H (1972) Systematics of carbon and sulfur isotopes in hydrothermal ore deposits. Econ Geol 67: 551-578
Palmer SE, Baker EW (1978) Copper porphyrins in deep-sea sediments: a possible indicator of oxidized terrestrial organic matter. Science 201: 49-51
Perch-Nielsen K, Supko PR et al (1977) Initial reports of the DSDP, vol 39. US Government Printing Office, Washington
Presley BJ, Kolodny Y, Nissenbaum A, Kaplan IR (1972) Early diagenesis in a reducing fjord, Saanich Inlet, British Columbia - II. Trace element distribution in interstitial water and sediment. Geochim Cosmochim Acta 36: 1073-1090
Ricke W (1964) Präparation von Schwefeldioxid zur massenspektrometrischen Bestimmung des S-Isotopenverhältnisses in natürlichen S-Verbindungen. Fresenius Z Anal Chem 199: 401-413
Riley JP, Taylor D (1972) The concentrations of cadmium, copper, iron, manganese, molybdenum, nickel, vanadium and zinc in part of the tropical northeast Atlantic Ocean. Deep Sea Res 19: 307-317
Ryan WBF, Cita MB (1977) Ignorance concerning episodes of oceanwide stagnation. Mar Geol 23: 197-215
Rhyter JH (1969) Photosynthesis and fish production in the sea. Science 166: 72-76
Sakai H (1968) Isotopic properties of sulfur compounds in hydrothermal processes. Geochem J 2: 29-49
Schidlowski M, Junge CE, Pietrek H (1977) Sulfur isotope variations in marine sulfate evaporites and the Phanerozoic oxygen budget. J Geophys Res 82: 2557-2565
Schutz DF, Turekian KK (1965) The investigation of the geographical and vertical distribution of several trace elements in sea water using neutron activation analysis. Geochim Cosmochim Acta 29: 259-313
Sclater FR, Boyle E, Edmond JM (1976) On the marine geochemistry of nickel. Earth Planet Sci Lett 31: 119-128
Silva JA, Bremner JM (1966) Determination and isotope-ratio analysis of different forms of nitrogen in soils: 5. Fixed ammonium. Soil Sci Soc Am Proc 30: 587-594
Stevenson FJ, Cheng CN (1972) Organic geochemistry of the Argentine Basin sediments: C-N relationships and Quaternary correlations. Geochim Cosmochim Acta 36: 653-671
Sugawara K, Naito H, Yamada S (1956) Geochemistry of vanadium in natural waters. J Earth Sci Nagoya Univ 4: 44
Sugawara K, Okabe S, Tanaka M (1961) Geochemistry of molybdenum in natural waters (II.). J Earth Sci Nagoy Univ 9: 114
Thode HG, Harrison AG, Monster J (1960) Sulphur isotope fractionation in early diagenesis of recent sediments of Northeast Venezuela. Am Assoc Petrol Geol Bull 44: 1809-1817
Tucholke BE, Vogt PR et al. (1979) Initial reports of the DSDP, vol 43. US Government Printing Office, Washington
Vine JD, Tourtelot EB (1970) Geochemistry of black shale deposits - a summary report. Econ Geol 65: 253-272

Vinogradova ZA, Kovalskiy UV (1962) Elemental composition of the Black Sea plankton. Dokl Acad Nauk SSSR 147:1458-1460 (in Russian)

von Stackelberg U (1972) Faziesverteilung in Sedimenten des indisch-pakistanischen Kontinentalrandes (Arabisches Meer). "Meteor" Forsch Ergebn C9:1-73

Wedepohl KH (1964) Untersuchungen am Kupferschiefer in Nordwestdeutschland: Ein Beitrag zur Deutung der Genese bituminöser Sedimente. Geochim Cosmochim Acta 28: 305-364

Wedepohl KH (1970) Environmental influences on the chemical composition of shales and clays. In: Ahrens LH, Press F, Runcorn SK, Urey HC (eds) Physics and chemistry of the earth, vol 8. Pergamon Press, Oxford New York, pp 305-333

Wedepohl KH (1974) Lead. In: Wedepohl KH (ed) Handbook of geochemistry, vol II/4. Springer, Berlin Heidelberg New York

Wedepohl KH (1980) The geochemistry of the Kupferschiefer bed in Central Europe. Eur Copper Depos Belgrade: 129-135

Yamamoto T, Fujita T (1966) A summary of chemical abundance in seaweeds. Bull hyoto Univ Educ Ser B 29:9-12

28 Geochemistry and Petrography of Organic Matter in Northwest African Continental Margin Sediments: Quantity, Provenance, Depositional Environment and Temperature History

Jürgen Rullkötter[1], Christopher Cornford[2], and Dietrich H. Welte[1]

ABSTRACT

This report synthesises the results of organic geochemical and organic petrographic analyses of sediments from deep sea drilling sites on the Northwest African continental margin. The investigation of the quantity, type, and composition of the organic matter and of the depositional environment leads to the recognition of a number of organofacies. Based on the organofacies, the sediments can be classified according to their differing hydrocarbon potentials, although the organic matter in all the sediments encountered has been determined to be immature, the onset of thermal hydrocarbon generation having not been reached at present burial depths.

1. Introduction

Deep sea drilling in the eastern North Atlantic off Northwest Africa (Fig. 1) has penetrated a large variety of clastic, hemipelagic and pelagic sediments in the continental slope and rise. Studying the organic matter in these sediments is of intrinsic interest to the organic geochemical community, has economic importance in terms of oil and gas potential of the sediments, and has helped in elucidating the general geological history of this area. The composition of organic matter down to the molecular level is sensitive to environmental conditions such as sediment provenance, climate, and water oxygen level and circulation patterns. Organic geochemical studies within the Deep Sea Drilling Project (DSDP) have considerably advanced our understanding of the relations between the organic matter that survives in the sediments and the organisms contributing to them. The examination of the nature of organic matter in the sediments, using microscopic, chemical and stable isotope methods, allows the recognition of characteristic assemblages of organic remains termed "organofacies". The assemblage of an organofacies is controlled by the rates of supply of organic matter of different origins, e.g., marine versus terrigenous material, its preservation (or destruction) under reducing or oxidizing conditions and the mechanisms operating during sedimentation.

The temperature sensitivity of organic matter, kerogen as well as small molecules, can be used to study both diagenetic and catagenetic processes, and define the organic maturity in marine sediments. Maturational changes, which range from microscopically determined parameters such as vitrinite reflectance and spore colour to molecular rearrangements and transformations, combined with mathematical methods, allow a reconstruction of the thermal history of the sediments. Together with the determination of organic facies and richness, the determination of maturity finally leads to an assessment of the hydrocarbon potential of deep sea sediments.

The following is a synthesis of organic geochemical and petrographic investigations of sediment cores from DSDP Legs 41 (Site 367, Cape Verde Basin; Site 368, Cape Verde Rise; Site 369, off Cape Bojador; Site 370, Morocco Basin), 47a (Site 397, off Cape Bojador) and 50 (Sites 415 and 416, Morocco Basin). An attempt is made to reinterpret the data taking into consideration the very recent results on organic geochemistry of deep sea sediments.

1 Institut für Erdöl und Organische Geochemie, Kernforschungsanlage Jülich GmbH, 5170 Jülich, FRG

2 British National Oil Corporation, Glasgow, Scotland

2. Quantity of Organic Matter

A compilation of data from the early phase of deep sea drilling (DSDP Legs 1 - 31) showed deep sea sediments to have a mean organic carbon content of 0.3%, with a median value of 0.1% (McIver, 1975). However, the range of samples was certainly not representative, and the organic carbon contents are most probably biased towards higher values. In the general case, sedimentation and preservation of organic matter is not favoured under deep water conditions. Extreme but typical examples, at least in terms of the sediment surface in the world's oceans, are the distal sites with little influx from terrestrial sources, like the Daito Ridge and Basin in the Philippine Sea, where an organic carbon content of 0.05% or less is the rule rather than the exception (Rullkötter et al., 1980). Similarly low organic carbon contents were observed in pelagic sediments at the Sierra Leone Rise, off West Africa, DSDP Site 366 (Lancelot, Seibold et al., 1978a).

Welte et al. (1979) have attempted a partial carbon mass balance for a typical hemipelagic sediment. Using data from the Atlantic ocean they note that typical near-surface water (< 400 m depth) contains about 30 - 750 μg/l of particulate matter (PM) of which some 25 - 300 μg/l is particulate organic carbon (POC). Typical figures for deep ocean waters are 10 - 25 μg PM/l and 5 - 10 μg POC/l, respectively. Thus, while particulate matter falling through the deep water column contains some 5% - 50% organic carbon, the underlying sediments typically contain 0.2% - 0.5% total organic carbon. After having taken differential settling into account, Welte et al. (1979) conclude that there is at least a tenfold reduction in organic carbon relative to mineral particles at and immediately above the deep sea floor (i.e., within the nepheloid layer).

2.1 Tertiary Sediments

At DSDP Site 397 (south of the Canary Islands) on the NW African continental margin, the Miocene and post-Miocene sections of the interval down to 690 m sub-sea floor are an example of continuous deposition of upper continental rise hemipelagic sediments. The organic carbon contents of the sediments are low (Fig. 1) and average 0.17% for the Miocene to Pliocene and 0.45% for the Pliocene to Recent intervals, respectively (von Rad, Ryan et al., 1979). As will be shown later in more detail, most of this organic matter is comprised of rather inert allochthonous terrigenous material, which apparently has undergone long-distance transportation. Similar observations were made for most of the other organic-lean sediments at the continental margin of NW Africa (e.g., Tertiary at Site 367, Fig. 1) confirming the generally low preservation rates of the easily degradable marine organic matter in deep ocean waters.

In contrast to this, some other Tertiary sediments contain unusually high amounts of organic matter (Fig. 1). At Site 397, organic carbon values up to nearly 5% were encountered mainly in the Lower Miocene sections. The sedimentology of the cores showed that the sediments were deposited from slumps and turbidites. The type of microfauna present indicated that original deposition occurred in the outer shelf or upper slope environment, slumping and turbidite transport leading to final deposition of the organic-carbon-rich Miocene mudstones in the oxidizing waters of the upper continental rise (Cornford, 1979; Cornford et al., 1979). The contrasting modes of deposition were indicated by the light-coloured organic-carbon-lean hemipelagic sediments interbedded with dark grey to olive organic-carbon-rich allochthonous units. At the shelf edge or upper slope, upwelling together with an associated mid-water oxygen-depleted layer led to the accumulation of the darker, olive-coloured organic-carbon-rich sediments. The low oxygen concentration reduced the degree of degradation in the water column and of benthonic reworking at the sea floor. Rapid burial after slumping or turbidite transport prevented extensive degradation of the organic matter in the oxidizing waters at the continental rise. Slumping was also observed in the pelagic sediments recovered at DSDP Site 369, off Cape Bojador. Rapid burial may also explain the relatively high organic carbon contents in the pre-Lower Miocene sediments at this site (Fig. 1). However, no detailed organic geochemical data are available for these samples.

2.2 Cretaceous Sediments

Cretaceous black shales found at DSDP Site 367, Cape Verde Basin, and 368, Cape Verde Rise, represent the other extreme of organic matter richness in the sediments of the NW African continental margin. Values greater than 10% organic carbon were observed (Fig. 1). Preservation of these large amounts of organic matter is a typical feature of highly reducing conditions in the sediments. In the deep sea, reducing environments can occur in the water column overlying the sediment in areas of restricted water circulation, as in the Black Sea. However, high bioproductivity associated with a high sedimentation rate may

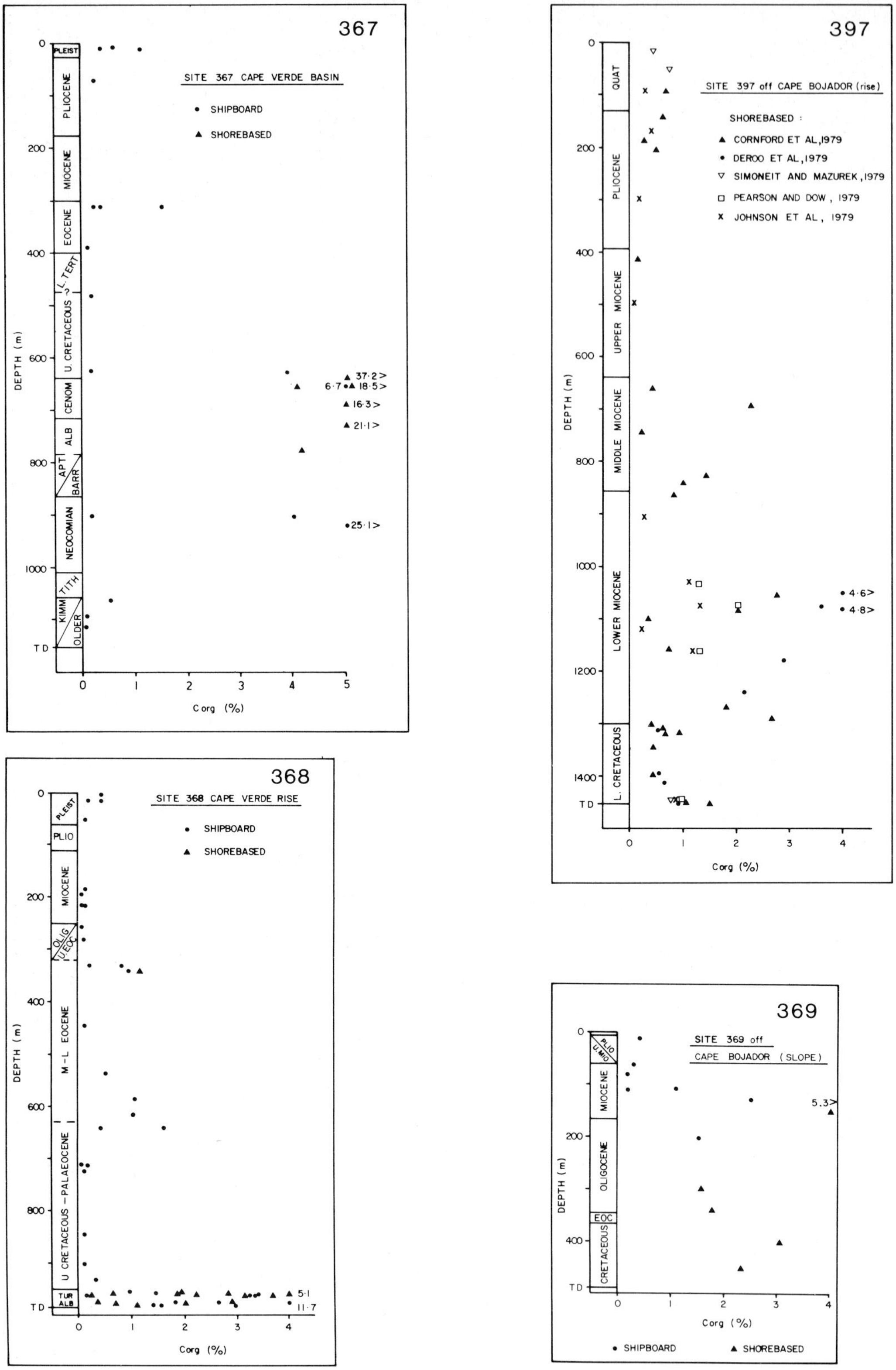

367
SITE 367 CAPE VERDE BASIN
• SHIPBOARD
▲ SHOREBASED
PLEIST
PLIOCENE
MIOCENE
EOCENE
L. TERT
U. CRETACEOUS ?
CENOM
ALB
APT
BARR
NEOCOMIAN
TITH
KIMM
OLDER
DEPTH (m)
Corg (%)
37·2>
6·7
18·5>
16·3>
21·1>
25·1>
397
SITE 397 off CAPE BOJADOR (rise)
SHOREBASED :
▲ CORNFORD ET AL,1979
• DEROO ET AL,1979
▽ SIMONEIT AND MAZUREK,1979
□ PEARSON AND DOW, 1979
x JOHNSON ET AL, 1979
QUAT
PLIOCENE
UPPER MIOCENE
MIDDLE MIOCENE
LOWER MIOCENE
L. CRETACEOUS
4·6>
4·8>
368
SITE 368 CAPE VERDE RISE
• SHIPBOARD
▲ SHOREBASED
PLEIST
PLIO
MIOCENE
OLIG
U. EOC
M-L EOCENE
U CRETACEOUS - PALAEOCENE
TUR
ALB
5·1
11·7
369
SITE 369 off
CAPE BOJADOR (SLOPE)
PLIO
U. MIO
MIOCENE
OLIGOCENE
EOC
CRETACEOUS
5.3>
• SHIPBOARD
▲ SHOREBASED

Fig. 1

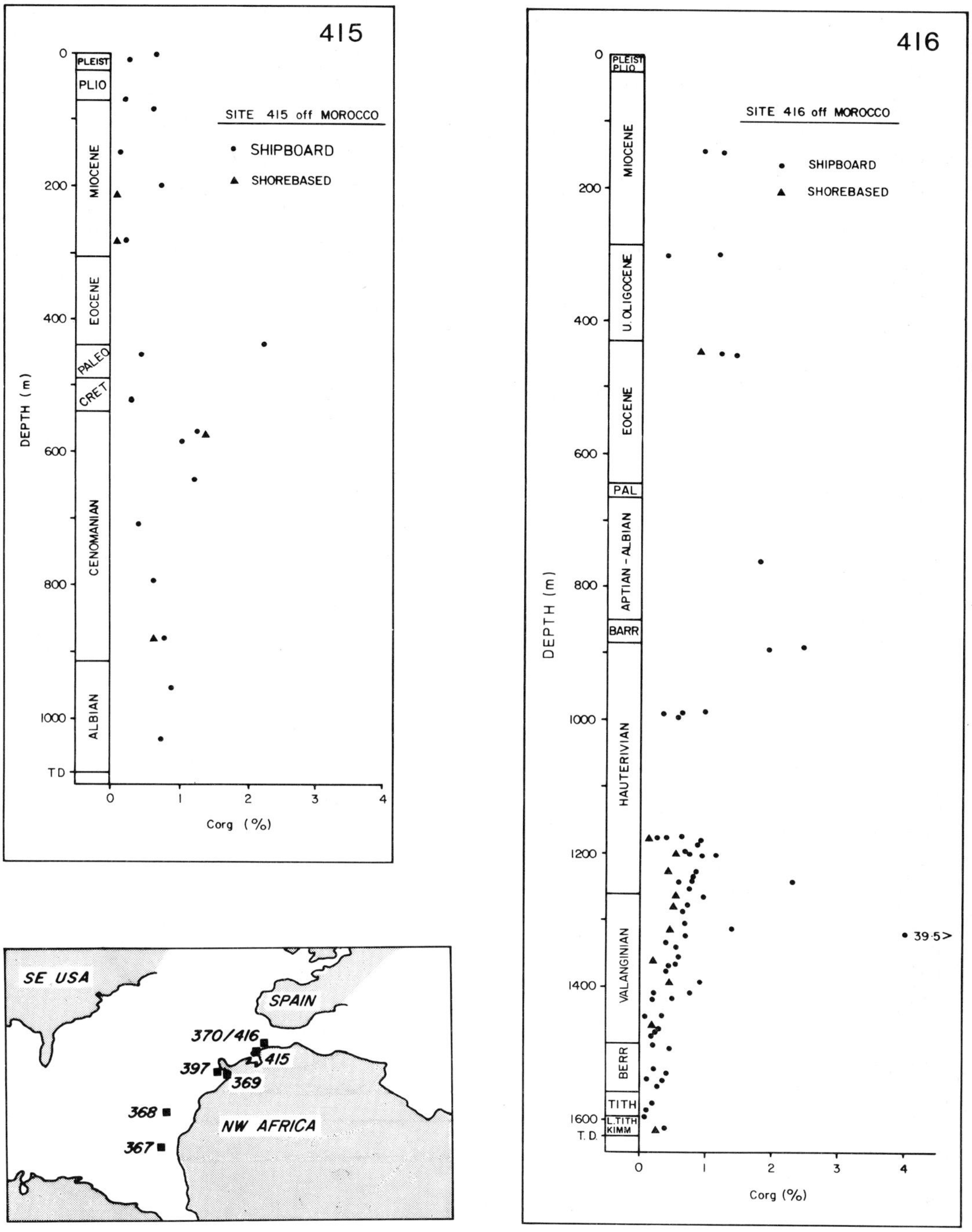

◄ Fig. 1
DSDP drill site locations on the NW African continental margin and total organic carbon versus depth profiles for each site. Unless stated otherwise, organic carbon data are taken from the corresponding site chapters in the Initial Reports of the Deep Sea Drilling Project: Lancelot, Seibold et al., 1978b,c,d (Sites 367, 368, and 369, respectively); Lancelot, Winterer et al., 1980a,b (Sites 415 and 416, respectively)

also reduce the oxygen level in the water column and lead to high preservation rates of organic matter.

The deposition of Cretaceous black shales in the southern part of the central North Atlantic (DSDP Sites 367 and 368) has been explained by Tissot et al. (1980) in terms of oxygen-depleted ocean waters in these areas due to restricted circulation in Cretaceous times. This restricted circulation also explains why corresponding facies were not encountered at the northern Sites 370, 397, 415, and 416. Here, the Cretaceous sections are dominated by turbidite sediments in prodelta, continental rise or deep-sea fan settings giving rise to the transport of terrestrial matter into these continental margin sites. Organic matter contents in most cases range from 0.3% to 1.0% organic carbon and rapid burial may have led to preservation of at least some of the autochthonous organic matter. A special case appears in the Cretaceous sediments of Site 369 where laminated hemipelagic sediments fairly rich in organic matter (Fig. 1) were deposited under conditions of low oxygen concentration in an area isolated from the mass transports of terrigenous matter.

3. Type and Composition of Organic Matter

Characterisation of the nature of organic matter preserved in sediments includes an evaluation of the type and source of organic matter as well as an assessment of the environmental conditions during deposition. A combined evaluation of these data has led to the recognition of different organofacies (Cornford et al., 1981). This concept allows a prediction of the fate of organic matter under conditions of increasing depth of burial and temperatures, i.e., in terms of hydrocarbon generation (Rogers, 1980). A number of techniques can be used to distinguish terrestrial from aquatic organic matter and the extent of bacterial and/or benthonic reworking. The four major methods are visual kerogen discription by microscopy, interpretation of pyrolysis yields, investigation of extractable organic matter by gas chromatography and combined gas chromatography/mass spectrometry, and stable carbon isotope measurements. Results of such techniques are summarized in Table 1 for a series of organofacies encountered in the NW African continental margin sediments.

3.1 Kerogen Microscopy, Pyrolysis, n-Alkane Distributions

As mentioned in the previous section, the hemipelagic and pelagic sediments that drape on the continental slope and rise, as exemplified by the autochthonous Tertiary sediments at Site 397 (Fig. 1), were found to contain dominantly detrital terrigenous organic matter in small quantities (<0.5% organic carbon). Vitrinite, altered to lesser or greater degree during long-distance transport, and inertinite are the dominant kerogen components identified under the optical microscope (Cornford et al., 1979). Occasional well-preserved algal bodies suggest chance preservation of the planktonic component. Spores and cuticles are consistent minor components. Simoneit and Mazurek (1979) have suggested a significant eolian contribution to these sediments on the basis of n-alkane distributions reaching a maximum at n-C_{31} with a high dominance of odd carbon-number

Table 1. Summary of kerogen type and hydrocarbon potential for DSDP sites off the Northwest African continental margin

Site/Interval	Source of organic matter	Mean C_{org}	Hydrocarbon potential	Comments
367, 368 (Albian, Aptian, Cenomanian)	Aquatic, planktonic and sedimentary bacteria	> 5%	Very high for oil and gas	Possibly thin unit at Site 369
397 (allochthonous Miocene slumps)	Mixed terrestrial and marine planktonic	1%-5%	High for oil and gas	Derives from outer shelf/ upper slope oxygen minimum
397, 369 415, 416/370 (Cretaceous)	Terrestrial with subsidiary aquatic	0.5%-1%	Moderate to low for gas only	Oxidation during transport reduces hydrocarbon potential
397, 415 (hemi-pelagic Tertiary)	Residual terrestrial	< 0.5%	Nil	

molecules and the presence of long-chain fatty acids and diterpenoid hydrocarbons. Pelagic nannofossil and foraminiferal oozes were encountered in the Neogene at Site 415 (Fig. 1), where analysis again indicates the preferential survival of a dominantly terrigenous type of organic matter (Boutefeu, 1980; Claypool and Baysinger, 1980; Schorno and Erdman, 1980).

In contrast to the terrigenous kerogens of the hemipelagic sediments, organic-carbon-rich Miocene slumps and turbidite flows at Site 397 (Fig. 1) contain a highly variable kerogen type (Cornford, 1979) with significant amounts of algal and amorphous liptinites of marine origin intermixed with terrestrial higher plants debris (spores, pollen, vitrinite, inertinite) (Cornford et al., 1979). The richly heterogeneous nature of the kerogen was confirmed by pyrolysis results and H/C ratios indicating a mixture of type III (vitrinite) and type II (spore/ pollen/algal) kerogen (Deroo et al., 1979), by bimodal n-alkane distributions (maxima at n-C_{17} to n-C_{20} and n-C_{27}) with a high percentage of branched and cyclic non-aromatic hydrocarbons (Cornford et al., 1979), and by carbon isotope values fluctuating in the range -20.1‰ to -22.5‰ $\delta^{13}C$ (Erdman and Schorno, 1979). The $\delta^{13}C$ value is defined as the carbon isotope ratio ($^{13}C/^{12}C$) of a sample relative to a standard, commonly the belemnite from the Pedee formation in South Carolina (e.g., Hunt, 1979):

$$\delta^{13}C = \left[\frac{(^{13}C/^{12}C)_{sample}}{(^{13}C/^{12}C)_{standard}} - 1\right] \times 1000$$

Two distinct organofacies were identified in the Cretaceous sediments penetrated in the continental rise and slope of the eastern North Atlantic. At the two most southerly locations (Sites 367 and 368) the Cretaceous consisted of an upper Cretaceous claystone section with an organic-carbon-rich black shale facies developed in the Late Aptian - Cenomanian at Site 367 and in the Turonian - Albian at Site 368. Kerogen at Site 367 was demonstrated by pyrolysis results to be hydrogen-rich (type II), while the n-alkane distributions indicate at least some admixture of terrigenous organic matter (Deroo et al., 1978). Phytane is dominant over pristane (ratios up to 3:1) indicating a highly reducing environment of deposition at the low level of maturity (Deroo et al., 1978; Simoneit, 1978). Dow (1978a) and Kendrick et al. (1978) describe a single black shale sample from Site 367 containing a dominance of amorphous kerogen but with subsidiary amounts of humic matter of terrestrial origin.

At Site 368 the black shale section was intruded by a series of diabase sills. The heat from these sills affected the maturity locally (see below); the kerogen type was described by Dow (1978a) as amorphous with subsidiary humic and liptinitic components. A dominantly marine algal origin was also indicated by pyrolysis results (Deroo et al., 1978). The n-alkane distributions were in part affected by the intrusions (see below) but indicated a mixed terrigenous and marine source.

Of the five sites further north (Sites 369, 397, 370, 415, and 416; Fig. 1), Sites 369, 370, 415, and 416 penetrated the lateral equivalents of the more southerly black shale facies. However, at all sites lower quantities of organic matter composed mainly of terrigenous kerogen were found generally associated with turbidite or pro-delta facies. Kendrick et al. (1978) reported on a single Albian sample from Site 369 with an organic carbon content of 2.3% and having a dominance of amorphous kerogen. This sample may represent a thin northerly extension of the black shale facies to the Site 369/397 area off Cape Bojador. The Upper Cretaceous and post-Hauterivian Lower Cretaceous sections are missing at Site 397, having been cut out by a Miocene to Hauterivian unconformity.

The dominantly terrigenous organic debris found in the Cretaceous sediments of the more northerly sites is characterized by vitrinitic kerogen with subsidiary spores and pollen (Cornford et al., 1979; Cornford, 1980). Pyrolysis yields indicate a type III terrigenous kerogen (Boutefeu, 1980). N-alkane distributions are dominated by long-chain molecules in the n-C_{27} to n-C_{31} range (Cornford et al., 1979; Galimov et al., 1980; Brassell et al., 1980). These sediments are probably the distal remains of the system of Lower Cretaceous deltas bounding the North Atlantic region.

The deltaic facies can also be characterized at the molecular level. Figure 2 shows the capillary column chromatograms of the non-aromatic hydrocarbon fractions of the dichloromethane extracts from two Cretaceous sediment samples from Sites 415 and 416, respectively. The lower trace (Fig. 2b) represents the n-alkane distribution of a composite Valanginian/Berriasian turbidite sample (0.16% and 0.18% C_{org}). The dominance of the long-chain n-alkanes indicates an origin from terrestrial higher plant waxes (Eglinton and Hamilton, 1963). The extremely low amounts of branched and cyclic components in the same fraction, together with the dominance of pristane over phytane, supports the idea of a long-distant transport through oxidizing waters, where only the most resistant n-alkanes survived.

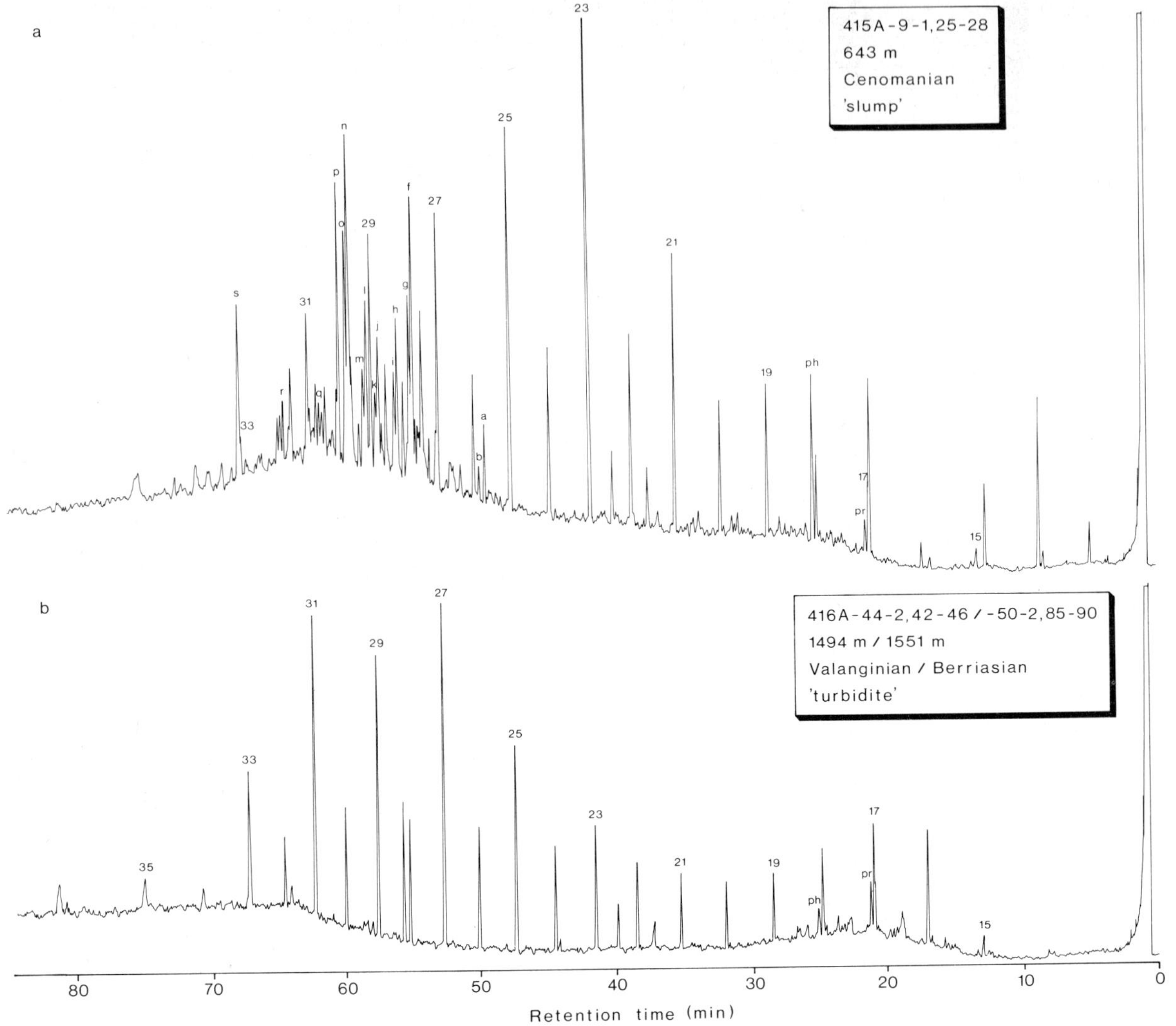

Fig. 2
Capillary column chromatograms of non-aromatic hydrocarbon fractions of two Cretaceous sediment samples from DSDP Sites 415 and 416. Branched and cyclic compounds identified are listed in Table 2; numbers indicate n-alkanes. Conditions: Siemens L 402 gas chromatograph with FID detector; 20 m x 0.3 mm i.d. SE 54 glass capillary column (Jaeggi); carrier gas: hydrogen; temperature programme: 80°C (2 min) - 80°C to 250°C (3°C/min) - 250°C (20 min)

In contrast to this observation a greater preservation of more labile organic matter is evident in the non-aromatic hydrocarbon fraction (Fig. 2a) obtained from a Cenomanian sediment sample within a slump interval (0.94% C_{org}). Numerous branched and cyclic hydrocarbons are present in addition to wax alkanes that appear to be affected to some extent by microbial alteration as indicated by the maximum at n-C_{23} (e.g., Simoneit et al., 1979; and references therein). Furthermore, the low pristane/phytane ratio indicates that the organic matter in this case was not exposed to oxygen-rich waters for extended periods.

Evidence from vitrinite reflectance histograms has also been used to show that much of the terrigenous organic matter has been transported long distances and underwent oxidation prior to deposition (Cornford et al., 1981). Broad and multi-modal distributions (Fig. 3) in the absence of evidence of reworked older and more mature kerogens or coals (e.g., bi- and trimacerites, reworked spores, anisotropy of reflectance), can be interpreted to result from oxidation during transport. The following interpretations can be given:

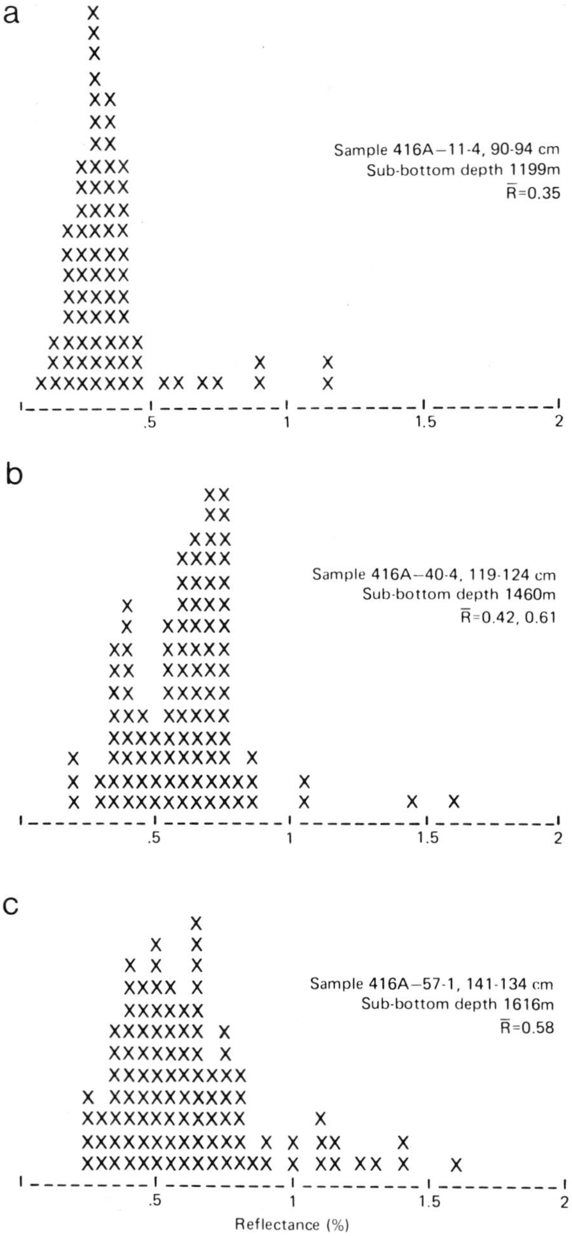

Fig. 3
Vitrinite reflectance histograms for three samples from DSDP Site 416 showing a unimodal, b bimodal, and c broad unstructured distributions. (After Cornford, 1980)

- Broad high reflectance peak and low organic carbon content (e.g., $< 0.5\%$): extensive destructive oxidation (Fig. 3c).
- Multi-modal reflectance peaks with higher organic carbon contents: partial oxidation of abundant land plant material (Fig. 3b).
- Single low reflectance peak: proximal source of land plant debris or protection from oxidation in a reducing environment, possibly associated with high sedimentation rate (Fig. 3a).

3.2 Detailed Composition of Extractable Organic Matter

A detailed study of the extractable organic matter in sediments helps to define the source, the depositional environment, and the degree of diagenetic and catagenetic alteration of the organic matter. Important progress has been made recently in relating the organic compounds of a wide variety of lipid classes within the extracts to different biosystems (Brassell et al., 1981). Taxonomic clues are drawn from biological markers which are not altered after incorporation into the sediment or altered to such a slight extent only that the original characteristics of the molecular structures are maintained. For the sediments so far recovered from the NW African continental margin, however, only a limited number of more detailed extract analyses are available.

The non-aromatic hydrocarbons are the most widely investigated extract fraction although they only represent a small portion of the total extracts. As an example, Figure 4 shows the capillary column chromatogram of the non-aromatic hydrocarbons of an Albian black shale from DSDP Site 367. Long-chain n-alkanes with a maximum at n-C_{29} indicate the presence of terrestrial higher plant material (Eglinton and Hamilton, 1963). The clear predominance of molecules with odd-numbered carbon atoms points to a low stage of diagenetic evolution. A second n-alkane maximum at n-C_{25} is most probably due to an element of microbial alteration (Simoneit et al., 1979; and references therein). Among the polycyclic non-aromatic hydrocarbons a number of major components could be identified by combined gas chromatography/mass spectrometry (Table 2). High concentrations of hopane-type pentacyclic triterpenoids are supporting evidence that major amounts of microbial biomass were incorporated into the sediment (Ourisson et al., 1979). The presence of 17β(H)-hopanes (VIII; cf. Fig. 5 for molecular structures) and the nearly complete absence of 17α(H)-hopanes again indicates the low maturity level of the organic matter (van Dorsselaer et al., 1977) as does the degree of unsaturation of the steroid hydrocarbons. The sterenes (II, III) and steradienes (IV, V) are early diagenetic alteration products of sterols (Dastillung and Albrecht, 1977). The abundance and type of distribution that have similarly been found in the slump sequences at Site 397 (Cornford et al., 1979) and Site 415 (Fig. 2a) and in an early Aptian turbidite sample at Site 416 (Rullkötter, unpub-

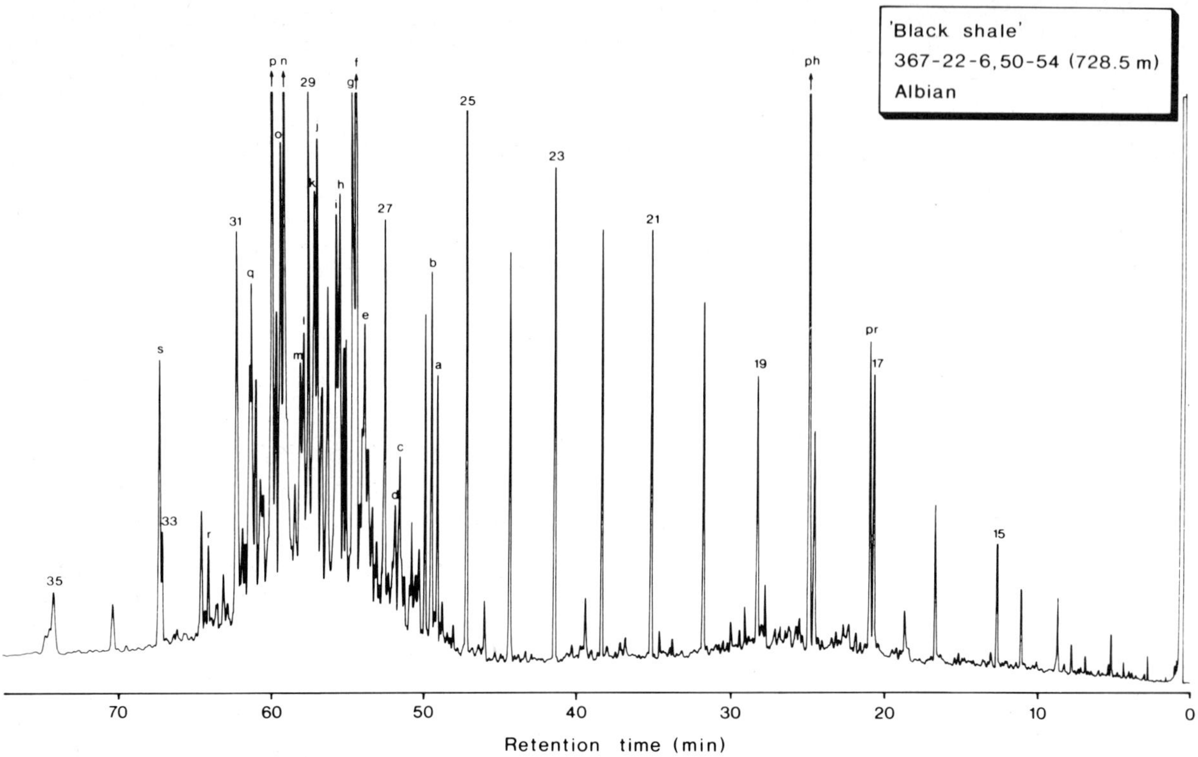

Fig. 4
Capillary column chromatogram of the non-aromatic hydrocarbon fraction of an Albian black shale sample from DSDP Site 367. Branched and cyclic compounds identified are listed in Table 2; numbers indicate n-alkanes. Conditions as given for Fig. 2

Table 2. Branched and cyclic compounds identified in the non-aromatic hydrocarbon fractions of a slump sediment from DSDP Site 415 (cf. Fig. 2) and a black shale sample from DSDP Site 367 (cf. Fig. 4). For structural formulae see Fig. 5

	Compound	Structure
pr	pristane	
ph	phytane	
a	C_{27}-diasterene	I, R = H
b	C_{27}-diasterene	I, R = H
c	C_{28}-diasterene	I, R = CH_3
d	C_{28}-diasterene	I, R = CH_3
e	C_{29}-diasterene	I, R = C_2H_5
f	cholest-4-ene	II, R = H
g	cholest-5-ene	III, R = H
h	24-methylcholesta-4,22-diene	IV, R = CH_3
i	24-methylcholesta-5,22-diene	V, R = CH_3
j	24-methylcholest-4-ene	II, R = CH_3
k	24-methylcholest-5-ene	III, R = CH_3
l	24-ethylcholesta-4,22-diene	IV, R = C_2H_5
m	24-ethylcholesta-5,22-diene	V, R = C_2H_5
n	24-ethylcholest-4-ene	II, R = C_2H_5
o	24-ethylcholest-5-ene	III, R = C_2H_5
p	hop-17(21)-ene	VI
q	neohop-13(18)-ene	VII
r	17β(H)-hopane	VIII, R = CH_3
s	17β(H)-homohopane	VIII, R = C_2H_5

lished), are interpreted as indication of favoured preservation of marine organic matter under anoxic conditions. The dominance of phytane over pristane (Fig. 4) gives supporting evidence for anoxic conditions during sedimentation. Diasterenes (or rearranged sterenes, I) found in the black shale sample (Fig.4) are rearrangement products of regular sterenes (II, III) and point to an elevated acidity in the sediment matrix (Rubinstein et al., 1975).

In a Late Cretaceous silty clay with interbedded black shale about 100 m further up in the same hole (Section 367-17-2) these diasterenes were not detected although the sterene/steradiene distribution was very similar to the one in the black shale shown in Fig. 4 (Rullkötter, unpublished). Diasterenes were found as major steroidal hydrocarbons, however, in Valanginian turbidite sediment samples from Site 416 by Brassell et al. (1980). An unusually high concentration of a C_{28}-triterpane, subsequently determined to be 17α(H),18α(H),21β(H)-28,30-bisnorhopane (Seifert et al., 1978), was detected by Cornford et al. (1979) in a Lower Miocene sediment sample (1241 m) from Site 397. The significance of this compound, its source, and the

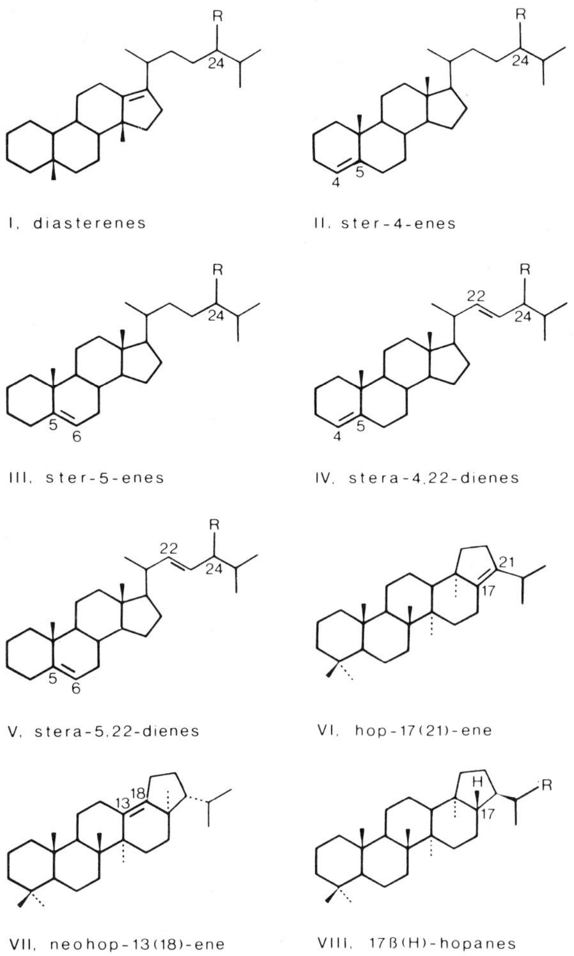

Fig. 5
Structural formulae of polycyclic non-aromatic hydrocarbons mentioned in the text

reason for its abnormal abundance are not at present known. Finally, aliphatic and aromatic diterpenoid hydrocarbons are excellent markers of terrigenous organic matter, and there are indications of an eolian transport of these compounds to Sites 367, 369 (Simoneit, 1978) and 397 (Simoneit and Mazurek, 1979).

Among the non-hydrocarbon compounds in the extracts, only fatty acids and methyl-n-alkylketones have been analysed in a wider range of samples, and long-chain members have been interpreted as indicators of the influx of terrigenous organic matter (Simoneit, 1978; Simoneit and Mazurek, 1979).

A more comprehensive survey in terms of lipid classes has been performed by Brassell et al. (1980) on three sediment cores from Site 416. They found a high degree of similarity of the extractable lipids in all three turbidite samples and suggested fairly uniform contributions of organic matter and constant depositional conditions during the Valanginian at this site. There were no significant differences in the stage of diagenesis in these samples covering the depth interval from about 1230 m to 1360 m.

4. Maturity of Organic Matter

Maturation of organic matter is a function of temperature and time. Increasing temperature as a result of deeper burial due to basin subsidence and exposure to elevated temperature for a given time leads to chemical and physical changes in organic matter. These changes can be observed by different analytical methods.

With the exception of local heating by intrusives, the organic matter in all sediment samples recovered by the Deep Sea Drilling Project along the NW African continental margin is immature with respect to thermal hydrocarbon generation. There are a number of reasons for this result, including the intermediate geothermal gradients encountered, the limited drilling capacity of the Glomar Challenger, and the specific avoidance of drilling into hydrocarbon-prone strata. This latter restriction derives from the safety and pollution concept which prohibits drilling into zones where the absence of a riser system makes a "blowout" possible.

Nevertheless, even in the immature sediment sections encountered, a number of trends indicating early maturation processes were observed. Table 3 contains a summary of data for the NW African continental margin, concentrating on three different maturation parameters, namely vitrinite reflectance, temperature of maximum pyrolysis yield, and the carbon preference index of n-alkanes found in the extractable organic matter.

4.1 Vitrinite Reflectance

The measurement of vitrinite reflectance by reflected-light microscopy is the most widely applicable method used to determine the maturity of organic matter (Bostick, 1979). Care has to be taken, however, that only primary vitrinite particles are considered. In samples where eroded, recycled (reworked) or oxidized organic matter is incorporated into sediments, vitrinite reflectance is higher than expected from the general downhole trend or the burial history of the site under investigation. This situation occurs at several sites on the NW African continental margin. Figure 3 shows vitrinite reflectance histograms for three samples from Site 416

Table 3. Summary of quantitative assessment of organic matter maturation. (Data taken from trends, ignoring extreme or dubious values)

	Vitrinite reflectance (% $\bar{R}_o$)	Pyrolysis (T_{max}, °C)	CPI (n-alkanes)
Site 367	0.18(649 m)[1]	396(639 m)-431(778 m)[2]	4.0(67 m)-2.9(649 m)[3]
Site 368, normal	0.22(388 m)-0.44(980 m)[1] 0.19(638 m)-0.30(748 m)[4]	423(939 m)-436(978 m)[2]	
Site 368, near diabase sill (depth below sill)	3.33(0.2 m)[4] 2.60(1.9 m)[5] 0.50(8,8 m)[4]		1.15(1.9 m)[5]
Site 369	0.21(444 m)[1]		3.6(12 m)[3]
Site 370	0.24(695 m)-0.26(879 m)[1] 0.19(446 m)-0.35(1088 m)[6]	400(104 m)-428(1169 m)[7]	
Site 397, Tertiary	0.20(660 m)-0.37(1288 m)[8] 0.32(51 m)-0.35(1159 m)[9]	414(1066 m)-423(1241 m)[10]	2.91(130 m)-1.29(1100 m)[8] 1.7(16.2 m)[11]
Site 397, L. Cretaceous	0.37(1310 m)-0.40(1453 m)[8] 0.42(1444 m)[9]	430(1449 m)[10]	4.42(1450 m)[8] 2.6(1444 m)[11]
Site 415	0.25(577 m)-0.33(881 m)[12]	390(0.5 m)-422(1044 m)[7] 440(210 m)[13]	2.9(580 m)-1.8(960 m)[14] 4.6(210 m)[13]
Site 416	0.29(452 m)-0.47(1616 m)[12] 0.21(302 m)-0.52(1615 m)[6]	417(148 m)-429(1615 m)[7] 438-464(1180-1532 m, no trend)[13] 424 ± 1(1180-1532 m)[15]	3.0(1180 m)-2.7(1532 m)[13] 1.5(300 m)-1.3(1500 m), no systematic trend[14]

[1] Kendrick et al., 1978; [2] Deroo et al., 1978; [3] Simoneit, 1978; [4] Baker et al., 1978; [5] Dow, 1978a; [6] Boutefeu et al., 1980; [7] Boutefeu, 1980; [8] Cornford et al., 1979; [9] Kendrick et al., 1979; [10] Deroo et al., 1979; [11] Simoneit and Mazurek, 1979; [12] Cornford, 1980; [13] Claypool and Baysinger, 1980; [14] Galimov et al., 1980; [15] Deroo et al., 1980.

(Morocco Basin) with a narrow reflectance range (Fig. 3a; Sample 416A-11-4, 1199 m), a bimodal distribution (Fig. 3b; Sample 416A-40-4, 1460 m) and a complex distribution from which no clear distinction between primary and altered vitrinite particles is possible and in which the total averaged reflectance may be too high a value (Fig. 3c; Sample 416A-57-1, 1616 m). Similar complex vitrinite reflectance distributions were found for the organic particles in DSDP Site 397 sediments (Cornford et al., 1979). Nevertheless, trends showing increasing maturation with depth (Fig. 6) were obtained for both sites (Cornford et al., 1979; Cornford, 1980). The maximum reflectance values measured for primary vitrinites at both sites nearly reach 0.5% $\bar{R}_o$, i.e., they are just approaching the commonly accepted range for the onset of thermal hydrocarbon generation from type II organic matter (Tissot and Welte, 1978). Data from the other holes, which are all somewhat shallower, confirm the immature nature of the organic matter in each case (Table 3).

4.2 Pyrolysis

Rock-Eval pyrolysis, in addition to information about the type of organic matter, provides evidence for the maturation level of the organic matter based on the temperature of the maximum pyrolysis yield, T_{max} (Espitalié et al., 1977). During Rock-Eval pyrolysis, a rock sample is gradually heated up in an oven under a stream of helium, and the amount of organic compounds released is monitored with a flame ionization detector. Commonly, two peaks are recorded, the first one corresponding to mobile organic compounds present in the rock and the second one, at higher temperature, corresponding to kerogen breakdown products. Although the temperature of the second peak maximum (T_{max}) may depend on the type of organic matter as well as maturity, trends of increasing T_{max} values were observed downhole in most of the sections drilled off NW Africa. A closely spaced set of samples from Sites 370, 415 and 416 have been analysed by Boutefeu (1980), showing that the increase of T_{max} is fairly rapid down to about 1000 m (corresponding to about 0.35% vitrinite reflectance), and then only a slight increase is observed down to 1600 m (∿0.5% R_o). A similar observation was made by Deroo et al. (1980) who pyrolysed samples from 1180 m to 1532 m from Site 416 and found a T_{max} value of 424 ± 1°C. Higher absolute values reported by Claypool and Baysinger (1980) are possibly due to a different method of pyrolysis.

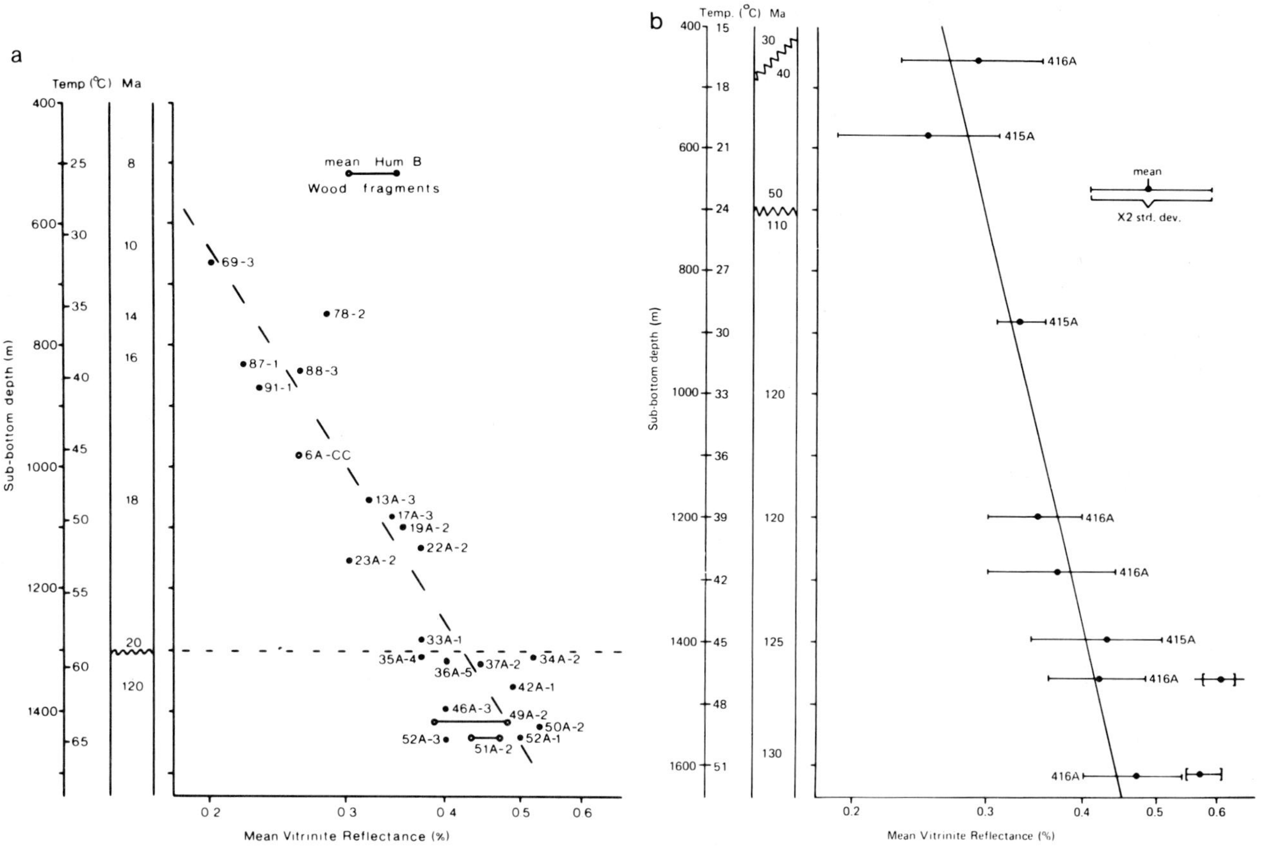

Fig. 6
Trends of increasing vitrinite reflectance with depth for a Site 397. (After Cornford et al., 1979); b Sites 415/416. (Cornford, 1980)

4.3 Carbon Preference Index

The carbon preference index (CPI) of the n-alkanes (also listed in Table 3) as a maturation parameter is very much dependent on the type of organic matter. The carbon preference index is defined as

$$CPI = \frac{\% \, C_{25} - C_{33} \text{ odd} + \% \, C_{23} - C_{31} \text{ odd}}{2 \, (\% \, C_{24} - C_{32} \text{ even})}$$

(Bray and Evans, 1961), although sometimes different carbon number ranges are used. Terrestrial higher plant waxes contain long-chain n-alkanes with a high predominance of the odd-numbered carbon molecules (Eglinton and Hamilton, 1963). Marine organisms normally contain low amounts of n-alkanes without any regular odd-over-even or even-over-odd predominance over a wider range. Thus, because the main phase of hydrocarbon generation is not reached in the sediments encountered, essentially only the n-alkanes indigenous to the living organisms are found in these deep sea sediments. However, small amounts of n-alkanes formed during early diagenesis/catagenesis have been reported. Systematic trends were observed, e.g., by Simoneit (1978) for samples from Site 367 with a decrease of the CPI (n-C_{14} to n-C_{36}) from 4.0 at 67 m to 2.9 at 649 m and by Cornford et al. (1979) for the Tertiary sediments from Site 397 (2.91 at 130 m, 1.29 at 1100 m; n-C_{24} to n-C_{31}), whereas no systematic trend is obvious from the data of Galimov et al. (1980) for the samples from Site 416 (n-C_{15} to n-C_{33}; Table 3).

4.4 Igneous Intrusion

The only exception from the series of immature sediments are the black shale layers affected by a diabase sill intrusion at Site 368. A number of investigators studied the effect of this sudden high heat flow on the composition and maturity of the organic matter.

An extremely high vitrinite reflectance of 3.33% $\bar{R}_o$ was measured at 0.2 m below the sill (Baker et al., 1978). This reflectance decreased to 2.60% $\bar{R}_o$ at a distance of 1.9 m (Dow, 1978a), and no further effect was noticed at 8.8 m below the sill ($\bar{R}_o$ = 0.50%; Baker et al., 1978). In agreement with this, a very low n-alkane carbon preference index of 1.15 was measured at 1.9 m below the sill (Dow, 1978a). The effect on the organic matter of the high temperature of the intrusive sill extends over a greater distance at Site 368 than it does in similar examples from cores in the Gulf of California (Einsele et al., 1980) where the effect disappeared at a distance of about 10 cm. This difference may be due to the fact that the intrusion at Site 368 occurred at a time when the sediments were already relatively more consolidated than were the sediments in the Gulf of California; the lower interstitial water content at Site 368 possibly reduced the dissipation of heat.

4.5 Calculation of Temperature History

A successful attempt has been made recently to simulate the development of a sedimentary basin by means of a deterministic three-dimensional computer model with the aim to enable exploration geologists to make quantitative exploration predictions for hydrocarbon generation and accumulation (Welte and Yükler, 1980a, b). In an early stage during the development of this concept, a one-dimensional form of this model was applied to the sediments encountered in DSDP Holes 397/397A during DSDP Leg 47a (Yükler et al., 1979). The study determines the mass and energy transport in the sedimentary sequences taking into account the major mechanisms operating during sedimentation and compaction. Initial data consist of heat flux, initial physical and thermodynamic properties of sediments, paleobathymetric estimations, and the sedimentation rate curve. The model then computes the two unknowns, namely pressure and temperature, as a function of time and vertical distance. The physical and thermodynamic parameters are evaluated from the pressure and temperature calculations. These values are then put into fluid and heat flow equations.

Figure 7 shows some of the results of the application of this model to data of the Tertiary sediments from DSDP Holes 397/397A (Yükler et al., 1979). A comparison of computed and observed porosity curves as a function of depth is given in Fig. 7a. Considerable scatter in the measured porosities reflects the heterogeneous nature of the sediments at this site and possibly some drilling disturbance. The computed and observed porosity trends show close correlation both in slope and in detailed points of inflection (average error ± 2% porosity). Reduced porosities in the sediments deposited on the plane of the disconformities at 235 m and 650 m are superimposed on the normal reduction in porosity with depth in Fig. 7a. A similar feature is observed at the base of the section where underlying Cretaceous sediments were modelled as an impermeable basement. Secondly, a reduction in porosity is also seen at and below the slump horizons (e.g., between 420 m and 480 m). Slumping appears to squeeze the underlying sediments.

Figure 7b shows the effect of geothermal gradient (G) on the temperature profile. The middle line represents the computed temperature profile with a geothermal gradient of 43.8°C/km from 0 m to 410 m and 35.6°C/km from 410 m to 1300 m. The computed temperature curve matches well with measured values at 360 m and 447 m depth. The curves to the left and right then demonstrate the effect of temperature on the physical properties of the sediments. The heat flux was varied such that a change of ± 5°C/km was observed in the geothermal gradients. The model run with the lower heat flux results in 1400 m of sediment, i.e., 100 m of sediment more than actually encountered; the run with the higher heat flux resulted in 1236 m of sediment, 64 m of sediment less than actually found. The results clearly indicate the effect of heat flow on compaction.

5. Hydrocarbon Potential

There is no a priori reason why hydrocarbon source beds should not exist in continental slope and rise sediment as they do in sediment on continental shelves (Dow, 1978b). As has been pointed out earlier, there are a number of mechanisms which preserve organic matter deposited in deep water environments. A high total organic carbon content alone, however, is not always an indication of a high hydrocarbon potential. Incorporation of major quantities of recycled or oxidized organic matter, which is of a dominantly resistant and inert nature, may lead to sediments rich in organic carbon, but, because of the low hydrogen content, the hydrocarbon generation potential is negligible. Kendrick et al. (1978, 1979) tried to determine experimentally the amount of organic carbon that is convertible to petroleum by measuring "effective organic carbon" using a pyrolysis-FID method. As an example, for samples from DSDP Site 397 they determined a C_{eff} to C_{org} ratio of about 0.25 for the Miocene slump sections, which are rich in marine algal organic matter, and a ratio of only 0.07 for a

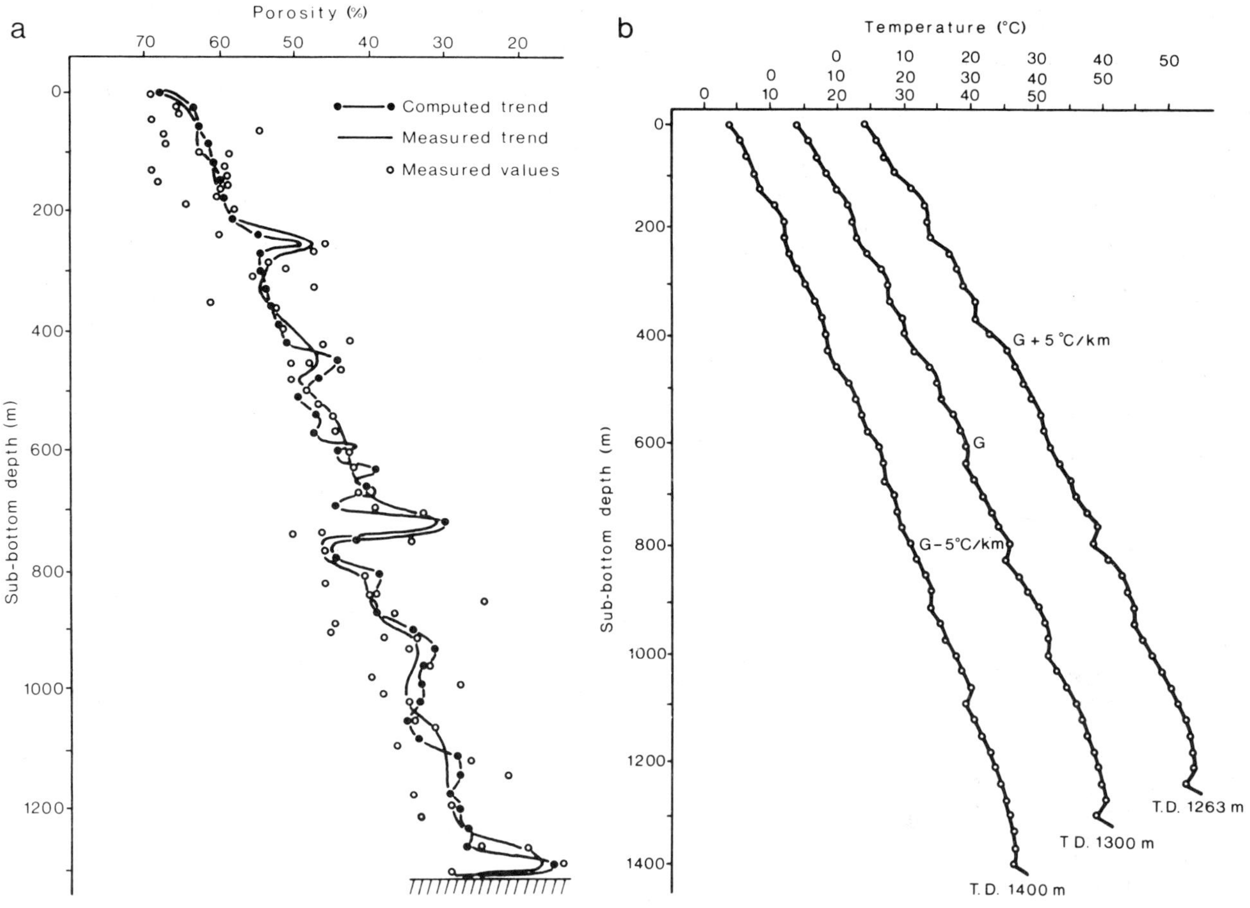

Fig. 7
Computer modelling of temperature history at DSDP Site 397. (After Yükler et al., 1979): a Comparison of measured and computed sediment porosities; b effect of different geothermal gradients (G) on sediment compaction

Lower Cretaceous sediment sample dominated by allochthonous terrigenous organic matter (Kendrick et al., 1979).

Table 1 summarizes the hydrocarbon potential of the different organofacies encountered during deep sea drilling on the NW African continental margin. A high potential for liquid and gaseous hydrocarbon generation may be ascribed to the organic matter in the Albian - Cenomanian sediments at Sites 367 and 368 and the Miocene slump sequences at Site 397. In both cases the amount of marine planktonic organic matter preserved is relatively high. A moderate to low gas potential is characteristic of the mostly terrigenous kerogen of all the other Cretaceous sediments on the NW African continental margin, while the autochthonous hemipelagic Tertiary sediments at Sites 397 and 415 have no hydrocarbon potential since the organic carbon contents are low and most of the organic matter consists of residual, inert terrigenous material.

As noted before during discussion of the maturity of the organic matter, all the sediment sections investigated so far have not yet reached the onset of thermal hydrocarbon generation. If hydrocarbon accumulations are to be found on the NW African continental margin, it must be assumed that facies described as oil- or gas-prone (Table 1) are more deeply buried at other locations or that other sediments with similar hydrocarbon potential exist in this area. Pearson and Dow (1979) extrapolate the measured maturity data for Site 397 and predict the onset of effective liquid hydrocarbon generation at 1870 m below the sea floor, assuming a minimum of 0.6% vitrinite reflectance for lipid-rich organic matter. These authors also predict that a significant wet gas formation (above 0.8% vitrinite reflectance) would not start before 2300 m depth below sea floor. It has to be emphasized, however, that the minimum vitrinite reflectance values indicating the onset of hydrocarbon generation are still subject of controversary discussion.

Calculation of the minimum temperature necessary for oil generation at Site 397 considering the geothermal heat flow and effective heating time (Connan, 1974; Hood et al., 1975) has been attempted by Arthur et al. (1979). For the Lower Miocene sediments they estimated a temperature of 100°C which has to be compared with the projected temperature of 65°C at total depth at Site 397 (1453 m). Similar calculations for the Hauterivian sediments, which because of their predominantly terrigenous organic matter content are gas-prone and thus need a higher maturity for hydrocarbon generation, are complicated by the major hiatus of about 100 m.y. between the Lower Miocene and the Hauterivian. In addition, the geothermal gradient (40°C/km), which since the Tertiary is related to the Canary Islands volcanism, was probably lower in the Cretaceous (Arthur et al., 1979).

Finally, suitable migration pathways and sealed porous reservoir rocks are an additional prerequisite for hydrocarbon accumulation. The situation off Cape Bojador (Sites 369 and 397) has been extensively discussed by Arthur et al. (1979). At the continental rise (Site 397), well-sorted distal turbidites were found to be relatively poor reservoir beds because of their low porosities. Their reservoir properties were in some cases further reduced by secondary calcite cementation. However, some of the thicker turbidites (up to 3 m) derived from the Cape Bojador slope exhibit reasonable reservoir properties. The best porosities (25% to 50%) were found at the continental rise in volcaniclastic sandstones which are derived from Fuerteventura and pinchout against the foot of the continental slope (Arthur et al., 1979). Since there is very little structural disturbance in this foot-of-slope setting, stratigraphic traps might be most important for the potential retention of hydrocarbons; the only major structure which might have trapped hydrocarbons is the "slope anticline" (Arthur et al., 1979).

The continental slope sediments are less likely to contain suitable reservoir beds, because sandy turbidity currents usually bypass the slopes, travelling more commonly through submarine canyons from the continental shelf to the continental rise and abyssal plains (Emery, 1980). This is supported by the observation made at the continental slope off Cape Bojador (Site 369) where no coarse terrigenous sediments were penetrated (Gardner et al., 1978).

Determination of most of these parameters, however, is beyond the scope of present deep sea drilling with the Glomar Challenger.

6. Conclusions

Typical deep sea sediments contain very little organic matter, because most of it is destroyed during settling through oxygen-rich oceanic water. Exceptional geological situations may, however, yield organic-carbon-rich sediments. The main factors affecting the organic matter content of shelf, slope, and rise sediments in terms of source and preservation of organic matter, and of redistribution of deposited material are schematical-

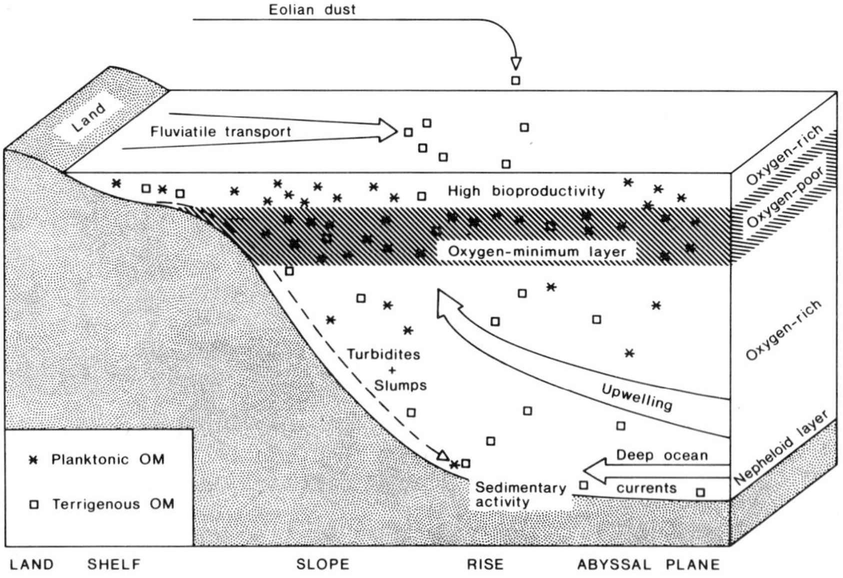

Fig. 8
Factors affecting the organic matter content of shelf, slope, and rise sediments. (After Cornford, 1979, and Welte et al., 1979)

ly summarized in Figure 8. These sediments can be described in terms of different organofacies according to the quantity, type, and composition of organic matter, and to the related hydrocarbon potential:

- Accumulation of the remains from high planktonic productivity associated with areas of upwelling, where the resultant reduced oxygen concentration allows a high survival rate of planktonic organic matter, may result in deposition of sediments with hydrogen-rich kerogen, which have a high hydrocarbon generation potential (e.g., black shale facies).
- The rapid influx and sedimentation of terrigenous organic matter with associated slumps and mass flows on a steep continental margin can preserve significant amounts of organic matter in continental rise and abyssal plain sediments with a limited hydrocarbon generation potential (e.g., Lower Miocene at DSDP Site 397 and Cretaceous turbidites at DSDP Sites 415/416).
- Incorporation of major quantities of recycled or oxidized terrigenous organic matter will yield sediments where the dominant kerogen component is of a chemically resistant and inert nature (e.g., hemipelagic Tertiary sediments at DSDP Sites 397 and 415).

Vitrinite reflectance measurements and other maturity indicators show that all sediments drilled so far at the Northwest African continental margin are immature with respect to hydrocarbon generation. To satisfy an interest in economic hydrocarbon accumulations, more deeply buried potential source rocks have to be investigated, as well as possibilities for migration and reservoir formation.

Acknowledgements. We would like to thank Dr. L.J. Bogoliubova (Academy of Science, Moscow) for providing total extracts of two sediment samples from DSDP Site 367 for organic geochemical investigation. We are indebted to Dr. Keith A. Kvenvolden (U.S. Geological Survey, Menlo Park) for carefully reviewing the manuscript and for his valuable comments. Financial support by the Deutsche Forschungsgemeinschaft (Bonn), grant No. We 346/21, is gratefully acknowledged.

REFERENCES

Arthur MA, von Rad U, Cornford C, McCoy FW, Sarnthein M (1979) Evolution and sedimentary history of the Cape Bojador continental margin, Northwestern Africa. In: von Rad U, Ryan WBF et al, Initl Repts DSDP. US Govt Printing Office, Washington, 47, part 1, pp 773-816

Baker EW, Huang WY, Rankin JG, Castaño JRG, Guinn JR, Fuex AN (1978) Electron paramagnetic resonance study of thermal alteration of kerogen in deep-sea sediments by basaltic sill intrusion. In: Lancelot Y, Seibold E et al, Initl Repts DSDP. US Govt Printing Office, Washington, 41, pp 839-847

Bostick NH (1979) Microscopic measurement of the level of catagenesis of solid organic matter in sedimentary rocks to aid exploration for petroleum and to determine former burial temperatures - a review. Soc Econ Paleont Mineral Spec Publ 26: 17-43.

Boutefeu A (1980) Pyrolysis study of organic matter from Deep Sea Drilling Project Sites 370 (Leg 41), 415, and 416 (Leg 50). In: Lancelot Y, Winterer EL et al, Initl Repts DSDP. US Govt Printing Office, Washington, 50, pp 555-566

Boutefeu A, Leplat P, Somers Y (1980) Preliminary results of petrographic and electron-spin-resonance studies of organic matter from Deep Sea Drilling Project Sites 370 and 416. In: Lancelot Y, Winterer EL et al, Initl Repts DSDP. US Govt Printing Office, Washington, 50, pp 567-573

Brassell SC, Comet PA, Eglinton G, McEvoy J, Maxwell JR, Quirke JME, Volkman JK (1980) Preliminary lipid analyses of cores 14, 18 and 28 from Deep Sea Drilling Project Hole 416A. In: Lancelot Y, Winterer EL et al, Initl Repts DSDP. US Govt Printing Office, Washington, 50, pp 647-664

Brassell SC, Comet PA, Eglinton G, Isaacson PJ, McEvoy J, Maxwell JR, Thomson ID, Tibbetts PJC, Volkman JK (1981) The origin and fate of lipids in the Japan trench. In: Douglas AG, Maxwell JR (eds) Adv Org Geochem - 1979, Prog Phys Chem of the Earth Series. Pergamon Press, Oxford, in press

Bray EE, Evans ED (1961) Distribution of n-paraffins as a clue to recognition of source bed. Geochim Cosmochim Acta 22: 2-15

Claypool GE, Baysinger JP (1980) Analysis of organic matter in sediment cores from the Morocco Basin, Deep Sea Drilling Project Sites 415 and 416. In: Lancelot Y, Winterer EL et al, Initl Repts DSDP. US Govt Printing Office, Washington, 50, pp 605-608

Connan J (1974) Time-temperature relation in oil genesis. Am Assoc Pet Geol Bull 58: 2516-2521

Cornford C (1979) Organic deposition at a continental rise: organic geochemical interpretation and synthesis at DSDP Site 397, eastern North Atlantic. In: von Rad U, Ryan WBF et al, Initl Repts DSDP. US Govt Printing Office, Washington, 47, part 1, pp 503-510

Cornford C (1980) Petrology of organic matter, Deep Sea Drilling Project Sites 415 and 416, Moroccan Basin, eastern North Atlantic. In: Lancelot Y, Winterer EL et al, Initl Repts DSDP. US Govt Printing Office, Washington, 50, 609-614

Cornford C, Rullkötter J, Welte DH (1979) Organic geochemistry of DSDP Leg 47a, Site 397 eastern North Atlantic: organic petrography and extractable hydrocarbons. In: von Rad U, Ryan WBF et al, Initl Repts DSDP. US Govt Printing Office, Washington, 47, part 1, pp 511-522

Cornford C, Rullkötter J, Welte DH (1981) A synthesis of organic petrographic and geochemical results from DSDP sites in the eastern central North Atlantic. In: Douglas AG, Maxwell JR (eds) Adv Org Geochem - 1979, Prog Phys Chem of the Earth Series. Pergamon Press, Oxford, in press

Dastillung M, Albrecht P (1977) Δ^2-Sterenes as diagenetic intermediates in sediments. Nature (London) 269: 678-679

Deroo G, Herbin JP, Roucaché J, Tissot B, Albrecht P, Schaeffle J (1978) Organic geochemistry of some Cretaceous black shales from Sites 367 and 368; Leg 41, eastern North Atlantic. In: Lancelot Y, Seibold E et al, Initl Repts DSDP. US Govt Printing Office, Washington, 41, pp 865-873

Deroo G, Herbin JP, Roucaché J, Tissot B (1979) Organic geochemistry of some organic-rich shales from DSDP Site 397, Leg 47a, eastern North Atlantic. In: von Rad U, Ryan WBF et al, Initl Repts DSDP. US Govt Printing Office, Washington, 47, part 1, pp 523-529

Deroo G, Herbin JP, Roucaché J, Tissot B (1980) Organic geochemistry of some Lower Cretaceous shales from Deep Sea Drilling Project Site 416, eastern North Atlantic. In: Lancelot Y, Winterer EL et al, Initl Repts DSDP. US Govt Printing Office, Washington, 50, pp 629-635

Dow WG (1978a) Contact metamorphism of kerogen in sediments from Leg 41: Cape Verde Rise and Basin. In: Lancelot Y, Seibold E et al, Initl Repts DSDP. US Govt Printing Office, Washington, 41, pp 821-824

Dow WG (1978b) Petroleum source beds on continental slopes and rises. In: Geological and geophysical investigations of continental margins. Am Assoc Pet Geol Mem 29: 423-442

Eglinton G, Hamilton RJ (1963) The distribution of alkanes. In: Swain T (ed) Chemical plant taxonomy. Academic Press, London New York, pp 187-217

Einsele G, Gieskes JM, Curray J, Moore DM, Aguayo E, Aubry M-P, Fornari D, Guerrero J, Kastner M, Kelts K, Lyle M, Matoba Y, Molina-Cruz A, Niemitz J, Rueda J, Saunders A, Schrader H, Simoneit B, Vacquier V (1980) Intrusion of basaltic sills into highly porous sediments, and resulting hydrothermal activity. Nature (London) 283: 441-445

Emery KO (1980) Continental margins - classification and petroleum prospects. Am Assoc Pet Geol Bull 64: 297-315

Erdman JG, Schorno KS (1979) Geochemistry of carbon: Deep Sea Drilling Project, Legs 47a and b. In: Sibuet J-C, Ryan WBF et al, Initl Repts DSDP. US Govt Printing Office, Washington, 47, part 2, pp 553-559

Espitalié J, Laporte JL, Madec M, Marquis F, Leplat P, Paulet J, Boutefeu A (1977) Méthode rapide de caractérization des roches-mères, de leur potentiel pétrolier et de leur degré d'évolution. Rev Inst Fr Pét 32: 23-42

Gardner JV, Dean WE, Jansa L (1978) Sediments recovered from the Northwest African continental margin, Leg 41, Deep Sea Drilling Project. In: Lancelot Y, Seibold E et al, Initl Repts DSDP. US Govt Printing Office, Washington, 41, pp 1121-1134

Galimov EM, Kodina LA, Shirinsky VG, Drozdova TV, Generalova VN, Bogachova MP, Chinyonor VA, Bannikova LA, Vernadsky VI (1980) A study of organic matter from deep sea oceanic bore holes, Deep Sea Drilling Project Sites 415 and 416, in the Moroccan Basin. In: Lancelot Y, Winterer EL et al, Initl Repts DSDP. US Govt Printing Office, Wahington, 50, pp 575-603

Hood A, Gutjahr CCM, Heacock RL (1975) Organic metamorphism and the generation of petroleum. Am Assoc Pet Geol Bull 59: 986-996

Hunt JM (1979) Petroleum geochemistry and gegeology. Freeman, San Francisco, pp 23-27

Johnson DL, McIver RD, Rogers MA (1979) Insoluble organic matter bitumens in Leg 47 samples. In: Sibuet J-C, Ryan WBF et al, Initl Repts DSDP. US Govt Printing Office, Washington, 47, part 2, pp 543-546

Kendrick JW, Hood A, Castaño JR (1978) Petroleum-generating potential of sediments from Leg 41, Deep Sea Drilling Project. In: Lancelot Y, Seibold E et al, Initl Repts DSDP. US Govt Printing Office, Washington, 41, pp 817-819

Kendrick JW, Hood A, Castaño JR (1979) Petroleum-generating potential of sediments from Leg 47, Deep Sea Drilling Project. In: Sibuet J-C, Ryan WBF et al, Initl Repts DSDP. US Govt Printing Office, Washington, 47, part 2, pp 547-551

Lancelot Y, Seibold E, Shipboard Party, Bukry D (1978a) Site 366: Sierra Leone Rise. In: Lancelot Y, Seibold E et al, Initl Repts DSDP. US Govt Printing Office, Washington, 41, pp 21-161

Lancelot Y, Seibold E, Shipboard Party, Bukry D (1978b) Site 367: Cape Verde Basin. In: Lancelot Y, Seibold E et al, Initl Repts DSDP. US Govt Printing Office, Washington, 41, pp 163-232

Lancelot Y, Seibold E, Shipboard Party, Bukry D (1978c) Site 368: Cape Verde Rise. In: Lancelot Y, Seibold E et al, Initl Repts DSDP. US Govt Printing Office, Washington, 41, pp 233-326

Lancelot Y, Seibold E, Shipboard Party, Bukry D (1978d) Site 369: Continental slope off Cape Bojador, Spanish Sahara. In: Lancelot Y, Seibold E et al, Initl Repts DSDP. US Govt Printing Office, Washington, 41, pp 327-420

Lancelot Y, Winterer EL, Shipboard Party (1980a) Site 415, Agadir Canyon, Deep Sea Drilling Project Leg 50. In: Lancelot Y, Winterer EL et al, Initl Repts DSDP. US Govt Printing Office, Washington, 50, pp 31-113

Lancelot Y, Winterer EL, Shipboard Party (1980b) Site 416, in the Moroccan Basin, Deep Sea Drilling Project Leg 50. In: Lancelot Y, Winterer et al, Initl Repts DSDP. US Govt Printing Office, Washington, 50, pp 115-301

McIver R (1975) Hydrocarbon occurrences from JOIDES Deep Sea Drilling Project. Proc 9th World Pet Congr (Tokyo). Appl Science Publ, Barking, 2, pp 269-280

Ourisson G, Albrecht P, Rohmer M (1979) The hopanoids. Palaeochemistry and biochemistry of a group of natural products. Pure Appl Chem 50: 709-729

Pearson B, Dow WG (1979) Geochemical analysis of samples from Sites 397 and 398. In: Sibuet J-C, Ryan WBF et al, Initl Repts DSDP. US Govt Printing Office, Washington, 47, part 2, pp 533-541

Rogers MA (1980) Application of organic facies concepts to hydrocarbon source rock evaluation. Proc 10th World Pet Congr (Bucharest). Heyden, London, 2, pp 23-30

Rubinstein I, Sieskind O, Albrecht P (1975) Rearranged sterenes in a shale: occurrence and simulated formation. J Chem Soc Perkin Trans I: 1833-1836

Rullkötter J, Flekken P, Welte DH (1980) Organic petrography and extractable hydrocarbons of sediments from the northern Philippine Sea, Deep Sea Drilling Project Leg 58. In: de Vries Klein G, Kobayashi K et al, Initl Repts DSDP. US Govt Printing Office, Washington, 58, pp 755-762

Schorno KS, Erdman JG (1980) Geochemistry of carbon: Deep Sea Drilling Project Sites 415 and 416. In: Lancelot Y, Winterer EL et al, Initl Repts DSDP. US Govt Printing Office, Washington, 50, pp 625-627

Seifert WK, Moldowan JM, Smith GW, Whitehead EV (1978) First proof of structure of a C_{28}-pentacyclic triterpane in petroleum. Nature (London) 271: 436-437

Simoneit BRT (1978) Leg 41 sediment lipids - Search for eolian organic matter in recent samples and examination of a black shale. In: Lancelot Y, Seibold E et al, Initl Repts DSDP. US Govt Printing Office, Washington, 41, pp 855-858

Simoneit BRT, Mazurek MA (1979) Search for eolian lipids in the Pleistocene off Cape Bojador and lipid geochemistry of a Cretaceous mudstone, DSDP/IPOD Leg 47a. In: von Rad U, Ryan WBF et al, Initl Repts DSDP. US Govt Printing Office, Washington, 47, part 1, pp 541-545

Simoneit BRT, Mazurek MA, Brenner S, Crisp PT, Kaplan IR (1979) Organic geochemistry of recent sediments from Guaymas Basin, Gulf of California. Deep-Sea Res 26A:879-891

Tissot BP, Welte DH (1979) Petroleum formation and occurrence. Springer, Berlin Heidelberg New York

Tissot B, Demaison G, Masson P, Delteil JR, Combaz A (1980) Paleoenvironment and petroleum potential of the mid-Cretaceous black shales in the Atlantic basins. Am Assoc Pet Geol Bull 64: 2051-2063

van Dorsselaer A, Albrecht P, Ourisson G (1977) Identification of novel 17α(H)-hopanes in shales, coals, lignites, sediments and petroleum. Bull Soc Chim Fr: 165-170

von Rad U, Ryan WBF, Shipboard Party (1979) Site 397. In: von Rad U, Ryan WBF et al, Initl Repts DSDP. US Govt Printing Office, Washington, 47, part 1, pp 17-217

Welte DH, Yükler A (1980a) Evolution of sedimentary basins from the standpoint of petroleum origin and accumulation - an approach for a quantitative basin study. Org Geochem 2: 1-8

Welte DH, Yükler A (1980b) Evolution of sedimentary basins from the standpoint of petroleum origin and accumulation - a three-dimensional quantitative basin study. Soc Econ Paleont Mineral Short Course on "How to assess maturation and paleotemperatures", AAPG-SEPM-EMD Annu Conv, Denver, June 1980

Welte DH, Cornford C, Rullkötter J (1979) Hydrocarbon source rocks in deep sea sediments. Proc 11th Annu Offshore Tech Conf (Houston), 1, pp 457-464

Yükler MA, Cornford C, Welte DH (1979) Simulation of geologic, hydrodynamic, and thermodynamic development of a sediment basin - a quantitative approach. In: von Rad U, Ryan WBF et al, Initl Repts DSDP. US Govt Printing Office, Washington, 47, part 1, pp 761-771